Islets of Langerhans

Md. Shahidul Islam
Editor

Islets of Langerhans

Volume 2

Second Edition

With 170 Figures and 34 Tables

Springer Reference

Editor
Md. Shahidul Islam
Department of Clinical Sciences and Education
Sodersjukhuset, Karolinska Institutet
Stockholm, Sweden

Department of Internal Medicine
Uppsala University Hospital
Uppsala, Sweden

ISBN 978-94-007-6685-3   ISBN 978-94-007-6686-0 (eBook)
ISBN 978-94-007-6687-7 (print and electronic bundle)
DOI 10.1007/978-94-007-6686-0
Springer Dordrecht Heidelberg New York London

Library of Congress Control Number: 2014950662

© Springer Science+Business Media Dordrecht 2010, 2015
This work is subject to copyright. All rights are reserved by the Publisher, whether the whole or part of the material is concerned, specifically the rights of translation, reprinting, reuse of illustrations, recitation, broadcasting, reproduction on microfilms or in any other physical way, and transmission or information storage and retrieval, electronic adaptation, computer software, or by similar or dissimilar methodology now known or hereafter developed. Exempted from this legal reservation are brief excerpts in connection with reviews or scholarly analysis or material supplied specifically for the purpose of being entered and executed on a computer system, for exclusive use by the purchaser of the work. Duplication of this publication or parts thereof is permitted only under the provisions of the Copyright Law of the Publisher's location, in its current version, and permission for use must always be obtained from Springer. Permissions for use may be obtained through RightsLink at the Copyright Clearance Center. Violations are liable to prosecution under the respective Copyright Law.
The use of general descriptive names, registered names, trademarks, service marks, etc. in this publication does not imply, even in the absence of a specific statement, that such names are exempt from the relevant protective laws and regulations and therefore free for general use.
While the advice and information in this book are believed to be true and accurate at the date of publication, neither the authors nor the editors nor the publisher can accept any legal responsibility for any errors or omissions that may be made. The publisher makes no warranty, express or implied, with respect to the material contained herein.

Printed on acid-free paper

Springer is part of Springer Science+Business Media (www.springer.com)

*"Dedicated to the living memory of
Matthias Braun, M.D., Ph.D. 1966–2013"*

Foreword

The tiny islets of Langerhans receive an extraordinary amount of attention from a variety of interested parties, many of whom will enthusiastically welcome publication of the second edition of the "Islets of Langerhans," ably edited by Md. Shahidul Islam, M.D., Ph.D., of the Karolinska Institute, Stockholm, Sweden. The amount of attention paid to islets is well deserved because the failure of their β cells to produce sufficient amounts of insulin results in diabetes, with its climbing prevalence worldwide and devastating complications. In type 1 diabetes the β cells are almost completely decimated by the vicious process of autoimmunity. With the far more common type 2 diabetes, the insulin resistance associated with obesity and our sedentary life style is linked to reduced β cell mass and function. The simplest view is that the β cells die because they are stressed by overwork, resulting in reduction of insulin secretion, which allows glucose levels to rise enough to cause further impairment of secretion through a process called glucotoxicity. Thus there is a loss of both β cell mass and function, resulting in the concept of decreased functional mass. Most people with insulin resistance never develop type 2 diabetes, which leads to the conclusion the β cell failure is the sine qua non for the development of the diabetic state.

Following from the above, the premise that β cell failure is the root cause of diabetes is conceptually very simple, which leads to the conclusion that the diabetic state should be reversed by administering insulin with injections, restoring β cell function with medication or by replenishment of the β cell deficit with transplantation or regeneration. Indeed, the all important proof-of-principle was achieved in the 1990s with the demonstration that both types 1 and 2 diabetes could be reversed with islet transplantation either as isolated islets placed in the liver or as whole organ pancreas transplants.

This second edition of "The Islets of Langerhans" is very timely, because in spite of the seeming simplicity of the basis of diabetes and progress with β cell replacement, we are still too far from our goal of providing these treatments for those in need. We need to understand islets on the most basic level so that preclinical therapeutic approaches can be explored and then taken to patients. The 49 chapters in "The Islets of Langerhans" provide up-to-date information on a carefully selected range of topics.

Important Unsolved Islet Puzzles

Knowing full well there are many opinions about which unsolved islet questions are most important, I will briefly mention a selection of issues that have captured my attention.

The islet as an organ The anatomy of islets is high organized with its cellular arrangements and islet-acinar portal blood flow. We know that β cell secretion has a major influence on glucagon secretion, but we have much to learn about the other interactions between beta, alpha and delta cells and how secretion from all of these influences downstream acinar cell development and maintenance. The role of the pancreatic polypeptide (PP) cells remains very much a mystery.

The mystery of glucose-stimulated insulin secretion (GSIS) For years we have had some understanding of the so-called K_{ATP} pathway of GSIS, yet we have little understanding of the quantitatively important K_{ATP}-independent pathway. This remains a major unsolved problem in β cell biology.

Finding new pharmacologic targets for insulin secretion Many of the chapters focus on the cell biology of insulin secretion, and there is much to be learned about these very basic facets, such is glucose and fat metabolism, ion and other transporters, mitochondrial function, calcium handling, phosphorylation reactions, insulin biosynthesis and more. A key question is how much more insulin secretion can we get out of a β cell? Simply put, if the cell is depolarized and fully stimulated by cyclic AMP, what approaches can be used to generate more insulin secretion?

Dedifferentiation of β cells and islet cell plasticity The phenotype of β cells in the diabetic state is deranged and accompanied by dysfunctional insulin secretion, with evidence pointing to glucotoxicity as the major driving force responsible for these changes. Restoration of normal glucose levels reverse these changes, but questions remain as to whether these β cells dedifferentiate toward a pluripotent progenitor state or some other distinct phenotype. The field is now swirling with the concept of islet cell plasticity, such as the potential alpha and delta cells being converted to β cells. There is also a big question about the alpha cell hyperplasia seen when glucagon action is inhibited: what is the signal of alpha cell growth?

The need for more β cells The β cell deficiency of diabetes could be restored by regeneration of new β cells in the pancreas or by transplanting β cells from some other source. As described in several chapters, this is one of the main priorities in diabetes research. Adult human β cells replicate very slowly but there has been great progress in understanding cell cycle mechanisms, which could somehow be exploited. Exciting progress has also be made with making mature β cells from human embryonic stem cells and from induced pluripotent stem cells. There have also been advances in exploiting the potential of exocrine multipotent progenitor cells and in bioengineering. Porcine cells also remain on the list.

Why do β cells die and how can this be prevented? We know that β cells in type 1 diabetes are killed by the immune system, and have watched impressive advances in defining the interactions among effector T cells, T regulatory cells, B cells, and the innate immune system. The process is very aggressive and there is a great need to control it with minimal or no immunosuppression. The Holy Grail is restoration of tolerance. An old approach receiving renewed attention is encapsulation of islets to protect them from immune killing. The new biomaterials and approaches are exciting but we cannot yet be confident about its eventual value. In the context of type 2 diabetes much has been written about how β cells die, with mechanisms receiving the most attention being oxidative stress, endoplasmic reticulum stress, toxicity from IAPP oligomers, and the general concept of "overwork." The reality is that the death rate is very low and we have little idea about which mechanisms are the most important.

Of course there are many other important questions, but this sampling fits well with the contents of this valuable new edition of "The islets of Langerhans." Its chapters contain important information about these key questions, which make it likely that hours spent reading this book should help our field connect the critical dots that will result in new treatments for people with diabetes.

Gordon C. Weir, M.D.
Co-Head Section on Islet Cell & Regenerative Biology
Diabetes Health and Wellness Foundation Chair
Joslin Diabetes Center
Professor of Medicine
Harvard Medical School

On Becoming an Islet Researcher

At the time of this writing, I have spent a quarter of a century in islet research, but the purpose of this article is not to share my journey with you. I do not want to bore you with anecdotes from my experiences, but it is impossible that my views will not be subjective.

Be Clear About Your Goal

Irrespective of whether you have started islet research recently, or you have spent almost a whole life in islet research, it is worthwhile to reflect upon your goals. Here is the big picture. About 194 million people in the world are suffering from some form of islet failure, and by 2025 this number may increase to 333 million. The β-cells of many young people and children are dead. To live a normal life, they need to take insulin injections daily, and they need to prick their finger tips for testing plasma glucose concentration, numerous times. In others, overwork of the β-cells caused by overeating leads to the failure, and eventually to the death of these cells. If you want to see the burden of islet failure, do not hesitate to visit a nearby diabetes clinic. This may open your eyes, or give you a much needed insight.

Your goal is to contribute to the discovery of something, so that this huge human tragedy can somehow be prevented, treated, or cured. Your goal is not primarily to publish papers or just to do some experiments solely to satisfy your own intellectual curiosity. Your goal is not just counting the numbers of your publications, and their impact factors, and not to secure a promotion, advance your own career, or receive prizes. You have a bigger goal, which you may or may not reach, within your lifetime, but if you are conscious of your ultimate goal, you are better prepared to work steadily towards that goal. You may then become the islet researcher that you dream to be.

If you wish not to have a clear goal and prefer to see your scientific journey as the goal, then it is up to you. I think it is important to have visions and goals, perhaps some small goals, if not a big goal to start with.

Become the Finest Islet Researcher

The making of an islet researcher is not easy. Becoming a good islet researcher can be a long process. Educate yourself, keeping in mind that it is never too early or too late to start learning anything new. Through a choice of an unconventional path of education, you may become a specialist in more than one subject, and may thus be better prepared. You may first become a molecular biologist, and then educate yourself as a chemist. Numerous other combinations are possible. Enrich yourself with the necessary knowledge, and the skills from whatever source you need to. You may need to move to the environments that promote creativity, that have better infrastructures, and traditions for good research. To do this, you may need to leave your home country, and then struggle hard to adapt yourself to the new environments.

You almost certainly need to acquire a broad base of knowledge before you focus on some special areas. At the same time, you must also be able to filter out as much unnecessary information and distractions as possible. In an age of information pollution, your ability to decide what to filter out, and to filter those out effectively, may determine how intelligent you are. Clearly, you will not be able to do many things, at least not at the same time.

Start with asking one of the most important questions in the field of islet research, keeping in mind that you are expected to discover things that you are not aware of beforehand. Do not waste time in rediscovering the wheel. If you are not asking an important question, then it does not matter how sophisticated instrument or advanced method you are using.

Identify your strength, strength of your institution, and that of your network if you wish to. Once you have identified the strengths, use those. Do what you think is the right thing to do without fear of being judged by others, but resist the temptation to work on many projects at the same time; take the one you have started to completion. You do not need to compare yourself with others. You do not need to think that you are less talented than others. Do not give up when the going forward seems tough. Dig as deep as possible or change the direction based on your sound judgment. See mistakes as valuable learning experiences. If you have time, get inspirations by reading the life histories of other great scientists. From such readings, you may get important insights about how to develop your own intuition and creativity, and about how to get clues about the so called "unknown unknowns."

Depending on your question and the nature of the project, you may find it useful to work alone or with a small dedicated team, or you may need to network personally with a handful of scientists, including some who are not conventional islet researches. You may benefit more if you attend meetings that do not deal with islet research or if you read papers that do not deal with islet research. If you can bring a small piece of new knowledge from the fields that are very distant from the contemporary islet research, and apply that knowledge to solve some of the common questions in the field of islet research, that may contribute to a breakthrough.

Islet research is not just about science, it is a way of life. You have to make difficult choices during your journey. You are sincere about your purpose in life. At times you may have to juggle with too many bolls in the air. It will affect your social life and your relationships with your near and dear ones. Set your priorities right. You have decided to spend your life for the benefit of people who have islet failure. You are not after money, fame, glamour or festivities. You are a genuine islet researcher.

The Ecosystem of Islet Research

Unfortunately, it is not enough that you have developed yourself as one of the finest islet researchers, and that you have clear visions and goals. The chances of breakthroughs in islet research will depend on what we can call the ecosystem of islet research. The ecosystem of islet research will determines the growth, survival, and creativity of the type of islet researchers that I have alluded to. Important components of this ecosystem include the educational and research enterprises, the funding agencies, the governments and policy makers, industries, publishers, and last but not the least, the patient organizations. The ecosystem of islet research, as well as the ecosystem of research in general, has changed over the past decades, and it will keep changing. For an individual islet researcher, it may be difficult to track these changes, and it may be impossible for them to adapt to the changes that are taking place rather rapidly. At first sight, it may appear that the ecosystem has worked well, and has ensured important discoveries at a steady rate. Islet researchers are not supposed to question the ecosystem; the only thing expected of them is to adapt to the changes for their own survival and earn their bread and butter.

Survival of the islets researchers depends on their ability to write grant applications, and their ability to convince the people who read those applications that their ideas are excellent and the goals are achievable. Islet researchers spend enormous amount of time, money, and energy on writing grants and in about 80 % of the cases, the applications are rejected. It is impossible to assess who is the most talented islet researcher. Since talent cannot be measured, an opportunistic way is to measure what islet researchers have published in the past and how many times those publications have been cited. Even if one is able to identify the most talented islet researchers based on their performances in the past, it is impossible that these selected islet researchers will perform equally well in the future. Some scientists think that the system we have is counter-productive, and wasteful of time and energy (Garwood 2011).

The ecosystem of research, in general, seems to have changed in such a way that it is possible for some academic psychopaths to fool the system. They will write in their grant applications whatever is needed, and they will do whatever else is necessary to manipulate the system in their favor. One of the most talented scientists in the world published in one of the world's most luxurious journal, one of the most exciting breakthroughs in stem cell research that turned out to be bogus (Normile 2009). In one investigation, a bogus manuscript, written by some bogus authors, from some bogus universities was accepted for publication by many scientific journals (Bohannon 2013). The system has become so corrupted that it is apparently possible for some scientists to publish without doing any experiment (Hvistendahl 2013). Don Poldermans published more than 300 papers some of which were fraudulent. Changes in clinical practice based on these papers has caused death of numerous people (Chopra and Eagle 2012). In islet research also, data included in many papers published in elegant journals cannot be reproduced. Many islet researchers are putting their names on papers written by their students, colleagues, and friends with minimal intellectual contributions.

It is possible that the altered ecosystem of islet research is supporting the proliferation of a group of islet researchers who are aggressive bullies, and academic psychopaths, and it is leading to the extinction of the finest islet researchers, who are genuinely talented and sincere, but are unable to survive in the ecosystem which is perceived as unsupportive and hostile.

Final Remarks

There is no take home message in this article. I have been partially able to write part of what I have thought, and if you have read this, then I have perhaps been able to transfer my thoughts to you.

Md. Shahidul Islam, M.D., Ph.D.
Karolinska Institutet
Department of Clinical Sciences
and Education
Stockholm
Uppsala University Hospital
Uppsala, Sweden

References

Bohannon J (2013) Who's afraid of peer review? Science 342:60–65
Chopra V, Eagle KA (2012) Perioperative mischief: the price of academic misconduct. Am J Med 125:953–955
Garwood J (2011) The heart of research is sick. Lab Times 2:24–31
Hvistendahl M (2013) China's publication bazaar. Science 342:1035–1039
Normile D (2009) Scientific misconduct. Hwang convicted but dodges jail; stem cell research has moved on. Science 326:650–651

Contents

Volume 1

1. The Comparative Anatomy of Islets 1
   R. Scott Heller

2. Microscopic Anatomy of the Human Islet of Langerhans 19
   Peter In't Veld and Silke Smeets

3. Basement Membrane in Pancreatic Islet Function 39
   Eckhard Lammert and Martin Kragl

4. Approaches for Imaging Pancreatic Islets: Recent Advances
   and Future Prospects 59
   Xavier Montet, Smaragda Lamprianou, Laurent Vinet, Paolo Meda,
   and Alfredo Fort

5. Mouse Islet Isolation 83
   Simona Marzorati and Miriam Ramirez-Dominguez

6. Regulation of Pancreatic Islet Formation 109
   Manuel Carrasco, Anabel Rojas, Irene Delgado,
   Nadia Cobo Vuilleumier, Juan R. Tejedo, Francisco J. Bedoya,
   Benoit R. Gauthier, Bernat Soria, and Franz Martín

7. (Dys)Regulation of Insulin Secretion by Macronutrients 129
   Philip Newsholme, Kevin Keane, Celine Gaudel, and
   Neville McClenaghan

8. Physiology and Pathology of the Anomeric Specificity for the
   Glucose-Induced Secretory Response of Insulin-, Glucagon-,
   and Somatostatin-Producing Pancreatic Islet Cells 157
   Willy J. Malaisse

9. Physiological and Pathophysiological Control of Glucagon
   Secretion by Pancreatic α-Cells 175
   Patrick Gilon, Rui Cheng-Xue, Bao Khanh Lai, Hee-Young Chae,
   and Ana Gómez-Ruiz

10	Electrophysiology of Islet Cells	249
	Gisela Drews, Peter Krippeit-Drews, and Martina Düfer	
11	ATP-Sensitive Potassium Channels in Health and Disease	305
	Peter Proks and Rebecca Clark	
12	β Cell Store-Operated Ion Channels	337
	Colin A. Leech, Richard F. Kopp, Louis H. Philipson, and Michael W. Roe	
13	Anionic Transporters and Channels in Pancreatic Islet Cells	369
	Nurdan Bulur and Willy J. Malaisse	
14	Chloride Channels and Transporters in β-Cell Physiology	401
	Mauricio Di Fulvio, Peter D. Brown, and Lydia Aguilar-Bryan	
15	Electrical, Calcium, and Metabolic Oscillations in Pancreatic Islets	453
	Richard Bertram, Arthur Sherman, and Leslie S. Satin	
16	Exocytosis in Islet β-Cells	475
	Haruo Kasai, Hiroyasu Hatakeyama, Mitsuyo Ohno, and Noriko Takahashi	
17	Zinc Transporters in the Endocrine Pancreas	511
	Mariea Dencey Bosco, Chris Drogemuller, Peter Zalewski, and Patrick Toby Coates	
18	High-Fat Programming of β-Cell Dysfunction	529
	Marlon E. Cerf	
19	Exercise-Induced Pancreatic Islet Adaptations in Health and Disease	547
	Sabrina Grassiolli, Antonio Carlos Boschero, Everardo Magalhães Carneiro, and Cláudio Cesar Zoppi	
20	Molecular Basis of cAMP Signaling in Pancreatic β Cells	565
	George G. Holz, Oleg G. Chepurny, Colin A. Leech, Woo-Jin Song, and Mehboob A. Hussain	
21	Calcium Signaling in the Islets	605
	Md. Shahidul Islam	
22	Role of Mitochondria in β-Cell Function and Dysfunction	633
	Pierre Maechler, Ning Li, Marina Casimir, Laurène Vetterli, Francesca Frigerio, and Thierry Brun	
23	IGF-1 and Insulin-Receptor Signalling in Insulin-Secreting Cells: From Function to Survival	659
	Susanne Ullrich	

24	Circadian Control of Islet Function	687
	Jeongkyung Lee, Mousumi Moulik, and Vijay K. Yechoor	

Volume 2

25	Wnt Signaling in Pancreatic Islets	707
	Joel F. Habener and Zhengyu Liu	
26	Islet Structure and Function in the GK Rat	743
	Bernard Portha, Grégory Lacraz, Audrey Chavey, Florence Figeac, Magali Fradet, Cécile Tourrel-Cuzin, Françoise Homo-Delarche, Marie-Héléne Giroix, Danièle Bailbé, Marie-Noëlle Gangnerau, and Jamileh Movassat	
27	β-Cell Function in Obese-Hyperglycemic Mice (*ob/ob* Mice)	767
	Per Lindström	
28	Role of Reproductive Hormones in Islet Adaptation to Metabolic Stress	785
	Ana Isabel Alvarez-Mercado, Guadalupe Navarro, and Franck Mauvais-Jarvis	
29	The β-Cell in Human Type 2 Diabetes	801
	Lorella Marselli, Mara Suleiman, Farooq Syed, Franco Filipponi, Ugo Boggi, Piero Marchetti, and Marco Bugliani	
30	Pancreatic β Cells in Metabolic Syndrome	817
	Marcia Hiriart, Myrian Velasco, Carlos Manlio Diaz-Garcia, Carlos Larqué, Carmen Sánchez-Soto, Alondra Albarado-Ibañez, Juan Pablo Chávez-Maldonado, Alicia Toledo, and Neivys García-Delgado	
31	Apoptosis in Pancreatic β-Cells in Type 1 and Type 2 Diabetes	845
	Tatsuo Tomita	
32	Mechanisms of Pancreatic β-Cell Apoptosis in Diabetes and Its Therapies	873
	James D. Johnson, Yu Hsuan Carol Yang, and Dan S. Luciani	
33	Clinical Approaches to Preserving β-Cell Function in Diabetes	895
	Bernardo Léo Wajchenberg and Rodrigo Mendes de Carvalho	
34	Role of NADPH Oxidase in β Cell Dysfunction	923
	Jessica R. Weaver and David A. Taylor-Fishwick	
35	The Contribution of Reg Family Proteins to Cell Growth and Survival in Pancreatic Islets	955
	Qing Li, Xiaoquan Xiong, and Jun-Li Liu	

36	Inflammatory Pathways Linked to β Cell Demise in Diabetes 989
	Yumi Imai, Margaret A. Morris, Anca D. Dobrian, David A. Taylor-Fishwick, and Jerry L. Nadler
37	Immunology of β-Cell Destruction 1047
	Åke Lernmark and Daria LaTorre
38	Current Approaches and Future Prospects for the Prevention of β-Cell Destruction in Autoimmune Diabetes 1081
	Carani B. Sanjeevi and Chengjun Sun
39	In Vivo Biomarkers for Detection of β Cell Death 1115
	Simon A. Hinke
40	Proteomics and Islet Research 1131
	Meftun Ahmed
41	Advances in Clinical Islet Isolation 1165
	Andrew R. Pepper, Boris Gala-Lopez, and Tatsuya Kin
42	Islet Isolation from Pancreatitis Pancreas for Islet Autotransplantation 1199
	A. N. Balamurugan, Gopalakrishnan Loganathan, Amber Lockridge, Sajjad M. Soltani, Joshua J. Wilhelm, Gregory J. Beilman, Bernhard J. Hering, and David E. R. Sutherland
43	Human Islet Autotransplantation 1229
	Martin Hermann, Raimund Margreiter, and Paul Hengster
44	Successes and Disappointments with Clinical Islet Transplantation 1245
	Paolo Cravedi, Piero Ruggenenti, and Giuseppe Remuzzi
45	Islet Xenotransplantation: An Update on Recent Advances and Future Prospects 1275
	Rahul Krishnan, Morgan Lamb, Michael Alexander, David Chapman, David Imagawa, and Jonathan R. T. Lakey
46	Islet Encapsulation 1297
	Jonathan R. T. Lakey, Lourdes Robles, Morgan Lamb, Rahul Krishnan, Michael Alexander, Elliot Botvinick, and Clarence E. Foster
47	Stem Cells in Pancreatic Islets 1311
	Erdal Karaöz and Gokhan Duruksu

48	**Generating Pancreatic Endocrine Cells from Pluripotent Stem Cells**...	1335
	Blair K. Gage, Rhonda D. Wideman, and Timothy J. Kieffer	
49	**Pancreatic Neuroendocrine Tumors**.......................	1375
	Apostolos Tsolakis and George Kanakis	
	Index...	1407

Contributors

Lydia Aguilar-Bryan Pacific Northwest Diabetes Research Institute, Seattle, WA, USA

Meftun Ahmed Department of Internal Medicine, Uppsala University Hospital, Uppsala, Sweden

Department of Physiology, Ibrahim Medical College, University of Dhaka, Dhaka, Bangladesh

Alondra Albarado-Ibañez Department of Neurodevelopment and Physiology, Instituto de Fisiología Celular, Universidad Nacional Autónoma de México, Mexico DF, Mexico

Michael Alexander Department of Surgery, University of California Irvine, Orange, CA, USA

Ana Isabel Alvarez-Mercado Division of Endocrinology and Metabolism, Tulane University Health Sciences Center, School of Medicine, New Orleans, LA, USA

Danièle Bailbé Laboratoire B2PE, Unité BFA, Université Paris-Diderot et CNRS EAC4413, Paris Cedex13, France

A. N. Balamurugan Islet Cell Laboratory, Cardiovascular Innovation Institute, Department of Surgery, University of Louisville, Louisville, KY, USA

Francisco J. Bedoya CIBERDEM, Barcelona, Spain

Andalusian Center of Molecular Biology and Regenerative Medicine (CABIMER), Seville, Andalucía, Spain

Gregory J. Beilman Department of Surgery, Schulze Diabetes Institute, University of Minnesota, Minneapolis, MN, USA

Richard Bertram Department of Mathematics, Florida State University, Tallahassee, FL, USA

Ugo Boggi Department of Translational Research and New Technologies, University of Pisa, Pisa, Italy

Antonio Carlos Boschero Department of Structural and Functional Biology, State University of Campinas, Campinas, Sao Paulo, Brazil

Mariea Dencey Bosco Basil Hetzel Institute at The Queen Elizabeth Hospital, Centre for Clinical and Experimental Transplantation Laboratory, Discipline of Medicine, University of Adelaide, Adelaide, SA, Australia

Elliot Botvinick Department of Surgery, Biomedical Engineering, University of California Irvine, Orange, CA, USA

Peter D. Brown Faculty of Life Sciences, Manchester University, Manchester, UK

Thierry Brun Department of Cell Physiology and Metabolism, University of Geneva Medical Centre, Geneva Switzerland

Marco Bugliani Department of Clinical and Experimental Medicine, Pancreatic Islet Laboratory, University of Pisa, Pisa, Italy

Nurdan Bulur Laboratory of Experimental Medicine, Université Libre de Bruxelles, Brussels, Belgium

Everardo Magalhães Carneiro Department of Structural and Functional Biology, State University of Campinas, Campinas, Sao Paulo, Brazil

Manuel Carrasco CIBERDEM, Barcelona, Spain

Andalusian Center of Molecular Biology and Regenerative Medicine (CABIMER), Seville, Andalucía, Spain

Marina Casimir Department of Cell Physiology and Metabolism, University of Geneva Medical Centre, Geneva Switzerland

Marlon E. Cerf Diabetes Discovery Platform, Medical Research Council, Cape Town, South Africa

Hee-Young Chae Institut de Recherche Expérimentale et Clinique, Université Catholique de Louvain, Pôle d'Endocrinologie, Diabète et Nutrition (EDIN), Brussels, Belgium

David Chapman Department of Experimental Surgery/Oncology, University of Alberta, Edmonton, AB, Canada

Audrey Chavey Laboratoire B2PE, Unité BFA, Université Paris-Diderot et CNRS EAC4413, Paris Cedex13, France

Juan Pablo Chávez-Maldonado Department of Neurodevelopment and Physiology, Instituto de Fisiología Celular, Universidad Nacional Autónoma de México, Mexico DF, Mexico

Rui Cheng-Xue Institut de recherche expérimentale et clinique, Université Catholique de Louvain, Pôle d'Endocrinologie, Diabète et Nutrition (EDIN), Brussels, Belgium

Oleg G. Chepurny Department of Medicine, SUNY Upstate Medical University, Syracuse, NY, USA

Rebecca Clark Department of Physiology, Anatomy and Genetics, University of Oxford, Oxford, UK

Patrick Toby Coates Centre for Clinical and Experimental Transplantation (CCET), University of Adelaide, Royal Adelaide Hospital, Australian Islet Consortium, Adelaide, SA, Australia

Paolo Cravedi IRCCS – Istituto di Ricerche Farmacologiche Mario Negri, Bergamo, Italy

Rodrigo Mendes de Carvalho Clinical Endocrinologist and Diabetologist, Rio de Janeiro, Brazil

Irene Delgado CIBERDEM, Barcelona, Spain

Andalusian Center of Molecular Biology and Regenerative Medicine (CABIMER), Seville, Andalucía, Spain

Mauricio Di Fulvio Pharmacology and Toxicology, Boonshoft School of Medicine, Wright State University, Dayton, OH, USA

Carlos Manlio Diaz-Garcia Department of Neurodevelopment and Physiology, Instituto de Fisiología Celular, Universidad Nacional Autónoma de México, Mexico DF, Mexico

Anca D. Dobrian Department of Physiological Sciences, Eastern Virginia Medical School, Norfolk, VA, USA

Gisela Drews Department of Pharmacology, Toxicology and Clinical Pharmacy, Institute of Pharmacy, University of Tübingen, Tübingen, Germany

Chris Drogemuller Centre for Clinical and Experimental Transplantation (CCET), University of Adelaide, Royal Adelaide Hospital, Australian Islet Consortium, Adelaide, SA, Australia

Martina Düfer Department of Pharmacology, Institute of Pharmaceutical and Medical Chemistry, University of Münster, Münster, Germany

Gokhan Duruksu Kocaeli University, Center for Stem Cell and Gene Therapies Research and Practice, Institute of Health Sciences, Stem Cell Department, Kocaeli, Turkey

Florence Figeac Laboratoire B2PE, Unité BFA, Université Paris-Diderot et CNRS EAC4413, Paris Cedex13, France

Franco Filipponi Department of Surgery, University of Pisa, Pisa, Italy

Alfredo Fort Department of Cell Physiology and Metabolism, Geneva University, Switzerland

Clarence E. Foster Department of Surgery, Biomedical Engineering, University of California Irvine, Orange, CA, USA

Department of Transplantation, University of California Irvine, Orange, CA, USA

Magali Fradet Laboratoire B2PE, Unité BFA, Université Paris-Diderot et CNRS EAC4413, Paris Cedex 13, France

Francesca Frigerio Department of Cell Physiology and Metabolism, University of Geneva Medical Centre, Geneva Switzerland

Blair K. Gage Department of Cellular and Physiological Sciences, Laboratory of Molecular and Cellular Medicine, University of British Columbia Vancouver, Vancouver, BC, Canada

Boris Gala-Lopez Clinical Islet Transplant Program, University of Alberta, Edmonton, AB, Canada

Marie-Noëlle Gangnerau Laboratoire B2PE, Unité BFA, Université Paris-Diderot et CNRS EAC4413, Paris Cedex 13, France

Neivys García-Delgado Department of Neurodevelopment and Physiology, Instituto de Fisiología Celular, Universidad Nacional Autónoma de México, Mexico DF, Mexico

Celine Gaudel INSERM U1065, Centre Mediterraneen de Medicine Moleculaire, C3M, Batiment Archimed, Nice, Cedex 2, France

Benoit R. Gauthier Andalusian Center of Molecular Biology and Regenerative Medicine (CABIMER), Seville, Spain

Patrick Gilon Institut de Recherche Expérimentale et Clinique, Université Catholique de Louvain, Pôle d'Endocrinologie, Diabète et Nutrition (EDIN), Brussels, Belgium

Marie-Héléne Giroix Laboratoire B2PE, Unité BFA, Université Paris-Diderot et CNRS EAC4413, Paris Cedex 13, France

Ana Gómez-Ruiz Institut de Recherche Expérimentale et Clinique, Université Catholique de Louvain, Pôle d'Endocrinologie, Diabète et Nutrition (EDIN), Brussels, Belgium

Sabrina Grassiolli Department of General Biology, State University of Ponta Grossa, Ponta Grossa, Brazil

Joel F. Habener Laboratory of Molecular Endocrinology, Massachusetts General Hospital and Harvard Medical School, Boston, MA, USA

Hiroyasu Hatakeyama Faculty of Medicine, Laboratory of Structural Physiology, Center for Disease Biology and Integrative, The University of Tokyo, Hongo, Tokyo, Japan

R. Scott Heller Histology and Imaging Department, Novo Nordisk, Måløv, Denmark

Paul Hengster Daniel Swarovski Laboratory, Department of Visceral-, Transplant- and Thoracic Surgery, Center of Operative Medicine, Innsbruck Medical University, Innsbruck, Austria

Bernhard J. Hering Department of Surgery, Schulze Diabetes Institute, University of Minnesota, Minneapolis, MN, USA

Martin Hermann Department of Anaesthesiology and Critical Care Medicine, Medical University of Innsbruck, Innsbruck, Austria

Simon A. Hinke Department of Pharmacology, University of Washington, Seattle, WA, USA

Current Address: Janssen Research & Development, Spring House, PA, USA

Marcia Hiriart Department of Neurodevelopment and Physiology, Instituto de Fisiología Celular, Universidad Nacional Autónoma de México, Mexico DF, Mexico

George G. Holz Departments of Medicine and Pharmacology, SUNY Upstate Medical University, Syracuse, NY, USA

Françoise Homo-Delarche Laboratoire B2PE, Unité BFA, Université Paris-Diderot et CNRS EAC4413, Paris Cedex13, France

Mehboob A. Hussain Departments of Pediatrics, Medicine, and Biological Chemistry, Johns Hopkins University School of Medicine, Baltimore, MD, USA

David Imagawa Department of Surgery, University of California Irvine, Orange, CA, USA

Yumi Imai Department of Internal Medicine, Eastern Virginia Medical School, Strelitz Diabetes Center, Norfolk, VA, USA

Peter In't Veld Department of Pathology, Vrije Universiteit Brussel, Brussels, Belgium

Md. Shahidul Islam Department of Clinical Sciences and Education, Södersjukhuset, Karolinska Institutet, Stockholm, Sweden

Department of Internal Medicine, Uppsala University Hospital, Uppsala, Sweden

James D. Johnson Diabetes Research Group, Department of Cellular and Physiological Sciences, University of British Columbia, Vancouver, BC, Canada

George Kanakis Department of Pathophysiology, University of Athens Medical School, Athens, Greece

Erdal Karaöz Kocaeli University, Center for Stem Cell and Gene Therapies Research and Practice, Institute of Health Sciences, Stem Cell Department, Kocaeli, Turkey

Haruo Kasai Faculty of Medicine, Laboratory of Structural Physiology, Center for Disease Biology and Integrative, The University of Tokyo, Hongo, Tokyo, Japan

Kevin Keane School of Biomedical Sciences, CHIRI Biosciences, Curtin University, Perth, WA, Australia

Timothy J. Kieffer Department of Cellular and Physiological Sciences, Laboratory of Molecular and Cellular Medicine, University of British Columbia Vancouver, Vancouver, BC, Canada

Department of Surgery, Life Sciences Institute, University of British Columbia Vancouver, Vancouver, BC, Canada

Tatsuya Kin Clinical Islet Laboratory, University of Alberta, Edmonton, AB, Canada

Richard F. Kopp Department of Medicine, State University of New York Upstate Medical University, Syracuse, NY, USA

Martin Kragl Institute of Metabolic Physiology, Heinrich-Heine University Düsseldorf, Düsseldorf, Germany

Peter Krippeit-Drews Department of Pharmacology, Toxikology and Clinical Pharmacy, Institute of Pharmacy, University of Tübingen, Tübingen, Germany

Rahul Krishnan Department of Surgery, University of California Irvine, Orange, CA, USA

Grégory Lacraz Laboratoire B2PE, Unité BFA, Université Paris-Diderot et CNRS EAC4413, Paris Cedex 13, France

Bao Khanh Lai Institut de Recherche Expérimentale et Clinique, Université Catholique de Louvain, Pôle d'Endocrinologie, Diabète et Nutrition (EDIN), Brussels, Belgium

Jonathan R. T. Lakey Department of Surgery, Biomedical Engineering, University of California Irvine, Orange, CA, USA

Morgan Lamb Department of Surgery, University of California Irvine, Orange, CA, USA

Eckhard Lammert Institute of Metabolic Physiology, Heinrich-Heine University Düsseldorf, Düsseldorf, Germany

Smaragda Lamprianou Department of Cell Physiology and Metabolism, Geneva University, Switzerland

Carlos Larqué Department of Neurodevelopment and Physiology, Instituto de Fisiología Celular, Universidad Nacional Autónoma de México, Mexico DF, Mexico

Daria LaTorre Department of Clinical Sciences, Lund University, CRC, University Hospital MAS, Malmö, Sweden

Jeongkyung Lee Department of Medicine, Division of Diabetes, Baylor College of Medicine, Endocrinology and Metabolism, Houston, TX, USA

Colin A. Leech Department of Medicine, SUNY Upstate Medical University, Syracuse, NY, USA

Åke Lernmark Department of Clinical Sciences, Lund University, CRC, University Hospital MAS, Malmö, Sweden

Ning Li Department of Cell Physiology and Metabolism, University of Geneva Medical Centre, Geneva Switzerland

Qing Li Fraser Laboratories for Diabetes Research, Department of Medicine, McGill University Health Centre, Montreal, QC, Canada

Per Lindström Department of Integrative Medical Biology, Section for Histology and Cell Biology, Umeå University, Umeå, Sweden

Jun-Li Liu Fraser Laboratories for Diabetes Research, Department of Medicine, McGill University Health Centre, Montreal, QC, Canada

Zhengyu Liu US Biopharmaceutical Regulatory Affairs Group, Sandoz Inc., A Novartis Company, Princeton, NY, USA

Amber Lockridge Department of Surgery, Schulze Diabetes Institute, University of Minnesota, Minneapolis, MN, USA

Gopalakrishnan Loganathan Department of Surgery, Schulze Diabetes Institute, University of Minnesota, Minneapolis, MN, USA

Dan S. Luciani Diabetes Research Program, Child & Family Research Institute, University of British Columbia, Vancouver, BC, Canada

Pierre Maechler Department of Cell Physiology and Metabolism, University of Geneva Medical Centre, Geneva, Switzerland

Willy J. Malaisse Laboratory of Experimental Medicine, Université Libre de Bruxelles, Brussels, Belgium

Piero Marchetti Department of Clinical and Experimental Medicine, Pancreatic Islet Laboratory, University of Pisa, Pisa, Italy

Raimund Margreiter Daniel Swarovski Laboratory, Department of Visceral-, Transplant- and Thoracic Surgery, Center of Operative Medicine, Innsbruck Medical University, Innsbruck Austria

Lorella Marselli Department of Clinical and Experimental Medicine, Pancreatic Islet Laboratory, University of Pisa, Pisa, Italy

Franz Martín CIBERDEM, Barcelona, Spain

Andalusian Center of Molecular Biology and Regenerative Medicine (CABIMER), Seville, Andalucía, Spain

Simona Marzorati β Cell Biology Unit, S. Raffaele Scientific Institute, Diabetes Research Institute-DRI, Milan, Italy

Franck Mauvais-Jarvis Division of Endocrinology & Metabolism, Tulane University Health Sciences Center, School of Medicine, New Orleans, LA, USA

Division of Endocrinology, Metabolism and Molecular Medicine, Northwestern University, Feinberg School of Medicine, Chicago, IL, USA

Neville McClenaghan School of Biomedical Sciences, University of Ulster, Coleraine, Londonderry, Northern Ireland

Paolo Meda Department of Cell Physiology and Metabolism, Geneva University, Switzerland

Xavier Montet Division of Radiology, Geneva University Hospital, Switzerland

Margaret A. Morris Departments of Internal Medicine and Microbiology and Molecular Cell Biology, Eastern Virginia Medical School, Strelitz Diabetes Center, Norfolk, VA, USA

Mousumi Moulik Department of Pediatrics, Division of Pediatric Cardiology, University of Texas Health Sciences Center at Houston, Houston, TX, USA

Jamileh Movassat Laboratoire B2PE, Unité BFA, Université Paris-Diderot et CNRS EAC4413, Paris Cedex 13, France

Jerry L. Nadler Department of Internal Medicine, Eastern Virginia Medical School, Strelitz Diabetes Center, Norfolk, VA, USA

Guadalupe Navarro Division of Endocrinology, Metabolism and Molecular Medicine, Northwestern University, Feinberg School of Medicine, Chicago, IL, USA

Philip Newsholme School of Biomedical Sciences, CHIRI Biosciences, Curtin University, Perth, WA, Australia

Mitsuyo Ohno Faculty of Medicine, Laboratory of Structural Physiology, Center for Disease Biology and Integrative, The University of Tokyo, Hongo, Tokyo, Japan

Andrew R. Pepper Clinical Islet Transplant Program, University of Alberta, Edmonton, AB, Canada

Louis H. Philipson Department of Medicine, University of Chicago, Chicago, IL, USA

Bernard Portha Laboratoire B2PE, Unité BFA, Université Paris-Diderot et CNRS EAC4413, Paris Cedex 13, France

Peter Proks Department of Physiology, Anatomy and Genetics, University of Oxford, Oxford, UK

Miriam Ramirez-Dominguez Department of Pediatrics, Faculty of Medicine and Odontology, University of the Basque Country, UPV/EHU, University Hospital Cruces, Leioa, País Vasco, Spain

Giuseppe Remuzzi IRCCS – Istituto di Ricerche Farmacologiche Mario Negri, Bergamo, Italy

Unit of Nephrology, Azienda Ospedaliera Papa Giovanni XXIII, Bergamo, Italy

Lourdes Robles Department of Surgery, University of California Irvine, Orange, CA, USA

Michael W. Roe Department of Medicine, State University of New York Upstate Medical University, Syracuse, NY, USA

Anabel Rojas CIBERDEM, Barcelona, Spain

Andalusian Center of Molecular Biology and Regenerative Medicine (CABIMER), Seville, Andalucía, Spain

Piero Ruggenenti IRCCS – Istituto di Ricerche Farmacologiche Mario Negri, Bergamo, Italy

Unit of Nephrology, Azienda Ospedaliera Papa Giovanni XXIII, Bergamo, Italy

Carmen Sánchez-Soto Department of Neurodevelopment and Physiology, Instituto de Fisiología Celular, Universidad Nacional Autónoma de México, Mexico DF, Mexico

Carani B. Sanjeevi Department of Medicine (Solna), Center for Molecular Medicine, Karolinska University Hospital, Solna, Stockholm, Sweden

Leslie S. Satin Department of Pharmacology and Brehm Diabetes Center, University of Michigan Medical School, Ann Arbor, MI, USA

Arthur Sherman Laboratory of Biological Modeling, National Institutes of Health, Bethesda, MD, USA

Silke Smeets Department of Pathology, Vrije Universiteit Brussel, Brussels, Belgium

Sajjad M. Soltani Department of Surgery, Schulze Diabetes Institute, University of Minnesota, Minneapolis, MN, USA

Woo-Jin Song Department of Pediatrics, Johns Hopkins University School of Medicine, Baltimore, MD, USA

Bernat Soria CIBERDEM, Barcelona, Spain

Andalusian Center of Molecular Biology and Regenerative Medicine (CABIMER), Seville, Andalucía, Spain

Mara Suleiman Department of Clinical and Experimental Medicine, Pancreatic Islet Laboratory, University of Pisa, Pisa, Italy

Chengjun Sun Department of Medicine (Solna), Center for Molecular Medicine, Karolinska University Hospital, Solna, Stockholm, Sweden

David E. R. Sutherland Department of Surgery, Schulze Diabetes Institute, University of Minnesota, Minneapolis, MN, USA

Farooq Syed Department of Clinical and Experimental Medicine, Pancreatic Islet Laboratory, University of Pisa, Pisa, Italy

Noriko Takahashi Faculty of Medicine, Laboratory of Structural Physiology, Center for Disease Biology and Integrative, The University of Tokyo, Hongo, Tokyo, Japan

David A. Taylor-Fishwick Department of Microbiology and Molecular Cell Biology, Department of Medicine, Eastern Virginia Medical School, Strelitz Diabetes Center, Norfolk, VA, USA

Juan R. Tejedo CIBERDEM, Barcelona, Spain

Andalusian Center of Molecular Biology and Regenerative Medicine (CABIMER), Seville, Andalucía, Spain

Alicia Toledo Department of Neurodevelopment and Physiology, Instituto de Fisiología Celular, Universidad Nacional Autónoma de México, Mexico DF, Mexico

Tatsuo Tomita Departments of Integrative Biosciences and Pathology and Oregon National Primate Center, Oregon Health and Science University, Portland, OR, USA

Cécile Tourrel-Cuzin Laboratoire B2PE, Unité BFA, Université Paris-Diderot et CNRS EAC4413, Paris Cedex13, France

Apostolos Tsolakis Department of Medical Sciences, Section of Endocrine Oncology, Uppsala University, Uppsala, Sweden

Susanne Ullrich Department of Internal Medicine, Clinical Chemistry and Institute for Diabetes Research and Metabolic Diseases of the Helmholtz Center Munich, University of Tübingen, Tübingen, Germany

Myrian Velasco Department of Neurodevelopment and Physiology, Instituto de Fisiología Celular, Universidad Nacional Autónoma de México, Mexico DF, Mexico

Laurène Vetterli Department of Cell Physiology and Metabolism, University of Geneva Medical Centre, Geneva, Switzerland

Laurent Vinet Department of Cell Physiology and Metabolism, Geneva University, Switzerland

Nadia Cobo Vuilleumier Andalusian Center of Molecular Biology and Regenerative Medicine (CABIMER), Sevilla, Spain

Bernardo Léo Wajchenberg Endocrine Service and Diabetes and Heart Center of the Heart Institute, Hospital, Clinicas of The University of São Paulo Medical School, São Paulo, Brazil

Jessica R. Weaver Department of Microbiology and Molecular Cell Biology, Eastern Virginia Medical School, Norfolk, VA, USA

Rhonda D. Wideman Department of Cellular and Physiological Sciences, Laboratory of Molecular and Cellular Medicine, University of British Columbia Vancouver, Vancouver, BC, Canada

Joshua J. Wilhelm Department of Surgery, Schulze Diabetes Institute, University of Minnesota, Minneapolis, MN, USA

Xiaoquan Xiong Fraser Laboratories for Diabetes Research, Department of Medicine, McGill University Health Centre, Montreal, QC, Canada

Yu Hsuan Carol Yang Diabetes Research Group, Department of Cellular and Physiological Sciences, University of British Columbia, Vancouver, BC, Canada

Vijay K. Yechoor Department of Medicine, Division of Diabetes, Baylor College of Medicine, Endocrinology and Metabolism, Houston, TX, USA

Peter Zalewski Department of Medicine, Basil Hetzel Institute at the Queen Elizabeth Hospital, University of Adelaide, Adelaide, SA, Australia

Cláudio Cesar Zoppi Department of Structural and Functional Biology, State University of Campinas, Campinas, Sao Paulo, Brazil

Wnt Signaling in Pancreatic Islets

25

Joel F. Habener and Zhengyu Liu

Contents

The Diabetes Problem	708
Wnt Signaling Pathways	710
The Canonical Wnt Signaling Pathway	711
Noncanonical Wnt Signaling	713
Wnt Signaling in Pancreas Development and Regeneration	713
Role of Wnt Signaling in β Cell Growth and Survival	716
Roles of Non-Wnt Hormonal Ligands in the Activation of the Wnt Signaling Pathway in Islets	716
Downstream Wnt Signaling Requirement for GLP-1-Induced Stimulation of β Cell Proliferation	717
Downstream Wnt Signaling Requirement for SDF-1-Induced Promotion of β Cell Survival	718
Potential Mechanisms by Which GLP-1 and SDF-1 May Act Cooperatively on Wnt Signaling to Enhance β Cell Growth and Survival	720
Type 2 Diabetes Genes	721
Genes Associated with Islet Development/Function and Wnt Signaling	721
Future Directions	732
References	733

Abstract

The Wnt signaling pathway is critically important not only for stem cell amplification, but also for the differentiation and migration and for organogenesis and the development of the body plan. β-catenin/TCF7L2-dependent

---

J.F. Habener (✉)
Laboratory of Molecular Endocrinology, Massachusetts General Hospital and Harvard Medical School, Boston, MA, USA
e-mail: jhabener@partners.org

Z. Liu
US Biopharmaceutical Regulatory Affairs Group, Sandoz Inc., A Novartis Company, Princeton, NY, USA
e-mail: zhengyu.liu@sandoz.com

Wnt signaling (the canonical pathway) is involved in pancreas development, islet function, and insulin production and secretion. The glucoincretin hormone glucagon-like peptide-1 and the chemokine stromal cell-derived factor-1 modulate canonical Wnt signaling in β cells which is obligatory for their mitogenic and cytoprotective actions.

Genome-wide association studies have uncovered approximately 90 gene loci that confer susceptibility for the development of type 2 diabetes (Marchetti P, Syed F, Suleiman M, Bugliani M, Marselli L, Islets 4:323–332, 2012). The majority of these diabetes risk alleles encode proteins that are implicated in islet growth and functioning (Marchetti P, Syed F, Suleiman M, Bugliani M, Marselli L, Islets 4:323–332, 2012, Ahlqvist E, Ahluwalia TS, Groop L, Clin Chem 57:241–254, 2011). At least 20 of the type 2 diabetes genes that affect islet functions are either components of or known target genes for Wnt signaling. The transcription factor TCF7L2 is particularly strongly associated with risk for diabetes and appears to be fundamentally important in both canonical Wnt signaling and β cell functioning. Experimental loss of TCF7L2 function in islets and polymorphisms in TCF7L2 alleles in humans impair glucose-stimulated insulin secretion suggesting that perturbations in the Wnt signaling pathway may contribute substantially to the susceptibility for, and pathogenesis of, type 2 diabetes. This review focuses on considerations of the hormonal regulation of Wnt signaling in islets and implications for mutations in components of the Wnt signaling pathway as a source for risk alleles for type 2 diabetes.

Keywords

β cell regeneration • Cell proliferation • Cytoprotection • Glucagon-like peptide-1 • Stromal cell-derived factor-1 • Tcf7L2

The Diabetes Problem

The prevalence of diabetes mellitus and its accompanying complications is increasing in populations throughout the world (American Diabetes Association, http://www.diabetes.org/about-diabetes.jsp). Diabetes results from a deficiency of the β cells of the islets of Langerhans to produce insulin in amounts sufficient to meet the body's needs, either absolute deficiency (type 1 diabetes) or relative deficiency (type 2 diabetes). In both forms of diabetes, the remaining β cells are placed under stress by (1) being forced to overproduce insulin to compensate for the lost β cells, (2) insulin resistance, and (3) by the glucotoxic effects of prolonged, sustained hyperglycemia. In the USA, 20 million individuals are currently afflicted with some form of diabetes, while an estimated 12 million additional people in the USA have diabetes but do not know it yet (Juvenile Diabetes Research Foundation, http://www.jdrf.org/index.cfm?fuseaction=home.viewPage&page_id=71927021-99EA-4D04-92E8463E607C84E1). Worldwide, an estimated 347 million people have the disease, and this global figure is expected to continue to increase in epidemic proportions (World Health Organization fact sheet #312, March 2013; http://who.int/topics/diabetes_mellitus/en/). Type 2 diabetes is the most prevalent form of diabetes comprising >90 % of all diabetes. Most individuals who develop type 2 diabetes do so in association with obesity (Jin and Patti 2009). Because a common feature of both type 1 and type 2 diabetes is a reduction in β cell mass, understanding the factors and the cellular mechanisms that govern β cell growth and survival may lead to new effective treatments for diabetes.

In adult rats and mice the entire mass of the β cells in the pancreas turns over approximately every 50 days (2–3 % per day) by processes of apoptosis counterbalanced by replication from existing β cells and neogenesis from progenitor cells believed to be located in the pancreatic ducts and possibly within the islets (Bonner-Weir and Sharma 2006). The adult pancreas of rodents, including the endocrine islets, has a substantial capacity for regeneration (Jensen et al. 2005). Rodent models of pancreatic injuries are followed by partial to nearly complete regeneration of the exocrine and endocrine pancreas. Such models of pancreas regeneration include partial pancreatectomy (Pauls and Bancroft 1950), streptozotocin-mediated ablation of the β cells (Cheta 1998; Rees and Alcolado 2005), duct ligation, and caerulein treatments (Sakaguchi et al. 2006).

Whether or not endocrine stem/progenitor cells exist in the adult pancreas and can give rise to new β cells has been a topic of debate (Bonner-Weir and Sharma 2006; Dor et al. 2004). Lineage tracing studies in mice during pancreas regeneration following partial pancreatectomy found that the majority of new β cells derive from preexisting insulin-expressing cells without evidence of regeneration from stem/progenitor cells (Dor et al. 2004; Cano et al. 2008). Compelling new evidence, however, indicates that stem-like progenitor cells reside in the duct linings of adult mouse pancreas (Xu et al. 2008; Collombat et al. 2009; Al-Hasani et al. 2013; Courtney et al. in press). Pancreas injury induced by partial duct ligation promotes the appearance and the expansion of a population of cells within the duct lining that express the transcription factor neurogenin-3 (Ngn3), a modulator of Wnt signaling (Gradwohl et al. 2000) and a major determinant of endocrine lineage commitment during pancreas development (Serafimidis et al. 2008). In adult mouse models with manipulations of the functions of the transcription factors Pax4 (paired homeobox-4) and Arx (Aristalis homeobox), the Ngn3-expressing cells differentiate into α cells and then transdifferentiate into β cells in response to metabolic stress and/or injuries of β cells (Collombat et al. 2009; Al-Hasani et al. 2013; Courtney et al. in press). Moreover, the β cells of mice undergoing metabolic stresses, such as glucotoxicity, dedifferentiate into Ngn3-expressing cells, many of which differentiate into α cells (Talchai et al. 2012).

Genome-wide scans of several large populations of diabetic cohorts have uncovered 70 or more of the genes associated with type 2 diabetes (Lyssenko et al. 2008; Florez 2008a; Van Hoek et al. 2008; Florez 2008b; Ahlqvist et al. 2011; Marchetti et al. 2012). Of note, the majority of the diabetes genes identified thus far appear to be involved in islet development and functions and, most notably, the functions of the insulin-producing β cells in the islets (Ahlqvist et al. 2011; Marchetti et al. 2012). Furthermore, as discussed later in this chapter, several of these genes appear to be involved in the Wnt signaling pathway, either components of the Wnt signaling system itself or target genes for downstream Wnt signaling by β-catenin and TCF7L2. The Wnt signaling pathway may be involved in the dysfunction of β cells in type 2 diabetes (Lee et al. 2008). Attention is directed to recent reviews on the role of Wnt signaling in pancreas development and function (Welters and Kulkarni 2008; Murtaugh 2008; Jin 2008) and the importance of the transcription factor TCF7L2 in pancreatic islet function and diabetes (Marchetti et al. 2012; Lyssenko 2008; Jin and Liu 2008; Perry and Frayling 2008; Cauchi and Froguel 2008; Bordonaro 2009; Hattersley 2007; Weedon 2007; Chiang et al. 2012; Florez 2007; Xiong et al. 2012; Owen and McCarthy 2007a; Smith 2007). In this review evidence is considered for the regulation of islet β cell functions by β-catenin/TCF7L2 induced by glucagon-like peptide-1 and stromal cell-derived factor-1, and speculations are presented on the potential involvement of the Wnt signaling pathway in the genetic predisposition to type 2 diabetes.

Wnt Signaling Pathways

The Wnt signaling cascade controls several cellular functions, including differentiation, proliferation, and migration (Kikuchi et al. 2006; Willert and Jones 2006; Moon et al. 2004; Nelson and Nusse 2004; Logan and Nusse 2004; Gordon and Nusse 2006; Nusse 2008). Useful brief summaries of the Wnt signaling pathways are provided in references MacDonald et al. 2007 and Semenov et al. 2007. The Wnt proteins form a large family of cell-secreted factors that control diverse aspects of development and organogenesis. Wnt proteins exert their effect by binding to cell-surface G-protein-coupled Frizzled (Fz) receptors and the lipoprotein receptor-like proteins, LRP5/6 co-receptors, and modulate the expression of various target genes through a series of intracellular processes ultimately leading to the regulation of transcription. There are several recognized Wnt signaling pathways: the β-catenin dependent, so-called canonical Wnt pathway that is dependent on the activation of the transcriptional complex of proteins consisting of β-catenin and TCF/LEF (Fig. 1), and several (at least nine) distinct and complex β-catenin, TCF/LEF-independent, noncanonical pathways, two of which are illustrated (Logan and Nusse 2004 and Fig. 2). In addition to Wnt signaling pathways activated by Wnt interactions with frizzed receptors, certain hormones, such as glucagon-like peptide-1 (GLP-1) and stromal cell-derived factor-1 (CXCL12) that activate G-protein-coupled receptors, can activate downstream Wnt signaling pathways via unique mechanisms culminating in the formation of transactivating complexes consisting of β-catenin and the DNA-binding proteins of the TCF/LEF family (see below) (Xiong et al. 2012; Liu and Habener 2008, 2009).

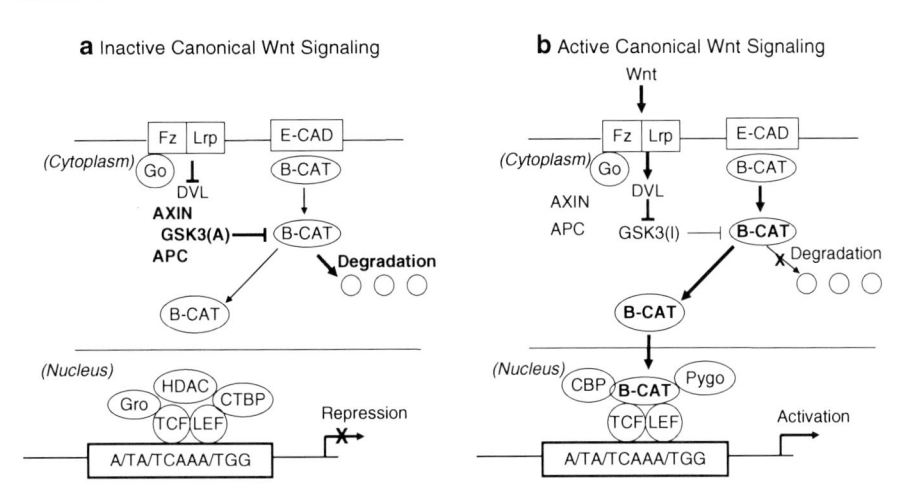

Fig. 1 Models depicting the canonical, β-catenin/TCF/LEF-dependent Wnt signaling pathway in inactive and active states. (**a**). Inactive Wnt signaling. In the absence of Wnt ligand-mediated activation of its receptor Frizzled (Fz), β-catenin in the cytoplasm is phosphorylated by the protein kinases glycogen synthase kinase-3β (GSK3β) and casein kinase 1 α (CK1α) leading to its degradation by proteosome complexes. GSK3β and CK1α are constitutively activated by the cofactors adenomatous polyposis coli (APC) and axin, which along with GSK3β and CK1α are known as the destruction complex. In the absence of sufficient levels of cytosolic β-catenin, nuclear levels are depleted and the DNA-binding transcription factors TCF and LEF act as repressors of gene transcription by the recruitment of corepressors such as Groucho and CtBP. (**b**). Active Wnt signaling. In the presence of Wnt ligands, Fz is activated via G-protein Gαi/o and small GTPases leading to the activation of disheveled (DVL) that disrupts the destruction complex composed of GSK3, CK1, APC, and axin, thereby inhibiting the activities of GSK3 and CK1. In the absence of phosphorylation, unphosphorylated β-catenin is stabilized and translocated to the nucleus where it non-covalently associates with TCF/LEF DNA-binding proteins and recruits co-activators such as CBP and Pygo resulting in the activation of gene transcription

The Canonical Wnt Signaling Pathway

The downstream canonical Wnt signaling pathway is defined as the pathway that ends in the formation of active, productive transcriptional, transactivation complexes composed of β-catenin and the DNA-binding proteins TCF (T-cell factor) and LEF (lymphocyte enhancer factor) (Fig. 1). It involves β-catenin that when stabilized translocates to the nucleus where it associates with the TCF/LEF family of transcription factors to regulate the expression of canonical Wnt target genes. In the absence of a Wnt signal, β-catenin is efficiently captured by the scaffold protein axin, which is present within a protein complex (referred to as the destruction complex) that also harbors adenomatous polyposis coli (APC), glycogen synthase kinase (GSK)-3, and casein kinase 1 (CSNK1) (Fig. 1a). The resident CSNK1 and GSK3 protein kinases sequentially phosphorylate conserved serine and threonine residues in the N-terminus of β-catenin subsequently targeting it for ubiquitination and degradation. The efficient suppression of β-catenin levels

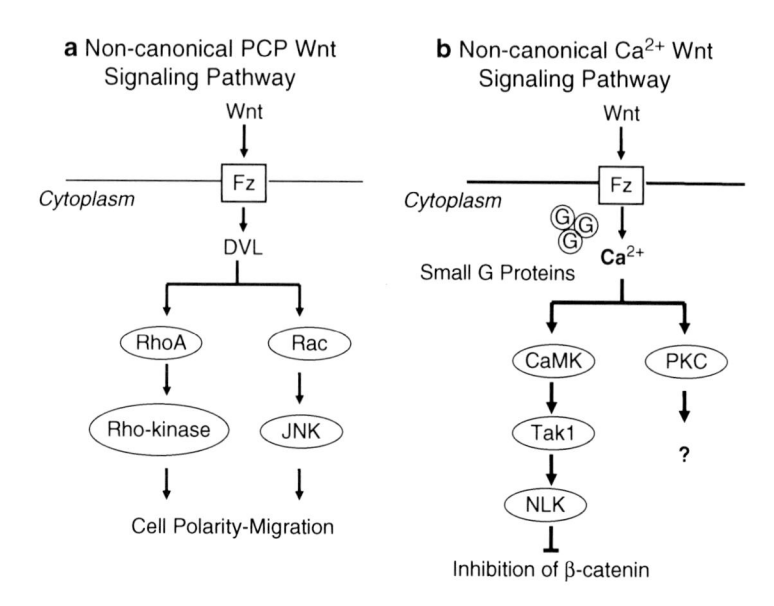

Fig. 2 Two models depicting noncanonical β-catenin-independent Wnt signaling pathways. (**a**). The planar cell polarity (PCP) pathway. The activation of Fz by Wnts leads to the activation of DVL and small G-proteins such as RhoA and Rac and the kinases Rho-kinase and Jun kinase (JNK). Through as yet undefined pathways, Rho-kinase and JNK modulate changes in the cytoskeleton involved in cell migration and polarity. (**b**). The Ca^{2+} pathway. Wnt ligands such as Wnt 5a activate Ca^{2+}-activated calmodulin kinases. CaMK and downstream kinases TAK1 and NLK. This pathway inhibits the canonical β-catenin-dependent Wnt signaling pathway and is active during gastrulation. The Ca^{2+} pathway also activates protein kinase C (PKC)

ensures that Groucho (TLE) proteins are free to bind members of the lymphocyte enhancer factor (LEF)/T-cell factor (TCF) family of transcription factors occupying the promoters and enhancers of Wnt target genes in the nucleus. These transcriptionally repressive complexes actively suppress the Wnt target genes such as c-Myc and cyclin D1, thereby silencing an array of biological responses, including cell proliferation. Rapid activation of the canonical pathway occurs when Wnt proteins interact with specific receptor complexes comprising members of the Frizzled family of proteins and the low-density lipid co-receptor LRP5 or LRP6 (Fig. 1b). The ligand-receptor binding activates the intracellular protein, disheveled (Dvl), which inhibits APC-GSK3β-axin activity and subsequently blocks degradation of β-catenin. This stabilization of β-catenin allows it to accumulate and translocate to the nucleus where it forms a transcriptionally active complex with the DNA-binding TCF transcription factors to activate the expression of Wnt signaling target genes.

In pancreatic β cells TCF7L2 is a major form of TCF involved in downstream Wnt signaling responsible for the activation of growth-promoting and survival (anti-apoptosis) genes in response to glucagon-like peptide-1 (GLP-1) agonists (Liu and Habener 2008; Shu et al. 2008; Boutant et al. 2012; Heller et al. 2011). Notably, as mentioned above and discussed in detail later, TCF7L2 is a major susceptibility factor for the development of T2D manifested by diminished insulin production (Jin 2008; Lyssenko 2008; Hattersley 2007; Chiang et al. 2012; Florez 2007; Schafer et al. 2007).

Noncanonical Wnt Signaling

Wnt signaling via Frizzled receptors can also lead to the activation of noncanonical pathways that are independent of β-catenin and TCF/LEF complexes (Semenov et al. 2007). Two of the several recognized (Semenov et al. 2007) β-catenin-independent pathways are considered (Fig. 2). One such noncanonical pathway consists of the release of intracellular calcium. Other intracellular second messengers associated with this pathway include heterotrimeric G-proteins, phospholipase C (PLC), and protein kinase C (PKC). The Wnt/Ca^{2+} pathway is important for cell adhesion and cell movements during gastrulation (Komiya and Habas 2008). Wnt/Ca^{2+} pathway is also known to control cell migration and involved in regulating endothelial cell migration. Interestingly, The Wnt/Ca^{2+} pathway may antagonize the canonical Wnt/β-catenin pathway. The canonical and noncanonical Wnt pathways are likely to have opposing effect on endothelial cells and probably antagonize each other in order to finely balance endothelial cell growth.

The Wd/planar cell polarity (PCP) signaling pathway is a second noncanonical Wnt signaling pathway (Komiya and Habas 2008; Wada and Okamoto 2009; Veeman et al. 2003). PCP controls tissue polarity and cell movement through the activation of RhoA, c-Jun N-terminal kinase (JNK), and nemo-like kinase (NLK) signaling cascades. In the planar cell polarity pathway, Wnt signaling through Frizzled receptors mediates asymmetric cytoskeletal organization and polarization of cells by inducing modifications to the actin cytoskeleton.

Wnt Signaling in Pancreas Development and Regeneration

Expression of components of the Wnt signaling pathway, including Wnt ligand family members and various Frizzled receptors, is well documented in the developing mouse, rat, chick, fish, and human pancreas (Heller et al. 2002, 2003; Pedersen and Heller 2005; Kim et al. 2005; Wang et al. 2006). A description of the subsets of the dozen or so Wnt ligands, Frizzled receptors, and the Wnt/FZ regulators, secreted frizzle-related proteins, and Dickkopfs is provided in Heller et al. (Liu and Habener 2008). Endogenous Wnt signaling also occurs in mouse and rat β cell lines (Liu and Habener 2008). Detailed information on the cellular distributions of expression of the various Wnt ligands, receptors, and regulators is not available. From the findings of Heller et al. (2002), it is clear that Wnt signaling factors are expressed both in epithelium and in mesenchyme. Several studies confirm that functional Wnt signaling is active in islets throughout development. A Wnt reporter strain of mice, in which lacZ was inserted into the locus of the Wnt target gene conductin/axin2, expressed β-galactosidase, the product of the *lacZ* gene, throughout the islets (Dessimoz et al. 2005). Expression of the conductin gene is transcriptionally activated by the canonical Wnt pathway via TCF binding sites in its promoter. Furthermore, the β-galactosidase (lacZ) reporter activity is maintained in islets of mice up to 6 weeks after birth. A monoclonal antibody specific for the non-phosphorylated form of β-catenin revealed a strong immunoreactivity in the pancreatic epithelium of the mouse at embryonic day 13 (Papadopoulou and Edlund 2005). Taken together, human and rodent islets and rodent β cell lines are known to express members of the Wnt ligand and Frizzled receptor families, along with modulators of Wnt signaling, the LRP co-receptors, and secreted Dkk (Dickkopf) proteins. Another source of Wnt ligands is adipose tissue (Schinner et al. 2008). Adipocytes secrete a wide range of signaling molecules including Wnt proteins. Fat-cell-conditioned media from human adipocytes increases the proliferation of INS-1 β cells and induces Wnt signaling, which could contribute to the β cell hyperplasia that occurs in humans and rodents in response to obesity. Interestingly inhibitory noncanonical Wnt ligand Wnt5b gene is associated strongly with obesity and type 2 diabetes (Schinner et al. 2008). Expression of Wnt5b in preadipocytes increases adipogenesis and expression of adipokine genes through the inhibition of canonical Wnt signaling (Schinner et al. 2008). Thusly, alterations in Wnt5b levels in humans could alter adipogenesis and, consequently, affect the risk of diabetes onset.

Wnt signaling loss-of-function studies. Following early pancreas specification, Wnt signaling appears to be indispensable for pancreas development, although its precise role remains controversial. The majority of studies have shown that Wnt signaling is essential in the development of the exocrine pancreas. Disruption of the Wnt signaling pathway results in an almost complete lack of exocrine cells (Dessimoz et al. 2005; Papadopoulou and Edlund 2005; Murtaugh et al. 2005; Wells et al. 2007). However, its role in endocrine cell development is still uncertain. Several studies in which Wnt signaling is abolished by conditional β-catenin knockout in the developing mouse pancreas have revealed that the endocrine component of the pancreas develops normally and is functionally intact. In the studies of Murtaugh et al. (2005) and Wells et al. (2007) in which the β-catenin gene in the epithelium of the pancreas and duodenum was specifically deleted, pancreatic islets are intact and contain all lineages of endocrine cells. In contrast, using a different β-catenin knockout approach, Dessimoz et al. (2005) found a reduction in endocrine islet numbers. Selective deletion of β-catenin in β cells of mice resulted in defective insulin secretion, reduced endocrine tissue, and perinatal mortality (Dabernat et al. 2009). Knockdown of β-catenin by administration of antisense RNA to normal and diabetic rats altered the normal and compensatory growth of β cells, mainly through the inhibition of proliferation (Figeac et al. 2010). It is worth noting that knockout studies should be interpreted with some caution because of the potential occurrence of adaptive compensatory mechanisms that could alter the phenotype. Furthermore, the use of different strains of mice expressing PDX-Cre, which have different recombination efficiencies, is expressed at different stages of development and is shown to have mosaic expression in the pancreata of transgenic mice (Heiser et al. 2006). It seems possible that β-catenin and Wnt signaling have several different roles throughout the development of the pancreas. Since the timing of the activation or inactivation of Wnt signaling is crucial for its effects on pancreas development, the currently available Cre-based recombinant technology might not be adequate to fully explore the role of Wnt signaling. Collectively, the loss-of-function studies have not yet provided a definitive role for β-catenin in the development and/or maintenance of function of adult islets. Nonetheless, these results underscore the possible dual nature of Wnt signaling in pancreas growth and development. Excessive Wnt signaling activation prevents proper differentiation and expansion of early pancreatic progenitor cells during early, first transition, specification. During the second transition, β-catenin acts as a pro-proliferative cue that induces gross enlargement of the exocrine and/or endocrine pancreas.

Wnt signaling gain-of-function studies. Gain-of-function experiments suggest an inhibitory role for Wnt pathway in pancreas specification, a stage when cells at the appropriate regions of the foregut begin to form a bud. Heller et al. (2002) showed that forced mis-expression of Wnt1 driven by PDX-1 promoter in mice induces a block in the expansion and differentiation of PDX-1 positive cells and causes ensuing reduction in endocrine cell number and a lack of organized islet formation. Excessive Wnt signaling in the epithelia limits the expansion of both the mesenchyme and epithelium and inhibits growth of the pancreas and islets. Using a different approach, the Heiser et al. (2006) study reached a similar conclusion. The conditional knockin of stable β-catenin in early pancreatic development of mice using PDX-1-driven Cre recombinase efficiently targets all three pancreatic lineages – the endocrine, exocrine, and duct – and results in upregulation of Hedgehog and leads to a loss of PDX-1 expression in early pancreatic progenitor cells (Heiser et al. 2006).

This genetic model of forced overexpression of β-catenin prevents normal formation of the exocrine and endocrine compartments of the pancreas. Using a Xenopus model, McLin et al. (2007) found forced Wnt/β-catenin signaling in the anterior endoderm, between gastrula and early somite stages, inhibits foregut development. By contrast, blocking β-catenin activity in the posterior endoderm is sufficient to initiate ectopic pancreas development (Heiser et al. 2006). These genetic manipulations of Wnt signaling in mice suggest a contribution of both inhibitory and facilitating roles of Wnt signaling during pancreas development. The gain-of-function studies by Dessimoz et al. (2005) show a distinctive role of Wnt signaling in endocrine development. Wnt3a induces the proliferation of islet and MIN-6 cells (Rulifson et al. 2007). The addition of the soluble Wnt inhibitor, Fz 8-cysteine-rich domain (Fz8-CRD), eliminated this stimulatory effect of Wnt3a on cell proliferation (Rulifson et al. 2007). The treatment of islets with Wnt3a significantly increased mRNA levels of cyclin D1, cyclin D2, and CDK4, all of which have Wnt responsive elements in the promoter regions of their genes (Wang et al. 2006). Conditional knockin of active β-catenin in mice promotes the expansion of functional β cells (Heiser et al. 2006), whereas the conditional knock-in of the Wnt inhibitor, axin, impaired proliferation of neonatal β cells (Rulifson et al. 2007).

Surprisingly, recent studies found that Wnt signaling may play a role in regulating the secretory function of mature β cells (Fujino et al. 2003). The Wnt co-receptor, LRP5, is required for glucose-induced insulin secretion from the pancreatic islets. The knockout of LRP5 in mice resulted in glucose intolerance (Fujino et al. 2003). Treatment of isolated mouse islets with purified Wnt3a and Wnt5a ligands causes potentiation of glucose-stimulated insulin secretion. Thus, LRP5 together with Wnt proteins appears to modulate glucose-induced insulin secretion. Furthermore, Schinner et al. (2008) reported that activating Wnt signaling increases insulin secretion in primary mouse islets and activates transcription of the glucokinase gene in both islets and INS-1 cells. The consummate evidence came in isolated mouse and human islets, in which reducing levels of TCF7L2 by siRNA decreases glucose-stimulated insulin secretion, expression of insulin and PDX-1, and insulin content (Shu et al. 2008; Loder et al. 2008; da Silva et al. 2009).

Role of Wnt Signaling in β Cell Growth and Survival

In addition to its potential role in regulating glucose-stimulated insulin secretion, the Wnt pathway is involved in β cell growth and survival. The activation of Wnt signaling in β cell lines or primary mouse islets results in an expansion of the functional β cell mass, with findings consistent with the upregulation of pro-proliferative genes including cyclin D1 and D2 (Liu and Habener 2008). Furthermore, the mis-expression of a negative regulator of Wnt signaling, axin, impairs the proliferation of neonatal β cells, demonstrating a requirement for Wnt signaling during β cell expansion (Rulifson et al. 2007). Axin expression impaired normal expression of islet cyclin D2 and pitx2, a transcriptional activator that directly associates with promoter regions of the cyclin D2 gene. The inhibition of GSK3β activity in normal and diabetic rats by the administration of antisense oligonucleotides, or lithium chloride, increased the stability of β-catenin and enhanced β cell proliferation (Figeac et al. 2010). Shu et al. (2008) provide further evidence in support of a role for Wnt signaling in β cell growth and survival in both mouse and human islets. Depletion of TCF7L2 in human islets causes a decrease in β cell proliferation, an increase in levels of apoptosis, and a decline in levels of active Akt, an important β cell survival factor (Liu and Habener 2008). Similarly, in INS-1 cells, expression of dominant-negative TCF7L2 decreases proliferation rates (Liu and Habener 2008). Furthermore, overexpression of TCF7L2 in both mouse and human islets protects β cells against glucotoxicity or cytokine-induced apoptosis (Liu and Habener 2009).

Roles of Non-Wnt Hormonal Ligands in the Activation of the Wnt Signaling Pathway in Islets

Several hormones and growth factors, such as insulin, insulin-like growth factor-1, platelet-derived growth factor, parathyroid hormone, and prostaglandins, are known to activate the canonical and noncanonical Wnt signaling pathways. However, these observations have been made in non-islet tissues such as the intestine, cancer cell lines, osteoblasts, and fibroblasts (Yi et al. 2008). It has been proposed that a primary function of Wnt signaling is to maintain stem cells in a pluripotent state and that growth factors such as FGF and EGF augment their proliferation (Nusse 2008). Very little is known, however, about the hormonal activation of Wnt signaling in pancreatic islets. Recent studies of glucagon-like peptide-1 (GLP-1) and stromal cell-derived factor-1 (SDF-1) actions on islet β cells demonstrate that both hormones activate downstream Wnt signaling via β-catenin/TCF7L2-regulated gene transcription and that downstream Wnt signaling is required for the pro-proliferative actions of GLP-1 (Liu and Habener 2008; Boutant et al. 2012) and the anti-apoptotic actions of SDF-1 (Liu and Habener 2009).

Downstream Wnt Signaling Requirement for GLP-1-Induced Stimulation of β Cell Proliferation

Glucagon-like peptide-1 (GLP-1) is a glucoincretin hormone released from the intestines in response to meals and stimulates glucose-dependent insulin secretion from pancreatic β cells (Kieffer and Habener 1999; Drucker 2006). GLP-1 also stimulates both the growth and the survival of β cells. GLP-1 is produced in the enteroendocrine L-cells that reside within the crypts of the intestinal mucosa by selective posttranslational enzymatic cleavages of the prohormonal polypeptide, proglucagon, the protein product of the expression of the glucagon gene (Gcg). Notably, the same proglucagon expressed from Gcg in the α cells of the pancreas is alternatively cleaved to yield the hormone glucagon in mature, fully differentiated α cells and is induced to produce GLP-1 and glucagon in response to β cell injuries or loss of glucagon signaling (Liu et al. 2011; Habener and Stanojevic 2012, 2013). Glucagon functions as an insulin counter-regulatory hormone to stimulate hepatic glucose production and thereby to maintain blood glucose levels in the postabsorptive, fasted state.

Genes expressed in Wnt signaling in β cells were examined using a focused Wnt signaling gene microarray and the clonal β cell line INS-1 (Liu and Habener 2008). Of the 118 probes represented on the Wnt signaling gene array, 37 were expressed above background in cultured INS-1 cells. Exposure of the cells to GLP-1 enhanced the expression of 14 of the genes, including cyclin D1 and c-Myc, strongly suggesting that GLP-1 agonists activate components and target genes of the Wnt signaling pathway. GLP-1 agonists activate β-catenin- and TCF7L2-dependent Wnt signaling in isolated mouse islets and INS-1 β cells, and antagonism of β-catenin by siRNAs and of TCF7L2 by a dominant-negative form of TCF7L2 inhibited GLP-1-induced proliferation (Liu and Habener 2008). These findings suggest that Wnt signaling is required for GLP-1-stimulated proliferation of β cells. Although INS-1 cells maintain high basal levels of Wnt signaling via Wnt ligands and Frizzled receptors, GLP-1 agonists specifically enhance Wnt signaling through their binding to the GLP-1 receptor (GLP-1R), a G-protein-coupled receptor coupled to GαS and the activation of cAMP-dependent protein kinase A (PKA).

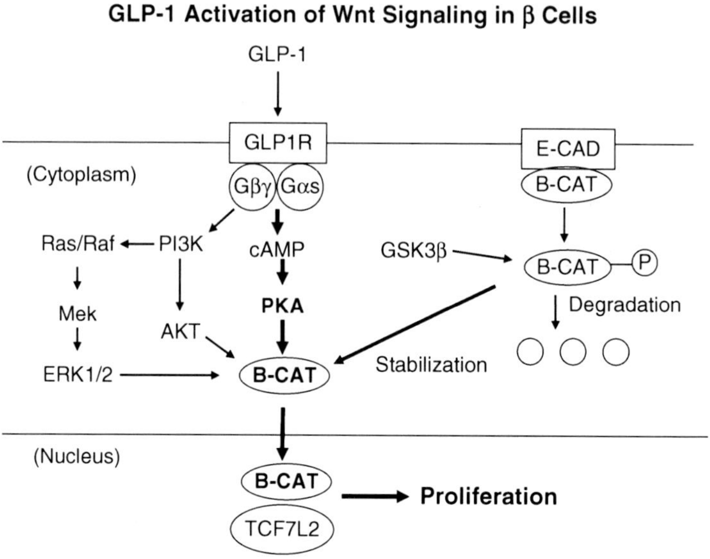

GLP-1 Activation of Wnt Signaling in β Cells

Fig. 3 Diagram summarizing the signaling pathway in pancreatic β cells by which GLP-1 actions couple to the downstream Wnt signaling pathway (Weedon 2007). The interaction of GLP-1 with the GLP-1 receptor (GLP-1R) activates G-protein α S (GαS) resulting in cAMP formation and activation of the cAMP-dependent protein kinase A (PKA). Remarkably, by the GLP-1-activated pathway, β-catenin is stabilized by direct phosphorylation by PKA, rendering it resistant to degradation in response to phosphorylations by GSK3β. This stabilization of β-catenin by PKA-mediated phosphorylation is a distinct departure from the canonical Wnt pathway in which phosphorylation of β-catenin by GSK3β results in its degradation. β-catenin thusly stabilized by PKA-mediated phosphorylation is resistant to degradation in response to phosphorylation by GSK3β, accumulates in the cytoplasm, and is translocated to the nucleus where it associates with TCF7L2 to form a productive transcriptional activation complex. β-catenin/TCF7L2 complexes activate the expression of target genes involved in β cell proliferation

Although PKA is not involved in maintaining basal levels of Wnt signaling, it is essential for the enhancement of Wnt signaling by GLP-1 (Liu and Habener 2008). In addition, the prosurvival protein kinase Akt, along with active MEK/ERK signaling, is required for maintaining both basal and GLP-1-induced Wnt signaling (Liu and Habener 2008; Fig. 3). In summary, both β-catenin and TCF7L2 appear to be required for GLP-1-mediated transcriptional responses and cell proliferation.

Downstream Wnt Signaling Requirement for SDF-1-Induced Promotion of β Cell Survival

SDF-1 is a chemokine originally identified as a bone marrow (BM) stromal cell-secreted factor and now recognized to be expressed in stromal tissues in multiple organs (Burger and Kipps 2006; Kucia et al. 2005; Ratajczak et al. 2006; Kryczek et al. 2007).

SDF-1 Activation of Wnt Signaling in β Cells

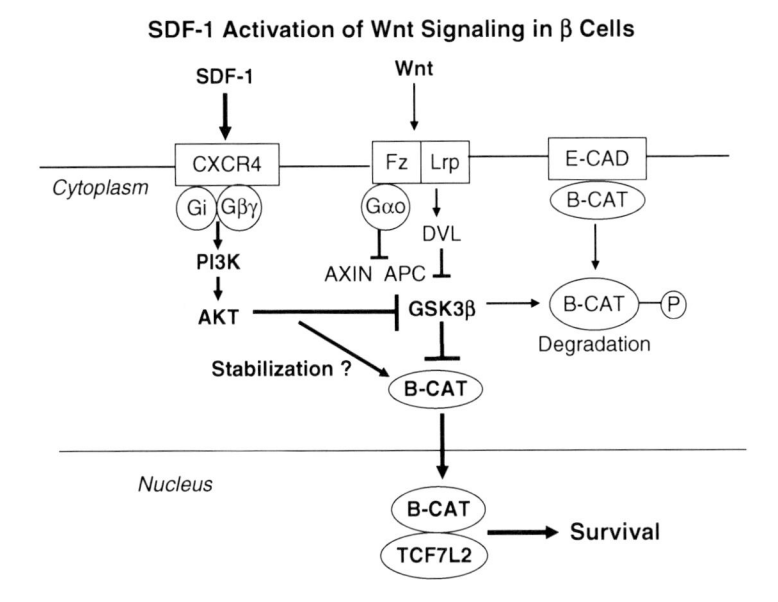

Fig. 4 Schematic model of signaling pathways utilized by SDF-1/CXCR4 in the activation of β catenin-/TCF7L2-mediated transcriptional expression of genes involved in β cell survival. Interactions of SDF-1 with its G-protein-coupled receptor CXCR4 activates G-protein i/o that activates the phosphoinositol kinase 3 (PI3K) and the downstream prosurvival kinase Akt. Akt is a potent inhibitor of the Wnt signaling destruction complex composed of axin, APC, and GSK3β. Inhibition of GSK3β by Akt results in the inhibition of phosphorylation of β-catenin by GSK3, prevents the degradation of β-catenin, and thereby results in the stabilization of β-catenin which accumulates in the cytoplasm and enters the nucleus, where it associates with TCF7L2. The β-catenin/TCF7L2 forms a transcriptional activation complex that activates the expression of genes that promote β cell survival. A direct action of Akt on the stabilization of β-catenin remains conjectural

The most extensively studied function of the SDF-1/receptor CXCR4 axis is that of chemoattraction involved in leukocyte trafficking and stem cell homing in which local tissue gradients of SDF-1 attract circulating stem/progenitor cells. SDF-1/CXCR4 signaling in the pancreas remains relatively unexplored. Kayali and co-workers reported expression of SDF-1 and CXCR4 in the fetal mouse pancreas and CXCR4 in the proliferating duct epithelium of the regenerating pancreas of the nonobese diabetic mouse (Kayali et al. 2003). The cross talk between the SDF-1-CXCR4 axis and Wnt signaling pathway was first demonstrated by Luo et al. (2006) in studies of rat neural progenitor cells. Transgenic mice expressing SDF-1 in their β cells (RIP-SDF-1 mice) are protected against streptozotocin-induced diabetes through activation of the prosurvival protein kinase Akt and resulting downstream prosurvival, antiapoptotic signaling pathways (Yano et al. 2007). An examination of SDF-1-activated Wnt signaling in both isolated islets and INS-1 cells using a β-catenin-/TCF-activated reporter gene assay revealed enhanced Wnt signaling through the Gαi/o-PI3K-Akt axis, suppression of GSK3β, and stabilization of β-catenin (Liu and Habener 2009; Fig. 4). Phosphorylation of GSK3 by Akt represses its phosphorylating activities on β-catenin and thereby reduces the degradation of β-catenin. Moreover, SDF-1 signaling in INS-1 β cells stimulates the accumulation of β-catenin mRNA, likely due to an enhancement in the transcription of the β-catenin gene (Liu and Habener 2009). Recent evidence also suggests that active Wnt signaling mediates, and is required for, the cytoprotective, survival actions of SDF-1 on β cells (Liu and Habener 2009).

Potential Mechanisms by Which GLP-1 and SDF-1 May Act Cooperatively on Wnt Signaling to Enhance β Cell Growth and Survival

There appear to be differences in the mechanisms of the interactions of SDF-1/CXCR4 signaling and GLP-1/GLP-1R signaling with the Wnt signaling pathway in β cells. Although both SDF-1 and GLP-1 activate the downstream pathway of Wnt signaling, consisting of β-catenin-/TCF7L2-mediated gene expression, they do so by way of different pathways of interactions with the more upstream components of the Wnt signaling pathway. These proposed different upstream pathways of signaling utilized by GLP-1 and SDF-1 raise the possibility of additive or synergistic effects on downstream Wnt signaling in the promotion of β cell growth and survival. SDF-1 inhibits the destruction complex of the canonical Wnt signaling pathway consisting of axin, APC, and the protein kinases, glycogen synthase kinase-3 (GSK3) and casein kinase-1 (CSNK1). This inhibition of GSK3 and CSNK1 by SDF-1 is likely mediated by the well-known actions of Akt to inhibit these kinases, resulting in the stabilization and accumulation of β-catenin. In marked contrast to the actions of SDF-1 on β cells, GLP-1 activates β-catenin/TCF7L2 complexes via the stabilization of β-catenin by a different mechanism involving the phosphorylation and stabilization of β-catenin by the cAMP-dependent protein kinase A (PKA). PKA activated by GLP-1/GLP-1R phosphorylates β-catenin on serine-675 resulting in its stabilization and accumulation. Thusly, unlike SDF-1, GLP-1-induced activation of gene expression by β-catenin/TCF7L2 in β cells occurs independently of the destruction box and the activities of GSK3. It also remains possible that β-catenin may be stabilized by its direct phosphorylation by Akt.

β-catenin is the activation domain, and TCF7L2 is the DNA-binding domain of the transactivator. It is tempting to speculate that different phosphorylations of β-catenin provided by SDF-1 signaling versus GLP-1 signaling result in different conformations of β-catenin. When different conformers of β-catenin interact with TCF7L2, they confer different conformations to the DNA-binding domains of TCF7L2 resulting in differing affinities of TCF7L2 for its cognate enhancer binding sites on the promoters of various Wnt signaling target genes. Such a combinatorial mechanism could account for the difference in genes regulated by β-catenin/TCF7L2 in β cells in response to SDF-1 compared to GLP-1. Wnt signaling may be a final downstream pathway for both SDF-1 and GLP-1 signaling in β cells. However, gene expression targets diverge so that SDF-1 predominately regulates genes involved in cell survival, whereas GLP-1 regulates genes involved in cell cycle control (proliferation). If this circumstance proves to be valid, our findings raise the possibility of a dual therapeutic approach for increasing β cell mass. GLP-1 is predominantly pro-growth and SDF-1 is predominantly prosurvival. Thereby the two peptides may act synergistically to promote both the growth and survival of β cells and to conserve, or even enhance, β cell mass in response to injury.

Type 2 Diabetes Genes

Genome-wide scans in several large populations have uncovered associations of specific genetic loci with the development of type 2 diabetes (Lyssenko et al. 2008; Florez 2008a; Van Hoek et al. 2008; Florez 2008b; Ahlqvist et al. 2011; Marchetti et al. 2012; Zeggini et al. 2007; Saxena et al. 2007; Sladek et al. 2007; Scott et al. 2007; Grarup et al. 2007; Hayes et al. 2007; Cauchi et al. 2008; Ruchat et al. 2009; Owen and McCarthy 2007b; Moore et al. 2008; Steinthorsdottir et al. 2007; Palmer et al. 2008). At least 90 genes have associations with diabetes that are consistent among various population studies (Table 1). Of note, the majority of these genes (55 of 99) are expressed in pancreatic β cells. Further, several of the genes (20) appear to be involved in the Wnt signaling pathway. TCF7L2, the DNA-binding component of the downstream transcription factor complex, appears to have a particularly strong association with type 2 diabetes. Brief descriptions and specific references to the 20 genes implicated in Wnt signaling in the pancreas are given below. References to the remaining genes in Table 1 can be found in the recent review articles (References #25 and #26).

Genes Associated with Islet Development/Function and Wnt Signaling

TCF7L2 (transcription factor 7-like 2). Grant and co-workers provided the index report on an association of polymorphisms in TCF7L2 with type 2 diabetes (Grant et al. 2006). Epidemiology studies from Icelandic, Danish, and US cohorts reported that the inheritance of specific single nucleotide polymorphisms (SNPs), at the region DG10S478, within the intron 3 region of TCF7L2 gene is related to an increased risk of type 2 diabetes (Lyssenko 2008; Jin and Liu 2008; Perry and Frayling 2008; Cauchi and Froguel 2008; Bordonaro 2009; Hattersley 2007; Weedon 2007; Chiang et al. 2012; Florez 2007; Xiong et al. 2012; Owen and McCarthy 2007a; Smith 2007). Then two other SNPs within introns 4 and 5 of TCF7L2, namely, rs12255372 and rs7903146, were found in strong linkage disequilibrium with DG10S478 and showed similarly robust associations with type 2 diabetes patients with glucose intolerance. In Asian populations, the frequencies of SNPs rs7903146 and rs12255372 are quite low, but two novel SNPs – rs290487 and rs11196218 – are associated with the risk of type 2 diabetes in a Chinese population.

Table 1 Type 2 diabetes genes identified by candidate gene approach and genome-wide association studies (99)

Genes associated with islet development/function and Wnt signaling

Gene symbol	Name(s)	β cell functions	Wnt signaling
TCF7L2	HMG transcription factor-7L2	β cell proliferation and survival, insulin secretion, impaired GLP-1 responses, increased hepatic glucose production	Canonical Wnt signaling, regulates target genes in association with β-catenin
FTO	Fatso. Fused toes locus. Includes FTS, FTM	Pancreas development, obesity	FTS, a target gene for Wnt signaling
NOTCH 2	Delta/notch signaling	Pancreas development	Wnt signaling interaction via phosphorylation by GSK3
IGF2BP2	Insulin growth factor 2 mRNA-binding protein 2	Islet growth	Expression induced by β-catenin and TCF7L2
HHEX	Hematopoietic homeobox transcription factor	Early pancreas development	Repressed by β-catenin and TCF7L2
CDKN2A/N2B	Cyclin-dependent kinase inhibitor, P16, INK4A	Islet regeneration regulates CDK4 in β cells	Cross talk with Wnt signaling, induced by β-catenin
HNF1B	HNF1 homeobox B	MODY 5, TCF2, islet development	β catenin, Wnt signaling
PROX1	Prospero homeobox 1	Pancreas development, islet endocrine development	Target Tcf/Lef/Ctnnb1 neural
PTPRD	Protein tyrosine phosphatase receptor type D	Growth, autism, diabetes	Target in PC12 cells
TLE4	Transducin-like enhancer of split 4, Xgrg4, Groucho	α cell development Nkx2.2	Hex β-catenin, co-repressor
ZBED3	Zinc finger BED-type containing protein 3	β cell function	Axin-interacting protein
TSPAN8/LGR5/GPR49	Tetraspanin 8. Leucine-rich G-protein-coupled receptor 5. G-protein-coupled receptor 49	Impaired GSIS	Wnt signaling target gene in intestinal crypt stem cells
TP53INP1	Tumor protein p53 inducible nuclear protein 1	β cell cytoprotection	Wnt signaling? Tcf7L2 regulated
WFS1	Wolfram syndrome 1 transmembrane protein	Insulin secretion, endoplasmic reticulum, protein trafficking	Strong target locus for TCF7L2 binding [Zhao 2010]
CDKAL1	Cyclin-dependent kinase 5 homolog inhibitor	Islet glucotoxicity, impaired insulin secretion	Strong target locus for TCF7L2 binding [Zhao 2010]
ADAMTS9	Metallopeptidase with thrombospondin 9	β cell survival and cytoprotection	Strong target locus for TCF7L2 binding [Zhao 2010]

(*continued*)

Table 1 (continued)

GLIS3	Kruppel zinc finger protein	Impaired GSIS	Regulates Ngn3 in β cells
GRK5	G-protein receptor kinase 5	Elevated plasma insulin levels	Regulates LRP6 Wnt signaling
SPRY1	Sprouty 1	Influences insulin secretion	Target for β-catenin/LEF
HNF4A	Hepatocyte nuclear factor 4	Impaired insulin secretion (MODY 1)	Liver zonation, intestinal development – Wnt signaling

Genes associated with islet development/function, Wnt signaling unknown (Bordonaro 2009)

Gene symbol	Name(s)	β cell functions
CDC123/CAMK1D	Cell division cycle 123/calcium/calmodulin dependent	β cell apoptosis? Impaired GSIS
PPARgamma	Peroxisome proliferator receptor gamma	Insulin resistance, insulin secretion
KCNJ11	Inward-rectifying potassium channel	Regulates insulin secretion and Sur1 (ABCC8)
SLC30A8	Solute carrier 30a8 zinc transporter	Insulin granules, secretion
KCNQ1	Potassium channel	Insulin secretion
MTNR1B	Melatonin receptor 1b increased in response to GLP-1	Insulin secretion
JAZF1	Nuclear zinc finger transcriptional repressor	β cell apoptosis? Impaired GSIS
ADAM30	ADAM metallopeptidase domain 30	β cell oxidative stress?
ADCY5	Adenylate cyclase 5	Increases conversion of proinsulin to insulin
AIF1	Allograft inflammatory factor 1	Modulates insulin production and GSIS. Role in T1D
ARAP1/CENTD2	ArfGAP with RhoGAP domain, ankyrin repeat with PH domain 1	Modulates GSIS
THADA	Thyroid adenoma associated	Lower β cell response to GLP-1, arginine, decreased β cell mass?
BCL11A	β-cell CLL/lymphoma zinc finger protein 11A	Altered GSIS
CAPN10	Calpain 10 Ca^{++}-activated neutral proteinase	β cell exocytosis apoptosis, metabolism increased in T2D β cells
DGKB/DGKG	Diacylglycerol kinases	Insulin secretion and β cell metabolism

(continued)

Table 1 (continued)

ETV5	Ets variant 5	Mediates mesenchymal to epithelial signaling in pancreas development
GCK	Glucokinase hexokinase 4	MODY 2, impaired insulin secretion
HNF1A	HNF1 homeobox A	Insulin secretion, MODY 3
RASGRP1	RAS guanyl nucleotide-releasing protein	Influences insulin secretion
GCKR	Glucokinase regulator	Insulin resistance, obesity, metabolic syndrome
TMEM195	Transmembrane protein 195	Reduced GSIS
GIPR	Gastric inhibitory peptide receptor	Increased proinsulin production, blunted insulin response
C2CD4B	C2 calcium-dependent domain-containing 4B	Impaired GSIS
FADS1	Fatty acid desaturase-1	Abnormal early insulin secretion
G6PC2	Glucose-6-phosphatase catalytic 2	Abnormal early insulin secretion
MADD	MAP kinase-activating death domain protein	Abnormal insulin processing
CRY2	Cryptochrome 2	Associated with fasting insulin
SLC2A2	Solute carrier family 2 Member 2, glucose carrier	Impaired GSIS
VPS13C	Vacuolar protein sorting 13 homolog C	Abnormal insulin processing and impaired secretion
ADRA2A	α 2A adrenergic receptor	Impaired GSIS
ANK1	Ankyrin 1, erythrocytic	Integral membrane protein
BCL2	β-cell CLL/lymphoma 2	Anti-apoptosis protein
ST6GAL1	ST6 β-galactosamine α-like 1	Associated with β cell function
CMIP	C-Maf-inducing protein	Modulates Maf and PI3K signaling
COBLL1-GRB14	Cordon bleu-like 1 growth factor receptor bound 14	Actin nucleator

Genes not known to be involved in either islet development/function or Wnt signaling (Willert and Jones 2006)

Gene symbol	Name(s)	Cell functions
DUSP9	Dual specificity phosphatase 9	Altered proinsulin conversion to insulin?

(*continued*)

Table 1 (continued)

BCDIN3D	BCDIN3 domain containing	BMI, obesity
CHCHD9	Coiled-coil-helix-coiled-coil-helix domain 9	Unknown
FAIM2	Fas apoptotic inhibitory molecule 2	Obesity related
GNPDA2	Glucosamine-6-P-deaminase 2	Obesity related
HHCA2	YY-associated protein1 HCCA2	Chromatin, oncogene insulin resistance
HMGA2	High-mobility group AT hook 2	Oncoprotein
IRS1	Insulin receptor substrate 1	Increased insulin resistance
KLF14	Kruppel-like factor 14	Increased insulin resistance
MCR4	Melanocortin 4 receptor	Obesity, hypothalamic
MTCH2	Mitochondrial carrier homolog 2	Obesity, mitochondrial death pathways
NCR3	Natural cytotoxicity triggering receptor 3	Autoimmunity modulation
NEGR1	Neuronal growth regulator 1	Obesity
PRC1	Protein regulator of cytokinesis 1	Cancer
RBMS1	RNA-binding motif single-stranded-interacting protein 1	DNA synthesis and gene transcription
SFRS10	Transformer 2 β homolog	Obesity, oxidative stress
SH2B1	SH2B adaptor protein 1	Obesity
SSR	Serine racemase	D-serine synthesis from L-serine
TMEM18	Transmembrane protein 18	Obesity DNA-binding protein
ZFAND6	Zinc finger AN1-type domain	Peroxisome biogenesis
HK1	Hexokinase 1 enzyme	Tissue ubiquitous, high-affinity glucose-phosphorylating enzyme
PEPD	Peptidase D, prolidase	Cytosolic dipeptidase, hydrolyzes dipeptides with proline or hydroxyproline at the carboxy terminus
UBE2E2	Ubiquitin-conjugating enzyme E2E2	Accepts ubiquitin and catalyzes its covalent attachment to proteins

(*continued*)

Table 1 (continued)

IGF-1	Insulin-like growth factor 1	Endocrine growth-promoting hormone and paracrine regulator. Akt activator and apoptosis inhibitor
FN3K	Fructosamine-3-kinase	Phosphorylates fructosamines resulting in deglycation of glycated proteins
SPTA1	Spectrin α erythrocytic 1	Actin cross-linking and molecular scaffold protein
ATP11A	ATPase, type 11A	Integral membrane ATPase
GATAD2A	GATA zinc finger domain-containing A	Transcriptional repressor. Enhances MBD2-mediated repression
SREBF1	Sterol regulatory element-binding transcription factor	Insulin-regulated transcription factor. Glucose and lipid production
TH/INS	Tyrosine hydroxylase/insulin	Obesity related to polymorphisms in the chromosome 11p15 locus containing TH/INS/IGF2
PCSK1	Prohormone convertase subtilisin kexin, type 1	Involved in the cleavages of proinsulin and proglucagon to the active hormones insulin and GLP-1
LARP6	La ribonucleoprotein domain, family, member 6	Translational regulator. Associated with fasting glucose traits, type 2 diabetes, and obesity
SGSM2	Small G-protein signaling modulator 2	Modulates small G-protein (RAP and RAB)-mediated signaling pathway. Associated with type 2 diabetes
SNX7	Sorting nexin 7	Contains a phox (PX) domain phosphoinositide-binding domain involved in intracellular trafficking
VPS26A	Vacuolar sorting protein 26A	Retrograde transport of proteins from endosomes to the trans-Golgi network
HMG20A	High-mobility group 20A	Transcription factor induces expression of neuronal genes
AP3S2	Adaptor-related protein complex	Facilitates protein trafficking from Golgi vesicles

(*continued*)

Table 1 (continued)

GRB14	Growth factor receptor-bound protein 14	Adaptor protein inhibits tyrosine kinase and insulin receptor signaling
ADIPOQ	Adiponectin, CTQ, and collagen domain containing	Adipokine involved in the control of fat metabolism and insulin sensitivity
ZNF664	Zinc finger protein 664	Transcription factor associated with obesity and diabetes
GNL3	Guanine nucleotide-binding protein-like 3	Nucleolar protein involved in stem cell proliferation. Stabilizes MDM2
LYPLAL1	Lysophospholipase-like 1	Hydrolyzes fatty acids from G-proteins and HRAS
PPP1R3B	Protein phosphatase 1 regulatory subunit 3B	Increases basal and insulin-stimulated glycogen synthesis
UHRF1BP1	UHRF1-binding protein 1	Ubiquitin-like containing PHD and ring finger domains 1-binding protein Function unknown

The most likely candidate is the rs7903146 single nucleotide polymorphism that has a strong association with type 2 diabetes (Gloyn et al. 2009). This polymorphism resides in an intronic, noncoding region of the gene, and the mechanisms for its effects on TCF7L2 expression are unclear. The TT risk allele in humans with T2D impairs β cell functions manifested in reduced glucose-stimulated insulin secretion (GSIS) and glucoincretin actions and increased hepatic glucose production (Schafer et al. 2007; Nauck and Meier 2007; Lyssenko et al. 2007).

The mechanisms involved in the regulation of β cell functions by Tcf7l2 have proved to be complex. In initial studies mRNA levels for Tcf7l2 were increased by 5-fold in donor islets obtained from T2D subjects, particularly in carriers of the TT genotype, and overexpression of Tcf7l2 in isolated human islets reduced GSIS (Lyssenko et al. 2007). However, depleting Tcf7l2 mRNA by siRNA in human islets resulted in a decrease in GSIS, proliferation, and an increase in apoptosis (Shu et al. 2008; da Silva et al. 2009; Le Bacquer et al. 2011) attributed in part to a reduction in the expression of glucoincretin (GLP-1 and GIP) receptors (Shu et al. 2009). Analyses of Tcf7l2 transcripts and protein levels in islets obtained from diabetic donors uncovered several alternatively spliced mRNAs and that the Tcf7l2 mRNAs and translated proteins are unstable (Le Bacquer et al. 2011). A role for Tcf7l2 in β cell proliferation and regeneration is provided by findings of an association of increased expression of Tcf7l2 with pancreatic duct proliferation and neogenesis of endocrine cells (Shu et al. 2012). Evidence obtained from genetic manipulations of the expression of Tcf7l2 in mice appears to be contradictory. Loss of Tcf7l2 function by deletion of Tcf7l2 in β cells impairs GSIS and β cell expansion in response to HFD (da Silva et al. 2012), whereas hemizygous

(Yang et al. 2012) Tcf7l2 mice and mice with a loss-of-function allele (Savic et al. 2011) improves glucose tolerance. Further, mice overexpressing Tcf7l2 in β cells display glucose intolerance (Savic et al. 2011). Notwithstanding the contradictory findings in mice, a major characteristic of islets of diabetic individuals harboring polymorphisms in Tcf7l2 is high Tcfl2 mRNA and low protein levels (Lyssenko et al. 2007; Le Bacquer et al. 2011; Shu et al. 2009), associated with impaired glucose-stimulated and glucoincretin-potentiated insulin secretion (Schafer et al. 2007; Nauck and Meier 2007; Lyssenko et al. 2007).

It is tempting to speculate that the discrepancies observed in mice with Tcf7l2 gene knockout or knockdown of expression, or forced overexpression of Tcf7l2, might reflect a necessary requirement for precise amounts of Tcfl2 protein within nuclei of cells to optimally activate and modulate gene transcription. Different-sized transcripts of the Tcf7l2 gene might translate different isoforms of Tcf7l2 protein with differing transactivation activities (activation and/or suppression), depending on its association with cofactors such as β-catenin. Since the polymorphisms, including the T mutation, are located in intronic regions of the gene locus, alternative RNA splicing (Le Bacquer et al. 2011) and/or intron-encoded microRNAs could regulate mRNA stability and translation efficiency.

The glucoincretin hormone GLP-1 appears to be involved in the pathogenesis of diabetes in individuals who carry TCF7L2 risk alleles. These carriers of TCF7L2 risk alleles have impaired insulin secretion as a major contributor to impaired glucose tolerance or diabetes (Lyssenko 2008; Jin and Liu 2008; Perry and Frayling 2008; Cauchi and Froguel 2008; Bordonaro 2009; Hattersley 2007; Weedon 2007; Chiang et al. 2012; Florez 2007; Xiong et al. 2012; Owen and McCarthy 2007a; Smith 2007). Glucose clamp studies on a large cohort of carriers of TCF7L2 polymorphisms revealed both reduced insulin secretion in response to oral glucose tolerance tests and impaired GLP-1-induced insulin secretion (Schafer et al. 2007). However, in these studies plasma GLP-1 levels were not influenced by the TCF7L2 variants (Schafer et al. 2007). These findings are of interest because two pathogenetic mechanisms involving GLP-1 have been proposed: impaired GLP-1 production in the intestine (Bordonaro 2009; Yi et al. 2008) and impaired GLP-1 actions on pancreatic β cells (Liu and Habener 2008). The studies of Schafer et al. (2007) suggest that the defect in the enteroinsular axis in individuals with defective TCF7L2 functions lies at the level of impaired actions of GLP-1 on insulin secretion from pancreatic β cells, rather than the level of impaired production of GLP-1 by intestinal L-cells. Evidence is reported from studies in vitro that support an important role for β-catenin-/TCF7L2-mediated Wnt signaling in both the expression of the proglucagon gene in intestinal cells (Korinek Barker et al. 1998) and in the regulation of insulin secretion (Shu et al. 2008; Loder et al. 2008; da Silva et al. 2009) and β cell proliferation (Liu and Habener 2008). Interestingly, there is some reported evidence that TCF7L2 may be expressed at low levels (Korinek Barker et al. 1998; Barker et al. 2007), or not at all (Yi et al. 2005) in β cells. These reports conflict with those of the Rutter (da Silva et al. 2009) and Maeder (Shu et al. 2008) laboratories, and our own observations

(Liu and Habener 2008). Based on the findings currently available, the contributions of TCF7L2 functions to the enteroinsular axis may occur at the levels of both the production of GLP-1 by intestinal L-cells and the actions of GLP-1 on pancreatic β cells. The two levels of involvement of TCF7L2 actions are not necessarily mutually exclusive.

FTO (Fat Mass and Obesity-Associated Protein). FTO encodes a protein that is homologous to the DNA repair AlkB family of proteins that are involved in the repair of alkylated nucleobases in DNA and RNA (Jia et al. 2008). The FTO gene is upregulated in orexigenic neurons in the feeding center of the hypothalamus (Frederiksson et al. 2008). Genetic variants in FTO result in excessive adiposity and insulin resistance as well as a markedly increased predisposition to the development of diabetes (Do et al. 2008). A 1.6 mb deletion mutation in the mouse results in the deletion of a locus containing FTO, FTS (fused toes), FTM, and three members of the Iroquois gene family, Irx3, Irx5, and Irx6 (Anselme et al. 2007), resulting in multiple defects in the patterning of the body plan during development (Anselme et al. 2007; Peters et al. 2002). The Irx (Iroquois) proteins are homeodomain transcription factors. The FTO, FTS, IRX locus is implicated in Wnt signaling. FTS is a small ubiquitin-like protein with conjugating protein ligase activity that is known to interact with the protein kinase Akt, a potent inhibitor of GSK3β activity in the Wnt signaling pathway. Moreover, Wnt signaling is reported to induce the expression of Irx3 (Braun et al. 2003). Irx1 and Irx2 are expressed in the endocrine pancreas of the mouse under the control of neurogenin-3 (Ngn3) expression (Petri et al. 2006).

Notch2. The delta/notch signaling pathway is an important cell-cell interactive signaling pathway (lateral inhibition) involved in embryonic stem cell amplification, differentiation, and determination of organogenesis. Notch2 is expressed in pancreatic ductal progenitor cells and may be involved in early branching morphogenesis of the pancreas (Lee et al. 2005). The conditional ablation of Notch2 signaling in mice moderately disturbed the proliferation of epithelial cells during early pancreas development (Nakhai et al. 2008). Evidence is presented linking Notch2 to Wnt signaling (Espinosa et al. 2003). GSK3β phosphorylates Notch2 thereby inhibiting the activation of Notch target genes.

IGF2BP2 (Insulin-Like Growth Factor 2-Binding Protein 2). IGF2BP2 is a paralog of IGF2BP1, which binds to the 5' UTR of the insulin-like growth factor 2 (IGF2) mRNA and regulates IGF2 translation (Nielsen et al. 1999). IGF2 is a member of the insulin family of polypeptide growth factors involved in development, growth, and stimulation of insulin action. Wnt1 is reported to induce the expression of IGF2 in preadipocytes (Longo et al. 2002).

Hhex (Hematopoietically Expressed Homeobox). Hhex is a homeodomain protein that regulates cell proliferation and tissue specification underlying vascular, pancreatic, and hepatic differentiation (Bort et al. 2004, 2006; Hallaq et al. 2004). Variants in the Hhex gene manifest in impaired β cell function (Pascoe et al. 2007). Hhex is associated with Wnt signaling during pancreas development as it acts with β-catenin to serve as a corepressor of Wnt signaling (Foley and Mercola 2005; Zamparnini et al. 2006).

HNF1β (Hepatocyte Nuclear Factor 1 β, TCF2, and MODY 5 Gene). Tcf2 is a critical regulator of a transcriptional network that controls the specification, growth, and differentiation of the embryonic pancreas (Maestro et al. 2007a). Mutations in the TCF2 gene result in hypoplasia of the pancreas resulting in exocrine pancreas dysfunction to varying degrees (Maestro et al. 2007a; Haldorsen et al. 2008; Haumaitre et al. 2006). Some mutations manifest as a form of maturity onset diabetes of the young (MODY 5).

CDKN2A/B (Cyclin-Dependent Kinase Inhibitor 2A/B, ARF, and p16INK4a). The CDKN2A/B gene generates several transcript variants which differ in their first exons. CDKN2A is a known tumor suppressor, and its product, p16INK4a, inhibits CDK4 (cyclin-dependent kinase 4), a powerful regulator of pancreatic β cell replication (Rane et al. 1999; Mettus and Rane 2003; Marzo et al. 2004). Overexpression of CDKN2A leads to decreased islet proliferation in ageing mice (Krishnamurthy et al. 2006). CDKN2B overexpression is also causally related to islet hypoplasia and diabetes in murine models (Moritani et al. 2005). p16(IND4a) is linked to the Wnt signaling pathway as stabilized β-catenin silences the p16(INK4a) promoter in melanoma cells (Delmas et al. 2007).

PROX1 (Prospero-Related Homeobox Gene). Prospero-related homeobox gene (Prox1) is a target of β catenin-TCF/LEF signaling involved in the differentiation of neural stem cells towards the neuronal lineage in hippocampal neurogenesis (Karalay et al. 2011).

PTPRD (Protein Tyrosine Phosphatase Receptor D). PTPRD (Pcp-2) is a protein tyrosine phosphatase that inhibits β-catenin signaling in the Wnt pathway (Yan et al. 2006). PTPRD prevents tyrosine phosphorylation and release of β-catenin bound to E-cadherin in the plasma membrane. By this action β-catenin remains sequestered and cannot translocate to the nucleus and partner with TCF/LEF DNA-binding proteins and activate transcription. The role of β-catenin bound to E-cadherin is to stabilize intercellular membrane adhesions and to maintain epithelial integrity.

TLE4/Groucho (Gro). TLE4/Groucho (Gro) is a member of a family of transcriptional corepressors that bind to TCF/LEF factors and prevent their activation by β-catenin (Hanson et al. 2012; Zamparini et al. 2006).

ZBED3 (Zinc Finger BED Domain-Containing Protein 3). ZBED3 is a zinc finger BED domain-containing protein that interacts with axin and activates β-catenin/Wnt signaling (Chen et al. 2009a). The interaction of ZBED3 with axin disrupts the destruction box and prevents phosphorylation and degradation of β-catenin, augmenting the activation of Wnt signaling.

TSPAN8/LGR5/GPR49. The protein encoded by this gene is a member of the transmembrane 4 superfamily, also known as the tetraspanin family. Most of these members are cell-surface proteins that have a role in the regulation of cell development, activation, growth, and motility. LGR5/GPR49 is a leucine-rich repeat-containing G-protein-coupled receptor. A role for TSPAN8 in the pancreas is as yet unknown. However, Tspan8/LGR5 is a recognized Wnt signaling target gene in small intestinal and colonic stem cells (Barker et al. 2007).

TP53INP1. TP53-induced nuclear protein-1 is induced by TP53 in response to depletion of Tcf7l2 and is involved in apoptosis (Zhou et al. 2012). Inhibition of TP53INP1 protects β cells from Tcf7l2 depletion-induced apoptosis suggesting that TP53INP1 is at least partially responsible for activating proapoptotic pathways in β cells deficient in Tcf7l2.

WFS1 (Wolfram Syndrome 1). WFS1 encodes a transmembrane protein of 890 amino acids that is highly expressed in the endoplasmic reticulum of neurons and pancreatic β cells (Takeda et al. 2001). Mutations in WFS1 result in Wolfram syndrome, an autosomal recessive neurodegenerative disorder. Disruption of the WFS1 gene in mice causes progressive β cell loss and impaired stimulus-secretion coupling in insulin secretion (Ishihara et al. 2004). The reduction in β cell mass is likely a consequence of enhanced endoplasmic reticulum stress resulting in the apoptosis of β cells (Riggs et al. 2005; Yamada et al. 2006; Fonseca et al. 2005). Impaired proinsulin processing to insulin and insulin transport through the secretory pathway may also be involved in the impaired insulin secretion. To date no information is available on the mechanisms that regulate WFS1 expression or of an involvement of Wnt signaling in its expression.

CDKAL1 (CDK5 Regulatory Subunit-Associated Protein-1-Like (1). CDKAL1 encodes a protein of unknown functions. However, the protein is similar to CDK5 regulatory subunit-associated protein-1 (encoded by CDK5RAP1), expressed in neuronal tissues. CDKAL1 inhibits cyclin-dependent kinase 5 (CDK5) activity by binding to the CDK5 regulatory subunit p35 (Ching et al. 2002). Variants in the CDKAL1 gene in humans are associated with decreased pancreatic β cell functioning (Pascoe et al. 2007). CDK5 has a role in the loss of β cell function in response to glucotoxicity as the inhibition of the CDK5/p35 complex prevents a decrease of insulin gene expression that results from glucotoxicity (Ubeda et al. 2006). Therefore, it seems possible that CDKAL1 may have a role in the inhibition of the CDK5/p35 complex in pancreatic β cells similar to that of CDK5RAP1 in neuronal tissue. One may conjecture that a reduced expression and inhibitory function of CDKAL1 or reduced inhibitory function could exacerbate β cell impairment in response to glucotoxicity.

ADAMTS9. The ADAMTS9 gene encodes a member of the ADAMTS (a disintegrin and metalloproteinase with thrombospondin motifs) protein family (http://www.ncbi.nlm.nih.gov/sites/entrez?Db=gene&Cmd=retrieve&dopt=full_report&list_uids=56999&log$=databasead&logdbfrom=protein). Members of the ADAMTS family have been implicated in the cleavage of proteoglycans, the control of organ shape during development, and the inhibition of angiogenesis. ADAMTS8 is widely expressed during mouse embryo development (Jungers et al. 2005). Functions for ADAMTS9 in the pancreas or in Wnt signaling are heretofore unrecognized.

GLIS3. GLIS3 is a Kruppel-like zinc finger transcription factor that regulates neurogenin-3 (Ngn3) through its distal promoter (Kim et al. 2012). GLIS3 and hepatic nuclear factor-6 (HNF6) bind close together to the distal promoter of Ngn3 and activate transcription, suggesting that cross talk between GLIS3 and HNF6 might be involved in the regulation of Ngn3 during pancreatic endocrine cell specification and development.

GRK5 (G-protein Coupled Receptor Kinase-5). G-protein-coupled receptor kinase-5, known to phosphorylate and inhibit the activities of G-protein-coupled receptors, also is found to phosphorylate the Wnt receptor LRP6 resulting in the disruption of the destruction box, stabilization of β-catenin, and activation of Wnt signaling (Chen et al. 2009b).

SPRY1 (Sprouty 1). Sprouty 1, a Wnt signaling target gene (Colli et al. 2013), is a member of conserved proteins involved in the modulation of receptor tyrosine kinases (RTKs) and branching morphogenesis during development (Edwin et al. 2009; Guy et al. 2009). The inducible expression of SPRY4 in pancreatic β cells of mice during development results in a reduction in islet size, an increased number of α cells, and an impaired islet cell type segregation (Jäggi et al. 2008).

HNF4A (Hepatic Nuclear Factor 4 α, MODY 1). Hepatic nuclear factor 4 α is a transcription factor involved in Wnt signaling by its interactions with lymphocyte enhancer factor-1 (LEF1) in controlling the zonation of hepatocytes in the liver (Colletti et al. 2009). HNF4A deficiency in mice causes abnormal insulin secretion and impaired expansion of β cell mass during pregnancy (Maestro et al. 2007b). Mutations in the HNF4A gene can lead to MODY 1, one of the several genetic causes of maturity onset diabetes of the young (McDonald and Ellard 2013) and, rarely, congenital hyperinsulinemia (James et al. 2009).

Future Directions

Continued studies of the involvement of the Wnt signaling pathway in islet development and function may reveal novel factors important in β cell growth and survival. A prerequisite for understanding the potential importance of Wnt signaling in islets is the identification of the specific Wnt signaling factors that are expressed in islets. Identification of these factors may provide opportunities for development of small molecules that target specific components of the pathways to promote growth and survival. Ongoing high-throughput screening studies of hundreds of thousands of compounds using islet tissues containing fluorescence reporter genes and growth or apoptosis-responsive promoters may uncover such small molecules.

Antidiabetogenic therapies consisting of combinations of GLP-1 and SDF-1 agonists may provide additive benefits in promoting both the growth and survival of β cells, thereby preserving or enhancing β cell mass. Recent findings suggest that both the pro-proliferative actions of GLP-1 and the anti-apoptosis actions of SDF-1 are mediated by the activation of β-catenin and TCF7L2 in β cells. Although the GLP-1/GLP-1R and SDF-1/CXCR4 axes both converge on downstream Wnt signaling at the level of the formation of transcriptionally productive complexes of β-catenin/TCF7L2, the target genes activated by GLP-1 and by SDF-1 differ. GLP-1-mediated activation of β-catenin/TCF7L2 results in the expression of genes involved in the cell division cycle, whereas SDF-1 actions result in the activation of the expression of genes engaged in cell survival. Furthermore, downstream β-catenin/TCF7L2 activation is a requisite for the pro-proliferative actions of GLP-1 and the anti-apoptotic actions of SDF-1. The two hormones, GLP-1 and SDF-1, acting together may provide additive benefits in promoting the regeneration and maintenance of β cell mass in diabetes.

Genome-wide association studies in search of risk alleles for type 2 diabetes are just beginning. It is estimated that 80–90 % of the human genome remains yet to be explored for the existence of diabetes-associated genes in the population. Predictably, further genome-wide scans in the future will uncover even more than the current 19 genes, and many will likely be involved in islet and β cell development and functions. It is tempting to speculate that the additional risk genes for type 2 diabetes that remain to be discovered in the future will include genes encoding components of the Wnt signaling pathway.

Intriguing current evidence warrants further investigations of Wnt ligands and Wnt signaling in the cross talk between adipose tissue and islets. Possibilities arise suggesting that Wnt ligands produced and secreted by adipocytes act on β cells to stimulate Wnt signaling.

Acknowledgments We thank Michael Rukstalis and Melissa Thomas for their helpful comments on this chapter and Sriya Avadhani, Violeta Stanojevic, and Karen McManus for their expert experimental assistance. Effort was supported in part by grants from the US Public Health Service, the American Diabetes Association, and the Juvenile Diabetes Research Foundation.

References

Ahlqvist E, Ahluwalia TS, Groop L (2011) Genetics of type 2 diabetes. Clin Chem 57:241–254, PMID: 21119033

Al-Hasani K, Pfeifer A, Courtney M, Ben-Othman N, Gjernes E, Vieira A, Druelle N, Avolio F, Ravassard P, Leuckx G, Lacas-Gervais S, Ambrosetti D, Benizri E, Hecksher-Sorensen J, Gounon P, Ferrer J, Gradwohl G, Heimberg H, Mansouri A, Collombat P (2013) Adult duct-lining cells can reprogram into β-like cells able to counter repeated cycles of toxin-induced diabetes. Dev Cell 26:86–100

Anselme I, Lacief C, Lanaud M, Ruther U, Schneider-Maunoury S (2007) Defects in brain patterning and head morphogenesis in the mouse mutant fused toes. Dev Biol 304:208–220

Barker N et al (2007) Identification of stem cells in small intestine and colon by marker gene Lgr5. Nature 449:1003–1007

Bonner-Weir S, Sharma A (2006) Are their pancreatic progenitor cells from which new islets form after birth? Nat Clin Pract Endocrinol Metab 2:240–241

Bordonaro M (2009) Role of Wnt signaling in the development of type 2 diabetes. Vitam Horm 80:563–581

Bort R, Martinez-Barbera JP, Beddington RS, Zaret KS (2004) Hex homeobox gene-dependent tissue positioning is required for organogenesis of the ventral pancreas. Development 131:797–806

Bort R, Signore M, Tremblay K, Martinez-Barbera JP, Zaret KS (2006) Hex homeobox gene controls the transition of the endoderm to a pseudostratified, cell emergent epithelium for liver bud development. Dev Biol 290:44–56

Boutant M, Ramos OH, Tourrel-Cuzin C, Movassat J, Ilias A, Vallois D, Planchais J, Pégorier JP, Schuit F, Petit PX, Bossard P, Maedler K, Grapin-Botton A, Vasseur-Cognet M (2012) COUP-TFII controls mouse pancreatic β-cell mass through GLP-1-β-catenin signaling pathways. PLoS One 7:e30847

Braun MM, Etheridge A, Bernard A, Robertson CP, Roelink H (2003) Wnt signaling is required at distinct stages of development for the induction of the posterior forebrain. Development 130:5579–5587

Burger JA, Kipps TJ (2006) CXCR4 a key receptor in the crosstalk between tumor cells and their microenvironment. Blood 107:1761–1767

Cano DA, Rulifson IC, Heiser PW, Swigart LB, Pelengaris S, German M, Evan GI, Bluestone JA, Hebrok M (2008) Regulated β-cell regeneration in the adult mouse pancreas. Diabetes 57:958–966

Cauchi S, Froguel P (2008) TCF7L2 genetic defect and type 2 diabetes. Curr Diab Rep 8:149–155, Review

Cauchi S et al (2008) Post genome-wide association studies of novel genes associated with type 2 diabetes show gene-gene interaction and high predictive value. PLoS One 3:e2031

Chen T, Li M, Ding Y, Zhang LS, Xi Y, Pan WJ, Tao DL, Wang JY, Li L (2009a) Identification of zinc-finger BED domain-containing 3 (Zbed3) as a novel Axin-interacting protein that activates Wnt/β-catenin signaling. J Biol Chem 284:6683–6689

Chen M, Philipp M, Wang J, Premont RT, Garrison TR, Caron MG, Lefkowitz RJ, Chen W (2009b) G Protein-coupled receptor kinases phosphorylate LRP6 in the Wnt pathway. J Biol Chem 284:35040–35048

Cheta D (1998) Animal models of type 1 (insulin-dependent) diabetes mellitus. J Pediatr Endocrinol Metab 11:11–19

Chiang YT, Ip W, Jin T (2012) The role of the Wnt signaling pathway in incretin hormone production and function. Front Physiol 12(3):273

Ching YP, Pang AS, Lam WH, Qi RZ, Wang JH (2002) Identification of a neuronal Cdk5 activator-binding protein as Cdk5 inhibitor J. Biol Chem 277:15237–15240

Colletti M, Cicchini C, Conigliaro A, Santangelo L, Alonzi T, Pasquini E, Tripodi M, Amicone L (2009) Convergence of Wnt signaling on the HNF4α-driven transcription in controlling liver zonation. Gastroenterology 137:660–672

Colli LM, Saggioro F, Serafini LN, Camargo RC, Machado HR, Moreira AC, Antonini SR, de Castro M (2013) Components of the canonical and non-canonical Wnt pathways are not mis-expressed in pituitary tumors. PLoS One 8:e62424

Collombat P, Xu X, Ravassard P, Sosa-Pineda B, Dussaud S, Billestrup N, Madsen OD, Serup P, Heimberg H, Mansouri A (2009) The ectopic expression of Pax4 in the mouse pancreas converts progenitor cells into α and subsequently β cells. Cell 138:449–462

Courtney M, Gjernes E, Druelle N, Ravaud C, Vieira A, Ben-Othman N, Pfeifer A, Avolio F, Leuckx G, Lacas-Gervais S, Hecksher-Sorensen J, Ravassard P, Heimberg H Mansouri A, Patrick Collombat P. The inactivation of Arx in pancreatic α-cells triggers their neogenesis and conversion into functional β-like cells. PLoS Genetics 9(10):e1003934

da Silva XG, Loder MK, McDonald A, Tarasov AI, Carzaniga R, Kronenberger K, Barg S, Rutter GA (2009) TCF7L2 regulates late events in insulin secretion from pancreatic islet β cells. Diabetes 58(4):894–905, Jan 23 ahead of print

da Silva XG, Mondragon A, Sun G, Chen L, McGinty JA, French PM, Rutter GA (2012) Abnormal glucose tolerance and insulin secretion in pancreas-specific Tcf7l2-null mice. Diabetologia 55:2667–2676

Dabernat S, Secrest P, Peuchant E, Moreau-Gaudry F, Dubus P, Sarvetnick N (2009) Lack of β-catenin in early life induces abnormal glucose homeostasis in mice. Diabetologia 52:1608–1617

Delmas V, Beermann F, Martinozzi S, Carreira S, Ackermann J, Kumasaka M, Denat L, Goodall J, Luciani F, Viros A, Demirkan N, Bastian BC, Goding CR, Larue L (2007) β-catenin induces immortalization of melanocytes by suppressing p16INK4a expression and cooperates with N-Ras in melanoma development. Genes Dev 21:2923–2935

Dessimoz J, Bonnard C, Huelsken J, Grapin-Botton A (2005) Pancreas-specific deletion of β-catenin reveals Wnt-dependent and Wnt-independent functions during development. Curr Biol 15:1677–1683

Do R, Bailey SD, Desbiens K, Belisle A, Montpetite, Bouchard C, Perusse L, Vohl MC, Engert JC (2008) Genetic variants of FTO influence adiposity, insulin sensitivity, leptin levels, and resting metabolic rate in the Quebec family diabetes. Diabetes 57:1147–1150

Dor Y, Brown J, Martinez OI, Melton DA (2004) Adult pancreatic β cells are formed by self-duplication rather than stem-cell differentiation. Nature 429:41–46

Drucker DJ (2006) The biology of incretin hormones. Cell Metab 3:153–165

Edwin F, Anderson K, Ying C, Patel TB (2009) Intermolecular interactions of Sprouty proteins and their implications in development and disease. Mol Pharmacol 76:679–691

Espinosa L, Engles-Esteve J, Aguilera C, Bigas A (2003) Phosphorylation by glycogen synthase kinase 3 β down-regulates Notch activity, a link for Notch and Wnt pathways. J Biol Chem 278:32227–32235

Figeac F, Uzan B, Faro M, Chelali N, Portha B, Movassat J (2010) Neonatal growth and regeneration of β-cells are regulated by the Wnt/β-catenin signaling in normal and diabetic rats. Am J Physiol Endocrinol Metab 298(2):E245–E256

Florez JC (2007) The new type 2 diabetes gene TCF7L2. Curr Opin Clin Nutr Metab Care 10:391–396

Florez J (2008a) Clinical review: the genetics of type 2 diabetes: a realistic appraisal in 2008. J Clin Endocrinol Metab 93:4633–4642

Florez J (2008b) Newly identified loci highlight β cell dysfunction as a key cause of type 2 diabetes: where are the insulin resistance genes? Diabetologia 51:1100–1110

Foley AC, Mercola M (2005) Heart induction by Wnt antagonists depends on the homeodomain transcription factor Hex. Genes Dev 19:387–396

Fonseca SG, Fukuma M, Lipson KL et al (2005) WFS1 is a novel component of the unfolded protein response and maintains homeostasis of the endoplasmic reticulum in pancreatic β cells. J Biol Chem 280:39609–39615

Frederiksson R, Hagglund M, Olszewski PK, Sstephansson O, Jacobsson JA, Olszewska AM, Levine AS, Lindblom J, Schioth HB (2008) The obesity gene, FTO, is of ancient origin, up-regulated during food deprivation and expressed in neurons of feeding-related nuclei of the brain. Endocrinology 149:2062–2071

Fujino T et al (2003) Low-density lipoprotein receptor-related protein 5 (LRP5) is essential for normal cholesterol metabolism and glucose-induced insulin secretion. Proc Natl Acad Sci USA 100:229–234

Gloyn AL, Braun M, Rorsman P (2009) Type 2 diabetes susceptibility gene TCF7L2 and its role in β cell function. Diabetes 58:832–834

Gordon MD, Nusse R (2006) Wnt signaling: multiple pathways, multiple receptors, and multiple transcription factors. J Biol Chem 281:22429–22433

Gradwohl G, Dierich A, LeMeur M, Guillemot F (2000) Neurogenin3 is required for the development of the four endocrine cell lineages of the pancreas. Proc Natl Acad Sci USA 9:1607–1611

Grant SF et al (2006) Variant of transcription factor 7-like 2 (TCF7L2) gene confers risk of type 2 diabetes. Nature Genet 38:320–323

Grarup N et al (2007) Studies of variants near the HHEX, CDKN2A/B, and IGF2BP2 genes with type 2 diabetes and impaired insulin release in 10,705 Danish subjects: validation and extension of genome-wide association studies. Diabetes 56:3105–3111

Guy GR, Jackson RA, Yusoff P, Chow SY (2009) Sprouty proteins: modified modulators, matchmakers or missing links? J Endocrinol 203:191–202

Habener JF, Stanojevic V (2012) α-cell role in β-cell generation and regeneration. Islets 4(3):188–198

Habener JF, Stanojevic V (2013) α cells come of age. Trends Endocrinol Metab 24:153–163

Haldorsen IS, Vesterhus M, Raeder H, Jensen DK, Sovik O, Molven A, Njelstad PR (2008) Lack of pancreatic body and tail in HNF1B mutation carriers. Diabet Med 25:782–787

Hallaq H et al (2004) A null mutation of Hhex results in abnormal cardiac development, defective vasculogenesis and elevated Vegfa levels. Development 131:5197–5209

Hanson AJ, Wallace HA, Freeman TJ, Beauchamp RD, Lee LA, Lee E (2012) XIAP monoubiquitylates Groucho/TLE to promote canonical Wnt signaling. Mol Cell 45(5):619–628

Hattersley AT (2007) Prime suspect: the TCF7L2 gene and type 2 diabetes risk. J Clin Invest 117:2077–2079

Haumaitre C, Fabre M, Cormier S, Baumann C, Delezoide AL, Ceereghini S (2006) Severe pancreas hypoplasia and multicystic renal dysplasia in two human fetuses carrying novel HNF1β/MODY5 mutations. Hum Mol Genet 15:2363–2375

Hayes MG et al (2007) Identification of type 2 diabetes genes in Mexican Americans through genome-wide association studies. Diabetes 56:3033–3044

Heiser PW, Lalu J, Taketo MM, Herrera PL, Hebrok M (2006) Stabilization of β-catenin impacts pancreatic growth. Development 133:2023–2033

Heller RS, Dichmann DS, Jensen J, Miller C, Wong G, Madsen OD, Serup P (2002) Expression patterns of Wnts, Frizzleds, sFRPs and misexpression in transgenic mice suggesting a role for Wnts in pancreas and foregut pattern formation. Dev Dyn 225:260–270

Heller RS, Klein T, Ling Z, Heimberg H, Katoh M, Madsen OD, Serup (2003) Expression of Wnt, Frizzled, sFRP, and DKK genes in adult human pancreas. Gene Expr 11:141–147

Heller C, Kühn MC, Mülders-Opgenoorth B, Schott M, Willenberg HS, Scherbaum WA, Schinner S (2011) Exendin-4 upregulates the expression of Wnt-4, a novel regulator of pancreatic β-cell proliferation. Am J Physiol Endocrinol Metab 301:E864–E872

Ishihara H, Takeda S, Tamura A et al (2004) Disruption of the WFS1 gene in mice causes progressive β cell loss and impaired stimulus-secretion coupling in insulin secretion. Hum Mol Genet 13:1159–1170

Jäggi F, Cabrita MA, Perl AK, Christofori G (2008) Modulation of endocrine pancreas development but not β-cell carcinogenesis by Sprouty4. Mol Cancer Res 6:468–482

James C, Kapoor RR, Ismail D, Hussain K (2009) The genetic basis of congenital hyperinsulinism. J Med Genet 46:289–299

Jensen JM, Cameron E, Baray MV, Starkev TW, Gianani R, Jensen J (2005) Recapitulation of elements on embryonic development in adult mouse pancreatic regeneration. Gastroenterology 128:728–741

Jia G, Yano CG, Yang S, Jian X, Yi C, Zhou ZA, He C (2008) Oxidative demethylation of 3-methylthymidine and 3-methyluracil in single-stranded DNA and RNA by mouse and human FTO. FEBS Lett 582:331319

Jin T (2008) The WNT, signalling pathway and diabetes mellitus. Diabetologia 51:1771–1780

Jin T, Liu L (2008) The Wnt signaling pathway effector TCF7L2 and type 2 diabetes mellitus. Mol Endocrinol 22:2383–2392

Jin W, Patti ME (2009) Genetic determinants and molecular pathways in the pathogenesis of type 2 diabetes. Clin Sci 116:99–111

Jungers KA, Le Goff C, Sommerville RP, Apte SS (2005) Adamts9 is widely expressed during mouse embryo development. Gene Expr Patterns 5:609–617

Karalay O, Doberauer K, Vadodaria KC, Knobloch M, Berti L, Miquelajauregui A, Schwark M, Jagasia R, Taketo MM, Tarabykin V, Lie DC, Jessberger S (2011) Prospero-related homeobox 1 gene (Prox1) is regulated by canonical Wnt signaling and has a stage-specific role in adult hippocampal neurogenesis. Proc Natl Acad Sci USA 108:5807–5812

Kayali AG, Van Gunst K, Campbell IL, Stotland A, Kritzik M, Liu G, Flodstrom-Tullberg M, Zhang YQ, Sarvetnick N (2003) The stromal cell-derived factor-1α/CXCR4 ligand-receptor axis is critical for progenitor survival and migration in the pancreas. J Cell Biol 163:859–869

Kieffer TJ, Habener JF (1999) The glucagon-like peptides. Endocr Rev 20:876–913

Kikuchi A, Kishido S, Yamamoto H (2006) Regulation of Wnt signaling by protein-protein interaction and post-translational modifications. Exp Mol Med 38:1–10

Kim HJ, Schieffarth JB, Jessurun J, Sumanas S, Petryk A, Lin S, Ekker SC (2005) Wnt5 signaling in vertebrate pancreas development. BMC Biol 24:3–23

Kim YS, Kang HS, Takeda Y, Hom L, Song HY, Jensen J, Jetten AM (2012) Glis3 regulates neurogenin 3 expression in pancreatic β-cells and interacts with its activator, Hnf6. Mol Cells 34:193–200

Komiya Y, Habas R (2008) Wnt signal transduction pathways. Organogenesis 4:68–75

Korinek V, Barker N, Moerer P, van Donselaar E, Huls G, Peters PJ, Clevers H (1998) Depletion of epithelial stem-cell compartments in the small intestine of mice lacking Tcf-4. Nat Genet 19:379–383

Krishnamurthy J, Ramsey MR, Ligon KL, Torrice C, Koh A, Bonner-Weir S, Sharpless NE (2006) p16INK4a induces an age-dependent decline in islet regenerative potential. Nature 443:453–457

Kryczek I, Wei S, Keller E, Liu R, Zou W (2007) Stroma-derived factor (SDF-1/CXCL12) and human tumor pathogenesis. Am J Physiol Cell Physiol 292:C987–C995

Kucia M, Ratajczak J, Ratajczak MZ (2005) Bone marrow as a source of circulating CXCR4$^+$ tissue-committed stem cells. Biol Cell 97:133–146

Le Bacquer O, Shu L, Marchand M, Neve B, Paroni F, Kerr Conte J, Pattou F, Froguel P, Maedler K (2011) TCF7L2 splice variants have distinct effects on β-cell turnover and function. Hum Mol Genet 20:1906–1915

Lee KM, Yasuda H, Hollingsworth MA, Ouellette MM (2005) Notch2-positive progenitors with the intrinsic ability to give rise to pancreatic duct cells. Lab Invest 85:1003–1012

Lee SH, Demeterco C, Geron I, Abrahamsson A, Levine F, Itkin-Ansari P (2008) Islet specific Wnt activation in human type 2 diabetes. Exp Diabetes Res 2008:728–763

Liu Z, Habener JF (2008) Glucagon-like peptide-1 activation of TCF7L2-dependent Wnt signaling enhances pancreatic β cell proliferation. J Biol Chem 283:8723–8735

Liu Z, Habener JF (2009) Stromal cell-derived factor-1 promotes survival of pancreatic β cells by the stabilisation of β-catenin and activation of transcription factor 7-like 2 (TCF7L2). Diabetologia 52:1589–1598

Liu Z, Stanojevic V, Avadhani S, Yano T, Habener JF (2011) Stromal cell-derived factor-1 (SDF-1)/chemokine (C-X-C motif) receptor 4 (CXCR4) axis activation induces intra-islet glucagon-like peptide-1 (GLP-1) production and enhances β cell survival. Diabetologia 54:2067–2076

Loder MK, da Silva XG, McDonald A, Rutter GA (2008) TCF7L2 controls insulin gene expression and insulin secretion in mature pancreatic β cells. Biochem Soc Trans 36:357–359

Logan CY, Nusse R (2004) The Wnt signaling pathway in development and disease. Annu Rev Cell Dev Biol 20:781–810

Longo KA, Kennell JA, Ochocinska MJ, Ross SE, Wright WS, McDougald OA (2002) Wnt signaling protects 3 T3-L1 preadipocytes from apoptosis through induction of insulin-like growth factors. J Biol Chem 277:38239–38244

Luo Y, Cai J, Xue H, Mattson MP, Rao MS (2006) SDF-1α/CXCR4 signaling stimulates β-catenin transcriptional activity in rat neural progenitors. Neurosci Lett 398:291–295

Lyssenko V (2008) The transcription factor 7-like 2 gene and increased risk of type 2 diabetes: an update. Curr Opin Clin Nutr Metab Care 11:385–392

Lyssenko V, Lupi R, Marchetti P, Del Guerra S, Orho-Melander M, Almgren P, Sjögren M, Ling C, Eriksson KF, Lethagen AL, Mancarella R, Berglund G, Tuomi T, Nilsson P, Del Prato S, Groop L (2007) Mechanisms by which common variants in the TCF7L2 gene increase risk of type 2 diabetes. J Clin Invest 117:2155–2163

Lyssenko V, Jonsson A, Almgren P, Pulizzi N, Isomaa B, Tusomi T, Gerglund G, Altshuler D, Nisson P, Groop L (2008) Clinical risk factors, DNA variants, and the development of type 2 diabetes. N Engl J Med 359:2220–2232

MacDonald BT, Semenov MV, He X (2007) SnapShot: Wnt/β-catenin signaling. Cell 131:1204

Maestro MA, Cardaida C, Boj SF, Luco RF, Servitja JM, Ferrer J (2007a) Distinct roles of HNF1β, HNF1α, and HNF4α in regulating pancreas development, β cell function, and growth. Endocr Rev 12:33–45

Maestro MA, Cardalda C, Boj SF, Luco RF, Servitja JM, Ferrer J (2007b) Distinct roles of HNF1β, HNF1α, and HNF4α in regulating pancreas development, β-cell function and growth. Endocr Dev 12:33–45

Marchetti P, Syed F, Suleiman M, Bugliani M, Marselli L (2012) From genotype to human β cell phenotype and beyond. Islets 4:323–332

Marzo N, Mora C, Fabregat ME, Martín J, Usac EF, Franco C, Barbacid M, Gomis R (2004) Pancreatic islets from cyclin-dependent kinase 4/R24C (Cdk4) knockin mice have significantly increased β cell mass and are physiologically functional, indicating that Cdk4 is a potential target for pancreatic β cell mass regeneration in Type 1 diabetes. Diabetologia 47:686–694

McDonald TJ, Ellard S (2013) Maturity onset diabetes of the young: identification and diagnosis. Ann Clin Biochem 50:403–415

McLin VA, Rankin SA, Zorn AM (2007) Repression of Wnt/β-catenin signaling in the anterior endoderm is essential for liver and pancreas development. Development 134:2207–2217

Mettus RV, Rane SG (2003) Characterization of the abnormal pancreatic development, reduced growth and infertility in Cdk4 mutant mice. Oncogene 22:8413–8422

Moon RT, Kohn AD, De Ferrari GV, Kaykas A (2004) WNT and β-catenin signalling: diseases and therapies. Nat Rev Genet 5:691–701

Moore AF et al (2008) Extension of type 2 diabetes genome-wide association scan results in the diabetes prevention program. Diabetes 57:2503–2510

Moritani M, Yamasaki S, Kagami M, Suzuki T, Yamaoka T, Sano T, Hata J, Itakura M (2005) Hypoplasia of endocrine and exocrine pancreas in homozygous transgenic TGF-β1. Mol Cell Endocrinol 229:175–184

Murtaugh LC (2008) The what, where, when and how of Wnt/β-catenin signaling in pancreas development. Organogenesis 4:81–86

Murtaugh LC, Law AC, Dor Y, Melton DA (2005) β-catenin is essential for pancreatic acinar but not islet development. Development 132:4663–4674

Nakhai H et al (2008) Conditional ablation of Notch signaling in pancreatic development. Development 135:2757–2765

Nauck MA, Meier JJ (2007) The enteroinsular axis may mediate the diabetogenic effects of TCF7L2 polymorphisms. Diabetologia 50:2413–2416

Nelson WJ, Nusse R (2004) Convergence of Wnt, β-catenin, and cadherin pathways. Science 303:1483–1487

Nielsen J, Christiansen J, Lykke-Andersen J, Johnsen AH, Wewer UM, Nielsen FC, Elsen J et al (1999) A family of insulin-like growth factor II mRNA-binding proteins represses translation in late development. Mol Cell Biol 19:1262–1269

Nusse R (2008) Wnt signaling and stem cell control. Cell Res 18:523–527

Owen KR, McCarthy MI (2007a) Genetics of type 2 diabetes. Curr Opin Genet Dev 17:239–244

Owen KR, McCarthy MI (2007b) Genetics of type 2 diabetes. Curr Opin Genet Develop 17:239–244

Palmer ND et al (2008) Quantitative trait analysis of type 2 diabetes susceptibility loci identified from whole genome association studies in the insulin resistance atherosclerosis family study. Diabetes 57:1093–1100

Papadopoulou S, Edlund H (2005) Attenuated Wnt signaling perturbs pancreatic growth but not pancreatic function. Diabetes 54:2844–2851

Pascoe L, Tura A, Patel SK, Ibrahim IM, Ferrannini E, Zeggini E, Weedon MN, Mari A, Hattersley AT, McCarthy MI, Frayling TM, Walker M (2007) RISC Consortium; U.K. Type 2 Diabetes Genetics Consortium. Common variants of the novel type 2 diabetes genes CDKAL1 and HHEX/IDE are associated with decreased pancreatic β cell function. Diabetes 56:3101–3104

Pauls F, Bancroft RW (1950) Production of diabetes in the mouse by partial pancreatectomy. Am J Physiol 160:103–106

Pedersen AH, Heller RS (2005) A possible role for the canonical Wnt pathway in endocrine cell development in chicks. Biochem Biophys Res Commun 333:961–968

Perry JR, Frayling TM (2008) New gene variants alter type 2 diabetes risk predominantly through reduced β cell function. Curr Opin Clin Nutr Metab Care 11:371–377

Peters T, Ausmeier K, Dildrop R, Ruther U (2002) The mouse Fused toes (Ft) mutation is the result of a 1.6 Mb deletion including the entire Iroquois B gene locus. Mamm Genome 13:186–188

Petri A, Anfelt-Ronne J, Fredericksen RS, Edwards DG, Madsen B, Serup P, Fleckner J, Heller RS (2006) The effect of neurogenin3 deficiency on pancreatic gene expression in embryonic mice. J Mol Endocrinol 37:301–316

Ruchat SM et al (2009) Association between insulin secretion, insulin sensitivity and type 2 diabetes susceptibility variants identified in genome-wide association studies. Acta Diabetol. 46:217–226

Rane SG, Dubus P, Mettus RV, Galbreath EJ, Boden G, Reddy EP, Barbacid M (1999) Loss of Cdk4 expression causes insulin-deficient diabetes and Cdk4 activation results in β-islet cell hyperplasia. Nat Genet 22:44–52

Ratajczak MZ, Zuba-Surma E, Kucia M, Reca R, Wojakowski W, Ratajczak J (2006) The pleiotropic effects of the SDF-1-CXCR4 axis in organogenesis, regeneration and tumorigenesis. Leukemia 20:1915–1924

Rees DA, Alcolado JC (2005) Animal models of diabetes mellitus. Diabet Med 22:359–370

Riggs AC, Bernal-Mizrachi E, Ohsugi M et al (2005) Mice conditionally lacking the Wolfram gene in pancreatic islet β cells exhibit diabetes as a result of enhanced endoplasmic reticulum stress and apoptosis. Diabetologia 48:2313–2321

Rulifson JC, Karnik SK, ten Heiser PW, Berge D, Chen H, Gu X, Taketo MM, Nusse R, Hebrok M, Kim SK (2007) Wnt signaling regulates pancreatic β cell proliferation. Proc Natl Acad Sci USA 104:6247–6252

Sakaguchi Y, Inaba M, Kusafuka K, Okazaki K, Ikehara S (2006) Establishment of animal models for three types of pancreatic and analyses of regeneration mechanisms. Pancreas 33:371–381

Savic D, Ye H, Aneas I, Park SY, Bell GI, Nobrega MA (2011) Alterations in TCF7L2 expression define its role as a key regulator of glucose metabolism. Genome Res 21(9):1417–1425

Saxena R et al (2007) Genome-wide association analysis identifies loci for type 2 diabetes and triglyceride levels. Science 316:1332–1336

Schafer SA, Tschritter O, Machicao F, Thamer C, Stefan N, Gallwitz B, Holst JJ, Dekker JM, 't Hart LM, Nipeis G, van Haeften TW, Haring HU, Fritsche A (2007) Impaired glucagon-like peptide-1-induced insulin secretion in carriers of transcription factor 7-like 2 (TCF7L2) gene polymorphisms. Diabetologia 59:2443–2450

Schinner S, Ulgen F, Papewalis C, Schott M, Woelk A, Vidal-Puig A, Scherbaurm WA (2008) Regulation of insulin secretion, glucokinase gene transcription and β cell proliferation by adipocyte-derived Wnt signalling molecules. Diabetologia 51:147–154

Scott LJ et al (2007) A genome-wide association study of type 2 diabetes in France detects multiple susceptibility variants. Science 316:1341–1345

Semenov MV, Habas R, Macdonald BT, He X (2007) SnapShot: Noncanonical Wnt signaling pathways. Cell 131:1738

Serafimidis I, Rakatzi I, Episkopou V, Gouti M, Gavalas A (2008) Novel effectors of directed and Ngn3-mediated differentiation of mouse embryonic stem cells into endocrine pancreas progenitors. Stem Cells 26:3–16

Shu L, Sauter NS, Schulthess FT, Matvevenko AV, Oberholzer J, Maedler K (2008) Transcription factor 7-like 2 regulates β cell survival and function in human pancreatic islets. Diabetes 57:645–653

Shu L, Matveyenko AV, Kerr-Conte J, Cho JH, McIntosh CH, Maedler K (2009) Decreased TCF7L2 protein levels in type 2 diabetes mellitus correlate with downregulation of GIP- and GLP-1 receptors and impaired β-cell function. Hum Mol Genet 18:2388–2399

Shu L, Zien K, Gutjahr G, Oberholzer J, Pattou F, Kerr-Conte J, Maedler K (2012) TCF7L2 promotes β cell regeneration in human and mouse pancreas. Diabetologia 55:3296–3307

Sladek R et al (2007) A genome-wide association study identifies novel risk loci for type 2 diabetes. Nature 445:881–885

Smith U (2007) TCF7L2 and diabetes – what we Wnt to know. Diabetologia 50:5–7
Steinthorsdottir V et al (2007) CDKAL1 influences insulin response and risk of type 2 diabetes. Nature Genet 39:770–775
Takeda K, Inoue H, Tanizawa Y et al (2001) WFS1 (Wolfram syndrome 1) gene product: predominant subcellular localization to endoplasmic reticulum in cultured cells and neuronal expression in rat brain. Hum Mol Genet 10:477–484
Talchai C, Xuan S, Lin HV, Sussel L, Accili D (2012) Pancreatic β cell dedifferentiation as a mechanism of diabetic β cell failure. Cell 150:1223–1234
Ubeda M, Rukstalis JM, Habener JF (2006) Inhibition of cyclin-dependent kinase 5 activity protects pancreatic β cells from glucotoxicity. J Biol Chem 281:28858–28864
Van Hoek M, Dehghan A, Witteman JC, Van Dulin CM, Utterfinden AG, Oostra BA, Hofman A, Sijbrands BA, Janssens AC (2008) Predicting type 2 diabetes based on polymorphisms from genome-wide association studies: a population-based study. Diabetes 57:3122–3128
Veeman MT, Axelrod JD, Moon RT (2003) A second canon. Functions and mechanisms of β-catenin-independent Wnt signaling. Dev Cell 5:367–377
Wada H, Okamoto H (2009) Roles of planar cell polarity pathway genes for neural migration and differentiation. Dev Growth Differ 51:233–240
Wang OM, Zhang Y, Yang KM, Zhou HY, Yano HJ (2006) Wnt/β-catenin signaling pathway is active in pancreatic development of rat embryo. World J Gastroenterol 12:2615–2619
Weedon MN (2007) The importance of TCF7L2. Diabet Med 24:1062–1066
Wells JM, Esni F, Bolvin GP, Aronow BJ, Stuart W, Combs C, Sklenka A, Leach SD, Lowy AM (2007) Wnt/β-catenin signaling is required for development of the exocrine pancreas. BMC Dev Biol 7:4
Welters HJ, Kulkarni RN (2008) Wnt signaling: relevance to β cell biology and diabetes. Trends Endocrinol Metab 19:349–355
Willert K, Jones KA (2006) Wnt signaling: is the party in the nucleus? Genes Dev 20:1394–1404
Xiong X, Shao W, Jin T (2012) New insight into the mechanisms underlying the function of the incretin hormone glucagon-like peptide-1 in pancreatic β-cells: the involvement of the Wnt signaling pathway effector β-catenin. Islets 4:359–365
Xu X, D'Hoker J, Stangé G, Bonné S, De Leu N, Xiao X, Van de Casteele M, Mellitzer G, Ling Z, Pipeleers D, Bouwens L, Scharfmann R, Gradwohl G, Heimberg H (2008) β cells can be generated from endogenous progenitors in injured adult mouse pancreas. Cell 132:197–207
Yamada T, Ishihara H, Tamura A et al (2006) WFS1-deficiency increases endoplasmic reticulum stress, impairs cell cycle progression and triggers the apoptotic pathway specifically in pancreatic β cells. Hum Mol Genet 15:1600–1609
Yan HX, Yang W, Zhang R, Chen L, Tang L, Zhai B, Liu SQ, Cao HF, Man XB, Wu HP, Wu MC, Wang HY (2006) Protein-tyrosine phosphatase PCP-2 inhibits β-catenin signaling and increases E-cadherin-dependent cell adhesion. J Biol Chem 281:15423–15433
Yang H, Li Q, Lee JH, Shu Y (2012) Reduction in Tcf7l2 expression decreases diabetic susceptibility in mice. Int J Biol Sci 8(6):791–801
Yano T, Liu Z, Donovan J, Thomas MK, Habener JF (2007) Stromal cell derived factor-1 (SDF-1)/CXCL12 attenuates diabetes in mice and promotes pancreatic β cell survival by activation of the prosurvival kinase Akt. Diabetes 56:2946–2957
Yi F, Brubaker PL, Jin T (2005) TCF-4 mediates cell type-specific regulation of proglucagon gene expression by β-catenin and glycogen synthase kinase 3 β. J Biol Chem 280:1457–1464
Yi F, Sun J, Lim GE, Fantus IG, Brubaker PL, Jin T (2008) Cross talk between the insulin and Wnt signaling pathways: evidence from intestinal endocrine L cells. Endocrinology 149:2341–2351
Zamparini AL, Watts T, Gardner CE, Tomlinson SR, Johnston GI, Brickman JM (2006) Hex acts with β-catenin to regulate anteroposterior patterning via a Groucho-related co-repressor and Nodal. Development 133:3709–3722
Zamparnini AL, Watts T, Gardner CE, Tomlinson SR, Johnston GI, Brickman JM (2006) Hex acts with β-catenin to regulate anteroposterior patterning via a Groucho-related co-repressor and Nodal. Development 133:3709–3722

Zeggini et al (2007) Replication of genome-wide association signals in UK samples reveals risk loci for type 2 diabetes. Science 316:1336–1341

Zhao J, Schug J, Li M, Kaestner KH, Grant SF (2010) Disease-associated loci are significantly over-represented among genes bound by transcription factor 7-like 2 (TCF7L2) in vivo. Diabetologia 53:2340–2346

Zhou Y, Zhang E, Berggreen C, Jing X, Osmark P, Lang S, Cilio CM, Göransson O, Groop L, Renström E, Hansson O (2012) Survival of pancreatic β cells is partly controlled by a TCF7L2-p53-p53INP1-dependent pathway. Hum Mol Genet 21:196–207

Islet Structure and Function in the GK Rat 26

Bernard Portha, Grégory Lacraz, Audrey Chavey,
Florence Figeac, Magali Fradet, Cécile Tourrel-Cuzin,
Françoise Homo-Delarche, Marie-Héléne Giroix, Danièle Bailbé,
Marie-Noëlle Gangnerau, and Jamileh Movassat

Contents

The Goto–Kakizaki (GK) Wistar Rat as Model of Spontaneous T2D	744
A Perturbed Islet Architecture, with Signs of Progressive Fibrosis, Inflammatory Microenvironment, Microangiopathy, and Increased Oxidative Stress	745
Less β-Cells Within the Pancreas with Less Replicative Activity but Intact Survival Capacity	748
Which Etiology for the β-Cell Mass Abnormalities?	749
Multiple β-Cell Functional Defects Mostly Targeting Insulin Release	750
Insulin Biosynthesis Is Grossly Preserved	750
Glucose-Induced Activation of Insulin Release Is Lost	751
Insulin Secretion Amplifying Mechanisms Are Altered	752
Insulin Exocytotic Machinery Is Abnormal	755
Secretory Response to Non-glucose Stimuli Is Partly Preserved	755
Islet ROS Scavenging Capacity Is Increased	756
Which Etiology for the Islet Functional Abnormalities?	756
Cross-References	759
References	759

Abstract

Type 2 diabetes mellitus (T2D) arises when the endocrine pancreas fails to secrete sufficient insulin to cope with the metabolic demand because of β-cell secretory dysfunction and/or decreased β-cell mass. Defining the nature of the pancreatic islet defects present in T2D has been difficult, in part because human islets are inaccessible for direct study. This review is aimed to illustrate to what extent the Goto–Kakizaki rat, one of the best characterized animal models of spontaneous T2D, has proved to be a valuable tool offering sufficient

---

B. Portha (✉) • G. Lacraz • A. Chavey • F. Figeac • M. Fradet • C. Tourrel-Cuzin •
F. Homo-Delarche • Marie-Héléne Giroix • D. Bailbé • M.-N. Gangnerau • J. Movassat
Laboratoire B2PE, Unité BFA, Université Paris-Diderot et CNRS EAC4413, Paris Cedex13, France
e-mail: portha@univ-paris-diderot.fr

commonalities to study this aspect. A comprehensive compendium of the multiple functional GK islet abnormalities so far identified is proposed in this perspective. The pathogenesis of defective β-cell number and function in the GK model is also discussed. It is proposed that the development of T2D in the GK model results from the complex interaction of multiple events: (i) several susceptibility loci containing genes responsible for some diabetic traits (distinct loci encoding impairment of β-cell metabolism and insulin exocytosis, but no quantitative trait locus for decreased β-cell mass); (ii) gestational metabolic impairment inducing an epigenetic programming of the offspring pancreas (decreased β-cell neogenesis and proliferation) transmitted over generations; and (iii) loss of β-cell differentiation related to chronic exposure to hyperglycemia/hyperlipidemia, islet inflammation, islet oxidative stress, islet fibrosis, and perturbed islet vasculature.

Keywords

Type 2 diabetes • GK rat • Islet cells • β-cell development • Differentiation and survival • Insulin release

The Goto–Kakizaki (GK) Wistar Rat as Model of Spontaneous T2D

Type 2 diabetes (T2D) arises when the endocrine pancreas fails to secrete sufficient insulin to cope with the metabolic demand (Butler et al. 2003; Donath and Halban 2004) because of β-cell secretory dysfunction and/or decreased β-cell mass. Hazard of invasive sampling and lack of suitable noninvasive methods to evaluate β-cell mass and β-cell functions are strong limitations for studies of the living pancreas in human. In such a perspective, appropriate rodent models are essential tools for identification of the mechanisms that increase the risk of abnormal β-cell mass/function and of T2D. Some answers to these major questions are available from studies using the endocrine pancreas of the Goto–Kakizaki (GK) rat model of T2D. It is the aim of this chapter to review the common features that make studies of the GK β-cell so compelling.

The GK line was established by repeated inbreeding from Wistar (W) rats selected at the upper limit of normal distribution for glucose tolerance (Goto et al. 1975, 1988; Portha et al. 2001, 2007; Östenson 2001; Portha 2005).

Until the end of the 1980s, GK rats were bred only in Sendai (Goto et al. 1975). Colonies were then initiated with breeding pairs from Japan, in Paris, France (GK/Par) (Portha et al. 1991); Dallas, TX, USA (GK/Dal) (Ohneda et al. 1993); Stockholm, Sweden (GK/Sto) (Östenson 2001); Cardiff, UK (GK/Card) (Lewis et al. 1996); Coimbra, Portugal (GK/Coi) (Duarte et al. 2004); and Tampa, USA (GK/Tamp) (Villar-Palasi and Farese 1994). Some other colonies existed for shorter periods during the 1990s in London, UK (GK/Lon) (Hughes et al. 1994); Aarhus, Denmark; and Seattle, USA (GK/Sea) (Metz et al. 1999). There are also GK rat colonies derived from Paris in Oxford, UK (GK/Ox) (Wallis et al. 2004), and Brussels, Belgium (GK/Brus) (Sener et al. 2001). Also, GK rats are available commercially from Japanese breeders Charles River Japan, Yokohama; Oriental Yeast, Tokyo; Clea Japan Inc, Osaka (GK/Clea); Japan SLC, Shizuoka (GK/SLC); and Takeda Lab Ltd, Osaka (GK/Taked) and from Taconic, USA (GK/Mol/Tac).

In our colony (GK/Par subline) maintained since 1989, the adult GK/Par body weight is 10–30 % lower than that of age- and sex-matched control animals. In male GK/Par rats, non-fasting plasma glucose levels are typically 10–14 mM (6–8 mM in age-matched Wistar (W) outbred controls). Despite the fact that GK rats in the various colonies bred in Japan and outside over 20 years have maintained rather stable degree of glucose intolerance, other characteristics such as β-cell number, insulin content, and islet metabolism and secretion have been reported to differ between some of the different colonies, suggesting that different local breeding environments and/or newly introduced genetic changes account for contrasting phenotypic properties.

Presently it is not clear whether the reported differences are artifactual or true. Careful and extensive identification of GK phenotype within each local subline is therefore necessary when comparing data from different GK sources. For further details concerning the pathogenic sequence culminating in the chronic hyperglycemia at adult age in the GK/Par rat, please refer to recent reviews (Östenson 2001; Portha 2005; Portha et al. 2007).

A Perturbed Islet Architecture, with Signs of Progressive Fibrosis, Inflammatory Microenvironment, Microangiopathy, and Increased Oxidative Stress

The adult GK/Par pancreas exhibits two different populations of islets in situ: large islets with pronounced fibrosis (Portha et al. 2001) and heterogeneity in the staining of their β-cells and small islets with heavily stained β-cells and normal architecture. One striking morphologic feature of GK rat islets is the occurrence of these big islets characterized by connective tissue separating strands of endocrine cells (Goto et al. 1988; Suzuki et al. 1992; Guenifi et al. 1995). Accordingly, the mantle of glucagon and somatostatin cells is disrupted and these cells are found intermingled between β-cells. These changes increase in prevalence with ageing (Suzuki et al. 1992).

No major alteration in pancreatic glucagon content, expressed per pancreatic weight, has been demonstrated in GK/Sto rats (Abdel-Halim et al. 1993), although the total α-cell mass was decreased by about 35 % in adult GK/Par rats (Movassat et al. 1997). The peripheral localization of glucagon-positive cells in W islets was replaced in GK/Sto rats with a more random distribution throughout the core of the islets (Guest et al. 2002). Pancreatic somatostatin content was slightly but significantly increased in GK/Sto rats (Abdel-Halim et al. 1993).

Chronic inflammation at the level of the GK/Par islet has recently received demonstration, and it is now considered as a pathophysiological contributor in

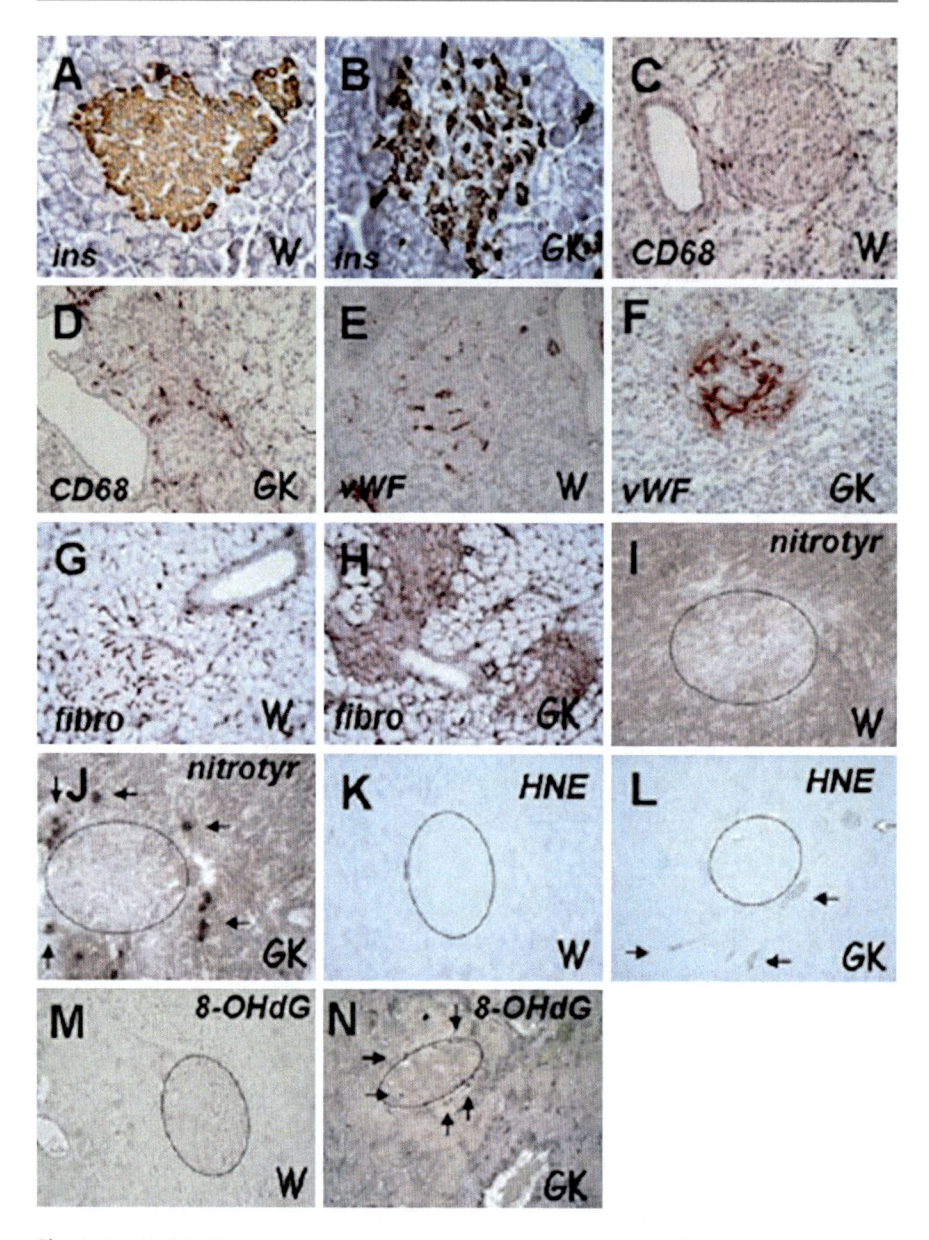

Fig. 1 Insulin labelling demonstrates the concomitant presence of large fibrotic islets (*B*) in adult GK/Par pancreas as compared with age-matched control Wistar (W) pancreas (*A*) (×500). Fibrosis is extensive in large GK/Par islets, as shown by fibronectin (*H*) labelling (×250). Small intact islets coexist with large fibrotic islets (not shown). Inflammatory cells infiltrate the islets of adult GK/Par rats. Compared with adult W rats, numerous macrophages are present in/around GK/Par islets, as shown by CD68 (*D* vs. *C*; ×500) and MHC class II (not shown) labellings. The concomitant presence of macrophages and granulocytes together with the quasi-absence of T and B cells and ED3 macrophages that are involved in autoimmune reaction (data not shown)

type 2 diabetes (Ehses et al. 2007a, b). Using an Affymetrix microarray approach to evaluate islet gene expression in freshly isolated adult GK/Par islets, we found that 34 % of the 71 genes found to be overexpressed belong to inflammatory/immune response gene family and 24 % belong to extracellular matrix (ECM)/cell adhesion gene family (Homo-Delarche et al. 2006) Numerous macrophages ($CD68^+$ and MHC class II^+) and granulocytes were found in/around adult GK/Par islets (Homo-Delarche et al. 2006). Upregulation of the MHC class II gene was also reported in a recent study of global expression profiling in GK/Takonic islets (Ghanaat-Pour et al. 2007). Immunolocalization with anti-fibronectin and anti-vWF antibodies indicated that ECM deposition progresses from intra- and peri-islet vessels, as it happens in microangiopathy (Homo-Delarche et al. 2006). These data demonstrate that a marked inflammatory reaction accompanies GK/Par islet fibrosis and suggest that islet alterations develop in a way reminiscent of microangiopathy (Ehses et al. 2007b). The previous reports by our group and others that increased blood flow and altered vascularization are present in the GK/Par and GK/Sto models (Atef et al. 1994; Svensson et al. 1994, 2000) are consistent with such a view. The increased islet blood flow in GK rats may be accounted for by an altered vagal nerve regulation mediated by nitric oxide, since vagotomy as well as inhibition of NO synthase normalized GK/Sto islet flow (Svensson et al. 1994). In addition, islet capillary pressure was increased in GK/Sto rats (Carlsson et al. 1997); this defect was reversed after 2 weeks of normalization of glycemia by phlorizin treatment. The precise relationship between islet microcirculation and β-cell secretory function remains to be established.

\Immunohistochemistry on diabetic GK/Par pancreases (Fig. 1) showed, unlike Wistar islets, the presence of nitrotyrosine and HNE labellings, which identify ROS and lipid peroxidation, respectively. Marker-positive cells were predominantly localized at the GK/Par islet periphery or along ducts and were accompanied by inflammatory infiltrates. Intriguingly, no marker-positive cell was detected within the islets in the same GK/Par pancreases (Lacraz et al. 2009). Such was not apparently the case in GK/Taked pancreases, as 8-OHdG and HNE-modified proteins accumulation was described within the islets. In this last study, the animals were older as compared to our study and accumulation of markers was correlated to hyperglycemia duration (Ihara et al. 1999). This suggests that the lack of OS-positive cells within islets as found in the young adult diabetic GK/Par is only transient and represents an early stage for a time-dependent evolutive islet adaptation.

Fig. 1 (continued) suggests a pure inflammatory process. Islet vascularization is altered in adult GK/Par rats. Labelling for vWF, a factor known to be produced by endothelial cells, shows the normal organization of islet vascularization in adult W rats (*E*). Islet vascularization differs markedly in age-matched GK/Par rats and appears to be hypertrophied (*F*) (×500). Nitrotyrosine, 4-hydroxy-2-nonenal (HNE)-modified proteins, and 8-hydroxy-2'-deoxyguanosine (8-OHdG) accumulate in the peri-islet vascular and inflammatory compartments of the adult GK/Par pancreas. Immunolabelling of nitrotyrosine, HNE adducts, or 8-OHdG (*arrows*) in pancreatic tissues of GK/Par (*J, L, N*) and W rats (*I, K, M*). An islet is encircled in each image (×250)

Less β-Cells Within the Pancreas with Less Replicative Activity but Intact Survival Capacity

In the adult hyperglycemic GK/Par rats (males or females), total pancreatic β-cell mass is decreased (by 60 %) (Portha et al. 2001; Movassat et al. 1997). This alteration of the β-cell population cannot be ascribed to increased β-cell apoptosis but is related, at least partly, to significantly decreased β-cell replication as measured in vivo, in situ (Portha et al. 2001). The islets isolated by standard collagenase procedure from adult GK/Par pancreases show limited decreased β-cell number (by 20 % only) and low insulin content compared with control islets (Giroix et al. 1999). The islet DNA content was decreased to a similar extent, consistent with our morphometric data, which indicates that there is no major change in the relative contribution of β-cells to total endocrine cells in the GK islets. In addition, the insulin content, when expressed relative to DNA, remains lower in GK islets than in control (inbred W/Par) islets, which supports some degranulation in the β-cells of diabetic animals (Giroix et al. 1999). Electron microscopy observation of β-cell in GK/SLC pancreas revealed that the number of β granules is decreased and that of immature granules increased. The Golgi apparatus was developed and the cisternae of the rough endoplasmic reticulum were dilated, indicating cell hyperfunction (Momose et al. 2006).

The distribution of various GK islet cell types appears to differ between some of the GK rat colonies. Thus, in the Stockholm colony, β-cell density and relative volume of insular cells were alike in adult GK/Sto rats and control W rats (Östenson 2001; Guenifi et al. 1995; Abdel-Halim et al. 1993). Similar results were reported in the Dallas colony (GK/Dal) (Ohneda et al. 1993). Reduction of adult β-cell mass, to an extent similar to that we reported in GK/Par rats, was however mentioned in GK rats from Sendai original colony (Goto et al. 1988), in GK/Taked (Koyama et al. 1998), in GK/Clea (Goda et al. 2007), and in GK/Coi (Seiça et al. 2003). Another element of heterogeneity between the different GK sources is related to the time of appearance of significant β-cell mass reduction when it is observed: It varies from fetal age in GK/Par to neonatal age in GK/Coi (Duarte et al. 2004) or young adult age (8 weeks) in GK/Taked (Koyama et al. 1998) and GK/SLC (Momose et al. 2006). The reason for such discrepancies in the onset and the severity of the β-cell mass reduction among colonies is not identified, but can be ascribed to differences in islet morphometric methodologies and/or characteristics acquired within each colony and arising from different nutritional and environmental conditions.

A meaningful set of data from our group (Movassat and Portha 1999; Miralles and Portha 2001; Plachot et al. 2001; Calderari et al. 2007) suggest that the permanently reduced β-cell mass in the GK/Par rat reflects a limitation of β-cell neogenesis during early fetal life and thereafter. Follow-up of the animals after delivery revealed that GK/Par pups become overtly hyperglycemic for the first time after 3–4 weeks of age only (i.e., during the weaning period). Despite normoglycemia, total β-cell mass was clearly decreased (by 60 %) in the GK/Par pups when compared with age-related W pups (Movassat et al. 1997). Since this early β-cell growth retardation in the prediabetic GK/Par rat pups can be ascribed neither to decreased β-cell replication nor to increased apoptosis (Movassat et al. 1997), we postulated that the recruitment of new β-cells from the precursor pool (β-cell neogenesis) was defective in the young prediabetic GK/Par rat. A comparative study of the development of GK/Par and W pancreases indicates that the β-cell deficit (reduced by more than 50 %) starts as early as fetal age 16 days (E16) (Miralles and Portha 2001). During the time window E16–E20, we detected an unexpected anomaly of proliferation and apoptosis of undifferentiated ductal cells in the GK/Par pancreatic rudiments (Miralles and Portha 2001; Calderari et al. 2007). Therefore, the decreased cell proliferation and survival in the ductal compartment of the pancreas, where the putative endocrine precursor cells localize, suggest that the impaired development of the β-cell in the GK/Par fetus could result from the failure of the proliferative and survival capacities of the endocrine precursor cells. Data from our group indicate that defective signaling through the IGF2/IGF1-R pathway is involved in this process at this stage. Importantly this represents a primary anomaly since Igf2 and IGF1-R protein expressions are already decreased within the GK/Par pancreatic rudiment at E13.5, at a time when β-cell mass (first wave of β-cell expansion) is in fact normal (Calderari et al. 2007). Low levels of pancreatic Igf2 associated with β-cell number deficiency are maintained thereafter in the GK/Par fetuses until delivery (Serradas et al. 2002). We have also published data illustrating a poor proliferation and/or survival of the endocrine precursors also during neonatal and adult life (Movassat and Portha 1999; Plachot et al. 2001). Altogether these arguments support the notion that an impaired capacity of β-cell neogenesis (either primary in the fetus or compensatory in the newborn and the adult) results from the permanently decreased pool of endocrine precursors in the GK/Par pancreas (Movassat et al. 2007).

Which Etiology for the β-Cell Mass Abnormalities?

During the last few years, some important information concerning the determinants (morbid genes vs. environment impact) for the low β-cell mass in the GK/Par model has been supplied. Hyperglycemia experienced during the fetal and/or early postnatal life may contribute to programming of the endocrine pancreas (Simmons 2006). Such a scenario potentially applies to the GK/Par rat, as GK/Par mothers are slightly hyperglycemic through their gestation and during the suckling period (Serradas et al. 1998). We have preliminary data using an embryo transfer strategy first described by Gill-Randall et al. (2004) suggesting that GK/Par embryos transferred in the uterus of euglycemic W mother still develop deficiency of β-cell mass when adults, to the same extent as the GK/Par rats from our stock colony (Chavey et al. 2008). While this preliminary conclusion rather favors a major role for inheritance of morbid genes, additional studies are needed to really eliminate the option that the gestational diabetic pattern of the GK/Par mothers does not contribute to establish and/or maintain the transmission of endocrine pancreas programming from one GK/Par generation to the next one. Moreover, studies on the offspring in crosses between GK/Par and W rats demonstrated that F1 hybrid fetuses, regardless of whether the mother was a GK or a W rat, exhibit decreased β mass and glucose-induced insulin secretion closely resembling those in

GK/GK fetuses (Serradas et al. 1998). This finding indicates that conjunction of GK genes from both parents is not required for defective β-cell mass to be fully expressed. We have also shown that to have one GK parent is a risk factor for a low β-cell mass phenotype in young adults, even when the other parent is a normal W rat (Calderari et al. 2006).

Search for identification of the morbid genes using a quantitative trait locus (QTL) approach has led to identification of six independently segregating loci containing genes regulating fasting plasma glucose and insulin levels, glucose tolerance, insulin secretion, and adiposity in GK/Par rats (Gauguier et al. 1996). The same conclusion was drawn by Galli et al. (1996) using GK/Sto rats. This established the polygenic inheritance of diabetes-related parameters in the GK rats whatever their origin. Both studies found the strongest evidence of linkage between glucose tolerance and markers spanning a region on rat chromosome 1, called Niddm1 locus. Recent works using congenic technology have identified a short region on the Niddm1i locus of GK/Sto rats that may contribute to defective insulin secretion (Lin et al. 2001). It has been recently reported that β-cell mass is intact in Niddm1i subcongenics (Granhall et al. 2006). These results are however inconsistent with the enhanced insulin release and increased islet size described in a GK/Ox congenic strain targeting a similar short region of the GK QTL Niddm1 (Wallis et al. 2008). Finally, no QTL association with β-cell mass or β-cell size could be found in the GK/Par rat (Ktorza and Gauguier, personal communication of unpublished data). Therefore, the likelihood that a genotype alteration directly contributes to the low β-cell mass phenotype in the GK/Par rat is reduced. The raised question to be answered now is whether or not epigenetic perturbation of gene expression occurs in the developing GK/Par pancreas and programs a durable alteration of the β-cell mass as seen in the adult. *igf2* and *igf1r* genes are good candidates for such a perspective.

Finally, since the loss of GK/Taked β-cells was mitigated by in vivo treatment with the α-glucosidase inhibitor voglibose (Koyama et al. 2000) or miglitol (Goda et al. 2007) or enhanced when the animals are fed with sucrose (Koyama et al. 1998; Mizukami et al. 2008), pathological progression (β-cell number, fibrosis) of the GK β-cell mass is also dependent on the metabolic (glycemic) control.

Multiple β-Cell Functional Defects Mostly Targeting Insulin Release

Insulin Biosynthesis Is Grossly Preserved

As for total pancreatic β-cell mass, there is some controversy regarding the content of pancreatic hormones in GK rats. In the adult hyperglycemic GK/Par rats, total pancreatic insulin stores are decreased by 60–40 % (Portha et al. 2001). In other GK rat colonies (Takeda, Stockholm, Seattle), total insulin store values have been found similarly or more moderately decreased, compared with control rats (Metz et al. 1999; Abdel-Halim et al. 1993; Keno et al. 1994; Östenson et al. 1993a;

Suzuki et al. 1997; Salehi et al. 1999). The islets isolated by standard collagenase procedure from adult GK/Par pancreases show lower insulin content compared with control islets (Giroix et al. 1999). In addition, when expressed relative to DNA, the GK/Par islet insulin content remains lower (by 30 %) than in that of the control (inbred W/Par) islets, therefore supporting some degranulation in the diabetic β-cells (Giroix et al. 1999).

Glucose-stimulated insulin biosynthesis in freshly isolated GK/Par, GK/Jap, or GK/Sto islets has been reported grossly normal (Guest et al. 2002; Giroix et al. 1993a; Nagamatsu et al. 1999). The rates of biosynthesis, processing, and secretion of newly synthesized (pro)insulin were comparable (Guest et al. 2002). This is remarkable in the face of markedly lower prohormone convertase PC2 immunoreactivity and expression in the GK/Sto islets, while the expression patterns of insulin, PC1, and carboxypeptidase E (CPE) remained normal (Guest et al. 2002). Circulating insulin immunoreactivity in GK/Sto rats was predominantly insulin 1 and 2 in the expected normal ratios with no (pro)insulin evident. The finding that proinsulin biosynthesis and processing of proinsulin appeared normal in adult GK rats suggests that the defective insulin release by β-cells does not arise from a failure to recognize glucose as an activator of prohormone biosynthesis and granule biogenesis. Rather it points to an inability of the β-cell population as a whole to meet the demands on insulin secretion imposed by chronic hyperglycemia in vivo. Although basal circulating GK insulin levels were similar or slightly elevated as compared to W rats, they were always inappropriate for the level of glycemia, indicative of a secretory defect.

Glucose-Induced Activation of Insulin Release Is Lost

Impaired glucose-stimulated insulin secretion has been repeatedly demonstrated in GK rats (whatever the colony), in vivo (Portha et al. 1991; Gauguier et al. 1994, 1996; Galli et al. 1996; Salehi et al. 1999), in the perfused isolated pancreas (Östenson 2001; Portha et al. 1991; Abdel-Halim et al. 1993, 1996; Östenson et al. 1993a; Kimura et al. 1982), and in freshly isolated islets (Hughes et al. 1994; Östenson et al. 1993a; Giroix et al. 1993a, b). A number of alterations or defects have been shown in the stimulus secretion coupling for glucose in GK islets. GLUT2 is underexpressed, but not likely to the extent that it could explain the impairment of insulin release (Ohneda et al. 1993). This assumption is supported by the fact that glucokinase/hexokinase activities are normal in GK rat islets (Östenson et al. 1993b; Tsuura et al. 1993; Giroix et al. 1993c). In addition, glycolysis rates in GK rat islets are unchanged or increased compared with control islets (Hughes et al. 1994; Östenson et al. 1993a; Giroix et al. 1993a, b, c; Ling et al. 1998, 2001; Fradet et al. 2008). Furthermore, oxidation of glucose has been reported decreased (Giroix et al. 1993a), unchanged (Hughes et al. 1994, 1998; Koyama et al. 2000; Östenson et al. 1993a; Giroix et al. 1993c; Fradet et al. 2008), or even enhanced (Ling et al. 1998). There exists however a common message between these data: The ratio of oxidized to glycolysed glucose was always reduced in GK islets compared to W islets. Also, lactate dehydrogenase gene expression (Lacraz et al. 2009) and lactate production (Ling et al. 1998) are increased and pyruvate dehydrogenase activity is decreased (Zhou et al. 1995) in GK rat islets. In GK/Par islets, we showed that mitochondria exhibit a specific decrease in the activities of FAD-dependent glycerophosphate dehydrogenase (Giroix et al. 1993a, c) and branched-chain ketoacid dehydrogenase (Giroix et al. 1999). Similar reduction of the FAD-linked glycerol phosphate dehydrogenase activity was reported in GK/Sto islets (Östenson et al. 1993b; MacDonald et al. 1996). These enzymatic abnormalities could work in concert to depress glucose oxidation. An inhibitory influence of islet fatty acid oxidation on glucose oxidation can be eliminated since the islet triglyceride content was found normal and etomoxir, an inhibitor of fatty acid oxidation, failed to restore glucose-induced insulin release in GK/Sto islets (Zhou et al. 1995).

We also found that the β-cells of adult GK/Par rats had a significantly smaller mitochondrial volume compared to control β-cells (Serradas et al. 1995). No major deletion or restriction fragment polymorphism could be detected in mtDNA from adult GK/Par islets (Serradas et al. 1995); however, they contained markedly less mtDNA than control islets. The lower islet mtDNA was paralleled by decreased content of some islet mt mRNAs such as cytochrome b (Serradas et al. 1995). In accordance with this, insufficient increase in ATP generation in response to high glucose was shown by our group (Giroix et al. 1993c). This supports the hypothesis that the defective insulin response to glucose in GK islet is accounted for by an impaired ATP production, closure of the ATP-regulated K^+ channels (Tsuura et al. 1993), and impaired elevation of intracellular $[Ca^{2+}]$ (Hughes et al. 1998; Marie et al. 2001; Dolz et al. 2005).

Such a view validated in the GK/Par β-cell is however contradictory to the reports in GK/Sto and GK/Sea islets that the rate of ATP production is unimpaired (Metz et al. 1999; Ling et al. 1998). Other energy metabolism defects identified in GK/Sto islets include increased glucose cycling due to increased glucose-6-phosphatase activity (Östenson et al. 1993a; Ling et al. 1998) and decreased pyruvate carboxylase activity (MacDonald et al. 1996). It is possible that these alterations may affect ATP concentrations locally. However, the enzyme dysfunctions were restored by normalization of glycemia in GK/Sto rats (MacDonald et al. 1996; Ling et al., unpublished observations), but with only partial improvement of glucose-induced insulin release. Hence, it is likely that these altered enzyme activities result from a glucotoxic effect rather than being primary causes behind the impaired secretion. Also, lipotoxic effects leading to defective insulin release have been observed in GK rats on high-fat diet (Shang et al. 2002; Briaud et al. 2002), possibly mediated by a mechanism partly involving modulation of UCP-2 expression.

Insulin Secretion Amplifying Mechanisms Are Altered

Phosphoinositide (Dolz et al. 2005) and cyclic AMP metabolism (Dolz et al. 2005, 2006) are also affected in GK/Par islets. While carbachol was able to promote normal inositol generation in GK/Par islets, high glucose failed to increase inositol phosphate accumulation. The inability of glucose to stimulate IP production is not related to defective phospholipase C activity per se (total activity in islet homogenates is normal) (Dolz et al. 2005). It is rather linked to abnormal targeting of the phosphorylation of phosphoinositides: The activity of phosphatidylinositol kinase, which is the first of the two phosphorylating activities responsible for the generation of phosphatidylinositol biphosphate, is clearly reduced (5, Giroix, unpublished data). Moreover, deficient calcium handling and ATP supply in response to glucose probably also contribute to abnormal activation of PI kinases and phospholipase C. A marked decrease in SERCA3 expression has also been described in the GK/Sto islets (Váradi et al. 1996).

Concerning cAMP, it is remarkable that its intracellular content is very high in GK/Par β-cells already at low glucose (Dolz et al. 2005). This is related to increased expression (mRNA) of the adenylyl cyclase isoforms 2 and 3, and of the GαS and Gαolf, while AC8 and phosphodiesterases PDE3B and PDE1C isoforms remain normal (Lacraz, unpublished data 2009). Furthermore, cAMP is not further enhanced at increasing glucose concentrations (at variance with the situation in normal β-cells) (Dolz et al. 2005, 2006). This suggests that there exists a block in the steps linking glucose metabolism to activation of adenylate cyclase in the GK/Par β-cell. In the GK/Sto rat, it has been shown that increased AC3 is due to functional mutations in the promoter region of the Ac3 gene (Abdel-Halim et al. 1998). We do not retain this hypothesis in the GK/Par islet since we found that the expressions (mRNA) of AC 2 and AC 3, and of GαS and Gαolf, are not increased in the prediabetic GK/Par islets (Lacraz, unpublished data 2009).

The increased cAMP production has also offered the possibility to fully restore the β-cell secretory competence to glucose in GK/Par as well as GK/Sto islets (Abdel-Halim et al. 1996; Dolz et al. 2006) with a clear biphasic response (Dolz et al. 2006). This also proves that the glucose incompetence of the GK/Par β-cell is not irreversible and emphasizes the usefulness of GLP-1 as a therapeutic agent in T2D. Also, cholinergic stimulation has been demonstrated to restore glucose-induced insulin secretion from GK/Sto as well as GK/Par islets (Dolz et al. 2005; Guenifi et al. 2001). We have proposed that such a stimulation is not mediated through activation of the PKC pathway, but via a paradoxical activation of the cAMP/PKA pathway to enhance Ca^{2+}-stimulated insulin release in the GK/Par β-cell (Dolz et al. 2005).

Other intriguing aspects of possible mechanisms behind defective glucose-induced insulin release in GK/Sto rat islets are the findings of dysfunction of islet lysosomal enzymes (Salehi et al. 1999), as well as excessive NO generation (Mosén et al. 2008; Salehi et al. 2008) or marked impairment of the glucose–heme oxygenase–carbon monoxide signaling pathway (Mosén et al. 2005). Islet activities of classical lysosomal enzymes, such as acid phosphatase, N-acetyl-β-D-glucosaminidase, β-glucuronidase, and cathepsin D, were reduced by 20–35 % in the GK rat. In contrast, the activities of the lysosomal α-glucosidehydrolases (acid glucan-1,4-α-glucosidase and acid α-glucosidase) were increased by 40–50 %. Neutral α-glucosidase (endoplasmic reticulum) was unaffected. Comparative analysis of liver tissue did not display such a difference. Since no sign of an acarbose effect on GK α-glucosidehydrolase activity (contrarily to Wistar islet) was seen, it was proposed that dysfunction of the islet lysosomal/vacuolar system participates to impairment of glucose-induced insulin release in the GK/Sto rat (Salehi et al. 1999). An abnormally increased NO production in the GK/Sto islets might also be an important factor in the pathogenesis of β-cell dysfunction, since it was associated with abnormal iNOS expression in insulin and glucagon cells, increased ncNOS activity, impaired glucose-stimulated insulin release, glucagon hypersecretion, and impaired glucose-induced glucagon suppression. Moreover, pharmacological blockade of islet NO production by the NOS inhibitor NG-nitro-L-arginine methyl ester greatly improved hormone secretion from GK/Sto islets, and GLP-1 suppressed iNOS and ncNOS expression and activity with almost full restoration of insulin release and partial restoration of glucagon release (Mosén et al. 2008; Salehi et al. 2008).

Also carbon monoxide (CO) derived from β-cell heme oxygenase (HO) might be involved in the secretory dysfunction. GK/Sto islets displayed a markedly decreased HO activity measured as CO production and immunoblotting revealed a 50 % reduction of HO-2 protein expression (Mosén et al. 2005). Furthermore, a prominent expression of inducible HO (HO-1) was found in GK/Sto (Mosén et al. 2005) as well as GK/Par (Lacraz et al. 2009) islets. The glucose-stimulated CO production and the glucose-stimulated insulin response were considerably reduced in GK/Sto islets. Since addition of the HO activator hemin or gaseous CO to incubation media brought about a normal amplification of glucose-stimulated insulin release in GK/Sto islets, it was proposed that distal steps in the HO–CO signaling pathway are not affected (Mosén et al. 2005).

A diminished pattern of expression and glucose-stimulated activation of several PKC isoenzymes (α, θ, and ζ) has been reported in GK/Sto islets, while the novel isoenzyme PKC ε not only showed a high expression level but also lacked glucose activation (Warwar et al. 2006; Rose et al. 2007). Since broad-range inhibition of the translocation of PKC isoenzymes by BIS increased the exocytotic efficacy of Ca^{2+} to trigger secretion in isolated GK/Sto β-cells (Rose et al. 2007), perturbed levels and/or activation of some PKC isoforms may be part of the defective signals downstream to glucose metabolism, involved in the GK insulin secretory lesion.

Peroxovanadium, an inhibitor of islet protein tyrosine phosphatase (PTP) activities, was shown to enhance glucose-stimulated insulin secretion from GK/Sto islets (Abella et al. 2003; Chen and Ostenson 2005). One possible target for this effect could be PTP sigma that is overexpressed in GK/Sto islets (Östenson et al. 2002). At present it is not known which exocytosis-regulating proteins are affected by the increased PTPase activity. In addition, defects in islet protein histidine phosphorylation have been proposed to contribute to impaired insulin release in GK/Sea islets (Kowluru 2003).

Lastly, an increased storage and secretion of amylin relative to insulin was found in the GK/Sto rat (Leckström et al. 1996), and GLP1 treatment in vivo was recently reported to exert a beneficial effect on the ratio of amylin to insulin mRNA in GK pancreas besides improvement of glucose-induced insulin release

(Weng et al. 2008). This is consistent with hypersecretion of amylin being one of the factors contributing to the impairment of glucose-induced insulin release.

Insulin Exocytotic Machinery Is Abnormal

In addition to these upstream abnormalities, important defects reside late in signal transduction, i.e., in the exocytotic machinery. Indeed, glucose-stimulated insulin secretion was markedly impaired in GK/Taked, GK/Sto, GK/Sea, and GK/Par islets also when the islets were depolarized by a high concentration of potassium chloride and the ATP-regulated K^+ channels kept open by diazoxide (Metz et al. 1999; Abdel-Halim et al. 1996; Okamoto et al. 1995, Szkudelski and Giroix, unpublished data). Similar results were obtained when insulin release was induced by exogenous calcium in electrically permeabilized GK/Jap islets (Okamoto et al. 1995). In fact, markedly reduced expressions of the SNARE complex proteins (α-SNAP, SNAP-25, syntaxin-1, Munc13-1, Munc18-1, N-ethylmaleimide-sensitive fusion protein and synaptotagmin 3) have been demonstrated in GK/Sto and GK/Taked islets (Nagamatsu et al. 1999; Gaisano et al. 2002; Zhang et al. 2002). We also recently found similar results in the GK/Par islets (Tourrel-Cuzin, unpublished data 2009). Thus, a reduced number of docking granules may account for impaired β-cell secretion (Ohara-Imaizumi et al. 2004), and this defect should partly be related to glucotoxicity (Gaisano et al. 2002). Actin cytoskeleton has also been implicated in regulated exocytosis. It has been proposed that in secretory cells, actin network under the plasma membrane acts as a physical barrier preventing the access of secretory granules to the membrane. However, the role of the subcortical actin is certainly more complex as it is also required for final transport of vesicles to the sites of exocytosis. The level of total actin protein evaluated by western blotting has been found similar in GK/Par and W islets (Movassat et al. 2005), at variance with reports in other GK rat lines (Nagamatsu et al. 1999; Gaisano et al. 2002). However, confocal analysis of the distribution of phalloidin-stained cortical actin filaments revealed a higher density of the cortical actin web nearby the plasma membrane in GK/Par islets as compared to W. Moreover, preliminary functional results suggest that the higher density of actin cortical web in the GK/Par islets contributes to the defects in glucose-induced insulin secretion exhibited by GK islets (Movassat et al. 2005).

Secretory Response to Non-glucose Stimuli Is Partly Preserved

Among the non-glucidic insulin stimulators, arginine has been shown to induce a normal or even augmented insulin response from perfused pancreases or isolated islets of GK/Clea, GK/Par, GK/Sto, and GK/Lon (Portha et al. 1991; Hughes et al. 1994; Kimura et al. 1982). Since preperfusion for 50–90 min in the absence of glucose reduced the insulin response to arginine in the GK/Par but not in the control pancreas (Portha et al. 1991), it is likely that previous exposure to glucose in vivo or during the perfusion experiment potentiates arginine-induced insulin secretion. Insulin responses to another amino acid, leucine, and its metabolite, ketoisocaproate (KIC), were diminished in GK/Par and GK/Sto rats (Östenson 2001; Giroix et al. 1993a, 1999). This was attributed to defective mitochondrial oxidative decarboxylation of KIC operated by the branched-chain 2-ketoacid dehydrogenase (BCKDH) complex (Giroix et al. 1999). However, in GK/Lon and GK/Taked islets, KIC induced normal insulin responses (Hughes et al. 1994; Tsuura et al. 1993). Finally it is of interest that GK islets are duly responsive to non-nutrient stimuli such as the sulfonylureas gliclazide (GK/Par) (Giroix et al. 1993a) and mitiglinide (GK/Sto) (Kaiser et al. 2005), the combination of Ba^{2+} and theophylline (GK/Par) (Giroix et al. 1993a), or high external K^+ concentrations (GK/Lon, GK/Sto, GK/Seat, GK/Par) (Metz et al. 1999; Abdel-Halim et al. 1996; Katayama et al. 1995, Dolz and Portha, unpublished data). However, this does not support the assumption from the molecular biology data that there exists a defect in the late steps of insulin secretion. As a tentative to elucidate this apparent contradiction, exocytosis assessment with high time-resolution membrane capacitance measurement in GK/Sto pancreatic slices showed a decreased efficacy of depolarization-evoked Ca^{2+} influx to trigger rapid vesicle release, contrasting with a facilitation of vesicle release in response to strong sustained Ca^{2+} stimulation (Rose et al. 2007).

Islet ROS Scavenging Capacity Is Increased

Considerable interest has recently been focused on the putative role of oxidative stress (OS) upon deterioration of β-cell function/survival in diabetes. Recent data from our group indicate that paradoxically GK/Par islets revealed protected against OS since they maintained basal ROS accumulation similar to or even lower than nondiabetic islets. Remarkably, GK/Par insulin secretion also exhibited strong resistance to the toxic effect of exogenous H_2O_2 or endogenous ROS exposures. Such adaptation was associated to both high glutathione content and overexpression of a large set of genes encoding antioxidant proteins as well as UCP2 (Portha et al. 2007; Lacraz et al. 2009).

Figure 2 illustrates a compendium of the abnormal intracellular sites so far identified in the diabetic GK islets from the different sources.

Which Etiology for the Islet Functional Abnormalities?

There are several arguments indicating that the GK β-cell secretory failure is, at least partially, related to the abnormal metabolic environment (gluco-lipotoxicity). When studied under in vitro static incubation conditions, islets isolated from normoglycemic (prediabetic) GK/Par pups amplified their secretory response to high glucose, leucine, or leucine plus glutamine to the same extent as age-related W islets (Portha et al. 2001). This suggests that there does not exist a major intrinsic secretory defect in the prediabetic GK/Par β-cells which can be considered as

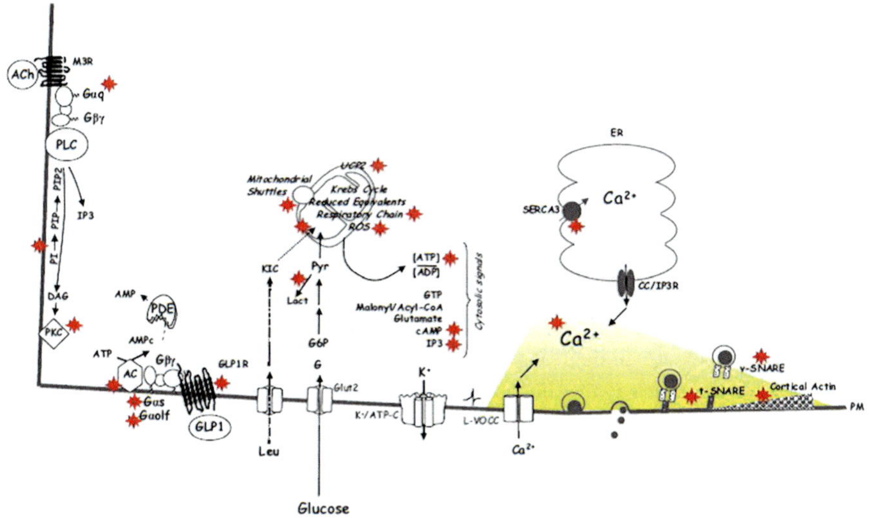

Fig. 2 Model for defective glucose-induced insulin release and the abnormal intracellular sites so far identified in the β-cell of the diabetic GK rats from the different sources (mostly the GK/Par and the GK/Sto sources). Where data are available, the impaired sites in the β-cell are indicated with the symbol ✱. Abbreviations: *Glut2* glucose transporter isoform 2, *Leu* leucine, *KIC* ketoisocaproate, *AC* adenylate cyclase isoforms, *Gαs, Gαolf, Gαq* α subunits of heterotrimeric G proteins, *Gβγ* β and γ subunits of heterotrimeric G proteins, *PI, PIP, PIP2* phosphoinositides, *PLC* phospholipase C, *PKC* protein kinase C, *DAG* diacylglycerol, *IP3* inositol-3-phosphate, *UCP2* uncoupling protein 2, *ROS* reactive oxygen species, *tSNARE, v-SNARE* SNARE proteins (syntaxin-1A, SNAP-25, VAMP-2, Munc-18), *SERCA-3* endoplasmic reticulum Ca^{2+}-ATPase isoform 3, *L-VOCC* L-type calcium channel modulated by the membrane polarization, *CC/IP3R* calcium channel modulated by receptor to IP3, K^+/ATP-*C* potassium channel modulated by the ATP/ADP ratio, *Ach* acetylcholine, *M3-R* muscarinic receptor isoform 3, *GLP-1* glucagon-like peptide 1, *GLP1-R* GLP1 receptor, *PDE* cAMP-dependent phosphodiesterase isoforms

normally glucose competent at this stage, at least when tested in vitro. In the GK/Par rat, basal hyperglycemia and normal to very mild hypertriglyceridemia are observed only after weaning (Portha et al. 2001). The onset of a profound alteration in glucose-stimulated insulin secretion by the GK/Par β-cell (after weaning) is time correlated with the exposure to the diabetic milieu. These changes in islet function could be ascribed, at least in part, to a loss of differentiation of β-cells chronically exposed to even mild chronic hyperglycemia and elevated plasma non-esterified fatty acids. This view is supported by the reports that chronic treatments of adult GK rats with phlorizin (Portha et al. 2007; Nagamatsu et al. 1999; Ling et al. 2001; Gaisano et al. 2002), T-1095 (Yasuda et al. 2002), glinides (Kaiser et al. 2005; Kawai et al. 2008), glibenclamide (Kawai et al. 2008), gliclazide (Dachicourt et al. 1998), JTT-608 (Ohta et al. 1999, 2003), voglibose (Ishida et al. 1998), or insulin (Kawai et al. 2008) partially improved glucose-induced insulin release, while hyperlipidemia induced by high-fat feeding markedly impaired their insulin secretion (Briaud et al. 2002).

The recent identification of TCF7L2 as a major predisposition gene for T2D and the predominant association of TCF7L2 variants with impaired insulin secretion have highlighted the importance of Wnt signaling in glucose homeostasis. In fact, two studies in human diabetic islets have reported that the expression of TCF7L2 is increased at mRNA (Lyssenko et al. 2007; Lee et al. 2008) and at protein levels (Lee et al. 2008), and it has been found that TCF7L2 overexpression in pancreatic β-cells is associated with reduced insulin secretion (Lyssenko et al. 2007). Islet TCF7L2 mRNA and protein levels revealed higher in GK/Par islets (Tourrel-Cuzin and Movassat, unpublished). Similar observation was reported in GK/Sto islets (Granhall et al. 2006). The functional link between the upregulation of TCF7L2 and the impairment of β-cell growth and function in the GK model remains to be uncovered.

Besides, there are indications in the GK/Sto rat that two distinct loci encode separately defects in β-cell glucose metabolism and insulin exocytosis (Granhall et al. 2006). Generation of congenic rat strains harboring different parts of GK/Sto-derived Niddm1i has recently enabled fine mapping of this locus. Congenic strains carrying the GK genotype distally in Niddm1i displayed reduced insulin secretion in response to both glucose and high potassium, as well as decreased single-cell exocytosis. By contrast, the strain carrying the GK genotype proximally in Niddm1i exhibited both intact insulin release in response to high potassium and intact single-cell exocytosis, but insulin secretion was suppressed when stimulated by glucose. Islets from this strain also failed to respond to glucose by increasing the cellular ATP to ADP ratio. Since the congenics had not developed overt hyperglycemia and their β-cell mass was found normal, it was concluded that their functional defects in glucose metabolism and insulin exocytosis were encoded by two distinct loci within Niddm1i (Granhall et al. 2006). These results in the GK/Sto are however inconsistent as previously mentioned (see Section 21.4), with the conclusion of a similarly designed congenic study indicating that the corresponding short region of the QTL Nidd/gk1 in GK/Ox congenics contributes to enhanced (and not decreased) insulin release (Wallis et al. 2008). Interestingly, the gene encoding for transcription factor TCF7L2 is also located in this locus and has recently been identified as a candidate gene for T2D in humans (Grant et al. 2006). However, Tcf7l2 RNA levels were not different in the GK/Sto congenics displaying reduced insulin secretion compared with controls (Granhall et al. 2006).

In conclusion, taking into account the diverse information so far available from the GK model through its different phenotype variants, it is proposed that the reduction of GK β-cell number and function reflects the complex interactions of different pathogenic items: multiple morbid genes causing impaired insulin secretion, early epigenetic programming of the pancreas by gestational diabetes (decreased β-cell neogenesis and/or proliferation) which is transmitted from one generation to the other, and acquired loss of β-cell differentiation due to chronic exposure to hyperglycemia/hyperlipidemia, inflammatory mediators, and oxidative stress and to perturbed islet microarchitecture. Last but not least, careful comparison of the alterations so far detected in the diabetic GK β-cell population and those found in the T2D human β-cell population put into the front stage a number of striking commonalities (Portha et al. 2009). Of course, the GK β-cell is not a blueprint for the diseased β-cell in human. There are however sufficient similarities with high value, to justify more efforts to understand the etiopathogenesis of T2D in this rat model now widely used and, more specifically, the central role played by the GK islet cells.

Acknowledgments The GK/Par studies done at Lab B2PE, BFA Unit, have been funded by the CNRS, the French ANR (programme Physio 2006 – Prograbeta), the EFSD/MSD European Foundation, MERCK-SERONO, French Diabetes Association, and NEB Research Foundation. G. Lacraz and F. Figeac received a doctoral fellowship from the Ministere de l'Education Nationale, de l'Enseignement Supérieur et de la Recherche (Ecole Doctorale 394, Physiologie/Physiopathologie). A. Chavey was the recipient of a CNRS postdoctoral fellowship and a NESTLE-France grant.

Cross-References

▶ Apoptosis in Pancreatic β-Islet Cells in Type 1 and Type 2 Diabetes
▶ β-Cell Function in Obese-hyperglycaemic Mice [ob/ob mice]
▶ The β-Cell in Human Type 2 Diabetes

References

Abdel-Halim SM, Guenifi A, Efendic S, Östenson CG (1993) Both somatostatin and insulin responses to glucose are impaired in the perfused pancreas of the spontaneously non-insulin dependent diabetic GK (Goto-Kakizaki) rat. Acta Physiol Scand 148:219–226

Abdel-Halim SM, Guenifi A, Khan A, Larsson O, Berggren PO, Östenson CG, Efendic S (1996) Impaired coupling of glucose signal to the exocytotic machinery in diabetic GK rats; a defect ameliorated by cAMP. Diabetes 45:934–940

Abdel-Halim SM, Guenifi A, He B, Yang B, Mustafa M, Höjeberg B, Hillert J, Bakhiet M, Efendic S (1998) Mutations in the promoter of adenylyl cyclase (AC)-III gene, overexpression of AC-III mRNA, and enhanced cAMP generation in islets from the spontaneously diabetic GK rat model of type 2 diabetes. Diabetes 47:498–504

Abella A, Marti L, Camps M, Claret M, Fernández-Alvarez J, Gomis R, Guma A, Viguerie N, Carpéné C, Palacin M, Testar X, Zorzan A (2003) Semicarbazide-sensitive amine oxidase/vascular adhesion protein-1 activity exerts an antidiabetic action in Goto–Kakizaki rats. Diabetes 52:1004–1013

Atef N, Portha B, Penicaud L (1994) Changes in islet blood flow in rats with NIDDM. Diabetologia 37:677–680

Briaud I, Kelpe CL, Johnson LM, Tran PO, Poitout V (2002) Differential effects of hyperlipidemia on insulin secretion in islets of Langerhans from hyperglycemic versus normoglycemic rats. Diabetes 51:662–668

Butler AE, Janson J, Bonner-Weir S, Ritzel R, Rizza RA, Butler PC (2003) β-cell deficit and increased β-cell apoptosis in humans with type 2 diabetes. Diabetes 52:102–110

Calderari S, Gangnerau MN, Meile MJ, Portha B, Serradas P (2006) Is defective pancreatic β-cell mass environmentally programmed in Goto Kakizaki rat model of type 2 diabetes: insights from cross breeding studies during suckling period. Pancreas 33:412–417

Calderari S, Gangnerau MN, Thibault M, Meile MJ, Kassis N, Alvarez C, Kassis N, Portha B, Serradas P (2007) Defective IGF-2 and IGFR1 protein production in embryonic pancreas precedes β cell mass anomaly in Goto-Kakizaki rat model of type 2 diabetes. Diabetologia 50:1463–1471

Carlsson PO, Jansson L, Östenson CG, Källskog O (1997) Islet capillary blood pressure increase mediated by hyperglycemia in NIDDM GK rats. Diabetes 46:947–952

Chavey A, Gangnerau MN, Maulny L, Bailbe D, Movassat J, Renard JP, Portha B (2008) Intrauterine programming of β-cell development and function by maternal diabetes. What tell us embryo-transfer experiments in GK/Par rats? (Abstract). Diabetologia 51[Suppl 1]:A151

Chen J, Ostenson CG (2005) Inhibition of protein-tyrosine phosphatases stimulates insulin secretion in pancreatic islets of diabetic Goto–Kakizaki rats. Pancreas 30:314–317

Dachicourt N, Bailbé D, Gangnerau MN, Serradas P, Ravel D, Portha B (1998) Effect of gliclazide treatment on insulin secretion and β-cell mass in non-insulin dependent diabetic Goto-Kakizaki rats. Eur J Pharmacol 361:243–251

Dolz M, Bailbé D, Giroix MH, Calderari S, Gangnerau MN, Serradas P, Rickenbach K, Irminger JC, Portha B (2005) Restitution of defective glucose-stimulated insulin secretion in diabetic GK rat by acetylcholine uncovers paradoxical stimulatory effect of β cell muscarinic receptor activation on cAMP production. Diabetes 54:3229–3237

Dolz M, Bailbé D, Movassat J, Le Stunff H, Kassis K, Giroix MH, Portha B (2006) Pivotal role of cAMP in the acute restitution of defective glucose-stimulated insulin release in diabetic GK rat by GLP-1. Diabetes 55(Suppl 1):A371

Donath MY, Halban PA (2004) Decreased β-cell mass in diabetes: significance, mechanisms and therapeutic implications. Diabetologia 47:581–589

Duarte AI, Santos MS, Seiça R, Oliveira CR (2004) Oxidative stress affects synaptosomal γ-aminobutyric acid and glutamate transport in diabetic rats. The role of insulin. Diabetes 53:2110–2116

Ehses JA, Perren A, Eppler E, Ribaux P, Pospisilik JA, Maor-Cahn R, Gueripel X, Ellingsgaard H, Schneider MKJ, Biollaz G, Fontana A, Reinecke M, Homo-Delarche F, Donath MY (2007a) Increased number of islet-associated macrophages in type 2 diabetes. Diabetes 56:2356–2370

Ehses JA, Calderari S, Irminger JC, Serradas P, Giroix MH, Egli A, Portha B, Donath MY, Homo-Delarche F (2007b) Islet Inflammation in type 2 diabetes (T2D): from endothelial to β-cell dysfunction. Curr Immunol Rev 3:216–232

Fradet M, Giroix MH, Bailbé D, El Bawab S, Autier V, Kergoat M, Portha B (2008) Glucokinase activators modulate glucose metabolism and glucose-stimulated insulin secretion in islets from diabetic GK/Par rats (Abstract). Diabetologia 51 [Suppl 1]:A198–9

Gaisano HY, Östenson CG, Sheu L, Wheeler MB, Efendic S (2002) Abnormal expression of pancreatic islet exocytotic soluble N-ethylmaleimide-sensitive factor attachment protein receptors in Goto-Kakizaki rats is partially restored by phlorizin treatment and accentuated by high glucose treatment. Endocrinology 143:4218–4226

Galli J, Li LS, Glaser A, Östenson CG, Jiao H, Fakhrai-Rad H, Jacob HJ, Lander E, Luthman H (1996) Genetic analysis of non-insulin dependent diabetes mellitus in the GK rat. Nat Genet 12:31–37

Gauguier D, Nelson I, Bernard C, Parent V, Marsac C, Cohen D, Froguel P (1994) Higher maternal than paternal inheritance of diabetes in GK rats. Diabetes 43:220–224

Gauguier D, Froguel P, Parent V, Bernard C, Bihoreau MT, Portha B, James MR, Penicaud L, Lathrop M, Ktorza A (1996) Chromosomal mapping of genetic loci associated with non-insulin dependent diabetes in the GK rat. Nat Genet 12:38–43

Ghanaat-Pour H, Huang Z, Lehtihet M, Sjöholm A (2007) Global expression profiling of glucose-regulated genes in pancreatic islets of spontaneously diabetic Goto-Kakizaki rats. J Mol Endocrinol 39:135–150

Gill-Randall R, Adams D, Ollerton RL, Lewis M, Alcolado JC (2004) Type 2 diabetes mellitus – genes or intrauterine environment? An embryo transfer paradigm in rats. Diabetologia 47:1354–1359

Giroix MH, Vesco L, Portha B (1993a) Functional and metabolic perturbations in isolated pancreatic islets from the GK rat, a genetic model of non-insulin dependent diabetes. Endocrinology 132:815–822

Giroix MH, Sener A, Portha B, Malaisse WJ (1993b) Preferential alteration of oxidative relative to total glycolysis in pancreatic islets of two rats models of inherited or acquired type 2 (non-insulin dependent) diabetes mellitus. Diabetologia 36:305–309

Giroix MH, Sener A, Bailbé D, Leclercq-Meyer V, Portha B, Malaisse WJ (1993c) Metabolic, ionic and secretory response to D-glucose in islets from rats with acquired or inherited non-insulin dependent diabetes. Biochem Med Metab Biol 50:301–321

Giroix MH, Saulnier C, Portha B (1999) Decreased pancreatic islet response to L-leucine in the spontaneously diabetic GK rat: enzymatic, metabolic and secretory data. Diabetologia 42:965–977

Goda T, Suruga K, Komori A, Kuranuki S, Mochizuki K, Makita Y, Kumazawa T (2007) Effects of miglitol, an α-glucosidase inhibitor, on glycaemic status and histopathological changes in islets in non-obese, non-insulin-dependent diabetic Goto-Kakizaki rats. Br J Nutr 98:702–710

Goto Y, Kakizaki M, Masaki N (1975) Spontaneous diabetes produced by selective breeding of normal Wistar rats. Proc Jpn Acad 51:80–85

Goto Y, Suzuki KI, Sasaki M, Ono T, Abe S (1988) GK rat as a model of nonobese, noninsulin-dependent diabetes. Selective breeding over 35 generations. In: Shafrir E, Renold AE (eds) Lessons from animal diabetes. Libbey, London, pp 301–303

Granhall C, Rosengren AH, Renstrom E, Luthman H (2006) Separately inherited defects in insulin exocytosis and β-cell glucose metabolism contribute to type 2 diabetes. Diabetes 55:3494–3500

Grant SF, Thorleifsson G, Reynisdottir I, Benediktsson R, Manolescu A, Sainz J, Helgason A, Stefansson H, Emilsson V, Helgadottir A, Styrkarsdottir U, Magnusson KP, Walters GB, Palsdottir E, Jonsdottir T, Gudmundsdottir T, Gylfason A, Saemundsdottir J, Wilensky RL, Reilly MP, Rader DJ, Bagger Y, Christiansen C, Gudnason V, Sigurdsson G, Thorsteinsdottir U, Gulcher JR, Kong A, Stefansson K (2006) Variant of transcription factor 7-like 2 (TCF7L2) gene confers risk of type 2 diabetes. Nat Genet 38:320–323

Guenifi A, Abdel-Halim SM, Höög A, Falkmer S, Östenson CG (1995) Preserved β-cell density in the endocrine pancreas of young, spontaneously diabetic Goto–Kakizaki (GK) rats. Pancreas 10:148–153

Guenifi A, Simonsson E, Karlsson S, Ahren B, Abdel-Halim SM (2001) Carbachol restores insulin release in diabetic GK rat islets by mechanisms largely involving hydrolysis of diacylglycerol and direct interaction with the exocytotic machinery. Pancreas 22:164–171

Guest PC, Abdel-Halim SM, Gross DJ, Clark A, Poitout V, Amaria R, Östenson CG, Hutton JC (2002) Proinsulin processing in the diabetic Goto–Kakizaki rat. J Endocrinol 175:637–647

Homo-Delarche F, Calderari S, Irminger JC, Gangnerau MN, Coulaud J, Rickenbach K, Dolz M, Halban P, Portha B, Serradas S (2006) Islet inflammation and fibrosis in a spontaneous model of type 2 diabetes, the GK rat. Diabetes 55:1625–1633

Hughes SJ, Suzuki K, Goto Y (1994) The role of islet secretory function in the development of diabetes in the GK Wistar rat. Diabetologia 37:863–870

Hughes SJ, Faehling M, Thorneley CW, Proks P, Ashcroft FM, Smith PA (1998) Electrophysiological and metabolic characterization of single β-cells and islets from diabetic GK rats. Diabetes 47:73–81

Ihara Y, Toyokuni S, Uchida K, Odaka H, Tanaka T, Ikeda H, Hiai H, Seino Y, Yamada Y (1999) Hyperglycemia causes oxidative stress in pancreatic β-cells of GK rats, a model of type 2 diabetes. Diabetes 48:927–932

Ishida H, Kato S, Nishimura M, Mizuno N, Fujimoto S, Mukai E, Kajikawa M, Yamada Y, Odaka H, Ikeda H, Seino Y (1998) Beneficial effect of long-term combined treatment with vogli-bose and pioglitazone on pancreatic islet function of genetically diabetic GK rats. Horm Metab Res 30:673–678

Kaiser N, Nesher R, Oprescu A, Efendic S, Cerasi E (2005) Characterization of the action of S21403 (mitiglinide) on insulin secretion and biosynthesis in normal and diabetic β-cells. Br J Pharmacol 146:872–881

Katayama N, Hughes SJ, Persaud SJ, Jones PM, Howell SL (1995) Insulin secretion from islets of GK rats is not impaired after energy generating steps. Mol Cell Endocrinol 111:125–128

Kawai J, Ohara-Imaizumi M, Nakamichi Y, Okamura T, Akimoto Y, Matsushima S, Aoyagi K, Kawakami H, Watanabe T, Watada H, Kawamori R, Nagamatsu S (2008) Insulin exocytosis in Goto-Kakizaki rat β-cells subjected to long-term glinide or sulfonylurea treatment. Biochem J 412:93–101

Keno Y, Tokunaga K, Fujioka S, Kobatake T, Kotani K, Yoshida S, Nishida M, Shimomura I, Matsuo T, Odaka H, Tarui S, Matsuzawa Y (1994) Marked reduction of pancreatic insulin content in male ventromedial hypothalamic-lesioned spontaneously non-insulin-dependent diabetic (Goto-Kakizaki) rats. Metabolism 43:32–37

Kimura K, Toyota T, Kakizaki M, Kudo M, Takebe K, Goto Y (1982) Impaired insulin secretion in the spontaneous diabetes rats. Tohoku J Exp Med 137:453–459

Kowluru A (2003) Defective protein histidine phosphorylation in islets from the Goto– Kakizaki diabetic rat. Am J Physiol 285:E498–E503

Koyama M, Wada R, Sakuraba H, Mizukami H, Yagihashi S (1998) Accelerated loss of islet β-cells in sucrose-fed Goto-Kakizaki rats, a genetic model of non-insulin-dependent diabetes mellitus. Am J Pathol 153:537–545

Koyama M, Wada R, Mizukami H, Sakuraba H, Odaka H, Ikeda H, Yagihashi S (2000) Inhibition of progressive reduction of islet β-cell mass in spontaneously diabetic Goto-Kakizaki rats by α-glucosidase inhibitor. Metabolism 49:347–352

Lacraz G, Figeac F, Movassat J, Kassis N, Coulaud J, Galinier A, Leloup C, Bailbé D, Homo-Delarche F, Portha B (2009) Diabetic Rat β-cells can achieve self-protection against oxidative stress through an adaptive up-regulation of their antioxidant defenses. PLoS ONE 4(8):e6500

Leckström A, Östenson CG, Efendic S, Arnelo U, Permert J, Lundquist I, Westermark P (1996) Increased storage and secretion of islet amyloid polypeptide relative to insulin in the spontaneously diabetic GK rat. Pancreas 13:259–267

Lee SH, Demeterco C, Geron I, Abrahamsson A, Levine F, Itkin-Ansari P (2008) Islet specific Wnt activation in human type 2 diabetes. Exp Diabetes Res Article ID 728763, doi: 10.1155/2008/728763

Lewis BM, Ismail IS, Issa B, Peters JR, Scanlon MF (1996) Desensitisation of somatostatin, TRH and GHRH responses to glucose in the diabetic Goto-Kakizaki rat hypothalamus. J Endocrinol 151:13–17

Lin JM, Ortsäter H, Fakhraid-Ra H, Galli J, Luthman H, Bergsten P (2001) Phenotyping of individual pancreatic islets locates genetic defects in stimulus secretion coupling to Niddm1i within the major diabetes locus in GK rats. Diabetes 50:2737–2743

Ling ZC, Efendic S, Wibom R, Abdel-Halim SM, Östenson CG, Landau BR, Khan A (1998) Glucose metabolism in Goto–Kakizaki rat islets. Endocrinology 139:2670–2675

Ling ZC, Hong-Lie C, Östenson CG, Efendic S, Khan A (2001) Hyperglycemia contributes to impaired insulin response in GK rat islets. Diabetes 50(Suppl 1):108–112

Lyssenko V, Lupi R, Marchetti P, Del Guerra S, Orho-Melander M, Almgren P, Sjögren M, Ling C, Eriksson KF, Lethagen AL, Mancarella R, Berglund G, Tuomi T, Nilsson P, Del Prato S, Groop L (2007) Mechanisms by which common variants in the TCF7L2 gene increase risk of type 2 diabetes. J Clin Invest 117:2155–2163

MacDonald MJ, Efendic S, Östenson CG (1996) Normalization by insulin treatment of low mitochondrial glycerol phosphate dehydrogenase and pyruvate carboxylase in pancreatic islets of the GK rat. Diabetes 45:886–890

Marie JC, Bailbé D, Gylfe E, Portha B (2001) Defective glucose-dependent cytosolic Ca^{2+} handling in islets of GK and nSTZ rat models of type2 diabetes. J Endocrinol 169:169–176

Metz SA, Meredith M, Vadakekalam J, Rabaglia ME, Kowluru A (1999) A defect late in stimulus secretion coupling impairs insulin secretion in Goto-Kakizaki diabetic rats. Diabetes 48:1754–1762

Miralles F, Portha B (2001) Early development of β-cells is impaired in the GK rat model of type 2 diabetes. Diabetes 50(Suppl 1):84–88

Mizukami H, Wada R, Koyama M, Takeo T, Suga S, Wakui M, Yagihashi S (2008) Augmented β-cell loss and mitochondrial abnormalities in sucrose-fed GK rats. Virchows Arch 452:383–392

Momose K, Nunomiya S, Nakata M, Yada T, Kikuchi M, Yashiro T (2006) Immunohistochemical and electron-microscopic observation of β-cells in pancreatic islets of spontaneously diabetic Goto-Kakizaki rats. Med Mol Morphol 39:146–153

Mosén H, Salehi A, Alm P, Henningsson R, Jimenez-Feltström J, Östenson CG, Efendic S, Lundquist I (2005) Defective glucose-stimulated insulin release in the diabetic Goto-Kakizaki (GK) rat coincides with reduced activity of the islet carbon monoxide signaling pathway. Endocrinology 146:1553–1558

Mosén H, Östenson CG, Lundquist I, Alm P, Henningsson R, Jimenez-Feltstrom J, Guenifi A, Efendic S, Salehi A (2008) Impaired glucose-stimulated insulin secretion in the GK rat is associated with abnormalities in islet nitric oxide production. Regul Pept 151:139–146

Movassat J, Portha B (1999) β-cell growth in the neonatal Goto-Kakizaki rat and regeneration after treatment with streptozotocin at birth. Diabetologia 42:1098–1106

Movassat J, Saulnier C, Serradas P, Portha B (1997) Impaired development of pancreatic β-cell mass is a primary event during the progression to diabetes in the GK rat. Diabetologia 40:916–925

Movassat J, Deybach C, Bailbé D, Portha B (2005) Involvement of cortical actin cytoskeleton in the defective insulin secretion in Goto-Kakizaki rats (Abstract). Diabetes 54:A1

Movassat J, Calderari S, Fernández E, Martín MA, Escrivá F, Plachot C, Gangneraw MN, Serradas P, Álvarez C, Portha B (2007) Type 2 diabetes – a matter of failing β-cell neogenesis? Clues from the GK rat model. Diabetes Obes Metab 9(Suppl 2):187–195

Nagamatsu S, Nakamichi Y, Yamamura C, Matsushima S, Watanabe T, Azawa S, Furukawa H, Ishida H (1999) Decreased expression of t-SNARE, syntaxin 1, and SNAP-25 in pancreatic β-cells is involved in impaired insulin secretion from diabetic GK rat islets: restoration of decreased t-SNARE proteins improves impaired insulin secretion. Diabetes 48:2367–2373

Ohara-Imaizumi M, Nishiwaki C, Kikuta T, Nagai S, Nakamichi Y, Nagamatsu S (2004) TIRF imaging of docking and fusion of single insulin granule motion in primary rat pancreatic β-cells: different behaviour of granule motion between normal and Goto–Kakizaki diabetic rat β-cells. Biochem J 381:13–18

Ohneda M, Johnson JH, Inman LR, Chen L, Suzuki KI, Goto Y, Alam T, Ravazzola M, Orci L, Unger RH (1993) GLUT2 expression and function in β-cells of GK rats with NIDDM. Diabetes 42:1065–1072

Ohta T, Furukawa N, Komuro G, Yonemori F, Wakitani K (1999) JTT-608 restores impaired early insulin secretion in diabetic Goto-Kakizaki rats. Br J Pharmacol 126:1674–1680

Ohta T, Miyajima K, Komuro G, Furukawa N, Yonemori F (2003) Antidiabetic effect of chronic administration of JTT-608, a new hypoglycaemic agent, in diabetic Goto-Kakizaki rats. Eur J Pharmacol 476:159–166

Okamoto Y, Ishida H, Tsuura Y, Yasuda K, Kato S, Matsubara H, Nishimura M, Mizuno N, Ikeda H, Seino Y (1995) Hyperresponse in calcium-induced insulin release from electrically permeabilized pancreatic islets of diabetic GK rats and its defective augmentation by glucose. Diabetologia 38:772–778

Ostenson CG (2001) The Goto-Kakizaki rat. In: Sima AAF, Shafrir E (eds) Animal models of diabetes: a primer. Harwood Academic, Amsterdam, pp 197–211

Östenson CG, Khan A, Abdel-Halim SM, Guenifi A, Suzuki K, Goto Y, Efendic S (1993a) Abnormal insulin secretion and glucose metabolism in pancreatic islets from the spontaneously diabetic GK rat. Diabetologia 36:3–8

Östenson CG, Abdel-Halim SM, Rasschaert J, Malaisse-Lagae F, Meuris S, Sener A, Efendic S, Malaisse WJ (1993b) Deficient activity of FAD-linked glycerophosphate dehydrogenase in islets of GK rats. Diabetologia 36:722–726

Östenson CG, Sandberg-Nordqvist AC, Chen J, Hallbrink M, Rotin D, Langel U, Efendic S (2002) Overexpression of protein tyrosine phosphatase PTP sigma is linked to impaired glucose-induced insulin secretion in hereditary diabetic Goto-Kakizaki rats. Biochem Biophys Res Commun 291:945–950

Plachot C, Movassat J, Portha B (2001) Impaired β-cell regeneration after partial pancreatectomy in the adult Goto-Kakizaki rat, a spontaneous model of type 2 diabetes. Histochem Cell Biol 116:131–139

Portha B (2005) Programmed disorders of β-cell development and function as one cause for type 2 diabetes? The GK rat paradigm. Diabetes Metab Res Rev 21:495–504

Portha B, Serradas P, Bailbé D, Suzuki KI, Goto Y, Giroix MH (1991) β-Cell insensitivity to glucose in the GK rat, a spontaneous nonobese model for type II diabetes. Dissociation between reductions in glucose transport and glucose-stimulated insulin secretion. Diabetes 40:486–491

Portha B, Giroix MH, Serradas P, Gangnerau MN, Movassat J, Rajas F, Bailbé D, Plachot C, Mithieux G, Marie JC (2001) β-cell function and viability in the spontaneously diabetic GK rat. Information from the GK/Par colony. Diabetes 50(Suppl 1):89–93

Portha B, Lacraz G, Dolz M, Giroix MH, Homo-Delarche F, Movassat J (2007) Issues surrounding β-cells and their roles in type 2 diabetes. What tell us the GK rat model. Expert Rev Endocrinol Metab 2:785–795

Portha B, Lacraz G, Kergoat M, Homo-Delarche F, Giroix MH, Bailbé D, Gangnerau MN, Dolz M, Tourrel-Cuzin C, Movassat J (2009) The GK rat β-cell: a prototype for the diseased human β-cell in type 2 diabetes? Mol Cell Endocrinol 297:73–85

Rose T, Efendic S, Rupnik M (2007) Ca^{2+} -secretion coupling is impaired in diabetic Goto Kakizaki rats. J Gen Physiol 129:493–508

Salehi A, Henningsson R, Mosén H, Östenson CG, Efendic S, Lundquist I (1999) Dysfunction of the islet lysosomal system conveys impairment of glucose-induced insulin release in the diabetic GK rat. Endocrinology 140:3045–3053

Salehi A, Meidute Abaraviciene S, Jimenez-Feltstrom J, Östenson CG, Efendic S, Lundquist I (2008) Excessive islet NO generation in type 2 diabetic GK rats coincides with abnormal hormone secretion and is counteracted by GLP1. PLoS ONE 3(5):e2165

Seiça R, Martins MJ, Pessa PB, Santos RM, Rosario LM, Suzuki KI, Martins MI (2003) Morphological changes of islet of Langerhans in an animal model of type 2 diabetes. Acta Med Port 16:381–388

Sener A, Ladrière L, Malaisse WJ (2001) Assessment by D-[(3)H]mannoheptulose uptake of β-cell density in isolated pancreatic islets from Goto-Kakizaki rats. Int J Mol Med 8:177–180

Serradas P, Giroix M-H, Saulnier C, Gangnerau MN, Borg LAH, Welsh M, Portha B, Welsh N (1995) Mitochondrial deoxyribonucleic acid content is specifically decreased in adult, but not fetal, pancreatic islets of the Goto-Kakizaki rat, a genetic model of non insulin-dependent diabetes. Endocrinology 136:5623–5631

Serradas P, Gangnerau MN, Giroix MH, Saulnier C, Portha B (1998) Impaired pancreatic β cell function in the fetal GK rat. Impact of diabetic inheritance. J Clin Invest 101:899–904

Serradas P, Goya L, Lacorne M, Gangnerau MN, Ramos S, Alvarez C, Pascual-Leone AM, Portha B (2002) Fetal insulin-like growth factor-2 production is impaired in the GK rat model of type 2 diabetes. Diabetes 51:392–397

Shang W, Yasuda K, Takahashi A, Hamasaki A, Takehiro M, Nabe K, Zhou H, Naito R, Fujiwara H, Shimono D, Ueno H, Ikeda H, Toyoda K, Yamada Y, Kurose T (2002) Effect of high dietary fat on insulin secretion in genetically diabetic Goto-Kakizaki rats. Pancreas 25:393–399

Simmons R (2006) Developmental origins of adult metabolic disease. Endocrinol Metab Clin N Am 35:193–204

Suzuki KI, Goto Y, Toyota T (1992) Spontaneously diabetic GK (Goto-Kakizaki) rats. In: Shafrir E (ed) Lessons from animal diabetes. Smith–Gordon, London, pp 107–116

Suzuki N, Aizawa T, Asanuma N, Sato Y, Komatsu M, Hidaka H, Itoh N, Yamauchi K, Hashizume K (1997) An early insulin intervention accelerates pancreatic β-cell dysfunction in young Goto-Kakizaki rats, a model of naturally occurring noninsulin-dependent diabetes. Endocrinology 138:1106–1110

Svensson AM, Östenson CG, Sandler S, Efendic S, Jansson L (1994) Inhibition of nitric oxide synthase by NG-nitro-L-arginine causes a preferential decrease in pancreatic islet blood flow in normal rats and spontaneously diabetic GK rats. Endocrinology 135:849–853

Svensson AM, Östenson CG, Jansson L (2000) Age-induced changes in pancreatic islet blood flow: evidence for an impaired regulation in diabetic GK rats. Am J Physiol Endocrinol Metab 279:E1139–E1144

Tsuura Y, Ishida H, Okamoto Y, Kato S, Sakamoto K, Horie M, Ikeda H, Okada Y, Seino Y (1993) Glucose sensitivity of ATP-sensitive K^+ channels is impaired in β-cells of the GK rat. A New Genet Model NIDDM Diabetes 42:1446–1453

Váradi A, Molnár E, Östenson CG, Ashcroft SJ (1996) Isoforms of endoplasmic reticulum Ca^{2+}–ATPase are differentially expressed in normal and diabetic islets of Langerhans. Biochem J 319:521–527

Villar-Palasi C, Farese RV (1994) Impaired skeletal muscle glycogen synthase activation by insulin in the Goto-Kakizaki (G/K) rat. Diabetologia 37:885–888

Wallis RH, Wallace KJ, Collins SC, Mc Ateer M, Argoud K, Bihoreau MT, Kaisaki PJ, Gauguier D (2004) Enhanced insulin secretion and cholesterol metabolism in congenic strains of the spontaneously diabetic (type 2) Goto-Kakizaki rat are controlled by independent genetic loci in rat chromosome 8. Diabetologia 47:1096–1106

Wallis RH, Collins SC, Kaisaki PJ, Argoud K, Wilder SP, Wallace KJ, Ria M, Ktorza A, Rorsman P, Bihoreau MT, Gauguier D (2008) Pathophysiological, genetic and gene expression features of a novel rodent model of the cardio-metabolic syndrome. PLoS ONE 3(8):e2962

Warwar N, Efendic S, Östenson CG, Haber EP, Cerasi E, Nesher R (2006) Dynamics of glucose-induced localization of PKC isoenzymes in pancreatic β-cells. Diabetes-related changes in the GK rat. Diabetes 55:590–599

Weng HB, Gu Q, Liu M, Cheng NN, Li D, Gao X (2008) Increased secretion and expression of amylin in spontaneously diabetic Goto-Kakizaki rats treated with rhGLP1(7–36). Acta Pharmacol Sin 29:573–579

Yasuda K, Okamoto Y, Nunoi K, Adachi T, Shihara N, Tamon A, Suzuki N, Mukai E, Fujimoto S, Oku A, Tsuda K, Seino Y (2002) Normalization of cytoplasmic calcium response in pancreatic β-cells of spontaneously diabetic GK rat by the treatment with T-1095, a specific inhibitor of renal Na^+–glucose co-transporters. Horm Metab Res 34:217–221

Zhang W, Khan A, Östenson CG, Berggren PO, Efendic S, Meister B (2002) Down-regulated expression of exocytotic proteins in pancreatic islets of diabetic GK rats. Biochem Biophys Res Commun 291:1038–1044

Zhou YP, Östenson CG, Ling ZC, Grill V (1995) Deficiency of pyruvate dehydrogenase activity in pancreatic islets of diabetic GK rats. Endocrinology 136:3546–3551

β-Cell Function in Obese-Hyperglycemic Mice (*ob/ob* Mice) 27

Per Lindström

Contents

The *ob/ob* Mouse	768
Discovery of Leptin	769
Insulin Resistance and Absence of Leptin	770
Pancreatic Islets	771
Oscillatory Insulin Release	772
β-Cell Mass	773
Glucotoxicity and Lipotoxicity	774
Incretins	775
Conclusions	776
Cross-References	776
References	776

Abstract

This review summarizes key aspects of what has been learned about β-cell physiology from studies in *ob/ob* mice. *Ob/ob* mice lack functional leptin. They are grossly overweight and hyperphagic particularly at young ages and develop severe insulin resistance with hyperglycemia and hyperinsulinemia. *Ob/ob* mice have large pancreatic islets. The β-cells respond adequately to most stimuli, and *ob/ob* mice have been used as a rich source of pancreatic islets with high insulin release capacity. Depending on the genetic background, *ob/ob* mice can be described as a model for a constant prediabetic state or as a model for β-cell events leading to overt type 2 diabetes. The large capacity for islet growth and insulin release makes *ob/ob* mice from the C57Bl/6J or Umeå *ob/ob* strain a good model for studies on how β-cells can cope with prolonged functional stress.

P. Lindström
Department of Integrative Medical Biology, Section for Histology and Cell Biology, Umeå University, Umeå, Sweden
e-mail: per.lindstrom@histocel.umu.se

> **Keywords**
> Mouse • Pancreatic islet • β-cell • Leptin

The *ob/ob* Mouse

The *ob/ob* syndrome was found in 1949 in an outbred mouse colony at Roscoe B. Jackson Memorial Laboratory, Bar Harbor, Maine (Ingalls et al. 1950) and was transferred to the already well-characterized C57Bl mice colony that had been established during the 1930s. The discovery that the *ob/ob* mouse syndrome is caused by a defective adipocytokine leptin opened a whole new era of metabolic studies and understanding of the endocrine functions of adipose tissue. Obesity is the most obvious characteristic of *ob/ob* mice. They are also hyperphagic, hyperinsulinemic, and hyperglycemic and have reduced metabolic rate and a low capacity for thermogenesis (Mayer et al. 1953; Garthwaite et al. 1980). The pancreatic islets are large and contain a high proportion of insulin-producing β-cells. It was soon discovered that *ob/ob* mice have a number of other traits except obesity. They are, e.g., infertile and have impaired immune functions.

The *ob/ob* syndrome varies considerably depending on the genetic background (Mayer and Silides 1953; Coleman and Hummel 1973). In this presentation *ob/ob* mice refer to 6J or Umeå *ob/ob* mice unless otherwise stated. On a 6J or Umeå *ob/ob* background, hyperglycemia is relatively mild particularly at old age and glycosuria is usually not present in the fasting state. They represent a mouse model for obesity and "diabetes" with moderate hyperglycemia, high insulin release capacity, and marked adiposity (Shafrir et al. 1999). On a KsJ or BTBR background, the mice have a higher food intake (Stoehr et al. 2004) and become overtly diabetic with a reduced life expectancy (Coleman 1978; Ranheim et al. 1997). On a 6J background the mice have a large lipogenic capacity in the liver (Clee et al. 2005), which may render them less susceptible to lipotoxic effects. β-Cells from *ob/ob* mice accumulate fat but only a small lipid increase is observed in β-cells from *ob/ob* mice on a 6J background (Garris and Garris 2004), which is in keeping with the better-preserved function. The importance of a high insulin release capacity was evident from studies where the *ob* trait was transferred to DBA mice (Chua et al. 2002). Mice with large islets and a high insulin release capacity maintained adiposity, whereas mice with lower serum insulin levels had diminished adiposity and a more severe diabetes (Chua et al. 2002). There are also differences between individual mice from the same colony of 6J and Umeå *ob/ob* strains with regard to hyperglycemia and other aspects of a "diabetes-like" condition. This can be used to select subgroups of animals within the same strain for metabolic studies. The fact that genetic background can have such a profound influence on the consequences of leptin deficiency has prompted a series of recent studies in search for factors involved in the development of diabetes. A comparison between BTBR and 6J *ob/ob* mouse islets indicates that the Alzheimer gene *App* may be one of the top candidates for the regulation of insulin release capacity (Tu et al. 2012). Cell cycle regulatory genes are differentially expressed in islets from C57Bl/6J and BTBR mice (Davis et al. 2010), and genes important for cell survival are overexpressed in C57Bl/6J *ob/ob* mice islets when compared with diabetes prone BTBR mice (Keller et al. 2008; Singh et al. 2013). Calcium channel blockers increase β-cell survival and insulin sensitivity in BTRB *ob/ob* mice (Xu et al. 2012). The adaptive unfolded protein response is better developed in islets from *ob/ob* mice compared with db/db mice (Chan et al. 2013), and the capacity for adaptive unfolded protein response is reduced in *ob/ob* mouse β-cells if they become autophagy deficient (Quan et al. 2012).

Ob/ob mice are indistinguishable from their lean littermates at birth, but within 2 weeks they become heavier and develop hyperinsulinemia. The syndrome becomes much more pronounced after weaning and overt hyperglycemia is observed during the fourth week. The blood glucose rises to reach a peak after 3–5 months when the mice also have a very high food intake and a rapid growth (Westman 1968a; Edvell and Lindström 1995, 1999). After that, blood glucose values decrease and eventually become nearly normal at old age. Serum insulin levels are also very high and peak at a higher age than blood glucose values (Westman 1968a). The animals remain insulin resistant but impaired glucose tolerance and glycosuria after a glucose load are observed mostly in the post-weaning period of rapid growth (Westman 1968a; Herberg et al. 1970; Danielsson et al. 1968; Edvell and Lindström 1998).

Discovery of Leptin

Elegant parabiosis experiments showed that *ob/ob* mice lack but are very sensitive to a circulating factor produced by their normal siblings (Coleman 1973, 1978). By extensive positional cloning experiments, this factor could be identified in 1994 by Friedman and co-workers as leptin produced in adipose tissue (Friedman et al. 1991; Zhang et al. 1994). The *ob/ob* syndrome can be reversed almost completely even in adult animals by exogenous leptin or transfection with the leptin gene (Larcher et al. 2001; Pelleymounter et al. 1995; Halaas et al. 1995). There are cases with leptin deficiency in obese humans, but this is uncommon; so *ob/ob* mice do not present a good model for the etiology of human obesity (Clement 2006). It has not been clarified if hyperglycemia and insulin resistance depend on the adiposity or are a consequence of leptin deficiency. However, the discovery of leptin has widened our understanding of the regulation of food intake, metabolic turnover, and obesity. We also have learned a lot more about the interrelationship between metabolism and other functions such as reproduction and the immune system. Much of what we know about the physiology of leptin has been achieved through studies in *ob/ob* mice but also from observations in animal models with leptin receptor defects such as *db/db* mice and *fa/fa* rats (Chehab et al. 2004; Unger and Orci 2001).

Insulin Resistance and Absence of Leptin

Ob/ob mice have severe insulin resistance. Peripheral insulin resistance induces hyperglycemia and worsens the functional load on the β-cells. *Ob/ob* mouse β-cells are insulin resistant from an early age (Loreti et al. 1974; Zawalich et al. 2002). Insulin inhibits insulin release, and insulin resistance coupled to reduced PI3K-dependent signaling may result in disinhibition of glucose-induced insulin release (Zawalich et al. 2002). Insulin resistance can therefore be beneficiary for β-cell function.

β-Cells have full-length leptin receptors and leptin inhibits insulin release and insulin biosynthesis in most studies (Kieffer et al. 1996; Emilsson et al. 1997; Melloul et al. 2002). Lack of leptin effects may enhance β-cell function and explain some of the functional differences between *ob/ob* mice and normal mice. The main signaling pathways for leptin are the JAK/STAT transduction cascade, the mitogen-activated protein kinase (MAPK) cascade, the phosphoinositide 3-kinase (PI3K), IRS, and the 5′-AMP-activated protein kinase (AMPK) pathways (Sweeney 2002; Frühbeck 2006; Cirillo et al. 2008). The role of these signal mediators in β-cell function has not been entirely clarified but the majority of findings suggest that AMPK (Rutter et al. 2003) and p38 MAPK (Cuenda and Nebreda 2009; Sumara et al. 2009) inhibit glucose-induced insulin release. There are different isoforms of the leptin receptor. The full-length leptin receptor present in pancreatic β-cells is required for the JAK/STAT response, and activation is accompanied also by a rise in suppressor of cytokine signaling (SOCS) (Seufert 2004). A shorter receptor form, which activates PI3K, is predominant in skeletal muscle (Dyck et al. 2006) but PI3K activation is found also in β-cells (Seufert 2004). Leptin signaling pathways may interact with insulin signaling at several points including JAKs, PI3K, and MAPK (Lulu Strat et al. 2005). This interaction between insulin and leptin is complex, but studies in *ob/ob* mice clearly indicate that the net effect of leptin is to increase insulin sensitivity (Lulu Strat et al. 2005; Rattarasarn 2006) and that leptin resistance worsens insulin resistance. Absence of leptin can therefore be one of the causes of insulin resistance in *ob/ob* mice. Obese individuals are usually both insulin resistant and leptin resistant. However, the total absence of leptin signaling already from the onset of obesity in *ob/ob* mice is in sharp contrast to obesity in humans, and the cross talk between the cellular effects of insulin and leptin is obviously absent.

Parasympathetic cholinergic axons are numerous and widespread throughout mouse islets, whereas sympathetic neurons are more abundant in the islet periphery (Rodriguez-Diaz et al. 2011). Leptin is a central mediator of autonomic nervous function and may inhibit islet function through activation of sympathetic neurons (Hinoi et al. 2008; Tentolouris et al. 2008). β-Cells from *ob/ob* mice are more sensitive than lean mouse β-cells to the stimulatory effect of acetylcholine and the inhibitory effect of noradrenalin on glucose-induced insulin release (Tassava et al. 1992). This could be because of sympathetic disinhibition due to the lack of leptin. However, there is an age dependence for these effects of neurotransmitters. Islets from young *ob/ob* mice have an increased β-cell responsiveness to cholinergic stimulation already from 10 to 12 days of age (Chen and Romsos 1995). The sensitivity to acetylcholine is reduced at old age, whereas the sensitivity to vagal neuropeptides may be increased (Persson-Sjögren and Lindström 2004; Persson-Sjögren et al. 2006). A reduced cholinergic activity at old age paralleling improved glycemic control is consistent with the finding that M3 receptor knockout in *ob/ob* mice reduces the severity of most of the phenotype (Gautam et al. 2008). Much of the effects of leptin are exerted at the level of the central nervous system, and it has been found that hypothalamic endocannabinoid signaling can be very important for the large insulin release capacity of leptin-deficient *ob/ob* mice (Li et al. 2013).

Ob/ob mouse islets have a rich supply of small vessels but a lower blood flow than lean mouse islets when calculated on the basis of islet size (Carlsson et al. 1996). *Ob/ob* mouse islet vessels are also more sensitive to sympathetic inhibition of islet circulation (Rooth and Täljedal 1987). This suggests that they have a reduced capacity to increase blood flow to meet metabolic demands (Carlsson et al. 1996), and this can increase β-cell stress. The increased demand for blood supply of *ob/ob* mouse β-cells may be met by vascular dilatation rather than by angiogenesis (Dai et al. 2013). Amyloid deposits surrounding islet cells is observed in most islets from type 2 diabetics (Butler et al. 2003) and may be part of the pathogenesis for β-cell damage. Mice do not normally form islet amyloid deposits, but *ob/ob* mice have high serum levels of the islet amyloid polypeptide (IAPP) (Leckström et al. 1999), and the islet content of IAPP increases during *ob/ob* syndrome development (Takada et al. 1996). The interaction between leptin and IAPP has not been much studied in *ob/ob* mouse islets but leptin inhibits IAPP release in lean mice (Karlsson et al. 1998). Leptin deficiency could therefore increase IAPP content in *ob/ob* mice. IAPP inhibits insulin and glucagon release (Ahrén and Sörhede Winzell 2008), and it has been suggested that IAPP also induces insulin resistance (Nyholm et al. 1998).

Pancreatic Islets

The islet volume is up to ten times higher in *ob/ob* mice than in normal mice (Bleisch et al. 1952; Gepts et al. 1960), and insulin-producing β-cells are by far the most numerous (Westman 1968a, b; Gepts et al. 1960; Baetens et al. 1978). The architecture and size distribution have been reevaluated in a comparative study of islets from different species (Kim et al. 2009). It was found that the distribution of islet sizes closely overlaps between species. Markedly large islets are found in *ob/ob* mice but also in humans and monkeys and in pregnant mice.

Both stimulatory and inhibitory effects of leptin have been found with regard to β-cell proliferation and survival (Marroquí et al. 2012). The islet hyperplasia is probably not caused by a primary abnormality in the islets due to leptin deficiency although this can contribute; the size and form of islets found in *ob/ob* mice are probably characteristic of islets able to adapt to increasing demands and not a feature specific for leptin deficiency. The growth may be triggered by hyperglycemia but also by other blood borne factors and nerve stimulation and is evident from the fourth week (Edvell and Lindström 1999). Partial pancreatectomy in *ob/ob* mice in a phase of rapid growth and severe hyperglycemia results in a huge expansion of islet area and islet number (Chen et al. 1989). The islet growth normally continues for more than 6 months and is paralleled by reduced insulin content per islet volume during conditions of free access to food (Tomita et al. 1992). The large islets with many insulin-producing β-cells are in contrast to the decreased β-cell mass found in diabetes (Wajchenberg 2007).

Intranuclear rodlets containing the cytoskeletal protein class III β-tubulin were first described in neurons (Milman et al. 2010). These rodlets are also found in β-cells from humans and normal mice but are in very low numbers in β-cells from *ob/ob* mice (Woulfe and Munoz 2000). The functional significance of intranuclear rodlets is not known. Advanced glycation end products are formed intracellularly in β-cells as a consequence of long-standing hyperglycemia and may be involved in β-cell damage leading to insufficient insulin release and overt type 2 diabetes (Han et al. 2013). The receptor for glycation products was upregulated in *ob/ob* mouse β-cells when compared with controls, but levels were lower than in *db/db* mouse β-cells (Han et al. 2013).

The cellular mechanisms for glucose-induced insulin release is not the subject of this article but islets isolated from *ob/ob* mice respond adequately to stimulators and inhibitors of insulin release in most experimental conditions (Hahn et al. 1974; Hellman et al. 1974), and they have been used in several hundred papers as a rich source of β-cells in studies of islet function. After an overnight fast the blood glucose is nearly normalized and *ob/ob* mouse islets release larger quantities of insulin after fasting when compared with normal mouse islets (Lavine et al. 1977). However, transplantation of coisogenic (+/+) islets to *ob/ob* mice lowered blood glucose values to nearly normal for 1 month (Barker et al. 1977).

The persistent hyperglycemia can therefore be a sign of insufficient β-cell function despite the high capacity to secrete insulin, and the *ob/ob* mouse can perhaps be described to be in a constant prediabetic state. The threshold for glucose-induced insulin release occurs at a lower glucose concentration than in lean mouse islets (Lavine et al. 1977; Chen et al. 1993). The mechanisms for this may in part be similar to the glucose hypersensitivity observed after prolonged exposure to elevated glucose in islets from normoglycemic animals and involve both metabolic and ionic events (Ling and Pipeleers 1996; Khaldi et al. 2004).

Oscillatory Insulin Release

Serum insulin shows diurnal oscillations, and it is thought that the effect of insulin on target organs is improved when insulin is delivered in a pulsatile manner. We know little about the periodicity of serum insulin in *ob/ob* mice, but serum insulin levels vary considerably in the same mouse also when sampled under tightly controlled conditions (Lindström unpublished). The oscillations can be triggered by several mechanisms including variations in cytosolic calcium and metabolic oscillations (Heart and Smith 2007). Variation in cAMP levels is also a likely candidate as evidenced from studies in *ob/ob* mouse β-cells (Grapengiesser et al. 1991) and β-cell lines (Dyachok et al. 2006). *Ob/ob* mouse islets have a reduced capacity to accumulate cAMP (Black et al. 1986, 1988a), but they are more sensitive to a rise in cAMP for stimulation of insulin release (Black et al. 1986). The β-cells have an increased Na/K-ATPase activity (Elmi 2001) and may be more sensitive to voltage-dependent events (Fournier et al. 1990) perhaps due to a reduced activation of K_{ATP} channels (Seufert 2004). However, the function of voltage-dependent Ca^{2+} channels is impaired (Black et al. 1988b), and there is a disturbed pattern of cytoplasmic calcium changes after glucose stimulation (Ravier et al. 2002). *Ob/ob* mouse β-cells also do not show the same type of cell-specific Ca^{2+} responses from individual cells that are found in lean mouse islets (Gustavsson et al. 2006). There is an excessive firing of cytoplasmic Ca^{2+} transients when *ob/ob* mouse β-cells are stimulated with glucagon (Ahmed and Grapengiesser 2001). This effect could be a direct consequence of leptin deficiency because it was reduced when leptin was also added. Ryanodine receptors in the endoplasmic reticulum may be involved in β-cell calcium regulation and stimulation of insulin release, but the precise role is controversial (Islam 2002; Bruton et al. 2003). In one study it was reported that β-cells from *ob/ob* mice have less ryanodine receptor activation than β-cells from lean mice (Takasawa et al. 1998).

An increased sensitivity to cAMP could also have other effects. Uncoupling protein-2 (UCP-2) was demonstrated in β-cells a decade ago, and it has been suggested that UCP-2 is important as a negative regulator of glucose-induced insulin release and protection against oxidative stress. cAMP was found to reduce the inhibitory effect of a rise in UCP-2 through PKA-mediated inhibition of the K_{ATP} channel (McQuaid et al. 2006). *Ob/ob* mouse β-cells have increased activity of UCP-2 (Saleh et al. 2006; Zhang et al. 2001) when compared with lean mice from the same background. Inhibition of UCP-2 improved glucose tolerance (De Souza et al. 2007), but knockdown of UCP-2 expression had no effect on glucose-induced insulin release in *ob/ob* mouse islets (Saleh et al. 2006). ACTH receptor activation is coupled to a rise in cAMP (Enyeart 2005). Leptin stimulates both CRF and ACTH secretion (Malendowicz et al. 2007), but *ob/ob* mice show signs of increased ACTH activity. Serum ACTH levels are high and islets from *ob/ob* mice respond with a larger increase in insulin release after stimulation with ACTH 1-39 (Bailey and Flatt 1987).

β-Cell Mass

One of the features of *ob/ob* mice is that they have large pancreatic islets consisting of mostly β-cells, and *ob/ob* mice have been used in studies of β-cell proliferation. β-Cell growth is probably directly or indirectly stimulated by hyperglycemia. There is a good correlation between the level of hyperglycemia and islet cell replication in rat (Bonner-Weir et al. 1989) and obese hyperglycaemic mice (Andersson et al. 1989), and the morphology of *ob/ob* mice islets reaggregated in vitro depends on the glucose concentration (Norlund et al. 1987). The source of β-cell expansion is still controversial. It has been suggested that cells recruited from the bone marrow increase the insulin release capacity in *ob/ob* mice (Kojima et al. 2004). Duct progenitor cells can also be involved in the expansion of the β-cell mass, but mitotic figures have been demonstrated in β-cells from *ob/ob* mice (Edvell and Lindström 1999; Gepts et al. 1960), and cells within existing islets can be important for expansion of the total islet mass (Bock et al. 2003). *Ob/ob* mice have a growth-promoting environment for β-cells depending on (extra) pancreatic factors (Tyrberg et al. 2001; Flier et al. 2001) perhaps including insulin (Lee and Nielsen 2009). Oncogenes stimulate *ob/ob* mice β-cell replication as a sign that they can be manipulated extrinsically (Welsh et al. 1988). Blood-borne factors probably include NPY (Cho and Kim 2004) and GLP-1 (Edvell and Lindström 1999; Stoffers et al. 2000; Blandino-Rosano et al. 2008) which both stimulate *ob/ob* mouse β-cell replication. Interestingly, NPY inhibits insulin release and *ob/ob* mouse islets have reduced expression of NPY receptors (Imai et al. 2007). Obesity probably also induces an indirect neuronal signal emanating from the liver which is important for stimulation of islet growth in *ob/ob* mice (Imai et al. 2008). Cytokines and growth hormone may be important mitogens for β-cells (Black et al. 1988b; Lindberg et al. 2005). Intracellular signaling for GH receptors includes JAK/STAT activation, and this is inhibited by SOCS that inhibit cytokine signaling (Black et al. 1988b; Gysemans et al. 2008). Inhibition of cytokine signaling by SOCS may prevent β-cell death induced by several cytokines such as IL-1β, TNFα, and IFNγ (Gysemans et al. 2008). Leptin activates both JAK/STAT and SOCS, and it is possible that the net effect of leptin deficiency is to stimulate β-cell growth through a lowering of SOCS. Low-grade inflammation may be important for increased adiposity and for the pathogenesis of type 2 diabetes (Hill et al. 2009; Donath et al. 2008). Leptin stimulates the immune system and is involved in macrophage activation and release of cytokines (Lam and Lu 2007). This could be part of the explanation why leptin deficiency may prevent β-cell death. Few studies have specifically addressed the effect of cytokines in *ob/ob* mouse islets, but they respond normally to cytokine activators and inhibitors (Prieto et al. 1992; Zaitseva et al. 2006; Peterson et al. 2008).

Glucotoxicity and Lipotoxicity

Glucotoxicity caused by long-standing hyperglycemia may be one factor inducing β-cell death in the development of type 2 diabetes. The toxicity may be caused by induction of reactive oxygen species and by inducing endoplasmic reticulum (ER) stress (Wajchenberg 2007). ER stress is probably an important cause of β-cell dysfunction in diabetes (Eizirik et al. 2008). ER stress can be an important cause of leptin resistance (Ozcan et al. 2009), but this may not be relevant to *ob/ob* mice since they lack leptin. However, *ob/ob* mice show clear signs of hepatocyte ER stress (Marí et al. 2006; Yang et al. 2007; Sreejayan et al. 2008), and it is likely that also the β-cells have ER stress because of the increased demands for protein synthesis.

Ob/ob mice are living proof that prolonged hyperglycemia is not necessarily deleterious. It is possible that the insulin resistance and leptin absence protects the β-cells from the damage that constant glucose stimulation would otherwise cause. However, only few differences from lean mice have been reported with regard to β-cell metabolic signaling and enzyme activities. The mitochondrial enzyme FAD-linked glycerophosphate dehydrogenase (m-GDH) is thought to play a key role in the glucose-sensing mechanism of the insulin-producing β-cell. It catalyzes a rate-limiting step of the glycerol phosphate shuttle, but there was no difference between islets in enzyme activity between normal and *ob/ob* mice (Sener et al. 1993). Perhaps *ob/ob* mouse β-cells are protected because they have an increased glucose cycling through glucose-6-phosphatase (Khan et al. 1995). They also have lower levels of the glucose transporter (GLUT2) (Jetton et al. 2001). A reduced glucokinase activity could lessen β-cell stress. There are conflicting data as to whether glucokinase is lower (Jetton et al. 2001) or higher (Hinoi et al. 2008) than in lean mouse islets, but glucokinase activation increases the insulin release capacity (Park et al. 2013).

Elevated serum levels of free fatty acids in the presence of hyperglycemia and aberrant lipoprotein profiles could cause lipotoxic damage to β-cells. *Ob/ob* mouse islets show signs of a reduced fatty acid oxidation in the presence of high glucose (Berne 1975) which could lead to toxic effects of lipids. However, β-cells from *ob/ob* mice lack lipoprotein lipase (Nyrén et al. 2012) and show low levels of hormone-sensitive lipase (Khan et al. 2001). Leptin treatment restores hormone levels to those found in normal islets (Nyrén et al. 2012; Khan et al. 2001). The lack of intracellular lipase may be important for the balance between glucose and lipid metabolism in *ob/ob* mouse islets making them less vulnerable to diabetic insult. *Ob/ob* mice have low serum VLDL levels and high HDL levels (Camus et al. 1988), and this can also be protective. It is likely that the large capacity to accumulate fat in adipose tissue protects *ob/ob* mice against β-cell lipotoxicity (Flowers et al. 2007).

Incretins

The incretins GLP-1 and GIP are released in response to food ingestion and play an important role in stimulating insulin release when blood glucose levels are elevated. The half-life in circulation is short because of enzymatic digestion through dipeptidyl peptidase-4 (DPP-4). GLP-1 and GLP-1 analogues stimulate β-cell proliferation (Edvell and Lindström 1999; Stoffers et al. 2000; Blandino-Rosano et al. 2008) and glucose-induced insulin release in *ob/ob* mice (Cullinan et al. 1994; Young et al. 1999; Rolin et al. 2002), and inhibition of DPP-4 improves β-cell function (Moritoh et al. 2008). On the other hand, chemical ablation of the GIP receptors causes normalization of hyperglycemia, serum insulin, insulin sensitivity, glucose tolerance, and islet hypertrophy in *ob/ob* mice (Gault et al. 2005; Irwin et al. 2007). This indicates that different incretins can have both beneficiary and adverse effects in obesity-related hyperglycemia and insulin resistance. Glucagon levels are high in *ob/ob* mice (Dubuc et al. 1977). It was early hypothesized that elevated glucagon secretion contributes to the altered metabolism of *ob/ob* mice (Mayer 1960) and immunoneutralization of endogenous glucagon improves metabolic control (Sorensen et al. 2006). There is a correlation between serum glucagon levels and hepatic glucose output in type 2 diabetic patients (Gastaldelli et al. 2000), and reduction of serum glucagon may be a target for diabetes treatment.

Peroxisome proliferator-activated receptors (PPARs) are ligand-activated transcription factors that belong to the nuclear hormone receptor superfamily. PPAR-γ and PPAR-α exert profound effects on lipid handling. PPAR-γ directs lipid toward adipose tissue and PPAR-α activation predominantly stimulates lipid oxidation. PPAR agonists have been used in the treatment of type 2 diabetes to reduce insulin resistance and improve β-cell function.

The adaptation of *ob/ob* mouse islets to insulin resistance and hyperglycemia is dependent on intact PPAR-$γ_2$ signaling (Medina-Gomez et al. 2009; Vivas et al. 2011), and treatment with both PPAR-γ agonists (Diani et al. 2004) and PPAR-α agonists (Lalloyer et al. 2006) improved glucose-stimulated insulin release in *ob/ob* mice. This is another indication that *ob/ob* mouse β-cells are normally under functional stress.

Conclusions

Ob/ob mouse islets are large and contain a high proportion of insulin-producing β-cells. They respond adequately to most stimulators and inhibitors of insulin release and have been used as a rich source of β-cells for in vitro studies of islet function. *Ob/ob* mouse β-cells show insulin resistance and other signs of leptin deficiency. The lack of leptin must always be taken into account when using *ob/ob* mice as a model. Nevertheless, *ob/ob* mice represent an excellent model for studies on how β-cells can adapt to increased demand and maintain a high insulin release capacity during prolonged functional stress.

Cross-References

▶ (Dys)Regulation of Insulin Secretion by Macronutrients
▶ Apoptosis in Pancreatic β-Islet Cells in Type 1 and Type 2 Diabetes
▶ Islet Structure and Function in the GK Rat
▶ The β-Cell in Human Type 2 Diabetes

References

Ahmed M, Grapengiesser E (2001) Pancreatic β-cells from obese-hyperglycemic mice are characterized by excessive firing of cytoplasmic Ca^{2+} transients. Endocrine 15:73–78

Ahrén B, Sörhede Winzell M (2008) Disturbed α-cell function in mice with β-cell specific overexpression of human islet amyloid polypeptide. Exp Diabetes Res 2008:304–313

Andersson A, Korsgren O, Naeser P (1989) DNA replication in transplanted and endogenous pancreatic islets of obese-hyperglycemic mice at different stages of the syndrome. Metabolism 38:974–978

Baetens D, Stefan Y, Ravazzola M, Malaisse-Lagae F, Coleman DL, Orci L (1978) Alteration of islet cell populations in spontaneously diabetic mice. Diabetes 27:1–7

Bailey CJ, Flatt PR (1987) Insulin releasing effects of adrenocorticotropin (ACTH 1-39) and ACTH fragments (1-24 and 18-39) in lean and genetically obese hyperglycaemic (ob/ob) mice. Int J Obes 11:175–181

Barker CF, Frangipane LG, Silvers WK (1977) Islet transplantation in genetically determined diabetes. Ann Surg 186:401–410

Berne C (1975) The metabolism of lipids in mouse pancreatic islets. The oxidation of fatty acids and ketone bodies. Biochem J 152:661–666

Black MA, Heick HM, Begin-Heick N (1986) Abnormal regulation of insulin secretion in the genetically obese (ob/ob) mouse. Biochem J 238:863–869

Black MA, Heick HM, Begin-Heick N (1988a) Abnormal regulation of cAMP accumulation in pancreatic islets of obese mice. Am J Physiol 255:E833–E838

Black MA, Fournier LA, Heick HM, Begin-Heick N (1988b) Different insulin secretory responses to calcium-channel blockers in islets of lean and obese (ob/ob) mice. Biochem J 249:401–407

Blandino-Rosano M, Perez-Arana G, Mellado-Gil JM, Segundo C, Aguilar- Diosdado M (2008) Anti-proliferative effect of pro-inflammatory cytokines in cultured β cells is associated with extracellular signal-regulated kinase 1/2 pathway inhibition: protective role of glucagon-like peptide -1. J Mol Endocrinol 41:35–44

Bleisch VR, Mayer J, Dickie MM (1952) Familiar diabetes mellitus in mice associated with insulin resistance, obesity, and hyperplasia of the islands of Langerhans. Am J Pathol 28:369–385

Bock T, Pakkenberg B, Buschard K (2003) Increased islet volume but unchanged islet number in ob/ob mice. Diabetes 52:1716–1722

Bonner-Weir S, Deery D, Leahy JL, Weir GC (1989) Compensatory growth of pancreatic β-cells in adult rats after short-term glucose infusion. Diabetes 38:49–53

Bruton JD, Lemmens R, Shi CL, Persson-Sjögren S, Westerblad H, Ahmed M, Pyne NJ, Frame M, Furman BL, Islam MS (2003) Ryanodine receptors of pancreatic β cells mediate a distinct context-dependent signal for insulin secretion. FASEB J 17:301–303

Butler AE, Janson J, Bonner-Weir S, Ritzel R, Rizza RA, Butler PC (2003) β-Cell deficit and increased β-cell apoptosis in humans with type 2 diabetes. Diabetes 52:102–110

Camus MC, Aubert R, Bourgeois F, Herzog J, Alexiu A, Lemonnier D (1988) Serum lipoprotein and apolipoprotein profiles of the genetically obese ob/ob mouse. Biochim Biophys Acta 961:53–64

Carlsson PO, Andersson A, Jansson L (1996) Pancreatic islet blood flow in normal and obese-hyperglycemic (ob/ob) mice. Am J Physiol 271:E990–E995

Chan JY, Luzuriaga J, Bensellam M, Biden TJ, Laybutt DR (2013) Failure of the adaptive unfolded protein response in islets of obese mice is linked with abnormalities in β-cell gene expression and progression to diabetes. Diabetes 62:1557–1568

Chehab FF, Qiu J, Ogus S (2004) The use of animal models to dissect the biology of leptin. Recent Prog Horm Res 59:245–266

Chen NG, Romsos DR (1995) Enhanced sensitivity of pancreatic islets from preobese 2-week-old ob/ob mice to neurohormonal stimulation of insulin secretion. Endocrinology 13:505–511

Chen L, Komiya I, Inman L, McCorkle K, Alam T, Unger RH (1989) Molecular and cellular responses of islets during perturbations of glucose homeostasis determined by in situ hybridization histochemistry. Proc Natl Acad Sci USA 86:1367–1371

Chen NG, Tassava TM, Romsos DR (1993) Threshold for glucose-stimulated insulin secretion in pancreatic islets of genetically obese (ob/ob) mice is abnormally low. J Nutr 123:1567–1574

Cho YR, Kim CW (2004) Neuropeptide Y promotes β-cell replication via extracellular signal-regulated kinase activation. Biochem Biophys Res Commun 314:773–780

Chua S Jr, Liu SM, Li Q, Yang L, Thassanapaff VT, Fisher P (2002) Differential β cell responses to hyperglycaemia and insulin resistance in two novel congenic strains of diabetes (FVB- Lepr (db)) and obese (DBA- Lep (ob)) mice. Diabetologia 45:976–990

Cirillo D, Rachiglio AM, la Montagna R, Giordano A, Normanno N (2008) Leptin signaling in breast cancer: an overview. J Cell Biochem 10:956–964

Clee SM, Nadler ST, Attie AD (2005) Genetic and genomic studies of the BTBR *ob/ob* mouse model of type 2 diabetes. Am J Ther 12:491–498

Clement K (2006) Genetics of human obesity. C R Biol 329:608–622

Coleman DL (1973) Effects of parabiosis of obese with diabetes and normal mice. Diabetologia 9:294–298

Coleman DL (1978) Obese and diabetes: two mutant genes causing diabetes obesity syndromes in mice. Diabetologia 14:141–148

Coleman DL, Hummel KP (1973) The influence of genetic background on the expression of the obese (ob) gene in the mouse. Diabetologia 9:287–293

Cuenda A, Nebreda AR (2009) p38 delta and PKD1: kinase switches for insulin secretion. Cell 136:209–210

Cullinan CA, Brady EJ, Saperstein R, Leibowitz MD (1994) Glucose-dependent alterations of intracellular free calcium by glucagon-like peptide-1(7-36amide) in individual *ob/ob* mouse β-cells. Cell Calcium 15:391–400

Dai C, Brissova M, Reinert RB, Nyman L, Liu EH, Thompson C, Shostak A, Shiota M, Takahashi T, Powers AC (2013) Pancreatic islet vasculature adapts to insulin resistance through dilation and not angiogenesis. Diabetes 62:4144–4153 [Epub ahead of print]

Danielsson Å, Hellman B, Täljedal I-B (1968) Glucose tolerance in the period preceding the appearance of the manifest obese-hyperglycemic syndrome in mice. Acta Physiol Scand 72:81–84

Davis DB, Lavine JA, Suhonen JI, Krautkramer KA, Rabaglia ME, Sperger JM, Fernandez LA, Yandell BS, Keller MP, Wang IM, Schadt EE, Attie AD (2010) FoxM1 is up-regulated by obesity and stimulates β-cell proliferation. Mol Endocrinol 24:1822–1834

De Souza CT, Araújo EP, Stoppiglia LF, Pauli JR, Ropelle E, Rocco SA, Marin RM, Franchini KG, Carvalheira JB, Saad MJ, Boschero AC, Carneiro EM, Velloso LA (2007) Inhibition of UCP2 expression reverses diet-induced diabetes mellitus by effects on both insulin secretion and action. FASEB J 21:1153–1163

Diani AR, Sawada G, Wyse B, Murray FT, Khan M (2004) Pioglitazone preserves pancreatic islet structure and insulin secretory function in three murine models of type 2 diabetes. Am J Physiol 286:E116–E122

Donath MY, Schumann DM, Faulenbach M, Ellingsgaard H, Perren A, Ehses JA (2008) Islet inflammation in type 2 diabetes: from metabolic stress to therapy. Diabetes Care 31(Suppl 2):S161–S164

Dubuc PU, Mobley PW, Mahler RJ, Ensinck JW (1977) Immunoreactive glucagon levels in obese-hyperglycemic (*ob/ob*) mice. Diabetes 26:841–846

Dyachok O, Isakov Y, Sågetorp J, Tengholm A (2006) Oscillations of cyclic AMP in hormone-stimulated insulin-secreting β-cells. Nature 439:349–352

Dyck DJ, Heigenhauser GJ, Bruce CR (2006) The role of adipokines as regulators of skeletal muscle fatty acid metabolism and insulin sensitivity. Acta Physiol 186:5–16

Edvell A, Lindström P (1995) Development of insulin secretory function in young obese hyperglycemic mice (Umeå *ob/ob*). Metabolism 44:906–913

Edvell A, Lindström P (1998) Vagotomy in young obese hyperglycemic mice: effects on syndrome development and islet proliferation. Am J Physiol 274:E1034–E1039

Edvell A, Lindström P (1999) Initiation of increased pancreatic growth in young normoglycemic mice (Umeå +/?). Endocrinology 140:778–783

Eizirik DL, Cardozo AK, Cnop M (2008) The role for endoplasmic reticulum stress in diabetes mellitus. Endocr Rev 29:42–61

Elmi A (2001) Increased number of Na^+/K^+ ATPase enzyme units in *Ob/Ob* mouse pancreatic islets. Pancreas 23:113–115

Emilsson V, Liu YL, Cawthorne MA, Morton NM, Davenport M (1997) Expression of the functional leptin receptor mRNA in pancreatic islets and direct inhibitory action of leptin on insulin secretion. Diabetes 46:313–316

Enyeart JJ (2005) Biochemical and ionic signaling mechanisms for ACTH-stimulated cortisol production. Vitam Horm 70:265–279

Flier SN, Kulkarni RN, Kahn CR (2001) Evidence for a circulating islet cell growth factor in insulin-resistant states. Proc Natl Acad Sci USA 98:7475–7480

Flowers JB, Rabaglia ME, Schueler KL, Flowers MT, Lan H, Keller MP, Ntambi JM, Attie AD (2007) Loss of stearoyl-CoA desaturase-1 improves insulin sensitivity in lean mice but worsens diabetes in leptin-deficient obese mice. Diabetes 5:1228–1239

Fournier LA, Heick HM, Begin-Heick N (1990) The influence of K^+-induced membrane depolarization on insulin secretion in islets of lean and obese (*ob/ob*) mice. Biochem Cell Biol 68:243–248

Friedman JM, Leibel RL, Siegel DS, Walsh J, Bahary N (1991) Molecular mapping of the mouse ob mutation. Genomics 11:1054–1062

Frühbeck G (2006) Intracellular signalling pathways activated by leptin. Biochem J 393:7–20

Garris DR, Garris BL (2004) Cytochemical analysis of pancreatic islet hypercytolipidemia following diabetes (db/db) and obese (*ob/ob*) mutation expression: influence of genomic background. Pathobiology 71:231–240

Garthwaite TL, Martinson DR, Tseng LF, Hagen TC, Menahan LA (1980) A longitudinal hormonal profile of the genetically obese mouse. Endocrinology 107:671–676

Gastaldelli A, Baldi S, Pettiti M, Toschi E, Camastra S, Natali A, Landau BR, Ferrannini E (2000) Influence of obesity and type 2 diabetes on gluconeogenesis and glucose output in humans: a quantitative study. Diabetes 49:1367–1373

Gault VA, Irwin N, Green BD, McCluskey JT, Greer B, Bailey CJ, Harriott P, O'harte FP, Flatt PR (2005) Chemical ablation of gastric inhibitory polypeptide receptor action by daily (Pro3)GIP administration improves glucose tolerance and ameliorates insulin resistance and abnormalities of islet structure in obesity-related diabetes. Diabetes 54:2436–2446

Gautam D, Jeon J, Li JH, Han SJ, Hamdan FF, Cui Y, Lu H, Deng C, Gavrilova O, Wess J (2008) Metabolic roles of the M3 muscarinic acetylcholine receptor studied with M3 receptor mutant mice: a review. J Recept Signal Transduct Res 28:93–108

Gepts W, Christophe J, Mayer J (1960) Pancreatic islets in mice with the obese hyperglycemic syndrome: lack of effect of carbutamide. Diabetes 9:63–69

Grapengiesser E, Gylfe E, Hellman B (1991) Cyclic AMP as a determinant for glucose induction of fast Ca^{2+} oscillations in isolated pancreatic β-cells. J Biol Chem 266:12207–12210

Gustavsson N, Larsson-Nyren G, Lindström P (2006) Cell specificity of the cytoplasmic Ca^{2+} response to tolbutamide is impaired in β-cells from hyperglycemic mice. J Endocrinol 190:461–470

Gysemans C, Callewaert H, Overbergh L, Mathieu C (2008) Cytokine signalling in the β-cell: a dual role for INFγ. Biochem Soc Trans 36:328–333

Hahn HJ, Hellman B, Lernmark Å, Sehlin J, Täljedal I-B (1974) The pancreatic β-cell recognition of insulin secretogogues. Influence of neuraminidase treatment on the release of insulin and the islet content of insulin, sialic acid, and cyclic adenosine 3':5'- monophosphate. J Biol Chem 249:5275–5284

Halaas JL, Gajiwala KS, Maffei M, Cohen SL, Chait BT, Rabinowitz D, Lallone RL, Burley SK, Friedman JM (1995) Weight-reducing effects of the plasma protein encoded by the obese gene. Science 269:543–546

Han D, Yamamoto Y, Munesue S, Motoyoshi S, Saito H, Win MT, Watanabe T, Tsuneyama K, Yamamoto H (2013) Induction of receptor for advanced glycation end products by insufficient leptin action triggers pancreatic β-cell failure in type 2 diabetes. Genes Cells 18:302–314

Heart E, Smith PJ (2007) Rhythm of the β-cell oscillator is not governed by a single regulator: multiple systems contribute to oscillatory behavior. Am J Physiol 292:E1295–E1300

Hellman B, Idahl L-Å, Lernmark Å, Sehlin J, Täljedal I-B (1974) The pancreatic β-cell recognition of insulin secretogogues. Comparisons of glucose with glyceraldehyde isomers and dihydroxyacetone. Arch Biochem Biophys 162:448–457

Herberg L, Major E, Hennings U, Grüneklee D, Freytag G, Gries FA (1970) Differences in the development of the obese-hyperglycemic syndrome in obob and NZO mice. Diabetologia 6:292–299

Hill MJ, Metcalfe D, McTernan PG (2009) Obesity and diabetes: lipids, 'nowhere to run to'. Clin Sci 116:113–123

Hinoi E, Gao N, Jung DY, Yadav V, Yoshizawa T, Myers MG Jr, Chua SC Jr, Kim JK, Kaestner KH, Karsenty G (2008) The sympathetic tone mediates leptin's inhibition of insulin secretion by modulating osteocalcin bioactivity. J Cell Biol 183:1235–1242

Imai Y, Patel HR, Hawkins EJ, Doliba NM, Matschinsky FM, Ahima RS (2007) Insulin secretion is increased in pancreatic islets of neuropeptide Y-deficient mice. Endocrinology 148:5716–5723

Imai J, Katagiri H, Yamada T, Ishigaki Y, Suzuki T, Kudo H, Uno K, Hasegawa Y, Gao J, Kaneko K, Ishihara H, Niijima A, Nakazato M, Asano T, Minokoshi Y, Oka Y (2008) Regulation of pancreatic β cell mass by neuronal signals from the liver. Science 322:1250–1254

Ingalls AM, Dickie MM, Snell GD (1950) Obese, a new mutation in the house mouse. J Hered 41:317–318

Irwin N, McClean PL, O'Harte FP, Gault VA, Harriott P, Flatt PR (2007) Early administration of the glucose-dependent insulinotropic polypeptide receptor antagonist (Pro3)GIP prevents the development of diabetes and related metabolic abnormalities associated with genetically inherited obesity in *ob/ob* mice. Diabetologia 50:1532–1540

Islam MS (2002) The ryanodine receptor calcium channel of β-cells: molecular regulation and physiological significance. Diabetes 51:1299–1309

Jetton TL, Liang Y, Cincotta AH (2001) Systemic treatment with sympatholytic dopamine agonists improves aberrant β-cell hyperplasia and GLUT2, glucokinase, and insulin immunoreactive levels in *ob/ob* mice. Metabolism 50:1377–1384

Karlsson E, Stridsberg M, Sandler S (1998) Leptin regulation of islet amyloid polypeptide secretion from mouse pancreatic islets. Biochem Pharmacol 56:1339–1346

Keller MP, Choi Y, Wang P, Davis DB, Rabaglia ME, Oler AT, Stapleton DS, Argmann C, Schueler KL, Edwards S, Steinberg HA, Chaibub Neto E, Kleinhanz R, Turner S, Hellerstein MK, Schadt EE, Yandell BS, Kendziorski C, Attie AD (2008) A gene expression network model of type 2 diabetes links cell cycle regulation in islets with diabetes susceptibility. Genome Res 18:706–716

Khaldi MZ, Guiot Y, Gilon P, Henquin JC, Jonas JC (2004) Increased glucose sensitivity of both triggering and amplifying pathways of insulin secretion in rat islets cultured for 1 wk in high glucose. Am J Physiol 287:E207–E217

Khan A, Hong-Lie C, Landau BR (1995) Glucose-6-phosphatase activity in islets from *ob/ob* and lean mice and the effect of dexamethasone. Endocrinology 136:1934–1938

Khan A, Narangoda S, Ahren B, Holm C, Sundler F, Efendic S (2001) Long-term leptin treatment of *ob/ob* mice improves glucose-induced insulin secretion. Int J Obes Relat Metab Disord 25:816–821

Kieffer TJ, Heller RS, Habener JF (1996) Leptin receptors expressed on pancreatic β-cells. Biochem Biophys Res Commun 224:522–527

Kim A, Miller K, Jo J, Kilimnik G, Wojcik P, Hara M (2009) Islet architecture: a comparative study. Islets 1:129–136

Kojima H, Fujimiya M, Matsumura K, Nakahara T, Hara M, Chan L (2004) Extrapancreatic insulin-producing cells in multiple organs in diabetes. Proc Natl Acad Sci USA 101:2458–2463

Lalloyer F, Vandewalle B, Percevault F, Torpier G, Kerr-Conte J, Oosterveer M, Paumelle R, Fruchart JC, Kuipers F, Pattou F, Fiévet C, Staels B (2006) Peroxisome proliferator-activated receptor α improves pancreatic adaptation to insulin resistance in obese mice and reduces lipotoxicity in human islets. Diabetes 55:1605–1613

Lam QJ, Lu L (2007) Role of leptin in immunity. Cell Mol Immunol 4:1–13

Larcher F, Del Rio M, Serrano F, Segovia JC, Ramirez A, Meana A, Page A, Abad JL, Gonzalez MA, Bueren J, Bernad A, Jorcano JL (2001) A cutaneous gene therapy approach to human leptin deficiencies: correction of the murine *ob/ob* phenotype using leptin targeted keratinocyte grafts. FASEB J 15:1529–1538

Lavine RL, Voyles N, Perrino PV, Recant L (1977) Functional abnormalities of islets of Langerhans of obese hyperglycemic mouse. Am J Physiol 233:E86–E90

Leckström A, Lundquist I, Ma Z, Westermark P (1999) Islet amyloid polypeptide and insulin relationship in a longitudinal study of the genetically obese (*ob/ob*) mouse. Pancreas 18:266–273

Lee YC, Nielsen JH (2009) Regulation of β cell replication. Mol Cell Endocrinol 297:18–27

Li Z, Schmidt SF, Friedman JM (2013) Developmental role for endocannabinoid signaling in regulating glucose metabolism and growth. Diabetes 62:2359–2367

Lindberg K, Rønn SG, Tornehave D, Richter H, Hansen JA, Rømer J, Jackerott M, Billestrup N (2005) Regulation of pancreatic β-cell mass and proliferation by SOCS-3. J Mol Endocrinol 35:231–243

Ling Z, Pipeleers DG (1996) Prolonged exposure of human β cells to elevated glucose levels results in sustained cellular activation leading to a loss of glucose regulation. J Clin Invest 98:2805–2812

Loreti L, Dunbar JC, Chen S, Foà PP (1974) The autoregulation of insulin secretion in the isolated pancreas islets of lean (obob) and obese-hyperglycemic (obob) mice. Diabetologia 10:309–315

Lulu Strat A, Kokta TA, Dodson MV, Gertler A, Wu Z, Hill RA (2005) Early signaling interactions between the insulin and leptin pathways in bovine myogenic cells. Biochim Biophys Acta 1744:164–175

Malendowicz LK, Rucinski M, Belloni AS, Ziolkowska A, Nussdorfer GG (2007) Leptin and the regulation of the hypothalamic-pituitary-adrenal axis. Int Rev Cytol 263:63–102

Marí M, Caballero F, Colell A, Morales A, Caballeria J, Fernandez A, Enrich C, Fernandez-Checa JC, García-Ruiz C (2006) Mitochondrial free cholesterol loading sensitizes to TNF- and Fas-mediated steatohepatitis. Cell Metab 4:185–198

Marroquí L, Gonzalez A, Ñeco P, Caballero-Garrido E, Vieira E, Ripoll C, Nadal A, Quesada I (2012) Role of leptin in the pancreatic β-cell: effects and signaling pathways. J Mol Endocrinol 49:R9–R17

Mayer J (1960) The obese hyperglycaemic syndrome of mice as an example of "metabolic" obesity. Am J Clin Nutr 8:712–718

Mayer J, Silides N (1953) A quantitative method of determination of the diabetogenic activity of growth hormone preparations. Endocrinology 52:54–56

Mayer J, Russel E, Bates MV, Dickie MM (1953) Metabolic, nutritional and endocrine studies of the hereditary obesity-diabetes syndrome of mice and mechanisms of its development. Metabolism 2:9–21

McQuaid TS, Saleh MC, Joseph JW, Gyulkhandanyan A, Manning-Fox JE, MacLellan JD, Wheeler MB, Chan CB (2006) cAMP-mediated signaling normalizes glucose stimulated insulin secretion in uncoupling protein-2 overexpressing β-cells. J Endocrinol 190:669–680

Medina-Gomez G, Yetukuri L, Velagapudi V, Campbell M, Blount M, Jimenez-Linan M, Ros M, Oresic M, Vidal-Puig A (2009) Adaptation and failure of pancreatic β cells in murine models with different degrees of metabolic syndrome. Dis Model Mech 2:582–592

Melloul D, Marshak S, Cerasi E (2002) Regulation of insulin gene transcription. Diabetologia 45:309–326

Milman P, Fu A, Screaton RA, Woulfe JM (2010) Depletion of intranuclear rodlets in mouse models of diabetes. Endocr Pathol 21:230–235

Moritoh Y, Takeuchi K, Asakawa T, Kataoka O, Odaka H (2008) Chronic administration of alogliptin, a novel, potent, and highly selective dipeptidyl peptidase-4 inhibitor, improves glycemic control and β-cell function in obese diabetic *ob/ob* mice. Eur J Pharmacol 588:325–332

Norlund R, Norlund L, Täljedal I-B (1987) Morphogenetic effects of glucose on mouse islet-cell re-aggregation in culture. Med Biol 65:209–216

Nyholm B, Fineman MS, Koda JE, Schmitz O (1998) Plasma amylin immunoreactivity and insulin resistance in insulin resistant relatives of patients with noninsulin-dependent diabetes mellitus. Horm Metab Res 30:206–212

Nyrén R, Chang CL, Lindström P, Barmina A, Vorrsjö E, Ali Y, Juntti-Berggren L, Bensadoun A, Young SG, Olivecrona T, Olivecrona G (2012) Localization of lipoprotein lipase and GPIHBP1 in mouse pancreas: effects of diet and leptin deficiency. BMC Physiol 12:14

Ozcan L, Ergin AS, Lu A, Chung J, Sarkar S, Nie D, Myers MG Jr, Ozcan U (2009) Endoplasmic reticulum stress plays a central role in development of leptin resistance. Cell Metab 9:35–51

Park K, Lee BM, Kim YH, Han T, Yi W, Lee DH, Choi HH, Chong W, Lee CH (2013) Discovery of a novel phenylethyl benzamide glucokinase activator for the treatment of type 2 diabetes mellitus. Bioorg Med Chem Lett 23:537–542

Pelleymounter MA, Cullen MJ, Baker MB, Hecht R, Winters D, Boone T, Collins F (1995) Effects of the obese gene product on body weight regulation in *ob/ob* mice. Science 269:540–543

Persson-Sjögren S, Lindström P (2004) Effects of cholinergic m-receptor agonists on insulin release in islets from obese and lean mice of different ages: the importance of bicarbonate. Pancreas 29:90–99

Persson-Sjögren S, Forsgren S, Lindström P (2006) Vasoactive intestinal polypeptide and pituitary adenylate cyclase activating polypeptide: effects on insulin release in isolated mouse islets in relation to metabolic status and age. Neuropeptides 40:283–290

Peterson SJ, Drummond G, Kim DH, Li M, Kruger AL, Ikehara S, Abraham NG (2008) L-4F treatment reduces adiposity, increases adiponectin levels, and improves insulin sensitivity in obese mice. J Lipid Res 49:1658–1669

Prieto J, Kaaya EE, Juntti-Berggren L, Berggren PO, Sandler S, Biberfeld P, Patarroyo M (1992) Induction of intercellular adhesion molecule-1 (CD54) on isolated mouse pancreatic β cells by inflammatory cytokines. Clin Immunol Immunopathol 65:247–253

Quan W, Hur KY, Lim Y, Oh SH, Lee JC, Kim KH, Kim GH, Kim SW, Kim HL, Lee MK, Kim KW, Kim J, Komatsu M, Lee MS (2012) Autophagy deficiency in β cells leads to compromised unfolded protein response and progression from obesity to diabetes in mice. Diabetologia 55:392–403

Ranheim T, Dumke C, Schueler KL, Cartee GD, Attie AD (1997) Interaction between BTBR and c57Bl/6J genomes produces an insulin resistance syndrome in [BTBNR x C57Bl/6J] F1 mice. Arterioscler Thromb Vasc Biol 17:3286–3293

Rattarasarn C (2006) Physiological and pathophysiological regulation of regional adipose tissue in the development of insulin resistance and type 2 diabetes. Acta Physiol 186:87–101

Ravier MA, Sehlin J, Henquin JC (2002) Disorganization of cytoplasmic Ca^{2+} oscillations and pulsatile insulin secretion in islets from *ob/ob* mice. Diabetologia 45:1154–1163

Rodriguez-Diaz R, Abdulreda MH, Formoso AL, Gans I, Ricordi C, Berggren PO, Caicedo A (2011) Innervation patterns of autonomic axons in the human endocrine pancreas. Cell Metab 14:45–54

Rolin B, Larsen MO, Gotfredsen CF, Deacon CF, Carr RD, Wilken M, Knudsen LB (2002) The long-acting GLP-1 derivative NN2211 ameliorates glycemia and increases β-cell mass in diabetic mice. Am J Physiol 283:E745–E752

Rooth P, Täljedal I-B (1987) Vital microscopy of islet blood flow: catecholamine effects in normal and *ob/ob* mice. Am J Physiol 252:E130–E135

Rutter GA, Da Silva Xavier G, Leclerc I (2003) Roles of 5′-AMP-activated protein kinase (AMPK) in mammalian glucose homoeostasis. Biochem J 375:1–16

Saleh MC, Wheeler MB, Chan CB (2006) Endogenous islet uncoupling protein-2 expression and loss of glucose homeostasis in *ob/ob* mice. Endocrinology 190:659–667

Sener A, Anak O, Leclercq-Meyer V, Herberg L, Malaisse WJ (1993) FAD glycerophosphate dehydrogenase activity in pancreatic islets and liver of *ob/ob* mice. Biochem Mol Biol Int 30:397–402

Seufert J (2004) Leptin effects on pancreatic β-cell gene expression and function. Diabetes 53(Suppl 1):S152–S158

Shafrir E, Ziv E, Mosthaf L (1999) Nutritionally induced insulin resistance and receptor defect leading to β-cell failure in animal models. Ann NY Acad Sci 892:223–246

Singh H, Farouk M, Bose BB, Singh P (2013) Novel genes underlying β cell survival in metabolic stress. Bioinformation 9:37–41

Sorensen H, Brand CL, Neschen S, Holst JJ, Fosgerau K, Nishimura E, Shulman GI (2006) Immunoneutralization of endogenous glucagon reduces hepatic glucose output and improves long-term glycemic control in diabetic *ob/ob* mice. Diabetes 55:2843–2848

Sreejayan N, Dong F, Kandadi MR, Yang X, Ren J (2008) Chromium alleviates glucose intolerance, insulin resistance, and hepatic ER stress in obese mice. Obesity 16:1331–1337

Stoehr JP, Byers JE, Clee SM, Lan H, Boronenkov OIV, Schueler KL, Yandell BS, Attie AD (2004) Identification of major quantitative trait loci controlling body weight variation in *ob/ob* mice. Diabetes 53:245–249

Stoffers DA, Kieffer TJ, Hussain MA, Drucker DJ, Bonner-Weir S, Habener JF, Egan JM (2000) Insulinotropic glucagon-like peptide 1 agonists stimulate expression of homeodomain protein IDX-1 and increase islet size in mouse pancreas. Diabetes 49:741–747

Sumara G, Formentini I, Collins S, Sumara I, Windak R, Bodenmiller B, Ramracheya R, Caille D, Jiang H, Platt KA, Meda P, Aebersold R, Rorsman P, Ricci R (2009) Regulation of PKD by the MAPK p38delta in insulin secretion and glucose homeostasis. Cell 136:235–248

Sweeney G (2002) Leptin signaling. Cell Signal 14:655–663

Takada K, Kanatsuka A, Tokuyama Y, Yagui K, Nishimura M, Saito Y, Makino H (1996) Islet amyloid polypeptide/amylin contents in pancreas change with increasing age in genetically obese and diabetic mice. Diabetes Res Clin Pract 33:153–158

Takasawa S, Akiyama T, Nata K, Kuroki M, Tohgo A, Noguchi N, Kobayashi S, Kato I, Katada T, Okamoto H (1998) Cyclic ADP-ribose and inositol 1,4,5-trisphosphate as alternate second messengers for intracellular Ca^{2+} mobilization in normal and diabetic β-cells. J Biol Chem 273:2497–2500

Tassava TM, Okuda T, Romsos DR (1992) Insulin secretion from *ob/ob* mouse pancreatic islets: effects of neurotransmitters. Am J Physiol 262:E338–E343

Tentolouris N, Argyrakopoulou G, Katsilambros N (2008) Perturbed autonomic nervous system function in metabolic syndrome. Neuromol Med 10:169–178

Tomita T, Doull V, Pollock HG, Krizsan D (1992) Pancreatic islets of obese hyperglycemic mice (*ob/ob*). Pancreas 7:367–375

Tu Z, Keller MP, Zhang C, Rabaglia ME, Greenawalt DM, Yang X, Wang IM, Dai H, Bruss MD, Lum PY, Zhou YP, Kemp DM, Kendziorski C, Yandell BS, Attie AD, Schadt EE, Zhu J (2012) Integrative analysis of a cross-loci regulation network identifies App as a gene regulating insulin secretion from pancreatic islets. PLoS Genet 8:e1003107

Tyrberg B, Ustinov J, Otonkoski T, Andersson A (2001) Stimulated endocrine cell proliferation and differentiation in transplanted human pancreatic islets: effects of the ob gene and compensatory growth of the implantation organ. Diabetes 50:301–307

Unger RH, Orci L (2001) Diseases of liporegulation: new perspective on obesity and related disorders. FASEB J 15:312–321

Vivas Y, Martínez-García C, Izquierdo A, Garcia-Garcia F, Callejas S, Velasco I, Campbell M, Ros M, Dopazo A, Dopazo J, Vidal-Puig A, Medina-Gomez G (2011) Early peroxisome proliferator-activated receptor gamma regulated genes involved in expansion of pancreatic β cell mass. BMC Med Genomics 4:86

Wajchenberg BL (2007) β-cell failure in diabetes and preservation by clinical treatment. Endocr Rev 28:187–218

Welsh M, Welsh N, Nilsson T, Arkhammar P, Pepinsky RB, Steiner DF, Berggren PO (1988) Stimulation of pancreatic islet β-cell replication by oncogenes. Proc Natl Acad Sci USA 85:116–120

Westman S (1968a) Development of the obese-hyperglycemic syndrome in mice. Diabetologia 4:141–149

Westman S (1968b) The endocrine pancreas of old obese-hyperglycemic mice. Acta Med Upsal 73:81–89

Woulfe J, Munoz D (2000) Tubulin immunoreactive neuronal intranuclear inclusions in the human brain. Neuropathol Appl Neurobiol 26:161–171

Xu G, Chen J, Jing G, Shalev A (2012) Preventing β-cell loss and diabetes with calcium channel blockers. Diabetes 61:848–856

Yang L, Jhaveri R, Huang J, Qi Y, Diehl AM (2007) Endoplasmic reticulum stress, hepatocyte CD1d and NKT cell abnormalities in murine fatty livers. Lab Invest 87:927–937

Young AA, Gedulin BR, Bhavsar S, Bodkin N, Jodka C, Hansen B, Denaro M (1999) Glucose-lowering and insulin-sensitizing actions of exendin-4: studies in obese diabetic (*ob/ob*, db/db) mice, diabetic fatty Zucker rats, and diabetic rhesus monkeys (Macaca mulatta). Diabetes 48:1026–1034

Zaitseva II, Sharoyko V, Størling J, Efendic S, Guerin C, Mandrup-Poulsen T, Nicotera P, Berggren PO, Zaitsev SV (2006) RX871024 reduces NO production but does not protect against pancreatic β-cell death induced by proinflammatory cytokines. Biochem Biophys Res Commun 347:1121–1128

Zawalich WS, Tesz GJ, Zawalich KC (2002) Inhibitors of phosphatidylinositol 3-kinase amplify insulin release from islets of lean but not obese mice. J Endocrinol 174:247–258

Zhang Y, Proenca R, Maffei M, Barone M, Leopold L, Friedman JM (1994) Positional cloning of the mouse obese gene and its human homologue. Nature 372:425–432

Zhang CY, Baffy G, Perret P, Krauss S, Peroni O, Grujic D, Hagen T, Vidal-Puig AJ, Boss O, Kim YB, Zheng XX, Wheeler MB, Shulman GI, Chan CB, Lowell BB (2001) Uncoupling protein-2 negatively regulates insulin secretion and is a major link between obesity, β cell dysfunction, and type 2 diabetes. Cell 105:745–755

Role of Reproductive Hormones in Islet Adaptation to Metabolic Stress

28

Ana Isabel Alvarez-Mercado, Guadalupe Navarro, and Franck Mauvais-Jarvis

Contents

Introduction	786
Estrogens	786
Androgens	790
Progesterone	791
Lactogens	791
Conclusion	794
Cross-References	795
References	795

Abstract

There is an interaction between reproduction and energy stores. Both the production and secretion of insulin by pancreatic islet β-cells must adapt to the metabolic demands of various environmental stresses related to reproduction and energy status. These adaptations must occur in a sex-specific manner.

A.I. Alvarez-Mercado
Division of Endocrinology and Metabolism, Tulane University Health Sciences Center, School of Medicine, New Orleans, LA, USA
e-mail: aalvare4@tulane.edu; analvarezmercado@gmail.com

G. Navarro
Division of Endocrinology, Metabolism and Molecular Medicine, Northwestern University, Feinberg School of Medicine, Chicago, IL, USA
e-mail: guadalupenavarro2012@u.northwestern.edu

F. Mauvais-Jarvis (✉)
Division of Endocrinology and Metabolism, Tulane University Health Sciences Center, School of Medicine, New Orleans, LA, USA

Division of Endocrinology, Metabolism and Molecular Medicine, Northwestern University, Feinberg School of Medicine, Chicago, IL, USA
e-mail: fmauvais@tulane.edu

It is therefore conceivable that reproductive hormones play a role in β-cell adaptation to environmental stresses. This review explores the roles of estrogen, androgen, progesterone, and lactogens in pancreatic β-cell mass and function under conditions of metabolic stress such as pregnancy, obesity, and diabetes.

Keywords

Islets • β-cell • Metabolic stress • Reproductive hormones • Estrogen • Androgen • Progesterone • Lactogens • Proliferation • β-cell survival • Insulin secretion

Introduction

There is an interaction between reproduction and energy metabolism under conditions of disrupted energy homeostasis such as obesity and cachexia, with both conditions negatively impacting fertility (Mauvais-Jarvis 2011). The pancreatic β-cell of the islets of Langerhans is critical for producing insulin, the main hypoglycemic and anabolic hormone. Insulin is essential for promotion and maintenance of cellular energy stores. It acts by stimulating the storage of glucose in the form of glycogen and triglycerides. Production and secretion of insulin by the β-cells must adapt to changes in metabolic demand associated with various environmental stresses, including those related to reproduction and energy status. It is therefore conceivable that reproductive hormones play sex-specific roles in β-cell adaptation to these environmental stresses. This review focuses on the roles of male and female reproductive hormones including estrogen, androgen, progesterone, prolactin, and placental lactogen as they relate to changes in pancreatic β-cell mass and function in conditions of metabolic stresses such as pregnancy, obesity, and diabetes.

Estrogens

In rodent models, treatment with 17 β-estradiol, the main circulating estrogen in females (E2), protects pancreatic β-cells against various diabetic injuries. These injuries include oxidative stress, amyloid polypeptide toxicity, lipotoxicity, and apoptosis (Tiano and Mauvais-Jarvis 2012a). Three ERs – ERα, ERβ, and the G protein-coupled ER (GPER) – have been identified in both rodent and human β-cells. Unlike the classical nuclear ER that acts as a ligand-activated transcription factor in breast and uterine cells, β-cell ERs reside mainly in extranuclear locations. They exert their effects via cytosolic interactions with kinases such as Src, ERK, and AMPK or via transcription factors of the STAT family (Tiano and Mauvais-Jarvis 2012a, b; Tiano et al. 2011; Wong et al. 2010). Activation of ERα enhances glucose-stimulated insulin biosynthesis (Wong et al. 2010; Alonso-Magdalena et al. 2008) through a pathway involving Src and ERK, and stimulation of nuclear translocation and binding to the insulin promoter of NeuroD1, an insulinotropic transcription factor (Wong et al. 2010). This action may assist islets in adapting to the increased metabolic demands of pregnancy by enhancing insulin biosynthesis. Activation of ERα reduces islet excess de novo synthesis of fatty acids and lipogenesis as well as accumulation of toxic lipid intermediates (Tiano et al. 2011; Tiano and Mauvais-Jarvis 2012b, c). This anti-lipogenic action involves at least two pathways. First, an extranuclear ERα activates and promotes the nuclear translocation of STAT3. This leads to inhibition of the master regulator of lipogenesis, the liver X receptor LXRβ, as well as its transcriptional targets, the sterol regulatory element-binding protein 1c (SREBP1c) and the carbohydrate response element-binding protein (ChREBP). Suppression of LXRβ and SREBP1c mRNA may be mediated by a pool of ERα associated with the plasma membrane that activates Src and leads to STAT3 activation (Tiano and Mauvais-Jarvis 2012b). In β-cells, chronic LXR activation leads to excess lipogenesis, which, in turn, is associated with lipotoxicity and apoptosis (Choe et al. 2007). Thus, ERα suppression of LXR mRNAs in β-cells may account for the inhibition of lipogenesis and prevention of islet lipotoxicity. In the second pathway, activation of ERα induces AMP kinase to suppress SREBP-1c gene and protein expression (Tiano and Mauvais-Jarvis 2012b). Together, ERα extranuclear actions in β-cells via STAT3 and AMPK lead to decreased expression and activity of the master effector of fatty acid (FA) synthesis under conditions of glucose surplus – fatty acid synthase (FAS). This converts malonyl-CoA into saturated long-chain FA that can then undergo β-oxidation or esterification to MAG, DAG, and TG (Tiano et al. 2011). Activation of ERα also promotes β-cell survival from most proapoptotic stimuli associated with diabetes (Le May et al. 2006; Liu et al. 2009, 2013; Liu and Mauvais-Jarvis 2009). These anti-apoptotic mechanisms involve a combination of rapid actions that are independent of nuclear events and that potentially lead to alteration in protein phosphorylation (Liu et al. 2009, 2013) as well as a more classical genomic mechanism that induces an anti-inflammatory cascade via expression of the liver receptor homolog-1(LRH-1), NR5A5 (Baquie et al. 2011). Activation of ERβ seems to preferentially enhance glucose-stimulated insulin secretion (Soriano et al. 2009, 2012) via a membrane pathway that leads to activation of the ANF receptor and closure of K_{ATP} channels (Soriano et al. 2009). Activation of GPER, however, protects β-cells from lipid accumulation (Tiano et al. 2011; Tiano and Mauvais-Jarvis 2012b), thereby promoting their survival (Liu et al. 2009; Balhuizen et al. 2010; Kumar et al. 2011). Activation of GPER also enhances glucose-stimulated insulin secretion (Balhuizen et al. 2010; Sharma and Prossnitz 2011) via activation of the epidermal growth factor receptor and ERK (Sharma and Prossnitz 2011), although it has no effect on insulin biosynthesis (Wong et al. 2010). However, it has been proposed that GPER induces expression of ERα36, a short isoform of the classical long isoform of ERα, ERα66 (Kang et al. 2010). Both ERα66 and ERα36 are expressed in β-cells (Tiano et al. 2011). Thus, it is unclear whether GPER-mediated effects in β-cells are due to intrinsic GPER actions or if GPER is interacting with ERα36 at the membrane level. Importantly, ERα, ERβ, and GPER are expressed in human β-cells, and the beneficial effects of ER ligands on β-cell survival, function, and nutrient homeostasis that are described above are all observed in human islets (Tiano et al. 2011; Tiano and Mauvais-Jarvis 2012b; Liu et al. 2009; Kumar et al. 2011; Contreras et al. 2002).

Perhaps the most translational prospect of E2 therapy for β-cell protection involves pancreatic islet transplantation (PIT). Fertile women with T1D exhibit E2 deficiency relative to healthy women (Salonia et al. 2006). Therefore, women with T1D undergoing islet transplantation may have lost part of their endogenous E2-related islet protection and could benefit from short-term E2 supplementation. To explore this hypothesis, we used a T1D model with xenotransplantation of a marginal dose of human islets in nude mice rendered insulin deficient by streptozotocin. In this model, a transient 4-week E2 treatment protected functional β-cell mass and enhanced islet revascularization and engraftment (Liu et al. 2013). E2 effects were retained in the presence of immunosuppression and persisted after discontinuation of E2 treatment. E2 treatment produced acutely decreased hypoxic damage and oxidative stress of the islet graft and suppressed graft β-cell apoptosis. Interestingly, E2 also acutely suppressed hyperglucagonemia without altering insulin secretion. These results suggest that transient E2 treatment in women could provide an immediate therapeutic alternative to improve PIT and also achieve insulin independence with fewer islets. This therapeutic approach could be developed long before other surrogate islet β-cell sources or β-cell regeneration therapy can be developed and therefore warrants further investigation.

From a therapeutic point of view, the risk of hormone-dependent cancer precludes use of general estrogen therapy as a chronic treatment for β-cell failure in diabetes. To preferentially target E2 to β-cells without the undesirable effect of general estrogen therapy, we created novel fusion peptides that combine glucagon-like peptide-1 (GLP-1) and E2 in a single molecule (Finan et al. 2012). By combining the pharmacologic properties of GLP-1 and E2, we postulated synergistic actions on β-cell function and survival resulting from the combined insulinotropic and anti-apoptotic activities on pancreatic β-cells that express ER and GLP-1R. Two conjugates were synthesized with E2 stably linked to GLP-1 to avoid E2 release in the circulation and to maximize E2 delivery at target cells: a GLP-1 agonist stably linked to E2 (aGLP1-E2) and an inactive GLP-1 stably linked to E2 (iGLP1-E2). The second conjugate binds GLP-1R normally, but is pharmacologically incapable of activating GLP-1R signaling and used to direct E2 to β-cells. Tiano et al. tested the efficiency of GLP1-E2 conjugates in preventing insulin-deficient diabetes in a model of β-cell destruction induced by multiple low-dose injections of streptozotocin (STZ). They observed that the iGLP1-E2 conjugate prevented STZ-induced insulin-deficient diabetes, thereby demonstrating that, in vivo, the inactive GLP-1 was able to bind the GLP-1R and to direct E2 to β-cells for protection. Most importantly, the aGLP1-E2 conjugate was more potent than either the GLP-1 agonist or the iGLP1-E2 individually in preventing STZ-induced diabetes. All conjugates were devoid of E2 gynecological effects compared to general E2 therapy (Tiano et al. 2012). These observations provide proof of concept that combining GLP-1 and E2 in a single molecule results in synergies for protection of β-cell function without the side effects associated with general estrogen therapy.

E2 might also promote islet β-cell proliferation under specific physiological and experimental conditions. An effect of estrogen on islet regeneration was initially suggested by Houssay et al. who observed that subtotal pancreatectomy followed by implantation of an estrogen pellet in the remaining pancreas induced regeneration of surrounding islets (Houssay et al. 1954). Further, the stimulatory effect of estrogen on islet and β-cell regeneration was also observed in the alloxan-induced diabetic rat model (Goodman and Hazelwood 1974) and in rat pancreatic islets damaged by streptozotocin (Yamabe et al. 2010). E2 also increases cultured rat islet cell proliferation (Sorenson et al. 1993). However, in these studies, estrogen was used at pharmacological concentrations, so the relevance of these observations to physiology is unclear. Nonetheless, in one study, physiological doses of estrogen have been reported to increase β-cell proliferation and restore the decrease in β-cell mass observed in ovariectomized rodents with subtotal pancreatectomy. This effect was associated with an increase in expression of IRS-2 and Pdx1 proteins via activation of the cAMP response element-binding protein (Choi et al. 2005). Thus, in classical models of β-cell regeneration or at high doses, E2 can induce β-cell proliferation. Regardless, in most of our studies, E2 – used at doses leading to physiological serum concentrations – has never induced significant β-cell proliferation in either male or female rodent models of diabetes with β-cell apoptosis induced by streptozotocin or lipotoxicity (Tiano et al. 2011; Le May et al. 2006; Liu et al. 2013).

Interestingly, GPER has recently been implicated in β-cell proliferation. Pregnancy is associated with an expansion of functional β-cell mass as a means to adapt to increased metabolic demand. In rodents, GPER expression is markedly upregulated during pregnancy. In addition, expansion of β-cell mass during pregnancy was associated with decreased expression of the islet microRNA, miR-338-3p. In rodents, downregulation of this small noconding RNA promoted β-cell proliferation and protected β-cells against apoptosis. In contrast, miR-338-3p upregulation triggered β-cell apoptosis and was associated with a decreased β-cell mass (Jacovetti et al. 2012). In liver cancer cells miR-338-3p expression was downregulated, and restoration of its expression produced a suppression of the invasive potential of cancer cells (Huang et al. 2011). In isolated rat islets, exposure to E2 or the GPER agonist G1 decreased miR-338-3p to levels observed in gestation, a level that was associated with increased β-cell proliferation. These E2 effects depend on cAMP and protein kinase A. However, under these conditions there is no proliferation of human β-cells (Liu et al. 2013). Nonetheless, E2 exposure reduces the level of miR-338-3p in human islet cells (Jacovetti et al. 2012). However, neither E2 nor silencing of miR-338-3p elicited replication of human β-cells in culture. Thus, the impressive effect of E2, GPER, and miR-338-3p observed in rodent β-cell proliferation is not observed in human β-cells. Finally, E2 was reported to promote the proliferation and inhibit the differentiation of adult human islet-derived precursor cells via ERα (Ren et al. 2010). Thus, in classical rodent models of β-cell regeneration and at pharmacological doses, E2 can induce β-cell proliferation. Still, further studies are needed to determine the validity of these findings in human β-cells.

Androgens

Although the role of the major male androgen, testosterone, in β-cell biology is poorly understood, aging men with testosterone deficiency exhibit increased T2D risk (Mauvais-Jarvis 2011; Zitzmann 2009). In addition, men who are on androgen depletion therapy for prostate cancer are also at high risk of T2D (Keating et al. 2012). Although the impact of testosterone deficiency on development of visceral obesity and insulin resistance (IR) in men is established (Mauvais-Jarvis 2011; Zitzmann 2009; Basaria et al. 2006; Khaw and Barrett-Connor 1992; Pitteloud et al. 2005; Zitzmann et al. 2006), the role of testosterone deficiency in β-cell dysfunction remains unknown. Nonetheless, low testosterone levels have been implicated in the pathogenesis of T2D (Haffner et al. 1996; Oh et al. 2002), raising the possibility that testosterone deficiency may predispose to β-cell failure. Early studies reported that in male mice in which β-cell destruction was induced by streptozotocin, testosterone accelerates hyperglycemic decompensation via a pathway involving the AR (Maclaren et al. 1980; Paik et al. 1982). By contrast, it was also reported that testosterone protects early apoptotic damage induced by streptozotocin in male rat pancreas via an AR-dependent mechanism (Morimoto et al. 2005; Palomar-Morales et al. 2010). In the latter study, however, the effect of testosterone on diabetes incidence was not reported. We have generated a β-cell-specific AR knockout mouse to examine the direct role of AR in male β-cell physiology (βARKO$^{-/y}$) (Navarro and Mauvais-Jarvis 2013). Male βARKO$^{-/y}$ mice exhibit decreased glucose-stimulated insulin secretion (GSIS) leading to glucose intolerance. The decreased GSIS is reproduced in cultured male βARKO$^{-/y}$ islets and in human islets treated with flutamide, an AR antagonist. This suggests that AR is a physiological regulator of male β-cell function, a finding that has important implications for prevention of T2D in aging men. A previous report suggested that testosterone stimulates islet insulin mRNA and content in culture and in vivo (Morimoto et al. 2001), but we found no evidence of AR involvement in insulin synthesis.

Women with hyperandrogenemia display β-cell dysfunction. Women with functional hyperandrogenism have significantly higher basal insulin secretory rates and attenuated secretory responses to meals (O'Meara et al. 1993). Women with polycystic ovary syndrome (PCOS) have been reported to show inadequate acute insulin release to the degree of insulin resistance (Dunaif and Finegood 1996) or an exaggerated early insulin response to glucose. These are not accounted for by insulin resistance and are closely associated with hyperandrogenicity (Holte et al. 1994). In these PCOS women, there is a robust relationship between β-cell function and bioavailable testosterone, raising the possibility that excess testosterone in women leads to insulin hypersecretion (Goodarzi et al. 2005). Thus, women with hyperandrogenism display β-cell hyperfunction which may predispose to secondary failure. Consistent with this hypothesis, in female mice, testosterone accelerates hyperglycemic decompensation in experimental models of insulin-dependent diabetes in which β-cell destruction is induced by oxidative stress or inflammation (Maclaren et al. 1980; Liu et al. 2010). In addition, hyperandrogenemia in women with PCOS is accompanied by systemic oxidative stress (Gonzalez et al. 2006), and we showed that excess testosterone in female mice induces systemic oxidative stress (Liu et al. 2010). Further, in the presence of a prior β-cell injury induced by streptozotocin, female mice exposed to excess testosterone are predisposed to β-cell failure via an AR-dependent mechanism (Liu et al. 2010). We also reported that female mice exposed to chronic androgen excess exhibit an islet failure to compensate for high-fat-feeding-induced insulin resistance that leads to T2D (Navarro et al. 2011). Androgen-excess-induced insulin resistance and hyperglycemia is eliminated in female βARKO$^{-/-}$ mice. Thus, excess AR activation in β-cells (and other tissues) may predispose to the β-cell dysfunction observed in women with androgen excess.

Progesterone

The presence of progesterone receptors in the human endocrine pancreas suggests a direct role of progesterone on pancreatic islet function (Doglioni et al. 1990). In vivo progesterone treatment of intact male and female mice stimulates islet α- and β-cell proliferation (Nieuwenhuizen et al. 1999). However, this effect is not observed in gonadectomized mice, suggesting that progesterone requires intact gonadal function to induce islet cell proliferation. Indeed, this effect of progesterone is not observed in cultured rat islet cells (Sorenson et al. 1993). In contrast, female mice deficient in the progesterone receptor (PR) have enhanced glucose tolerance related to improved β-cell function (Picard et al. 2002). The improved β-cell function in these female mice is attributed to increased β-cell mass with enhanced proliferation. The increased β-cell proliferation is not associated with differences in islet expression levels of the cell cycle regulators p21, p27, cyclin D1, cyclin B1, or cyclin E. In contrast, the protein levels of the tumor suppressor p53 were markedly decreased in PR-deficient islets, which may enhance islet proliferation. Progesterone did not affect miR-338-3p levels in cultured INS cells (Jacovetti et al. 2012). In MIN6 β-cells, progesterone was reported to enhance basal- and glucose-stimulated insulin secretion, in part by increasing glucokinase activity and amplifying cAMP levels (Shao et al. 2004).

Lactogens

Lactogens – prolactin (PRL) and placental lactogen (PL) – mediate their biological responses through a common prolactin receptor (PRLR) (Goffin et al. 1999). In situ hybridization analysis has revealed the presence of both the short and long forms of the PRLR mRNA in the endocrine pancreas (Moldrup et al. 1993; Ouhtit et al. 1994). This was confirmed by immunohistochemical staining of PRLR in islet β-cells (Brelje et al. 2002). Lactogens play a role in regulation of normal islet development. Male and female PRLR-deficient mice exhibit reduced islet mass and density that is observed as early as 3 weeks of age (Freemark et al. 2002).

There is a blunted insulin secretory response to glucose in male PRLR-deficient mice and in isolated cultured islets from PRLR-deficient mice of both sexes.

Lactogens are especially important during pregnancy, when islets are exposed to metabolic stress and need to adapt to the increased metabolic demands of the fetus (Newberna and Freemark 2011). Physiological changes associated with pregnancy include β-cell proliferation, lowering of the glucose threshold for insulin release, increased glucose-stimulated insulin secretion, and islet β-cell coupling (Terra et al. 2011). This suggests that prolactin may increase β-cell sensitivity to glucose. Indeed, during pregnancy, increased islet glucokinase and GLUT2 glucose transporters are associated with prolactin-induced augmentation of glucose-stimulated insulin secretion, as well as increases in β-cell proliferation, increased glucose metabolism, gap-junctional coupling among β-cells, and c-AMP signaling (Sorenson and Brelje 1997). There is evidence to suggest that PRL and placental lactogen are instrumental in inducing these changes and promoting β-cell expansion and insulin production. The increase in serum prolactin and placental lactogen levels parallels increases in β-cell mass (Hughes and Huang 2011; Vetere and Wagner 2012), and expression of PRLR in pancreatic β-cells increases during pregnancy (Hughes and Huang 2011). Further, the onset of placental lactogen secretion during pregnancy occurs at the same time that the earliest changes in β-cell division and insulin secretion are detected (Huang et al. 2009). In vivo and in vitro rodent studies comparing the effect of lactogenic hormones with those observed in pregnancy on islets revealed that lactogens induce the same changes in islets as those observed during pregnancy. This led to the hypothesis that β-cell PRLR is central to the mechanisms islets use to adapt to pregnancy (Sorenson and Brelje 2009). During pregnancy, female PRLR-deficient mice exhibit decreased islet mass and lower rates of β-cell proliferation leading to impaired glucose tolerance (Huang et al. 2009). PRL-induced increases in insulin secretion and β-cell proliferation are also mediated by the JAK2/STAT5 pathway (Brelje et al. 2002; Fujinaka et al. 2007). Increases in proliferation require PRLR signaling through Jak2, Akt, menin/p18, and p21 (Hughes and Huang 2011). Transgenic-specific overexpression of PL β-cells in mice produces a marked increase in islet mass due to augmented islet size and number. These mice are also resistant to STZ-induced β-cell death, suggesting that PL may promote β-cell survival during pregnancy (Vasavada et al. 2000). PRL has been reported to improve islet graft function, and pretreatment of human islets for transplantation with PRL promotes islet viability and survival (Yamamoto et al. 2008; Terra et al. 2011). PRL-treated transplanted islets have increased revascularization and improved function (Johansson et al. 2009). Lactogen-mediated β-cell protection is also mediated by JAK2/STAT5 signaling pathway (Fujinaka et al. 2007; Kondegowda et al. 2012) and involves the upregulation of the anti-apoptotic mediator Bcl-X_L (Fujinaka et al. 2007). PRL upregulates expression of cell cycle regulators (D-type cyclins, CDK4) as well as genes involved in cell proliferation (members of the MAPK signaling pathway) (Bordin et al. 2004). The enhanced compensatory glucose-stimulated insulin secretion during pregnancy also involves the IRS/PI3K and SHC/ERK pathway (Amaral et al. 2004).

In humans, prolactin and placental lactogen play central roles in insulin production (Lombardo et al. 2011). Although limited, studies on human islets indicate that lactogen treatment increases insulin secretion and islet cell proliferation and survival (Yamamoto et al. 2008; Terra et al. 2011; Lombardo et al. 2011). Thus, it is quite likely that in humans as well as in rodents, lactogenic hormones are at least partly responsible for islet adaptation to the metabolic stress of pregnancy (Sorenson and Brelje 2009). The molecular pathway involved in lactogen transduction in human islets involves JAK2/STAT5, IRS-1 and IRS-2, PI3 kinase, MAPKs, as well as the signal transduction mechanisms activated by PL in β-cells in pregnancy (Lombardo et al. 2011). Evidence suggests that during pregnancy, lactogens induce proliferation via serotonin release. PL induces a marked rise of serotonin production in islets in a subpopulation of β-cells (Schraenen et al. 2010). Inhibition of serotonin synthesis blocks β-cell expansion and induces glucose intolerance in pregnant mice without affecting insulin sensitivity (Kim et al. 2010).

Thus, lactogens promote β-cell survival and increase insulin secretion and islet cell proliferation. Further studies are needed to determine how these findings can be translated to therapeutic avenue to protect β-cells at the onset of T2D.

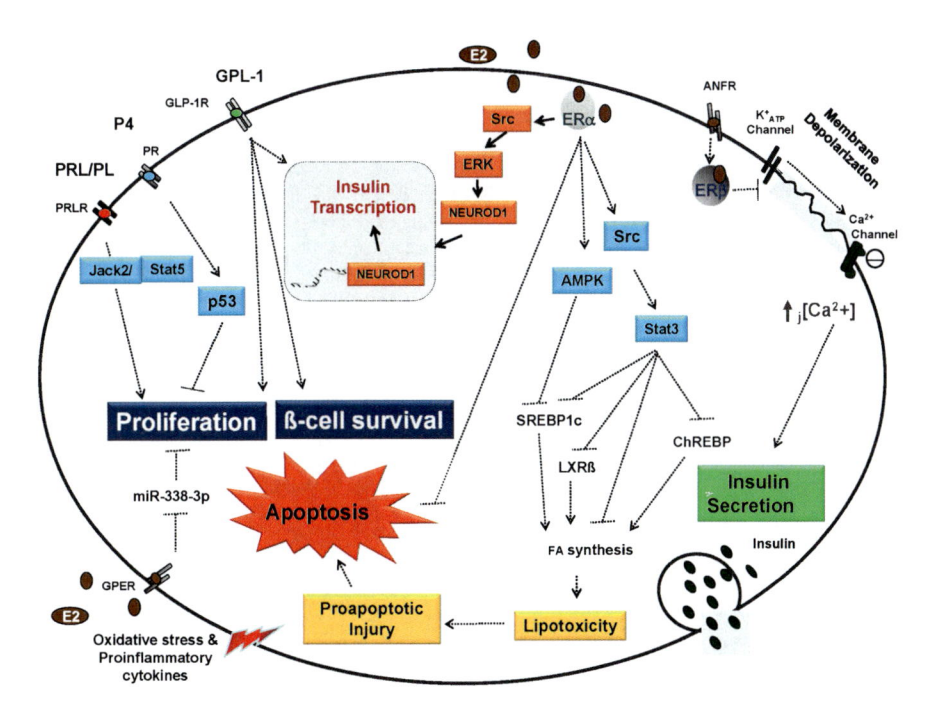

Fig. 1 Summary of reproductive hormone effects in the islet adaptation to metabolic stress. In females, reproductive hormones are important to the β-cell adaptation to the metabolic stress of pregnancy. Thus, placental lactogen (*PL*) and prolactin (*PRL*) promote β-cell expansion to adapt to the increase in insulin demand. This synergizes with 17β-estradiol (*E2*) that increases insulin production and secretion and promotes β-cell survival. Progesterone (*P4*) could act to limit β-cell expansion in vivo. In males, physiological levels of testosterone enhance β-cell function

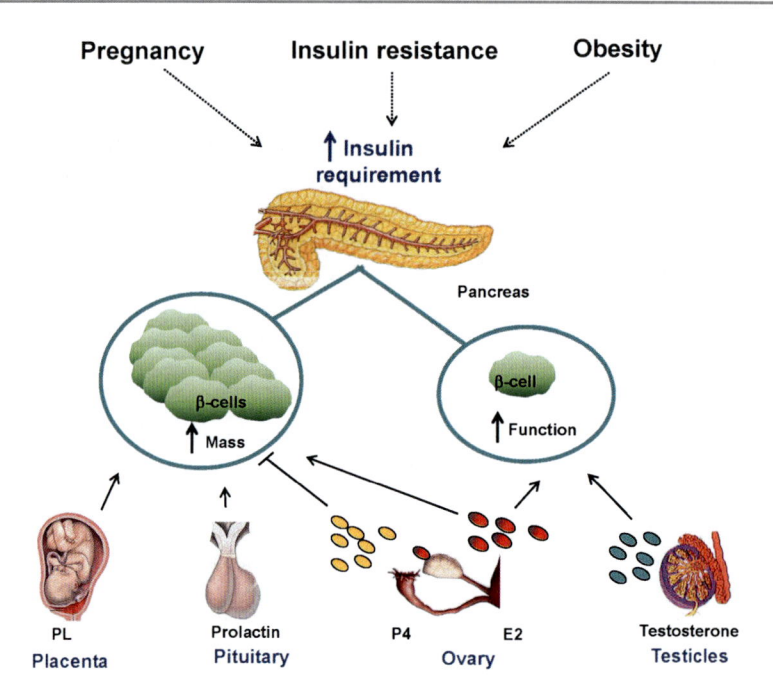

Fig. 2 Molecular pathways used by female reproductive hormones in the islet adaptation to metabolic stress. Prolactin (*PRL*) and placental lactogen (*PL*) promote β-cell proliferation via the lactogen-prolactin receptor (*PRLR*) and the Janus kinase 2/signal transducer and activator of transcription 5 (*JAK2/STAT5*) pathway. E2 signals in β-cells via three different receptors, ERα, ERβ, and the G protein coupled ER (*GPER*). ERα enhances glucose-stimulated insulin biosynthesis by enhancing NeuroD1 nuclear translocation to the insulin promoter. ERα suppresses lipogenesis through (1) a Src-signal transducer and activator of transcription 3 (*STAT3*)-dependent pathway inhibiting liver X receptor β (*LXRβ*) mRNA expression and eventually sterol regulatory element-binding protein 1c (*SREBP1c*) and carbohydrate response element-binding protein (*ChREBP*) expression and (2) an 5′ adenosine monophosphate-activated protein kinase (*AMPK*)-dependent pathway directly inhibiting SREBP1c expression. Altogether this prevents lipotoxicity. Activation of ERβ enhances glucose-stimulated insulin secretion (*GSIS*) via activation of the atrial natriuretic peptide receptor (*ANFR*) and closure of K_{ATP} channels. GPER stimulates β-cell mass during pregnancy via a decreased expression of the islet microRNA, miR-338-3p. Finally, progesterone (*P4*) suppresses β-cell proliferation by inducing the expression of tumor suppressor p53

Conclusion

Pancreatic β-cells located in islets of Langerhans have the ability to adapt to the increased insulin requirements by increasing their function, mass, or both. During pregnancy, reproductive hormones play a critical role in this adaptive process by increasing β-cell function, growth, and survival via β-cell receptors. These effects and their mechanisms are summarized in Figs. 1 and 2. A better understanding of the mechanisms of action of reproductive hormones in β-cells promises to yield therapeutic avenues to protect functional β-cell mass in diabetes.

Acknowledgments This work was supported by grants from NIH RO1 DK074970, the Juvenile Diabetes Research Foundation (1-2006-837) and the March of Dimes (6-FY7-312).

Cross-References

- Glucose-Induced Apoptosis in Pancreatic Islets
- Prevention of β-Cell Destruction in Autoimmune Diabetes: Current Approaches and Future Prospects
- The β-Cell in Human Type 2 Diabetes
- β-Cell Function in Obese-Hyperglycemic Mice

References

Alonso-Magdalena P, Ropero AB, Carrera MP, Cederroth CR, Baquie M et al (2008) Pancreatic insulin content regulation by the estrogen receptor ER α. PLoS One 3:e2069

Amaral ME, Cunha DA, Anhe GF, Ueno M, Carneiro EM et al (2004) Participation of prolactin receptors and phosphatidylinositol 3-kinase and MAP kinase pathways in the increase in pancreatic islet mass and sensitivity to glucose during pregnancy. J Endocrinol 183:469–476

Balhuizen A, Kumar R, Amisten S, Lundquist I, Salehi A (2010) Activation of G protein-coupled receptor 30 modulates hormone secretion and counteracts cytokine-induced apoptosis in pancreatic islets of female mice. Mol Cell Endocrinol 320:16–24

Baquie M, St-Onge L, Kerr-Conte J, Cobo-Vuilleumier N, Lorenzo PI et al (2011) The liver receptor homolog-1 (LRH-1) is expressed in human islets and protects β-cells against stress-induced apoptosis. Hum Mol Genet 20:2823–2833

Basaria S, Muller DC, Carducci MA, Egan J, Dobs AS (2006) Hyperglycemia and insulin resistance in men with prostate carcinoma who receive androgen-deprivation therapy. Cancer 106:581–588

Bordin S, Amaral ME, Anhe GF, Delghingaro-Augusto V, Cunha DA et al (2004) Prolactin-modulated gene expression profiles in pancreatic islets from adult female rats. Mol Cell Endocrinol 220:41–50

Brelje TC, Svensson AM, Stout LE, Bhagroo NV, Sorenson RL (2002) An immunohistochemical approach to monitor the prolactin-induced activation of the JAK2/STAT5 pathway in pancreatic islets of Langerhans. J Histochem Cytochem 50:365–383

Choe SS, Choi AH, Lee JW, Kim KH, Chung JJ et al (2007) Chronic activation of liver X receptor induces β-cell apoptosis through hyperactivation of lipogenesis: liver X receptor-mediated lipotoxicity in pancreatic β-cells. Diabetes 56:1534–1543

Choi SB, Jang JS, Park S (2005) Estrogen and exercise may enhance β-cell function and mass via insulin receptor substrate 2 induction in ovariectomized diabetic rats. Endocrinology 146:4786–4794

Contreras JL, Smyth CA, Bilbao G, Young CJ, Thompson JA et al (2002) 17β-Estradiol protects isolated human pancreatic islets against proinflammatory cytokine-induced cell death: molecular mechanisms and islet functionality. Transplantation 74:1252–1259

Doglioni C, Gambacorta M, Zamboni G, Coggi G, Viale G (1990) Immunocytochemical localization of progesterone receptors in endocrine cells of the human pancreas. Am J Pathol 137:999–1005

Dunaif A, Finegood DT (1996) β-cell dysfunction independent of obesity and glucose intolerance in the polycystic ovary syndrome. J Clin Endocrinol Metab 81:942–947

Finan B, Yang B, Ottaway N, Stemmer K, Muller TD et al (2012) Targeted estrogen delivery reverses the metabolic syndrome. Nat Med 18:1847–1856

Freemark M, Avril I, Fleenor D, Driscoll P, Petro A et al (2002) Targeted deletion of the PRL receptor: effects on islet development, insulin production, and glucose tolerance. Endocrinology 143:1378–1385

Fujinaka Y, Takane K, Yamashita H, Vasavada RC (2007) Lactogens promote β cell survival through JAK2/STAT5 activation and Bcl-XL upregulation. J Biol Chem 282:30707–30717

Goffin V, Binart N, Clement-Lacroix P, Bouchard B, Bole-Feysot C et al (1999) From the molecular biology of prolactin and its receptor to the lessons learned from knockout mice models. Genet Anal 15:189–201

Gonzalez F, Rote NS, Minium J, Kirwan JP (2006) Increased activation of nuclear factor kappaB triggers inflammation and insulin resistance in polycystic ovary syndrome. J Clin Endocrinol Metab 91:1508–1512

Goodarzi MO, Erickson S, Port SC, Jennrich RI, Korenman SG (2005) β-Cell function: a key pathological determinant in polycystic ovary syndrome. J Clin Endocrinol Metab 90:310–315

Goodman MN, Hazelwood RL (1974) Short-term effects of oestradiol benzoate in normal, hypophysectomized and alloxan-diabetic male rats. J Endocrinol 62:439–449

Haffner SM, Laakso M, Miettinen H, Mykkanen L, Karhapaa P et al (1996) Low levels of sex hormone-binding globulin and testosterone are associated with smaller, denser low density lipoprotein in normoglycemic men. J Clin Endocrinol Metab 81:3697–3701

Holte J, Bergh T, Berne C, Berglund L, Lithell H (1994) Enhanced early insulin response to glucose in relation to insulin resistance in women with polycystic ovary syndrome and normal glucose tolerance. J Clin Endocrinol Metab 78:1052–1058

Houssay BA, Foglia VG, Rodriguez RR (1954) Production or prevention of some types of experimental diabetes by oestrogens or corticosteroids. Acta Endocrinol (Cph) 17:146–164

Huang C, Snider F, Cross J (2009) Prolactin receptor is required for normal glucose homeostasis and modulation of β-cell mass during pregnancy. Endocrinology 150:1618–1626

Huang X, Chen J, Wang Q, Chen X, Wen W et al (2011) miR-338-3p suppresses invasion of liver cancer cell by targeting smoothened. J Pathol 225:463–472

Hughes E, Huang C (2011) Participation of Akt, menin, and p21 in pregnancy-induced β-cell proliferation. Endocrinology 152:847–855

Jacovetti C, Abderrahmani A, Parnaud G, Jonas JC, Peyot ML et al (2012) MicroRNAs contribute to compensatory β cell expansion during pregnancy and obesity. J Clin Invest 122:3541–3551

Johansson M, Olerud J, Jansson L, Carlsson PO (2009) Prolactin treatment improves engraftment and function of transplanted pancreatic islets. Endocrinology 150:1646–1653

Kang L, Zhang X, Xie Y, Tu Y, Wang D et al (2010) Involvement of estrogen receptor variant ER-α36, not GPR30, in nongenomic estrogen signaling. Mol Endocrinol 24:709–721

Keating NL, O'Malley A, Freedland SJ, Smith MR (2012) Diabetes and cardiovascular disease during androgen deprivation therapy: observational study of veterans with prostate cancer. J Natl Cancer Inst 104(19):1518–1523

Khaw KT, Barrett-Connor E (1992) Lower endogenous androgens predict central adiposity in men. Ann Epidemiol 2:675–682

Kim H, Toyofuku Y, Lynn FC, Chak E, Uchida T et al (2010) Serotonin regulates pancreatic β cell mass during pregnancy. Nat Med 16:804–U106

Kondegowda N, Mozar A, Chin C, Otero A, Garcia-Ocaña A et al (2012) Lactogens protect rodent and human β cells against glucolipotoxicity-induced cell death through Janus kinase-2 (JAK2)/signal transducer and activator of transcription-5 (STAT5) signalling. Diabetologia 55:1721–1732

Kumar R, Balhuizen A, Amisten S, Lundquist I, Salehi A (2011) Insulinotropic and antidiabetic effects of 17β-estradiol and the GPR30 agonist G-1 on human pancreatic islets. Endocrinology 152:2568–2579

Le May C, Chu K, Hu M, Ortega CS, Simpson ER et al (2006) Estrogens protect pancreatic β-cells from apoptosis and prevent insulin-deficient diabetes mellitus in mice. Proc Natl Acad Sci U S A 103:9232–9237

Liu S, Mauvais-Jarvis F (2009) Rapid, nongenomic estrogen actions protect pancreatic islet survival. Islets 1:273–275

Liu S, Le May C, Wong WP, Ward RD, Clegg DJ et al (2009) Importance of extranuclear estrogen receptor-α and membrane G protein-coupled estrogen receptor in pancreatic islet survival. Diabetes 58:2292–2302

Liu S, Navarro G, Mauvais-Jarvis F (2010) Androgen excess produces systemic oxidative stress and predisposes to β-cell failure in female mice. PLoS One 5:e11302

Liu S, Kilic G, Meyers MS, Navarro G, Wang Y et al (2013) Oestrogens improve human pancreatic islet transplantation in a mouse model of insulin deficient diabetes. Diabetologia 56:370–381

Lombardo MF, De Angelis F, Bova L, Bartolini B, Bertuzzi F et al (2011) Human placental lactogen (hPL-A) activates signaling pathways linked to cell survival and improves insulin secretion in human pancreatic islets. Islets 3:250–258

Maclaren NK, Neufeld M, McLaughlin JV, Taylor G (1980) Androgen sensitization of streptozotocin-induced diabetes in mice. Diabetes 29:710–716

Mauvais-Jarvis F (2011) Estrogen and androgen receptors: regulators of fuel homeostasis and emerging targets for diabetes and obesity. Trends Endocrinol Metab 22:24–33

Moldrup A, Petersen ED, Nielsen JH (1993) Effects of sex and pregnancy hormones on growth hormone and prolactin receptor gene expression in insulin-producing cells. Endocrinology 133:1165–1172

Morimoto S, Cerbón MA, Alvarez-Alvarez A, Romero-Navarro G, Díaz-Sánchez V (2001) Insulin gene expression pattern in rat pancreas during the estrous cycle. Life Sci 68(26):2979–2985

Morimoto S, Mendoza-Rodriguez CA, Hiriart M, Larrieta ME, Vital P et al (2005) Protective effect of testosterone on early apoptotic damage induced by streptozotocin in rat pancreas. J Endocrinol 187:217–224

Navarro G, Mauvais-Jarvis F (2013) The role of the Androgen Receptor in β-cell function in male mice. Diabetes 62(Suppl 1):A571

Navarro G, Suhuan Liu P, De Gendt K, Verhoeven G, Mauvais-Jarvis F (2011) Importance of the β – cell Androgen Receptor in type 2 diabetes. Endocr Rev 32:OR22–OR23

Newberna N, Freemark M (2011) Placental hormones and the control of maternal metabolism and fetal growth. Curr Opin Endocrinol Diabetes Obes 18:409–416

Nieuwenhuizen AG, Schuiling GA, Liem SM, Moes H, Koiter TR et al (1999) Progesterone stimulates pancreatic cell proliferation in vivo. Eur J Endocrinol 140:256–263

O'Meara NM, Blackman JD, Ehrmann DA, Barnes RB, Jaspan JB et al (1993) Defects in β-cell function in functional ovarian hyperandrogenism. J Clin Endocrinol Metab 76:1241–1247

Oh JY, Barrett-Connor E, Wedick NM, Wingard DL (2002) Endogenous sex hormones and the development of type 2 diabetes in older men and women: the Rancho Bernardo study. Diabetes Care 25:55–60

Ouhtit A, Kelly PA, Morel G (1994) Visualization of gene expression of short and long forms of prolactin receptor in rat digestive tissues. Am J Physiol 266:G807–G815

Paik SG, Michelis MA, Kim YT, Shin S (1982) Induction of insulin-dependent diabetes by streptozotocin. Inhibition by estrogens and potentiation by androgens. Diabetes 31:724–729

Palomar-Morales M, Morimoto S, Mendoza-Rodriguez CA, Cerbon MA (2010) The protective effect of testosterone on streptozotocin-induced apoptosis in β cells is sex specific. Pancreas 39:193–200

Picard F, Wanatabe M, Schoonjans K, Lydon J, O'Malley BW et al (2002) Progesterone receptor knockout mice have an improved glucose homeostasis secondary to β-cell proliferation. Proc Natl Acad Sci U S A 99:15644–15648

Pitteloud N, Mootha VK, Dwyer AA, Hardin M, Lee H et al (2005) Relationship between testosterone levels, insulin sensitivity, and mitochondrial function in men. Diabetes Care 28:1636–1642

Ren Z, Zou C, Ji H, Zhang YA (2010) Oestrogen regulates proliferation and differentiation of human islet-derived precursor cells through oestrogen receptor α. Cell Biol Int 34:523–530

Salonia A, Lanzi R, Scavini M, Pontillo M, Gatti E et al (2006) Sexual function and endocrine profile in fertile women with type 1 diabetes. Diabetes Care 29:312–316

Schraenen A, Lemaire K, de Faudeur G, Hendrickx N, Granvik M et al (2010) Placental lactogens induce serotonin biosynthesis in a subset of mouse β cells during pregnancy. Diabetologia 53:2589–2599

Shao J, Qiao L, Friedman JE (2004) Prolactin, progesterone, and dexamethasone coordinately and adversely regulate glucokinase and cAMP/PDE cascades in MIN6 β-cells. Am J Physiol Endocrinol Metab 286:E304–E310

Sharma G, Prossnitz ER (2011) Mechanisms of estradiol-induced insulin secretion by the G protein-coupled estrogen receptor GPR30/GPER in pancreatic β-cells. Endocrinology 152:3030–3039

Sorenson RL, Brelje TC (1997) Adaptation of islets of langerhans to pregnancy: β-cell growth, enhanced insulin secretion and the role of lactogenic hormones. Horm Metab Res 29:301–307

Sorenson RL, Brelje TC (2009) Prolactin receptors are critical to the adaptation of islets to pregnancy. Endocrinology 150:1566–1569

Sorenson RL, Brelje TC, Roth C (1993) Effects of steroid and lactogenic hormones on islets of Langerhans: a new hypothesis for the role of pregnancy steroids in the adaptation of islets to pregnancy. Endocrinology 133:2227–2234

Soriano S, Ropero AB, Alonso-Magdalena P, Ripoll C, Quesada I et al (2009) Rapid regulation of K_{ATP} channel activity by 17β-estradiol in pancreatic β-cells involves the estrogen receptor β and the atrial natriuretic peptide receptor. Mol Endocrinol 23:1973–1982

Soriano S, Alonso-Magdalena P, Garcia-Arevalo M, Novials A, Muhammed SJ et al (2012) Rapid insulinotropic action of low doses of bisphenol-A on mouse and human islets of Langerhans: role of estrogen receptor β. PLoS One 7:e31109

Terra LF, Garay-Malpartida MH, Wailemann RA, Sogayar MC, Labriola L (2011) Recombinant human prolactin promotes human β cell survival via inhibition of extrinsic and intrinsic apoptosis pathways. Diabetologia 54:1388–1397

Tiano JP, Mauvais-Jarvis F (2012a) Importance of oestrogen receptors to preserve functional β-cell mass in diabetes. Nat Rev Endocrinol 8:342–351

Tiano JP, Mauvais-Jarvis F (2012b) Molecular mechanisms of estrogen receptors' suppression of lipogenesis in pancreatic β-cells. Endocrinology 153:2997–3005

Tiano J, Mauvais-Jarvis F (2012c) Selective estrogen receptor modulation in pancreatic β-cells and the prevention of type 2 diabetes. Islets 4:173–176

Tiano JP, Delghingaro-Augusto V, Le May C, Liu S, Kaw MK et al (2011) Estrogen receptor activation reduces lipid synthesis in pancreatic islets and prevents β cell failure in rodent models of type 2 diabetes. J Clin Invest 121:3331–3342

Tiano J, Finan B, DiMarchi R, Mauvais-Jarvis F (2012) A Glucagon-like peptide-1-estrogen fusion peptide shows enhanced efficacy in preventing insulin-deficient diabetes in mice. Endocr Rev 33:OR21–OR26

Vasavada RC, Garcia-Ocana A, Zawalich WS, Sorenson RL, Dann P et al (2000) Targeted expression of placental lactogen in the β cells of transgenic mice results in β cell proliferation, islet mass augmentation, and hypoglycemia. J Biol Chem 275:15399–15406

Vetere A, Wagner BK (2012) Chemical methods to induce β-cell proliferation. Int J Endocrinol 2012:925143

Wong WP, Tiano JP, Liu S, Hewitt SC, Le May C et al (2010) Extranuclear estrogen receptor-α stimulates NeuroD1 binding to the insulin promoter and favors insulin synthesis. Proc Natl Acad Sci USA 107:13057–13062

Yamabe N, Kang KS, Zhu BT (2010) Beneficial effect of 17β-estradiol on hyperglycemia and islet β-cell functions in a streptozotocin-induced diabetic rat model. Toxicol Appl Pharmacol 249:76–85

Yamamoto T, Ricordi C, Mita A, Miki A, Sakuma Y et al (2008) β-Cell specific cytoprotection by prolactin on human islets. Transplant Proc 40:382–383

Zitzmann M (2009) Testosterone deficiency, insulin resistance and the metabolic syndrome. Nat Rev Endocrinol 5:673–681

Zitzmann M, Faber S, Nieschlag E (2006) Association of specific symptoms and metabolic risks with serum testosterone in older men. J Clin Endocrinol Metab 91:4335–4343

The β-Cell in Human Type 2 Diabetes   29

Lorella Marselli, Mara Suleiman, Farooq Syed, Franco Filipponi,
Ugo Boggi, Piero Marchetti, and Marco Bugliani

Contents

Introduction	802
β-Cell Mass Defects	803
β-Cell Functional Defects	805
Molecular Changes	806
The Role of Genetic and Acquired Factors	808
Reversal of β-Cell Damage in Type 2 Diabetes	810
Conclusions	812
Cross-References	812
References	812

Abstract

β-cell dysfunction is central to the onset and progression of type 2 diabetes. Reduced islet number and/or diminished β-cell mass/volume in the pancreas of type 2 diabetic subjects have been reported by many authors, mainly due to increased apoptosis not compensated for by adequate regeneration. In addition, ultrastructural analysis has shown reduced insulin granules and morphological changes in several β-cell organelles, including mitochondria and endoplasmic

---

L. Marselli (✉) • M. Suleiman • F. Syed • P. Marchetti • M. Bugliani
Department of Clinical and Experimental Medicine, Pancreatic Islet Laboratory, University of Pisa, Pisa, Italy
e-mail: lorella.marselli@med.unipi.it; mara.suleiman@gmail.com; farooqnobel@gmail.com; piero.marchetti@med.unipi.it; m.bugliani@ao-pisa.toscana.it

F. Filipponi
Department of Surgery, University of Pisa, Pisa, Italy
e-mail: franco.filipponi@med.unipi.it

U. Boggi
Department of Translational Research and New Technologies, University of Pisa, Pisa, Italy
e-mail: u.boggi@med.unipi.it

reticulum. Several quantitative and qualitative defects of β-cell function have been described in human type 2 diabetes using isolated islets, including alterations in early phase and glucose-stimulated insulin release. These survival and functional changes are accompanied by modifications of islet gene and protein expression. The impact of genotype in affecting β-cell function and survival has been addressed in a few studies, and a number of gene variants have been associated with β-cell dysfunction. Among acquired factors, the role of glucotoxicity and lipotoxicity could be of particular importance, due to the potential deleterious impact of elevated levels of glucose and/or free fatty acids in the natural history of β-cell damage. More recently, it has been proposed that inflammation might also play a role in the dysfunction of the β-cell in type 2 diabetes. Encouraging, although preliminary, data show that some of these defects might be directly counteracted, at least in part, by appropriate in vitro pharmacological intervention.

Keywords

β-cell volume • β-cell mass • Insulin secretion • Apoptosis • Regeneration • Mitochondria • Endoplasmic reticulum • Gene polymorphisms • Gene expression • Protein expression • Glucotoxicity • Lipotoxicity • Inflammation

Introduction

β-cell dysfunction is central to the development and progression of type 2 diabetes (American Diabetes Association 2008; Stumvoll et al. 2005; Kahn 2003). Reduced β-cell functional mass in diabetes and other categories of glucose intolerance has been described in patients, and decreased islet and/or β-cell volume in the pancreas of type 2 diabetic patients has been consistently observed (Marchetti et al. 2008; Wajchenberg 2007; Meier 2008). These findings are in agreement with the results obtained with healthy humans who underwent hemipancreatectomy for the purpose of organ donation, and 43 % of cases developed impaired fasting glucose, impaired glucose tolerance, or diabetes on 3–10 years of follow-up (Kumar et al. 2008). In addition, studies in patients and the use of isolated islets have shown both quantitative and qualitative defects of glucose-stimulated insulin secretion in type 2 diabetes (Porte 1991; Ferrannini and Mari 2004; Kahn et al. 2008). The importance of β-cell function (in the absence of obvious reduction of β-cell mass) is supported by the MODY2 type of diabetes, due to mutations of the enzyme glucokinase, leading to decreased glycolytic flux in the β-cell (Vaxillaire and Froguel 2008). In this chapter, we describe the mass and functional defects of β-cells in type 2 diabetes and discuss the accompanying molecular alterations. Then, the role of a few genetic and acquired factors affecting the β-cell is briefly discussed, followed by the description of the beneficial effects that some compounds directly have on the diabetic β-cell.

β-Cell Mass Defects

Early work reported that total islet number was approximately 30 % lower in pancreatic histology samples from type 2 diabetic subjects as compared to those from nondiabetic individuals (Saito et al. 1978). The reduction in total islet volume in diabetic vs. nondiabetic pancreata (1.01 ± 0.12 vs. 1.60 ± 0.16 cm^3) was confirmed (Westermark and Wilander 1978) and resulted even more marked when corrected for the presence of amyloid (Westermark and Wilander 1978). Successively, it was found that β-cell volume was 30–40 % reduced in type 2 diabetic islets (Saito et al. 1979). In the following years, although a few authors were not able to find differences in β-cell amount in diabetic versus nondiabetic pancreas specimens (Stefan et al. 1982; Rahier et al. 1983), several studies have consistently shown that β-cell mass is reduced in type 2 diabetes (Clark et al. 1988; Sakuraba et al. 2002; Yoon et al. 2003; Butler et al. 2003; Rahier et al. 2008). Clark and colleagues studied the pancreas of 15 type 2 diabetic and ten control subjects and observed 24 % β-cell area reduction in the diabetic samples (Clark et al. 1988). More recently, it has been reported that islet β-cell volume density and total β-cell mass were significantly lower (~30 %) in pancreatic specimens from Japanese type 2 diabetic patients in comparison with those obtained from nondiabetic individuals (Sakuraba et al. 2002). Accordingly, when pancreas samples following surgical removal were studied (Yoon et al. 2003), it was found that in the nondiabetic cases β-cell volume was 1.94 ± 0.7 %, whereas specimens from type 2 diabetic patients contained a lower β-cell volume (1.37 ± 1.0 %). In addition, in the diabetic samples, no correlation was found between β-cell volume and diabetes duration (Yoon et al. 2003). Pancreatic autoptic samples from type 2 diabetic patients, subjects with impaired fasting glycemia (IFG), and nondiabetic individuals (the groups were subdivided into lean or obese according to BMI) have been studied lately (Butler et al. 2003). In normoglycemic cases, obesity was associated with 50 % higher β-cell volume, as compared to nonobese individuals. However, obese subjects with IFG or diabetes had 40–60 % reduction in β-cell volume in comparison to BMI-matched, nondiabetic cases. This was due to β-cell number decrease, rather than changes in islet size. In the nonobese group, diabetes was associated with 41 % reduction in the volume of the β-cells. A detailed study has been published very recently (Rahier et al. 2008). The authors analyzed autoptic samples from 57 type 2 diabetic and 52 nondiabetic European subjects and confirmed that β-cell mass was lower (around 30 %) in the former (Fig. 1). However, there was marked intersubject variability and large overlap between the two groups (Fig.1). No difference was found between diabetic patients treated with oral agents and insulin, whereas β-cell mass increased with BMI values and decreased with duration of diabetes (Rahier et al. 2008). Finally, a reduced number of β-cells in islets from type 2 diabetic subjects has been demonstrated by electron microscopy as well, which also showed that volume density of mature insulin granules was lower in type 2 diabetic than in nondiabetic β-cells (Marchetti et al. 2004).

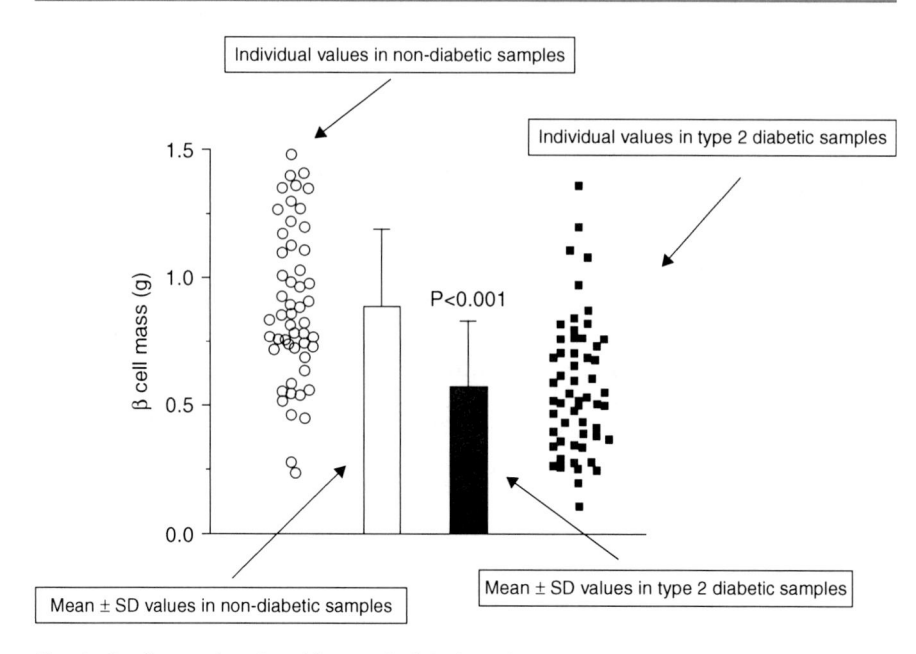

Fig. 1 β-cell mass is reduced in type 2 diabetic patients, as compared to nondiabetic controls, although there is a marked intersubject variability and clear overlap between the two groups (Adapted from Rahier et al. (2008))

It is generally assumed that β-cell loss in type 2 diabetes is mainly due to increased β-cell apoptosis (Butler et al. 2003; Marchetti et al. 2007). As a matter of fact, in autoptic samples, apoptosis was shown to be three- and tenfold higher in obese and lean type 2 diabetic samples, respectively, than in BMI-matched, normoglycemic individuals (Butler et al. 2003), and increased β-cell apoptosis in diabetic islets has been reported following electron microscopy analysis (Marchetti et al. 2007). In addition, by assessing cytoplasmic histone-associated DNA fragments, it has been observed that there is a twofold higher amount of islet cell death with isolated diabetic islets, as compared to nondiabetic islets (Marchetti et al. 2004) (Fig.2). This was accompanied by a significant increase in the activity of caspase-3 and caspase-8, key molecules in the apoptotic pathway (Marchetti et al. 2004) (Fig.2). Several factors can contribute to cause β-cell apoptosis (see below), and intracellular organelles, including the endoplasmic reticulum, are likely to be actively involved (Marchetti et al. 2007). On the other hand, the enhanced β-cell death rate does not seem to be adequately compensated for by regenerative phenomena in diabetic islets. In autoptic specimens, it has been reported that the relative rate of new islet formation, estimated by fraction of duct cells positive for insulin, and the frequency of β-cell replication, assessed by Ki67 staining, were substantially similar in type 2 diabetic and control pancreata (Butler et al. 2003).

Therefore, current evidence shows a reduced β-cell amount in human type 2 diabetes, possibly due to increased apoptosis without adequate regeneration.

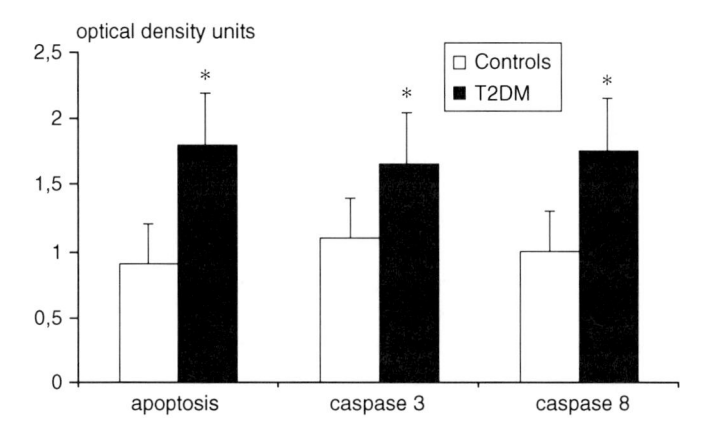

Fig. 2 Isolated type 2 diabetic (*T2DM*) islets show increased apoptosis and enhanced caspase-3 and caspase-8 activities, as compared to nondiabetic controls. Death was measured by ELISA methods evaluating cytoplasmic histone-associated DNA fragments, and caspase activity was determined using a colorimetric assay. *$p < 0.05$ vs. controls (Adapted from Marchetti et al. (2004))

However, the loss of β-cell appears to be 30 % on average, which is unlikely to lead to overt diabetes, unless a defect in β-cell function is present as well.

β-Cell Functional Defects

Several functional properties of the pancreatic β-cells in type 2 diabetes have been directly evaluated ex vivo following islet isolation from the human pancreas. Earlier work showed that the release of insulin evoked by glucose was lower in type 2 diabetic than in nondiabetic islets (Fernandez-Alvarez et al. 1994). However, the secretory response to the combination of l-leucine and l-glutamine appeared less severely altered (Fernandez-Alvarez et al. 1994). In a detailed study by Deng and colleagues, islets isolated from eight diabetic and nine normal donors were evaluated by in vitro islet perifusion experiments (Deng et al. 2004). Basal insulin secretion was similar for both normal and diabetic islets. However, the islets from diabetic donors released less total insulin in response to glucose and also exhibited an elevated threshold for insulin secretion triggering. In addition, it was observed that in comparison with normal islets, an equivalent amount of type 2 diabetic islets did not fully reverse the hyperglycemic condition when transplanted into diabetic mice (Deng et al. 2004). In another study, when insulin secretion was measured in response to glucose, arginine, and glibenclamide in isolated nondiabetic and type 2 diabetic islets, again no significant difference as for insulin release in response to 3.3 mmol/l glucose was observed (Del Guerra et al. 2005). However, when challenged with 16.7 mmol/l glucose, diabetic islets secreted significantly less insulin than did nondiabetic cells. Insulin secretion during arginine and glibenclamide stimulation was also lower from diabetic islets than from control islets; however, type 2 diabetic islets released a significantly higher amount of insulin in response to arginine and glibenclamide than in response to glucose. In addition, when perifusion experiments were performed, glucose stimulation did not elicit any apparent increase in the early insulin secretion phase from diabetic islets, which however promptly released insulin when challenged with arginine or sulfonylurea (Del Guerra et al. 2005). Consistent with the observation that β-cell insulin secretion defects in type 2 diabetes β-cells are more selective for glucose-induced stimulation, it has been observed that in type 2 diabetic islets glucose oxidation is reduced, as compared to nondiabetic islets (Fernandez-Alvarez et al. 1994; Del Guerra et al. 2005). This has led to the speculation that mitochondria might be involved in causing β-cell dysfunction in type 2 diabetes. In this regard, the morphology and the function of mitochondria in human type 2 diabetic β-cells have been studied (Anello et al. 2005). By electron microscopy, mitochondria in type 2 diabetes β-cells appeared round shaped, hypertrophic, and with higher density volume when compared to control β-cells. When adenine nucleotide content was measured, it was found that islets from diabetic subjects were not able to increase their ATP content in the presence of acute glucose stimulation (Fig.3). As a consequence, the ATP/ADP ratio was approximately 40 % lower in diabetic than in control islets, which could contribute to the blunted or absent glucose-stimulated insulin release in the former (Anello et al. 2005) (Fig.3).

In summary, insulin secretion defects in human type 2 diabetic islets have been described by several authors, and data show more marked changes in insulin release in response to glucose, as compared to other fuel and nonfuel stimuli. This suggests that type 2 diabetic β-cells may have alterations in some steps of glucose metabolism, including those at the mitochondria level, leading to reduced ATP production.

Molecular Changes

Changes at the gene and protein expression levels have been reported in type 2 diabetic pancreatic islets by several authors. Using oligonucleotide microarrays of pancreatic islets isolated from humans with type 2 diabetes vs. normal glucose tolerant controls, Gunton et al. found that 370 genes were differently expressed in the two groups (243 upregulated and 137 downregulated) (Gunton et al. 2005). Quantitative RT-PCR studies were performed on selected genes, which confirmed changes in the expression of genes known to be important in β-cell function, including major decreases in the expression of HNF4 α, insulin receptor, IRS2, Akt2, and several glucose-metabolic-pathway genes. There was also a 90 % decrease in the expression of the transcription factor ARNT/HIF1β (hydrocarbon nuclear receptor translocator/hypoxia-inducible factor 1β) (Gunton et al. 2005). Successively, several genes encoding for the following proteins were found to be downregulated in type 2 diabetic islets by real-time RT-PCR: insulin, glucose transporter 1, glucose transporter 2, glucokinase, and molecules involved in insulin granules exocytosis (Del Guerra et al. 2005; Ostenson et al. 2006). Conversely, several genes implicated in differentiation and proliferation pathways have been reported to be increased in diabetic islets, including PDX-1, Foxo-1, Pax-4,

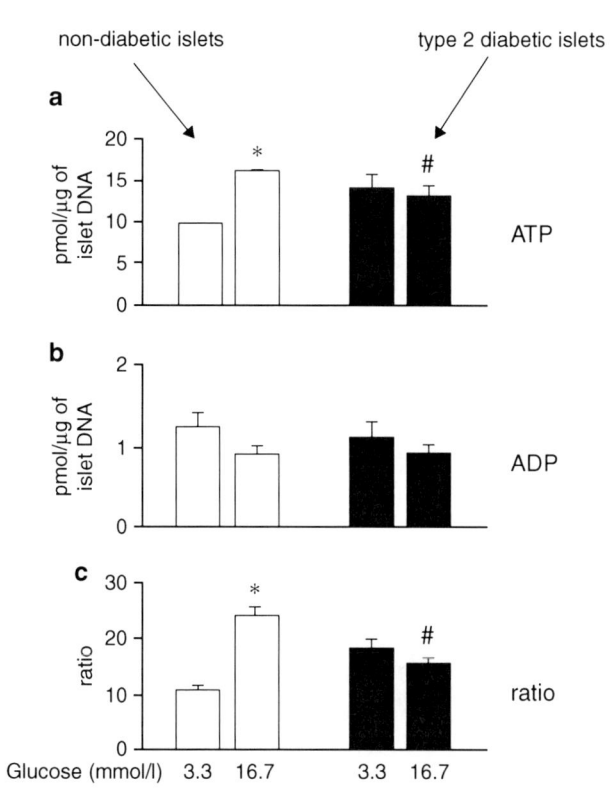

Fig. 3 ATP production and ATP/ADP ratio increase in nondiabetic but not in type 2 diabetic islets following exposure to 3.3–16.7 mmol/l glucose concentration.*: significantly higher vs.3.3 mmol/l glucose; #: significantly lower vs. nondiabetic islets at16.7 mmol/l glucose (Adapted from Anello et al. (2005))

andTCF7L2 (Del Guerra et al. 2005; Brun et al. 2008; Lyssenko et al. 2007). Furthermore, changes at the level of the expression of genes involved in regulating cell redox balance have been shown (Marchetti et al. 2004). As a matter of fact, mRNA expression of NADPH oxidase has been found to be increased and that of manganese and copper/zinc superoxide dismutases to be decreased in diabetic islets, together with enhanced expression of catalase and GSH peroxidase (Marchetti et al. 2004). In a recent paper, the expression of several genes associated with the function of the endoplasmic reticulum (in particular, those encoding for immunoglobulin heavy chain-binding protein, BiP, and X-box binding protein 1, XBP-1) has been described to be induced by exposure to high glucose in type 2 diabetic islets but not in control islets (Marchetti et al. 2007). When β-cell-enriched preparations obtained by the laser capture microdissection technique were studied (Marselli et al. 2007), transcript to some analysis preliminarily performed on four type 2 diabetic and four samples showed that in diabetic samples, there were 1,532 upregulated and 528 downregulated genes (Marselli et al. 2007).

Some information is also available as for protein expression in type 2 diabetic islets. The amount of insulin has been reported to be decreased 30–40 % in diabetic islet cells (Marchetti et al. 2004; Ostenson et al. 2006). The expression of AMP-activated kinase, IRS-2, PDX-1 (this latter at odds with gene expression data), and proteins involved in exocytosis was also found to be decreased in type 2 diabetic islets in comparison to nondiabetic samples (Marchetti et al. 2004; Ostenson et al. 2006). Preliminary data on type 2 diabetic islet protein profiling have been reported recently (Nyblom et al. 2007). The results showed that although considerable variability existed within the individuals, 31 differentially expressed peaks were detected, and the intensities of some of them were significantly correlated with ex vivo islet insulin release (Nyblom et al. 2007).

Whereas many defects at the gene and protein expression level have been described in islet cells from type 2 diabetic subjects, at present it is not possible to distinguish between primary β-cell molecular changes (leading to diabetes) and those occurring as a consequence of the unfavorable microenvironment associated with the diabetic conditions (see below). Since prospective studies in this regard are not feasible for obvious reasons, it would be of interest to compare the molecular properties of β-cells from individuals at different stages of disease.

The Role of Genetic and Acquired Factors

Type 2 diabetes is a polygenic disease, and in the past few years, link age studies, candidate-gene approaches, and genome-wide association studies have identified several gene variants which associate with this form of diabetes (Jafar-Mohammadi and McCarthy 2008; Owen and McCarthy 2007; Groop and Lyssenko 2008; Parikh and Groop 2004; Vaxillaire and Froguel 2008; Hattersley and Pearson 2006). The majority of these genes are involved in β-cell function and survival, and for some of them the description is available as for their direct effects on some β-cell features in humans. The common Gly(972) → Arg amino acid polymorphism of insulin receptor substrate 1, Arg(972) IRS-1, has been found to be associated with functional and morphological alterations of isolated human islets, including increased susceptibility to apoptosis, diminished glucose-stimulated insulin secretion, and lower amount of insulin granules (Marchetti et al. 2002; Federici et al. 2001). Similarly, the E23K variant of KCNJ11 gene, encoding the pancreatic β-cell adenosine 5′-triphosphate-sensitive potassium channel subunit Kir6.2 and associated with an increased risk of secondary failure to sulfonylurea in patients with type 2 diabetes (Sesti et al. 2006), has been shown to be associated with impairment of glibenclamide-induced insulin release following 24-h exposure to high glucose concentration. However, those studies were performed on islets isolated from nondiabetic subjects. More recently, genetic variants in the gene encoding for transcription factor-7-like 2 (TCF7L2) have been associated with type 2 diabetes and impaired β-cell function (Cauchi and Froguel 2008). It has been shown that the CT/TT genotypes of SNP rs7903146 strongly predicted future diabetes in independent cohorts of patients and that TCF7L2 expression in human islets was increased fivefold in type 2 diabetes, particularly in carriers of the TT genotype (Lyssenko et al. 2007). In this study, overexpression of TCF7L2 in human islets reduced glucose-stimulated insulin secretion. However, in another report, depleting TCF7L2 by siRNA resulted in decreased glucose-stimulated insulin release, increased β-cell apoptosis, and decreased β-cell proliferation in human islets (Shu et al. 2008). In contrast, overexpression of TCF7L2 protected islets from glucose and cytokine-induced apoptosis and impaired function (Shu et al. 2008). It cannot be excluded that in the presence of diabetes, phenotypic changes occurring independent of the genotype may render the overall picture less clear.

Several acquired factors can affect β-cell survival and function (Stumvoll et al. 2005; Kahn 2003; Marchetti et al. 2008; Wajchenberg 2007; Meier 2008). In particular, the effects of glucotoxicity and lipotoxicity (terms used to indicate the deleterious effects induced on tissues and cells by prolonged exposure to increased glucose or free fatty acid concentrations) have been studied with isolated islets. Both conditions can lead to increased apoptosis, reduced glucose-stimulated insulin release, and molecular changes (Marchetti et al. 2008). Unfortunately, very little information is available on gluco- and/or lipotoxicity on human type 2 diabetic islets. In a recently published study (Marchetti et al. 2007), several features of β-cell endoplasmic reticulum were investigated in islets from nondiabetic and type 2 diabetic subjects. Whereas signs of endoplasmic reticulum stress were found in diabetic β-cells, it was also reported that when the islets were cultured for 24 h in 11.1 mmol/l glucose, there was the induction of immunoglobulin heavy chain-binding protein (BiP) and X-box binding protein 1(XBP-1) in the type 2 diabetic islets (Marchetti et al. 2007) (Fig.4). Obviously, more work is needed on these issues.

The mechanisms mediating the deleterious effects of acquired factors are being actively investigated, with increased oxidative stress probably playing an important role (Poitout and Robertson 2002). As a matter of fact, when the presence of 8-hydroxy-2′-deoxyguanosine (a marker of oxidative stress-induced DNA damage) and 4-hydroxy-2-nonenal modified proteins (a marker of lipid peroxidation products) was determined by immunostaining in islets of type 2 diabetic patients, both markers significantly increased as compared with nondiabetic individuals (Sakuraba et al. 2002). In addition, reduced staining of Cu/Zn superoxide dismutase was observed in the diabetic islet cells (Sakuraba et al. 2002). Similar findings were reported in a study performed with isolated type 2 diabetic islets (Marchetti et al. 2004), which showed increased content of nitrotyrosine and 8-hydroxy-2′-deoxyguanosine and reduced expression of Cu/Zn- and Mn superoxide dismutase. All this may contribute to produce a proinflammatory soil, which has been proposed to lead to β-cell damage in type 2 diabetes (Böni-Schnetzler et al. 2008; Ehses et al. 2008). Pancreatic islets may respond to metabolic stress by producing inflammatory factors, such as IL-1, and macrophage infiltration has been found in human type 2 diabetic islets. It is however possible that some of these pathways may be activated in subgroups of patients (Welsh et al. 2005).

Dealing with all the information continuously and rapidly coming from genetic studies is not an easy task, but the assessment of the relationships between β-cell

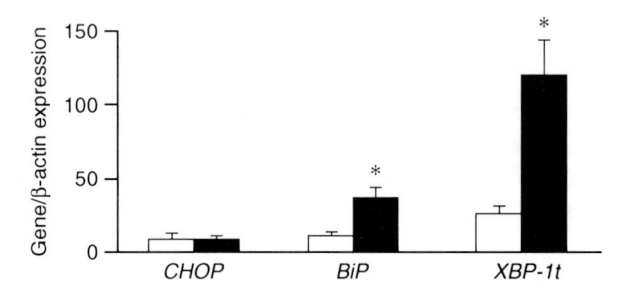

Fig. 4 When isolated type 2 diabetic islets were exposed for 24 h at increased glucose concentration (see text for details), a significant induction of genes involved in endoplasmic reticulum stress (BiP and XBP-1t) was observed, as measured by quantitative RT-PCR. The expression of another gene (CHOP) did not change (Adapted from Marchetti et al. (2007))

genotype and phenotype is crucial to understand why the β-cell fails in type 2 diabetes and in which way it is affected by acquired factors.

Reversal of β-Cell Damage in Type 2 Diabetes

The possibility that pancreatic β-cell damage induced by acquired factors can be prevented has been demonstrated in isolated nondiabetic islets exposed to different metabolic perturbations (Marchetti et al. 2008). More importantly, a few studies have shown that β-cell dysfunction in type 2 diabetes may be reversible. Exposure of isolated type 2 diabetic islets to antioxidants has led to improved glucose-stimulated insulin secretion and normalized expression of a few ROS scavenging enzymes (Del Guerra et al. 2005; Lupi et al. 2007). As mentioned above, a study showed that isolated type 2 diabetic islets were characterized by reduced insulin content, decreased amount of mature insulin granules, impaired glucose-induced insulin secretion, reduced insulin mRNA expression, and increased apoptosis with enhanced caspase-3 and caspase-8 activities (Marchetti et al. 2004). These alterations were associated with increased oxidative stress, as shown by higher nitrotyrosine concentrations, increased expression of protein kinase C-β2 and NADH oxidase, and changes in mRNA expression of Mn superoxide dismutase, Cu/Zn superoxide dismutase, catalase, and glutathione peroxidase (Marchetti et al. 2004). When these islets were incubated for 24 h in the presence of therapeutic concentration of metformin, insulin content and the number of mature insulin granules increased (Fig.5), and glucose-induced insulin release improved, with induction of insulin mRNA expression. Moreover, apoptosis was reduced, with concomitant decrease of caspase-3 and -8 activities. These changes were accompanied by reduction or normalization of markers of oxidative stress (Marchetti et al. 2004). Recently, the role of incretins (GLP-1, glucose-dependent insulinotropic polypeptide [GIP], and some of their analogs) in the therapy of diabetes has received much attention, mainly because of the beneficial actions of

Fig. 5 The amount of insulin granules in type 2 diabetic β-cells increases following preexposure for 24 h with therapeutic concentration of metformin. Electron microscopy evaluation, magnification × 160,000 (Reproduced with modifications from Marchetti et al. (2004)

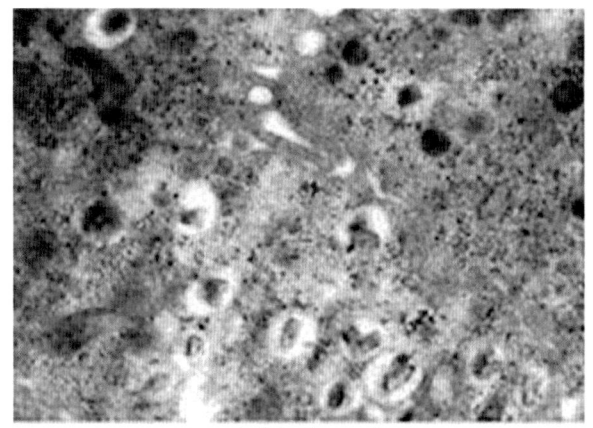

type 2 diabetic β-cell

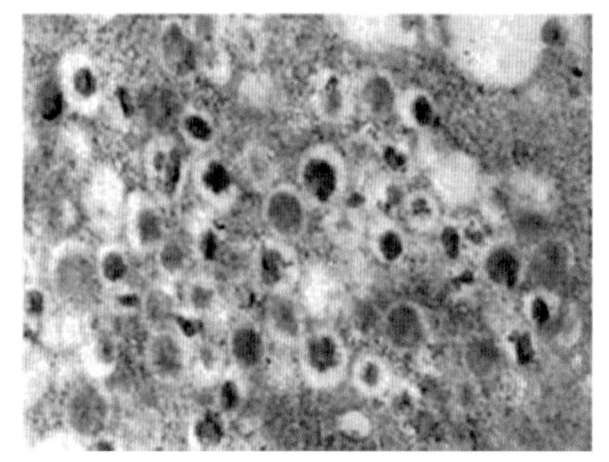

type 2 diabetic β-cell pre-exposed to metformin

these molecules (GLP-1 in particular) on the β-cell (Drucker and Nauck 2006). In a recent study (Lupi et al. 2008), pancreatic islets were prepared from the pancreas of nondiabetic and type 2 diabetic donors and then incubated in the presence of 5.5 mmol/l glucose, with or without the addition of exendin-4 (a long-acting GLP-1 mimetic). Insulin secretion from the type 2 diabetic islets improved after incubation with exendin-4, which also induced a significantly higher expression of insulin, glucose transporter 2, glucokinase, and some β-cell regeneration and differentiation factors, including pancreas duodenum homeobox-1(Pdx-1).

Therefore, acting directly at the β-cell level to prevent damage or restore functional and survival competence is feasible in vitro. Strategies need to be developed to deliver the appropriate treatment to the β-cell in vivo, to be combined with therapies aiming to limit the negative impact on the islets of acquired conditions such as glucotoxicity and lipotoxicity (see above).

Table 1 Main defects of pancreatic β-cells in human type 2 diabetes

β-cell mass
Increased apoptosis
Not sufficient proliferation
Not sufficient neogenesis
β-cell function
Reduced glucose-stimulated insulin secretion
Blunted or absent early phase insulin secretion
Increased proinsulin/insulin ratio
Altered pulsatility of insulin release
Molecular features
Altered expression of genes involved in β-cell function and survival
Altered expression of proteins involved in β-cell function and survival
Increased production of reactive oxygen and nitrogen species

Conclusions

Pancreatic β-cells in type 2 diabetes have several defects (Table 1). Decreased β-cell mass is due to increased apoptosis not compensated for by adequate β-cell regeneration. Insulin secretion defects are more marked in response to glucose, suggesting that handling of this fuel by the β-cell is defective somewhere along the road leading to ATP production. These alterations are accompanied by several molecular defects, possibly due, at least in part, to genetic variations and acquired factors, which still need to be set in a more comprehensive picture. The observation that β-cell defects may be reversible supports the concept that β-cell dysfunction in human type 2 diabetes could not be relentless.

Acknowledgments Supported in part by the Italian Ministry of University and Research (PRIN2007–2008).

Cross-References

▶ (Dys)Regulation of Insulin Secretion by Macronutrients
▶ Apoptosis in Pancreatic β-Islet Cells in Type 1 and Type 2 Diabetes
▶ Islet Structure and Function in the GK Rat
▶ Microscopic Anatomy of the Human Islet of Langerhans

References

American Diabetes Association (2008) Diagnosis and classification of diabetes mellitus. Diabetes Care 31(Suppl 1):S55–S60

Anello M, Lupi R, Spampinato D, Piro S, Masini M, Boggi U, Del Prato S, Rabuazzo AM, Purrello F, Marchetti P (2005) Functional and morphological alterations of mitochondria in pancreatic β cells from type 2 diabetic patients. Diabetologia 48:282–289

Böni-Schnetzler M, Thorne J, Parnaud G, Marselli L, Ehses JA, Kerr-Conte J, Pattou F, Halban PA, Weir GC, Donath MY (2008) Increased interleukin IL-1β messenger ribonucleic acid expression in β-cells of individuals with type 2 diabetes and regulation of IL-1β inhuman islets by glucose and autostimulation. J Clin Endocrinol Metab 93:4065–4074

Brun T, Hu He KH, Lupi R, Boehm B, Wojtusciszyn A, Sauter N, Donath M, Marchetti P, Maedler K, Gauthier BR (2008) The diabetes-linked transcription factor Pax4 is expressed in human pancreatic islets and is activated by mitogens and GLP-1. Hum Mol Genet 17:478–489

Butler AE, Janson J, Bonner-Weir S, Ritzel R, Rizza RA, Butler PC (2003) β-cell deficit and increased β-cell apoptosis in humans with type 2 diabetes. Diabetes 52:102–110

Cauchi S, Froguel P (2008) TCF7L2 genetic defect and type 2 diabetes. Curr Diabetes Rep 8:149–155

Clark A, Wells CA, Buley ID et al (1988) Islet amyloid, increased α-cells, reduced β-cells and exocrine fibrosis: quantitative changes in the pancreas in type 2 diabetes. Diabetes Res 9:151–159

Del Guerra S, Lupi R, Marselli L, Masini M, Bugliani M, Sbrana S, Torri S, Pollera M, Boggi U, Mosca F, Del Prato S, Marchetti P (2005) Functional and molecular defects of pancreatic islets inhuman type 2 diabetes. Diabetes 54:727–735

Deng S, Vatamaniuk M, Huang X, Doliba N, Lian MM, Frank A et al (2004) Structural and functional abnormalities in the islets isolated from type 2 diabetic subjects. Diabetes 53:624–632

Drucker DJ, Nauck MA (2006) The incretin system: glucagon-like peptide-1 receptor agonists and dipeptidyl peptidase-4 inhibitors in type 2 diabetes. Lancet 368:1696–1705

Ehses JA, Boni-Schnetzler M, Faulenbach M, Donath MY (2008) Macrophages, cytokines and β cell death in type 2 diabetes. Biochem Soc Trans 36:340–342

Federici M, Hribal ML, Ranalli M, Marselli L, Porzio O, Lauro D, Borboni P, Lauro R, Marchetti P, Melino G, Sesti G (2001) The common Arg972 polymorphism in insulin receptorsubstrate-1 causes apoptosis of human pancreatic islets. FASEB J 15:22–24

Fernandez-Alvarez J, Conget I, Rasschaert J, Sener A, Gomis R, Malaisse WJ (1994) Enzymatic, metabolic and secretory patterns in human islets of type 2 (non-insulin-dependent) diabetic patients. Diabetologia 37:177–181

Ferrannini E, Mari A (2004) β-cell function and its relation to insulin action in humans: a critical appraisal. Diabetologia 47:943–956

Groop L, Lyssenko V (2008) Genes and type 2 diabetes mellitus. Curr Diabetes Rep 8:192–197

Gunton JE, Kulkarni RN, Yim S, Okada T, Hawthorne WJ, Tseng YH, Roberson RS, Ricordi C, O'Connell PJ, Gonzalez FJ, Kahn CR (2005) Loss of ARNT/HIF1β mediates altered gene expression and pancreatic-islet dysfunction in human type 2 diabetes. Cell 122:337–349

Hattersley AT, Pearson ER (2006) Minireview: pharmacogenetics and beyond: the interaction of therapeutic response, β-cell physiology, and genetics in diabetes. Endocrinology 147:2657–2663

Jafar-Mohammadi B, McCarthy MI (2008) Genetics of type 2 diabetes mellitus and obesity-a review. Ann Med 40:2–10

Kahn SE (2003) The relative contribution of insulin resistance and β-cell dysfunction to the pathophysiology of type 2 diabetes. Diabetologia 46:3–19

Kahn SE, Carr DB, Faulenbach MV, Utzschneider KM (2008) An examination of β-cell function measures and their potential use for estimating β-cell mass. Diabetes Obes Metab 10(Suppl 4):63–76

Kumar AF, Gruessner RW, Seaquist ER (2008) Risk of glucose intolerance and diabetes in hemipancreatectomized donors selected for normal preoperative glucose metabolism. Diabetes Care 31:1639–1643

Lupi R, Del Guerra S, Mancarella R, Novelli M, Valgimigli L, Pedulli GF, Paolini M, Soleti A, Filipponi F, Mosca F, Boggi U, Del Prato S, Masiello P, Marchetti P (2007) Insulin secretion defects of human type 2 diabetic islets are corrected in vitro by a new reactive oxygen species scavenger. Diabetes Metab 33:340–345

Lupi R, Mancarella R, Del Guerra S (2008) Effects of exendin-4 on islets Bugliani M, Del PratoS, Boggi U, Mosca F, Filipponi F, Marchetti P. from type 2 diabetes patients. Diabetes Obes Metab 10:515–519

Lyssenko V, Lupi R, Marchetti P, Del Guerra S, Orho-Melander M, Almgren P, Sjogren M, Ling C, Eriksson KF, Lethagen UL, Mancarella R, Berglund G, Tuomi T, Nilsson P, Del Prato S, Groop L (2007) Mechanisms by which common variants in the TCF7L2 gene increase risk of type2 diabetes. J Clin Invest 117:2155–2163

Marchetti P, Lupi R, Federici M, Marselli L, Masini M, Boggi U, Del Guerra S, Patane G, Piro S, Anello M, Bergamini E, Purrello F, Lauro R, Mosca F, Sesti G, Del Prato S (2002) Insulin secretory function is impaired in isolated human islets carrying the Gly(972) → Arg IRS-1polymorphism. Diabetes 51:1419–1424

Marchetti P, Del Guerra S, Marselli L, Lupi R, Masini M, Pollera M (2004) Pancreatic islets from type 2 diabetic patients have functional defects and increased apoptosis that are ameliorated by metformin. J Clin Endocrinol Metab 89:5535–5541

Marchetti P, Bugliani M, Lupi R, Marselli L, Masini M, Boggi U, Filipponi F, Weir GC, Eizirik DL, Cnop M (2007) The endoplasmic reticulum in pancreatic β cells of type 2 diabetes patients. Diabetologia 50:2486–2494

Marchetti P, Dotta F, Lauro D, Purrello F (2008) An overview of pancreatic β-cell defects inhuman type 2 diabetes: implications for treatment. Regul Pept 146:4–11

Marselli L, Sgroi DC, Thorne J, Dahiya S, Torri S, Omer A, Del Prato S, Towia L, Out HH, Sharma A, Bonner-Weir S, Marchetti P, Weir GC (2007) Evidence of inflammatory markers in β-cells of type 2 diabetic subjects. Diabetologia 50(Suppl 1):S178

Meier JJ (2008) β cell mass in diabetes: a realistic therapeutic target? Diabetologia 51:703–713

Nyblom HK, Bugliani M, Marchetti P, Bergsten P (2007) Islet protein expression from type 2 diabetic donors correlating with impaired secretory response. Diabetologia 50(Suppl 1):S178

Ostenson CG, Gaisano H, Sheu L, Tibell A, Bartfai T (2006) Impaired gene and protein expression of exocytotic soluble N-ethylmaleimide attachment protein receptor complex proteins in pancreatic islets of type 2 diabetic patients. Diabetes 55:435–440

Owen KR, McCarthy MI (2007) Genetics of type 2 diabetes. Curr Opin Genet Dev 17:239–244

Parikh H, Groop L (2004) Candidate genes for type 2 diabetes. Rev Endocr Metab Disord 5:151–176

Poitout V, Robertson RP (2002) Minireview: secondary β-cell failure in type 2 diabetes – a convergence of glucotoxicity and lipotoxicity. Endocrinology 143:339–342

Porte D Jr (1991) Banting lecture 1990. β-cells in type II diabetes mellitus. Diabetes 40:166–180

Rahier J, Goebbels RM, Henquin JC (1983) Cellular composition of the human diabetic pancreas. Diabetologia 24:366–371

Rahier J, Guiot Y, Goebbels RM, Sempoux C, Henquin JC (2008) Pancreatic β-cell mass in European subjects with type 2 diabetes. Diabetes Obes Metab 10(Suppl 4):32–42

Saito K, Takahashi T, Yaginuma N, Iwama N (1978) Islet morphometry in the diabetic pancreas of man. Tohoku J Exp Med 125:185–197

Saito K, Yaginuma N, Takahashi T (1979) Differential volumetry of A, B and D cells in the pancreatic islets of diabetic and non diabetic subject. Tohoku J Exp Med 129:273–283

Sakuraba H, Mizukami H, Yagihashi N, Wada R, Hanyu C, Yagihashi S (2002) Reduced β cell mass and expression of oxidative stress related DNA damage in the islets of Japanese type 2diabetic patients. Diabetologia 45:85–96

Sesti G, Laratta E, Cardellini M, Andreozzi F, Del Guerra S, Irace C, Gnasso A, Grupillo M, Lauro R, Hribal ML, Perticone F, Marchetti P (2006) The E23K variant of KCNJ11 encoding the pancreatic β-cell K_{ATP} channel subunits Kir6.2 is associated with an increased risk of secondary failure to sulfonylurea in patients with type 2 diabetes. J Clin Endocrinol Metab 91:2334–2339

Shu L, Sauter NS, Schulthess FT, Matveyenko AV, Oberholzer J, Maedler K (2008) Transcription factor 7-like 2 regulates β-cell survival and function in human pancreatic islets. Diabetes 57:645–653

Stefan Y, Orci L, Malaisse-Lagae F et al (1982) Quantitation of endocrine cell content in the pancreas of nondiabetic and diabetic humans. Diabetes 31:694–700

Stumvoll M, Goldstein BJ, van Haeften TW (2005) Type 2 diabetes: principles of pathogenesis and therapy. Lancet 365:1333–13

Vaxillaire M, Froguel P (2008) Monogenic diabetes in the young, pharmacogenetics and relevance to multifactorial forms of type 2 diabetes. Endocr Rev 29:254–264

Wajchenberg BL (2007) β-cell failure in diabetes and preservation by clinical treatment. Endocr Rev 28:187–218

Welsh N, Cnop M, Kharroubi I, Bugliani M, Lupi R, Marchetti P, Eizirik DL (2005) Is there a role for locally produced interleukin-1 in the deleterious effects of high glucose or the type 2 diabetes milieu to human pancreatic islets? Diabetes 54:3238–3244

Westermark P, Wilander E (1978) The influence of amyloid deposits on the islet volume in maturity onset diabetes mellitus. Diabetologia 15:417–421

Yoon KH, Ko SH, Cho JH et al (2003) Selective β-cell loss and α-cell expansion in patients with type 2 diabetes in Korea. J Clin Endocrinol Metab 88:2300–2308

Pancreatic β Cells in Metabolic Syndrome 30

Marcia Hiriart, Myrian Velasco, Carlos Manlio Diaz-Garcia, Carlos Larqué, Carmen Sánchez-Soto, Alondra Albarado-Ibañez, Juan Pablo Chávez-Maldonado, Alicia Toledo, and Neivys García-Delgado

Contents

Introduction	819
Insulin Secretion	820
Glucose-Stimulated Insulin Secretion (GSIS)	820
Amino Acid-Stimulated Insulin Secretion (AASIS)	823
Nonesterified Fatty Acids (NEFAs)	823
Pancreatic Hormones	824
Incretins: A Crosstalk Between the gut and the Endocrine Pancreas	827
Neurotransmitters	828
Insulin Functions and Signaling	829
Metabolic Syndrome and β Cells	831
Adipose Tissue and Obesity	831
Metabolic Effects of Cytokines	832
Hyperinsulinemia and Insulin Resistance	834
β-Cell Exhaustion and T2DM	835
Conclusions	838
Cross-References	839
References	839

Abstract

Obesity is considered a major public health problem worldwide. Metabolic syndrome is a cluster of signs that increases the risk of developing cardiovascular disease and type 2 diabetes mellitus (T2DM). The main characteristics of metabolic syndrome are central obesity, dyslipidemia, hypertension, hyperinsulinemia, and insulin resistance. It is clear that the progression of

---

M. Hiriart (✉) • M. Velasco • C.M. Diaz-Garcia • C. Larqué • C. Sánchez-Soto • A. Albarado-Ibañez • J.P. Chávez-Maldonado • A. Toledo • N. García-Delgado
Department of Neurodevelopment and Physiology, Instituto de Fisiología Celular, Universidad Nacional Autónoma de México, Mexico DF, Mexico
e-mail: mhiriart@ifc.unam.mx

metabolic syndrome to T2DM depends on the environment and the genetic traits of individuals.

Pancreatic β cells are fundamental for nutrient homeostasis. They are the unique cells in the organisms that produce and secrete insulin. The actions of insulin are anabolic, stimulating glucose entry to adipose tissue and skeletal muscle, and promoting nutrient storage.

However, insulin receptors are present in every mammalian cell, and not all the physiological effects of this hormone are completely understood. Nutrients, other hormones, and neurotransmitters regulate insulin secretion, and the main ones will be discussed in this chapter. We will summarize how metabolic changes modify β-cell physiology and the actions of insulin in metabolic syndrome, eventually leading to the development of T2DM.

Keywords

β-cell exhaustion • Insulin resistance • Obesity • Ion channels • Cytokines • β-cell dysfunction • Insulin hypersecretion • Hyperinsulinemia

Glossary

ACh	Acetylcholine
acyl-CoA	Acyl coenzyme A
ADP	Adenosine diphosphate
AMPK	5′ adenosine monophosphate-activated protein kinase
AR	Adrenoreceptor
ATP	Adenosine triphosphate
BDNF	Brain-derived neurotrophic factor
BMI	Body mass index
cAMP	Cyclic adenosine monophosphate
CAP	Cbl-associated protein
Cbl	Casitas B-lineage lymphoma proto-oncogene
Cytokine R	R Cytokine receptor
DAG	Diacylglycerol
DPP-4	Enzyme dipeptidylpeptidase-4
ER	Endoplasmic reticulum
ERK	Extracellular signal-regulated kinase
GK	Glucokinase
GLP-1	Glucagon-like peptide-1
GLUT2	Glucose transporters type 2
GSIS	Glucose-stimulated insulin secretion
IGF1R	Insulin-like growth factor 1 receptor
IKK	Kinase of IKB (inhibitor of KB)
IL-6	Interleukin-6
IR	Insulin receptor
IRS	Insulin receptor substrate

JAKs	Kinases of the Janus family
JNK	c-Jun N-terminal kinase
K_{ATP}	ATP-sensitive potassium channel
MAPK	Mitogen-activated protein kinase
MODY	Maturity onset diabetes of the young
MS	Metabolic syndrome
mTOR	Mammalian target of rapamycin
NEFAs	Nonesterified fatty acids
NGF	Nerve growth factor
PDK	Phosphoinositide-dependent kinase
PHHI	Persistent hypoglycemic hyperinsulinemia of the infancy
PI3K	Phosphoinositol-3 kinase
PKA	Protein kinase A
PKB/Akt	Protein kinase B
PKC	Protein kinase C
PPAR gamma	Peroxisome-proliferation-activated receptor gamma
PTPs	Protein tyrosine phosphatases
Ras	Rat sarcoma protein family
RBP4	Retinol binding protein-4
ROS	Reactive oxygen species
SNARE	Soluble NSF attachment protein receptor
SOCS	Suppressor of cytokine signaling
SREBP	Sterol regulatory element-binding protein
T2DM	Type 2 diabetes mellitus
TLR4	Toll-like receptor 4
TNFa	Tumor necrosis factor α
TNFR	Tumor necrosis factor receptor
TrkA	Tyrosine kinase receptor A
TRP	Transient receptor channels
WAT	White adipose tissue

Introduction

Insulin secretion is fundamental for nutrient homeostasis. Insulin is an anabolic hormone that regulates the storage of carbohydrates, lipids, and proteins in the liver, muscle, and adipose tissue but also glucose uptake in the last two. Although insulin receptor is ubiquitously expressed in mammalian cells, the physiological effects of this hormone are not completely understood in every tissue. The autoimmune destruction of pancreatic β cells leads to a chronic insulin deprivation, a pathologic condition called type 1 diabetes mellitus. On the other hand, type 2 diabetes mellitus (T2DM) results from a combination of insulin resistance and an impaired insulin secretion, which is not enough to maintain euglycemia.

The metabolic syndrome (MS) is a cluster of signs including central obesity, dyslipidemia, hypertension, hyperinsulinemia, and insulin resistance that increase the risk to develop cardiovascular disease and diabetes mellitus (Hunt et al. 2004), reviewed by Larqué et al. (2011). The progression from MS to T2DM depends on individual genetic traits.

According to the World Health Organization (2013), diabetes is one of the leading causes of death worldwide. It is clear that some risk factors like obesity and metabolic syndrome are associated with T2DM; for instance, in North America 60 % of the population exhibits overweight or obesity. In 2010, it has been estimated that 79 million people around the world presented a prediabetic diagnosis. It is mandatory to understand how obesity and MS contribute to β-cell exhaustion and finally to diabetes. Here, we review the β-cell physiology and the physiopathological changes in MS.

Insulin Secretion

Pancreatic β-cells may be considered fuel sensors. They are continuously monitoring and responding to changes in the concentration of circulating nutrients and thus regulating their homeostasis. Insulin secretion is stimulated by several secretagogues like metabolized nutrients, neurotransmitters, hormones, and drugs that bind to membrane receptors (Fig. 1).

Glucose-Stimulated Insulin Secretion (GSIS)

The best characterized mechanism of insulin secretion is the stimulated by glucose (Fig. 1) (see also chapter "▶ (Dys)Regulation of Insulin Secretion by Macronutrients"). When its level rises, glucose enters the β cell mainly via glucose transporters: GLUT2 in rodents (SLC2A2) or GLUT1 in humans (SLC2A1). This is followed by glucose phosphorylation catalyzed by glucokinase (GK) (Prentki and Matschinsky 1987; Prentki et al. 2013). The glycolytic flux is regulated by a combination of the glucose flux and GK activity. Studies of GK mutations in animals, as well as in some human conditions, have led to similar conclusions. Mutations that increase GK activity promote persistent hypoglycemic hyperinsulinemia of the infancy (PHHI). A reduction in the GK catalytic activity has been documented in patients with maturity onset diabetes of the youth (MODY) and type 2 diabetes (Huypens et al. 2012). Excellent reviews on the importance of glucose metabolism have been published elsewhere, and we will not review them in this chapter (Jensen et al. 2008; Nolan and Prentki 2008).

It is well established that GSIS is dependent on β-cell electrical activity (Drews et al. 2010; Henquin and Meissner 1984) (see also chapter "▶ Electrophysiology of Islet Cells"). At glucose concentrations below 3 mM, the cell is electrically silent with a resting membrane potential of about -70 mV. Rising external glucose produces a slow depolarization, which is dependent on the sugar concentration.

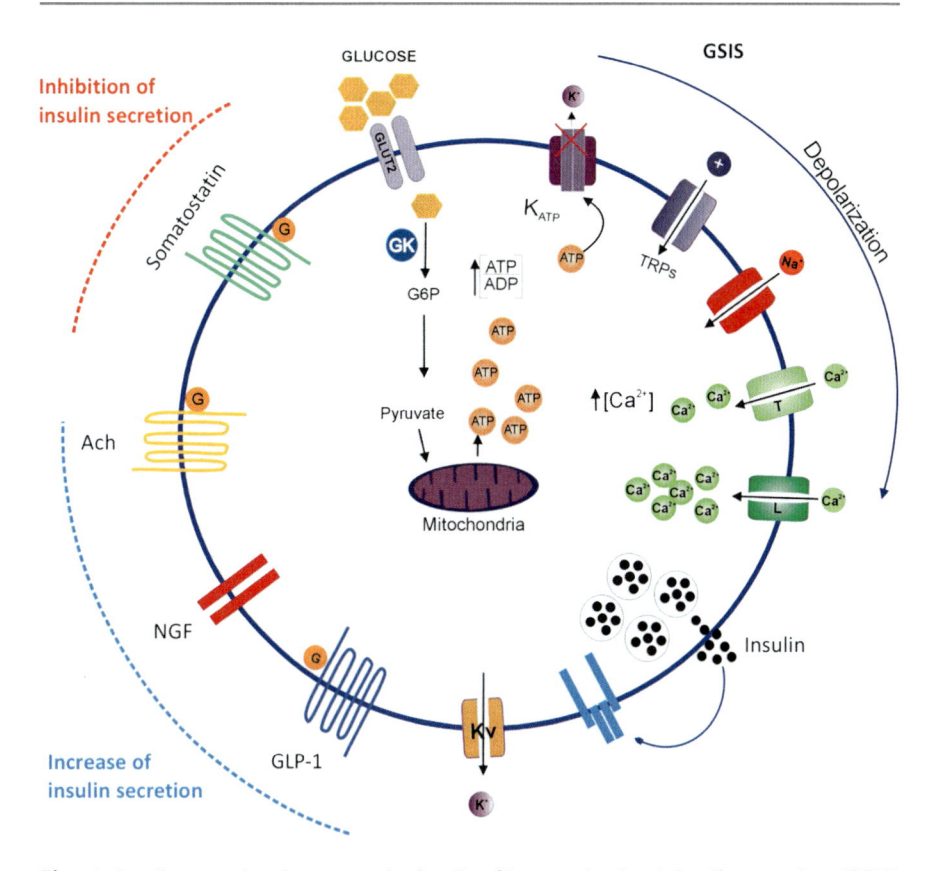

Fig. 1 Insulin secretion in pancreatic β-cells. Glucose-stimulated insulin secretion (GSIS) depends on the activation of different types of ionic channels placed on the cell membrane. Insulin secretion is regulated by other hormones, neurotransmitters, growth factors and incretins

When plasma glucose rises, the rate of metabolism by the cell is stimulated. Therefore, the intracellular ATP/ADP ratio increases, which leads to the closure of K_{ATP} channels and membrane depolarization (see also chapter "▶ ATP-Sensitive Potassium Channels in Health and Disease"). This step is also dependent on the activity of TRP channels, which are nonselective cationic channels. Membrane depolarization results in the activation of voltage-dependent sodium and low-threshold calcium channels, which accentuate it. Consequently, high-threshold voltage-dependent calcium channels (mainly L-type channels) are activated and action potentials start from a plateau potential. The Ca^{2+} influx through these channels increases the intracellular Ca^{2+} concentration, which triggers insulin exocytosis from secretory granules (see also chapter "▶ Exocytosis in Islet β-Cells"; Hiriart and Aguilar-Bryan 2008).

Delayed rectifier voltage-dependent K^+ channels and high-conductance calcium- dependent K^+ channels (K_{Ca}) repolarize the membrane to the resting potential (Drews et al. 2010). The process of ATP synthesis, closure of K_{ATP}

channels, and raising of intracellular calcium levels is repeated in subsequent cycles (see also chapter "▶ Electrical, Calcium, and Metabolic Oscillations in PancreaticIslets").

The membrane potential of β cell begins to oscillate at a suprathreshold glucose concentration. It is known that lasting of both, the depolarized burst phases with superimposed spikes and the silent hyperpolarized inter-burst phases are glucose dependent. As the glucose concentration increases, burst phases are longer and inter-burst phases are shortened until continuous activity is achieved at a glucose concentration above 25 mM (Drews et al. 2010).

It is well established that triggering Ca^{2+} signals is essential for both first and second phases of GSIS. However, glucose also induces insulin secretion by an amplifying pathway, sensitizing the exocytotic machinery to cytosolic calcium concentration changes, independently of K_{ATP} and membrane potential. This pathway is strongly dependent on metabolism and is only mimicked by other metabolizable secretagogues (see also chapter "▶ Electrical, Calcium, and Metabolic Oscillations in Pancreatic Islets"; Henquin 2011).

It is discussed in the literature that amplifying signals produced by glucose are only involved in the second phase of insulin secretion. However, dynamic studies of insulin secretion in mouse islets with various glucose concentrations point to a role also in the first phase that could account for no less than 50 % of the response. The rate of insulin secretion relies on the intracellular Ca^{2+} signals (see also chapter "▶ Calcium Signaling in the Islets"), while the magnitude of insulin pulses increases with the glucose concentration, reflecting the participation of the amplifying pathway (Henquin 2011).

Furthermore, the distribution of insulin functional pools must be well controlled to guarantee an optimal physiological response to glucose (Rorsman and Renstrom 2003). The initial exocytotic response (see also chapter "▶ Exocytosis in Islet β-Cells;") may result from stimulus that increases intracellular Ca^{2+}. However, sustained secretion phase can only be elicited by metabolizable secretagogues as glucose, which also promotes the translocation of granules to the membrane, from a reservoir pool, followed by docking and priming (Ohara-Imaizumi et al. 2007). Proteins dependent on GTP/GDP cycling play an important role in vesicle recruitment in this second phase (Henquin 2000; Straub et al. 2004). The fusion of an exocytotic vesicle with the plasma membrane is mediated by SNARE proteins (Soluble NSF attachment protein receptor).

The SNARE core complex includes the plasma membrane proteins syntaxin and SNAP-23/25 and the vesicular protein VAMP (synaptobrevin). This complex is regulated by the Sec1/Munc18 (SM) family of proteins, which selectively bind with high affinity to their syntaxin isoforms (Jewell et al. 2010). Other proteins, such as synaptotagmin III, V, VII, VIII, and IX, act as Ca^{2+} sensors for exocytosis in β cells. The interaction of various SNARE proteins with the voltage-dependent calcium channels (VDCCs) mediates a tight coupling of Ca^{2+} entry and exocytosis of insulin vesicles (reviewed by MacDonald et al. 2005). In summary, glucose-sensing mechanisms are important for amplifying and maintaining insulin secretion.

Amino Acid-Stimulated Insulin Secretion (AASIS)

The role of amino acids in insulin secretion has long been recognized, but their mechanisms of action are still unclear (see also chapter "▶ (Dys)Regulation of Insulin Secretion by Macronutrients"). Individually, amino acids, such as alanine, asparagine, glycine, glutamate, phenylalanine, and tryptophan, or mixtures of all amino acids are poor secretagogues, and a relatively small number of them promote or synergistically enhance GSIS from β cells. Leucine is an exception, since its stimulatory actions in perfused rat pancreas are eradicated in the presence of glucose (Zhang and Li 2013).

Amino acids that are cotransported with Na^+ depolarize the membrane, causing Ca^{2+} influx and triggering insulin secretion (Newsholme et al. 2007). Alanine and leucine also increase ATP production, promoting the K_{ATP} channel closure.

In normal humans, amino acids also potentiate the GSIS, which is the basis for the application of combinatory therapies of both secretagogues in order to achieve a maximal insulin secretory responsiveness (Li et al. 2012). Furthermore, an important role for amino acids in insulin secretion comes from recent discoveries related to clinical disorders associated with amino acid-sensitive hypoglycemia in children with congenital hyperinsulinism (Zhang and Li 2013).

Glutamine exerts amplifying effects on GSIS, although the precise mechanism is still unknown. Studies in SUR1-KO mice suggested that its effect is mediated by the cAMP-dependent pathways, analogous to the GLP-1 receptor (Zhang and Li 2013).

During GSIS the intracellular levels of many amino acids change, probably via transamination reactions between some amino acids and the intermediaries of the tricarboxylic acid cycle. For example, the intracellular aspartate is reduced, while alanine increases (Li et al. 2012).

Nonesterified Fatty Acids (NEFAs)

It has been demonstrated that normal blood levels of NEFAs are necessary to maintain GSIS in β cells (see also chapter "▶ (Dys)Regulation of Insulin Secretion by Macronutrients"; Stein et al. 1997). The effects of NEFAs are time and concentration dependent. A transient elevation in the NEFA levels potentiates GSIS, while a long-term increase impairs this process (reviewed by Graciano et al. 2011; Nolan et al. 2006).

Several mechanisms had been proposed to explain the physiological role of NEFAs in GSIS. During fasting state, circulating NEFAs are oxidized, increasing ATP production, β-cell depolarization, and insulin secretion. On the fed state, β-cell NEFAs, associated to the glucose metabolism, regulate the mitochondrial anaplerotic and cataplerotic pathways and thus insulin secretion. Moreover, when glucose levels rise, NEFAs oxidation decreases, leading to long-chain acyl-CoA accumulation which facilitates the fusion of insulin vesicles to the β-cell membrane. In addition, long-chain fatty acids may be esterified increasing intracellular levels of complex lipids such as triglycerides, DAG, or phospholipids. GPR40 is a G protein-coupled receptor activated by NEFAs that amplifies insulin exocytosis by DAG, PKC, IP3, increased calcium influx, and reduced outward voltage-gated potassium channel conductance. Finally, β-oxidation of NEFAs generates electrons that activate the respiratory chain and promotes mitochondrial production of ROS through complex I (NADH dehydrogenase) and complex III (cytochrome bc1), which are the major producing sites of superoxide, also enhancing insulin secretion (reviewed by Graciano et al. 2011; Nolan et al. 2006).

It has been demonstrated that long-term high blood levels of NEFAs impair insulin secretion. In MS high intracellular levels of NEFAs increase the mitochondrial transmembrane potential and shift the redox state of complex III components toward reduced values favoring superoxide production. In addition, NEFAs may inhibit electron transport by interacting with components of the respiratory chain and promote superoxide production by transferring electrons to complex I. Furthermore, peroxisome metabolism of long-chain fatty acids leads to the generation of hydrogen peroxide due to the O_2 participation as the electron acceptor (reviewed by Graciano et al. 2011; Nolan et al. 2006).

Pancreatic Hormones

Glucagon

Glucagon is secreted by pancreatic α-cells (see also chapter "▶ Physiological and Pathophysiological Control of Glucagon Secretion by Pancreatic α-Cells"), which comprise nearly 15 % of islet volume (reviewed by Taborsky 2010). In rodents, the α cells are localized in the periphery of the islet together with delta cells, while the β cells are concentrated at the core of the islet (Cabrera et al. 2006). Blood flow direction inside the islet impacts the regulation of the different intraislet hormones. Two blood flow pattern have been proposed in mice: (a) the most accepted in the literature, in which the direction is inner to outer, allowing β-cell hormones to reach and regulate α and delta cells physiology, and (b) top to bottom, in which the blood perfuses from one side to the other (Nyman et al. 2008). The endocrine cells in human islets have a distinct distribution, where cells appear to be randomly distributed within the islet. In human islets, 70 % of β cells are in contact with α and delta cells (Gromada et al. 2007). According to cellular arrangement and blood flow, it has been proposed that glucagon can increase insulin secretion by a paracrine mechanism and insulin and gamma-aminobutyric acid (GABA) secreted by β cells may decrease glucagon secretion (Bagger et al. 2011).

Glucagon receptor (GlucR) is expressed in the α and β cells. GlucRs are proteins with seven transmembrane domains, coupled to G proteins (Ahren 2009), adenylyl cyclase (AC), cAMP production, and protein kinase A (PKA), and finally promote calcium-dependent exocytosis (reviewed by Gromada et al. 2007; Koh et al. 2012).

Somatostatin

Somatostatin is secreted by pancreatic δ-cells and inhibits insulin and glucagon secretion. There are five types of somatostatin receptors (SSTRs), which are tissue specific. SSTRs are G protein-coupled receptors with seven transmembrane domains. Human and rat β-cells express type 1 and 5 SSTRs, while human, mouse, and rat α-cells express type 2 SSTRs (Koh et al. 2012).

The inhibitory effectors of somatostatin signaling are potassium channels, promoting membrane potential hyperpolarization and decreasing calcium channel activity, which mediates exocytosis. Moreover somatostatin may inhibit adenylyl cyclase activity resulting in a reduction of intracellular cAMP levels and a decreased activity of PKA. Furthermore, it has been proposed that in high glucose concentrations, somatostatin may modulate insulin secretion by interacting with cholinergic receptors (Youos 2011).

Autocrine Effects of Insulin

β-cell secretion is also regulated by hormones, neurotransmitters, and other proteins that are secreted by themselves or their neighbors causing autocrine and paracrine effects respectively. The autocrine effects of insulin on β cells through insulin autoreceptors are a matter of controversy. However, it has been shown that autocrine regulation of insulin release plays an important role in apoptosis, proliferation, and gene transcription (see also chapter "▶ IGF-1 and Insulin Receptor Signalling in Insulin-Secreting Cells: From Function to Survival"; reviewed by (Leibiger et al. 2008; Rhodes et al. 2013).

Amylin

Amylin, also known as islet amyloid polypeptide protein (IAPP), is another small protein, member of the calcitonin family peptides that is synthesized and co-secreted with insulin by β cells and delta cells. In the literature, the possible role of amylin in insulin secretion is controversial because some articles affirm that it increases insulin secretion, while others describe no effects or significant inhibition of GSIS by physiological doses of this hormone. Other extrapancreatic effects of amylin described throughout the literature are suppression of glucagon secretion, inhibition of glucose release from the liver, decreasing gastric emptying, and stimulating satiety (reviewed by Gebre-Medhin et al. 2000; Pillay and Govender 2013).

Human IAPP aggregates and forms toxic fibrils in T2DM; it is considered that the latter may cause islet dysfunction and apoptosis of β cells (Lorenzo et al. 1994). Rodents under experimental diabetes express more amylin than insulin; however the structure of amylin in these animals is different, and it does not form fibrils or plaques (Chakraborty et al. 2013).

Amylin receptors are formed by a complex of calcitonin receptor and a receptor-modifying protein (RAMP). The affinity of amylin to the receptor is higher in the complex form. Moreover, different subtypes of receptors are formed by splice variants of the calcitonin receptor and different RAMPs, although these aspects have not been fully characterized in β cells (Abedini and Schmidt 2013).

When large amounts of insulin are secreted, there is also an elevated amylin secretion, which eventually could form aggregates. Important information on the impact of amylin aggregation come from transgenic animals for human IAPP, for example, it has been observed that high carbohydrates or fat diets promote amyloid fibril formation in human-amylin transgenic mice (Pillay and Govender 2013; Westermark et al. 2011).

The complete role of amylin in the pathogenesis of T2DM is not yet resolved, but considering its possible peripheral roles, it is now used as a complementary treatment in T1DM.

Gamma-aminobutyric Acid (GABA)

Pancreatic β cells produce and secrete the neurotransmitter GABA. It could be present in the cytoplasm, contained in synaptic-like microvesicles, or stored in a subpopulation of insulin-containing granules. In addition, β cells express both $GABA_A$ and $GABA_B$ receptors, indicating a probable autocrine role of the neurotransmitter (Reetz et al. 1991). It has been demonstrated that glucose induces GABA secretion in a Ca^{2+}-dependent manner (Smismans et al. 1997). However, some evidence suggests that GABA could also be secreted through a non-vesicular and glucose-independent pathway (Braun et al. 2004). The G-protein-coupled $GABA_B$ receptor has been involved in the inhibition of insulin secretion by decreasing the exocytotic process probably through the modulation of voltage-gated calcium and G-coupled inwardly-rectifying potassium channels. Finally, the activation of the ligand-gated Cl^- channel $GABA_A$ receptor may lead to membrane depolarization thus promoting insulin secretion (Braun et al. 2010; Dong et al. 2006).

Nerve Growth Factor

Nerve growth factor (NGF), initially identified as a soluble factor enhancing growth and differentiation of sympathetic ganglia, exerts many other effects on the physiology of neuronal and nonneuronal tissues (Aloe 2011). Pancreatic β cells secrete NGF (Rosenbaum et al. 1998) and also express both high- (TrkA) and low-affinity (p75) receptors for this neurotrophin (Polak et al. 1993). Some of the effects of this neurotrophin on insulin-secreting cells include morphological changes (i.e., neurite-like processes), improved survival, and increased excitability (Hiriart et al. 2001).

The regulatory role of NGF on ion channels that participate in insulin secretion has been extensively studied by our research group. Today, it is well known that a long-term treatment with NGF increases the synthesis of voltage-sensitive Na^+ channels and consequently insulin secretion (Vidaltamayo et al. 2002). The currents through L-type Ca^{2+} channels are also upregulated by NGF due to an increase in the density of ion channels at the plasma membrane and their activity through a direct modulation (Rosenbaum et al. 2001, 2002). The latter supports a role of NGF as an autocrine modulator of β-cell physiology. Nerve growth factor improves the responsiveness of immature β-cells from neonate rats increasing the translocation of voltage-gated Ca^{2+} channels to the plasma membrane (Navarro-Tableros et al. 2007). Furthermore, it has been suggested that NGF guides the vasculature and sympathetic innervation of pancreatic islets during development (Cabrera-Vasquez et al. 2009). Moreover, NGF deprivation causes a reduction in insulin levels and secretion, as well as an exacerbated apoptosis, which are more prominent in β cells from hyperglycemic rats (Gezginci-Oktayoglu et al. 2012). Indeed, NGF shapes the endocrine physiopathology from islet architecture to the ultrastructure of single β cells.

Interestingly, β cells increase transcription and secretion of NGF after streptozotocin injury, probably as a protective strategy that counteracts apoptosis (Larrieta et al. 2006). Circulating NGF is also associated with diabetes mellitus and metabolic syndrome. It is known that NGF levels are reduced in patients with diabetic neuropathy and correlated with an impairment of motor nerve conduction velocity (Faradji and Sotelo 1990).

NGF participates in the etiology of metabolic syndrome and neuroendocrine disorders (Chaldakov et al. 2010; Chaldakov et al. 2009). In fact, during early stages of MS, hypersecretion of NGF causes alterations of neurotransmitters levels in the brain and leads to vegetodystonia. In the early stages, the vagal tone increases, which also enhances insulin secretion. Moreover, NGF also causes an increased brain expression of the orexigenic neuropeptide Y, as well as the hyperactivation of the hypothalamic–pituitary–adrenal axis that contributes to obesity. The resulting long-term hormonal imbalance and the exhaustion of compensatory mechanisms finally establish hypo-neurotrophinemia in the generalized MS (Hristova and Aloe 2006).

Incretins: A Crosstalk Between the gut and the Endocrine Pancreas

Some intestinal hormones cause trophic effects in β cells and link insulin secretion to nutrient ingestion. Glucagon-like peptide-1 (GLP-1) and the glucose insulinotropic polypeptide (GIP) are named incretins because they account for the higher insulin secretion after an oral challenge of glucose, compared with an intravenous bolus of glucose. Moreover, both incretins decrease apoptosis and potentiate insulin release and β-cell proliferation (Phillips and Prins 2011).

The way GLP-1 affects insulin release involves its binding to a membrane receptor and PKA activation, which in turn increases β-cell excitability and Ca^{2+} entry, in part through cationic nonselective currents (Togashi et al. 2006). Interestingly, application of the GLP-1 mimetic exendin-4 prevents β-cell loss and augments NGF and p75 expression in islets from streptozotocin-treated rodents (Gezginci-Oktayoglu and Bolkent 2009). GLP-1 could also induce cardio- and neuroprotective effects as well as weight loss, which contribute to the beneficial effects in therapeutics of GLP-1 analogs and also the inhibitors of its degrading enzyme dipeptidylpeptidase-4 (DPP-4) (Phillips and Prins 2011).

It is worth to mention that overexpression of GIP in a transgenic mouse model ameliorates diet-induced obesity while improving glucose tolerance and insulin sensitivity (Kim et al. 2012). Moreover, administration of a GLP-1-derived nonapeptide showed similar effects on mice fed with a high-fat diet, indicating its potential in the treatment of MS (Tomas et al. 2011). Recently, it has been suggested that GLP-1 could physiologically modulate insulin secretion by activating the pancreatic vagal innervation, because vagotomized mice show less insulin secretion in response to intraportal administration of GLP-1 with respect to a control group of animals (Nishizawa et al. 2013).

Controversial results are reported regarding the mechanisms of incretin deregulation in diabetes mellitus. Apparently, an impairment of β-cell response to incretins is more likely than a reduced secretion of these hormones (Ahren 2012). It has been shown that exenatide, a GLP-1 receptor agonist, decreases the postprandial glucose and promotes insulin secretion with a higher efficacy than sitagliptin (a DPP-4 inhibitor) in patients with T2DM (DeFronzo et al. 2008). Interestingly, patients with MS manifested a higher improvement of their metabolic status when treated with exenatide versus those treated with DPP-4 inhibitors, overcoming the GLP-1 resistance (Fadini et al. 2011).

Neurotransmitters

The pancreas is richly innervated by both parasympathetic and sympathetic branches of the autonomic nervous system. These fibers do not form classic synapses with endocrine cells but have release sites near the islet cells (reviewed by Osundiji and Evans 2013). The overall effect of parasympathetic stimulation is an increase in pancreatic insulin release, whereas the net effect of sympathetic nerve stimulation is to lower the plasma insulin concentration (Rodriguez-Diaz and Caicedo 2013).

Sympathetic nerve stimulation is carried out by adrenaline and noradrenaline. The presence of both α2- and β2-adrenergic receptors (AR) have been reported on pancreatic islet cells (Ullrich and Wollheim 1985), and the selective activation of α2- and β-receptors results in inhibition and stimulation of insulin secretion, respectively.

The predominant physiological effect on insulin secretion of the sympathetic system is inhibitory, due to activation of potassium channels, which hyperpolarizes the membrane; another target is the inhibition of adenylyl cyclases activity, preventing the stimulation of insulin release by cAMP, or to the actions of the G protein β-γ subunits, blocking the interaction of the calcium sensor synaptotagmin with the proteins involved in exocytosis (SNAREs) (Straub and Sharp 2012).

The blockade of α2-adrenoceptors with specific antagonists (i.e., phentolamine) (Ahren and Lundquist 1981), silencing of receptor expression, or blockade of downstream effectors reverses this effect. Mice lacking α2A- or α2C-AR show hyperinsulinemia, reduced blood glucose levels, and improved glucose tolerance (Ruohonen et al. 2012). A recent study also supported the role of the α2A-AR in the regulation of β-cell function and glucose homeostasis. In agreement with the above, overexpression of α2-AR in diabetic rats causes impaired insulin-granule membrane docking and less GSIS in β cells (Rosengren et al. 2010).

On the other hand, acetylcholine (ACh) release by parasympathetic innervation stimulates insulin secretion by activating muscarinic receptors (AChRs), linked to Gq proteins signaling pathways that increase the intracellular calcium concentration and activate different PKC isoforms (Gautam et al. 2006). Studies in KO mice showed that M3 receptor subtype is responsible for regulating insulin release (de Azua et al. 2012). Muscarinic M3 receptor in β cells also triggers PKD1 activation and induces insulin secretion (Sumara et al. 2009).

In vivo studies in M3 AChRs KO mice show impaired glucose tolerance and modest insulin release after glucose administration, while transgenic mice with β-cell M3 AChRs overexpression show the opposite metabolic phenotype (Rodriguez-Diaz et al. 2011).

Insulin Functions and Signaling

Insulin increases glucose uptake in adipose tissue and muscle and also inhibits hepatic glucose production. In addition, it stimulates cell growth and differentiation and promotes storage of nutrients by enhancing lipogenesis, glycogenesis, glycolysis, and protein synthesis and by inhibiting lipolysis, gluconeogenesis, glycogenolysis, and protein breakdown. Insulin resistance results in important deregulation of glucose and lipid metabolism (Saltiel and Kahn 2001).

Normal insulin signaling and effects have been extensively reviewed elsewhere (Fig. 2; Saltiel and Kahn 2001; Cheng et al. 2010; Leavens and Birnbaum 2011; Taniguchi et al. 2006; Wang and Jin 2009; White 2003). Briefly, the pathway includes the insulin receptor (IR), which is a prototype of integral membrane proteins with tyrosine kinase (TK) activity, and is formed by two α-β dimers. Insulin binds to the extracellular domain of the α subunit of the receptor and produces a conformational change that promotes the β subunit TK activity. This induces the IR transphosphorylation and phosphorylation of other IR-recruited proteins. However, the binding properties depend on insulin receptor structure, which may vary due to tissue-specific alternative splicing and alternative posttranslational assembly with IGF-1 receptor (White 2003).

Once IR is activated, several proteins are recruited including insulin receptor substrate (IRS) proteins and scaffold proteins such as SHC, CBL, APS, SH2B, GABs, and DOCKs. It has been suggested that IRS proteins mediate most of the intracellular insulin signaling. These proteins can couple to IR and other activated receptors, such as growth hormone receptor, integrins, and some interleukin receptors, through an NH2-terminal pleckstrin homology (PH) domain and a phosphotyrosine-binding (PTB) domain. In addition, IR and IRSs contain other tyrosine and serine/threonine phosphorylation sites capable of binding to numerous effectors or adapter proteins (White 2003). Their phosphorylation could modify the TK activity and interaction with other proteins, thus impairing downstream signaling. IR serine phosphorylation is observed during hyperinsulinemia, resulting from the activation of PKC or AMPK (Vollenweider 2003).

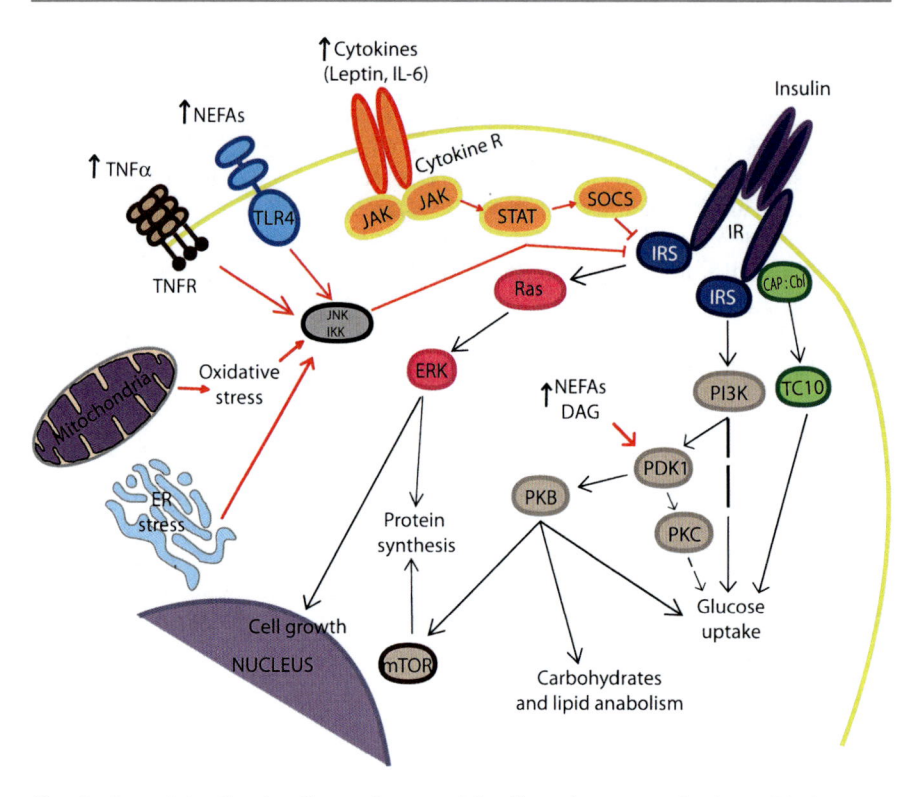

Fig. 2 General insulin signaling pathway and insulin resistance mechanisms. *Black arrows* denote activation processes within insulin signaling pathway. *Red arrows* denote activation processes in other signaling pathways. *Red blocked arrows* represent inhibition processes impairing insulin signaling. *Discontinuous lines* represent activation of intermediate mediators

Phosphorylation of IRS proteins activates three main intracellular downstream signaling pathways. First, the PI3K-PKC-phosphoinositide-dependent kinase (PDK) and PKB/Akt pathway is mainly involved in the metabolic actions of insulin. This cascade modulates glucose-related metabolic enzymes through glycogen synthase kinase 3 (GSK3) and lipid metabolic enzymes, regulating mammalian target of rapamycin (mTOR)-related protein synthesis, phosphorylating the forkhead box O1 (FOXO1), and activating sterol regulatory element-binding protein (SREBP) transcription factors, which regulate metabolism and partially mediate the translocation of GLUT-4-containing vesicles to the cell membrane (Taniguchi et al. 2006). In addition, the Ras-mitogen-activated protein kinase (MAPK) pathway regulates the expression of cell growth- and differentiation-related genes (Taniguchi et al. 2006). Finally, the proto-oncogene Cbl and the Cbl-associated protein (CAP) are released from the IR complex after being activated. They accumulate in the caveolar membrane domains due to an interaction with the protein flotillin. These discrete plasma membrane regions are enriched in lipid-modified signaling proteins, glycophosphatidylinositol-anchored proteins, glycolipids, sphingolipids, and cholesterol. The Cbl/CAP-activated complex can recruit SH2-containing proteins, such as CRK II and the guanine nucleotide exchange factor C3G. Furthermore, it has been suggested that this complex activates the rho family protein TC10, which partially mediates the translocation of GLUT-4-containing vesicles to the membrane (Vollenweider 2003).

The extent of tyrosine phosphorylation of the IR and downstream proteins reflects a balance between autophosphorylation, in addition to the TK activity of the IR, and their dephosphorylation mediated by protein tyrosine phosphatases (PTPs). The PTB1B is a well studied phosphatase that interacts with the phosphorylated activated IR (Vollenweider 2003; Asante-Appiah and Kennedy 2003), directly dephosphorylating it and finishing the intracellular insulin signaling. Moreover, PTP1B KO mice are resistant to diet-induced obesity, probably due to enhanced hepatic and muscular insulin sensitivity (Asante-Appiah and Kennedy 2003).

The phosphatase SHP-2/PTP1D also binds to ligand-activated receptors with tyrosine kinase activity through their SH2 domains (Vollenweider 2003) and acts as a negative modulator of insulin signaling through dephosphorylation of IR and IRSs (Asante-Appiah and Kennedy 2003).

Metabolic Syndrome and β Cells

The diagnostic criteria for MS are summarized in Table 1, according to the World Health Organization (WHO) and the Adult Treatment Program III (ATPIII) of the National Cholesterol Education Program (NCEP). The presence of at least three factors of Table 1 is required for a positive MS diagnosis. NCEP criteria do not include inflammatory or homeostatic variables. However, they are used because of their clinical feasibility (Expert Panel on Detection, Evaluation, and Treatment of High Blood Cholesterol in Adults 2002). Although the pathogenesis of MS is not completely clear, it is known that a genetic predisposition, coupled to a sedentary lifestyle and a high-energy diet, contributes to the development of this syndrome (Rask-Madsen and Kahn 2012).

Insulin resistance is defined as a defect in insulin ability to decrease plasma glucose levels, due in part to an impaired signaling of this hormone in sensitive tissues. The impairment of insulin-dependent biological actions on glucose and lipid metabolism results in fasting and postprandial hyperinsulinemia and increased glucose levels.

Adipose Tissue and Obesity

Obesity is a risk factor to develop insulin resistance, T2DM, and cardiovascular disease. It is well established that hypertrophy and hyperplasia of visceral adipocytes are components of central obesity. Therefore, abdominal circumference,

Table 1 Criteria to diagnose MS according to WHO and NCEP ATP III

Risk factor	WHO	NCEP ATP III
Abdominal girth	Waist to hip ratio > 0.9 in men, > 0.85 in women and/or BMI >30 kg/m^2	> 35 in. (88 cm) in women >40 in. (102 cm) in men
Triglycerides	>1.7mmol/l	>150 mg/dl (1.69 mmol/l) or drug elevated triglycerides
HDL cholesterol	Men < 0.09 mmol/l Women <1.0 mmol/l	Men <40 mg/dl; women <50 mg/dl or drug treatment for reduced HDL-C
Blood pressure	>140/90 mmHg	>130/85 mmHg or drug treatment for elevated tryglicerides
Fasting glucose		>110 mg/dl or drug treatment for elevated glucose
Microalbuminuria	Urinary albumin excretion rate >20 μg/min or Albumin: creatine ratio >30 mg/g	

rather than body mass index (BMI), has become the most important parameter to identify central obesity (Alberti et al. 2009).

Adipose tissue is not only metabolically active and plays an important role in glucose and lipid metabolism, sequestering fat in triglycerides stores, which attenuates the effects of NEFAs, but also secretes hormones called adipokines that could be related to the pathogenesis of various pathological processes including the metabolic syndrome (Hotamisligil 2006).

Interestingly, nearly 25–30 % of morbidly obese humans (BMI > 35 kg/m^2) are insulin sensitive, and the insulin-resistant individuals show a decrease in AMPK activity and increased oxidative stress compared to the insulin-sensitive counterparts (Xu et al. 2012).

The metabolic effects from the abdominal fat are more adverse than those from the fat tissue in other locations, being associated with ectopic fat accumulation in the liver and nonalcoholic fatty liver disease (NAFLD). However, patients with lipodystrophy may also develop NAFLD, independent of obesity, indicating that it is adipose tissue function per se which is important for developing this pathology. Fatty acid fluxes from adipose tissue to the liver increase the availability of NEFAs, particularly in sedentary subjects, which leads to DAG synthesis that contributes to insulin resistance in the liver (Byrne 2012).

Metabolic Effects of Cytokines

Adipose tissue as an endocrine organ is involved in the whole-body energetic balance, by producing and secreting adipokines, such as leptin, adiponectin, resistin, interleukin-6 (IL-6), retinol binding protein-4 (RBP4), NGF, and brain-derived neurotrophic factor (BDNF). A mixed cytokine excess in MS leads to inflammation and disrupts β-cell functions (see also chapter "▶ Inflammatory Pathways Linked to β Cell Demise in Diabetes").

Leptin is mainly secreted by adipocytes, and its plasma levels correlate with the total body fat mass (Maffei et al. 1995). Leptin inhibits insulin action in white adipose tissue (WAT) and decreases insulin synthesis and secretion in β cells (Denroche et al. 2012). The classical animal models of leptin deficiency exhibit obesity, hyperglycemia, and hyperinsulinemia. The most commonly used are the leptin-deficient *ob/ob* (see also chapter "▶ β-Cell Function in Obese-Hyperglycemic Mice") and the leptin receptor-deficient db/db mice, as well as some rat models (fa/fa, Zucker, etc., reviewed by Larqué et al. 2011). Another mouse model with a disrupted signaling domain of leptin receptor in β cells and hypothalamus is characterized by overweight, insulin resistance, impaired β-cell functions, and glucose intolerance (Covey et al. 2006). Recently, it was observed that a variant of this model, deficient in leptin signaling in β cells by a Cre-loxP recombination, shows insulin hypersecretion under basal conditions and a deficient secretion in response to glucose (Tuduri et al. 2013).

Adiponectin is exclusively secreted by adipocytes and its plasma levels show a negative correlation with the risk of obesity, insulin resistance, T2DM, and cardiovascular disease (Harwood 2012). Reports about the direct effects of adiponectin on insulin secretion are variable and inconsistent between animal models and humans and require further clarification (Lee et al. 2011).

Circulating levels of tumor necrosis factor α (TNFα) are also increased in nondiabetic obese and individuals with T2DM (Bays et al. 2004), but the correlation between insulin resistance and plasma TNFα levels is weak in both cases (Hotamisligil et al. 1995). The primary source of TNFα is macrophages from the stromal vascular fraction of adipose tissue; however, differentiated white adipocytes also produce this cytokine.

Little is known about the direct effect of TNFα in the islets. However, plasma levels of insulin and leptin are decreased in TNFα$^{-/-}$ mice (Romanatto et al. 2009) and their islets are pronouncedly infiltrated by inflammatory cells. Paradoxically, insulitis and diabetes are not well correlated, as observed in the NOD/WEHI strain of NOD mice, which shows insulitis without diabetes. These observations are consistent with the idea that insulitis may be necessary but not sufficient to induce diabetes (Charlton et al. 1989). Furthermore, TNFα induces the synthesis of amylin in β cells, without affecting insulin expression in the MIN6 cell line and islets from mice (Cai et al. 2011).

Interleukin-6 (IL-6) is a cytokine with pleiotropic biological effects that is secreted by adipocytes and the stromal fraction of adipose tissue. IL-6 expression has been correlated with insulitis and β-cell destruction in NOD mice. However, this is not the case for human islets. In fact, it has been observed in other models that IL-6 may protect β-cells from other inflammatory insults; at least, it is clear that it is not cytotoxic without other cytokines. Interestingly, several studies indicate that IL-6 inhibits insulin secretion in rodent β cells; in contrast, human islets are not affected by this cytokine (Kristiansen and Mandrup-Poulsen 2005). We have also observed that IL-6 plasma levels are increased in Wistar rats with MS, developed by sucrose ingestion in drinking water (unpublished observation).

Brain-derived neurotrophic factor (BDNF) is a member of the neurotrophin (NT) family of factors, which play important roles in the development of both the central and peripheral nervous systems. BDNF is produced by adipocytes and may be altered in MS and diabetes mellitus (Chaldakov 2011; Sornelli et al. 2009). Evidence in humans indicates that BDNF could play an important role in the body weight control and energy homeostasis (Rosas-Vargas et al. 2011). It has been suggested that BDNF plasma levels are decreased in obesity, showing an inverse correlation with body mass index. However, in T2DM, BDNF levels decrease, independently of obesity. Plasma BDNF is also inversely associated with fasting plasma glucose, but not with insulin, suggesting a contribution to glucose metabolism (Krabbe et al. 2007).

Direct effects of BDNF on β cells have not been extensively explored. It was described that chronic BDNF administration in db/db mice prevents exhaustion of β cells by maintaining the cellular organization in the islets and restoring the levels of insulin-secreting granules (Yamanaka et al. 2006).

Hyperinsulinemia and Insulin Resistance

Insulin resistance is a major sign in metabolic syndrome and an important warning for T2DM development, due to dysregulation of glucose and lipid metabolism (Saltiel and Kahn 2001). Mechanisms involved in the insulin-signaling elements that could be affected in insulin resistance include the phosphorylation–dephosphorylation processes, interaction with insulin-signaling negative regulator proteins, interaction with non-insulin-signaling proteins, interaction with lipid metabolites or oxidative stress mediators, proteolytic processes, or susceptibility to transcriptional modulation by other signaling pathways, among others. Moreover, several elements of the insulin-signaling network may play potential crosstalk roles with other signaling pathways.

Interaction of proteins from the insulin pathway with other signaling molecules such as SOCS is suggested as an important mechanism of insulin resistance. TNF-α, IL6, and even the insulin cascade enhance the transcription of SOCS genes. These proteins are suggested to play different roles in insulin resistance, including interaction with IR and competition with IRS proteins, inhibition of Janus kinase involved in insulin signaling, and targeting IRSs to proteasomal degradation (Howard and Flier 2006; Mlinar et al. 2007).

Evidence has showed that ROS, such as hydrogen peroxide, are involved in the normal regulation of cellular processes. However, chronic or increased production of ROS or a reduced capacity for their elimination could lead to impairment of intracellular signaling and result either in inflammation or in insulin resistance. As mentioned before, IKKβ and JNK proteins have been proposed as oxidative stress-sensitive kinases and may play a role in the attenuation of insulin signaling. The resulting effect of activating oxidative stress-sensitive kinases is the serine/threonine phosphorylation of proteins such as IR and IRSs (Evans et al. 2005).

During MS, the β cells are continuously stimulated. In this condition, they secrete high amounts of insulin and eventually become exhausted and incapable to secrete sufficient amounts of hormone to maintain glucose homeostasis, leading to T2DM development. (see also chapter " ▶ The β-Cell in Human Type 2 Diabetes").

The MS has been extensively studied in different animal models. We recently reported that adult male *Wistar* rats, treated for 2 or 6 months with 20 % sucrose in drinking water, develop MS, characterized by obesity, hypertriglyceridemia, hyperinsulinemia, hypertension, and insulin resistance.

After 2 months of treatment, the short-term/early effects of MS on β cells include an increase in the total amount and expression level of the GLUT2 protein (Larqué et al. 2011) and a discrete increment of calcium currents (unpublished observations). We consider that these cellular changes play an important role in hyperinsulinemia.

We also analyzed the long-term effects of MS on β-cell electrical activity after a 6-month treatment (Fig. 3). Electrophysiological studies of K_{ATP} single-channel activity demonstrated that the channel conductance was not modified. However, in isolated patches of membrane, the ATP sensitivity of K_{ATP} channels in MS rats increased with respect to control rats. The change in the Kd for ATP indicates that the channel closes at lower intracellular ATP concentrations in the MS β cells (Velasco et al. 2012).

Furthermore, we found three subpopulations of β cells, according to their calcium currents (Fig. 3). One half of the cells had low current density (we named them MS1), 35 % of the cells showed large currents (MS2), and 15 % of them showed no IBa^{2+} at all. We observed that during the metabolic syndrome, MS2 cells showed an increase (92 %) in the maximum peak of barium current density, which could explain hyperinsulinemia. However, MS1 cells begin to get exhausted, by showing less current and probably less insulin secretion (Velasco et al. 2012).

β-Cell Exhaustion and T2DM

Cytokines in obesity affect islet health and potentially contribute to islet inflammation in T2DM (Fig. 4). β cells show a low antioxidant defense and are more vulnerable to oxidative stress caused by high glucose and NEFAs levels (see also chapter "▶ Role of NADPH Oxidase in β Cell Dysfunction"). This can cause ER

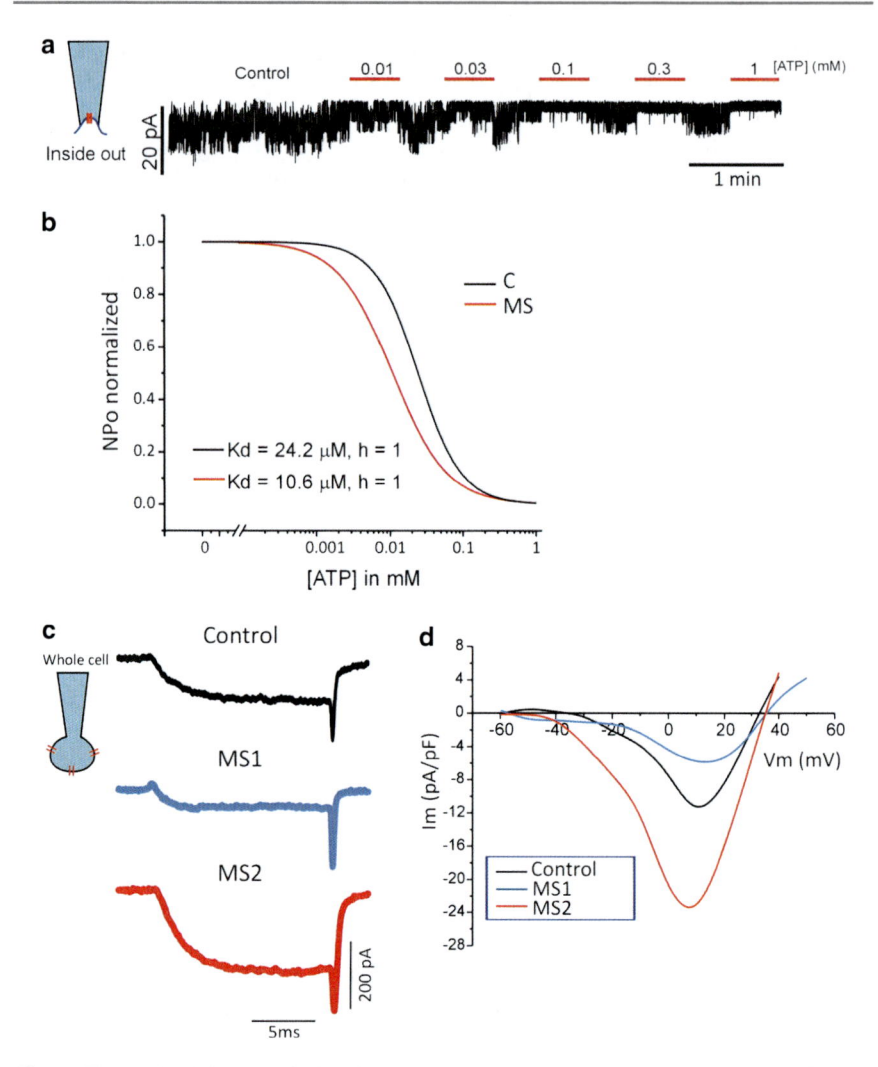

Fig. 3 Electrophysiological changes in β cells during metabolic syndrome. (**a**) Time course of K_{ATP} channels activity in metabolic syndrome registered with inside-out configuration in a cell membrane patch. K_{ATP} channel blockade in a concentration-dependent fashion is observed. (**b**) ATP concentration-response curve. (**c**) Barium currents observed at a +10 mV potential in normal β-cells, and in β-cells during metabolic syndrome. Two β-cell populations, named MS1 and MS2 were classified according to their current amplitude compared to controls. (**d**) Current to voltage relationship obtained from cells in Section C

stress and amplify cell dysfunction (Imai et al. 2013). A decrease in β-cell mass by 60–75 % is observed in T2DM (see also chapters "▶ Mechanisms of Pancreatic β-Cell Apoptosis in Diabetes" and "▶ The β-Cell in Human Type 2 Diabetes"). However, it is difficult to calculate this parameter in vivo. Also, the β-cell mass function of the surviving β cells is also important for the development of T2DM.

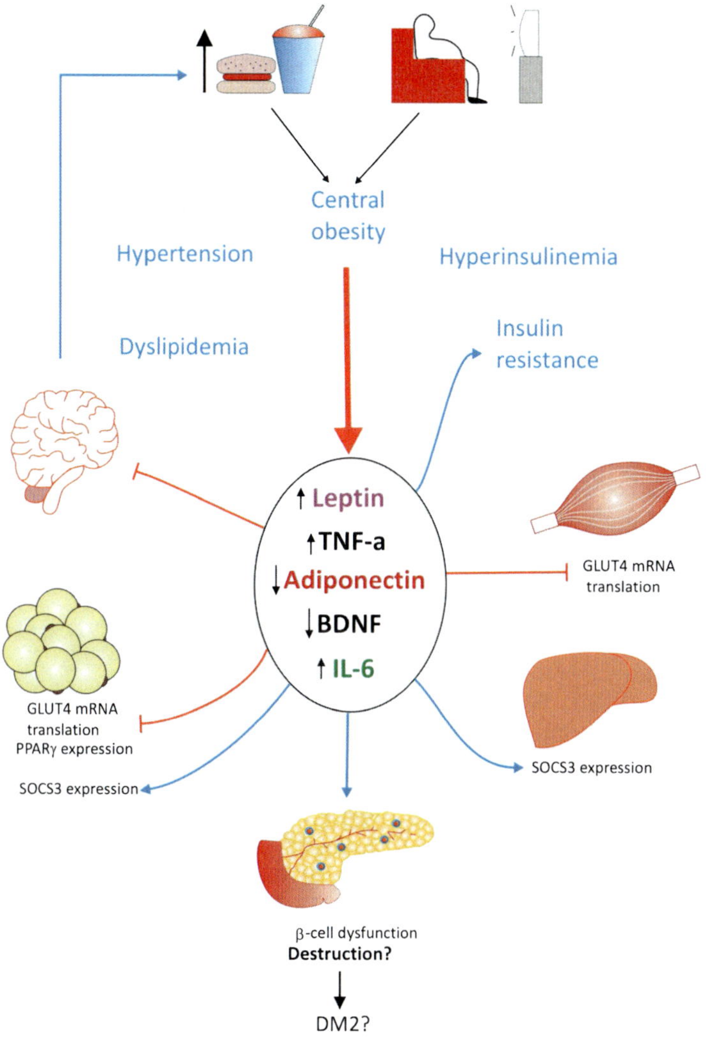

Fig. 4 Cytokines contribution to insulin resistance and β-cell dysfunction in the metabolic syndrome

β-cell exhaustion is not unlikely in a long-term metabolic disease considering the increased levels of ROS production (see also chapter "▶ Role of Mitochondriain β-cell Function and Dysfunction"), circulating pro-inflammatory cytokines, lipids, and hyperglycemic hyperinsulinemia. The oxidative stress is an inherent part of insulin secretion, signaling, and regulation, but β-cell dysfunction could arise in a glycolipotoxic context, along with a dysregulated ROS production as occurs in T2DM (Jezek et al. 2012; Somesh et al. 2013). Indeed, β cells face a challenging environment in MS, which is plenty of stimulating signals that keep their metabolism activated and could cause a Ca^{2+} overload (see also chapter "▶ Calcium Signaling in the Islets").

Both the hyperexcitability and the exacerbation of several signaling pathways simultaneously could cause β-cell failure, because of desensitizing mechanisms and/or proapoptotic processes. Hypoglycemic agents, such as glibenclamide, which faces insulin resistance by increasing insulin secretion, eventually fail to control glucose levels in diabetic patients (Matthews et al. 1998). In vitro, they also impair β-cell performance and survival (Sawada et al. 2008) indicating that a chronic exposure to these agents could lead to β-cell failure or even death. Moreover, several of the signs that MS shares with diabetes could converge into toxic pathways and synergize their deleterious effects, as has been suggested for TRPA1 activation (Diaz-Garcia 2013).

Some new factors have recently been described with therapeutic potential in β-cell decrement in prediabetic and diabetic states. Betatrophin is a hormone secreted by the mouse liver that selectively promotes β-cell mass expansion, which could improve glucose tolerance (Yi et al. 2013). However, more basic research in β cells is needed to fully understand metabolic syndrome and β-cells failure but also how to prevent it.

Conclusions

- Pancreatic β-cell physiology is medullar for glucose homeostasis. β cells are sensitive to metabolic changes in the system; they are also especially vulnerable to ROS damage, and their replacement is low.
- Insulin signaling is important for the metabolism of carbohydrates, lipids, and proteins. When insulin resistance is developed, the nutrient homeostasis is affected.
- MS affects β-cell function even at early stages, and when inflammation and NEFAs levels rise, they promote, among other factors, an increase in ROS. One of the signs of MS is hyperinsulinemia, which is maintained until, depending on individual genetic traits, β cells are no longer capable of secreting enough insulin to maintain glucose levels in normal values and T2DM develops.
- The complete scenario of the individual presentation of MS is not well understood, and the time needed for β-cell exhaustion is not known, under this condition. Moreover, many of the effects of the different cytokines and adipokines on β cells are incompletely understood. Clearly more research is needed to have the whole picture of β-cell response to MS.
- Finally, the time and conditions needed to revert the pathophysiological changes of MS at different levels, including those in β cells, are not known. However, the main challenge for humans in this moment is to change their lifestyles, by having a moderate and healthy food consumption and by increasing their exercise habits. It will be difficult, but it is worth trying.

Cross-References

▶ (Dys)Regulation of Insulin Secretion by Macronutrients
▶ Electrophysiology of Islet Cells
▶ High-Fat Programming of β-Cell Dysfunction
▶ β-Cell function in obese-hyperglycaemic mice [*ob/ob* mice]
▶ The β-Cell in Human Type 2 Diabetes

References

Abedini A, Schmidt AM (2013) Mechanisms of islet amyloidosis toxicity in type 2 diabetes. FEBS Lett 587(8):1119–1127

Ahren B (2009) Islet G protein-coupled receptors as potential targets for treatment of type 2 diabetes. Nat Rev Drug Discov 8(5):369–385

Ahren B (2012) Islet nerves in focus – defining their neurobiological and clinical role. Diabetologia 55(12):3152–3154

Ahren B, Lundquist I (1981) Effects of selective and non-selective β-adrenergic agents on insulin secretion in vivo. Eur J Pharmacol 71(1):93–104

Alberti KG, Eckel RH et al (2009) Harmonizing the metabolic syndrome: a joint interim statement of the International Diabetes Federation Task Force on Epidemiology and Prevention; National Heart, Lung, and Blood Institute; American Heart Association; World Heart Federation; International Atherosclerosis Society; and International Association for the Study of Obesity. Circulation 120(16):1640–1645

Aloe L (2011) Rita Levi-Montalcini and the discovery of NGF, the first nerve cell growth factor. Arch Ital Biol 149(2):175–181

Asante-Appiah E, Kennedy BP (2003) Protein tyrosine phosphatases: the quest for negative regulators of insulin action. Am J Physiol Endocrinol Metab 284(4):E663–E670

Bagger JI, Knop FK et al (2011) Glucagon antagonism as a potential therapeutic target in type 2 diabetes. Diabetes Obes Metab 13(11):965–971

Bays H, Mandarino L et al (2004) Role of the adipocyte, free fatty acids, and ectopic fat in pathogenesis of type 2 diabetes mellitus: peroxisomal proliferator-activated receptor agonists provide a rational therapeutic approach. J Clin Endocrinol Metab 89(2):463–478

Braun M, Ramracheya R et al (2010) Gamma-aminobutyric acid (GABA) is an autocrine excitatory transmitter in human pancreatic β-cells. Diabetes 59(7):1694–1701

Braun M, Wendt A et al (2004) Regulated exocytosis of GABA-containing synaptic-like microvesicles in pancreatic β-cells. J Gen Physiol 123(3):191–204

Byrne CD (2012) Dorothy Hodgkin Lecture 2012: non-alcoholic fatty liver disease, insulin resistance and ectopic fat: a new problem in diabetes management. Diabet Med 29(9):1098–1107

Cabrera O, Berman DM et al (2006) The unique cytoarchitecture of human pancreatic islets has implications for islet cell function. Proc Natl Acad Sci USA 103(7):2334–2339

Cabrera-Vasquez S, Navarro-Tableros V et al (2009) Remodelling sympathetic innervation in rat pancreatic islets ontogeny. BMC Dev Biol 9:34

Cai K, Qi D et al (2011) TNF-α acutely upregulates amylin expression in murine pancreatic β cells. Diabetologia 54(3):617–626

Chakraborty S, Mukherjee B et al (2013) Pinpointing proline substitution to be responsible for the loss of amyloidogenesis in IAPP. Chem Biol Drug Des 82(4):446–452

Chaldakov G (2011) The metabotrophic NGF and BDNF: an emerging concept. Arch Ital Biol 149(2):257–263

Chaldakov GN, Fiore M et al (2010) Neuroadipology: a novel component of neuroendocrinology. Cell Biol Int 34(10):1051–1053

Chaldakov GN, Tonchev AB et al (2009) NGF and BDNF: from nerves to adipose tissue, from neurokines to metabokines. Riv Psichiatr 44(2):79–87

Charlton B, Bacelj A et al (1989) Cyclophosphamide-induced diabetes in NOD/WEHI mice. Evidence for suppression in spontaneous autoimmune diabetes mellitus. Diabetes 38(4):441–447

Cheng Z, Tseng Y et al (2010) Insulin signaling meets mitochondria in metabolism. Trends Endocrinol Metab 21(10):589–598

Covey SD, Wideman RD et al (2006) The pancreatic β cell is a key site for mediating the effects of leptin on glucose homeostasis. Cell Metab 4(4):291–302

DeFronzo RA, Okerson T et al (2008) Effects of exenatide versus sitagliptin on postprandial glucose, insulin and glucagon secretion, gastric emptying, and caloric intake: a randomized, cross-over study. Curr Med Res Opin 24(10):2943–2952

Denroche HC, Huynh FK et al (2012) The role of leptin in glucose homeostasis. Journal of Diabetes Investigation 3(2):115–129

Diaz-Garcia CM (2013) The TRPA1 channel and oral hypoglycemic agents: Is there complicity in β-cell exhaustion? Channels (Austin) 7(6):420–422

Dong H, Kumar M et al (2006) Gamma-aminobutyric acid up- and downregulates insulin secretion from β cells in concert with changes in glucose concentration. Diabetologia 49(4):697–705

Drews G, Krippeit-Drews P et al (2010) Electrophysiology of islet cells. Adv Exp Med Biol 654:115–163

Evans JL, Maddux BA et al (2005) The molecular basis for oxidative stress-induced insulin resistance. Antioxid Redox Signal 7(7–8):1040–1052

Expert Panel on Detection, E., and Treatment of High Blood Cholesterol in Adults (2002) Third report of the National Cholesterol Education Program (NCEP) expert panel on detection, evaluation, and treatment of high blood cholesterol in adults (Adult Treatment Panel III) final report. Circulation 106(25):3143–3421

Fadini GP, de Kreutzenberg SV et al (2011) The metabolic syndrome influences the response to incretin-based therapies. Acta Diabetol 48(3):219–225

Faradji V, Sotelo J (1990) Low serum levels of nerve growth factor in diabetic neuropathy. Acta Neurol Scand 81(5):402–406

Gautam D, Han SJ et al (2006) A critical role for β cell M3 muscarinic acetylcholine receptors in regulating insulin release and blood glucose homeostasis in vivo. Cell Metab 3(6):449–461

Gebre-Medhin S, Olofsson C et al (2000) Islet amyloid polypeptide in the islets of Langerhans: friend or foe? Diabetologia 43(6):687–695

Gezginci-Oktayoglu S, Bolkent S (2009) Exendin-4 exerts its effects through the NGF/p75NTR system in diabetic mouse pancreas. Biochem Cell Biol 87(4):641–651

Gezginci-Oktayoglu S, Karatug A et al (2012) The relation among NGF, EGF and insulin is important for triggering pancreatic β cell apoptosis. Diabetes Metab Res Rev 28(8):654–662

Graciano MF, Valle MM et al (2011) Regulation of insulin secretion and reactive oxygen species production by free fatty acids in pancreatic islets. Islets 3(5):213–223

Gromada J, Franklin I et al (2007) α-cells of the endocrine pancreas: 35 years of research but the enigma remains. Endocr Rev 28(1):84–116

Harwood HJ Jr (2012) The adipocyte as an endocrine organ in the regulation of metabolic homeostasis. Neuropharmacology 63(1):57–75

Henquin JC (2000) Triggering and amplifying pathways of regulation of insulin secretion by glucose. Diabetes 49(11):1751–1760

Henquin JC (2011) The dual control of insulin secretion by glucose involves triggering and amplifying pathways in β-cells. Diabetes Res Clin Pract 93(Suppl 1):S27–S31

Henquin JC, Meissner HP (1984) Significance of ionic fluxes and changes in membrane potential for stimulus-secretion coupling in pancreatic β-cells. Experientia 40(10):1043–1052

Hiriart M, Aguilar-Bryan L (2008) Channel regulation of glucose sensing in the pancreatic β-cell. Am J Physiol Endocrinol Metab 295(6):E1298–E1306

Hiriart M, Vidaltamayo R et al (2001) Nerve and fibroblast growth factors as modulators of pancreatic β cell plasticity and insulin secretion. Isr Med Assoc J 3(2):114–116

Hotamisligil GS (2006) Inflammation and metabolic disorders. Nature 444(7121):860–807

Hotamisligil GS, Arner P et al (1995) Increased adipose tissue expression of tumor necrosis factor-α in human obesity and insulin resistance. J Clin Invest 95(5):2409–2415

Howard JK, Flier JS (2006) Attenuation of leptin and insulin signaling by SOCS proteins. Trends Endocrinol Metab 17(9):365–371

Hristova M, Aloe L (2006) Metabolic syndrome – neurotrophic hypothesis. Med Hypotheses 66(3):545–549

Hunt KJ, Resendez RG et al (2004) National Cholesterol Education Program versus World Health Organization metabolic syndrome in relation to all-cause and cardiovascular mortality in the San Antonio Heart Study. Circulation 110(10):1251–1257

Huypens PR, Huang M et al (2012) Overcoming the spatial barriers of the stimulus secretion cascade in pancreatic β-cells. Islets 4(1):1–9

Imai Y, Dobrian AD et al (2013) Islet inflammation: a unifying target for diabetes treatment? Trends Endocrinol Metab 24(7):351–360

Jensen MV, Joseph JW et al (2008) Metabolic cycling in control of glucose-stimulated insulin secretion. Am J Physiol Endocrinol Metab 295(6):E1287–E1297

Jewell JL, Oh E et al (2010) Exocytosis mechanisms underlying insulin release and glucose uptake: conserved roles for Munc18c and syntaxin 4. Am J Physiol Regul Integr Comp Physiol 298(3):R517–R531

Jezek P, Dlaskova A et al (2012) Redox homeostasis in pancreatic β cells. Oxid Med Cell Longev 2012:932838

Kim SJ, Nian C et al (2012) GIP-overexpressing mice demonstrate reduced diet-induced obesity and steatosis, and improved glucose homeostasis. PLoS One 7(7):e40156

Koh DS, Cho JH et al (2012) Paracrine interactions within islets of Langerhans. J Mol Neurosci 48(2):429–440

Krabbe KS, Nielsen AR et al (2007) Brain-derived neurotrophic factor (BDNF) and type 2 diabetes. Diabetologia 50(2):431–438

Kristiansen OP, Mandrup-Poulsen T (2005) Interleukin-6 and diabetes: the good, the bad, or the indifferent? Diabetes 54(Suppl 2):S114–S124

Larqué C, Velasco M et al (2011) Early endocrine and molecular changes in metabolic syndrome models. IUBMB Life 63(10):831–839

Larrieta ME, Vital P et al (2006) Nerve growth factor increases in pancreatic β cells after streptozotocin-induced damage in rats. Exp Biol Med (Maywood) 231(4):396–402

Leavens KF, Birnbaum MJ (2011) Insulin signaling to hepatic lipid metabolism in health and disease. Crit Rev Biochem Mol Biol 46(3):200–215

Lee YH, Magkos F et al (2011) Effects of leptin and adiponectin on pancreatic β-cell function. Metabolism 60(12):1664–1672

Leibiger IB, Leibiger B et al (2008) Insulin signaling in the pancreatic β-cell. Annu Rev Nutr 28:233–251

Li C, Matschinsky FM et al (2012) Amino acid-stimulated insulin secretion: the role of the glutamine-glutamate-α-ketoglutarate axis. Monogenic hyperinsulinemic hypoglycemia disorders. S. C. A. and D. L. D. D., Philadelphia, p 21

Lorenzo A, Razzaboni B et al (1994) Pancreatic islet cell toxicity of amylin associated with type-2 diabetes mellitus. Nature 368(6473):756–760

MacDonald PE, Wheeler MB (2003) Voltage-dependent K^+ channels in pancreatic β cells: role, regulation and potential as therapeutic targets. Diabetologia 46(8):1046–1062

Maffei M, Halaas J et al (1995) Leptin levels in human and rodent: measurement of plasma leptin and ob RNA in obese and weight-reduced subjects. Nat Med 1(11):1155–1161

Matthews DR, Cull CA et al (1998) UKPDS 26: sulphonylurea failure in non-insulin-dependent diabetic patients over six years. UK Prospective Diabetes Study (UKPDS) Group. Diabet Med 15(4):297–303

Mlinar B, Marc J et al (2007) Molecular mechanisms of insulin resistance and associated diseases. Clin Chim Acta 375(1–2):20–35

Navarro-Tableros V, Fiordelisio T et al (2007) Nerve growth factor promotes development of glucose-induced insulin secretion in rat neonate pancreatic β cells by modulating calcium channels. Channels (Austin) 1(6):408–416

Newsholme P, Bender K et al (2007) Amino acid metabolism, insulin secretion and diabetes. Biochem Soc Trans 35(Pt 5):1180–1186

Nishizawa M, Nakabayashi H et al (2013) Intraportal GLP-1 stimulates insulin secretion predominantly through the hepatoportal-pancreatic vagal reflex pathways. Am J Physiol Endocrinol Metab 305(3):E376–E387

Nolan CJ, Leahy JL et al (2006) β cell compensation for insulin resistance in Zucker fatty rats: increased lipolysis and fatty acid signalling. Diabetologia 49(9):2120–2130

Nolan CJ, Prentki M (2008) The islet β-cell: fuel responsive and vulnerable. Trends Endocrinol Metab 19(8):285–291

Nyman LR, Wells KS et al (2008) Real-time, multidimensional in vivo imaging used to investigate blood flow in mouse pancreatic islets. J Clin Invest 118(11):3790–3797

Ohara-Imaizumi M, Fujiwara T et al (2007) Imaging analysis reveals mechanistic differences between first- and second-phase insulin exocytosis. J Cell Biol 177(4):695–705

Osundiji MA, Evans ML (2013) Brain control of insulin and glucagon secretion. Endocrinol Metab Clin North Am 42(1):1–14

Phillips LK, Prins JB (2011) Update on incretin hormones. Ann N Y Acad Sci 1243:E55–E74

Pillay K, Govender P (2013) Amylin uncovered: a review on the polypeptide responsible for type II diabetes. Biomed Res Int 2013:826706

Polak M, Scharfmann R et al (1993) Nerve growth factor induces neuron-like differentiation of an insulin-secreting pancreatic β cell line. Proc Natl Acad Sci U S A 90(12):5781–5785

Prentki M, Matschinsky FM (1987) Ca^{2+}, cAMP, and phospholipid-derived messengers in coupling mechanisms of insulin secretion. Physiol Rev 67(4):1185–248

Prentki M, Matschinsky FM et al (2013) Metabolic signaling in fuel-induced insulin secretion. Cell Metab 18(2):162–85

Rask-Madsen C, Kahn CR (2012) Tissue-specific insulin signaling, metabolic syndrome, and cardiovascular disease. Arterioscler Thromb Vasc Biol 32(9):2052–2059

Reetz A, Solimena M et al (1991) GABA and pancreatic β-cells: colocalization of glutamic acid decarboxylase (GAD) and GABA with synaptic-like microvesicles suggests their role in GABA storage and secretion. EMBO J 10(5):1275–1284

Rhodes CJ, White MF et al (2013) Direct autocrine action of insulin on β-cells: does it make physiological sense? Diabetes 62(7):2157–2163

Rodriguez-Diaz R, Caicedo A (2013) Novel approaches to studying the role of innervation in the biology of pancreatic islets. Endocrinol Metab Clin North Am 42(1):39–56

Rodriguez-Diaz R, Dando R et al (2011) α cells secrete acetylcholine as a non-neuronal paracrine signal priming β cell function in humans. Nat Med 17(7):888–8892

Romanatto T, Roman EA et al (2009) Deletion of tumor necrosis factor-α receptor 1 (TNFR1) protects against diet-induced obesity by means of increased thermogenesis. J Biol Chem 284(52):36213–36222

Rorsman P, Renstrom E (2003) Insulin granule dynamics in pancreatic β cells. Diabetologia 46(8):1029–1045

Rosas-Vargas H, Martinez-Ezquerro JD et al (2011) Brain-derived neurotrophic factor, food intake regulation, and obesity. Arch Med Res 42(6):482–494

Rosenbaum T, Castanares DT et al (2002) Nerve growth factor increases L-type calcium current in pancreatic β cells in culture. J Membr Biol 186(3):177–184

Rosenbaum T, Sanchez-Soto MC et al (2001) Nerve growth factor increases insulin secretion and barium current in pancreatic β-cells. Diabetes 50(8):1755–1762

Rosenbaum T, Vidaltamayo R et al (1998) Pancreatic β cells synthesize and secrete nerve growth factor. Proc Natl Acad Sci U S A 95(13):7784–7788

Rosengren AH, Jokubka R et al (2010) Overexpression of α2A-adrenergic receptors contributes to type 2 diabetes. Science 327(5962):217–220

Ruiz de Azua I, Gautam D et al (2012) Critical metabolic roles of β-cell M3 muscarinic acetylcholine receptors. Life Sci 91(21–22):986–991

Ruohonen ST, Ruohonen S et al (2012) Involvement of α2-adrenoceptor subtypes A and C in glucose homeostasis and adrenaline-induced hyperglycaemia. Neuroendocrinology 96(1):51–59

Saltiel AR, Kahn CR (2001) Insulin signalling and the regulation of glucose and lipid metabolism. Nature 414(6865):799–806

Sawada F, Inoguchi T et al (2008) Differential effect of sulfonylureas on production of reactive oxygen species and apoptosis in cultured pancreatic β-cell line, MIN6. Metabolism 57(8):1038–1045

Smismans A, Schuit F et al (1997) Nutrient regulation of gamma-aminobutyric acid release from islet β cells. Diabetologia 40(12):1411–1415

Somesh BP, Verma MK et al (2013) Chronic glucolipotoxic conditions in pancreatic islets impair insulin secretion due to dysregulated calcium dynamics, glucose responsiveness and mitochondrial activity. BMC Cell Biol 14:31

Sornelli F, Fiore M et al (2009) Adipose tissue-derived nerve growth factor and brain-derived neurotrophic factor: results from experimental stress and diabetes. Gen Physiol Biophys 28 Spec No:179–183

Stein DT, Stevenson BE et al (1997) The insulinotropic potency of fatty acids is influenced profoundly by their chain length and degree of saturation. J Clin Invest 100(2):398–403

Straub SG, Shanmugam G et al (2004) Stimulation of insulin release by glucose is associated with an increase in the number of docked granules in the β-cells of rat pancreatic islets. Diabetes 53(12):3179–3183

Straub SG, Sharp GW (2012) Evolving insights regarding mechanisms for the inhibition of insulin release by norepinephrine and heterotrimeric G proteins. Am J Physiol Cell Physiol 302(12):C1687–C1698

Sumara G, Formentini I et al (2009) Regulation of PKD by the MAPK p38delta in insulin secretion and glucose homeostasis. Cell 136(2):235–248

Taborsky GJ Jr, Ahren B et al (2002) Autonomic mechanism and defects in the glucagon response to insulin-induced hypoglycaemia. Diabetes Nutr Metab 15(5):318–322, discussion 322–323

Taniguchi CM, Emanuelli B et al (2006) Critical nodes in signalling pathways: insights into insulin action. Nat Rev Mol Cell Biol 7(2):85–96

Togashi K, Hara Y et al (2006) TRPM2 activation by cyclic ADP-ribose at body temperature is involved in insulin secretion. EMBO J 25(9):1804–1815

Tomas E, Wood JA et al (2011) Glucagon-like peptide-1(9–36)amide metabolite inhibits weight gain and attenuates diabetes and hepatic steatosis in diet-induced obese mice. Diabetes Obes Metab 13(1):26–33

Tuduri E, Bruin JE et al (2013) Impaired Ca^{2+} Signaling in β-Cells Lacking Leptin Receptors by Cre-loxP Recombination. PLoS One 8(8):e71075

Ullrich S, Wollheim CB (1985) Expression of both α 1- and α 2-adrenoceptors in an insulin-secreting cell line. Parallel studies of cytosolic free Ca^{2+} and insulin release. Mol Pharmacol 28(2):100–106

Velasco M, Larque C et al (2012) Metabolic syndrome induces changes in K_{ATP}-channels and calcium currents in pancreatic β-cells. Islets 4(4):302–311

Vidaltamayo R, Sanchez-Soto MC et al (2002) Nerve growth factor increases sodium channel expression in pancreatic β cells: implications for insulin secretion. FASEB J 16(8):891–892

Vollenweider P (2003) Insulin resistant states and insulin signaling. Clin Chem Lab Med 41(9):1107–1119

Wang Q, Jin T (2009) The role of insulin signaling in the development of β-cell dysfunction and diabetes. Islets 1(2):95–101

Westermark P, Andersson A et al (2011) Islet amyloid polypeptide, islet amyloid, and diabetes mellitus. Physiol Rev 91(3):795–826

White MF (2003) Insulin signaling in health and disease. Science 302(5651):1710–1711

Xu XJ, Gauthier MS et al (2012) Insulin sensitive and resistant obesity in humans: AMPK activity, oxidative stress, and depot-specific changes in gene expression in adipose tissue. J Lipid Res 53(4):792–801

Yamanaka M, Itakura Y et al (2006) Protective effect of brain-derived neurotrophic factor on pancreatic islets in obese diabetic mice. Metabolism 55(10):1286–1292

Yi P, Park JS et al (2013) Betatrophin: a hormone that controls pancreatic β cell proliferation. Cell 153(4):747–758

Youos JG (2011) The role of α-, δ- and F cells in insulin secretion and action. Diabetes Res Clin Pract 93(Suppl 1):S25–S26

Zhang T, Li C (2013) Mechanisms of amino acid-stimulated insulin secretion in congenital hyperinsulinism. Acta Biochim Biophys Sin (Shanghai) 45(1):36–43

Apoptosis in Pancreatic β-Cells in Type 1 and Type 2 Diabetes

31

Tatsuo Tomita

Contents

Introduction	846
Apoptosis in Pancreatic β-Cells by Immunological and Immunocytochemical Study	849
Amyloid Theory on β-Cell Apoptosis in Type 2 Diabetes	856
Hyperglycemia-Induced Apoptosis in β-Cells	859
Recent Studies on β-Cell Apoptosis	862
Cross-References	867
References	868

Abstract

Apoptosis plays an important role in the pathophysiology of both type 1 and type 2 diabetes. In type 1 diabetes, β-cell death by apoptosis following autoimmune insulitis causes an absolute insulin deficiency triggered by an extrinsic receptor-mediated pathway, which activates a cascade of caspase family reaction. The etiology of type 2 diabetes is multifactorial, including obesity-associated insulin resistance, defective insulin secretion, and loss of β-cell mass through β-cell apoptosis. β-cell apoptosis is mediated through a milliard of caspase family cascade machinery in both type 1 and type 2 diabetes. The glucose-induced insulin secretion is the principle pathophysiology of diabetes and insufficient insulin secretion results in chronic hyperglycemia and diabetes. Recently, hyperglycemia-induced β-cell apoptosis has been extensively studied with regard to the balance of pro-apoptotic genes (Bad, Bid, and Bik) and the anti-apoptotic Bcl family toward apoptosis in in vitro isolated islets. Apoptosis can only occur when the concentration of pro-apoptotic Bcl-2 exceeds that of anti-apoptotic proteins at the mitochondrial membrane of the intrinsic pathway.

T. Tomita (✉)
Departments of Integrative Biosciences and Pathology and Oregon National Primate Center, Oregon Health and Science University, Portland, OR, USA
e-mail: tomitat@ohsu.edu

The bulk of recent research on hyperglycemia-induced apoptosis on β-cells unveiled complex details of glucose toxicity on β-cells at a molecular level coupled with cell membrane potential by the K^+ and Ca^{2+} channels opening and closing. Further, animal models using knockout mice will shed light on our basic understanding of the pathophysiology of diabetes as a glucose metabolic disease complex, and on the balance of the anti-apoptotic Bcl family and pro-apoptotic genes. The cumulative knowledge will provide a better understanding of the metabolic control of glucose metabolism at a molecular level and will lead to eventual prevention and therapeutic application for type 1 and type 2 diabetes.

Keywords

Amyloid • Anti-apoptotic genes and proteins • Apoptosis • β-cells • Bcl family • Ca^{++} channel • Caspase • Glucokinase • Glucose-induced insulin secretion • Glucose toxicity • Immunocytochemistry • Insulin • Islet amyloid polypeptide • TUNEL assay • Type 1 and type 2 diabetes mellitus

Introduction

There is increasing evidence to support the notion that both type 1 and type 2 diabetes (DM) is modulated through β-cell apoptosis. Apoptosis is a complex biological phenomenon characterized by cell shrinkage, chromatic condensation, internucleosomal DNA fragmentation, and disassembly into membrane-encircled vesicles (apoptotic bodies) (Kerr et al. 1972). This programmed cell death is implicated in the remodeling of the normal endocrine pancreas after birth, and plays an important role in the development of final β-cell mass (Finegood et al. 1995).

The role of apoptosis in the physiology of normal pancreatic development has been demonstrated in neonatal pancreas, which has a threefold higher frequency of apoptotic cells than adult animals (Scaglia et al. 1997). Pancreatic β-cell apoptosis is also a pathological feature that is common to both type 1 and type 2 diabetes mellitus (DM). In type 1 DM (T1DM), β-cells are selectively destroyed by insulitis of T-lymphocytic infiltration in islets, and autoimmune destruction follows after the appearance of islet autoantibody, and results in insulin deficiency (Urusova et al. 2004). In type 2 DM (T2DM), insulin resistance with obesity leads to a glucose toxicity effect, which contributes to β-cell death by apoptosis (Mandrup-Poulsen 2001).

Defects in apoptotic regulatory machinery are implicated in a variety of pathological statuses: excess apoptosis is the underlying cause for β-cell loss for both T1DM and T2DM and inadequate apoptosis may contribute to the oncogenesis of pancreatic endocrine tumors (Lee and Pervaiz 2007). Of the two apoptosis pathways, the extrinsic (receptor-mediated) and the intrinsic (mitochondria-driven) pathways, it is the extrinsic pathway that is activated upon ligation of the cells'

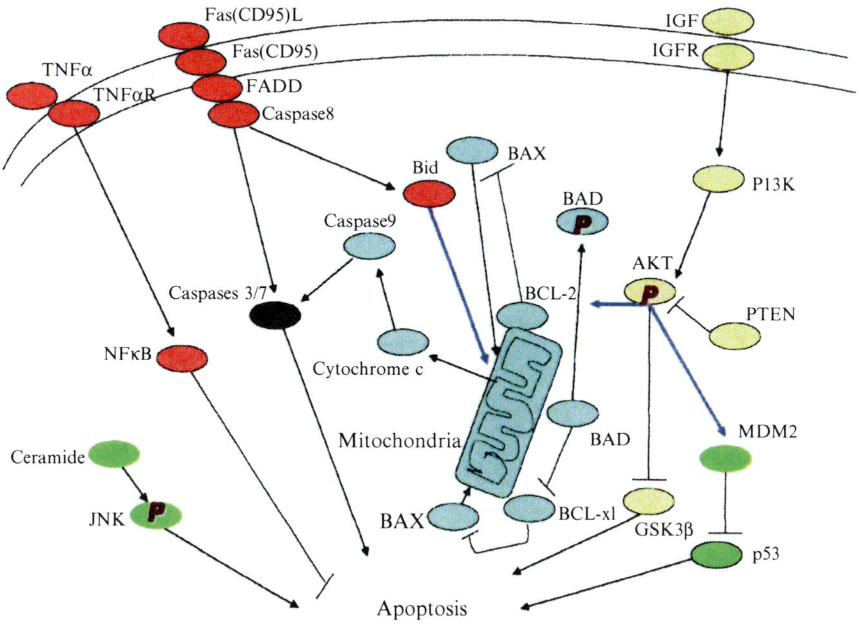

31 Apoptosis in Pancreatic β-Cells in Type 1 and Type 2 Diabetes

Fig. 1 There are the extrinsic (receptor-mediated, *red*) and intrinsic (mitochondria-driven, *blue*) apoptosis pathways as opposed to the survival proteins such as the P13/Akt signaling circuitry (*yellow*) (From Lee and Pervaiz (2007))

surface death receptor(s), which in turn activate(s) a downstream effector mechanism orchestrated by the caspase family of cysteine proteases (Fig. 1; Finegood et al. 1995; Lee and Pervaiz 2007; Green 2005; Emamaullee and Shapiro 2006). The prototype example of death signaling via the extrinsic pathway is the Fas death receptor, which instigates assembly of the death-inducing signaling complex (DISC), a multi-protein complex comprising the cytoplasmic aspects of the Fas receptor, the adaptor protein FADD (Fas-associated death domain containing protein), and procaspase-8 (Fig. 1; Finegood et al. 1995; Lee and Pervaiz 2007; Green 2005; Emamaullee and Shapiro 2006).

Caspase-3 is a converging point of the apoptotic pathway (Fig. 2; Emamaullee and Shapiro 2006) and its peptide inhibitors have been shown to prevent islet apoptosis and improve islet graft function (Nakano et al. 2004; Brandhorst et al. 2006; Cheng et al. 2008). Caspases are cysteine-containing aspartic acid-specific proteases that exist as zymogens in the soluble cytoplasm, endoplasmic reticulum, mitochondrial intermembrane space, and nuclear matrix (Nicholson and Thornberry 1997). Apoptosis induced by ligation of cell surface receptors like Fas (CD 95) or TNF receptors, "death receptors," represents a pathway controlled by caspases (Chandra et al. 2001). Ligand binding of the receptor causes assembly of a series of proteins of the death-inducing signaling complex, which then activates an apical caspase, procaspase-8 (Peter and Krammer 1998). The resulting events

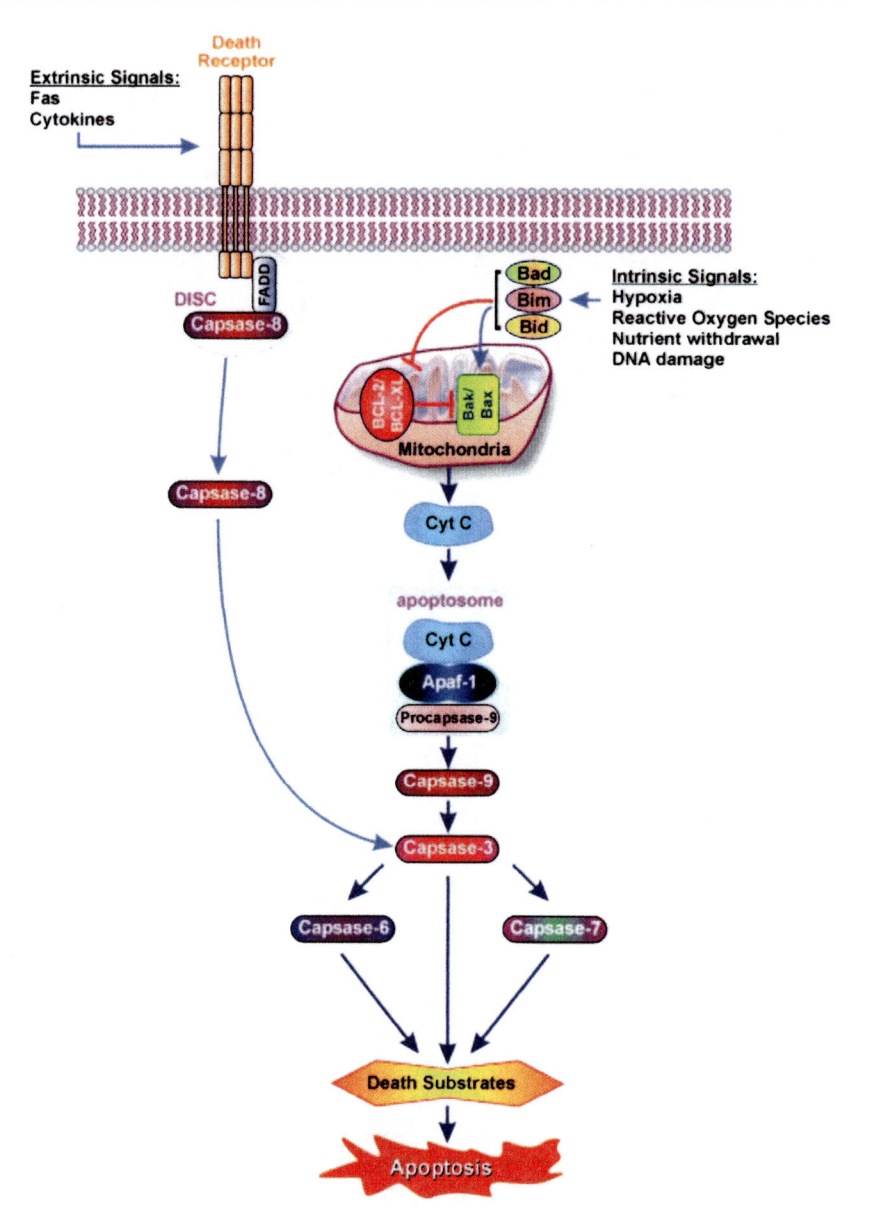

Fig. 2 Extrinsic and intrinsic pathways lead to apoptosis via cytochrome c and "apoptosome". Extrinsic pathway: Fas-Fas L binding leads to the death-inducing signaling complexes (DISC) where the DISC-caspase-8 complex is activated, leading to caspase-3 activation. Intrinsic pathway: pro-apoptotic proteins (Bad, Bid, Bik, Bim) become activated and translocate to the mitochondria, where they bind or inactivate Bcl proteins or form pores in the mitochondrial membrane, which facilitates the release of cytochrome c into the cytosol. Once cytochrome c accumulates in the cytosol, it forms a complex with apocaspase-9 and Apaf-1 to make the "apoptosome", which in turn activates caspase-3. From Emamaulee and Shapiro (2006)

proceed in cascades that caspase-8 induces activation of caspase-3 (Chandra et al. 2001). One of these proteins is a caspase-dependent endonuclease (CAD), which is freed from its inhibitor (ICAD) by caspase-3 and subsequently cuts DNA into oligonucleosomal (180-bp) fragments (Peter and Krammer 1998; Sakahira et al. 1998).

Apoptosis manifests in two major execution programs, downstream of the death signal, the caspase pathway (Gross et al. 1999), and upstream of irreversible cellular damage reside the Bcl family members, which are proteins with both pro-apoptotic and anti-apoptotic properties, playing a pivotal role in the life and death of cells (Figs. 1 and 2) (Green 2005; Emamaullee and Shapiro 2006; Federici et al. 2001). Anti-apoptotic members of the Bcl family, including Bcl-2 and Bcl-xL, blunt intrinsic death signaling by blocking the recruitment of pro-apoptotic members to the mitochondria (Green 2005).

The cumulative data support the notion that high glucose might modulate the balance of the pro-apoptotic caspase family and anti-apoptotic Bcl proteins toward apoptosis, thus leading to β-cell death (Emamaullee and Shapiro 2006; Federici et al. 2001).

Apoptosis in Pancreatic β-Cells by Immunological and Immunocytochemical Study

Apoptosis is a cause of absolute β-cell deficiency in T1DM and relative β-cell deficiency in T2DM. In T1DM, a rapid β-cell loss occurs by T-cell mediated autoimmunity through the interaction of helper (CD 4^+) and cytotoxic (CD8^+) cells under the influence of MHC loci such as HLA DR-3 and HLA DR-4, and non-MHC determinants (Lee and Pervaiz 2007). "Insulitis" is surrounded by peri-islet inflammation associated with pro-inflammatory cytokines (Il-β, TNF-σ, interferon-ϒ) release by monocytes, Fas-ligand (CD 178) and autoreactive T-cells, leading to destruction of the β-cells and the onset of hyperglycemia (Hui et al. 2004). β-cells from newly diagnosed T1DM patients have increased cell surface expression of Fas (CD 95), which is then delivered via the Fas L(CD178) expressed on the infiltrating T lymphocytes (Stassi et al. 1997). Expression of dominant-negative Fas or neutralizing antibody to Fas significantly blocks apoptosis, manifests adequate β-cell function, blocks the adoptive transfer of diabetes by primed T-cells, and impedes the course of T1DM development (Willcox et al. 2008). With immunocytochemical staining of 29 cases of pancreata from the subjects with T1DM who died within 18 months of diagnosis, Willcox et al. (2008) showed the following findings: CD 8^+ cytotoxic T cells were the most abundant population in insulitis; CD 68^+ macrophages were also present during insulitis, whereas CD 4^+ cells were present in the islet infiltrates, but were less numerous than CD 8^+ cells or CD 68^+ cells. Both CD 8^+ and CD 68^+ cells may contribute to β-cell death during early insulitis, whereas CD 20^+ cells increased in number during late insulitis, since CD 20^+ cells were recruited late during insulitis (Allison et al. 2005). It is presumed that β-cell damage and destruction are mediated by both CD 8^+ and CD 4^+ (Eisenbarth and Kotzin 2003). Natural killer cells

(NK cells) do not appear to be required for β-cell death (Willcox et al. 2008). Dysregulation of apoptosis is a central defect in diverse murine autoimmune diseases, including the NOD mouse model for human T1DM (Willcox et al. 2008). Mutations in Fas (CD 95) or Fas L (CD 178) have been identified to render lymphoid cells resistant to apoptosis (Hayashi and Faustman 2003). The MHC region of the genome contains immune response genes, which are important for T-cell education and for antigen presentation by both MHC class I and II molecules. Studies of both humans and rodents have suggested that the centrally located MHC class II genes confer the greatest statistical risk for autoimmune diseases. Cellular abnormalities in the expression of maturation markers or in antigen presentation have been detected in NOD mice and human T1DM (Hayashi and Faustman 2003; Rabinovitch and Skyler 1998). The defects include reduced expression of the maturation antigen CD 45 and reduced the abundance of conformational correct complexes of MHC class I molecules and self-peptides on the cell surface (Faustman et al. 1989). Human autoimmune diseases are associated with impairment of antigen processing controlled by the MHC. Cytosolic extracts of lymphocytes from humans with T1DM exhibit altered patterns of the cleavage of test substrates by the proteasome (Hayashi and Faustman 2003). These results are followed by the generation of peptides, which are poorly suited for assembly with MHC class I molecules (Faustman et al. 1991). Clinical studies have shown that the antigen presentation defect correlates with disease expression in identical twins with T1DM (Faustman et al. 1991; Jansen et al. 1995). The certain MHC class II haplotypes, HLA-DR3 or -DR4, or both, are positively associated with 95 % of T1DM cases compared with 20 % of the general population, whereas HLA-DR2 is negatively associated with T1DM (Cucca et al. 2001).

T2DM is characterized by insulin resistance, defective insulin secretion, loss of β-cell mass with increased β-cell apoptosis and islet amyloid deposits (Haataja et al. 2008). In T2DM, insulin resistance with obesity precedes insulin deficiency and plays a considerable role (Reaven 1988), followed by a failure of β-cell insulin production against the progressive insensitivity to insulin. β-cell mass fluctuates according to the body's need for insulin:
1. β-cell mass can increase during insulin resistance
2. Progressive β-cell loss is present in T2DM
3. β-cell deficiency correlates with glucose intolerance
4. β-cell death may directly lead to insulin deficiency when loss of 60 % or more is accompanied by the presence of insulin resistance with obesity (Butler et al. 2003a)

B-cell mass is regulated by a balance of β-cell replication and apoptosis, and islet hyperplasia and new islet formation from exocrine pancreatic ducts (Bonner-Weir 2000a; Leonardi et al. 2003; Marchetti et al. 2004). Elevated caspase-3 and -8 are activated in β-cells from T2DM subjects, which can be inhibited by the anti-diabetic agent metform (Marchetti et al. 2004). Hyperglycemia-induced β-cell apoptosis has been implicated and has been studied mainly in T2DM (Butler et al. 2003a). Butler et al. extensively studied 124 cases of pancreata from autopsy, including 91 in obese patients (BMI > 27 kg/m^2 : 41 cases – T2DM, 15 cases– impaired fasting glucose, and 35 cases – non-DM cases) and 33 lean patients (BMI <25 kg/m^2 : 16 cases – type 2 DM, 17 cases – non-diabetic cases). The authors measured relative β-cell mass volume using Image-Pro Plus software (Media Cybermetric, Silber Springs, MD, USA), the frequency of β-cell apoptosis by TUNEL (terminal deoxynucleotydyl transferase-mediated dUTP nick-end labeling), and the replication index using Ki-67 immunocytochemical staining (Butler et al. 2003a). With the use of TUNEL staining, only discernible cells with TUNEL-positive nuclei were included as positive cells (Butler et al. 2003a). Obese patients with impaired fasting glucose and T2DM subjects showed 40 % and 63 % less β-cell volume compared with non-diabetic obese and lean controls respectively (Butler et al. 2003a). The frequency of β-cell replication was very low at 0.04–0.06 % of β-cell mass, but the frequency of β-cell apoptosis by TUNEL was increased tenfold in lean patients with DM (0.47 % of the β-cell area) and threefold in obese patients with DM (0.31 % of the β-cell area) compared with the respective non-diabetic control subjects (Butler et al. 2003a). It appears that β-cell replication by Ki-67 is underestimated and that β-cell apoptosis by TUNEL is overestimated, since the replication and apoptosis rates should be about the same in order to maintain the β-cell mass in a delicate balance. The authors conclude in T2DM that β-cell mass is decreased and that the mechanism underlying the β-cell loss is increased β-cell apoptosis (Butler et al. 2003a).

Another Immunocytochemical marker for apoptosis is cleaved caspase-3. Each caspase family protease becomes active when the precursor is cleaved into a large subunit with a molecular mass of ~20 kDa and a small subunit with a molecular mass of ~10 kDa, which then forms a tetramer consisting of two large and two small units (Hirata et al. 1998; Tewari et al. 1995). One of these cleaved caspases is present on activated caspase-3, a ubiquitously distributed caspase that is a main effector caspase of the apoptotic cascade within cells (Marchetti et al. 2004; Hirata et al. 1998). The commercially available polyclonal anti-cleaved caspase-3 detects endogenous levels of the large (17/19 kDa) cleaved caspase-3 resulting from cleavage adjacent to Asp 175 and does not recognize the full length or other cleaved caspases (Cleaved caspase-3 (Asp 175) antibody 2006; Gown and Willingham 2002). Recently, involvement of caspase-3 in both T1DM and T2DM was implicated: in T1DM, Fas(CD 95)-Fas L (CD 178) may be critical for β-cell destruction, as apoptosis in the β-cell clone expressing the human Fas β-cell line is mediated by elevated caspase-3-like activity in tissue culture (Martin and Green 1995) and the frequency of β-cell apoptosis in T2DM pancreatic tissues from autopsy was increased using TUNEL, as described before (Butler et al. 2003a). Our group studied 16 cases of T2DM pancreata compared with 10 control pancreata using rabbit antihuman cleaved caspase-3 (Cell Signaling Technology, Beverly, MA, USA) for immunocytochemicalstaining: the control islets revealed that 4.7 % of total islet cells were cleaved caspase-3-positive islet cells, with large and small islets positive at 4.1 % and 7.0 % respectively, whereas type 2 DM islets showed a higher level of positive cells at 8.7 % of total islet cells, with large and small islets positive at 7.7 % and 12 % respectively, about twice the control values (Table 1; Tomita 2010a). A double immunostaining for insulin and cleaved caspase-3 supports the notion that β-cell nuclei in the degranulated cytoplasm were positive for cleaved caspase-3 (Tomita 2010a). Cleaved caspase-3-positive islet cells were more frequent in the islets containing fewer amyloid deposits than in the islet cells containing more amyloid deposits; the latter corresponded to the end-stage T2DM islets, which have completed the apoptosis process (Tomita 2010a). Thus, the higher number of cleaved caspase-3 positive islets from T2DM subjects may implicate an accelerated apoptotic cascade, accompanied by increasing amyloid deposits, before proceeding to ultimate β-cell death by overwhelming interstitial amyloid deposits (Tomita 2010a).

We performed cleaved caspase-3 immunocytochemical staining in 8 cases of T1DM pancreata compared with 8 controls. T1DM islets showed higher amounts of cleaved caspase-3-positive cells at 16 % of the total islet cells, with large and small islets positive at 14 %and 17 % respectively, at 3.3, 3.6 and 2.4 times that of the corresponding control values (Table 2; Tomita 2010b). The T1DM islets were a mixture of major small-sized islets consisting of β-cell-poor and σ-cell-rich islets with more caspase-3-positive cells and occasional large islets, consisting of non-β-cells and σ-, δ-, and PP-rich islets with moderately increased caspase-3-positive cells (Fig. 3a, b; Table 2; Tomita 2010b). These increased caspase-3 positive islet cells in T1DM pancreas may correspond to a more accelerated apoptosis cascade than in T2DM islets before entirely exhausting the β-cell mass by apoptosis (Tomita 2010b).

In adult islets, β-cells have an estimated life-span of about 60 days (Bonner-Weir 2000b). Under normal conditions, 0.5 % of control adult β-cells were reportedly to undergo apoptosis (Rhodes 2005). By cleaved caspase-3 immunostaining, about 5 % of β-cells are positive for this apoptosis marker (Martin and Green 1995). Thus, there is a wide range of 0.5–5 % of apoptotic islet cells in the control islets according to TUNEL and cleaved caspase immunocytochemical staining (Butler et al. 2003a; Tomita 2010a; b).

Each of the two immunocytochemical staining methods for apoptosis has advantages and disadvantages. The TUNEL assay is very sensitive and widely used, but it is prone to some pitfalls. The TUNEL technique can label non-apoptotic nuclei showing signs of active gene transcription (Tomita 2010a; Bonner-Weir 2000b; Rhodes 2005; Barett et al. 2001). Tumor necrosis (Sava et al. 2001) and autolysis generate a significantly higher number of DNA ends, which can be positively labeled under certain conditions (Duan et al. 2003). The technical problem of TUNEL is mostly related to DNA strand breaks associated with excessive levels of protease digestion, fixation, and processing procedures (Duan et al. 2003). Therefore, techniques that detect DNA fragmentation are not specific to apoptosis and frequently generate erroneously higher results (Duan et al. 2003; Butler et al. 2003b). Duan et al. carefully studied apoptosis in histological sections of prostatic cancer cell line PC-3 using both TUNEL and cleaved caspase-3

Table 1 Cleaved capase-3 immunostaining in type 2 diabetic islets (From Tomita (Tomita 2010a), with permission)

Diabetic subjects, case no.	Age/sex, history[a]	Large islets			Small islets			Total islets		
		Positive cells	Islet cells	Positive[b] %	Positive cells	Islet cells	Positive[b] %	Positive cells	Islet cells	Positive[b] %
1	49/M$_1$	3.5	61.4	5.7 (18)	3.1	26.0	11.9 (12)	3.3	87.3	7.1 (30)
2	54/M$_2$	3.1	65.7	4.8 (14)	2.7	23.4	11.0 (16)	2.9	43.2	6.7 (30)
3	57/M$_2$	13.8	53.9	25.6 (17)	7.3	24.4	29.9 (13)	11.0	41.1	26.8 (30)
4	62/M$_3$	5.9	69.0	8.5 (15)	3.1	25.3	12.4 (15)	4.5	47.2	9.5 (30)
5	62/F$_3$	4.8	89.2	5.4 (18)	2.4	26.3	9.2 (12)	4.0	41.1	9.5 (30)
6	62/F$_2$	3.5	45.2	5.1 (20)	2.0	24.0	8.3 (10)	3.0	53.2	5.1 (30)
7	62/F$_2$	3.9	62.9	6.2 (17)	2.4	21.5	11.2 (13)	3.2	45.4	7.0 (30)
8	63/M$_3$	2.4	57.6	4.2 (16)	2.1	26.3	7.8 (14)	2.3	43.0	5.3 (30)
9	63/F$_2$	3.4	58.9	5.8 (21)	2.2	26.6	8.4 (9)	3.1	59.2	6.2 (30)
10	64/M$_2$	5.9	56.7	10.4 (12)	2.4	21.4	11.1 (18)	3.8	35.8	10.7 (30)
11	66/M$_3$	2.9	73.1	4.0 (20)	2.3	23.5	10.1 (10)	2.7	36.3	4.8 (30)
12	71/F$_4$	3.4	71.4	4.8 (18)	2.8	22.7	12.5 (12)	3.2	51.9	6.2 (30)
13	72/F$_3$	3.6	72.4	5.5 (17)	2.3	21.2	10.6 (13)	3.0	47.1	6.4 (30)
14	75/F$_4$	6.1	59.0	10.3 (16)	2.9	22.2	12.9 (14)	4.6	41.8	10.9 (30)
15	76/F$_4$	5.6	56.1	10.0 (18)	2.9	19.1	15.1 (12)	4.0	40.5	10.9 (30)
16	77/1M$_3$	4.4	68.9	6.3 (19)	2.1	23.8	8.8 (11)	3.5	52.4	6.7 (30)
	Mean	4.8	63.8	7.7[c]	2.8	23.6	12.0[c]	3.9	49.8	8.7 (30)[c]
	SE	0.67	2.55	1.31	0.31	0.55	1.29	0.50	3.97	1.31 (30)
Controls ($n = 10$)										
	Mean	2.9	71.1	4.1[c]	1.7	24.4	7.0[c]	2.6	55.7	4.7 (30)[c]
	SE	0.23	4.30	0.28	0.10	1.10	0.36	0.18	2.78	0.33 (30)

[a]History of diabetes: 1, <5 years; 2, 5–10 years; 3, 11–15 years; 4, > 15 years
[b]Numbers in parentheses are the numbers of islets examined
[c]p values calculated with the paired type 2 diabetic and corresponding control values: $p < 0.001$
F female, M male, SE standard error

Table 2 Cleaved caspase-3 immunostaining in type 1 diabetic islets (From Tomita (Tomita 2010b), with permission)

Diabetic subjects		Large islets			Small islets			Total islets		
Case	Age/sex	Positive cells	Islet cells	Positive %	Positive cells	Islet cells	Positive %	Positive cells	Islet cells	Positive %
1	18/M	7.3	37.3	19.9 (12)	5.4	22.7	23.7 (18)	6.1	28.6	21.3 (30)
2	35/F	9.2	62.1	14.8 (18)	3.3	38.3	8.7 (12)	6.8	48.6	14.0 (30)
3	43/F	2.8	44.9	6.3 (13)	2.3	14.1	16.3 (17)	2.5	30.3	8.4 (30)
4	50/F	21.8	73.3	29.7 (20)	6.4	27.2	23.5 (10)	16.7	57.9	28.8 (30)
5	61/F	7.7	43.1	18.0 (15)	5.4	27.2	19.9 (15)	6.6	35.1	18.7 (30)
6	75/F	3.6	50.6	7.1 (9)	5.4	37.4	14.5 (21)	5.5	41.4	13.2 (30)
7	75/F	5.1	51.6	9.8 (18)	3.7	28.1	13.1 (12)	4.5	42.2	10.7 (30)
8	75/F	4.4	44.3	9.9 (15)	3.5	29.4	11.9 (15)	3.9	39.4	9.9 (30)
	Mean	7.7[a]	50.9	14.4[a]	4.5[a]	26.8	16.5[a]	6.6[a]	40.4[a]	15.7 (30)[a]
	SE	2.1	3.75	2.80	0.50	2.60	1.94	1.50	3.40	2.45 (30)
Controls ($n = 8$)										
	Mean	2.9[a]	72.4	4.0[a]	1.7[a]	25.0	6.8[a]	2.7[a]	58.5[a]	4.7 (30)[a]
	SE	0.29	5.32	0.39	0.13	1.20	0.43	0.19	2.91	0.40 (30)

Numbers in parenthesis are the numbers of islets examined
[a]p values calculated with the corresponding control values are: $p < 0.001$

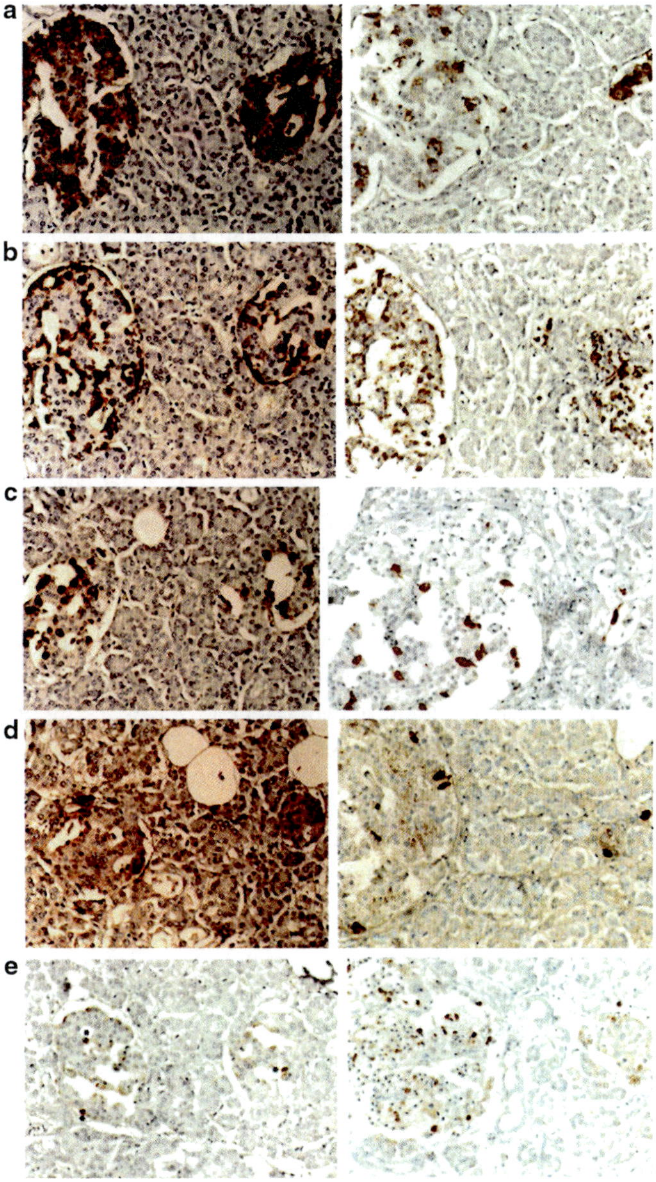

Fig. 3 Normal islets (*left column*), and β-cell-less and σ-cell-rich type 1 diabetic islets (*right column*) by cleaved caspase-3 immunostaining. (*Left column*) Normal islets consist of major β-cells (80 %), next major σ-cells (10 %) and minor δ-cells (<7 %), and PP cells (<5 %). There is about 5 % cleaved caspase-3-positive nuclear staining for normal islet cell nuclei. (*Right column*) β-cell-poor (3–20 %) and σ-cell-rich (80 %) type 1 diabetic islets are the major islet cells with relatively increased δ-cells (>7 %) and PP cells (>5 %). About 17 % of islet cell nuclei are positive for cleaved caspase-3. (**a**) Insulin, (**b**) glucagon, (**c**) SRIF, (**d**) pancreatic polypeptide, (**e**) cleaved caspase −3 immunostained (From Tomita (2010b), with permission)

immunostaining (Duan et al. 2003). TUNEL staining depends on the optimal concentration of terminal deoxynucleotydyl transferase (TdT), with which less diluted (1:7) solution positively stained the majority (>90 %) of the transplanted cancer cells compared with an optimally diluted solution (1:16) stained adequate numbers (< 2 %) of TUNEL-positive cancer cells (Duan et al. 2003). By comparing TUNEL and cleaved caspase-3 immunostaining, the authors concluded that cleaved caspase-3 immunostaining was an easy, sensitive, and reliable method for detecting and quantifying apoptosis and that a good correlation of apoptotic indices existed between caspase-3 immunostaining and the TUNEL assay (Duan et al. 2003). We agree with Duan et al. that cleaved caspase-3 immunostaining is an easier and more reliable immunostaining method than TUNEL, although the former is not as commonly used as the latter. Cleaved caspase-3 immunostaining also has its own pitfalls, as does immunocytochemical staining in general: good fixation with optimal tissue preservation and proper tissue processing are mandatory using an adequate concentration of antibody for optimal cleaved caspase-3 immunostaining, and the stained sections have to be critically evaluated for discernible nuclear positive staining by excluding false-positive staining in tissue debris and autolytic tissues (Sava et al. 2001; Duan et al. 2003; Butler et al. 2003b; Tomita 2009).

Our group studied cleaved caspase-3 immunostaining in 37 cases of pancreatic endocrine tumors (PETs) (Tomita 2009): among 15 cases of insulinomas, 5 cases were positive and 10 cases were negative for cleaved caspase-3 (67 %). Among non-β cell PETs, 2 out of 2 glucagonomas (100 %), 6 out of 9 pancreatic polypeptidomas (67 %), 10 out of 12 gastrinomas (83 %), and 3 out of 3 non-functioning PETs (100 %) were negative for cleaved caspase-3, with a total of 21 out of 24 non-β cell tumors (88 %) being negative. This results supports the notion that 88 % of non-β cell PETs are potentially malignant according to the absence of cleaved caspase-3 immunostaining (Tomita 2009) and that negative cleaved caspa-3 immunostaining may be a candidate for being a malignant marker for PETs (Tomita 2009). So far, positive activated caspase-3 immunocytochemical staining appears to be a good marker for β-cell apoptosis.

Amyloid Theory on β-Cell Apoptosis in Type 2 Diabetes

The amyloid deposit in pancreatic islets is a characteristic finding for T2DM (Ehrlich and Ratner 1961; Hoppener et al. 2000; Hull et al. 2004). The chief constituent of amyloid deposits is islet amyloid polypeptide (IAPP) (Butler et al. 2003b; Ehrlich and Ratner 1961; Hoppener et al. 2000; Hull et al. 2004; Westermark et al. 1986; Cooper et al. 1987). IAPP is concomitantly co-secreted with insulin from β-cells into the blood in response to glucose-induced insulin secretion (Cooper et al. 1987; Kahn et al. 1999). IAPP hyposecretion is well established in T1DM and insulin-requiring T2DM (Cooper et al. 1987; Kahn et al. 1999):a synthetic IAPP, pramlintide (Cucca et al. 2001; Haataja et al. 2008;

Reaven 1988) (pro-hIAPP) has been used for treating both T1DM and insulin-requiring T2DM, together with insulin injection, for better glycemic control (Kruger et al. 1999; Weyer et al. 2001; Buse et al. 2002; Fineman et al. 2002). We studied the transformation of soluble IAPP in β-cell granules to insoluble amyloid deposits in islet stroma by using immunocytochemical staining for IAPP (Tomita 2012): a ratio of IAPP-positive cells to insulin-positive cells was 43 % in control islets compared with 25 % inT2DM islets, in support of the decreased IAPP serum levels in T2DM, since the source of serum IAPP is the β-cell granule (Tomita 2012). Pancreatic extracts from normal humans contain less IAPP, at 10 % that of insulin, and the fasting serum IAPP level in non-obese controls is 2.0 μmol/l, 5 % that of the insulin level of 48 μmol/l (Clark and Nilsson 2004). As IAPP-positive β-cell cytoplasm decreased, stromal islet amyloid deposits increased in T2DM islets (Tomita 2012). In advanced stages of islets with amyloid deposits, weakening IAPP-immunostaining in the residual β-cells co-existed with the adjacent fine amyloid fibrils, which were positively immunostained for IAPP (Fig. 4f; Tomita 2012). This finding suggests that disappearing soluble IAPP from β-granules are transformed to amyloid fibrils in the adjacent islet stroma. Freshly prepared IAPP oligomers can form non-selective iron permeable membrane pores, leading to increased Ca^{2+} concentration, endoplasmic reticulum stress, and apoptosis (Haataja et al. 2008). In early stages of islet amyloidosis, round to sickle-shaped β-cell cytoplasm without a nucleus was strongly immunopositive for IAPP (Fig. 4f; Tomita 2012). This cytoplasm probably represents an early fibrillar form of amyloidogenic β-cell cytoplasmic proteins consisting of smaller IAPP polymers, which may subsequently form extracellular amyloid β-sheets. IAPP has the propensity to form membrane-permeable toxic oligomers, but it remains unclear why amyloidogenic proteins form oligomers in vivo, what their exact structure is, and to what extent these oligomers play a role in the cytotoxicity that is now often called unfolded protein disease (Haataja et al. 2008; Ehrlich and Ratner 1961; Clark and Nilsson 2004; Clark et al. 1988; Mirzabekov et al. 1996; O'Brien et al. 2010; Jurgens et al. 2011).

According to the toxic oligomer theory, β-cells in T2DM are killed through IAPP-induced damage of the β-cell membrane (Clark et al. 1988; Mirzabekov et al. 1996; O'Brien et al. 2010; Jurgens et al. 2011; Anguiano et al. 2002; Jansen et al. 1999; Engel et al. 2008). These toxic oligomers (not monomers or mature amyloid fibrils) eventually form the end product of β-sheets containing IAPP, Aβ, synuclein, transthyretin, and prion protein, and share a common epitope (Kayed et al. 2003; Tomita 2005). Amyloid p is a universal immunological and immunocytochemical marker for all forms of amyloidosis (Kayed et al. 2003; Tomita 2005). Freshly prepared intermediate IAPP polymers (25–6,000 IAPP molecules) have a toxic effect on β-cells, but do not exhibit a toxic effect on σ, δ, and PP islet cells (Haataja et al. 2008; Ehrlich and Ratner 1961; Tomita 2012; Clark and Nilsson 2004). Water-soluble IAPP with low molecular weight in β-cell granules is readily and densely immunostained for IAPP, whereas water-insoluble amyloid fibrils containing IAPP polymers are only weakly immunostained (Ehrlich and Ratner 1961; Kahn et al. 1999). The cause of this lack of strong IAPP immunostaining for

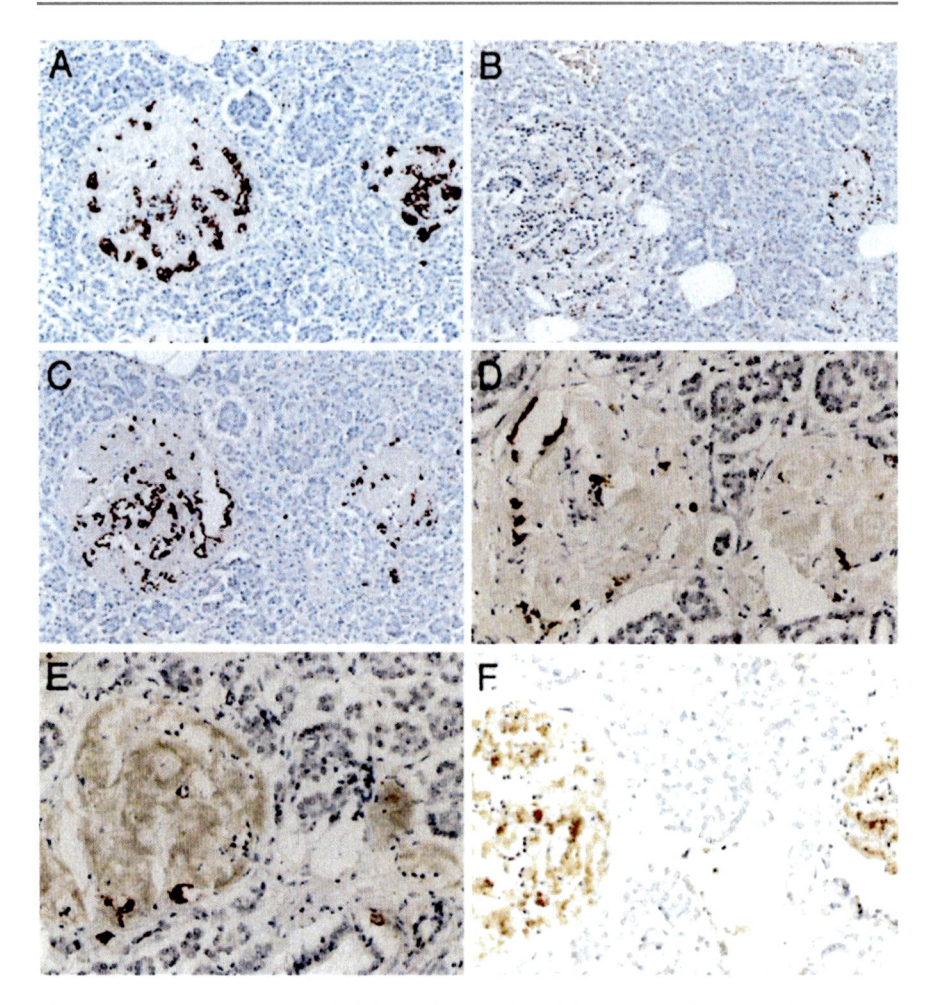

Fig. 4 Amyloid deposits in type 2 diabetic islets. Diabetic islets are occupied by amyloid deposits in 95 % (**a–c**) and are also occupied by > 99 % (**d–f**). The residual β-cells with plump cytoplasm (**a**) are minor cells and σ-cells with small and compact cytoplasm (**c**) are major islet cells. IAPP immunostaining with 1:400 diluted IAPP antibody reveals weak staining in the islets occupied by amyloid at 95 % (**b**) whereas islets occupied by amyloid > 99 % reveal stronger IAPP immunostaining with round to sickle-shaped positive immunostaining in the β-cell cytoplasm (**f**). Thus, IAPP immunostaining revealed various immunostaining for IAPP-depleted islets (**b**) to a mixture of dense IAPP-positive cytoplasm and weaker interstitial fibers to darker interstitial fibers (**e**) in the same patient. (**a**) and (**d**) Insulin, (**c**) glucagon, (**b**), (**e**), and (**f**) IAPP antibody at 1:400 immunostained for IAPP (From Tomita (2012a), with permission)

amyloid fibrils is not clear, but one likely reason is that the unexposed epitope of IAPP within the water-insoluble amyloid fibrils forms β-sheet conformation, into which IAPP antibody cannot penetrate to bind (Anguiano et al. 2002). Antibodies generated to this epitope using toxic $A\beta_{1-40}$ bind toxic oligomers generated from the other amyloidogenic proteins in cell culture, and block the cytotoxic effects of each of these diverse oligomers (Kayed et al. 2003; Ritzel and Butler 2003).

Butler and his associates tested the hypothesis that β-cells are preferably vulnerable to hIAPP-induced apoptosis with isolated human islets in tissue culture: apoptotic cells by TUNEL were increased fivefold after incubation with 40 μmol/l hIAPP compared with control islets (Ritzel and Butler 2003). Further, in T2DM islets, the apoptotic cells in islets were adjacent to each other and contained two separate nuclei, suggesting that cells might undergo apoptosis shortly after mitosis (Ritzel and Butler 2003). More studies are needed to answer the question whether amyloid deposits are the cause or the result of T2DM or amyloid deposits are both the cause and the results of T2DM.

Hyperglycemia-Induced Apoptosis in β-Cells

Glucose is the main fuel that stimulates insulin secretion and mechanisms of glucose-induced insulin secretion is the fundamental principle for the pathophysiology of insulin secretion (Tomita et al. 1974; Tomita and Scarpelli 1977), and chronic hyperglycemia causes β-cell glucose toxicity and eventually leads to β-cell apoptosis (Federici et al. 2001). Glucose-induced insulin secretion with isolated perifused islets typically presents both an initial insulin secretion within several minutes and a second larger sustained secretion in about 20 min after exposure to a high glucose medium, in which the initial small secretion occurs before glucose is metabolized (Tomita et al. 1974; Tomita and Scarpelli 1977). This early peak of insulin secretion has brought about the glucose sensor theory by Matschinsky et al., who proposed glucokinase (GK, hexokinase IV) as a glucose sensor (Begoya et al. 1986; Shimizu et al. 1988; Matschinsky 1990, 1995; Matschinsky et al. 1993). GK constitutes a key component of mammalian glucose-sensing machinery (Matschinsky 1990, 1995; Matschinsky et al. 1993). In the liver, GK controls glycogen synthesis and glucose output, whereas in pancreatic islets it regulates insulin secretion (McDonald et al. 2005). Subsequent studies showed that glucose-sensing mechanisms in the β-cells are divided into the two components: proximal events of glucose entry and metabolism, and the distal mechanisms of insulin secretion, spanning from mitochondrial signal generation and initiation of electrical activity to the ultimate effectors of β-granule exocytosis (McDonald et al. 2005). The proximal sensing and metabolic signal generation includes the following:
1. Glucose equilibrates rapidly across the β-cell membrane owing to expression of the high-capacity, low-affinity glucose-transporter-2.
2. After glucose has entered the β-cells, it is phosphorylated to glucose-6-phosphate by the high K_MGK, which constitutes the flux-determining step of glycolysis and is considered a glucose sensor.
3. Once the glucose is phosphorylated, it is metabolized by glycolysis to pyruvate, NADH, and ATP (Innedian 1993; De Vos et al. 1995).

Pyruvate is the main end product of glycolysis in β-cells and is essential for mitochondrial ATP synthesis, and is suggested to be an important modulator of insulin secretion (McDonald et al. 2005). In the mitochondrial matrix, pyruvate is oxidized by pyruvate dehydrogenase to form acetyl-CoA. Acetyl-CoA enters the TCA cycle to undergo additional oxidation steps, generating CO_2 and reducing equivalents, $FADH_2$ and NADH (Innedian 1993; De Vos et al. 1995; Newgard and McGarry 1995). Oxidation of reducing equivalents by the respiratory chain is coupled with the extrusion of protons from the matrix to the outside of the mitochondria, thereby establishing the electrochemical gradient across the inner mitochondrial membrane (Fridlyand and Philson 2010). The final electron acceptor of these reactions is molecular oxygen (Fridlyand and Philson 2010). The distal sensing of metabolic signals includes the following:

1. In the absence of stimulatory glucose (<5 mmol/l) rodent β-cells are electrically silent, with a resting membrane potential of ~70 mV.
2. Reduction of the resting K^+ conductance by stimulated glucose leads to membrane depolarization and initiation of electrical activity characterized by slow wave depolarization.
3. ATP-sensitive K^+ channels set the β-cell membrane potential and closure of these leads to depolarization.
4. Membrane depolarization triggers action potential firing and opening voltage-dependent Ca^{2+} channels, leading to Ca^{2+} influx, which triggers β-granule exocytosis.
5. Action potential is terminated by the opening of voltage-dependent K^+ channels, which limit Ca^{2+} entry and thus insulin release (McDonald et al. 2005; Innedian 1993; De Vos et al. 1995; Newgard and McGarry 1995; Fridlyand and Philson 2010).

A comprehensive mathematical model of β-cell sensitivity to glucose predicts the special role of the mitochondrial control mechanism in insulin secretion and reactive oxygen species (ROS) generation in the β-cells (McDonald et al. 2005; Newgard and McGarry 1995; Fridlyand and Philson 2010). A failure of the insulin secretory machine results in insulin deficiency and subsequent hyperglycemia. Mutations in the GK gene lead to impaired insulin secretion in maturity onset diabetes of the young T2DM (MODY2), an autosomal-dominant DM, and this fact further supports the role of GK as a glucose sensor in diabetes (Porter and Barette 2005; Rossetti et al. 1990; DeFrenzo 1999). The individuals with mutations within the GCK gene on chromosome 7p (GCK-MODY2) have an abnormal glucose sensing with stable mild hyperglycemia and do not predispose to complications of diabetes (Froguel et al. 1992). The bulk of detailed information on hyperglycemia-induced apoptosis has been derived from controlled cultured human and rodent islets in vitro (Federici et al. 2001; Leahy et al. 1992). In order to study glucose toxicity as a deleterious effect of chronic hyperglycemia on β-cells in vitro, Federici et al. cultured 400 isolated human islets per batch for 5 days in a low-glucose (5.5 mmol/L) and a high-glucose medium (16.6 mmol/L) for studying any possible high glucose effects on Bcl-2 (β-cell lymphoma 2) family gene expression by RT-PCR (Federici et al. 2001).

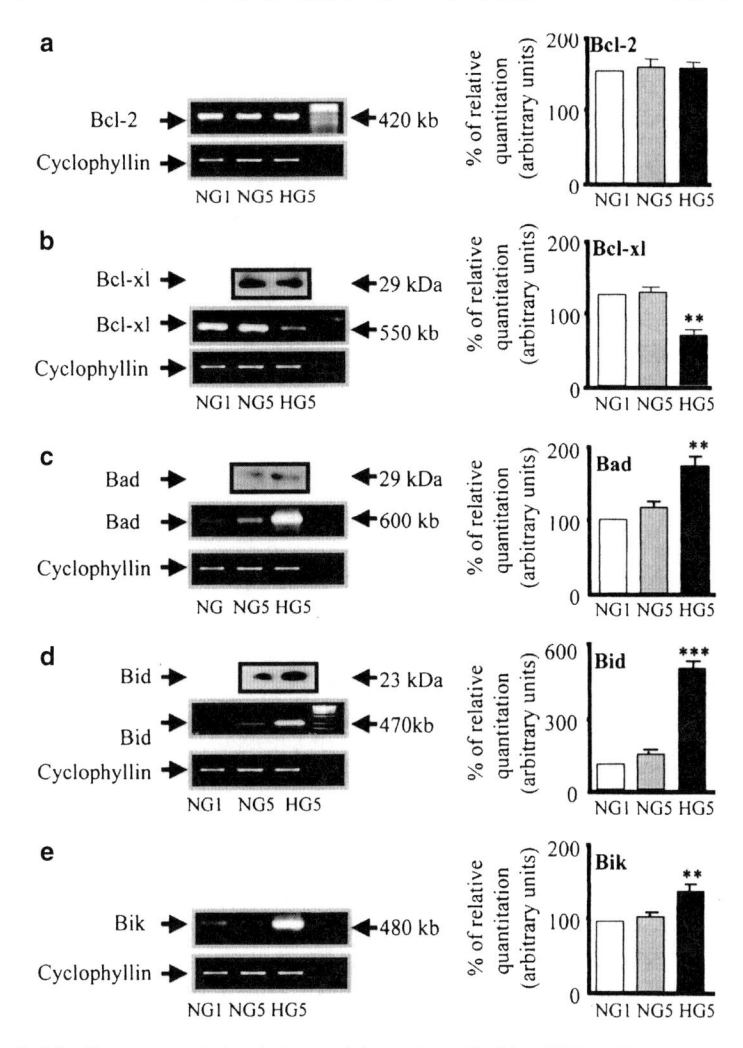

Fig. 5 Bcl family gene regulation in human islets cultured in high (HG, 16.7 mM) versus normal glucose (NG, 5.5 mM). (**a**) Bcl-2, (**b**) Bcl-xL, (**c**) Bad, (**d**) Bid, and (**e**) Bik RNA. *NG1* incubated in NG for 1 day, *NG5* incubated in NG for 5 days, *HG5* incubated in HG for 5 days (From Federici et al.(2001))

1. Bcl-2 was highly expressed in both low- and high-glucose media and expression did not change between a high- and a low-glucose condition (Fig. 5a) (Federici et al. 2001).
2. However, Bcl-xL was reduced by 45 % in high-glucose cultured islets compared with low-glucose cultured islets (Fig. 5b; Federici et al. 2001), supporting the notion that a reduction in Bcl-xL protein expression was due to high glucose exposure (Fig. 5b; Federici et al. 2001).
3. Bad, Bid, and Bik were expressed in low-glucose medium at low levels. Bad gene expression was greatly increased with a high-glucose medium and the

Bad protein level increased 80 % as well compared with low-level glucose (Fig. 5c–e; Federici et al. 2001).

Bid gene expression was markedly increased with high-glucose medium and so was Bik protein (Fig. 5d–e; Federici et al. 2001). Thus, anti-apoptotic Bcl-2 was unaffected by high glucose, but the pro-apoptotic genes, Bad, Bid, and Bik markedly increased in high-glucose cultured islets (Fig. 5c–e; Federici et al. 2001). These data support the notion that chronic high-glucose incubation in vitro modulates the balance of pro-apoptotic and anti-apoptotic Bcl proteins toward apoptosis, thus leading to eventual β-cell death (Federici et al. 2001).

Recent Studies on β-Cell Apoptosis

The main studies on apoptotic Bcl proteins were performed in artificially forced overexpression experiments in vitro (Zhou et al. 2000; Ou et al. 2005; Saldeen 2000; Chan and Yu 2004; Hengartner 2000). Caspases are activated in a hierarchy order, in which initiator caspases (caspase-8 and -10) function to cleave effector caspases (caspase-3 and -7), the latter in turn degrade a number of intercellular protein substrates and lead to the classical morphological changes of apoptosis (Figs. 1 and 2). Extracellular events present during the inflammatory response through the release of cytokines, including INF-σ, Il-1β, and interferon-ϒ by infiltrating leukocytes or direct cytotoxic T-cell engagement, can initiate apoptosis (Emamaullee and Shapiro 2006). These intrinsic cues function via surface molecules in the death receptor pathway, where specific ligand-receptor binding such as TNF–TNF receptor binding, Fas (CD 95)-Fas L (CD 178) binding lead to receptor clustering, adaptor molecule recruitment, and formation of DISC) (Fig. 2; Emamaullee and Shapiro 2006). Caspase-8 associates with DISC complex, where it is activated, released, and leads to effector activation for caspase-3 (Boatright and Salvesen 2003). Intracellular cues such as DNA damage, hypoxia, nutrient deprivation or reactive oxygen species (ROS) function via the mitochondrial pathway, which is tightly modulated by the Bcl-2 proteins (Fig. 2; Emamaullee and Shapiro 2006). In healthy cells, pro-apoptotic Bcl-2 proteins (Bim, Bid, Bad, Bax, and Bak) are present in the mature form, while anti-apoptotic Bcl-2 proteins (Bcl-2 and Bcl-xL) are constitutively active and reside in the outer membrane of mitochondria (Fig. 2; Emamaullee and Shapiro 2006; Hui et al. 2004; Boatright and Salvesen 2003). Following an intrinsic cue, pro-apoptotic Bcl proteins become activated and translocate to the mitochondria, where they result in inactivation of anti-apoptotic Bcl-2 proteins or form pores in the mitochondrial membrane, which facilitates the release of cytochrome c into the cytosol (Fig. 2; Emamaullee and Shapiro 2006). When cytochrome c accumulates in the cytosol, it forms a complex with procaspase-9 and Apaf-1 to form the "apoptosome," which in turn activates caspase-3 (Fig. 2; Emamaullee and Shapiro 2006). Both intrinsic and extrinsic signaling cascades converge at the point of caspase-3 activation, which is often considered the "point of no return" in apoptosis (Emamaullee and Shapiro 2006).

Apoptosis can only occur when the concentration of pro-apoptotic Bcl proteins exceeds that of anti-apoptotic proteins at the mitochondrial membrane of the intrinsic pathway (Emamaullee and Shapiro 2006).

Recent studies unveiled new roles for certain Bcl-2 family members in other physiological pathways, including glucose metabolism (Hengartner 2000), Ca^{2+} homeostasis (Daniel et al. 2003, 2008; Karbowski et al. 2006), and mitochondrial morphology (Daniel et al. 2008; Karbowski et al. 2006). BAD nucleates a core complex at the mitochondria-containing GK, the product of the gene associated with MODY2 (Shimizu et al. 1988). BAD resides in a GK-containing complex that regulates glucose-driven mitochondrial respiration (Daniel et al. 2008). Daniel et al. studied new insights into the role of BAD in glucose-stimulated insulin secretion by β-cells from $Bad^{-/-}$ compared with $Bad^{+/+}$ islets (Fig. 6) (Daniel et al. 2008): perifused islets from $Bad^{-/-}$ mice secreted significantly less insulin in response to 25 mM of glucose compared with $Bad^{+/+}$ mouse islets, at about 40 % for the first-phase secretion (0–15 min) and 60 % for the second-phase secretion (15–40 min) (Fig. 6a, b; Daniel et al. 2008); however, the total insulin secretion by 30 mM KCl (Fig. 6a; Daniel et al. 2008), glucose-induced (25 mM) changes in the ATP/ADP ratio (Fig. 6c; Daniel et al. 2008) and insulin secretion in response to 10 mM KIC and 25 mM tolbutamide (Fig. 6d; Daniel et al. 2008) were compatible in both $Bad^{+/+}$ and $Bad^{-/-}$ islets. GK activity in homogenates of $Bad^{-/-}$ islets was about 25 % that of $Bad^{+/+}$ islets (Fig. 6e; Daniel et al. 2008). Insulin secretion by $Bad^{-/-}$ islets was considerably lower than that of $Bad^{+/+}$ islets in response to glucose concentration of 15–25 mM (Fig. 6f; Daniel et al. 2008). A signature of β-cell dysfunction associated with impaired GK activity is a loss on glucose sensing (Fig. 6e, f; Daniel et al. 2008) and $Bad^{-/-}$ islets require more glucose to secrete insulin than wild-type islets (Fig. 6d, f; Daniel et al. 2008). Glucose-induced changes in mitochondrial membrane potential are significantly reduced in $Bad^{-/-}$ β-cells (Daniel et al. 2008). The reduction does not cause a global impairment of mitochondrial respiratory chains, as both genotypes ($Bad^{+/+}$ and $Bad^{-/-}$) show comparable changes in membrane potential to KIC (Daniel et al. 2008). Basal $[Ca^{2+}]_i$ at 3 mM glucose is compatible for both genotypes, indicating that basic control mechanisms for Ca^{2+} handling are presented in $Bad^{-/-}$ cells (Daniel et al. 2008). $Bad^{-/-}$ islets do not present a stepwise increase in insulin secretion when exposed to incrementally increased glucose concentration (Fig. 6f; Daniel et al. 2008). The efficiency of glucose and other fuel secretagogues to stimulate insulin secretion correlates with their capacity to hyperpolarize the mitochondrial potential (Karbowski et al. 2006; Antinozzi et al. 2002). In β-cells, the characteristic features of glucose-driven mitochondrial respiration correspond to those of glucose phosphorylation by GK (Liang et al. 1996; Berggren and Larson 1994). Glucose-induced changes in the mitochondrial membrane potential were significantly reduced in $Bad^{-/-}$ β-cells and the average $[Ca^{2+}]_i$ response to 11 mM glucose was significantly lower in $Bad^{-/-}$ β-cells (Daniel et al. 2008). The BH3 domain of BAD is an amphipathic σ-helix, which binds to Bcl-2 and Bcl-xL and neutralizes the apoptotic activity of BAD (Daniel et al. 2008). An intact BH3 domain is required for glucose-stimulated insulin secretion by its binding to Bcl-2 and Bcl-xL. Treatment with BAD $SAHB_A$

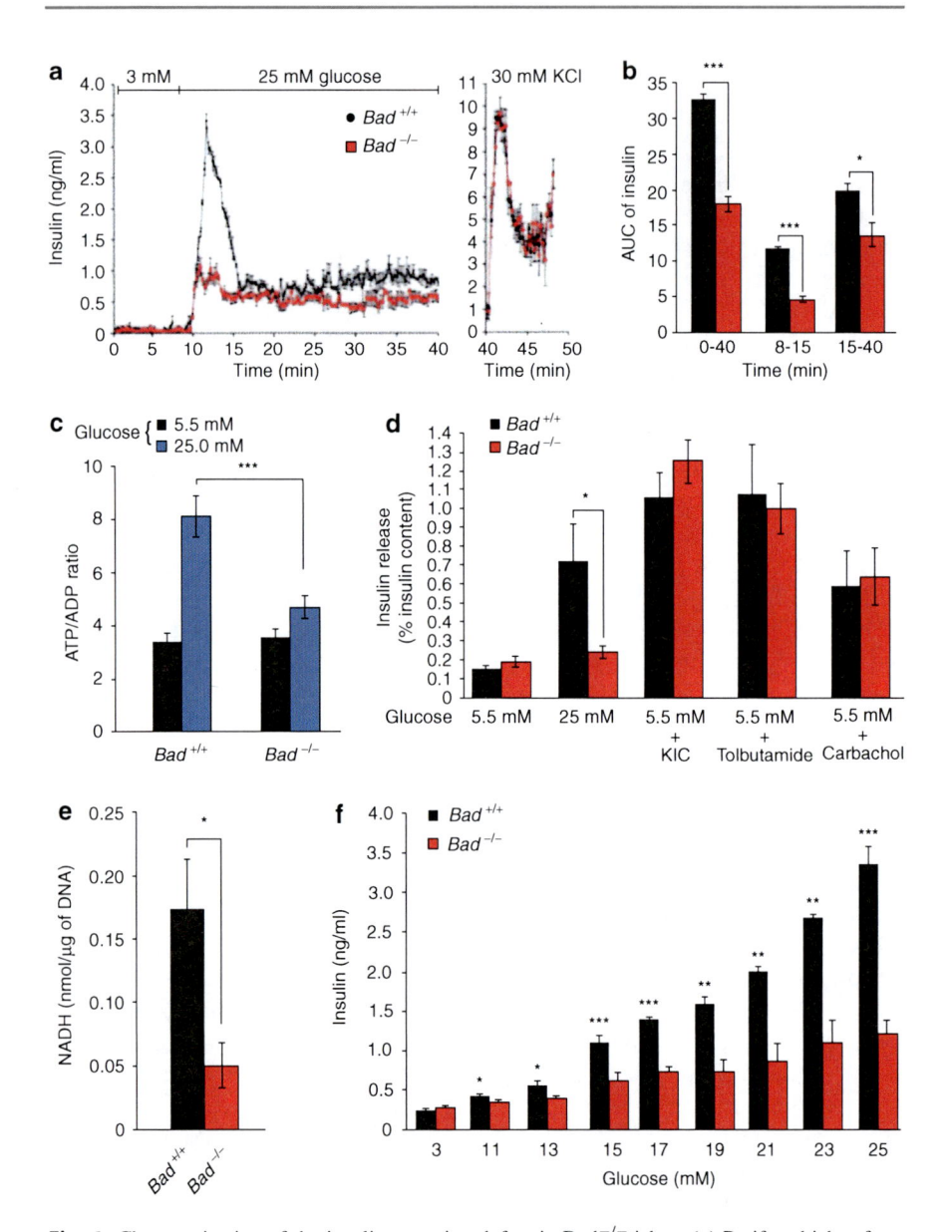

Fig. 6 Characterization of the insulin secretion defect in Bad$^{-/-}$ islets. (**a**) Perifused islets from Bad$^{-/-}$ mice (*red*) with 25 mM glucose secreted significantly less insulin compared with Bad$^{+/+}$ islets (*black*). (**b**) Insulin secretion throughout the perifusion (0–40 min), first phase (8–15 min) and second phase (15–40 min). (**c**) Glucose-induced changes in the ATP/ADP ratio in Bad$^{+/+}$ and Bad$^{-/-}$ islets – 5.5 mM (*black*), 25 mM (*blue*). (**d**) Insulin secretion in response to glucose 5.5 mM and 25 mM, 10 mM σ-ketoisocaproate (KIC), 0.25 mM tolbutamide and carbachol. (**e**) GK activity in homogenates of primary islets isolated from Bad$^{+/+}$ (*black*) and Bad$^{-/-}$ mice (*red*). (**f**) Insulin secretion by Bad$^{+/+}$ (*black*) and Bad$^{-/-}$ (*red*) islets perifused with incrementally increasing concentrations of glucose (From Daniel et al. (2003))

(stabilized σ-helix of Bcl-2 domain) restored the secretion defect in Bad$^{-/-}$ islets (Daniel et al. 2008), underscoring the sequence specificity of the BAD SAHB effect. Mutating the conserved leucine and aspartic acid residues of the BAD BH3 sequence (BAD SAHB$_{A(L,D \rightarrow A)}$) abrogated its effect on insulin secretion (Daniel et al. 2008). In their extensive elegant study, Daniel et al. identified GK as a novel and direct physiological target of the BAD BH3 domain in β-cells and that phosphorylation within the BH3 domain drives the metabolic functionality of BAD and serves as a physiological switch of its apoptotic and metabolic effects (Daniel et al. 2008). They demonstrated that Bad plays a physiological role in β-cells, aside from its role in β-cell apoptosis, and specifically that Bad phosphorylated at serine 155 promotes glucose-stimulated insulin secretion via interactions with GK. The therapeutic application of BAD SAHB$_s$ and other BAD BH3 may be applied in restoring insulin secretion. A phosphorylated BAD SAHB that activates glucose-stimulated insulin secretion, but does not affect the survival function of Bcl-xL, may serve as a prototype therapeutic in diabetes and islet transplantation (Daniel et al. 2003, 2008; Karbowski et al. 2006; Antinozzi et al. 2002; Liang et al. 1996; Berggren and Larson 1994).

Recently, Luciani et al. have demonstrated that chemical and genetic loss-of-function of anti-apoptotic Bcl-2 and Bcl-xL significantly augments glucose-dependent metabolic and Ca^{2+} signals by extending the role of Bcl-2 and Bcl-xL through suppressing glucose signaling in pancreatic β-cells (Luciani et al. 2013). Prolonged Bcl antagonism induced dose- and time-dependent cell death in human and mouse islet cells (Fig. 7; Luciani et al. 2013). The enhancement of β-cell glucose response by Bcl-2 was studied using genetically ablated mice as a genetic loss-of-function approach. Real-time PCR confirmed the loss of Bcl-2$^{-/-}$ β-cells and Bcl-2$^{+/-}$ islets compared with Bcl$^{+/+}$ control islets with no compensatory increase in Bcl-xL (Fig. 7a; Luciani et al. 2013). Bcl-2$^{-/-}$ and Bcl$^{+/-}$ β-cells showed enhanced sensibility to glucose (Fig. 7b, c; Luciani et al. 2013). Intact islets from Bcl-2$^{-/-}$ mice also showed increased Ca^{2+} and metabolic NAD(P)H response to glucose (Fig. 7e, f; Luciani et al. 2013). Perifused Bcl-2$^{-/-}$ islets to glucose revealed significantly increased insulin secretion compared with Bcl-2$^{+/+}$ islets (Fig. 7g; Luciani et al. 2013). Loss of Bcl-2 had no effect on responses to depolarization with KCl (Fig. 7d, h; Luciani et al. 2013). The immediate augmentation of glucose-induced Ca^{2+} responses in Bcl-2 heterozygous β-cells indicates that effects were dependent on gene dosage (Fig. 7b, c; Luciani et al. 2013). Inducible deletion of Bcl-xL mouse β-cells also increased glucose-stimulated NAD(H)P and Ca^{2+} responses and resulted in an improved in vivo glucose tolerance in the Bcl-xL knockout mice (Luciani et al. 2013). These results suggest that prosurvival Bcl proteins normally dampen the β-cell response to glucose and physiology of β-cells (Luciani et al. 2013). Bcl proteins directly affect mitochondrial proteins in the β-cells. The study of anti-apoptotic activates of Bcl-2 and Bcl-xL has shown that they interact with mitochondria via their BH4 domain (Luciani et al. 2013; Real et al. 2004). A cell-permeable Bcl-xL BH4 domain peptide triggers cytosolic and mitochondrial Ca^{2+} fluctuations in β-cells, which may result from direct mitochondrial actions of the BH4 domain and endoplasmic reticulum Ca^{2+} release

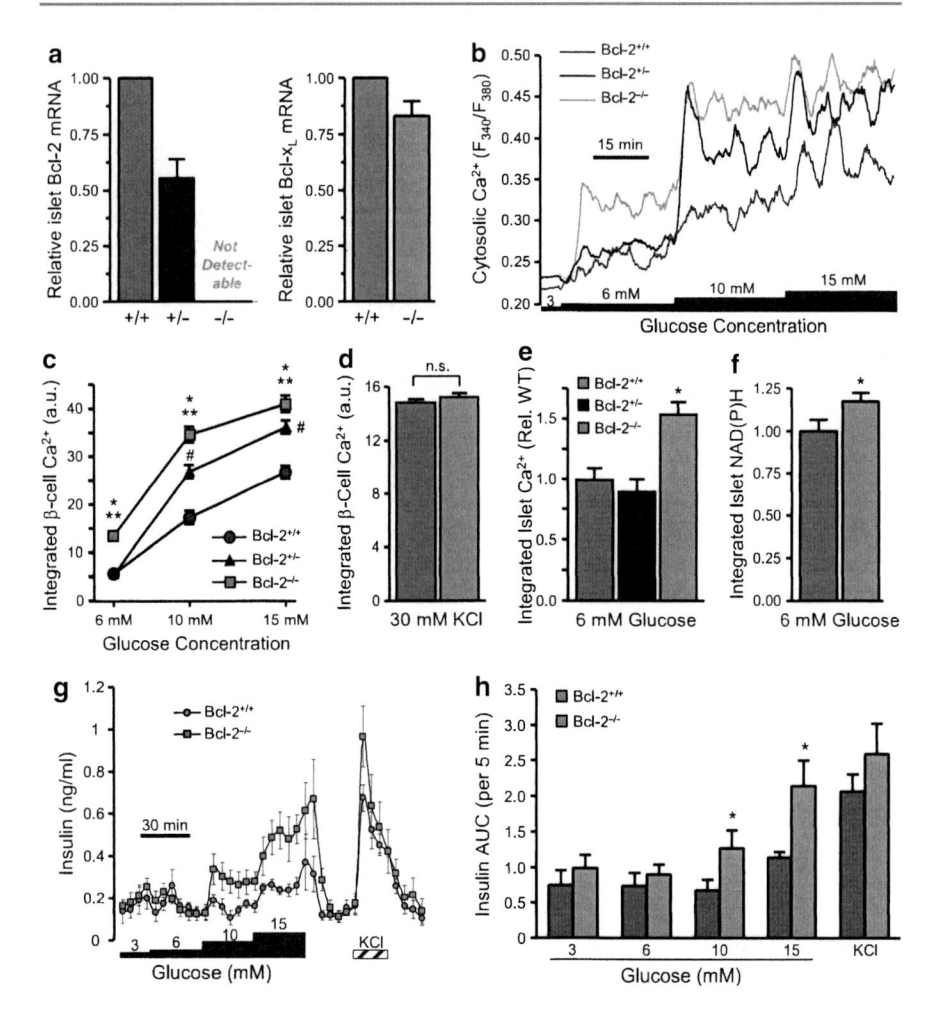

Fig. 7 Loss of Bcl-2 enhances β-cell glucose responses. (**a**) Quantitative PCR quantification of Bcl-2 and Bcl-xL mRNA levels in islets from Bcl-2$^{+/-}$ and Bcl-2$^{-/-}$ islets compared with Bcl-2$^{+/+}$ islets. (**b**) Average cytosolic Ca^{2+} levels of dispersed islet cells from Bcl-2$^{+/+}$, Bcl-2$^{+/-}$, and Bcl-2$^{-/-}$ islets. (**c**) Incremental area under the curve of Ca^{2+} responses by Bcl-2$^{+/+}$, Bcl-2$^{+/-}$, and Bcl-2$^{-/-}$ islets. (**d**) Integrated cytosolic Ca^{2+} responses of Bcl-2$^{-/-}$ and Bcl-2$^{+/+}$ β-cells depolarized with 30 mM KCl. (**e**) and (**f**) Integrated Ca^{2+} and NAD(P)H autofluorescence increases of intact islet cells, normalized Bcl-2$^{+/+}$ control islet cells. (**g**) Insulin secretion profiles of perifused islets from Bcl-2$^{+/+}$ and Bcl-2$^{-/-}$ islets. (**h**) Quantified area under the curve of insulin secretion profiles by Bcl-2$^{+/+}$ and Bcl-2$^{-/-}$ islets (From Luciani et al. (2013))

(Real et al. 2004). Bcl-xL can lower acetyl-CoA levels independently of Bax and Bak (Yi et al. 2011). Anti-apoptotic Bcl-2 family proteins can modulate β-cell function, and thus have implications for the pathophysiology of DM. The reduction in Bcl-2 and Bcl-xL under prediabetic conditions can affect β-cell function and insulin hypersecretion is an early marker of human DM (Tsujimoto and Shimizu 2007;

Rong et al. 2009; Simonson 1990). These results suggest that endogenous Bcl-2 and Bcl-xL suppress the β-cell response to glucose. The emerging evidence places Bcl family proteins at the intersection of β-cell function and survival. The involvement of apoptosis-regulating proteins provides fertile ground for future insights into the pathophysiology of diabetes and other diseases. GK is a novel and direct physiological target of the Bad BH3 domain in β-cells and genetic evidence combined with the pharmacological activity of novel stapled Bad BH3 peptides indicates that phosphorylation within the BH3 domain drives the metabolic function of Bad and serves as a physiological switch between its apoptotic and metabolic effects (Real et al. 2004). The molecular targeting of GK activation holds therapeutic promise and is leading to the development of several GK activator compounds; further, there is a therapeutic application of Bad, SAHBs, and other Bad BH3 mimics in restoring insulin secretion (Luciani et al. 2013).

The recent reports on β-cell apoptosis have come from well-funded laboratories with many researchers of different disciplines to cooperate in this competitive area of research. The current and ongoing research is focusing on insulin secretion coupled with glucose metabolism using isolated islets from genetically ablated mice. These studies explore the effects of Bcl-2 and Bcl-xL proteins versus anti-apoptotic proteins, Bim, Bid, Bad, and Bik as pro-apoptotic proteins (Daniel et al. 2008). Another genetic approach by McKenzie showed that in wild-type mice, islets were exposed to 33.3 mmol/L glucose or 50 mmol/L ribose in tissue culture for 5 days, which showed increased DNA fragmentation and cytochrome c release. The pan caspase-inhibitor qVD.oph significantly inhibited both ribose- and glucose-induced islet cell DNA fragmentation and cytochrome c release, supporting the notion that islet cell killing occurred through a caspase-dependent apoptotic process (McKenzie et al. 2010). The islets from RIP Bcl-2 transgenic mice did not reveal increased DNA fragmentation and cytochrome c release. These results implicate the Bcl-2 regulated apoptotic pathway in glucose-induced β-cell killing (McKenzie et al. 2010).

Glucose induces insulin secretion through its glucose sensor and metabolism, and insulin secretion in turn modulates glucose metabolism through its myriad of glycolytic reactions in mitochondria and K^+ and Ca^{2+} channel opening and closing at the β-cell membrane. Chronic hyperglycemia also induces a myriad of reactions causing glucose toxicity in β-cells. How to prevent and reverse β-cell apoptosis by modulating the balance of the Bcl family and apoptotic genes against apoptosis may lead to prevention and therapeutic application for both T1DM and T2DM (Luciani et al. 2013).

Cross-References

- ▶ Immunology of β-Cell Destruction
- ▶ Islet Structure and Function in the GK Rat
- ▶ Mechanisms of Pancreatic β-Cell Apoptosis in Diabetes and Its Therapies

- Microscopic Anatomy of the Human Islet of Langerhans
- The β-Cell in Human Type 2 Diabetes

References

Allison J, Thomas HE, Catterall T, Catterall T, Kay TW, Strasser A (2005) Transgenic expression of dominant-negative Fas-associated death domain protein in β cells protects against Fas ligand-induced apoptosis and reduces spontaneous diabetes in nonobese diabetic mice. J Immunol 175:293–301

Anguiano M, Nowak RJ, Lansbury PT Jr (2002) Protofibrillar islet amyloid polypeptide permeabilizes synthetic vesicles by a pore-like mechanisms that may be relevant to type II diabetes. Biochemistry 41:11338–11343

Antinozzi PA, Ishihara H, Newgard CB, Wollheim CB (2002) Mitochondrial metabolism sets the maximal limit of fuel-stimulated insulin secretion in a model pancreatic β cell: a survey of four fuel secretagogues. J Biol Chem 277:11746–11755

Barett KL, Willingham JM, Garvin AJ, Willingham MC (2001) Advances in cytochemical methods for detection of apoptosis. J Histochem Cytochem 49:821–832

Begoya FJ, Matschinsky F, Shimizu T, O'Neil JJ, Appel MC (1986) Differential regulation of glucokinase activity in pancreatic islets. J Biol Chem 261:10760–10764

Berggren PO, Larson D (1994) Ca^{2+} and pancreatic β-cell function. Biochem Soc Trans 22:12–18

Boatright KM, Salvesen GS (2003) Mechanisms of caspase activation. Curr Opin Cell Biol 15:725–731

Bonner-Weir S (2000a) Islet growth and development in the adult. J Mol Endocrinol 24:297–302

Bonner-Weir S (2000b) Life and death of the pancreatic β cells. Trends Endocrinol Metab 11:375–378

Bonner-Weir S, O'Brien TD (2008) Islets in type 2 diabetes in honor of Dr Robert C Turner. Diabetes 57:2899–2904

Brandhorst D, Kumarasamy V, Maatoui A, Aht A, Bretzel RG, Brandhorst H et al (2006) Porcine islet graft function is affected by pretreatment with a caspase-3 inhibitor. Cell Transplant 15:311–317

Buse JB, Weyer C, Maggs DC (2002) Amylin replacement with Pramlintide in type 1 and type 2 diabetes: a physiological approach to overcome a barrier with insulin therapy. Clin Diabetes 20:137–144

Butler AE, Janson J, Bonner-Weir S, Ritzel R, Rizza RA, Butler PC (2003a) β cell deficit and increased β-cell apoptosis in humans with type 2 diabetes. Diabetes 52:102–110

Butler AE, Janson J, Soeller WC, Butler PC (2003b) Increased β-cell apoptosis prevents adoptive increase in β-cell mass in mouse model of type 2 diabetes: evidence for role of islet amyloid formation rather than direct action of amyloid. Diabetes 52:2304–2314

Chan SI, Yu VC (2004) Proteins of the Bcl-2 family in apoptosis signaling: from mechanic insights to therapeutic opportunities. Clin Exp Pharmacol Physiol 31:119–128

Chandra J, Zhivotovsky B, Zaitsev S, Juntti-Berggren H, Berggren PO, Orreni S (2001) Role of apoptosis in pancreatic β-cell death in diabetes. Diabetes 50(Suppl 1):S44–S47

Cheng G, Ahu L, Mahato RI (2008) Caspase gene silencing for inhibiting apoptosis in insulin cells and human islets. Mol Pharmacol 5:1093–1102

Clark A, Nilsson MR (2004) Islet amyloid: a complication of islet dysfunction or an aetiological factor in type 2 diabetes? Diabetologia 47:157–169

Clark A, Wells CA, Buley ID, Cruickshank JK, Vanhegan RI et al (1988) Islet amyloid, increased α-cells, reduced β-cells and exocrine fibrosis quantitative changes in the pancreas in type 2 diabetes. Diabetes Res 9:151–159

Cleaved caspase-3 (Asp 175) antibody (2006). vol 9661. Cell Signaling Technology, Beverly, MA, Cat. # 9661 pp 1–7

Cooper GJ, Willis AC, Clark A, Tumer RC, Sim RB, Reid KB (1987) Purification and characterization of a peptide from amyloid-rich pancreas of type 2 diabetic pancreas. Proc Natl Acad Sci U S A 84:8628–8632

Cucca F, Lampis R, Congia M, Angius E, Nutland S et al (2001) A correlation between the relative predisposition of MHC class II alleles to type 1 diabetes and their structure of their proteins. Hum Mol Gen 10:2025–2037

Daniel NN, Gramm CF, Scorrano L, Zhang CY, Kraus S et al (2003) BAD and glucokinase reside in a mitochondrial complex that integrates glycolysis and apoptosis. Nature 424:952–956

Daniel NN, Walensky LD, Zhang CY, Choi CS, Fisher JK et al (2008) Dual role of proapoptotic BAD in insulin secretion and β cell survival. Nat Med 14:144–153

De Vos A, Heimberg H, Quatier E, Huypens P, Bouwens L et al (1995) Human and rat β cells differ in glucose transporter but not in glucokinase gene expression. J Clin Invest 96:2489–2495

DeFrenzo RA (1999) Pharmacologic therapy for type 2 diabetes mellitus. Ann Int Med 131:281–303

Duan WR, Garmer DS, Williams SD, Funkes-Shippy C, Spath I, Blomme EAG (2003) Comparison of immunohistochemistry of activated caspase-3 and cleaved cytokeratin 18 with the TUNEL method for quantification of apoptosis in histological section of PC-3 subcutaneous xenografts. J Pathol 199:221–228

Ehrlich JC, Ratner IM (1961) Amyloidosis of the islets of Langerhans. A restudy of islet hyaline in diabetic and non-diabetic individuals. Am J Pathol 38:49–59

Eisenbarth GS, Kotzin BL (2003) Enumerating autoreactive T cells in peripheral blood: a big step in diabetes prediction. J Clin Invest 111:179–181

Emamaullee JA, Shapiro AM (2006) Interventional strategies to prevent β-cell apoptosis in islet transplantation. Diabetes 55:1907–1914

Engel MFM, Khemtemourian I, Kleijer CC, Meeklik HJ, Jacobs J et al (2008) Membrane damage by human islet amyloid polypeptide through fibril growth at the membrane. Proc Natl Acad Sci U S A 105:6033–6038

Faustman D, Eisenbarth G, Daley J, Beitmyer J (1989) Abnormal T lymphocytes subsets in type 1 diabetes mellitus. Analysis with anti-2H4 andanti-4B4 antibodies. Diabetes 38:1462–1468

Faustman DL, Li X, Lim HY, Eisenbarth G, Avruch J, Guo J (1991) Linkage of family major histocompatibility complex class I to autoimmune diseases. Science 254:1756–1761

Federici M, Hrivbal M, Perego L, Ranalli M, Caradonna Z et al (2001) High glucose causes apoptosis in cultured human pancreatic islets of Langerhans. A potential role for regulation of specific Bcl family genes toward an apoptotic cell death program. Diabetes 50:1290–1301

Finegood D, Scaglia L, Bonner-Weir S (1995) Dynamics of β-cell mass in the growing rat pancreas: estimation with a simple mathematical model. Diabetes 44:249–256

Fineman M, Weyer C, Maggs DG, Strobel S, Koltermano G (2002) The human amylin analog, pramlintide, reduces postprandial hyperglucagonemia in patients with type 2 diabetes mellitus. Horm Metab Res 34:504–508

Fridlyand L, Philson LH (2010) Glucose sensing in the pancreatic β cells: a computational systems analysis. Theor Biol Med Model 7:1–44

Froguel P, Vaxillaire M, Sun F, Valho G, Zouali H et al (1992) Close linkage of glucokinase locus on chromosome 7p to early-onset non-insulin-dependent diabetes mellitus. Nature 356:162–164

Gown AM, Willingham MC (2002) Improved detection of apoptotic cells in archival paraffin sections: immunohistochemistry using antibodies to cleaved caspase3. J Histochem Cytochem 50:449–454

Green D (2005) Apoptotic pathways: ten minutes to death. Cell 121:671–674

Gross A, McDonnell JM, Korsmeyer SJ (1999) Bcl-2 family members and the mitochondria in apoptosis. Genes Dev 13:1899–1911

Haataja L, Gurlo T, Huang CJ, Butler PC (2008) Islet amyloid in type 2 diabetes and the toxic oligomer hypothesis. Endocr Rev 29:303–316

Hayashi T, Faustman DL (2003) Role of defective apoptosis in type 1 diabetes and other autoimmune diseases. Rec Prog Horm Res 58:131–153

Hengartner MO (2000) The biochemistry of apoptosis. Nature 407:770–776

Hirata H, Takahashi A, Kobayashi S, Yonehara S, Sawai H et al (1998) Caspases are activated in a branched protease cascade and control distinct down-stream processes in Fas-induced apoptosis. J Exp Med 187:587–600

Hoppener JW, Ahren B, Lips CJ (2000) Islet amyloid and type 2 diabetes mellitus. N Eng J Med 343:411–419

Hui H, Dotta F, De Maria U, Perfetti R (2004) Role of caspases in the regulation of apoptotic pancreatic islet β-cell death. J Cell Physiol 200:177–200

Hull RI, Westermark GT, Westermark P, Kahn SE (2004) Islet amyloid: a critical entity in the pathogenesis of type 2diabetes. J Clin Endocrinol Metab 89:3629–3643

Innedian BB (1993) Mammalian glucokinase and its gene. Biochem J 293:1–13

Jansen A, Van Hagen M, Drexhage HA (1995) Defective mutation and function of antigen-presenting cells in type1 diabetes. Lancet 345:491–492

Jansen J, Ashley RH, Harrison D, McIntyre S, Butler PC (1999) The mechanisms of islet amyloid polypeptide toxicity is membrane disruption by intermediate-sized toxic amyloid particle. Diabetes 48:491–498

Jurgens CA, Toukatly MN, Flinger CL, Udayasankar J (2011) Subramanian, SL et al: β-cell loss and β-cell apoptosis inhuman type2 diabetes are related to islet amyloid deposition. Am J Pathol 178:2632–2640

Kahn SE, Andrikopoulos S, Verchere CBN (1999) Islet amyloid: a long-recognized but underappreciated pathological feature of type 2 diabetes. Diabetes 48:241–253

Karbowski M, Norris KL, Cleland NM, Jeong SY, Youle RJ (2006) Role of Bax and Bak in mitochondrial morphogenesis. Nature 443:658–662

Kayed R, Head E, Thompson JI, McIntyre TM, Milton SC et al (2003) Common structure of soluble amyloid oligomers implies common mechanism of pathogenesis. Science 300:486–489

Kerr JP, Wyllie AH, Currie AR (1972) Apoptosis: a basic biological phenomenon with wide-range implications in tissue kinetics. Br J Cancer 26:239–257

Kruger DF, Gatcomb PM, Owen SK (1999) Clinical implication of amylin and amylin deficiency. Diabetes Educ 25:389–397

Leahy JL, Bonner-Weir S, Weir CC (1992) β-cell dysfunction induced by chronic hyperglycemia: current ideas on mechanisms of impaired glucose-induced insulin secretion. Diabetes Care 15:442–455

Lee SC, Pervaiz S (2007) Apoptosis in pathophysiology of diabetes mellitus. Int J Biochem 39:497–504

Leonardi O, Mints G, Hussain MA (2003) β-cell apoptosis in the pathogenesis of human type 2 diabetes mellitus. Eur J Endocrinol 149:99–102

Liang Y, Bai N, Doliba C, Wang L, Barner DK, Matschinsky FM (1996) Glucose metabolism and insulin release in mouse HC9 cells, as model of wild-type pancreatic β cells. Am J Physiol 270: E846–E857

Luciani DS, White SA, Widenmaier SB, Saran VV, Taghizadeh F et al (2013) Bcl-2 and Bcl-xL suppress glucose signaling in pancreatic β-cells. Diabetes 62:170–182

Mandrup-Poulsen T (2001) β-cell apoptosis: stimuli and signaling. Diabetes 50(Suppl 1):S58–S63

Marchetti P, Guerra S, Marselli L, Marselli L, Lupi R, Masini M et al (2004) Pancreatic islets from type 2 diabetic patients have functional defects and increased apoptosis that are ameliorated by metformin. J Clin Endocrinol Metab 89:5535–5541

Martin SJ, Green DR (1995) Protease activation during apoptosis: death by a thousand cuts? Cell 82:349–452

Matschinsky FM (1990) Glucokinase as glucose sensor and metabolic signal generator in pancreatic β-cell and hepatocytes. Diabetes 39:647–652

Matschinsky FM (1995) Banting lecture. A lesson in metabolic regulation inspired by the glucokinase glucose sensor paradigm. Diabetes 1995(45):223–241

Matschinsky F, Liang Y, Kesavan P (1993) Glucokinase as pancreatic β cell sensor and diabetes. J Clin Invest 92:2092–2098

McDonald PE, Joseph JW, Rorsman P (2005) Glucose-sensing mechanisms in pancreatic β-cells. Phil Trans R Soc 360:2211–2225

McKenzie MD, Jamieson E, Jansen ES, Scott CL, Huang DCS et al (2010) Glucose induces pancreatic islet cell apoptosis that requires the BH3-only proteins Bim and Puma and multi-BH domain protein Bax. Diabetes 59:644–652

Mirzabekov TA, Lin MC, Kagan BL (1996) Pore formation by the cytotoxic islet amyloid peptide amylin. J Biol Chem 27:1988–1992

Nakano M, Matsumoto I, Sawada T, Ansite J, Oberbroeckling J et al (2004) Caspase-3 inhibitor prevents apoptosis of human islets immediately after isolation and improves islet graft function. Pancreas 29:104–109

Newgard CB, McGarry JD (1995) Metabolic coupling factors in pancreatic β-cell signal transduction. Ann Rev Biochem 64:689–719

Nicholson DW, Thornberry NA (1997) Caspases: killer proteases. Trends Biochem Sci 22:299–306

O'Brien TD, Glabe CG, Butler PC (2010) Evidence for proteotoxicity in β cells in type 2 diabetes: toxic islet amyloid oligomers from intracellularly in the secretory pathway. Am J Pathol 176:861–869

Ou D, Wang X, Metzger DL, James RFL, Pozzilli P et al (2005) Synergetic inhibition of tumor necrosis factor-related apoptosis-inducing ligand-induced apoptosis inhuman pancreatic β cells by Bcl-2 and X-linked inhibitor of apoptosis. Hum Immunol 66:274–284

Peter ME, Krammer PH (1998) Mechanisms of CD95 (APO-1/Fas)-mediated apoptosis. Curr Opin Immunol 10:545–551

Porter JR, Barette TG (2005) Monogenic syndromes of abnormal glucose homeostasis: clinical review and relevance to the understanding of the pathology of insulin resistance and β-cell failure. J Med Genet 42:893–902

Rabinovitch A, Skyler JS (1998) Prevention of type 1 diabetes. Med Clin North Am 82:739–755

Real PJ, Cao Y, Wang R, Nikolovska-Coleska Z, Sanz-Ortiz J et al (2004) Breast cancer cells can evade apoptosis-mediated selective killing by a novel small molecule inhibitor of Bcl-2. Cancer Res 64:7947–7953

Reaven G (1988) Role of insulin resistance in human disease. Diabetes 37:1595–1607

Rhodes CJ (2005) Type 2 diabetes – a matter of β-cell life and death? Science 307:380–384

Ritzel RA, Butler PC (2003) Replication increases β-cell vulnerability to human islet amyloid polypeptide-induced apoptosis. Diabetes 52:1701–1708

Rong YP, Bultynck G, Aromolaran AS, Zhong F, Parys JB et al (2009) The BH4 domain of Bcl-2 inhibits ER calcium release and apoptosis by binding the regulatory and coupling domain of the IP3 receptor. Proc Natl Acad Sci U S A 106:14397–14402

Rossetti I, Giaccari A, DeFrenzo RA (1990) Glucose toxicity. Diabetes Care 13:610–630

Sakahira H, Enari M, Nagata S (1998) Cleavage of CAD inhibitor in CAD activation and DNA degradation during apoptosis. Nature 391:96–99

Saldeen J (2000) Cytokines induce both necrosis and apoptosis via a common Bcl-2-inhibitable pathway in rat insulin-producing cells. Endocrinology 141:2003–2010

Sava V, Caates JP, Hall PA (2001) Analysis of apoptosis in tissue sections. Endocr Mol Biol 174:347–359

Scaglia L, Cahill CJ, Finegood DT (1997) Apoptosis participates in the remodeling of endocrine pancreas in the neonatal rat. Endocrinology 138:1736–1741

Shimizu T, Knowles BB, Matschinsky FM (1988) Control of glucose phosphorylation and glucose usage in clonal insulinoma cells. Diabetes 37:563–568

Simonson DC (1990) Hyperinsulinemia and its sequence. Horm Metab Res 22(Suppl):17–25

Stassi GD, De Maria R, Trucco G, Rudert W, Testi R et al (1997) Nitric oxide primes pancreatic β cells for Fas (CD95)-mediated destruction in insulin-dependent diabetes mellitus. J Exp Med 186:1193–1200

Tewari M, Quan LT, O'Rourke K, Dosmoyers K, Zeng Z, Beidler DR et al (1995) Yama/cp32β, a mammalian homolog of CED-3, is a CrmA-inhibitable protease that cleaves the death substrate poly(ADP-ribose) polymerase. Cell 81:801–809

Tomita T (2005) Amyloidosis of pancreatic islets in primary amyloidosis (AL type). Pathol Int 55:223–227

Tomita T (2009) Cleaved caspase-3 immunocytochemical staining for pancreatic islets and pancreatic endocrine tumors. A potential marker for biological malignancy. Islets 2:82–88

Tomita T (2010a) Immunocytochemical localisation of cleaved caspase-3 in pancreatic islets from type 2 diabetic subjects. Pathology 42:432–437

Tomita T (2010b) Immunocytochemical localization of cleaved casspase-3 in pancreatic islets from type 1 diabetic subjects. Islets 2:24–29

Tomita T (2012) Islet amyloid polypeptide in pancreatic islets from type 2 diabetic subjects. Islets 4:223–232

Tomita T, Scarpelli DG (1977) Interaction of cyclic AMP and alloxan on insulin secretion in isolated rat islets perifused in vitro. Endocrinology 100:1327–1333

Tomita T, Lacy PE, Matschinsky FM, McDaniel M (1974) Effect of alloxan on insulin secretion in isolated rat islets perifused in vitro. Diabetes 23:517–524

Tsujimoto Y, Shimizu S (2007) VDAC regulation by the Bcl-2 family of proteins. Cell Death Differ 7:1174–1181

Urusova IA, Farila L, Hui H, D'Amico E, Perfetti R (2004) GLP-inhibition of pancreatic islet cell apoptosis. Trends Endocrinol Metab 15:27–33

Westermark P, Westermark C, Wilander E, Sletten K (1986) A novel peptide in the calcitonin gene related peptide family as an amyloid fibril protein in the endocrine pancreas. Biochem Biophys Res Commun 140:827–831

Weyer C, Maggs DG, Young AA, Kohlman OG (2001) Amylin replacement with pramlintide as an adjunct to insulin therapy in type 1 and type 2 diabetes mellitus: a physiological approach toward improved metabolic control. Curr Pharm Des 7:1353–1373

Willcox A, Richardson SJ, Bone AJ, Foulis AK, Morgan NG (2008) Analysis of islet inflammation in human type 1 diabetes. Clin Exp Immunol 155:173–181

Yi CH, Pan H, Seebacher J, Jang F, Hyberts SG, Hoffron GJ et al (2011) Metabolic regulation of protein N-α acetylation by Bcl-xL promotes cell survival. Cell 146:607–620

Zhou YP, Pena JC, Roe MW, Mittal A, Levisetti M et al (2000) Overexpression of Bcl-xL in β cells prevents cell death but impairs mitochondrial signal for insulin secretion. Am J Physiol Endocrinol Metab 278:E340–E351

Mechanisms of Pancreatic β-Cell Apoptosis in Diabetes and Its Therapies

32

James D. Johnson, Yu H.C. Yang, and Dan S. Luciani

Contents

Introduction to β-Cell Death	874
Increased β-Cell Death as a Trigger and Mediator of Type 1 Diabetes	875
Pancreatic β-Cell Death as a Complication of Diabetes: Glucose Toxicity	877
Programmed β-Cell Death as a Contributing Factor in Type 2 Diabetes	878
Mechanisms of β-Cell Apoptosis in Type 2 Diabetes: ER Stress	879
Mechanisms of β-Cell Apoptosis in Type 2 Diabetes: Lipotoxicity	880
Mechanisms of β-Cell Apoptosis in Type 2 Diabetes: Proinflammatory Cytokines	881
Genetic Factors Affecting β-Cell Apoptosis in Type 2 Diabetes	881
The Role of β-Cell Apoptosis in Rare Forms of Diabetes	882
Islet Engraftment and β-Cell Death in Islet Transplantation	883
Survival Factors that Prevent β-Cell Apoptosis	884
The Role of Other β-Cell Death Modalities: Beyond β-Cell Apoptosis	885
β-Cell Apoptosis as a Therapeutic Target in Diabetes: Future Directions	888
Cross-References	889
References	889

Abstract

Diabetes occurs when β-cells no longer function properly or have been mostly destroyed. Pancreatic β-cell loss by apoptosis and other modes of death contributes to both autoimmune type 1 diabetes and type 2 diabetes. Programmed pancreatic β-cell death can be induced by multiple stresses in both major types of diabetes. There are also several rare forms of diabetes, including Wolcott-Rallison

---

J.D. Johnson (✉) • Y.H.C. Yang
Diabetes Research Group, Department of Cellular and Physiological Sciences, University of British Columbia, Vancouver, BC, Canada
e-mail: james.d.johnson@ubc.ca; carol_yh_yang@hotmail.com

D.S. Luciani
Diabetes Research Program, Child & Family Research Institute, University of British Columbia, Vancouver, BC, Canada
e-mail: dluciani@cfri.ca

syndrome, Wolfram syndrome, as well as some forms of maturity onset diabetes of the young that are caused by mutations in genes that may play important roles in β-cell survival. The use of islet transplantation as a treatment for diabetes is also limited by excessive β-cell death. Mechanistic insights into the control of multiple modes of β-cell death are therefore important for the prevention and treatment of diabetes. Indeed, a substantial quantity of research has been dedicated to this area over the past decade. In this chapter, we will review the factors that influence the propensity of β-cells to die and the mechanisms of programmed cell death involved in the initiation and progression of diabetes.

Keywords

Clinical islet transplantation • Autoimmune diabetes • Glucotoxicity and lipotoxicity • Endoplasmic reticulum stress • Gene-environment interactions • Mitochondrial death pathway

List of Abbreviations

GLP-1   Glucagon-like peptide 1
MODY   Maturity onset diabetes of the young
NOD   Nonobese diabetic
UPR   Unfolded protein response
VNTR   Variable number of tandem repeats

Introduction to β-Cell Death

A person's functional β-cell mass determines, to a large extent, their glucose homeostasis and susceptibility to diabetes. Functional β-cell mass is the product of β-cell number, β-cell size, and the ability of individual β-cells to secrete mature insulin in a correct manner (Fig. 1)(Szabat et al. 2012).

It has become increasingly evident that β-cell apoptosis contributes to the development of both type 1 diabetes (autoimmune diabetes) and type 2 diabetes (adult-onset diabetes), as well as to more rare forms of the disease such as the various types of maturity onset diabetes of the young (MODY) (Oyadomari et al. 2002; Mathis et al. 2001; Leonardi et al. 2003; Johnson 2007; Donath and Halban 2004). Basal β-cell apoptosis also plays a role in the remodeling and development of the normal endocrine pancreas. For example, β-cells undergo a wave of apoptosis around the time of birth (Mathis et al. 2001), which is followed by a proliferation-driven postnatal expansion of β-cell mass (Meier et al. 2008). At all stages of life, β-cell replication and death are tightly controlled by intrinsic and extrinsic factors that control how β-cell mass adjusts to meet metabolic demand (Bell and Polonsky 2001). Only when a combination of genetic and environmental influences causes this balance to fail does diabetes develop. Despite major advances in recent years, the nature of the gene-environment interactions that promote β-cell apoptosis in diabetes remains unclear, as do many aspects of the apoptotic pathways

Fig. 1 Factors that dictate the functional β-cell mass

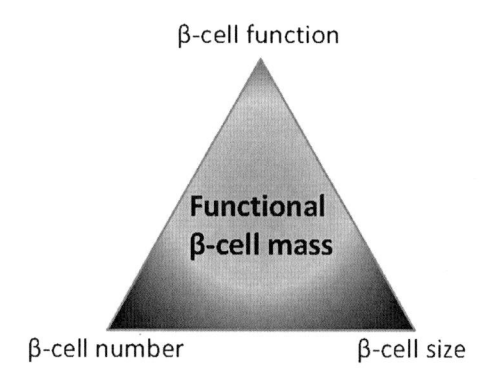

β-cell function

Functional β-cell mass

β-cell number   β-cell size

involved. In the following article, we will review some of the central mechanisms that have been implicated in the control β-cell apoptosis to date, as well as current therapeutic efforts that target these pathways.

Increased β-Cell Death as a Trigger and Mediator of Type 1 Diabetes

Type 1 diabetes is an autoimmune disease in which the pancreatic β-cells are gradually destroyed, but the initial trigger for this destruction and the exact mechanisms of β-cell death remain enigmatic. Like necrosis, excessive apoptosis is capable of initiating an immune response in susceptible individuals. It has been suggested that a perinatal wave of β-cell apoptosis may promote the presentation of β-cell autoantigens and thus provoke an autoimmune response against β-cells (Mathis et al. 2001; Liadis et al. 2005; Trudeau et al. 2000).

Clues to the cause and pathobiology of type 1 diabetes also come from the analysis of its genetics. In most cases, genes linked to type 1 diabetes are known to play specific roles in the immune system. IDDM1 is the human leukocyte antigen system superlocus containing the major histocompatibility complex genes. This region of the human and nonobese diabetic (NOD) mouse genome confers the majority of the risk for type 1 diabetes. Interestingly, the insulin gene itself (IDDM2) is the second most significant type 1 diabetes gene in humans. The genetic alterations are not in the coding sequence of insulin, but in an upstream regulatory region called the "variable number of tandem repeats" or VNTR (Pugliese et al. 1997; Bennett and Todd 1996; Barratt et al. 2004). At-risk alleles appear to reduce the expression of the insulin gene in the thymus where it is thought to play a role in tolerance (Pugliese et al. 1997). At the same time, VNTR sequences that confer diabetes risk increase insulin mRNA in the islets. High doses of insulin can have deleterious effects on the survival of β-cells under some culture conditions (Johnson et al. 2006a; Guillen et al. 2008). If this was also the case in vivo, one might expect that the VNTR could increase type 1 diabetes risk via direct effects on β-cell apoptosis. Recent in vivo evidence has demonstrated that insulin gene dose can

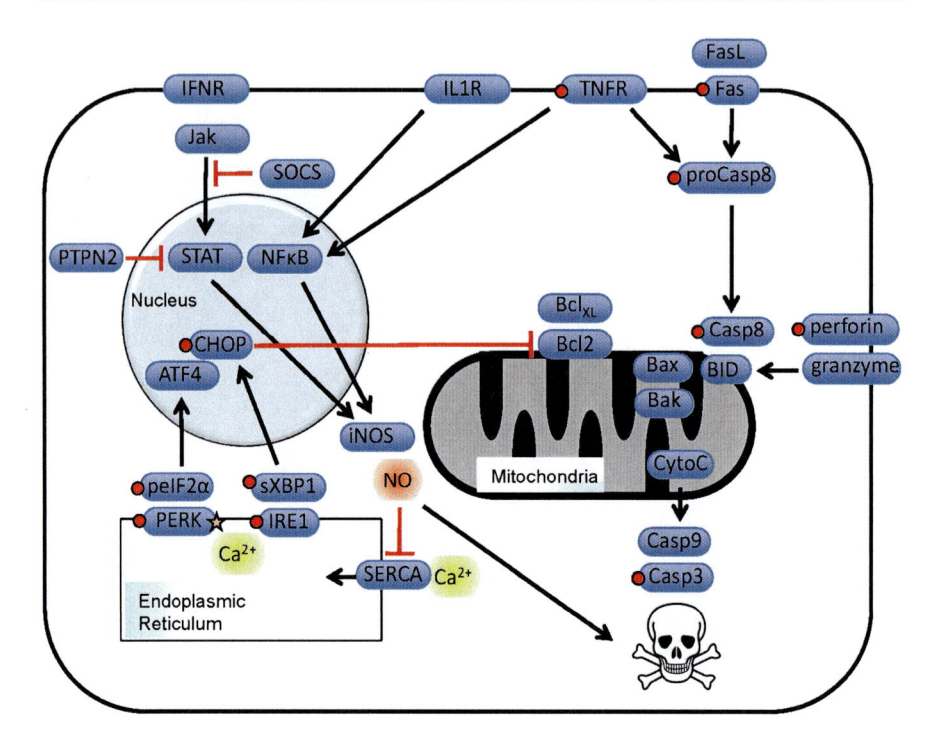

Fig. 2 Molecular mechanisms controlling β-cell apoptosis in type 1 diabetes. Shown is a partial description of signaling cascades that modulate β-cell survival in type 1 diabetes. Protein products of genes that are linked to human diabetes are denoted with a *star*. Genes that have been implicated in β-cell apoptosis or β-cell mass using in vivo or molecular loss-of-function experiments (i.e., knockout mice) are denoted with a *dot*

modulate β-cell mass (Mehran et al. 2012) and mathematical modeling has supported the feasibility of anti-apoptotic autocrine/paracrine insulin action (Wang et al. 2013).

Genome-wide association studies have identified several single nucleotide polymorphisms that contribute to the risk of type 1 diabetes. While it had long been assumed that most of these genes are involved exclusively in immune cells, it is now clear that virtually all of the currently known type 1 diabetes susceptibility genes are expressed in pancreatic β-cells as well (Eizirik et al. 2012), where they could theoretically modulate cell survival. For example, PTPN2 modulates β-cell apoptosis via effects on BIM and the transcription factor STAT1 (Moore et al. 2009; Santin et al. 2011; Colli et al. 2010). GLIS3 is unique in that it may modulate β-cell fate in the context of both type 1 diabetes and type 2 diabetes (Nogueira et al. 2013). Thus, genes that confer risk to type 1 diabetes may also affect β-cell death directly.

The mechanisms by which β-cells are selectively killed by the immune system have been studied extensively and appear to involve multiple pathways (Fig. 2).

One mechanism is the activation of "death receptors", Fas and tumor necrosis factor receptor, by their respective ligands. Interestingly, Fas expression is negligible in normal β-cells and it may be upregulated by cytokines such as IL-1

(Thomas et al. 2009). Activation of Fas by FasL converts pro-caspase-8 to active caspase-8 (Mathis et al. 2001). Caspase-8 then acts via the proapoptotic BH3-only Bcl family member Bid to promote mitochondrial outer membrane permeabilization and cytochrome c release (McKenzie et al. 2008). Bid may do so by interacting directly with the proapoptotic effector Bcl protein Bax and activate its channel forming functions in the outer mitochondrial membrane (Lovell et al. 2008).

Another pathway of β-cell apoptosis in type 1 diabetes involves perforin and granzyme B, cytotoxic components released by $CD8^+$ T cells. Mouse models suggest $CD8^+$ T cells to be major effectors of immune mediated β-cell death and perforin knockout mice on a NOD background have reduced diabetes incidence compared with NOD controls (Thomas et al. 2009). Granzyme B cleaves multiple substrates in the target cell, including Bid and studies with islets from Bid knockout mice demonstrate that Bid is also key in this β-cell death cascade (Estella et al. 2006).

The involvement of other Bcl family members in type 1 diabetes and its animal models is less clear. Pancreatic islets isolated from Bax knockout mice are partially protected from death receptor-triggered β-cell apoptosis, in agreement with Bax being the downstream effector of mitochondrial outer membrane permeabilization following Bid activation (McKenzie et al. 2008). Efforts to block diabetes using transgenic mice overexpressing Bcl-2 under the control of the rat insulin promoter provided mixed results (Thomas et al. 2009). To date, no in vivo loss-of-function experiments have demonstrated an essential role for anti-apoptotic Bcl-2 or Bcl-x_L in basal β-cell survival (Luciani et al. 2013). Interestingly, Bcl family proteins also play key roles in β-cell metabolic function (Luciani et al. 2013; Danial et al. 2008), making studies into the joint role of these proteins especially important.

While the Bcl proteins collectively control mitochondrial outer membrane permeabilization and cytochrome c release, the majority of the β-cell "execution" steps are triggered by the activation of effector caspases, such as caspase-3. These proteases also coordinate the semi-ordered disassembly of β-cells with members of the calpain family of calcium-activated proteases. Pancreatic β-cell apoptosis is promoted by caspase-3 and caspase-9, essential mediators in the intrinsic pathway of apoptosis. In cell culture models, β-cell death can be abrogated with inhibitors of caspase-3 activity (Yamada et al. 1999). In vivo, mice lacking caspase-3 in their β-cells are protected from type 1 diabetes (Liadis et al. 2005). Interestingly, isolated islets from β-cell-specific caspase-8 knockout mice are protected from Fas-induced apoptosis, but have increased "basal" apoptosis and glucose intolerance in the absence of frank diabetes (Liadis et al. 2007). These results suggest that the action of caspases can be context-dependent in the β-cell.

Pancreatic β-Cell Death as a Complication of Diabetes: Glucose Toxicity

Pancreatic β-cells are exquisitely sensitive to metabolic stress, since they must transduce changes in blood glucose levels into insulin release via glycolytic and mitochondrial ATP production (Bell and Polonsky 2001). Since hyperglycemia and hyperlipidemia both are hallmarks of the diabetic state, β-cell apoptosis is also likely to be an important complication of diabetes. This downward spiral likely plays a significant role in the rapid reduction in functional β-cell mass that precipitates the onset of both type 1 and type 2 diabetes. Chronically elevated glucose induces β-cell apoptosis via multiple mechanisms, including modulating the gene expression of multiple Bcl family members (Federici et al. 2001). Toxic high levels of reactive oxygen species are produced by hyperactive mitochondria and β-cells contain relatively low levels of some key antioxidant proteins (Federici et al. 2001; Kaneto et al. 2005; Robertson et al. 2004). Moreover, elevated Ca^{2+} levels associated with overworked β-cells are toxic to the cells (Maedler et al. 2004; Efanova et al. 1998). This excitotoxicity may be the cause of the eventual clinical failure of long-term sulfonylurea treatment, which depolarizes β-cells by directly closing K_{ATP} channels (Maedler et al. 2005). Prolonged hyperglycemia may also activate Fas-mediated β-cell apoptosis (Maedler and Donath 2004) and pathways controlled by the proapoptotic protein TXNIP (Chen et al. 2008). Moreover, chronic hyperglycemia increases secretory demand, which has been speculated to cause ER stress due to the increased requirement for protein synthesis and processing (see below).

Programmed β-Cell Death as a Contributing Factor in Type 2 Diabetes

It is established that pancreatic β-cell death is a key event in type 1 diabetes, but evidence has only recently emerged supporting an important role for β-cell apoptosis in the pathobiology of type 2 diabetes (Leonardi et al. 2003; Donath and Halban 2004; Rhodes 2005; Shu et al. 2008; Jeffrey et al. 2008; Cnop et al. 2005; Johnson et al. 2004; Butler et al. 2003a, b) (Fig. 3).

Type 2 diabetes is a disease of gene-environment interactions, with obesity and hyperlipidemia being the main manifestations of the "environment." Obesity is associated with inflammation and insulin resistance in a multitude of key metabolic tissues, including liver, fat, and muscle (Ozcan et al. 2004, 2006). In the majority of obese people, an expansion of β-cell mass and workload can effectively compensate for the increased insulin secretory demand (Rhodes 2005; Butler et al. 2003a, b). However, if this compensatory increase in β-cell mass and function fails, the obese individual will progress to frank type 2 diabetes. Compared to weight-matched controls, patients with type 2 diabetes have been reported to exhibit a 60 % reduction in β-cell mass associated with significantly increased β-cell apoptosis and ER stress (Butler et al. 2003a, b). A disruption of islet architecture and an accumulation of amyloid deposits are also associated with type 2 diabetes (Haataja et al. 2008). It is clear that much work remains to be done to distinguish cause and effect relationships between these pathologies. Moreover, the concepts of β-cell health and β-cell dedifferentiation, intermediates between dysfunction and death, are gaining increased attention.

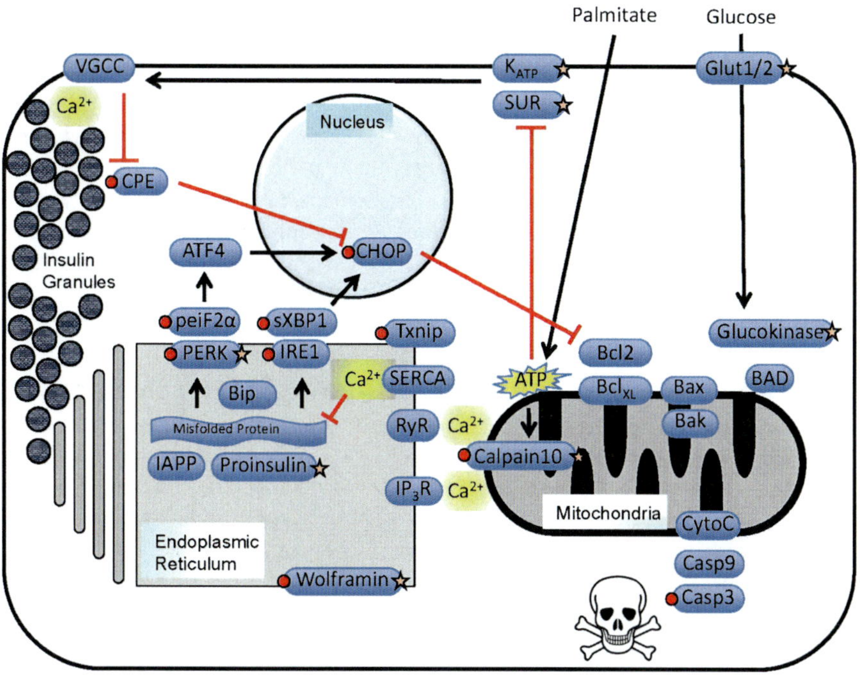

Fig. 3 Molecular mechanisms controlling lipid- and glucose-induced β-cell apoptosis in type 2 diabetes. Shown is a partial description of signaling cascades that modulate β-cell survival. Protein products of genes that are linked to human diabetes are denoted with a *star*. Genes that have been implicated in β-cell apoptosis or β-cell mass using in vivo loss-of-function experiments (i.e., knockout mice) are denoted with a *dot*

Mechanisms of β-Cell Apoptosis in Type 2 Diabetes: ER Stress

Pancreatic β-cells are the body's only source of blood-borne insulin and therefore must produce and secrete large amounts of this hormone as well as other hormones such as amylin. This high secretory demand makes them susceptible to secretory pathway stress, especially if demand is increased by insulin resistance. Elevated protein flux through the ER and Golgi can result in misfolded proteins and activation of the unfolded protein response (UPR) (Oyadomari et al. 2002; Eizirik et al. 2008; Harding and Ron 2002). Three main ER-resident signaling molecules, PERK, ATF6, and IRE1, act as sensors to trigger cellular adaptation responses or ultimately β-cell apoptosis if the stress is not alleviated. Important components of the initial "rescue response" are the PERK-triggered and eIF2α-mediated regulation of protein translation as well as an increased ER-associated degradation of misfolded proteins. When these rescue efforts fail, apoptosis is triggered. The relative sensitivity of β-cells to ER stress-induced cell death is illustrated by humans and mice with mutations in PERK, since other cells in the body can be largely unaffected (Harding and Ron 2002). The transcription factor CHOP is a major mediator of ER stress-induced apoptosis downstream of PERK and ATF6. Mice lacking the CHOP gene are resistant to β-cell apoptosis following ER stress and are protected from developing diabetes under these conditions (Eizirik et al. 2008; Song et al. 2008). Importantly, there is now increasing evidence of ER stress in islets of human type 2 diabetes patients (Eizirik et al. 2008; Laybutt et al. 2007), suggesting that ER stress does in fact contribute to β-cell apoptosis during the progression of type 2 diabetes. It is less clear whether β-cell ER stress can initiate β-cell death prior to the onset of type 2 diabetes.

Mechanisms of β-Cell Apoptosis in Type 2 Diabetes: Lipotoxicity

Obesity is thought to trigger type 2 diabetes by causing hyperlipidemia and insulin resistance. These events impose increased secretory demand on individual β-cells, which can activate the UPR, as outlined above. Moreover, elevated fatty acids, such as palmitate, have direct toxic effects on the β-cell via activation of a number of relatively separate apoptosis-inducing events, including the generation of ceramide and excessive reactive oxygen species. Palmitate activates the caspase-3-dependent mitochondrial apoptosis pathway (Jeffrey et al. 2008). Some investigators have shown that the activation of caspase-3 by palmitate is synergistic with the detrimental effects of high glucose (El-Assaad et al. 2003; Prentki and Nolan 2006), but it also triggers β-cell apoptosis in the absence of elevated glucose levels (Jeffrey et al. 2008). Palmitate also decreases the expression of the anti-apoptotic Bcl-2 protein (Lupi et al. 2002). The type 2 diabetes susceptibility gene, calpain-10, is also implicated in palmitate-induced β-cell death, since islets lacking calpain-10 have ~30% less apoptosis and mice with transgenic overexpression of calpain-10 are more susceptible to palmitate toxicity (Johnson et al. 2004). Moreover, palmitate has been demonstrated to directly act on the distal components of the insulin processing machinery of the β-cell. Specifically, palmitate induces a rapid, Ca^{2+}-dependent degradation of carboxypeptidase E, the final enzyme required for the conversion of proinsulin into mature insulin (Jeffrey et al. 2008). Carboxypeptidase E is also reduced in high fat-fed mice and the transgenic MKR mouse model of insulin resistance (Lu et al. 2008). A decrease in carboxypeptidase E is sufficient to induce CHOP-dependent ER stress and β-cell apoptosis in vivo and in vitro. It is unclear how reduced carboxypeptidase E modulates β-cell apoptosis, but two possibilities can be considered. In one scenario, a backlog of unprocessed insulin induces the UPR from inside the cell. It is also possible that a reduction in local release of mature insulin could impair β-cell survival. Substantial evidence suggests local insulin levels at the right concentration may help protect β-cells against ER stress and apoptosis (Johnson et al. 2006a, b; Mehran et al. 2012; Johnson and Alejandro 2008; Martinez et al. 2008), and islets from patients with type 2 diabetes exhibit reductions in several critical insulin signaling components (Gunton et al. 2005). To add to the complexity, insulin may modulate the levels of CPE

(Chu et al. 2011). Fatty acids, including palmitate, also modulate secretory pathway stress by partially depleting ER Ca^{2+} stores (Gwiazda et al. 2009). Although an incomplete ER Ca^{2+} reduction alone is not sufficient to induce ER stress, this event activates PERK and it is likely that this could potentiate ER stress induced by other factors (Gwiazda et al. 2009). Other systems involved in β-cell lipotoxicity include de novo lipogenesis, the ubiquitin proteasome system, lipophagy, and subcellular lipid distribution/flux (Chu et al. 2010, 2012; Pearson et al. 2014; Boslem et al. 2011, 2012, 2013; Preston et al. 2009). This is an intense area of investigation with new breakthroughs on the horizon.

Mechanisms of β-Cell Apoptosis in Type 2 Diabetes: Proinflammatory Cytokines

There is emerging evidence that proinflammatory cytokines and immune cell infiltration of the islet are common factors in type 1 diabetes and type 2 diabetes. The type 2 diabetic milieu of increased hyperglycemia and hyperlipidemia appears to stimulate the production of IL-1β from islets themselves. This has been suggested to have local inflammatory effects and advance subsequent islet infiltration by macrophages to promote apoptosis in type 2 diabetes (Ehses et al. 2007). There is evidence that proapoptotic cytokines (IL-1β, TNFα, IFNγ) can act through nitric oxide to decrease the expression of the SERCA pumps that load Ca^{2+} into the ER, which in turn impairs Ca^{2+}-dependent protein processing and promotes ER stress-induced β-cell apoptosis (Eizirik et al. 2008; Cardozo et al. 2005). This is in addition to changes in ER Ca^{2+}-release channels seen in the diabetic state (Lee et al. 1999). Cytokines might thus promote similar types of β-cell apoptosis in type 1 and type 2 diabetes, but the extent to which overlapping pathways are involved has been questioned (Cnop et al. 2005). Interested readers are referred to some excellent reviews on this topic (Eizirik et al. 2013; Eizirik and Grieco 2012).

Genetic Factors Affecting β-Cell Apoptosis in Type 2 Diabetes

Type 2 diabetes is a polygenic disease, with dozens of genes being implicated via both candidate studies and unbiased genome-wide approaches. Some of the first gene candidates studied for their role in type 2 diabetes were ones that play important roles in β-cell function. These included the components of the ATP-sensitive potassium channels (KCNJ11, ABCC8). PPARγ was also linked to type 2 diabetes risk, and recent experiments point to a role for PPARγ in β-cell apoptosis (Lin et al. 2005). The first type 2 diabetes susceptibility gene discovered by unbiased linkage mapping was calpain-10 (Horikawa et al. 2000), although this association is not seen in all populations. In the β-cell, calpain-10 likely plays a proapoptotic role in addition to a role promoting insulin secretion (Marshall et al. 2005). Additional in vivo studies are required to determine the detailed roles of the calpain-10 gene, which encodes for 8 splice variants, in the maintenance of glucose homeostasis.

Newer genome-wide association studies have found about 20 single nucleotide polymorphisms that show significant and reproducible associations with type 2 diabetes (Lyssenko and Groop 2009). The susceptibilities conferred by these loci are greater than those of the candidate genes or calpain-10. Most of these genes are expressed in the endocrine pancreas, suggesting β-cells can be considered the main target of the genetic component in type 2 diabetes. In European populations the strongest association is associated with TCF7L2, a transcription factor involved in the development and survival of islet cells and enteroendocrine cells of the gut. In vitro studies implicate TCF7L2 in β-cell apoptosis associated with increased caspase-3 cleavage and decreased Akt activity (Shu et al. 2008). Pancreatic β-cell function is also reduced in patients with TCF7L2 polymorphisms (Lyssenko et al. 2008). It is important to realize that each of the top 20 diabetes-linked genes has minimal effects on their own and that their combined effects are not synergistic. Also, their net contribution cannot explain the apparent heredity of type 2 diabetes, suggesting either that heredity has been overestimated or that epigenetic factors are dominant in the development of type 2 diabetes. The epigenetics of β-cell death in type 2 diabetes will be an important area for investigation in the future, given the persistent effects of fetal and early nutrition on β-cell function and survival.

The Role of β-Cell Apoptosis in Rare Forms of Diabetes

Although the common forms of type 1 and type 2 diabetes are polygenic, several rare forms of diabetes are caused by mutations in single genes. In most cases, these genes are important for β-cell survival or function. Monogenic causes of diabetes include mutations in proinsulin that prevent its proper folding, cause ER stress and β-cell death, and result in early-onset diabetes (Stoy et al. 2007). Wolcott-Rallison syndrome is caused by mutations in the ER stress-sensing protein PERK (Eizirik et al. 2008). ER stress-induced β-cell apoptosis may also be the cause of diabetes in Wolfram syndrome (Riggs et al. 2005). Several of the six MODY genes may also influence β-cell survival. The prime example here appears to be Pdx-1 (MODY4). Mice lacking one allele of Pdx-1 have increased β-cell apoptosis, caspase-3 activation, a reduction in the Bcl-x_L to Bax ratio, and 50% decrease in β-cell mass evident at 1 year of age (Johnson et al. 2003). This increase in apoptosis might reflect the fact that full expression of Pdx-1 is required for the pro-survival effects of insulin and incretin hormones in the β-cell (Johnson et al. 2006a; Li et al. 2005). Other MODY genes have also been linked to β-cell apoptosis, including HNF1a (Johnson 2007; Wobser et al. 2002). Pancreatic β-cells expressing a dominant-negative HNF1a exhibit caspase-3- and Bcl-x_L-dependent apoptosis (Wobser et al. 2002b). Collectively, the genes implicated in monogenic diabetes illustrate the critical importance of β-cell function and survival in human glucose homeostasis.

Islet Engraftment and β-Cell Death in Islet Transplantation

Islet transplantation is severely limited by β-cell death at several stages of this clinical treatment. Since islets are isolated from cadaveric donors, a number of factors reduce the viability of islets even before they are isolated, including the age and health status of the donor as well as organ ischemia and the time from donor death to islet harvest. The process of islet isolation itself also causes significant β-cell death, by both necrosis and apoptosis. Islets are then cultured, typically at high density, and this is associated with a 2–20 % apoptosis rate, which is markedly higher than what is observed in vivo (Dror et al. 2007, 2008a). The implantation of islets into the liver is associated with rapid β-cell death, with only a fraction of islets engrafting with sufficient microvasculature. During and after the process of engraftment, β-cells also experience toxicity from the immunosuppressant drugs that are currently required to prevent allo- and auto-rejection of the transplant. A side-by-side comparison of three clinically significant immunosuppressant drugs revealed distinct differences in the mechanisms by which they impair β-cell function and survival (Johnson et al. 2009). Clinically relevant doses of rapamycin and mycophenolate mofetil increased caspase-3-dependent apoptosis and CHOP-dependent ER stress in human islets, but did not have direct effects on glucose-stimulated insulin secretion. On the other hand, FK506, which had direct deleterious effects on insulin secretion, but caused relatively modest induction of caspase-3 activation and ER stress, resulted in the worst graft function in vivo when transplanted into STZ-diabetic NOD/scid mice. Treating islet cultures with the glucagon-like peptide 1 (GLP-1) agonist exenatide ameliorated the effects of these drugs on human β-cell function and survival (Johnson et al. 2009).

Thus, islet transplantation is associated with a cluster of related stresses including hypoxia and nutrient deprivation. The specific mechanisms that mediate β-cell death from hypoxia remain to be fully elucidated, but likely involves hypoxia-inducible factors (HIF)(Miao et al. 2006). Interestingly, von Hippel-Lindau factor and HIF1b have also been implicated in β-cell function (Gunton et al. 2005; Zehetner et al. 2008). Pancreatic β-cells can undergo programmed cell death under hypoglycemic conditions, and this environment appears to regulate the expression of HIF1b (Dror et al. 2008b). Interestingly, the RyR2 Ca^{2+} channel and calpain-10 appear to be involved in β-cell death in hypoglycemia as well (Johnson et al. 2004). In adult islets, these genes form a network that also includes presenilin, notch, neurogenin-3, and Pdx-1. This gene network appears to influence the basal rate of apoptosis, specifically under low glucose conditions (Dror et al. 2007, 2008a). Whether hypoglycemia, nutrient deprivation, or hypoxia is involved in the progression of diabetes is not well understood. Such a scenario might occur under conditions where genetic or acquired defects in the extensive intra-islet vascular network restrict the delivery of oxygen and nutrients to the β-cells (Clee et al. 2006).

Survival Factors that Prevent β-Cell Apoptosis

A large number of endogenous and exogenous growth factors have been shown to promote β-cell survival, in vitro or in vivo. Some of the key factors will be discussed here (Fig. 4).

Examples of such anti-apoptotic signaling cascades are those activated by the gut hormones GLP-1 and glucose-dependent insulinotropic polypeptide (GIP), which were first investigated for their positive effects on glucose-stimulated insulin secretion. The new diabetes drug Byetta acts by mimicking GLP-1 and has been shown to protect rodent β-cells from apoptosis when administered at high doses (Li et al. 2005). It is likely that other hormones that increase cAMP and activate RyR Ca^{2+} channels would also have anti-apoptotic effects on β-cells. It has also been suggested that inhibiting dipeptidyl peptidase-4, an enzyme that degrades GLP-1, might increase β-cell mass by preventing apoptosis

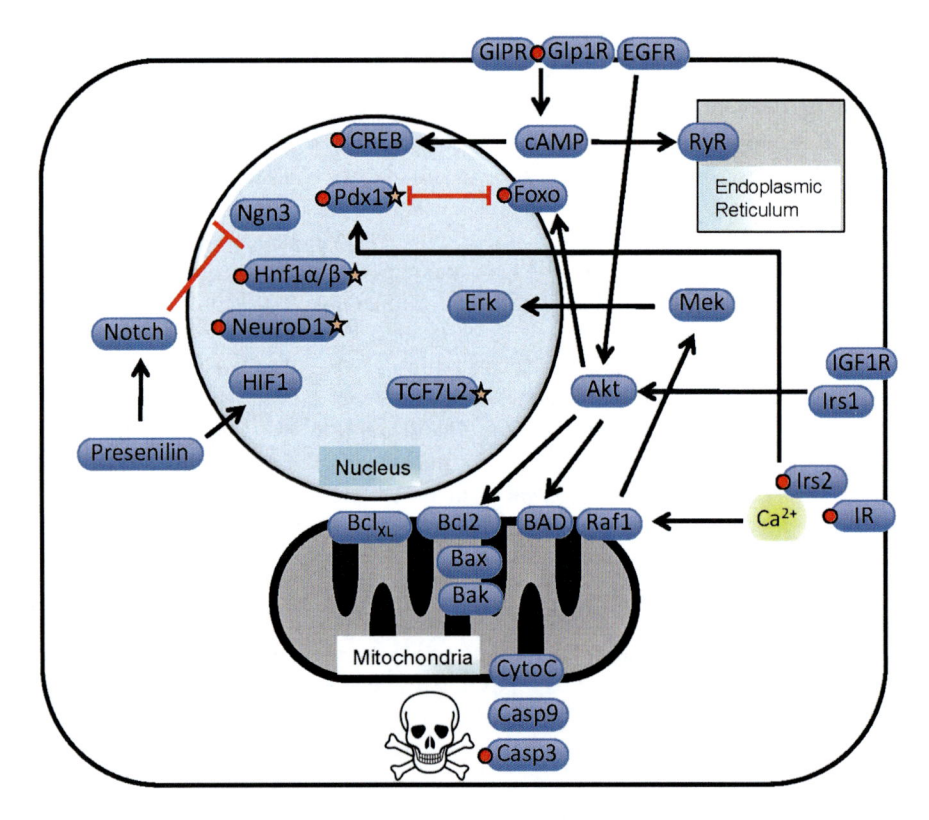

Fig. 4 Molecular mechanisms controlling basal β-cell apoptosis and survival factor signaling cascades. Shown is a partial description of signaling cascades that modulate β-cell survival. Protein products of genes that are linked to human diabetes are denoted with a *star*. Genes that have been implicated in β-cell apoptosis or β-cell mass using in vivo loss-of-function experiments (i.e., knockout mice) are denoted with a *dot*

and increasing proliferation. Nevertheless, caution is critical since this ubiquitous enzyme has many targets.

Many other β-cell growth factor systems, including hepatocyte growth factor, fibroblast growth factors, parathyroid hormone-related protein, gastrin, delta/notch, netrin, and Slit-Robo also promote β-cell survival (Dror et al. 2007; Garcia-Ocana et al. 2001; Yang et al. 2011, 2013). The Slit-Robo system is interesting as it is one of the only examples discovered to date of a local secreted factor that is required for β-cell survival (Yang et al. 2013). One of the most important local endogenous β-cell growth factors appears to be insulin itself (Johnson et al. 2006a, b; Johnson and Alejandro 2008; Alejandro and Johnson 2008; Beith et al. 2008; Ohsugi et al. 2005; Kulkarni et al. 1999a; Hennige et al. 2003; Otani et al. 2004; Ueki et al. 2006; Okada et al. 2007). Based on knockout mouse studies, the insulin receptor even appears more important than the IGF-1 receptor (Ueki et al. 2006). Insulin acts via a complex series of signaling events, including both the PI3-kinase/Akt pathway and the Raf-1/Erk pathway (Johnson et al. 2006a; Johnson and Alejandro 2008; Alejandro and Johnson 2008; Ueki et al. 2006). We have recently shown that 14-3-3 proteins play an important role in coordinating β-cell survival signaling (Lim et al. 2013). Akt acts on multiple downstream targets, including Bad. In addition to stimulating Erk, Raf-1 can also dephosphorylate and inactivate Bad at the mitochondria. Interestingly, signaling through IRS-2 rather than IRS-1 appears to play a role in β-cell survival (Kulkarni et al. 1999b). While constitutive insulin signaling seems to be essential for β-cell survival under stressful conditions, excessive concentrations of insulin may be deleterious (Johnson and Alejandro 2008). Further work is needed to understand the ideal way to harness this and other endogenous anti-apoptotic signaling pathways. Ongoing studies have employed high-throughput imaging methodologies to identify generation and stress-specific β-cell survival factors from large libraries of endogenous biologic factors.

The Role of Other β-Cell Death Modalities: Beyond β-Cell Apoptosis

Numerous intrinsic and extrinsic signals are required for the maintenance of functional β-cell mass by providing pro-survival and pro-death signals. Mechanistic studies on the initiation and progression of β-cell death can make significant contributions to the prevention and treatment of type 1 diabetes and type 2 diabetes, in addition to improving the success of islet transplantations. Programmed cell death via apoptosis has been well characterized as an important mechanism of β-cell death (Mathis et al. 2001; Cnop et al. 2005). However, there are other commonly characterized forms of cell death distinguished by morphological and biochemical features, including necrosis and autophagy (Fig. 5) (Kroemer et al. 2009; Galluzzi et al. 2009, 2012).

The redundancy of signaling molecules involved in the temporal cascade of events leading to β-cell apoptosis, autophagy, and necrosis has not been well

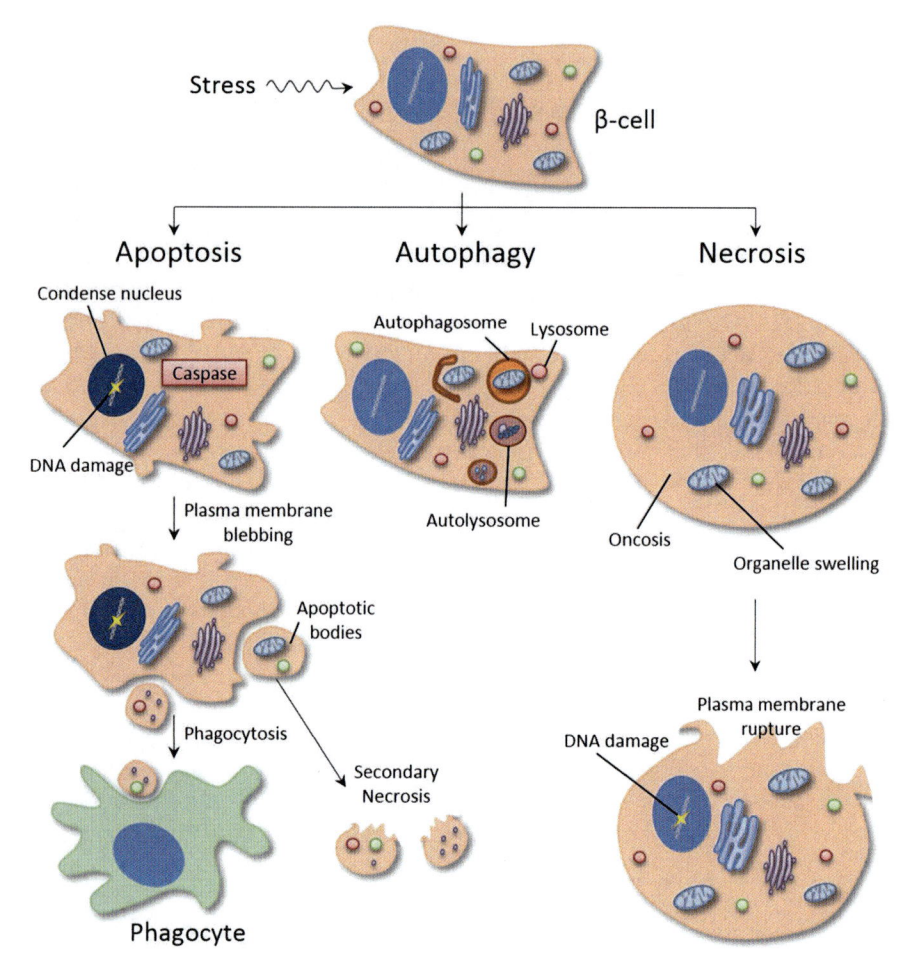

Fig. 5 Schematic of common β-cell death pathways. Prolonged exposure to stress conditions, including cytokine exposure hyperglycemia, hyperlipidemia, oxidative stress, and ER stress, can lead to β-cell death. Apoptosis, autophagy-mediated, and necrosis are the most well-characterized modes of cell death

characterized on a single cell level, resulting in the underappreciation of non-apoptotic forms of cell death (Yang et al. 2013b). Understanding the redundancy and exclusiveness of different mechanisms of cell death has important implications for the detection and therapeutic manipulation of cell death.

Autophagy is a catabolic process often favoring cell survival under conditions of nutrient deprivation, hypoxia, ER stress, pathogen infection, and DNA damage (Kroemer et al. 2009, 2010; Fleming et al. 2011; Levine and Yuan 2005). These conditions are relevant to the initiation and progression and diabetes. Also, in islet transplantation, islet cells are exposed to hypoxia and nutrient deprivation prior to vascularization and engraftment. The formation of double membrane vacuoles that sequester damaged organelles and harmful cytoplasmic contents, termed autophagosomes, is a defining feature of autophagy, which concludes with the delivery of the contents to the lysosome for degradation and recycling (Kroemer et al. 2010; Levine and Yuan 2005). Autophagic cell death is simply characterized by the lack of chromatin condensation and accumulation of autophagosomes and does not necessarily implicate autophagy as the cause of cell death (Kroemer et al. 2009). Ablation of free fatty acid-induced autophagy leads to a lack of compensatory β-cell hyperplasia and impaired glucose tolerance (Ebato et al. 2008). Diminished maintenance of functional β-cell mass by autophagy may increase the susceptibility to β-cell death under basal and stressed conditions and consequently affect diabetes initiation and progression (Ebato et al. 2008; Jung et al. 2008; Levine and Kroemer 2008). The mutual inhibition between apoptosis and autophagy further supports the involvement of autophagy in maintaining β-cell health (Kroemer et al. 2010; Kang et al. 2011).

β-Cell death via necrosis has also been implicated in the pathogenesis of diabetes (Fujimoto et al. 2010; Steer et al. 2006). Necrosis is often defined as cell death lacking the characteristics of apoptosis or autophagy (Yang et al. 2013b). In addition, key morphological features of necrosis include plasma membrane rupture and swelling of cytoplasmic organelles (Kroemer et al. 2009; Golstein and Kroemer 2007). Although initially believed to be an uncontrolled form of cell death leading to the release of inflammatory cellular contents, there is accumulating evidence supporting the notion that necrotic cell death is regulated by a defined set of signaling events induced by oxidative stress, loss of Ca^{2+} homeostasis, or ischemia (Kroemer et al. 2009; Golstein and Kroemer 2007; Fink and Cookson 2005). In fact, apoptosis and necrosis may share common signaling pathways involving mitochondrial membrane permeabilization through activation of proapoptotic Bcl-2 family members (Golstein and Kroemer 2007; Kim et al. 2003). Receptor-interacting protein kinase 1 (RIP1)-dependent necrosis is a well-characterized case of regulated necrosis that can be activated upon binding of tumor necrosis factor α (TNFα) to TNF receptor 1 in the absence of caspase-8 activity (Galluzzi et al. 2012; Vandenabeele et al. 2010). Consequently, RIP1 is deubiquitinated and associates with RIP3 to activate necrotic cell death. Upon exposure to stress, inhibition of the apoptotic signaling cascade by direct inhibition of caspase activation or depletion of ATP (which is required for caspase activation) can favor necrotic cell death (Kim et al. 2003; Leist et al. 1997; Eguchi et al. 1997). This suggests that multiple modes of cell death can coexist within the same cell and they have the potential to substitute for each other. The complex interplay between different modes of cell death further complicates the development of therapeutics for preventing β-cell death.

Understanding the molecular processes behind cell death may reveal novel therapeutic targets. In addition to the complex interplay between apoptosis, necrosis, and autophagy, other pathways can also proceed. The diversity of the molecular pathways mediating cell death has led to the characterization of new modalities of cell death that sometimes share similar features. Mitotic catastrophe is initiated by aberrant mitosis leading to cell death during mitosis or interphase via apoptosis or necrosis (Kroemer et al. 2009). Anoikis is an intrinsic apoptotic response of adherent cells to the detachment from extracellular matrix interactions (Galluzzi et al. 2012; Frisch and Francis 1994). Parthanatos is a caspase-independent cell death pathway involving DNA damage induced by over-activation of poly-ADP-ribose polymerases (PARPs), which can further result in ATP and NAD^+ depletion, PAR accumulation, loss of mitochondrial membrane potential, and subsequently AIF release (David et al. 2009; Luo and Kraus 2012). Pyroptosis is a caspase-1-mediated cell death pathway that exhibits morphological features of apoptosis and/or necrosis. The activation of caspase-1 leads to the mature processing of inflammatory cytokines interleukin-1β (IL-1β) and IL-18 (Fink and Cookson 2005). It remains controversial whether these new modalities constitute unique cell death subroutines or whether they represent specific cases of apoptosis and/or necrosis. Elucidating the context-dependent distribution of various mechanisms of cell death may determine the success of targeted therapeutic interventions to control β-cell death. It is conceivable that therapies for promoting β-cell survival may require inhibition of all forms of cell death through targeted inhibition of upstream events or combinatorial therapies.

β-Cell Apoptosis as a Therapeutic Target in Diabetes: Future Directions

The protection of existing β-cells and the regeneration of new ones is a major goal in diabetes research. Therapeutic strategies to protect β-cells could have an immediate impact on clinical islet transplantation, where close to half of the islets transplanted into the liver die before becoming engrafted. In future years we also expect drugs may be developed that improve endogenous β-cell survival in vivo. These treatments would theoretically slow the progression of, or perhaps reverse, type 1 diabetes or type 2 diabetes. Once the exact molecular defects are better known, specific components of the β-cell apoptosis system could be targeted more selectively. For diabetes caused by β-cell ER stress, so-called molecular chaperones might be useful to decrease unfolded proteins in the ER. In cases where diabetes is associated with apoptosis controlled by cellular metabolism, we expect that direct interventions at the level of β-cell mitochondria might be of benefit. Since islet amyloid formation can be found in type 2 diabetes and in transplantation, chemical inhibitors of this process might have therapeutic potential (Potter et al. 2009). A thorough understanding of survival signaling pathways induced by endogenous β-cell growth factors will hopefully provide new targets for intervention, based on the β-cell's own defenses. Moreover, unbiased and high-throughput methods promise to accelerate the pace at which we discover the mechanisms of β-cell apoptosis and treatments that target β-cell apoptosis in diabetes.

Cross-References

Interested readers are encouraged to view other articles on this platform. For example, there is an entire chapter devoted to the effects of glucose, insulin signaling, and immune cells on β-cell survival:
► IGF-1 and insulin-receptor signaling in insulin secreting cells: from function to survival
► Immunology of β-Cell Destruction
► Inflammatory Pathways Linked to β Cell Demise in Diabetes

There is also information on β-cell dysfunction in:
► (Dys)Regulation of Insulin Secretion by Macronutrients
► High Fat Programming of β-Cell Dysfunction
► Role of Mitochondria in β-Cell Function and Dysfunction

A discussion on possible approaches to detect and treat β-cell death can be found in:
► Clinical Approaches to Preserve β-Cell Function in Diabetes
► Current Approaches and Future Prospects for the Prevention of β-Cell Destruction in Autoimmune Diabetes
► In Vivo Biomarkers for Detection of β Cell Death

References

Alejandro EU, Johnson JD (2008) Inhibition of raf-1 alters multiple downstream pathways to induce pancreatic β-cell apoptosis. J Biol Chem 283(4):2407–2417

Barratt BJ et al (2004) Remapping the insulin gene/IDDM2 locus in type 1 diabetes. Diabetes 53 (7):1884–1889

Beith JL, Alejandro EU, Johnson JD (2008) Insulin stimulates primary β-cell proliferation via Raf-1 kinase. Endocrinology 149(5):2251–2260

Bell GI, Polonsky KS (2001) Diabetes mellitus and genetically programmed defects in β-cell function. Nature 414(6865):788–791

Bennett ST, Todd JA (1996) Human type 1 diabetes and the insulin gene: principles of mapping polygenes. Annu Rev Genet 30:343–370

Boslem E et al (2011) A lipidomic screen of palmitate-treated MIN6 β-cells links sphingolipid metabolites with endoplasmic reticulum (ER) stress and impaired protein trafficking. Biochem J 435(1):267–276

Boslem E, Meikle PJ, Biden TJ (2012) Roles of ceramide and sphingolipids in pancreatic β-cell function and dysfunction. Islets 4(3):177–187

Boslem E et al (2013) Alteration of endoplasmic reticulum lipid rafts contributes to lipotoxicity in pancreatic β-cells. J Biol Chem 288(37):26569–26582

Butler AE et al (2003a) Increased β-cell apoptosis prevents adaptive increase in β-cell mass in mouse model of type 2 diabetes: evidence for role of islet amyloid formation rather than direct action of amyloid. Diabetes 52(9):2304–2314

Butler AE et al (2003b) β-cell deficit and increased β-cell apoptosis in humans with type 2 diabetes. Diabetes 52(1):102–110

Cardozo AK et al (2005) Cytokines downregulate the sarcoendoplasmic reticulum pump Ca^{2+} ATPase 2b and deplete endoplasmic reticulum Ca^{2+}, leading to induction of endoplasmic reticulum stress in pancreatic β-cells. Diabetes 54(2):452–461

Chen J et al (2008) Thioredoxin-interacting protein: a critical link between glucose toxicity and β-cell apoptosis. Diabetes 57(4):938–944

Chu KY et al (2010) ATP-citrate lyase reduction mediates palmitate-induced apoptosis in pancreatic β cells. J Biol Chem 285(42):32606–32615

Chu KY et al (2011) Differential regulation and localization of carboxypeptidase D and carboxypeptidase E in human and mouse β-cells. Islets 3(4):155–165

Chu KY et al (2012) Ubiquitin C-terminal hydrolase L1 is required for pancreatic β cell survival and function in lipotoxic conditions. Diabetologia 55(1):128–140

Clee SM et al (2006) Positional cloning of Sorcs1, a type 2 diabetes quantitative trait locus. Nat Genet 38(6):688–693

Cnop M et al (2005) Mechanisms of pancreatic β-cell death in type 1 and type 2 diabetes: many differences, few similarities. Diabetes 54(Suppl 2):S97–S107

Colli ML et al (2010) MDA5 and PTPN2, two candidate genes for type 1 diabetes, modify pancreatic β-cell responses to the viral by-product double-stranded RNA. Hum Mol Genet 19(1):135–146

Danial NN et al (2008) Dual role of proapoptotic BAD in insulin secretion and β cell survival. Nat Med 14(2):144–153

David KK et al (2009) Parthanatos, a messenger of death. Front Biosci 14:1116–1128

Donath MY, Halban PA (2004) Decreased β-cell mass in diabetes: significance, mechanisms and therapeutic implications. Diabetologia 47(3):581–589

Dror V et al (2007) Notch signalling suppresses apoptosis in adult human and mouse pancreatic islet cells. Diabetologia 50(12):2504–2515

Dror V et al (2008a) Glucose and Endoplasmic Reticulum Calcium Channels Regulate HIF-1β via Presenilin in Pancreatic β-Cells. J Biol Chem 283(15):9909–9916

Dror V et al (2008b) Glucose and endoplasmic reticulum calcium channels regulate HIF-1β via presenilin in pancreatic β-cells. J Biol Chem 283(15):9909–9916

Ebato C et al (2008) Autophagy is important in islet homeostasis and compensatory increase of β cell mass in response to high-fat diet. Cell Metab 8(4):325–332

Efanova IB et al (1998) Glucose and tolbutamide induce apoptosis in pancreatic β-cells. A process dependent on intracellular Ca^{2+} concentration. J Biol Chem 273(50):33501–33507

Eguchi Y, Shimizu S, Tsujimoto Y (1997) Intracellular ATP levels determine cell death fate by apoptosis or necrosis. Cancer Res 57(10):1835–1840

Ehses JA et al (2007) Increased number of islet-associated macrophages in type 2 diabetes. Diabetes 56(9):2356–2370

Eizirik DL, Grieco FA (2012) On the immense variety and complexity of circumstances conditioning pancreatic β-cell apoptosis in type 1 diabetes. Diabetes 61(7):1661–1663

Eizirik DL, Cardozo AK, Cnop M (2008) The role for endoplasmic reticulum stress in diabetes mellitus. Endocr Rev 29(1):42–61

Eizirik DL et al (2012) The human pancreatic islet transcriptome: expression of candidate genes for type 1 diabetes and the impact of pro-inflammatory cytokines. PLoS Genet 8(3):e1002552

Eizirik DL, Miani M, Cardozo AK (2013) Signalling danger: endoplasmic reticulum stress and the unfolded protein response in pancreatic islet inflammation. Diabetologia 56(2):234–241

El-Assaad W et al (2003) Saturated fatty acids synergize with elevated glucose to cause pancreatic β-cell death. Endocrinology 144(9):4154–4163

Estella E et al (2006) Granzyme B-mediated death of pancreatic β-cells requires the proapoptotic BH3-only molecule bid. Diabetes 55(8):2212–2219

Federici M et al (2001) High glucose causes apoptosis in cultured human pancreatic Islets of Langerhans – a potential role for regulation of specific Bcl family genes toward an apoptotic cell death program. Diabetes 50(6):1290–1301

Fink SL, Cookson BT (2005) Apoptosis, pyroptosis, and necrosis: mechanistic description of dead and dying eukaryotic cells. Infect Immun 73(4):1907–1916

Fleming A et al (2011) Chemical modulators of autophagy as biological probes and potential therapeutics. Nat Chem Biol 7(1):9–17

Frisch SM, Francis H (1994) Disruption of epithelial cell-matrix interactions induces apoptosis. J Cell Biol 124(4):619–626

Fujimoto K et al (2010) Loss of Nix in Pdx1-deficient mice prevents apoptotic and necrotic β cell death and diabetes. J Clin Invest 120(11):4031–4039

Galluzzi L et al (2009) Guidelines for the use and interpretation of assays for monitoring cell death in higher eukaryotes. Cell Death Differ 16(8):1093–1107

Galluzzi L et al (2012) Molecular definitions of cell death subroutines: recommendations of the Nomenclature Committee on Cell Death 2012. Cell Death Differ 19(1):107–120

Garcia-Ocana A et al (2001) Using β-cell growth factors to enhance human pancreatic Islet transplantation. J Clin Endocrinol Metab 86(3):984–988

Golstein P, Kroemer G (2007) Cell death by necrosis: towards a molecular definition. Trends Biochem Sci 32(1):37–43

Guillen C et al (2008) Biphasic effect of insulin on β cell apoptosis depending on glucose deprivation. FEBS Lett 582(28):3855–3860

Gunton JE et al (2005) Loss of ARNT/HIF1β mediates altered gene expression and pancreatic-islet dysfunction in human type 2 diabetes. Cell 122(3):337–349

Gwiazda KS et al (2009) Effects of palmitate on ER and cytosolic Ca^{2+} homeostasis in β-cells. Am J Physiol Endocrinol Metab 296(4):E690–E701

Haataja L et al (2008) Islet amyloid in type 2 diabetes, and the toxic oligomer hypothesis. Endocr Rev 29(3):303–316

Harding HP, Ron D (2002) Endoplasmic reticulum stress and the development of diabetes: a review. Diabetes 51(Suppl 3):S455–S461

Hennige AM et al (2003) Upregulation of insulin receptor substrate-2 in pancreatic β cells prevents diabetes. J Clin Invest 112(10):1521–1532

Horikawa Y et al (2000) Genetic variation in the gene encoding calpain-10 is associated with type 2 diabetes mellitus. Nat Genet 26(2):163–175

Jeffrey KD et al (2008) Carboxypeptidase E mediates palmitate-induced β-cell ER stress and apoptosis. Proc Natl Acad Sci U S A 105(24):8452–8457

Johnson JD (2007) Pancreatic β-cell apoptosis in maturity onset diabetes of the young. Can J Diabetes 31(1):001–008

Johnson JD, Alejandro EU (2008) Control of pancreatic β-cell fate by insulin signaling: the sweet spot hypothesis. Cell Cycle 7(10):1343–1347

Johnson JD et al (2003) Increased islet apoptosis in $Pdx1^{+/-}$ mice. J Clin Invest 111(8):1147–1160

Johnson JD et al (2004) RyR2 and calpain-10 delineate a novel apoptosis pathway in pancreatic islets. J Biol Chem 279(23):24794–24802

Johnson JD et al (2006a) Insulin protects islets from apoptosis via Pdx1 and specific changes in the human islet proteome. Proc Natl Acad Sci U S A 103(51):19575–19580

Johnson JD et al (2006b) Suppressed insulin signaling and increased apoptosis in CD38-null islets. Diabetes 55(10):2737–2746

Johnson JD, et al. (2009) Different effects of FK506, rapamycin, and mycophenolate mofetil on glucose-stimulated insulin release and apoptosis in human islets. Cell Transpl 18(8):833–845

Jung HS et al (2008) Loss of autophagy diminishes pancreatic β cell mass and function with resultant hyperglycemia. Cell Metab 8(4):318–324

Kaneto H et al (2005) Oxidative stress and pancreatic β-cell dysfunction. Am J Ther 12(6):529–533

Kang R et al (2011) The Beclin 1 network regulates autophagy and apoptosis. Cell Death Differ 18(4):571–580

Kim JS, He L, Lemasters JJ (2003) Mitochondrial permeability transition: a common pathway to necrosis and apoptosis. Biochem Biophys Res Commun 304(3):463–470

Kroemer G et al (2009) Classification of cell death: recommendations of the Nomenclature Committee on Cell Death 2009. Cell Death Differ 16(1):3–11

Kroemer G, Marino G, Levine B (2010) Autophagy and the integrated stress response. Mol Cell 40(2):280–293

Kulkarni RN et al (1999a) Tissue-specific knockout of the insulin receptor in pancreatic β cells creates an insulin secretory defect similar to that in type 2 diabetes. Cell 96(3):329–339

Kulkarni RN et al (1999b) Altered function of insulin receptor substrate-1-deficient mouse islets and cultured β-cell lines. J Clin Invest 104(12):R69–R75

Laybutt DR et al (2007) Endoplasmic reticulum stress contributes to β cell apoptosis in type 2 diabetes. Diabetologia 50(4):752–763

Lee B et al (1999) Glucose regulates expression of inositol 1,4,5-trisphosphate receptor isoforms in isolated rat pancreatic islets. Endocrinology 140(5):2173–2182

Leist M et al (1997) Intracellular adenosine triphosphate (ATP) concentration: a switch in the decision between apoptosis and necrosis. J Exp Med 185(8):1481–1486

Leonardi O, Mints G, Hussain MA (2003) β-cell apoptosis in the pathogenesis of human type 2 diabetes mellitus. Eur J Endocrinol 149(2):99–102

Levine B, Kroemer G (2008) Autophagy in the pathogenesis of disease. Cell 132(1):27–42

Levine B, Yuan J (2005) Autophagy in cell death: an innocent convict? J Clin Invest 115(10):2679–2688

Li Y et al (2005) β-Cell Pdx1 expression is essential for the glucoregulatory, proliferative, and cytoprotective actions of glucagon-like peptide-1. Diabetes 54(2):482–491

Liadis N et al (2005) Caspase-3-dependent β-cell apoptosis in the initiation of autoimmune diabetes mellitus. Mol Cell Biol 25(9):3620–3629

Liadis N et al (2007) Distinct in vivo roles of caspase-8 in β-cells in physiological and diabetes models. Diabetes 56(9):2302–2311

Lim GE, Piske M, Johnson JD (2013) 14-3-3 proteins are essential signalling hubs for β cell survival. Diabetologia 56(4):825–837

Lin CY et al (2005) Activation of peroxisome proliferator-activated receptor-gamma by rosiglitazone protects human islet cells against human islet amyloid polypeptide toxicity by a phosphatidylinositol 3′-kinase-dependent pathway. J Clin Endocrinol Metab 90(12):6678–6686

Lovell JF et al (2008) Membrane binding by tBid initiates an ordered series of events culminating in membrane permeabilization by Bax. Cell 135(6):1074–1084

Lu H et al (2008) The identification of potential factors associated with the development of type 2 diabetes: a quantitative proteomics approach. Mol Cell Proteomics 7(8):1434–1451

Luciani DS et al (2013) Bcl-2 and Bcl-xL suppress glucose signaling in pancreatic β-cells. Diabetes 62(1):170–182

Luo X, Kraus WL (2012) On PAR with PARP: cellular stress signaling through poly(ADP-ribose) and PARP-1. Genes Dev 26(5):417–432

Lupi R et al (2002) Prolonged exposure to free fatty acids has cytostatic and pro-apoptotic effects on human pancreatic islets: evidence that β-cell death is caspase mediated, partially dependent on ceramide pathway, and Bcl-2 regulated. Diabetes 51(5):1437–1442

Lyssenko V, Groop L (2009) Genome-wide association study for type 2 diabetes: clinical applications. Curr Opin Lipidol 20(2):87–91

Lyssenko V et al (2008) Clinical risk factors, DNA variants, and the development of type 2 diabetes. N Engl J Med 359(21):2220–2232

Maedler K, Donath MY (2004) β-cells in type 2 diabetes: a loss of function and mass. Horm Res 62(Suppl 3):67–73

Maedler K et al (2004) Glucose- and interleukin-1β-induced β-cell apoptosis requires Ca^{2+} influx and extracellular signal-regulated kinase (ERK) 1/2 activation and is prevented by a sulfonylurea receptor 1/inwardly rectifying K^+ channel 6.2 (SUR/Kir6.2) selective potassium channel opener in human islets. Diabetes 53(7):1706–1713

Maedler K et al (2005) Sulfonylurea induced β-cell apoptosis in cultured human islets. J Clin Endocrinol Metab 90(1):501–506

Marshall C et al (2005) Evidence that an isoform of calpain-10 is a regulator of exocytosis in pancreatic β-cells. Mol Endocrinol 19(1):213–224

Martinez SC et al (2008) Inhibition of Foxo1 protects pancreatic islet β-cells against fatty acid and endoplasmic reticulum stress-induced apoptosis. Diabetes 57(4):846–859

Mathis D, Vence L, Benoist C (2001) β-cell death during progression to diabetes. Nature 414(6865):792–798

McKenzie MD et al (2008) Proapoptotic BH3-only protein Bid is essential for death receptor-induced apoptosis of pancreatic β-cells. Diabetes 57(5):1284–1292

Mehran AE et al (2012) Hyperinsulinemia drives diet-induced obesity independently of brain insulin production. Cell Metab 16(6):723–737

Meier JJ et al (2008) β-cell replication is the primary mechanism subserving the postnatal expansion of β-cell mass in humans. Diabetes 57(6):1584–1594

Miao G et al (2006) Dynamic production of hypoxia-inducible factor-1α in early transplanted islets. Am J Transplant 6(11):2636–2643

Moore F et al (2009) PTPN2, a candidate gene for type 1 diabetes, modulates interferon-gamma-induced pancreatic β-cell apoptosis. Diabetes 58(6):1283–1291

Nogueira TC et al (2013) GLIS3, a susceptibility gene for type 1 and type 2 diabetes, modulates pancreatic β cell apoptosis via regulation of a splice variant of the BH3-only protein Bim. PLoS Genet 9(5):e1003532

Ohsugi M et al (2005) Reduced expression of the insulin receptor in mouse insulinoma (MIN6) cells reveals multiple roles of insulin signaling in gene expression, proliferation, insulin content, and secretion. J Biol Chem 280(6):4992–5003

Okada T et al (2007) Insulin receptors in β-cells are critical for islet compensatory growth response to insulin resistance. Proc Natl Acad Sci U S A 104(21):8977–8982

Otani K et al. (2004) Reduced β-cell mass and altered glucose sensing impair insulin-secretory function in βIRKO mice. Am J Physiol Endocrinol Metab 286(1):E41–9

Oyadomari S, Araki E, Mori M (2002) Endoplasmic reticulum stress-mediated apoptosis in pancreatic β-cells. Apoptosis 7(4):335–345

Ozcan U et al (2004) Endoplasmic reticulum stress links obesity, insulin action, and type 2 diabetes. Science 306(5695):457–461

Ozcan U et al (2006) Chemical chaperones reduce ER stress and restore glucose homeostasis in a mouse model of type 2 diabetes. Science 313(5790):1137–1140

Pearson GL et al (2014) Lysosomal acid lipase and lipophagy are constitutive negative regulators of glucose-stimulated insulin secretion from pancreatic β cells. Diabetologia 57(1):129–139

Potter KJ et al (2009) Amyloid inhibitors enhance survival of cultured human islets. Biochim Biophys Acta 1790(6):566–574

Prentki M, Nolan CJ (2006) Islet β cell failure in type 2 diabetes. J Clin Invest 116(7):1802–1812

Preston AM et al (2009) Reduced endoplasmic reticulum (ER)-to-Golgi protein trafficking contributes to ER stress in lipotoxic mouse β cells by promoting protein overload. Diabetologia 52(11):2369–2373

Pugliese A et al (1997) The insulin gene is transcribed in the human thymus and transcription levels correlated with allelic variation at the INS VNTR-IDDM2 susceptibility locus for type 1 diabetes. Nat Genet 15(3):293–297

Rhodes CJ (2005) Type 2 diabetes-a matter of β-cell life and death? Science 307(5708):380–384

Riggs AC et al (2005) Mice conditionally lacking the Wolfram gene in pancreatic islet β cells exhibit diabetes as a result of enhanced endoplasmic reticulum stress and apoptosis. Diabetologia 48(11):2313–2321

Robertson RP et al (2004) β-cell glucose toxicity, lipotoxicity, and chronic oxidative stress in type 2 diabetes. Diabetes 53(Suppl 1):S119–S124

Santin I et al (2011) PTPN2, a candidate gene for type 1 diabetes, modulates pancreatic β-cell apoptosis via regulation of the BH3-only protein Bim. Diabetes 60(12):3279–3288

Shu L et al (2008) Transcription factor 7-like 2 regulates β-cell survival and function in human pancreatic islets. Diabetes 57(3):645–653

Song B et al (2008) Chop deletion reduces oxidative stress, improves β cell function, and promotes cell survival in multiple mouse models of diabetes. J Clin Invest 118(10):3378–3389

Steer SA et al (2006) Interleukin-1 stimulates β-cell necrosis and release of the immunological adjuvant HMGB1. PLoS Med 3(2):e17

Stoy J et al (2007) Insulin gene mutations as a cause of permanent neonatal diabetes. Proc Natl Acad Sci U S A 104(38):15040–15044

Szabat M et al (2012) Maintenance of β-cell maturity and plasticity in the adult pancreas: developmental biology concepts in adult physiology. Diabetes 61(6):1365–1371

Thomas HE et al (2009) β cell apoptosis in diabetes. Apoptosis 14(12):1389–1404

Trudeau JD et al (2000) Neonatal β-cell apoptosis: a trigger for autoimmune diabetes? Diabetes 49 (1):1–7

Ueki K et al (2006) Total insulin and IGF-I resistance in pancreatic β cells causes overt diabetes. Nat Genet 38(5):583–588

Vandenabeele P et al (2010) Molecular mechanisms of necroptosis: an ordered cellular explosion. Nat Rev Mol Cell Biol 11(10):700–714

Wang M et al (2013) Is dynamic autocrine insulin signaling possible? A mathematical model predicts picomolar concentrations of extracellular monomeric insulin within human pancreatic islets. PLoS ONE 8(6):e64860

Wobser H et al (2002) Dominant-negative suppression of HNF-1 α results in mitochondrial dysfunction, INS-1 cell apoptosis, and increased sensitivity to ceramide-, but not to high glucose-induced cell death. J Biol Chem 277(8):6413–6421

Yamada K et al (1999) Essential role of caspase-3 in apoptosis of mouse β-cells transfected with human Fas. Diabetes 48(3):478–483

Yang YH, Johnson JD (2013) Multi-parameter, single-cell, kinetic analysis reveals multiple modes of cell death in primary pancreatic β-cells. J Cell Sci 126(Pt 18):4286–95

Yang YH et al (2011) Paracrine signalling loops in adult human and mouse pancreatic islets: netrins modulate β cell apoptosis signalling via dependence receptors. Diabetologia 54 (4):828–842

Yang YH et al (2013) Intraislet SLIT-ROBO signaling is required for β-cell survival and potentiates insulin secretion. Proc Natl Acad Sci U S A 110(41):16480–16485

Zehetner J et al (2008) PVHL is a regulator of glucose metabolism and insulin secretion in pancreatic β cells. Genes Dev 22(22):3135–3146

Clinical Approaches to Preserving β-Cell Function in Diabetes

33

Bernardo Léo Wajchenberg and Rodrigo Mendes de Carvalho

Contents

Introduction	897
Clinical Impact of Therapies Aimed at β-Cell Preservation	899
Short-Term Improvement of β-Cell Insulin Secretion	899
Long-Term Improvement of β-Cell Insulin Secretion	899
Short-Term Intensive Insulin Therapy of Newly Diagnosed DM2	899
Long-Term Intensive Insulin Therapy in Newly Diagnosed DM2	900
Glitazones	902
Indirect Effects with Amelioration of Insulin Sensitivity	902
Direct Effects via PPARγ Activation in Pancreatic Islands	902
Incretin Mimetics	903
Exenatide BID	906
Liraglutide	906
Exenatide Once-Weekly	908
Exenatide OW Safety and Tolerability	909
Incretin Enhancers (DPP-4 Inhibitors)	910
Pharmacological Differences Among the DPP-4 Inhibitors	911
Liver Impairment	911
DPP-4 Inhibitors and CV Protection	912
Long-Term Clinical Efficiency	913
Mechanism of Action of DPP-4 Inhibitors	913
Cross-References	916
References	916

B.L. Wajchenberg (✉)
Endocrine Service and Diabetes and Heart Center of the Heart Institute, Hospital, Clinicas of The University of São Paulo Medical School, São Paulo, Brazil
e-mail: bernarwaj@globo.com

R.M. de Carvalho
Clinical Endocrinologist and Diabetologist, Rio de Janeiro, Brazil
e-mail: rodrigomendesdecarvalho@hotmail.com

Abstract

In type 2 diabetes (DM2) there is progressive deterioration of β-cell function and mass. It was found that islet function was about 50 % of normal at the time of diagnosis and there was a reduction of β-cell mass of about 60 % at necropsy (accelerated apoptosis). Among the interventions to preserve the β-cells, those that lead to short-term improvement of β-cell secretion are weight loss, metformin, sulfonylureas, and insulin. Long-term improvement was demonstrated with short-term insulin therapy of newly diagnosed DM2. Besides, long-term intensive insulin therapy plus metformin or triple oral therapy (metformin + glyburide + pioglitazone) for 3.5 years enabled β-cell function to be preserved for at least that period of time. Furthermore, long-term improvement was also shown with the isolated use of anti-apoptotic drugs such as glitazones, and the use of glucagon-like peptide-1 receptor agonists (GLP-1 mimetics), not inactivated by the enzyme dipeptidyl peptidase-4, and/or to inhibit that enzyme (GLP-1 enhancers). The incretin hormones are released from the gastrointestinal tract in response to nutrient ingestion to enhance glucose-dependent insulin secretion from the pancreas and overall maintenance of glucose homeostasis. Of the incretins, only GLP-1 mimetics or enhancers can be used for the treatment of DM2. Although incretin-based medications maintain β-cell function, there is no evidence that they increase β-cell mass.

Abbreviations

A1c = HbA1c	Glycated hemoglobin
aGLP-1	Active glucagon-like peptide-1
ALT	Alanine aminotransferase
AST	Aspartate aminotransferase
BID	Twice a day
BMI	Body mass index
CV	Cardiovascular
CVD	Cardiovascular disease
DBP	Diastolic blood pressure
DM2	Type 2 diabetes mellitus
DPP-4	Dipeptidyl peptidase-4
ER	Endoplasmic reticulum
FA	Fatty acid
FFA	Free fatty acid
FPG	Fasting plasma glucose
GFR	Glomerular filtration rate
GIP	Glucose-dependent insulinotropic polypeptide
GLP-1	Glucagon-like peptide-1
GLP-1R	Glucagon-like peptide-1 receptor
GLP-2	Glucagon-like peptide-2
HOMA	homeostasis model assessment

HOMA-β or B	HOMA of β-cell function
IFG	Impaired fasting glucose
IGT	Impaired glucose tolerance
OW	Once a week
PI/IRI ratio	Proinsulin to total immunoreactive insulin ratio
PPARγ	Peroxisome proliferator-activated receptor γ
PPG	Postprandial plasma glucose
ROS	Reactive oxygen species
SBP	Systolic blood pressure
tGLP-1	Total GLP-1

Introduction

Type 2 diabetes (DM2) is caused by insufficient insulin secretion usually in the context of resistance of the peripheral tissues to the action of the hormone and characterized by progressive deterioration of the β-cell function over time. The deterioration occurs regardless of therapy allocation, albeit conventional (mainly diet), insulin, sulfonylureas, or sensitizers such as glitazones and metformin (UK Prospective Diabetes Study (UKPDS) Group 1998; Holman 2006). DM2 subjects show both quantitative and qualitative disturbances in plasma insulin levels (loss of acute insulin response to glucose loss of the first phase; impaired insulin oscillations during the sustained second phase of glucose-induced insulin secretion, and defects in proinsulin processing at the β-cell level, resulting in an increase in the proinsulin to insulin ratio (Wajchenberg 2007)). Associated with reduced β-cell function, which was found to be about 50 % of the normal level at the time of diagnosis, independent of the degree of insulin resistance and probably commencing 10–12 years before diagnosis, as well as being aggravated by increasing fasting plasma glucose levels (Holman 1998), a reduction in the β-cell mass of about 60 % has been observed at necropsy. The underlying mechanism was found to be increased β-cell apoptosis, while new islet formation and β-cell replication (normalized to relative β-cell volume) remained normal or increased (Butler et al. 2003).

While there is consensus that hyperglycemia develops in the context of insulin resistance only if insulin secretion is insufficient, the question remains whether this insufficiency reflects functional abnormalities in each β-cell or too few appropriately functioning β-cells, usually referred to as a low β-cell mass (Henquin et al. 2008). As indicated by Rahier et al. (2008), sub-optimal β-cell function leads to a higher risk of developing DM2 if there is also a low β-cell mass while the slow decrease in β-cell mass with duration of diabetes could, at least in part, be a secondary phenomenon caused by exposure to a metabolically abnormal environment: glucolipotoxicity (Poitout and Robertson 2008).

Paradoxically, it has also been proposed that an important mechanism contributing to β-cell failure in DM2 is the ability to hypersecrete insulin

(Aston-Mourney et al. 2008). Hypersecretion, a characteristic in the early stages of the disease, is beneficial in maintaining normal glucose tolerance, which may also be an important factor in the progression of β-cell failure (Aston-Mourney et al. 2008). A state of hyperinsulinemia can be caused by increased insulin demand (insulin resistance), a genetic abnormality leading to hypersecretion (as in persistent hyperinsulinemic hypoglycemia of infancy) or the use of insulin secretory drugs (sulfonylureas such as glibenclamide). The increased demands for insulin production could overload the endoplasmic reticulum (ER), resulting in ER stress and inducing the unfolded protein response. Furthermore, apoptosis of the β cells has been shown to be a result of the activation of an ER stress response (Laybutt et al. 2007). Alternatively, the increased glycolytic flux required for increased insulin secretion could result in oxidative stress. In individuals with a genetic predisposition, the increased ER stress could lead to β cell failure and subsequent diabetes. The treatment of diabetes with insulin secretory drugs could further promote insulin hypersecretion, leading to worsening of β-cell function.

Besides glucotoxicity, lipotoxicity, and glucolipotoxicity, which are secondary phenomena that play a role in β-cell dysfunction, other factors could contribute to the progressive loss of β-cell function in DM2 (Wajchenberg 2007).

In conclusion, drawing on all the information available, it can be suggested that the link between reduced β-cell mass and impaired function could be due to an increased demand on residual β-cells per se, leading to changes in function (ER stress or other mechanisms), or related to the hyperglycemia resulting from decreased β-cell mass, driving the impairment in β-cell function. In vitro and in vivo studies in rodents (not in humans, as shown previously) have indicated that persistently high glucose levels play a central role among those factors (free fatty acids [FFA], lipoproteins, leptin, and cytokines), contributing to β-cell demise.

Understanding the mechanisms of β-cell death and thus decreased β-cell mass, at least in rodents, and impaired function has provided the basis of β-cell preservation, especially when one considers that the impaired β-cell function and possibly β-cell mass appear to be reversible to a certain degree, particularly during the early stages of the disease, where the threshold for reversibility of decreased β-cell mass has probably not been passed. Therefore, any therapeutic intervention aimed at preserving β-cell activity should improve function and prevent further reduction in mass.

It is known that humans have a much lower capacity for β-cell regeneration than rodents, which declines even further with aging. Data from studies with human autopsy and human donor pancreata suggest that, after the age of 15–20 years β cell growth by replication (probably the primary mechanism for regeneration) is minimal (Ritzel 2009). Accordingly, in a small series of 10 deceased patients (aged 17–74 years) who had received thymidine analogs 8 days to 4 years before death, Perl et al. (2010) found that human β-cell turnover is limited to the first three decades of life. Since most humans requiring antidiabetic therapy are over 20 it will require long-term studies to identify which treatment option might induce β-cell regeneration in the clinical setting.

Clinical Impact of Therapies Aimed at β-Cell Preservation

Short-Term Improvement of β-Cell Insulin Secretion

The current diabetes treatment options that lead to short-term improvement of β-cell secretion include weight loss and antidiabetic medications: oral insulin secretagogues and insulin:

Weight loss improves insulin secretion in obese DM2 patients (Gumbiner et al. 1990). Among the oral antidiabetic drugs, metformin improves glucose levels before and after meals without significant changes in insulin secretion and levels, indicating improved glucose sensitivity (Wu et al. 1990). The sulfonylureas and glinides are commonly used to stimulate insulin secretion in DM2 patients, enhancing β-cell responsiveness to glucose (Shapiro et al. 1989). Several studies have shown that treatment with sulfonylureas is not associated with any change in the decay curve of β-cell function with time (UK Prospective Diabetes Study 16 1995; Kahn et al. 2006). Moreover, these compounds have been shown to cause apoptosis and therefore loss of β-cell mass (Maedler et al. 2005). Finally, short-term intensive insulin therapy in patients with DM2 has been shown to improve endogenous β-cell function and insulin resistance (Garvey et al. 1985; Glaser et al. 1988). However, prolonged benefit has rarely been demonstrated, with virtually all patients becoming hyperglycemic again after a few weeks (Gormley et al. 1986). Until recently, it was unknown whether such outcomes pertained to new-onset DM2, although patients having failed with diet therapy can show a good response to a short period of intensive insulin therapy by continuous subcutaneous insulin infusion (CSII), as initially demonstrated by Ilkova et al. (1997).

Long-Term Improvement of β-Cell Insulin Secretion

Treatments that may lead to long-term improvement in β-cell insulin secretion include short- and long-term intensive insulin therapy of newly diagnosed DM2 and the use of oral insulin sensitizers: glitazones and incretin mimetics (GLP-1 mimetics and enhancers), which have shown clinical evidence of effects on human β-cell function, the latter drugs having demonstrated, at least in rodents, that they are associated with expansion of β-cell mass via stimulation of β-cell proliferation, promotion of islet cell neogenesis, and inhibition of β-cell apoptosis (Wajchenberg 2007; Xu et al. 1999; Li et al. 2003; Baggio and Drucker 2006).

Short-Term Intensive Insulin Therapy of Newly Diagnosed DM2

Insulin therapy is the most effective antidiabetic therapy and has a variety of effects that may protect against the progression of β-cell dysfunction as suggested by the clinical studies to be outlined later. First, correcting hyperglycemia with insulin may alleviate glucolipotoxicity. Preclinical studies also suggest that insulin has anti-apoptotic effects via its action on IRS (insulin receptor substrate) proteins and may promote β-cell growth (Tseng et al. 2002). Numerous in vitro and clinical studies have also demonstrated that insulin therapy has potential anti-inflammatory benefits, independent of its ability to lower blood glucose levels (Dandona et al. 2007). Further investigation is needed to determine the clinical implications of the anti-inflammatory properties of insulin in the progression of DM2.

Optimal metabolic control, especially early intensive glycemic control, plays a role in the prevention of progressive β-cell dysfunction and possibly destruction of the β-cells with worsening of diabetes, as will be presented below (Retnakaran and Drucker 2008).

Table 1 shows that in the available studies, early implementation of a short course of intensive insulin therapy, either by continuous subcutaneous insulin infusion or multiple daily injections can induce sustained euglycemia, in patients with DM2 (Ryan et al. 2004; Li et al. 2004; Weng et al. 2008; Chen et al. 2008; Wen et al. 2009), while off any antidiabetic therapy. The remission of DM2 achieved in these studies persisted for 1 year after cessation of insulin therapy in about 46 % of patients. In the small series of patients treated for 1 year, after a short-term bout of intensive insulin therapy, accompanied by Chen et al. (2008), HbA1c levels were significantly lower in the insulin group than in the oral hypoglycemic agent(s) group at 6 months, and after 1 year the glycated hemoglobin level remained lower in the insulin group. Furthermore, Li and colleagues (2004), as well as Weng et al. (2008) and Chen et al. (2008), reported that patients who maintained euglycemia while off oral antidiabetic therapy for 1 year showed greater recovery of β-cell function than their counterparts. In a set of patients studied by Wen et al. out of 84 newly diagnosed DM2 patients post-treatment with continuous sub-cutaneous insulin infusion for 2 weeks, and followed for 2 years, remission was observed in 53 % of the subjects studied (Wen et al. 2009).

It was suggested that an improvement in β-cell function, especially restoration of the first-phase insulin secretion, might be responsible for the ability of intensive insulin therapy to induce sustained euglycemia. Furthermore, proinsulin decreased highly significantly as did the proinsulin to total immunoreactive insulin (PI/IRI) ratio, indicating an improvement in the quality of insulin secretion (Li et al. 2004; Weng et al. 2008).

It should be noticed that in all series of patients, except in that from Ryan et al. (2004), the mean BMI was within or slightly above the normal range, which is infrequent in western countries where the majority of the patients are obese at admission. It could be suggested, at least for the Asian patients, that they presented a different phenotype of the disease with predominant β-cell failure and much less insulin resistance.

Long-Term Intensive Insulin Therapy in Newly Diagnosed DM2

Intensive insulin therapy (plus metformin) at the time of diagnosis of DM2, treatment-naïve, with a mean BMI of 36 kg/m^2, was followed by random assignment to insulin + metformin or triple oral therapy, with metformin, glyburide, and

Table 1 Intensive insulin therapy in newly diagnosed type 2 diabetes

Author	n	Mean age (years)	Mean BMI (kg/m²)	Baseline HbA₁c (%)	Therapy Type	Therapy Duration (days)	Patients c/ englycemia c/ therapy (%)	Patients c/ englycemia At 6 months (%)	At 12 month (%)
Ilkova et al. (1997)	13	50	26.9	11.0	CSII	14	92	69	N/A
Ryan et al. (2004)	16	52	30.8	11.8	MDI	14–21	88	N/A	44
Li et al. (2004)	138	49	25.0	10.9	CSII	14	91	67	47
Weng et al. (2008)	133	50	25.1	9.8	CSII	14–35	97	N/A	51
	118	51	24.4	9.7	MDI	14–35	95	N/A	45
	101	52	25.1	9.5	OHA	14–35	84	N/A	27
Chen et al. (2008)	22	59	27.7	11.7	MDI[a]	1 year	N/A	65[b]	55[b]
	8	56	26.6	11.3	OHA[a]	1 year	N/A	35[b]	32[b]
Wen et al. (2009)	84	48	25.3	9.9	CSII	14	N/A	74	62[b] (24 mo:53%)

Modified, with permission, from Retnakaran and Drucker (2008)

N/A not available, *CSII* continuus subcutaneous insulin infusion, *MDI* multiple daily injections, *OHA* oral hypoglycemic agents

[a] After 10–14 days of intensive (MDI) therapy

[b] Patients (%) with HbA1c<6.5

pioglitazone, for 3.5 years, with preservation of both glycemic control and β-cell function, which was assessed by a mixed-meal challenge test at 6, 12, 18, 30, and 42 months, measuring C-peptide. β-cell function was preserved for at least 3.5 years with either therapy (Harrison et al. 2012).

Glitazones

Indirect Effects with Amelioration of Insulin Sensitivity

The glitazones are agonists of PPARγ, a nuclear receptor that regulates transcription genes involved in lipid and glucose metabolism. Although predominantly expressed in adipose tissue, PPARγ is present in other insulin-sensitive tissues, including the pancreatic islet cells (Dubois et al. 2001). The development of small, insulin-sensitive adipocytes, enhances glucose uptake and decreases hepatic glucose output, improving glycemic control as well as lowering plasma FFAs in DM2. Improving insulin sensitivity in the periphery may improve the glucose-sensing ability of β-cells and preserve β-cell function by reducing the demand on these cells. It has been postulated that the improvement in β-cell function, particularly the normalization of the asynchronous insulin secretion that characterizes β-cell failure, could be related to a reduction in glucolipotoxicity due to improved glycemic control and/or improved insulin sensitivity seen with glitazones. This could suggest an increased ability of the β-cell to sense and respond to glucose changes within the physiological range after glitazone treatment (Wajchenberg 2007).

Direct Effects via PPARγ Activation in Pancreatic Islands

Preclinical data in rodents have suggested that glitazones might decrease β-cell apoptosis, maintaining β-cell neogenesis and preventing islet amyloidosis. Various mechanisms of action have been proposed to explain these effects (Wajchenberg 2007).

In humans, as a class effect, glitazones consistently improve basal β-cell function, as measured by the HOMA model and observed during glitazone monotherapy and combination therapy. Further evidence of the beneficial effects on β-cells originates from other studies, in which treatment with glitazones alone or added to maximal doses of sulfonylurea and metformin or insulin, restored the first-phase insulin response to an intravenous glucose tolerance test (Ovalle and Bell 2004). In all studies, the beneficial effect of glitazones on β-cell function was independent of glucose control (as suggested by a similar reduction in HbA1c, with no improvement in β-cell function found in the insulin-treated group), indicating that glitazones can promote recovery of β-cell function independently of the amelioration of insulin sensitivity (Wajchenberg 2007).

Furthermore, extension studies with glitazones indicate that improvements in β-cell function are sustained over time in some individuals, both as monotherapy and

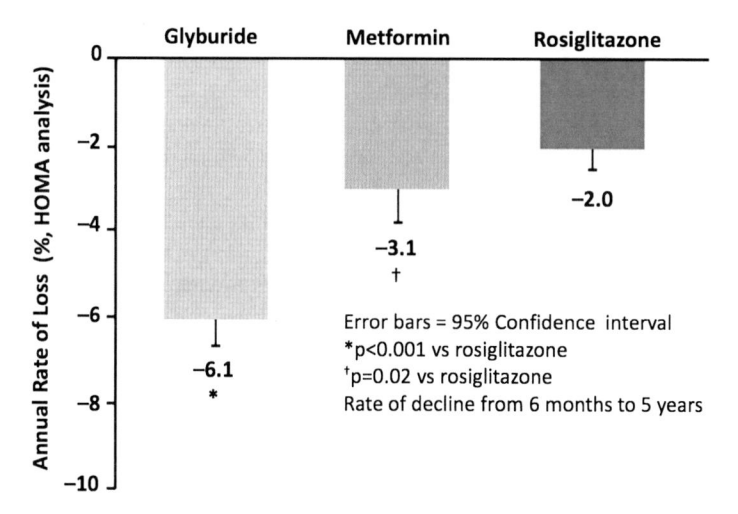

Fig. 1 ADOPT: Rosiglitazone reduces the rate of loss of β-cell function (Analysis includes only patients continuing on monotherapy (Adapted from Kahn et al. (2006))

in combination with metformin and/or sulfonylurea (Campbell 2004; Bell and Ovalle 2002). Another study evaluated the durability of the efficacy of rosiglitazone, metformin, and glyburide (glibenclamide) treatment for recently diagnosed DM2, in maintaining long-term glycemic control along with their effects on insulin sensitivity and β-cell function in 4,360 patients (Kahn et al. 2006). In this study, the cumulative incidence of monotherapy failure at 5 years was 15 % with rosiglitazone, 21 % with metformin, and 34 % with glyburide ($p < 0.001$; for both comparisons with rosiglitazone). During the first 6 months, levels of β-cell function (as evaluated by HOMA-β) increased more in the glyburide group than in either the rosiglitazone or the metformin groups. Thereafter, levels of β-cell function declined in all three groups. The annual rate of decline after 6 months was 6.1 % with glyburide, 3.1 % with metformin and 2 % with rosiglitazone ($p < 0.001$ vs. glyburide and $p = 0.02$ vs. metformin; Fig. 1) (Kahn et al. 2006). In conclusion, the study showed that the efficacy of glitazones, compared with other oral glucose-lowering medications, in maintaining long-term glycemic control in DM2.

Incretin Mimetics

Incretin hormones are released by the gastrointestinal tract in response to nutrient ingestion to enhance insulin secretion and aid in the maintenance of glucose homeostasis. The two major incretins are GLP-1 and GIP, which are released by enteroendocrine L cells located in the distal ileum and the colon, and by the K cells in the duodenum respectively (Drucker 2003). They provide the additional stimulus to insulin secretion during oral ingestion not provided by IV glucose infusion.

These incretins increase insulin secretion in a glucose-dependent manner through activation of their specific receptors in β-cells. In newly-diagnosed DM2 with relatively good glycemic control (HbA1c ~6.9 %), both GIP and GLP-1 secretion in response to glucose and mixed-meal challenges are the same or even increased compared with healthy subjects (Theodorakis et al. 2006; Vollmer et al. 2008). However, in long-standing DM2 with poor glycemic control (HbA1c ~8–9 %) the GLP-1 response is decreased whereas GIP secretion is unchanged (Chia and Egan 2008). In addition, insulin response to exogenous GLP-1 is three to fivefold lower in DM2. However, acute GLP-1 administration is able to increase insulin secretion to normal levels and to lower plasma glucose effectively (Kjems et al. 2003). In contrast, exogenous GIP, even at supraphysiological doses, has markedly reduced insulinotropic action with little or no glucose-lowering effects in DM2 (Nauck et al. 1993).

Thus, deterioration of glucose homeostasis can develop in the absence of any impairment in GLP-1 levels. This could suggest that the defects in GLP-1 concentrations previously described in patients with long-standing DM2 are likely to be secondary to other hormonal and metabolic alterations, such as fasting hyperglucagonemia and body weight, which were negatively associated with GLP-1 levels, as assessed by the incremental areas under the curves, after oral glucose and meal ingestion (Vollmer et al. 2008). Conversely, there is a positive relationship between GLP-1 and increasing age and a negative association with higher BMI levels. However, these associations were stronger after oral glucose ingestion than after mixed meal ingestion. Accordingly, another study found that obesity and glucose tolerance each attenuate the incretin effect (i.e., the gain in β-cell function after oral glucose versus intravenous glucose) on β-cell function and GLP-1 response (Muscelli et al. 2008). In both studies it was concluded that GIP and GLP-1 appeared to be regulated by different factors and are independent of each other (Vollmer et al. 2008; Muscelli et al. 2008).

Therefore, therapeutic strategies for DM2 within the incretin field focused on the use of GLP-1, GLP-1 analogs (GLP-1 receptor [GLP-1R] agonists or GLP-1 mimetics) and GLP-1 enhancers, but not GIP.

GLP-1 at pharmacological doses also has other non-insulinotropic effects beneficial for treating DM2: suppression of glucagon secretion in the presence of hyperglycemia and euglycemia, but not hypoglycemia, leading to improved hepatic insulin resistance and glycemic control; slowing of gastric emptying and gut motility, causing delayed nutrition, absorption, and dampened postprandial glucose excursion; and increasing the duration of postprandial satiety, leading to lower food intake, weight loss, and improved insulin resistance (Drucker 2003). More importantly, acute GLP-1 infusion normalized fasting plasma glucose in patients with long-standing uncontrolled DM2 who were no longer responsive to sulfonylureas or metformin (Willms et al. 1998). One major drawback of GLP-1 treatment is its short half-life (1–2 min), since it is rapidly degraded by dipeptidyl peptidase-4 (DPP-4), which cleaves the N-terminal dipeptides (His 7-Ala 8) from GLP-1 (Rahier et al. 2008; Aston-Mourney et al. 2008; Ovalle and Bell 2004).

Modifications in the GLP-1 molecule to prevent degradation by DPP-4 have resulted in the development of long-acting GLP-1 receptor agonists (GLP-1 mimetics) for the management of DM2.

1. Exendin-4-based: exenatide is a synthetic version of exendin-4, which is a 39-amino-acid peptide produced in the salivary glands of the lizard "Gila monster" with 53 % homology to full-length GLP-1. It binds more avidly to GLP-1R than GLP-1 and exendin-4 is not a substrate for DPP-4 because it has Gly8 in place of Ala8 (Deacon 2007). Available are exenatide with a short-half life (~2–4 h) has to be given at least twice daily (exenatide BID) and exenatide long-acting release (LAR) given once-weekly (OW). The other exendin-4 based GLP-1 receptor agonist, lixisenatide, given twice daily, is in phase 3 clinical trials.
2. Human GLP-1-based: liraglutide is a long-acting GLP-1 analog has 97 % homology with GLP-1 and resists DPP-4 degradation by fatty acid acylation and albumin binding, with a half-life of 12–14 h, allowing for single daily-dose administration (Table 2) (Agerso et al. 2002). The other two human GLP-1 analogs, albiglutide and dulaglutide, are in phase 3 trials and are to be given once a week.

The acute effect of exogenous GLP-1 or GLP-1 R agonists on β-cells in rodent models of diabetes and in cultured β-cells is the stimulation of glucose-dependent insulin release, whereas the subacute effect is enhancing insulin biosynthesis and stimulation of insulin gene transcription. Their chronic action is stimulation of β-cell proliferation, induction of islet neogenesis from precursor ductal cells, and inhibition of β-cell apoptosis, thus promoting an expansion of the β-cell mass. This

Table 2 Incretin mimetics: exenatide BID versus liraglutide versus exenatide OW

	Exenatide BID	Liraglutide	Exenatide OW[a]
Administration	Injection	Injection	Injection
Half life [h]	≈2–4	≈12–14	2 weeks
Frequency of inj.	twice daily	once daily	once weekly
Dose per injection	5–10μg	up to 1.8mg	2mg weekly
DPP-4 subtrate?	no	no	no
Insulin secretion[b]	↑	↑	↑
Glucagon secretion[b]	↓	↓	↓
Fasting glucose	↓↓	↓↓↓	↓↓↓
Postprandial glucose	↓↓↓	↓↓	↓↓
↓HbA 1c	≈0.6–1.4	≈1.0–1.8	≈1.3–1.9
Gastric emptying	↓	(↓)	yes
Antibody production	yes (≈45 %, ↓ c/therapy)	no	yes (≈22 %, ↓ c/therapy)

Partial reproduction with permission from Wajchenberg BL. Endocine Reviews: β-cell failure in diabetes and preservation by clinical treatment. vol 28, no 2, April 2007: 187–218. Copyright [2007]. The Endocrine Society
[a]Hydrolyzable polymer microspheres
[b]Glucose-dependent

was also demonstrated in human islets freshly isolated from three cadaveric donors treated with liraglutide (Wajchenberg 2007). These effects have major implications for the treatment of DM2 because they directly address one of the fundamental defects in DM2, i.e., β-cell failure.

Exenatide BID

Clinical trials in DM2 patients who have not achieved adequate glycemic control on metformin and/or sulfonylurea, metformin and/or TZD, as well as comparative trials with insulin glargine and biphasic insulin aspart (30/70), are available in the literature (Drucker and Nauck 2006; Amori et al. 2007). With exenatide BID, 10 μg twice daily as adjuvant therapy to oral hypoglycemic agents, a significant proportion of patients (32–62 %) achieved HbA1c of 7 % or less compared with placebo (7–13 %), glargine (48 %), and biphasic insulin aspart (24 %). HbA1c reductions of 0.8–1.1 % were sustained over 3 years when added to metformin and/or sulfonylurea, resulting in significant and sustained improvements in glycemic control: HbA1c (-1.0 ± 0.1 %; $p < 0.0001$) and FPG (-23.5 ± 3.8 mg/dl; $p < 0.0001$), and improvement in HOMA-β as a surrogate of β-cell function (Klonoff et al. 2008).

There were reports of pancreatitis during the exenatide BID development program and the postmarketing period, which were passed to the FDA. Out of 30 cases 90 % reported one or more possible contributory factors, including the concomitant use of medications that list pancreatitis among the reported adverse effects in product labeling, or confounding conditions, such as obesity, gallstones, severe hypertriglyceridemia, and alcohol use (De Vries 2009).

It should be mentioned that a retrospective cohort study with a large US healthcare claims database found that the cohort with DM2 were at a 2.8-fold greater risk of acute pancreatitis compared with the cohort without diabetes (Noel et al. 2009).

Liraglutide

In the 52-week monotherapy trial, liraglutide was investigated versus the sulfonylurea glimepiride (Liraglutide Effect and Action in Diabetes [LEAD-3] trial), in which participants in the liraglutide groups lost weight, independent of the presence of nausea, up to 2.45 kg by end of study, compared with the weight gain of about 1.12 kg on glimepiride. HOMA-IR and fasting plasma glucagon showed significant decreases with liraglutide, but mean increases with glimepiride. The proinsulin to insulin ratio and HOMA-B showed no significant differences between treatments (Garber et al. 2009).

When liraglutide was used in combination trials with other oral agents and insulin, significant improvements were also demonstrated in β-cell function, as measured by HOMA-B and the proinsulin to insulin ratio. As with the monotherapy trial, liraglutide resulted in substantial weight loss in the combination trials – up to 2.8 kg after 24 weeks (LEAD-2) – and also demonstrated clinically meaningful reductions in systolic blood pressure (up to an average of 6.7 mmHg in LEAD-4). Again, liraglutide was associated with a low rate of hypoglycemic events (Marre et al. 2009; Nauck et al. 2009; Zinman et al. 2008; Russell-Jones et al. 2008).

Meta-Analyses of the LEAD (Liraglutide Effect and Action in Diabetes) Trials
- **Factors affecting glycemic control with liraglutide**
  1. Adding liraglutide to existing therapy was more effective in lowering HbA1c than substituting an existing therapy with liraglutide.
  2. Liraglutide induced clinically relevant reductions in HbA1c across the continuum of disease progression, but achieved the greatest reductions in patients with poorer initial glycemic control (higher HbA1c).
  3. Liraglutide reduced HbA1c independent of concomitant weight loss.
- **Systolic Blood Pressure**
  1. Liraglutide significantly and sustainably reduced systolic blood pressure (SBP), with greatest improvements observed with elevated SBP at baseline.
  2. Mean SBP was significantly reduced by up to 2.6 mmHg from baseline within 2 weeks of liraglutide treatment, before any weight loss had occurred, and reductions were sustained up to 26 weeks (duration of the study).
  3. Patients with the highest quartile of baseline SBP (>140 mmHg to 190 mmHg) displayed the greatest reduction in SBP from baseline, independent of concomitant treatment with antihypertensive medication.
- **Lipids and Cardiovascular markers**
  Liraglutide significantly improved lipids and cardiovascular (CV) risk factors from baseline after 26 weeks of treatment:
  1. Total cholesterol, LDL cholesterol, FFA, and triglycerides all decreased significantly from baseline with liraglutide treatment ($p < 0.01$ for all).
  2. Liraglutide treatment also significantly reduced brain natriuretic peptide and high-sensitivity C-reactive protein, both markers of CV risk.

In relation to the use of liraglutide and the development of pancreatitis, acute pancreatitis occurred at a rate of 1.6 cases per 1,000 subject-years of liraglutide exposure and 0.6 cases per 1,000 subject-years of glimepiride exposure; chronic pancreatitis occurred at a rate of 0.6 cases per 1,000 subject-years of liraglutide exposure. Of the patients recruited for the LEAD program, 24 had previous histories of acute (n = 17) or chronic (n = 7) pancreatitis. Pancreatitis did not recur in any of these patients during liraglutide treatment (Liraglutide scientific synopsis 2010).

Data from the liraglutide phase 2 and 3 clinical trials, plus open-label controlled extension periods involving 6,638 subjects, 4,257 of whom had been exposed to liraglutide, provided consistent results across trials and did not indicate a signal of increased CV event (death, myocardial infarct or stroke) associated with liraglutide treatment, from the retrospective major adverse cardiovascular event (MACE) analyses (Liraglutide scientific synopsis 2010).

Exenatide Once-Weekly

The difference between exenatide BID and once-weekly (OW) is related to the continuous availability of the drug over 24 h compared with exposure only at mealtime with exenatide BID and deterioration of control during the night and at lunch time. Therefore, greater reduction of HbA1c with exenatide OW compared with BID is an indication that continuous glycemic control with once-weekly doses, not only at mealtime, but whenever blood glucose levels are elevated, thus resulting in the powerful efficacy of OW medication (Bergenstal et al. 2010; Drucker et al. 2008; Blevins et al. 2011; Buse et al. 2010, 2013; Diamant et al. 2010, 2012; Russell-Jones et al. 2012).

The core design of exenatide OW clinical trials corresponded to that of the so-called DURATION (Diabetes Therapy Utilization: Researching Changes in A1c, Weight and Other Factors Through Intervention With Exenatide Once Weekly) trials as follows: patients with DM2 and HbA1c from 7.1 % to 11 %, received either exenatide OW or a comparative agent(s) for 24–30 weeks, with optional OW extension. The primary endpoint was the change in HbA1c from baseline to endpoint. Secondary endpoints were the change in body weight, blood pressure, cardiovascular risk factors, safety, and tolerability from baseline to endpoint (Bergenstal et al. 2010; Drucker et al. 2008; Blevins et al. 2011; Buse et al. 2010, 2013; Diamant et al. 2010, 2012; Russell-Jones et al. 2012).

Relative to CV risk factors, patients treated with exenatide OW for 52 weeks (Buse et al. 2010) experienced improvements in the following factors: clinically significant blood pressure improvements were observed in patients treated with exenatide OW from baseline, SBP −6.2 mmHg and DBP −2.8 mmHg, and in patients switching from exenatide BID to OW, SBP −3.8 mmHg and DBP −1.8 mmHg. The majority (84 %) of the 154 patients who had been using an antihypertensive medication at screening did not change their dose after completing 52 weeks. Improvements in serum lipid profiles were demonstrated in both treatment groups, with clinically significant reductions in total cholesterol (−9.6 mg/dl vs. −9 mg/dl), LDL cholesterol (−3.4 mg/dl vs. −2.8 mg/dl), HDL cholesterol (−0.7 mg/dl vs. −1.6 mg/dl), triglycerides (−15 % vs. −13 %), exenatide OW vs. exenatide BID respectively.

Regarding baseline-to-end changes in cardiometabolic parameters and β-cell functions after 84 weeks of treatment with exenatide OW, greater reductions were observed compared with insulin glargine for waist and hip circumference ($p < 0.001$) and SBP −4.2 mmHg vs. −0.8 after insulin ($p = 0.027$), but there was no difference in DBP (−1.5 mmHg after exenatide OW vs. −1.4 mmHg after glargine; $p = 0.879$). While the heart rate increased after exenatide OW +1.97 it decreased to 0.79 beats/min after insulin ($p = 0.0034$). No differences were found in total cholesterol or triglycerides. C-reactive protein, urinary albumin to creatinine ratio decreased similarly after both treatments. HOMA-β values improved to a greater extent in the exenatide OW group than in the glargine group (Diamant et al. 2012).

Exenatide OW Safety and Tolerability

Durability of Glycemic Control with GLP-1 Receptor Agonists
There are three clinical trials with open-label extension:
1. Exenatide BID extension for over 3 years when added to metformin and/or sulphonylurea, with sustained improvement in glycemic control: A1c −1.0 % and FPG −23.5 mg/dl (Klonoff et al. 2008).
2. Exenatide OW extension for 3 years. This study followed the 30-week controlled trial (DURATION-1) in which, as already indicated, a more robust glucose lowering effect occurred than with the BID formulation of exenatide. The controlled period of the trial was followed by an open-label period in which all patients either continued with exenatide OW treatment or switched from exenatide BID to OW for 52 weeks (Diamant et al. 2012). Approximately 66 % (n = 194) of the original 295 patients completed 3 years of treatment. Baseline mean: HbA1c 8.2 %, FPG 167 mg/dl; weight: 101 kg; therapy at screening: metformin (33 %), metformin + sulphonylurea (29 %), and metformin + TZD (9 %). Significant HbA1c improvement was observed with 3 years of treatment (−1.6 %), 57 % achieving HbA1c ≤ 7.0 %. Significant improvements were observed in mean FPG (−33 mg/dl) and mean weight (−2.3 kg). There was also an improvement in CV risk factors: SBP, LDL cholesterol, and triglycerides (MacConell et al. 2011).
3. Liraglutide for 2 years. In this study, participants were randomized to receive once-daily liraglutide 1.2 mg, liraglutide 1.8 mg or glimepiride 8 mg. For patients completing 2 years of therapy, HbA1c reductions were −0.6 % with glimepiride vs. −0.9 % with liraglutide 1.2 mg (difference: −0.37 %, $p = 0.0376$) and −1.1 % with liraglutide 1.8 mg (difference: −0.55 %, $p = 0.0016$). Liraglutide was more effective in reducing HbA1c, FPG, and weight. Rates of minor hypoglycemia were significantly lower with liraglutide 1.2 mg and 1.8 mg compared with glimepiride ($p < 0.0001$) (Garber et al. 2011).

Regarding the important question of whether GLP-1 and GLP-1 mimetics have an effect on β-cell mass in humans, even though they have favorable effects on β cell function, such as first-phase insulin secretion and homeostasis assessment, the β-cell index (HOMA β), as seen with chronic exenatide BID use up to 3 years (Klonoff et al. 2008), this improvement in function may be due to the restoration of glucose competence to β-cells and the insulinotropic glucose-lowering and weight-loss effects of exenatide, and perhaps not because of any direct effect of exenatide on β-cell mass, as previously indicated. At present there is no strong evidence that incretin mimetics and DPP-4 inhibitors can expand or at least maintain β-cell mass in humans, and as such be able to delay the progression of the disease (Salehi et al. 2008).

Another question yet to be elucidated is the mechanism by which GLP-1 and GLP-1 mimetics lower glucagon secretion from α-cells. The ability of GLP-1 and incretin mimetics to lower glucagon levels in DM2 patients, in whom they are high throughout the day, contributes to the overall glucose-lowering effect. By enhancing endogenous insulin secretion with suppression of glucagon secretion, a more physiological insulin to glucagon ratio in the portal vein should be established, resulting in better suppression of hepatic glucose output. The mechanism(s) by which GLP-1 and GLP-1 mimetics lower glucagon secretion remain(s) unclear and discussion of this issue is beyond the scope of this publication. According to Dunning and Gerich (Dunning and Gerich 2007) in a review of published studies, the defect(s) in α-cell function that occur(s) in type 2 diabetes reflect(s) impaired glucose sensing. Because local insulin is a key regulator of glucagon secretion and defective β-cell glucose sensing in DM2 is indisputable, many, if not all, of the characteristic defects in the α-cell function may be secondary to β-cell dysfunction. It is interesting to note that attenuated and delayed glucagon suppression in DM2 occurs after oral ingestion of glucose, while isoglycemic intravenous administration of glucose results in normal suppression of glucagon, possibly because of the glucagonotropic action of GIP and GLP-1 after oral glucose. This phenomenon contributes both to the glucose intolerance and to the reduced incretin effect observed in DM2 patients (Knop et al. 2007; Meier et al. 2007).

Incretin Enhancers (DPP-4 Inhibitors)

Preclinical studies have demonstrated that DPP-4 inhibitors, which prevent the degradation of native GLP-1 by inhibiting the activity of the DPP-4 enzyme and thus increasing endogenous GLP-1 (and GIP) levels, may promote β-cell proliferation and neogenesis and inhibit apoptosis (Wajchenberg 2007). Thus, they have emerged as a therapeutic strategy for enhancing GLP-1 action in vivo. However, there are suggestions that mediators other than GLP-1 might contribute to the therapeutic effect of DPP-4 inhibition (Nauck and El-Ouaghlidi 2005). Alternatively, there are indications that GLP-1 may work indirectly through activation of the autonomic nervous system (Ahrén 2004).

DPP-4 inhibitors were developed to augment biologically active, endogenously secreted plasma GLP-1. In humans, sitagliptin, a DPP-4 inhibitor, both after a single dose and a once-daily dose for 10 days, resulted in an approximately twofold increase in active GLP-1 (aGLP-1) after meals. Besides, sitagliptin decreased total GLP-1 (tGLP-1) in the presence of increased aGLP-1 (Chia and Egan 2008). Whether the twofold increase in aGLP-1 is sufficient to explain the glucose-lowering effect with a reduction of HbA1c remains a matter of controversy.

If DPP-4 inhibitors lower blood glucose as a direct consequence of increased aGLP-1 levels, plasma insulin should also increase. However, fasting and postprandial plasma insulin and C-peptide levels did not differ before and after 10 days of DPP-4 inhibition in both healthy and DM2 subjects (Bergman et al. 2006; Ahrén et al. 2004). Indeed, infusions of GLP-1 that result in comparable plasma aGLP-1 attained by DPP-4 inhibition do not induce insulin secretion, but the same amount of insulin is secreted at a lower glucose level, or the insulinogenic index is improved.

DPP-4 inhibition results in lower postprandial plasma glucagon levels (Drucker and Nauck 2006). However, the reduced glucagon secretion is not evident in the fasting state when it would be most beneficial to decreasing nocturnal hepatic glucose output. The postprandial glucagon suppressive effects of DPP-4 inhibitors, although significantly different from placebo, are weak and short-lived, while levels are much higher than in nondiabetic subjects, and therefore are unlikely to account for the full antihyperglycemic effect (Chia and Egan 2008).

Currently, four DPP-4 inhibitors have been approved for treatment of DM2: sitagliptin, saxagliptin, linagliptin, and vildagliptin. The latter is not available in the USA. Another DPP-4 inhibitor is undergoing clinical development, namely alogliptin.

Pharmacological Differences Among the DPP-4 Inhibitors

While all DPP-4 inhibitors share a common mechanism of action, there are clinically important differences within the group considering their structural heterogeneity, which may account for variations in the pharmacokinetic profile among the DPP-4 inhibitors (Gerich 2010).

The different DPP-4 inhibitors are distinctive in their metabolism (saxagliptin and vildagliptin are metabolized in the liver, whereas sitagliptin and linagliptin are not), their excretion (linagliptin is excreted mostly unchanged by the liver, in contrast to other DPP-4 inhibitors, which are mainly excreted via the kidneys) and the potential of cytochrome-mediated drug–drug interactions (observed only with saxagliptin). These differences may be clinically relevant in patients with renal or hepatic impairment (Scheen 2012).

Liver Impairment

Regarding the safety of the DPP-4 inhibitors the risk of acute pancreatitis with DPP-4 therapy remains controversial. A recently published study that analyzed a large administrative database in the USA from 2005 to 2008, of DM2 patients aged 18–64 years, identified 1,269 hospitalized cases with acute pancreatitis and 1,269 control subjects matched for age, category, sex, enrollment pattern, and diabetes complications. Cases were significantly more likely than controls to have hypertriglyceridemia, alcohol use, gallstones, tobacco abuse, obesity, and biliary and pancreatic cancer (2.8 % vs. 0 %) and any neoplasm (20.94 % vs. 18.05 %). The conclusion of this administrative database study of US adults with DM2, treatment with GLP-1-based therapies, sitagliptin and exenatide BID, was associated with increased odds of hospitalization for acute pancreatitis (Singh et al. 2013). However, the limitations of observational claims-based analyses cannot exclude the possibility of an increased risk of acute pancreatitis (Garg et al. 2010).

As indicated by Scheen (2012) "further investigation is needed and long-term careful postmarketing surveillance is mandatory. Indeed, various experimental data in animal model suggested that there are grounds for concern that GLP-1 class of drugs may induce asymptomatic pancreatitis and perhaps over time, in some individuals, induce pancreatic cancer (Butler et al. 2010)."

Butler et al. examining the pancreata from 20 age-matched organ donors with DM2 treated with incretin therapy (n = 8) for 1 year or more, 7 having received sitagliptin and 1 exenatide BID. The remaining 12 did not receive GLP-1 drugs. Pancreata were also obtained from 14 nondiabetic controls matched for age, sex, and BMI. The study revealed a ~40 % increased pancreatic mass in DM2 patients treated with incretin therapy, with both increased exocrine cell proliferation ($p < 0.0001$) and dysplasia (increased pancreatic intraepithelial neoplasia, $p < 0.01$). Pancreas in DM2 patients treated with incretin therapy presented α-cell hyperplasia and glucagon microadenomas (3/8) and a neuroendocrine tumor. β-cell mass was reduced by approximately 60 % in those with DM2, yet a sixfold increase was observed in incretin-treated subjects, although diabetes persisted. Endocrine cells co-staining for insulin and glucagon were increased in diabetics compared with nondiabetic controls ($p < 0.05$) and markedly further increased with incretin therapy ($p < 0.05$). The authors concluded that, in humans, incretin therapy resulted in a marked expansion of the exocrine and endocrine pancreatic compartments, the former being accompanied by increased proliferation and dysplasia, the latter by α-cell hyperplasia with the potential for evolution into neuroendocrine tumors (Butler et al. 2013). The US Food and Drug Administration (FDA) is evaluating the reported findings by Butler et al. (2013). "These findings were based on examination of a small number of pancreatic tissue specimens taken from patients after they died from unspecified causes. FDA has asked the researchers to provide the methodology used to collect and study these specimens and to provide the tissue samples so the Agency can further investigate potential pancreatic toxicity associated with the incretin mimetics. FDA will communicate its final conclusions and recommendations when its review is complete or when the Agency has additional information to report" (Drug Safety and Availability. FDA Drug Safety Communication 2013).

At this time, the FDA and The Endocrine Society advise that patients should continue to take their medicine as directed until they talk to their health care professional, and health care professionals should continue to follow the prescribing recommendations on the drug labels (FDA Drug Safety Communication 2013).

The European Medicines Agency´s Committee for Medicinal Products for Human Use (CHMP) has finalized a review of GLP-1-based diabetes therapies. The Committee concluded that presently available data do not confirm recent concerns over an increased risk of pancreatic adverse events with these medicines (European Medicines Agency (EMA) 2013).

DPP-4 Inhibitors and CV Protection

A potential benefit of incretin-based therapies is their effect on CVD. In a meta-analysis of 41 randomized controlled trials (RCTs), 32 published (9 unpublished), performed in type 2 diabetic patients with DPP-4 inhibitors, with a duration >12 weeks, at the time of publication, the risk of CV events and all-cause death was 0.76 (0.46–1.28) and 0.78 (0.40–1.51) respectively (Monami et al. 2010).

There is some plausibility based on the influences of GLP-1-based drugs on CV risk factors: cardioprotection (rodents); mimicking of cardiac pre- and post-conditioning (Rahmi et al. 2013); improved myocardial survival in ischemic heart disease; improved myocardial performance in non-ischemic heart failure; reduced systolic blood pressure by 2–5 mmHg, mechanistically explained improved endothelial function and vasodilation, enhanced natriuresis and fluid excretion; lipid profiles are modestly improved; there is weight loss (incretin mimetics) or weight neutrality (DPP-4 inhibitors) (Ussher and Drucker 2012).

Long-Term Clinical Efficiency

Studies in humans with DM2 showed improvement of islet-cell function, in fasting and post-prandial states, and these beneficial effects were sustained in studies of up to 2 years' duration. However, there is at present no evidence in humans to suggest that DPP-4 inhibition has durable effects on β-cell function after cessation of therapy, as previously indicated with GLP-1 analogs (Scheen 2012; Gomis et al. 2012). The duration of these trials was too short to draw any definite conclusions. There are no long-term controlled trials to evaluate the effects of DPP-4 inhibitors on β-cell function in humans and the durability of the glucose-lowering effect of gliptins, as opposed to the escape phenomenon observed with sulfonylureas. Perhaps the analysis of large ongoing clinical trials with CV outcomes will provide additional information regarding the durability of glucose control with gliptins.

Mechanism of Action of DPP-4 Inhibitors

The contributions of increased insulin secretion and inhibition of glucagon secretion in the glucose-lowering effects in both fasting and post-prandial states still need to be better explored (Scheen 2012). Unresolved issues, such as the effects of GLP-1 mimetics and DPP-4 inhibitors on β-cell mass in humans, the mechanism by which GLP-1 mimetics lower glucagon levels from α-cells, the modest increase in active GLP-1 levels as a possible sole modulator of glycemia using DPP-4 inhibitors, and exactly how DPP-4 inhibition leads to a decline in plasma glucose levels without an increase in insulin secretion, need to be further evaluated (Chia and Egan 2008).

The measurement of islet function and glucose utilization was performed with the DPP-4 inhibitor, vildagliptin, given to DM2 patients for 6 weeks versus the same period of time with placebo. Vildagliptin increased postprandial GLP-1 and GIP, after oral glucose, by three- and twofold respectively, reduced FPG and PPG significantly (both $p < 0.01$), and improved the glucose responsiveness of insulin secretion by 50 % ($p < 0.01$). Vildagliptin lowered postprandial glucagon by 16 % ($p < 0.01$). Examined using a hyperinsulinemic euglycemic clamp, insulin sensitivity and glucose clearance improved after the administration of the DPP-4 inhibitor ($p < 0.01$). This was due to an increase in the glucose oxidation rate at the expense of fat oxidation and was also associated with a decrease in fasting lipolysis. Decreasing fasting lipolysis over 6 weeks is predicted to decrease stored triglycerides in non-fat tissues and may explain the increased glucose oxidation during use of the clamp at the expense of lipid oxidation. Thus, it was demonstrated that vildagliptin improves islet function in DM2 and glucose metabolism in peripheral tissues (Azuma et al. 2008).

To assess the effect of a DPP4-inhibitor, sitagliptin, on β-cell function in patients with DM2, a C-peptide minimal model was applied to extensive blood sampling of a nine-point meal tolerance test performed at baseline and at the end of treatment for 18–24 weeks of sitagliptin 100 mg q.d. as an add-on to metformin therapy or as monotherapy. In this model-based analysis, sitagliptin improved β-cell function relative to placebo in both fasting and postprandial states in the patients with DM2. The disposition indices (DIs), which assess insulin secretion in the context of changes in insulin sensitivity, for all measures performed, were significantly ($p < 0.05$) increased with sitagliptin treatment compared with placebo. The AA concluded that in their model-based analysis, sitagliptin improved β-cell function relative to placebo in both fasting and postprandial states in patients with DM2 (Xu et al. 2008).

The effects of saxagliptin on β-cell function of DM2 were assessed at baseline and at week 12 using an intravenous hyperglycemic clamp (fasting state) and an intravenous–oral hyperglycemic clamp (postprandial state) following oral ingestion of 75 g glucose. DPP-4 inhibition improved β-cell function in both postprandial (increased insulin secretion by 18.5 % adjusted difference vs. placebo, $p = 0.04$, associated with increased plasma concentrations of GLP-1 and GIP) and fasting states (increased insulin secretion by 27.9 % adjusted difference vs. placebo, $p = 0.02$). Saxagliptin also improved the glucagon area under the curve in the postprandial state (adjusted difference –21.8 % vs. placebo, $p = 0.03$). The AA indicated that given the magnitude of insulin response in the fasting state, further study into the effect of DPP-4 inhibition on the β-cell is warranted (Henry et al. 2011).

To quantify the incretin effect by comparing insulin secretory responses with oral as well as isoglycemic intravenous glucose infusions in patients with DM2 with and without the administration of a DPP-4 inhibitor, Vardarli et al. (2011) assessed the incretin effect after treatment with the DPP-4 inhibitor vildagliptin or placebo in patients previously treated with metformin. Vildagliptin augmented insulin secretory responses both after oral glucose (accompanied by the release of incretin hormones: GLP-1 and GIP) and during the intravenous infusion of glucose (without a major incretin response). Thus, against expectations, according to the AA, the incretin effect is not enhanced by DPP-4 inhibitor treatment, mainly because of a "surprising" augmentation of insulin secretory responses, even with intravenous glucose infusions. Thus, slight variations in basal incretin levels may be more important than previously thought. Another possibility is that the DPP-4 inhibitor-induced change in the incretin-related environment of islets might persist overnight, augmenting insulin secretory responses to intravenous glucose as well. Alternatively, as yet unidentified mediators of DPP-4 inhibition may have caused these effects (Vardarli et al. 2011).

Regarding the effect of alogliptin on pancreatic β-cell function, there are several studies of short duration (26 weeks) showing modest or nonsignificant increases in the proinsulin/insulin ratio and a trend toward increased homeostasis model assessment of β-cell function (HOMA-β) (DeFronzo et al. 2008; Pratley et al. 2009), but long-term studies will be needed to prove that alogliptin, like other DPP-4 inhibitors, preserves β-cell function.

Table 3 shows the differences between GLP-1 agonists and DPP-4 inhibitors (Wajchenberg 2007).

In conclusion, the DPP-4 inhibitors make an important contribution to the treatment of DM2, providing effective glucose control with a low risk for hypoglycemia, a neutral effect on body weight, and a general lack of gastrointestinal and other side effects differentiate DPP-4 inhibitors from some other classes of oral antidiabetic drugs. Experimental and particularly clinical studies suggest that DPP-4 inhibitors could preserve the progressive destruction of β-cells and the loss of insulin secretory capacity characteristic of DM2 (Gerich 2010).

As indicated by Nauck (personal communication), generally speaking, the optimism that incretin-based medications clinically improve β-cell function has fallen away, especially since it has been shown that in old rodents, GLP-1 and its derivatives do not increase β-cell replication and mass. Also, much of the "improved β-cell function" has been measured during treatment with GLP-1 receptor agonists, which acutely stimulate insulin secretion (which is not "improved β-cell function"); thus, some of the information is misleading.

Table 3 GLP-1 agonists versus DPP 4 inhibitors

	GLP-1 R	DPP-4 Inhibitors
Administration	Injection	Orally available
GLP-1 concentrations	Pharmacological	Physiological
Mechanism of actions	GLP-1	GLP-1 + GIP
Activation of portal glucose sensor	No	Yes
↑ Insulin secretion	+++	+
↓ Glucagon secretion	++	++
Gastric emptying	Inhibited	+/−
Weight loss	Yes	No
Expansion of β-cell mass in preclinical studies	Yes	Yes
Nausea and vomiting	Yes	No
Potential immunogenicity	Yes	No

Reproduced with permission from Wajchenberg BL. Endocrine Reviews: β-cell failure in diabetes and preservation by clinical treatment. vol 28, no 2, April 2007: 187–218. Copyright [2007]. The Endocrine society

Cross-References

▶ Mechanisms of Pancreatic β-Cell Apoptosis in Diabetes and Its Therapies
▶ Physiological and Pathophysiological Control of Glucagon Secretion by Pancreatic α Cells
▶ The β-Cell in Human Type 2 Diabetes

References

Agerso H, Jensen LB, Elbrond B, Rolan P, Zdravkovic M (2002) The pharmacokinetic, pharmacodynamics, safety and tolerability of NN2211, a long-acting GLP-1 derivative, in healthy men. Diabetologia 45:195–202

Ahrén B (2004) Sensory nerves contribute to insulin secretion by glucagon-like peptide-1(GLP-1) in mice. Am J Physiol Regul Integr Comp Physiol 286:R269–R272

Ahrén B, Landin-Olsson M, Jansson PA, Svensson M, Holmes D, Schweizer A (2004) Inhibition of dipeptidyl peptidase-4 reduces glycemia, sustain insulin levels, and reduces glucagon levels in type 2 diabetes. J Clin Endocrinol Metab 89:2078–2084

Amori RE, Lau J, Pittas AG (2007) Efficacy and safety of incretin therapy in type 2 diabetes: systematic review and meta-analysis. JAMA 298:194–206

Aston-Mourney K, Proietto J, Morahan G, Andrikopoulos S (2008) Too much of a good thing: why it is bad to stimulate the β cell to secrete insulin. Diabetologia 51:540–545

Azuma K, Rádiková Z, Mancino J, Toledo FGS, Thomas E, Kangani C, Dalla Man C, Cobelli C, Holst JJ, Deacon CF, He Y, Ligueros-Saylan M, Serra D, Foley JF, Kelley DE (2008) Measurements of islet function and glucose metabolism with the dipeptidyl peptidase 4 inhibitor vildagliptin in patients with type 2 diabetes. J Clin Endocrinol Metab 93:459–464

Baggio LL, Drucker DJ (2006) Incretin hormones in the treatment of type 2 diabetes: therapeutic applications of DPP-IV inhibitors. Medscape Diabetes Endocrinol 8:1–5

Bell DSH, Ovalle F (2002) Long-term efficacy of triple oral therapy for type 2 diabetes mellitus. Endocr Pract 8:271–275

Bergenstal RM, Wyshm C, MacConell L, Malloy J, Walsh B, Yan P, Wilhelm K, Malone J, Porter LE, DURATION-2 Study Group (2010) Efficacy and Safety of exenatide once weekly versus sitagliptin or pioglitazone as an adjunct to metformin for treatment of type 2 diabetes (DURATION-2): a randomised trial. Lancet 376:431–439

Bergman AJ, Stevens C, Zhou Y, Yi B, Laethem M, De Smet M, Snyder K, Hilliard D, Tanaka W, Zeng W, Tanen M, Wang AQ, Chen L, Winchell G, Davies MJ, Ramael S, Wagner JA, Herman GA (2006) Pharmacokinetic and pharmacodynamic properties of multiple oral doses of sitagliptin, a dipeptidyl peptidase-IV inhibitor: a double-blind, randomized, placebo-controlled study in healthy male volunteers. Clin Ther 28:55–72

Blevins T, Pullman J, Malloy J, Yen P, Taylor K, Schulteis C, Trautmann M, Porter L (2011) DURATION-5: exenatide once weekly resulted in greater improvements in glycemic control compared with exenatide twice daily in patients with type 2 diabetes. J Clin Endocrinol Metab 96:1301–1310

Buse JB, Drucker DJ, Taylor KL, Kim T, Walsh B, Hu H, Wilhelm K, Trautmann M, Shen LZ, Porter LE, for the DURATION-1 Study Group (2010) DURATION-1: exenatide once weekly produces sustained glycemic control and weight loss over 52 weeks. Diabetes Care 33:1255–1261

Buse JB, Nauck M, Forst T, Sheu WH, Shenouda SK, Heilmann CR, Hoogwerf BJ, Gao A, Boardman MK, Fineman M, Porter L, Schernthaner G (2013) Exenatide once weekly versus liraglutide once daily in patients with type 2 diabetes (DURATION-6): a randomized, open-label study. Lancet 381:117–124

Butler AE, Janson J, Bonner-Weir S, Ritzel R, Rizza RA (2003) Butler PC: β-cell deficit and increased β-cell apoptosis in humans with type 2 diabetes. Diabetes 52:102–110

Butler PC, Matveyenko AV, Dry S, Bhushan A, Elashoff R (2010) Glucagon-like peptide-1 therapy and the exocrine pancreas: innocent bystander or friendly fire? Diabetologia 53:1–6

Butler AE, Campbell-Thompson M, Gurlo T, Dawson DW, Atkinson M, Butler PC (2013) Marked expansion of exocrine and endocrine pancreas with incretin therapy in humans with increased exocrine pancreas dysplasia and the potential for glucagon-producing neuroendocrine tumors. Diabetes 62:2595–2604

Campbell JW (2004) Long-term glycemic control with pioglitazone in patients with type 2 diabetes. Int J Clin Pract 58:192–200

Chen H-S, Wu TE, Jap TS, Hsiao LC, Lee SH, Lin HD (2008) Beneficial effects of insulin on glycemic control and β-cell function in newly diagnosed type 2 diabetes with severe hyperglycemia after short-term intensive insulin therapy. Diabetes Care 31:1927–1932

Chia CW, Egan JM (2008) Incretin-based therapies in type 2 diabetes mellitus. J Clin Endocrinol Metab 93:3703–3716

Dandona P, Chaudhuri A, Mohanty P, Ghanim H (2007) Anti-inflammatory effects of insulin. Curr Opin Clin Nutr Metab Care 10:511–517

De Vries JH (2009) Is pancreatitis an adverse drug effect of GLP-1 receptor agonists? In: Advances in Glucagon-like peptides for the treatment of type 2 diabetes, vol 3. pp 9–13. http://cme.medscape.com/viewprogram/19082_pnt

Deacon CF (2004) Circulation and degradation of GIP and GLP-1. Horm Metab Res 36:761–765

Deacon CF (2007) Incretin-based treatment of type 2 diabetes: glucagon-like-peptide-1 receptor agonists and dipeptidyl peptidase-4 inhibitors. Diabetes Obes Metab 9(Suppl 1):23–31

DeFronzo RA, Fleck PR, Wilson CA, Mekki Q, Alogliptin Study 010 Group (2008) Efficacy and safety on the dipeptidyl peptidase-4 inhibitor alogliptin in patients with type 2 diabetes and inadequate glycemic control: a randomized, double-blind, placebo-controlled study. Diabetes Care 31:2315–2317

Diamant M, Van Gaal L, Stranks S, Northrup J, Cao D, Taylor K, Trautmann M (2010) Once weekly exenatide compared with insulin glargine titrated to target in patients with type 2 diabetes (DURATION-3): an open-label randomized trial. Lancet 375:2234–2243

Diamant M, Van Gaal L, Stranks S, Guerci B, MacConell L, Heber H, Scism-Bacon J, Trautmann M (2012) Safety and efficacy of once-weekly exenatide compared with insulin glargine titrated to target in patients with type 2 diabetes over 84 weeks. Diabetes Care 35:683–689

Drucker DJ (2003) Enhancing incretin action for the treatment of type 2 diabetes. Diabetes Care 26:2928–2940

Drucker DJ, Nauck MA (2006) The incretin system: glucagon-like peptide-1 receptor agonists and dipeptidyl-peptidase-4 inhibitors in type 2 diabetes. Lancet 368:1696–1705

Drucker DJ, Buse JB, Taylor K, Kendall DM, Trautmann M, Zhuang D, Porter L, DURATION-1 Study Group (2008) Exenatide once weekly versus twice daily for treatment of Type 2 diabetes: a randomised, open-label, non-inferiority study. Lancet 372:1240–1250

Drug Safety and Availability. FDA Drug Safety Communication (2013) FDA investigating reports of possible increased risk of pancreatitis and pre-cancerous findings of the pancreas from incretin mimetic drugs for type 2 diabetes, 14 March 2013

Dubois M, Pattou F, Kerr-Conte J, Gmyr V, Vanderwalle B, Desreumaux P, Auswers J, Schoonjans K, Lefebvre J (2001) Expression of peroxisome-proliferator-activated receptor γ (PPARγ) in normal human pancreatic islet cells. Diabetologia 43:1165–1169

Dunning BE, Gerich JE (2007) The role of α-cell dysregulation in fasting and postprandial hyperglycemia in type 2 diabetes and therapeutic implications. Endocr Rev 28:253–283

European Medicines Agency (EMA) (2013) Investigation into GLP-1 based diabetes therapies concluded. 26 July 2013

FDA Drug Safety Communication (2013) FDA investigating reports of possible increased risk of pancreatitis and pre-cancerous findings of the pancreas from incretin mimetic drugs for type 2 diabetes. Endocrine Daily Briefing, The Endocrine Society, 15 March 2013

Garber A, Henry R, Ratner R, Rodriguez-Pattzi H, Olvera-Alvarez I, Hale PM, Zdravkovic M, Bode B, Garcia-Hernandez PA, for the LEAD-3 (Mono) Study Group (2009) Liraglutide versus glimepiride monotherapy for type 2 diabetes (LEAD-3 Mono): a randomized, 52-week, phase III, double-blind, parallel-treatment trial. Lancet 373:473–481

Garber A, Henry RR, Ratner R, Hale P, Chang CT, Bode B, LEAD-3 (Mono) Study Group (2011) Liraglutide, a once-daily human glucagon-like peptide 1 analogue, provides sustained improvements in glycaemic control and weight for 2 years as monotherapy compared with glimepiride in patients with type 2 diabetes. Diabetes Obes Metab 13:348–356

Garg R, Chen W, Pendergrass M (2010) Acute pancreatitis in type 2 diabetes treated with exenatide or sitagliptin. A retrospective observational pharmacy claims analysis. Diabetes Care 33:2349–2354

Garvey WT, Olefsky JM, Griffin J, Hamman RF, Kolterman OG (1985) The effect of insulin treatment on insulin secretion and insulin action in type II diabetes mellitus. Diabetes 34:222–234

Gerich J (2010) DPP-4 inhibitors: what may be the clinical differentiators? Diabetes Res Clin Practice 90:131–140. Review article

Glaser B, Leibovich G, Nesher R, Hartling S, Binder C, Cerasi E (1988) Improved β-cell function after intensive insulin treatment in severe non-insulin-dependent diabetes. Acta Endocrinol (Copenh) 118:365–373

Gomis R, Owens OR, Taskinen MR, Del Prato S, Patel S, Pivovarova A, Schlosser A, Woerle HJ (2012) Long-term safety and efficacy of linagliptin as monotherapy or in combination with other glucose-lowering agents in 2121 subjects with type 2 diabetes: up to 2 years exposure in 24-week phase III trials followed by a 78-week open-label extension. Int J Clin Pract 66:731–740

Gormley MJ, Hadden DR, Woods R, Sheridan B, Andrews WJ (1986) One month's insulin treatment of type II diabetes: the early and medium-term effects following insulin withdrawal. Metabolism 35:1029–1036

Gumbiner B, Polonsky KS, Beltz WF, Griver K, Wallace P, Brechtel G, Henry RR (1990) Effects of weight loss and reduced hyperglycemia on the kinetics of insulin secretion in obese non-insulin dependent diabetes mellitus. J Clin Endocrinol Metab 70:1594–1602

Harrison LB, Adams-Huet B, Raskin P, Ligvay I (2012) β-cell function preservation after 3.5 years after 3.5 years of intensive diabetes therapy. Diabetes Care 35:1406–1412

Henquin J-C, Cerasi E, Efendic S, Steiner DF, Boitard C (2008) Pancreatic β-cell mass or β-cell function? Editorial. Diabetes Obes Metab 10(Suppl 4):1–4

Henry RR, Smith SR, Schwartz SL, Mudaliar SR, Deacon CF, Holst JJ, Duan RY, Chen RS, List JF (2011) Effects of saxagliptin on β-cell stimulation and insulin secretion in patients with type 2 diabetes. Diabetes Obes Metab 13:850–858

Holman RR (1998) Assessing the potential for α-glucosidase inhibitors in prediabetic states. Diabetes Res Clin Pract 40(Suppl):S21–S25

Holman RR (2006) Long-term efficacy of sulfonylureas: a United Kingdom prospective diabetes study perspective. Metabolism 55(Suppl 1):S2–S5

Ilkova H, Glaser B, Tunckale A, Bagriacik N, Cerasi E (1997) Induction of long-term glycemic control in newly diagnosed type 2 diabetic patients by transient intensive insulin treatment. Diabetes Care 20:1353–1356

Kahn SE, Haffner SM, Heise MA, Herman WH, Holman RR, Jones NP, Kravitz BG, Lachin JM, O'Neill C, Zinman B, Viberti G, for the ADOPT (A Diabetes Outcome Progression Trial) Study Group (2006) Glycemic durability of rosiglitazone, metformin, or glyburide monotherapy. N Engl J Med 355:2427–2443

Kjems LL, Holst JJ, Volund A, Madsbad S (2003) The influence of GLP-1 on glucose-stimulated insulin secretion: effects on β-cell sensitivity in type 2 and nondiabetic subjects. Diabetes 52:380–386

Klonoff DC, Buse JB, Nielsen LI, Guan X, Bowlus CL, Holcombe JH, Wintle ME, Maggs DG (2008) Exenatide effects on diabetes, obesity, cardiovascular risk factors and hepatic biomarkers in patients with type 2 diabetes treated for at least 3 years. Curr Med Res Opin 24:275–286

Knop FK, Visboll T, Madsbad S, Holst JJ, Krarup T (2007) Inappropriate suppression of glucagon during OGTT but not during isoglycemic i.v. glucose infusion contributes to the reduce incretin effect in type 2 diabetes mellitus. Diabetologia 50:797–805

Laybutt DR, Preston AM, Akerfeldt MC, Kench JG, Bush AK, Biankin AV, Biden TJ (2007) Endoplasmic reticulum stress contributes to β cell apoptosis in type 2 diabetes. Diabetologia 50:752–763

Li Y, Hansotia Y, Yusta B, Ris F, Halban PA, Drucker DJ (2003) Glucagon-like-peptide-1 receptor signaling modulates β-cell apoptosis. J Biol Chem 278:471–478

Li Y, Xu W, Liao Z, Yao B, Chen X, Huang Z, Hu G, Weng JP (2004) Induction of long-term glycemic control in newly diagnosed type diabetic patients is associated with improvement of β-cell function. Diabetes Care 27:2597–2602

Liraglutide scientific synopsis. Novo Nordisk A/S, june 2010

MacConell L, Walsh B, Li Y, Pencek R, Maggs D (2013) Exenatide once weekly: sustained improvement in glycemic control and weight loss through 3 years. Diabetes, Metabolic Syndrome and Obesity: Targets and Therapy 6: 31–41

Maedler K, Carr RD, Bosco D, Zuellig RA, Berney T, Donath MY (2005) Sulfonylurea induced β-cell apoptosis in cultured human islets. J Clin Endocrinol Metab 90:501–506

Marre M, Shaw J, Brandle M, Bebakar WM, Kamaruddin NA, Strand J, Zdravkovic M, Le-Thi TD, Colagiuri S, LEAD-1 SU Study Group (2009) Liraglutide, a once-daily human GLP-1 analogue, added to a sulphonylurea over 26 weeks produces greater improvements on glycaemic and weight control compared with adding rosiglitazone or placebo in subjects with Type 2 diabetes (LEAD-1 SU). Diabet Med 26:268–278

Meier JJ, Deacon CF, Schmidt WE, Holst JJ, Nauck MA (2007) Suppression of glucagon secretion is lower after oral glucose administration than during intravenous glucose administration in human subjects. Diabetologia 50:806–813

Monami M, Iacomelli I, Marchionni N, Mannucci E (2010) Dipeptydil peptidase-4 inhibitors in type 2 diabetes: a meta-analysis of randomized clinical trials. Nutr Metab Cardiovasc Dis 20:224–235

Muscelli E, Mari A, Casolaro A, Camastra S, Seghieri G, Gastaldelli A, Holst JJ, Ferrannini E (2008) Separate impact of obesity and glucose tolerance on the incretin effect in normal subjects and type 2 diabetic patients. Diabetes 57:1340–1348

Nauck MA, El-Ouaghlidi A (2005) The therapeutic actions of DPP-IV inhibition are not mediated by glucagon-like peptide-1. Diabetologia 48:608–611

Nauck MA, Heimesaat MM, Orskow C, Holst JJ, Ebert R, Creutzfeldt W (1993) Preserved incretin activity of glucagon-like peptide 1[17-36 amide] but not of synthetic human gastric inhibitory polypeptide in patients with type 2 diabetes mellitus. J Clin Invest 91:301–307

Nauck M, Frid A, Hermansen K, Shah NS, Tankova T, Mitha IH, Zdravkovic M, During M, Matthews DR, for the LEAD-2 Study Group (2009) Efficacy and safety comparison of Liraglutide, glimepiride, and placebo, all in combination with metformin, in type 2 diabetes. Diabetes Care 32:84–90

Noel RA, Braun DK, Patterson RE, Bloomgren G (2009) Increased risk of acute pancreatitis and biliary disease observed in patients with type 2 diabetes: a retrospective, cohort study. Diabetes Care 32:834–838

Ovalle F, Bell DSH (2004) Effect of rosiglitazone versus insulin on the pancreatic β-cell function of subjects with type 2 diabetes. Diabetes Care 27:2585–2589

Perl S, Kushner JA, Buchholz BA, Meeker AK, Stein GM, Hsieh M, Kirby M, Pechhold S, Liu EH, Harlan DM, Tisdale JF (2010) Significant human β-cell turnover is limited to the first three decades of life as determined by in vivo thymidine analog incorporation and radiocarbon dating. J Clin Endocrinol Metab 95:E234–E239

Poitout V, Robertson RP (2008) Glucolipotoxicity: fuel excess and β-Cell dysfunction. Endocr Rev 29:351–366

Pratley RE, Kipnes MS, Fleck PR, Wilson C, Mekki Q, Alogliptin Study 007 Group (2009) Efficacy and safety of the dipeptidyl peptidase-4 inhibitor alogliptin in patients with type2 diabetes inadequately controlled by glyburide monotherapy. Diabetes Obes Metab 11:167–176

Rahier J, Guiot Y, Goebbels RM, Sempoux C, Henquin JC (2008) Pancreatic β-cell mass in European subjects with type 2 diabetes. Diabetes Obes Metab 10(Suppl 4):32–42

Rahmi RM, Uchida AH, Rezende PC, Lima EG, Garzillo CL, Favarato D, Strunz CMC, Takiuti M, Girardi P, Hueb W, Kalil Filho R, Ramires JAF (2013) Effect of hypoglycemic agents on ischemic preconditioning in patients with type 2 diabetes and symptomatic coronary artery disease. Diabetes Care 36:1654–1659

Retnakaran R, Drucker DJ (2008) Intensive insulin therapy in newly diagnosed type 2 diabetes. Lancet 371:1725–1726

Ritzel RA (2009) Therapeutic approaches based on β-cell mass preservation and/or regeneration. Front Biosci 14:1835–1850

Russell-Jones D, Vaag A, Schmitz O, Sethi BK, Lalic N, Antic S, Zdravkovic M, Ravn GM, Simo R (2008) Significantly better glycemic control and weight reduction with liraglutide, a once-daily human GLP-1 analog, compared with insulin glargine: all as add-on to metformin and a sulfonylurea in type 2 diabetes. Diabetes 57(Suppl 1):A 159

Russell-Jones D, Cuddihy RM, Hanefeld M, Kumar A, Gonzalez JG, Chan M, Wolka AM, Boardman MK, on behalf of the DURATION-4 Study Group (2012) Efficacy and safety of exenatide once weekly versus metformin, pioglitozone and sitagliptin used as monotherapy in drug-naive patients with Type 2 diabetes (DURATION-4): a 26-weekly double-blind study. Diabetes Care 35:252–258

Ryan EA, Imes S, Wallace C (2004) Short-term intensive insulin therapy in newly diagnosed type 2 diabetes. Diabetes Care 27:1028–1032

Salehi M, Aulinger BA, D'Alessio DA (2008) Targeting β-cell mass in type 2 diabetes: promise and limitations of new drugs based on incretins. Endocr Rev 29:357–379

Scheen AJ (2012) A review of gliptins in 2011. Expert Opin Pharmacother 13:81–99

Shapiro ET, Van Cauter E, Tillil H, Given BD, Hirsch L, Beebe C, Rubenstein AH, Polonsky KS (1989) Glyburide enhances the responsiveness of the β-cell to glucose but does not correct the abnormal patterns of insulin secretion in non-insulin-dependent diabetes mellitus. J Clin Endocrinol Metab 69:571–576

Singh S, Chang HY, Richards TM, Weiner JP, Clark JM, Segal JB (2013) Glucagon-like peptide-1 based therapies and risk of hospitalization for acute pancreatitis in type 2 diabetes mellitus: a population-based matched cases-control study. JAMA Intern Med 173:534–539

Theodorakis MJ, Carlson O, Michopoulos S, Doyle ME, Juhaszova M, Petraki K, Egan JM (2006) Human duodenal enteroendocrine cells: source of both incretin peptides, GLP-1 and GIP. Am J Physiol Endocrinol Metab 290:E550–E559

Tseng YH, Ueki K, Kriauciunas KM, Kahn CR (2002) Differential roles of insulin receptor substrates in the anti-apoptotic function of insulin-like growth factor-1 and insulin. J Biol Chem 277:31601–31611

UK Prospective Diabetes Study (UKPDS) Group (1998) Intensive blood-glucose control with sulphonylureas or insulin compared with conventional treatment and risk of complications in patients with type 2 diabetes (UKPDS 33). Lancet 352:837–853

UK Prospective Diabetes Study 16 (1995) Overview of 6 years' therapy of type II diabetes: a progressive disease. Diabetes 44:1249–1258

Ussher JR, Drucker DJ (2012) Cardiovascular biology of the incretin system. Endocr Rev 33:187–215

Vardarli I, Nauck MA, Köthe LD, Deacon CE, Holst JJ, Schweizer A, Foley JE (2011) Inhibition of DPP-4 with vildagliptin improved insulin secretion in response to oral as well as "isoglycemic" intravenous glucose without numerically changing the incretin effect in patients with type 2 diabetes. J Clin Endocrinol Metab 96:945–954

Vollmer K, Holst JJ, Baller B, Elrichmann M, Nauck MA, Schmidt WE, Meyer JJ (2008) Predictors of incretin concentrations in subjects with normal, impaired, and diabetic glucose tolerance. Diabetes 57:678–687

Wajchenberg BL (2007) β-cell failure in diabetes and preservation by clinical treatment. Endocr Rev 28:187–218

Wen XU, Li Y-B, Deng W-P, Hao Y-T, Weng J-P (2009) Remission of hyperglycemia following intensive insulin therapy in newly diagnosed type 2 diabetic patients: a long-term follow-up study. Chin Med J 122:2554–2559

Weng JP, Li Y, Xu W, Shi L, Zhang Q, Xhu D, Hu Y, Zhou Z, Yan X, Tian H, Ran X, Luo Z, Xian J, Yan L, Li F, Zeng L, Chen Y, Yang L, Yan S, Liu J, Li M, Fu Z, Cheng H (2008) Effect of intensive insulin therapy on β-cell function and glycaemic control in patients with newly diagnosed type 2 diabetes: a multicentre randomised parallel-group trial. Lancet 371:1753–1760

Willms B, Idowu K, Holst JJ, Creutzfeldt W, Nauck MA (1998) Overnight GLP-1 normalizes fasting but not daytime plasma glucose levels in NIDDM patients. Exp Clin Endocrinol Diabetes 106:103–107

Wu MS, Johnston P, Sheu WH, Hollenbeck CB, Jeng CY, Goldfine ID, Chen YD, Reaven GM (1990) Effect of metformin on carbohydrate and lipoprotein metabolism in NIDDM patients. Diabetes Care 13:1–8

Xu G, Stoffers DA, Habener JF, Bonner-Weir S (1999) Exendin-4 stimulates both β-cell replication and neogenesis, resulting in increased β-cell mass and improved glucose tolerance in diabetic rats. Diabetes 48:2270–2276

Xu L, Man CD, Charbonnel B, Meninger G, Davies MJ, Williams-Herman D, Cobelli C, Stein PP (2008) Effect of sitagliptin, a dipeptidyl peptidase-4 inhibitor, on β-cell function in patients with type 2 diabetes. Diabetes Obes Metab 10:1212–1220

Zinman B, Gerich J, Buse J, Lewin A, Schwartz SL, Raskin P, Hale PM, Zdravkovic M, Blonde L (2008) Effect of the GLP-1 analog liraglutide on glycemic control and weight reduction in patients on metformin and rosiglitazone: a randomized double-blind placebo-controlled trial. Diabetologia 51(Suppl 1):A 898

Role of NADPH Oxidase in β Cell Dysfunction

34

Jessica R. Weaver and David A. Taylor-Fishwick

Contents

Introduction	924
NADPH Oxidase Family	925
NADPH Oxidase Family Members	927
NOX-2	927
NOX-1	930
NOX-3	931
NOX-4	931
NOX-5	932
DUOX-1 and DUOX-2	932
NADPH Oxidases and β Cells	933
NOX and Glucose-Stimulated Insulin Secretion	933
NOX and β Cell Dysfunction	934
NOX-1 and β Cell Function	937
NOX-4 and Insulin Signaling	938
Redox Signaling	939
NOX-1 and Second Messenger Src-Kinase Signaling	939
NOX-2 and AngII Signaling	940
NOX-4 and Renal Signaling	940
Redox Signaling and Activators of Transcription	941
NOX Inhibition	942
Conclusions	945
Cross-References	945
References	945

J.R. Weaver
Department of Microbiology and Molecular Cell Biology, Eastern Virginia Medical School, Norfolk, VA, USA
e-mail: weaverjr@evms.edu

D.A. Taylor-Fishwick (✉)
Department of Microbiology and Molecular Cell Biology, Department of Medicine, Eastern Virginia Medical School, Strelitz Diabetes Center, Norfolk, VA, USA
e-mail: taylord@evms.edu

Abstract

Dysfunction of pancreatic β cells and loss of β cell mass is a major factor in the development of diabetes. Currently there is no cure for diabetes, and available therapies do not focus on halting or reversing the loss of β cell function. New strategies to preserve β cells in diabetes are needed. Conferring protection to the β cells against the effects of sustained intracellular reactive oxygen species (ROS) presents a novel approach to preserve β cell mass. Important contributors to increases in intercellular ROS in β cells are nicotinamide adenine dinucleotide phosphate (NADPH) oxidase enzymes. Discussed in this review are the roles of NADPH oxidases in the β cell, their contribution to β cell dysfunction, and new emerging selective inhibitors of NADPH oxidase.

Keywords

β cell • Diabetes • Inhibitors • Islet • NADPH oxidase • NOX • Proinflammatory cytokines • ROS

Introduction

An underlying feature of diabetes is a loss of functional β cell capacity. The therapeutic potential to cure diabetes by restoring functional β cell mass has been best illustrated in islet transplantation studies. These proof-of-principle efforts reversed insulin dependence. As a therapeutic strategy, islet transplantation is not currently viable (Taylor-Fishwick et al. 2008). However, these studies validate efforts to seek approaches to preserve and protect β cell mass in diabetes. Fundamental to this is a better appreciation of key events that drive β cell failure. Sensitivity of β cells to reactive oxygen species and oxidative stress is well recognized (Lenzen 2008). Relative to other cells, pancreatic β cells are vulnerable to sustained elevation in intracellular reactive species (Lenzen 2008). This is due, in part, to the low activity of free-radical detoxifying enzymes such as catalase, superoxide dismutase, and glutathione peroxidase in β cells (Grankvist et al. 1981; Lenzen et al. 1996; Tiedge et al. 1997; Modak et al. 2007). Islets also exhibit poor ability to rectify oxidative damage to DNA (Modak et al. 2007). Consequently, β cells are easily overwhelmed by elevated reactive oxygen species (ROS) and enter a state of oxidative stress (Lenzen 2008). Under oxidative stress conditions, ROS in addition to oxidizing proteins, lipids, and DNA also activate stress-sensitive second messengers such as p38-mitogen-activated protein kinase (MAPK), c-Jun N-terminal kinase (JNK), and protein kinase C (PKC) (Koya and King 1998; Purves et al. 2001). For example, a consequence of JNK activation is translocation of the homeodomain transcription factor *pdx-1* from the nucleus to the cytoplasm. *Pdx-1* is a key transactivator of the insulin gene in the nucleus (Ohneda et al. 2000). An outcome of cytoplasmic translocation of *pdx-1* is defective insulin expression that contributes to β cell dysfunction. As *pdx-1* transactivates its own expression, the impact of a cytoplasmic relocation for *pdx-1* is amplified

(Kawamori et al. 2003). It seems a paradox then that while β cells are susceptible to sustained elevations in ROS, transient increases in ROS are required for function and the generation of intracellular signaling (Goldstein et al. 2005; Pi et al. 2007; Newsholme et al. 2009). The source of intracellular ROS in β cells is a fertile area of research and reflects recognition for the key role elevated ROS can play in β cell pathophysiology. Intracellular events such as mitochondrial stress and/or endoplasmic reticulum stress do contribute to elevated intracellular ROS in the β cells (reviewed in Newsholme et al. (2009), Volchuk and Ron (2010)). Peroxisome metabolism of fatty acids elevates cellular ROS, and this ROS elevation has been linked to β cell lipotoxicity and dysfunction (Gehrmann et al. 2010; Elsner et al. 2011; Fransen et al. 2012). More recently, recognition has been made for an important contribution of the ROS-generating NADPH oxidase enzyme family in β cell pathophysiology. Reviewed in this chapter are the roles that NADPH oxidases have in the β cell with an emphasis on contributions to β cell dysfunction. Initially, the family of NADPH oxidase enzymes is introduced with a description of common features for this family in terms of structure, function, and expression. Specific features of each member of the family are considered relative to structure, function, expression, and regulation. The contribution of NADPH oxidase enzymes to β cell function and dysfunction is considered by describing the members and subunits that are expressed and functional in β cells. The activity of these NADPH oxidase enzymes is regulated by stimuli known to be elevated in diabetes. The consequence of stimulating NADPH oxidase activity is relevant for both β cell physiology and β cell pathophysiology. The role of NADPH oxidases in insulin secretion, β cell signaling, and induced β cell dysfunction is explored. Newly recognized are the contributions of intracellular ROS to discrete activation of second messengers. Thus, action of NADPH oxidase in terms of influencing the β cell ranges from subtle control to frank cell destruction. Lastly, major advances in the development of new selective inhibitors of NADPH oxidases are reviewed. These inhibitors represent candidate agents to advance the field and propose new therapeutic strategies to preserve and protect β cell mass in diabetes.

NADPH Oxidase Family

NADPH oxidases are multi-protein complexes that generate reactive oxygen species, including superoxide ($O_2^{\cdot-}$) and hydroxyl radical ($^{\cdot}OH$). Characteristically, the family shares a common basic structure but differ by their membrane localization and protein components (illustrated in Fig. 1).

There are seven identified members of the NADPH oxidase family. In humans, all seven enzymes are expressed. NADPH oxidases contain the following characteristic elements: (1) a COOH terminus, (2) six conserved transmembrane domains, (3) a flavin adenine dinucleotide (FAD)-binding site, (4) a NADPH-binding site at the COOH terminus, and (5) generally two heme-binding histidines in the third transmembrane domain and two in the fifth transmembrane domain, although the location of these sites vary (Rotrosen et al. 1992; Cross et al. 1995; Finegold

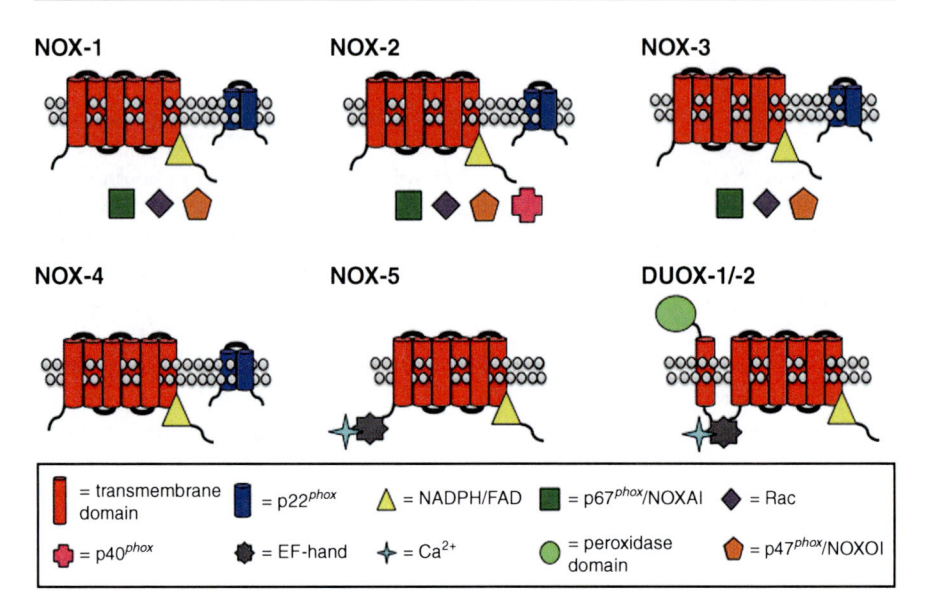

Fig. 1 NADPH oxidase (*NOX*) family members and their structures. Each NOX member contains six conserved transmembrane domains and a NADPH/FAD site at the COOH terminus. Additional structures needed for an active complex for some of the NOXs include $p22^{phox}$, $p67^{phox}$/NOXA1, $p47^{phox}$, Rac, and $p40^{phox}$. NOX-5 and DUOX-1/-2 have an EF-hand where Ca^{2+} binds for activation. DUOX-1/-2 have an extra-transmembrane domain with an extracellular peroxidase domain

et al. 1996; Biberstine-Kinkade et al. 2001; Lambeth et al. 2007). Some family members have additional domains, such as EF hands, or an additional NH_2-terminal transmembrane domain. As a class, NADPH oxidases are members of the flavocytochrome family. Functional enzymes require association with protein subunits to form complexes. Lambeth et al. reviewed subunit and domain interaction in NADPH oxidase enzymes (Lambeth et al. 2007).

Functionally, NADPH oxidase complexes traverse membranes and convert oxygen to a reactive radical via electron transfer across plasma or organelle membranes. NADPH serves as the single electron donor. NADPH donates a single electron to FAD which is located in the cytoplasmic end of the NOX enzyme (Nisimoto et al. 1999). The electron is passed between heme groups before reacting with oxygen to create superoxide (Finegold et al. 1996). Superoxide can then be converted to hydrogen peroxide (H_2O_2), a principle stable product.

NADPH oxidase family members are found primarily in eukaryotic cells of animals, plants, and fungi. Members of the NADPH oxidase family are implicated in numerous biological functions in animals. For example, phagocyte NADPH oxidase plays a critical role in host immune defense by activating the respiratory burst associated with neutrophil-mediated innate immunity (Rada et al. 2008). NADPH oxidases function in thyroid hormone synthesis and help regulate inner ear function (Caillou et al. 2001; Paffenholz et al. 2004; Ris-Stalpers 2006). Maintenance of the vascular system, hypertensive tone, and vascular wall function are influenced by NADPH oxidase activity (reviewed in Manea (2010)). Indeed, reflective of the central influence of ROS in many biological processes, NADPH oxidases are associated with physiological or pathophysiological processes of most mammalian organ systems (reviewed (Bedard and Krause 2007; Bedard et al. 2007; Lambeth et al. 2007)). Additionally, an elevation of intracellular ROS and association of NADPH oxidase activity are implicated in numerous diseases involving these mammalian organ systems. In plants, NADPH oxidase plays a role in plant defenses and development (reviewed in Marino et al. (2012)) (Torres and Dangl 2005; Gapper and Dolan 2006; Sagi and Fluhr 2006). NADPH oxidase regulates root growth by cell expansion (Foreman et al. 2003). Fungi express several different isoforms of NOXs (reviewed in Tudzynski et al. (2012)). Filamentous fungi express two subfamilies of NADPH oxidase that are similar to mammalian phagocyte NADPH oxidase. Homologous subunits to other mammalian NADPH oxidases have also been reported in fungi (Aguirre et al. 2005). The NADPH oxidase subfamilies present in fungi play a role in growth and development including germination (Lara-Ortiz et al. 2003; Malagnac et al. 2004; Cano-Dominguez et al. 2008). Evidence to suggest that fungal NADPH oxidases may play a role in pathogenesis of cellulose degradation has also been reported (Brun et al. 2009).

NADPH Oxidase Family Members

Phagocyte NADPH oxidase, named for its expression in phagocytes, especially neutrophils, was the first identified NADPH oxidase and plays a critical role in the respiratory burst response associated with innate immune defense (Babior et al. 1973, 2002). Subsequent gene discoveries have identified a family of genes/ proteins that are distinct enzymes but homologous to the catalytic core subunit of phagocyte NADPH oxidase. This family of NADPH oxidases is termed NOX/ DUOX with the nomenclature of the seven members being termed NOX-1, NOX-2, NOX-3, NOX-4, NOX-5, DUOX-1, and DUOX-2. Under this nomenclature the core catalytic subunit of phagocyte NADPH oxidase (previously termed $gp91^{phox}$) is called NOX-2. A brief overview of each family member is provided starting with the archetypal phagocyte NADPH oxidase (NOX-2) before consideration of the role of NADPH oxidase in β cells (Table 1).

NOX-2

The structure of the functional NOX-2 enzyme is made up of five protein subunits that are located in the plasma membrane or translocated to the plasma membrane from the cytosol. Based upon the original name, phagocyte NADPH oxidase, the naming convention of subunits is gpX^{phox}, where gpX indicates glycoprotein of "X" molecular weight. Activation of phagocyte NADPH oxidase (NOX-2) occurs through a complex series of protein interactions. The core catalytic component of NADPH oxidase, $gp91^{phox}$, is stabilized in the membrane by $p22^{phox}$. The adaptor

Table 1 The members of the NADPH oxidase family have distinct structures, functions, locations of expression, and specific regulators. NOX-1 and NOX-2 have roles in β cell function with respect to insulin signaling and secretion

Complex	Structure	Function	Expression	Regulation	β cell role
NOX-1	gp91^{phox} NOXO1 NOXA1 p22^{phox} Rac	Immune modulator Cell proliferation	Colon epithelium Pancreas Vascular smooth muscle	AngII Interferon PMA EGF 12-LO	Increases intracellular ROS Uncouples GSIS Induces apoptosis
NOX-2	gp91^{phox} p47^{phox} p67^{phox} p22^{phox} p40^{phox} Rac	Immune defense	Phagocytes Pancreas B lymphocytes Neurons Cardiomyocytes	AngII Endothelin-1 Growth factors Cytokines LPS	Increases intracellular ROS Uncouples GSIS Induces apoptosis
NOX-3	p22^{phox} NOXO1 NOXA1 Rac p47^{phox}	Balance Gravity	Inner ear Fetal kidney Fetal spleen Skull bone Brain		
NOX-4	p22^{phox} Rac	Cell migration Apoptosis Cell survival Cell differentiation Insulin signaling	Kidney Pancreas Vascular endothelial cells Osteoclasts	Constitutively active TGF-β Insulin IGF-1	

NOX-5	EF-hand domain	Cell proliferation Cell migration Angiogenesis Cancer growth Sperm maturation Pregnancy Fetal development	Lymphoid tissue Testis Spleen Endothelium Uterus Placenta Fetal tissue	Calcium Thrombin Platelet-derived growth factor AngII Endothelin-1
DUOX-1	EF-hand domain Peroxidase domain	Thyroid hormone synthesis Immune defense Inflammation	Thyroid Respiratory tract Prostate	Calcium Forskolin IL-4 IL-13
DUOX-2	EF-hand domain Peroxidase domain	Thyroid hormone synthesis Heme peroxidase Immune defense Inflammation Antibacterial response	Thyroid Respiratory tract	Calcium Interferon gamma Flagellin Phorbol esters

NOX: Nicotinamide adenine dinucleotide phosphate-oxidase enzyme; DUOX: Dual oxidase; NOXO1: NOX Organizer 1; gp: glycoprotein; NOXA1: NOX Activator 1; AngII: Angiotensin II; PMA: Phorbol 12-myristate 13-acetate; EGF: Epidermal growth factor; 12-LO: 12-Lipoxygenase; LPS: Lipopolysaccharide; TGF-β: Transforming growth factor-β; IGF-1: Insulin-like growth factor 1; IL-4: Interleukin-4; IL-13: Interleukin-13; ROS: Reactive oxygen species; GSIS: Glucose-stimulated-insulin-secretion

protein p47^{phox} is activated by phosphorylation. Recruitment of p47^{phox} facilitates addition to the complex of p40^{phox}, p67^{phox}, and Rac (small GTP-binding protein). The latter two appear to regulate catalysis (Abo et al. 1991; Ando et al. 1992; Wientjes et al. 1993; Heyworth et al. 1994; Hordijk 2006; Orient et al. 2007; Guichard et al. 2008).

NOX-2 in phagocytes functions in the host defense by providing a respiratory burst response that serves to kill bacteria enclosed within the phagosome through production of hypochlorous acid. NOX-2 also activates inflammatory and immune responses (Rada et al. 2008). Genetic defects in NOX-2 activity arising from mutations in the genes encoding gp91^{phox}, p22^{phox}, p47^{phox}, and p67^{phox} are associated with chronic granulomatous disease (Heyworth et al. 2003). Generation of ROS by NOX-2 has also been shown to activate pathways involved in angiogenesis (as reviewed in Ushio-Fukai (2006)).

In addition to phagocytes, NOX-2 expression has been reported in pancreas, B lymphocytes, neurons, cardiomyocytes, endothelium, skeletal muscle, hepatocytes, smooth muscle, and hematopoietic stem cells (Bedard and Krause 2007).

NOX-2 is regulated by angiotensin II (AngII), endothelin-1, growth factors such as vascular endothelial growth factor (VEGF), cytokines such as tumor necrosis factor α (TNF-α), interferon gamma (IFN-γ), lipopolysaccharide, mechanical forces, and hyperlipidemia (Cassatella et al. 1990; Dworakowski et al. 2008).

NOX-1

The first homolog of the NOX-2 catalytic subunit (p91^{phox}) to be described was NOX-1. Similar to functional NOX-2, the functional NOX-1 enzyme is a multi-protein complex. The subunits that bind NOX-1 are NOXO1 (NOX organizer 1) and NOXA1 (NOX activator 1), which while being distinct proteins are homologs of p47^{phox} and p67^{phox}, respectively (Takeya et al. 2003; Uchizono et al. 2006). For a functional complex, active NOX-1 requires the NOX-1 catalytic core bound to the membrane subunit p22^{phox}, NOXO1, and NOXA1 along with the GTPase Rac subunit. NOXO1 is predominantly associated with membrane-bound p22^{phox} (Sumimoto 2008). NOXA1 and Rac translocate to the membrane, in a phosphorylation-dependent process, and are required for NOX-1 activation (Miyano et al. 2006). Transfection experiments have shown that NOX-1 can use the p47^{phox} and p67^{phox} subunits of NOX-2, raising the possibility of a dynamic interaction amongst NOX isotypes where subunits are interchangeable between the different NOX family members (Banfi et al. 2003).

NOX-1 activity is associated in immune defense in inflammatory bowel disease and cancer (Rokutan et al. 2006, 2008). Studies have linked NOX-1 activity to colon cancer by affecting cell proliferation of colon carcinoma cell lines and controlling cell migration of colon adenocarcinoma cells (de Carvalho et al. 2008; Sadok et al. 2008). NOX-1 also functions in the vasculature by regulating smooth muscle growth, migration (cell movement), and blood pressure (Cave et al. 2006; Gavazzi et al. 2006; Garrido and Griendling 2009). NOX-1 helps regulate neuronal differentiation by negatively affecting excessive neurite outgrowth and influences pain sensitivity during inflammation (Ibi et al. 2006, 2008).

NOX-1 is expressed in the colon epithelium and at lower levels in the pancreas, vascular smooth muscle, endothelium, uterus, placenta, prostate, osteoclasts, and retinal pericytes (Nisimoto et al. 2008). Patterns of expression for NOX-1 are related to species. For example, the expression of NOX-1 is found in rodent stomach but not in human stomach (Kawahara et al. 2005; Kusumoto et al. 2005; Rokutan et al. 2008).

Regulators of NOX-1 include AngII, IFN-γ, protein kinase C (PKC) activation (PMA: 4β-phorbol 12-myristate 13-acetate), and epidermal growth factor receptor ligation (Suh et al. 1999; Lassegue et al. 2001; Wingler et al. 2001; Katsuyama et al. 2002; Seshiah et al. 2002; Touyz et al. 2002; Geiszt et al. 2003a; Takeya et al. 2003; Fan et al. 2005). In the liver cell line FaO, NOX-1 regulation involves an autoregulatory feedforward loop involving second messenger activation of Src kinase and extracellular-signal-regulated kinase (ERK) (Fan et al. 2005; Adachi et al. 2008; Sancho and Fabregat 2010).

NOX-3

Activity of NOX-3 is dependent on $p22^{phox}$. Additional cytosolic subunits NOXO1, NOXA1, $p47^{phox}$, or Rac are not needed for basal NOX-3 activity. The activity of NOX-3 is however significantly increased following association with cytosolic subunits (Banfi et al. 2004a; Cheng et al. 2004; Ueyama et al. 2006).

A major site for NOX-3 expression is the inner ear where its function is to assist biogenesis of otoconia and regulation of balance and gravity (Paffenholz et al. 2004). Expression of NOX-3 has additionally been described in fetal kidney, fetal spleen, skull bone, brain, and lung endothelial cells where it is associated with the development of emphysema (Banfi et al. 2004a; Zhang et al. 2006; Bedard and Krause 2007).

NOX-4

NOX-4 activity requires association with the $p22^{phox}$ subunit, but unlike NOX-2, other subunits are not essential (Martyn et al. 2006). It is unresolved whether the Rac subunit is required for NOX-4 activity (Gorin et al. 2003). Unlike the other NOX family members that initially produce superoxide, NOX-4 produces H_2O_2 (Martyn et al. 2006; Serrander et al. 2007).

NOX-4 function has been associated with cell migration, apoptosis, cell survival, cell differentiation, insulin signaling, cell migration, the unfolded protein response, and differentiation (Mahadev et al. 2004; Pedruzzi et al. 2004; Vaquero et al. 2004; Cucoranu et al. 2005; Li et al. 2006; Meng et al. 2008; Pendyala et al. 2009; Santos et al. 2009). It is a major source of oxidative stress in the failing heart (Kuroda et al. 2010).

NOX-4 is expressed mainly in the kidney but is also detected in vascular endothelial cells, osteoclasts, endothelium, smooth muscle, hematopoietic stem cells, fibroblasts, keratinocytes, melanoma cells, neurons, pancreas, and adipocytes (Geiszt et al. 2000; Cheng et al. 2001; Shiose et al. 2001; Ago et al. 2004; Bedard and Krause 2007).

NOX-4 is constitutively active and levels of mRNA directly correlate with enzyme function (Serrander et al. 2007). The activity of NOX-4 can be upregulated by tumor growth factor β (TGF-β) in cardiac fibroblasts, pulmonary artery smooth muscle cells, and lungs (Cucoranu et al. 2005; Sturrock et al. 2006, 2007). Insulin activates NOX-4 in adipocytes, and it can be activated by insulin-like growth factor-1 (IGF-1) in vascular smooth muscle cells (VSMCs) (Mahadev et al. 2004; Meng et al. 2008; Schroder et al. 2009).

NOX-5

Distinct to the other NOX enzymes, activation of NOX-5 is calcium dependent. At its amino terminal, NOX-5 has a calmodulin-like domain that has four calcium-binding EF-hand domains (Lambeth 2007). NOX-5 does not require $p22^{phox}$ or other phox subunits for activity. Several splice variant forms of NOX-5 exist (Banfi et al. 2001, 2004b; BelAiba et al. 2007; Pandey et al. 2012).

NOX-5 activity has functional relevance in endothelial cell proliferation, migration, and angiogenesis (BelAiba et al. 2007; Jay et al. 2008; Schulz and Munzel 2008). NOX-5 activity is linked to cancer, including prostate cancer growth, esophageal adenocarcinoma, breast cancer, and hairy cell leukemia (Brar et al. 2003; Kamiguti et al. 2005; Fu et al. 2006; Kumar et al. 2008; Juhasz et al. 2009). NOX-5 exerts a role in fetal development and sperm maturation (reviewed Bedard et al. (2012)).

NOX-5 is expressed mainly in the lymphoid tissue, testis, and spleen (Banfi et al. 2001; Cheng et al. 2001). Expression of NOX-5 in endothelium, smooth muscle, pancreas, placenta, ovary, uterus, stomach, certain prostate cancers, and various fetal tissues has been reported (Banfi et al. 2001; Cheng et al. 2001; Bedard and Krause 2007; BelAiba et al. 2007).

Calcium, thrombin, platelet-derived growth factor (PDGF), AngII, and endothelin-1 are factors reported to regulate NOX-5 (Montezano et al. 2010, 2011). IFN-γ has also been shown to activate NOX-5 in smooth muscle cells possibly through the release of intracellular calcium stores (Manea et al. 2012).

DUOX-1 and DUOX-2

DUOX-1 and DUOX-2 do not require subunits to form an active enzyme complex. Like NOX-5, calcium is required to activate to each DUOX enzyme (Selemidis et al. 2008; Rigutto et al. 2009). Structurally, DUOX enzymes differ to NOX enzymes by having an additional transmembrane domain at the amino terminal and an extracellular peroxidase domain (De Deken et al. 2000). DUOX-1/DUOX-2 locate to the plasma membrane from the ER with the help of maturation factors DUOXA1 and DUOXA2 (Grasberger and Refetoff 2006; Morand et al. 2009).

DUOX-1 and DUOX-2 are highly expressed in the thyroid and directly produce H_2O_2 that is used during thyroid hormone T4 synthesis (Dupuy et al. 1989, 1999; De Deken et al. 2000; Caillou et al. 2001; Ris-Stalpers 2006). While current knowledge suggests the major role for DUOX is in the synthesis of thyroid hormone, expression of DUOX has been described in other sites including the respiratory tract, prostate, testis, pancreas, colon, and heart (Edens et al. 2001; Geiszt et al. 2003b; Harper et al. 2005; Allaoui et al. 2009; Gattas et al. 2009).

In addition to elevation in calcium, activation of DUOXs can be regulated by forskolin, interleukin IL-4, IL-13, IFN-γ, phorbol esters, and insulin (Morand et al. 2003; Harper et al. 2005, 2006; Rigutto et al. 2009).

NADPH Oxidases and β Cells

Select NOX family members are expressed in pancreatic β cells. NOX-1, NOX-2, NOX-4, NOXO1 (homolog of $p47^{phox}$), NOXA1 (homolog of $p67^{phox}$), and $p40^{phox}$ have been described in a variety of pancreatic and islet studies, including isolated rat β cells (Oliveira et al. 2003; Nakayama et al. 2005; Lupi et al. 2006; Shao et al. 2006; Uchizono et al. 2006; Rebelato et al. 2012). Expression of NOX-5, DUOX-1 and DUOX-2, as determined by RT-PCR, has additionally been described in the pancreas, though the functional relevance of their expression has yet to be determined (Cheng et al. 2001; Edens et al. 2001). Historically, expression of NADPH oxidase family members in β cells has been associated with regulation of glucose-stimulated insulin secretion (Morgan et al. 2007; Pi et al. 2007; Morgan et al. 2009). More recently, activity of NOX enzymes has been linked to β cell dysfunction. β cell damage likely arises from generation of intracellular ROS stimulating redox signaling pathways and, more chronically, oxidative stress. In the following sections, the roles of the specific NOXs expressed in the β cell are discussed.

NOX and Glucose-Stimulated Insulin Secretion

Several key observations have linked NADPH oxidase activity with regulation of insulin secretion. The product of NADPH oxidase activity, generation of H_2O_2, is required for insulin secretion (Pi et al. 2007). Elevated glucose leads to an increase in H_2O_2 generation, thus linking NADPH oxidase activity to regulation of insulin secretion (Morgan et al. 2007, 2009). Inhibition of NADPH oxidase by the general inhibitor, diphenyleneiodonium (DPI), led to a decrease in H_2O_2 production and also impaired insulin secretion (Imoto et al. 2008). Antisense-mediated decrease in the expression of $p47^{phox}$, an important subunit for the active NOX enzyme, reduced glucose-stimulated insulin secretion (Morgan et al. 2009). Activation and translocation of p47^{phox} is required for an active NADPH complex. Translocation of proteins to membranes or subunits is supported since inhibitors of protein prenylation or protein farnesyltransferase significantly decrease NOX-2-induced ROS generation and decrease glucose-stimulated insulin secretion (Syed et al. 2011b; Matti et al. 2012). These results suggest NOX activity is necessary for the transient increase in ROS that is needed for glucose-stimulated insulin secretion.

Which NOX family isotypes are involved in insulin secretion and what exactly their role is remains an unanswered question. A major limitation to resolving this question has been the lack of isoform-specific inhibitors of the NOX enzymes.

NOX and β Cell Dysfunction

Associated with a diabetic state is an increase in the serum levels of proinflammatory cytokines, free-fatty acids (FFA), and glucose. These serum mediators have been shown to elevate the expression and activity of NADPH oxidases (Morgan et al. 2007). Additionally, deposition of fibrillar human islet amyloid polypeptide (IAPP) in the β cell line RIN5mF cells increases NADPH oxidase activity and intracellular lipid peroxidation (Janciauskiene and Ahren 2000). Accumulation of amyloid in islets is a pathogenic state associated with type 2 diabetes (Marzban and Verchere 2004). In a small sample group, we showed that NOX-1 expression is elevated in islets from human type 2 diabetic donors (Weaver et al. 2012). Animal models of type 2 diabetes have also reported an increase in NOXs. For example, the role of NOX-2 in β cell dysfunction has been explored in the Zucker diabetic fatty rat (Syed et al. 2011a). The Zucker diabetic fatty (ZDF) rat is a model of type 2 diabetes where rats become obese and develop hyperinsulinemia, hyperglycemia, and β cell dysfunction. Examination of islets isolated from ZDF rats showed an increase in intracellular ROS levels that corresponded with elevated expression of the NOX enzyme subunits p47^{phox}, gp91^{phox}, and Rac1 (Syed et al. 2011a). Exposure of islets from Wistar rats to the free-fatty acid palmitate resulted in elevation of p47^{phox} protein and an increase in the mRNA levels of p22^{phox}, gp91^{phox}, p47^{phox}, proinsulin, and the G protein-coupled protein receptor 40, a signaling receptor that activates the phospholipase C signaling pathway (Graciano et al. 2011). With global inhibition of NADPH oxidase activity, β cells are protected from the effects of cytokine or FFA treatment (Michalska et al. 2010).

Advanced glycation end products (AGE) contribute to oxidative stress and the development of diabetes (Kaneto et al. 1996; Hofmann et al. 2002; Peppa et al. 2003; Cai et al. 2008; Zhao et al. 2009; Coughlan et al. 2011). AGEs form when carbohydrates, such as glucose, react nonenzymatically (e.g., glycation and oxidation) with amino groups. Binding of AGE to its receptor, RAGE (receptor for advanced glycation end products), generates ROS leading to oxidative stress in β cells. Evidence suggests that AGE may be increasing ROS generation through NADPH oxidase. In isolated rat islets, NADPH-dependent superoxide generation in homogenates increased after treatment with high glucose plus glycolaldehyde (Costal et al. 2013); addition of the NADPH oxidase inhibitor DPI decreased superoxide production. VAS2870, a NADPH oxidase inhibitor, also decreased intracellular superoxide production in islets treated with glucose plus glycolaldehyde (Costal et al. 2013). The increase in superoxide preceded apoptosis. Pancreatic β cell lines including INS-1, MIN6, and BTC-6 cells and isolated primary rat islets treated with AGE showed an increase in ROS generation that was followed by apoptosis (Lim et al. 2008). INS-1 β cells exposed to varying concentrations of AGE showed a time-dependent increase in intracellular ROS and apoptosis that was dependent upon NADPH oxidase (Lin et al. 2012). Since superoxide production in vascular smooth muscle cells exposed to AGE has been linked to an increase in the transcription of NOX-1, a similar pathway linking AGE and NADPH oxidase-1 in β cells may occur (San Martin et al. 2007).

NOX-2 and β Cell Function

(a) *NOX-2 and Regulation of Insulin Secretion*

Of the NOX family members expressed in β cells, NOX-2 has most closely been associated with regulation of insulin secretion. In an attempt to address critical roles of NOX enzymes in insulin secretion and discern isotype function, Li et al. performed studies using knockout mice, an approach that uses genetic gene depletion. Knockout mice that were genetically depleted of the core catalytic unit of NOX-1, NOX-2, or NOX-4 were evaluated to determine the role of each NADPH oxidase member in physiological insulin secretion. In an unanticipated outcome, none of the knockout mice showed a decrease in insulin secretion when challenged for glucose-stimulated insulin secretion (Li et al. 2012). This raises the question of whether the NADPH oxidase family is necessary for physiological glucose-stimulated insulin secretion. These results differ from a previous report in isolated rat islets. When treated with the general NADPH oxidase inhibitor, diphenyleneiodonium (DPI) chloride, a decrease in glucose-stimulated insulin secretion was observed (Uchizono et al. 2006). The discrepancy between the outcomes of the two studies has been explained by the nonspecific inhibitory action of DPI. As a nonspecific inhibitor of flavoenzymes, DPI inhibits numerous flavoproteins in addition to NOX enzymes. However, glucose-stimulated insulin secretion was reduced in isolated rat islets following depletion of $p47^{phox}$ using antisense oligonucleotide (Morgan et al. 2009). This effect was not seen with control (scrambled) oligonucleotide. The $p47^{phox}$ subunit is a required element for activity in certain NOX isoforms. This experimental approach is not subject to the arguments of nonspecificity attributed to the use of the chemical inhibitor of flavoenzymes, DPI. Depletion of $p47^{phox}$ was associated with a reduction in intracellular ROS (H_2O_2) production supporting a functional decrease in NOX activity. The results for inhibition of glucose-stimulated insulin secretion reported by Morgan et al. using depletion of $p47^{phox}$ were matched with their evaluation of DPI. The role of NOX enzymes in physiological insulin secretion may not yet be resolved. Clarity will be better achieved with discovery of selective isoform-specific inhibitors of NOX enzymes. An additional consideration is the possibility of the cross-use of isotype subunits between the NOX enzymes ensuring homeostatic regulation. It is possible that adaptation in genetic deletion (transgenic knockout) studies may occur or be overcome by functional compensation. This was addressed in the studies by Li et al., which looked in each NOX-knockout mouse for a homeostatic upregulation in remaining NOX enzymes. None was reported nor was physiological insulin secretion affected following a transient knockdown of NOX-2 (Li et al. 2012). Li et al. reported that NOX-2 played a role in insulin secretion through cyclic adenosine monophosphate (cAMP)/protein kinase A (PKA) signaling. In isolated wild-type islets, an increase in NOX-2 activity and generation of intracellular ROS results in negative modulation of the insulin secretory response and reduction in adenylate cyclase/cAMP/PKA signaling. Overall, these results show a link between NOX-2 and insulin regulation, though the mechanism may be more complex than first assumed.

(b) *NOX-2 and Regulation of ROS Generation*

Importantly, the β cell has to traverse a delicate balance in ROS generation. Acutely, a transient increase in ROS is a physiological requirement for insulin secretion. However, chronic sustained ROS generation negatively regulates insulin secretion and promotes β cell dysfunction (Morgan et al. 2007; Pi et al. 2007; Morgan et al. 2009). The activity of NADPH oxidase, whose primary enzymatic function is the production of ROS, is a logical candidate for a sustained chronic increase in intracellular ROS in the β cell. In addition to direct activation of NOX, mitochondrial activity has been related to NOX-2 activity in β cells (Syed et al. 2011a, b; Matti et al. 2012). Glucose or specific mitochondrial fuels (monomethyl succinate and α-ketoisocaproate) lead to an increase in NOX-2 activity and generation of ROS. In both the isolated rat islets and the homogeneous INS-1 β cell line, stimulation of NOX-2 by hyperglycemia and hyperlipidemia results in ROS generation and activation of the JNK1/2 signaling pathway. Activation of the JNK pathway precedes mitochondrial dysfunction and an increase in caspase-3 activity (Syed et al. 2011a). NOX enzymes can be upregulated in response to pathogenic stimuli, including proinflammatory cytokines, elevated FFAs, and high glucose. These serum factors are recognized as promoting β cell dysfunction, presumably by sustained activation of NAPDH oxidase and elevation of ROS.

(c) *NOX-2 and β Cell Dysfunction/Survival*

In NOX-2-deficient mice, protection to β cell destruction associated with streptozotocin (STZ) exposure was observed (Xiang et al. 2010). NOX-2 expression in the β cell line INS 832/13 is increased in response to hyperglycemic conditions, a condition known to cause β cell dysfunction (Mohammed and Kowluru 2013). Upon exposure of β cells to FFA or low-density lipoprotein, NOX-2 activity was linked to β cell dysfunction (Yuan et al. 2010b; Jiao et al. 2012). Decreased glucose-stimulated insulin secretion and apoptosis occur in pancreatic NIT-1 cells following exposure to very low-density lipoprotein (VLDL). This β cell dysfunction is co-associated with an increase in NOX-2 generated ROS and a decrease in expression and secretion of insulin (Jiao et al. 2012). NIT-1 cells showed only expression of NOX-2 and its subunits, and no expression of any of the other NADPH oxidase family members (Yuan et al. 2010a, b). In contrast, when NIT-1 cells were treated with VLDL plus siRNA-NOX-2, β cell function was preserved. Similar results were found when NIT-1 cells were treated with palmitate or oleate (Yuan et al. 2010b). FFAs induced β cell dysfunction and increased apoptosis through elevated ROS generation that arises from an increase in NOX-2 activity. When siRNA-NOX-2 was used to knock down NOX-2 protein, FFA-treated cells responded like control (untreated) cells with preserved β cell function and negligible apoptosis.

NOX-1 and β Cell Function

(a) *NOX-1 and Inflammation*

Our own studies have explored the regulation of NOX enzyme expression in β cells following stimulation with inflammatory cytokines. The cocktail combination of inflammatory cytokines used (TNF-α, IL-1β, IFN-γ) is widely reported to induce β cell dysfunction. We have shown that acute treatment of primary human islets, mouse islets, or homogeneous murine β cell lines with this inflammatory cytokine cocktail (termed PICs) induces gene expression, leads to a loss of glucose-stimulated insulin secretion, and induces apoptosis (Weaver et al. 2012; Weaver and Taylor-Fishwick 2013). In these model systems, expression of NOX-1 is selectively upregulated relative to other NOX isotypes, following stimulation with PICs (Weaver et al. 2012). The upregulation of NOX-1 co-associates with elevated intracellular ROS, loss of glucose-stimulated insulin secretion, and induction of β cell apoptosis (Weaver et al. 2012; Weaver and Taylor-Fishwick 2013). Significantly, inhibition of NOX-1 protected β cells from the damaging effects resulting from PIC stimulation, suggesting NOX-1 may be an important target for β cell preservation in diabetes (Weaver and Taylor-Fishwick 2013).

(b) *NOX-1 and 12-Lipoxygenase*

In terms of intracellular regulators of NOX-1 expression, the lipid-metabolizing enzyme, 12-lipoxygenase (12-LO), has been shown to induce NOX-1 expression in β cells (Weaver et al. 2012). 12-LO and one of its major bioactive lipid products, 12-HETE, are key mediators of β cell dysfunction and inflammation (as detailed in the chapter "▶ Inflammatory pathways linked to β cell demise in diabetes" by Imai et al.). The 12-LO pathway is associated with inflammation and activation of the transcription factor STAT4 (signal transducer and activator of transcription 4) and cytokines IL-12 and IFN-γ. The significance of the 12-LO pathway for diabetes has been shown in several mouse models. Mice with a deletion in 12-LO are resistant to diabetes induced by low-dose streptozotocin (Bleich et al. 1999). The non-obese diabetic (NOD) mouse model of type 1 diabetes was also protected from spontaneous diabetes development when 12-LO was deleted (McDuffie et al. 2008). When β cells were treated with 12-HETE, a product of 12-LO activity, or proinflammatory cytokines, there was an induction of NOX-1 expression (Weaver et al. 2012). Conversely, selective inhibition of 12-LO activity by selective small molecules (Kenyon et al. 2011) reduced proinflammatory cytokine-induced NOX-1 expression (Weaver et al. 2012). These data integrate inflammation with induction of 12-LO activity and NOX-1 expression in a pathway regulating β cell dysfunction.

(c) *NOX-1 and Feedforward Regulation in β Cells*

Our studies have additionally provided evidence for a feedforward regulation of NOX-1 in β cells. An autoregulatory feedback loop amplifies NOX-1 upregulation in β cells. This work parallels similar observations in a liver cell line (Sancho and Fabregat 2010). In the β cell, where relatively limited defense mechanisms exist to counter a sustained increase in ROS, oxidative stress arising from feedforward regulation of NOX-1 could be a significant event. Identifying and inhibiting such regulation could prove important in developing new strategies for preservation and protection of functional β cell mass in diabetes. Our studies demonstrated that induced expression of NOX-1 by PIC stimulation of β cells was abrogated with inhibitors of NADPH oxidase activity. Assuming NOX activity is subsequent to induced gene expression, the simplest explanation of the data is that NOX activity upregulates NOX-1 gene expression. The resultant elevation of intracellular ROS was implicated. General antioxidants, which neutralize cellular ROS, inhibited NOX-1 expression induced by PIC stimulation. In contrast, pro-oxidants that directly elevate cellular ROS in the absence of other stimuli induced NOX-1 expression. Redox-sensitive signaling pathways (discussed below) were shown to mediate this feedforward regulation of NOX-1 in β cells. It will be interesting to evaluate if feedforward regulation of NADPH oxidase in β cells is a phenomenon restricted to NOX-1 or has a more broad relevance to other NOX enzymes functional in β cells. Selective inhibition of NOX-1 or key redox signaling events arising from NOX-1 activity in β cells may offer therapeutic opportunities.

In summary, NADPH oxidases play important roles in regulation of β cell biology. This is in terms of both regulation of physiological insulin secretion and mediation of β cell pathophysiology. The latter may result in β cell dysfunction arising from uncontrolled oxidative stress or more discrete modulation of function mediated by redox signaling initiated events. In terms of diabetes, NADPH oxidase activity is stimulated by several diabetes-associated stimuli.

NOX-4 and Insulin Signaling

NOX- 4 activity has additionally been implicated in the regulation of insulin signaling (Mahadev et al. 2004). NOX-4 is expressed in insulin-sensitive adipose cells (Mahadev et al. 2004). It has been previously shown that 3T3-L1 adipocytes produce H_2O_2 in response to insulin (Krieger-Brauer and Kather 1995). Dominant negative deletion constructs (missing either the NADPH-binding domain or the FAD/NADPH domains of NOX-4) were expressed in differentiated 3T3-L1 adipocyte cells. These cells expressing a deregulated/mutated NOX-4 showed a decrease in the generation of H_2O_2 when stimulated with insulin and a decrease in tyrosine phosphorylation of both the insulin receptor and insulin receptor substrate-1 (IRS-1) (Mahadev et al. 2004). Intracellular events associated with activation of the insulin signaling pathway, such as activation of ERK1/2 and glucose uptake, were also inhibited in the cells expressing the NOX-4 mutants. These studies highlight a link between NOX-4, ROS generation, and changes in insulin signaling (Mahadev et al. 2004).

Redox Signaling

A consequence of NADPH oxidase activation is production of ROS. Sustained increase in ROS results in oxidative stress that can be destructive to cells, resulting in dysfunction and cell death. Transient increases in ROS are, however, part of normal physiological processes. There is increasing recognition that intracellular signaling pathways are sensitive to changes in redox levels (Goldstein et al. 2005). Thus, in addition to global oxidative stress, NADPH oxidase activity is likely to also play an important role in regulation of discrete signaling pathways, kinase activation in particular. Upregulation of ROS, especially superoxide and hydrogen peroxide, can lead to activation of specific signaling pathways. Several signaling pathways are regulated by changes in intracellular ROS (reviewed in Goldstein et al. (2005), Mittler et al. (2011)). As NADPH oxidase is upregulated, there is an increase in ROS generation. For example, overexpression of NOX-1 in NIH 3T3 cells resulted in elevation in intracellular ROS and activation of signaling kinases JNK and ERK1/2 (Go et al. 2004). Adjustments to intracellular ROS levels likely lead to conformational changes in kinases and access to phosphorylation sites. Modification of kinase phosphorylation can result in either activation or inhibition of subsequent signals in a specific pathway. Known signaling pathways regulated by cellular redox state include kinases in the mitogen-activated protein kinase (MAPK) family, ERK, JNK, p38 kinase, and the Src-kinase family (Giannoni et al. 2005). Some transcription factors, such as NF-κB, AP-1, Nrf2, and c-Jun, are also sensitive to redox signaling. Elevated ROS produced from NADPH oxidase activity also inactivate phosphatases, including protein-tyrosine phosphatase 1B (PTP1B) (Mahadev et al. 2004).

NOX-1 and Second Messenger Src-Kinase Signaling

The expression of NOX-1 in the INS-1 β cell line is regulated through a feedforward loop in which NOX-mediated ROS generation affects second messengers resulting in a signal to upregulate NOX-1 protein expression. The second messengers involved in this pathway include activation of Src kinase (Weaver and Taylor-Fishwick 2013). When INS-1 cells were treated with proinflammatory cytokines, elevation in intracellular ROS and NADPH oxidase activity led to an increase in NOX-1 expression. This feedforward regulation was blocked by the selective Src-kinase inhibitor, PP2. Importantly, PP3, the structural chemical analog of PP2 that is inactive for Src-kinase inhibition, did not block the upregulation of NOX-1. Signaling pathways associated with NOX-1 activation include p38MAPK, Akt, and Src kinase (Gianni et al. 2008; Sancho and Fabregat 2010). Both p38MAPK and Akt are downstream mediators of NOX-1-activated ROS elevation in vascular smooth muscle cells (Lassegue et al. 2001).

NOX-2 and AngII Signaling

In addition to NOX-2 activation of cAMP/PKA in β cells, stimulation of NOX-2 by AngII also mediates JNK and Janus kinase (JAK)/STAT activation (Alves et al. 2012; Li et al. 2012). Alves et al. found that when rat islets were stimulated with AngII, NOX-2 was activated and ROS was generated. This elevation in intracellular ROS led to phosphorylation of JAK/STAT and JNK proteins (Alves et al. 2012). Islets isolated from the type 2 diabetic-like animal model, ZDF rat, have increased expression and phosphorylation of the NADPH oxidase $p47^{phox}$ subunit. They also have increased expression of the $gp91^{phox}$ subunit and increased activation of Rac (Syed et al. 2011a). Increase in NOX subunit expression and phosphorylation was correlated with activation of JNK1/2 and a decrease in activation of ERK1/2 in isolated ZDF islets. This study therefore linked the increase in NOX-2 subunits with second messenger activation and β cell dysfunction (Syed et al. 2011a). Supportive data was additionally presented in the INS-1 rat β cell line. Upon treatment with high glucose or the FFA palmitate, JNK1/2 phosphorylation increased, ERK1/2 phosphorylation decreased, and caspase-3 was active (Syed et al. 2011a). Both high glucose and palmitate are effective activators of NADPH oxidase in β cells. Studies in human islets have provided similar results. Islets isolated from human type 2 diabetic donors show an increase in Rac expression, JNK1/2 activation, and caspase-3 degradation, results that were analogous to the ZDF model. Exposure of human islets to high glucose activated Rac (Syed et al. 2011a). Collectively, these data indicate that exposure of β cells to stimuli (high glucose, cytokines, FFAs) results in an NADPH oxidase-mediated activation of second messengers. Activation of these second messengers is associated with β cell dysfunction.

NOX-4 and Renal Signaling

Redox signaling involving NOX-4 activity is an important contributor to renal dysfunction. High glucose leads to an upregulation of NOX-4 and an associated increase in ROS generation in the type 2 diabetic-like mouse model db/db. Elevated

ROS results in phosphorylation of p38MAPK and increased expression of TGF-β1/2 and fibronectin (Sedeek et al. 2010). NOX-4 generated ROS in adipose cells inhibits PTP1B by blocking its catalytic activity and affecting insulin signaling (Mahadev et al. 2004).

Redox Signaling and Activators of Transcription

In terms of signaling mediators, several transcription factors are associated with NAPDH oxidase activation. Both NOX-1 and NOX-2 were shown to activate NF-κB in vascular smooth muscle cells and MCF-7 cells (human mammary epithelial cells). Generation of H_2O_2 by endosomal NOX-2 facilitates formation of an active TNF receptor 1 (TNFR1) complex which is required for NF-κB activation. These are redox-dependent events (Li et al. 2009). Similarly, in vascular smooth muscle cells, ligation of receptors for IL-1β or TNF-α triggers receptor-ligand internalization into an endosomal compartment containing NOX-1. The elevation in ROS that is mediated by NOX-1 results in NF-κB activity (Miller et al. 2007). AngII activates the transcription factors NF-κB and AP-1 in arterial smooth muscle cells. This redox signaling-mediated event is dependent on NOX-1 activity and leads to cell migration and proliferation (Valente et al. 2012).

Adenoviral expression of NOX-4 results in *GATA-4* gene transcription in pluripotent progenitor cells (Murray et al. 2013). Signaling is mediated via ROS generation and activation of c-Jun. NOX-4 is implicated in the regulation of Smad2/3 and differentiation of cardiac fibroblasts into myofibroblasts. TGF-β activates the transcription factors Smad2/3 (Cucoranu et al. 2005). The knockdown of NOX-4 with siRNA blocked TGF-β-stimulated Smad2/3, highlighting the key role of NOX-4 in this signaling pathway. These studies, described in cardiomyocytes, may have relevance to β cells. TGF-β signaling activates Smad3 in β cells, which regulates insulin gene transcription (Lin et al. 2009). Smad3-deficient mice developed moderate hyperinsulinemia and mild hypoglycemia (Lin et al. 2009).

The contribution of redox signaling to β cell function has likely been underappreciated. NADPH oxidase family members NOX-1, NOX-2, and NOX-4 play significant roles in elevating intracellular ROS and therefore regulating redox signaling events in the β cell. NOX enzymes clearly regulate signaling outcomes in non-β cell systems. Whether this influence on cell regulation by NADPH oxidase isotypes is cell type-specific or translates also to regulation of the β cell will be revealed with further study. The activation of specific kinases, phosphatases, and transcription factors can be dependent upon changes in intracellular ROS. Induction of these pathways has influence on major physiological and pathophysiological responses by the β cell. Mapping redox signaling responses in the β cell will help characterize the relative importance and therapeutic potential of redox signaling events in terms of β cell survival and preservation.

NOX Inhibition

Early identified inhibitors of NADPH oxidase activity have helped in defining the important contribution of NADPH oxidase enzymes to biological processes and disease. The resulting identification of pathophysiology associated with NADPH oxidase and elevated NADPH oxidase products has stimulated the search for improved inhibitors. In addition to enhanced efficacy, the focus has been to identify inhibitors that are selective to different NOX isoforms, thereby establishing the relative contribution of each NOX enzyme to disease processes and facilitating development of a therapeutic strategy to control, treat, or reverse the disease.

Historically accepted inhibitors of NADPH oxidase include apocynin and diphenyleneiodonium (DPI). Apocynin was first identified in the 1800s from plant root and was recognized and widely used as a NADPH oxidase inhibitor from the mid-1900s. Apocynin (4′-hydroxy-3′-methoxyacetophenone) also known as acetovanillone is a naturally occurring methoxy-substituted catechol. It is not considered selective for NOX subunits. Apocynin is a pro-drug, being converted to an active form by peroxidase. Marketed as an inhibitor of NADPH oxidase with limited adverse effects *in vivo*, apocynin is reported to exhibit off-target effects (Lafeber et al. 1999). Significantly, peroxide-deficient cells are sensitive to apocynin inhibition, and ROS production in non-phagocytes has been associated with apocynin treatment (Vejrazka et al. 2005). Apocynin demonstrated antioxidant effects in endothelial and vascular smooth muscle cells (Heumuller et al. 2008). Reported mechanisms of action for apocynin include a block in the membrane translocation of p67^{phox} and p47^{phox}, sequestration of H_2O_2, and interaction of an apocynin radical with NOX thiol groups (Stolk et al. 1994; Johnson et al. 2002; Ximenes et al. 2007). Diphenyleneiodonium (DPI) is an inhibitor of flavoprotein dehydrogenases and has been extensively described in the research literature as an inhibitor of NADPH oxidase activity. DPI inhibits electron transporters in flavoenzymes. Thus, in addition to inhibition of NADPH oxidase, DPI also inhibits other flavin-dependent enzymes including nitric oxide synthase, xanthine oxidase, NADPH dehydrogenase, glucose phosphate dehydrogenase, mitochondrial complex I, and cytochrome *P*-450 reductase. Other nonspecific NOX inhibitors have been reported including AEBSF (pefabloc, 4-(2-aminoethyl)-benzenesulfonyl fluoride), which is an irreversible serine protease inhibitor and blocks complex assembly by inhibiting binding of p47^{phox} (Diatchuk et al. 1997).

Discovery of endogenous peptide inhibitors of NOX revealed candidate peptide sequences that could inhibit enzyme complex assembly by blocking targetable interacting domains (Kleinberg et al. 1990; Nauseef et al. 1993; DeLeo et al. 1995; Uhlinger et al. 1995; Shi et al. 1996). Exploitation of peptide inhibition has prompted efforts to design peptides with selectivity to the NOX enzyme isoforms. Peptide construct gp91ds-tat combines a peptide sequence from the NOX-2 B-loop that interacts with the p47^{phox} subunit and a nine amino acid ds-tat sequence. While designed to be specific, concern of inhibition of other NOX enzymes has been raised. This was driven by the sequence homology between NOX-2 and, to a lesser extent, NOX-4 (Jackson et al. 2010). Reconstitution studies that express enzyme components in an endogenously devoid cell however support specificity for NOX-2 (Csanyi et al. 2011). This 18 amino acid peptide (which has later been termed NOX-2ds-tat) inhibited angiotensin II-mediated ROS in vascular smooth muscle cells and attenuated vascular superoxide and systolic blood pressure in mice (Rey et al. 2001; Yang et al. 2005). The concept of peptide-based inhibition of complex assembly remains attractive due to the ability to target specific regions offering the potential for specificity and fewer off-target effects. Reviewed by Dahan and Pick is a detailed consideration of parameters to consider in peptide-based inhibition of NOX (Dahan and Pick 2012). Other peptide regions of NOX-2 have been targeted along with peptide regions in $p22^{phox}$, $p47^{phox}$, and accessory molecules (Rotrosen et al. 1990; Nauseef et al. 1993; Dahan et al. 2002). While offering the prospect of specificity, peptide-based therapeutics are limited by poor oral bioavailability, being readily inactivated in the gastrointestinal tract. It is possible that this limitation could be overcome by alternative delivery routes, including injection, patches, or inhalation, as has been explored for insulin administration. Further, advances in nanoparticle encapsulation and gene therapy/DNA therapeutics could help facilitate delivery of peptide-based strategies.

Current limitations in bioavailability of peptide approaches to selectively inhibit NOX enzymes paired with the emerging realization of the therapeutic potential of inhibitors targeted to NOX isotypes have renewed investigation of small molecular weight compound inhibitors. In recent years, significant investment in high-throughput screening of small molecule libraries has resulted in new emergent inhibitors with promising selectivity profiles. Initial evidenced inhibitors of neutrophil NADPH oxidase were VAS2870 and its second generation relative VAS3947. Both arose from high-throughput screens. These compounds display micromolar potency for NOX inhibition and have widely reported efficacy in assays dependent on NOX-induced ROS. Unlike earlier compounds, these molecules offer selectivity to NADPH oxidase and do not have activity against other flavoproteins, e.g., xanthine oxidase (ten Freyhaus et al. 2006; Wind et al. 2010). Although screened against neutrophil NADPH oxidase (NOX-2), the efficacy of these compounds in a variety of assays suggests they are selective pan-NOX inhibitors and lack isoform selectivity for individual NOX enzymes. Other high-throughput screening approaches have used NOX isoform-specific assays. These approaches have identified new compounds that preferentially inhibit NOX isoforms. Screening for ROS production in a cell-free assay of human NOX-4 identified a series of first-in-class pyrazolopyridine dione inhibitors with nanomolar potency (Laleu et al. 2010). Subsequent enhancement in chemical structure led to identification of GKT136901 and GKT137831, orally active potent inhibitors. The pyrazolopyridine dione inhibitors, while screened for NOX-4, also exhibit equipotent activity for NOX-1 inhibition. The lead compound has demonstrated a high degree of potency in several in vitro and in vivo assays (Laleu et al. 2010). Further, GKT136901 is 20-fold less potent at inhibiting NOX-2 and thus can be considered as a selective dual NOX-1/NOX-4 inhibitor. As an alternative approach, a cell-based high-throughput screen of NOX-1 expressing colon cancer cell line identified a chemical hit in a distinct chemical class. Studies of the structure-activity relationship of the trifluoromethyl-phenothiazine hit led to characterization of ML171 (2-acetylphenothiazine), a nanomolar inhibitor of NOX-1. Moreover, ML171 displays isoform selectivity for NOX-1 inhibition, having a potency for NOX-1 inhibition 20-fold lower than NOX-2, NOX-3, and NOX-4. Inhibition of NOX-1 by ML171 is evidenced in human colon cancer cells that only express NOX-1 plus NOXO1 and NOXA1. Overexpression of NOX-1 reverses ML171 inhibition supporting a selective inhibitory profile for ML171 (Gianni et al. 2010).

Identification of NADPH oxidase and ROS efficacy in pathways promoting pathogenesis in the β cell fosters the potential utility of selective NOX inhibition as a strategy to preserve β cell mass in diabetes. The progress of a range of new selective inhibitors, either small molecule, peptide, or peptidomimetic, that target NOX isoforms greatly facilitates the discovery of the role of NADPH oxidases in islet biology and diabetes. GKT136901 has been implicated in mitigation of oxidative stress associated in diabetic nephropathy, stroke, and neurodegeneration (Schildknecht et al. 2013). Our own unpublished data adds support since selective NOX-1 inhibition with ML171 confers β cells with protection in an inflammatory environment. Clearly an underlying concern, even with highly efficacious and selective inhibitors, will be off-target effects since NOX enzymes are expressed and function in several tissues. The prospect of functional redundancy for physiological actions of NADPH oxidases, and a general theme of overactive NAPDH oxidases in pathological conditions, offers encouragement for the therapeutic potential of selective NOX inhibitors. Studies in vivo with newer NOX inhibitors have been well tolerated (Laleu et al. 2010). Evolution of these investigations is expected to identify the following: (1) new investigative tools, (2) a better understanding of β cell vulnerability, and (3) original therapeutic strategies that can be applied to diabetes (Fig. 2).

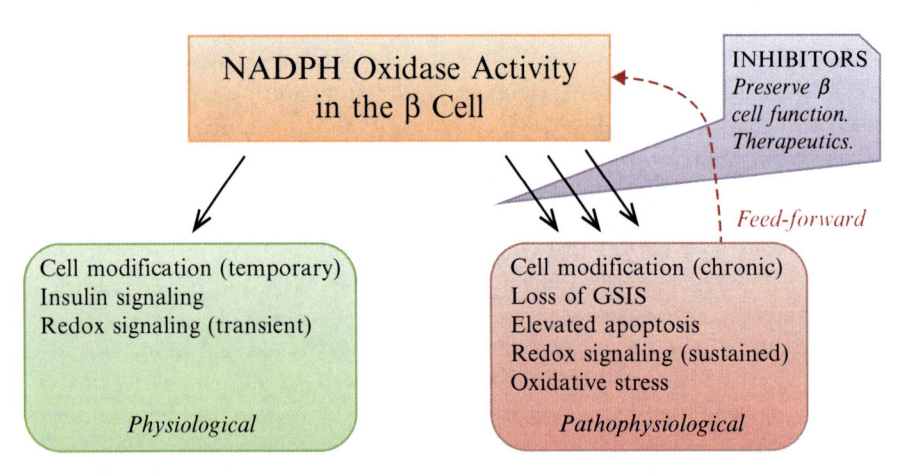

Fig. 2 NADPH oxidase (*NOX*) activity and β cell function. Physiologic responses involve transient NADPH oxidase activity and changes in ROS (*single arrow*). Chronic sustained NADPH oxidase activity in β cells (*multiple arrows*) is associated with pathophysiologic responses. This can be amplified with a feedforward regulation of NADPH oxidase activity. Inhibition of NOX could help protect islets and preserve β cell function

Conclusions

A major contributor to loss of functional β cell mass in diabetes is a sustained elevation in intracellular ROS. There are several sources of ROS in the β cell, including NADPH oxidase activity. In terms of β cell dysfunction, NADPH oxidases may have a greater role than previously appreciated. Upon chronic activation of NADPH oxidase activity and sustained elevation of intracellular ROS, β cell function and viability decreases. An increased appreciation for the role of NADPH oxidase activity in regulating β cell biology is driving the quest to understand the contribution of each NADPH oxidase isoform to β cell biology. The ability to selectively inhibit each isoform is a key goal. Selective inhibition of NOX enzymes may potentially reveal the factors and pathways needed to preserve a healthy β cell pool and offer new approaches in the treatment of diabetes.

Cross-References

▶ Current Approaches and Future Prospects for the Prevention of β-Cell Destruction in Autoimmune Diabetes
▶ Immunology of β-Cell Destruction
▶ Inflammatory Pathways Linked to β Cell Demise in Diabetes

References

Abo A, Pick E et al (1991) Activation of the NADPH oxidase involves the small GTP-binding protein p21rac1. Nature 353(6345):668–670

Adachi Y, Shibai Y et al (2008) Oncogenic Ras upregulates NADPH oxidase 1 gene expression through MEK-ERK-dependent phosphorylation of GATA-6. Oncogene 27(36):4921–4932

Ago T, Kitazono T et al (2004) Nox4 as the major catalytic component of an endothelial NAD(P)H oxidase. Circulation 109(2):227–233

Aguirre J, Rios-Momberg M et al (2005) Reactive oxygen species and development in microbial eukaryotes. Trends Microbiol 13(3):111–118

Allaoui A, Botteaux A et al (2009) Dual oxidases and hydrogen peroxide in a complex dialogue between host mucosae and bacteria. Trends Mol Med 15(12):571–579

Alves ES, Haidar AA et al (2012) Angiotensin II-induced JNK activation is mediated by NAD(P)H oxidase in isolated rat pancreatic islets. Regul Pept 175(1–3):1–6

Ando S, Kaibuchi K et al (1992) Post-translational processing of rac p21s is important both for their interaction with the GDP/GTP exchange proteins and for their activation of NADPH oxidase. J Biol Chem 267(36):25709–25713

Babior BM, Kipnes RS et al (1973) Biological defense mechanisms. The production by leukocytes of superoxide, a potential bactericidal agent. J Clin Invest 52(3):741–744

Babior BM, Lambeth JD et al (2002) The neutrophil NADPH oxidase. Arch Biochem Biophys 397(2):342–344

Banfi B, Molnar G et al (2001) A Ca^{2+}-activated NADPH oxidase in testis, spleen, and lymph nodes. J Biol Chem 276(40):37594–37601

Banfi B, Clark RA et al (2003) Two novel proteins activate superoxide generation by the NADPH oxidase NOX1. J Biol Chem 278(6):3510–3513

Banfi B, Malgrange B et al (2004a) NOX3, a superoxide-generating NADPH oxidase of the inner ear. J Biol Chem 279(44):46065–46072

Banfi B, Tirone F et al (2004b) Mechanism of Ca^{2+} activation of the NADPH oxidase 5 (NOX5). J Biol Chem 279(18):18583–18591

Bedard K, Krause KH (2007) The NOX family of ROS-generating NADPH oxidases: physiology and pathophysiology. Physiol Rev 87(1):245–313

Bedard K, Lardy B et al (2007) NOX family NADPH oxidases: not just in mammals. Biochimie 89(9):1107–1112

Bedard K, Jaquet V et al (2012) NOX5: from basic biology to signaling and disease. Free Radic Biol Med 52(4):725–734

BelAiba RS, Djordjevic T et al (2007) NOX5 variants are functionally active in endothelial cells. Free Radic Biol Med 42(4):446–459

Biberstine-Kinkade KJ, DeLeo FR et al (2001) Heme-ligating histidines in flavocytochrome b(558): identification of specific histidines in gp91(phox). J Biol Chem 276(33):31105–31112

Bleich D, Chen S et al (1999) Resistance to type 1 diabetes induction in 12-lipoxygenase knockout mice. J Clin Invest 103(10):1431–1436

Brar SS, Corbin Z et al (2003) NOX5 NAD(P)H oxidase regulates growth and apoptosis in DU 145 prostate cancer cells. Am J Physiol Cell Physiol 285(2):C353–C369

Brun S, Malagnac F et al (2009) Functions and regulation of the Nox family in the filamentous fungus Podospora anserina: a new role in cellulose degradation. Mol Microbiol 74(2):480–496

Cai W, He JC et al (2008) Oral glycotoxins determine the effects of calorie restriction on oxidant stress, age-related diseases, and lifespan. Am J Pathol 173(2):327–336

Caillou B, Dupuy C et al (2001) Expression of reduced nicotinamide adenine dinucleotide phosphate oxidase (ThoX, LNOX, Duox) genes and proteins in human thyroid tissues. J Clin Endocrinol Metab 86(7):3351–3358

Cano-Dominguez N, Alvarez-Delfin K et al (2008) NADPH oxidases NOX-1 and NOX-2 require the regulatory subunit NOR-1 to control cell differentiation and growth in Neurospora crassa. Eukaryot Cell 7(8):1352–1361

Cassatella MA, Bazzoni F et al (1990) Molecular basis of interferon-gamma and lipopolysaccharide enhancement of phagocyte respiratory burst capability. Studies on the gene expression of several NADPH oxidase components. J Biol Chem 265(33):20241–20246

Cave AC, Brewer AC et al (2006) NADPH oxidases in cardiovascular health and disease. Antioxid Redox Signal 8(5–6):691–728

Cheng G, Cao Z et al (2001) Homologs of gp91phox: cloning and tissue expression of Nox3, Nox4, and Nox5. Gene 269(1–2):131–140

Cheng G, Ritsick D et al (2004) Nox3 regulation by NOXO1, p47phox, and p67phox. J Biol Chem 279(33):34250–34255

Costal F, Oliveira E et al (2013) Dual effect of advanced glycation end products in pancreatic islet apoptosis. Diabetes Metab Res Rev 29(4):296–307

Coughlan MT, Yap FY et al (2011) Advanced glycation end products are direct modulators of β-cell function. Diabetes 60(10):2523–2532

Cross AR, Rae J et al (1995) Cytochrome b-245 of the neutrophil superoxide-generating system contains two nonidentical hemes. Potentiometric studies of a mutant form of gp91phox. J Biol Chem 270(29):17075–17077

Csanyi G, Cifuentes-Pagano E et al (2011) Nox2 B-loop peptide, Nox2ds, specifically inhibits the NADPH oxidase Nox2. Free Radic Biol Med 51(6):1116–1125

Cucoranu I, Clempus R et al (2005) NAD(P)H oxidase 4 mediates transforming growth factor-β1-induced differentiation of cardiac fibroblasts into myofibroblasts. Circ Res 97(9):900–907

Dahan I, Pick E (2012) Strategies for identifying synthetic peptides to act as inhibitors of NADPH oxidases, or "all that you did and did not want to know about Nox inhibitory peptides". Cell Mol Life Sci 69(14):2283–2305

Dahan I, Issaeva I et al (2002) Mapping of functional domains in the p22(phox) subunit of flavocytochrome b(559) participating in the assembly of the NADPH oxidase complex by "peptide walking". J Biol Chem 277(10):8421–8432

de Carvalho DD, Sadok A et al (2008) Nox1 downstream of 12-lipoxygenase controls cell proliferation but not cell spreading of colon cancer cells. Int J Cancer 122(8):1757–1764

De Deken X, Wang D et al (2000) Cloning of two human thyroid cDNAs encoding new members of the NADPH oxidase family. J Biol Chem 275(30):23227–23233

DeLeo FR, Nauseef WM et al (1995) A domain of p47phox that interacts with human neutrophil flavocytochrome b558. J Biol Chem 270(44):26246–26251

Diatchuk V, Lotan O et al (1997) Inhibition of NADPH oxidase activation by 4-(2-aminoethyl)-benzenesulfonyl fluoride and related compounds. J Biol Chem 272(20):13292–13301

Dupuy C, Kaniewski J et al (1989) NADPH-dependent H2O2 generation catalyzed by thyroid plasma membranes. Studies with electron scavengers. Eur J Biochem 185(3):597–603

Dupuy C, Ohayon R et al (1999) Purification of a novel flavoprotein involved in the thyroid NADPH oxidase. Cloning of the porcine and human cdnas. J Biol Chem 274(52):37265–37269

Dworakowski R, Alom-Ruiz SP et al (2008) NADPH oxidase-derived reactive oxygen species in the regulation of endothelial phenotype. Pharmacol Rep 60(1):21–28

Edens WA, Sharling L et al (2001) Tyrosine cross-linking of extracellular matrix is catalyzed by Duox, a multidomain oxidase/peroxidase with homology to the phagocyte oxidase subunit gp91phox. J Cell Biol 154(4):879–891

Elsner M, Gehrmann W et al (2011) Peroxisome-generated hydrogen peroxide as important mediator of lipotoxicity in insulin-producing cells. Diabetes 60(1):200–208

Fan C, Katsuyama M et al (2005) Transactivation of the EGF receptor and a PI3 kinase-ATF-1 pathway is involved in the upregulation of NOX1, a catalytic subunit of NADPH oxidase. FEBS Lett 579(5):1301–1305

Finegold AA, Shatwell KP et al (1996) Intramembrane bis-heme motif for transmembrane electron transport conserved in a yeast iron reductase and the human NADPH oxidase. J Biol Chem 271 (49):31021–31024

Foreman J, Demidchik V et al (2003) Reactive oxygen species produced by NADPH oxidase regulate plant cell growth. Nature 422(6930):442–446

Fransen M, Nordgren M et al (2012) Role of peroxisomes in ROS/RNS-metabolism: implications for human disease. Biochim Biophys Acta 1822(9):1363–1373

Fu X, Beer DG et al (2006) cAMP-response element-binding protein mediates acid-induced NADPH oxidase NOX5-S expression in Barrett esophageal adenocarcinoma cells. J Biol Chem 281(29):20368–20382

Gapper C, Dolan L (2006) Control of plant development by reactive oxygen species. Plant Physiol 141(2):341–345

Garrido AM, Griendling KK (2009) NADPH oxidases and angiotensin II receptor signaling. Mol Cell Endocrinol 302(2):148–158

Gattas MV, Forteza R et al (2009) Oxidative epithelial host defense is regulated by infectious and inflammatory stimuli. Free Radic Biol Med 47(10):1450–1458

Gavazzi G, Banfi B et al (2006) Decreased blood pressure in NOX1-deficient mice. FEBS Lett 580 (2):497–504

Gehrmann W, Elsner M et al (2010) Role of metabolically generated reactive oxygen species for lipotoxicity in pancreatic β-cells. Diabetes Obes Metab 12(Suppl 2):149–158

Geiszt M, Kopp JB et al (2000) Identification of renox, an NAD(P)H oxidase in kidney. Proc Natl Acad Sci U S A 97(14):8010–8014

Geiszt M, Lekstrom K et al (2003a) NAD(P)H oxidase 1, a product of differentiated colon epithelial cells, can partially replace glycoprotein 91phox in the regulated production of superoxide by phagocytes. J Immunol 171(1):299–306

Geiszt M, Witta J et al (2003b) Dual oxidases represent novel hydrogen peroxide sources supporting mucosal surface host defense. FASEB J 17(11):1502–1504

Gianni D, Bohl B et al (2008) The involvement of the tyrosine kinase c-Src in the regulation of reactive oxygen species generation mediated by NADPH oxidase-1. Mol Biol Cell 19(7):2984–2994

Gianni D, Taulet N et al (2010) A novel and specific NADPH oxidase-1 (Nox1) small-molecule inhibitor blocks the formation of functional invadopodia in human colon cancer cells. ACS Chem Biol 5(10):981–993

Giannoni E, Buricchi F et al (2005) Intracellular reactive oxygen species activate Src tyrosine kinase during cell adhesion and anchorage-dependent cell growth. Mol Cell Biol 25(15):6391–6403

Go YM, Gipp JJ et al (2004) H2O2-dependent activation of GCLC-ARE4 reporter occurs by mitogen-activated protein kinase pathways without oxidation of cellular glutathione or thioredoxin-1. J Biol Chem 279(7):5837–5845

Goldstein BJ, Mahadev K et al (2005) Redox paradox: insulin action is facilitated by insulin-stimulated reactive oxygen species with multiple potential signaling targets. Diabetes 54(2):311–321

Gorin Y, Ricono JM et al (2003) Nox4 mediates angiotensin II-induced activation of Akt/protein kinase B in mesangial cells. Am J Physiol Renal Physiol 285(2):F219–F229

Graciano MF, Santos LR et al (2011) NAD(P)H oxidase participates in the palmitate-induced superoxide production and insulin secretion by rat pancreatic islets. J Cell Physiol 226(4):1110–1117

Grankvist K, Marklund SL et al (1981) CuZn-superoxide dismutase, Mn-superoxide dismutase, catalase and glutathione peroxidase in pancreatic islets and other tissues in the mouse. Biochem J 199(2):393–398

Grasberger H, Refetoff S (2006) Identification of the maturation factor for dual oxidase. Evolution of an eukaryotic operon equivalent. J Biol Chem 281(27):18269–18272

Guichard C, Moreau R et al (2008) NOX family NADPH oxidases in liver and in pancreatic islets: a role in the metabolic syndrome and diabetes? Biochem Soc Trans 36(Pt 5):920–929

Harper RW, Xu C et al (2005) Differential regulation of dual NADPH oxidases/peroxidases, Duox1 and Duox2, by Th1 and Th2 cytokines in respiratory tract epithelium. FEBS Lett 579(21):4911–4917

Harper RW, Xu C et al (2006) Duox2 exhibits potent heme peroxidase activity in human respiratory tract epithelium. FEBS Lett 580(22):5150–5154

Heumuller S, Wind S et al (2008) Apocynin is not an inhibitor of vascular NADPH oxidases but an antioxidant. Hypertension 51(2):211–217

Heyworth PG, Bohl BP et al (1994) Rac translocates independently of the neutrophil NADPH oxidase components p47phox and p67phox. Evidence for its interaction with flavocytochrome b558. J Biol Chem 269(49):30749–30752

Heyworth PG, Cross AR et al (2003) Chronic granulomatous disease. Curr Opin Immunol 15(5):578–584

Hofmann SM, Dong HJ et al (2002) Improved insulin sensitivity is associated with restricted intake of dietary glycoxidation products in the db/db mouse. Diabetes 51(7):2082–2089

Hordijk PL (2006) Regulation of NADPH oxidases: the role of Rac proteins. Circ Res 98(4):453–462

Ibi M, Katsuyama M et al (2006) NOX1/NADPH oxidase negatively regulates nerve growth factor-induced neurite outgrowth. Free Radic Biol Med 40(10):1785–1795

Ibi M, Matsuno K et al (2008) Reactive oxygen species derived from NOX1/NADPH oxidase enhance inflammatory pain. J Neurosci 28(38):9486–9494

Imoto H, Sasaki N et al (2008) Impaired insulin secretion by diphenyleneiodium associated with perturbation of cytosolic Ca^{2+} dynamics in pancreatic β-cells. Endocrinology 149(11):5391–5400

Jackson HM, Kawahara T et al (2010) Nox4 B-loop creates an interface between the transmembrane and dehydrogenase domains. J Biol Chem 285(14):10281–10290

Janciauskiene S, Ahren B (2000) Fibrillar islet amyloid polypeptide differentially affects oxidative mechanisms and lipoprotein uptake in correlation with cytotoxicity in two insulin-producing cell lines. Biochem Biophys Res Commun 267(2):619–625

Jay DB, Papaharalambus CA et al (2008) Nox5 mediates PDGF-induced proliferation in human aortic smooth muscle cells. Free Radic Biol Med 45(3):329–335

Jiao J, Dou L et al (2012) NADPH oxidase 2 plays a critical role in dysfunction and apoptosis of pancreatic β-cells induced by very low-density lipoprotein. Mol Cell Biochem 370(1–2): 103–113

Johnson DK, Schillinger KJ et al (2002) Inhibition of NADPH oxidase activation in endothelial cells by ortho-methoxy-substituted catechols. Endothelium 9(3):191–203

Juhasz A, Ge Y et al (2009) Expression of NADPH oxidase homologues and accessory genes in human cancer cell lines, tumours and adjacent normal tissues. Free Radic Res 43(6):523–532

Kamiguti AS, Serrander L et al (2005) Expression and activity of NOX5 in the circulating malignant B cells of hairy cell leukemia. J Immunol 175(12):8424–8430

Kaneto H, Fujii J et al (1996) Reducing sugars trigger oxidative modification and apoptosis in pancreatic β-cells by provoking oxidative stress through the glycation reaction. Biochem J 320 (Pt 3):855–863

Katsuyama M, Fan C et al (2002) NADPH oxidase is involved in prostaglandin F2α-induced hypertrophy of vascular smooth muscle cells: induction of NOX1 by PGF2α. J Biol Chem 277 (16):13438–13442

Kawahara T, Kohjima M et al (2005) Helicobacter pylori lipopolysaccharide activates Rac1 and transcription of NADPH oxidase Nox1 and its organizer NOXO1 in guinea pig gastric mucosal cells. Am J Physiol Cell Physiol 288(2):C450–C457

Kawamori D, Kajimoto Y et al (2003) Oxidative stress induces nucleo-cytoplasmic translocation of pancreatic transcription factor PDX-1 through activation of c-Jun NH(2)-terminal kinase. Diabetes 52(12):2896–2904

Kenyon V, Rai G et al (2011) Discovery of potent and selective inhibitors of human platelet-type 12- lipoxygenase. J Med Chem 54(15):5485–5497

Kleinberg ME, Malech HL et al (1990) The phagocyte 47-kilodalton cytosolic oxidase protein is an early reactant in activation of the respiratory burst. J Biol Chem 265 (26):15577–15583

Koya D, King GL (1998) Protein kinase C activation and the development of diabetic complications. Diabetes 47(6):859–866

Krieger-Brauer HI, Kather H (1995) Antagonistic effects of different members of the fibroblast and platelet-derived growth factor families on adipose conversion and NADPH-dependent H2O2 generation in 3 T3 L1-cells. Biochem J 307(Pt 2):549–556

Kumar B, Koul S et al (2008) Oxidative stress is inherent in prostate cancer cells and is required for aggressive phenotype. Cancer Res 68(6):1777–1785

Kuroda J, Ago T et al (2010) NADPH oxidase 4 (Nox4) is a major source of oxidative stress in the failing heart. Proc Natl Acad Sci U S A 107(35):15565–15570

Kusumoto K, Kawahara T et al (2005) Ecabet sodium inhibits Helicobacter pylori lipopolysaccharide-induced activation of NADPH oxidase 1 or apoptosis of guinea pig gastric mucosal cells. Am J Physiol Gastrointest Liver Physiol 288(2):G300–G307

Lafeber FP, Beukelman CJ et al (1999) Apocynin, a plant-derived, cartilage-saving drug, might be useful in the treatment of rheumatoid arthritis. Rheumatology 38(11):1088–1093

Laleu B, Gaggini F et al (2010) First in class, potent, and orally bioavailable NADPH oxidase isoform 4 (Nox4) inhibitors for the treatment of idiopathic pulmonary fibrosis. J Med Chem 53 (21):7715–7730

Lambeth JD (2007) Nox enzymes, ROS, and chronic disease: an example of antagonistic pleiotropy. Free Radic Biol Med 43(3):332–347

Lambeth JD, Kawahara T et al (2007) Regulation of Nox and Duox enzymatic activity and expression. Free Radic Biol Med 43(3):319–331

Lara-Ortiz T, Riveros-Rosas H et al (2003) Reactive oxygen species generated by microbial NADPH oxidase NoxA regulate sexual development in *Aspergillus nidulans*. Mol Microbiol 50(4):1241–1255

Lassegue B, Sorescu D et al (2001) Novel gp91(phox) homologues in vascular smooth muscle cells : nox1 mediates angiotensin II-induced superoxide formation and redox-sensitive signaling pathways. Circ Res 88(9):888–894

Lenzen S (2008) Oxidative stress: the vulnerable β-cell. Biochem Soc Trans 36(Pt 3):343–347

Lenzen S, Drinkgern J et al (1996) Low antioxidant enzyme gene expression in pancreatic islets compared with various other mouse tissues. Free Radic Biol Med 20(3):463–466

Li J, Stouffs M et al (2006) The NADPH oxidase NOX4 drives cardiac differentiation: role in regulating cardiac transcription factors and MAP kinase activation. Mol Biol Cell 17(9):3978–3988

Li Q, Spencer NY et al (2009) Endosomal Nox2 facilitates redox-dependent induction of NF-kappaB by TNF-α. Antioxid Redox Signal 11(6):1249–1263

Li N, Li B et al (2012) NADPH oxidase NOX2 defines a new antagonistic role for reactive oxygen species and cAMP/PKA in the regulation of insulin secretion. Diabetes 61(11):2842–2850

Lim M, Park L et al (2008) Induction of apoptosis of β cells of the pancreas by advanced glycation end-products, important mediators of chronic complications of diabetes mellitus. Ann N Y Acad Sci 1150:311–315

Lin HM, Lee JH et al (2009) Transforming growth factor-β/Smad3 signaling regulates insulin gene transcription and pancreatic islet β-cell function. J Biol Chem 284(18):12246–12257

Lin N, Zhang H et al (2012) Advanced glycation end-products induce injury to pancreatic β cells through oxidative stress. Diabetes Metab 38(3):250–257

Lupi R, Del Guerra S et al (2006) The direct effects of the angiotensin-converting enzyme inhibitors, zofenoprilat and enalaprilat, on isolated human pancreatic islets. Eur J Endocrinol 154(2):355–361

Mahadev K, Motoshima H et al (2004) The NAD(P)H oxidase homolog Nox4 modulates insulin-stimulated generation of H2O2 and plays an integral role in insulin signal transduction. Mol Cell Biol 24(5):1844–1854

Malagnac F, Lalucque H et al (2004) Two NADPH oxidase isoforms are required for sexual reproduction and ascospore germination in the filamentous fungus Podospora anserina. Fungal Genet Biol 41(11):982–997

Manea A (2010) NADPH oxidase-derived reactive oxygen species: involvement in vascular physiology and pathology. Cell Tissue Res 342(3):325–339

Manea A, Manea SA et al (2012) Positive regulation of NADPH oxidase 5 by proinflammatory-related mechanisms in human aortic smooth muscle cells. Free Radic Biol Med 52(9):1497–1507

Marino D, Dunand C et al (2012) A burst of plant NADPH oxidases. Trends Plant Sci 17(1):9–15

Martyn KD, Frederick LM et al (2006) Functional analysis of Nox4 reveals unique characteristics compared to other NADPH oxidases. Cell Signal 18(1):69–82

Marzban L, Verchere CB (2004) The role of islet amyloid polypeptide in type 2 diabetes. Can J Diabetes 28(4):39–47

Matti A, Kyathanahalli C et al (2012) Protein farnesylation is requisite for mitochondrial fuel-induced insulin release: further evidence to link reactive oxygen species generation to insulin secretion in pancreatic β-cells. Islets 4(1):74–77

McDuffie M, Maybee NA et al (2008) Nonobese diabetic (NOD) mice congenic for a targeted deletion of 12/15-lipoxygenase are protected from autoimmune diabetes. Diabetes 57(1):199–208

Meng D, Lv DD et al (2008) Insulin-like growth factor-I induces reactive oxygen species production and cell migration through Nox4 and Rac1 in vascular smooth muscle cells. Cardiovasc Res 80(2):299–308

Michalska M, Wolf G et al (2010) Effects of pharmacological inhibition of NADPH oxidase or iNOS on pro-inflammatory cytokine, palmitic acid or H2O2-induced mouse islet or clonal pancreatic β-cell dysfunction. Biosci Rep 30(6):445–453

Miller FJ Jr, Filali M et al (2007) Cytokine activation of nuclear factor kappa B in vascular smooth muscle cells requires signaling endosomes containing Nox1 and ClC-3. Circ Res 101(7):663–671

Mittler R, Vanderauwera S et al (2011) ROS signaling: the new wave? Trends Plant Sci 16(6):300–309

Miyano K, Ueno N et al (2006) Direct involvement of the small GTPase Rac in activation of the superoxide-producing NADPH oxidase Nox1. J Biol Chem 281(31):21857–21868

Modak MA, Datar SP et al (2007) Differential susceptibility of chick and mouse islets to streptozotocin and its co-relation with islet antioxidant status. J Comp Physiol B 177(2):247–257

Mohammed AM, Kowluru A (2013) Activation of apocynin-sensitive NADPH oxidase (Nox2) activity in INS-1 832/13 cells under glucotoxic conditions. Islets 5(3):129–131

Montezano AC, Burger D et al (2010) Nicotinamide adenine dinucleotide phosphate reduced oxidase 5 (Nox5) regulation by angiotensin II and endothelin-1 is mediated via calcium/calmodulin-dependent, rac-1-independent pathways in human endothelial cells. Circ Res 106(8):1363–1373

Montezano AC, Burger D et al (2011) Novel Nox homologues in the vasculature: focusing on Nox4 and Nox5. Clin Sci 120(4):131–141

Morand S, Dos Santos OF et al (2003) Identification of a truncated dual oxidase 2 (DUOX2) messenger ribonucleic acid (mRNA) in two rat thyroid cell lines. Insulin and forskolin regulation of DUOX2 mRNA levels in FRTL-5 cells and porcine thyrocytes. Endocrinology 144(2):567–574

Morand S, Ueyama T et al (2009) Duox maturation factors form cell surface complexes with Duox affecting the specificity of reactive oxygen species generation. FASEB J 23(4):1205–1218

Morgan D, Oliveira-Emilio HR et al (2007) Glucose, palmitate and pro-inflammatory cytokines modulate production and activity of a phagocyte-like NADPH oxidase in rat pancreatic islets and a clonal β cell line. Diabetologia 50(2):359–369

Morgan D, Rebelato E et al (2009) Association of NAD(P)H oxidase with glucose-induced insulin secretion by pancreatic β-cells. Endocrinology 150(5):2197–2201

Murray TV, Smyrnias I et al (2013) NADPH oxidase 4 regulates cardiomyocyte differentiation via redox activation of c-Jun protein and the cis-regulation of GATA-4 gene transcription. J Biol Chem 288(22):15745–15759

Nakayama M, Inoguchi T et al (2005) Increased expression of NAD(P)H oxidase in islets of animal models of Type 2 diabetes and its improvement by an AT1 receptor antagonist. Biochem Biophys Res Commun 332(4):927–933

Nauseef WM, McCormick S et al (1993) Functional domain in an arginine-rich carboxyl-terminal region of p47phox. J Biol Chem 268(31):23646–23651

Newsholme P, Morgan D et al (2009) Insights into the critical role of NADPH oxidase(s) in the normal and dysregulated pancreatic β cell. Diabetologia 52(12):2489–2498

Nisimoto Y, Motalebi S et al (1999) The p67(phox) activation domain regulates electron flow from NADPH to flavin in flavocytochrome b(558). J Biol Chem 274(33):22999–23005

Nisimoto Y, Tsubouchi R et al (2008) Activation of NADPH oxidase 1 in tumour colon epithelial cells. Biochem J 415(1):57–65

Ohneda K, Mirmira RG et al (2000) The homeodomain of PDX-1 mediates multiple protein-protein interactions in the formation of a transcriptional activation complex on the insulin promoter. Mol Cell Biol 20(3):900–911

Oliveira HR, Verlengia R et al (2003) Pancreatic β-cells express phagocyte-like NAD(P)H oxidase. Diabetes 52(6):1457–1463

Orient A, Donko A et al (2007) Novel sources of reactive oxygen species in the human body. Nephrol Dial Transplant 22(5):1281–1288

Paffenholz R, Bergstrom RA et al (2004) Vestibular defects in head-tilt mice result from mutations in Nox3, encoding an NADPH oxidase. Genes Dev 18(5):486–491

Pandey D, Patel A et al (2012) Expression and functional significance of NADPH oxidase 5 (Nox5) and its splice variants in human blood vessels. Am J Physiol Heart Circ Physiol 302(10):H1919–H1928

Pedruzzi E, Guichard C et al (2004) NAD(P)H oxidase Nox-4 mediates 7-ketocholesterol-induced endoplasmic reticulum stress and apoptosis in human aortic smooth muscle cells. Mol Cell Biol 24(24):10703–10717

Pendyala S, Gorshkova IA et al (2009) Role of Nox4 and Nox2 in hyperoxia-induced reactive oxygen species generation and migration of human lung endothelial cells. Antioxid Redox Signal 11(4):747–764

Peppa M, He C et al (2003) Fetal or neonatal low-glycotoxin environment prevents autoimmune diabetes in NOD mice. Diabetes 52(6):1441–1448

Pi J, Bai Y et al (2007) Reactive oxygen species as a signal in glucose-stimulated insulin secretion. Diabetes 56(7):1783–1791

Purves T, Middlemas A et al (2001) A role for mitogen-activated protein kinases in the etiology of diabetic neuropathy. FASEB J 15(13):2508–2514

Rada B, Hably C et al (2008) Role of Nox2 in elimination of microorganisms. Semin Immunopathol 30(3):237–253

Rebelato E, Mares-Guia TR et al (2012) Expression of NADPH oxidase in human pancreatic islets. Life Sci 91(7–8):244–249

Rey FE, Cifuentes ME et al (2001) Novel competitive inhibitor of NAD(P)H oxidase assembly attenuates vascular $O(2)(-)$ and systolic blood pressure in mice. Circ Res 89(5):408–414

Rigutto S, Hoste C et al (2009) Activation of dual oxidases Duox1 and Duox2: differential regulation mediated by camp-dependent protein kinase and protein kinase C-dependent phosphorylation. J Biol Chem 284(11):6725–6734

Ris-Stalpers C (2006) Physiology and pathophysiology of the DUOXes. Antioxid Redox Signal 8(9–10):1563–1572

Rokutan K, Kawahara T et al (2006) NADPH oxidases in the gastrointestinal tract: a potential role of Nox1 in innate immune response and carcinogenesis. Antioxid Redox Signal 8(9–10):1573–1582

Rokutan K, Kawahara T et al (2008) Nox enzymes and oxidative stress in the immunopathology of the gastrointestinal tract. Semin Immunopathol 30(3):315–327

Rotrosen D, Kleinberg ME et al (1990) Evidence for a functional cytoplasmic domain of phagocyte oxidase cytochrome b558. J Biol Chem 265(15):8745–8750

Rotrosen D, Yeung CL et al (1992) Cytochrome b558: the flavin-binding component of the phagocyte NADPH oxidase. Science 256(5062):1459–1462

Sadok A, Bourgarel-Rey V et al (2008) Nox1-dependent superoxide production controls colon adenocarcinoma cell migration. Biochim Biophys Acta 1783(1):23–33

Sagi M, Fluhr R (2006) Production of reactive oxygen species by plant NADPH oxidases. Plant Physiol 141(2):336–340

San Martin A, Foncea R et al (2007) Nox1-based NADPH oxidase-derived superoxide is required for VSMC activation by advanced glycation end-products. Free Radic Biol Med 42(11):1671–1679

Sancho P, Fabregat I (2010) NADPH oxidase NOX1 controls autocrine growth of liver tumor cells through up-regulation of the epidermal growth factor receptor pathway. J Biol Chem 285(32):24815–24824

Santos CX, Tanaka LY et al (2009) Mechanisms and implications of reactive oxygen species generation during the unfolded protein response: roles of endoplasmic reticulum oxidoreductases, mitochondrial electron transport, and NADPH oxidase. Antioxid Redox Signal 11(10):2409–2427

Schildknecht S, Weber A et al (2013) The NOX1/4 inhibitor GKT136901 as selective and direct scavenger of peroxynitrite. Curr Med Chem 21:365–376

Schroder K, Wandzioch K et al (2009) Nox4 acts as a switch between differentiation and proliferation in preadipocytes. Arterioscler Thromb Vasc Biol 29(2):239–245

Schulz E, Munzel T (2008) NOX5, a new "radical" player in human atherosclerosis? J Am Coll Cardiol 52(22):1810–1812

Sedeek M, Callera G et al (2010) Critical role of Nox4-based NADPH oxidase in glucose-induced oxidative stress in the kidney: implications in type 2 diabetic nephropathy. Am J Physiol Renal Physiol 299(6):F1348–F1358

Selemidis S, Sobey CG et al (2008) NADPH oxidases in the vasculature: molecular features, roles in disease and pharmacological inhibition. Pharmacol Ther 120(3):254–291

Serrander L, Cartier L et al (2007) NOX4 activity is determined by mRNA levels and reveals a unique pattern of ROS generation. Biochem J 406(1):105–114

Seshiah PN, Weber DS et al (2002) Angiotensin II stimulation of NAD(P)H oxidase activity: upstream mediators. Circ Res 91(5):406–413

Shao J, Iwashita N et al (2006) Beneficial effects of candesartan, an angiotensin II type 1 receptor blocker, on β-cell function and morphology in db/db mice. Biochem Biophys Res Commun 344(4):1224–1233

Shi J, Ross CR et al (1996) PR-39, a proline-rich antibacterial peptide that inhibits phagocyte NADPH oxidase activity by binding to Src homology 3 domains of p47 phox. Proc Natl Acad Sci U S A 93(12):6014–6018

Shiose A, Kuroda J et al (2001) A novel superoxide-producing NAD(P)H oxidase in kidney. J Biol Chem 276(2):1417–1423

Stolk J, Hiltermann TJ et al (1994) Characteristics of the inhibition of NADPH oxidase activation in neutrophils by apocynin, a methoxy-substituted catechol. Am J Respir Cell Mol Biol 11(1):95–102

Sturrock A, Cahill B et al (2006) Transforming growth factor-β1 induces Nox4 NAD(P)H oxidase and reactive oxygen species-dependent proliferation in human pulmonary artery smooth muscle cells. Am J Physiol Lung Cell Mol Physiol 290(4):L661–L673

Sturrock A, Huecksteadt TP et al (2007) Nox4 mediates TGF-β1-induced retinoblastoma protein phosphorylation, proliferation, and hypertrophy in human airway smooth muscle cells. Am J Physiol Lung Cell Mol Physiol 292(6):L1543–L1555

Suh YA, Arnold RS et al (1999) Cell transformation by the superoxide-generating oxidase Mox1. Nature 401(6748):79–82

Sumimoto H (2008) Structure, regulation and evolution of Nox-family NADPH oxidases that produce reactive oxygen species. FEBS J 275(13):3249–3277

Syed I, Kyathanahalli CN et al (2011a) Increased phagocyte-like NADPH oxidase and ROS generation in type 2 diabetic ZDF rat and human islets: role of Rac1-JNK1/2 signaling pathway in mitochondrial dysregulation in the diabetic islet. Diabetes 60(11):2843–2852

Syed I, Kyathanahalli CN et al (2011b) Phagocyte-like NADPH oxidase generates ROS in INS 832/13 cells and rat islets: role of protein prenylation. Am J Physiol Regul Integr Comp Physiol 300(3):R756–R762

Takeya R, Ueno N et al (2003) Novel human homologues of p47phox and p67phox participate in activation of superoxide-producing NADPH oxidases. J Biol Chem 278(27):25234–25246

Taylor-Fishwick DA, Pittenger GL et al (2008) Transplantation and beyond. Drug Dev Res 69:165–176

ten Freyhaus H, Huntgeburth M et al (2006) Novel Nox inhibitor VAS2870 attenuates PDGF-dependent smooth muscle cell chemotaxis, but not proliferation. Cardiovasc Res 71(2):331–341

Tiedge M, Lortz S et al (1997) Relation between antioxidant enzyme gene expression and antioxidative defense status of insulin-producing cells. Diabetes 46(11):1733–1742

Torres MA, Dangl JL (2005) Functions of the respiratory burst oxidase in biotic interactions, abiotic stress and development. Curr Opin Plant Biol 8(4):397–403

Touyz RM, Chen X et al (2002) Expression of a functionally active gp91phox-containing neutrophil-type NAD(P)H oxidase in smooth muscle cells from human resistance arteries: regulation by angiotensin II. Circ Res 90(11):1205–1213

Tudzynski P, Heller J et al (2012) Reactive oxygen species generation in fungal development and pathogenesis. Curr Opin Microbiol 15(6):653–659

Uchizono Y, Takeya R et al (2006) Expression of isoforms of NADPH oxidase components in rat pancreatic islets. Life Sci 80(2):133–139

Ueyama T, Geiszt M et al (2006) Involvement of Rac1 in activation of multicomponent Nox1- and Nox3-based NADPH oxidases. Mol Cell Biol 26(6):2160–2174

Uhlinger DJ, Tyagi SR et al (1995) On the mechanism of inhibition of the neutrophil respiratory burst oxidase by a peptide from the C-terminus of the large subunit of cytochrome b558. Biochemistry 34(2):524–527

Ushio-Fukai M (2006) Redox signaling in angiogenesis: role of NADPH oxidase. Cardiovasc Res 71(2):226–235

Valente AJ, Yoshida T et al (2012) Angiotensin II enhances AT1-Nox1 binding and stimulates arterial smooth muscle cell migration and proliferation through AT1, Nox1, and interleukin-18. Am J Physiol Heart Circ Physiol 303(3):H282–H296

Vaquero EC, Edderkaoui M et al (2004) Reactive oxygen species produced by NAD(P)H oxidase inhibit apoptosis in pancreatic cancer cells. J Biol Chem 279(33):34643–34654

Vejrazka M, Micek R et al (2005) Apocynin inhibits NADPH oxidase in phagocytes but stimulates ROS production in non-phagocytic cells. Biochim Biophys Acta 1722(2):143–147

Volchuk A, Ron D (2010) The endoplasmic reticulum stress response in the pancreatic β-cell. Diabetes Obes Metab 12(Suppl 2):48–57

Weaver JR, Taylor-Fishwick DA (2013) Regulation of NOX-1 expression in β cells: a positive feedback loop involving the Src-kinase signaling pathway. Mol Cell Endocrinol 369 (1–2):35–41

Weaver JR, Holman TR et al (2012) Integration of pro-inflammatory cytokines, 12-lipoxygenase and NOX-1 in pancreatic islet β cell dysfunction. Mol Cell Endocrinol 358(1):88–95

Wientjes FB, Hsuan JJ et al (1993) p40phox, a third cytosolic component of the activation complex of the NADPH oxidase to contain src homology 3 domains. Biochem J 296(Pt 3):557–561

Wind S, Beuerlein K et al (2010) Comparative pharmacology of chemically distinct NADPH oxidase inhibitors. Br J Pharmacol 161(4):885–898

Wingler K, Wunsch S et al (2001) Upregulation of the vascular NAD(P)H-oxidase isoforms Nox1 and Nox4 by the renin-angiotensin system in vitro and in vivo. Free Radic Biol Med 31 (11):1456–1464

Xiang FL, Lu X et al (2010) NOX2 deficiency protects against streptozotocin-induced β-cell destruction and development of diabetes in mice. Diabetes 59(10):2603–2611

Ximenes VF, Kanegae MP et al (2007) The oxidation of apocynin catalyzed by myeloperoxidase: proposal for NADPH oxidase inhibition. Arch Biochem Biophys 457(2):134–141

Yang M, Foster E et al (2005) Insulin-stimulated NAD(P)H oxidase activity increases migration of cultured vascular smooth muscle cells. Am J Hypertens 18(10):1329–1334

Yuan H, Lu Y et al (2010a) Suppression of NADPH oxidase 2 substantially restores glucose-induced dysfunction of pancreatic NIT-1 cells. FEBS J 277(24):5061–5071

Yuan H, Zhang X et al (2010b) NADPH oxidase 2-derived reactive oxygen species mediate FFAs-induced dysfunction and apoptosis of β-cells via JNK, p38 MAPK and p53 pathways. PLoS One 5(12):e15726

Zhang X, Shan P et al (2006) Toll-like receptor 4 deficiency causes pulmonary emphysema. J Clin Invest 116(11):3050–3059

Zhao Z, Zhao C et al (2009) Advanced glycation end products inhibit glucose-stimulated insulin secretion through nitric oxide-dependent inhibition of cytochrome c oxidase and adenosine triphosphate synthesis. Endocrinology 150(6):2569–2576

The Contribution of Reg Family Proteins to Cell Growth and Survival in Pancreatic Islets

35

Qing Li, Xiaoquan Xiong, and Jun-Li Liu

Contents

Introduction	956
An Overview of the Regenerating Gene Family	958
Classification of Reg Proteins Based on Protein Sequence	958
Expression Pattern During Development in Human and Rodents	961
Reg Protein Receptor(s)	962
The Roles in Promoting Cell Replication and Preventing Apoptosis of Adult β-Cells	963
Reg1 [hReg1A and hReg1B]	963
Reg2	966
Reg3α [hReg3G]	967
Reg3β (hReg3A)	969
Reg3δ (INGAP)	971
Antibacterial Reg3γ	972
Reg4	972
Possible Role of Reg Proteins on β-Cell Neogenesis	973
General and Isoform-Specific Functions of Reg Proteins	975
Molecular Factors Regulating the Expression of Reg Family Genes	976
Perspective	978
Cross-References	978
References	978

Abstract

In 2008, we have reviewed Reg family proteins which have been found and characterized in several systems including cell growth and regeneration in the pancreas. Since then the research scope has expanded significantly to the (patho-)physiology of the liver, intestine, immunity, and cancer. More importantly, in communicating our research findings, we feel the need of further classification in the family of seven independent genes and among key species. A more

Q. Li • X. Xiong • J.-L. Liu (✉)
Fraser Laboratories for Diabetes Research, Department of Medicine, McGill University Health Centre, Montreal, QC, Canada
e-mail: qing.li3@mail.mcgill.ca; hhyyxxq@gmail.com; jun-li.liu@mcgill.ca

uniformed terminology should help us to understand their isoform-specific functions and/or mode of activation.

Keywords

INGAP • Reg1, Reg2, Reg3, Reg4 • Orthology • EXTL3/EXTR1 • Regeneration • Transdifferentiation

Abbreviations

HIP	Gene expressed in hepatocellular carcinoma-intestine-pancreas
PAP	Pancreatitis-associated protein
PSP	Pancreatic stone protein
PTP	Pancreatic thread protein
RELP	Regenerating protein-like protein

Introduction

The mammalian pancreas is composed of three main cell types: the exocrine acini, endocrine islets, and ducts. The endocrine islets constitute about 5 % of the volume and consist of α, β, δ, ε, and PP cells that produce glucagon, insulin, somatostatin, ghrelin, and pancreatic polypeptide, respectively (Gu et al. 2003; Harbeck et al. 1996). Insulin and glucagon are two hormones with opposing roles working together to maintain the balance of glucose storage and utilization (Moses et al. 1996). Somatostatin and pancreatic polypeptide exert inhibitory effects on both pancreatic endocrine and exocrine secretions (D'Ercole 1999; Bonner-Weir 2000a). Ghrelin regulates insulin secretion and expression of genes essential for β-cell biology, promotes β-cell proliferation and survival, and inhibits β-cell apoptosis (Bonner-Weir 2000b). The exocrine cells that are organized into acini constitute about 85 % of the pancreas. They secrete digestive enzymes, such as amylase, elastase, trypsinogen, into the pancreatic ducts, a branched network of tubules formed by epithelial duct cells. These enzymes, along with bicarbonate and other electrolytes secreted by ductal cells, constitute the pancreas juice and are drained into the duodenum through the main duct (Gu et al. 2003). Also, increasing interest has been placed on the interactions between the exocrine and endocrine portions of pancreas in the structures and functions.

In both normal and pathophysiological states, β-cell mass is determined by changes in the rate of replication and neogenesis, individual cell volume, and cell death rate (Dheen et al. 1997). High rate β-cell replication has been observed in the embryos of late gestation and in newborns. A similar increase in the activity of cellular apoptosis also occurs during the newborn and postpartum period in mother. The dynamic changes in replication and apoptosis may contribute to the remodeling of β-cell mass during these periods (Dheen et al. 1997; Zhou et al. 2000; Cheng et al. 2000; Mauras et al. 2000). The rates of replication and apoptosis of β-cells are both reduced significantly beyond 3 months of life and remain low except in response to physiological/pathological changes. Low-rate β-cell replication lasts throughout the lifespan, which is closely correlated to the body weight increment. In young ages, increases of both β-cell size and number contribute to the increase of β-cell mass, but in old animals the increase of β-cell size is mainly responsible for the increase in mass (Mauras et al. 2000). For in-depth discussion on apoptosis, please refer to the chapter entitled "▶ Mechanisms of Pancreatic β-Cell Apoptosis in Diabetes and Its Therapies."

A common pathology of diabetes is the loss of functional β-cells. Pancreatic islet regeneration is now an attractive alternative for diabetes cell therapy. There are three major efforts to achieve this goal:

1. To promote the replication and reverse the destruction of existing β-cells. Increasing interest has been focused on the initiation of growth and the role of proliferation factors in pancreatic islet, such as insulin-like growth factor (IGF)-I, hepatocyte growth factor (HGF), and glucagon-like peptide (GLP)-1 (Suarez-Pinzon et al. 2008).
2. To directly differentiate stem cells or pancreatic progenitor cells into β-cells. It is generally accepted that pancreatic ducts and ductular progenitor cells can differentiate to β-cells, as evidenced by the observation of islet budding from ductal structures during embryogenesis or postnatal growth (Scharfmann et al. 1989). Although replication is the major source of β-cell renewal, around 30 % of new β-cells can arise from neogenesis from non-β-cell precursors in adult rats (Bonner-Weir et al. 2004). A differentiation process that converts human embryonic stem cells (ESCs) to endocrine cells has been developed by using transcriptional factors at different stages (D'Amour et al. 2006).
3. To transdifferentiat from other endogenous pancreatic cells. Cells in the periphery of islets in the neonatal pancreas strongly express ductal markers CK19 and CK20 and may serve as islet progenitors (Bouwens et al. 1994). Shortly after birth, CK19 expression expands to the whole pancreas, which is turned off in differentiated islet cells (Billestrup and Nielsen 1991). This transient expression of CK19 suggests that new islets may arise from ductal tissues. Neogenesis of β-cells was also observed in response to various stresses, including 90 % pancreatectomy (Terazono et al. 1988) and partial obstruction of the pancreas (Sieradzki et al. 1988).

Recent studies also showed that adult ductal and acinar cells could be dedifferentiated into a progenitor state and then re-differentiated into β-like cells using a series of transcriptional factors, including Pdx1, neurogenin (Ngn)3, and Maf A (Liu et al. 2009). Transfection of Ngn3 into pancreatic ductal cells can also induce islet neogenesis (Xu et al. 2008). What should be pointed out is that there is a distinction between the terms islet regeneration and neogenesis. "Islet regeneration describes an increase in β-cell mass in general, regardless of the mechanism, while islet neogenesis refers specifically to an increase in β-cell mass via the transdifferentiation of adult pancreatic stem cells, putatively found in the ductal epithelium or acinar tissue, into functioning, physiologically regulated tissue" (Pittenger et al. 2009a). For additional consideration, please refer to the chapter entitled "▶ Stem Cells in Pancreatic Islets."

A number of growth factors have been reported to promote β-cell expansion in animal models, including IGF-I (George et al. 2002; Smith et al. 1991), gastrin

(Hansson et al. 1996; Hansson and Thoren 1995), transforming growth factor (TGF)-α (Song et al. 1999; Wang et al. 1993; Sandgren et al. 1990), GLP-1 (Pospisilik et al. 2003), exendin-4 (Gedulin et al. 2005; Xu et al. 1999; Tourrel et al. 2001; Xu et al. 2006), and Reg family proteins Reg1 (Terazono et al. 1990, 1988) and INGAP (Reg3δ) (Rosenberg et al. 2004). Transgenic mice with β-cell-specific overexpression of IGF-I displayed increased β-cell mass in parallel with a higher rate of neogenesis and β-cell replication, hence better recovery from the hyperglycemia and hypoinsulinemia induced by streptozotocin, compared to control animals (George et al. 2002). In the pancreatic regeneration model stimulated by duct ligation, gastrin expression was strongly induced in the ligated part at both mRNA and protein levels, shortly after surgery (Wang et al. 1997). Gastrin alone, or in combination with epidermal growth factor (EGF), increased the expression of Pdx1 and insulin in isolated CK19-positive human ductal cells (Hansson et al. 1996). Overexpression of TGF-α upregulates the Pdx1-expressing epithelium characterized by the expression of Pax6 and initiates islet neogenesis (Song et al. 1999). Mice that overexpress both gastrin and TGF-α showed significant increase in islet cell mass, suggesting a synergistic effect of the two factors on stimulating islet cell growth (Wang et al. 1993). Exendin-4, an agonist of GLP-1 with a longer half-life, facilitates β-cell neogenesis in rat and human pancreatic ducts (Xu et al. 2006). GLP-1 or exendin-4 treatment increased pancreatic insulin content and β-cell mass and decreased basal plasma glucose in streptozotocin-treated neonatal rats in both short and long terms (Tourrel et al. 2001). This chapter will focus on the contribution of Reg family proteins to cell growth and survival in pancreatic islets.

An Overview of the Regenerating Gene Family

Reg and Reg-related genes constitute a family within the C-type lectin superfamily (Lasserre et al. 1994; Chakraborty et al. 1995; Hartupee et al. 2001). In the last two decades, over 29 Reg genes have been discovered in several different species (Table 1). These secretory proteins share structural and functional properties associated with tissue injury, inflammation, diabetes, carcinogenesis, and cell proliferation or differentiation in the pancreas, liver, neurons, and gastrointestinal tract (Dusetti et al. 1994; Christa et al. 1996; Hill et al. 1999; He et al. 2010; Nishimune et al. 2000). Ever since Reg1 was discovered, special attention has been paid to the therapeutic potential of Reg proteins for the regeneration of pancreatic islets and treatment of diabetes (Okamoto 1999).

Classification of Reg Proteins Based on Protein Sequence

In the mouse, seven unique Reg genes have been discovered, all of which are located on chromosome 6C except Reg4. Consequently, five Reg genes in rat and

Table 1 *Members of the Reg family proteins in the mouse, rat, and human.* Literatures have used a few other names, such as HIP, gene expressed in hepatocellular carcinoma-intestine-pancreas; PAP, pancreatitis-associated protein; PSP, pancreatic stone protein; PTP, pancreatic thread protein; and RELP, regenerating protein-like protein. Reg stands for regenerating islet derived. The data is mostly based on NCBI collections. Based on the degree of sequence identity, mouse Reg1 seems to correspond to two human genes, Reg1A and Reg1B; and Reg3α and Reg3β to two unique rat genes and two human genes, respectively. When identical proteins were repeatedly submitted to Genbank, NCBI, or UniProtKB/Swiss-Prot, "=" is used to list identical protein IDs and orthologies in a single cell

Mouse genes	Orthology	Transcript	Polypeptide
Reg1	Reg, PTP, PSP, lithostathine	NM_009042	NP_033068 =P43137
	Rat Reg1/1α (Bimmler et al. 1999; Rouquier et al. 1991)	NM_012641	NP_036773 (Terazono et al. 1988; Li et al. 2013a) =P10758
	Human Reg1A, PSP, PTP, lithostathine	NM_002909	NP_002900 (Fujishiro et al. 2012; Rouimi et al. 1988) =P05451
	Human Reg1B, RegL (Bartoli et al. 1993), PSP2	NM_006507	NP_006498.1 =P48304.1
Reg2	PTP2, PSP2, lithostathine 2	NM_009043	NP_033069.1 (Luo et al. 2013; Unno et al. 1993) =Q08731
Reg3α	PAP2, PAP II, Reg IIIα	NM_011259	NP_035389 (Narushima et al. 1997; Lai et al. 2012) =O09037
	Rat Reg3α (REG 3A), Rat Reg III	NM_001145846 NM_172077.2	NP_001139318 =NP_742074.2 (Li et al. 2013a; Frigerio et al. 1993a)
	Rat PAP II	L10229	AAA02980.1 =P35231.1 (Frigerio et al. 1993a; Suzuki et al. 1994)
	Human Reg3G, PAPIB, Reg III	NM_198448 =NM_001008387.2 AB161037 AY428734	NP_940850.1 (Lee et al. 2012; Nata et al. 2004) =Q6UW15.1 =NP_001008388.1 =BAD51394.1 =AAR88147.1
Reg3β	PAP, PAP1, PAP I, HIP, Reg IIIβ	NM_011036	NP_035166 (Luo et al. 2013; Itoh and Teraoka 1993) =P35230
	Rat Reg3β, PAP	NM_053289 M98049	NP_445741 (Li et al. 2013a; Iovanna et al. 1991) =P25031.1 =AAA16341.1 (Iovanna et al. 1993)
	Rat Reg-2, Reg2	S43715	AAB23103.1 (Kamimura et al. 1992; Lieu et al. 2006)

(*continued*)

Table 1 (continued)

Mouse genes	Orthology	Transcript	Polypeptide
	Human Reg3A, HIP, PAP, INGAP (Rafaeloff et al. 1997)	NM_002580.2 NM_138937.2 NM_138938.2 BC036776	NP_002571.1 (Lai et al. 2012; Lasserre et al. 1992) =Q06141.1 =NP_620354.1 =NP_620355.1 =AAH36776.1
		M84337.1	AAA36415.1 (Orelle et al. 1992)
Reg3γ	PAP3, PAP III, Reg IIIgamma	NM_011260	NP_035390.1 (Narushima et al. 1997; Choi et al. 2013) =O09049
	Rat Reg3γ, PAP III	NM_173097	NP_775120.1 (Frigerio et al. 1993b; Konishi et al. 2013) =P42854.1
Reg3δ	INGAP, Reg3d, RegIII delta	NM_013893	NP_038921 (Sasahara et al. 2000; Skarnes et al. 2011) =Q9QUS9
		NM_001161741.1	NP_001155213.1 (Skarnes et al. 2011)
	INGAP-related protein	AB028625.1	BAA92141.1 (Sasahara et al. 2000)
Reg4	RELP, RegIV	NM_026328	NP_080604.2 (Hu et al. 2011; Kamarainen et al. 2003) =Q9D8G5
	Rat Reg4	NM_001004096.1	NP_001004096 (Namikawa et al. 2005) =Q68AX7.1
	Human Reg4	NM_032044 =NM_001159352.1	NP_114433.1 (Hartupee et al. 2001; Ying et al. 2013) =Q9BYZ8.1 =NP_001152824.1

five in human have been identified. The molecular relationships based on sequence comparison and alignment are summarized in Table 1. Based on the sequence homology, and the phylogenetic analysis conducted with blast method in the data retrieved from NCBI, Reg proteins can be divided into four groups: Reg1, Reg2, Reg3, and Reg4 (Okamoto 1999). With the exception of Reg4, all of the other Reg family genes are structured into six exons separated by five introns spanning about 3 kb. The first exon encodes the 5′-UTR, and the second encodes the remainder of the 5′-UTR, the ATG start codon, and the initial protein coding sequence. Exons 3–6 encode the body of the proteins with the 3′-UTR located in the sixth exon (Narushima et al. 1997). In Reg3 subfamily proteins, there is a common 5-aa insertion in the C-terminal regions (Narushima et al. 1997). Based on the high degrees of domain/sequences identities, the gene family is probably derived from the same ancestor gene by gene duplication events accumulated during evolution. Currently, Reg2 and Reg3δ/INGAP are only found in mice; more Reg3 isoforms should be discovered in rats and humans.

All members of the Reg family contain the typical C-type lectin-like domain (CTLD). They are subject to trypsin cleavage at the Arg-Ile bond located at 11th residue at the N-terminal, resulting in the formation of insoluble fibrils (Graf et al. 2006). The sensitive cleavage site is conserved in 18 Reg/Reg-related proteins from 6 different species (human, bovine, mouse, hamster, pig, and rat). Studies on rat Reg proteins (Reg1, Reg3α, Reg3β) showed that cleavage of the N-terminal undecapeptide produces peptides of 133–138 residues. Trypsin cleavage converts three of the soluble 16-kDa Reg proteins (Reg1, Reg3β, Reg3γ) into 14-kDa insoluble products that are completely resistant to trypsin and partially resistant to other proteases from pancreatic juice.

Reg3α is also processed into the 14-kDa form, but its product remained soluble and is only resistant to trypsin, but not to other proteases. How this cleavage affects the function of Reg proteins remains to be understood. It was found that trypsin-activated insoluble isoforms of Reg1, Reg3β, and Reg3γ polymerize into highly organized fibrillar structures with helical configurations (Graf et al. 2001). The C-terminal cleavage product of rat Reg1 spontaneously precipitates at a neutral pH (Schiesser et al. 2001). This insoluble form may play a key role in forming protein plugs in chronic pancreatitis. In more than one half of patients with pancreaticobiliary maljunction, Reg1, together with trypsinogen and activated trypsin, was detected in both the duct bile and the gallbladder bile, whereas none of the pancreatic enzymes or Reg1 was detected in the controls (Ochiai et al. 2004).

Several consensus transcriptional regulatory elements have been identified after examining the 5′-flanking sequences of Reg family genes. IL-6 response elements, mediating putative acute-phase responses, are located in the 5′-flanking region of all mouse Reg genes. Pan-1 motif sequences (CACCTG) are located in the promoter regions of mouse Reg3α and Reg3β, rat Reg1, and hamster INGAP genes. Pit-1 element, which mediates pituitary-specific transcription, is located in the promoter regions of mouse Reg3α, Reg3δ, and Reg3γ and rat Reg3β genes (Narushima et al. 1997). In addition to the IL-6 and Pit-1 response elements shared with other Reg3 genes, Reg3δ also contains consensus motifs for MyoD and Irf 1/Irf 2 binding sites in the promoter region, which suggests isoform-specific expression and response (Abe et al. 2000).

Expression Pattern During Development in Human and Rodents

The expression pattern of Reg genes during development may differ from each other depending upon the types of tissue and the age of development. The expressions of Reg1 and Reg2 genes have been investigated in mouse embryos at 8.5–12 days of the development, along with the expression of Ins1 and Ins2 genes. Reg1 mRNA became detectable at day E9, following the onset of Ins2 expression at day E8.5. Reg2 mRNA was not detectable until E12, when Ins1 transcription takes place (Perfetti et al. 1996a). This suggests that the two insulin genes and the two Reg genes are induced and expressed differentially during early development. In the human fetus, Reg1A expression was observed only in the pancreas, in contrast to its widespread expression in adults (Bartoli et al. 1998). The level of human pancreatic Reg1A transcript is low before 16 weeks of gestation, at which time they increase dramatically and reach a similar level as in the adult by 20-week gestation (Mally et al. 1994). Despite its early expression, the Reg1A/Reg1 gene might not be involved in β- or acinar cell growth during human and rat fetal development due to a lack of coordination between Reg mRNA levels and insulin gene expression (Moriscot et al. 1996; Smith et al. 1994). Human Reg1B transcript is present not only in the pancreas but also in the colon and brain of the fetus (Bartoli et al. 1998). Interestingly, expression of Reg1A was higher than Reg1B in human fetal pancreata, but the reversed expression pattern is observed in adult pancreas where Reg1B is higher than Reg1A (Sanchez et al. 2001). Reg3A/PAP mRNA expression displayed a broad distribution in the human fetus, being observed in the pancreas, stomach, jejunum, and colon and to a much lower level in the pituitary gland. The expression of Reg3A/PAP transcript in these tissues lasts throughout the adult lifespan, being especially high in the jejunum (Bartoli et al. 1998). Reg3A/PAP protein was first detectable at 8 weeks in endocrine nests co-stained with chromogranin A. The expression of Reg3A/PAP reached a level comparable with of the adult pancreas at 10 weeks of fetal life, being detected only in the glucagon-producing cells. In the meantime, no expression of Reg3A/PAP protein was detected in the pancreatic ducts or acinar cells of the fetal pancreas (Hervieu et al. 2006). Different from Reg3A/PAP in human, mouse INGAP/Reg3δ is present in cells that co-expressed insulin or somatostatin, but not glucagon, in the developing pancreatic bud of the embryo. Surprisingly, the colocalization of glucagon and Reg3δ/INGAP only occurred in the mouse islet cells to a significant level after birth (Hamblet et al. 2008).

Postnatal expression of the Reg proteins has only been systemically analyzed in rodents. After birth, total Reg1 and Reg2 expression in the pancreas showed an age-dependent decline, being decreased by 45 % at 30-month-old vs. 1-month-old in mice. While Reg1 mRNA level in the pancreas decreased progressively with age, Reg2 mRNA levels did not decline significantly, indicating that Reg1 and Reg2 expressions in the pancreas have differential age-dependent regulation (Perfetti et al. 1996b). Reg1 was also detectable in the duodenum and pancreas of newborn rats and dramatically increased at 3 weeks of age. Reg3β mRNA was undetectable in neonatal rat and displayed a sudden increase in the ileum around the time of weaning. A decline of Reg1 and Reg3β expressions in the ileum was observed in older rats (Chakraborty et al. 1995).

Reg Protein Receptor(s)

The putative interactions of Reg proteins and Reg receptor(s) have not been sufficiently elucidated to date. The Reg1α receptor cDNA was isolated from rat islets from a 2,760 bp open reading frame. The 919-amino-acid protein was suggested to be a type II transmembrane protein with a long extracellular domain (868 aa), a single transmembrane domain (residues 29–51), and a short N-terminal intracellular region (Kobayashi et al. 2000). The rat receptor is homologous to human EXTL3/EXTR1, a member of the EXT family, and can modulate NF-κB signaling upon stimulation by TNF-α (Nguyen et al. 2006). The mRNA of Rat Reg1α receptor was detected in normal pancreatic islets, regenerating islets, and insulinoma RINm5F cells. The receptor transcript was also expressed in a wide range of other tissues, including the liver, kidney, spleen, thymus, testis, adrenal gland, stomach, ileum, colon, pituitary gland, and brain, but not in the heart and jejunum (Kobayashi et al. 2000). Reg1α receptor-expressing RINm5F cells showed significant Reg1-dependent growth acceleration as indicated by BrdU incorporation. However, the expression of Reg1α receptor remained unchanged in regenerating islets as compared to normal ones, suggesting that both proliferation and apoptosis of pancreatic β-cells are primarily regulated by the expression of the Reg genes, but not the receptor (Kobayashi et al. 2000).

Beside the isolation of Reg1α receptor, the mechanism of Reg3δ/INGAP action on RINm5F cell proliferation was also explored (Petropavlovskaia et al. 2012). Both the full-length recombinant protein and bioactive peptide of INGAP (INGAP-P, a pentadecapeptide corresponding to amino acids 104–118) stimulated cell regeneration via binding to Gi protein-coupled receptor and by activating the Ras/Raf/Erk signaling pathway. Activation of ERK1/2 can be blocked by pertussis toxin, a reagent that can prevent the G proteins from interacting with corresponding receptors on the cell membrane. Further, PI3K/Akt pathway was also activated after INGAP administration. But to date, the sequence of this particular INGAP receptor has not been determined.

The Roles in Promoting Cell Replication and Preventing Apoptosis of Adult β-Cells

Reg1 [hReg1A and hReg1B]

Reg1/pancreatic stone protein (PSP) was first found in the study of pancreatitis, which was proposed to control the formation of calcium carbonate crystals in the pancreas (Multigner et al. 1983). However, in the diabetes field, the term Reg1 was used more often to describe its function on islet regeneration (Graf et al. 2006). The cDNA of Reg1 was first isolated in screening the regenerating islets-derived library from 90 % pancreatectomized rats in 1988 (Terazono et al. 1988). The Reg1 gene was found to be expressed in rat regenerating islets and induced by the administration of the poly (ADP-ribose) polymerase (PARP) inhibitor nicotinamide to depancreatized rats and also by the treatment with aurothioglucose, a drug that induced islet hyperplasia (Terazono et al. 1988). As human homologues of mouse Reg1, both Reg1A and Reg1B proteins consist of 166 amino acids and differ only by 22 (Moriizumi et al. 1994); their protein sequences share 87 % identities (Bartoli et al. 1993).

Under physiological condition, very low concentration of Reg1 protein is detected in the islets. Most of the protein are located in the acinar cells (Kimura et al. 1992) and increased significantly under inflammatory stress or other forms of pancreatic injury. Its mRNA level was significantly elevated in response to interleukin IL-6, interferon (IFN), or tumor necrosis factor (TNF)-α but decreased by dexamethasone (Zenilman et al. 1997). Expression of Reg1 mRNA in the pancreas was also increased by 12-fold after a 2-week 75 % high-protein diet (Rouquier et al. 1991). A consistent increase of Reg1 protein in the pancreatic juice was also detected within 2 weeks of an 82 % high-protein feeding (Bimmler et al. 1999). In addition, Reg1 is also normally expressed in the gastrointestinal tract, such as in the duodenum and jejunum, and gastric mucosa, with species variations (Terazono et al. 1988; Rouquier et al. 1991; Unno et al. 1993; Perfetti et al. 1996b; Watanabe et al. 1990). Reg1 expression is closely associated with pancreatic β-cell function. It is proposed to be a paracrine or autocrine factor in the proliferation and differentiation of cells in the digestive and endocrine systems (Acquatella-Tran Van Ba et al. 2012). In regenerating islets induced by the administration of PARP inhibitors to 90 % depancreatized rats, Reg1 was found to colocalize with insulin in secretory granules, suggesting that Reg1 is synthesized in and secreted from regenerating β-cells (Terazono et al. 1990). Reg1 mRNA levels were increased threefold within 2 days in the rat pancreas that received surgical wrapping, which correlated with ductular proliferation and emerging insulin staining within the ductular epithelia in the wrapped lobe. However, the induced Reg1 gene expression was localized to the exocrine tissue, suggesting that Reg1 may be involved in the maintenance of normal islet function through induction of new islet formation from precursors of ductal origins (Zenilman et al. 1996a).

Reg1 plays a role in the replication of β-cells. Using isolated rat islets, Reg1 protein stimulated β-cell replication by increasing [^{3}H] thymidine incorporation in a dose-dependent manner (Watanabe et al. 1994). Reg1 proteins isolated from human and bovine pancreas were mitogenic to both ARIP ductal and RIN β-cell lines in a dose-dependent manner but had no effect on AR42J acinar cells or isolated mature islets (Zenilman et al. 1996b). Isolated human and rat Reg1 proteins were also mitogenic to primary ductal cells and may modulate the expansion of the pancreatic ductal population during islet regeneration (Zenilman et al. 1998). It suggests that Reg1 can potentiate proliferation of ductal cells and islet β-cells, but not acinar cells. (More discussion related to islet isolation can be found in chapter entitled "▶ Isolation of Rodent Islets of Langerhans.") Administration of rat Reg1 protein to depancreatized rats ameliorated the surgical diabetes after 2 months, as evidenced by decreased blood glucose and preserved insulin-producing capacity (Watanabe et al. 1994). Diabetic NOD mice treated with recombinant human Reg1A protein showed increased β-cell mass and decreased mortality rate than untreated animals. This was interpreted as a result of Reg1A-induced maturation of β-cell precursors in NOD mice (Gross et al. 1998). The expression of Reg1 may be directly stimulated by gastrin, since the induced mRNA abundance was diminished by gastrin/cholecystokinin B antagonist (O'Hara et al. 2013; Ashcroft et al. 2004). In murine pancreatic tumors with acinar-specific overexpression of the gastrin receptor (CCK2R), expressions of Reg1 and Reg3α proteins were strongly upregulated in duct-like cells in preneoplastic lesions, or in the periphery of tumors and adjacent acini.

Moreover, the CCK2R transgenic mice showed improved glucose tolerance, increased insulin secretion, and doubled insulin contents compared to control animals (Gigoux et al. 2008), which indirectly indicated Reg1 could promote pancreatic cell proliferation and improve glucose tolerance. The rate of [^{3}H] thymidine incorporation was low in cultured pancreatic islets from Reg1-deficient mice (Unno et al. 2002) but high in those from β-cell-specific Reg1-overexpressing (Ins-Reg) mice, indicating that Reg1 protein was secreted from the islets which stimulated DNA synthesis through an autocrine mechanism. The Reg1-deficient mice had significantly smaller β-cell mass than control animals following gold thioglucose treatment, a drug inducing hyperplasic islets, suggesting that Reg1 might be essential for the cell cycle progression in pancreatic β-cells. The NOD mice carrying the Reg1 transgene showed a delayed onset of diabetes, which coincided with a threefold increase in islet cell volume. These data further support the notion that Reg1 promotes the regeneration of β-cells, which, as a consequence, compensates for the β-cell loss and delays the onset of autoimmune diabetes (Unno et al. 2002). It is thus conceivable that Reg1A level in the serum of both T1D and T2D patients was significantly elevated (Astorri et al. 2010). However, controversies existed in an early study on transgenic mice overexpressing Reg1 protein in the islets, i.e., the mice became diabetic as a result of increased β-cell apoptosis, as well as the development of various tumors (Yamaoka et al. 2000), as we have assessed before (Liu et al. 2008). Whether Reg1 promote islet cell proliferation and/or differentiation from other types of pancreatic cells needs to be studied further.

The identification of a putative receptor for Reg1, EXTL3, supports its direct effect on islet proliferation. The mechanism of how Reg1 stimulates DNA synthesis in islet β-cells appeared to involve phosphoinositide 3-kinase (PI3K) and downstream targets of transcriptional factor ATF-2 and cyclin D1 (Takasawa et al. 2006). Consequently, in Reg1 knockout islets, the levels of phospho-ATF-2, cyclin D1 and phospho-retinoblastoma protein (pRb), and the rate of DNA synthesis were all decreased. Cyclin D1 is established to promote cell cycle progression by inactivating retinoblastoma protein through cyclin-dependent kinases (cdks) and stimulating cell proliferation. Alternatively, ERK1/2 pathway was activated during mitogenesis of ductal and β-cell lines triggered by Reg1 overexpression or treatment with recombinant protein (Wang et al. 2011). Using cDNA microarray, significant elevations of mitogen-activated protein kinase phosphatases (MKP-1) and cyclins were detected in Reg1-treated cells.

In addition, Reg1 effects exhibited a dose-dependent manner. Endogenously expressed Reg1 in high concentration may form a complex with EXTL3, bind to MKP-1, and inactivate JNK, leading to cell apoptosis or differentiation into other cells (Mueller et al. 2008), a pathway that seemed to exist in both ductal and β-cells. In this experiment, Reg1 protein displayed a dual action on cell proliferation under low-dose administration, while high-dose Reg1 or endogenous overexpression induced cell apoptosis. In fact, it has been reported that high extracellular level of Reg1 over 100 nM could inhibit cell growth (Jung and Kim 2002). With overexpression of Reg1, more differentiated state was observed in rat insulinoma cells. It was thought that under low dose of Reg1, the protein could bind to its receptor and activate MAPK-cyclin D1 pathway. When overexpressed within cells or cultured in high concentration, Reg1 can inhibit growth by binding to MKP-1, leading to differentiation into other types of cells.

Reg1 expression is associated with pancreatic pathology. In patients with cystic fibrosis (CF), in addition to its normal localization in acinar cells, Reg1 immunostaining was induced in the duct-like cells of the tubular complexes and dilated duct cells co-stained with the ductal marker CK19 (Sanchez et al. 2004). In vitro, Reg1 inhibited proliferation and migration of pancreatic stellate cells and stimulated fibrolysis by increasing the ratio of matrix metalloproteinases (MMPs) to tissue inhibitors of matrix metalloproteinases (TIMPs). Hence, it might rescue pancreatitis by promoting the resolution of fibrosis (Li et al. 2010). On the other hand, autoimmunity to Reg1 may be associated with the development of diabetes. A significant increase in anti-Reg1 autoantibodies was found in both T1D and T2D patients compared with healthy subjects. Serum from diabetic patients with Reg1 autoantibodies demonstrated significantly attenuated BrdU incorporation induced by Reg1, while nondiabetic serum without the autoantibodies had little effect (Shervani et al. 2004). It supports that Reg1 protein can be used as a replacement therapy for diabetes.

The action of Reg1 is not restricted to the pancreas. Reg1 knockout mice had a greater number of severe lesions in the small intestines induced by indomethacin, a nonsteroidal anti-inflammatory drug. These intestinal injuries were rescued by the administration of Reg1 protein, indicating a physiological role for Reg1 in maintaining the intercellular integrity in the small intestine (Pittenger et al. 2009b). In addition, Reg1 regulates cell growth that is required for the maintenance of the villous structure of the small intestine (Ose et al. 2007). In rat regenerating liver, after 2-acetylaminofluorene administration and subject to 70 % partial hepatectomy (2-AAF/PH), Reg1 was significantly induced, with increasing formation of bile ductules (Wilding Crawford et al. 2008). It suggested that Reg1 is closely related to the cell regeneration in the liver through activation of the stem cell compartment. Hence, the level of Reg1 expression is closely associated with the regeneration of the small intestine and liver.

Reg2

In addition to the pancreas, Reg2 is normally expressed in the mouse liver, duodenum, small intestine, and colon (Unno et al. 1993; Perfetti et al. 1996b). There have been different opinions on the precise cellular source of Reg2 in the pancreas. Sanchez et al. reported that the expression of Reg2 mRNA and protein was restricted to the exocrine tissue regardless of the age and the presence of insulitis and/or diabetes (Sanchez et al. 2000); we only detected Reg2 immunostaining in the peri-islet acinar cells of normal mice (Luo et al. 2013; Spak et al. 2010) using a specific antibody from R&D System. However, Gurr et al. demonstrated expression of Reg2 in the endocrine cells of NOD mice (Gurr et al. 2007).

There have been numerous reports on the induction of Reg2 gene as part of islet protection or regeneration. During islet regeneration, 5 days after 50 % pancreatectomy, Reg2, Reg3β, and Reg3γ were the most abundantly induced (>10-fold) transcripts in the pancreas (Rankin and Kushner 2010). Exendin-4 and INGAP-P, which stimulate β-cell replication and/or neogenesis, increase insulin production and partially reverse insulitis in diabetic mice also increased Reg2 gene expression. These evidences suggest its role in islet survival and/or regeneration (Huszarik et al. 2010). In the NOD mice, Reg2 expression was increased more compared to Reg1, irrespective of sex or state of the diabetes, suggesting a possible difference in the physiological functions of the two proteins (Baeza et al. 1997). Following mycobacterial adjuvant treatment that made a partial recovery of β-cell mass, Reg2 expression was significantly increased, which correlated with an increase in the number of newly formed small islets and improved glucose tolerance in NOD and in streptozotocin-induced diabetic mice (Huszarik et al. 2010). Similar to Reg3β, Reg2 was suggested to be a β-cell-derived autoantigen in NOD mice since vaccination with the C-terminal fragment of Reg2 delayed the onset of T1D (Gurr et al. 2007). We have reported that Reg2 overexpression protected MIN6 insulinoma cells from streptozotocin-induced mitochondrial disruption and cell apoptosis, by attenuating the activation of caspase-3 and cleavage of PARP. These changes correlated with a persistent suppression of JNK phosphorylation by streptozotocin and a clear reversal by Reg2. These data demonstrate that Reg2 protects insulin-producing cells against streptozotocin-induced apoptosis by interfering with its cytotoxic signaling upstream of the intrinsic pro-apoptotic events by preventing its ability to inactivate JNK (Liu et al. 2010). Our recent data suggested Reg2, as well as Reg3β, can be activated by glucocorticoids and IL-6 in pancreatic acinar and islet cells and may serve a role in response to inflammation during pancreatitis (Luo et al. 2013). The direct effect of Reg2 on islet regeneration has not been tested.

Reg3α [hReg3G]

Different from other Reg family members, the Reg3 proteins (Reg3a, Reg3b, Reg3g, and Reg3δ) are characterized by the extra five amino acids close to the C-terminus in their primary structure. Reg3α belongs to a subfamily with pancreatitis-associated protein (PAP/PAP I/Reg3β/peptide 23/HIP), which was first discovered in pancreatic juice and homogenate of rat pancreatitis in 1984, but not normal pancreas, and differed from Reg1 (pancreatic stone protein, PTP) (Keim et al. 1984; Closa et al. 2007). Because of 74 % amino acid identity with Reg3β, rat Reg3α (PAP II) was first discovered in 1993 (Frigerio et al. 1993a); the human homology is Reg3G/RegIII/PAPIB which shares 66 % identity to mouse Reg3α in protein sequence, while shares 65 % identity to another mouse homology, Reg3γ (Lee et al. 2012; Nata et al. 2004; Lasserre et al. 1992; Laurine et al. 2005). Normally in rats it is hardly detectable but can be drastically induced after pregnancy and pancreatitis, indicating its involvement in functional adaption (Frigerio et al. 1993a; Bimmler et al. 2004; Honda et al. 2002). In the endocrine pancreas, it was reported to be expressed by mouse islet α-cells (Gurr et al. 2007). Reg3α protein was also detected in the small intestine, the proximal colon, and the pancreatic primordium (Hervieu et al. 2006).

Some indirect evidences support its effects on promoting islet proliferation. Acinar-specific overexpression of gastrin/CCK2 receptor induced carcinogenesis and the level of Reg3α/RegIII protein, which is proposed to be involved in the adaptive and regenerative responses of the endocrine tissues (Gigoux et al. 2008). C-Myc is a potent driver of β-cell proliferation (Pelengaris et al. 2004); activation of Myc transcription factor in mouse islets significantly promoted cell cycle progression, with steady induction of the mRNAs of Reg2, Reg3α, Reg3β, and Reg3γ up to twofold within 24 h, supporting their role in cell cycle progression or islet cell transformation. In NOD mice, pancreatic Reg3α mRNA level (but not other isoforms) was increased 1.5-fold after onset of insulitis leading to T1D (GEO profile ID 15736194, 15738507). Among the changes brought by diabetes, hyperglycemia seemed to have specific effect on Reg3α. In primary rat islets, increasing glucose concentrations from 2 to 10 mM caused no change in its expression. However, further increase to 30 mM which is known to be detrimental or proliferative, significantly doubled Reg3α mRNA level (GEO profile ID 59898598). This response was isoform specific and did not occur to other Reg proteins.

Reg3α overexpression in vitro: in order to explore whether Reg3α can directly stimulate islet β-cell replication, similar to Reg1 (Takasawa et al. 2006), we overexpressed its cDNA in stably transfected MIN6 cells (Cui et al. 2009). Using real-time PCR and Western blots, Reg3α expression was barely detectable in vector-transfected cells; in contrast, the levels of its mRNA and protein in pcDNA-Reg3α-expressing clones were increased 10- and 6-fold, respectively. Western blots also revealed Reg3α protein being released into the culture medium, which is consistent with its detection in patients' serum (Astorri et al. 2010) and supports its endocrine action. In MTT cell viability assay, Reg3α-overexpression caused ~2-fold higher rate of growth vs. vector-transfected cells. In order to investigate possible intracellular mechanisms, we detected an average 1.8-fold increase in Akt phosphorylation and 2.2- and 2.5-fold increase in the levels of cyclin D1 and cdk-4 in these cells vs. vector-transfected cells. These effects were not revealed when Reg2 or Reg3β gene was transfected, indicating isoform specificity. It is well established that β-cell replication is associated with increased cyclin D1 and cdk-4 levels (Cozar-Castellano et al. 2004), deficiency in cdk-4 or cyclin D2 results in β-cell loss and diabetes (Rane et al. 1999; Georgia and Bhushan 2004), and both Reg1 and Reg3δ cause PI3K-mediated increases in cyclin D1 and cdk-4 levels (Takasawa et al. 2006; Jamal et al. 2005; Sherr 2001; Diehl et al. 1998; Rane and Reddy 2000). Our result thus suggests that Reg3α stimulate β-cell replication, by activating Akt kinase and increasing the levels of cyclin D1/cdk-4 (Cui et al. 2009). The identical effect of Rat Reg3α in proliferation has since been confirmed in human pancreatic carcinoma Panc-1 cells, with increased expression of the islet transcription factors NeuroD, Nkx6.1, and Pax6 (Choi et al. 2010).

Reg3β (hReg3A)

In 1986, another Reg protein was found in the rat and named pancreatitis-associated protein (PAP) due to its induction during experimental pancreatitis and correlation with the severity of pancreatitis (Keim and Loffler 1986). In fact, Peptide-23 gene was first detected in primary cultures of rat pituitary and was then proved to be identical with Reg3β (Chakraborty et al. 1995). Pancreatic expression of Reg3β was strong in glucagon-producing islet cells and acinar cells close to the islets, but quite low in other endocrine cells located in the center of islets or other cells located in their outer rim (Hervieu et al. 2006; Baeza et al. 2001). This suggests Reg3β may act as a paracrine modulator of β-cell function. Reg3β is also expressed in the ileum and to a lesser extent in the jejunum and duodenum (Waelput et al. 2000). Surprisingly, its homology in human is hReg3A/HIP, with 70 % identities. In a panel of 36 adult human tissues, human Reg3A was only expressed in the pancreas and small intestines, with 100-fold higher levels than others (profile ID 10124642, Gene Expression Omnibus (GEO), www.ncbi.nlm.nih.gov/geo).

Pathophysiological roles of Reg3β in diabetes have been proposed. Reg3β protein can be released into the culture medium from primary islets taken from a T1D patient, and this release was stimulated by IL-6, indicating that Reg3β is involved in a local inflammatory response in diabetic islets (Gurr et al. 2002). The pancreatic mRNA level of Reg3β in NOD mice was significantly higher than in control IOPS-OF1 mice (Baeza et al. 2001). And the protein was also expressed in the islets and ductal epithelium in pancreata of prediabetic and diabetic NOD mice, in contrast to its restricted expression in acinar cells and peri-islet cells in nondiabetic controls. The lymphocytes from islets infiltrates and pancreatic lymph nodes of 7-week-old NOD mice showed a strong proliferative response to Reg3β, suggesting a possible role as an autoantigen (Gurr et al. 2002).

In order to test whether it can promote islet cell growth or survival against experimental damage, we established a pancreatic islet-specific overexpression of Reg3β mouse model using rat insulin I promoter and evaluated the changes in normal islet function, gene expression profiles, and the response to streptozotocin-induced diabetes. Significant and specific overexpression of Reg3β was achieved in the pancreatic islets of RIP/Reg3β mice, which exhibited normal islet histology, β-cell mass, and insulin secretion in response to high glucose yet were slightly hyperglycemic and low in islet GLUT2 level. Upon streptozotocin treatment, in contrast to wild-type littermates that became hyperglycemic in 3 d and lost 15 % weight, RIP/Reg3β mice were significantly protected from hyperglycemia and weight loss. To identify specific targets affected by Reg3β overexpression, cDNA microarray on islet RNA isolated from the transgenic mice revealed that more than 45 genes were either up- or downregulated significantly. Among them, islet-protective osteopontin/SPP1 and acute responsive nuclear protein p8/NUPR1 were significantly induced. These results were further confirmed by real-time PCR, Western blots, and immunohistochemistry (Xiong et al. 2011). This suggests that, compared to the regenerating effects of other Reg proteins on islets, Reg3β is more likely an islet-protective factor in response to stress and inflammation.

As regards the effect of human Reg3A in pancreatic regeneration, it is still being debated whether the regeneration of pancreatic β-cells is achieved by self-replication (Meier et al. 2008; Teta et al. 2007; Dor et al. 2004), neogenesis, or both (Seaberg et al. 2004). Based on rodent experimental evidence, the adult pancreas may harbor a small progenitor population, perhaps resident among centroacinar or ductal cells, which can be activated by injury and inflammation and give rise to new islet cells (Reichert and Rustgi 2011). Transcription factors and extracellular regulators control this process of islet neogenesis. Direct evidence was demonstrated by a 5-d injection of a 15-aa peptide based on human Reg3A (Human proIslet Peptide: IGLHDPTQGTEPNGE). It increased the volume of small extra-islets, insulin-positive clusters 1.5-fold in NMRI mouse pancreas and showed a tendency of increased Ngn3 and Nkx6.1 expression in IHC (Kapur et al. 2012). This was the first report to indicate that human Reg3A peptide has the bioactivity in vivo to promote new islet formation by elevation of transcription factors. Earlier, a similar 16-aa peptide based on Reg3A (WIGLHDPTQGTEPNGE) prevented streptozotocin diabetes by increasing the islet cell mass in mice (Levetan et al. 2008, 2010).

Reg3β also seems to be a tumor promoter as its deficiency caused increased tumor cell apoptosis, decreased tumor growth, and impaired angiogenesis in pancreatic cancer (Gironella et al. 2013). Moreover, Reg3β is a crucial mitogenic and antiapoptotic factor for the liver as its knockout caused cellular apoptosis and impaired liver regeneration (Lieu et al. 2006) (mislabeled as Reg2 in this publication; see our editorial correspondence (Liu and Cui 2007)). The mechanisms of Reg3β-mediated antiapoptotic and anti-inflammatory effects were further explored. Reg3β knockout mice displayed increased apoptosis in acinar cells in pancreatitis, as shown by elevated levels of caspase-3 and cleaved PARP; pretreatment with Reg3β protein reversed those effects and protected the pancreas (Gironella et al. 2007). This knockout pancreas also showed more neutrophil infiltration and higher levels of inflammatory cytokines, including TNF-α, IL-6, and IL-1β; both parameters were significantly reduced when the mice were pretreated with Reg3β, further supporting an anti-inflammatory role (Gironella et al. 2007). Conversely, antisense knockdown and antibody neutralization of Reg3β worsened the symptoms of pancreatitis induced by sodium taurocholate in rats (Viterbo et al. 2009).

Thus, Reg3β is antiapoptotic by inhibiting caspase-3 and PARP activation and anti-inflammatory by controlling cytokine output. Transcription factors of NF-κB family are crucial in controlling the inflammatory responses and cell survival. Pancreatic-specific deletion of RelA/p65 and thus NF-*k*B signaling abolished Reg3β induction and resulted in more severe pancreatitis, suggesting that Reg3β protects the pancreatic acinar cells (Algul et al. 2007). So far other Reg proteins, including Reg2 and Reg3β on pancreatic islets, have not been evaluated using similar knockout approaches. The mechanism of Reg3β action in pancreatic acinar cells (which seems distinct from that of Reg1/INGAP) involves MAPK (Ferrés-Masó et al. 2009) and/or cytokine receptor-mediated activation of JAK and STAT family of transcription factors. In acinar cell line AR42J, Reg3β activated JAK and caused the phosphorylation and nuclear translocation of STAT3 and the induction of suppressor of cytokine signaling 3 (SOCS3). Meanwhile, it was shown to induce JAK-dependent NF-*k*B inhibition, pointing to a cross talk between JAK/STAT and NF-*k*B signaling pathways (Folch-Puy et al. 2006).

Reg3δ (INGAP)

Reg3δ was first identified and purified in hamsters after partial obstruction of the pancreatic duct in 1997 (Rafaeloff et al. 1997). It was originally known as islet neogenesis-associated protein (INGAP) since it was identified as the local pancreatic factor that reversed streptozotocin-induced diabetes presumably by the induction of islet neogenesis. Normally, it is expressed in the stomach and duodenum and in the glucagon-producing islets and pancreatic ductal cells (Abe et al. 2000; Borelli et al. 2005; Taylor-Fishwick et al. 2008). The expression level of INGAP increased significantly in acinar cells of cellophane-wrapped pancreata, but not in pancreatic islet cells (Rafaeloff et al. 1997). Pancreatic transcription factors, such as Pdx1, Ngn3, NeuroD, and Isl-1, can directly activate the INGAP promoter individually or in combination (Hamblet et al. 2008).

To demonstrate a direct effect, the administration of 15-amino-acid INGAP-P to nondiabetic and streptozotocin-induced diabetic mice or rats caused increased islet cell number and mass and new islet formation (small foci of islet-like cells budding from intralobular and terminal ductules), thus capable of hyperglycemia reversal in diabetic animals (Rosenberg et al. 2004; Lipsett et al. 2007a). Intramuscular injection of INGAP stimulated islet neogenesis in healthy dogs (Pittenger et al. 2007). Petropavlovskaia M et al. then explored the mechanisms of INGAP on the proliferation of RINm5F cells (Petropavlovskaia et al. 2012). Both the recombinant protein and INGAP-P stimulated cell regeneration via binding to Gi protein-coupled receptor and activating the Ras/Raf/Erk (Petropavlovskaia et al. 2012) and PI3K/Akt pathways. Ex vivo studies showed that INGAP-P peptide could enhance glucose- and amino acid-stimulated insulin secretion from both adult and neonatal rat islets without affecting the islet survival rate or the relative proportion of the islet cells. A significant increase in β-cell size was observed in the cultured islets in the presence of INGAP-P peptide compared to controls (Borelli et al. 2005). A microarray analysis of INGAP-P peptide-treated rat neonatal islets shows many genes that are upregulated, especially those related to islet metabolism, insulin secretion, β-cell mass, and islet neogenesis. They include hepatocyte nuclear factor 3β (Hnf3β), upstream stimulatory factor 1 (Usf1), K^+ channel proteins (Sur1 and Kir6.2), Ins1, glucagon, MAPK1, Snap-25 that may regulate insulin exocytosis, and Pdx1 (Barbosa et al. 2006).

Overexpression of INGAP in pancreatic acinar cells also caused a significant increase in both the β-cell mass and pancreatic insulin content, which was mainly contributed by increased number of small islets. These mice were resistant to β-cell destruction, hyperglycemia, or hypoinsulinemia following streptozotocin treatment and had a markedly preserved islet structure (Taylor-Fishwick et al. 2006a). Meanwhile, targeted expression of INGAP to pancreatic β-cells (IP-INGAP) in mice enhanced glucose tolerance and significantly delayed the development of hyperglycemia caused by streptozotocin. Isolated islets from these INGAP-overexpressing mice displayed increased insulin release in response to glucose stimulation in the presence of streptozotocin. This is partially due to a decreased induction of apoptosis and oxidative stress in the islets of the transgenic mice, indicated by lower levels of caspase 3 and NADPH oxidase-1 (NOX1), respectively (Hashimoto et al. 2006).

The application of INGAP peptide in diabetic intervention has been explored in a report on the 65th Scientific Sessions of the American Diabetes Association (2005). Ratner RE et al. reported a double-blind, placebo-controlled trial on INGAP peptide therapy which induced islet neogenesis and improved insulin secretion both in T1D and T1D patients, which suggests that INGAP can be used as an effective therapy alone or in combination with other antidiabetic drugs. The detailed experiment was further published, indicating INGAP peptide (600 mg/d) increased C-peptide secretion in T1D and reduced HbA1c levels in T2D patients (Dungan et al. 2009). However, similar result has not been peer-reviewed and reported.

Antibacterial Reg3γ

Consistent to the role of most Reg proteins, Reg3γ expression was activated in the skeletal muscles and innervating nerves after different models of tissue injury in rat (Klasan et al. 2013). In the intestines, Reg3γ is produced together with lysozyme and cryptdin by Paneth cells, which constitute innate intestinal immunity. In the ileal tissue after 2-day food deprivation, Reg3γ mRNA level was decreased significantly, and the decrease in protein content and localization were confirmed by Western blot and immunohistochemistry. This decrease was associated with increased bacterial translocation into the mesenteric lymph nodes (Hodin et al. 2011). Similar to Reg3β, Reg3γ secreted by specialized epithelial cells is involved in limiting the epithelial contact with bacteria in the small intestine. In Reg3γ-deficient mice, more bacterial reach the small intestinal epithelium; Reg3γ specifically affects Gram-positive bacteria (Johansson and Hansson 2011; Vaishnava et al. 2011). But so far, there is no direct evidence indicating the involvement of Reg3γ in islet cells proliferation.

Reg4

Reg4 is a distinct isoform in the family, starting from its structure and chromosomal location to its high expression in colon cancer. Human Reg4, also named RELP (regenerating protein-like protein), was discovered from a high-throughput screening of inflammatory large bowel disease library. The Reg4 cDNA consists of 7 exons rather than the 6 found in other Reg family genes (Kamarainen et al. 2003). The protein differs in the five-amino-acid insertion (P-N/D-G-E/D-G) present in Reg3 proteins and the six residues (S/A-Q-T-E-L-P) near the N-terminus found in all other Reg family proteins. Its chromosome location also differs from the rest, e.g., chromosome 1 vs. 2p12 in the human and chromosome 3 vs. 6C in the mouse. Reg4 is normally expressed in the prostate, testes, stomach, duodenum, jejunum, ileum, and colon (Hartupee et al. 2001; Oue et al. 2005). Cells positive for Reg4 are mostly enteroendocrine and mucin-producing goblet cells, e.g., immunofluorescent dual labeling demonstrated its colocalization with chromogranin A in the neuroendocrine cells of the duodenal epithelium (Violette et al. 2003).

Reg4 expression in normal islets has not been determined. In a single report, positive Reg4 staining was shown in mouse insulin-producing islet cells (Oue et al. 2005), but we could not confirm it in normal or malignant pancreas (data not shown), nor was the Human Protein Atlas (www.proteinatlas.org). In islet-derived neuroendocrine tumor, Reg4 is totally negative, while a distinct positive staining for Hath1 (helix-loop-helix transcription factor, regulating differentiation of neural and intestinal secretory cells) was found in peri-islet cells. The same pattern was followed in liver and lymph node metastases from islets (Heiskala et al. 2010). So far, there is no evidence showing that Reg4 is related to cell proliferation of the pancreatic islets.

Reg4 expression is increased in various types of human diseases, especially cancer. In the Atlas, Reg4 is clearly demonstrated in the glandular cells of the small intestines, colon, and rectum. The protein is strongly expressed in the cryptal epithelium of ulcerative colitis, and to a lesser extent in the parietal cells of the gastric corpus mucosa. It was also upregulated in the goblet cells of the glands representing intestinal metaplasia in the esophagus and the gastric antrum (Kamarainen et al. 2003). Reg4 overexpression in human gastric cancer cells caused an increased number and size of tumors and worsened survival of nude mice with peritoneal metastasis, whereas Reg4 knockdown improved survival (Heald et al. 2006). Consequently, Reg4 is highly expressed in drug-resistant colon cancers vs. drug-sensitive ones. Compared to low or no expression of other Reg genes (Reg1A, Reg1B, Reg3), Reg4 is expressed in 71 % colorectal tumors (Violette et al. 2003; Zhang et al. 2003). Reg4 positive colorectal cancer patients have a significantly worse prognosis than those negative for it (Maake and Reinecke 1993). Reg4 also plays important roles in tumor progression and deterioration of prostate cancer (Ohara et al. 2008). High serum Reg4 level in patients with pancreatic cancer was associated with poor response to chemoradiotherapy (Quaife et al. 1989). Applications of specific Reg4 antibodies or small interfering RNAs against Reg4 resulted in increased cell apoptosis and decreased proliferation, leading to decreased tumor growth and increased host animal survival (Bishnupuri et al. 2010). These data indicate that Reg4 may play a role in tumor formation, diagnosis, and/or treatment. More discussion on pancreatic tumors can be found in the chapter entitled "▶ Pancreatic Neuroendocrine Tumors."

Possible Role of Reg Proteins on β-Cell Neogenesis

The stem cells have the ability of differentiating from one precursor cell to multiple specialized terminal types, whereas progenitor cells can only be differentiated into their (one and only) direct targets. In adult pancreas, β-cells preserve a limited ability to replicate and generate new cells under stress or injury. Therefore, identifying those progenitor cells and factors influencing their differentiating outcome represents a promising avenue toward rescuing diabetes. Recent studies revealed that progenitor cells located in the ductal epithelium can be induced to express Ngn3 and become new β-cells after partial duct ligation in mouse pancreas (Xu et al. 2008).

Reg1 expression was detected in mouse embryonic stem cells (ESCs) and can be activated by Wnt/β-catenin signaling pathway; the latter is important for the maintenance of ESC in an undifferentiated state (Parikh et al. 2012). It indicates that Reg1 might play an important role during embryonic development. Attempts have been made to assess the effect of Reg1 protein on ESC differentiation by adding recombinant protein and overexpressing Reg1 gene. Unfortunately, no significant effect was observed in cell growth compared with those untreated. Nevertheless, the potential effect of Reg1 on stem cell proliferation should be reexamined. If it has protective and/or proliferative effects on stem cells, it can be used along with other transcriptional factors to facilitate the differentiation into β-cells.

In β-cell differentiation, INGAP seems to play a more active role than other Reg family proteins. INGAP promoter activations driven by pancreatic protein 1 (Pan-1), phorbol myristate acetate (PMA), or leukemia inhibitory factor (LIF) can be inhibited by Pdx1 through direct promoter binding (Taylor-Fishwick et al. 2006b). Further studies revealed that the repressing effect of Pdx1 is dependent on its interaction with Pan-1/NeuroD (Taylor-Fishwick et al. 2010); Pdx1 may be part of a negative feedback mechanism to control islet expansion. Similar to Reg1, INGAP-positive cells were also present in mouse embryonic pancreatic buds (Hamblet et al. 2008). Based on that INGAP and some growth factors are essential for islet development, a strategy was proposed by using a combination of INGAP and growth factors including EGF and GLP-1 to cause the expansion of embryonic stem cells and pancreatic progenitor cells. Moreover, administration of INGAP peptide intraperitoneally to hamsters, which received multiple low doses of streptozotocin, stimulated the growth of new endocrine cells with mature islets appearance. The INGAP peptide can also normalize blood glucose and insulin levels; the mechanism of this protective effect seems to include increased expression of Pdx1 in ductal and islet cells (Rosenberg et al. 2004). As for other members of Reg family, Reg2 expression was highly induced in HFD-fed mice or after 70 % pancreatectomy (16, 139, 171); Reg1 and Reg3β exhibited similar changes. Whether they can be considered for islet neogenesis or differentiation needs to be assessed.

Transdifferentiation of other endogenous pancreatic cells into β-cells is now considered as another mechanism of islet regeneration. In response to appropriate stimuli, ductal epithelial and acinar cells can be induced to become regenerating islet cells (Pittenger et al. 2009a). Both direct acinar/ductal to islet transformation and indirect acinar to ductal and then to islet transformation have been proposed to increase β-cell numbers and mass in vitro (Bonner-Weir et al. 2008; Rooman et al. 2002; Schmied et al. 2001). In the early stages of streptozotocin-induced diabetes, both Reg1 expression and BrdU incorporation can be induced in residual β-cells, indicating a role in β-cell proliferation (Anastasi et al. 1999). In the meantime, co-expression of Reg1 and cytokeratin 19 in acini-ductal cells suggests that Reg1 may participate in the transdifferentiation of acinar and/or ductal cells to islet cells (Tezel et al. 2004).

Compared to other isoforms, the role of INGAP in islet transdifferentiation is better established. In acinar-derived, duct-like cells cultured with a mixture of gastrin, HGF, and INGAP, the mass of islet-like clusters was significantly increased (Lipsett et al. 2007b). Several transcriptional factors, which were crucial for β-cell differentiation, were introduced in order to promote the transdifferentiation of ductal cells into islet-like clusters (ILCs) (Li et al. 2009). A 4-step strategy using nicotinamide, exendin-4, TGF-β, and INGAP peptide was tested. The mass of ILCs was much larger in the group treated with INGAP than that with a scrambled peptide, with increased expression of Pdx1, insulin, and glucagon. Furthermore, the ILCs with INGAP treatment secreted higher levels of insulin and C-peptide during the differentiation process, illustrating a gain in the secreting capacity. INGAP or INGAP peptide can stimulate the proliferation of ductal cells, thereby maintaining a pool of possible precursors of islet cells. In another study, short-term incubation of the primitive duct-like structures derived from quiescent adult human islets with INGAP peptide induced a re-differentiation back to islet-like structures. Those newly generated islets resembled freshly isolated islets with respect to the number and topological arrangement of cell types within an islet and the capacity of glucose-stimulated insulin secretion. Furthermore, these islet-like structures also express the islet-specific transcription factors Pdx1, ISL-1, and Nkx-2.2 and the islet-specific proteins GLUT2 and C-peptide to levels comparable to freshly isolated islets (Rafaeloff et al. 1997). Thus, limited in vitro evidence supports the role of INGAP and Reg1 not only in β-cell proliferation but, more interestingly, also in new β-cell formation, especially through a process of transdifferentiation.

General and Isoform-Specific Functions of Reg Proteins

Different isoforms also share some similar biological effects in the pancreatic development and pathophysiological conditions due to structural resemblances. Both Reg1 and INGAP contribute to pancreatic cell protection against apoptosis during oxidative stress. The level of Reg1 and INGAP transcripts was increased significantly in hamsters with chronic diabetes induced by streptozotocin and was slightly decreased upon administration of the antioxidant probucol. The latter caused an increase in number of insulin-positive cells in the pancreata of diabetic hamsters (Takatori et al. 2003). As well, both Reg1 and INGAP were suggested to play a role in the differentiation of stem cells into the islet cells and transdifferentiating endogenous pancreatic cells to islet cells.

However, members of the Reg family also possess isoform-specific properties. For example, several Reg proteins can be induced in the pancreas in response to injury, such as pancreatectomy or caerulein-induced acute pancreatitis, in an isoform-specific fashion (Graf et al. 2002; De Leon et al. 2006). Upon caerulein or sodium taurocholate treatment, the level of Reg3 proteins, especially Reg3β, showed an acute increase in the first day after the induction of pancreatitis that rapidly returned to baseline, while Reg1 exhibited a persistent elevation and did not return to normal level after 35 days of caerulein injection. Post-injury level of Reg3β, but not Reg1, showed the highest induction and was significantly correlated with the severity of the pancreatic injury and the mortality rate (Graf et al. 2002; Zenilman et al. 2000). Administration of anti-Reg1 and/or anti-Reg3α antibody to rats with established pancreatitis increased pancreatic wet weight, indicating worsened tissue inflammation and cell necrosis (Viterbo et al. 2009). Thus, endogenous Reg1 and Reg3α, but not Reg3β, seem to be protective against the onset of acute pancreatitis.

Both the synthesis and secretion of Reg1 and Reg3 proteins (Reg3α, Reg3β, Reg3δ) were increased after 9 weeks of age in the chronic pancreatic WBN/Kob rat and peaked at 6 months. Elevations of these proteins correlated with the disease progression and coincided with increased cell apoptosis and tissue fibrosis (Bimmler et al. 2004). Using immunoreactivity, both Reg1 and Reg3 proteins were increased and colocalized to the same areas of pancreatic acinar cells displaying active inflammation and fibrosis. The immunogold technique revealed the intracellular localization of Reg1 and Reg3 proteins in the secretory apparatus. However, despite the anti-inflammatory and antiapoptotic effect of Reg3 proteins, the acinar damage in the WBN/Kob rat might be attributed more to Reg1, as supported by the elongated structures with fibrillar contents formed from fusion of Reg1-positive zymogen granules (Meili et al. 2003).

In a subset of gastric cancers, both Reg1A and Reg4 were overexpressed, suggesting their involvement in gastric carcinogenesis. Reg1A expression was closely related to the venous invasion and tumor stage, whereas Reg4 showed no such clear relationship (Yamagishi et al. 2009). Although several Reg proteins have been demonstrated to promote islet cell growth, survival, and/or function, there are signs of isoform specificities. Their pattern of activation in the rat intestine following antidiabetic duodenal-jejunal bypass was isoform and segment specific (Li et al. 2013b). For isoforms of Reg3, we and others have found that Reg3α promoted cell proliferation in vitro (Cui et al. 2009); overexpressed Reg3β and Reg3δ (INGAP) protected the β-cells in vivo against streptozotocin-induced diabetes (Taylor-Fishwick et al. 2006a; Xiong et al. 2011; Chang et al. 2011); although the expression of Reg3γ in regenerating pancreas was significantly induced, its direct effect on β-cells has not been studied (Rankin and Kushner 2010; De Leon et al. 2006).

Molecular Factors Regulating the Expression of Reg Family Genes

The molecular mechanisms regulating Reg protein expression are not fully understood yet. Regulating factors and their interactions may differ depending on the tissue and specific isoform under study. Reg proteins are induced by several inflammatory cytokines, including IL-1β, IL-6, and TNF-α (Dusetti et al. 1995). Reg1 mRNA expression is induced by the combination of IL-6 and dexamethasone, but not by the treatment with an individual pro- or anti-inflammatory factor, such as IL-1β, TNF-α, IFNγ, dexamethasone, or IL-6. Induction of Reg1 mRNA levels by the combined IL-6/dexamethasone is further enhanced by the addition of nicotinamide or 3-aminobenzamide, the inhibitors for PARP which normally binds to a 12-bp cis element on Reg1 promoter (TGCCCCTCCCAT) and inhibits the formation of protein/DNA complex (Akiyama et al. 2001). Thus, PARP is a negative regulator of Reg1 transcription in the β-cells. The highly conserved element at -81-bp/-70-bp region of the Reg1 promoter has been proven essential for the activations by both IL-6/dexamethasone and IL-6/dexamethasone/nicotinamide treatments (Akiyama et al. 2001).

Both human Reg1A and Reg3A proteins are downstream targets of the Wnt/β-catenin pathway during liver tumorigenesis (Cavard et al. 2006). Upregulation of Reg1A and Reg3A gene expression was confirmed in the liver by Northern blot analysis and immunohistochemistry. In the adenoma and hepatocellular carcinoma, a strong immunoreactivity was detected for Reg3A and a less pronounced signal for Reg1A. Using the Huh7 hepatoma cell line, the Reg3A gene was upregulated upon the activation of Wnt/β-catenin signaling by stabilizing β-catenin in its unphosphorylated form with lithium chloride (LiCl). This induction was abolished by inhibition of β-catenin signaling with siRNA (Cavard et al. 2006). The overexpression of Reg1A and Reg3A and the activation of the Wnt/β-catenin pathway were also detected in colon adenoma from familial adenomatous polyposis, but not in the pediatric liver tumor hepatoblastoma (Cavard et al. 2006).

Interestingly, Reg3β gene expression can be induced during both TNF-α induced cell apoptosis and Cdx1-induced cell proliferation. Long-term incubation of acinar AR42J cells with TNF-α induced cell apoptosis. In the meantime, Reg3β gene expression was induced by a MAPK/MEK1-mediated pathway which antagonized TNF-α-induced cell death (Harrison et al. 1998). Reg3β prevents the activation of macrophages by TNF-α, probably through inhibition of the nuclear translocation of NF-κB (Vasseur et al. 2004). In intestinal epithelial cells, Reg3β gene expression was increased at both mRNA and protein levels by Cdx1 overexpression. Cdx1 may directly activate the Reg3β promoter through a putative Cdx1-response element. Overexpression of the Reg3β gene or in vitro administration of Reg3β protein significantly increased the proliferation of intestinal cells (Zhao et al. 1997).

In addition, IL-22 also caused a robust induction of Reg3β mRNA in an acinar cell line, but not in a β-cell line, presumably through activation of STAT3 and resulting changes in gene transcription. Ex vivo incubation of isolated acinar cells with IL-22 also induced a substantial upregulation of Reg3β mRNA and a modest effect on the expression of osteopontin (OPN), a proliferative glycoprotein that can be induced by Reg3β itself (Xiong et al. 2011). Deficiency of the IL-22 receptor in mice resulted in loss of Reg3β mRNA upregulation upon IL-22 injection (Aggarwal et al. 2001). The activation of toll-like receptor 2 (TLR2) is required for *Yersinia*-induced expression of Reg3β and the subsequent clearance of the bacterial load in Peyer's patches or aggregated lymphoid nodules (Dessein et al. 2009). Reg3β can also increase its own transcription by increasing the binding of the nuclear factors C/EBPβ, P-CREB, P-ELK1, EGR1, STAT3, and ETS2 to its own promoters. This self-activation seems to involve MAPK signaling and the activation of p44/p42, p38, and JNK (Ferrés-Masó et al. 2009).

The implantation of Reg3β-expressing hepatocytes into SCID mice enhanced liver regeneration following hepatectomy, probably by modulating the effects of TNF-α, IL-6, and STAT3, thus shortening the cell cycles of hepatocytes (Lieu et al. 2005). In Reg3β-deficient mice, there was a delayed ERK and AKT signaling but persistent activation of TNF-α/IL-6/STAT3 pathway during liver regeneration following hepatectomy, which may induce delayed liver regeneration and persistently inflammatory condition (Lieu et al. 2006).

Perspective

After two decades of research, more Reg family proteins have been characterized with relevance to the function of endocrine and exocrine pancreas. We remain hopeful that some isoforms of Reg proteins under certain circumstances will be proved to promote islet cell growth, regeneration, or survival against harsh conditions. There are plenty of indirect evidences showing associations of the changes in Reg proteins with islet regeneration or protection; more transgenic and knockout mice are being developed; however, in vitro direct incubations with the islet cells and recombinant proteins need to be carefully performed. We hope one day to demonstrate that Reg proteins are secreted extracellularly and act on cell membrane receptor(s); however, the only evidence of Reg1 receptor is still thin. The classification of Reg family proteins in Table 1 and our preliminary analysis on their regulation, effects, and isoform specificity should help to expand our research toward establishing the biological relevance of Reg proteins.

Acknowledgments Our research activity was supported by the Canadian Diabetes Association (OG-3-11-3469-JL) and the China Scholarship Council (201208370055).

Cross-References

► Mechanisms of Pancreatic β-Cell Apoptosis in Diabetes and Its Therapies
► Pancreatic Neuroendocrine Tumors
► Stem Cells in pancreatic Islets

References

Abe M et al (2000) Identification of a novel Reg family gene, Reg IIIdelta, and mapping of all three types of Reg family gene in a 75 kilobase mouse genomic region. Gene 246 (1–2):111–122

Acquatella-Tran Van Ba I et al (2012) Regenerating islet-derived 1α (Reg-1α) protein is new neuronal secreted factor that stimulates neurite outgrowth via exostosin Tumor-like 3 (EXTL3) receptor. J Biol Chem 287(7):4726–4739

Aggarwal S et al (2001) Acinar cells of the pancreas are a target of interleukin-22. J Interferon Cytokine Res 21(12):1047–1053

Akiyama T et al (2001) Activation of Reg gene, a gene for insulin-producing β-cell regeneration: poly(ADP-ribose) polymerase binds Reg promoter and regulates the transcription by autopoly (ADP-ribosyl)ation. Proc Natl Acad Sci USA 98(1):48–53

Algul H et al (2007) Pancreas-specific RelA/p65 truncation increases susceptibility of acini to inflammation-associated cell death following cerulein pancreatitis. J Clin Invest 117 (6):1490–1501

Anastasi E et al (1999) Expression of Reg and cytokeratin 20 during ductal cell differentiation and proliferation in a mouse model of autoimmune diabetes. Eur J Endocrinol 141(6):644–652

Ashcroft FJ et al (2004) Control of expression of the lectin-like protein Reg-1 by gastrin: role of the Rho family GTPase RhoA and a C-rich promoter element. Biochem J 381(Pt 2):397–403

Astorri E et al (2010) Circulating Reg1α proteins and autoantibodies to Reg1α proteins as biomarkers of β-cell regeneration and damage in type 1 diabetes. Horm Metab Res 42 (13):955–960

Baeza N et al (1997) Specific reg II gene overexpression in the non-obese diabetic mouse pancreas during active diabetogenesis. FEBS Lett 416(3):364–368

Baeza N et al (2001) Pancreatitis-associated protein (HIP/PAP) gene expression is upregulated in NOD mice pancreas and localized in exocrine tissue during diabetes. Digestion 64(4):233–239

Barbosa H et al (2006) Islet Neogenesis Associated Protein (INGAP) modulates gene expression in cultured neonatal rat islets. Regul Pept 136(1–3):78–84

Bartoli C et al (1993) A gene homologous to the reg gene is expressed in the human pancreas. FEBS Lett 327(3):289–293

Bartoli C et al (1998) Expression of peptide-23/pancreatitis-associated protein and Reg genes in human pituitary and adenomas: comparison with other fetal and adult human tissues. J Clin Endocrinol Metab 83(11):4041–4046

Billestrup N, Nielsen JH (1991) The stimulatory effect of growth hormone, prolactin, and placental lactogen on β-cell proliferation is not mediated by insulin-like growth factor-I. Endocrinology 129(2):883–888

Bimmler D et al (1999) Regulation of PSP/reg in rat pancreas: immediate and steady-state adaptation to different diets. Pancreas 19(3):255–267

Bimmler D et al (2004) Coordinate regulation of PSP/reg and PAP isoforms as a family of secretory stress proteins in an animal model of chronic pancreatitis. J Surg Res 118 (2):122–135

Bishnupuri KS et al (2010) Reg IV regulates normal intestinal and colorectal cancer cell susceptibility to radiation-induced apoptosis. Gastroenterology 138(2):616.e1–626.e2

Bonner-Weir S (2000a) Life and death of the pancreatic β cells. Trends Endocrinol Metab 11 (9):375–378

Bonner-Weir S (2000b) Perspective: postnatal pancreatic β cell growth. Endocrinology 141 (6):1926–1929

Bonner-Weir S et al (2004) The pancreatic ductal epithelium serves as a potential pool of progenitor cells. Pediatr Diabetes 5:16–22

Bonner-Weir S et al (2008) Transdifferentiation of pancreatic ductal cells to endocrine β-cells. Biochem Soc Trans 36(Pt 3):353–356

Borelli MI et al (2005) INGAP-related pentadecapeptide: its modulatory effect upon insulin secretion. Regul Pept 131(1–3):97–102

Bouwens L et al (1994) Cytokeratins as markers of ductal cell differentiation and islet neogenesis in the neonatal rat pancreas. Diabetes 43(11):1279–1283

Cavard C et al (2006) Overexpression of regenerating islet-derived 1 α and 3 α genes in human primary liver tumors with β-catenin mutations. Oncogene 25(4):599–608

Chakraborty C et al (1995) Age-related changes in peptide-23/pancreatitis-associated protein and pancreatic stone protein/reg gene expression in the rat and regulation by growth hormone-releasing hormone. Endocrinology 136(5):1843–1849

Chang TJ et al (2011) Targeted expression of INGAP to β cells enhances glucose tolerance and confers resistance to streptozotocin-induced hyperglycemia. Mol Cell Endocrinol 335(2):104–109

Cheng CM et al (2000) Insulin-like growth factor 1 regulates developing brain glucose metabolism. Proc Natl Acad Sci USA 97(18):10236–10241

Choi JH et al (2010) Isolation of genes involved in pancreas regeneration by subtractive hybridization. Biol Chem 391(9):1019–1029

Choi SM et al (2013) Innate Stat3-mediated induction of the antimicrobial protein Reg3gamma is required for host defense against MRSA pneumonia. J Exp Med 210(3):551–561

Christa L et al (1996) HIP/PAP is an adhesive protein expressed in hepatocarcinoma, normal Paneth, and pancreatic cells. Am J Physiol 271(6 Pt 1):G993–G1002

Closa D, Motoo Y, Iovanna JL (2007) Pancreatitis-associated protein: from a lectin to an anti-inflammatory cytokine. World J Gastroenterol 13(2):170–174

Cozar-Castellano I et al (2004) Induction of β-cell proliferation and retinoblastoma protein phosphorylation in rat and human islets using adenovirus-mediated transfer of CDK4 and cyclin D1. Diabetes 53(1):149–159

Cui W et al (2009) Overexpression of Reg3α increases cell growth and the levels of cyclin D1 and CDK4 in insulinoma cells. Growth Factors 27(3):195–202

D'Amour KA et al (2006) Production of pancreatic hormone-expressing endocrine cells from human embryonic stem cells. Nat Biotechnol 24(11):1392–1401

D'Ercole AJ (1999) Actions of IGF system proteins from studies of transgenic and gene knockout models. In: Rosenfeld RG, Roberts J (eds) The IGF system: molecular biology, physiology, and clinical applications. Humana Press, Totowa, pp 545–576

De Leon DD et al (2006) Identification of transcriptional targets during pancreatic growth after partial pancreatectomy and exendin-4 treatment. Physiol Genomics 24(2):133–143

Dessein R et al (2009) Toll-like receptor 2 is critical for induction of Reg3β expression and intestinal clearance of *Yersinia pseudotuberculosis*. Gut 58(6):771–776

Dheen ST, Rajkumar K, Murphy LJ (1997) Islet cell proliferation and apoptosis in insulin-like growth factor binding protein-1 in transgenic mice. J Endocrinol 155(3):551–558

Diehl JA et al (1998) Glycogen synthase kinase-3β regulates cyclin D1 proteolysis and subcellular localization. Genes Dev 12(22):3499–3511

Dor Y et al (2004) Adult pancreatic β-cells are formed by self-duplication rather than stem-cell differentiation. Nature 429(6987):41–46

Dungan KM, Buse JB, Ratner RE (2009) Effects of therapy in type 1 and type 2 diabetes mellitus with a peptide derived from islet neogenesis associated protein (INGAP). Diabetes Metab Res Rev 25(6):558–565

Dusetti NJ et al (1994) Molecular cloning, genomic organization, and chromosomal localization of the human pancreatitis-associated protein (PAP) gene. Genomics 19(1):108–114

Dusetti NJ et al (1995) Pancreatitis-associated protein I (PAP I), an acute phase protein induced by cytokines. Identification of two functional interleukin-6 response elements in the rat PAP I promoter region. J Biol Chem 270(38):22417–22421

Ferrés-Masó M et al (2009) PAP1 signaling involves MAPK signal transduction. Cell Mol Life Sci 66(13):2195–2204

Folch-Puy E et al (2006) Pancreatitis-associated protein I suppresses NF-kappa B activation through a JAK/STAT-mediated mechanism in epithelial cells. J Immunol 176(6):3774–3779

Frigerio JM et al (1993a) Identification of a second rat pancreatitis-associated protein. Messenger RNA cloning, gene structure, and expression during acute pancreatitis. Biochemistry 32(35):9236–9241

Frigerio JM et al (1993b) The pancreatitis associated protein III (PAP III), a new member of the PAP gene family. Biochim Biophys Acta 1216(2):329–331

Fujishiro M et al (2012) Regenerating gene (REG) 1 α promotes pannus progression in patients with rheumatoid arthritis. Mod Rheumatol 22(2):228–237

Gedulin BR et al (2005) Exenatide (exendin-4) improves insulin sensitivity and β-cell mass in insulin-resistant obese fa/fa Zucker rats independent of glycemia and body weight. Endocrinology 146(4):2069–2076

George M et al (2002) β cell expression of IGF-I leads to recovery from type 1 diabetes. J Clin Invest 109(9):1153–1163

Georgia S, Bhushan A (2004) β cell replication is the primary mechanism for maintaining postnatal β cell mass. J Clin Invest 114(7):963–968

Gigoux V et al (2008) Reg genes are CCK2 receptor targets in ElasCCK2 mice pancreas. Regul Pept 146(1–3):88–98

Gironella M et al (2007) Experimental acute pancreatitis in PAP/HIP knock-out mice. Gut 56(8):1091–1097

Gironella M et al (2013) Reg3β deficiency impairs pancreatic tumor growth by skewing macrophage polarization. Cancer Res 73(18):5682–5694

Graf R et al (2001) A family of 16-kDa pancreatic secretory stress proteins form highly organized fibrillar structures upon tryptic activation. J Biol Chem 276(24):21028–21038

Graf R et al (2002) Coordinate regulation of secretory stress proteins (PSP/reg, PAP I, PAP II, and PAP III) in the rat exocrine pancreas during experimental acute pancreatitis. J Surg Res 105(2):136–144

Graf R et al (2006) Exocrine meets endocrine: pancreatic stone protein and regenerating protein–two sides of the same coin. J Surg Res 133(2):113–120

Gross DJ et al (1998) Amelioration of diabetes in nonobese diabetic mice with advanced disease by linomide-induced immunoregulation combined with Reg protein treatment. Endocrinology 139(5):2369–2374

Gu G, Brown JR, Melton DA (2003) Direct lineage tracing reveals the ontogeny of pancreatic cell fates during mouse embryogenesis. Mech Dev 120(1):35–43

Gurr W et al (2002) A Reg family protein is overexpressed in islets from a patient with new-onset type 1 diabetes and acts as T-cell autoantigen in NOD mice. Diabetes 51(2):339–346

Gurr W et al (2007) RegII is a β-cell protein and autoantigen in diabetes of NOD mice. Diabetes 56(1):34–40

Hamblet NS et al (2008) The Reg family member INGAP is a marker of endocrine patterning in the embryonic pancreas. Pancreas 36(1):1–9

Hansson A, Thoren M (1995) Activation of MAP kinase in Swiss 3T3 fibroblasts by insulin-like growth factor-I. Growth Regul 5(2):92–100

Hansson A, Hehenberger K, Thoren M (1996) Long-term treatment of Swiss 3T3 fibroblasts with dexamethasone attenuates MAP kinase activation induced by insulin-like growth factor- I (IGF-I). Cell Biochem Funct 14(2):121–129

Harbeck MC et al (1996) Expression of insulin receptor mRNA and insulin receptor substrate 1 in pancreatic islet β-cells. Diabetes 45(6):711–717

Harrison M et al (1998) Growth factor protection against cytokine-induced apoptosis in neonatal rat islets of Langerhans: role of Fas. FEBS Lett 435(2–3):207–210

Hartupee JC et al (2001) Isolation and characterization of a cDNA encoding a novel member of the human regenerating protein family: Reg IV. Biochim Biophys Acta 1518(3):287–293

Hashimoto N et al (2006) Ablation of PDK1 in pancreatic β cells induces diabetes as a result of loss of β cell mass. Nat Genet 38(5):589–593

He S-Q et al (2010) Inflammation and nerve injury induce expression of pancreatitis-associated protein-II in primary sensory neurons. Mol Pain 6:23

Heald A, Stephens R, Gibson JM (2006) The insulin-like growth factor system and diabetes–an overview. Diabet Med 23(Suppl 1):19–24

Heiskala K et al (2010) Expression of Reg IV and Hath1 in neuroendocrine neoplasms. Histol Histopathol 25(1):63–72

Hervieu V et al (2006) HIP/PAP, a member of the reg family, is expressed in glucagon-producing enteropancreatic endocrine cells and tumors. Hum Pathol 37(8):1066–1075

Hill DJ et al (1999) Insulin-like growth factors prevent cytokine-mediated cell death in isolated islets of Langerhans from pre-diabetic non-obese diabetic mice. J Endocrinol 161(1):153–165

Hodin CM et al (2011) Starvation compromises Paneth cells. Am J Pathol 179(6):2885–2893

Honda H, Nakamura H, Otsuki M (2002) The elongated PAP II/Reg III mRNA is upregulated in rat pancreas during acute experimental pancreatitis. Pancreas 25(2):192–197

Hu G et al (2011) Reg4 protects against acinar cell necrosis in experimental pancreatitis. Gut 60(6):820–828

Huszarik K et al (2010) Adjuvant immunotherapy increases β cell regenerative factor Reg2 in the pancreas of diabetic mice. J Immunol 185(9):5120–5129

Iovanna J et al (1991) Messenger RNA sequence and expression of rat pancreatitis-associated protein, a lectin-related protein overexpressed during acute experimental pancreatitis. J Biol Chem 266(36):24664–24669

Iovanna JL et al (1993) PAP, a pancreatic secretory protein induced during acute pancreatitis, is expressed in rat intestine. Am J Physiol 265(4 Pt 1):G611–G618

Itoh T, Teraoka H (1993) Cloning and tissue-specific expression of cDNAs for the human and mouse homologues of rat pancreatitis-associated protein (PAP). Biochim Biophys Acta 1172(1–2):184–186

Jamal AM et al (2005) Morphogenetic plasticity of adult human pancreatic islets of Langerhans. Cell Death Differ 12:702–712

Johansson ME, Hansson GC (2011) Microbiology. Keeping bacteria at a distance. Science 334(6053):182–183

Jung EJ, Kim CW (2002) Interaction between chicken protein tyrosine phosphatase 1 (CPTP1)-like rat protein phosphatase 1 (PTP1) and p60(v-src) in v-src-transformed Rat-1 fibroblasts. Exp Mol Med 34(6):476–480

Kamarainen M et al (2003) RELP, a novel human REG-like protein with up-regulated expression in inflammatory and metaplastic gastrointestinal mucosa. Am J Pathol 163(1):11–20

Kamimura T, West C, Beutler E (1992) Sequence of a cDNA clone encoding a rat Reg-2 protein. Gene 118(2):299–300

Kapur R et al (2012) Short-term effects of INGAP and Reg family peptides on the appearance of small β-cells clusters in non-diabetic mice. Islets 4(1):40–48

Keim V, Loffler HG (1986) Pancreatitis-associated protein in bile acid-induced pancreatitis of the rat. Clin Physiol Biochem 4(2):136–142

Keim V et al (1984) An additional secretory protein in the rat pancreas. Digestion 29(4):242–249

Kimura N et al (1992) Expression of human regenerating gene mRNA and its product in normal and neoplastic human pancreas. Cancer 70(7):1857–1863

Klasan GS et al (2013) Reg3G gene expression in regenerating skeletal muscle and corresponding nerve. Muscle Nerve 49(1):61–68

Kobayashi S et al (2000) Identification of a receptor for reg (regenerating gene) protein, a pancreatic β-cell regeneration factor. J Biol Chem 275(15):10723–10726

Konishi H et al (2013) N-terminal cleaved pancreatitis-associated protein-III (PAP-III) serves as a scaffold for neurites and promotes neurite outgrowth. J Biol Chem 288(15):10205–10213

Lai Y et al (2012) The antimicrobial protein REG3A regulates keratinocyte proliferation and differentiation after skin injury. Immunity 37(1):74–84

Lasserre C et al (1992) A novel gene (HIP) activated in human primary liver cancer. Cancer Res 52(18):5089–5095

Lasserre C et al (1994) Structural organization and chromosomal localization of a human gene (HIP/PAP) encoding a C-type lectin overexpressed in primary liver cancer. Eur J Biochem 224(1):29–38

Laurine E et al (2005) PAP IB, a new member of the Reg gene family: cloning, expression, structural properties, and evolution by gene duplication. Biochim Biophys Acta 1727(3):177–187

Lee KS et al (2012) Helicobacter pylori CagA triggers expression of the bactericidal lectin REG3gamma via gastric STAT3 activation. PLoS One 7(2):e30786

Levetan CS et al (2008) Discovery of a human peptide sequence signaling islet neogenesis. Endocr Pract 14(9):1075–1083

Levetan CS et al (2010) Human Reg3a gene protein as a novel islet neogenesis therapy for reversal of type 1 and 2 diabetes. In: American Diabetes Association, 70th scientific sessions, Orlando

Li J et al (2009) Islet neogenesis-associated protein-related pentadecapeptide enhances the differentiation of islet-like clusters from human pancreatic duct cells. Peptides 30(12):2242–2249

Li L et al (2010) PSP/reg inhibits cultured pancreatic stellate cell and regulates MMP/TIMP ratio. Eur J Clin Invest 41(2):151–158

Li B et al (2013a) Intestinal adaptation and Reg gene expression induced by antidiabetic duodenal-jejunal bypass surgery in Zucker fatty rats. Am J Physiol Gastrointest Liver Physiol 304(7):G635–G645

Li B et al (2013b) Intestinal adaptation and Reg gene expression induced by anti-diabetic duodenal-jejunal bypass surgery in Zucker fatty rats. Am J Physiol Gastrointestin Liver Physiol 304(7):G635–G645

Lieu HT et al (2005) HIP/PAP accelerates liver regeneration and protects against acetaminophen injury in mice. Hepatology 42(3):618–626

Lieu HT et al (2006) Reg2 inactivation increases sensitivity to Fas hepatotoxicity and delays liver regeneration post-hepatectomy in mice. Hepatology 44(6):1452–1464

Lipsett M et al (2007a) The role of islet neogenesis-associated protein (INGAP) in islet neogenesis. Cell Biochem Biophys 48(2–3):127–137

Lipsett MA, Castellarin ML, Rosenberg L (2007b) Acinar plasticity: development of a novel in vitro model to study human acinar-to-duct-to-islet differentiation. Pancreas 34(4):452–457

Liu JL, Cui W (2007) Which gene, Reg2 or Reg3β, was targeted that affected liver regeneration? Hepatology 45(6):1584–1585

Liu JL et al (2008) Possible roles of reg family proteins in pancreatic islet cell growth. Endocr Metab Immune Disord Drug Targets 8(1):1–10

Liu Y et al (2009) β-Cells at the crossroads: choosing between insulin granule production and proliferation. Diab Obes Metab 11:54–64

Liu L, Liu JL, Srikant CB (2010) Reg2 protects mouse insulinoma cells from streptozotocin-induced mitochondrial disruption and apoptosis. Growth Factors 28(5):370–378

Luo C et al (2013) Transcriptional activation of Reg2 and Reg3β genes by glucocorticoids and interleukin-6 in pancreatic acinar and islet cells. Mol Cell Endocrinol 365(2):187–196

Maake C, Reinecke M (1993) Immunohistochemical localization of insulin-like growth factor 1 and 2 in the endocrine pancreas of rat, dog, and man, and their coexistence with classical islet hormones. Cell Tissue Res 273(2):249–259

Mally MI et al (1994) Developmental gene expression in the human fetal pancreas. Pediatr Res 36(4):537–544

Marselli L et al (2010) Gene expression profiles of β-cell enriched tissue obtained by laser capture microdissection from subjects with Type 2 diabetes. PLoS One 5(7):e11499

Mauras N et al (2000) Recombinant human insulin-like growth factor I has significant anabolic effects in adults with growth hormone receptor deficiency: studies on protein, glucose, and lipid metabolism. J Clin Endocrinol Metab 85(9):3036–3042. [MEDLINE record in process]

Meier JJ et al (2008) β-cell replication is the primary mechanism subserving the postnatal expansion of β-cell mass in humans. Diabetes 57(6):1584–1594

Meili S et al (2003) Secretory apparatus assessed by analysis of pancreatic secretory stress protein expression in a rat model of chronic pancreatitis. Cell Tissue Res 312(3):291–299

Moriizumi S et al (1994) Isolation, structural determination and expression of a novel reg gene, human regI β. Biochim Biophys Acta 1217(2):199–202

Moriscot C et al (1996) Absence of correlation between reg and insulin gene expression in pancreas during fetal development. Pediatr Res 39(2):349–353

Moses A et al (1996) Recombinant human insulin-like growth factor I increases insulin sensitivity and improves glycemic control in type II diabetes. Diabetes 45(1):91–100

Mueller CM, Zhang H, Zenilman ME (2008) Pancreatic reg I binds MKP-1 and regulates cyclin D in pancreatic-derived cells. J Surg Res 150(1):137–143

Multigner L et al (1983) Pancreatic stone protein, a phosphoprotein which inhibits calcium carbonate precipitation from human pancreatic juice. Biochem Biophys Res Commun 110(1):69–74

Namikawa K et al (2005) Expression of Reg/PAP family members during motor nerve regeneration in rat. Biochem Biophys Res Commun 332(1):126–134

Narushima Y et al (1997) Structure, chromosomal localization and expression of mouse genes encoding type III Reg, RegIII α, RegIII β, RegIII γ. Gene 185(2):159–168

Nata K et al (2004) Molecular cloning, expression and chromosomal localization of a novel human REG family gene, REG III. Gene 340(1):161–170

Nguyen KT et al (2006) Essential role of Pten in body size determination and pancreatic β-cell homeostasis in vivo. Mol Cell Biol 26(12):4511–4518

Nishimune H et al (2000) Reg-2 is a motoneuron neurotrophic factor and a signalling intermediate in the CNTF survival pathway. Nat Cell Biol 2(12):906–914

O'Hara A et al (2013) The role of proteasome β subunits in gastrin-mediated transcription of plasminogen activator inhibitor-2 and regenerating protein1. PLoS One 8(3):e59913

Ochiai K et al (2004) Activated pancreatic enzyme and pancreatic stone protein (PSP/reg) in bile of patients with pancreaticobiliary maljunction/choledochal cysts. Dig Dis Sci 49(11–12):1953–1956

Ohara S et al (2008) Reg IV is an independent prognostic factor for relapse in patients with clinically localized prostate cancer. Cancer Sci 99(8):1570–1577

Okamoto H (1999) The Reg gene family and Reg proteins: with special attention to the regeneration of pancreatic β-cells. J Hepatobiliary Pancreat Surg 6(3):254–262

Orelle B et al (1992) Human pancreatitis-associated protein. Messenger RNA cloning and expression in pancreatic diseases. J Clin Invest 90(6):2284–2291

Ose T et al (2007) Reg I-knockout mice reveal its role in regulation of cell growth that is required in generation and maintenance of the villous structure of small intestine. Oncogene 26(3):349–359

Oue N et al (2005) Expression and localization of Reg IV in human neoplastic and non-neoplastic tissues: Reg IV expression is associated with intestinal and neuroendocrine differentiation in gastric adenocarcinoma. J Pathol 207(2):185–198

Parikh A, Stephan AF, Tzanakakis ES (2012) Regenerating proteins and their expression, regulation and signaling. Biomol Concept 3(1):57–70

Pelengaris S et al (2004) Brief inactivation of c-Myc is not sufficient for sustained regression of c-Myc-induced tumours of pancreatic islets and skin epidermis. BMC Biol 2:26

Perfetti R et al (1996a) Regenerating (reg) and insulin genes are expressed in prepancreatic mouse embryos. J Mol Endocrinol 17(1):79–88

Perfetti R et al (1996b) Differential expression of reg-I and reg-II genes during aging in the normal mouse. J Gerontol A Biol Sci Med Sci 51(5):B308–B315

Petropavlovskaia M et al (2012) Mechanisms of action of islet neogenesis-associated protein: comparison of the full-length recombinant protein and a bioactive peptide. Am J Physiol Endocrinol Metab 303(7):E917–E927

Pittenger GL et al (2007) Intramuscular injection of islet neogenesis-associated protein peptide stimulates pancreatic islet neogenesis in healthy dogs. Pancreas 34(1):103–111

Pittenger GL, Taylor-Fishwick D, Vinik AI (2009a) The role of islet neogenesis-associated protein (INGAP) in pancreatic islet neogenesis. Curr Protein Pept Sci 10(1):37–45

Pittenger GL, Taylor-Fishwick D, Vinik AI (2009b) A role for islet neogenesis in curing diabetes. Diabetologia 52(5):735–738

Pospisilik JA et al (2003) Dipeptidyl peptidase IV inhibitor treatment stimulates β-cell survival and islet neogenesis in streptozotocin-induced diabetic rats. Diabetes 52(3):741–750

Quaife CJ et al (1989) Histopathology associated with elevated levels of growth hormone and insulin-like growth factor I in transgenic mice. Endocrinology 124(1):40–48

Rafaeloff R et al (1997) Cloning and sequencing of the pancreatic islet neogenesis associated protein (INGAP) gene and its expression in islet neogenesis in hamsters. J Clin Invest 99 (9):2100–2109

Rane SG, Reddy EP (2000) Cell cycle control of pancreatic β cell proliferation. Front Biosci 5:D1–D19

Rane SG et al (1999) Loss of Cdk4 expression causes insulin-deficient diabetes and Cdk4 activation results in β-islet cell hyperplasia. Nat Genet 22(1):44–52

Rankin MM, Kushner JA (2010) Aging induces a distinct gene expression program in mouse islets. Islets 2(6):4–11

Reichert M, Rustgi AK (2011) Pancreatic ductal cells in development, regeneration, and neoplasia. J Clin Invest 121(12):4572–4578

Rooman I, Lardon J, Bouwens L (2002) Gastrin stimulates β-cell neogenesis and increases islet mass from transdifferentiated but not from normal exocrine pancreas tissue. Diabetes 51 (3):686–690

Rosenberg L et al (2004) A pentadecapeptide fragment of islet neogenesis-associated protein increases β-cell mass and reverses diabetes in C57BL/6J mice. Ann Surg 240(5):875–884

Rouimi P et al (1988) The disulfide bridges of the immunoreactive forms of human pancreatic stone protein isolated from pancreatic juice. FEBS Lett 229(1):171–174

Rouquier S et al (1991) Rat pancreatic stone protein messenger RNA. Abundant expression in mature exocrine cells, regulation by food content, and sequence identity with the endocrine reg transcript. J Biol Chem 266(2):786–791

Sanchez D et al (2000) Overexpression of the reg gene in non-obese diabetic mouse pancreas during active diabetogenesis is restricted to exocrine tissue. J Histochem Cytochem 48 (10):1401–1410

Sanchez D et al (2001) Preferential expression of reg I β gene in human adult pancreas. Biochem Biophys Res Commun 284(3):729–737

Sanchez D et al (2004) Implication of Reg I in human pancreatic duct-like cells in vivo in the pathological pancreas and in vitro during exocrine dedifferentiation. Pancreas 29 (1):14–21

Sandgren EP et al (1990) Overexpression of TGF α in transgenic mice: induction of epithelial hyperplasia, pancreatic metaplasia, and carcinoma of the breast. Cell 61(6):1121–1135

Sasahara K et al (2000) Molecular cloning and tissue-specific expression of a new member of the regenerating protein family, islet neogenesis-associated protein-related protein. Biochim Biophys Acta 1500(1):142–146

Scharfmann R, Corvol M, Czernichow P (1989) Characterization of insulin-like growth factor I produced by fetal rat pancreatic islets. Diabetes 38(6):686–690

Schiesser M et al (2001) Conformational changes of pancreatitis-associated protein (PAP) activated by trypsin lead to insoluble protein aggregates. Pancreas 22(2):186–192

Schmied BM et al (2001) Transdifferentiation of human islet cells in a long-term culture. Pancreas 23(2):157–171

Seaberg RM et al (2004) Clonal identification of multipotent precursors from adult mouse pancreas that generate neural and pancreatic lineages. Nat Biotechnol 22(9):1115–1124

Sherr CJ (2001) The INK4a/ARF network in tumour suppression. Nat Rev Mol Cell Biol 2 (10):731–737

Shervani NJ et al (2004) Autoantibodies to REG, a β-cell regeneration factor, in diabetic patients. Eur J Clin Invest 34(11):752–758

Sieradzki J et al (1988) Stimulatory effect of insulin-like growth factor-I on [^{3}H]thymidine incorporation, DNA content and insulin biosynthesis and secretion of isolated pancreatic rat islets. J Endocrinol 117(1):59–62

Skarnes WC et al (2011) A conditional knockout resource for the genome-wide study of mouse gene function. Nature 474(7351):337–342

Smith FE et al (1991) Enhanced insulin-like growth factor I gene expression in regenerating rat pancreas. Proc Natl Acad Sci USA 88(14):6152–6256

Smith FE et al (1994) Pancreatic Reg/pancreatic stone protein (PSP) gene expression does not correlate with β-cell growth and regeneration in rats. Diabetologia 37(10):994–999

Song SY et al (1999) Expansion of Pdx1-expressing pancreatic epithelium and islet neogenesis in transgenic mice overexpressing transforming growth factor α. Gastroenterology 117(6):1416–1426

Spak E et al (2010) Changes in the mucosa of the Roux-limb after gastric bypass surgery. Histopathology 57(5):680–688

Suarez-Pinzon WL et al (2008) Combination therapy with glucagon-like peptide-1 and gastrin restores normoglycemia in diabetic NOD mice. Diabetes 57(12):3281–3288

Suzuki Y et al (1994) Structure and expression of a novel rat RegIII gene. Gene 144(2):315–316

Takasawa S et al (2006) Cyclin D1 activation through ATF-2 in Reg-induced pancreatic β-cell regeneration. FEBS Lett 580(2):585–591

Takatori A et al (2003) Protective effects of probucol treatment on pancreatic β-cell function of SZ-induced diabetic APA hamsters. Exp Anim 52(4):317–327

Taylor-Fishwick DA et al (2006a) Islet neogenesis associated protein transgenic mice are resistant to hyperglycemia induced by streptozotocin. J Endocrinol 190(3):729–737

Taylor-Fishwick DA et al (2006b) PDX-1 can repress stimulus-induced activation of the INGAP promoter. J Endocrinol 188(3):611–621

Taylor-Fishwick DA et al (2008) Pancreatic islet immunoreactivity to the Reg protein INGAP. J Histochem Cytochem 56(2):183–191

Taylor-Fishwick DA et al (2010) Pdx-1 regulation of the INGAP promoter involves sequestration of NeuroD into a non-DNA-binding complex. Pancreas 39(1):64–70

Terazono K et al (1988) A novel gene activated in regenerating islets. J Biol Chem 263(5):2111–2114

Terazono K et al (1990) Expression of reg protein in rat regenerating islets and its co-localization with insulin in the β cell secretory granules. Diabetologia 33(4):250–252

Teta M et al (2007) Growth and regeneration of adult β cells does not involve specialized progenitors. Dev Cell 12(5):817–826

Tezel E et al (2004) REG I as a marker for human pancreatic acinoductular cells. Hepatogastroenterology 51(55):91–96

Tourrel C et al (2001) Glucagon-like peptide-1 and exendin-4 stimulate β-cell neogenesis in streptozotocin-treated newborn rats resulting in persistently improved glucose homeostasis at adult age. Diabetes 50(7):1562–1570

Unno M et al (1993) Structure, chromosomal localization, and expression of mouse reg genes, reg I and reg II. A novel type of reg gene, reg II, exists in the mouse genome. J Biol Chem 268(21):15974–15982

Unno M et al (2002) Production and characterization of Reg knockout mice: reduced proliferation of pancreatic β-cells in Reg knockout mice. Diabetes 51(Suppl 3):S478–S483

Vaishnava S et al (2011) The antibacterial lectin RegIIIgamma promotes the spatial segregation of microbiota and host in the intestine. Science 334(6053):255–258

Vasseur S et al (2004) p8 improves pancreatic response to acute pancreatitis by enhancing the expression of the anti-inflammatory protein pancreatitis-associated protein I. J Biol Chem 279(8):7199–7207

Violette S et al (2003) Reg IV, a new member of the regenerating gene family, is overexpressed in colorectal carcinomas. Int J Cancer 103(2):185–193

Viterbo D et al (2009) Administration of anti-Reg I and anti-PAPII antibodies worsens pancreatitis. JOP 10(1):15–23

Waelput W et al (2000) Identification and expression analysis of leptin-regulated immediate early response and late target genes. Biochem J 348(Pt 1):55–61

Wang TC et al (1993) Pancreatic gastrin stimulates islet differentiation of transforming growth factor α-induced ductular precursor cells. J Clin Invest 92(3):1349–1356

Wang RN et al (1997) Expression of gastrin and transforming growth factor-α during duct to islet cell differentiation in the pancreas of duct-ligated adult rats. Diabetologia 40(8):887–893

Wang F et al (2011) Identification of RegIV as a novel GLI1 target gene in human pancreatic cancer. PLoS One 6(4):e18434

Watanabe T et al (1990) Complete nucleotide sequence of human reg gene and its expression in normal and tumoral tissues. The reg protein, pancreatic stone protein, and pancreatic thread protein are one and the same product of the gene. J Biol Chem 265(13):7432–7439

Watanabe T et al (1994) Pancreatic β-cell replication and amelioration of surgical diabetes by Reg protein. Proc Natl Acad Sci USA 91(9):3589–3592

Wilding Crawford L et al (2008) Gene expression profiling of a mouse model of pancreatic islet dysmorphogenesis. PLoS One 3(2):e1611

Xiong X et al (2011) Pancreatic islet-specific overexpression of Reg3β protein induced the expression of pro-islet genes and protected mice against streptozotocin-induced diabetes. Am J Physiol Endocrinol Metab 300:E669–E680

Xu G et al (1999) Exendin-4 stimulates both β-cell replication and neogenesis, resulting in increased β-cell mass and improved glucose tolerance in diabetic rats. Diabetes 48(12):2270–2276

Xu G et al (2006) GLP-1/exendin-4 facilitates [β]-cell neogenesis in rat and human pancreatic ducts. Diabetes Res Clin Pract 73(1):107–110

Xu X et al (2008) β cells can be generated from endogenous progenitors in injured adult mouse pancreas. Cell 132(2):197–207

Yamagishi H et al (2009) Expression profile of REG family proteins REG I[α] and REG IV in advanced gastric cancer: comparison with mucin phenotype and prognostic markers. Mod Pathol 22(7):906–913

Yamaoka T et al (2000) Diabetes and tumor formation in transgenic mice expressing Reg I. Biochem Biophys Res Commun 278(2):368–376

Ying LS et al (2013) Enhanced RegIV expression predicts the intrinsic 5-fluorouracil (5-FU) resistance in advanced gastric cancer. Dig Dis Sci 58(2):414–422

Zenilman ME et al (1996a) Pancreatic regeneration (reg) gene expression in a rat model of islet hyperplasia. Surgery 119(5):576–584

Zenilman ME et al (1996b) Pancreatic thread protein is mitogenic to pancreatic-derived cells in culture. Gastroenterology 110(4):1208–1214

Zenilman ME et al (1997) Pancreatic reg gene expression is inhibited during cellular differentiation. Ann Surg 225(3):327–332

Zenilman ME, Chen J, Magnuson TH (1998) Effect of reg protein on rat pancreatic ductal cells. Pancreas 17(3):256–261

Zenilman ME et al (2000) Comparison of reg I and reg III levels during acute pancreatitis in the rat. Ann Surg 232(5):646–652

Zhang Y et al (2003) Reg IV, a differentially expressed gene in colorectal adenoma. Chin Med J (Engl) 116(6):918–922

Zhao AZ et al (1997) Attenuation of insulin secretion by insulin-like growth factor 1 is mediated through activation of phosphodiesterase 3B. Proc Natl Acad Sci USA 94(7):3223–3228

Zhou J, Bievre M, Bondy CA (2000) Reduced GLUT1 expression in $Igf1^{-/-}$ null oocytes and follicles. Growth Horm IGF Res 10(3):111–117

Inflammatory Pathways Linked to β Cell Demise in Diabetes

36

Yumi Imai, Margaret A. Morris, Anca D. Dobrian, David A. Taylor-Fishwick, and Jerry L. Nadler

Contents

Introduction	993
Autoimmunity-Based Inflammation in T1D	993
General Introduction	993
Development of Autoimmunity Against β Cells	996
Pathway by Which Cytokines Mediate β Cell Death	999
Metabolic Stress and Islet Inflammation: Implication in Islet Dysfunction Associated with T2D	1001
T2D: The Most Common Form of Diabetes	1001
Does Metabolic Stress Cause Islet Inflammation in T2D?	1002
Evidence of Islet Inflammation in T2D	1007
The Effects of Metabolic Alteration in Other Organs on Islet Function and Islet Inflammation	1010
T1D and T2D: Are They a Continuum or Distinct Entities?	1017
Does Insulin Resistance Modify the Development of T1D?	1018
Does Autoimmunity Contribute to Islet Dysfunction in T2D?	1018

Y. Imai (✉) • J.L. Nadler
Department of Internal Medicine, Eastern Virginia Medical School, Strelitz Diabetes Center, Norfolk, VA, USA
e-mail: imaiy@evms.edu; nadlerjl@evms.edu

M.A. Morris
Departments of Internal Medicine and Microbiology and Molecular Cell Biology, Eastern Virginia Medical School, Strelitz Diabetes Center, Norfolk, VA, USA
e-mail: morrisma@evms.edu

A.D. Dobrian
Department of Physiological Sciences, Eastern Virginia Medical School, Norfolk, VA, USA
e-mail: dobriaad@evms.edu

D.A. Taylor-Fishwick
Department of Microbiology and Molecular Cell Biology, Eastern Virginia Medical School, Norfolk, VA, USA
e-mail: taylord@evms.edu

Inflammatory Pathways as a Therapeutic Target for Diabetes 1019
　　Effort to Halt β Cell Loss Through Anti-inflammatory Therapy in T1D 1019
　　Therapeutics Targeting IL-1 Pathway 1021
Concluding Remarks 1027
Cross-References 1028
References 1028

Abstract

Inflammation is proposed to play a key role in the development of both type 1 diabetes (T1D) and type 2 diabetes (T2D). It is well established that autoimmunity against β cells is responsible for massive loss of β cells in T1D. Recently, it has been recognized that chronic low-grade inflammation is not limited to insulin target organs but is also seen in islets in T2D. In T1D, T lymphocytes are primed to destroy the β cells. However, the process that leads to the development of self-reactive T lymphocytes remains elusive and is an area of intense research. A complex interplay between genetic and environmental factors, β cells, and immune cells is likely involved in the process. Immunomodulatory therapies have been attempted with some promises in animal models of T1D but have not yielded satisfactory effects on humans. Recent initiatives to evaluate T1D pathology in human donor pancreata hold promise to increase our knowledge in the coming years. In T2D, overnutrition results in metabolic stress including glucolipotoxicity, endoplasmic reticulum stress, and oxidative stress that potentially trigger an inflammatory response in the islets. Substantial evidence exists for an increase in humoral inflammatory mediators and an accumulation of macrophages in the islets of T2D subjects. Anti-inflammatory therapies targeting IL-1β and NFkB have shown improvement in β cell functions in small short term studies of T2D humans, providing a proof of principle for targeting islet inflammation in T2D. However, the nature of islet inflammation in T2D needs better characterization to tailor anti-inflammatory therapies that are effective and durable for T2D.

Keywords

Type 1 diabetes • Type 2 diabetes • Cytokines • Immune cells • 12-Lipoxygenase

Abbreviation

12/15LO	12/15-Lipoxygenase
12-HETE	12-Hyrdroxyeicosatetraenoic acid
12LO	12-Lipoxygenase
A1c	Hemoglobin A1c
AAb	Autoantibody
ACE	Angiotensin-converting enzyme
AdipoR1	Adiponectin receptor 1
AMPK	AMP-activated kinase
ANGPTL8	Angiopoietin-like 8

AP-1	Activator protein-1
ARb	Angiotensin receptor blockers
AT	Adipose tissue
ATF6	Activating transcription factor-6
CANTOS	Canakinumab Anti-inflammatory Thrombosis Outcomes Study
CB1	Cannabinoid receptor type 1
CCL	Chemokine (C-C motif) ligand
CDKAL1	CDK5 regulatory subunit-associated protein-1-like
CHOP	C/EBP homology protein
CLECL1	C-type leptin-like 1
CVD	Cardiovascular disease
CXCL	Chemokine(C-X-C motif) ligand
DAISY	Diabetes Association in Support of Youth
DC	Dendritic cells
DEXI	Dexamethasone induced
DHA	Docosahexaenoic acid
DMT1	Divalent metal transporter 1
DPPIV	Dipeptidyl peptidase-IV
EIF2AK3	Eukaryotic translation initiation factor 2α kinase-3
ER	Endoplasmic reticulum
ERK	Extracellular signal-regulated kinases
FasL	Fas ligand
GAD	Glutamic acid decarboxylase
GIP	Gastric inhibitory polypeptide
GK rat	Goto-Kakizaki rat
GLIS3	Gli-similar 3
GLP	Glucagon-like peptide
GWAS	Genome-wide association study
HLA	Human leukocyte antigen
HOMA	Homeostasis Model Assessment
hsCRP	High sensitivity C-reactive protein
IA2	Islet antigen 2
IAA	Insulin autoantibody
IAPP	Islet amyloid polypeptide
ICA	Islet cell antibody
IFN	Interferon
IL	Interleukin
IL-18RAP	IL-18R accessory protein
IL-1Ra	IL-1 receptor antagonist
iNKT	Invariant natural killer T cells
IRE1	Inositol requiring enzyme-1
IRS	Insulin receptor substrate
JDRF	Juvenile Diabetes Research Foundation
JNK	c-Jun N-terminal kinase

KIR	Killer immunoglobulin-like receptor
LADA	Latent autoimmune diabetes of adults
Lp	*Lactobacillus plantarum*
LPS	Lipopolysaccharide
MΦ	Macrophages
MAPK	Mitogen-activated protein kinase
MHC	Major histocompatibility complex
MIP	Macrophage Inflammatory Proteins
NADPH	Nicotinamide adenine dinucleotide phosphate
NEFA	Non-esterified fatty acids
NFκB	Nuclear factor kappa light chain enhancer of activated B cells
NK	Natural killer
NLRP3	NLR family, pyrin domain containing 3
NOD	Nonobese diabetic
Non-DM	Nondiabetic
nPOD	Network for the Pancreatic Organ Donor with Diabetes
OM	Omental
PA	Palmitic acids
PERK	PKR-like eukaryotic initiation factor 2α kinase
PEVNET	Persistent Virus Infection in Diabetes Network
PI3K	Phosphoinositide 3-kinase
PKB	Protein kinase B
PPAR	Peroxisome proliferator-activated receptor
PTB1B	Protein tyrosine phosphatase 1B
PTEN	Phosphatase and tensin homolog
qRT-PCR	Real-time reverse transcription polymerase chain reaction
ROS	Reactive oxygen species
SC	Subcutaneous
SLE	Systemic lupus erythematosus
SOCS3	Suppressor of cytokine signaling 3
STAT	Signal transducer and activator of signal transduction
STZ	Streptozotocin
SUOX	Sulfite oxidase
SVF	Seminal vesicle fluid
T1D	Type 1 diabetes
T2D	Type 2 diabetes
Th	T helper
TINSAL-T2D	Targeting-Inflammation Using Salsalate in Type 2 Diabetes
TLR4	Toll receptor 4
TNF	Tumor necrosis factor
TXNIP	Thioredoxin-interacting protein
UKPDS	United Kingdom Prospective Diabetes Study
UPR	Unfolded protein response
WFS1	Wolfram syndrome 1

Introduction

The incidence of diabetes, both T1D and T2D, is on the rise globally, creating an urgent need to understand their pathogenesis and to identify effective therapeutic targets. There is a substantial volume of evidence supporting the idea that inflammatory responses play an active role in the development of T1D and T2D. Inflammation serves as a defense system to protect a host by removing pathogens, foreign bodies, toxins, and damaged cells/tissues and promoting healing of tissues. Damaged or infected host cells produce humoral factors that communicate with immune cells to orchestrate inflammatory responses to achieve the removal of "harm" and to restore the normal architecture and function of cells. Innate immune responses involve immune cells without specified antigens, while adaptive immune responses target specified antigens through recognition by lymphocytes. It is well established that β cell loss in T1D results from an adaptive immune response against β cells. However, cross talk between β cells and variety of leukocytes, from both innate and adaptive immune responses, is required to establish self-reactive lymphocytes.

T2D was originally regarded as nonimmune-mediated diabetes. However, inflammation involving both innate and adaptive immune cells has been shown in insulin target tissues and peripheral circulation in T2D. More recently, signs of inflammation have been shown in human islets affected by T2D. Here, we will discuss the nature of islet inflammation in T1D and T2D, the communication of islets with other metabolic organs in T1D and T2D, and the current status of anti-inflammatory therapy for T1D and T2D.

Autoimmunity-Based Inflammation in T1D

General Introduction

Although the exact cause of T1D remains unknown, a great number of contributing factors have been appreciated. Many of these factors play into the development of autoimmunity and contribute to the inflammatory responses that result in the loss of β cells in T1D. Especially of note are the roles played by certain genetic haplotypes, potential environmental triggers (including viral infections, interactions of the immune system with the gut microbiome, and diet), immune cells (Mϕ, dendritic cells, B cells, T cells, and NK cells), and defects in the pancreatic islets themselves. This section will provide an overview of the natural progression of the disease as it is seen in the clinic, as well as the involvement of key cell types as we currently understand.

T1D is an inflammatory autoimmune disease in which the body's T cells are primed to destroy the β cells of the pancreatic islets. This autoimmune destruction leads to the ultimate reliance on exogenous sources of insulin for survival.

There are many facets to this disease, including altered T cell development (Rosmalen et al. 2002; Tanaka et al. 2010), the development of AAb (Ziegler and Nepom 2010), increased pancreatic inflammation (Arif et al. 2004; Saha and Ghosh 2012; Willcox et al. 2009), and deficiencies in regulatory T cells (Lindley et al. 2005). However, T cells alone are not solely responsible for diabetes development. There is also ample evidence to support the idea that the β cells themselves have defects in coping with ER stress (Marhfour et al. 2012; Tersey et al. 2012b). Yet more evidence suggests that β cells normally undergo apoptosis without autoimmunity arising, but in the presence of inflammatory insults, this apoptosis initiates the autoimmune cascade (Carrington et al. 2011; Roggli et al. 2012). When combined, all of these processes can lead to autoimmunity against the islets and a lifetime reliance on exogenous sources of insulin.

Perhaps the greatest indicator of potential diabetes development is the expression of certain susceptibility genes. Genetic susceptibility to T1D is defined by a multitude of genes, although the most important are those in the HLA haplotype. These genes determine the interactions of the T cell with MHC, which include the antigenic peptides that bind within the grooves of the MHC molecules. It appears that in the case of T1D, certain haplotypes lead to less stringent negative selection of T cells in the thymus, thereby increasing the number of circulating mature T cells that recognize self-antigens (Thorsby and Ronningen 1993). The HLA alleles most strongly associated with disease development are DRB1, DQA1, and DQB1, which encode HLA class II proteins involved in $CD4^+$ T cell recognition of antigens (Todd 2010).

For many years, it has been proposed that T1D is initiated by viral infection in at least a portion of those who develop T1D. Indeed, it is quite clear that environmental factors play a role in T1D development, as the concordance rate between monozygotic twins is only 30–50 %, indicating that there is a significant environmental contribution above genetic susceptibility (Redondo et al. 1999). The group of viruses most commonly associated with T1D development is that of the enteroviruses, which are members of the *Picornaviridae* family. Enteroviruses are RNA viruses able to form four structural and seven functional proteins (Jaidane and Hober 2008). There are several lines of evidence correlating infection with Coxsackie B viruses and the development of T1D (Dotta et al. 2007; Roivainen et al. 1998; Ylipaasto et al. 2012). Based on these data, the nPOD-Virus Group, a consortium involving 35 investigators worldwide, has formed in order to address this question with a multitude of complementary approaches that will be applied to the joint study of human pancreas from T1D patients identified by the JDRF nPOD. It is anticipated that the study will provide conclusive data regarding the type of virus strongly associated with T1D development. Many of the same investigators are also involved with the PEVNET, or Persistent Virus Infection in Diabetes Network, which also seeks answers to this question.

Beyond the mounting evidence that viruses may play a role in the development of T1D, it has also become evident that alterations in the gut microbiome are associated with disease progression. This will be discussed in further detail in a subsequent section of this chapter. A future goal will be to identify biomarkers of risk for progression to T1D.

T1D is on the rise, especially in children younger than 5 years of age. A recent prospective study (named the DAISY study) focused on infant diet to determine the risk of developing T1D (Frederiksen et al. 2013). Genetically susceptible children are at greater risk for developing T1D when either the father or siblings are T1D. Genetically susceptible patients showed a significant correlation with disease progression based upon their exposure to certain foods in infancy. Infants with relatives affected by T1D were studied for a history of breastfeeding, formula feeding, and timing of exposure to solid foods recorded. Both early (before 4 months of age) and late exposures (after 6 months of age) to any cereal led to an increased risk of T1D. Late exposure to non-gluten cereals specifically increased the risk to T1D, but early exposure did not. Gluten-containing cereals did not increase the risk when separated out from non-gluten-containing cereals. Additionally, early exposure to fruits increased the risk of T1D. The age at first exposure to cow's milk did not correlate with diabetes risk. Overall, the study recommends that new food antigens be introduced concurrently with breastfeeding between 4 and 6 months of age to minimize the risk of developing T1D, as breastfeeding provides protection from disease development.

To date, the best indicator of disease progression to overt T1D is the expression of AAb. These are antibodies produced against antigens related directly to β cells and insulin production. In another prospective study, 13,377 children were assessed for seroconversion, or development of AAb against insulin, GAD65, and IA2, and the risk of progression to T1D (Ziegler et al. 2013). 7.9 % of the children seroconverted, while 92.1 % remained free of islet AAb. It is not uncommon for children at high risk to develop AAb prior to developing full-blown diabetes. This occurs faster in children that develop multiple AAb prior to 3 years of age, with the 10-year risk being 74.9 % versus 60.9 % in those children who develop AAb after the age of 3. The development of multiple AAb is, by itself, predictive of the development of T1D in children, although the time to disease onset varied greatly. It has been suggested that islet inflammation develops by the deposition of autoantigens binding with islet AAb to form pro-inflammatory immune complexes.

Beyond the production of AAb, there is no good way of detecting T1D clinically before most patients present with hyperglycemia. Unless a family history dictates closer evaluation of potential disease progression, most patients are not tested for their HLA genotypes, production of AAb, or loss of the first-phase insulin response in a glucose tolerance test. From laboratory research, scientists have begun to piece together the events leading up to this clinical presentation.

Development of Autoimmunity Against β Cells

While it is still unclear what initiates autoimmunity against β cells, there are certainly key events that must occur in order for diabetes to fully develop. For example, it is understood that T cells are needed in order for T1D to develop, but it is unclear what recruits the autoimmune T cells to the pancreas in order to react against the β cells. Recent evidence suggests that there are metabolic defects within the pancreatic islets that might lead to increased susceptibility to autoimmune-mediated destruction (Tersey et al. 2012b). Perhaps stress signals supplied to the resident Mϕ and dendritic cells by the inherently stressed islets are sufficient to cause upregulation of inflammatory cytokine and chemokine production, which then leads to recruitment of additional immune effectors.

Research in this area is complicated by the fact that there are limited means of studying human disease. While our understanding has been greatly increased by the formation of the nPOD and PEVNET consortia, we still cannot study the disease as it develops within a single patient, and we are limited to snapshots of the disease or extrapolation of in vitro data in hopes that the same scenarios play out in vivo. Research in this area would benefit greatly from the further development of imaging modalities that allow us to study what is going on in vivo in real time. Some strides have been made in this arena (Antkowiak et al. 2013; Brom et al. 2010); however, in the absence of sufficient means to fully study human disease, we must use alternative models to test certain hypotheses. The following section provides details of our understanding of both rodent and human disease, and how they compare to one another.

Rodent Models

The use of mouse and rat models affords some insight into human disease. However, these models are not by any means perfect substitutes for the development of T1D in humans. We know that Mϕs and dendritic cells are recruited to the islets early in the process (between 3 and 6 weeks of age) of diabetes progression in NOD mouse model, the most frequently used model of T1D (Charre et al. 2002). The 12/15LO, a mouse gene for leukocyte 12LO implicated in inflammation (discussed in detail below), has been linked to β cell damage and T1D. Mϕ levels are significantly decreased in the absence of the 12/15LO enzyme (a mouse isoform of 12 LO), which metabolizes arachidonic acid to pro-inflammatory mediators (McDuffie et al. 2008). The reduction of Mϕ was associated with decrease in diabetes incidence in NOD mice with 12/15LO deletion. Additionally, NOD mice have increased numbers of pro-inflammatory circulating Mϕ (Nikolic et al. 2005). Evidence of aberrant dendritic cell development and function in NOD mice suggests one mechanism for the downstream development of autoimmune T cells in the periphery of these mice (Boudaly et al. 2002). Both types of antigen-presenting cells are potent cytokine producers and therefore contribute substantially to the development of β cell autoimmunity. The role of cytokines will be reviewed in greater detail in subsequent sections.

Additional early responders include NK cells, which are capable of producing large amounts of IFN-γ (Brauner et al. 2010). However, a recent study suggests that while more numerous in diseased mice, NK cells are dispensable in disease progression of NOD mice, as depleting NOD NK cells do not alter the disease course (Beilke et al. 2012). In non-autoimmune strains, NK cells are present in the pancreas and thought to function as sentinel cells rather than effectors of autoimmunity.

T cells begin to appear in the NOD pancreas by 6–8 weeks of age, with a mixture of $CD4^+$, $CD8^+$, and $Foxp3^+$ T cells making up the bulk of the insulitis. While regulatory $Foxp3^+$ T cells make an appearance in the pancreas, they are unable to stop the ensuing autoimmunity (D'Alise et al. 2008). In NOD models, the effector T cells ($CD4^+$ and $CD8^+$ cells) cause invasive and rapid destruction of most islets in affected mice. One significant difference between rodent and human research is the level of insulitis present around the islets during the autoimmune response. Murine T cells completely overwhelm the islet, while examples of human insulitis are much more subdued (Atkinson et al. 2013).

In the NOD model, it has been suggested that $CD4^+$ T cells, but not $CD8^+$ T cells, contribute to the development of autoimmunity by producing IFN-γ. $CD8^+$ T cells, however, tend to use either the granzyme/perforin pathway or the Fas/FasL pathway (Scott et al. 2010; Varanasi et al. 2012). Increased levels of granzyme A and Fas are able to overcome a granzyme B deficiency in the NOD model, such that disease incidence is not reduced or slowed (Kobayashi et al. 2013). Progression of research in diabetes has enabled us to more accurately assess which cells play a pathogenic role in the development of T1D. Recently, CD27-γδ T cells have been implicated in the development of diabetes in NOD mice by contributing to IL-17 production and playing a role in the development of islet inflammation (Markle et al. 2013b).

Humans

The appearance of islet-specific AAb occurs early in the progression of T1D relative to clinical evidence of β cell failure. While antibodies serve as a marker of autoimmunity, their role in the disease progression is not fully understood. As the antigens they recognize are not cell surface antigens, it does not appear that islet AAb cause direct destruction of the pancreas. Perhaps the AAb serve merely as indicators of underlying autoimmunity (Skyler and Sosenko 2013); however, some believe that they could form immune complexes that promote islet inflammation (Achenbach et al. 2004; Bonifacio et al. 1990; Orban et al. 2009). There is evidence that some diabetic patients develop secondary vascular complications based on the formation of immune complexes (Nicoloff et al. 2004).

Concurrently with AAb production, some patients begin to display impaired glucose tolerance characterized by loss of first-phase insulin secretion, indicating metabolic deficiencies present in the islets (Scheen 2004). Unfortunately, due to a lack of imaging modalities, we are unable to determine if these deficiencies correlate with increased T cell mediated damage in the islets.

Examination of peripheral blood cells has shown that plasmacytoid dendritic cell balance is significantly skewed in recent onset T1D patients (Allen et al. 2009). These cells are able to enhance the development of the autoimmune T cell response to the islet autoantigens. Dendritic cells show some promise as therapeutic targets in tolerizing protocols in which treatment with mammalian glycolipids increases IL-10 production, reduces IL-12 production, and reduces autoreactive T cell development by dendritic cells (Buschard et al. 2012).

Many have studied monocytes and Mϕ in order to discern their role in T1D development in humans. $CD68^+$ Mϕ has been detected in the islets of diabetic patients at both early and late time points in disease progression (Willcox et al. 2009). In one study, Bradshaw et al. found that monocytes from T1D patients were more likely to produce cytokines that stimulated the development of Th17 cells (Bradshaw et al. 2009). Another study has shown that IL-1β expression due to signaling through toll-like receptors was increased in monocytes of AAb + patients as compared to AAb- controls (Alkanani et al. 2012), implicating alterations in innate immune pathways in the development of T1D (Meyers et al. 2010). Furthermore, there are several candidate genes expressed by monocytes and monocyte-derived cells that have been suggested as contributors to diabetes pathogenesis, including CD226, CLECL1, DEXI, and SUOX (Wallace et al. 2012).

There is evidence that the IL-18R accessory protein (IL-18RAP) is altered on the surface of NK cells of T1D patients, which can lead to increased IFN-γ production by NK cells in some patients, which may stimulate downstream T cell responses (Myhr et al. 2013). In children diagnosed before the age of 5, NK cell phenotyping based on KIR gene and HLA class I gene haplotypes has shown that there is an increased representation of genotypes predictive of NK cell activation, which may indicate a role for these cells in some cases of T1D (Mehers et al. 2011). Additional evidence supports the idea that NK cell frequency and activation states are altered in both classical T1D and LADA patients (Akesson et al. 2010; Rodacki et al. 2007). There may be functional differences in NK cells due to expression of particular alleles and how they interact with HLA-C1 molecules (Ramos-Lopez et al. 2009).

Downstream of AAb production, T cells are recruited to the islets, and it is thought that $CD8^+$ T cells use cytotoxic mechanisms to destroy the pancreatic β cells. This destruction appears to occur mainly through cytotoxic degranulation involving perforin and granzymes, although maximal destruction is dependent upon the strength of the signal through the T cell receptor (Knight et al. 2013).

It is clear that pro-inflammatory cytokines can directly affect islet health (Eizirik and Mandrup-Poulsen 2001). In healthy humans, there is a subset of IL-10 producing T cells that appear to regulate pro-inflammatory cytokine production, thereby limiting the damaging effects of these cytokines (Tree et al. 2010). Below, we review our knowledge about cytokines involved in β cell demise.

Pathway by Which Cytokines Mediate β Cell Death

Elevated inflammatory cytokines have been reported for both T1D and T2D (Al-Maskari et al. 2010; Catalan et al. 2007; Eizirik and Mandrup-Poulsen 2001; Igoillo-Esteve et al. 2010; Jorns et al. 2005; Kang et al. 2010; Steinberg 2007; Su et al. 2010; Tilg and Moschen 2008). Acute exposure of ex vivo islets and/or β cell lines to a variety of inflammatory cytokines that are elevated in diabetes induces β cell dysfunction. Key inflammatory cytokine interactions include those mediated by IL-1β, IFN-γ, or TNF-α. Paired or triple combinations of these cytokines that include IL-1β are widely reported to induce β cell failure and promote apoptosis. Single cytokine treatment of the islets does not induce β cell dysfunction. A synergy in intracellular signaling pathways mediated by cytokines is required (Eizirik and Mandrup-Poulsen 2001; Rabinovitch et al. 1990).

Ex vivo studies on human or mouse primary islets show that a brief (6 h) exposure to three inflammatory cytokines (TNF-α, IL-1β, IFN-γ) is sufficient to result in loss of glucose-stimulated insulin secretion, increased inflammatory gene expression, and induction of apoptosis (Taylor-Fishwick et al. 2013; Weaver et al. 2012). Dissecting the intracellular events that mediate the transition of a β cell from functional to dysfunctional in an inflammatory environment may offer new approaches to preserve β cells exposed to inflammation. This strategic approach would be relevant to slowing diabetes progression as a monotherapy or reversing diabetes as a combinatorial approach with an islet regeneration strategy, either endogenous stimulation or exogenous repopulation (cell transplantation, encapsulation, xenotransplantation, etc.) (Taylor-Fishwick and Pittenger 2010; Taylor-Fishwick et al. 2008).

Literature reports describe many gene changes in β cells or islets that result from exposure to pro-inflammatory cytokines (Eizirik et al. 2012). Our own studies have identified changes in gene expression and enzyme activation that co-associate with pro-inflammatory cytokine-induced β cell dysfunction. The development of targeted inhibitors will allow description of a framework to integrate these pathways.

12LO enzymes are associated with the development of T1D. Genetic deletion of 12/15LO (a mouse 12LO) in the diabetes-prone NOD mouse confers protection to T1D onset relative to control (wild type) NOD mice (McDuffie et al. 2008). Islets from 12/15LO-deficient mice are resistant to β cell dysfunction induced by pro-inflammatory cytokines (Bleich et al. 1999). 12LO oxidizes cellular polyunsaturated acids to form lipid mediators termed eicosanoids. By definition, 12LO oxidizes carbon-12 of arachidonic acid. There are several isozymes that catalyze the reactions including leukocyte 12LO and platelet 12LO (Imai et al. 2013b). A major stable bioactive metabolite of 12LO activity is 12-HETE. Stimulation of human donor islets with pro-inflammatory cytokines increases the gene expression for 12LO. The active lipid product of 12LO activity, 12-HETE, reproduces, in part, β cell dysfunction mediated by pro-inflammatory cytokine stimulation (Ma et al. 2010). Direct addition of 12-HETE results in a loss of glucose-stimulated insulin secretion and promotes β cell apoptosis. Newly described small-molecular-weight compounds selectively inhibit the activity of 12LO (Kenyon et al. 2011).

These selective inhibitors of 12LO preserve islets/β cell function in an inflammatory cytokine environment and inhibit apoptosis (Taylor-Fishwick and Nadler unpublished). These data connect pro-inflammatory cytokine stimulation with 12LO activation placing 12LO as one mediator of inflammatory cytokine-induced β cell dysfunction. Pro-inflammatory cytokines are potent inducers of cellular ROS, as are other serum conditions associated with the diabetic state, high free fatty acids, and elevated glucose (Cunningham et al. 2005; Inoguchi and Nawata 2005; Janciauskiene and Ahren 2000; Michalska et al. 2010; Morgan et al. 2007; Nakayama et al. 2005; Oliveira et al. 2003; Uchizono et al. 2006). Emerging contributors to sustained elevation in β cell ROS are NADPH oxidase enzymes. NADPH oxidase activity is regulated by 12LO activation (Weaver et al. 2012). A detailed description of NADPH oxidase enzymes in relation to β cell function is provided in the chapter by Taylor-Fishwick, "Role of NADPH Oxidase in β Cell Dysfunction."

Our studies provide an additional component to a molecular framework of intracellular β cell changes following inflammatory exposure. These discoveries support an emerging concept of an interactive interface between the islets and the immune system. IL-12 production and function has been described in β cells (Taylor-Fishwick et al. 2013). Expression of IL-12 gene and protein are observed following inflammatory cytokine stimulation (TNF-α, IL-1β, IFN-γ) of primary human and mouse islets, or β cell lines. Homogeneous β cell lines are devoid of any potential immune cell "contamination." While this is a concept that challenges established immune-based sources of IL-12, local production of IL-12 in the β cell could play a signsificant role in targeting the immune mediator recruitment. Separately, support for a paracrine function of IL-12 in β cells is also provided (Taylor-Fishwick et al. 2013). β cells, including human β cells, express the receptor for the IL-12 ligand and are responsive to IL-12-ligand/IL-12-receptor ligation. Administration of exogenous IL-12 directly mediated β cell dysfunction.

Observed were induction of apoptosis and disruption of glucose-stimulated insulin secretion (Taylor-Fishwick et al. 2013). Exogenous IL-12 induced a dose-dependent expression of IFN-γ in β cell lines suggesting a functional IL-12/STAT4/IFN axis. Previous studies have provided evidence of an active STAT4 signaling pathway in islet β cells (Yang et al. 2003, 2004). Importantly, neutralization of IL-12 with an IL-12 antibody blocked the β cell dysfunction induced by inflammatory cytokine stimulation (Taylor-Fishwick et al. 2013). A small molecule inhibitor of IL-12 prevented STAT4 activation, prevented T1D in NOD, and, in combination with exendin-4 or a β cell growth factor, reversed established diabetes (Yang et al. 2002, 2006). The functional defects mediated by IL-12 corresponded to those seen with inflammatory cytokine stimulation. Both receptor and ligand for IL-12 are upregulated in β cells exposed to inflammatory cytokine stimulation (TNF-α, IL-1β, IFN-γ). These studies suggest that pro-inflammatory cytokines induce local IL-12 expression that may be a mediator of inflammatory cytokine-induced β cell dysfunction. Lastly, selective inhibitors of 12LO

(Kenyon et al. 2011) suppress the induction of IL-12 ligand in islets and β cells exposed to inflammatory cytokine stimulation (Taylor-Fishwick and Nadler, unpublished). Inflammatory cytokine stimulation of 12LO activity is implied in the regulation of β cell IL-12 expression.

Beyond the key pro-inflammatory cytokines, there are others that likely affect disease progression of diabetes. The particular role of IL-17 in the development of T1D is rather controversial, as some have shown that it does not play a role (Joseph et al. 2012), while others maintain that it plays an important role in disease progression. Still others see a limited role for IL-17 in diabetes progression (Saxena et al. 2012). It has been suggested that IL-17 enhances the apoptotic response of β cells to TNF-α, IL-1β, and IFN-γ (Arif et al. 2011). The presence of Th22 (IL-22-producing helper T cells) and Th17 cells correlated strongly in T1D patients versus controls; however, it is unclear what the role of the Th22 cells is at this point in time (Xu et al. 2014). There is evidence in mice that IL-22 is increased under protective circumstances due to modulations in gut bacteria (Kriegel et al. 2011).

A conclusion from these observations is that β cells are not merely passive in terms of immune interactions. There appears a growing body of evidence that β cells actively interact with the immune system. Transcriptome analysis of islets exposed to pro-inflammatory cytokines further serves to reinforce the concept of an active dialog between the pancreatic islets and the immune system in T1D (Eizirik et al. 2012).

Metabolic Stress and Islet Inflammation: Implication in Islet Dysfunction Associated with T2D

T2D: The Most Common Form of Diabetes

T2D affects 347 million individuals in a recent worldwide study and has seen astonishing doubling of prevalence in less than 3 decades (Danaei et al. 2011). T2D is the most common form of diabetes that is characterized by insufficient insulin secretion to overcome insulin resistance. Extreme hyperglycemia itself could cause acute illness such as nonketotic hyperosmolar status. In addition, persistent hyperglycemia, even at modest degree, increases risks of macrovascular complications (represented by cardiovascular disease and stroke) and microvascular complications (retinopathy, nephropathy, and neuropathy) (Stratton et al. 2000). T2D is a leading cause of end-stage renal disease and blindness in developed countries and increases the risk of heart attack to two to four times. Thus, financial and emotional burdens of the disease are significant for affected individuals, as well as for society at large.

The recent rise in T2D incidence is largely attributed to environmental factors, including a sedentary lifestyle and excessive energy intake, which result in insulin resistance commonly seen in obesity (Danaei et al. 2011). In order to maintain normoglycemia, the body will compensate for insulin resistance by increasing the islet mass. However, a fraction of insulin-resistant subjects fails to produce enough insulin and develops T2D, highlighting the critical role of the islets in the pathogenesis of T2D (Kahn et al. 2006). Indeed, many of the T2D susceptibility genes identified from GWAS are related to the regulation of insulin secretion, rather than insulin resistance (Petrie et al. 2011). Also, the natural history of T2D is characterized by gradual worsening of the disease due to progressive islet dysfunction, rather than a continuous increase in insulin resistance (Kahn et al. 2006; Prentki and Nolan 2006). Both a reduction in islet function and an eventual reduction in mass are noted in T2D. One early sign of islet dysfunction is the impairment of first-phase insulin secretion, which is an acute rise in insulin in response to glucose (Kahn et al. 2009). Forty to fifty percent of islet mass is already lost by the time T2D subjects present with hyperglycemia. An increase in β cell death and possibly dedifferentiation of β cells is considered to be responsible for the loss of islet mass (reviewed in Weir and Bonner-Weir (2013)). Here, we will discuss the potential mechanisms that lead to reduction of functional β cell mass and the evidence supporting involvement of inflammatory pathways in the islets during the development of T2D.

Does Metabolic Stress Cause Islet Inflammation in T2D?

Insulin resistance and the associated metabolic stress have profound effects on the health of the islets, which is discussed in a chapter, "Pancreatic β Cells in Metabolic Syndrome," and in a recent review (Imai et al. 2013a). Here, we will discuss the potential mechanisms by which metabolic stress provokes inflammatory responses in the islets.

Hyperglycemia and Dyslipidemia

Obesity, the most common cause of insulin resistance, is characterized by excessive energy intake from carbohydrate and lipids. Thus, hyperglycemia and dyslipidemia are potentially responsible for the loss of islet mass and function in T2D. In experimental conditions, it is well documented that prolonged exposure to elevated levels of glucose impairs β cell function and survival through oxidative stress, ER stress, enhancement of the hexosamine pathway, and others (Bensellam et al. 2012; Poitout and Robertson 2008). The activation of ER stress and oxidative stress could provoke inflammatory responses in the islets (discussed below); however, there are conflicting reports as to whether high levels of glucose directly induce the islet production of IL-1β, a cytokine that is proposed to play a key role in chronic inflammation associated with T2D (Bensellam et al. 2012). It has been widely accepted that exposure of β cell lines or islets to saturated fatty acids, especially PA, in culture conditions negatively affects insulin secretion and viability of β cells (Poitout and Robertson 2008). Increased production of ROS and ceramides, alterations in the integrity of ER, and aberrant protein palmitoylation are proposed mechanisms by which fatty acids elicit deleterious effects on the islets (Baldwin et al. 2012; Boslem et al. 2012; Cnop et al. 2010; Poitout and Robertson 2008). Fatty acids have been shown to provoke inflammatory responses directly in the islets in experimental settings. Treatment of human islets with PA increased the expression of multiple cytokines including IL-1β, TNF-α, IL-6, IL-8, CXCL1, and CCL2 (Igoillo-Esteve et al. 2010). In this study, the blockade of IL-1β in human islets treated with PA prevented the rise of inflammatory cytokines but did not prevent apoptosis, indicating that inflammatory responses may not be solely responsible for the cytotoxic effects of PA in human islets ex vivo (Igoillo-Esteve et al. 2010). Interestingly, an in vivo study that raised serum PA levels by infusion in C57Bl/6 mice caused islet dysfunction, which seemed to result from cross talk between β cells and pro-inflammatory Mϕs that were recruited to the islets (Eguchi et al. 2012). A similar result was obtained in prediabetic, diabetes-prone BioBreeding rats, a model for T1D. A 48-h infusion of Intralipid (lipid emulsion), used to increase serum fatty acids, induced CCL2, IL-1β, TNF-α, IFN-γ, and IL-10 expression in the islets, caused mononuclear cell accumulation in islets, and impaired insulin secretion in prediabetic, diabetes-prone BioBreeding rats, but not in nondiabetes-prone rats (Tang et al. 2013). Therefore, recruitment and activation of Mϕ into the islets may play an important role in islet dysfunction in the setting of hyperlipidemia in rodent models. However, it still remains controversial whether fatty acids directly impair islet function and survival in humans in vivo, as the increase in serum fatty acids causes insulin resistance, which complicates the interpretation of islet function (Giacca et al. 2011). Finally, we need to bear in mind that cytotoxic effects of saturated fatty acids are not specific to the islets but seen in a wide range of cells. These effects are substantially lessened when cells are treated with a mixture of saturated and unsaturated fatty acids that simulates the typical lipid profile in vivo (van Raalte and Diamant 2011).

Oxidative Stress

The association of oxidative stress with islet failure in T2D was implicated in a study that demonstrated differential expression of genes related to oxidative stress in the islets obtained by laser microdissection of human pancreata from T2D donors (Marselli et al. 2010). Although ROS serves as an intracellular signal at a physiological level, excessive ROS is detrimental to the β cells (Pi and Collins 2010), the cells that are vulnerable to oxidative stress due to low antioxidant capacity (Lenzen 2008). It has been shown that high glucose, fatty acids, and IAPP (discussed below) result in an increased production of ROS by mitochondria and NADPH oxidase in the islets (Koulajian et al. 2013; Lightfoot et al. 2012; Rolo and Palmeira 2006; Zraika et al. 2009). Mitochondria are critically important for glucose-stimulated insulin secretion and a highly active organelle in β cells. Defects in their function and morphology are commonly seen in T2D and result in oxidative stress (Supale et al. 2012). A role of NADPH oxidase in oxidative stress is further discussed in chapter "NADPH Oxidase in β Cell Dysfunction." As ROS is a strong signal to provoke pro-inflammatory response in immune cells and nonimmune cells, ROS is a potential pathway connecting metabolic stress and inflammation in T2D islets through activation of JNK, p38-MAPK, NFkB, and AP-1 and subsequent production of cytokines including IL-1β, TNF-α, and CCL2 (Lamb and Goldstein 2008; Padgett et al. 2013).

ER Stress

The ER is an intracellular organelle with key roles in proper peptide synthesis and maturation (folding), in Ca^{2+} homeostasis, and in the regulation of carbohydrate and lipid metabolism (Hotamisligil 2010). The UPR is a programmed response to protect the integrity of the ER from accumulation of misfolded proteins and metabolic stress by orchestrating an adaptive response or to produce cell death signals due to overwhelming pressures on the ER (Oslowski and Urano 2010). Three ER membrane-bound proteins, PERK, IRE1*a*, and ATF6, function as sensors of ER stress and coordinator of the UPR response (Oslowski and Urano 2010; Scheuner and Kaufman 2008). The importance of UPR response for the maintenance of β cell health is well known through studies of humans and mice with rare mutations. Misfolding of insulin due to gene mutations results in diabetes due to ER stress (Scheuner and Kaufman 2008). Similarly, Wolfram syndrome and Wolcott-Rallison syndrome both result from mutations affecting ER homeostasis and lead to the development of diabetes: the former due to defects in the WFS1 protein that regulates Ca^{2+} store in the ER and the latter due to defects in PERK (EIF2AK3) (Scheuner and Kaufman 2008). Interestingly, a polymorphism of CDKAL1, which is associated with T2D in GWAS, may introduce an insulin mutation at translation level, providing another example in humans in which misfolding of insulin may contribute to T2D development (Wei et al. 2011).

Aside from mutations, the insulin-resistant status associated with T2D could provoke ER stress in the β cells through several pathways. The increased demand for insulin secretion under insulin-resistant status necessitates upregulation of insulin production, which could account for 30–50 % of protein synthesis in the β cells (Scheuner and Kaufman 2008). In addition, high glucose levels, fatty acids, oxidative stress, mitochondrial dysfunction, and IAPP (see below) all induce ER stress (Hotamisligil 2010; Scheuner and Kaufman 2008). Indeed, a contribution of ER stress to islet failure has been implicated in both human and mouse studies. CHOP, a proapoptotic transcription factor upregulated in ER stress, was increased in the β cells of pancreata from human T2D donors (Huang et al. 2007). *Ob/ob* mice and db/db mice are both insulin resistant and obese due to lack of leptin signaling but show different capacities for β cell compensation. Interestingly, the expression of genes related to adaptive UPR was increased in *ob/ob* mice that compensate for insulin resistance well, but those genes failed to increase in db/db mice that progressively lose the β cells due to failure to alleviate ER stress (Chan et al. 2013).

Once the UPR, a programmed response to ER stress, is activated to a certain level, a member of UPR proteins initiates inflammatory responses in many cells including the β cells (Eizirik et al. 2013; Garg et al. 2012; Zhang and Kaufman 2008). In brief, PERK reduces translation of I*k*B, while IRE1*a* promotes degradation of I*k*B, both resulting in activation of NF*k*B through removal of inhibition by I*k*B (Zhang and Kaufman 2008). PERK and IRE1*a* also induces TXNIP (also discussed in below) expression that could increase IL-1β secretion through an interaction of TXNIP with NLRP3 (Lerner et al. 2012; Oslowski et al. 2012).

A Potential Role of TXNIP in Islet Dysfunction and Inflammation in T2D

TXNIP regulates cellular redox status by inhibiting thioredoxin in a redox state-dependent manner (Spindel et al. 2012). A series of studies has revealed that TXNIP may be one of the mediators of β cell demise in diabetes. The expression of TXNIP is upregulated by glucose and increased in diabetic islets. TXNIP provokes apoptosis and impairs islet function through several pathways, including induction of mitochondrial apoptosis pathway and regulation of insulin transcription via microRNA (Saxena et al. 2010; Xu et al. 2013). Interestingly, as discussed in above, TXNIP is also induced by ER stress and contributes to ER stress-induced inflammation by activating an inflammasome (Lerner et al. 2012; Oslowski et al. 2012). Therefore, TXNIP may initiate β cell demise in response to elevated glucose, ROS, and ER stress through a multitude of pathways including inflammatory pathway.

A Potential Role of IAPP in Islet Dysfunction and Inflammation in T2D

One of the distinct characteristics of human islets affected by T2D is the accumulation of amyloid that was first noted in 1900 (Weir and Bonner-Weir 2013). Aggregates derived from IAPP, which is co-secreted with insulin from the β cells, are the major component of amyloid found in T2D islets. IAPP has physiological functions in the regulation of gastric emptying and satiety (Cao et al. 2013). Although human IAPP is soluble in monomeric form, it typically forms amyloid fibrils through a multistep self-driven reaction, a process not seen in mouse IAPP that does not possess an amyloidogenic sequence (Westermark et al. 2011). Several transgenic mice that express human IAPP have been created to determine whether human IAPP actively participates in β cell demise in T2D or is a by-product of β cell death. In a recent review, cytotoxic effects of human IAPP in transgenic mice were noted to be most apparent when overexpression of human IAPP is combined with some stressors such as high-fat diet (Montane et al. 2012). In humans, the deposition of islet amyloid correlates with β cell apoptosis indicating a close association between islet amyloid and β cell death (Jurgens et al. 2011).

Rapid accumulation of amyloid is also seen in human islet transplants and considered to contribute a graft failure (Westermark et al. 2011). Although the mechanisms by which IAPP elicits toxicity are not fully understood and could be multifactorial, the involvement of inflammatory pathways has been implicated. The activation of JNK pathway was shown to play a key role in inducing apoptosis in the islets overexpressing human IAPP cultured in high glucose (Subramanian et al. 2012). Furthermore, it has been shown that the islets exposed to human IAPP fibril increase the expression of CCL2 and CXCL1, which can potentially initiate recruitment of leukocytes (Westwell-Roper et al. 2011). Interestingly, human IAPP fibrils also induce the activation of inflammasomes and the production of IL-1β and other cytokines from bone marrow derived-monocytes (Masters et al. 2010; Westwell-Roper et al. 2011). Indeed, mouse islets overexpressing human IAPP transplanted into STZ-induced diabetic NOD.scid mice showed increased recruitment of F4/80-positive Mφ, supporting the idea that human IAPP fibril provokes an inflammatory response in the islets (Westwell-Roper et al. 2011).

Other Pathways Implicated in T2D Pathogenesis That Potentially Induce Islet Inflammation

There are several other factors that may contribute to islet dysfunction in T2D and may provoke inflammatory responses in the islets. Autophagy, or degradation and recycling of cytoplasmic components, is upregulated in the β cells under insulin resistance and appears to be important for the adaptive increase seen in β cell mass in mouse studies (Fujitani et al. 2010). Recently, a role for autophagy in chronic inflammation has gained attention, as defects in proteins regulating autophagy are associated with Crohn's disease, SLE, and others (Levine et al. 2011). Future studies are needed to determine whether or not human T2D is associated with defects in autophagy in the islets. If present, it needs to be addressed whether defects in autophagy modulate inflammatory responses in the islets. The alteration in gut microbiota is another pathway implicated in T2D pathogenesis, and bacteria may produce inflammatory signals for the islets, such as LPS. This is discussed further in below.

Role of Inflammation as an Amplifier of Metabolic Stress

As we discussed, there are multiple pathways that could generate inflammatory signals under metabolic stress in the islets (Fig. 1). Importantly, once induced, islet inflammation, in turn, can exacerbate oxidative stress, ER stress, and mitochondrial dysfunction and further increase islet inflammation in T2D, as well as in T1D. Pro-inflammatory cytokines including IL-1β, TNF-α, and IFN-γ impair ER homeostasis through nitric oxide-dependent and nitric oxide-independent mechanisms (Eizirik et al. 2013). 12LO is another pro-inflammatory mediator that potentially amplifies islet inflammation (Imai et al. 2013b). 12LO, a lipoxygenase upregulated in the islets of an animal model of T2D (Imai et al. 2013b) and in some humans with T2D (Galkina and Imai, unpublished), leads to the production of pro-inflammatory lipid metabolites, such as 12-HETE. 12-HETE, in turn, will induce additional pro-inflammatory cytokines such as IL-12, ER stress, and oxidative stress (Cole et al. 2012a; Weaver et al. 2012). Recently DMT1, an iron transporter, was shown to mediate oxidative stress in response to IL-1β in the islets (Hansen et al. 2012b). DMT1 is upregulated by IL-1β in the islets, and increases iron-catalyzed formation of ROS providing another pathway to amplify an inflammatory response (Hansen et al. 2012b). Importantly, deletion of both DMT1 and 12LO in mice was protective against development of diabetes in both T1D and T2D models, supporting the significant contribution of these key molecules that may serve as amplifiers of inflammatory responses in the islets (Hansen et al. 2012b; Imai et al. 2013b).

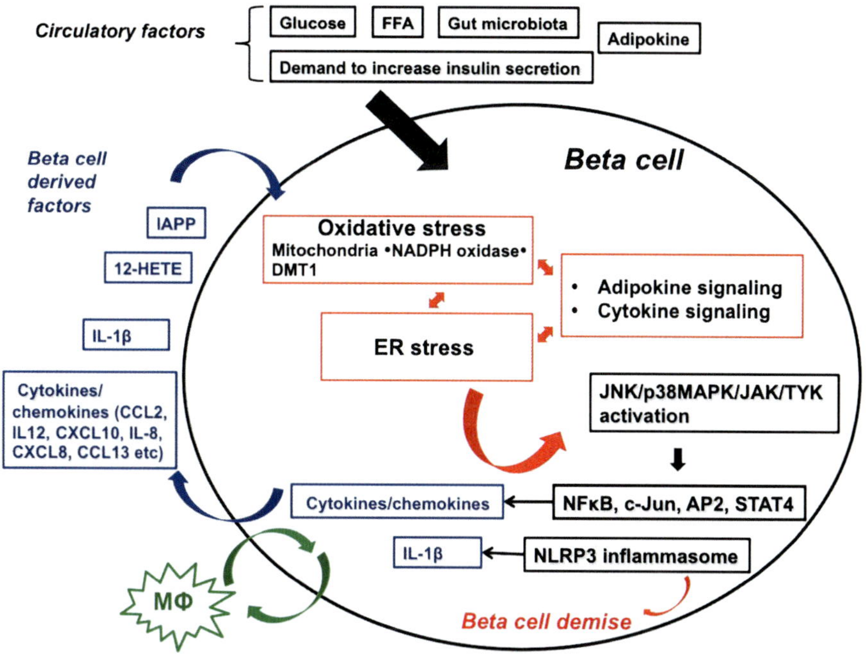

Fig. 1 Major inflammatory mediators and signaling pathways implicated in β cell demise in type 2 diabetes. Metabolic stress is considered to trigger inflammation in islets in type 2 diabetes. Increased glucose, excess saturated fatty acids (*FFA*), and higher demand for insulin secretion result in ER stress and oxidative stress. The alteration in adipokines and gut microbiota may also provoke islet inflammation. Cytokines produced locally or from circulation activate inflammatory pathways such as JAK/TYK, p38MAPK, and JNK via specific cytokine receptors in β cells. Oxidative stress and ER stress also result in the activation of JAK/TYK, p38MAPK, and JNK. In β cells under metabolic stress, there is extensive cross talk between signaling molecules involved in oxidative stress, ER stress, and inflammatory responses. Ultimately the activation of downstream transcription factors NFκB, c-Jun, and AP2 (downstream of JNK) will increase the production of pro-inflammatory mediators such as CCL2, IL-12, IL-8, CXCL10, CXCL8, and CCL13. The production of 12-HETE (via 12LO activation) and IL-1β (via Nlrp3 inflammasome) will be also increased in β cells under stress. These pro-inflammatory mediators produced by β cells will further increase oxidative stress, ER stress, and cytokine signaling creating a feedforward loop to exacerbate the inflammatory response. IL-1β increases the expression of NADPH oxidase and DMT1, which increases iron transport; both lead to an increase in reactive oxygen species and, together with IAPP, contribute to increased oxidative stress. 12-HETE will induce the production of IL-12, oxidative stress, and ER stress. IL-12 activates STAT4 signaling. Chemokine produced by β cells will also recruit macrophages (Mϕ) into the islets that further augment inflammatory response in islets

Evidence of Islet Inflammation in T2D

Humoral Mediators of Inflammation in the Islets of T2D

An increase in several humoral mediators of inflammation has been reported in human T2D islets. Gene arrays comparing the islets obtained by laser capture microdissection of frozen pancreata sections in 9 non-DM and 10 T2D donors found upregulation of IL-1β and IL-8 in the T2D group. While the expression of IL-1β determined by qRT-PCR was barely detectable in the islets from non-DM donors, six out of ten T2D donors showed high IL-1β expression in the islets (56-fold increase over the controls) (Boni-Schnetzler et al. 2008). There was a trend of increased IL-8 qRT-PCR expression in the same study (Boni-Schnetzler et al. 2008). Reanalysis of the gene array above performed using the same cohorts plus one additional non-DM showed upregulation of CCL2 and CCL13 (Igoillo-Esteve et al. 2010). No difference was noted in IFN-γ. In both studies, reflecting the heterogeneity of the T2D population, the expression levels of cytokines varied widely in the T2D cohort compared with non-DM control.

CXCL10 is a chemokine that has been shown to play an important role in early insulitis associated with T1D. Both mRNA levels, determined by qRT-PCR, and secretion of CXCL10 were increased in human islets from three T2D donors (Schulthess et al. 2009). The protein levels of heterodimeric IL-12 and the phosphorylation of downstream signaling molecule, STAT4, were increased in human islets from three T2D donors implicating the activation of IL-12 pathways (Taylor-Fishwick et al. 2013). More recently, microarray gene expression data from 48 human donors including 10 T2D donors were analyzed using a principle of weighted gene co-expression network to identify a group of genes that are associated with T2D traits. The study found that a group containing IL-1 related genes correlates with A1c (Mahdi et al. 2012). Collectively, human islets affected by T2D tend to show an increase in inflammatory humoral factors, which potentially contributes to a reduced islet function and viability. It has been shown that IL-12 and CXCL10 impair insulin secretion in human islets ex vivo (Schulthess et al. 2009; Taylor-Fishwick et al. 2013). The effects of IL-1β on the islets have been studied extensively, as it has been considered one of the major mediators of insulitis in T1D (Donath and Shoelson 2011).

The activation of IL-1β results in the production of additional pro-inflammatory cytokines in the islets and impairs islet functions (Donath and Shoelson 2011). In addition to direct toxicity, CCL2 and CXCL10 are known chemoattractants that may amplify inflammation through recruitment of Mϕ, lymphocytes, and other leukocytes. In a transgenic mouse model, CCL2 overexpression (20- to 700-fold greater than wild-type control) resulted in the accumulation of monocytes and DC in the islets and caused hyperglycemia (Martin et al. 2008). On the other hand, the overexpression of CCL2 in NOD mice was protective against the development of diabetes due to increased tolerogenic CD11b + CD11c + DC cells, indicating the complexity of immune regulation by CCL2 in the islets (Kriegel et al. 2012). Overall, the actual contributions of these factors in apoptosis and the impairment of islet function in T2D need to be further clarified.

Cellular Mediators of Inflammation in the Islets of T2D

Several rodent models of T2D show an increased accumulation of innate immune cells during the development of hyperglycemia. The GK rat, which is also discussed in chapter "Islet Structure and Function in GK Rat," shows an increase in MΦs and granulocytes in and around the islets (Homo-Delarche et al. 2006). An increase in

CD68⁺ MΦs in the islets was also noted in a histological study of C57BL/6 J mice on a high-fat diet and in db/db mice (Ehses et al. 2007).

Elevation of serum PA over the course of a 14-h infusion led to the recruitment of CD11b + Ly-6C + M1-type pro-inflammatory Mϕs into the islets of C57Bl/6 mice, indicating that metabolic alterations may be sufficient to initiate Mϕ accumulation (Eguchi et al. 2012). In humans, two studies have reported an increase in CD68⁺ Mϕs in the islets of pancreata from T2D donors (Ehses et al. 2007; Richardson et al. 2009). With the exception of one study that tested the effects of acute infusion of fatty acids (Eguchi et al. 2012), the majority of the aforementioned studies are histological analyses, leaving detailed characterization of Mϕ subtypes for future studies. Recently, an intriguing study demonstrated that Mϕ infiltration indeed may contribute to the impairment of β cell function in Zucker diabetic fatty rat, a rat model of T2D (Fig. 1). In this model, endocannabinoid that increases insulin resistance through CB1 receptor was also shown to impair the islet function by activating NLRP3 inflammasome in Mϕ that is recruited to the islets (Jourdan et al. 2013).

T2D might not show the classical criteria for autoimmune disease including loss of tolerance to self-tissue antigen, such as the β cells in case of T1D, and disease phenotypes transferable through pathological antibodies or immune cells (Velloso et al. 2013); however, immune cells classically associated with adaptive immune responses are altered in insulin target tissues and peripheral circulation of T2D patients and implicated in the development of insulin resistance (DeFuria et al. 2013; Nikolajczyk et al. 2012). Importantly, at least in rodent models, modulations that target these cells ameliorate insulin resistance (Nikolajczyk et al. 2012). Since islets are under metabolic stress that is similar to insulin target tissues, the involvement of lymphocytes and DC in islet dysfunction of T2D cannot be ruled out. The most intriguing data supporting the role of adaptive immune cells in islet dysfunction of T2D is a series of works that demonstrate T cells reactive to islet extract in the peripheral circulation of clinically defined T2D subjects (Brooks-Worrell et al. 2011, 2012; Brooks-Worrell and Palmer 2013). Currently, antigens recognized by these peripheral T cells are not defined, and the number of subjects studied is limited. If this phenomenon is to be detected in a larger population of T2D, it will support the involvement of lymphocytes in islet dysfunction of T2D.

Beyond peripheral lymphocytes, only limited information is currently available regarding the involvement of adaptive immune cells locally in the islets of T2D. The heterogeneity of islet DC has recently evaluated in detail in mice (Yin et al. 2012), but little is known about the profiles of intra-islet DC in human islets of healthy controls or T2D patients. Interestingly, significant lymphocyte infiltration consisting of CD3⁺ T cells was noted in C57BL/6 mice placed on a high-fat diet at an advanced age (Omar et al. 2013). Although C57Bl/6 mice on a high-fat diet is a commonly used model of T2D, this inbred strain of mice is known to develop autoimmune lesions in multiple organs with advanced age, making any extrapolations of this mouse model to pathology of human T2D difficult (Hayashi et al. 1989). From histological studies, the infiltration of lymphocytes is considered to be relatively limited in human T2D in general (Ehses et al. 2007). As in the case for Mϕs, detailed analysis of subset and functional status of lymphocytes and DC may demonstrate their contribution in T2D, especially the slow progressive decline in islet mass and function seen in T2D that may result from accumulation of subtle damage.

The Effects of Metabolic Alteration in Other Organs on Islet Function and Islet Inflammation

Glucose and energy homeostasis are regulated by cross talk between several key organs including the pancreatic islets, the AT, the liver, the brain, the skeletal muscle, and the gut. Here, we focus on how the AT, the liver, and the gut affect the function and health of the pancreatic islets.

The Regulation of Islet Function by AT in Health and Disease

AT is an active endocrine organ with a wide, biologically active secretome capable of controlling energy homeostasis through peripheral and central regulation. AT is also endowed with remarkable plasticity being able to respond in face of nutrient overload by more than doubling its mass and storage capacity. However, this metabolic adaptation to chronically positive energy balance leads to many pathogenic changes. In particular, glucotoxicity, lipotoxicity, and innate and adaptive immune responses in AT all lead to production of pro-inflammatory cytokines and a perturbed secretome reflected in quantitative changes of the various adipokines (Deng and Scherer 2010; Piya et al. 2013). These changes lead to the subclinical inflammation in AT that is a key pathogenic contributor to obesity-mediated insulin resistance and diabetes (Hotamisligil 2006; Ouchi et al. 2011). In this section, we will discuss the interaction between AT dysregulation and pancreatic islet inflammation leading to β cell functional demise. Since the interaction between the two organs relies on secretory factors produced by AT that both directly or indirectly affect the pancreatic islets, the focus will be on a causal relationship between metabolic dysregulation and changes in the adipocytokine production and how the latter could in turn influence β cell functional failure.

Metabolic Responses in AT Responsible for a Dysregulated Secretome

The AT secretome is composed of molecules commonly referred to as adipokines or adipocytokines. The list of these biologically active molecules is growing, as both novel and existing molecules secreted by adipocytes or AT are being discovered. It is becoming increasingly clear that adipokines form an important part of an adipo-insular axis, dysregulation of which may have key roles in β cell failure and the development of T2D (Dunmore and Brown 2013; Imai et al. 2013a).

The majority of the adipokines are peptides/proteins with hormone-like properties, including cytokines that are either specific to adipocytes or are also produced by other cells or endocrine organs. In this context, NEFA can also be considered adipokines. Phenotypically, an enlarged AT is associated with dysregulated adipokine secretion (Wellen and Hotamisligil 2003, 2005). However, the

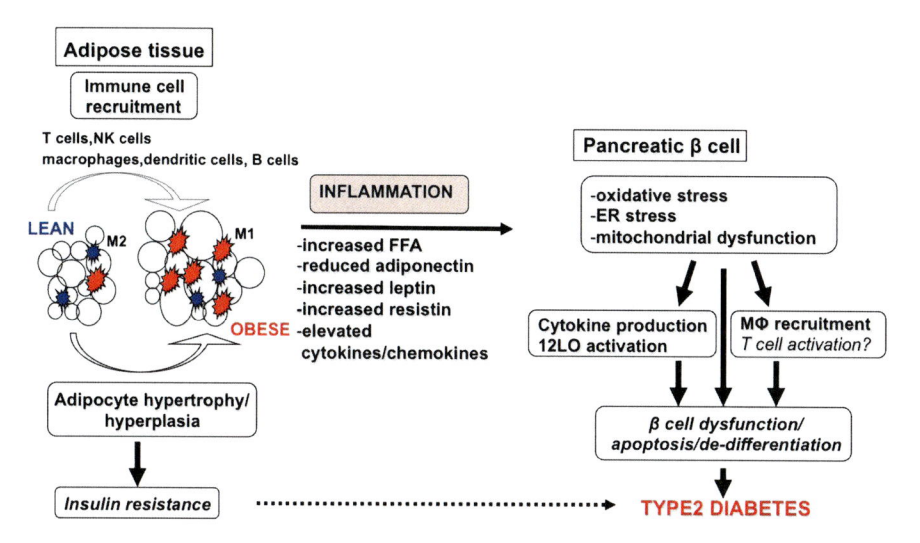

Fig. 2 Dysfunctional adipose tissue secretome in obesity drives β cell inflammation. Excess caloric excess in obesity leads to adipocyte hypertrophy and some degree of hyperplasia. Also, immune cell recruitment takes place in the adipose tissue and a macrophage (Mϕ) phenotypic shift occurs from the M2 anti-inflammatory (in *blue*) to M1 pro-inflammatory (in *red*) Mϕ. As a result adipocytes will generate a pro-inflammatory secretome: increased adipocyte leptin and resistin, reduced adiponectin, and increased chemokines and cytokine. In adipose tissue, M1 Mϕs, Th1CD4+, CD8+, and NK cells together will increase pro-inflammatory cytokines and chemokines. These circulatory factors will result in cellular stresses leading to increased production of pro-inflammatory cytokine and 12LO activation in the β cells. Also Mϕ seems to be recruited into the islets in response to cellular stresses. T cell activation may also take place in the islets under stress. The progressive inflammatory insult combined with cellular stresses will cause loss of β cell function and mass and result in type 2 diabetes when combined with insulin resistance

mechanisms leading to this dysregulation are complex and only partly understood. Both the hyperplastic adipocytes and other cells residing in the AT are significant contributors.

In Fig. 2 we illustrate some of the established key players and the effects on islet dysfunction. Adipocytes themselves can produce cytokines and chemokines in response to changes in their cell size (Skurk et al. 2007). Also, in response to increased saturated fatty acids, adipocytes activate TLR4 and NF*k*B, leading to downstream adipokine production (Youssef-Elabd et al. 2012). Hypoxia, associated with a reduced angiogenic response in the growing AT, induces pro-inflammatory responses in adipocytes including formation of adipokines such as CCL5 (Skurk et al. 2009). Other cytokines were reportedly produced by adipocytes in obesity including IL-18, CXCL10, and MIP-1 (Herder et al. 2007; Skurk et al. 2005a, b). These locally produced chemokines can contribute to the elevation of the circulating chemokine pool and hence may affect distant sites, including the pancreatic islets, the liver, and the muscle. Interestingly, epidemiologic studies showed significant correlations between systemic chemokine concentration and glucose intolerance and T2D (Herder et al. 2005). Also, adipocyte chemokine secretion may contribute to the recruitment of immune cells into the inflamed AT.

Cell components of both innate and adaptive immunity populate AT in obesity and may themselves represent an important source of inflammatory cytokines (Lumeng and Saltiel 2011). Mφs are abundantly present in AT in obesity in both rodents and humans (Chawla et al. 2011; Osborn and Olefsky 2012). In particular, the pro-inflammatory "M1-like" Mφ subset accumulates in visceral fat in obesity due to progressive lipid accumulation (Nguyen et al. 2007; Prieur et al. 2011) (Fig. 2). These Mφs, whose differentiation is promoted by lipopolysaccharide or IFN-γ, produce pro-inflammatory mediators such as TNF-α, IL-6, IL-1β, IL-12, etc. Alternatively, the activated "M2-like" Mφ, the main subset present in the lean AT, secretes anti-inflammatory molecules such as IL-10 and IL-1 receptor antagonist (Osborn and Olefsky 2012). There are many cellular mechanisms and pathways responsible for the increased inflammation in both Mφ and adipocytes in obesity. Some of these include the activation of TLR- or NLRP3-dependent pathways via NEFA or ceramides produced mainly in adipocytes (Nguyen et al. 2007; Vandanmagsar et al. 2011), as well as ER stress and autophagy (Martinez et al. 2013). Also, recent reports emphasized the pro-inflammatory roles of the incretin hormone GIP in adipocyte via increased production of cytokines and free fatty acids via lipolysis (Nie et al. 2012; Timper et al. 2013). Besides Mφ, several other immune cells populate AT in obesity such as neutrophils, mast cells, CD8[+] and CD4[+] T cells, NK cells, and B cells (reviewed in (Mathis 2013)). All of these cells may contribute to production of pro-inflammatory cytokines and chemokines that could also affect islet function via their release in circulation. The mechanisms by which various cytokines and chemokines induce islet inflammation and β cell dysfunction were described earlier in this chapter and in a few recent reviews (Donath et al. 2013; Imai et al. 2013a).

Effects of AT Adipokines on Pancreatic β Cell Dysfunction

Besides cytokines, which can be produced by adipocytes and other cells, of particular interest are the adipokines that are uniquely produced by the adipocytes such as leptin, adiponectin, omentin, resistin, and visfatin, which can also act in an endocrine manner and affect the function or dysfunction of the islet cells (Fig. 2). Ample evidence shows that the adipokine balance is significantly altered in obesity, both in rodent models and in humans, showing an increase of leptin, resistin, and TNF-α and a reduction of adiponectin and visfatin.

Leptin is known to be key in the regulation of glucose homeostasis via central and peripheral actions. Human and rodent β cells express the long form of the leptin receptor that is required for intracellular signaling, as well as the truncated forms. In obesity, excess leptin production occurs due to an increase in adipose mass; however, the peripheral and central signaling of leptin is attenuated due to the phenomenon of leptin resistance (Patel et al. 2008). It has been proposed that the leptin resistance in the β cell may lead to dysfunction and could contribute to T2D in human obesity (Morioka et al. 2007). Although the leptin signaling in the β cells is similar to other leptin-responding tissues (Marroqui et al. 2012), the mechanisms responsible for leptin resistance have not been yet studied. It is possible that increases in SOCS3 and PTP1B, reportedly associated with obesity and responsible for inhibition of leptin signaling in AT and hypothalamus, could be key (Myers et al. 2010). Additional effects of leptin on β cell mass generated contradictory results. As opposed to some rodent models, leptin in humans appears to have an antiproliferative effect on β cells that involves the PI3K pathway and inhibition of the protein and lipid phosphatase PTEN, which ultimately leads to increased activation of the K_{ATP} channels and inhibition of glucose-stimulated insulin secretion (Wang et al. 2010). Also, in human islets, leptin exerted a U-shape response on insulin secretion, with lower concentrations inhibiting insulin release and higher concentrations stimulating it (Dunmore and Brown 2013). Therefore, it is possible that in the early stages of insulin resistance, an increase in circulating leptin prior to the development of defects in signaling mechanism may be responsible for the compensatory hyperinsulinemia. In severe obesity, which is characterized by impaired leptin signaling, the latter may reduce insulin secretion and contribute to hyperglycemia and progression to T2D.

Another adipokine secreted exclusively by the adipocytes is adiponectin. Adiponectin improves insulin sensitivity and vascular function. Obesity is characterized by reduced adiponectin secretion by the hypertrophic adipocytes. Both of the adiponectin receptors are expressed in primary and clonal β cells with AdipoR1 expressed at significantly higher levels. Signaling involves AMPK, PPARγ, PPARα, and p38MAPK (Wijesekara et al. 2010). A wealth of evidence suggests a beneficial role of adiponectin in β cell function and survival. Increased adiponectin in both a GK model following gastric bypass and the clonal β cell lines BRIN BD11 and MIN1 following in vitro palmitate treatment (Brown et al. 2010a) was associated with a reduction of β cell apoptosis (Chai et al. 2011). In vivo mouse studies indicated that many of the adiponectin actions are mediated via activation of ceramidase activity, which leads to generation of sphingosine-1-phosphate (Holland et al. 2011). Furthermore, a recent report showed that adiponectin caused a significant increase in insulin content and secretion via a PPARγ-mediated effect and also had a proliferative effect independent of PPARγ in MIN6 cells (Rao et al. 2012). This latter effect apparently requires generation of reactive oxygen species that may have a beneficial role. Therefore, the combined antiapoptotic and proliferative effects of adiponectin suggest an important contribution of adiponectin to preserve β cell mass and function. The reduced levels of adiponectin found in obesity and T2D may negate these beneficial effects.

Another important adipokine that is increased in obesity and T2D is TNF-α. TNF-α can induce β cell apoptosis via the NFκB pathway and may have direct effects on insulin secretion (Ortis et al. 2012). Recently, TNF-α reportedly increased IAPP expression in β cells with no concurrent expression of proinsulin, potentially leading to amyloid production, subsequent β cell death, and systemic increase in insulin resistance (Cai et al. 2011).

Recent findings showed that DPPIV is produced and secreted by human adipocytes (Lamers et al. 2011) and therefore may reduce the half-life of GLP-1 with important implications on the insulinotropic effects of this gastric hormone on the β cell. Although it is unclear if obesity is associated with increased levels of DPPIV, inhibition of the latter by sitagliptin in a rodent model of obesity and insulin resistance reduced inflammatory cytokine and chemokine production both in the pancreatic islets and in AT and improved the glucose-stimulated insulin secretion in the pancreatic islets in vitro (Dobrian et al. 2010c).

Resistin is an adipocyte-specific-secreted molecule shown to induce insulin resistance in rodents but not in humans (Steppan et al. 2001). However, human islets express resistin and its expression is upregulated in T2D (Al-Salam et al. 2011). In clonal β cells, resistin downregulated insulin receptor expression, decreased cell viability, induced insulin resistance in the pancreatic islets, and caused a subsequent reduction in glucose-stimulated insulin secretion (Nakata et al. 2007).

One of the more recently described adipokines, visfatin, is a phosphoribosyl-transferase that is secreted from AT via a nonclassical pathway. A recent meta-analysis study indicated that visfatin can be used as a predictor for the development of insulin resistance and diabetes and is positively associated with obesity and T2D (Chang et al. 2011). Interestingly, visfatin can increase insulin secretion and directly induce activation of the β cell insulin receptors by increasing their phosphorylation (Brown et al. 2010b; Revollo et al. 2007). In a recent study, visfatin was shown to stimulate β cell proliferation and to reduce palmitate-induced β cell apoptosis via ERK1/2 and PI3K/PKB-mediated pathways (Cheng et al. 2011). Although most of the in vitro studies suggest a positive effect of visfatin on β cell function, a recent study suggests that the effect is concentration dependent with potentially deleterious effects at pathologically high concentrations (Brown et al. 2010b). Future research is needed to refine our understanding on the effects of visfatin on islet functional health.

In conclusion, several pro-inflammatory mechanisms in AT may influence pancreatic β cell inflammation, survival, and functions (Fig. 2). Additional data will be valuable in understanding whether various adipokines play a causative role in the development of diabetes and whether they contribute to the progression of T2D as well. Also, investigating the cross talk between various adipokines will refine our understanding of the delicate balance between the overall beneficial vs. deleterious effects of the AT secretome on pancreatic β cell apoptosis, proliferation, and insulin secretion. The information will provide new approaches to target the "adipo-insular" axis for more efficient future therapies.

Influence of the Gut Microbiome

There is unquestionably a role for the gut microbiome in the development of both T1D and T2D. We are increasingly appreciative of the effects of our symbiotic counterparts in the gut and how they can alter the levels of inflammation in our system by changing gut permeability to antigens and pathogens as well as altering how nutritional intake is processed. All of these factors combined can affect downstream immune responses, which may protect or exacerbate the inflammatory processes related to diabetes development and β cell demise.

T1D

Early postnatal time period plays a large role in the development of the autoimmune response leading to T1D. The antibiotic vancomycin may deplete deleterious bacteria and support *Akkermansia muciniphila*, which appears to be protective in NOD mice (Hansen et al. 2012a). It is thought that *Akkermansia* may stimulate gram-negative bacterial interactions with gut wall, leading to TLR4 stimulation. Segmented filamentous bacteria have also been shown to protect against diabetes development in NOD mice by directing immune responses in the small intestine lamina propria cells (Kriegel et al. 2011). Furthermore, it has been determined that sex differences play a role in the composition of the gut microbiome and may contribute to female predominance of diabetes development in NOD mice (Markle et al. 2013a). Gut microbiota from male NOD mice can confer protection to female mice and do so by increasing the testosterone levels in female mice.

In a rat model of virus-induced T1D, protection against diabetes development was afforded to virus-infected rats treated with trimethoprim and sulfamethoxazole (Sulfatrim), as the viral infection led to increases in *Bifidobacterium* and *Clostridium* (Hara et al. 2012). While the infection increased the level of inflammation due to innate immune cells in the Peyer's patches and pancreatic draining lymph nodes, Sulfatrim treatment decreased the level of inflammation. These data point to a role of the gut microbiota in the development of virus-induced T1D.

An additional rat model showed that supplementing diabetes-prone BioBreeding rats with *Lactobacillus johnsonii* afforded protection from diabetes development, in part by altering the expression of pro-inflammatory cytokines (TNF-α and IFN-γ) that are known to lead to β cell toxicity (Valladares et al. 2010).

Enteric bacterial pathogens may disrupt the gut barrier, causing acceleration of insulitis (Lee et al. 2010), as this is one mechanism that increases insulitis in the NOD mouse strain. $CD8^+$ T cells are activated by the loss of intestinal barrier and pathogenic bacteria. Additionally, there are several lines of evidence that suggest increased intestinal permeability in human T1D patients (Bosi et al. 2006; Secondulfo et al. 2004).

Studies of AAb + vs. AAb- children matched for HLA genotype, age, sex, and feeding history suggest that *Bifidobacterium* (Actinobacteria, decreased), *Bacteroides* (increased), and butyrate-producing (decreased) species are altered in those with β cell autoimmunity (de Goffau et al. 2013). These differential profiles of microbiota might affect the intestinal epithelial barrier function and thereby modify the levels of inflammation. The authors of this work discussed the advantage of performing fecal microbe profiling before diagnosis to eliminate the possible influence of diabetic status on the gut microbiota of individuals and thereby justifying the analysis of those that were AAb + vs. AAb-.

Several studies have shown that T1D patients exhibit skewed gut microbiota as compared to controls (Brown et al. 2011; Giongo et al. 2011). Once diagnosed with T1D, patients appear to have decreased *Bifidobacterium* and increased *Bacteroides*, indicating that this trend continues beyond disease onset (Murri et al. 2013).

In this study, patients were also matched by age, sex, dietary habits, race, mode of delivery at birth, and duration of breastfeeding. Furthermore, the *Bacteroides* genus appears to be associated with T1D patients, while the enterotype 2 groups (*Prevotella* genus) are more strongly affiliated with healthy patients.

It is certainly possible that the gut microbiome of T1D patients is directly affected by viral infections (i.e., enterovirus), leading to altered immunity and the development of T1D (Hara et al. 2012).

T2D

Obesity and T2D are physiological states shown to have altered gut microbiota (Larsen et al. 2010). Both states show significant decreases in *Akkermansia muciniphila*, which degrades the mucus layer of the gut. Supplementing obese and T2D mice with this prebiotic correlates with improved metabolic profile, including decreased AT inflammation, decreased insulin resistance, and increased levels of the gut barrier, gut peptide secretion, and endocannabinoids, which control inflammation (Everard et al. 2013). Perhaps treatment to increase this colonizer will aid in treating obesity and metabolic disorders. Additionally, selectively increasing *Bifidobacterium* via prebiotics in *ob/ob* mice appears to increase the GLP-2 production, which decreases intestinal permeability and inflammation that is associated with obesity (Cani et al. 2009). Furthermore, rodents on a high-fat diet are more likely to exhibit decreases in *Lactobacillus* strains and *Bifidobacterium*. Decreases in these two strains correlate inversely with plasma glucose levels, insulin sensitivity, and inflammation (Chen et al. 2011; Sakai et al. 2013). These two groups, along with others, have shown that dietary supplementation with certain *Lactobacillus* strains and *Bifidobacterium* strains can alleviate high blood glucose levels (Honda et al. 2012).

Another attempt to alter the composition of the gut microbiome toward a less inflammatory phenotype includes treating mice to decrease phenolic acids by prebiotic supplementation with green tea powder and Lp (Axling et al. 2012). This study also corroborates the idea that increased levels of *Akkermansia* are complicit in blunting inflammatory responses. The authors conclude that green tea powder and Lp exert their effects on different portions of the gut, with the green tea powder standardizing the flora in the small intestine and the Lp standardizing flora in the cecum. Among the parameters tested, green tea powder lowered the HOMA index and increased insulin sensitivity in treated mice, but did not improve oral glucose tolerance.

In human patients, compositional alterations have been noted in the gut microbiome of those who have normal, impaired, and diabetic glucose control (Karlsson et al. 2013). This study suggests that the risks for diabetes development can be assessed more accurately by studying the gut microbiome and metagenomic clusters of patients.

The overarching theme in this arena is that inflammation resulting from obesity can be downplayed by altering the endogenous gut microbiome. It remains unclear which change occurs first: are gut microbiota affecting metabolic markers, or are the changes in gut microbiota result from metabolic alterations? Furthermore, it appears that the changes in the gut microbiome of T1D and T2D patients do not necessarily undergo the same kinds of changes. Little is known whether the gut microbiome directly contributes to islet dysfunction in T2D. This is an area for active future investigation.

The Communication Between the Liver and Islets

The liver plays the critical role in glucose and energy homeostasis and has close bidirectional relationship with pancreatic islets. Insulin being secreted from the islets is first delivered to the liver via the portal vein, and then the liver will take up 80 % of insulin to regulate the metabolic function of the liver (Meier et al. 2005). Indeed, pulsatility of insulin delivery in the liver was shown to be an important determinant of hepatic insulin sensitivity in an experimental setting (Matveyenko et al. 2012). Thus, islet dysfunction may be a proximal cause for insulin resistance in the liver. However, it remains to be determined whether the impairment of insulin secretion contributes to the development of hepatosteatosis or steatohepatitis in humans.

On the other hand, recent studies demonstrate that the liver regulates insulin secretion and β cell mass through neuronal and humoral mechanisms. The activation of ERK in response to insulin resistance (*ob/ob* leptin-deficient mouse) or insulin deficiency (STZ or Akita mouse) was shown to trigger an increase in β cell mass through a neuronal relay, connecting the liver and pancreas by involving afferent splanchnic and efferent pancreatic vagal nerves (Imai et al. 2008). The unique contribution of the liver in the determination of islet mass was also implicated by mice with tissue-targeted deletion of insulin receptors. The genetic deletion of insulin receptors from the liver, but not from the muscle or AT, results in an increase in islet mass (El Ouaamari et al. 2013). Interestingly, the blockade of insulin signaling, either pharmacologically or genetically, results in the increase of β cell proliferation at least partly through a humoral pathway (El Ouaamari et al. 2013; Yi et al. 2013). Recently, an effort to identify a factor responsible for humoral mitotic signal leads to betatrophin (also known as ANGPTL8, lipasin), a 22 kDa peptide secreted from the liver (Yi et al. 2013). Currently, little is known regarding the pathway by which betatrophin promotes β cell proliferation. Furthermore, translational value of betatrophin requires additional studies, as it is associated with an increase in serum triglycerides and hepatic cancer (Quagliarini et al. 2012). Also, it remains an important area of research to determine whether or not humoral and neuronal regulations of islet mass by the liver are impaired when integrity of the liver is affected, such as in nonalcoholic fatty liver disease.

T1D and T2D: Are They a Continuum or Distinct Entities?

T2D subjects include a heterogeneous population with variable degrees of insulin resistance and β cell dysfunction. On the other hand, T1D is regarded as an independent entity of disease characterized by autoimmune destruction of the β cells (Atkinson et al. 2013). However, the question has arisen whether these 2 forms of diabetes are completely different, as the chronic inflammation is not only limited to insulin target tissues but also seen in the islets in T2D. Indeed, interventions targeting inflammatory pathways, such as IL-1β, DMT1, 12LO, and others, have shown efficacy in animal models of both T1D and T2D (Dobrian et al. 2011; Ehses et al. 2009; Hansen et al. 2012b; Tersey 2012a; Thomas et al. 2004), highlighting the possibility that particular anti-inflammatory therapy may become a unifying therapeutic tactic for both forms of diabetes (Imai et al. 2013a). To consider a continuum for the two forms of diabetes, there are two major questions to be addressed: whether insulin resistance modifies development of autoimmunity in T1D (Wilkin 2009) and whether the alteration in β cells during T2D development shares any commonalities with autoimmune insulitis associated with T1D.

Does Insulin Resistance Modify the Development of T1D?

One of the strongest arguments supporting the role of insulin resistance in T1D development comes from epidemiology. The recent global rise in obesity has been followed by increase in T1D incidence in many areas of the world (Atkinson et al. 2013). Meta-analysis based on five clinical studies supported association between obesity and subsequent development of T1D (pooled odds ratio of 1.25 with 95 % CI 1.04–1.51) (Verbeeten et al. 2011). Ex vivo and animal studies indicate that AT inflammation, lipotoxicity, ER stress, mitochondrial dysfunction, oxidative stress, and other pathways activated by obesity and insulin resistance could trigger inflammatory processes in the islets. It is intriguing to postulate that these insults may trigger or amplify autoimmune process in T1D. A recent report revealed signs of ER stress and NFκB activation along with early insulitis at a prediabetic stage of NOD, indicating that these cell distress pathways may interact and amplify autoimmune insulitis (Tersey et al. 2012b). Although not directly relevant to the pathogenesis of β cell demise in T1D, it is noteworthy that environmental pressure and advancement in insulin therapy have seen the substantial proportion of T1D subjects becoming insulin resistant as adults. The recent review of "double diabetes" (T1D exhibiting insulin resistance) provides excellent insight into the contribution of insulin resistance to cardiovascular risks in T1D population (Cleland et al. 2013).

Does Autoimmunity Contribute to Islet Dysfunction in T2D?

Increases in Mφs and inflammatory cytokines localized in affected islets provide strong support for the involvement of innate immune responses in T2D development (Imai et al. 2013a). In contrast, the adaptive immune response against the islets has been considered to play a little role in T2D (Velloso et al. 2013). UKPDS and other large population-based studies have shown low levels of positivity for islet autoantibodies, including GADA, ICA, or IA-2A, in clinically defined T2D, UKPDS study being 11.6 % positive (Davis et al. 2005; Lohmann et al. 2001).

Those positive likely represent subjects with autoimmune diabetes or LADA, which shares pathogenesis with T1D but initially presents as non-insulin-requiring diabetes in adulthood (Rolandsson and Palmer 2010). However, a series of studies by Brooks-Worrell et al. indicate that a significant proportion of clinically defined T2D patients may harbor autoimmunity against the islets (Brooks-Worrell and Palmer 2012; Brooks-Worrell et al. 2011). They detected islet antibodies, including ICA, GADA, IAA, and IA-2A, along with T cell responses against human islet extracts, in 36 phenotypical T2D of less than 5-year duration, and A1C below 8 % on one non-insulin, diabetic medication. They found 11/36 cases were positive for at least one AAb and 22/36 cases harbored T cells in peripheral circulation that showed proliferative responses against human islet extracts. The glucagon-stimulated C-peptide response was significantly lower in those with T cell responses against the islets, implicating an association of T cell responses and β cell demise. Of 11 antibody positive subjects, only 4 were positive for GADA, an antibody commonly used to diagnose LADA (Brooks-Worrell et al. 2011). Together with islet-reactive T cells, the study proposes the presence of adaptive immune responses in a significant population of clinically defined T2D who do not fit into the classic classification of autoimmune diabetes or LADA. Further studies will be required to test whether similar frequencies are seen in large populations of clinically defined adult T2D, especially those with circulatory islet-reactive T cells, as these have not been analyzed widely (Velloso et al. 2013).

In closing, it needs to be stressed that there are many distinctions between typical T1D and T2D that justified the classification of diabetes. Age of onset, rate of progression for β cell destruction and insulin deficiency, profiles of AAb, genetic susceptibility, and histology of the islets are clearly different between typical T1D and T2D (Brooks-Worrell et al. 2012; Cnop et al. 2005; Igoillo-Esteve et al. 2010). Additionally, genetic susceptibility foci identified by GWAS are mostly distinct between T1D and T2D. GLIS3, a Kruppel-like zinc finger protein important for the development of the pancreas and normal insulin secretion from adult β cells, is exceptional, being associated with both T1D and T2D (Nogueira et al. 2013). Also, neither T2D risk loci identified by GWAS nor body weight led to T1D progression, at least in the first-degree relatives of T1D, cautioning against attributing the increase in T1D incidence seen in recent years to the increase in the incidence of obesity (Winkler et al. 2012). Therefore, the field needs to continue to evaluate commonality and distinction between two forms of diabetes, so as to obtain more insights into the pathways involved in β cell demise in both forms of diabetes.

Inflammatory Pathways as a Therapeutic Target for Diabetes

Effort to Halt β Cell Loss Through Anti-inflammatory Therapy in T1D

The remission of newly diagnosed T1D by cyclosporine originally provided support for immunomodulatory therapy for T1D in 1980s (Staeva et al. 2013). As cyclosporine is not suitable for long-term therapy due to toxicity and side effects, many scientists have made an effort to treat and/or prevent T1D in humans by targeting the immune pathways. Thus far, little has come of these efforts despite the many treatments that cure disease in NOD mice (reviewed in (Shoda et al. 2005)). More recent work has studied novel targets shown to be involved in the pathogenesis of T1D. The dipeptidyl peptidase-IV inhibitor, MK-626, has beneficial effects on β cell area, insulitis, and regulatory T cell populations when paired with the histone deacetylase inhibitor, vorinostat, in NOD mice (Cabrera et al. 2013). In an effort to stimulate regulatory T cells in vivo, but preventing the exacerbation of autoimmunity, the combination of rapamycin and IL-2 was tested in mice (Baeyens et al. 2013). While both show positive results alone, these results showed unexpected deleterious effects by stimulating NK cells and directly affecting β cell health. In humans, the β cell function was only transiently affected, while both regulatory T cells and NK cells were augmented in a persistent manner (Long et al. 2012).

One goal of new therapeutics to treat T1D is aimed at reducing the amount of inflammation locally in the pancreas of T1D patients, thus limiting toxicity to other organs. One means of accomplishing this goal is to use improved delivery systems that specifically target the pancreas. In one instance, nanoparticles have aided drug delivery in a targeted fashion (Ghosh et al. 2012). Others are working on the generation of tolerized autologous cells to squelch the adaptive immune response. Regulatory T cells have been the cells of choice for many years (Putnam et al. 2009). More recent studies are honing in on dendritic cells. In vitro development of vitamin D3-dexamethasone-modulated dendritic cells has significant hurdles to overcome before it could be considered as a potential therapy for T1D (Kleijwegt et al. 2013). The current studies have tolerized $CD8^+$ T cells to themselves, which works well with naïve $CD8^+$ T cells, but not with memory T cells.

A comprehensive list of human trials and outcomes was recently reviewed by von Herrath et al. (von Herrath et al. 2013). For the most part, few of these trials have shown any positive outcomes. The results are not fully available for all of the studies covered to date. Several studies concentrated on anti-inflammatory treatments, which might help protect β cells prior to disease onset (i.e., DHA treatment in infants at high risk (Miller et al. 2010, 2011)), or controlling diet in infants (removing gluten has minimal effect (Frederiksen et al. 2013; Hummel et al. 2011), but removal of bovine insulin shows decreased levels of islet AAb (Hummel et al. 2011; Vaarala et al. 2012)). Many studies looked at the effect of reducing the IL-1β levels; however, these treatments do not appear to halt disease progression in T1D (Moran et al. 2013).

Results from human islet transplantation studies have demonstrated the potential to reverse T1D by replacement of the functional β cell pool (Ryan et al. 2002; Shapiro et al. 2000). As a curative strategy, islet transplantation is currently limited by a number of factors including donor islet availability, donor-recipient ratio, and lack of long-term islet survival (Taylor-Fishwick et al. 2008). Dysfunction of the islet graft results, in part, from inadequate vascularization and inflammation-induced damage. The role of inflammation in β cell damage is recognized, but the cellular events that mediate the effects have been less well defined. Consequently, research with foci on pathogenic mechanisms resulting in β cell dysfunction and reduced β cell survival will improve the outcome of islet transplant.

A central contribution of inflammation and inflammatory cytokines to β cell dysfunction is anticipated to apply to T1D development and must be addressed in strategies to slow, halt, or reverse diabetes progression. Description of an integrated pathway associated with T1D development with discrete points for therapeutic intervention is encouraging when considering the future for T1D therapeutics. However, the restoration of β cell mass remains as the major challenge, since substantial β cell mass is already lost when T1D subjects present with hyperglycemia. The greatest promise for an effective therapy against T1D would be one that confers protection to β cells from the damaging effects of inflammation that is combined with a strategy to enhance β cell mass.

Therapeutics Targeting IL-1 Pathway

Several clinical trials have tested the efficacy of IL-1 pathway for treatment of both T1D and T2D. It is not surprising since the IL-1 pathway has been strongly implicated in both forms of diabetes. As several IL-1-targeted compounds are already in clinical use for systemic inflammatory diseases such as rheumatoid arthritis (anakinra), neonatal onset multisystem inflammatory disease (anakinra), cryopyrin-associated periodic syndromes (canakinumab), and systemic juvenile idiopathic arthritis (canakinumab), the availability of agents facilitated the initiation of clinical trials for their application to diabetes. As current status of IL-1-targeted therapy for T1D is discussed above, here, we focus on its application to T2D.

In 2007, Larsen et al. reported the first proof-of-concept study that showed the efficacy of targeting the IL-1 pathway to improve glycemic control in T2D (Larsen et al. 2007). Anakinra is a synthetic analog of the naturally occurring antagonist for IL-1Ra that blocks the action of IL-1α and IL-1β. It reduced A1c by 0.46 % in T2D who had pretreatment A1c of 8.7 % in a 13-week placebo-controlled, double-blinded, randomized trial. The reduction in hsCRP and IL-6 supported the idea that anakinra elicited anti-inflammatory effect in the study cohort. The improvement in insulin secretion was considered to contribute to reduction of A1c based on the increase in C-peptide and the reduction of the proinsulin/insulin ratio in the absence of significant improvement in insulin sensitivity (Larsen et al. 2007). Thirty-nine-week follow-up of the same cohort after cessation of anakinra showed that the improvement in proinsulin/insulin ratio was maintained, along with the reduction in hsCRP and IL-6. The increase in C-peptide was not maintained when all T2D were combined. However, a subgroup who initially responded to anakinra by reduction in A1c after 13 weeks of treatment continued to maintain higher C-peptide compared with nonresponders after termination of anakinra (Larsen et al. 2009).

IL-1β antibodies (gevokizumab (XOMA-052), canakinumab, LY2189102) are another class of therapeutics targeting the IL-1 pathway. IL-1β antibodies have benefit of a longer half-life compared with recombinant IL-1Ra, thereby allowing less frequent injections, and they block IL-1β without blocking IL-1α or other cytokines. Gevokizumab administered to T2D with average A1c of ~9 % reduced A1c by 0.85 % and increased both C-peptide and insulin sensitivity after 3 months (Cavelti-Weder et al. 2012). Twelve weeks of phase II trial of LY2189102 in T2D showed ~0.3 % reduction in A1c along with the reduction in glycemia and inflammatory markers (Sloan-Lancaster et al. 2013). Diacerein is an anti-inflammatory agent marketed for osteoarthritis in some countries that also reduces IL-1β. In a 2-month randomized double-blinded, placebo-controlled trial, diacerein increased insulin secretion without changing insulin sensitivity in drug naïve T2D (Ramos-Zavala et al. 2011). Overall, IL-1-targeted therapies were well tolerated in diabetic subjects. However, it should be noted that studies so far include relatively small numbers of subjects treated for a limited period of time. Also, several of the IL-1-targeted trials did not show an improvement in glycemic control in T2D. There was no improvement in insulin sensitivity or β cell function following 4 weeks of treatment with anakinra in 19 prediabetic subjects (13 completed the study), despite the reduction in CRP and blood leukocyte counts (van Asseldonk et al. 2011). In another study, 4 weeks of treatment with canakinumab resulted in statistically significant improvement in insulin secretion upon meal challenge in subjects with impaired glucose tolerance; however, there were no statistically significant changes in subjects with well-controlled diabetes despite the reduction in hsCRP (Rissanen et al. 2012). It is apparent that a larger study with longer duration is needed to clarify the clinical benefits of IL-1-targeted therapy in T2D. CANTOS trial is a multinational, event-driven, intent-to-treat protocol enrolling 17,200 stable, post-MI subjects with persistent elevation of C-reactive protein (Ridker et al. 2011).

Although early reports following 4 months of canakinumab administration to relatively well-controlled T2D (A1c 7.4 %) from the CANTOS trial showed no improvement in A1c despite the reduction in CRP (Ridker et al. 2012), the trial will likely provide much needed data about the long-term efficacy of IL-1-targeted therapy on glycemic control. In addition, the trial monitors the prevalence of new-onset diabetes and provides the information regarding its efficacy in prevention of T2D. As the focus of trial is on cardiovascular outcomes, it will provide critical information about overall mortality and morbidity benefits for T2D. Drawbacks for IL-1-based therapies include the cost of currently available formulations and the fact that most, except for diacerein, need to be administered by injections.

From the basic research point of view, there also is some reservation for long-term benefit of IL-1 targeted therapy. PA, one of the major saturated fatty acids found in circulation in humans, provokes apoptosis of human islets in culture, a process considered to be involved in β cell demise in T2D. This process is accompanied by the production of multiple cytokines including IL-1β, TNF-α, CCL2, IL-6, CXCL1, and IL-8. Although IL-1β antagonism was able to abolish the upregulation of many cytokines upon PA treatment, it failed to prevent apoptosis (Igoillo-Esteve et al. 2010). Also IL-6, one of the cytokines induced by IL-1β in the islets, may be beneficial for the islets by increasing insulin secretion and islet function through GLP-1 secretion at least when tested in a noninflammatory condition (Ellingsgaard et al. 2011).

Nonacetylated Salicylate for T2D

Salicylic acid and its derivatives, salicylates, have been used to alleviate inflammatory conditions in humans for centuries. They are originally derived from plants where they act as hormones regulating growth, development, and defense against pathogens. The hypoglycemic effect of salicylates was documented in 1876 (reviewed in Goldfine et al. (2011)) but received little attention until its role as an inhibitor of the NFκB pathway attracted attention. Nonacetylated salicylates, such as salsalate, have very weak activity as cyclooxygenase inhibitors but suppress the NFκB pathway, which is known to be activated in animal models of obesity and T2D (Donath and Shoelson 2011). However, salicylates may elicit hypoglycemic effects through multiple mechanisms in addition to inhibition of NFκB pathway. Proposed targets of salicylates encompass a wide range of enzymes and transcription factors, including mitochondrial dehydrogenases involved in glucose metabolism (Hines and Smith 1964) and transcription factors implicated in immune cell activation (Aceves et al. 2004). Recently, the direct activation of AMPK was proposed as another mechanism responsible for at least part of the metabolic effects of salicylates (Hawley et al. 2012). Salicylates have also been shown to suppress the activation of 11-β-hydroxysteroid dehydrogenase in AT and may prevent glucocorticoid production and insulin resistance (Nixon et al. 2012). There were several clinical trials that tested efficacy of salsalate for glycemic control in T2D. TINSAL-T2D is a multicenter randomized clinical trial that tested the efficacy of salsalate for T2D. Stage I, involving 14 weeks of 3, 3.5, and 4 gm/day salsalate for T2D, showed a significant reduction of A1c ranging 0.3–0.5 % in all doses of salsalate tested (Goldfine et al. 2010).

For the stage II trial, 286 T2D patients were enrolled in double-blinded, randomized, prospective study of 3.5 gm salsalate daily for 48 weeks as an add-on to the diet or other therapies excluding injectables and thiazolidinedione. Although A1c reduction has been modest (0.37 %, 0.53–0.21 95 % CI, $p < 0.001$), there was a reduction in the use of diabetic medications. The treatment group showed increases in fasting insulin levels and a reduction in C-peptide. Reductions in uric acid, glucose, leukocytes, and triglycerides and an increase in adiponectin were noted, indicating an overall improvement in metabolic syndrome. Increases in LDL and body weight, however, were concerning. A reversible increase in urine microalbumin was also noted in the study (Goldfine et al. 2013b). Sixty drug naïve, newly diagnosed T2D treated for 12 weeks with 3 gm/kg of salsalate showed reductions in fasting glucose (6.3 ± 0.2 mmol/l to 5.4 ± 0.2 mmol/l), triglycerides, and WBC, and a 0.5 % decline in A1c. This was accompanied by increased fasting insulin (Faghihimani et al. 2013). A 12-week, randomized placebo-controlled study of salsalate in 71 subjects with impaired fasting glucose and/or impaired glucose tolerance showed 6 % reduction in fasting glucose along with 25 % reduction in triglyceride with 53 % increase in adiponectin (Goldfine et al. 2013a). Overall, the reductions in glucose and A1c were modest but were generally reproducible in multiple studies. Several mechanisms responsible for improvement in glycemia are proposed from clinical data. Fasting C-peptide was reduced, arguing against improvement in β cell function (Goldfine et al. 2013a).

In addition, euglycemic hyperinsulinemic clamp testing did not support the improvement in insulin sensitivity (Goldfine et al. 2013a). Several studies implicate salsalate in reducing the clearance of insulin, resulting in reduced glycemia in humans (Hundal et al. 2002; Koska et al. 2009). The short half-life of salsalate requires multiple doses per day, and mechanisms of action are not fully elucidated, especially at the high dose used in clinical trials. However, this low-cost and orally active compound holds potential to serve as T2D agent, especially if shown to have cardiovascular benefit as well. Currently, cardioprotective effects of salsalate are being evaluated in the ongoing TINSAL-CVD trial.

12LO as a Target for T1D and T2D

The role of 12LO in the development of T1D is gaining appreciation. Several models have shown that T1D pathogenesis is disrupted in the absence of 12/15LO (a mouse gene for 12LO) in mice (Bleich et al. 1999; McDuffie et al. 2008). Further research has shown that it is likely that both Mϕ and islet expression of 12LO contribute to disease development (Green-Mitchell et al. 2013). Ongoing studies will investigate the contribution of individual cells to diabetes development in the NOD mouse model. In humans, 12LO is expressed in the islets, although specific cellular expression has not yet been determined (Ma et al. 2010). 12LO has also been implicated in the pathogenesis of enteroaggregative E. coli infection (Boll et al. 2012), which leads to upregulation of inflammation and possible alterations of the flora in the gut. Therefore, it is possible that the 12LO pathway could be activated by changes in the gut microbiome. The collective works thus far have indicated that inhibitors of 12LO might suitably and selectively target a portion of the inflammatory pathway with relatively minor side effects, as knockout mice show no major ill effects. In vitro investigations have shown promise in protecting the islets from cytokine injury using newly developed inhibitors of the 12LO pathway (Weaver et al. 2012).

The role of 12LO is not limited to islet inflammation in T1D. 12LO and their products play important roles in many tissues and organs, including the vasculature, kidney, AT, brain, and pancreatic islet during the development of T2D and their complications (Dobrian et al. 2010a). There is emerging evidence that the various 12LO isoforms (platelet 12LO, 12/15LO) in both white and brown fats are expressed in multiple cell-type constituent of AT: adipocytes, vascular cells, macrophages, and pre-adipocytes. The presence and roles of 12LO in adipocytes have been extensively investigated by our group and by others in mouse models of obesity, insulin resistance, and T2D. C57BL/6 J mice that have been on a high-fat diet for 8–16 weeks exhibit increased expression of 12/15LO in isolated white adipocytes (Chakrabarti et al. 2009), and Zucker obese rats, a genetically induced rodent model of obesity and insulin resistance, also exhibit increased expression of 12/15LO in isolated white adipocytes compared to lean controls (Chakrabarti et al. 2011). Also, more recent data from our lab showed that in db/db mice, leukocyte and platelet 12LO in AT and pancreatic islets undergo expressional increases that coincide with the metabolic decline.

Addition of one of the primary 12LO metabolites, 12-HETE, to fully differentiated 3T3-L1 adipocyte cultures significantly induced pro-inflammatory gene expression and secretion of many pro-inflammatory cytokines, including TNF-α, CCL2, IL-6, and IL-12p40. Also, the anti-inflammatory adiponectin was significantly decreased under these conditions. Importantly, addition of the same metabolites led to an increase in activation of c-Jun, while insulin-mediated activation of key insulin signaling proteins such as Akt and IRS1 was decreased. Addition of PA to 3T3-L1 adipocytes increased 12/15LO expression with concomitant increased cytokine expression. Also, recent evidence from our lab demonstrates that 12LO is a novel inflammatory pathway that mediates ER stress in the adipocyte (Cole et al. 2012a). Interestingly, in a 12/15LO conditional knockout in AT, the inflammation in response to high dietary fat was reduced not only in adipocytes but also in the pancreatic islets, and the glucose intolerance was restored, indicating the important role of 12/15LO expressed in AT for the islet functional demise (Cole et al. 2012b).

Relevant for the translational value of the therapeutic potential of various 12LO inhibitors, important differences were emphasized in the 12LO pathway between rodents and humans (Dobrian et al. 2010a). In a recent publication, we reported human 12LO mRNA and protein expression in human AT with exclusive localization in the SVF both in the SC and in the OM fat (Dobrian et al. 2010b). Increased expression of all of the 12LO enzyme isoforms in OM vs. SC AT suggests that the pathway may contribute to the pro-inflammatory milieu prominently associated to visceral fat in obesity (Dobrian et al. 2010b).

Recent evidence suggests a pro-inflammatory role of 12LO pathway in humans. Gene array analysis of AT showed that arachidonic acid metabolism is the second most significantly upregulated pathway in human omental compared to subcutaneous AT in human obese subjects. The very limited information on 12LO functional roles and changes with different pathological conditions in human AT grants future studies to identify the roles of different isoforms and the lipid mediators that are key for regulation of inflammation in human obesity and T2D.

While understanding the mechanisms that control the regulation of the 12LO pathway require substantial future efforts, the contribution of this pathway to inflammation in AT in obesity is well established and may constitute a valuable future therapeutic target. Therefore, 12LO inhibitors hold promise in T2D therapy by reducing insulin resistance and restoring insulin secretion.

Anti-inflammatory Effects of Therapies Currently Available for Diabetic Patients

Some of the drugs that are currently used to treat hyperglycemia in T2D (incretins, PPARγ agonists) or various diabetic complications, such as hypertension (ACE inhibitors, angiotensin receptor blockers), have anti-inflammatory effects that may contribute to their overall positive outcomes. Also, bariatric surgery performed in morbidly obese subjects in particular the Roux-en-Y procedure has anti-inflammatory effects in subjects with T2D that may be in part independent of weight loss.

GLP-1 is a gut incretin hormone that reduces hyperglycemia primarily via insulinotropic effects and by reducing postprandial excursions of plasma glucose (Drucker 2006). Modulation of the GLP-1 system is a current therapeutic option for the treatment of T2D and involves the use of either long-acting GLP-1 analogs, mimetic peptides acting as GLP-1 receptor agonists (Estall and Drucker 2006), or inhibitors of the DPPIV (such as sitagliptin and analogs) that extend the endogenous GLP-1 half-life (Drucker 2007). Both approaches proved successful for the control of glycemia in T2D. However, recent evidence unveiled off-target anti-inflammatory effects of the GLP-1 analogs and DPPIV inhibitors that may account for their overall positive therapeutic outcome independent of glucoregulation. A recent report showed that in a cohort of obese T2D subjects, 8 weeks after starting GLP-1 analog therapy, soluble CD163, a molecule released from inflammatory macrophage activation, was significantly reduced in the cohort that received therapy independent of reduction in A1c. Also, GLP-1 therapy in the cohort of T2D patients reduced the basal levels of pro-inflammatory cytokines TNF-α and IL-6 while increasing adiponectin levels (Hogan et al. 2013).

Also, GLP-1 overexpression via adenoviral GLP-1 delivery to *ob/ob* mice was shown to improve insulin sensitivity and glucose tolerance (Lee et al. 2012). The effect may be attributed to the anti-inflammatory effects in adipocytes and AT reflected by reduced JNK-related signaling and a bias toward the M2 anti-inflammatory macrophage phenotype in AT. GLP-1 agonism may also influence local inflammation by modulating innate immunity. iNKT cells have a rather controversial role in the pathogenesis of obesity-related insulin resistance with reports showing either a pro- or an anti-inflammatory role. A recent report showed a significant improvement in psoriasis plaques in subjects with obesity and T2D treated with the GLP-1 agonist exenatide (Hogan et al. 2011). Interestingly, the effect was mediated via activation of iNKT cells that have been shown to have insulin-sensitizing effects in AT. This opens the possibility to modulate the NK-cell phenotype by using GLP-1 agonists which may enhance the insulin-sensitizing effects of the treatment. DPPIV inhibitors represent another class of drugs that emerged as potent anti-inflammatory agents in various tissues. DPPIV is expressed in the immune cells, vasculature, pancreatic islets, and AT. Aside from its enzymatic activity, DPPIV has multiple nonenzymatic effects including stimulating T cell proliferation and monocyte migration through the endothelium, cytokine production, and processing of various chemokines.

The multiplicity of functions and targets suggests that DPPIV may play a distinct role aside from its effects on the incretin axis (Zhong et al. 2013). Several recent studies emphasized anti-inflammatory actions of DPPIV inhibitors in various tissues. Our lab was among the first to report a reduction in pro-inflammatory cytokine production in the AT, adipocytes, and pancreatic islets of C67Bl6 mice on high-fat diet treated for 12 weeks with the DDPIV inhibitor sitagliptin (Dobrian et al. 2010c). Other studies also showed anti-inflammatory effects of vildagliptin in the islets and enhanced β cell function in an advanced-aged diet-induced obesity mouse model (Omar et al. 2013) as well as reversal of new-onset diabetes in NOD mice through modulation of inflammation in the islets and stimulation of β cell replication (Jelsing et al. 2012). Clinical trials are currently ongoing aiming to establish the beneficial anti-inflammatory effects of different DPPIV inhibitors on β cell function and on the cardiac and vascular function.

Hypertension is a common complication of obesity and T2D. Inhibition of the renin-angiotensin axis via ACE inhibitors or ARb represents the preferred approach in patients with obesity and T2D. Modulation of the renin-angiotensin system may also have anti-inflammatory effects aside from the blood pressure-lowering effect. A number of studies with ARbs found a reduction in systemic inflammation with reduced levels of TNF-α, IL-6, and C-reactive protein in hypertensive patients with T2D (Fliser et al. 2004; Pavlatou et al. 2011). The angiotensin receptor type 1 is expressed in several immune cells including T cells, monocytes, and macrophages. Not surprisingly, the ARb telmisartan was found to modulate AT macrophage polarization toward an M2-like anti-inflammatory phenotype in diet-induced obese mice (Fujisaka et al. 2011).

Importantly, several clinical trials have indicated that ACEs and ARbs reduce the incidence of new-onset T2D in high-risk populations. This effect was in part independent from the blood pressure-lowering activity and attributed to improved insulin signaling in the muscle and adipocyte (van der Zijl et al. 2012). Whether these effects are attributable to reduced inflammation remains to be determined. Additional anti-inflammatory drugs currently used for other conditions may have off-target anti-inflammatory effects and could be selected in the future keeping in mind the convincing evidence for the safety profiles in humans.

Concluding Remarks

Substantial progress has been made in recent years for the characterization of inflammatory response in the islets of both T1D and T2D. It is clear that the loss of tolerance to β cells in T1D is established through complex interaction between genetic and environmental factors, β cells, humoral mediators, and wide ranges of immune cells. Importantly, metabolic stress classically associated with T2D may modify immune response and may be an important contributing factor for the development of T1D. For T2D, metabolic stress such as glucolipotoxicity, ER stress, and oxidative stress is likely responsible for inflammatory response in the islets that in turn amplifies ER stress and oxidative stress, accelerating loss of β cell mass and function. Mϕs and humoral mediators such as IL-1β, CCL2, 12LO, IL-12, and CXCL10 are implicated in islet inflammation in T2D, but the role of adaptive immunity in the islets of T2D has not been firmly established. Islet functions and inflammatory response in the islets are likely under effects of metabolic status of other tissues such as the AT, the liver, and the gut. Further research is warranted in this area. Anti-inflammatory/immune therapy for T1D has been challenging despite the numerous efforts. The area may benefit from analyses of human donor pancreata to better understand pathogenesis of T1D in humans. Two major anti-inflammatory therapies tested for T2D to date are IL-1β-targeted therapy and salsalate. Both show promise and support the contribution of inflammatory response in β cell dysfunction of T2D. However, the most effective target for anti-inflammatory therapy to halt heterogeneous and the slow progressive decline in the islets in T2D is yet to be determined. Thus, islet inflammation will likely remain an important area of research for the coming years.

Acknowledgement Human studies in unpublished data were approved by the Institutional Review Board at EVMS. Human islets in unpublished data were provided by Integrated Islet Distribution Program (IIDP). Funding support for the authors includes Juvenile Research Foundation grant (Nadler, Taylor-Fishwick), American Diabetes Association (Morris), National Institutes of Health (R01-DK090490 to Imai, R15-HL114062 to Dobrian, R01-HL112605 to Nadler), Astra Zeneca (Dobrian, Nadler), and Congressionally Directed Medical Research Program, Department of Defense (PR093521 to Taylor-Fishwick).

Cross-References

- Clinical Approaches to Preserve β-Cell Function in Diabetes
- Current Approaches and Future Prospects for the Prevention of β-Cell Destruction in Autoimmune Diabetes
- High Fat Programming of β-Cell Dysfunction
- Immunology of β-Cell Destruction
- Islet Structure and Function in the GK Rat
- Mechanisms of Pancreatic β-Cell Apoptosis in Diabetes and Its Therapies
- Pancreatic β Cells in Metabolic Syndrome
- Role of Mitochondria in β-cell Function and Dysfunction
- Role of NADPH Oxidase in β Cell Dysfunction
- The β-Cell in Human Type 2 Diabetes

References

Aceves M, Duenas A, Gomez C, San Vicente E, Crespo MS, Garcia-Rodriguez C (2004) A new pharmacological effect of salicylates: inhibition of NFAT-dependent transcription. J Immunol 173:5721–5729

Achenbach P, Warncke K, Reiter J, Naserke HE, Williams AJ, Bingley PJ, Bonifacio E, Ziegler AG (2004) Stratification of type 1 diabetes risk on the basis of islet autoantibody characteristics. Diabetes 53:384–392

Akesson C, Uvebrant K, Oderup C, Lynch K, Harris RA, Lernmark A, Agardh CD, Cilio CM (2010) Altered natural killer (NK) cell frequency and phenotype in latent autoimmune diabetes in adults (LADA) prior to insulin deficiency. Clin Exp Immunol 161:48–56

Alkanani AK, Rewers M, Dong F, Waugh K, Gottlieb PA, Zipris D (2012) Dysregulated Toll-like receptor-induced interleukin-1β and interleukin-6 responses in subjects at risk for the development of type 1 diabetes. Diabetes 61:2525–2533

Allen JS, Pang K, Skowera A, Ellis R, Rackham C, Lozanoska-Ochser B, Tree T, Leslie RD, Tremble JM, Dayan CM, Peakman M (2009) Plasmacytoid dendritic cells are proportionally expanded at diagnosis of type 1 diabetes and enhance islet autoantigen presentation to T-cells through immune complex capture. Diabetes 58:138–145

Al-Maskari M, Al-Shukaili A, Al-Mammari A (2010) Pro-inflammatory cytokines in Omani type 2 diabetic patients presenting anxiety and depression. Iran J Immunol 7:124–129

Al-Salam S, Rashed H, Adeghate E (2011) Diabetes mellitus is associated with an increased expression of resistin in human pancreatic islet cells. Islets 3:246–249

Antkowiak PF, Stevens BK, Nunemaker CS, McDuffie M, Epstein FH (2013) Manganese-enhanced magnetic resonance imaging detects declining pancreatic β-cell mass in a cyclophosphamide-accelerated mouse model of type 1 diabetes. Diabetes 62:44–48

Arif S, Tree TI, Astill TP, Tremble JM, Bishop AJ, Dayan CM, Roep BO, Peakman M (2004) Autoreactive T cell responses show proinflammatory polarization in diabetes but a regulatory phenotype in health. J Clin Invest 113:451–463

Arif S, Moore F, Marks K, Bouckenooghe T, Dayan CM, Planas R, Vives-Pi M, Powrie J, Tree T, Marchetti P, Huang GC, Gurzov EN, Pujol-Borrell R, Eizirik DL, Peakman M (2011) Peripheral and islet interleukin-17 pathway activation characterizes human autoimmune diabetes and promotes cytokine-mediated β-cell death. Diabetes 60:2112–2119

Atkinson MA, Eisenbarth GS, Michels AW (2013) Type 1 diabetes. Lancet 383(9911):69–82

Axling U, Olsson C, Xu J, Fernandez C, Larsson S, Strom K, Ahrne S, Holm C, Molin G, Berger K (2012) Green tea powder and Lactobacillus plantarum affect gut microbiota, lipid metabolism and inflammation in high-fat fed C57BL/6J mice. Nutr Metab 9:105

Baeyens A, Perol L, Fourcade G, Cagnard N, Carpentier W, Woytschak J, Boyman O, Hartemann A, Piaggio E (2013) Limitations of IL-2 and rapamycin in immunotherapy of type 1 diabetes. Diabetes 62(9):3120–3131

Baldwin AC, Green CD, Olson LK, Moxley MA, Corbett JA (2012) A role for aberrant protein palmitoylation in FFA-induced ER stress and β-cell death. Am J Physiol Endocrinol Metab 302:E1390–E1398

Beilke JN, Meagher CT, Hosiawa K, Champsaur M, Bluestone JA, Lanier LL (2012) NK cells are not required for spontaneous autoimmune diabetes in NOD mice. PloS One 7:e36011

Bensellam M, Laybutt DR, Jonas JC (2012) The molecular mechanisms of pancreatic β-cell glucotoxicity: recent findings and future research directions. Mol Cell Endocrinol 364:1–27

Bleich D, Chen S, Zipser B, Sun D, Funk CD, Nadler JL (1999) Resistance to type 1 diabetes induction in 12-lipoxygenase knockout mice. J Clin Invest 103:1431–1436

Boll EJ, Struve C, Sander A, Demma Z, Krogfelt KA, McCormick BA (2012) Enteroaggregative *Escherichia coli* promotes transepithelial migration of neutrophils through a conserved 12-lipoxygenase pathway. Cell Microbiol 14:120–132

Bonifacio E, Bingley PJ, Shattock M, Dean BM, Dunger D, Gale EA, Bottazzo GF (1990) Quantification of islet-cell antibodies and prediction of insulin-dependent diabetes. Lancet 335:147–149

Boni-Schnetzler M, Thorne J, Parnaud G, Marselli L, Ehses JA, Kerr-Conte J, Pattou F, Halban PA, Weir GC, Donath MY (2008) Increased interleukin IL-1β messenger ribonucleic acid expression in β-cells of individuals with type 2 diabetes and regulation of IL-1β in human islets by glucose and autostimulation. J Clin Endocrinol Metab 93:4065–4074

Bosi E, Molteni L, Radaelli MG, Folini L, Fermo I, Bazzigaluppi E, Piemonti L, Pastore MR, Paroni R (2006) Increased intestinal permeability precedes clinical onset of type 1 diabetes. Diabetologia 49:2824–2827

Boslem E, Meikle PJ, Biden TJ (2012) Roles of ceramide and sphingolipids in pancreatic β-cell function and dysfunction. Islets 4:177–187

Boudaly S, Morin J, Berthier R, Marche P, Boitard C (2002) Altered dendritic cells (DC) might be responsible for regulatory T cell imbalance and autoimmunity in nonobese diabetic (NOD) mice. Eur Cytokine Netw 13:29–37

Bradshaw EM, Raddassi K, Elyaman W, Orban T, Gottlieb PA, Kent SC, Hafler DA (2009) Monocytes from patients with type 1 diabetes spontaneously secrete proinflammatory cytokines inducing Th17 cells. J Immunol 183:4432–4439

Brauner H, Elemans M, Lemos S, Broberger C, Holmberg D, Flodstrom-Tullberg M, Karre K, Hoglund P (2010) Distinct phenotype and function of NK cells in the pancreas of nonobese diabetic mice. J Immunol 184:2272–2280

Brom M, Andralojc K, Oyen WJ, Boerman OC, Gotthardt M (2010) Development of radiotracers for the determination of the β-cell mass in vivo. Curr Pharm Des 16:1561–1567

Brooks-Worrell B, Palmer JP (2012) Immunology in the Clinic Review Series; focus on metabolic diseases: development of islet autoimmune disease in type 2 diabetes patients: potential sequelae of chronic inflammation. Clin Exp Immunol 167:40–46

Brooks-Worrell BM, Palmer JP (2013) Attenuation of islet-specific T cell responses is associated with C-peptide improvement in autoimmune type 2 diabetes patients. Clin Exp Immunol 171:164–170

Brooks-Worrell BM, Reichow JL, Goel A, Ismail H, Palmer JP (2011) Identification of autoantibody-negative autoimmune type 2 diabetic patients. Diabetes Care 34:168–173

Brooks-Worrell B, Narla R, Palmer JP (2012) Biomarkers and immune-modulating therapies for type 2 diabetes. Trends Immunol 33:546–553

Brown JE, Conner AC, Digby JE, Ward KL, Ramanjaneya M, Randeva HS, Dunmore SJ (2010a) Regulation of β-cell viability and gene expression by distinct agonist fragments of adiponectin. Peptides 31:944–949

Brown JE, Onyango DJ, Ramanjaneya M, Conner AC, Patel ST, Dunmore SJ, Randeva HS (2010b) Visfatin regulates insulin secretion, insulin receptor signalling and mRNA expression of diabetes-related genes in mouse pancreatic β-cells. J Mol Endocrinol 44:171–178

Brown CT, Davis-Richardson AG, Giongo A, Gano KA, Crabb DB, Mukherjee N, Casella G, Drew JC, Ilonen J, Knip M, Hyoty H, Veijola R, Simell T, Simell O, Neu J, Wasserfall CH, Schatz D, Atkinson MA, Triplett EW (2011) Gut microbiome metagenomics analysis suggests a functional model for the development of autoimmunity for type 1 diabetes. PLoS One 6: e25792

Buschard K, Mansson JE, Roep BO, Nikolic T (2012) Self-glycolipids modulate dendritic cells changing the cytokine profiles of committed autoreactive T cells. PLoS One 7:e52639

Cabrera SM, Colvin SC, Tersey SA, Maier B, Nadler JL, Mirmira RG (2013) Effects of combination therapy with dipeptidyl peptidase-IV and histone deacetylase inhibitors in the non-obese diabetic mouse model of type 1 diabetes. Clin Exp Immunol 172:375–382

Cai K, Qi D, Wang O, Chen J, Liu X, Deng B, Qian L, Liu X, Le Y (2011) TNF-α acutely upregulates amylin expression in murine pancreatic β cells. Diabetologia 54:617–626

Cani PD, Possemiers S, Van de Wiele T, Guiot Y, Everard A, Rottier O, Geurts L, Naslain D, Neyrinck A, Lambert DM, Muccioli GG, Delzenne NM (2009) Changes in gut microbiota control inflammation in obese mice through a mechanism involving GLP-2-driven improvement of gut permeability. Gut 58:1091–1103

Cao P, Marek P, Noor H, Patsalo V, Tu LH, Wang H, Abedini A, Raleigh DP (2013) Islet amyloid: from fundamental biophysics to mechanisms of cytotoxicity. FEBS letters 587:1106–1118

Carrington EM, Kos C, Zhan Y, Krishnamurthy B, Allison J (2011) Reducing or increasing β-cell apoptosis without inflammation does not affect diabetes initiation in neonatal NOD mice. Eur J Immunol 41:2238–2247

Catalan V, Gomez-Ambrosi J, Ramirez B, Rotellar F, Pastor C, Silva C, Rodriguez A, Gil MJ, Cienfuegos JA, Fruhbeck G (2007) Proinflammatory cytokines in obesity: impact of type 2 diabetes mellitus and gastric bypass. Obes Surg 17:1464–1474

Cavelti-Weder C, Babians-Brunner A, Keller C, Stahel MA, Kurz-Levin M, Zayed H, Solinger AM, Mandrup-Poulsen T, Dinarello CA, Donath MY (2012) Effects of gevokizumab on glycemia and inflammatory markers in type 2 diabetes. Diabetes Care 35:1654–1662

Chai F, Wang Y, Zhou Y, Liu Y, Geng D, Liu J (2011) Adiponectin downregulates hyperglycemia and reduces pancreatic islet apoptosis after roux-en-y gastric bypass surgery. Obes Surg 21:768–773

Chakrabarti SK, Cole BK, Wen Y, Keller SR, Nadler JL (2009) 12/15-lipoxygenase products induce inflammation and impair insulin signaling in 3T3-L1 adipocytes. Obesity (Silver Spring) 17(9):1657–1663

Chakrabarti SK, Wen Y, Dobrian AD, Cole BK, Ma Q, Pei H, Williams MD, Bevard MH, Vandenhoff GE, Keller SR, Gu J, Nadler JL (2011) Evidence for activation of inflammatory lipoxygenase pathways in visceral adipose tissue of obese zucker rats. Am J Physiol Endocrinol Metab 300(1):E175–E187

Chan JY, Luzuriaga J, Bensellam M, Biden TJ, Laybutt DR (2013) Failure of the adaptive unfolded protein response in islets of obese mice is linked with abnormalities in β-cell gene expression and progression to diabetes. Diabetes 62:1557–1568

Chang YH, Chang DM, Lin KC, Shin SJ, Lee YJ (2011) Visfatin in overweight/obesity, type 2 diabetes mellitus, insulin resistance, metabolic syndrome and cardiovascular diseases: a meta-analysis and systemic review. Diabetes Metab Res Rev 27:515–527

Charre S, Rosmalen JG, Pelegri C, Alves V, Leenen PJ, Drexhage HA, Homo-Delarche F (2002) Abnormalities in dendritic cell and macrophage accumulation in the pancreas of nonobese diabetic (NOD) mice during the early neonatal period. Histol Histopathol 17:393–401

Chawla A, Nguyen KD, Goh YP (2011) Macrophage-mediated inflammation in metabolic disease. Nat Rev Immunol 11:738–749

Chen JJ, Wang R, Li XF, Wang RL (2011) Bifidobacterium longum supplementation improved high-fat-fed-induced metabolic syndrome and promoted intestinal Reg I gene expression. Exp Biol Med (Maywood) 236:823–831

Cheng Q, Dong W, Qian L, Wu J, Peng Y (2011) Visfatin inhibits apoptosis of pancreatic β-cell line, MIN6, via the mitogen-activated protein kinase/phosphoinositide 3-kinase pathway. J Mol Endocrinol 47:13–21

Cleland SJ, Fisher BM, Colhoun HM, Sattar N, Petrie JR (2013) Insulin resistance in type 1 diabetes: what is 'double diabetes' and what are the risks? Diabetologia 56:1462–1470

Cnop M, Welsh N, Jonas JC, Jorns A, Lenzen S, Eizirik DL (2005) Mechanisms of pancreatic β-cell death in type 1 and type 2 diabetes: many differences, few similarities. Diabetes 54(Suppl 2):S97–S107

Cnop M, Ladriere L, Igoillo-Esteve M, Moura RF, Cunha DA (2010) Causes and cures for endoplasmic reticulum stress in lipotoxic β-cell dysfunction. Diabetes Obes Metab 12(Suppl 2):76–82

Cole BK, Kuhn NS, Green-Mitchell SM, Leone KA, Raab RM, Nadler JL, Chakrabarti SK (2012a) 12/15-lipoxygenase signaling in the endoplasmic reticulum stress response. Am J Physiol Endocrinol Metab 302:E654–E665

Cole BK, Morris MA, Grzesik WJ, Leone KA, Nadler JL (2012b) Adipose tissue-specific deletion of 12/15-lipoxygenase protects mice from the consequences of a high-fat diet. Mediators Inflamm 2012:851798

Cunningham GA, McClenaghan NH, Flatt PR, Newsholme P (2005) L-alanine induces changes in metabolic and signal transduction gene expression in a clonal rat pancreatic β-cell line and protects from pro-inflammatory cytokine-induced apoptosis. Clin Sci (Lond) 109:447–455

D'Alise AM, Auyeung V, Feuerer M, Nishio J, Fontenot J, Benoist C, Mathis D (2008) The defect in T-cell regulation in NOD mice is an effect on the T-cell effectors. Proc Natl Acad Sci USA 105:19857–19862

Danaei G, Finucane MM, Lu Y, Singh GM, Cowan MJ, Paciorek CJ, Lin JK, Farzadfar F, Khang YH, Stevens GA, Rao M, Ali MK, Riley LM, Robinson CA, Ezzati M (2011) National, regional, and global trends in fasting plasma glucose and diabetes prevalence since 1980: systematic analysis of health examination surveys and epidemiological studies with 370 - country-years and 2.7 million participants. Lancet 378:31–40

Davis TM, Wright AD, Mehta ZM, Cull CA, Stratton IM, Bottazzo GF, Bosi E, Mackay IR, Holman RR (2005) Islet autoantibodies in clinically diagnosed type 2 diabetes: prevalence and relationship with metabolic control (UKPDS 70). Diabetologia 48:695–702

de Goffau MC, Luopajarvi K, Knip M, Ilonen J, Ruohtula T, Harkonen T, Orivuori L, Hakala S, Welling GW, Harmsen HJ, Vaarala O (2013) Fecal microbiota composition differs between children with β-cell autoimmunity and those without. Diabetes 62:1238–1244

DeFuria J, Belkina AC, Jagannathan-Bogdan M, Snyder-Cappione J, Carr JD, Nersesova YR, Markham D, Strissel KJ, Watkins AA, Zhu M, Allen J, Bouchard J, Toraldo G, Jasuja R, Obin MS, McDonnell ME, Apovian C, Denis GV, Nikolajczyk BS (2013) B cells promote inflammation in obesity and type 2 diabetes through regulation of T-cell function and an inflammatory cytokine profile. Proc Natl Acad Sci USA 110:5133–5138

Deng Y, Scherer PE (2010) Adipokines as novel biomarkers and regulators of the metabolic syndrome. Ann N Y Acad Sci 1212:E1–E19

Dobrian AD, Lieb DC, Cole BK, Taylor-Fishwick DA, Chakrabarti SK, Nadler JL (2010a) Functional and pathological roles of the 12- and 15-lipoxygenases. Prog Lipid Res 50:115–131

Dobrian AD, Lieb DC, Ma Q, Lindsay JW, Cole BK, Ma K, Chakrabarti SK, Kuhn NS, Wohlgemuth SD, Fontana M, Nadler JL (2010b) Differential expression and localization of 12/15 lipoxygenases in adipose tissue in human obese subjects. Biochem Biophys Res Commun 403:485–490

Dobrian AD, Ma Q, Lindsay JW, Leone KA, Ma K, Coben J, Galkina EV, Nadler JL (2010c) Dipeptidyl peptidase IV inhibitor sitagliptin reduces local inflammation in adipose tissue and in pancreatic islets of obese mice. Am J Physiol Endocrinol Metab 300:E410–E421

Dobrian AD, Lieb DC, Cole BK, Taylor-Fishwick DA, Chakrabarti SK, Nadler JL (2011) Functional and pathological roles of the 12- and 15-lipoxygenases. Prog Lipid Res 50:115–131

Donath MY, Shoelson SE (2011) Type 2 diabetes as an inflammatory disease. Nat Rev Immunol 11:98–107

Donath MY, Dalmas E, Sauter NS, Boni-Schnetzler M (2013) Inflammation in obesity and diabetes: islet dysfunction and therapeutic opportunity. Cell Metab 17:860–872

Dotta F, Censini S, van Halteren AG, Marselli L, Masini M, Dionisi S, Mosca F, Boggi U, Muda AO, Del Prato S, Elliott JF, Covacci A, Rappuoli R, Roep BO, Marchetti P (2007) Coxsackie B4 virus infection of β cells and natural killer cell insulitis in recent-onset type 1 diabetic patients. Proc Natl Acad Sci USA 104:5115–5120

Drucker DJ (2006) The biology of incretin hormones. Cell Metab 3:153–165

Drucker DJ (2007) Dipeptidyl peptidase-4 inhibition and the treatment of type 2 diabetes: preclinical biology and mechanisms of action. Diabetes Care 30:1335–1343

Dunmore SJ, Brown JE (2013) The role of adipokines in β-cell failure of type 2 diabetes. J Endocrinol 216:T37–T45

Eguchi K, Manabe I, Oishi-Tanaka Y, Ohsugi M, Kono N, Ogata F, Yagi N, Ohto U, Kimoto M, Miyake K, Tobe K, Arai H, Kadowaki T, Nagai R (2012) Saturated fatty acid and TLR signaling link β cell dysfunction and islet inflammation. Cell Metab 15:518–533

Ehses JA, Perren A, Eppler E, Ribaux P, Pospisilik JA, Maor-Cahn R, Gueripel X, Ellingsgaard H, Schneider MK, Biollaz G, Fontana A, Reinecke M, Homo-Delarche F, Donath MY (2007) Increased number of islet-associated macrophages in type 2 diabetes. Diabetes 56:2356–2370

Ehses JA, Lacraz G, Giroix MH, Schmidlin F, Coulaud J, Kassis N, Irminger JC, Kergoat M, Portha B, Homo-Delarche F, Donath MY (2009) IL-1 antagonism reduces hyperglycemia and tissue inflammation in the type 2 diabetic GK rat. Proc Natl Acad Sci USA 106:13998–14003

Eizirik DL, Mandrup-Poulsen T (2001) A choice of death – the signal-transduction of immune-mediated β-cell apoptosis. Diabetologia 44:2115–2133

Eizirik DL, Sammeth M, Bouckenooghe T, Bottu G, Sisino G, Igoillo-Esteve M, Ortis F, Santin I, Colli ML, Barthson J, Bouwens L, Hughes L, Gregory L, Lunter G, Marselli L, Marchetti P, McCarthy MI, Cnop M (2012) The human pancreatic islet transcriptome: expression of candidate genes for type 1 diabetes and the impact of pro-inflammatory cytokines. PLoS Genet 8:e1002552

Eizirik DL, Miani M, Cardozo AK (2013) Signalling danger: endoplasmic reticulum stress and the unfolded protein response in pancreatic islet inflammation. Diabetologia 56:234–241

El Ouaamari A, Kawamori D, Dirice E, Liew CW, Shadrach JL, Hu J, Katsuta H, Hollister-Lock J, Qian WJ, Wagers AJ, Kulkarni RN (2013) Liver-derived systemic factors drive β cell hyperplasia in insulin-resistant states. Cell Rep 3:401–410

Ellingsgaard H, Hauselmann I, Schuler B, Habib AM, Baggio LL, Meier DT, Eppler E, Bouzakri K, Wueest S, Muller YD, Hansen AM, Reinecke M, Konrad D, Gassmann M, Reimann F, Halban PA, Gromada J, Drucker DJ, Gribble FM, Ehses JA, Donath MY (2011)

Interleukin-6 enhances insulin secretion by increasing glucagon-like peptide-1 secretion from L cells and α cells. Nat Med 17:1481–1489

Estall JL, Drucker DJ (2006) Glucagon and glucagon-like peptide receptors as drug targets. Curr Pharm Des 12:1731–1750

Everard A, Belzer C, Geurts L, Ouwerkerk JP, Druart C, Bindels LB, Guiot Y, Derrien M, Muccioli GG, Delzenne NM, de Vos WM, Cani PD (2013) Cross-talk between Akkermansia muciniphila and intestinal epithelium controls diet-induced obesity. Proc Natl Acad Sci USA 110:9066–9071

Faghihimani E, Aminorroaya A, Rezvanian H, Adibi P, Ismail-Beigi F, Amini M (2013) Salsalate improves glycemic control in patients with newly diagnosed type 2 diabetes. Acta Diabetologica 50:537–543

Fliser D, Buchholz K, Haller H (2004) Antiinflammatory effects of angiotensin II subtype 1 receptor blockade in hypertensive patients with microinflammation. Circulation 110:1103–1107

Frederiksen B, Kroehl M, Lamb MM, Seifert J, Barriga K, Eisenbarth GS, Rewers M, Norris JM (2013) Infant exposures and development of type 1 diabetes mellitus: the Diabetes Autoimmunity Study in the Young (DAISY). JAMA Pediatr 167:808–815

Fujisaka S, Usui I, Kanatani Y, Ikutani M, Takasaki I, Tsuneyama K, Tabuchi Y, Bukhari A, Yamazaki Y, Suzuki H, Senda S, Aminuddin A, Nagai Y, Takatsu K, Kobayashi M, Tobe K (2011) Telmisartan improves insulin resistance and modulates adipose tissue macrophage polarization in high-fat-fed mice. Endocrinology 152:1789–1799

Fujitani Y, Ueno T, Watada H (2010) Autophagy in health and disease. 4. The role of pancreatic β-cell autophagy in health and diabetes. Am J physiol Cell Physiol 299:C1–C6

Garg AD, Kaczmarek A, Krysko O, Vandenabeele P, Krysko DV, Agostinis P (2012) ER stress-induced inflammation: does it aid or impede disease progression? Trends Mol Med 18:589–598

Ghosh K, Kanapathipillai M, Korin N, McCarthy JR, Ingber DE (2012) Polymeric nanomaterials for islet targeting and immunotherapeutic delivery. Nano Lett 12:203–208

Giacca A, Xiao C, Oprescu AI, Carpentier AC, Lewis GF (2011) Lipid-induced pancreatic β-cell dysfunction: focus on in vivo studies. Am J Physiol Endocrinol Metab 300:E255–E262

Giongo A, Gano KA, Crabb DB, Mukherjee N, Novelo LL, Casella G, Drew JC, Ilonen J, Knip M, Hyoty H, Veijola R, Simell T, Simell O, Neu J, Wasserfall CH, Schatz D, Atkinson MA, Triplett EW (2011) Toward defining the autoimmune microbiome for type 1 diabetes. ISME J 5:82–91

Goldfine AB, Fonseca V, Jablonski KA, Pyle L, Staten MA, Shoelson SE (2010) The effects of salsalate on glycemic control in patients with type 2 diabetes: a randomized trial. Ann Intern Med 152:346–357

Goldfine AB, Fonseca V, Shoelson SE (2011) Therapeutic approaches to target inflammation in type 2 diabetes. Clin Chem 57:162–167

Goldfine AB, Conlin PR, Halperin F, Koska J, Permana P, Schwenke D, Shoelson SE, Reaven PD (2013a) A randomised trial of salsalate for insulin resistance and cardiovascular risk factors in persons with abnormal glucose tolerance. Diabetologia 56:714–723

Goldfine AB, Fonseca V, Jablonski KA, Chen YD, Tipton L, Staten MA, Shoelson SE (2013b) Salicylate (salsalate) in patients with type 2 diabetes: a randomized trial. Ann Intern Med 159:1–12

Green-Mitchell SM, Tersey SA, Cole BK, Ma K, Kuhn NS, Cunningham TD, Maybee NA, Chakrabarti SK, McDuffie M, Taylor-Fishwick DA, Mirmira RG, Nadler JL, Morris MA (2013) Deletion of 12/15-lipoxygenase alters macrophage and islet function in NOD-Alox15 (null) mice, leading to protection against type 1 diabetes development. PLoS One 8:e56763

Hansen CH, Krych L, Nielsen DS, Vogensen FK, Hansen LH, Sorensen SJ, Buschard K, Hansen AK (2012a) Early life treatment with vancomycin propagates Akkermansia muciniphila and reduces diabetes incidence in the NOD mouse. Diabetologia 55:2285–2294

Hansen JB, Tonnesen MF, Madsen AN, Hagedorn PH, Friberg J, Grunnet LG, Heller RS, Nielsen AO, Storling J, Baeyens L, Anker-Kitai L, Qvortrup K, Bouwens L, Efrat S, Aalund M, Andrews NC, Billestrup N, Karlsen AE, Holst B, Pociot F, Mandrup-Poulsen T (2012b)

Divalent metal transporter 1 regulates iron-mediated ROS and pancreatic β cell fate in response to cytokines. Cell Metab 16:449–461

Hara N, Alkanani AK, Ir D, Robertson CE, Wagner BD, Frank DN, Zipris D (2012) Prevention of virus-induced type 1 diabetes with antibiotic therapy. J Immunol 189:3805–3814

Hawley SA, Fullerton MD, Ross FA, Schertzer JD, Chevtzoff C, Walker KJ, Peggie MW, Zibrova D, Green KA, Mustard KJ, Kemp BE, Sakamoto K, Steinberg GR, Hardie DG (2012) The ancient drug salicylate directly activates AMP-activated protein kinase. Science 336:918–922

Hayashi Y, Utsuyama M, Kurashima C, Hirokawa K (1989) Spontaneous development of organ-specific autoimmune lesions in aged C57BL/6 mice. Clin Exp Immunol 78:120–126

Herder C, Haastert B, Muller-Scholze S, Koenig W, Thorand B, Holle R, Wichmann HE, Scherbaum WA, Martin S, Kolb H (2005) Association of systemic chemokine concentrations with impaired glucose tolerance and type 2 diabetes: results from the Cooperative Health Research in the Region of Augsburg Survey S4 (KORA S4). Diabetes 54(Suppl 2):S11–S17

Herder C, Hauner H, Kempf K, Kolb H, Skurk T (2007) Constitutive and regulated expression and secretion of interferon-gamma-inducible protein 10 (IP-10/CXCL10) in human adipocytes. Int J Obes (Lond) 31:403–410

Hines WJ, Smith MJ (1964) Inhibition of dehydrogenases by salicylate. Nature 201:192

Hogan AE, Tobin AM, Ahern T, Corrigan MA, Gaoatswe G, Jackson R, O'Reilly V, Lynch L, Doherty DG, Moynagh PN, Kirby B, O'Connell J, O'Shea D (2011) Glucagon-like peptide-1 (GLP-1) and the regulation of human invariant natural killer T cells: lessons from obesity, diabetes and psoriasis. Diabetologia 54:2745–2754

Hogan AE, Lynch L, Gaoatswe G, Woods C, Jackson R, O'Connell J, Moynagh PN, O'Shea D (2013) Glucagon-like peptide-1 (GLP-1) analogue therapy impacts inflammatory macrophages and cytokine increasing evidence for its anti-inflammatory actions. Endocr Rev 34:FP13–FP15

Holland WL, Miller RA, Wang ZV, Sun K, Barth BM, Bui HH, Davis KE, Bikman BT, Halberg N, Rutkowski JM, Wade MR, Tenorio VM, Kuo MS, Brozinick JT, Zhang BB, Birnbaum MJ, Summers SA, Scherer PE (2011) Receptor-mediated activation of ceramidase activity initiates the pleiotropic actions of adiponectin. Nat Med 17:55–63

Homo-Delarche F, Calderari S, Irminger JC, Gangnerau MN, Coulaud J, Rickenbach K, Dolz M, Halban P, Portha B, Serradas P (2006) Islet inflammation and fibrosis in a spontaneous model of type 2 diabetes, the GK rat. Diabetes 55:1625–1633

Honda K, Moto M, Uchida N, He F, Hashizume N (2012) Anti-diabetic effects of lactic acid bacteria in normal and type 2 diabetic mice. J Clin Biochem Nutr 51:96–101

Hotamisligil GS (2006) Inflammation and metabolic disorders. Nature 444:860–867

Hotamisligil GS (2010) Endoplasmic reticulum stress and the inflammatory basis of metabolic disease. Cell 140:900–917

Huang CJ, Lin CY, Haataja L, Gurlo T, Butler AE, Rizza RA, Butler PC (2007) High expression rates of human islet amyloid polypeptide induce endoplasmic reticulum stress mediated β-cell apoptosis, a characteristic of humans with type 2 but not type 1 diabetes. Diabetes 56:2016–2027

Hummel S, Pfluger M, Hummel M, Bonifacio E, Ziegler AG (2011) Primary dietary intervention study to reduce the risk of islet autoimmunity in children at increased risk for type 1 diabetes: the BABYDIET study. Diabetes Care 34:1301–1305

Hundal RS, Petersen KF, Mayerson AB, Randhawa PS, Inzucchi S, Shoelson SE, Shulman GI (2002) Mechanism by which high-dose aspirin improves glucose metabolism in type 2 diabetes. J Clin Invest 109:1321–1326

Igoillo-Esteve M, Marselli L, Cunha DA, Ladriere L, Ortis F, Grieco FA, Dotta F, Weir GC, Marchetti P, Eizirik DL, Cnop M (2010) Palmitate induces a pro-inflammatory response in human pancreatic islets that mimics CCL2 expression by β cells in type 2 diabetes. Diabetologia 53:1395–1405

Imai J, Katagiri H, Yamada T, Ishigaki Y, Suzuki T, Kudo H, Uno K, Hasegawa Y, Gao J, Kaneko K, Ishihara H, Niijima A, Nakazato M, Asano T, Minokoshi Y, Oka Y (2008) Regulation of pancreatic β cell mass by neuronal signals from the liver. Science 322:1250–1254

Imai Y, Dobrian AD, Morris MA, Nadler JL (2013a) Islet inflammation: a unifying target for diabetes treatment? Trends Endocrinol Metab: TEM 24:351–360

Imai Y, Dobrian AD, Weaver JR, Butcher MJ, Cole BK, Galkina EV, Morris MA, Taylor-Fishwick DA, Nadler JL (2013b) Interaction between cytokines and inflammatory cells in islet dysfunction, insulin resistance, and vascular disease. Diabetes Obes Metab 15(Suppl 3):117–129

Inoguchi T, Nawata H (2005) NAD(P)H oxidase activation: a potential target mechanism for diabetic vascular complications, progressive β-cell dysfunction and metabolic syndrome. Curr Drug Targets 6:495–501

Jaidane H, Hober D (2008) Role of coxsackievirus B4 in the pathogenesis of type 1 diabetes. Diabetes Metab 34:537–548

Janciauskiene S, Ahren B (2000) Fibrillar islet amyloid polypeptide differentially affects oxidative mechanisms and lipoprotein uptake in correlation with cytotoxicity in two insulin-producing cell lines. Biochem Biophys Res Commun 267:619–625

Jelsing J, Vrang N, van Witteloostuijn SB, Mark M, Klein T (2012) The DPP4 inhibitor linagliptin delays the onset of diabetes and preserves β-cell mass in non-obese diabetic mice. J Endocrinol 214:381–387

Jorns A, Gunther A, Hedrich HJ, Wedekind D, Tiedge M, Lenzen S (2005) Immune cell infiltration, cytokine expression, and β-cell apoptosis during the development of type 1 diabetes in the spontaneously diabetic LEW.1AR1/Ztm-iddm rat. Diabetes 54:2041–2052

Joseph J, Bittner S, Kaiser FM, Wiendl H, Kissler S (2012) IL-17 silencing does not protect nonobese diabetic mice from autoimmune diabetes. J Immunol 188:216–221

Jourdan T, Godlewski G, Cinar R, Bertola A, Szanda G, Liu J, Tam J, Han T, Mukhopadhyay B, Skarulis MC, Ju C, Aouadi M, Czech MP, Kunos G (2013) Activation of the Nlrp3 inflammasome in infiltrating macrophages by endocannabinoids mediates β cell loss in type 2 diabetes. Nat Med 19:1132–1140

Jurgens CA, Toukatly MN, Fligner CL, Udayasankar J, Subramanian SL, Zraika S, Aston-Mourney K, Carr DB, Westermark P, Westermark GT, Kahn SE, Hull RL (2011) β-cell loss and β-cell apoptosis in human type 2 diabetes are related to islet amyloid deposition. Am J Pathol 178:2632–2640

Kahn SE, Hull RL, Utzschneider KM (2006) Mechanisms linking obesity to insulin resistance and type 2 diabetes. Nature 444:840–846

Kahn SE, Zraika S, Utzschneider KM, Hull RL (2009) The β cell lesion in type 2 diabetes: there has to be a primary functional abnormality. Diabetologia 52:1003–1012

Kang YS, Song HK, Lee MH, Ko GJ, Cha DR (2010) Plasma concentration of visfatin is a new surrogate marker of systemic inflammation in type 2 diabetic patients. Diabetes Res Clin Pract 89:141–149

Karlsson FH, Tremaroli V, Nookaew I, Bergstrom G, Behre CJ, Fagerberg B, Nielsen J, Backhed F (2013) Gut metagenome in European women with normal, impaired and diabetic glucose control. Nature 498:99–103

Kenyon V, Rai G, Jadhav A, Schultz L, Armstrong M, Jameson JB 2nd, Perry S, Joshi N, Bougie JM, Leister W, Taylor-Fishwick DA, Nadler JL, Holinstat M, Simeonov A, Maloney DJ, Holman TR (2011) Discovery of potent and selective inhibitors of human platelet-type 12-lipoxygenase. J Med Chem 54:5485–5497

Kleijwegt FS, Jansen DT, Teeler J, Joosten AM, Laban S, Nikolic T, Roep BO (2013) Tolerogenic dendritic cells impede priming of naive $CD8^+$ T cells and deplete memory $CD8^+$ T cells. Eur J Immunol 43:85–92

Knight RR, Kronenberg D, Zhao M, Huang GC, Eichmann M, Bulek A, Wooldridge L, Cole DK, Sewell AK, Peakman M, Skowera A (2013) Human β-cell killing by autoreactive preproinsulin-specific CD8 T cells is predominantly granule-mediated with the potency dependent upon T-cell receptor avidity. Diabetes 62:205–213

Kobayashi M, Kaneko-Koike C, Abiru N, Uchida T, Akazawa S, Nakamura K, Kuriya G, Satoh T, Ida H, Kawasaki E, Yamasaki H, Nagayama Y, Sasaki H, Kawakami A (2013) Genetic deletion of granzyme B does not confer resistance to the development of spontaneous diabetes in non-obese diabetic mice. Clin Exp Immunol 173:411–418

Koska J, Ortega E, Bunt JC, Gasser A, Impson J, Hanson RL, Forbes J, de Courten B, Krakoff J (2009) The effect of salsalate on insulin action and glucose tolerance in obese non-diabetic patients: results of a randomised double-blind placebo-controlled study. Diabetologia 52:385–393

Koulajian K, Desai T, Liu GC, Ivovic A, Patterson JN, Tang C, El-Benna J, Joseph JW, Scholey JW, Giacca A (2013) NADPH oxidase inhibition prevents β cell dysfunction induced by prolonged elevation of oleate in rodents. Diabetologia 56:1078–1087

Kriegel MA, Sefik E, Hill JA, Wu HJ, Benoist C, Mathis D (2011) Naturally transmitted segmented filamentous bacteria segregate with diabetes protection in nonobese diabetic mice. Proc Natl Acad Sci USA 108:11548–11553

Kriegel MA, Rathinam C, Flavell RA (2012) Pancreatic islet expression of chemokine CCL2 suppresses autoimmune diabetes via tolerogenic CD11c + CD11b + dendritic cells. Proc Natl Acad Sci USA 109:3457–3462

Lamb RE, Goldstein BJ (2008) Modulating an oxidative-inflammatory cascade: potential new treatment strategy for improving glucose metabolism, insulin resistance, and vascular function. Int J Clin Pract 62:1087–1095

Lamers D, Famulla S, Wronkowitz N, Hartwig S, Lehr S, Ouwens DM, Eckardt K, Kaufman JM, Ryden M, Muller S, Hanisch FG, Ruige J, Arner P, Sell H, Eckel J (2011) Dipeptidyl peptidase 4 is a novel adipokine potentially linking obesity to the metabolic syndrome. Diabetes 60:1917–1925

Larsen CM, Faulenbach M, Vaag A, Volund A, Ehses JA, Seifert B, Mandrup-Poulsen T, Donath MY (2007) Interleukin-1-receptor antagonist in type 2 diabetes mellitus. N Engl J Med 356:1517–1526

Larsen CM, Faulenbach M, Vaag A, Ehses JA, Donath MY, Mandrup-Poulsen T (2009) Sustained effects of interleukin-1 receptor antagonist treatment in type 2 diabetes. Diabetes Care 32:1663–1668

Larsen N, Vogensen FK, van den Berg FW, Nielsen DS, Andreasen AS, Pedersen BK, Al-Soud WA, Sorensen SJ, Hansen LH, Jakobsen M (2010) Gut microbiota in human adults with type 2 diabetes differs from non-diabetic adults. PLoS One 5:e9085

Lee AS, Gibson DL, Zhang Y, Sham HP, Vallance BA, Dutz JP (2010) Gut barrier disruption by an enteric bacterial pathogen accelerates insulitis in NOD mice. Diabetologia 53:741–748

Lee YS, Park MS, Choung JS, Kim SS, Oh HH, Choi CS, Ha SY, Kang Y, Kim Y, Jun HS (2012) Glucagon-like peptide-1 inhibits adipose tissue macrophage infiltration and inflammation in an obese mouse model of diabetes. Diabetologia 55:2456–2468

Lenzen S (2008) Oxidative stress: the vulnerable β-cell. Biochem Soc Trans 36:343–347

Lerner AG, Upton JP, Praveen PV, Ghosh R, Nakagawa Y, Igbaria A, Shen S, Nguyen V, Backes BJ, Heiman M, Heintz N, Greengard P, Hui S, Tang Q, Trusina A, Oakes SA, Papa FR (2012) IRE1α induces thioredoxin-interacting protein to activate the NLRP3 inflammasome and promote programmed cell death under irremediable ER stress. Cell Metab 16:250–264

Levine B, Mizushima N, Virgin HW (2011) Autophagy in immunity and inflammation. Nature 469:323–335

Lightfoot YL, Chen J, Mathews CE (2012) Oxidative stress and β cell dysfunction. Methods Mol Biol 900:347–362

Lindley S, Dayan CM, Bishop A, Roep BO, Peakman M, Tree TI (2005) Defective suppressor function in CD4$^+$CD25$^+$ T-cells from patients with type 1 diabetes. Diabetes 54:92–99

Lohmann T, Kellner K, Verlohren HJ, Krug J, Steindorf J, Scherbaum WA, Seissler J (2001) Titre and combination of ICA and autoantibodies to glutamic acid decarboxylase discriminate two clinically distinct types of latent autoimmune diabetes in adults (LADA). Diabetologia 44:1005–1010

Long SA, Rieck M, Sanda S, Bollyky JB, Samuels PL, Goland R, Ahmann A, Rabinovitch A, Aggarwal S, Phippard D, Turka LA, Ehlers MR, Bianchine PJ, Boyle KD, Adah SA, Bluestone JA, Buckner JH, Greenbaum CJ (2012) Rapamycin/IL-2 combination therapy in patients with type 1 diabetes augments Tregs yet transiently impairs β-cell function. Diabetes 61:2340–2348

Lumeng CN, Saltiel AR (2011) Inflammatory links between obesity and metabolic disease. J Clin Invest 121:2111–2117

Ma K, Nunemaker CS, Wu R, Chakrabarti SK, Taylor-Fishwick DA, Nadler JL (2010) 12-Lipoxygenase Products Reduce Insulin Secretion and β-Cell Viability in Human Islets. J Clin Endocrinol Metab 95:887–893

Mahdi T, Hanzelmann S, Salehi A, Muhammed SJ, Reinbothe TM, Tang Y, Axelsson AS, Zhou Y, Jing X, Almgren P, Krus U, Taneera J, Blom AM, Lyssenko V, Esguerra JL, Hansson O, Eliasson L, Derry J, Zhang E, Wollheim CB, Groop L, Renstrom E, Rosengren AH (2012) Secreted frizzled-related protein 4 reduces insulin secretion and is overexpressed in type 2 diabetes. Cell Metab 16:625–633

Marhfour I, Lopez XM, Lefkaditis D, Salmon I, Allagnat F, Richardson SJ, Morgan NG, Eizirik DL (2012) Expression of endoplasmic reticulum stress markers in the islets of patients with type 1 diabetes. Diabetologia 55:2417–2420

Markle JG, Frank DN, Mortin-Toth S, Robertson CE, Feazel LM, Rolle-Kampczyk U, von Bergen M, McCoy KD, Macpherson AJ, Danska JS (2013a) Sex differences in the gut microbiome drive hormone-dependent regulation of autoimmunity. Science 339:1084–1088

Markle JG, Mortin-Toth S, Wong AS, Geng L, Hayday A, Danska JS (2013b) Gammadelta T cells are essential effectors of type 1 diabetes in the nonobese diabetic mouse model. J Immunol 190:5392–5401

Marroqui L, Gonzalez A, Neco P, Caballero-Garrido E, Vieira E, Ripoll C, Nadal A, Quesada I (2012) Role of leptin in the pancreatic β-cell: effects and signaling pathways. J Mol Endocrinol 49:R9–R17

Marselli L, Thorne J, Dahiya S, Sgroi DC, Sharma A, Bonner-Weir S, Marchetti P, Weir GC (2010) Gene expression profiles of β-cell enriched tissue obtained by laser capture microdissection from subjects with type 2 diabetes. PloS One 5:e11499

Martin AP, Rankin S, Pitchford S, Charo IF, Furtado GC, Lira SA (2008) Increased expression of CCL2 in insulin-producing cells of transgenic mice promotes mobilization of myeloid cells from the bone marrow, marked insulitis, and diabetes. Diabetes 57:3025–3033

Martinez J, Verbist K, Wang R, Green DR (2013) The relationship between metabolism and the autophagy machinery during the innate immune response. Cell Metab 17:895–900

Masters SL, Dunne A, Subramanian SL, Hull RL, Tannahill GM, Sharp FA, Becker C, Franchi L, Yoshihara E, Chen Z, Mullooly N, Mielke LA, Harris J, Coll RC, Mills KH, Mok KH, Newsholme P, Nunez G, Yodoi J, Kahn SE, Lavelle EC, O'Neill LA (2010) Activation of the NLRP3 inflammasome by islet amyloid polypeptide provides a mechanism for enhanced IL-1β in type 2 diabetes. Nat Immunol 11:897–904

Mathis D (2013) Immunological goings-on in visceral adipose tissue. Cell Metab 17:851–859

Matveyenko AV, Liuwantara D, Gurlo T, Kirakossian D, Dalla Man C, Cobelli C, White MF, Copps KD, Volpi E, Fujita S, Butler PC (2012) Pulsatile portal vein insulin delivery enhances hepatic insulin action and signaling. Diabetes 61:2269–2279

McDuffie M, Maybee NA, Keller SR, Stevens BK, Garmey JC, Morris MA, Kropf E, Rival C, Ma K, Carter JD, Tersey SA, Nunemaker CS, Nadler JL (2008) Nonobese diabetic (NOD) mice congenic for a targeted deletion of 12/15-lipoxygenase are protected from autoimmune diabetes. Diabetes 57:199–208

Mehers KL, Long AE, van der Slik AR, Aitken RJ, Nathwani V, Wong FS, Bain S, Gill G, Roep BO, Bingley PJ, Gillespie KM (2011) An increased frequency of NK cell receptor and HLA-C group 1 combinations in early-onset type 1 diabetes. Diabetologia 54:3062–3070

Meier JJ, Veldhuis JD, Butler PC (2005) Pulsatile insulin secretion dictates systemic insulin delivery by regulating hepatic insulin extraction in humans. Diabetes 54:1649–1656

Meyers AJ, Shah RR, Gottlieb PA, Zipris D (2010) Altered Toll-like receptor signaling pathways in human type 1 diabetes. J Mol Med (Berl) 88:1221–1231

Michalska M, Wolf G, Walther R, Newsholme P (2010) The effects of pharmacologic inhibition of NADPH oxidase or iNOS on pro-inflammatory cytokine, palmitic acid or H2O2 -induced mouse islet or clonal pancreatic β cell dysfunction. Biosci Rep 30:445–453

Miller MR, Seifert J, Szabo NJ, Clare-Salzler M, Rewers M, Norris JM (2010) Erythrocyte membrane fatty acid content in infants consuming formulas supplemented with docosahexaenoic acid (DHA) and arachidonic acid (ARA): an observational study. Matern Child Nutr 6:338–346

Miller MR, Yin X, Seifert J, Clare-Salzler M, Eisenbarth GS, Rewers M, Norris JM (2011) Erythrocyte membrane omega-3 fatty acid levels and omega-3 fatty acid intake are not associated with conversion to type 1 diabetes in children with islet autoimmunity: the Diabetes Autoimmunity Study in the Young (DAISY). Pediatr Diabetes 12:669–675

Montane J, Klimek-Abercrombie A, Potter KJ, Westwell-Roper C, Verchere CB (2012) Metabolic stress, IAPP and islet amyloid. Diabetes Obes Metab 14:68–77

Moran A, Bundy B, Becker DJ, DiMeglio LA, Gitelman SE, Goland R, Greenbaum CJ, Herold KC, Marks JB, Raskin P, Sanda S, Schatz D, Wherrett DK, Wilson DM, Krischer JP, Skyler JS, Pickersgill L, de Koning E, Ziegler AG, Boehm B, Badenhoop K, Schloot N, Bak JF, Pozzilli P, Mauricio D, Donath MY, Castano L, Wagner A, Lervang HH, Perrild H, Mandrup-Poulsen T, Pociot F, Dinarello CA (2013) Interleukin-1 antagonism in type 1 diabetes of recent onset: two multicentre, randomised, double-blind, placebo-controlled trials. Lancet 381:1905–1915

Morgan D, Oliveira-Emilio HR, Keane D, Hirata AE, Santos da Rocha M, Bordin S, Curi R, Newsholme P, Carpinelli AR (2007) Glucose, palmitate and pro-inflammatory cytokines modulate production and activity of a phagocyte-like NADPH oxidase in rat pancreatic islets and a clonal β cell line. Diabetologia 50:359–369

Morioka T, Asilmaz E, Hu J, Dishinger JF, Kurpad AJ, Elias CF, Li H, Elmquist JK, Kennedy RT, Kulkarni RN (2007) Disruption of leptin receptor expression in the pancreas directly affects β cell growth and function in mice. J Clin Invest 117:2860–2868

Murri M, Leiva I, Gomez-Zumaquero JM, Tinahones FJ, Cardona F, Soriguer F, Queipo-Ortuno MI (2013) Gut microbiota in children with type 1 diabetes differs from that in healthy children: a case-control study. BMC Med 11:46

Myers MG Jr, Leibel RL, Seeley RJ, Schwartz MW (2010) Obesity and leptin resistance: distinguishing cause from effect. Trends Endocrinol Metab: TEM 21:643–651

Myhr CB, Hulme MA, Wasserfall CH, Hong PJ, Lakshmi PS, Schatz DA, Haller MJ, Brusko TM, Atkinson MA (2013) The autoimmune disease-associated SNP rs917997 of IL18RAP controls IFNgamma production by PBMC. J Autoimmun 44:8–12

Nakata M, Okada T, Ozawa K, Yada T (2007) Resistin induces insulin resistance in pancreatic islets to impair glucose-induced insulin release. Biochem Biophys Res Commun 353:1046–1051

Nakayama M, Inoguchi T, Sonta T, Maeda Y, Sasaki S, Sawada F, Tsubouchi H, Sonoda N, Kobayashi K, Sumimoto H, Nawata H (2005) Increased expression of NAD(P)H oxidase in islets of animal models of type 2 diabetes and its improvement by an AT1 receptor antagonist. Biochem Biophys Res Commun 332:927–933

Nguyen MT, Favelyukis S, Nguyen AK, Reichart D, Scott PA, Jenn A, Liu-Bryan R, Glass CK, Neels JG, Olefsky JM (2007) A subpopulation of macrophages infiltrates hypertrophic adipose tissue and is activated by free fatty acids via Toll-like receptors 2 and 4 and JNK-dependent pathways. J Biol Chem 282:35279–35292

Nicoloff G, Blazhev A, Petrova C, Christova P (2004) Circulating immune complexes among diabetic children. Clin Dev Immunol 11:61–66

Nie Y, Ma RC, Chan JC, Xu H, Xu G (2012) Glucose-dependent insulinotropic peptide impairs insulin signaling via inducing adipocyte inflammation in glucose-dependent insulinotropic peptide receptor-overexpressing adipocytes. Faseb J 26:2383–2393

Nikolajczyk BS, Jagannathan-Bogdan M, Denis GV (2012) The outliers become a stampede as immunometabolism reaches a tipping point. Immunol Rev 249:253–275

Nikolic T, Bouma G, Drexhage HA, Leenen PJ (2005) Diabetes-prone NOD mice show an expanded subpopulation of mature circulating monocytes, which preferentially develop into macrophage-like cells in vitro. J Leukoc Biol 78:70–79

Nixon M, Wake DJ, Livingstone DE, Stimson RH, Esteves CL, Seckl JR, Chapman KE, Andrew R, Walker BR (2012) Salicylate downregulates 11β-HSD1 expression in adipose tissue in obese mice and in humans, mediating insulin sensitization. Diabetes 61:790–796

Nogueira TC, Paula FM, Villate O, Colli ML, Moura RF, Cunha DA, Marselli L, Marchetti P, Cnop M, Julier C, Eizirik DL (2013) GLIS3, a susceptibility gene for type 1 and type 2 diabetes, modulates pancreatic β cell apoptosis via regulation of a splice variant of the BH3-only protein Bim. PLoS Genet 9:e1003532

Oliveira HR, Verlengia R, Carvalho CR, Britto LR, Curi R, Carpinelli AR (2003) Pancreatic β-cells express phagocyte-like NAD(P)H oxidase. Diabetes 52:1457–1463

Omar BA, Vikman J, Winzell MS, Voss U, Ekblad E, Foley JE, Ahren B (2013) Enhanced β cell function and anti-inflammatory effect after chronic treatment with the dipeptidyl peptidase-4 inhibitor vildagliptin in an advanced-aged diet-induced obesity mouse model. Diabetologia 56:1752–1760

Orban T, Sosenko JM, Cuthbertson D, Krischer JP, Skyler JS, Jackson R, Yu L, Palmer JP, Schatz D, Eisenbarth G (2009) Pancreatic islet autoantibodies as predictors of type 1 diabetes in the Diabetes Prevention Trial-Type 1. Diabetes Care 32:2269–2274

Ortis F, Miani M, Colli ML, Cunha DA, Gurzov EN, Allagnat F, Chariot A, Eizirik DL (2012) Differential usage of NF-kappaB activating signals by IL-1β and TNF-α in pancreatic β cells. FEBS Lett 586:984–989

Osborn O, Olefsky JM (2012) The cellular and signaling networks linking the immune system and metabolism in disease. Nat Med 18:363–374

Oslowski CM, Urano F (2010) The binary switch between life and death of endoplasmic reticulum-stressed β cells. Curr Opin Endocrinol Diabetes Obes 17:107–112

Oslowski CM, Hara T, O'Sullivan-Murphy B, Kanekura K, Lu S, Hara M, Ishigaki S, Zhu LJ, Hayashi E, Hui ST, Greiner D, Kaufman RJ, Bortell R, Urano F (2012) Thioredoxin-interacting protein mediates ER stress-induced β cell death through initiation of the inflammasome. Cell Metab 16:265–273

Ouchi N, Parker JL, Lugus JJ, Walsh K (2011) Adipokines in inflammation and metabolic disease. Nat Rev Immunol 11:85–97

Padgett LE, Broniowska KA, Hansen PA, Corbett JA, Tse HM (2013) The role of reactive oxygen species and proinflammatory cytokines in type 1 diabetes pathogenesis. Ann N Y Acad Sci 1281:16–35

Patel SB, Reams GP, Spear RM, Freeman RH, Villarreal D (2008) Leptin: linking obesity, the metabolic syndrome, and cardiovascular disease. Curr Hypertens Rep 10:131–137

Pavlatou MG, Mastorakos G, Margeli A, Kouskouni E, Tentolouris N, Katsilambros N, Chrousos GP, Papassotiriou I (2011) Angiotensin blockade in diabetic patients decreases insulin resistance-associated low-grade inflammation. Eur J Clin Invest 41:652–658

Petrie JR, Pearson ER, Sutherland C (2011) Implications of genome wide association studies for the understanding of type 2 diabetes pathophysiology. Biochem Pharmacol 81:471–477

Pi J, Collins S (2010) Reactive oxygen species and uncoupling protein 2 in pancreatic β-cell function. Diabetes Obes Metab 12(Suppl 2):141–148

Piya MK, McTernan PG, Kumar S (2013) Adipokine inflammation and insulin resistance: the role of glucose, lipids and endotoxin. J Endocrinol 216:T1–T15

Poitout V, Robertson RP (2008) Glucolipotoxicity: fuel excess and β-cell dysfunction. Endocr Rev 29:351–366

Prentki M, Nolan CJ (2006) Islet β cell failure in type 2 diabetes. J Clin Invest 116:1802–1812

Prieur X, Mok CY, Velagapudi VR, Nunez V, Fuentes L, Montaner D, Ishikawa K, Camacho A, Barbarroja N, O'Rahilly S, Sethi JK, Dopazo J, Oresic M, Ricote M, Vidal-Puig A (2011) Differential lipid partitioning between adipocytes and tissue macrophages modulates macrophage lipotoxicity and M2/M1 polarization in obese mice. Diabetes 60:797–809

Putnam AL, Brusko TM, Lee MR, Liu W, Szot GL, Ghosh T, Atkinson MA, Bluestone JA (2009) Expansion of human regulatory T-cells from patients with type 1 diabetes. Diabetes 58:652–662

Quagliarini F, Wang Y, Kozlitina J, Grishin NV, Hyde R, Boerwinkle E, Valenzuela DM, Murphy AJ, Cohen JC, Hobbs HH (2012) Atypical angiopoietin-like protein that regulates ANGPTL3. Proc Natl Acad Sci USA 109:19751–19756

Rabinovitch A, Sumoski W, Rajotte RV, Warnock GL (1990) Cytotoxic effects of cytokines on human pancreatic islet cells in monolayer culture. J Clin Endocrinol Metab 71:152–156

Ramos-Lopez E, Scholten F, Aminkeng F, Wild C, Kalhes H, Seidl C, Tonn T, Van der Auwera B, Badenhoop K (2009) Association of KIR2DL2 polymorphism rs2756923 with type 1 diabetes and preliminary evidence for lack of inhibition through HLA-C1 ligand binding. Tissue Antigens 73:599–603

Ramos-Zavala MG, Gonzalez-Ortiz M, Martinez-Abundis E, Robles-Cervantes JA, Gonzalez-Lopez R, Santiago-Hernandez NJ (2011) Effect of diacerein on insulin secretion and metabolic control in drug-naive patients with type 2 diabetes: a randomized clinical trial. Diabetes Care 34:1591–1594

Rao JR, Keating DJ, Chen C, Parkington HC (2012) Adiponectin increases insulin content and cell proliferation in MIN6 cells via PPARgamma-dependent and PPARgamma-independent mechanisms. Diabetes Obes Metab 14:983–989

Redondo MJ, Rewers M, Yu L, Garg S, Pilcher CC, Elliott RB, Eisenbarth GS (1999) Genetic determination of islet cell autoimmunity in monozygotic twin, dizygotic twin, and non-twin siblings of patients with type 1 diabetes: prospective twin study. BMJ 318:698–702

Revollo JR, Korner A, Mills KF, Satoh A, Wang T, Garten A, Dasgupta B, Sasaki Y, Wolberger C, Townsend RR, Milbrandt J, Kiess W, Imai S (2007) Nampt/PBEF/Visfatin regulates insulin secretion in β cells as a systemic NAD biosynthetic enzyme. Cell Metab 6:363–375

Richardson SJ, Willcox A, Bone AJ, Foulis AK, Morgan NG (2009) Islet-associated macrophages in type 2 diabetes. Diabetologia 52:1686–1688

Ridker PM, Thuren T, Zalewski A, Libby P (2011) Interleukin-1β inhibition and the prevention of recurrent cardiovascular events: rationale and design of the Canakinumab Anti-inflammatory Thrombosis Outcomes Study (CANTOS). Am Heart J 162:597–605

Ridker PM, Howard CP, Walter V, Everett B, Libby P, Hensen J, Thuren T (2012) Effects of interleukin-1β inhibition with canakinumab on hemoglobin A1c, lipids, C-reactive protein, interleukin-6, and fibrinogen: a phase IIb randomized, placebo-controlled trial. Circulation 126:2739–2748

Rissanen A, Howard CP, Botha J, Thuren T (2012) Effect of anti-IL-1β antibody (canakinumab) on insulin secretion rates in impaired glucose tolerance or type 2 diabetes: results of a randomized, placebo-controlled trial. Diabetes Obes Metab 14:1088–1096

Rodacki M, Svoren B, Butty V, Besse W, Laffel L, Benoist C, Mathis D (2007) Altered natural killer cells in type 1 diabetic patients. Diabetes 56:177–185

Roggli E, Gattesco S, Pautz A, Regazzi R (2012) Involvement of the RNA-binding protein ARE/poly(U)-binding factor 1 (AUF1) in the cytotoxic effects of proinflammatory cytokines on pancreatic β cells. Diabetologia 55:1699–1708

Roivainen M, Knip M, Hyoty H, Kulmala P, Hiltunen M, Vahasalo P, Hovi T, Akerblom HK (1998) Several different enterovirus serotypes can be associated with prediabetic autoimmune episodes and onset of overt IDDM. Childhood Diabetes in Finland (DiMe) Study Group. J Med Virol 56:74–78

Rolandsson O, Palmer JP (2010) Latent autoimmune diabetes in adults (LADA) is dead: long live autoimmune diabetes! Diabetologia 53:1250–1253

Rolo AP, Palmeira CM (2006) Diabetes and mitochondrial function: role of hyperglycemia and oxidative stress. Toxicol Appl Pharmacol 212:167–178

Rosmalen JG, van Ewijk W, Leenen PJ (2002) T-cell education in autoimmune diabetes: teachers and students. Trends Immunol 23:40–46

Ryan EA, Lakey JR, Paty BW, Imes S, Korbutt GS, Kneteman NM, Bigam D, Rajotte RV, Shapiro AM (2002) Successful islet transplantation: continued insulin reserve provides long-term glycemic control. Diabetes 51:2148–2157

Saha SS, Ghosh M (2012) Antioxidant and anti-inflammatory effect of conjugated linolenic acid isomers against streptozotocin-induced diabetes. Br J Nutr 108:974–983

Sakai T, Taki T, Nakamoto A, Shuto E, Tsutsumi R, Toshimitsu T, Makino S, Ikegami S (2013) Lactobacillus plantarum OLL2712 regulates glucose metabolism in C57BL/6 mice fed a high-fat diet. J Nutr Sci Vitaminol 59:144–147

Saxena G, Chen J, Shalev A (2010) Intracellular shuttling and mitochondrial function of thioredoxin-interacting protein. J Biol Chem 285:3997–4005

Saxena A, Desbois S, Carrie N, Lawand M, Mars LT, Liblau RS (2012) Tc17 $CD8^+$ T cells potentiate Th1-mediated autoimmune diabetes in a mouse model. J Immunol 189:3140–3149

Scheen AJ (2004) Pathophysiology of insulin secretion. Ann Endocrinol (Paris) 65:29–36

Scheuner D, Kaufman RJ (2008) The unfolded protein response: a pathway that links insulin demand with β-cell failure and diabetes. Endocr Rev 29:317–333

Schulthess FT, Paroni F, Sauter NS, Shu L, Ribaux P, Haataja L, Strieter RM, Oberholzer J, King CC, Maedler K (2009) CXCL10 impairs β cell function and viability in diabetes through TLR4 signaling. Cell Metab 9:125–139

Scott GS, Fishman S, Khai Siew L, Margalit A, Chapman A, Chervonsky AV, Wen L, Gross G, Wong FS (2010) Immunotargeting of insulin reactive CD8 T cells to prevent diabetes. J Autoimmun 35:390–397

Secondulfo M, Iafusco D, Carratu R, deMagistris L, Sapone A, Generoso M, Mezzogiomo A, Sasso FC, Carteni M, De Rosa R, Prisco F, Esposito V (2004) Ultrastructural mucosal alterations and increased intestinal permeability in non-celiac, type I diabetic patients. Dig Liver Dis: Off J Ital Soc Gastroenterol Ital Assoc Study Liver 36:35–45

Shapiro AM, Lakey JR, Ryan EA, Korbutt GS, Toth E, Warnock GL, Kneteman NM, Rajotte RV (2000) Islet transplantation in seven patients with type 1 diabetes mellitus using a glucocorticoid-free immunosuppressive regimen. N Engl J Med 343:230–238

Shoda LK, Young DL, Ramanujan S, Whiting CC, Atkinson MA, Bluestone JA, Eisenbarth GS, Mathis D, Rossini AA, Campbell SE, Kahn R, Kreuwel HT (2005) A comprehensive review of interventions in the NOD mouse and implications for translation. Immunity 23:115–126

Skurk T, Herder C, Kraft I, Muller-Scholze S, Hauner H, Kolb H (2005a) Production and release of macrophage migration inhibitory factor from human adipocytes. Endocrinology 146:1006–1011

Skurk T, Kolb H, Muller-Scholze S, Rohrig K, Hauner H, Herder C (2005b) The proatherogenic cytokine interleukin-18 is secreted by human adipocytes. Eur J Endocrinol 152:863–868

Skurk T, Alberti-Huber C, Herder C, Hauner H (2007) Relationship between adipocyte size and adipokine expression and secretion. J Clin Endocrinol Metab 92:1023–1033

Skurk T, Mack I, Kempf K, Kolb H, Hauner H, Herder C (2009) Expression and secretion of RANTES (CCL5) in human adipocytes in response to immunological stimuli and hypoxia. Horm Metab Res 41:183–189

Skyler JS, Sosenko JM (2013) The evolution of type 1 diabetes. JAMA: J Am Med Assoc 309:2491–2492

Sloan-Lancaster J, Abu-Raddad E, Polzer J, Miller JW, Scherer JC, De Gaetano A, Berg JK, Landschulz WH (2013) Double-blind, randomized study evaluating the glycemic and anti-inflammatory effects of subcutaneous LY2189102, a neutralizing IL-1β antibody, in patients with type 2 diabetes. Diabetes Care 36(8):2239–2246

Spindel ON, World C, Berk BC (2012) Thioredoxin interacting protein: redox dependent and independent regulatory mechanisms. Antioxid Redox Signal 16:587–596

Staeva TP, Chatenoud L, Insel R, Atkinson MA (2013) Recent lessons learned from prevention and recent-onset type 1 diabetes immunotherapy trials. Diabetes 62:9–17

Steinberg GR (2007) Inflammation in obesity is the common link between defects in fatty acid metabolism and insulin resistance. Cell Cycle 6:888–894

Steppan CM, Bailey ST, Bhat S, Brown EJ, Banerjee RR, Wright CM, Patel HR, Ahima RS, Lazar MA (2001) The hormone resistin links obesity to diabetes. Nature 409:307–312

Stratton IM, Adler AI, Neil HA, Matthews DR, Manley SE, Cull CA, Hadden D, Turner RC, Holman RR (2000) Association of glycaemia with macrovascular and microvascular complications of type 2 diabetes (UKPDS 35): prospective observational study. BMJ 321:405–412

Su SC, Pei D, Hsieh CH, Hsiao FC, Wu CZ, Hung YJ (2010) Circulating pro-inflammatory cytokines and adiponectin in young men with type 2 diabetes. Acta Diabetol 48:113–119

Subramanian SL, Hull RL, Zraika S, Aston-Mourney K, Udayasankar J, Kahn SE (2012) cJUN N-terminal kinase (JNK) activation mediates islet amyloid-induced β cell apoptosis in cultured human islet amyloid polypeptide transgenic mouse islets. Diabetologia 55:166–174

Supale S, Li N, Brun T, Maechler P (2012) Mitochondrial dysfunction in pancreatic β cells. Trends Endocrinol Metab: TEM 23:477–487

Tanaka S, Maeda S, Hashimoto M, Fujimori C, Ito Y, Teradaira S, Hirota K, Yoshitomi H, Katakai T, Shimizu A, Nomura T, Sakaguchi N, Sakaguchi S (2010) Graded attenuation of TCR signaling elicits distinct autoimmune diseases by altering thymic T cell selection and regulatory T cell function. J Immunol 185:2295–2305

Tang C, Naassan AE, Chamson-Reig A, Koulajian K, Goh TT, Yoon F, Oprescu AI, Ghanim H, Lewis GF, Dandona P, Donath MY, Ehses JA, Arany E, Giacca A (2013) Susceptibility to fatty acid-induced β-cell dysfunction is enhanced in prediabetic diabetes-prone biobreeding rats: a potential link between β-cell lipotoxicity and islet inflammation. Endocrinology 154:89–101

Taylor-Fishwick D, Pittenger GL (2010) Harnessing the pancreatic stem cell. Endocrinol Metab Clin North Am 39:763

Taylor-Fishwick D, Pittenger GL, Vinik AI (2008) Transplantation and beyond. Drug Dev Res 69:165–176

Taylor-Fishwick DA, Weaver JR, Grzesik W, Chakrabarti S, Green-Mitchell S, Imai Y, Kuhn N, Nadler JL (2013) Production and function of IL-12 in islets and β cells. Diabetologia 56:126–135

Tersey SA, Carter JD, Rosenberg L, Taylor-Fishwick DA, Mirmira RG, Nadler JL (2012a) Amelioration of type 1 diabetes following treatment of non-obese diabetic mice with INGAP and lisofylline. J Diabetes Mellitus 2:251

Tersey SA, Nishiki Y, Templin AT, Cabrera SM, Stull ND, Colvin SC, Evans-Molina C, Rickus JL, Maier B, Mirmira RG (2012b) Islet β-cell endoplasmic reticulum stress precedes the onset of type 1 diabetes in the nonobese diabetic mouse model. Diabetes 61:818–827

Thomas HE, Irawaty W, Darwiche R, Brodnicki TC, Santamaria P, Allison J, Kay TW (2004) IL-1 receptor deficiency slows progression to diabetes in the NOD mouse. Diabetes 53:113–121

Thorsby E, Ronningen KS (1993) Particular HLA-DQ molecules play a dominant role in determining susceptibility or resistance to type 1 (insulin-dependent) diabetes mellitus. Diabetologia 36:371–377

Tilg H, Moschen AR (2008) Inflammatory mechanisms in the regulation of insulin resistance. Mol Med 14:222–231

Timper K, Grisouard J, Sauter NS, Herzog-Radimerski T, Dembinski K, Peterli R, Frey DM, Zulewski H, Keller U, Muller B, Christ-Crain M (2013) Glucose-dependent insulinotropic polypeptide induces cytokine expression, lipolysis, and insulin resistance in human adipocytes. Am J Physiol Endocrinol Metab 304:E1–E13

Todd JA (2010) Etiology of type 1 diabetes. Immunity 32:457–467

Tree TI, Lawson J, Edwards H, Skowera A, Arif S, Roep BO, Peakman M (2010) Naturally arising human CD4 T-cells that recognize islet autoantigens and secrete interleukin-10 regulate proinflammatory T-cell responses via linked suppression. Diabetes 59:1451–1460

Uchizono Y, Takeya R, Iwase M, Sasaki N, Oku M, Imoto H, Iida M, Sumimoto H (2006) Expression of isoforms of NADPH oxidase components in rat pancreatic islets. Life Sci 80:133–139

Vaarala O, Ilonen J, Ruohtula T, Pesola J, Virtanen SM, Harkonen T, Koski M, Kallioinen H, Tossavainen O, Poussa T, Jarvenpaa AL, Komulainen J, Lounamaa R, Akerblom HK, Knip M (2012) Removal of bovine insulin from cow's milk formula and early initiation of β-cell autoimmunity in the FINDIA Pilot Study. Arch Pediatr Adolesc Med 166:608–614

Valladares R, Sankar D, Li N, Williams E, Lai KK, Abdelgeliel AS, Gonzalez CF, Wasserfall CH, Larkin J, Schatz D, Atkinson MA, Triplett EW, Neu J, Lorca GL (2010) Lactobacillus johnsonii N6.2 mitigates the development of type 1 diabetes in BB-DP rats. PLoS One 5: e10507

van Asseldonk EJ, Stienstra R, Koenen TB, Joosten LA, Netea MG, Tack CJ (2011) Treatment with Anakinra improves disposition index but not insulin sensitivity in nondiabetic subjects with the metabolic syndrome: a randomized, double-blind, placebo-controlled study. J Clin Endocrinol Metab 96:2119–2126

van der Zijl NJ, Moors CC, Goossens GH, Blaak EE, Diamant M (2012) Does interference with the renin-angiotensin system protect against diabetes? Evidence and mechanisms. Diabetes Obes Metab 14:586–595

van Raalte DH, Diamant M (2011) Glucolipotoxicity and β cells in type 2 diabetes mellitus: target for durable therapy? Diabetes Res Clin Pract 93(Suppl 1):S37–S46

Vandanmagsar B, Youm YH, Ravussin A, Galgani JE, Stadler K, Mynatt RL, Ravussin E, Stephens JM, Dixit VD (2011) The NLRP3 inflammasome instigates obesity-induced inflammation and insulin resistance. Nat Med 17:179–188

Varanasi V, Avanesyan L, Schumann DM, Chervonsky AV (2012) Cytotoxic mechanisms employed by mouse T cells to destroy pancreatic β-cells. Diabetes 61:2862–2870

Velloso LA, Eizirik DL, Cnop M (2013) Type 2 diabetes mellitus-an autoimmune disease? Nat Rev Endocrinol

Verbeeten KC, Elks CE, Daneman D, Ong KK (2011) Association between childhood obesity and subsequent Type 1 diabetes: a systematic review and meta-analysis. Diabet Med: J British Diabet Assoc 28:10–18

von Herrath M, Peakman M, Roep B (2013) Progress in immune-based therapies for type 1 diabetes. Clin Exp Immunol 172:186–202

Wallace C, Rotival M, Cooper JD, Rice CM, Yang JH, McNeill M, Smyth DJ, Niblett D, Cambien F, Tiret L, Todd JA, Clayton DG, Blankenberg S (2012) Statistical colocalization of monocyte gene expression and genetic risk variants for type 1 diabetes. Hum Mol Genet 21:2815–2824

Wang L, Liu Y, Yan Lu S, Nguyen KT, Schroer SA, Suzuki A, Mak TW, Gaisano H, Woo M (2010) Deletion of Pten in pancreatic β-cells protects against deficient β-cell mass and function in mouse models of type 2 diabetes. Diabetes 59:3117–3126

Weaver JR, Holman TR, Imai Y, Jadhav A, Kenyon V, Maloney DJ, Nadler JL, Rai G, Simeonov A, Taylor-Fishwick DA (2012) Integration of pro-inflammatory cytokines, 12-lipoxygenase and NOX-1 in pancreatic islet β cell dysfunction. Mol Cell Endocrinol 358:88–95

Wei FY, Suzuki T, Watanabe S, Kimura S, Kaitsuka T, Fujimura A, Matsui H, Atta M, Michiue H, Fontecave M, Yamagata K, Suzuki T, Tomizawa K (2011) Deficit of tRNA(Lys) modification by Cdkal1 causes the development of type 2 diabetes in mice. J Clin Invest 121:3598–3608

Weir GC, Bonner-Weir S (2013) Islet β cell mass in diabetes and how it relates to function, birth, and death. Ann N Y Acad Sci 1281:92–105

Wellen KE, Hotamisligil GS (2003) Obesity-induced inflammatory changes in adipose tissue. J Clin Invest 112:1785–1788

Wellen KE, Hotamisligil GS (2005) Inflammation, stress, and diabetes. J Clin Invest 115:1111–1119

Westermark P, Andersson A, Westermark GT (2011) Islet amyloid polypeptide, islet amyloid, and diabetes mellitus. Physiol Rev 91:795–826

Westwell-Roper C, Dai DL, Soukhatcheva G, Potter KJ, van Rooijen N, Ehses JA, Verchere CB (2011) IL-1 blockade attenuates islet amyloid polypeptide-induced proinflammatory cytokine release and pancreatic islet graft dysfunction. J Immunol 187:2755–2765

Wijesekara N, Krishnamurthy M, Bhattacharjee A, Suhail A, Sweeney G, Wheeler MB (2010) Adiponectin-induced ERK and Akt phosphorylation protects against pancreatic β cell apoptosis and increases insulin gene expression and secretion. J Biol Chem 285:33623–33631

Wilkin TJ (2009) The accelerator hypothesis: a review of the evidence for insulin resistance as the basis for type I as well as type II diabetes. Int J Obes (Lond) 33:716–726

Willcox A, Richardson SJ, Bone AJ, Foulis AK, Morgan NG (2009) Analysis of islet inflammation in human type 1 diabetes. Clin Exp Immunol 155:173–181

Winkler C, Raab J, Grallert H, Ziegler AG (2012) Lack of association of type 2 diabetes susceptibility genotypes and body weight on the development of islet autoimmunity and type 1 diabetes. PLoS One 7:e35410

Xu G, Chen J, Jing G, Shalev A (2013) Thioredoxin-interacting protein regulates insulin transcription through microRNA-204. Nat Med 19:1141–1146

Xu X, Zheng S, Yang F, Shi Y, Gu Y, Chen H, Zhang M, Yang T (2014) Increased Th22 cells are independently associated with Th17 cells in type 1 diabetes. Endocrine 46:90–98

Yang ZD, Chen M, Wu R, McDuffie M, Nadler JL (2002) The anti-inflammatory compound lisofylline prevents Type I diabetes in non-obese diabetic mice. Diabetologia 45:1307–1314

Yang Z, Chen M, Fialkow LB, Ellett JD, Wu R, Nadler JL (2003) The novel anti-inflammatory compound, lisofylline, prevents diabetes in multiple low-dose streptozotocin-treated mice. Pancreas 26:e99–e104

Yang Z, Chen M, Ellett JD, Fialkow LB, Carter JD, McDuffie M, Nadler JL (2004) Autoimmune diabetes is blocked in Stat4-deficient mice. J Autoimmun 22:191–200

Yang Z, Chen M, Carter JD, Nunemaker CS, Garmey JC, Kimble SD, Nadler JL (2006) Combined treatment with lisofylline and exendin-4 reverses autoimmune diabetes. Biochem Biophys Res Commun 344:1017–1022

Yi P, Park JS, Melton DA (2013) Betatrophin: a hormone that controls pancreatic β cell proliferation. Cell 153:747–758

Yin N, Xu J, Ginhoux F, Randolph GJ, Merad M, Ding Y, Bromberg JS (2012) Functional specialization of islet dendritic cell subsets. J Immunol 188:4921–4930

Ylipaasto P, Smura T, Gopalacharyulu P, Paananen A, Seppanen-Laakso T, Kaijalainen S, Ahlfors H, Korsgren O, Lakey JR, Lahesmaa R, Piemonti L, Oresic M, Galama J, Roivainen M (2012) Enterovirus-induced gene expression profile is critical for human pancreatic islet destruction. Diabetologia 55:3273–3283

Youssef-Elabd EM, McGee KC, Tripathi G, Aldaghri N, Abdalla MS, Sharada HM, Ashour E, Amin AI, Ceriello A, O'Hare JP, Kumar S, McTernan PG, Harte AL (2012) Acute and chronic saturated fatty acid treatment as a key instigator of the TLR-mediated inflammatory response in human adipose tissue, in vitro. J Nutr Biochem 23:39–50

Zhang K, Kaufman RJ (2008) From endoplasmic-reticulum stress to the inflammatory response. Nature 454:455–462

Zhong J, Rao X, Rajagopalan S (2013) An emerging role of dipeptidyl peptidase 4 (DPP4) beyond glucose control: potential implications in cardiovascular disease. Atherosclerosis 226:305–314

Ziegler AG, Nepom GT (2010) Prediction and pathogenesis in type 1 diabetes. Immunity 32:468–478

Ziegler AG, Rewers M, Simell O, Simell T, Lempainen J, Steck A, Winkler C, Ilonen J, Veijola R, Knip M, Bonifacio E, Eisenbarth GS (2013) Seroconversion to multiple islet autoantibodies and risk of progression to diabetes in children. JAMA: J Am Med Assoc 309:2473–2479

Zraika S, Hull RL, Udayasankar J, Aston-Mourney K, Subramanian SL, Kisilevsky R, Szarek WA, Kahn SE (2009) Oxidative stress is induced by islet amyloid formation and time-dependently mediates amyloid-induced β cell apoptosis. Diabetologia 52:626–635

Immunology of β-Cell Destruction

37

Åke Lernmark and Daria LaTorre

Contents

Background and Historical Perspectives 1049
Autoimmune β-Cell Destruction 1053
    Genetic Etiology 1053
    Immune Cells in Tolerance 1055
    What Happens in the Islet? 1058
    Antigen Presentation in Pancreatic Lymph Nodes 1061
    Homing of T-Cells to Islets 1062
    Insulitis and β-Cell Destruction 1063
    Is β-Cell Destruction Reflected in the Blood? 1064
Prediction of β-Cell Destruction 1067
Concluding Remarks 1068
Cross-References 1069
References 1069

Abstract

The pancreatic islet β cells are the target for an autoimmune process that eventually results in an inability to control blood glucose due to the lack of insulin. The different steps that eventually lead to the complete loss of the β cells are reviewed to include the very first step of a triggering event that initiates the development of β-cell autoimmunity to the last step of appearance of islet cell autoantibodies, which may mark that insulitis is about to form. The observations that the initial β-cell destruction by virus or other environmental factors triggers islet autoimmunity not in the islets but in the draining pancreatic lymph nodes are reviewed along with possible basic mechanisms of loss of tolerance to islet autoantigens. Once islet autoimmunity is established, the question is how β cells are progressively killed by autoreactive lymphocytes, which eventually results

Å. Lernmark (✉) • D. LaTorre
Department of Clinical Sciences, Lund University, CRC, University Hospital MAS, Malmö, Sweden
e-mail: ake.lernmark@med.lu.se

in chronic insulitis. These events have been examined in spontaneously diabetic mice or rats, but controlled clinical trials have shown that rodent observations cannot always be translated into mechanisms in humans. Attempts are therefore needed to clarify the step 1 triggering mechanisms and the step to chronic autoimmune insulitis to develop evidence-based treatment approaches to prevent type 1 diabetes.

Keywords
Antigen-presenting cells • Autoantigen • CD4$^+$ T-cells • CD8$^+$ T-cells • Dendritic cells • Insulitis • Islet autoantibodies • Islet autoimmunity • Prediction • Prevention • T regulatory cells

Abbreviations

APC	Antigen-presenting cells
BB	Biobreeding
BCR	β-cell receptor
CTLA-4	Cytolytic T-lymphocyte-associated antigen
cTreg	Conventional regulatory T
DC	Dendritic cells
Fas-L	Fas-Ligand
FOXP3	Forkhead–winged helix
GABA	Gamma-aminobutyric acid
GAD	Glutamic acid decarboxylase
HLA	Human leukocyte antigens
HSP	Heat-shock protein
IA-2	Insulinoma-associated antigen-2
IAA	Insulin autoantibodies
ICA	Islet cell antibodies
ICAM	Intercellular adhesion molecule
ICSA	Islet cell surface antibodies
IDO	Indoleamine 2 3-dioxygenase
IFN	Interferon
IL	Interleukin
iVEC	Islet vascular endothelial cells
LFA-1	Leukocyte function-associated antigen-1
NF	Nuclear factor
NK	Natural killer lymphocyte
NKT	Natural killer T-cell
NO	Nitric oxide
NOD	Nonobese diabetic
nTreg	Natural regulatory T
PBMC	Peripheral blood mononuclear cells
PD-1	Programmed death-1
pDC	Plasmacytoid dendritic cell
pLN	Pancreatic lymph node

pMHC	Peptide–MHC
PRR	Pattern recognition receptors
TCR	T-cell receptor
TEDDY study	The Environmental Determinants of Diabetes in the Young
TF	Transcription factor
TGF	Transforming growth factor
TLR	Toll-like receptor
TNF	Tumor necrosis factor
Treg	Regulatory T-cell
TSA	Tissue-specific antigen
VNTR	Variable nucleotide tandem repeat
ZnT8	Zinc transporter isoform-8

Background and Historical Perspectives

Immune-mediated selective destruction of the pancreatic islet β-cells is the hallmark of autoimmune (type 1) diabetes mellitus (T1D) formerly known as juvenile diabetes or insulin-dependent diabetes mellitus (Eisenbarth 1986; Atkinson et al. 2014; Bach 1997). The immunogenetic feature of the disease is a polygenic inheritance of susceptibility, which is reflected in a highly polyclonal autoimmune response targeting several β-cell autoantigens. The autoimmune response is associated with progressive β-cell destruction that eventually leads to overt clinical disease. As attested by prospective studies of children at genetic risk for T1D (DIPP, DAISY, and BabyDIAB), the appearance of specific islet autoantibodies marks the initiation of islet autoimmunity and may be detectable for months to years (Ziegler et al. 2013) during which time β-cell dysfunction proceeds asymptomatically (Vehik et al. 2011). T1D may therefore be viewed as a two-step disease. The first step is the initiation of islet autoimmunity. The second step is appearance of diabetes when islet autoimmunity has caused a major β-cell loss (>80 %) and insulin deficiency becomes clinically manifest (Bingley and Gale 2006).

At diagnosis the typical histological finding of affected islets, first described in short-duration diabetes patients at the beginning of the last century (Weichselbaum 1910) and termed "insulitis" (Von Meyenburg 1940; Gepts 1965), consists of an infiltrate of inflammatory cells associated with a loss of the β-cell endocrine subpopulation. The infiltrate consists of mononuclear cells (Gepts and De Mey 1978; Gepts and Lecompte 1981) and T and B lymphocytes (Itoh 1989; Bottazzo et al. 1985). Little is known about insulitis during the first step of the disease when subjects have preclinical islet autoimmunity. Recent studies suggest that the mere presence of a single islet autoantibody does not predict insulitis (In't Veld et al. 2007; Campbell-Thompson et al. 2013).

The understanding of T1D etiology and pathogenesis is complicated by the lack of epidemiological data on the first step of the disease. In contrast the epidemiological knowledge of T1D is developing rapidly through registers in many different countries. The incidence is different among age groups, highest among children

(EURODIAB ACE Study Group 2000; Patterson et al. 2009; Onkamo et al. 1999; Karvonen et al. 2000), but the disease may occur at any age (Todd and Farrall 1996).

Annual incidence shows geographical variation among different countries and ethnic groups, from 0.1 per 100,000 children in parts of Asia and South America to the highest rate in Finland (64.2 per 100,000) (Onkamo et al. 1999; Karvonen et al. 2000). The mode of inheritance is complex as 80–85 % of T1D is occurring sporadically (Dahlquist et al. 1989), and the risk of becoming diabetic is approximately 7 % for a sibling and 6 % for the children of T1D parents (Dahlquist et al. 1989; Akesson et al. 2005).

An autoimmune etiology for T1D was suspected approximately 40 years ago from the association between diabetes and other autoimmune diseases (Ungar et al. 1968; Nerup and Binder 1973; Barker 2006). The first attempt to identify an autoimmune reaction toward the endocrine pancreas dates back to 1973 when testing for leukocyte migration inhibition to islet antigens suggested that T1D patients might be sensitized to pancreatic antigens (Nerup et al. 1973). Nearly concomitantly T1D was reported to be correlated to histocompatibility antigens (HLA), previously shown to be associated with other autoimmune diseases (Nerup et al. 1974; 1979). Association studies have proved that the greatest contribution to genetic susceptibility to T1D is exerted by HLA class II alleles on chromosome 6 where the HLA-DQ haplotypes DQ2 and DQ8 confer the highest risk and DQ6.2 the highest protection (Owerbach et al. 1983; Todd et al. 1987; Todd 2010). The detailed mechanisms by which different HLA molecules provide either risk or resistance to T1D are not fully understood (Todd 2010; Thorsby and Lie 2005). It is possible that different conformations of the MHC molecules pocket yield different peptide-binding properties and influence antigen presentation by antigen-presenting cells (APC) to effector T-cells (Todd 2010; Thorsby and Lie 2005; Delli et al. 2012).

The HLA genetic factors are necessary but not sufficient for islet autoimmunity and T1D. Environmental factors will therefore play a major role in the penetrance of a susceptible genotype. Virus infections have figured prominently in T1D epidemiological investigations (Yoon et al. 1989; Jun and Yoon 1994). Recent studies indicate that virus infections may have multiple effects on β-cell destruction and T1D development. The first is that gestational infections may sensitize the offspring to the eventual development of T1D (Dahlquist et al. 1999a, b; Lindehammer et al. 2011). The second is that virus may trigger islet cell autoimmunity (Lee et al. 2013; Parikka et al. 2012). The third is that the progression in islet autoantibody-positive subjects may be accelerated by virus infection (Stene et al. 2010). The possible contribution of a virus infection to trigger islet autoimmunity (step 1 of the disease) or to affect the progression to clinical onset therefore needs to be sorted out. The contribution of dietary factors is equally controversial (Blom et al. 1991, 1989; Lamb et al. 2008; Virtanen et al. 2012; Landin-Olsson et al. 2013). Maternal factors (Blom et al. 1991; Dahlquist et al. 1990), vaccinations (Blom et al. 1991; Helmke et al. 1986), or toxins (Myers et al. 2003) have also been considered.

Table 1 β-cell autoantigens

Antigen	Mol weight (Da)	Autoantibody abbreviation	References
Glutamic acid decarboxylase	65,000	GAD65A; GADA	Uibo and Lernmark 2008
Insulin	6,000	IAA	Palmer et al. 1983
IA-2	40,000	IA-2A	Lan et al. 1996
IA-2-β (Phogrin)	37,000	IA-2βA	Kawasaki et al. 1996
Zinc transporter ZnT8 R/W/Q variants	41,000	ZnT8A	Wenzlau et al. 2007

Although the event that initiates the autoimmune process (step 1) is not yet understood, the fact that it specifically targets the β cells promoted the attempts to find β-cell-specific autoantigens. The interest was initially focused on autoantibodies as useful tools in identification of autoantigenic molecules (Table 1) and to clarify the pathological immune response. The first description of pancreatic islet autoantibodies was published in 1974 when indirect immunofluorescence on frozen human pancreas sections revealed circulating islet cell antibodies (ICA) in the serum of T1D patients with polyendocrine disease (Bottazzo et al. 1974). However, no specific attempt was made at the time to identify the autoantigen(s). A few years later islet cell surface antibodies (ICSA) were demonstrated in newly diagnosed T1D patients using dispersed cell preparations of rodent pancreatic islets (Lernmark et al. 1978; Dobersen et al. 1980). The molecular characteristics of islet autoantigens remained unknown until the demonstration in 1982 that sera from new-onset T1D patients had autoantibodies immunoprecipitating a 64 kDa protein in isolated human islets (Baekkeskov et al. 1982). The 64 kDa immunoprecipitate proved in 1990 to have gamma-aminobutyric acid (GABA)-synthesizing enzymatic activity (Baekkeskov et al. 1990). Molecular cloning of human islet glutamic acid decarboxylase (GAD) showed that the β cells expressed the unique human isoenzyme, GAD65 (Karlsen et al. 1991). GAD65 is expressed in several cell types, but, apart from some brain neurons, it is mainly localized to synaptic-like microvesicles in the β cells (Karlsen et al. 1992; Christgau et al. 1991). GAD65 is in part responsible for the β-cell-specific pattern of ICA (Marshall et al. 1994).

In 1983 autoantibodies reacting with insulin (insulin autoantibodies, IAA) were demonstrated in T1D patients uncorrelated to insulin administration (Palmer et al. 1983). In 1994 trypsin digestion of the 64 kDa immunoprecipitate revealed a 37/40 kDa autoantigen pair recognized by sera of T1D patients (Christie 1993). This observation eventually led to the identification of the insulinoma-associated antigen-2 (IA-2) (Lan et al. 1994) and IA-2 β (or phogrin) (Wasmeier and Hutton 1996; Kawasaki et al. 1996). IA-2 is a transmembrane molecule of islet secretory granules and is implicated in insulin secretion (Kubosaki et al. 2005). In 2007 autoantibodies to the zinc transporter isoform-8 (ZnT8) were reported (Wenzlau et al. 2007). The ZnT8 protein mediates Zn^{2+} cation transport into the insulin granules facilitating the formation of insulin crystals (Wenzlau et al. 2008; Lemaire et al. 2009). ZnT8 polymorphic variants (Wenzlau et al. 2008) represent not only targets of islet autoimmunity but also a genetic marker for type 2 diabetes (Chimienti et al. 2004, 2013).

Continued study of T1D sera has identified additional candidate targets of the humoral immune response. Autoantigens reported so far have different tissue expression patterns and subcellular localization and are referred to as minor autoantigens when the frequency of the respective autoantibodies is below 30 % in newly diagnosed T1D patients (Hirai et al. 2008). Potential minor autoantigens that either need confirmation or further studies include DNA topoisomerase II (Chang et al. 1996, 2004), heat-shock protein 60 (HSP60) (Ozawa et al. 1996), HSP-70 (Abulafia-Lapid et al. 2003), HSP-90 (Qin et al. 2003), vesicle-associated membrane protein-2 (VAMP2)73, neuropeptide Y (NPY)73, (Skarstrand et al. 2013a), carboxypeptidase H (CPH) (Castano et al. 1991), ICA69 (Kerokoski et al. 1999) (Pietropaolo et al. 1993), SOX13 (Torn et al. 2002), Glima38 (Aanstoot et al. 1996), and INS-IGF2 (Kanatsuna et al. 2013). Ganglioside such as the GM-2 ganglioside (Dotta et al. 1997) and sulfatides (Buschard et al. 1993) have also been considered. This wide array of islet autoantigens needs to be better defined. Autoantigens that trigger islet autoimmunity (step 1) needs to be separated from those representing autoantigen spreading during progression to clinical onset (step 2). It is critical that all autoantigen candidates are tested in samples from the past Diabetes Autoantibody Standardization Program (Torn et al. 2008; Yu et al. 2012) or the ongoing Islet Autoantibody Standardization Program (IASP). At this time, the relevance of minor autoantigens for the prediction of T1D is unclear. Despite their attested association with T1D, there is no evidence that islet autoantibodies directly contribute to β-cell damage. The role of B lymphocytes producing islet autoantibodies as APC is understudied (Pihoker et al. 2005; Lernmark and Larsson 2013).

Studies of the cellular arm in human T1D have detected $CD4^+$ and $CD8^+$ T-cells that recognize the same autoantigens as targeted by the humoral arm (Roep and Peakman 2012; Nepom and Buckner 2012). Cellular immunoreactivity to islet autoantigens is less easily assessed than the autoantibody response and is not yet applicable in the clinic. Most of the studies performed in the last decade to identify self-reactive T-cells in the peripheral blood of T1D patients are based on the indirect detection of T-cell presence through antigen-induced proliferation assays, tetramer stimulation, or cytokine release (ELISPOT) analysis (Nepom and Buckner 2012; Ziegler and Nepom 2010). The tetramer technique is highly specific for the HLA type and the peptide lodged in the MHC peptide-binding groove. As pointed out in a recent T-cell workshop, traditional in vitro proliferation assays suffer from methodological limitations (Mallone et al. 2011a; Nagata et al. 2004). As will be discussed later, these studies have produced inconsistent results and have globally failed to detect marked differences between T1D patients and controls.

The understanding of T1D has improved over the years since the rediscovery of insulitis in 1965, the realization that islet autoantibody -positive subjects may not have insulitis14 and that newly diagnosed T1D patients have β cells left and that these apparently dysfunctional cells tend to hyperexpress HLA class I heterodimers 15,

(Atkinson and Gianani 2009; Campbell-Thompson et al. 2012; Pociot et al. 2010). The recognition that T1D is a two-step disease characterized by a long prodrome of islet autoimmunity prior to clinical onset has allowed new hypotheses to be developed as to the initiation of the β-cell destructive process. The transition from islet autoimmunity to clinical T1D will also require a redefinition of the role of environmental factors triggering the disease. In this chapter we will review possible mechanisms of induction of β-cell autoimmunity and the role of environmental factors in this process.

Autoimmune β-Cell Destruction

Genetic Etiology

The major genetic factor for T1D is HLA-DQ on chromosome 620,27,32. Both sib-pair analyses and association studies in Caucasians have indicated that the HLA-DQ A1-B1 haplotypes A1*0301-B1*0302 (DQ8) and A1*0501-B1*0201 (DQ2), alone or in combination (DQ2/8), confer the highest risk for T1D. Nearly 90 % of newly diagnosed children carry DQ2/8 (about 30 %), DQ8, or DQ2 in combination with other haplotypes 33, (Sanjeevi et al. 1995). Among the many haplotypes, there are combinations, in particular with the DQA1*0201-B1*0602 (DQ6.2) haplotype, which is negatively associated (protective) with T1D. The effect is attenuated with increasing age (Graham et al. 2002). The rising incidence of T1D is puzzling as it is associated with a reduced overall contribution of high-risk HLA types in parallel with an increase in DQ8 and DQ2 combinations which did not confer risk 20 years ago (Resic-Lindehammer et al. 2008; Fourlanos et al. 2008). The mechanisms by which DQ8, DQ2, or both increase the risk for T1D are not fully clarified. The function of the DQ heterodimers to present antigenic peptides to the immune system is well understood. It remains to be determined why the DQ2/8 heterozygosity is associated with a young age at onset (Badenhoop et al. 2009). It has been speculated that the DQ2 and DQ8 molecules are important to maintain central or peripheral tolerance to the β-cell autoantigens GAD65, IA-2, insulin, or ZnT8. This possibility needs further exploration as it cannot be excluded that the primary association between T1D and HLA is the "step 1" part of the disease rather than the progression to clinical onset. This hypothesis is supported by the observation that the presence in healthy subjects of GAD65 autoantibodies is associated with DQ2 and IA-2 autoantibodies with DQ8 (Rolandsson et al. 1999).

Several investigations suggest that HLA contributes to about 60 % to the genetic risk of T1D (Todd et al. 1988; Concannon et al. 2005). Major efforts have therefore been made to identify non-HLA genetic risk factors for type 1 diabetes (Concannon et al. 2009). These studies have been highly rewarding as more than 40 genetic factors (see examples in Table 2) have been found to contribute (Pociot et al. 2010; Rich et al. 2009). Interestingly enough many of the genetic factors are important to the function of the immune system. PTPN22 is a regulator of T-cell function, and a

Table 2 Non-HLA genetic factors in type 1 diabetes

Gene (Syno.)	Name	Chromosome	Function	Association with other autoimmune diseases
PTPN22 (PEP, Lypl, Lyp2, LYP, PTPN8)	Protein tyrosine phosphatase, non-receptor type 22 (lymphoid)	1p13	Encodes tyrosine phosphatase and may be involved in regulating CBL function in the T-cell receptor signaling pathway	T1D and 22 other diseases
CTLA-4 (DDM12, CELIAC3)	Cytotoxic T-lymphocyte-associated protein 4	2q33	Possible involvement in regulating T-cell activation	T1DM and 99 other diseases
IFH1 (MDA5)	Interferon induced with helicase C domain 1	Chr.2q24	Proposed involvement in innate immune defense against viruses through interferon response	T1DM association
IL2 (lymphokine. TCGF)	Interleukin 2	Chr.4q27	Encodes a cytokine important for T- and β-cells proliferation. Stimulate β-cells, monocytes, and NK cells	T1DM and 39 other diseases
ITPR3 (IP3R3)	Inositol 1, 4, 5-triphosphate receptor 3	Chr.6p21.3	A second messenger that mediates the release of intracellular calcium	Strong T1DM association
BACH2 (BTB and CNC homology 1)	Basic leucine zipper transcription factor 2	Chr.6q15	Important roles in coordinating transcription activation and repression by MAFK (by similarity)	T1DM association
IL2RA (IDDM 10, CD25)	Interleukin-2 receptor α (chain)	Chr.10p15	Receptor for interleukin-2	Strong association with T1DM
INS VNTR (proinsulin, ILPR, MODY)	Insulin II; insulin 2; insulin	Chr.11p15	Regulating glucose metabolism through adjusting central tolerance to insulin	T1DM and 38 other diseases
TH (TYH, The)	Tyrosine hydroxylase	Chr.11p15	Encodes a protein that converts tyrosine to dopamine. Plays a key role in adrenergic neurons physiology	T1DM and 35 other diseases
ERBB3 (c-erbB3, HER3, LCCS2)	v-erb-b2 erythroblastic leukemia viral oncogene homolog 3	Chr.12p13	Encodes a member of the epidermal growth factor receptor (EGFR) family of receptor tyrosine kinases binds and is activated by neuregulins and NTAK	T1DM and multiple sclerosis
C12orf30 (C12orf51, KIAA0614)	Similar to KIAA0614 protein	Chr.12q24	Not yet determined	T1DM association

(*continued*)

Table 2 (continued)

Gene (Syno.)	Name	Chromosome	Function	Association with other autoimmune diseases
CLEC16A/ KIAA0350 (GoP-1)	C-type lectin domain family 16, member A	Chr.16p13	Unknown. Proposed to be related to immune modulation mechanisms	Strong association with T1DM
PTPN2	Protein tyrosine phosphatase, non-receptor type 2	Chr.18p11	Encode a PTP family protein and may be related to growth factor-mediated cell signaling	T1DM association
BASH3A (TULA, CLIP4)	Ubiquitin-associated and SH3 domain-containing protein A	Chr.21q22	Promotes accumulation of activated target receptors, such as T-cell receptors. EGFR and PDGFRB	T1DM association

genetic polymorphism results in a phosphatase variant that increases the risk not only for T1D but also for rheumatoid arthritis, juvenile rheumatoid arthritis, systemic lupus erythematosus, Graves' disease, generalized vitiligo, and other human autoimmune diseases (Gregersen 2005; Vang et al. 2007). The PTPN22 polymorphism seems in particular to affect progression from prediabetes to clinical disease (Hermann et al. 2006) also in individuals with lower-risk HLA genotypes (Maziarz et al. 2010).

The variable nucleotide tandem repeat in the promoter region of the insulin gene INS VNTR seems to contribute to T1D by the mechanisms of central tolerance (Pugliese et al. 1997). In newly diagnosed T1D patients, the presence of insulin autoantibodies was associated with the INS VNTR polymorphism 101. The many other genetic factors listed in Table 2 are all shown to be significantly associated with T1D 108. The function of these genes is understood individually, but it is not clear how these factors interact to increase the risk for the development of islet autoimmunity (step 1), T1D (step 2), or both. Since the majority of the genetic factors seem to be associated with the immune system, it is attractive to speculate that their contribution is related to the ability of the immune system to mount an autoimmune reaction specifically directed toward the islet β-cells.

Immune Cells in Tolerance

Epitope presentation to T- and β-cells is the key step in the generation of tolerance, in its early failure and during the maintenance of autoimmunity. The capacity to distinguish between self and nonself, which is the hallmark of a functional immune system, is lost when central and peripheral tolerization fail, leading to the development and expansion of autoreactive pathogenic effector cells. Central tolerance is induced at the site of lymphocyte development (the thymus and bone marrow, respectively, for T- and β-cells), while peripheral tolerance occurs at sites of antigen recognition, namely, in lymphoid and nonlymphoid tissues. Central to the function of tolerance is APC.

APC

The recognition by T and B lymphocytes of antigens presented in the context of MHC surface of an APC is the first step of the adaptive immune response. Macrophages and particularly DC are the most efficient APC as they show constitutive expression of MHC class II molecules, cytokine secretion, and migrating capacity. APC have a dual role: uptake, processing, and presentation of antigens to T-cells and regulating T-cell-driven responses through cytokine release. APC are involved in T-cell tolerance mechanisms at both central (clonal deletion) and peripheral level (clonal anergy). Negative selection of autoreactive clonotypes derived by random T-cell receptor (TCR) rearrangement is guided by T-cell affinity for self-peptide–MHC (pMHC) complexes presented in the thymus (Geenen et al. 2010). An inadequate binding affinity spares self-reactive T-cells from apoptosis. The thymic expression of tissue-specific antigens (TSA) is regulated by the autoimmune regulatory (AIRE) transcription factor. Insufficient level of expression and presentation of TSA-derived peptides is observed in subjects with a mutated AIRE gene. It is possible that some β-cell autoantigens are not present in the thymus at sufficient concentrations to induce negative selection. This mechanism may explain the correlation of T1D protection with the "long form" of INS VNTR (Bennett and Todd 1996; Walter et al. 2003; Nielsen et al. 2006). The number of "tandem repeats" modulate thymic expression of this autoantigen, and the "long variant" results in increased insulin mRNA within the thymus. This higher thymic insulin expression is thought to enhance the deletion of insulin-specific thymocytes and may account for the protective phenotype.

Transcriptional modifications due to alternative splicing have been proposed to explain IA-2 immunogenicity as IA-2 is not expressed full length in thymus but in an alternatively spliced transcript derived from the deletion of exon 13 (Pugliese et al. 2001). This may account for the escape of a subset of IA-2-reactive T-cells. Interestingly, several B- and T-cell epitopes map to IA-2 exon 13 (Diez et al. 2001). So far it is not clear to what extent central tolerance and thymic expression are important to antigen presentation of the other two major autoantigens.

Among APC, DC are peculiar, highly specialized effectors with ontogenic, morphologic, and functional heterogeneity and can be mainly divided into conventional or myeloid DC (mDC) and plasmacytoid DC (pDC), depending on superficial clusters of differentiation and secretive function (Liu 2005). pDC are potent productors of IFN-α and are connected to the innate immune system through the expression of toll-like receptors (TLR) specific for the detection of viral infections. Emerging evidence suggests a close relationship between pDC and autoimmune conditions. In healthy subjects, autoantigen-bearing DC are physiologically found in blood, peripheral lymphoid organs, and thymus where they are an important source of TSA (Hernandez et al. 2001). DC were also reported to display proinsulin epitopes through direct transcriptional events in a capture-independent way (Garcia et al. 2005). After the activation by an antigen, DC undergo maturation, express pMHC complexes, and promote antigen-specific T-cell clonal expansion. At this mature stage, DC are generally immunogenic and produce costimulatory molecules and several cytokines.

The physiology of β-cells as APC indicates that these cells are able to take up antigen at very low concentrations through their antigen-specific membrane-bound immunoglobulin and to present it to T-cells. The antigen presentation is enhanced in the presence of specific autoantibodies (Amigorena and Bonnerot 1998). β-cells may be important for the spreading of T- and β-cell determinants during the progression of the disease (Jaume et al. 2002; Steed et al. 2008). The minute amounts of antigen presented by β-cells may be important for the maintenance of autoimmune reactivity in the later phase once most of the target tissue has been destroyed (Uibo and Lernmark 2008; Steed et al. 2008). HLA-restricted B- and T-cell epitopes are in close proximity within the GAD65 molecule (Fenalti et al. 2008), and recently an overlap within T and B IA-2 epitopes has been described (Weenink et al. 2009). These observations suggest that antigen–antibody complexes may influence antigen presentation by APC and thereby T-cell reactivity. There are major gaps in our understanding of the possible importance of the T-β-cell synapse within the human islets of Langerhans.

There is wide evidence from studies on T1D pancreas with insulitis that MHC class II expression is increased on islet vascular endothelial cells (iVEC) (Hanninen et al. 1992; Greening et al. 2003). More recent data on iVEC suggest that these cells are capable to internalize, process, and present disease-relevant epitopes from GAD65 (Greening et al. 2003) and insulin (Savinov et al. 2003). The in vivo acquisition of these autoantigens by iVEC is not clearly established. Since iVEC are physiologically exposed to very high insulin concentration, it is likely that these cells take up insulin and process it into peptides through endosomal degradation rather than acquire peptides or pMHC complexes produced by β-cells (Savinov et al. 2003). The mechanism is even more unclear for non-secreted antigens. Although it is uncertain whether islet vascular endothelium has any prominent role in the priming of autoreactive T-cells, given the recognized importance of professional APC, it has been suggested that iVEC may be important for the trafficking of activated T-cells providing antigen-driven homing specificity (Savinov et al. 2003).

T-Cells

Recent progress in studying peripheral tolerance has highlighted the importance of immunoregulation by Treg, co-expressing CD4 and the α chain of the IL-2 receptor complex (CD 25) (Sakaguchi et al. 2008). Treg are potent suppressors of organ-specific autoimmunity. Natural Treg (nTreg) originate from intrathymic recognition of self-pMHC complexes [177, 178] and are functionally marked by the constitutive expression of forkhead–winged helix transcription factor (FOXP3) (Tang and Bluestone 2008), while conventional Treg (cTreg) differentiate from naïve $CD4^+$ T-cells in the periphery (Sakaguchi et al. 2008). Although FOXP3 plays a major role in Treg development and activity, as mutations in FOXP3 gene in humans determine severe multiorgan autoimmunity (IPEX syndrome) (Gambineri et al. 2003), Treg function is complex and involves other transcriptional signaling as TGF-β, IL-2, and possibly others. The possible dysregulation of IL-2 signaling in Treg suppressor activity is supported by the association of T1D and polymorphisms within the IL-2 receptor α gene region in humans (Bach 2003). Immunoregulation by Treg affects T-cells, β-cells, and APC antigen-specific cellular responses in different manners. These include production of anti-inflammatory cytokines (TGF-β, IL-10, and IL-35) and contact-dependent mechanisms possibly involving CTLA-4 and direct cytolysis (Zhou et al. 2009). Further studies in humans are complicated by the difficulties to obtain T-cells from the pancreatic islets let alone the pancreatic draining lymph nodes.

B-Cells

Little is known about self-tolerance mechanisms for β-cells (Shlomchik 2008). Immature β-cells in the bone marrow express a potentially polyreactive β-cell receptor (BCR), which results from stochastic gene recombination. It is thought that 20–50 % of autoreactive immature β-cells undergo rearrangement of immunoglobulin light chain genes. The remaining self-reactive β-cells undergo peripheral deletion or peripheral anergy. The extent to which deletion and anergy contribute to β-cell tolerance has not yet been determined. Although evidence of aberrant receptor editing has been associated with autoimmunity in mouse and human diseases (Wardemann and Nussenzweig 2007), it is still unclear to what extent these defects participate in the establishment of autoimmunity.

What Happens in the Islet?

It is presently unclear whether the initiation of autoimmune β-cell destruction in humans requires a set of autoreactive T-cells recognizing β-cell antigen peptides presented on HLA class I heterodimers. The best current evidence in a case report (Bottazzo et al. 1985) or in the nPOD pancreata (Atkinson and Gianani 2009; Campbell-Thompson et al. 2012) is the presence of $CD8^+$ T-cells that are thought to recognize specific autoantigens. The variability in reactivity to individual autoantigens may in turn be due to "epitope spreading" which consists of intramolecular shifting of the recognized epitopes with the progression of the autoimmune attack. In addition, autoantigen spreading may occur as the risk for clinical diagnosis is increasing with an increased number of islet autoantibodies (Sosenko et al. 2013) preceded by activation of new T-cell clonotypes. These events may provide an explanation for the widely diversified anti-islet immune response in T1D.

Priming of naïve $CD4^+$ T-cells by islet antigen-presenting APC would be the first event in initiating islet autoimmunity (step 1) and diabetogenesis (step 2) (Fig. 1). This event most likely takes place in the pancreatic lymph nodes (pLN). Islet antigen presentation in pLN in humans is unclear. What promotes the earliest event, namely, uptake of antigen by APC in the islets, is still a matter of debate.

37 Immunology of β-Cell Destruction

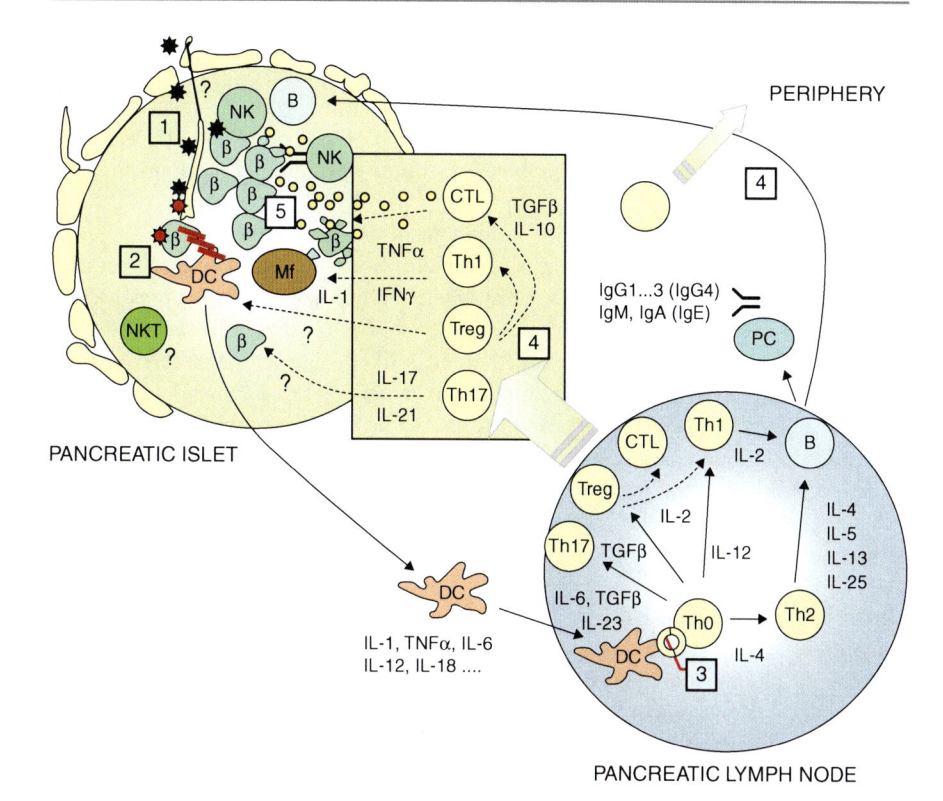

Fig. 1 Schematic view on possible immunopathogenesis of β-cell destruction. Steps of events: (Eisenbarth 1986) environmental factors are conditioning the relevant milieu by activation of dendritic cells (DC), macrophages (Mf), natural killer (NK) cells, and natural killer T-cells (NKT); (Atkinson et al. 2013) intake of antigens or cross-reactive peptides by dendritic cells (DC); (Bach 1997) presentation of peptides to naive T helper (Th0) cells and subsequent activation and proliferation of type 1 (Th1) and type 2 (Th2) helper cells, IL-17-producing helper cells (Th17), regulatory T (Treg)-cells, cytotoxic T-cells (CTL), β-cells and plasma cells (PC) and activation of different cell subsets by cytokines (Ziegler et al. 2013); migration of activated cells from pancreatic lymph node to the islets, cross talk with periphery; and β-cell destruction by cytokine- and perforin-/granzyme-mediated mechanisms. – environmental factor (virus, etc.); ✤ – islet autoantigens; ▬ – islet autoantigens or cross-reactive peptides; ○ – islet-specific T-cell; Ab – autoantibodies

Initial, still not fully characterized insults (virus infection or other external damage, e.g., environmental toxins) may elicit an innate immune response through the generation of exogenous or endogenous ligands for the pattern recognition receptors (PRR) on the β-cell surface. The activation of these receptors triggers intracellular responses including cytokine production, endoplasmic reticulum stress, and accumulation of misfolded proteins. β-cell apoptosis and local inflammation may ensue. Dying β-cells may release immunostimulatory "danger" signals physiologically aimed at eliminating the initial harmful factor. This requires a transfer to adaptive immune response mediated by the enrollment of APC and the establishment of a pro-inflammatory local environment (IFN, IL1β, and chemokines) to attract other immune cells. A defective resolution of the early inflammation results in a chronic destructive autoimmune reaction and may be dependent on the individual genetic background. For example, the DR3-DQ2 haplotype seems to be permissive of organ-specific autoimmunity (Lio et al. 1997).

DC have been demonstrated in human insulitis (Lernmark et al. 1995), but further studies of human pancreatic specimens from islet autoantibody-positive subjects or newly diagnosed T1D patients will be needed. DC may cross-present peptides derived from apoptotic cells directly onto MHC class I molecules without processing in the cytosol. Taken together, it is likely that antigens derived from β-cells dying upon external damage may be taken up by APC in the pancreatic islets and transported to pLN (Fig. 1). In humans this has not been possible to be fully demonstrated though expression of the β-cell autoantigens proinsulin, GAD65, and IA-2 has been detected in human peripheral DC. Currently our understanding of possible mechanisms of the very early events in islet autoimmunity relies on studies in animals.

Virus-Induced β-Cell Killing

Regardless of the numerous reports of TD1 onset following viral diseases, no conclusive pathogenic connection has been found between viral infection and human islet autoimmunity. The studies are complicated by the lack of data that distinguished triggering of islet autoimmunity (step 1) from virus infection affecting islets in subjects with islet autoantibodies or studied at the time of clinical diagnosis. Nevertheless, pancreatic islets showed the expression of innate PRR when infected by virus or exposed to virus-related cytokines as IFN and IL-1β (Hultcrantz et al. 2007). Virus antigen peptides need to be presented on MHC class I on the β-cell surface to be recognized by CTL. The critical question is to what extent a virus-infected β cell is copresenting viral and β-cell antigens on MHC class I molecules. Some viral antigen sequences are similar to self-peptides and may mislead T-cell responses. This phenomenon of "molecular mimicry" has been proposed between PC-2 antigen from Coxsackie B and GAD65 (Atkinson et al. 1994), between Rotavirus and IA-2 and for rubella (Honeyman et al. 1998). It is possible that these events are more relevant to the amplification of the autoimmune process and its maintenance after the resolution of the viral infection, than to the initial triggering of autoimmunity. As previously described, virus may activate β-cell intracellular signaling that induces altered expression of self-antigens on the β-cell surface ("neoantigens" or "cryptic antigens") and participates in the cascade leading to β-cell apoptosis and insulitis. Virus replication in the β cell may result in its necrosis and in release of previously sequestered cellular constituents ("hidden antigens") lacking induced thymic tolerance. The uptake and presentation of these self-antigens by APC to CD4[+] T-cells may eventually lead to the formation of specific autoantibodies. Coxsackie B4 has been isolated from the β cells of T1D new-onset patients (Dotta et al. 2007). In summary, there is wide evidence that virus infections may accelerate islet autoimmunity (step 2) leading to clinical onset of T1D. The mechanism may be an increase in insulin resistance or a boost in β-cell killing induced by the virus infection. The major question to be answered is whether a virus infecting and replicating in human β-cells induces islet autoimmunity. The DIPP study (Oikarinen et al. 2012) and the ongoing TEDDY (the Environmental Determinants of Diabetes in the Young) study may be able to answer this question (TEDDY Study Group 2008; Hagopian et al. 2011).

Cytotoxin-Induced β-Cell Killing

Alloxan, streptozotocin, and the rodenticide (Vacor) are well-known β-cell cytotoxic agents. It is important to note that both alloxan and streptozotocin are more toxic to rodent than human β cells. Other chemicals that may be potential human β-cell cytotoxins are nitrosamine derivatives as well as dietary microbial toxins. Epidemiological data suggest that an increase in nitrate-treated food items increases the risk for children to develop T1D (Dahlquist et al. 1990). Other compounds structurally related to streptozotocin or alloxan have been implicated as possible environmental agents contributing to human T1D. Most prominently these compounds include the rodenticide pyriminil (Vacor) that induces islet cell surface antibodies and confirms that β-cell destruction in humans may cause islet autoimmunity (Karam et al. 1980; Esposti et al. 1996).

In summary, several virus and chemical agents directly affecting islet cells may be causative in the initiation of the autoimmune β-cell destructive process. Alternatively these factors may potentiate a process initiated by other environmental factors, which are currently under scrutiny in the TEDDY study (TEDDY Study Group 2008; Hagopian et al. 2011). In individuals prone to develop T1D, environmental chemicals may play a detrimental role by repeat injuries to the pancreatic β cells over several years of life. This in combination with a poor regenerative capacity of the β cell and islet autoimmunity may eventually induce diabetes.

Antigen Presentation in Pancreatic Lymph Nodes

Although specific mechanisms in humans remain unclear, APC loaded with β-cell antigens migrate from the islets to the pLN, where the processed antigens are presented to naïve $CD4^+$ T-cells (Th0) (Fig. 1). In the pLN primed $CD4^+$ T-cells proliferate and differentiate into several subsets, as type 1 $CD4^+$ T-cells (Th1), IL-17-producing $CD4^+$ T-cells (Th17), and Treg cells, and activate naïve $CD8^+$ T- and β-cells into CTL and plasma cells, respectively. The expansion of $CD4^+$ T-cells toward lineages of pro-inflammatory subtype (Th1 and Th17) is mainly promoted by the cytokine milieu, through IL-6, IL-12, and IL-23, whereas a balance toward IL-4, IL-5, IL-13, and IL-25 would decrease the inflammation. Th1 cells release IFN-γ, which activates macrophages, TNF-α, IL-12, and IL-18. The recent discovery of Th17 cells that are potent inducers of tissue inflammation and autoimmunity is of interest as they may have a role in islet destruction.

Activation and differentiation of naïve CD8$^+$ T-cells to antigen-specific CTL are dependent on "cross-priming." This is the cognate recognition of the same antigen by the CD8$^+$ and the CD4$^+$ T-cells on the same APC. The interaction between CD40 on APC and CD154 on CD4$^+$ T-cells induces upregulation of costimulatory molecules for the activation of the CD8$^+$ T-cells and increases the local production of pro-inflammatory cytokines such as TNF-α and IL-12. Alternatively the IFN-γ produced by CD8$^+$ T-cells could enhance CD4$^+$ T-cell action (Viglietta et al. 2002). When cognate interaction occurs between β-cells and activated CD4$^+$ T-cells, the β-cells differentiate into plasma cells. They start to secrete immunoglobulins with the same specificity as the previous membrane-bound immunoglobulin upon stimulation of T-cell-released "Th2" cytokines IL-4 and IL-5.

Homing of T-Cells to Islets

Primed β-cell-specific effector T-cells gain access to peripheral nonlymphoid tissues, migrate to the pancreas, and reach the β cells (Fig. 1). The molecular basis for this directed migration (homing) of autoreactive T-cells to the islets and for endothelial transmigration is not clarified. The processes guiding islet autoantigen-specific T-cells into islets are not known. In pancreas transplantations between monozygotic twins without immunosuppression islets in the donor pancreas were infiltrated by CD8$^+$ T-cells in association with the loss of β-cell function (Sibley et al. 1985). These experiments demonstrate the immunologic memory of the recipient, as well as β-cell killing by CTL, indicating that autoreactive CTL are reactivated. The mechanism of reactivation is unclear. It has been proposed that T-cells can be programmed to a specific tissue tropism through a distinct "homing receptor pattern" acquired at the site of priming (Dudda and Martin 2004). Upon second contact with cognate antigen in the islet, CTL are retained inside the islet tissue and may initiate insulitis (Fig. 1). Any β-cell-specific CTL may recognize antigens expressed on MHC class I molecules. MHC class I overexpression on islet cells, previously described in pancreas with insulitis (Atkinson and Gianani 2009; Foulis et al. 1991), is likely to be involved. Although routine investigation of the early phase (step 1) is not feasible in humans, immunocytochemistry on pancreas biopsy specimens from new-onset T1D patients in Japan indicates the presence of CD8$^+$ T-cells and activated macrophages secreting inflammatory cytokines (Imagawa et al. 2001). The ongoing inflammatory islet milieu expands the recruitment of autoreactive CTL through the expression of chemokines and homing ligands from the β-cells. As physiological response to the inflammation, islet endothelium upregulates the expression of surface adhesion molecules that increase vascular permeability and facilitate the recruitment of effector cells. Adhesion and diapedesis of T-cells are possible through the interactions of T-cell surface molecules (integrins) such as leukocyte function-associated antigen-1 (LFA-1) and very late activation antigen-4 (VLA-4) with their counter ligands on VEC. These include intercellular adhesion molecules (ICAM) and junctional adhesion molecules (JAM-1). This hyperexpression of adhesion molecules is documented in new-onset diabetes pancreas (Uno et al. 2007). It is now proposed that after migration from pLN, activated T-cells require an additional upregulation of LFA-1 functional activity for the successful adhesion to VEC (Somoza et al. 1994). The hypothesis that iVEC may participate in T-cell-selective recruitment and adhesion in an antigen-specific fashion is intriguing. A recent study reported that GAD65 presentation by iVEC markedly promotes the in vitro transmigration of GAD65-autoreactive T-cells across iVEC monolayers in an LFA-1-dependent fashion (Somoza et al. 1994). In this process, $CD4^+$ T-cells may also intervene by secreting various lymphokines that attract and activate other cell types such as monocytes, eosinophils, and natural killer lymphocytes (NK). Whether islet-specific autoantibodies secreted by plasma cells take part in the islet destruction or are merely recruited upon the ongoing discharge of autoantigens is still a matter of debate since a defined pathogenetic effect has not been proven. Clinical evidence in humans does not support this hypothesis. It is still a matter of debate whether autoantibodies reacting to antigen-binding areas of autoantibodies (anti-idiotype) may be of relevance within the autoimmune process through the blockade of circulating self-autoantibodies (Oak et al. 2008; Ortqvist et al. 2010; Wang et al. 2012).

Insulitis and β-Cell Destruction

The progression from the initiating phase to an adaptive immune response is thought to take place very early during insulitis and determine the final outcome toward the generation of a prolonged devastating autoimmune reaction or the resolution of inflammation and preservation of islet integrity. Target-cell death further activates PRR that in turn promote the progression of insulitis perhaps through IFN-α-mediated upregulation of MHC class I molecules on pancreatic islet cells. IFN and other macrophage-derived cytokines prompt NK activation. These cells exert nonantigen-specific cytotoxicity through the release of perforin, after the activation of surface receptors, as NKG2D that recognizes viral products and other specific ligands. NKT cells on the other hand may be considered as innate-like lymphocytes as they may co-express NK cell surface markers including NK1.1 (human CD161) and TCR. Most NKT cells recognize glycolipid antigens presented on the MHC class I-like molecule CD1d (Kronenberg 2005). The possible role of NK in β-cell damage has not yet been clarified.

As the islet invasion progresses, chemokine-attracted macrophages contribute to the recruitment of other immune cells that also release multiple chemokines and pro-inflammatory cytokines. These inflammatory signals create an overall immunoactivatory environment that modifies DC phenotype and shifts $CD4^+$ T-cells toward "Th1-like" responses which promote the expansion of CTL and shelters them from peripheral tolerance. If this vicious circle is not interrupted, the maintenance and amplification of insulitis result in accumulation of immune cells and their cytotoxic mediators that may act synergistically to destroy the β cells (Fig. 1). In the later stages, the destructive process may be worsened in the course of β-cell failure as the hyperglycemic environment may locally enhance insulin or GAD65 epitope presentation (Skowera et al. 2008).

Further studies of early insulitis (step 2) in humans will be needed to fully appreciate the initiating mechanisms of infiltration of immune cells. Effective prevention of T1D may require a better understanding of the events of chronic insulitis. We speculate that a chronic insulitis, which includes APC-presenting islet autoantigens within the islets as opposed to the pLN, represents a refractory state to immunosuppression. This may explain why immunosuppression at the time of clinical diagnosis of T1D is ineffective. It cannot be excluded that immunosuppression therapy may be successful provided that the treatment is started prior to chronic insulitis.

Is β-Cell Destruction Reflected in the Blood?

Assaying the cells involved in β-cell damage may give insights about the induction and maintenance of islet autoimmune destruction. Several possible immunological alterations have been investigated in the peripheral blood of T1D patients and at-risk subjects to differentiate them from healthy subjects.

APC

An abnormal cytokine response by DC from T1D patients upon antigenic (Mollah et al. 2008) or nonantigenic stimulation was proposed (Allen et al. 2009). More robustly, phenotypic characterization suggests that DC from recent-onset T1D patients exhibit an immature phenotype and may have a decreased T-cell stimulatory capacity compared to controls (Vidard et al. 1992). DC may therefore indirectly participate to T1D autoimmunity through a reduced efficacy in stimulating Treg. This immature phenotype of T1D human DC may result from abnormal activation of the NF-kB pathway (Vidard et al. 1992). This is consistent with the strong involvement of this transcription factor in the induction of self-tolerance (Osugi et al. 2002). Studies investigating the peripheral DC count reported a reduction in absolute number of blood DC in T1D children and more recently a modest but significant increase in the relative frequency of pDC subset, strictly time related with disease onset (Allen et al. 2009). However, the present observations about DC in human diabetes rely upon studies on in vitro monocyte-generated DC that may not reflect the true in vivo situation.

T-Cells

Many studies on peripheral blood mononuclear cells (PBMC) of T1D patients are aimed at detecting the presence of islet-specific $CD4^+$ and $CD8^+$ T-cells upon stimulation with synthetic peptides from islet antigens. The immunogenic epitopes are selected among putative immunodominant regions within the multiple islet autoantigens. Many of these studies report a higher frequency of islet-specific self-reactive T-cells in T1D patients than in control subjects when T-cells are detected by either functional tests of antigen-induced proliferative (Endl et al. 1997), cytokine secretion (Dang et al. 2011; Faresjo et al. 2006; Herold et al. 2009), or tetramer staining (Nepom 2012; Reijonen et al. 2004).

CD8$^+$ and CD4$^+$ T-cells from T1D patients target a wide array of epitopes within GAD65 molecule (Endl et al. 1997; Wicker et al. 1996; Schloot et al. 1997; Harfouch-Hammoud et al. 1999; Peakman et al. 2001; Cernea and Herold 2010; Oling et al. 2012; Hjorth et al. 2011), insulin and proinsulin (Rudy et al. 1995; Geluk et al. 1998; Chen et al. 2001; Mannering et al. 2009; Mallone et al. 2011b; Abreu et al. 2012; Heninger et al. 2013; Tree et al. 2000; Kelemen et al. 2004), IA-2 (Hanninen et al. 2010; Velthuis et al. 2010; Peakman et al. 1999), IGRP (Velthuis et al. 2010; Alkemade et al. 2013; Unger et al. 2007), I-A2β (phogrin) (Tree et al. 2000; Kelemen et al. 2004; Achenbach et al. 2002), islet amyloid polypeptide (IAPP) (Velthuis et al. 2010), and glial fibrillary acidic protein (GFAP) (Standifer et al. 2006) as comprehensively summarized in a review updated in 2006 (Seyfert-Margolis et al. 2006; Roep and Peakman 2011). These investigations, mostly oriented toward epitope identification, provide evidence of multiple immunodominant β-cell regions targeted by CTL in human T1D but do not fully clarify the development of the T-cell-specific responses during the progression of the disease. In fact, no single epitope has proven to be discriminatory between health and disease. A hierarchy of T-cell responsiveness was proposed for proinsulin peptides (Arif et al. 2004). In some ways the choice of the epitope may also be misleading. Candidate sequences are usually selected on the basis of predicted TCR-pMHC-binding motifs (Rammensee et al. 1999) or affinity algorithms (Parker et al. 1994), whereas the strength of the TCR-pMHC complex interaction may inversely correlate with immunogenicity (Baker et al. 2008), in accordance with an insufficient negative thymic selection. This bias can be avoided through the analysis of multiepitope, multiantigen panels (Baker et al. 2008). Moreover, epitopes that have been proved of relevance in mice may guide the search efforts in humans, as recently done with IGRP peptides (Jarchum et al. 2008). At the present time, there is a lack of a precise, reproducible, and standardized method for detection and identification of β-cell specific autoreactive T-cells. Such a method is needed to reliably identify subjects with islet autoimmunity who may progress to clinical onset. Some authors report that the use of multiple epitopes achieves more diagnostic sensitivity and better discrimination of T1D from controls (van Endert et al. 2006). It is therefore still unclear to what extent all the data provided may be translated into evaluation of risk for islet autoimmunity or clinical onset. Moreover, autoreactive T-cells-specific responses for T1D self-antigens have been widely described in healthy individuals in stimulation assays with peptides from GAD65 (Danke et al. 2005; Monti et al. 2007) and insulin (Yang et al. 2008; Monti et al. 2009). Several differences have been proposed between self-reactive T CD4$^+$ T-cells from T1D and controls. Only GAD65-reactive T-cells from T1D subjects seem to be fully autoantigen-experienced in vivo and express the memory T-cell marker CD45RA (Danke et al. 2005) and are capable of activation in the absence of CD28/B7 costimulatorysignals (Viglietta et al. 2002). It was also recently proposed that CD4⁺ T-cells from T1D subjects may have a lower threshold of activation as compared to healthy controls (Yang et al. 2008).

In healthy individuals, self-reactive T-cells are probably present but quiescent for the immunosuppressive action of Treg. This is confirmed by the experimental observation that Treg in vitro depletion is followed by amplification of autoreactive T-cells only in samples from healthy individuals (Yang et al. 2008). The Treg pool in human T1D has also been extensively investigated, and a deficiency in Treg peripheral frequency has been reported in patients compared to controls (Tree et al. 2006). Subsequent investigations have failed to uniformly replicate these findings and have suggested that T1D nTreg may rather display an impaired immune suppressor function (Brusko et al. 2007; Lindley et al. 2005). Globally it seems that a simple deficiency in the peripheral Treg repertoire is not confirmed, but a local impairment of Treg activity at the site of inflammation cannot be excluded.

The peripheral blood from T1D patients may display an imbalance toward inflammation. Autoantigen-driven cytokine secretion by CD4⁺ T-cells from T1D patients may be polarized toward INF-γ, while HLA-matched healthy controls display IL-10⁺ c Treg-like responses (Arif et al. 2004). This "regulatory phenotype" skewed toward IL-10 has also been reported in association with later onset of TID (Arif et al. 2004) and better glycemic control (Sanda et al. 2008). Increased levels of "Th1 cell"-derived chemokines CC13, CC14, and CXC110 (Sanda et al. 2008; Nicoletti et al. 2002) and of adhesion molecules ICAM and L-selectin (CD62L) (Lampeter et al. 1992) have been found in serum of T1D patients. The NK population in the peripheral blood of T1D patients may be decreased, but these findings have not been universally replicated (Rodacki et al. 2006). A larger study confirmed a functional impairment of NK cells in T1D patients, i.e., reduced surface expression of activating receptors and low levels of IFN-γ and perforin, and suggested that these alterations may be a consequence of T1D, since they are evident exclusively in long-standing disease (Rodacki et al. 2007). It has also been reported that activated NK cells in T1D patients display a reduced expression of NKG2D receptor (Rodacki et al. 2006). It is possible that a downregulation of NKG2D receptor mediates the increased risk for T1D associated with polymorphisms of MHC class I chain-related (MIC) proteins that are NKG2D natural ligands (Gambelunghe et al. 2000).

B-Cells and Autoantibodies

The assessment of disorders of humoral immunity in T1D relies on the monitoring of circulating islet-reactive autoantibodies (Table 2). Autoantibodies against at least one of the islet cell autoantigens GAD65, IA-2, insulin, and ZnT8 represent in more than 95 % of T1D patients and in only 1–2 % of the general population (Andersson et al. 2011). Radio-binding assay of these autoantibodies has replaced the ICA assay. GAD65 antibodies are found in 70–75 % of T1D patients (Delli et al. 2012) and show a diagnostic sensitivity of 70–80 % and a diagnostic specificity of 98–99 % (Vaziri-Sani et al. 2010). The major antigenic epitopes of

GAD65 are the middle-(Padoa et al. 2003) and C-terminal (Hampe et al. 2000; Richter et al. 1993) region and are in close proximity to T-cell disease-relevant determinants (Fenalti and Buckle 2010). Differential epitope specificities, as identified by monoclonal antibodies to GAD65 epitopes within the C-terminal region, align with distinct autoimmune disease phenotypes, and the binding of N-terminal epitope is associated with slowly progressive β-cell failure (Kobayashi et al. 2003). It was recently suggested that the presence of GAD65 autoantibodies in T1D patients may be the result of an "unmasking' due to the lack of GAD65-anti-idiotypic antibodies (Oak et al. 2008). These anti-idiotypic antibodies are reported to highly discriminate T1D from healthy subjects and may be of some relevance in the pathogenesis of islet autoimmunity.

IAA are found in approximately 50–70 % of T1D patients (Delli et al. 2012; Hagopian et al. 1995) and are the first islet autoantibody to appear (Kukko et al. 2005) suggesting an involvement of insulin as primary autoimmune triggering antigen also in humans (Eisenbarth and Jeffrey 2008). Epitopes targeted by IAA are placed within A and B chains and are shared between insulin and proinsulin (Brooks-Worrell et al. 1999).

IA-2 antibodies are detected in 60–70 % of patients with new-onset T1D and tend to appear closer to the clinical onset (Andersson et al. 2011; 2013). Epitopes for IA-2 autoantibodies are found exclusively within the cytoplasmic region of the molecule and predominantly within the tyrosine phosphatase-like domain (Hatfield et al. 1997; Zhang et al. 1997).

Autoantibodies to ZnT8 are detected in 60–80 % of newly diagnosed T1D (Andersson et al. 2011; Andersson et al. 2013). The polymorphism at position 325 is a major target for ZnT8 autoantibodies demonstrating considerable binding specificity (Skarstrand et al. 2013b).

Prediction of β-Cell Destruction

Standardized methods have made islet autoantibodies the most useful marker for T1D and for enrollment of subjects into clinical preventiontrials (Torn et al. 2008). The number of islet autoantibodies is the best predictor for the risk of clinical onset (Sosenko et al. 2011a; Sosenko et al. 2011b; Elding Larsson et al. 2013). The predictive power is enhanced by the combination of multiple markers. The stepwise appearance of islet autoantibodies may signal autoantigen spreading and a worsening of the pathogenic process that eventually may lead to a major loss of β cells. The ongoing and escalating islet autoimmunity may also signal that the islets may eventually be infiltrated by mononuclear cells. Most critical will be the infiltration of $CD8^+$ T-cells that are recognizing islet autoantigens presented on HLA class I proteins on the β-cell surface.

The prediction power for T1D reaches 100 % in case of multiple positivity (Fig. 2). Importantly, in case of a single autoantibody, the correlation between islet

Fig. 2 Diagrammatic presentation of the effect of multiple islet autoantibodies on the risk of developing T1D in the Diabetes Prevention Trial – Type 1(DTP-1) (Courtesy of Jay Skyler)

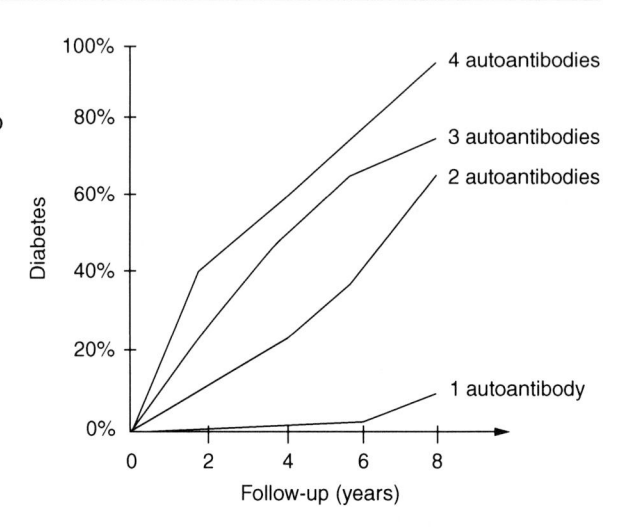

autoimmunity and histological evidence of insulitis is weak (In't Veld et al. 2007). Longitudinal studies investigating DC and T-cells in at-risk subjects are lacking. Some reports have found poor in vitro maturation and pro-inflammatory cytokine response in DC from children at genetic risk for TID.

Concluding Remarks

In conclusion, the β cell in T1D is the major target for an autoimmune process that takes place in two steps. The first step is the development of an autoimmune reaction directed toward specific β-cell autoantigens. This step results in the appearance of circulating autoantibodies to β-cell autoantigens including GAD65, IA-2, ZnT8, and insulin. While the triggering phase may be short (hours, days), the ensuing islet autoimmunity leading up to step 2 – onset of clinical diabetes – may be months to years. The number of autoantibodies predicts T1D risk. The second step, progression from islet autoimmunity to the clinical onset of T1D, is associated with a major loss of β cells due to insulitis, but recent data indicate that the function of residual β cells may be inhibited. Insulitis appears late in the autoimmune process and can be recapitulated in pancreas and islets transplantation. The immunological memory of β-cell autoantigen is chronic. Efforts are needed both to detect intra-islet events that precede the development of autoantibodies and to disclose when islet autoantibody positivity is marking that the β-cell destructive process of insulitis is about to be established. A better understanding of step one and two events will be necessary for the ultimate prevention of β-cell destruction and of T1D.

Acknowledgement The research in the authors laboratory has been supported by the National Institutes of Health (grant DK63861), Juvenile Diabetes research foundation, the Swedish Research Council, Diabetesfonden, Childhood Diabetes Fund, and Skåne County Council for Research and Development.

Cross-References

▶ Apoptosis in Pancreatic β-Islet Cells in Type 1 and Type 2 Diabetes
▶ Current Approaches and Future Prospects for the Prevention of β-Cell Destruction in Autoimmune Diabetes
▶ Inflammatory Pathways Linked to β Cell Demise in Diabetes
▶ Microscopic Anatomy of the Human Islet of Langerhans

References

Aanstoot HJ, Kang SM, Kim J et al (1996) Identification and characterization of glima 38, a glycosylated islet cell membrane antigen, which together with GAD65 and IA2 marks the early phases of autoimmune response in type 1 diabetes. J Clin Invest 97(12):2772–2783

Abreu JR, Martina S, Verrijn Stuart AA et al (2012) CD8 T cell autoreactivity to preproinsulin epitopes with very low human leucocyte antigen class I binding affinity. Clin Exp Immunol 170(1):57–65

Abulafia-Lapid R, Gillis D, Yosef O, Atlan H, Cohen IR (2003) T cells and autoantibodies to human HSP70 in type 1 diabetes in children. J Autoimmun 20(4):313–321

Achenbach P, Kelemen K, Wegmann DR, Hutton JC (2002) Spontaneous peripheral T-cell responses to the IA-2β (phogrin) autoantigen in young nonobese diabetic mice. J Autoimmun 19(3):111–116

Akesson K, Nystrom L, Farnkvist L, Ostman J, Lernmark A, Kockum I (2005) Increased risk of diabetes among relatives of female insulin-treated patients diagnosed at 15–34 years of age. Diabet Med 22(11):1551–1557

Alkemade GM, Clemente-Casares X, Yu Z et al (2013) Local autoantigen expression as essential gatekeeper of memory T-cell recruitment to islet grafts in diabetic hosts. Diabetes 62(3):905–911

Allen JS, Pang K, Skowera A et al (2009) Plasmacytoid dendritic cells are proportionally expanded at diagnosis of type 1 diabetes and enhance islet autoantigen presentation to T-cells through immune complex capture. Diabetes 58(1):138–145

Amigorena S, Bonnerot C (1998) Role of β-cell and Fc receptors in the selection of T-cell epitopes. Curr Opin Immunol 10(1):88–92

Andersson C, Larsson K, Vaziri-Sani F et al (2011) The three ZNT8 autoantibody variants together improve the diagnostic sensitivity of childhood and adolescent type 1 diabetes. Autoimmunity 44(5):394–405

Andersson C, Vaziri-Sani F, Delli A et al (2013) Triple specificity of ZnT8 autoantibodies in relation to HLA and other islet autoantibodies in childhood and adolescent type 1 diabetes. Pediatr Diabetes 14(2):97–105

Arif S, Tree TI, Astill TP et al (2004) Autoreactive T cell responses show proinflammatory polarization in diabetes but a regulatory phenotype in health. J Clin Invest 113(3):451–463

Atkinson MA, Gianani R (2009) The pancreas in human type 1 diabetes: providing new answers to age-old questions. Curr Opin Endocrinol Diabetes Obes 16(4):279–285

Atkinson MA, Bowman MA, Campbell L, Darrow BL, Kaufman DL, Maclaren NK (1994) Cellular immunity to a determinant common to glutamate decarboxylase and Coxsackie virus in insulin-dependent diabetes. J Clin Invest 94(5):2125–2129

Atkinson MA, Eisenbarth GS, Michels AW (2014) Type 1 diabetes. Lancet 383:69-82

Bach JF (1997) Autoimmunity and type I diabetes. Trends Endocrinol Metab 8(2):71–74

Bach JF (2003) Regulatory T, cells under scrutiny. Nat Rev Immunol 3(3):189–198

Badenhoop K, Kahles H, Seidl C et al (2009) MHC-environment interactions leading to type 1 diabetes: feasibility of an analysis of HLA DR-DQ alleles in relation to manifestation periods and dates of birth. Diabetes Obes Metab 11(Suppl 1):88–91

Baekkeskov S, Nielsen JH, Marner B, Bilde T, Ludvigsson J, Lernmark A (1982) Autoantibodies in newly diagnosed diabetic children immunoprecipitate human pancreatic islet cell proteins. Nature 298(5870):167–169

Baekkeskov S, Aanstoot HJ, Christgau S et al (1990) Identification of the 64 K autoantigen in insulin-dependent diabetes as the GABA-synthesizing enzyme glutamic acid decarboxylase. Nature 347(6289):151–156

Baker C, de Marquesini LGP, Bishop AJ, Hedges AJ, Dayan CM, Wong FS (2008) Human CD8 responses to a complete epitope set from preproinsulin: implications for approaches to epitope discovery. J Clin Immunol 28(4):350–360

Barker JM (2006) Clinical review: type 1 diabetes-associated autoimmunity: natural history, genetic associations, and screening. J Clin Endocrinol Metab 91(4):1210–1217

Bennett ST, Todd JA (1996) Human type 1 diabetes and the insulin gene: principles of mapping polygenes. Annu Rev Genet 30:343–370

Bingley PJ, Gale EA (2006) Progression to type 1 diabetes in islet cell antibody-positive relatives in the European Nicotinamide Diabetes Intervention Trial: the role of additional immune, genetic and metabolic markers of risk. Diabetologia 49(5):881–890

Blom L, Lundmark K, Dahlquist G, Persson LA (1989) Estimating children's eating habits. Validity of a questionnaire measuring food frequency compared to a 7-day record. Acta Paediatr Scand 78(6):858–864

Blom L, Nystrom L, Dahlquist G (1991) The Swedish childhood diabetes study. Vaccinations and infections as risk determinants for diabetes in childhood. Diabetologia 34(3):176–181

Bottazzo GF, Florin-Christensen A, Doniach D (1974) Islet-cell antibodies in diabetes mellitus with autoimmune polyendocrine deficiencies. Lancet 2(7892):1279–1283

Bottazzo GF, Dean BM, McNally JM, MacKay EH, Swift PG, Gamble DR (1985) In situ characterization of autoimmune phenomena and expression of HLA molecules in the pancreas in diabetic insulitis. N Engl J Med 313(6):353–360

Brooks-Worrell BM, Nielson D, Palmer JP (1999) Insulin autoantibodies and insulin antibodies have similar binding characteristics. Proc Assoc Am Physicians 111(1):92–96

Brusko T, Wasserfall C, McGrail K et al (2007) No alterations in the frequency of FOXP3$^+$ regulatory T-cells in type 1 diabetes. Diabetes 56(3):604–612

Buschard K, Josefsen K, Horn T, Fredman P (1993) Sulphatide and sulphatide antibodies in insulin-dependent diabetes mellitus. Lancet 342(8875):840

Campbell-Thompson M, Wasserfall C, Kaddis J et al (2012) Network for Pancreatic Organ Donors with Diabetes (nPOD): developing a tissue biobank for type 1 diabetes. Diabetes Metab Res Rev 28(7):608–617

Campbell-Thompson ML, Atkinson MA, Butler AE et al (2013) The diagnosis of insulitis in human type 1 diabetes. Diabetologia 56(11):2541–2543

Castano L, Russo E, Zhou L, Lipes MA, Eisenbarth GS (1991) Identification and cloning of a granule autoantigen (carboxypeptidase-H) associated with type I diabetes. J Clin Endocrinol Metab 73(6):1197–1201

Cernea S, Herold KC (2010) Monitoring of antigen-specific CD8 T cells in patients with type 1 diabetes treated with antiCD3 monoclonal antibodies. Clin Immunol 134(2):121–129

Chang YH, Hwang J, Shang HF, Tsai ST (1996) Characterization of human DNA topoisomerase II as an autoantigen recognized by patients with IDDM. Diabetes 45(4):408–414

Chang YH, Shiau MY, Tsai ST, Lan MS (2004) Autoantibodies against IA-2, GAD, and topoisomerase II in type 1 diabetic patients. Biochem Biophys Res Commun 320(3):802–809

Chen W, Bergerot I, Elliott JF et al (2001) Evidence that a peptide spanning the B-C junction of proinsulin is an early Autoantigen epitope in the pathogenesis of type 1 diabetes. J Immunol 167(9):4926–4935

Chimienti F (2013) Zinc, pancreatic islet cell function and diabetes: new insights into an old story. Nutr Res Rev 26(1):1–11

Chimienti F, Devergnas S, Favier A, Seve M (2004) Identification and cloning of a β-cell-specific zinc transporter, ZnT-8, localized into insulin secretory granules. Diabetes 53(9):2330–2337

Christgau S, Schierbeck H, Aanstoot HJ et al (1991) Pancreatic β cells express two autoantigenic forms of glutamic acid decarboxylase, a 65-kDa hydrophilic form and a 64-kDa amphiphilic form which can be both membrane-bound and soluble. J Biol Chem 266(34):23516

Christie MR (1993) Characterization of distinct islet protein autoantigens associated with type 1 diabetes. Autoimmunity 15(Suppl):1–3

Concannon P, Erlich HA, Julier C et al (2005) Type 1 diabetes: evidence for susceptibility loci from four genome-wide linkage scans in 1,435 multiplex families. Diabetes 54(10):2995–3001

Concannon P, Rich SS, Nepom GT (2009) Genetics of type 1A diabetes. N Engl J Med 360(16):1646–1654

Dahlquist G, Blom L, Tuvemo T, Nystrom L, Sandstrom A, Wall S (1989) The Swedish childhood diabetes study – results from a nine year case register and a one year case-referent study indicating that type 1 (insulin-dependent) diabetes mellitus is associated with both type 2 (non-insulin-dependent) diabetes mellitus and autoimmune disorders. Diabetologia 32(1):2–6

Dahlquist GG, Blom LG, Persson LA, Sandstrom AI, Wall SG (1990) Dietary factors and the risk of developing insulin dependent diabetes in childhood. BMJ 300(6735):1302–1306

Dahlquist GG, Patterson C, Soltesz G (1999a) Perinatal risk factors for childhood type 1 diabetes in Europe. The EURODIAB Substudy 2 Study Group. Diabetes Care 22(10):1698–1702

Dahlquist GG, Boman JE, Juto P (1999b) Enteroviral RNA and IgM antibodies in early pregnancy and risk for childhood-onset IDDM in offspring. Diabetes Care 22(2):364–365

Dang M, Rockell J, Wagner R et al (2011) Human type 1 diabetes is associated with T cell autoimmunity to zinc transporter 8. J Immunol 186(10):6056–6063

Danke NA, Yang J, Greenbaum C, Kwok WW (2005) Comparative study of GAD65-specific $CD4^+$ T cells in healthy and type 1 diabetic subjects. J Autoimmun 25(4):303–311

Delli AJ, Vaziri-Sani F, Lindblad B et al (2012) Zinc transporter 8 autoantibodies and their association with SLC30A8 and HLA-DQ genes differ between immigrant and Swedish patients with newly diagnosed type 1 diabetes in the better diabetes diagnosis study. Diabetes 61(10):2556–2564

Diez J, Park Y, Zeller M et al (2001) Differential splicing of the IA-2 mRNA in pancreas and lymphoid organs as a permissive genetic mechanism for autoimmunity against the IA-2 type 1 diabetes autoantigen. Diabetes 50(4):895–900

Dobersen MJ, Scharff JE, Ginsberg-Fellner F, Notkins AL (1980) Cytotoxic autoantibodies to β cells in the serum of patients with insulin-dependent diabetes mellitus. N Engl J Med 303(26):1493–1498

Dotta F, Falorni A, Tiberti C et al (1997) Autoantibodies to the GM2-1 islet ganglioside and to GAD-65 at type 1 diabetes onset. J Autoimmun 10(6):585–588

Dotta F, Censini S, van Halteren AG et al (2007) Coxsackie B4 virus infection of β cells and natural killer cell insulitis in recent-onset type 1 diabetic patients. Proc Natl Acad Sci USA 104(12):5115–5120

Dudda JC, Martin SF (2004) Tissue targeting of T cells by DCs and microenvironments. Trends Immunol 25(8):417–421

Eisenbarth GS (1986) Type I, diabetes mellitus. A chronic autoimmune disease. N Engl J Med 314(21):1360–1368

Eisenbarth GS, Jeffrey J (2008) The natural history of type 1A diabetes. Arq Bras Endocrinol Metabol 52(2):146–155

Elding Larsson H, Vehik K, Gesualdo P et al (2014) Children followed in the TEDDY study are diagnosed with type 1 diabetes at an early stage of disease. Pediatr Diabetes 15:118-26

Endl J, Otto H, Jung G et al (1997) Identification of naturally processed T cell epitopes from glutamic acid decarboxylase presented in the context of HLA-DR alleles by T lymphocytes of recent onset IDDM patients. J Clin Invest 99(10):2405–2415

Esposti MD, Ngo A, Myers MA (1996) Inhibition of mitochondrial complex I may account for IDDM induced by intoxication with the rodenticide Vacor. Diabetes 45(11):1531–1534

EURODIAB ACE Study Group (2000) Variation and trends in incidence of childhood diabetes in Europe. Lancet 355(9207):873–876

Faresjo MK, Vaarala O, Thuswaldner S, Ilonen J, Hinkkanen A, Ludvigsson J (2006) Diminished IFN-gamma response to diabetes-associated autoantigens in children at diagnosis and during follow up of type 1 diabetes. Diabetes Metab Res Rev 22(6):462–470

Fenalti G, Buckle AM (2010) Structural biology of the GAD autoantigen. Autoimmun Rev 9(3):148–152

Fenalti G, Hampe CS, Arafat Y et al (2008) COOH-terminal clustering of autoantibody and T-cell determinants on the structure of GAD65 provide insights into the molecular basis of autoreactivity. Diabetes 57(5):1293–1301

Foulis AK, McGill M, Farquharson MA (1991) Insulitis in type 1 (insulin-dependent) diabetes mellitus in man–macrophages, lymphocytes, and interferon-gamma containing cells. J Pathol 165(2):97–103

Fourlanos S, Varney MD, Tait BD et al (2008) The rising incidence of type 1 diabetes is accounted for by cases with lower-risk human leukocyte antigen genotypes. Diabetes Care 31(8): 1546–1549

Gambelunghe G, Ghaderi M, Cosentino A, Falorni A, Brunetti P, Sanjeevi CB (2000) Association of MHC Class I chain-related A (MIC-A) gene polymorphism with Type I diabetes. Diabetologia 43(4):507–514

Gambineri E, Torgerson TR, Ochs HD (2003) Immune dysregulation, polyendocrinopathy, enteropathy, and X-linked inheritance (IPEX), a syndrome of systemic autoimmunity caused by mutations of FOXP3, a critical regulator of T-cell homeostasis. Curr Opin Rheumatol 15(4):430–435

Garcia CA, Prabakar KR, Diez J et al (2005) Dendritic cells in human thymus and periphery display a proinsulin epitope in a transcription-dependent, capture-independent fashion. J Immunol 175(4):2111–2122

Geenen V, Mottet M, Dardenne O et al (2010) Thymic self-antigens for the design of a negative/tolerogenic self-vaccination against type 1 diabetes. Curr Opin Pharmacol 10(4):461–472

Geluk A, van Meijgaarden KE, Schloot NC, Drijfhout JW, Ottenhoff TH, Roep BO (1998) HLA-DR binding analysis of peptides from islet antigens in IDDM. Diabetes 47(10):1594–1601

Gepts W (1965) Pathologic anatomy of the pancreas in juvenile diabetes mellitus. Diabetes 14(10):619–633

Gepts W, De Mey J (1978) Islet cell survival determined by morphology. An immunocytochemical study of the islets of Langerhans in juvenile diabetes mellitus. Diabetes 27(Suppl 1):251–261

Gepts W, Lecompte PM (1981) The pancreatic islets in diabetes. Am J Med 70(1):105–115

Graham J, Hagopian WA, Kockum I et al (2002) Genetic effects on age-dependent onset and islet cell autoantibody markers in type 1 diabetes. Diabetes 51(5):1346–1355

Greening JE, Tree TI, Kotowicz KT et al (2003) Processing and presentation of the islet autoantigen GAD by vascular endothelial cells promotes transmigration of autoreactive T-cells. Diabetes 52(3):717–725

Gregersen PK (2005) Gaining insight into PTPN22 and autoimmunity. Nat Genet 37(12): 1300–1302

Hagopian WA, Sanjeevi CB, Kockum I et al (1995) Glutamate decarboxylase-, insulin-, and islet cell-antibodies and HLA typing to detect diabetes in a general population-based study of Swedish children. J Clin Invest 95(4):1505–1511

Hagopian WA, Erlich H, Lernmark A et al (2011) The Environmental Determinants of Diabetes in the Young (TEDDY): genetic criteria and international diabetes risk screening of 421 000 infants. Pediatr Diabetes 12(8):733–743

Hampe CS, Hammerle LP, Bekris L et al (2000) Recognition of glutamic acid decarboxylase (GAD) by autoantibodies from different GAD antibody-positive phenotypes. J Clin Endocrinol Metab 85(12):4671–4679

Hanninen A, Jalkanen S, Salmi M, Toikkanen S, Nikolakaros G, Simell O (1992) Macrophages, T cell receptor usage, and endothelial cell activation in the pancreas at the onset of insulin-dependent diabetes mellitus. J Clin Invest 90(5):1901–1910

Hanninen A, Soilu-Hanninen M, Hampe CS et al (2010) Characterization of $CD4^+$ T cells specific for glutamic acid decarboxylase (GAD65) and proinsulin in a patient with stiff-person syndrome but without type 1 diabetes. Diabetes Metab Res Rev 26(4):271–279

Harfouch-Hammoud E, Walk T, Otto H et al (1999) Identification of peptides from autoantigens GAD65 and IA-2 that bind to HLA class II molecules predisposing to or protecting from type 1 diabetes. Diabetes 48(10):1937–1947

Hatfield EC, Hawkes CJ, Payton MA, Christie MR (1997) Cross reactivity between IA-2 and phogrin/IA-2β in binding of autoantibodies in IDDM. Diabetologia 40(11):1327–1333

Helmke K, Otten A, Willems WR et al (1986) Islet cell antibodies and the development of diabetes mellitus in relation to mumps infection and mumps vaccination. Diabetologia 29(1):30–33

Heninger AK, Monti P, Wilhelm C et al (2013) Activation of islet autoreactive naive T cells in infants is influenced by homeostatic mechanisms and antigen-presenting capacity. Diabetes 62(6):2059–2066

Hermann R, Lipponen K, Kiviniemi M et al (2006) Lymphoid tyrosine phosphatase (LYP/PTPN22) Arg620Trp variant regulates insulin autoimmunity and progression to type 1 diabetes. Diabetologia 49(6):1198–1208

Hernandez J, Aung S, Redmond WL, Sherman LA (2001) Phenotypic and functional analysis of $CD8^+$ T cells undergoing peripheral deletion in response to cross-presentation of self-antigen. J Exp Med 194(6):707–717

Herold KC, Brooks-Worrell B, Palmer J et al (2009) Validity and reproducibility of measurement of islet autoreactivity by T-cell assays in subjects with early type 1 diabetes. Diabetes 58(11):2588–2595

Hirai H, Miura J, Hu Y et al (2008) Selective screening of secretory vesicle-associated proteins for autoantigens in type 1 diabetes: VAMP2 and NPY are new minor autoantigens. Clin Immunol 127(3):366–374

Hjorth M, Axelsson S, Ryden A, Faresjo M, Ludvigsson J, Casas R (2011) GAD-alum treatment induces GAD65-specific $CD4^+$ $CD25^{high}FOXP3^+$ cells in type 1 diabetic patients. Clin Immunol 138(1):117–126

Honeyman MC, Stone NL, Harrison LC (1998) T-cell epitopes in type 1 diabetes autoantigen tyrosine phosphatase IA-2: potential for mimicry with rotavirus and other environmental agents. Mol Med 4(4):231–239

Hultcrantz M, Huhn MH, Wolf M et al (2007) Interferons induce an antiviral state in human pancreatic islet cells. Virology 367(1):92–101

Imagawa A, Hanafusa T, Tamura S et al (2001) Pancreatic biopsy as a procedure for detecting in situ autoimmune phenomena in type 1 diabetes: close correlation between serological markers and histological evidence of cellular autoimmunity. Diabetes 50(6):1269–1273

In't Veld P, Lievens D, De Grijse J et al (2007) Screening for insulitis in adult autoantibody-positive organ donors. Diabetes 56(9):2400–2404

Itoh M (1989) Immunological aspects of diabetes mellitus: prospects for pharmacological modification. Pharmacol Ther 44(3):351–406

Jarchum I, Nichol L, Trucco M, Santamaria P, DiLorenzo TP (2008) Identification of novel IGRP epitopes targeted in type 1 diabetes patients. Clin Immunol 127(3):359–365

Jaume JC, Parry SL, Madec AM, Sonderstrup G, Baekkeskov S (2002) Suppressive effect of glutamic acid decarboxylase 65-specific autoimmune B lymphocytes on processing of T cell determinants located within the antibody epitope. J Immunol 169(2):665–672

Jun HS, Yoon JW (1994) Initiation of autoimmune type 1 diabetes and molecular cloning of a gene encoding for islet cell-specific 37kd autoantigen. Adv Exp Med Biol 347:207–220

Kanatsuna N, Taneera J, Vaziri-Sani F et al (2013) Autoimmunity against INS-IGF2 expressed in human pancreatic islets. J Biol Chem 288:29013–29023

Karam JH, Lewitt PA, Young CW et al (1980) Insulinopenic diabetes after rodenticide (Vacor) ingestion: a unique model of acquired diabetes in man. Diabetes 29(12):971–978

Karlsen AE, Hagopian WA, Grubin CE et al (1991) Cloning and primary structure of a human islet isoform of glutamic acid decarboxylase from chromosome 10. Proc Natl Acad Sci USA 88(19):8337–8341

Karlsen AE, Hagopian WA, Petersen JS et al (1992) Recombinant glutamic acid decarboxylase (representing the single isoform expressed in human islets) detects IDDM-associated 64,000-M(r) autoantibodies. Diabetes 41(10):1355–1359

Karvonen M, Viik-Kajander M, Moltchanova E, Libman I, LaPorte R, Tuomilehto J (2000) Incidence of childhood type 1 diabetes worldwide. Diabetes Mondiale (DiaMond) Project Group. Diabetes Care 23(10):1516–1526

Kawasaki E, Eisenbarth GS, Wasmeier C, Hutton JC (1996) Autoantibodies to protein tyrosine phosphatase-like proteins in type I diabetes. Overlapping specificities to phogrin and ICA512/IA-2. Diabetes 45(10):1344–1349

Kelemen K, Gottlieb PA, Putnam AL, Davidson HW, Wegmann DR, Hutton JC (2004) HLA-DQ8-associated T cell responses to the diabetes autoantigen phogrin (IA-2 β) in human prediabetes. J Immunol 172(6):3955–3962

Kerokoski P, Ilonen J, Gaedigk R et al (1999) Production of the islet cell antigen ICA69 (p69) with baculovirus expression system: analysis with a solid-phase time-resolved fluorescence method of sera from patients with IDDM and rheumatoid arthritis. Autoimmunity 29(4):281–289

Kobayashi T, Tanaka S, Okubo M, Nakanishi K, Murase T, Lernmark A (2003) Unique epitopes of glutamic acid decarboxylase autoantibodies in slowly progressive type 1 diabetes. J Clin Endocrinol Metab 88(10):4768–4775

Kronenberg M (2005) Toward an understanding of NKT cell biology: progress and paradoxes. Annu Rev Immunol 23:877–900

Kubosaki A, Nakamura S, Notkins AL (2005) Dense core vesicle proteins IA-2 and IA-2β: metabolic alterations in double knockout mice. Diabetes 54(Suppl 2):S46–S51

Kukko M, Kimpimaki T, Korhonen S et al (2005) Dynamics of diabetes-associated autoantibodies in young children with human leukocyte antigen-conferred risk of type 1 diabetes recruited from the general population. J Clin Endocrinol Metab 90(5):2712–2717

Lamb MM, Yin X, Barriga K et al (2008) Dietary glycemic index, development of islet autoimmunity, and subsequent progression to type 1 diabetes in young children. J Clin Endocrinol Metab 93(10):3936–3942

Lampeter ER, Kishimoto TK, Rothlein R et al (1992) Elevated levels of circulating adhesion molecules in IDDM patients and in subjects at risk for IDDM. Diabetes 41(12):1668–1671

Lan MS, Lu J, Goto Y, Notkins AL (1994) Molecular cloning and identification of a receptor-type protein tyrosine phosphatase, IA-2, from human insulinoma. DNA Cell Biol 13(5):505–514

Lan MS, Wasserfall C, Maclaren NK, Notkins AL (1996) IA-2, a transmembrane protein of the protein tyrosine phosphatase family, is a major autoantigen in insulin-dependent diabetes mellitus. Proc Natl Acad Sci USA 93(13):6367–6370

Landin-Olsson M, Hillman M, Erlanson-Albertsson C (2013) Is type 1 diabetes a food-induced disease? Med Hypotheses 81(2):338–342

Lee HS, Briese T, Winkler C et al (2013) Next-generation sequencing for viruses in children with rapid-onset type 1 diabetes. Diabetologia 56(8):1705–1711

Lemaire K, Ravier MA, Schraenen A et al (2009) Insulin crystallization depends on zinc transporter ZnT8 expression, but is not required for normal glucose homeostasis in mice. Proc Natl Acad Sci USA 106(35):14872–14877

Lernmark A, Larsson HE (2013) Immune therapy in type 1 diabetes mellitus. Nat Rev Endocrinol 9(2):92–103

Lernmark A, Freedman ZR, Hofmann C et al (1978) Islet-cell-surface antibodies in juvenile diabetes mellitus. N Engl J Med 299(8):375–380

Lernmark A, Kloppel G, Stenger D et al (1995) Heterogeneity of islet pathology in two infants with recent onset diabetes mellitus. Virchows Arch 425(6):631–640

Lindehammer SR, Hansson I, Midberg B et al (2011) Seroconversion to islet autoantibodies between early pregnancy and delivery in non-diabetic mothers. J Reprod Immunol 88(1):72–79

Lindley S, Dayan CM, Bishop A, Roep BO, Peakman M, Tree TI (2005) Defective suppressor function in CD4$^+$CD25$^+$ T-cells from patients with type 1 diabetes. Diabetes 54(1):92–99

Lio D, Candore G, Romano GC et al (1997) Modification of cytokine patterns in subjects bearing the HLA-B8, DR3 phenotype: implications for autoimmunity. Cytokines Cell Mol Ther 3(4):217–224

Liu YJ (2005) IPC: professional type 1 interferon-producing cells and plasmacytoid dendritic cell precursors. Annu Rev Immunol 23:275–306

Mallone R, Scotto M, Janicki CN et al (2011a) Immunology of Diabetes Society T-cell workshop: HLA class I tetramer-directed epitope validation initiative T-cell workshop report – HLA class I tetramer validation initiative. Diabetes Metab Res Rev 27(8):720–726

Mallone R, Brezar V, Boitard C (2011b) T cell recognition of autoantigens in human type 1 diabetes: clinical perspectives. Clin Dev Immunol 2011:513210

Mannering SI, Pang SH, Williamson NA et al (2009) The A-chain of insulin is a hot-spot for CD4$^+$ T cell epitopes in human type 1 diabetes. Clin Exp Immunol 156(2):226–231

Marshall MO, Hoyer PE, Petersen JS et al (1994) Contribution of glutamate decarboxylase antibodies to the reactivity of islet cell cytoplasmic antibodies. J Autoimmun 7(4):497–508

Maziarz M, Janer M, Roach JC et al (2010) The association between the PTPN22 1858C > T variant and type 1 diabetes depends on HLA risk and GAD65 autoantibodies. Genes Immun 11(5):406–415

Mollah ZU, Pai S, Moore C et al (2008) Abnormal NF-kappa B function characterizes human type 1 diabetes dendritic cells and monocytes. J Immunol 180(5):3166–3175

Monti P, Scirpoli M, Rigamonti A et al (2007) Evidence for in vivo primed and expanded autoreactive T cells as a specific feature of patients with type 1 diabetes. J Immunol 179(9):5785–5792

Monti P, Heninger AK, Bonifacio E (2009) Differentiation, expansion, and homeostasis of autoreactive T cells in type 1 diabetes mellitus. Curr Diab Rep 9(2):113–118

Myers MA, Hettiarachchi KD, Ludeman JP, Wilson AJ, Wilson CR, Zimmet PZ (2003) Dietary microbial toxins and type 1 diabetes. Ann N Y Acad Sci 1005:418–422

Nagata M, Kotani R, Moriyama H, Yokono K, Roep BO, Peakman M (2004) Detection of autoreactive T cells in type 1 diabetes using coded autoantigens and an immunoglobulin-free cytokine ELISPOT assay: report from the fourth immunology of diabetes society T cell workshop. Ann N Y Acad Sci 1037:10–15

Nepom GT (2012) MHC class II tetramers. J Immunol 188(6):2477–2482

Nepom GT, Buckner JH (2012) A functional framework for interpretation of genetic associations in T1D. Curr Opin Immunol 24(5):516–521

Nerup J, Binder C (1973) Thyroid, gastric and adrenal auto-immunity in diabetes mellitus. Acta Endocrinol (Copenh) 72(2):279–286

Nerup J, Andersen OO, Bendixen G et al (1973) Anti-pancreatic, cellular hypersensitivity in diabetes mellitus. Experimental induction of anti-pancreatic, cellular hypersensitivity and associated morphological β-cell changes in the rat. Acta Allergol 28(4):231–249

Nerup J, Platz P, Andersen OO et al (1974) HL-A antigens and diabetes mellitus. Lancet 2(7885):864–866

Nerup J, Christy M, Kromann H et al (1979) HLA and insulin-dependent diabetes mellitus. Postgrad Med J 55(Suppl 2):8–13

Nicoletti F, Conget I, Di Mauro M et al (2002) Serum concentrations of the interferon-gamma-inducible chemokine IP-10/CXCL10 are augmented in both newly diagnosed Type I diabetes mellitus patients and subjects at risk of developing the disease. Diabetologia 45(8):1107–1110

Nielsen LB, Mortensen HB, Chiarelli F et al (2006) Impact of IDDM2 on disease pathogenesis and progression in children with newly diagnosed type 1 diabetes: reduced insulin antibody titres and preserved β cell function. Diabetologia 49(1):71–74

Oak S, Gilliam LK, Landin-Olsson M et al (2008) The lack of anti-idiotypic antibodies, not the presence of the corresponding autoantibodies to glutamate decarboxylase, defines type 1 diabetes. Proc Natl Acad Sci USA 105(14):5471–5476

Oikarinen M, Tauriainen S, Oikarinen S et al (2012) Type 1 diabetes is associated with enterovirus infection in gut mucosa. Diabetes 61(3):687–691

Oling V, Reijonen H, Simell O, Knip M, Ilonen J (2012) Autoantigen-specific memory $CD4^+$ T cells are prevalent early in progression to Type 1 diabetes. Cell Immunol 273(2):133–139

Onkamo P, Vaananen S, Karvonen M, Tuomilehto J (1999) Worldwide increase in incidence of type 1 diabetes – the analysis of the data on published incidence trends. Diabetologia 42(12):1395–1403

Ortqvist E, Brooks-Worrell B, Lynch K et al (2010) Changes in GAD65Ab-specific antiidiotypic antibody levels correlate with changes in C-peptide levels and progression to islet cell autoimmunity. J Clin Endocrinol Metab 95(11):E310–E318

Osugi Y, Vuckovic S, Hart DN (2002) Myeloid blood $CD11c^+$ dendritic cells and monocyte-derived dendritic cells differ in their ability to stimulate T lymphocytes. Blood 100(8):2858–2866

Owerbach D, Lernmark A, Platz P et al (1983) HLA-D region β-chain DNA endonuclease fragments differ between HLA-DR identical healthy and insulin-dependent diabetic individuals. Nature 303(5920):815–817

Ozawa Y, Kasuga A, Nomaguchi H et al (1996) Detection of autoantibodies to the pancreatic islet heat shock protein 60 in insulin-dependent diabetes mellitus. J Autoimmun 9(4):517–524

Padoa CJ, Banga JP, Madec AM et al (2003) Recombinant Fabs of human monoclonal antibodies specific to the middle epitope of GAD65 inhibit type 1 diabetes-specific GAD65Abs. Diabetes 52(11):2689–2695

Palmer JP, Asplin CM, Clemons P et al (1983) Insulin antibodies in insulin-dependent diabetics before insulin treatment. Science 222(4630):1337–1339

Parikka V, Nanto-Salonen K, Saarinen M et al (2012) Early seroconversion and rapidly increasing autoantibody concentrations predict prepubertal manifestation of type 1 diabetes in children at genetic risk. Diabetologia 55(7):1926–1936

Parker KC, Bednarek MA, Coligan JE (1994) Scheme for ranking potential HLA-A2 binding peptides based on independent binding of individual peptide side-chains. J Immunol 152(1):163–175

Patterson CC, Dahlquist GG, Gyurus E, Green A, Soltesz G (2009) Incidence trends for childhood type 1 diabetes in Europe during 1989-2003 and predicted new cases 2005-20: a multicentre prospective registration study. Lancet 373(9680):2027–2033

Peakman M, Stevens EJ, Lohmann T et al (1999) Naturally processed and presented epitopes of the islet cell autoantigen IA-2 eluted from HLA-DR4. J Clin Invest 104(10):1449–1457

Peakman M, Tree TI, Endl J, van Endert P, Atkinson MA, Roep BO (2001) Characterization of preparations of GAD65, proinsulin, and the islet tyrosine phosphatase IA-2 for use in detection of autoreactive T-cells in type 1 diabetes: report of phase II of the Second International Immunology of Diabetes Society Workshop for Standardization of T-cell assays in type 1 diabetes. Diabetes 50(8):1749–1754

Pietropaolo M, Castano L, Babu S et al (1993) Islet cell autoantigen 69 kD (ICA69). Molecular cloning and characterization of a novel diabetes-associated autoantigen. J Clin Invest 92(1):359–371

Pihoker C, Gilliam LK, Hampe CS, Lernmark A (2005) Autoantibodies in diabetes. Diabetes 54 (Suppl 2):S52–S61

Pociot F, Akolkar B, Concannon P et al (2010) Genetics of type 1 diabetes: what's next? Diabetes 59(7):1561–1571

Pugliese A, Zeller M, Fernandez A Jr et al (1997) The insulin gene is transcribed in the human thymus and transcription levels correlated with allelic variation at the INS VNTR-IDDM2 susceptibility locus for type 1 diabetes. Nat Genet 15(3):293–297

Pugliese A, Brown D, Garza D et al (2001) Self-antigen-presenting cells expressing diabetes-associated autoantigens exist in both thymus and peripheral lymphoid organs. J Clin Invest 107(5):555–564

Qin HY, Mahon JL, Atkinson MA, Chaturvedi P, Lee-Chan E, Singh B (2003) Type 1 diabetes alters anti-hsp90 autoantibody isotype. J Autoimmun 20(3):237–245

Rammensee H, Bachmann J, Emmerich NP, Bachor OA, Stevanovic S (1999) SYFPEITHI: database for MHC ligands and peptide motifs. Immunogenetics 50(3–4):213–219

Reijonen H, Mallone R, Heninger AK et al (2004) GAD65-specific CD4$^+$ T-cells with high antigen avidity are prevalent in peripheral blood of patients with type 1 diabetes. Diabetes 53(8):1987–1994

Resic-Lindehammer S, Larsson K, Ortqvist E et al (2008) Temporal trends of HLA genotype frequencies of type 1 diabetes patients in Sweden from 1986 to 2005 suggest altered risk. Acta Diabetol 45(4):231–235

Rich SS, Akolkar B, Concannon P et al (2009) Overview of the type I diabetes genetics consortium. Genes Immun 10(Suppl 1):S1–S4

Richter W, Shi Y, Baekkeskov S (1993) Autoreactive epitopes defined by diabetes-associated human monoclonal antibodies are localized in the middle and C-terminal domains of the smaller form of glutamate decarboxylase. Proc Natl Acad Sci USA 90(7):2832–2836

Rodacki M, Milech A, de Oliveira JE (2006) NK cells and type 1 diabetes. Clin Dev Immunol 13(2–4):101–107

Rodacki M, Svoren B, Butty V et al (2007) Altered natural killer cells in type 1 diabetic patients. Diabetes 56(1):177–185

Roep BO, Peakman M (2011) Diabetogenic T lymphocytes in human Type 1 diabetes. Curr Opin Immunol 23(6):746–753

Roep BO, Peakman M (2012) Antigen targets of type 1 diabetes autoimmunity. Cold Spring Harb Perspect Med 2(4):a007781

Rolandsson O, Hagg E, Hampe C et al (1999) Glutamate decarboxylase (GAD65) and tyrosine phosphatase-like protein (IA-2) autoantibodies index in a regional population is related to glucose intolerance and body mass index. Diabetologia 42(5):555–559

Rudy G, Stone N, Harrison LC et al (1995) Similar peptides from two β cell autoantigens, proinsulin and glutamic acid decarboxylase, stimulate T cells of individuals at risk for insulin-dependent diabetes. Mol Med 1(6):625–633

Sakaguchi S, Yamaguchi T, Nomura T, Ono M (2008) Regulatory T cells and immune tolerance. Cell 133(5):775–787

Sanda S, Roep BO, von Herrath M (2008) Islet antigen specific IL-10$^+$ immune responses but not CD4$^+$ CD25$^+$ FoxP3$^+$ cells at diagnosis predict glycemic control in type 1 diabetes. Clin Immunol 127(2):138–143

Sanjeevi CB, Lybrand TP, DeWeese C et al (1995) Polymorphic amino acid variations in HLA-DQ are associated with systematic physical property changes and occurrence of IDDM. Members of the Swedish Childhood Diabetes Study. Diabetes 44(1):125–131

Savinov AY, Wong FS, Stonebraker AC, Chervonsky AV (2003) Presentation of antigen by endothelial cells and chemoattraction are required for homing of insulin-specific CD8$^+$ T cells. J Exp Med 197(5):643–656

Schloot NC, Roep BO, Wegmann DR, Yu L, Wang TB, Eisenbarth GS (1997) T-cell reactivity to GAD65 peptide sequences shared with coxsackie virus protein in recent-onset IDDM, post-onset IDDM patients and control subjects. Diabetologia 40(3):332–338

Seyfert-Margolis V, Gisler TD, Asare AL et al (2006) Analysis of T-cell assays to measure autoimmune responses in subjects with type 1 diabetes: results of a blinded controlled study. Diabetes 55(9):2588–2594

Shlomchik MJ (2008) Sites and stages of autoreactive B cell activation and regulation. Immunity 28(1):18–28

Sibley RK, Sutherland DE, Goetz F, Michael AF (1985) Recurrent diabetes mellitus in the pancreas iso- and allograft. A light and electron microscopic and immunohistochemical analysis of four cases. Lab Invest 53(2):132–144

Skarstrand H, Dahlin LB, Lernmark A, Vaziri-Sani F (2013a) Neuropeptide Y autoantibodies in patients with long-term type 1 and type 2 diabetes and neuropathy. J Diabetes Complications 27:609–617

Skarstrand H, Lernmark A, Vaziri-Sani F (2013b) Antigenicity and epitope specificity of ZnT8 autoantibodies in type 1 diabetes. Scand J Immunol 77(1):21–29

Skowera A, Ellis RJ, Varela-Calvino R et al (2008) CTLs are targeted to kill β cells in patients with type 1 diabetes through recognition of a glucose-regulated preproinsulin epitope. J Clin Invest 118(10):3390–3402

Somoza N, Vargas F, Roura-Mir C et al (1994) Pancreas in recent onset insulin-dependent diabetes mellitus. Changes in HLA, adhesion molecules and autoantigens, restricted T cell receptor V β usage, and cytokine profile. J Immunol 153(3):1360–1377

Sosenko JM, Skyler JS, Palmer JP et al (2011a) A longitudinal study of GAD65 and ICA512 autoantibodies during the progression to type 1 diabetes in Diabetes Prevention Trial-Type 1 (DPT-1) participants. Diabetes Care 34(11):2435–2437

Sosenko JM, Skyler JS, Mahon J et al (2011b) Validation of the Diabetes Prevention Trial-Type 1 Risk Score in the TrialNet Natural History Study. Diabetes Care 34(8):1785–1787

Sosenko JM, Skyler JS, Palmer JP et al (2013) The prediction of type 1 diabetes by multiple autoantibody levels and their incorporation into an autoantibody risk score in relatives of type 1 diabetic patients. Diabetes Care 36:2615–2620

Standifer NE, Ouyang Q, Panagiotopoulos C et al (2006) Identification of Novel HLA-A*0201-restricted epitopes in recent-onset type 1 diabetic subjects and antibody-positive relatives. Diabetes 55(11):3061–3067

Steed J, Gilliam LK, Harris RA, Lernmark A, Hampe CS (2008) Antigen presentation of detergent-free glutamate decarboxylase (GAD65) is affected by human serum albumin as carrier protein. J Immunol Methods 334(1–2):114–121

Stene LC, Oikarinen S, Hyoty H et al (2010) Enterovirus infection and progression from islet autoimmunity to type 1 diabetes: the Diabetes and Autoimmunity Study in the Young (DAISY). Diabetes 59(12):3174–3180

Tang Q, Bluestone JA (2008) The Foxp3$^+$ regulatory T cell: a jack of all trades, master of regulation. Nat Immunol 9(3):239–244

TEDDY Study Group (2008) The Environmental Determinants of Diabetes in the Young (TEDDY) study. Ann N Y Acad Sci 1150:1–13

Thorsby E, Lie BA (2005) HLA associated genetic predisposition to autoimmune diseases: genes involved and possible mechanisms. Transpl Immunol 14(3–4):175–182

Todd JA (2010) Etiology of type 1 diabetes. Immunity 32(4):457–467

Todd JA, Farrall M (1996) Panning for gold: genome-wide scanning for linkage in type 1 diabetes. Hum Mol Genet 5:1443–1448

Todd JA, Bell JI, McDevitt HO (1987) HLA-DQ β gene contributes to susceptibility and resistance to insulin-dependent diabetes mellitus. Nature 329(6140):599–604

Todd JA, Bell JI, McDevitt HO (1988) A molecular basis for genetic susceptibility to insulin-dependent diabetes mellitus. Trends Genet 4(5):129–134

Torn C, Shtauvere-Brameus A, Sanjeevi CB, Landin-Olsson M (2002) Increased autoantibodies to SOX13 in Swedish patients with type 1 diabetes. Ann N Y Acad Sci 958:218–223

Torn C, Mueller PW, Schlosser M, Bonifacio E, Bingley PJ (2008) Diabetes antibody standardization program: evaluation of assays for autoantibodies to glutamic acid decarboxylase and islet antigen-2. Diabetologia 51(5):846–852

Tree TI, O'Byrne D, Tremble JM et al (2000) Evidence for recognition of novel islet T cell antigens by granule-specific T cell lines from new onset type 1 diabetic patients. Clin Exp Immunol 121(1):100–105

Tree TI, Roep BO, Peakman M (2006) A mini meta-analysis of studies on $CD4^+$ $CD25^+$ T cells in human type 1 diabetes: report of the Immunology of Diabetes Society T Cell Workshop. Ann N Y Acad Sci 1079:9–18

Uibo R, Lernmark A (2008) GAD65 autoimmunity-clinical studies. Adv Immunol 100:39–78

Ungar B, Stocks AE, Martin FI, Whittingham S, Mackay IR (1968) Intrinsic-factor antibody, parietal-cell antibody, and latent pernicious anaemia in diabetes mellitus. Lancet 2(7565):415–417

Unger WW, Pinkse GG, Van der Kracht SM et al (2007) Human clonal CD8 autoreactivity to an IGRP islet epitope shared between mice and men. Ann N Y Acad Sci 1103:192–195

Uno S, Imagawa A, Okita K et al (2007) Macrophages and dendritic cells infiltrating islets with or without β cells produce tumour necrosis factor-α in patients with recent-onset type 1 diabetes. Diabetologia 50(3):596–601

van Endert P, Hassainya Y, Lindo V et al (2006) HLA class I epitope discovery in type 1 diabetes. Ann N Y Acad Sci 1079:190–197

Vang T, Miletic AV, Bottini N, Mustelin T (2007) Protein tyrosine phosphatase PTPN22 in human autoimmunity. Autoimmunity 40(6):453–461

Vaziri-Sani F, Oak S, Radtke J et al (2010) ZnT8 autoantibody titers in type 1 diabetes patients decline rapidly after clinical onset. Autoimmunity 43(8):598–606

Vehik K, Beam CA, Mahon JL et al (2011) Development of autoantibodies in the TrialNet Natural History Study. Diabetes Care 34(9):1897–1901

Velthuis JH, Unger WW, Abreu JR et al (2010) Simultaneous detection of circulating autoreactive $CD8^+$ T-cells specific for different islet cell-associated epitopes using combinatorial MHC multimers. Diabetes 59(7):1721–1730

Vidard L, Rock KL, Benacerraf B (1992) Heterogeneity in antigen processing by different types of antigen-presenting cells. Effect of cell culture on antigen processing ability. J Immunol 149(6):1905–1911

Viglietta V, Kent SC, Orban T, Hafler DA (2002) GAD65-reactive T cells are activated in patients with autoimmune type 1a diabetes. J Clin Invest 109(7):895–903

Virtanen SM, Nevalainen J, Kronberg-Kippila C et al (2012) Food consumption and advanced β cell autoimmunity in young children with HLA-conferred susceptibility to type 1 diabetes: a nested case-control design. Am J Clin Nutr 95(2):471–478

Von Meyenburg H (1940) Über "insulitis" bei diabetes. Schweitz Med Wochenschr 21:554–561

Walter M, Albert E, Conrad M et al (2003) IDDM2/insulin VNTR modifies risk conferred by IDDM1/HLA for development of type 1 diabetes and associated autoimmunity. Diabetologia 46(5):712–720

Wang X, Zhang Y, Liu Y et al (2012) Anti-idiotypic antibody specific to GAD65 autoantibody prevents type 1 diabetes in the NOD mouse. PLoS One 7(2):e32515

Wardemann H, Nussenzweig MC (2007) β-cell self-tolerance in humans. Adv Immunol 95:83–110

Wasmeier C, Hutton JC (1996) Molecular cloning of phogrin, a protein-tyrosine phosphatase homologue localized to insulin secretory granule membranes. J Biol Chem 271(30):18161–18170

Weenink SM, Lo J, Stephenson CR et al (2009) Autoantibodies and associated T-cell responses to determinants within the 831–860 region of the autoantigen IA-2 in Type 1 diabetes. J Autoimmun 33(2):147–154

Weichselbaum A (1910) Über die veranderungen des pankreas bei diabetes mellitus. Sitzungsber Akad Wiss Wien Math Naturw Klasse 119:73–281

Wenzlau JM, Juhl K, Yu L et al (2007) The cation efflux transporter ZnT8 (Slc30A8) is a major autoantigen in human type 1 diabetes. Proc Natl Acad Sci USA 104(43):17040–17045

Wenzlau JM, Liu Y, Yu L et al (2008) A common nonsynonymous single nucleotide polymorphism in the SLC30A8 gene determines ZnT8 autoantibody specificity in type 1 diabetes. Diabetes 57(10):2693–2697

Wicker LS, Chen SL, Nepom GT et al (1996) Naturally processed T cell epitopes from human glutamic acid decarboxylase identified using mice transgenic for the type 1 diabetes-associated human MHC class II allele, DRB1*0401. J Clin Invest 98(11):2597–2603

Yang J, Danke N, Roti M et al (2008) CD4$^+$ T cells from type 1 diabetic and healthy subjects exhibit different thresholds of activation to a naturally processed proinsulin epitope. J Autoimmun 31(1):30–41

Yoon JW, Ihm SH, Kim KW (1989) Viruses as a triggering factor of type 1 diabetes and genetic markers related to the susceptibility to the virus-associated diabetes. Diabetes Res Clin Pract 7 (Suppl 1):S47–S58

Yu L, Miao D, Scrimgeour L, Johnson K, Rewers M, Eisenbarth GS (2012) Distinguishing persistent insulin autoantibodies with differential risk: nonradioactive bivalent proinsulin/insulin autoantibody assay. Diabetes 61(1):179–186

Zhang B, Lan MS, Notkins AL (1997) Autoantibodies to IA-2 in IDDM: location of major antigenic determinants. Diabetes 46(1):40–43

Zhou X, Bailey-Bucktrout S, Jeker LT, Bluestone JA (2009) Plasticity of CD4$^+$ FoxP3$^+$ T cells. Curr Opin Immunol 21(3):281–285

Ziegler AG, Nepom GT (2010) Prediction and pathogenesis in type 1 diabetes. Immunity 32(4):468–478

Ziegler AG, Rewers M, Simell O et al (2013) Seroconversion to multiple islet autoantibodies and risk of progression to diabetes in children. JAMA 309(23):2473–2479

Current Approaches and Future Prospects for the Prevention of β-Cell Destruction in Autoimmune Diabetes

38

Carani B. Sanjeevi and Chengjun Sun

Contents

Introduction	1082
Antigen-Based Therapy	1084
Insulin	1084
Insulin and Cholera Toxin	1085
GAD65	1086
DiaPeP277	1087
Monoclonal Antibody-Based Therapy	1088
Anti-CD3 Antibodies	1088
Improving the Existing anti-CD3 Antibody Therapy	1089
Anti-CD20 Antibodies	1090
CTLA4 Immunoglobulin (CTLA4 Ig)	1090
IL-1 antagonist and anti-IL-1β Antibody	1091
Anti-TNF-α Antibody	1091
Others (Anti-CD52 Antibody and Anti-CD2 Antibody)	1091
Immunomodulating Therapy	1092
Autologous Hematopoietic Stem Cell Transplantation	1092
Stem Cell Therapies	1093
Other Cell Therapies	1094
IFN-α and IL-2	1094
Anti-T-Cell Globulin	1095
Other Forms of Therapy	1095
DNA Vaccination	1095
DNA Vaccination with GAD65	1095
Microsphere-Based Vaccine	1096
Anti-Inflammatory Agents	1096
Vitamin D	1096
N-3 Polyunsaturated Fatty Acids and Other Dietary Supplements	1097
Other Drugs	1097
Past Trials	1099

C.B. Sanjeevi (✉) • C. Sun
Department of Medicine (Solna), Center for Molecular Medicine, Karolinska University Hospital, Solna, Stockholm, Sweden
e-mail: sanjeevi.carani@ki.se; chengjsun@gmail.com

Immunosuppression Drugs	1099
Nicotinamide	1099
BCG	1099
Future Directions	1101
Strategies on Islet Expansion	1102
Islet Regeneration	1102
Probiotic Approach	1103
Promising Therapies	1103
Key Points	1103
Cross-References	1103
References	1104

Abstract

Type 1 diabetes (T1D) is an autoimmune disease resulting from the destruction of pancreatic β-cells. The main aim of treatment for T1D should be to prevent β-cell destruction and preserve existing β-cells in individuals with progressive autoimmunity. This can be achieved in several ways, and in this chapter, the authors have reviewed recent approaches that are currently being tested in animal models and human T1D patients under the following categories: (i) - antigen-based therapy, (ii) antibody-based therapy, (iii) immunomodulating therapy, and (iv) other form of therapies.

Keywords

Type 1 diabetes • β-cell • Antigen-based therapy • Biologics • Immunomodulation

Introduction

Type 1 diabetes mellitus (T1D) results from autoimmune destruction of pancreatic β-cells. Autoimmunity is thought to occur in genetically predisposed individuals after exposure to one or more environmental triggers such as dietary factors, viral infections, etc. Infiltrating T cells, B cells, and NK cells in pancreatic islets initiate the autoimmune response and progressively destroy the insulin-producing β-cells. The entire process of β-cell destruction can take anywhere from a few months to a few years, finally resulting in hyperglycemia. HLA-DQ8 and HLA-DQ2 have been associated with high risk to T1D, and 89 % of newly diagnosed children from Sweden are positive for these HLA alleles (Sanjeevi et al. 1994, 1995a, b, 1996). Association of these HLA alleles with T1D has been shown to be inversely proportional to age (Graham et al. 1999).

Latent autoimmune diabetes in adults (LADA) is a slowly progressive form of autoimmune diabetes. Patients initially diagnosed as classical type 2 diabetics are identified as LADA according to the following criteria of the Immunology of Diabetes Society: (i) adult age (>30 years) at onset of diabetes, (ii) the presence

Table 1 List of therapeutics used in prevention of β-cell death in autoimmune diabetes

Therapy	Reference
1. Antigen-based therapy	
Insulin (subcutaneous)	DPT-1 Study Group 2002
Insulin (oral)	Skyler et al. 2005; Barker et al. 2007
Insulin (intranasal)	Achenbach et al. 2008; Nanto-Salonen et al. 2008
Insulin (B chain with adjuvant)	Orban et al. 2010
alum-formulated GAD65	Agardh et al. 2005; Ludvigsson et al. 2008, 2011; Wherrett et al. 2011
DiaPep277	Raz et al. 2001; Elias et al. 2006; Lazar et al. 2007; Ziegler et al. 2010; Buzzetti et al. 2011
2. Monoclonal antibody-based therapy	
Anti-CD3 (otelixizumab)	Keymeulen et al. 2005, 2010
Anti-CD3 (teplizumab)	Herold et al. 2002, 2005, 2009
Anti-CD20 (rituximab)	Pescovitz et al. 2009
CTLA4 Ig (abatacept)	Orban et al. 2011
Anti-IL-1 (IL-1R antagonist, anakinra and anti-IL-1β antibody, canakinumab)	Moran et al. 2013
Anti-TNF-α (etanercept)	Mastrandrea et al. 2009
3. Immunomodulating therapy	
Autologous hematopoietic stem cell transplantation	Voltarelli et al. 2007; Couri et al. 2009
Umbilical cord blood transfusion	Haller et al. 2009
PBMCs educated by cord blood stem cell	Zhao et al. 2012
IFN-α (ingestion)	Brod and Burns 1994; Brod et al. 1997; Rother et al. 2009
Anti-T-cell globulin	Eisenbarth et al. 1985; Saudek et al. 2004
4. Miscellaneous	
Classic anti-inflammatory agents	Hundal et al. 2002; Tan et al. 2002; van de Ree et al. 2003
Vitamin D	Bizzarri et al. 2010; Walter et al. 2010; Gabbay et al. 2012
Atorvastatin	Strom et al. 2012
Diazoxide	Radtke et al. 2010
5. Past trials	
Immunosuppression drugs	Elliott et al. 1981; Harrison et al. 1985; Behme et al. 1988; Bougneres et al. 1988; Silverstein et al. 1988; Cook et al. 1989
Nicotinamide	Yamada et al. 1982; Gale et al. 2004
BCG vaccination	Allen et al. 1999; Huppmann et al. 2005

of circulating islet autoantibodies, and (iii) lack of a requirement for insulin for at least 6 months after diagnosis.

Considering this sequence of events, preventing β-cell destruction is vital to preserving the residual β-cells in individuals with progressive β-cell loss and those at risk of developing T1D and LADA (referred to as autoimmune diabetes in adults). Antigen-specific and antigen-nonspecific immune therapies that aim to reduce islet cell autoimmunity are in different stages of clinical development. Recent insights into the autoimmune process are elucidating the etiology of autoimmune diabetes, conceivably identifying therapeutic targets. Stand-alone and/or combinational therapies that reduce autoimmunity in islets, regenerate β-cells, and restore insulin secretion are ongoing and appear to be the future of autoimmune diabetes intervention. Aggressive autoimmunity appears significantly earlier than overt disease, and therefore, pursuing therapeutic strategies before disease presentation should be beneficial for susceptible patients. Early intervention before the autoimmunity is initiated is the best. Second best is intervention after autoimmunity is initiated but before the disease becomes insulin dependent. The detailed discussion of immunological β-cell destruction can be found here as follows: "▶ Immu nology of β-cell Destruction" and "▶ Inflammatory Pathways Linked to β Cell Demise in Diabetes".

Preservation of β-cells is advantageous in autoimmune diabetes as it may significantly reduce both short- and long-term complications (hypoglycemia, retinopathies, etc.) while at the same time stabilize blood glucose levels and improve quality of life. To this end, pharmaceuticals are being developed using the available knowledge to generate target antigen-specific immune response. Ideally, tolerance induction would be a short time course, leading to a long-lasting tolerant stage, without debilitating the capability of the immune system to mount effective immune response against invading pathogens. In the following sections, the authors have discussed recent strategies employed to prevent β-cell destruction and preserve residual β-cells in autoimmune diabetic patients in the following categories: (i) antigen-based therapy, (ii) antibody-based therapy, (iii) other forms of therapy, and (iv) failed therapies in the past (summarized in Table 1).

Antigen-Based Therapy

Insulin

Autoimmunity against insulin in T1D has long been observed since the 1980s that T1D patient had circulation insulin autoantibodies (IAA) before the insulin treatment (Palmer et al. 1983). As one of the major autoantigens in T1D, insulin is also among the earliest used antigens to induce immune tolerance in T1D patients (to preserve β-cells) as well as T1D relatives (to prevent the disease). The diabetes prevention trial 1 (DPT-1) was performed to access the capability of insulin administered as injections to prevent T1D among T1D relatives. The study, however, failed to demonstrate any beneficial preventive outcome (2002). The insignificant outcome led to subsequent change in insulin administration in similar trials.

Oral tolerance is a term used to describe the immune tolerance, which can be induced by the exogenous administration of antigen to the peripheral immune system via the gut. The active suppression of low doses of administered antigen appears to be mediated by the oral antigen-generating regulatory T cells that migrate to lymphoid organs and to organs expressing the antigen, thus conferring suppression via the secretion of downregulatory cytokines including IL-4, IL-10, and TGF-β. Since Oral administration is one of the easiest ways to induce immune tolerance. Another prevention approach in the DPT-1 was to administer insulin orally in first-degree relatives of T1D patients; however, the treatment failed to delay or prevent T1D (Skyler et al. 2005). At the same time, it was also found that in DPT-1 trial, oral administration did not alter IAA levels over time in those already positive for IAA at the start of treatment (Barker et al. 2007).

Similar to oral tolerance, immune tolerance could also be induced by administration of antigen to the respiratory tract (mucosal tolerance). At first, insulin administration through the respiratory tract was developed as an alternative to subcutaneous insulin injection. However, the inhaled insulin is associated with increased risk of lung cancer (Gatto et al. 2012). The use of inhaled insulin to prevent β-cell destruction requires further safety studies. While at the same time, the safety of nasal insulin administration is well documented. However, nasal insulin administration in children carrying high-risk HLA (for T1D) soon after detection of autoantibodies failed to prevent or delay the disease (Nanto-Salonen et al. 2008).

Exposure of the nasal mucous membranes to insulin may also cause act like a vaccine effect whereby protective immune cells are stimulated and then counteract the autoreactive immune cells which damage the β cells. There is an ongoing trial that aims to determine if intranasal insulin can protect β cells and stop progression to diabetes in individuals who are at risk (NCT00336674). The Pre-POINT (Primary Oral/intranasal Insulin Trial) is a dose-finding safety and immune efficacy pilot study aiming primary prevention in children genetically at risk to T1D, using oral or intranasal insulin (Achenbach et al. 2008). The results from Pre-POINT, in the future can give us more information on the effectiveness of insulin tolerance induction in T1D.

While most of the above trials end with insignificant outcomes, trials using insulin with modifications was then carried out to treat or prevent T1D. A recent published phase I trial used intramuscular human insulin B chain in incomplete Freund's adjuvant for T1D treatment (Orban et al. 2010). After 2-year follow-up, although there was no statistical difference in stimulated C-peptide responses between treated and untreated patients, a robust insulin-specific humoral and regulatory T-cell (Treg) response was developed in treated patients (Orban et al. 2010). Results from long-term follow-up or trials using upgraded insulin modification can probably give us more data on the treatment.

Insulin and Cholera Toxin

A mechanism of tolerance induction that is currently showing promise is oral insulin conjugated to β-subunit of the cholera toxin (CTB) (Bergerot et al. 1997). It has been shown recently that oral administration of microgram amounts of antigen coupled to the CTB subunit can effectively suppress systemic T-cell reactivity in animal models. Bergerot et al. report that feeding small amounts (2–20 μg) of human insulin conjugated to CTB can effectively suppress β-cell destruction and clinical diabetes in adult nonobese diabetic (NOD) mice (Bergerot et al. 1997). The protective effect could be transferred by T cells from CTB-insulin-treated animals and was associated with reduced lesions of insulitis. Furthermore, adoptive cotransfer experiments show concomitant reduction in islet cell infiltration. These results suggest that protection against autoimmune diabetes can be achieved by feeding minute amounts of a pancreas islet cell autoantigen linked to CTB and appears to involve the selective migration and retention of protective T cells into lymphoid tissues draining the site of organ injury.

CTB subunit carries the insulin to the intestine and helps in the transfer of the insulin molecule across the intestinal barrier. The CTB conjugation also helps in the reduction of the dosage of insulin that can be administered orally without causing hypoglycemia. Further, this approach has also been tried successfully by intranasal administration. Both approaches have prevented the development of diabetes in the NOD mouse model of the autoimmune disease. CTB-insulin β-chain fusion protein produced in silk worms has been shown to suppress insulitis in NOD mice (Gong et al. 2007).

GAD65

Glutamic acid decarboxylase isoform 65 (GAD65) is a major autoantigen in T1D. Studies in NOD mouse have shown that destruction of islet β-cells was associated with T cells recognizing GAD65. Kaufman et al. showed that in NOD mice, intravenously injection of GAD65 before diabetes onset effectively prevents autoimmune β-cell destruction and reduce the development of spontaneous diabetes (Kaufman et al. 1993).

Diamyd AB evaluated this by using alum-formulated human recombinant GAD65 in LADA patients. They selected diabetic patients of both sexes aged 30–70 years, diagnosed with type 2 diabetes (T2D) and positive for GAD65 antibodies in their phase I/II trial. These patients were treated with either diet or oral tablets. A total of 34 patients and 13 controls were tested with 4, 20, 100, and 500 μg dose. This was injected subcutaneously twice but 4 weeks apart. No serious adverse effects were reported. In the follow-up, the C-peptide level (both fasting and stimulated) was significantly elevated in the group receiving 20 μg dose compared to placebo. Likewise, the HbA1c and mean glucose levels were significantly lowered in the 20 μg dose compared to placebo. The $CD4^+CD25^+$ T cells which reflect the increase in regulatory T cells associated with nondestructive response to β-cell were elevated in the 20 μg dose but not in other doses. All these findings were relevant even after a follow-up period of 24 months and 5 years (www.diamyd.com; Agardh et al. 2005). It is thought that the prevention of β-cell destruction and β-cell recovery is due to shifting of immune response from destructive to nondestructive which is mediated by the Diamyd GAD65 vaccine.

Subsequent phase IIb trials in Swedish T1D patients with alum-formulated GAD65 showed significant preservation of β-cell function 30 months after the first 20 μg dose administrations. It also induced antigen-specific T-cell population, cytokines involved in regulation of the immune system, and a long-lasting B cell memory, suggesting that modulation of general immune responses to GAD65 can be helpful in preserving residual β-cells (Ludvigsson et al. 2008). In the extended evaluation of the Swedish phase IIb trials, it showed that the alum-formulated GAD65 was able to delay the progressive β-cell destruction (Ludvigsson et al. 2011). However, the phase IIb trial in Canada did not significantly differ C-peptide levels between alum-formulated GAD65 treated and control groups at 1 year (Wherrett et al. 2011). Whether it is due to the shorter follow-up duration or the immunological difference between populations is not clear. A phase III trial recently concluded to verify the previous observed effect of alum-formulated GAD65 did not meet the end point when the results were analyzed from the European sites. If Sweden and Finland were excluded in the analysis, the results in the rest of the European sites showed significant end point. The reason why Sweden and Finland sites showed nonsignificant end point was because all the children in the trial had taken state-recommended H1N1 vaccine, which were not a part of the inclusion criteria. Even in Sweden, if the results were analyzed in those children who completed the study before the H1N1 vaccine was recommended by the state, significant result was obtained. It is not clear what H1N1 does to the protective effect of the alum-formulated GAD65 vaccine. Meanwhile, few studies are in progress like the phase II trial which is ongoing in Norway to see whether the difference in the effect of alum-formulated GAD65 is in association with enterovirus infections (NCT01129232). In supplementation with other anti-inflammatory drugs, another phase II trial is ongoing in Sweden to see the effect of using alum-formulated GAD65, vitamin D, and ibuprofen (NCT01785108).

Alum-formulated GAD65 is the only antigen-based vaccine candidate which has been shown to be effective in LADA. LADA is often misdiagnosed as type 2 diabetes and treated accordingly. This may lead to additional stress on an already declining β-cell mass (due to autoimmune destruction). Hence, diagnosis and treatment of LADA are vital.

DiaPeP277

Heat shock protein 60 (hsp60) is a 60 kDa protein which is one of the self-antigens in T1D. DiaPeP277 is a 24-amino-acid peptide which comprises 24 residues (437–460) analog to hsp60 (www.develogen.com). Administration of DiaPep277 in NOD mice arrested diabetes (Elias et al. 1990). A randomized double-blind phase II trial using DiaPeP277 in human subjects with newly onset disease (<6 months) resulted in preservation of the endogenous insulin production compared to the placebo group (Raz et al. 2001). In a follow-up study (Elias et al. 2006), the findings were reiterated. In both studies, immunomodulation was observed and associated with downregulation of Th1 cells and upregulation of IL-10 producing

T cells. The immune responses were antigen specific as T-cell responses to bacterial antigens remained unaffected. However, studies performed in children did not show any improvement in the preservation of β-cell function or metabolic control (Lazar et al. 2007).

The preliminary results from phase III trials of DiaPep277 showed that T1D patients treated with DiaPep277 maintained C-peptide level better than in placebo group. Meanwhile, patients in the treated group had lower HbA1c level (Ziegler et al. 2010). Subsequent analysis found that T1D adults with low- and moderate-risk HLA genotypes would benefit the most from the intervention with DiaPep277 (Buzzetti et al. 2011). Additional phase III trial is ongoing for further evaluation on the efficacy/safety (NCT01103284, NCT01898086) and treatment effect (NCT01460251) of DiaPep277.

Monoclonal Antibody-Based Therapy

Anti-CD3 Antibodies

Experiments in the early 1990s in NOD mice demonstrated that hamster-derived anti-CD3 monoclonal antibodies reversed diabetes in hyperglycemic mice (Chatenoud et al. 1994, 1997). In order to increase safety in future clinical application, Fc-mutated (Fc-nonbinding) monoclonal anti-CD3 antibodies were engineered and were found to be less mitogenic, but were equally tolerogenic compared to functional Fc anti-CD3 antibodies (Chatenoud et al. 1994, 1997). These series of experiments demonstrated several unique features of the antibody therapy. First, continuous immunosuppression was not required, and second, the ability of the antibody to reverse disease after hyperglycemia has occurred was demonstrated. Treated NOD mice were resistant to transfer of diabetes by diabetogenic spleen cells, implying the involvement of active immune regulation preventing diabetes (Chatenoud et al. 1997).

Two of the earliest monoclonal antibody specific for the CD3 T-cell epitope that were tested in clinical trials are the ChAglyCD3 antibody (known after as otelixizumab), having a single mutation (Asn→Ala) at residue 297 in the Fc region that prevents glycosylation, derived from rat YTH 12.5 antibody (Routledge et al. 1995), and the hOKT3Ala-Ala antibody (known after as teplizumab), having two mutations at residues 234 (Lue→Ala) and 235 (Lue→Ala) in the Fc region. This antibody is derived from OKT3 (Bolt et al. 1993). A 6-day otelixizumab treatment in newly diagnosed T1D patients was found to be associated with lower requirement of insulin, however, susceptibility to infection at 1-year follow-up time (Keymeulen et al. 2005). When these T1D patients were followed longer (4 years), the treatment with otelixizumab after their T1D diagnosis was still able to suppress the rise in insulin requirements (Keymeulen et al. 2010). At the same time, teplizumab was also showed to change the ratio of $CD4^+$ T cells to $CD8^+$ T cells within 3 months and subsequently preserved insulin production up to 5 years (Herold et al. 2002, 2009) with better clinical parameters found in teplizumab-treated

T1D patients (Herold et al. 2005). A current ongoing trial is undergoing to evaluate the effect of teplizumab to prevent or delay the onset of T1D in relatives determined to be at very high risk for developing T1D (NCT01030816).

The FcR-nonbinding anti-CD3 antibody (anti-CD3 antibody with a mutated Fc portion) therapy was effective only if the immune response was primed and ongoing. Locally, they target autoreactive T cells, and the strength of the T-cell receptor (TCR)/CD3 is important in determining the efficacy. Thus, it can be hypothesized that though CD3 is expressed on all T cells, anti-CD3 antibodies mediate signaling depending on the functional stage of the target T cell whether it is naive or effector or memory. Administration of anti-CD3 antibodies induces depletion of effector T cells in the target tissue and lymphoid organs. In the pancreas-draining lymph nodes, apoptosis is induced in effector T cells compared to regulatory T cells and resting T cells (Hirsch et al. 1988; Wong and Colvin 1991; Carpenter et al. 2000). Apoptotic effector T cells are engulfed and digested by phagocytes (macrophages and immature dendritic cells [DC]). These phagocytes secrete large amounts of transforming growth factor (TGF-β) which creates a noninflammatory environment and also plays a major role in the maturation of DCs. TGF-β production has been suggested and experimental data demonstrate that TGF-β is central to the tolerance induced by FcR-nonbinding anti-CD3 antibodies. TGF-β-neutralizing antibodies are shown to completely neutralize the tolerogenicity induced by anti-CD3 therapy (Belghith et al. 2003). Local production of TGF-β has been shown to have the capability to convert a proinflammatory environment to a noninflammatory and tolerogenic environment (Li et al. 2006). A high concentration of TGF-β also promotes upregulation of inhibitory receptor ligands (programmed cell death ligand 1, ICOS ligand) and downregulation of MHC and costimulatory molecules on antigen-presenting cells (Li et al. 2006; Rutella et al. 2006). This in turn induces the induction or expansion of $CD4^+CD25^+FOXP3^+$ T-regulatory cells (Treg) (Rutella et al. 2006). From the available experimental data, it has been proposed that the FcR-nonbinding anti-CD3 antibody treatment triggers a massive local production of TGF-β, by phagocytes engulfing activated effector T cells (You et al. 2008).

Improving the Existing anti-CD3 Antibody Therapy

Administration of drugs which promote β-cell survival and growth (such as exendin-4) may increase the β-cell growth and replication in the "tolerant" environment. In NOD mice, combination of exendin-4 and anti-CD3 monoclonal antibodies led to effective reversal or the disease with increased insulin content of the β-cell as compared with individual exendin-4 or anti-CD3 monoclonal antibody treatment (Sherry et al. 2007). Frequent side effects because of interferences with the T-cell population in proximity with treatment periods and recurrent autoimmunity might be a problem in anti-CD3 antibody-treated individuals. Repetitive treatment can be a possible way out in such a situation but formation of anti-idiotypic antibodies should be taken into consideration.

Anti-CD20 Antibodies

B cells constitute about 60–70 % of the immune cells infiltrating the pancreatic islets (Green and Flavell 1999). Until recently B cells were thought to play an important role in priming T cells (Wong and Wen 2005). However, a recent study showed for the first time that B cells promote the survival of $CD8^+$ T cells in the islets and thereby promote the disease (Brodie et al. 2008). CD20 is a cell surface marker expressed on all mature B cells. Rituximab (Roche/Genentech), a humanized anti-CD20 monoclonal antibody (CD20 mAb), has been shown to successfully deplete human B cells from peripheral circulation via mechanisms involving Fc- and complement-mediated cytotoxicity and probably via proapoptotic signals (Rastetter et al. 2004; Martin and Chan 2006). Given the important role of B cells in the pathogenesis of T1D, depleting B cells is a very interesting therapeutic option. Transgenic NOD mice engineered to express human CD20 on B cells, when treated with a single dose of CD20 mAb, gave interesting results (Hu et al. 2007; Xiu et al. 2008). First, treatment of mice in the early stage of the disease (insulitis) prevented or delayed the progression to disease; second, clinical hyperglycemia could be reversed in over one-third of the experimental animals; third, B cell levels were restored to pre-depletion levels within 3 months of treatment, but the progression to T1D was delayed almost indefinitely. Recent published data from TrailNet anti-CD20 study group showed that four infusions of rituximab in the first month after diagnosis of T1D can preserve β-cell function at 1-year follow-up (Pescovitz et al. 2009). A recent clinical trial (NCT01280682) is ongoing to investigate further the immunomodulating role of rituximab in the treatment of T1D.

CTLA4 Immunoglobulin (CTLA4 Ig)

Cytotoxic T-lymphocyte antigen 4 (CTLA4) is expressed on the surface of T-helper (Th) cells. CTLA4 binds CD80 and CD86 on the antigen-presenting cells and blockades the most important second signal (costimulation signal) for full activation of T cells transduced by the binding between CD80/CD86 and CD28 on the T cells. Thus, CTLA4 functions as a negative regulator of T-cell activation, and the costimulation blockade has been proposed as a therapeutic modality for autoimmunity and transplantation (Bluestone et al. 2006; Teft et al. 2006). The CTLA4 Ig (or lately named abatacept) is a fusion protein of human CTLA4 and the Ig–Fc region designed to bind CD80/CD86 and block the T-cell costimulatory signal (Lenschow et al. 1992). A study in NOD mice showed that costimulatory blockade with CTLA4 Ig fusion protein prevented diabetes when administered before overt diabetes (Lenschow et al. 1995). A recent phase II trial showed that abatacept infusion regularly after T1D diagnosis can preserve β-cell function throughout the 2 years during the investigation (Orban et al. 2011). Further trial is ongoing (NCT01773707) to determine whether the infusion of abatacept can delay or prevent the development of T1D among T1D relatives at risk.

IL-1 antagonist and anti-IL-1β Antibody

Interleukin-1β (IL-1β) is a proinflammatory factor secreted by several cell types in response to tissue insult. It has been shown that IL-1β bound to pancreatic β-cell interleukin-1 (IL-1) type 1 receptors (IL-1R), IL-1 induces β-cell dysfunction and apoptosis through mitogen-activated protein kinase pathways (Mandrup-Poulsen et al. 2010). In addition, IL-1β enhances expansion and survival of T cells, promotes differentiation of T cells toward pathological phenotypes, and enables effector T cells to proliferate despite the presence of Tregs (Dinarello et al. 2012). These evidence makes blockade of IL-1β is an attractive therapeutic target. A recent report combined results from two independent studies blocking IL-1 (anakinra, IL-1 receptor antagonist, and canakinumab, anti-IL-1β mAb) showed that both treatment did not preserve C-peptide level at 9-month or 12-month time (Moran et al. 2013). Although future results of long follow-up duration might give us more information, the investigator questioned that the IL-1 blockade treatment should be given as prevention for T1D before T1D onset. However, up to date, there is no registered trial investigating the preventive effect of IL-1 blockade treatment in T1D. There are currently trials investigating the use of other IL-1 blockade in the treatment of T1D (rilonacept, IL-1 Trap, NCT00962026, and gevokizumab, anti-IL-1β antibody, NCT01788033 and NCT00998699).

Anti-TNF-α Antibody

It was found in NOD mouse that TNF-α mRNA is produced by $CD4^+$ T cells within inflamed islets during the development of diabetes (Held et al. 1990). In vitro models show that TNF-α potentiates the destruction of β-cells by other cytokines (Mandrup-Poulsen et al. 1987). Transgenic mice with increased β-cell expression of TNF-α have significant lymphocytic insulitis, which is abrogated in TNF receptor knockout mice (Herrera et al. 2000). It was also indicated clinically that anti-TNF therapy may induce IL-10-secreting regulatory cells with a consequent resolution of the inflammation in the pancreatic islets (Arif et al. 2010). Etanercept is a recombinant fusion protein of TNF receptor to the constant end of the IgG1 antibody that binds to TNF-α, thereby clearing the TNF-α from circulation. The usage of etanercept twice a week subcutaneously resulted in lower A1C and increased endogenous insulin production at 6-month follow-up (Mastrandrea et al. 2009). However, long-term effect of etanercept in T1D is currently unknown and awaiting future investigations.

Others (Anti-CD52 Antibody and Anti-CD2 Antibody)

Mycophenolate mofetil (MMF) is rapidly absorbed after oral administration and hydrolyzed to MPA, an inhibitor of inosine monophosphate dehydrogenase that inhibits guanosine nucleotide synthesis and thus inhibits T and B cell proliferation with no obvious effect on other cell types. It was found that MMF was effective in diabetic animal models (Hao et al. 1992, 1993). Recent study in DRBB rat model demonstrated a synergistic effect of MMF with daclizumab in the treatment of diabetes (Ugrasbul et al. 2008). Daclizumab is a humanized monoclonal antibody that binds to CD25, the α subunit of the interleukin-2 (IL-2) receptor on the surface of activated lymphocytes, functioning to arrest the proliferation of activated lymphocytes. However, clinical trial in human using MMF and daclizumab or alone did not have an effect on the loss of C-peptide in subjects with new-onset T1D (Gottlieb et al. 2010). In addition, the increased risk of virus infection after coadministration of MMF and daclizumab might impede further investigations (Loechelt et al. 2013). An ongoing trial using daclizumab alone (NCT00198146) is under investigation.

Alefacept is a soluble LFA3/IgG1 fusion protein that binds CD2 on T cells. Alefacept was found recently to be efficacious in treatment of chronic psoriasis (Ellis and Krueger 2001). Alefacept was found to downregulate circulating memory ($CD45RO^+$) T cells (da Silva et al. 2002). The trial using alefacept in the treatment of T1D (NCT00965458) is ongoing.

Immunomodulating Therapy

Autologous Hematopoietic Stem Cell Transplantation

It was shown as early as in the late 1980s that diabetes in NOD mice may be prevented by allogeneic transplantation of hematopoietic stem cells (HSC) from a non-disease-prone strain (LaFace and Peck 1989). In the 1990s, it was demonstrated that environmentally induced animal autoimmune diseases such as experimental autoimmune encephalomyelitis and adjuvant arthritis could be cured by syngeneic, autologous, or pseudo-autologous (using syngeneic animals in the same stage of disease as the recipient) HSC transplants (Burt et al. 1995; Kroger et al. 1998; Karussis et al. 1999). These studies in animal models indicated that HSC transplantation can probably reintroduce tolerance to autoantigens. Indeed, in a recent published uncontrolled clinical trial, HSC transplantation procedure in human showed that it is associated with more insulin independence and lower insulin usage and increased C-peptide level (Couri et al. 2009). Aside from classic HSC transplantation, autologous non-myeloablative hematopoietic stem cell transplantation also showed effectiveness in slowing C-peptide decrease, decreasing GAD65 antibody, and better diabetic metabolism measurements (Voltarelli et al. 2007). However, since earlier trials had shown that low-dose immunosuppression can induce a slow decline or even improvement in C-peptide level (see below in section "Immunosuppression Drugs"), it was argued whether the effect is due to repopulation of the immune cells or the high-dose immunosuppression itself before the transplantation. There are two ongoing trials in China (NCT01341899 and NCT00807651) that might provide more knowledge on the HSC transplantation in T1D.

While the transplantation of HSC was showed to halt the autoimmune destruction of β-cells, research and trials on the transplantation of islets heightened further the importance of transplantation in the treatment of T1D (discussed elsewhere: "▶ Human Islet Autotransplantation," "▶ Islet Encapsulation," "▶ Islet Xenotransplantation: Recent Advances and Future Prospect," and "▶ Successes and Disapointments with Clinical Islet Transplantation").

Stem Cell Therapies

Mesenchymal stem cells (MSCs) were found to be able to protect NOD mice from diabetes by induction of Tregs (Fiorina et al. 2009; Madec et al. 2009). On one side, it was shown that MSC and PBMC mixture is capable of switching T-cell response from Th1 to Th2 when it encounters autoantigen GAD65 (Zanone et al. 2010), and on the other hand, MSCs were found to be able to home to injured tissue and stimulate endogenous islet regeneration (Yagi et al. 2010; Bell et al. 2012). Being capable of shooting two birds with one stone, MSC attracts researchers' interest and there are several ongoing trials investigating different types of MSCs (NCT00690066, NCT01322789, NCT01374854, NCT01068951, NCT01219465, NCT01496339).

In addition to MSC, immature unprimed functional regulatory T cells (Tregs) were found abundant in the umbilical cord blood (Godfrey et al. 2005). These highly functional Tregs might limit inflammatory cytokine responses and anergize effector T cells in autoimmune processes (Fruchtman 2003). Thus, the umbilical cord blood has become a focus of researchers to design cell-based therapies for T1D patients. However, umbilical cord blood transfusion did not demonstrate efficacy in preserving C-peptide in the newly diagnosed after 1 year of transfusion (Haller et al. 2009). Larger randomized studies as well as 2-year post-infusion follow-up of this cohort are needed to determine whether autologous cord blood-based approaches can be used to slow the decline of endogenous insulin production in children with T1D is an idea source to be transplanted to patients with T1D. An ongoing trial is investigating the effectiveness of autologous cord blood transfusion in the treatment of T1D (NCT00989547). At the same time, the combination of MSC transplantation and umbilical cord blood transfusion is also under investigation (NCT01143168).

It has been reported in NOD mouse model that the autologous $CD4^+CD62L^+$ Tregs co-cultured with the human cord blood stem cells (CB-SC) can eliminate hyperglycemia and promote β-cell regeneration (Zhao et al. 2009). Although the underlying mechanism is not clear, it was proposed that CB-SC could suppress the proliferation of β-cell-specific autoreactive T cells (Zhao et al. 2007). The phase I trial was published recently using CB-SC-treated peripheral lymphocytes (stem cell educator) reinfusion to patients with existing T1D (Zhao et al. 2012). The stem cell educator showed its safety and capability of persistently improving metabolic control after a single treatment (Zhao et al. 2012). A phase II trial (NCT01350219) is currently ongoing to investigate the effectiveness of this method in the treatment of T1D.

The other usage of stem cell therapies prospered in the production of islets in vitro as well as in vivo. The in vitro islets generated from human embryonic stem cells and its potential clinical implementation is further discussed in "▶ Making Islets From Human Embryonic Stem Cells". Potential usage of in vivo stem cells is discussed in "▶ Stem Cells in Pancreatic Islets".

Other Cell Therapies

Dendritic cell (DC) is one of the most important antigen-presenting cells that also showed immunoregulatory characteristics (Hackstein et al. 2001). Studies in mouse models showed that treating NOD mice dendritic cells ex vivo with antisense oligonucleotides targeting the primary transcripts of CD40, CD80, and CD86 (costimulatory molecules) can downregulate those cell surface molecules, thus increasing $CD4^+CD25^+$ T cells (Tregs) in NOD recipients (Machen et al. 2004). This regulation effect in mice was then found to be mediated by IL-7 produced by the treated DCs (Harnaha et al. 2006). The trial using this method (NCT00445913) is ongoing and awaiting results.

Tregs (T-regulatory cells) can act through DC to prevent autoreactive T-cell differentiation, thus preventing or slowing down the progression of autoimmune diseases (Tang and Bluestone 2006). The shortage of Tregs can lead to the development and accumulation of Treg-resistant pathogenic T cells in patients with autoimmune diseases (Tang and Bluestone 2006). Thus, restoration of self-tolerance using Tregs in these patients will likely be able to control ongoing tissue injury. A recent report showed the feasibility of expanding Tregs isolated from patient with recent-onset T1D (Putnam et al. 2009). The ongoing trial using autologous expanded Tregs transfusion in T1D patients (NCT01210664) will investigate the effectiveness of this hypothesis.

IFN-α and IL-2

Early trial showed that parenteral IFN-α (interferon-α) provided no benefits in patients with newly diagnosed T1D patients (Koivisto et al. 1984). However, it was found later that ingested IFN-α has immunomodulatory effect in experimental autoimmune animal model and in multiple sclerosis in humans (Brod and Burns 1994; Brod et al. 1997). In a recent phase I trial using ingested human recombinant IFN-α in T1D, it was found that 5,000 units of IFN-α administered once daily in newly diagnosed T1D patients for 1 year could maintain more β-cell function (Rother et al. 2009). Future study is required to verify the results and pave the way for clinical use of IFN-α.

Interleukin 2 (IL-2) is a dependant cytokine for Tregs to maintain viability and function. IL-2 down signaling events activate Akt/Erk pathway and targets CTLA4 gene. At the same time, IL-2 also signals through STAT5 pathway and activates FOXP3 and CD25 genes (Hulme et al. 2012). Reduction in IL-2 in T1D may contribute to Treg decline. Aldesleukin (Proleukin) is a commercialized IL-2 and currently under the investigation of several trials (NCT01353833, NCT01827735, NCT01862120).

Anti-T-Cell Globulin

Early clinical experience on immunomodulating therapy using anti-T-cell globulin (ATG) in T1D showed effectiveness in delaying and lowering insulin requirement (Eisenbarth et al. 1985) A recent clinical trial using polyclonal ATG confirmed that short-term ATG therapy in newly diagnosed T1D contributes to the preservation of residual C-peptide production and to lower insulin requirements at 1 year after diagnosis (Saudek et al. 2004). A current ongoing trial (NCT00515099) is further investigating the therapeutic effect of ATG in T1D.

Other Forms of Therapy

DNA Vaccination

DNA vaccination involves administration of a gene that encodes the target antigen, instead of the antigen as in classical vaccination. Variety of vectors can be used to transfer the target gene as DNA or RNA, along with genes encoding immunomodulatory molecules. Several studies have been performed using administration of plasmids encoding antigens such as insulin B chain, GAD65, and immunoglobulin G–Fc fusion constructs in animal models. However, plasmids carry unmethylated CpG motifs (ISS, immunostimulatory sequences) which activate the innate immune system. Therefore, DNA vaccination against T1D should block or overcome the effect of such stimulatory elements. DNA vaccine hold good promise in treatment of autoimmune diseases as they have been used, in experimental models, to direct the immune response toward a Th1 or a Th2 response (Prud'homme 2003).

DNA Vaccination with GAD65

Intramuscular injections of plasmid containing GAD65 fused with IgG–Fc and IL-4 were reported to generate a GAD65-specific Th2 response, protecting NOD mice from developing T1D (Tisch et al. 2001). A study performed to evaluate two different modes of delivery of a plasmid coding for GAD65 reported the elicitation of IL-4 secreting T-cell response. Two methods of plasmid delivery, intramuscular and a novel gene gun method, were tested in this study. Intramuscular injections fail to stop the ongoing β-cell autoimmunity, whereas the gene gun method was successful in eliciting immunomodulation, significantly delaying the disease onset in NOD mice (Goudy et al. 2008).

Microsphere-Based Vaccine

Microparticulate carriers have the capability to shape the functional phenotype of dendritic cells (DC) (Waeckerle-Men et al. 2006; Yoshida and Babensee 2006). A nucleic acid-based vaccine using antisense oligonucleotides coated on microspheres, directed against CD40, CD80, and CD86 (costimulatory molecules important in DC maturation), has been shown to prevent T1D in NOD mice as well as reverse new-onset disease (Phillips et al. 2008). Microspheres administered are taken up by DCs by phagocytosis. Inflammation in the pancreatic islets associated with β-cell apoptosis is suggested to drive the antisense oligonucleotide-loaded DCs to acquire the β-cell antigen(s). This is followed by the accumulation of these DCs in the pancreatic lymph nodes, where they are hypothesized to interact with regulatory T cells inducing a β-cell-specific immune hyporesponsiveness or functional tolerance to β-cell antigens (Tarbell et al. 2006). The detailed mode of action of the microsphere-based vaccine is yet to be established and clinical trials in human subjects will decide the efficacy of this approach in the prevention of T1D.

Anti-Inflammatory Agents

Use of anti-inflammatory drugs such as aspirin (Hundal et al. 2002), statins (Tan et al. 2002), and glitazone (van de Ree et al. 2003) has been shown to be beneficial in type 2 diabetes. These drugs have been shown to have anti-inflammatory effect by affecting either the signaling pathways (such as NFkB signaling) or cytokines involved in inflammation. Such drugs can be vital in bringing down the overall islet inflammation and thereby creating a better islet environment which can respond to other forms of treatment.

Newly developed anti-inflammatory drugs have joined the tide against T1D recently. α-1 antitrypsin or α1-antitrypsin (AAT) is a naturally occurring anti-inflammatory glycoprotein; AAT is a protease inhibitor belonging to the serpin superfamily. AAT has been shown to facilitate Treg expansion in the NOD mice mode, protecting the mice from diabetes (Koulmanda et al. 2008). AAT alters CCR7 expression on DC surface; thus, promoting semimature DC migration to the lymph nodes subsequently activates Tregs (Ozeri et al. 2012). There are two trials now investigating AAT in the treatment of T1D (NCT01319331, NCT01661192).

Vitamin D

Vitamin D has been shown to suppress proinflammatory responses by suppressing enhanced activity of immune cells taking part in autoimmune processes. In NOD mice, vitamin D has been shown to prevent autoimmune diabetes (Mathieu et al. 1994). Supplementation of vitamin D has been shown to be protective in children against T1D. High dosage and the timing of the dose have also been shown to play a role. A randomized open-label, pilot trial is currently under way

(NCT00141986), where increased dose of vitamin D (2,000 IU/day instead of the current practice of 400 IU/day) is administered to children genetically at risk of developing T1D. However, trials with vitamin D in new-onset T1D have shown mixed results, with one showing benefit (Gabbay et al. 2012) while the other two did not (Bizzarri et al. 2010; Walter et al. 2010). The dosing and timing of the treatment using vitamin D require future studies.

N-3 Polyunsaturated Fatty Acids and Other Dietary Supplements

It has been shown early that synthesis of IL-1β, IL-1α, and tumor necrosis factor can be suppressed by dietary supplementation with long-chain n-3 fatty acids (Endres et al. 1989). An ongoing trial (NCT00333554) investigating the effect of long-chain n-3 fatty acids in the prevention of T1D might provide us more useful information.

In addition, study using streptozotocin-treated diabetic rats showed that chromium supplementation lowered blood levels of proinflammatory cytokines. Although there is no benefit in plasma glucose level from chromium supplement found in the animal study (Vinson 2007), the investigators believed T1D patients could benefit from chromium supplement. The effect of chromium supplement human T1D is currently under investigation (NCT01709123).

Other Drugs

Lansoprazole and other proton-pump inhibitors consistently elevated serum gastrin levels (Ligumsky et al. 2001). It has already been found as early as the mid-1950s that gastrin has the potential to increase new β-cell formation (Zollinger and Ellison 1955). Thus, it was hypothesized that lansoprazole could probably induce β-cell regeneration by increasing serum gastrin level. There are two current trials investigating the safety of coadministration of cyclosporin and lansoprazole among patients with existing T1D (NCT01762657) and newly diagnosed T1D (NCT01762644).

The lipid-lowering drug atorvastatin was also found to have immunomodulating effect in rheumatoid arthritis intervention trials (McCarey et al. 2004). Atorvastatin treatment was found to be able to preserve β-cell function in T1D patients with median inflammation mediator levels (Strom et al. 2012). There is an ongoing trial (NCT00529191) which might give further information on the effect of atorvastatin treatment in T1D. Similarly, another lipid-lowering drug simvastatin is also under investigation in a current trial (NCT 00441844).

Imatinib is originally designed as a specific inhibitor of Abl protein tyrosine kinases and used in the treatment of chronic myeloid leukemia (Druker et al. 1996). Recent studies showed that imatinib had a strong anti-inflammatory effect by inhibiting TNF-α production in macrophages (Wolf et al. 2005). A recent study showed that both imatinib and sunitinib treatments led to durable remission in NOD mice (Louvet et al. 2008). The underlying mechanism is probably due to the multikinase inhibiting characteristics of imatinib and sunitinib which inhibited platelet-derived growth factor receptor (PDGFR) (Louvet et al. 2008). The phase I trial using imatinib for the treatment of T1D (NCT01781975) is planned and awaiting participant recruitment.

Table 2 Ongoing trials

Therapy	NCT number
1. Antigen-based therapy	
Insulin (intranasal, prevention)	NCT00336674
alum-formulated GAD65	NCT01122446, NCT01129232, NCT01785108
DiaPep277	NCT01103284, NCT01898086, NCT01460251
2. Monoclonal antibody-based therapy	
Anti-CD3 (teplizumab, prevention)	NCT01030816
Anti-CD20 (rituximab)	NCT01280682
CTLA4 Ig (abatacept)	NCT01773707
Anti-IL-1 (IL-1 Trap, rilonacept)	NCT00962026
Anti-IL-1 (IL-1β antibody, gevokizumab)	NCT01788033, NCT00998699
Anti-CD25 (daclizumab)	NCT00198146
Anti-CD2 (alefacept)	NCT00965458
3. Immunomodulating therapy	
Autologous hematopoietic stem cell transplantation	NCT01341899, NCT00807651
Mesenchymal stem cells transfusion	NCT00690066, NCT01322789, NCT01374854, NCT01068951, NCT01219465, NCT01496339
Umbilical cord blood transfusion	NCT00989547, NCT01143168
PBMCs educated by cord blood stem cell	NCT01350219
Dendritic cell treated ex vivo	NCT00445913
Tregs expanded ex vivo	NCT01210664
IL-2	NCT01353833, NCT01827735, NCT01862120
Anti-T-cell globulin	NCT00515099
4. Miscellaneous	
Novel anti-inflammatory agents (AAT)	NCT01319331, NCT01661192
Vitamin D (prevention)	NCT00141986
N-3 polyunsaturated fatty acid (dietary supplement)	NCT00333554
Chromium (dietary supplement)	NCT01709123
Lansoprazole	NCT01762657, NCT01762644
Atorvastatin and simvastatin	NCT00529191 and NCT 00441844
Imatinib	NCT01781975

Diazoxide is a potassium channel activator that is frequently used in the treatment for hypertension. It was found that diazoxide can provide β-cell rest by reversibly suppressing glucose-induced insulin secretion through opening ATP-sensitive K^+ channels in the β-cell (Trube et al. 1986). Early trial using diazoxide in T1D showed that those treated T1D displayed higher residual insulin secretion than the placebo group (Bjork et al. 1996; Ortqvist et al. 2004). However, a recent trial did not observe the preservation effect from diazoxide, although better metabolic control was found among diazoxide-treated T1D patients (Radtke et al. 2010). More studies in the future with regard to diazoxide function may unveil more on the effect of diazoxide as well as T1D pathogenesis. Ongoing clinical trials for new therapy in managing type 1 diabetes is listed in Table 2.

Past Trials

Immunosuppression Drugs

Cyclosporin was one of the first immunosuppressive drugs used in treatment of T1D, which could delay the onset of the disease (Bougneres et al. 1988). However, cyclosporin achieved immunosuppression by targeting intracellular processes, which is nonspecific and unrelated to autoantigens involved in the disease. Withdrawal of the treatment resulted in invariable recurrence of the pathogenic immune response. Considering the nephrotoxic potential of the drug, it was not a choice of long-term treatment, and therefore, it was not considered for therapy (Behme et al. 1988).

Other early used drugs are prednisone (Elliott et al. 1981), azathioprine (Harrison et al. 1985; Cook et al. 1989), and coadministration of prednisone plus azathioprine (Silverstein et al. 1988). Although these drugs showed slow decline or even improvement in C-peptide level, they were not used in clinic afterward due to considerable side effects.

Nicotinamide

The European Nicotinamide Diabetes Intervention Trial (ENDIT) tested the efficacy of nicotinamide in preventing diabetes in human subjects. Previous studies in animal models demonstrated that the administration of nicotinamide can prevent T1D (Yamada et al. 1982). Nicotinamide is speculated to confer protection by inhibiting DNA repair enzyme poly-ADP-ribose polymerase and prevent the depletion of β-cell NAD. However, in the ENDIT, nicotinamide treatment did not result in successful prevention of T1D (Gale et al. 2004).

BCG

Bacille Calmette–Guerin (BCG) vaccination has been proposed as an adjuvant therapy to prevent T1D. A study reported that administration of BCG vaccination soon after T1D onset preserves β-cell function (Allen et al. 1999). However, this was not the case in the trials that followed. BCG vaccination could not prevent the development of T1D in children genetically at risk (Huppmann et al. 2005).

Ongoing Prediction Studies

Several international collaborative efforts are under way. These studies will identify potential population/risk groups who would benefit from various therapies for prevention of β-cell death.

Potential therapies aiming at prevention of β-cell death would directly benefit patients suffering from autoimmune diabetes (T1D/LADA). Successful therapies can also benefit prediabetics, first-degree relatives of T1D patients, and individuals at risk of developing autoimmune diabetes.

TEDDY: The Environmental Determinants of Diabetes in the Young (TEDDY) study are an effort to screen more than 360,000 children around the world to the environmental factors that might play a role in T1D pathogenesis (2008). Several genome-wide association scans have been completed and are under way, with an aim to identify the T1D risk loci across the human genome. Identification of environmental and genetic factors involved in the etiology of T1D can broaden the scope of therapeutic interventions.

TrailNet: It is an international consortium of clinical research centers working toward achieving prevention of T1D (Skyler et al. 2008).

TRIGR: Trial to reduce T1M in the genetically at-risk (TRIGR) study is another collaborative effort, which aims at testing the hypothesis that weaning to an extensively hydrolyzed infant formula will decrease the incidence of T1D in children who carry high-risk HLA and in those who have a first-degree relative with T1D (2007). Initial findings from TRIGR suggest that introduction of cow's milk at an early age in children with dysfunctional gut immune system might result in aberrant immune response, leading to T1D (Luopajarvi et al. 2008).

DAISY: The DAISY study (the Diabetes Autoimmunity Study in the Young) aims at elucidating the interaction between genes and the environment that can trigger T1D. Children who are genetically at risk or those who have a first-degree T1D relative are being studied and followed up.

BABY-DIAB: BABY-DIAB is (Roll et al. 1996) a prospective study conducted from birth among children of mothers with T1D or gestational diabetes or fathers with T1D to investigate the temporal sequence of antibody responses to islet cells (ICA), insulin (IAA), GAD65 (GADA), and the protein tyrosine phosphatase IA-2/ICA512 (IA-2A). A total of 78.6 % of children (17,055 out of 21,700) born in the southeast of Sweden were entered in the *ABIS* (All Babies in Southeast Sweden) study with an aim to study environmental factors affecting the development of immune-mediated diseases in children, with special focus on T1D (Ludvigsson et al. 2002).

DIPP: The DIabetes Prediction and Prevention Project was launched in 1994 in Finland. In the study, general population newborns are screened for increased genetic risk for T1D in the University Hospitals of Turku, Tampere, and Oulu.

BABYDIET: BABYDIET study (Schmid et al. 2004) is a primary prevention trial in Germany initiated to investigate whether delay of the introduction of dietary gluten can prevent the development of islet autoimmunity in newborns with a first-degree relative with T1D, who are at genetically high risk of T1D. However, the result showed that delaying gluten exposure until the age of 12 months is safe but does not substantially reduce the risk for islet autoimmunity in genetically at-risk children (Hummel et al. 2011).

TIRGR: The Trial to Reduce IDDM in the Genetically at Risk (TRIGR) study is an international randomized double-blind controlled intervention trial that was designed to establish whether weaning to a highly hydrolyzed formula in infancy reduces the risk of T1D later in childhood (Group 2007). A recent report from the study showed that dietary intervention using casein hydrolysate formula had a long-lasting effect on markers of β-cell autoimmunity (Knip et al. 2010). And the difference in fecal microbiota composition between children with β-cell autoimmunity and those without has been found (de Goffau et al. 2013). Further trial (NCT01735123) is ongoing to investigate whether extensively hydrolyzed casein formula is able to protect children at risk for type 1 diabetes.

Future Directions

Intervention/prevention of β-cell destruction in T1D is the final goal resulting in good metabolic control of blood glucose. Balancing the risks and benefits in intervention/prevention of T1D is very complicated. Individual response to a particular therapy might differ. Biomarkers which can identify individuals who would or would not respond to a particular therapy are the need of the hour. T1D is associated with end-organ complications. The number of adverse events in an individual undergoing a particular therapy might differ from another, depending upon the time and intensity of progress to end-organ complications. Therefore, identification of those at risk becomes important while considering therapy.

Disease diagnosis is another important factor. T1D is usually diagnosed when the existing β-cells fail to meet the insulin needs of the body and thereby insufficient metabolic control. Earlier identification of existing autoimmunity is very crucial.

Research in the past few decades has highlighted many ways in which this can be achieved. Several promising candidates (such as alum-formulated GAD65 and anti-CD3 antibodies) have also reached different stages of clinical trials. Alum-formulated GAD65 was tested in several phase III clinical trials, including a trial in nine countries in Europe apart from Norway and a 4-year follow-up in Sweden (www.diamyd.com). It is interesting to note that although alum-formulated GAD65 and anti-CD3 seem to show similar efficacy the alum-formulated GAD65 product has not been associated with any relevant side effects and moreover it is easy to administer. Considering the complex etiology of the disease, involving several susceptibility factors and immune cells, it is possible that multi-therapy, involving more than one therapeutic agent, may be of advantage. With the increasing insights into the etiology of the disease, more and more targets are being identified for prevention/intervention.

Strategies on Islet Expansion

Nutrient ingestion stimulates the gastrointestinal tract to secrete incretin hormones to enhance glucose-dependent insulin secretion, thereby maintaining glucose homeostasis. The success of several therapies in reversing islet cell autoimmunity has led to the search of agents that enhance β-cell preservation or restoration. The safety of the combined usage of the above strategies was under investigation (NCT00873925, NCT00064714) which might give exciting results in the near future.

Glucagon-like peptide-1 (GLP-1) is a gut hormone secreted from the intestinal L cells. GLP-1 has a very short circulating half-life due to rapid inactivation by the enzyme dipeptidyl peptidase IV (DPP-4). Since GLP-1 has been well identified as an insulin stimulator and glucagon inhibitor, both GLP-1 and DPP-4 inhibitor were widely tested in the treatment for T2D (Mari et al. 2005; Mu et al. 2006; Duttaroy et al. 2011). It was recently found that GLP-1 peptide could not only induce β-cell proliferation and neogenesis but also suppress β-cell apoptosis and delay the onset of T1D in mouse model (Hadjiyanni and Drucker 2007; Zhang et al. 2007; Hadjiyanni et al. 2008; Xue et al. 2010). Meanwhile, clinical trials showed that GLP-1 improve glucose control in T1D patients (Behme et al. 2003). Sitagliptin (a DPP-4 inhibitor) improved glucose control in T1D patients (Ellis et al. 2011). There are several ongoing trials investigating GLP-1 agonist (NCT01722227, NCT01722240 and NCT01879917), DPP-4 inhibitor (NCT00813228, NCT01159847, NCT01099618 and NCT01559025), and the co-application of both GLP-1 agonist and DPP-4 inhibitor (NCT01782261). The effect of GLP-1 in T1D might soon be revealed in the near future.

Islet Regeneration

When the treatment aimed to suppress autoimmunity is developing rapidly, treatment aimed to induce islet/β-cell regeneration is also under way. Currently, there are several stem cell therapies under different stages of investigation. An ongoing trial (NCT00465478) is using autologous bone marrow stem cell transplantation to stimulate islet stem cell regeneration. Similarly, another trial (NCT00703599) using autologous adipose-derived stem cells to stimulate islet stem cell regeneration is also under investigation. The combination of stem cell therapies is also interesting, a current trial (NCT01143168) combining bone marrow mononuclear cells and umbilical cord MSC for treating T1D patients is also ongoing.

Probiotic Approach

Identification of the role of environmental agents (viruses and more recently bacteria) and their potential use as therapeutics throws open a vast range of possibilities. Use of food supplements or even a probiotic yoghurt containing "friendly bacteria" in prevention of autoimmune diabetes has been suggested. The idea seems farfetched but considering the influx of information on the disease etiology, it is not completely impractical; however, such concepts should be approached with extreme caution. In conclusion, therapies aiming at preserving/preventing β-cell function should aim at providing safe, long-term, and clinically relevant improvements over standard insulin therapy.

Promising Therapies

- *Alum-formulated GAD65:* Specific modulation of long-lasting immune response to β-cells GAD65
- *Anti-CD3 antibodies:* Prevention of β-cell destruction by depletion of T cells
- *Anti-CD20 antibodies:* Prevention of β-cell destruction by depletion of B cells
- *DiaPep277:* Immunomodulation and shift from Th1 response to a Th2 response
- *Cell/stem cell therapies:* Reintroduce immune tolerance and induce islet regeneration

Key Points

- Therapeutic interventions can be beneficial to individuals identified at risk and to individuals with existing autoimmunity to prevent the damage to residual β-cells.
- Modern therapies aimed at reducing β-cell autoimmunity should ideally be short-term treatment which can induce long-lasting "tolerance," but does not debilitate the capacity of the immune system to fight pathogens.
- Successful intervention using autoantigen-specific therapies like alum-formulated GAD65 is the need of the hour.
- Combinatorial therapies can be very helpful in β-cell regeneration and arresting aggressive β-cell autoimmunity.
- The correct timing of immunomodulating therapies could be very important.
- New approaches such as DNA vaccines can be beneficial, but should be approached with caution.

Cross-References

▶ Generating Pancreatic Endocrine Cells from Pluripotent Stem Cells
▶ Human Islet Autotransplantation

- Immunology of β-Cell Destruction
- Inflammatory Pathways Linked to β Cell Demise in Diabetes
- Islet Encapsulation
- Islet Xenotransplantation: Recent Advances and Future Prospects
- Stem Cells in Pancreatic Islets
- Successes and Disappointments with Clinical Islet Transplantation

References

Achenbach P, Barker J, Bonifacio E (2008) Modulating the natural history of type 1 diabetes in children at high genetic risk by mucosal insulin immunization. Curr Diab Rep 8(2):87–93

Agardh CD, Cilio CM, Lethagen A, Lynch K, Leslie RD, Palmer M, Harris RA, Robertson JA, Lernmark A (2005) Clinical evidence for the safety of GAD65 immunomodulation in adult-onset autoimmune diabetes. J Diabetes Complications 19(4):238–246

Allen HF, Klingensmith GJ, Jensen P, Simoes E, Hayward A, Chase HP (1999) Effect of Bacillus Calmette-Guerin vaccination on new-onset type 1 diabetes. A randomized clinical study. Diabetes Care 22(10):1703–1707

Arif S, Cox P, Afzali B, Lombardi G, Lechler RI, Peakman M, Mirenda V (2010) Anti-TNFα therapy–killing two birds with one stone? Lancet 375(9733):2278

Barker JM, McFann KK, Orban T (2007) Effect of oral insulin on insulin autoantibody levels in the Diabetes Prevention Trial Type 1 oral insulin study. Diabetologia 50(8):1603–1606

Behme MT, Dupre J, Stiller CR (1988) Effect of cyclosporine on insulin binding to erythrocytes in type 1 diabetes mellitus of recent onset. Clin Invest Med 11(2):113–122

Behme MT, Dupre J, McDonald TJ (2003) Glucagon-like peptide 1 improved glycemic control in type 1 diabetes. BMC Endocr Disord 3(1):3

Belghith M, Bluestone JA, Barriot S, Megret J, Bach JF, Chatenoud L (2003) TGF-β-dependent mechanisms mediate restoration of self-tolerance induced by antibodies to CD3 in overt autoimmune diabetes. Nat Med 9(9):1202–1208

Bell GI, Broughton HC, Levac KD, Allan DA, Xenocostas A, Hess DA (2012) Transplanted human bone marrow progenitor subtypes stimulate endogenous islet regeneration and revascularization. Stem Cells Dev 21(1):97–109

Bergerot I, Ploix C, Petersen J, Moulin V, Rask C, Fabien N, Lindblad M, Mayer A, Czerkinsky C, Holmgren J, Thivolet C (1997) A cholera toxoid-insulin conjugate as an oral vaccine against spontaneous autoimmune diabetes. Proc Natl Acad Sci USA 94(9):4610–4614

Bizzarri C, Pitocco D, Napoli N, Di Stasio E, Maggi D, Manfrini S, Suraci C, Cavallo MG, Cappa M, Ghirlanda G, Pozzilli P, Group I (2010) No protective effect of calcitriol on β-cell function in recent-onset type 1 diabetes: the IMDIAB XIII trial. Diabetes Care 33(9):1962–1963

Bjork E, Berne C, Kampe O, Wibell L, Oskarsson P, Karlsson FA (1996) Diazoxide treatment at onset preserves residual insulin secretion in adults with autoimmune diabetes. Diabetes 45 (10):1427–1430

Bluestone JA, St Clair EW, Turka LA (2006) CTLA4Ig: bridging the basic immunology with clinical application. Immunity 24(3):233–238

Bolt S, Routledge E, Lloyd I, Chatenoud L, Pope H, Gorman SD, Clark M, Waldmann H (1993) The generation of a humanized, non-mitogenic CD3 monoclonal antibody which retains in vitro immunosuppressive properties. Eur J Immunol 23(2):403–411

Bougneres PF, Carel JC, Castano L, Boitard C, Gardin JP, Landais P, Hors J, Mihatsch MJ, Paillard M, Chaussain JL et al (1988) Factors associated with early remission of type I diabetes in children treated with cyclosporine. N Engl J Med 318(11):663–670

Brod SA, Burns DK (1994) Suppression of relapsing experimental autoimmune encephalomyelitis in the SJL/J mouse by oral administration of type I interferons. Neurology 44(6):1144–1148

Brod SA, Kerman RH, Nelson LD, Marshall GD Jr, Henninger EM, Khan M, Jin R, Wolinsky JS (1997) Ingested IFN-α has biological effects in humans with relapsing-remitting multiple sclerosis. Mult Scler 3(1):1–7

Brodie GM, Wallberg M, Santamaria P, Wong FS, Green EA (2008) β-cells promote intra-islet CD8$^+$ cytotoxic T-cell survival to enhance type 1 diabetes. Diabetes 57(4):909–917

Burt RK, Burns W, Ruvolo P, Fischer A, Shiao C, Guimaraes A, Barrett J, Hess A (1995) Syngeneic bone marrow transplantation eliminates V β 8.2 T lymphocytes from the spinal cord of Lewis rats with experimental allergic encephalomyelitis. J Neurosci Res 41(4):526–531

Buzzetti R, Cernea S, Petrone A, Capizzi M, Spoletini M, Zampetti S, Guglielmi C, Venditti C, Pozzilli P, Grp DT (2011) C-peptide response and HLA genotypes in subjects with recent-onset type 1 diabetes after immunotherapy with DiaPep277 an exploratory study. Diabetes 60(11):3067–3072

Carpenter PA, Pavlovic S, Tso JY, Press OW, Gooley T, Yu XZ, Anasetti C (2000) Non-Fc receptor-binding humanized anti-CD3 antibodies induce apoptosis of activated human T cells. J Immunol 165(11):6205–6213

Chatenoud L, Thervet E, Primo J, Bach JF (1994) Anti-CD3 antibody induces long-term remission of overt autoimmunity in nonobese diabetic mice. Proc Natl Acad Sci USA 91(1):123–127

Chatenoud L, Primo J, Bach JF (1997) CD3 antibody-induced dominant self tolerance in overtly diabetic NOD mice. J Immunol 158(6):2947–2954

Cook JJ, Hudson I, Harrison LC, Dean B, Colman PG, Werther GA, Warne GL, Court JM (1989) Double-blind controlled trial of azathioprine in children with newly diagnosed type I diabetes. Diabetes 38(6):779–783

Couri CE, Oliveira MC, Stracieri AB, Moraes DA, Pieroni F, Barros GM, Madeira MI, Malmegrim KC, Foss-Freitas MC, Simoes BP, Martinez EZ, Foss MC, Burt RK, Voltarelli JC (2009) C-peptide levels and insulin independence following autologous nonmyeloablative hematopoietic stem cell transplantation in newly diagnosed type 1 diabetes mellitus. JAMA 301(15):1573–1579

da Silva AJ, Brickelmaier M, Majeau GR, Li Z, Su L, Hsu YM, Hochman PS (2002) Alefacept, an immunomodulatory recombinant LFA-3/IgG1 fusion protein, induces CD16 signaling and CD2/CD16-dependent apoptosis of CD2$^+$ cells. J Immunol 168(9):4462–4471

de Goffau MC, Luopajarvi K, Knip M, Ilonen J, Ruohtula T, Harkonen T, Orivuori L, Hakala S, Welling GW, Harmsen HJ, Vaarala O (2013) Fecal microbiota composition differs between children with β-cell autoimmunity and those without. Diabetes 62(4):1238–1244

Diabetes Prevention Trial–Type 1 Diabetes Study Group (2002) Effects of insulin in relatives of patients with type 1 diabetes mellitus. N Engl J Med 346(22):1685–1691

Dinarello CA, Simon A, van der Meer JW (2012) Treating inflammation by blocking interleukin-1 in a broad spectrum of diseases. Nat Rev Drug Discov 11(8):633–652

Druker BJ, Tamura S, Buchdunger E, Ohno S, Segal GM, Fanning S, Zimmermann J, Lydon NB (1996) Effects of a selective inhibitor of the Abl tyrosine kinase on the growth of Bcr-Abl positive cells. Nat Med 2(5):561–566

Duttaroy A, Voelker F, Merriam K, Zhang X, Ren X, Subramanian K, Hughes TE, Burkey BF (2011) The DPP-4 inhibitor vildagliptin increases pancreatic β cell mass in neonatal rats. Eur J Pharmacol 650(2–3):703–707

Eisenbarth GS, Srikanta S, Jackson R, Rabinowe S, Dolinar R, Aoki T, Morris MA (1985) Anti-thymocyte globulin and prednisone immunotherapy of recent onset type 1 diabetes mellitus. Diabetes Res 2(6):271–276

Elias D, Markovits D, Reshef T, van der Zee R, Cohen IR (1990) Induction and therapy of autoimmune diabetes in the non-obese diabetic (NOD/Lt) mouse by a 65-kDa heat shock protein. Proc Natl Acad Sci USA 87(4):1576–1580

Elias D, Avron A, Tamir M, Raz I (2006) DiaPep277 preserves endogenous insulin production by immunomodulation in type 1 diabetes. Ann N Y Acad Sci 1079:340–344

Elliott RB, Crossley JR, Berryman CC, James AG (1981) Partial preservation of pancreatic β-cell function in children with diabetes. Lancet 2(8247):631–632

Ellis CN, Krueger GG (2001) Treatment of chronic plaque psoriasis by selective targeting of memory effector T lymphocytes. N Engl J Med 345(4):248–255

Ellis SL, Moser EG, Snell-Bergeon JK, Rodionova AS, Hazenfield RM, Garg SK (2011) Effect of sitagliptin on glucose control in adult patients with type 1 diabetes: a pilot, double-blind, randomized, crossover trial. Diabet Med 28(10):1176–1181

Endres S, Ghorbani R, Kelley VE, Georgilis K, Lonnemann G, van der Meer JW, Cannon JG, Rogers TS, Klempner MS, Weber PC et al (1989) The effect of dietary supplementation with n-3 polyunsaturated fatty acids on the synthesis of interleukin-1 and tumor necrosis factor by mononuclear cells. N Engl J Med 320(5):265–271

Fiorina P, Jurewicz M, Augello A, Vergani A, Dada S, La Rosa S, Selig M, Godwin J, Law K, Placidi C, Smith RN, Capella C, Rodig S, Adra CN, Atkinson M, Sayegh MH, Abdi R (2009) Immunomodulatory function of bone marrow-derived mesenchymal stem cells in experimental autoimmune type 1 diabetes. J Immunol 183(2):993–1004

Fruchtman S (2003) Stem cell transplantation. Mt Sinai J Med 70(3):166–170

Gabbay MA, Sato MN, Finazzo C, Duarte AJ, Dib SA (2012) Effect of cholecalciferol as adjunctive therapy with insulin on protective immunologic profile and decline of residual β-cell function in new-onset type 1 diabetes mellitus. Arch Pediatr Adolesc Med 166(7): 601–607

Gale EA, Bingley PJ, Emmett CL, Collier T (2004) European Nicotinamide Diabetes Intervention Trial (ENDIT): a randomised controlled trial of intervention before the onset of type 1 diabetes. Lancet 363(9413):925–931

Gatto NM, Koralek DO, Bracken MB, Duggan WT, Lem J, Klioze SS, Jackson NC (2012) Comparative lung cancer mortality with inhaled insulin or comparator: an observational follow-up study of patients previously enrolled in exubera controlled clinical trials (FUSE). Pharmacoepidemiol Drug Saf 21:26

Godfrey WR, Spoden DJ, Ge YG, Baker SR, Liu B, Levine BL, June CH, Blazar BR, Porter SB (2005) Cord blood $CD4^+CD25^+$-derived T regulatory cell lines express FoxP3 protein and manifest potent suppressor function. Blood 105(2):750–758

Gong Z, Jin Y, Zhang Y (2007) Suppression of diabetes in Non-Obese Diabetic (NOD) mice by oral administration of a cholera toxin B subunit-insulin B chain fusion protein vaccine produced in silkworm. Vaccine 25(8):1444–1451

Gottlieb PA, Quinlan S, Krause-Steinrauf H, Greenbaum CJ, Wilson DM, Rodriguez H, Schatz DA, Moran AM, Lachin JM, Skyler JS (2010) Failure to preserve β-cell function with mycophenolate mofetil and daclizumab combined therapy in patients with new- onset type 1 diabetes. Diabetes Care 33(4):826–832

Goudy KS, Wang B, Tisch R (2008) Gene gun-mediated DNA vaccination enhances antigen-specific immunotherapy at a late preclinical stage of type 1 diabetes in nonobese diabetic mice. Clin Immunol 129(1):49–57

Graham J, Kockum I, Sanjeevi CB, Landin-Olsson M, Nystrom L, Sundkvist G, Arnqvist H, Blohme G, Lithner F, Littorin B, Schersten B, Wibell L, Ostman J, Lernmark A, Breslow N, Dahlquist G (1999) Negative association between type 1 diabetes and HLA DQB1*0602-DQA1*0102 is attenuated with age at onset. Swedish Childhood Diabetes Study Group. Eur J Immunogenet 26(2–3):117–127

Green EA, Flavell RA (1999) Tumor necrosis factor-α and the progression of diabetes in non-obese diabetic mice. Immunol Rev 169:11–22

Group TS (2007) Study design of the Trial to Reduce IDDM in the Genetically at Risk (TRIGR). Pediatr Diabetes 8(3):117–137

Hackstein H, Morelli AE, Thomson AW (2001) Designer dendritic cells for tolerance induction: guided not misguided missiles. Trends Immunol 22(8):437–442

Hadjiyanni I, Drucker DJ (2007) Glucagon-like peptide 1 and type 1 diabetes: NOD ready for prime time? Endocrinology 148(11):5133–5135

Hadjiyanni I, Baggio LL, Poussier P, Drucker DJ (2008) Exendin-4 modulates diabetes onset in nonobese diabetic mice. Endocrinology 149(3):1338–1349

Haller MJ, Wasserfall CH, McGrail KM, Cintron M, Brusko TM, Wingard JR, Kelly SS, Shuster JJ, Atkinson MA, Schatz DA (2009) Autologous umbilical cord blood transfusion in very young children with type 1 diabetes. Diabetes Care 32(11):2041–2046

Hao L, Calcinaro F, Gill RG, Eugui EM, Allison AC, Lafferty KJ (1992) Facilitation of specific tolerance induction in adult mice by RS-61443. Transplantation 53(3):590–595

Hao L, Chan SM, Lafferty KJ (1993) Mycophenolate mofetil can prevent the development of diabetes in BB rats. Ann N Y Acad Sci 696:328–332

Harnaha J, Machen J, Wright M, Lakomy R, Styche A, Trucco M, Makaroun S, Giannoukakis N (2006) Interleukin-7 is a survival factor for $CD4^+$ $CD25^+$ T-cells and is expressed by diabetes-suppressive dendritic cells. Diabetes 55(1):158–170

Harrison LC, Colman PG, Dean B, Baxter R, Martin FI (1985) Increase in remission rate in newly diagnosed type I diabetic subjects treated with azathioprine. Diabetes 34(12):1306–1308

Held W, MacDonald HR, Weissman IL, Hess MW, Mueller C (1990) Genes encoding tumor necrosis factor α and granzyme A are expressed during development of autoimmune diabetes. Proc Natl Acad Sci USA 87(6):2239–2243

Herold KC, Hagopian W, Auger JA, Poumian-Ruiz E, Taylor L, Donaldson D, Gitelman SE, Harlan DM, Xu D, Zivin RA, Bluestone JA (2002) Anti-CD3 monoclonal antibody in new-onset type 1 diabetes mellitus. N Engl J Med 346(22):1692–1698

Herold KC, Gitelman SE, Masharani U, Hagopian W, Bisikirska B, Donaldson D, Rother K, Diamond B, Harlan DM, Bluestone JA (2005) A single course of anti-CD3 monoclonal antibody hOKT3gamma1(Ala-Ala) results in improvement in C-peptide responses and clinical parameters for at least 2 years after onset of type 1 diabetes. Diabetes 54(6):1763–1769

Herold KC, Gitelman S, Greenbaum C, Puck J, Hagopian W, Gottlieb P, Sayre P, Bianchine P, Wong E, Seyfert-Margolis V, Bourcier K, Bluestone JA, I. T. N. A. I. S. G. Immune Tolerance Network (2009) Treatment of patients with new onset Type 1 diabetes with a single course of anti-CD3 mAb Teplizumab preserves insulin production for up to 5 years. Clin Immunol 132 (2):166–173

Herrera PL, Harlan DM, Vassalli P (2000) A mouse CD8 T cell-mediated acute autoimmune diabetes independent of the perforin and Fas cytotoxic pathways: possible role of membrane TNF. Proc Natl Acad Sci USA 97(1):279–284

Hirsch R, Eckhaus M, Auchincloss H Jr, Sachs DH, Bluestone JA (1988) Effects of in vivo administration of anti-T3 monoclonal antibody on T cell function in mice. I. Immunosuppression of transplantation responses. J Immunol 140(11):3766–3772

Hu CY, Rodriguez-Pinto D, Du W, Ahuja A, Henegariu O, Wong FS, Shlomchik MJ, Wen L (2007) Treatment with CD20-specific antibody prevents and reverses autoimmune diabetes in mice. J Clin Invest 117(12):3857–3867

Hulme MA, Wasserfall CH, Atkinson MA, Brusko TM (2012) Central role for interleukin-2 in type 1 diabetes. Diabetes 61(1):14–22

Hummel S, Pfluger M, Hummel M, Bonifacio E, Ziegler AG (2011) Primary dietary intervention study to reduce the risk of islet autoimmunity in children at increased risk for type 1 diabetes: the BABYDIET study. Diabetes Care 34(6):1301–1305

Hundal RS, Petersen KF, Mayerson AB, Randhawa PS, Inzucchi S, Shoelson SE, Shulman GI (2002) Mechanism by which high-dose aspirin improves glucose metabolism in type 2 diabetes. J Clin Invest 109(10):1321–1326

Huppmann M, Baumgarten A, Ziegler AG, Bonifacio E (2005) Neonatal Bacille Calmette-Guerin vaccination and type 1 diabetes. Diabetes Care 28(5):1204–1206

Karussis D, Vourka-Karussis U, Mizrachi-Koll R, Abramsky O (1999) Acute/relapsing experimental autoimmune encephalomyelitis: induction of long lasting, antigen-specific tolerance by syngeneic bone marrow transplantation. Mult Scler 5(1):17–21

Kaufman DL, Clare-Salzler M, Tian J, Forsthuber T, Ting GS, Robinson P, Atkinson MA, Sercarz EE, Tobin AJ, Lehmann PV (1993) Spontaneous loss of T-cell tolerance to glutamic acid decarboxylase in murine insulin-dependent diabetes. Nature 366(6450):69–72

Keymeulen B, Vandemeulebroucke E, Ziegler AG, Mathieu C, Kaufman L, Hale G, Gorus F, Goldman M, Walter M, Candon S, Schandene L, Crenier L, De Block C, Seigneurin JM, De Pauw P, Pierard D, Weets I, Rebello P, Bird P, Berrie E, Frewin M, Waldmann H, Bach JF, Pipeleers D, Chatenoud L (2005) Insulin needs after CD3-antibody therapy in new-onset type 1 diabetes. N Engl J Med 352(25):2598–2608

Keymeulen B, Walter M, Mathieu C, Kaufman L, Gorus F, Hilbrands R, Vandemeulebroucke E, Van de Velde U, Crenier L, De Block C, Candon S, Waldmann H, Ziegler AG, Chatenoud L, Pipeleers D (2010) Four-year metabolic outcome of a randomised controlled CD3-antibody trial in recent-onset type 1 diabetic patients depends on their age and baseline residual β cell mass. Diabetologia 53(4):614–623

Knip M, Virtanen SM, Seppa K, Ilonen J, Savilahti E, Vaarala O, Reunanen A, Teramo K, Hamalainen AM, Paronen J, Dosch HM, Hakulinen T, Akerblom HK, Finnish TSG (2010) Dietary intervention in infancy and later signs of β-cell autoimmunity. N Engl J Med 363(20):1900–1908

Koivisto VA, Aro A, Cantell K, Haataja M, Huttunen J, Karonen SL, Mustajoki P, Pelkonen R, Seppala P (1984) Remissions in newly diagnosed type 1 (insulin-dependent) diabetes: influence of interferon as an adjunct to insulin therapy. Diabetologia 27(2):193–197

Koulmanda M, Bhasin M, Hoffman L, Fan Z, Qipo A, Shi H, Bonner-Weir S, Putheti P, Degauque N, Libermann TA, Auchincloss H Jr, Flier JS, Strom TB (2008) Curative and β cell regenerative effects of α1-antitrypsin treatment in autoimmune diabetic NOD mice. Proc Natl Acad Sci USA 105(42):16242–16247

Kroger N, Kruger W, Wacker-Backhaus G, Hegewisch-Becker S, Stockschlader M, Fuchs N, Russmann B, Renges H, Durken M, Bielack S, de Wit M, Schuch G, Bartels H, Braumann D, Kuse R, Kabisch H, Erttmann R, Zander AR (1998) Intensified conditioning regimen in bone marrow transplantation for Philadelphia chromosome-positive acute lymphoblastic leukemia. Bone Marrow Transplant 22(11):1029–1033

LaFace DM, Peck AB (1989) Reciprocal allogeneic bone marrow transplantation between NOD mice and diabetes-nonsusceptible mice associated with transfer and prevention of autoimmune diabetes. Diabetes 38(7):894–901

Lazar L, Ofan R, Weintrob N, Avron A, Tamir M, Elias D, Phillip M, Josefsberg Z (2007) Heat-shock protein peptide DiaPep277 treatment in children with newly diagnosed type 1 diabetes: a randomised, double-blind phase II study. Diabetes Metab Res Rev 23(4):286–291

Lenschow DJ, Zeng Y, Thistlethwaite JR, Montag A, Brady W, Gibson MG, Linsley PS, Bluestone JA (1992) Long-term survival of xenogeneic pancreatic islet grafts induced by CTLA4Ig. Science 257(5071):789–792

Lenschow DJ, Ho SC, Sattar H, Rhee L, Gray G, Nabavi N, Herold KC, Bluestone JA (1995) Differential effects of anti-B7-1 and anti-B7-2 monoclonal antibody treatment on the development of diabetes in the nonobese diabetic mouse. J Exp Med 181(3):1145–1155

Li MO, Wan YY, Sanjabi S, Robertson AK, Flavell RA (2006) Transforming growth factor-β regulation of immune responses. Annu Rev Immunol 24:99–146

Ligumsky M, Lysy J, Siguencia G, Friedlander Y (2001) Effect of long-term, continuous versus alternate-day omeprazole therapy on serum gastrin in patients treated for reflux esophagitis. J Clin Gastroenterol 33(1):32–35

Loechelt BJ, Boulware D, Green M, Baden LR, Gottlieb P, Krause-Steinrauf H, Weinberg A, G. Type 1 Diabetes TrialNet Daclizumab/Mycophenolic Acid Study (2013) Epstein-Barr and other herpesvirus infections in patients with early onset type 1 diabetes treated with daclizumab and mycophenolate mofetil. Clin Infect Dis 56(2):248–254

Louvet C, Szot GL, Lang J, Lee MR, Martinier N, Bollag G, Zhu S, Weiss A, Bluestone JA (2008) Tyrosine kinase inhibitors reverse type 1 diabetes in nonobese diabetic mice. Proc Natl Acad Sci USA 105(48):18895–18900

Ludvigsson J, Gustafsson-Stolt U, Liss PE, Svensson T (2002) Mothers of children in ABIS, a population-based screening for prediabetes, experience few ethical conflicts and have a positive attitude. Ann N Y Acad Sci 958:376–381

Ludvigsson J, Faresjo M, Hjorth M, Axelsson S, Cheramy M, Pihl M, Vaarala O, Forsander G, Ivarsson S, Johansson C, Lindh A, Nilsson NO, Aman J, Ortqvist E, Zerhouni P, Casas R (2008) GAD treatment and insulin secretion in recent-onset type 1 diabetes. N Engl J Med 359(18):1909–1920

Ludvigsson J, Hjorth M, Cheramy M, Axelsson S, Pihl M, Forsander G, Nilsson NO, Samuelsson BO, Wood T, Aman J, Ortqvist E, Casas R (2011) Extended evaluation of the safety and efficacy of GAD treatment of children and adolescents with recent-onset type 1 diabetes: a randomised controlled trial. Diabetologia 54(3):634–640

Luopajarvi K, Savilahti E, Virtanen SM, Ilonen J, Knip M, Akerblom HK, Vaarala O (2008) Enhanced levels of cow's milk antibodies in infancy in children who develop type 1 diabetes later in childhood. Pediatr Diabetes 9(5):434–441

Machen J, Harnaha J, Lakomy R, Styche A, Trucco M, Giannoukakis N (2004) Antisense oligonucleotides down-regulating costimulation confer diabetes-preventive properties to nonobese diabetic mouse dendritic cells. J Immunol 173(7):4331–4341

Madec AM, Mallone R, Afonso G, Abou Mrad E, Mesnier A, Eljaafari A, Thivolet C (2009) Mesenchymal stem cells protect NOD mice from diabetes by inducing regulatory T cells. Diabetologia 52(7):1391–1399

Mandrup-Poulsen T, Bendtzen K, Dinarello CA, Nerup J (1987) Human tumor necrosis factor potentiates human interleukin 1-mediated rat pancreatic β-cell cytotoxicity. J Immunol 139(12):4077–4082

Mandrup-Poulsen T, Pickersgill L, Donath MY (2010) Blockade of interleukin 1 in type 1 diabetes mellitus. Nat Rev Endocrinol 6(3):158–166

Mari A, Sallas WM, He YL, Watson C, Ligueros-Saylan M, Dunning BE, Deacon CF, Holst JJ, Foley JE (2005) Vildagliptin, a dipeptidyl peptidase-IV inhibitor, improves model-assessed β-cell function in patients with type 2 diabetes. J Clin Endocrinol Metab 90(8):4888–4894

Martin F, Chan AC (2006) B cell immunobiology in disease: evolving concepts from the clinic. Annu Rev Immunol 24:467–496

Mastrandrea L, Yu J, Behrens T, Buchlis J, Albini C, Fourtner S, Quattrin T (2009) Etanercept treatment in children with new-onset type 1 diabetes: pilot randomized, placebo-controlled, double-blind study. Diabetes Care 32(7):1244–1249

Mathieu C, Waer M, Laureys J, Rutgeerts O, Bouillon R (1994) Prevention of autoimmune diabetes in NOD mice by 1,25 dihydroxyvitamin D3. Diabetologia 37(6):552–558

McCarey DW, McInnes IB, Madhok R, Hampson R, Scherbakov O, Ford I, Capell HA, Sattar N (2004) Trial of Atorvastatin in Rheumatoid Arthritis (TARA): double-blind, randomised placebo-controlled trial. Lancet 363(9426):2015–2021

Moran A, Bundy B, Becker DJ, DiMeglio LA, Gitelman SE, Goland R, Greenbaum CJ, Herold KC, Marks JB, Raskin P, Sanda S, Schatz D, Wherrett DK, Wilson DM, Krischer JP, Skyler JS, G. Type 1 Diabetes TrialNet Canakinumab Study, Pickersgill L, de Koning E, Ziegler AG, Boehm B, Badenhoop K, Schloot N, Bak JF, Pozzilli P, Mauricio D, Donath MY, Castano L, Wagner A, Lervang HH, Perrild H, Mandrup-Poulsen T, Group AS, Pociot F, Dinarello CA (2013) Interleukin-1 antagonism in type 1 diabetes of recent onset: two multicentre, randomised, double-blind, placebo-controlled trials. Lancet 381(9881):1905–1915

Mu J, Woods J, Zhou YP, Roy RS, Li Z, Zycband E, Feng Y, Zhu L, Li C, Howard AD, Moller DE, Thornberry NA, Zhang BB (2006) Chronic inhibition of dipeptidyl peptidase-4 with a sitagliptin analog preserves pancreatic β-cell mass and function in a rodent model of type 2 diabetes. Diabetes 55(6):1695–1704

Nanto-Salonen K, Kupila A, Simell S, Siljander H, Salonsaari T, Hekkala A, Korhonen S, Erkkola R, Sipila JI, Haavisto L, Siltala M, Tuominen J, Hakalax J, Hyoty H, Ilonen J, Veijola R, Simell T, Knip M, Simell O (2008) Nasal insulin to prevent type 1 diabetes in children with HLA genotypes and autoantibodies conferring increased risk of disease: a double-blind, randomised controlled trial. Lancet 372(9651):1746–1755

Orban T, Farkas K, Jalahej H, Kis J, Treszl A, Falk B, Reijonen H, Wolfsdorf J, Ricker A, Matthews JB, Tchao N, Sayre P, Bianchine P (2010) Autoantigen-specific regulatory T cells induced in patients with type 1 diabetes mellitus by insulin B-chain immunotherapy. J Autoimmun 34(4):408–415

Orban T, Bundy B, Becker DJ, DiMeglio LA, Gitelman SE, Goland R, Gottlieb PA, Greenbaum CJ, Marks JB, Monzavi R, Moran A, Raskin P, Rodriguez H, Russell WE, Schatz D, Wherrett D, Wilson DM, Krischer JP, Skyler JS, G. Type 1 Diabetes TrialNet Abatacept Study (2011) Co-stimulation modulation with abatacept in patients with recent-onset type 1 diabetes: a randomised, double-blind, placebo-controlled trial. Lancet 378(9789):412–419

Ortqvist E, Bjork E, Wallensteen M, Ludvigsson J, Aman J, Johansson C, Forsander G, Lindgren F, Berglund L, Bengtsson M, Berne C, Persson B, Karlsson FA (2004) Temporary preservation of β-cell function by diazoxide treatment in childhood type 1 diabetes. Diabetes Care 27(9):2191–2197

Ozeri E, Mizrahi M, Shahaf G, Lewis EC (2012) α-1 antitrypsin promotes semimature, IL-10-producing and readily migrating tolerogenic dendritic cells. J Immunol 189(1):146–153

Palmer JP, Asplin CM, Clemons P, Lyen K, Tatpati O, Raghu PK, Paquette TL (1983) Insulin antibodies in insulin-dependent diabetics before insulin treatment. Science 222 (4630):1337–1339

Pescovitz MD, Greenbaum CJ, Krause-Steinrauf H, Becker DJ, Gitelman SE, Goland R, Gottlieb PA, Marks JB, McGee PF, Moran AM, Raskin P, Rodriguez H, Schatz DA, Wherrett D, Wilson DM, Lachin JM, Skyler JS, C. D. S. G. Type 1 Diabetes TrialNet Anti (2009) Rituximab, B-lymphocyte depletion, and preservation of β-cell function. N Engl J Med 361(22):2143–2152

Phillips B, Nylander K, Harnaha J, Machen J, Lakomy R, Styche A, Gillis K, Brown L, Lafreniere D, Gallo M, Knox J, Hogeland K, Trucco M, Giannoukakis N (2008) A microsphere-based vaccine prevents and reverses new-onset autoimmune diabetes. Diabetes 57(6):1544–1555

Prud'homme GJ (2003) Prevention of autoimmune diabetes by DNA vaccination. Expert Rev Vaccines 2(4):533–540

Putnam AL, Brusko TM, Lee MR, Liu W, Szot GL, Ghosh T, Atkinson MA, Bluestone JA (2009) Expansion of human regulatory T-cells from patients with type 1 diabetes. Diabetes 58(3):652–662

Radtke MA, Nermoen I, Kollind M, Skeie S, Sorheim JI, Svartberg J, Hals I, Moen T, Dorflinger GH, Grill V (2010) Six months of diazoxide treatment at bedtime in newly diagnosed subjects with type 1 diabetes does not influence parameters of β-cell function and autoimmunity but improves glycemic control. Diabetes Care 33(3):589–594

Rastetter W, Molina A, White CA (2004) Rituximab: expanding role in therapy for lymphomas and autoimmune diseases. Annu Rev Med 55:477–503

Raz I, Elias D, Avron A, Tamir M, Metzger M, Cohen IR (2001) β-cell function in new-onset type 1 diabetes and immunomodulation with a heat-shock protein peptide (DiaPep277): a randomised, double-blind, phase II trial. Lancet 358(9295):1749–1753

Roll U, Christie MR, Fuchtenbusch M, Payton MA, Hawkes CJ, Ziegler AG (1996) Perinatal autoimmunity in offspring of diabetic parents. The German multicenter BABY-DIAB study: detection of humoral immune responses to islet antigens in early childhood. Diabetes 45 (7):967–973

Rother KI, Brown RJ, Morales MM, Wright E, Duan Z, Campbell C, Hardin DS, Popovic J, McEvoy RC, Harlan DM, Orlander PR, Brod SA (2009) Effect of ingested interferon-α on β-cell function in children with new-onset type 1 diabetes. Diabetes Care 32(7):1250–1255

Routledge EG, Falconer ME, Pope H, Lloyd IS, Waldmann H (1995) The effect of aglycosylation on the immunogenicity of a humanized therapeutic CD3 monoclonal antibody. Transplantation 60(8):847–853

Rutella S, Danese S, Leone G (2006) Tolerogenic dendritic cells: cytokine modulation comes of age. Blood 108(5):1435–1440

Sanjeevi CB, Lybrand TP, Landin-Olsson M, Kockum I, Dahlquist G, Hagopian WA, Palmer JP, Lernmark A (1994) Analysis of antibody markers, DRB1, DRB5, DQA1 and DQB1 genes and modeling of DR2 molecules in DR2-positive patients with insulin-dependent diabetes mellitus. Tissue Antigens 44(2):110–119

Sanjeevi CB, Landin-Olsson M, Kockum I, Dahlquist G, Lernmark A (1995a) Effects of the second HLA-DQ haplotype on the association with childhood insulin-dependent diabetes mellitus. Tissue Antigens 45(2):148–152

Sanjeevi CB, Lybrand TP, DeWeese C, Landin-Olsson M, Kockum I, Dahlquist G, Sundkvist G, Stenger D, Lernmark A (1995b) Polymorphic amino acid variations in HLA-DQ are associated with systematic physical property changes and occurrence of IDDM. Members of the Swedish Childhood Diabetes Study. Diabetes 44(1):125–131

Sanjeevi CB, Hook P, Landin-Olsson M, Kockum I, Dahlquist G, Lybrand TP, Lernmark A (1996) DR4 subtypes and their molecular properties in a population-based study of Swedish childhood diabetes. Tissue Antigens 47(4):275–283

Saudek F, Havrdova T, Boucek P, Karasova L, Novota P, Skibova J (2004) Polyclonal anti-T-cell therapy for type 1 diabetes mellitus of recent onset. Rev Diabet Stud 1(2):80–88

Schmid S, Buuck D, Knopff A, Bonifacio E, Ziegler AG (2004) BABYDIET, a feasibility study to prevent the appearance of islet autoantibodies in relatives of patients with Type 1 diabetes by delaying exposure to gluten. Diabetologia 47(6):1130–1131

Sherry NA, Chen W, Kushner JA, Glandt M, Tang Q, Tsai S, Santamaria P, Bluestone JA, Brillantes AM, Herold KC (2007) Exendin-4 improves reversal of diabetes in NOD mice treated with anti-CD3 monoclonal antibody by enhancing recovery of β-cells. Endocrinology 148(11):5136–5144

Silverstein J, Maclaren N, Riley W, Spillar R, Radjenovic D, Johnson S (1988) Immunosuppression with azathioprine and prednisone in recent-onset insulin-dependent diabetes mellitus. N Engl J Med 319(10):599–604

Skyler JS, Krischer JP, Wolfsdorf J, Cowie C, Palmer JP, Greenbaum C, Cuthbertson D, Rafkin-Mervis LE, Chase HP, Leschek E (2005) Effects of oral insulin in relatives of patients with type 1 diabetes: the Diabetes Prevention Trial–Type 1. Diabetes Care 28(5):1068–1076

Skyler JS, Greenbaum CJ, Lachin JM, Leschek E, Rafkin-Mervis L, Savage P, Spain L (2008) Type 1 Diabetes TrialNet–an international collaborative clinical trials network. Ann N Y Acad Sci 1150:14–24

Strom A, Kolb H, Martin S, Herder C, Simon MC, Koenig W, Heise T, Heinemann L, Roden M, Schloot NC, Group DS (2012) Improved preservation of residual β cell function by atorvastatin in patients with recent onset type 1 diabetes and high CRP levels (DIATOR trial). PLoS One 7 (3):e33108

Tan KC, Chow WS, Tam SC, Ai VH, Lam CH, Lam KS (2002) Atorvastatin lowers C-reactive protein and improves endothelium-dependent vasodilation in type 2 diabetes mellitus. J Clin Endocrinol Metab 87(2):563–568

Tang Q, Bluestone JA (2006) Regulatory T-cell physiology and application to treat autoimmunity. Immunol Rev 212:217–237

Tarbell KV, Yamazaki S, Steinman RM (2006) The interactions of dendritic cells with antigen-specific, regulatory T cells that suppress autoimmunity. Semin Immunol 18(2):93–102

TEDDY Study Group (2008) The Environmental Determinants of Diabetes in the Young (TEDDY) study. Ann N Y Acad Sci 1150:1–13

Teft WA, Kirchhof MG, Madrenas J (2006) A molecular perspective of CTLA-4 function. Annu Rev Immunol 24:65–97

Tisch R, Wang B, Weaver DJ, Liu B, Bui T, Arthos J, Serreze DV (2001) Antigen-specific mediated suppression of β cell autoimmunity by plasmid DNA vaccination. J Immunol 166 (3):2122–2132

TRIGR Study Group (2007) Study design of the Trial to Reduce IDDM in the Genetically at Risk (TRIGR). Pediatr Diabetes 8(3):117–137

Trube G, Rorsman P, Ohno-Shosaku T (1986) Opposite effects of tolbutamide and diazoxide on the ATP-dependent K^+ channel in mouse pancreatic β-cells. Pflugers Arch 407(5):493–499

Ugrasbul F, Moore WV, Tong PY, Kover KL (2008) Prevention of diabetes: effect of mycophenolate mofetil and anti-CD25 on onset of diabetes in the DRBB rat. Pediatr Diabetes 9(6):596–601 van de Ree MA, Huisman MV, Princen HM, Meinders AE, Kluft C (2003) Strong decrease of high sensitivity C-reactive protein with high-dose atorvastatin in patients with type 2 diabetes mellitus. Atherosclerosis 166(1):129–135

Vinson JA (2007) So many choices, so what's a consumer to do?: a commentary on effect of chromium niacinate and chromium picolinate supplementation on lipid peroxidation, TNF-α, IL-6, CRP, glycated hemoglobin, triglycerides, and cholesterol levels in blood of streptozotocin-treated diabetic rats. Free Radic Biol Med 43(8):1121–1123

Voltarelli JC, Couri CE, Stracieri AB, Oliveira MC, Moraes DA, Pieroni F, Coutinho M, Malmegrim KC, Foss-Freitas MC, Simoes BP, Foss MC, Squiers E, Burt RK (2007) Autologous nonmyeloablative hematopoietic stem cell transplantation in newly diagnosed type 1 diabetes mellitus. JAMA 297(14):1568–1576

Waeckerle-Men Y, Allmen EU, Gander B, Scandella E, Schlosser E, Schmidtke G, Merkle HP, Groettrup M (2006) Encapsulation of proteins and peptides into biodegradable poly(D, L-lactide-co-glycolide) microspheres prolongs and enhances antigen presentation by human dendritic cells. Vaccine 24(11):1847–1857

Walter M, Kaupper T, Adler K, Foersch J, Bonifacio E, Ziegler AG (2010) No effect of the 1α,25-dihydroxyvitamin D3 on β-cell residual function and insulin requirement in adults with new-onset type 1 diabetes. Diabetes Care 33(7):1443–1448

Wherrett DK, Bundy B, Becker DJ, DiMeglio LA, Gitelman SE, Goland R, Gottlieb PA, Greenbaum CJ, Herold KC, Marks JB, Monzavi R, Moran A, Orban T, Palmer JP, Raskin P, Rodriguez H, Schatz D, Wilson DM, Krischer JP, Skyler JS, T. D. T. G. S. G (2011) Antigen-based therapy with Glutamic Acid Decarboxylase (GAD) vaccine in patients with recent-onset type 1 diabetes: a randomised double-blind trial. Lancet 378(9788):319–327

Wolf AM, Wolf D, Rumpold H, Ludwiczek S, Enrich B, Gastl G, Weiss G, Tilg H (2005) The kinase inhibitor imatinib mesylate inhibits TNF-α production in vitro and prevents TNF-dependent acute hepatic inflammation. Proc Natl Acad Sci USA 102(38):13622–13627

Wong JT, Colvin RB (1991) Selective reduction and proliferation of the $CD4^+$ and $CD8^+$ T cell subsets with bispecific monoclonal antibodies: evidence for inter-T cell-mediated cytolysis. Clin Immunol Immunopathol 58(2):236–250

Wong FS, Wen L (2005) B cells in autoimmune diabetes. Rev Diabet Stud 2(3):121–135

Xiu Y, Wong CP, Bouaziz JD, Hamaguchi Y, Wang Y, Pop SM, Tisch RM, Tedder TF (2008) B lymphocyte depletion by CD20 monoclonal antibody prevents diabetes in nonobese diabetic mice despite isotype-specific differences in Fc gamma R effector functions. J Immunol 180 (5):2863–2875

Xue S, Wasserfall C, Parker M, McGrail S, McGrail K, Campbell-Thompson M, Schatz DA, Atkinson MA, Haller MJ (2010) Exendin-4 treatment of nonobese diabetic mice increases β-cell proliferation and fractional insulin reactive area. J Diabetes Complications 24(3):163–167

Yagi H, Soto-Gutierrez A, Parekkadan B, Kitagawa Y, Tompkins RG, Kobayashi N, Yarmush ML (2010) Mesenchymal stem cells: mechanisms of immunomodulation and homing. Cell Transplant 19(6):667–679

Yamada K, Nonaka K, Hanafusa T, Miyazaki A, Toyoshima H, Tarui S (1982) Preventive and therapeutic effects of large-dose nicotinamide injections on diabetes associated with insulitis. An observation in nonobese diabetic (NOD) mice. Diabetes 31(9):749–753

Yoshida M, Babensee JE (2006) Molecular aspects of microparticle phagocytosis by dendritic cells. J Biomater Sci Polym Ed 17(8):893–907

You S, Candon S, Kuhn C, Bach JF, Chatenoud L (2008) CD3 antibodies as unique tools to restore self-tolerance in established autoimmunity their mode of action and clinical application in type 1 diabetes. Adv Immunol 100:13–37

Zanone MM, Favaro E, Miceli I, Grassi G, Camussi E, Caorsi C, Amoroso A, Giovarelli M, Perin PC, Camussi G (2010) Human mesenchymal stem cells modulate cellular immune response to islet antigen glutamic acid decarboxylase in type 1 diabetes. J Clin Endocrinol Metab 95 (8):3788–3797

Zhang J, Tokui Y, Yamagata K, Kozawa J, Sayama K, Iwahashi H, Okita K, Miuchi M, Konya H, Hamaguchi T, Namba M, Shimomura I, Miyagawa JI (2007) Continuous stimulation of human glucagon-like peptide-1 (7–36) amide in a mouse model (NOD) delays onset of autoimmune type 1 diabetes. Diabetologia 50(9):1900–1909

Zhao Y, Huang Z, Qi M, Lazzarini P, Mazzone T (2007) Immune regulation of T lymphocyte by a newly characterized human umbilical cord blood stem cell. Immunol Lett 108(1):78–87

Zhao Y, Lin B, Darflinger R, Zhang Y, Holterman MJ, Skidgel RA (2009) Human cord blood stem cell-modulated regulatory T lymphocytes reverse the autoimmune-caused type 1 diabetes in nonobese diabetic (NOD) mice. PLoS One 4(1):e4226

Zhao Y, Jiang Z, Zhao T, Ye M, Hu C, Yin Z, Li H, Zhang Y, Diao Y, Li Y, Chen Y, Sun X, Fisk MB, Skidgel R, Holterman M, Prabhakar B, Mazzone T (2012) Reversal of type 1 diabetes via islet β cell regeneration following immune modulation by cord blood-derived multipotent stem cells. BMC Med 10:3

Ziegler AG, Elias D, Dagan S, Raz I, Grp D-AS (2010) DIA-AID 1-An international phase III clinical study to evaluate the biological effect of DiaPep277 (R) in preservation of β cell function in newly diagnosed T1D patients. Diabetologia 53:S190

Zollinger RM, Ellison EH (1955) Primary peptic ulcerations of the jejunum associated with islet cell tumors of the pancreas. Ann Surg 142(4):709–723, discussion, 724–8

In Vivo Biomarkers for Detection of β Cell Death

Simon A. Hinke

Contents

Introduction 1116
Qualities of Good Biomarkers 1117
GAD65 1118
Doublecortin 1119
Protein Phosphatase Inhibitor 1 1121
Methylated Insulin Gene 1122
miR-375 1124
Summary and Future Perspectives 1125
Cross-References 1126
References 1126

Abstract

There is an immediate need for a noninvasive approach to detect β cell destruction during the initial stages of type I diabetes and late stages of type II diabetes in order to permit early preventative and interventional treatment strategies. The appearance of circulating biomarkers has been highly useful in diagnosing specific organ damage and early detection of some types of cancer. Hence, for nearly two decades, there have been efforts to identify analogous prognosticating factors in diabetes. To date, studies have identified several β cell-selective proteins which have recently been validated as biomarkers for detection of acute islet insulin cell death in vivo: glutamic acid decarboxylase 65 (GAD65), doublecortin (DCX), and protein phosphatase inhibitor 1 (PPP1R1A). More recently, the PCR detection of circulating genomic DNA originating from the β cell by its methylation fingerprint, as well as circulating β cell-specific microRNA, has offered more sensitive means to discern in vivo islet damage.

S.A. Hinke
Department of Pharmacology, University of Washington, Seattle, WA, USA

Current Address: Janssen Research & Development, Spring House, PA, USA
e-mail: shinke@uw.edu

This chapter surveys the current state of β cell biomarkers for real-time detection of cell death in vivo. While preclinical data suggests that we still have a long way to go, the successful translation of the biomarker approach to humans will revolutionize the way in which diabetes is treated.

Keywords

β cell mass • Streptozotocin • Islet transplantation • Necrosis • Apoptosis

Introduction

Both type 1 and type 2 diabetes mellitus are diseases of inadequate β cell mass. In type 1 diabetes (T1D), autoimmune T-cell-mediated destruction of islet β cells results in a progressive loss of β cell mass which presents itself clinically as hyperglycemia after >70 % of the insulin-secreting cells have been destroyed (Cnop et al. 2005; Eisenbarth 1986). In contrast, type 2 diabetic (T2D) patients suffer from insulin resistance that induces an initial compensatory expansion of β cell mass and hyperinsulinemia; subsequently as the disease progresses, hyperglycemia can be restrained in T2D by lifestyle modification (diet and exercise), insulin secretagogues (e.g., sulfonylureas, incretin mimetics) or insulin-sensitizing agents (e.g., biguanides, glitazones). Later in T2D, β cell apoptosis is also observed, and once patients become refractory to current pharmaceutical agents, they too become dependent upon injection of exogenous insulin to adequately control their blood glucose levels (Butler et al. 2003; Pan et al. 1997; Prentki and Nolan 2006; Tuomilehto et al. 2001).

The preclinical progression of both type 1 and type 2 diabetes may go undetected until hyperglycemia is diagnosed, and thus, a tool to monitor β cell mass in susceptible patient cohorts would permit interventional strategies at earlier time points to prevent ongoing β cell stress and injury (reviewed in the accompanying chapters: "▶ Clinical Approaches to Preserve β-Cell Function in Diabetes; "▶ Pre vention of β-Cell Destruction in Autoimmune Diabetes: Current Approaches and Future Prospects"). Currently, our capabilities for selecting candidates for prevention or intervention trials are based upon genetic susceptibility (including diabetes in first degree relatives) and the presence of one or more islet autoantibodies (Hinke 2011). While the presence of islet-reactive antibodies in the circulation is highly predictive of diabetes development (Wenzlau et al. 2007; Winter and Schatz 2011), the degree of β cell damage at the point of autoantibody detection is unclear. Other classical biomarkers of β cell function are primarily metabolic parameters: glycated hemoglobin (HbA1C), fasting plasma glucose, oral glucose tolerance testing, circulating proinsulin or proinsulin:insulin ratio, and C-peptide (Lebastchi and Herold 2012; Neutzsky-Wulff et al. 2012). However, these are end-stage markers, giving little or no therapeutic window for prevention or reversal of disease progression.

Biomarkers released upon cell death or injury have been used to rapidly detect or assess tissue damage, for example, troponin I in myocardial ischemia (Babuin and

Jaffe 2005), prostate-specific antigen for prostate cancer (Makarov et al. 2009), and alanine aminotransferase in liver injury (Schomaker et al. 2013). Extensive effort has been made to elucidate similar candidate biomarkers in β cells to permit real-time detection of β cell destruction, and recently several novel candidates have been identified. In addition to the real-time detection of progressive β cell destruction in vivo during diabetes, such biomarkers will allow greater optimization of islet transplantation, where there are still many inefficiencies in the procedure leading to significant β cell death during cadaveric isolation and culture, as much as 50 % β cell death during transplantation, and then varying degrees of engraftment.

Qualities of Good Biomarkers

One must consider a number of criteria when evaluating the utility of potential biomarkers for detection of β cell destruction. Furthermore, the qualities of several good examples of clinically useful biomarkers can be compared to putative novel biomarkers as a yardstick. Perhaps the most desirable traits of good biomarkers are tissue selectivity and a highly sensitive detection method. The former quality ensures that measurement of the circulating molecule is unequivocally detectable due to selective tissue damage and is not confounded by other pathological conditions. If this condition is not met, or if the biomarker is upregulated during tumorigenesis, it will require additional diagnostic confirmation. Importantly, it is undesirable to observe expression in tissues contributing the majority of plasma proteins, liver and muscle, as they could be the source of false positives. The expression level of a protein can also directly influence the ease of measurement, whether by mass spectrometry, a biochemical enzymatic spectrophotometric method, or antibody-based approaches such as radioimmunoassay (RIA), enzyme-linked immunosorbent assay (ELISA), or immunoprecipitation and immunoblotting. The method of detection, combined with the abundance of the biomarker, will determine the lower limit of detection, which in turn dictates the actual worth of the analyte in research or clinical diagnosis. A more recent advance in the biomarker field has been the finding that tissue-specific microRNA species and methylated genomic DNA can be released into the circulation and subsequently detected by PCR-based approaches in the days or weeks following tissue injury. While this methodology, by virtue of its signal amplification during detection, may result in a higher frequency of erroneous positive results, the ease and sensitivity of the detection method greatly favor this approach over protein-based biomarkers.

For β cell-specific biomarkers, proteins and miRNA species should likely not be metabolically active agents, be actively secreted by metabolic stimuli, nor their expression be metabolically regulated in order to properly interpret results. Generally, avoidance of transmembrane proteins would also be advisable. For both protein and nucleic acid biomarkers, a detailed pharmacokinetic profile should be characterized to help identify the detection window duration. Diabetes is a progressive disease, and thus, there is unlikely to be an acute insult causing a spike of biomarker appearance in the circulation – rather, a low level of biomarker release over a period of years (in contrast, during islet transplantation, acute necrosis during engraftment may show a rapid peak in released biomarker). The following sections examine the current successes and challenges reported in early stage biomarker validation studies in rodent models and human diabetes.

GAD65

Glutamic acid decarboxylase (GAD) 65 was one of the first candidate biomarkers for detection of β cell death in vitro and in vivo. GAD65 along with GAD67 are the two predominant forms of the enzyme responsible for generation of the inhibitory neurotransmitter γ-aminobutyric acid (GABA) (Hinke 2007). GAD expression is restricted to neuroendocrine cell types, with the human isoform of GAD65 being abundantly expressed in brain and α, β, and δ cells of the endocrine pancreas (Mally et al. 1996); rats express both GAD65 and GAD67 in islet cells, whereas mice appear to synthesize exclusively GAD67 (Chessler and Lernmark 2000). Studies on rat islet tissue indicate GAD65 to be localized to membrane structures via palmitoylation, while GAD67 is cytosolic (Dirkx et al. 1995). GAD65 has many of the characteristics of a decent biomarker, with respect to tissue selectivity and sufficient molar abundance, however, because of the prevalence of anti-GAD65 autoantibodies in the same population of patients in which use of GAD65 as a biomarker is sought, the potential use of this enzyme is limited by antibody interference.

The first in vitro evidence that GAD65 might serve as a biomarker for β cell death came from Smismans et al. (1996), showing enzyme release into the culture media following streptozotocin (STZ) toxin treatment of flow-sorted purified rat β cells. Biochemical assays detected a significant reduction in cellular GAD activity and GABA content in STZ-treated cells, with a concomitant appearance of GAD activity in concentrated media extracts. In a preclinical study examining auto- and allotransplantation of islets into dogs under temporary immunosuppression, an enzymatic assay for GAD achieved limited success in tracking islet graft rejection but left authors optimistic to the possibility of detection of islet-specific proteins prior clinical manifestation of hyperglycemia in diabetes (Shapiro et al. 2001). Development of more sensitive detection methods, sandwich ELISA, RIA, and time-resolved fluorescent immunoassay technologies (Hao et al. 1999; Rui et al. 2007; Schlosser et al. 1997), have shown promise for in vivo applications but were still limited by the relative abundance of GAD65 in islets, the scarce but vital organ, and the dilution of the enzyme in the circulation below the lower limits of detection of these assays. This has prompted subsequent investigators of similar biomarkers to use GAD65 abundance as an internal benchmark criterion, hypothesizing proteins with lower expression than GAD will prove to be too difficult to detect in the circulation.

In 2006, a new magnetic bead immunoassay to GAD65 using a modified dot blot chemiluminescent detection strategy was described by Waldrop et al. (2006). The authors were able to achieve a quantitation limit of 31 pg/mL, and more importantly, as the method employed a denaturation step prior to detection, it was not susceptible to autoantibody interference. The subsequent application of this assay to a proof-of-concept in vivo study in rats was the first to measure acute release of GAD65 into the circulation in response to β cell toxin administration (Waldrop et al. 2007). Remarkably, it was possible to temporally separate β cell death as being well prior to the onset of hyperglycemia. Two β cell-specific toxic compounds, STZ and alloxan, were used at high doses to severely damage β cells in Wistar rats; markers of β cell function were assessed over time. Either toxin treatment induced acute GAD65 appearance in the peripheral circulation or eventually elevated blood sugar – in the same animals, β cell death was evident by significantly reduced circulating insulin and C-peptide levels. Increasing doses of STZ administered to rats caused a concentration-dependent GAD65 shedding into the circulation within 6 h, while blood glucose remained close to basal values. By 24 h postinjection, overt hyperglycemia (>16 mM) and significantly elevated circulating GAD65 were observed at all STZ doses above 40 mg/kg. At a lower dose of 20 mg/kg STZ, there was no induction of β cell TUNEL positivity in rat pancreatic sections at the 24-h time point, nor was there hyperglycemia, yet GAD65 was elevated more than fourfold in the circulation (Waldrop et al. 2007). The latter result implies that it may be possible to detect mild insults to insulin-secreting cells by tracking GAD65 as a circulating biomarker.

These results show promise for use of this type of strategy to study the early stages of diabetes initiation in animal models and human subjects, possibly permitting elucidation of the triggering events during the etiology of diabetes prior to the onset of hyperglycemia. The major hurdle in this regard is the window of detection: Waldrop and colleagues estimated a circulating half-life of recombinant GAD65 of 2.9 h, and following acute toxin, the biomarker was detectable in the circulation for 24 h, but returned to baseline within 2 days (Waldrop et al. 2007). During human T1D, there is a gradual decline in β cell mass over a period of years and a persistence of a small number of β cells in later stages of diagnosis (Pipeleers et al. 2001). Mild acute STZ injury of β cells in rats did cause a progressive appearance of circulating GAD65 (Waldrop et al. 2007), but it is unclear if the prolonged shedding of GAD65 during the development of human autoimmune diabetes will be detectable. To date, no follow-up reports have been published to describe preclinical experiments in rodent models of spontaneous progressive β cell death, such as the BioBreeding rat model of autoimmune islet destruction or late-stage glucolipotoxic β cell death in the ZDF Fatty Zucker rat (Hinke 2007).

Doublecortin

Using the GAD65 proof-of-concept study as the established gold standard of in vivo β cell death biomarkers, Jiang and co-workers sought to improve on the diagnostic sensitivity by focusing on more abundant proteins in insulin-secreting cells. The Belgium-based group used a bottom-up LC-MS/MS unlabeled proteomic strategy, which is inherently biased towards proteins with higher expression, to identify candidate alternate biomarkers in flow-sorted primary rat β and α cell populations (Jiang et al. 2013b) (also refer to the chapter "▶ Proteomics and Islet Research" for greater detail on the identification of β cell-specific proteins). This approach provided a list of 521 proteins above the limit of detection that could be identified with confidence, of which 164 were specific to insulin-secreting cells and 54 to glucagon-enriched cells. This data set was cross-referenced to tissue comparative gene transcript array data from the same group on the same islet populations, as well as brain, pituitary, liver, muscle, and white adipose tissues (Martens et al. 2011), thus permitting candidate selection based upon tissue selectivity. As β cells share a similar neuroendocrine gene expression program to α cells, brain, and pituitary, the authors did not set exclusion criteria based on expression of transcripts in these tissues; however, it was possible to examine the relative levels of mRNAs corresponding to the mass spectrometry identified β cell-expressed proteins and narrow the putative candidate biomarkers to a field of 36 proteins with at least 1.5-fold greater mRNA expression than liver, muscle, and adipose. These candidates were then further restricted by excluding plasmalemmal and intravesicular proteins (Jiang et al. 2013b).

The cytoplasmic protein doublecortin (DCX) was chosen for further in vitro validation, as it showed similar β cell selectivity as islet amyloid polypeptide (IAPP) and prohormone convertase 1 (PC1) and was not detected in the glucagon-positive-enriched cell population (Jiang et al. 2013b). Doublecortin expression in insulin-positive cells had not been previously described, and this protein was primarily used as a marker of neurogenesis during migration of newly formed neurons into the neocortex (Couillard-Despres et al. 2005; Gleeson et al. 1999). Using quantitative immunoblotting, Jiang et al. estimated that the DCX content of human and rat β cells was 5 and 40 times more abundant that GAD65, respectively (Jiang et al. 2013b). The sensitivity of the LC-MS/MS proteomic approach was unable to detect GAD65 in tryptic digests of rat β cell proteins. Quantitative PCR and Western blotting confirmed DCX expression was restricted to islet and brain, with about 20-fold enrichment in β cells as compared to whole brain or pituitary lysates. Fluorescence microscopy localized DCX immunoreactivity to the cytosol of all insulin-positive cells and rare weak staining in some glucagon-positive cells (Jiang et al. 2013b).

In vitro validation studies to evaluate DCX as a biomarker of β cell necrosis were performed using STZ or H_2O_2 treatments and performed in parallel with GAD65 measurements. Using a magnetic bead-based quantitative immunoprecipitation method, a detection limit of 24 nM DCX was obtained, which was sufficient to measure the proportional appearance of DCX in the islet culture media and disappearance from cell extracts following toxin treatment (Jiang et al. 2013b). While this offered proof-of-principle, the rudimentary method to assay DCX prevented further validation in vivo due to lack of sensitivity. Regardless, it was possible to show that DCX was stable in plasma and pharmacokinetic data measuring the removal of recombinant DCX from the circulation indicated an in vivo half-life of around 3 h (remarkably similar to the $t_{1/2}$ calculated for GAD65, perhaps reflecting exclusion from renal filtration) (Jiang et al. 2013b; Waldrop et al. 2007). Additionally, no disease-related DCX autoreactivity was observed in human T1D samples, suggesting that antibody interference is unlikely to cause complications in biomarker detection (Jiang et al. 2013b). Hence, DCX appears to be a promising candidate biomarker for tracking β cell death, with several apparent advantages over GAD65 for this purpose. However, full in vivo validation is precluded pending development of more sensitive DCX detection assays.

Protein Phosphatase Inhibitor 1

A second candidate biomarker came out of the combined efforts of the Martens' group examining β cell-selective transcripts by gene array and abundant protein by label-free mass spectrometry. Protein phosphatase inhibitor 1 (PPP1R1A) fits the criteria for tracking β cell death in vivo based upon its high protein abundance in rat β cells, with relatively good tissue-specific expression (Jiang et al. 2013a). PPP1R1A is an endogenous regulatory inhibitor of protein phosphatase 1 (PP1); when PPP1R1A is phosphorylated by protein kinase A (PKA) on Thr^{35}, it potently inhibits PP1 activity with a K_i of 1.6 nM (Oliver and Shenolikar 1998). It has been studied primarily in regard to its role in PP1 regulation of skeletal muscle glycogen metabolism and contraction, synaptic plasticity, and cell growth. The presence of PPP1R1A in muscle tissue causes some concern for its application as a circulating biomarker, particularly considering its expression is elevated in some cancers (Wai et al. 2002). Nevertheless, Jiang et al. demonstrated expression in purified rat primary β cells and a transformed rat insulin-secreting cell line (INS-1) by Western blot, and protein expression of PPP1R1A was at least one order of magnitude greater in the islet-derived tissue than in rodent skeletal muscle or whole brain, consistent with other studies (Lilja et al. 2005; Martens et al. 2010, 2011; Nicolaou et al. 2009; Vander Mierde et al. 2007). Fluorescent immunostaining human pancreatic sections indicated cytoplasmic PPP1R1A co-staining only in insulin-positive cells (Jiang et al. 2013a).

Using a similar study design to in vitro validation of DCX as a putative biomarker (Jiang et al. 2013b), flow-sorted rat β cells were treated with STZ or cultured cryopreserved human islets were exposed to H_2O_2. PPP1R1A release into the culture media and remaining content in the damaged tissue was assessed by a combination of magnetic Dynabead immunoprecipitation and Western blot detection, followed by densitometry (Jiang et al. 2013a). STZ treatment caused 60–70 % β cell death by 24 h, and similarly, hydrogen peroxide exposure stimulated islet disintegration and 40–50 % cell necrosis at the same time point. In the case of rodent β cells, a 20-fold increase in immunoprecipitated PPP1R1A was observed in media following STZ administration relative to controls, with a reciprocal loss of cellular PPP1R1A protein content. Concordant results from human islet cultures were observed following H_2O_2 treatment (Jiang et al. 2013a).

For in vivo validation, rats were injected with a diabetogenic dose of STZ (60 mg/kg), and in addition to tracking blood glucose and plasma insulin, samples were obtained at 2-h intervals for immunohistochemical analysis, total pancreatic

PPP1R1A protein content, and immunoprecipitated circulating PPP1R1A. Pancreatic PPP1R1A protein expression began to drop within the first 2 h and fell to barely detectable levels after 4 h; likewise, fluorescent staining of PPP1R1A could no longer be visualized by 4 h (Jiang et al. 2013a). At the same time, Jiang and colleagues were able to show a rise in the protein phosphatase 1 inhibitor in the circulation within 2 h and a significant peak at 4 h post-STZ injection. Subsequently, the circulating PPP1R1A rapidly returned to baseline values and was undetectable beyond the 8-h time point (Jiang et al. 2013a). This latter finding contrasts to the in vivo validation results with GAD65, where the protein was undetectable in the circulation in the first 3 h, peaking at 6 h post-STZ and remaining detectable in the circulation for at least 24 h (Waldrop et al. 2007). While these disparate results are likely partially due to the sensitivity of the assays used to detect the circulating proteins, the circulating half-life of PPP1R1A was found to be only 15 min (Jiang et al. 2013a). Injection of recombinant PPP1R1A was cleared quite rapidly from the blood stream, most likely because protein is only approximately 19 kDa and thus freely filtered by the glomerulus. This highlights the need to fully characterize the pharmacokinetic properties of putative biomarkers as part of the validation process.

While this window for detection is quite narrow, it was possible to show in vivo substantiation of PPP1R1A as a biomarker using an alternative model system. During islet transplantation, as many as 50 % of portally injected islets fail to functionally engraft due to ischemic, inflammatory, and mechanical stress during the procedure (Emamaullee and Shapiro 2006; Piemonti et al. 2010) (also refer to chapter "▶ Successes and Disappointments with Clinical Islet Transplantation"). As such, similar to the preclinical canine islet transplantation study examining serum GAD activity (Shapiro et al. 2001), it was hoped that PPP1R1A could be measured in plasma samples of human islet recipients and give some index of the engraftment efficiency. Four islet transplant patients were studied, and 75 % had detectable circulating PPP1R1A biomarker that appeared to correlate with the number of islets transplanted (graft size) (Jiang et al. 2013a). In these graft recipients, the biomarker peaked within the first 15 min and then rapidly returned to baseline following similar kinetics to the pharmacokinetic experiment suggesting complete renal filtration. Control subject serum (kidney transplant, stroke, pancreatitis, type 2 diabetes) did not show any detectable PPP1R1A, nor were any autoantibodies to PPP1R1A measured in GAD autoantibody-positive sera from type 1 diabetics (Jiang et al. 2013a).

Methylated Insulin Gene

Epigenetic regulation of genes can control tissue-specific expression. DNA methylation typically favors a transcriptionally repressed state, whereas demethylation is transcriptionally permissive (Miranda and Jones 2007). Differential methylation of specific oncogenes has been demonstrated in specific cancers, and their detection in serum has been applied as a diagnostic biomarker in these patients

(Grady et al. 2001; Muller et al. 2003; Wallner et al. 2006). Two groups have searched for diabetic biomarkers using the same strategy. Initial efforts by Kuroda et al. (2009) examined the methylation state of the mouse (*Ins2*) and human (*INS*) insulin promoters to characterize the tissue-specific expression of insulin in β cells and during differentiation of stem cells to insulin-positive tissue. Three methylation sites were found in the mouse *Ins2* promoter and 9 in the human *INS* promoter, with the striking observation of all sites being methylated in non-islet tissues. Insulin promoter methylation was shown to directly repress gene transcription by inhibiting transcription factor binding to the upstream CRE site (Kuroda et al. 2009). In a follow-up study from the same group, researchers describe the development and thorough characterization of a quantitative, methylation-specific PCR assay for circulating β cell *Ins2* exon 2 DNA and apply it to a mouse model of T1D (Husseiny et al. 2012). It was estimated that the assay sensitivity achieved could detect the β cell-specific DNA from as few as one thousand cells in the circulation of a mouse. NOD/scid mice were injected with 50 mg/kg STZ on three consecutive days, and glycemia- and methylation-specific PCR were performed on bisulfite-converted gDNA from blood for 5 weeks post-STZ. Hyperglycemia was evident within 5–6 days, but a significant increase in demethylated β cell-specific *Ins2* DNA could be measured as early as 2 days following STZ (Husseiny et al. 2012). The sensitivity of this method and the ability to detect islet injury prior to hyperglycemia hopefully point towards the future ability to intervene in the progression of human diabetes to prevent further loss of β cell mass.

In a similar approach, the Herold lab examined the methylation state of the coding region of the mouse *Ins1* gene and identified seven methylation sites that were modified in a tissue-specific manner (Akirav et al. 2011). Optimization of a nested two-step methylation-specific PCR reaction was used to show more than 12-fold greater demethylated *Ins1* DNA in crude islet preparations compared to a panel of non-islet tissues. Using a single high dose of STZ (200 mg/kg) and analyzing circulating *Ins1* DNA, these researchers measured a significant 2.6-fold increase in the demethylation index at 8 h after STZ, which increased to 3.8-fold by 24 h. Notably, at the early time point, immunohistochemical indicators of β cell injury were present in the absence of elevated blood glucose (Akirav et al. 2011). Application of these methods to the NOD mouse pre-diabetic state showed a significant, but variable elevation of demethylated *Ins1* DNA in the circulation, which was inversely correlated with pancreatic insulin content. In extending this study to humans, Akirav et al. were able to show demethylation of *INS* DNA in sorted human β cells at eight putative sites. The most promising result shown was the detection of demethylated insulin DNA in the sera from five newly diagnosed type 1 diabetic patients (<1.5 years since diagnosis), and the proportion of demethylated *INS* DNA was significantly greater than in control subjects (Akirav et al. 2011).

Broader application of the two-step nested methylation-sensitive PCR method to human diabetes was reported by Lebastchi et al. (2013). Serum samples were obtained during a clinical trial (DELAY) at enrollment and 1 year later from 13 age-matched nondiabetic control patients, 43 patients recently diagnosed with type 1 diabetes, and a third group of 37 diabetic patients treated with anti-CD3 monoclonal antibody. At entry, newly diagnosed T1D patients were found to have significantly more circulating unmethylated insulin DNA compared to control subjects. Patients receiving the anti-CD3 intervention treatment had a reduced decline in C-peptide levels and reduced insulin requirement after 1 year, relative to placebo-treated diabetics; at the same time, a significant decline in abundance of unmethylated *INS* DNA was measured in the sera from the treatment group, suggesting that treatment was associated with less β cell loss than the placebo group (Lebastchi et al. 2013). Hence, proof-of-concept has been verified for use of demethylated insulin DNA as a circulating biomarker for β cell death and independently replicated by two laboratories. Possibly the most challenging aspect for the future of this technology in diabetes research will be the adaptation of the technique to facilitate larger clinical screening efforts.

miR-375

MicroRNAs (miRNAs) are single-stranded RNA molecules (approx. 21–23 nt in length) that have emerged as epigenetic regulators of gene expression and cellular function (Bartel 2009). Several β cell-specific miRNAs have been identified, of which the most abundant and best characterized is miR-375 (Fernandez-Valverde et al. 2011; Poy et al. 2004). These microRNAs have been shown to play a role in endocrine pancreas development, α and β cell mass, insulin production, and glucose-stimulated insulin release (El Ouaamari et al. 2008; Fernandez-Valverde et al. 2011; Kredo-Russo et al. 2012; Melkman-Zehavi et al. 2011; Poy et al. 2009). Kieffer's group examined the potential of miR-375 as marker of β cell death, as it was the first islet-specific miRNA identified and the most abundant (Landgraf et al. 2007; Poy et al. 2004). Furthermore, studies have recently shown some extracellular miRNAs to have remarkable stability due to complex formation (Koberle et al. 2013); the use of quantitative PCR to detect circulating miRNAs similarly avoids the limitations in sensitivity-hampering application of protein biomarkers.

Initially, results were confirmed for miR-375 showing expression in both human and mouse pancreas and enrichment in islet tissue. Although the tissue specificity was primarily restricted to pancreas, some lower level expression was observed in pulmonary tissue, testis, and the gastrointestinal tract (Erener et al. 2013). As miR-375 is elevated in some prostate cancers (Brase et al. 2011), caution should be used if it is applied as a single diagnostic test for a β cell biomarker. In vitro, when mouse islets were treated with either a cytotoxic cocktail of cytokines (IL-1β, TNF-α, and IFN-γ) or STZ, it was possible to detect miR-375 release into the media samples at 20–24 times the amount found in control samples. Additionally, the use of caspase or PARP inhibitors to block the cell death induced by cytokines or STZ showed a 2-to 4-fold reduction in miR-375 abundance in the culture media (Erener et al. 2013).

Two in vivo mouse models were employed to examine if miR-375 could be used to detect β cell death prior to appearance of overt diabetes. The first was similar to the approaches used for protein-based biomarkers: a single diabetogenic dose of STZ. In this model, a mild reduction in blood glucose was observed on day 1 (consistent with loss of β cell membrane integrity, thus releasing insulin), and hyperglycemia was observed 1 week following intraperitoneal injection of the β cell toxin (Erener et al. 2013). Peak plasma detection of miR-375 occurred within 2 h of STZ treatment (6.8-fold control values) and remained significantly elevated in one cohort of animals for a week. In a second cohort of mice, miR-375 was elevated on the first day of STZ injection, but fell back below baseline by 7 days; interestingly, a second phase of elevated circulating miR-375 appeared from day 14 that persisted to two months post-STZ. Treatment of the diabetic mice with subcutaneous insulin pellets appeared to suppress the level of circulating miR-375 in these animals. Parallel examination of miR-16.1 in this in vivo study indicated that it was unaffected by STZ treatment and hyperglycemia, and thus, this phenomenon was selective for miR-375 (Erener et al. 2013).

As human T1D manifests following a progressive decline in β cell mass over a period of years of autoimmune β cell destruction, Erener and colleagues turned to one of the most common rodent models of autoimmune diabetes, the nonobese diabetic (NOD) mouse. A subset of female NOD mice spontaneously develop a diabetic phenotype after 12 weeks of age, but insulitis can be observed during histological analysis of pancreata as early as 3 weeks (Delovitch and Singh 1997). In this experimental model, it is possible to segregate mice based upon whether or not they develop diabetes at 3–4 months of age, thus providing an internally controlled experiment. During examination of plasma miR-375 in female NOD mice, significantly elevated microRNA was detectable at least 2 weeks prior to the onset of hyperglycemia and remained raised until animals presented overt diabetes. In contrast, measurement of miR-375 in plasma from the six mice that did not develop any hyperglycemia indicated no change in circulating abundance of the microRNA (Erener et al. 2013). The potential sensitivity of miRNA detection in the circulation combined with the ease of quantitation seem to favor this class of biomarker over both protein-based and genomic DNA-based methods.

Summary and Future Perspectives

The silent and complex etiology of diabetes prior to clinical diagnosis poses an immense challenge to preventative and interventional therapeutic approaches. The use of late-stage biomarkers, such as detection of islet autoantibodies and alterations in β cell function parameters, has been the only criteria for selecting patients for immunomodulator therapy. The mixed results of clinical outcomes may stem from the late stage at which diabetics are enrolled in the trials; clearly, the earliest detection of the initial injury provides the largest window for therapy. This chapter has examined the current state of biomarkers that have been proposed to detect β cell injury in vivo. In general, all protein-based biomarkers had drawbacks in regard to sensitivity of the assays to detect them or limitations in the pharmacokinetic properties that narrow the window of detection. Nevertheless, the utility of GAD65, DCX, and PPP1R1A may primarily lie in the optimization of islet transplantation, where a clearer relationship between the proportions of protein detected in the circulation and the degree of β cell loss is directly related. In terms of sensitivity of detection, tissue-specific miRNA and methylated DNA appear to have a great advantage; however, amplification by PCR will likely generate more false positives, and aberrant expression during cancer or other nondiabetic maladies is a concern. Until it is possible to image β cell mass by noninvasive means (reviewed in another chapter in this book), identifying the best circulating biomarker that most accurately reflects ongoing β cell death is likely our best option for early disease intervention.

Cross-References

► Current Approaches and Future Prospects for the Prevention of β-Cell Destruction in Autoimmune Diabetes
► Immunology of β-Cell Destruction
► Inflammatory Pathways Linked to β Cell Demise in Diabetes

References

Akirav EM, Lebastchi J, Galvan EM, Henegariu O, Akirav M, Ablamunits V, Lizardi PM, Herold KC (2011) Detection of β cell death in diabetes using differentially methylated circulating DNA. Proc Natl Acad Sci U S A 108:19018–19023

Babuin L, Jaffe AS (2005) Troponin: the biomarker of choice for the detection of cardiac injury. CMAJ 173:1191–1202

Bartel DP (2009) MicroRNAs: target recognition and regulatory functions. Cell 136:215–233

Brase JC, Johannes M, Schlomm T, Falth M, Haese A, Steuber T, Beissbarth T, Kuner R, Sultmann H (2011) Circulating miRNAs are correlated with tumor progression in prostate cancer. Int J Cancer 128:608–616

Butler AE, Janson J, Bonner-Weir S, Ritzel R, Rizza RA, Butler PC (2003) β-cell deficit and increased β-cell apoptosis in humans with type 2 diabetes. Diabetes 52:102–110

Chessler SD, Lernmark A (2000) Alternative splicing of GAD67 results in the synthesis of a third form of glutamic-acid decarboxylase in human islets and other non-neural tissues. J Biol Chem 275:5188–5192

Cnop M, Welsh N, Jonas JC, Jorns A, Lenzen S, Eizirik DL (2005) Mechanisms of pancreatic β-cell death in type 1 and type 2 diabetes: many differences, few similarities. Diabetes 54(Suppl 2):S97–S107

Couillard-Despres S, Winner B, Schaubeck S, Aigner R, Vroemen M, Weidner N, Bogdahn U, Winkler J, Kuhn HG, Aigner L (2005) Doublecortin expression levels in adult brain reflect neurogenesis. Eur J Neurosci 21:1–14

Delovitch TL, Singh B (1997) The nonobese diabetic mouse as a model of autoimmune diabetes: immune dysregulation gets the NOD. Immunity 7:727–738

Dirkx R Jr, Thomas A, Li L, Lernmark A, Sherwin RS, De Camilli P, Solimena M (1995) Targeting of the 67-kDa isoform of glutamic acid decarboxylase to intracellular organelles is mediated by its interaction with the NH2-terminal region of the 65-kDa isoform of glutamic acid decarboxylase. J Biol Chem 270:2241–2246

Eisenbarth GS (1986) Type I diabetes mellitus. A chronic autoimmune disease. N Engl J Med 314:1360–1368
El Ouaamari A, Baroukh N, Martens GA, Lebrun P, Pipeleers D, van Obberghen E (2008) miR-375 targets 3'-phosphoinositide-dependent protein kinase-1 and regulates glucose-induced biological responses in pancreatic β-cells. Diabetes 57:2708–2717
Emamaullee JA, Shapiro AM (2006) Interventional strategies to prevent β-cell apoptosis in islet transplantation. Diabetes 55:1907–1914
Erener S, Mojibian M, Fox JK, Denroche HC, Kieffer TJ (2013) Circulating miR-375 as a biomarker of β-cell death and diabetes in mice. Endocrinology 154:603–608
Fernandez-Valverde SL, Taft RJ, Mattick JS (2011) MicroRNAs in β-cell biology, insulin resistance, diabetes and its complications. Diabetes 60:1825–1831
Gleeson JG, Lin PT, Flanagan LA, Walsh CA (1999) Doublecortin is a microtubule-associated protein and is expressed widely by migrating neurons. Neuron 23:257–271
Grady WM, Rajput A, Lutterbaugh JD, Markowitz SD (2001) Detection of aberrantly methylated hMLH1 promoter DNA in the serum of patients with microsatellite unstable colon cancer. Cancer Res 61:900–902
Hao W, Daniels T, Pipeleers DG, Smismans A, Reijonen H, Nepom GT, Lernmark A (1999) Radioimmunoassay for glutamic acid decarboxylase-65. Diabetes Technol Ther 1:13–20
Hinke SA (2007) Finding GAD: early detection of β-cell injury. Endocrinology 148:4568–4571
Hinke SA (2011) Inverse vaccination with islet autoantigens to halt progression of autoimmune diabetes. Drug Dev Res 72:788–804
Husseiny MI, Kuroda A, Kaye AN, Nair I, Kandeel F, Ferreri K (2012) Development of a quantitative methylation-specific polymerase chain reaction method for monitoring β cell death in type 1 diabetes. PLoS One 7:e47942
Jiang L, Brackeva B, Ling Z, Kramer G, Aerts JM, Schuit F, Keymeulen B, Pipeleers D, Gorus F, Martens GA (2013a) Potential of protein phosphatase inhibitor 1 as biomarker of pancreatic β-cell injury in vitro and in vivo. Diabetes 62:2683–2688
Jiang L, Brackeva B, Stange G, Verhaeghen K, Costa O, Couillard-Despres S, Rotheneichner P, Aigner L, Van Schravendijk C, Pipeleers D et al (2013b) LC-MS/MS identification of doublecortin as abundant β cell-selective protein discharged by damaged β cells in vitro. J Proteomics 80:268–280
Koberle V, Pleli T, Schmithals C, Augusto Alonso E, Haupenthal J, Bonig H, Peveling-Oberhag J, Biondi RM, Zeuzem S, Kronenberger B et al (2013) Differential stability of cell-free circulating microRNAs: implications for their utilization as biomarkers. PLoS One 8:e75184
Kredo-Russo S, Mandelbaum AD, Ness A, Alon I, Lennox KA, Behlke MA, Hornstein E (2012) Pancreas-enriched miRNA refines endocrine cell differentiation. Development 139:3021–3031
Kuroda A, Rauch TA, Todorov I, Ku HT, Al-Abdullah IH, Kandeel F, Mullen Y, Pfeifer GP, Ferreri K (2009) Insulin gene expression is regulated by DNA methylation. PLoS One 4:e6953
Landgraf P, Rusu M, Sheridan R, Sewer A, Iovino N, Aravin A, Pfeffer S, Rice A, Kamphorst AO, Landthaler M et al (2007) A mammalian microRNA expression atlas based on small RNA library sequencing. Cell 129:1401–1414
Lebastchi J, Herold KC (2012) Immunologic and metabolic biomarkers of β-cell destruction in the diagnosis of type 1 diabetes. Cold Spring Harb Perspect Med 2:a007708
Lebastchi J, Deng S, Lebastchi AH, Beshar I, Gitelman S, Willi S, Gottlieb P, Akirav EM, Bluestone JA, Herold KC (2013) Immune therapy and β-cell death in type 1 diabetes. Diabetes 62:1676–1680
Lilja L, Meister B, Berggren PO, Bark C (2005) DARPP-32 and inhibitor-1 are expressed in pancreatic β-cells. Biochem Biophys Res Commun 329:673–677
Makarov DV, Loeb S, Getzenberg RH, Partin AW (2009) Biomarkers for prostate cancer. Annu Rev Med 60:139–151
Mally MI, Cirulli V, Otonkoski T, Soto G, Hayek A (1996) Ontogeny and tissue distribution of human GAD expression. Diabetes 45:496–501

Martens GA, Jiang L, Verhaeghen K, Connolly JB, Geromanos SG, Stange G, Van Oudenhove L, Devreese B, Hellemans KH, Ling Z et al (2010) Protein markers for insulin-producing β cells with higher glucose sensitivity. PLoS One 5:e14214

Martens GA, Jiang L, Hellemans KH, Stange G, Heimberg H, Nielsen FC, Sand O, Van Helden J, Van Lommel L, Schuit F et al (2011) Clusters of conserved β cell marker genes for assessment of β cell phenotype. PLoS One 6:e24134

Melkman-Zehavi T, Oren R, Kredo-Russo S, Shapira T, Mandelbaum AD, Rivkin N, Nir T, Lennox KA, Behlke MA, Dor Y et al (2011) miRNAs control insulin content in pancreatic β-cells via downregulation of transcriptional repressors. EMBO J 30:835–845

Miranda TB, Jones PA (2007) DNA methylation: the nuts and bolts of repression. J Cell Physiol 213:384–390

Muller HM, Widschwendter A, Fiegl H, Ivarsson L, Goebel G, Perkmann E, Marth C, Widschwendter M (2003) DNA methylation in serum of breast cancer patients: an independent prognostic marker. Cancer Res 63:7641–7645

Neutzsky-Wulff AV, Andreassen KV, Hjuler ST, Feigh M, Bay-Jensen AC, Zheng Q, Henriksen K, Karsdal MA (2012) Future detection and monitoring of diabetes may entail analysis of both β-cell function and volume: how markers of β-cell loss may assist. J Transl Med 10:214

Nicolaou P, Rodriguez P, Ren X, Zhou X, Qian J, Sadayappan S, Mitton B, Pathak A, Robbins J, Hajjar RJ et al (2009) Inducible expression of active protein phosphatase-1 inhibitor-1 enhances basal cardiac function and protects against ischemia/reperfusion injury. Circ Res 104:1012–1020

Oliver CJ, Shenolikar S (1998) Physiologic importance of protein phosphatase inhibitors. Front Biosci 3:D961–D972

Pan XR, Li GW, Hu YH, Wang JX, Yang WY, An ZX, Hu ZX, Lin J, Xiao JZ, Cao HB et al (1997) Effects of diet and exercise in preventing NIDDM in people with impaired glucose tolerance. The Da Qing IGT and Diabetes Study. Diabetes Care 20:537–544

Piemonti L, Guidotti LG, Battaglia M (2010) Modulation of early inflammatory reactions to promote engraftment and function of transplanted pancreatic islets in autoimmune diabetes. Adv Exp Med Biol 654:725–747

Pipeleers D, Hoorens A, Marichal-Pipeleers M, Van de Casteele M, Bouwens L, Ling Z (2001) Role of pancreatic β-cells in the process of β-cell death. Diabetes 50(Suppl 1):S52–S57

Poy MN, Eliasson L, Krutzfeldt J, Kuwajima S, Ma X, Macdonald PE, Pfeffer S, Tuschl T, Rajewsky N, Rorsman P et al (2004) A pancreatic islet-specific microRNA regulates insulin secretion. Nature 432:226–230

Poy MN, Hausser J, Trajkovski M, Braun M, Collins S, Rorsman P, Zavolan M, Stoffel M (2009) miR-375 maintains normal pancreatic α- and β-cell mass. Proc Natl Acad Sci U S A 106:5813–5818

Prentki M, Nolan CJ (2006) Islet β cell failure in type 2 diabetes. J Clin Invest 116:1802–1812

Rui M, Hampe CS, Wang C, Ling Z, Gorus FK, Lernmark A, Pipeleers DG, De Pauw PE (2007) Species and epitope specificity of two 65 kDa glutamate decarboxylase time-resolved fluorometric immunoassays. J Immunol Methods 319:133–143

Schlosser M, Hahmann J, Ziegler B, Augstein P, Ziegler M (1997) Sensitive monoclonal antibody-based sandwich ELISA for determination of the diabetes-associated autoantigen glutamic acid decarboxylase GAD65. J Immunoassay 18:289–307

Schomaker S, Warner R, Bock J, Johnson K, Potter D, Van Winkle J, Aubrecht J (2013) Assessment of emerging biomarkers of liver injury in human subjects. Toxicol Sci 132:276–283

Shapiro AM, Hao EG, Lakey JR, Yakimets WJ, Churchill TA, Mitlianga PG, Papadopoulos GK, Elliott JF, Rajotte RV, Kneteman NM (2001) Novel approaches toward early diagnosis of islet allograft rejection. Transplantation 71:1709–1718

Smismans A, Ling Z, Pipeleers D (1996) Damaged rat β cells discharge glutamate decarboxylase in the extracellular medium. Biochem Biophys Res Commun 228:293–297

Tuomilehto J, Lindstrom J, Eriksson JG, Valle TT, Hamalainen H, Ilanne-Parikka P, Keinanen-Kiukaanniemi S, Laakso M, Louheranta A, Rastas M et al (2001) Prevention of type 2 diabetes mellitus by changes in lifestyle among subjects with impaired glucose tolerance. N Engl J Med 344:1343–1350

Vander Mierde D, Scheuner D, Quintens R, Patel R, Song B, Tsukamoto K, Beullens M, Kaufman RJ, Bollen M, Schuit FC (2007) Glucose activates a protein phosphatase-1-mediated signaling pathway to enhance overall translation in pancreatic β-cells. Endocrinology 148:609–617

Wai DH, Schaefer KL, Schramm A, Korsching E, Van Valen F, Ozaki T, Boecker W, Schweigerer L, Dockhorn-Dworniczak B, Poremba C (2002) Expression analysis of pediatric solid tumor cell lines using oligonucleotide microarrays. Int J Oncol 20:441–451

Waldrop MA, Suckow AT, Hall TR, Hampe CS, Marcovina SM, Chessler SD (2006) A highly sensitive immunoassay resistant to autoantibody interference for detection of the diabetes-associated autoantigen glutamic acid decarboxylase 65 in blood and other biological samples. Diabetes Technol Ther 8:207–218

Waldrop MA, Suckow AT, Marcovina SM, Chessler SD (2007) Release of glutamate decarboxylase-65 into the circulation by injured pancreatic islet β-cells. Endocrinology 148:4572–4578

Wallner M, Herbst A, Behrens A, Crispin A, Stieber P, Goke B, Lamerz R, Kolligs FT (2006) Methylation of serum DNA is an independent prognostic marker in colorectal cancer. Clin Cancer Res 12:7347–7352

Wenzlau JM, Juhl K, Yu L, Moua O, Sarkar SA, Gottlieb P, Rewers M, Eisenbarth GS, Jensen J, Davidson HW et al (2007) The cation efflux transporter ZnT8 (Slc30A8) is a major autoantigen in human type 1 diabetes. Proc Natl Acad Sci U S A 104:17040–17045

Winter WE, Schatz DA (2011) Autoimmune markers in diabetes. Clin Chem 57:168–175

Proteomics and Islet Research

40

Meftun Ahmed

Contents

Introduction 1132
Proteome and Proteomics 1132
Application of Proteomics in Islet Research 1138
   Protein Profiling of Pancreatic Islets 1139
   Comparative and Quantitative Islet Proteomics 1143
   Glucolipotoxicity and Islet Proteomics 1145
   Type 1 Diabetes and Islet Proteomics 1149
   Pharmacoproteomics and Pancreatic Islets 1152
Conclusion 1153
Cross-References 1154
References 1154

Abstract

Almost a decade has elapsed since the contemporary scientists, fascinated by the promising possibilities of proteomics, conducted extensive proteomic studies to unlock the secret of islet biology in the pathogenesis of diabetes. In recent years, proteomics has been revolutionized by the successful application of improved techniques such as 2D gel-based proteomics, mass spectrometric techniques, protein arrays, nanotechnology, and single-cell proteomics. These techniques have tremendous potential for biomarker development, target validation, diagnosis, prognosis, and optimization of treatment in medical care, especially in the field of islet and diabetes research. This chapter will highlight the contributions of proteomic technologies towards the dissection of complex network of signaling molecules regulating islet function, the identification of potential biomarkers, and the understanding of mechanisms involved in the pathogenesis of diabetes.

M. Ahmed
Department of Internal Medicine, Uppsala University Hospital, Uppsala, Sweden

Department of Physiology, Ibrahim Medical College, University of Dhaka, Dhaka, Bangladesh
e-mail: meftun.khandker@akademiska.se; meftun@hotmail.com

Keywords

Proteomics • Islets • Two-dimensional gel electrophoresis • Mass spectrometry • Proteome • Glucotoxicity • Lipotoxicity

Introduction

Pancreatic islets, the fascinating little magic box, because of their vital performance in blood glucose regulation have long been the central focus of diabetes research. The essential illusion of these magical islets is the β-cell, a "mysterious maiden" with bags full of insulin. The search for the understanding the β-cells has given rise new ideas, imagination, and creativity in a worldwide scientific community, but till now not a single phenomenon of the β-cell has been fully understood. Every new discovery tells a tale about the previous one – a little more, but the story it seems a never ending one. In this promising journey of biomedical discovery, the completion of the human genome project has facilitated the entry of the biomedical researchers into a new dimension – the post-genomic era. This era is marked by an explosion of terms containing the suffix "omics," like the word genomics, transcriptomics, and metabolomics. One of the very stylish and trendy labels carrying the appellation "omics" is *proteomics*, which attracted the attention of contemporary scientists and offered to fill the void left by the human genome project to gain an in-depth understanding of future disease prevention and innovation of novel drug targets. The contributions of proteomic technologies towards the insights of the pathophysiology of the pancreatic islet function will be discussed in the following sections.

Proteome and Proteomics

All cells in the human body have essentially the same genetic information, and the genes possess only the information which is sequentially encoded to construct the final products – the proteins. These proteins are dynamic in nature and considered as the molecular engineers for a cell; their composition in a cell may vary at its different stages of development, whereas the genes remain as the static component of a cell. A classic example is the caterpillar and its mature form, the butterfly; they have the same genetic makeup, whereas their protein composition is quite different and it is the protein which is responsible for different shape and form of the organism. The renaissance of proteomics is due to the fact that proteins are expressed in quantities and physical forms that cannot be predicted from DNA and mRNA analysis (Anderson and Anderson 1998; Gygi et al. 1999). In addition, the diseased cells often produce proteins that healthy cells do not have and vice versa. Hence, scientists are aiming towards creating a complete catalogue of all the human proteins with an intention to uncover their interactions with one another (Anderson and Anderson 1998; Gygi et al. 1999; Anderson et al. 2001).

Their definitive goal is to discover biomarkers and to devise better drugs with fewer side effects. Significant progress has already made in biomarker discovery where several groups have announced that using proteomic techniques it is highly possible to make an accurate early diagnosis for cancers including ovarian, breast, and prostate cancer (Adam et al. 2001; Petricoin and Liotta 2004; Maurya et al. 2007).

In general, proteomics includes cataloging all the proteins present in a cell or tissue type at a specific time under specific conditions, quantitation, and functional characterization of these proteins to elucidate their relationships (protein-protein interaction networks) and functional roles and ultimately outlining their precise three-dimensional structures in order to find where the drugs might turn their activity on or off – the "Achilles heels" (Celis et al. 1998; Cahill et al. 2001; Ezzell 2002). The term proteome was coined as a linguistic equivalent to the concept of genome and first used in 1994 at the Siena 2D electrophoresis meeting (Ezzell 2002; Williams and Hochstrasser 1997; Abbott 1999). It denotes the entire PROTEin complement to a genOME, expressed by a cell or tissue type, at a specific time in the development of the organism under specific conditions (Wasinger et al. 1995; Wilkins et al. 1996). While humans are estimated to have approximately 20,000–25,000 genes, alternate RNA splicing and posttranslational modification may lead to encoding as many as 250,000–1 million individual proteins or peptides. For example, more than 22 different isoforms of α-1-antitrypsin exist in human plasma (Hoogland et al. 1999). In addition, the proteome undergoes dynamic changes as it continuously responds to autocrine, paracrine, and endocrine factors as well as exposure to any pathogen, changes in external environment, and during time course of disease and drug treatment. Various gene products, including microRNA (Bartel 2009), as well as epigenetic factors (Strohman 1994) also influence the expression levels of genes and their transcripts. As a consequence, the proteome is far more complex than the genome. Thus, the scale of protein discovery task is challenging and very large indeed. And multiple specialists from different fields must collaborate to provide a range of sophisticated tools to analyze nature's tremendous complexity. However, proteomics is still in an early stage, and at the time when mRNA expression arrays are spreading like cell phones in industry and in academic institutions, systems for large-scale protein analysis are still novelties. The commonly available proteomic technologies to date are summarized in Table 1.

Over the years, proteomics has expanded to include profiling, quantitative, functional, and structural proteomics based on a broad range of technologies. *Protein profiling* involves identifying and making a list of the proteins present in a biological sample (Figeys 2003). *Quantitative proteomics* discovers molecular physiology at the protein level and allows comparisons between samples by measuring relative changes in protein expression in response to external stimuli (Molloy et al. 2003; Domon and Broder 2004). *Functional proteomics* attempts to identify proteins in a cell, tissue, or organism that undergo changes in abundance, localization, or modification in response to a specific biological condition and discover their functions based on the presence of specific functional groups or

Table 1 Currently used proteomic technologies

Proteome profiling	Emerging technologies
1D gel electrophoresis	SILAC
2D gel electrophoresis	Imaging mass spectrometry (IMS)
2D-DIGE	Molecular scanner
MS-based methods	iTRAQ
SELDI-TOF	ICAT
MALDI-TOF	HysTag
CE-ESI-MS	Label-free LC-MS/MS quantitation
LC-MS	Protein chips:
Protein identification	*Spotted array based tools*:
Mass spectrometry	Forward phase arrays (FPA), e.g.,
MudPIT or shotgun proteomics	antibody arrays, protein arrays
2D LC-MS/MS	Reverse phase arrays (RPA)
Protein function	*Microfluidic based tools*
Yeast two hybrid	Single-cell proteomics
Phase display	Nanoproteomics
Surface plasmon resonance analysis	
Immunoaffinity	
Structural proteomics	
X-ray crystallography	
NMR spectroscopy	
Electron tomography	
Immunoelectron microscopy	

2D-DIGE two-dimensional differential in-gel electrophoresis, *CE-ESI-MS* capillary electrophoresis electrospray ionization mass spectrometry, *ICAT* isotope-coded affinity tags, *iTRAQ* isobaric tagging for relative and absolute quantitation. *LC-MS* liquid chromatography mass spectrometry, *MudPIT* multidimensional protein identification technology, *MALDI-TOF* matrix-assisted laser desorption/ionization-time of flight, *NMR* nuclear magnetic resonance, *SELDI-TOF* surface-enhanced laser desorption/ionization-time of flight, *SILAC* stable isotope labeling by amino acids in cell culture

based on their involvement in protein-ligand interactions (Figeys 2003; Molloy et al. 2003; Domon and Broder 2004; Graves and Haystead 2003). Similarly, pathways can be characterized as a cascade of specific protein interactions required to activate cellular functions. Functional proteomics thus focuses on understanding part of the wiring diagram of a cell. *Structural proteomics* attempts to determine the three-dimensional structure of proteins, the structure of protein complexes, and the small-molecule/protein complexes. X-ray crystallography and NMR are its main approaches (Yee et al. 2002; Sali et al. 2003).

In the plethora of proteomic technologies, two-dimensional gel electrophoresis (2DGE) remains as a cornerstone of protein profiling (Lopez 2007; Gorg et al. 2004). The 2DGE separates proteins according to two independent parameters, isoelectric point (pI) in the first dimension and molecular mass (Mr) in the second dimension by coupling isoelectric focusing (IEF) and sodium dodecyl sulfate polyacrylamide gel electrophoresis (SDS-PAGE) (Klose 1975; O'Farrell 1975). Theoretically, 2DGE is capable of resolving up to 10,000 proteins simultaneously, with approximately 2,000 proteins being routine, and detecting and quantifying protein amounts of less than 1 ng per spot (Lopez 2007; Gorg et al. 2004). Despite the well-known limitations of the 2DGE approach, e.g., poor solubility of membrane proteins, limited dynamic range, difficulties in displaying and identifying low-abundant proteins, lack of reproducibility and automation, 2DGE will remain as a powerful and versatile tool for display and quantification of a majority of proteins in biological samples (Rogowska-Wrzesinska et al. 2013; Rabilloud 2012). The detailed technology, challenges, as well as the application potential and future of high-resolution 2DGE have been elegantly reviewed in several papers (Lopez 2007; Gorg et al. 2004, 2000; Vercauteren et al. 2006). However, gel-free high-throughput protein profiling techniques have leapt prominence and now become preferred method of choice including multidimensional protein identification technology (MudPIT) (Florens and Washburn 2006), molecular scanner (Binz et al. 2004), stable isotope labeling by amino acids in cell culture (SILAC) (Ong et al. 2002; Fenselau 2007), isotope-coded affinity tag (ICAT) (Gygi et al. 1999), isobaric tagging for relative and absolute quantitation (iTRAQ) (Aggarwal et al. 2006), protein microarrays (Zhu and Snyder 2003; Bertone and Snyder 2005; Cretich et al. 2006; Uttamchandani and Yao 2008), and HysTag reagent (Olsen et al. 2004). It should be noted that the use of these emerging techniques is limited to certain specialized and privileged laboratories. Also, the choice of a given proteomic approach depends on the type of biological question asked, since each proteomic technology is characterized by specific applications, technical advantages, and limitations. A typical gel-based proteomic workflow is schematically illustrated in Fig. 1. Peptide mass fingerprinting (PMF) and tandem mass spectrometry (peptide fragmentation to generate partial sequence; MS/MS) are commonly used for protein identification on 2D proteomic patterns (Thiede et al. 2005; Yates 1998; Aebersold and Goodlett 2001; Canas et al. 2006). The recent progress in the sensitivity of mass spectrometry analysis has significantly increased the applicability of proteomic technologies (Zhou and Veenstra 2008) as protein identification and profiling tool as well as determining protein interactions and the type and location of posttranslational modifications (Aebersold and Goodlett 2001; Aebersold and Mann 2003; Mann and Jensen 2003). Surface-enhanced laser desorption/ionization-time of flight (SELDI-TOF) is a suitable technique for high-throughput proteomic analysis of complex mixtures of proteins where proteins are retained on solid-phase chromatographic surfaces with specific properties and are subsequently ionized and detected by TOF MS (Merchant and Weinberger 2000; Issaq et al. 2002; Kiehntopf et al. 2007). However, this system is limited for profiling low-molecular-weight proteins (<20 kDa) (Issaq et al. 2002). In another protein profiling strategy, commonly referred as *bottom-up* or *shotgun proteomics* (multidimensional LC-MS/MS or MudPIT), complex protein mixtures are digested into peptides, followed by chromatographic separation of peptides prior to analysis by tandem mass spectrometry, and computer algorithms then map the peptides onto proteins to determine the original content of the mixture (Swanson and Washburn 2005). *Top-down proteomics* refers to the analysis of intact proteins, in contrast to bottom-up proteomics, which are not enzymatically digested prior to MS analysis (Angel et al. 2012; Messana et al. 2013). Applications of top-down

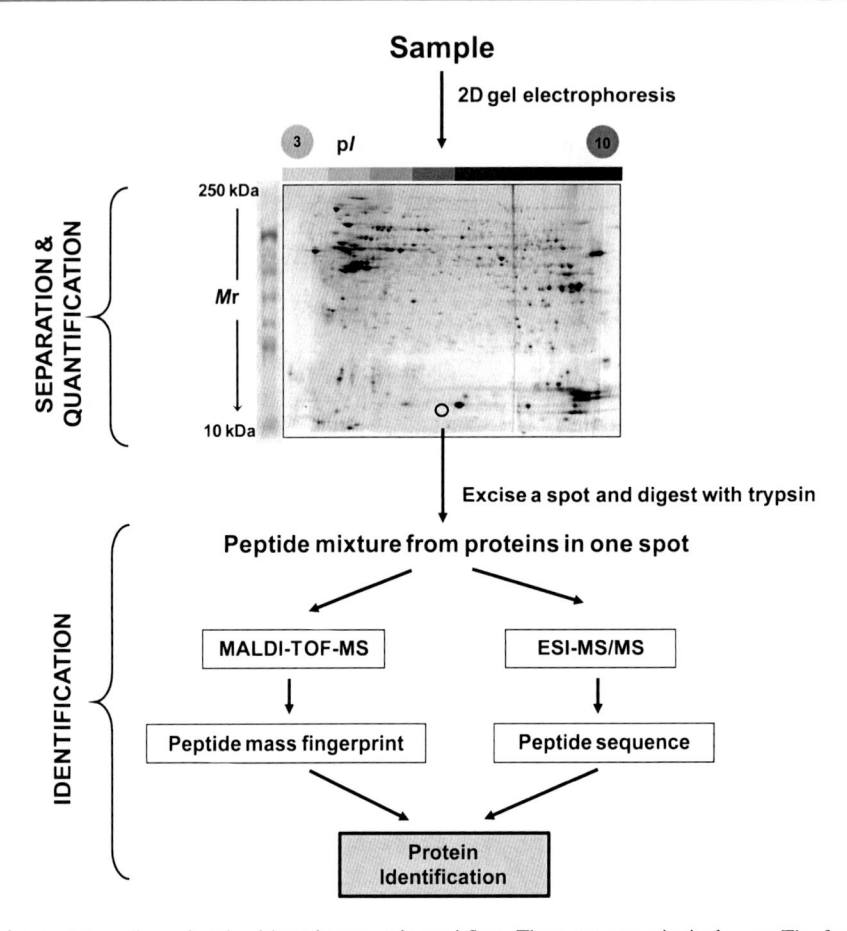

Fig. 1 A two-dimensional gel-based proteomic workflow. There are two principal steps. The *first* is separation and quantification of proteins in a sample using 2D gels. In the first dimension, proteins are separated in a pH gradient according to their molecular charge, known as isoelectric focusing. In the second dimension, the proteins are separated orthogonally by electrophoresis based on their molecular mass. The end result is a 2D gel with thousands of spots where individual spot represents a protein/peptide or a mixture. The *second* is identification of the separated proteins, typically using mass spectrometry (MS) techniques and bioinformatics. A protein spot can be excised from the 2D gel, digested with a protease and the peptides extracted. These peptides can then be analyzed using MS techniques such as matrix-assisted laser desorption/ionization-time of flight (MALDI-TOF) and electrospray ionization tandem MS (ESI-MS/MS)

proteomics include identification of protein isoforms arising from amino acid modifications, gene variants, transcript variation, and posttranslational modifications as well as proteolytic processing of proteins.

Microfluidic devices combine multiple sample preparation, purification, and separation steps in a single integrated device and emerged as an important tool for single-cell proteomics (Angel et al. 2012; Zare and Kim 2010; Chao and

Hansmeier 2013). For details of microfluidic chips, cell preparation, cell lysis, purification, and separations, readers are referred to the article by Chao and Hansmeier (2013). In a microfluidic device with microchip capillary electrophoresis (CE) platform, cells passing through an intersection are subjected to an abrupt change in solvent environment and electric field, which led to rapid lysis. Following a brief CE separation, the contents are ionized on chip and detected by MS. Several microfluidic technologies have recently been developed with the aim of creating a platform for MS-based single-cell analysis (Kelly et al. 2009, 2008; Sun et al. 2010). An emerging technology in the microfluidics field that shows perhaps the most promise for single-cell analysis involves the use of picoliter-sized aqueous droplets surrounded by an immiscible oil phase such that each droplet constitutes an individual sample vessel (Chiu 2010). While the single cell proteomics are at an early stage of development, flexibility in design of chip-based microfluidic devices, rapid analysis time, ability to automate, and successful integration of multiple functions within a single platform provide promises for high-throughput proteomic analysis at the single cell level. Nanotechnology has been used to study the dynamic concentration range of various proteins in complex biological samples, especially the low-abundance proteins (nanoproteomics) (Ray et al. 2010; Jia et al. 2013). Recently, nanomaterials have been employed for improvements of proteomic analysis especially when they are coupled with MALDI-TOF. Among the diverse classes of nanomaterials, gold nanoparticles, carbon nanotubes, silicon nanowires, and QDs are the most commonly used nanomaterials to offer several advantages in nanoproteomics such as ultralow detection, short assay time, high-throughput capability, and low sample consumption (Jia et al. 2013).

Quantifying changes in protein abundance between samples is a key goal of proteomics. Promising novel methods for high-throughput quantitation involve *label-free* approaches. Several studies have demonstrated that LC-MS peptide ions spectral peak intensities are directly proportional to the protein abundances in complex samples (America and Cordewener 2008). Another label-free method, termed spectral counting, compares the number of MS/MS spectra assigned to each protein (America and Cordewener 2008; Old et al. 2005). With controls for normalization between runs, label-free quantitation offers a simpler approach for analysis. Spectral sampling also enables ranking different proteins by their relative abundances, providing information that other methods cannot achieve (Resing and Ahn 2005).

In addition to the protein profiling and comparative proteomics, functional study of target proteins is essential in any successful proteomic study. Functional proteomic approaches are based on interactions of proteins or specific activities of proteins. Phage display is a powerful proteomic tool used to express proteins or domains of proteins (McCafferty et al. 1990; Jestin 2008). The system has played a pivotal role in mapping epitopes of monoclonal and polyclonal antibodies, defining amino acid substrate sequences, and identifying peptide ligands for drug research. Yeast two-hybrid system detects binary protein interactions by activating expression of a reporter gene upon direct binding between the two tested proteins (Fields 2005; Lalonde et al. 2008). SELDI-TOF MS has also been used to characterize protein-protein interaction (Issaq et al. 2002). Recently, for studying the functions and interactions of proteins, protein microarrays have been developed in analogy to DNA microarrays which can also be applied for comparative studies of expression of large sets of proteins (Schweitzer et al. 2003). There are two major types of protein microarrays – forward (FPA) and reverse phase array (RPA) (Espina et al. 2003; Kikuchi and Carbone 2007). In forward protein arrays, thousands of recombinant antibodies carrying the desired specificities are arrayed on glass slides, which make it very well suited for high-throughput screening of biological samples for specific disease markers (Kusnezow and Hoheisel 2002; Wingren and Borrebaeck 2006). The BD Clontech™ Ab Microarray 500 represents a significant step in that direction. With this array, over 500 specific proteins can be assayed to detect and compare expression level of both cytosolic and membrane-bound proteins representing a broad range of biological functions, including signal transduction, cell-cycle regulation, gene transcription, and apoptosis. In contrast to using chips with immobilized antibodies to detect specific proteins, protein chips carrying the proteome of a specific organism or cell type can be made by cloning and purification of these proteins (Zhu et al. 2001). This protein microarray can then be screened on the basis of the ability of the chip to bind specific ligands or interact with specific proteins. The human ProtoArray® protein microarray (Invitrogen™) contains more than 8,000 full-length human proteins purified under native conditions. This high-content discovery tool provides highly sensitive and reproducible results enabling rapid and easy profiling of thousands of biochemical interactions. In a reverse phase microarray, tissues (Speer et al. 2005), cell lysates (Geho et al. 2005), or serum samples (Janzi et al. 2005) are spotted on the surface and probed with one antibody per analyte for a multiplex readout. Thus, this analysis evaluates the expression level of given protein in multiple samples. Both forward and reverse phase protein microarrays are novel technologies in proteomics and offer great promise for use in clinical applications.

Application of Proteomics in Islet Research

The accomplishment of human genome sequences has conferred the islet scientists with immense errands to assess the relative levels of expression of these gene products including the proteins and their posttranslational modifications in pancreatic islets. In the post-genomic era, to clarify the molecular mechanism of islet function in both normal and disease states, it is important to understand the entire gene products which regulate the phenotypes of islet cells and their ability to differentiate and secrete specific hormones. An important advantage of global protein expression profiling compared with individual gene or protein regulation studies is the ability to monitor changes in several functional groups simultaneously. It should be kept in mind that proteomics per se is not a hypothesis-driven experimental approach, but rather a hypothesis-generating *fishing expedition* where one explores the proteins that are not a priori expected to be associated with any pathophysiological conditions, which allows discovering novel proteins and signaling networks opening new research avenues. Since its introduction in 1994, the proteomic booms continue and got considerable attention of the islet researchers as well. Improvements of the core technologies, especially advancement of protein identification by mass spectrometry and bioinformatics tools, have recently encouraged the application of proteomics to unlock the secret of islet pathophysiology. It is indeed interesting to note that the most widely used protein separation technique, the 2DGE, has been employed in 1982 for insulin granule protein profiling (Hutton et al. 1982). In those early days more than 150 protein/peptide spots were detected in a 2DG of insulin secretory granule, and some of the high-molecular-weight spots were presumed as glycoprotein. Lack of high-throughput protein identification method did not permit annotation of the granule proteins but provided an opportunity to study the functional properties of the insulin secretory granule and to dissect the molecular events of exocytosis. A similar proteomic approach has been utilized to explore the glucose-responsive granule proteins in ^{35}S-methionine-labeled rat islet and insulinoma cells, and the study showed that biosynthesis of 25 granule proteins was stimulated 15–30-fold by glucose (Guest et al. 1991). In a subsequent subproteomic study, almost after 25 years, Brunner et al. (2007) separated the INS-1E granule proteins by 1D SDS-PAGE and identified 130 different proteins by LC-MS/MS. Combining improved insulin secretory granule preparation and quantitative proteomics, Schvartz et al. (2012) have provided data on potential proteome changes during maturation of insulin secretory granule (ISG) from Golgi to the plasma membrane. They have elegantly demonstrated localization of a PC1 inhibitor, proSAAS protein in ISG, and glucose-induced modulation of proSAAS mRNA expression. It has been proposed that proSAAS may be involved in regulation of proinsulin processing in the β-cells and dysregulation of it can result in altered glucose-stimulated insulin secretion (GSIS).

Protein Profiling of Pancreatic Islets

A high-quality 2DGE reference map of the isolated pancreatic islets is essential for a 2DG-based comparative proteomic study and for generation of hypothesis. In the holy grail of protein profiling of pancreatic islets, Sanchez et al. (2001) did a pioneering work where they mapped 63 spots corresponding to 44 mouse islet protein entries. This protein map is available in the SWISS-2D database (http://us.expasy.org/ch2d/). Nicolls et al. (2003) identified 88 proteins in total from mouse islets of which 18 were already identified by Sanchez and coworkers. Continued attempts in "shooting at stars" generated another mouse islet 2DGE reference map where 124 spots corresponded to 77 distinct proteins (Ahmed and Bergsten 2005). A reference map of rat insulinoma-derived clonal INS-1E β-cell proteins has also been constructed (Fig. 2). This 2D map contains 686 valid spots, among which 118 spots corresponding to 63 different proteins have successfully identified by MALDI-TOF MS and a combination of liquid chromatography and electrospray tandem MS (LC-ESI-MS/MS). Using 2DGE and MALDI-TOF MS, the first protein map and database of human islets have been generated in 2005 where 130 spots

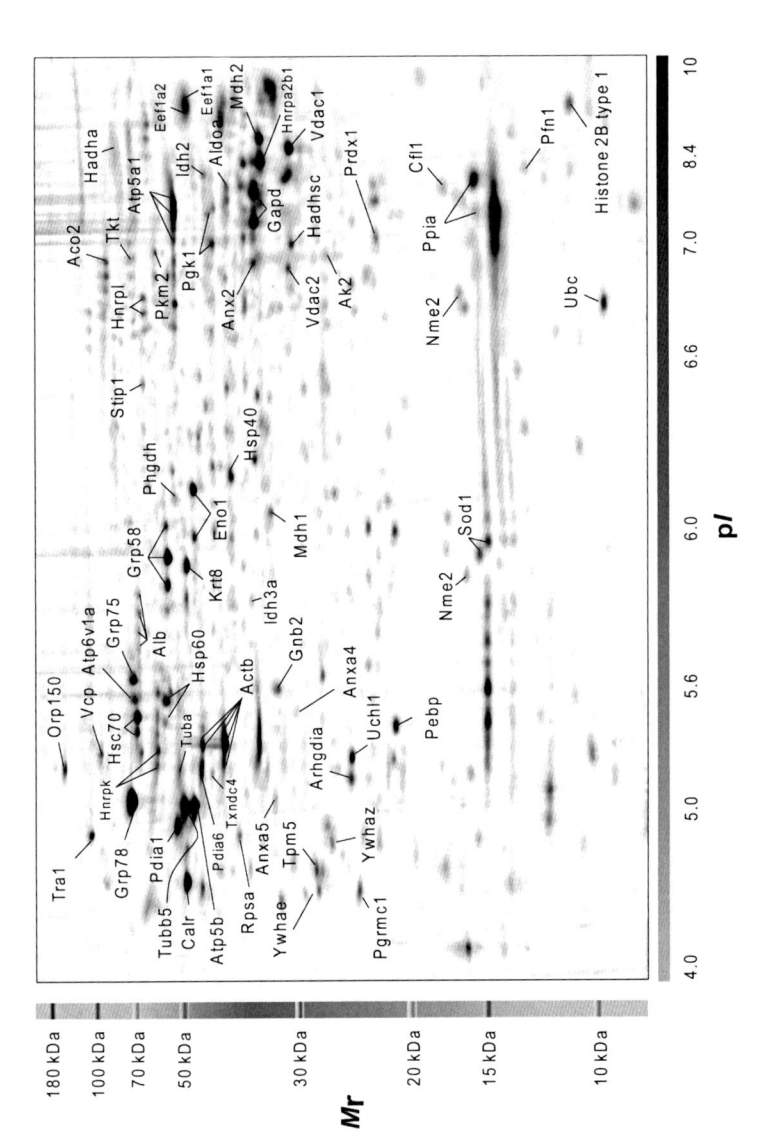

Fig. 2 2D PAGE image of INS-1E proteins. Proteins (200 μg) were loaded onto an IPG strip (pH 3–10 NL) and subsequently separated by mass on a gradient (8–16 %) SDS-PAGE gel. The gel was stained with colloidal Coomassie blue and the filtered image was generated by PDQuest software. Experimental masses and p*I*s are indicated. The gene names mark the location of the corresponding proteins on the gel. *Aco2* aconitate hydratase, mitochondrial, *Actb* β-actin, *Ak2* adenylate kinase isoenzyme 2, mitochondrial, *Alb* serum albumin, *Aldoa* fructose-bisphosphate aldolase A, *Anx2* annexin A2, *Anxa4* annexin A4, *Anxa5* annexin A5, *Arhgdia* Rho GDP dissociation inhibitor 1 (RhoGDI-α), *Atp5a1* ATP synthase subunit α, mitochondrial, *Atp6v1a* V-type proton ATPase catalytic

corresponding to 66 different protein entries were successfully identified (Ahmed et al. 2005a). A high level of reproducibility was reported among the gels, and a total of 744 protein spots were detected (Ahmed et al. 2005a). All the protein profiling studies (Sanchez et al. 2001; Nicolls et al. 2003; Ahmed and Bergsten 2005; Ahmed et al. 2005a) using 2DGE categorized the identified proteins according to cellular location and function. Any attempt to compare these studies renders déjà vu since a number of prevailing proteins were repeatedly reported and most proteins fell into the cytosolic category followed by mitochondrial and endoplasmic reticulum (reviewed by Sundsten and Ortsater 2008). In aforementioned studies a large part of the proteins have either chaperone (e.g., protein disulfide isomerase, PDI; calreticulin; 78 kDa glucose-regulated protein, GRP78; 58 kDa glucose-regulated protein, GRP58; endoplasmin) or metabolic (e.g., α-enolase, transketolase, pyruvate kinase, and hydroxyacyl-CoA dehydrogenase, SCHAD) functions. However, every laboratory blessed with the successful application of 2DGE has its own protocol for protein extraction, isoelectric focusing, and SDS-PAGE. Therefore, a reference map produced by one group cannot necessarily be useful for any other group interested in comparative islet proteomics. Moreover, since introduction in SWISS-2D database, the islet proteome map has not been updated assigning identification of more protein spots. Therefore, the technical hurdle remains for the laborious protein identification procedure even if one follows a similar protocol.

Recent advances in mass spectrometry techniques allowed the use of strong cation exchange fractionation coupled with reversed phase LC-MS/MS and

◂───

Fig. 2 (continued) subunit A, *Calr* calreticulin, *Cfl1* cofilin-1, *Eef1a1* elongation factor 1-α 1, *Eef1a2* elongation factor 1-α 2, *Eno1* α-enolase, *Gapd* glyceraldehyde-3-phosphate dehydrogenase, *Gnb2* guanine nucleotide-binding protein $G_i/G_s/G_t$ subunit β-2, *Grp58* protein disulfide-isomerase A3, *Grp75* stress-70 protein, mitochondrial (75 kDa glucose-regulated protein), *Grp78* 78 kDa glucose-regulated protein, *Hadha* trifunctional enzyme subunit α, mitochondrial, *Hadhsc* hydroxyacyl-coenzyme A dehydrogenase, mitochondrial, *Hnrpa2b1* heterogeneous nuclear ribonucleoproteins A2/B1, *Hnrpk* heterogeneous nuclear ribonucleoprotein K, *Hnrpl* heterogeneous nuclear ribonucleoprotein L, *Hsc70* Heat shock cognate 71 kDa protein (Hspa8), *Hsp40* DnaJ homolog subfamily B member 1 (heat shock 40 kDa protein 1), *Hsp60* 60 kDa heat shock protein, mitochondrial, *Idh3a* isocitrate dehydrogenase [NAD] subunit α, mitochondrial, *Ihd2* isocitrate dehydrogenase [NADP], mitochondrial, *Krt8* keratin, type II cytoskeletal 8, *Mdh1* malate dehydrogenase, cytoplasmic, *Mdh2* malate dehydrogenase, mitochondrial, *Nme2* nucleoside diphosphate kinase B, *Orp150* 150 kDa oxygen-regulated protein (hypoxia upregulated protein 1), *Pdia1* protein disulfide isomerase, *Pdia6* protein disulfide-isomerase A6, *Pebp* phosphatidylethanolamine-binding protein 1, *Pfn1* profilin-1, *Pgk1* phosphoglycerate kinase 1, *Pgrmc1* membrane-associated progesterone receptor component 1, *Phgd* D-3-phosphoglycerate dehydrogenase, *Pkm2* pyruvate kinase isozymes M1/M2, *Ppia* peptidyl-prolyl cis-trans isomerase A (cyclophilin A), *Prdx1* peroxiredoxin-1 (thioredoxin peroxidase 2), *Rpsa* 40S ribosomal protein SA, *Sod1* superoxide dismutase [Cu-Zn], *Stip1* stress-induced-phosphoprotein 1, *Tkt* transketolase, *Tpm5* tropomyosin α-3 chain, *Tra1* endoplasmin, *Tuba* tubulin α, *Tubb5* tubulin β-5 chain, *Txndc4* thioredoxin domain-containing protein 4, *Ubc* polyubiquitin, *Uchl1* ubiquitin carboxyl-terminal hydrolase isozyme L1, *Vcp* transitional endoplasmic reticulum ATPase, *Vdac* voltage-dependent anion-selective channel protein, *Ywhae* 14-3-3 protein ε, *Ywhaz* 14-3-3 protein ζ/δ (protein kinase C inhibitor protein 1)

characterization of 2,612 proteins in the mouse islet proteome (Petyuk et al. 2008). Using nano-UPLC coupled to ESI-MS/MS, more than thousand proteins have been identified in mouse islet (unpublished data). A 2D LC-MS/MS study of the human islets characterized 3,365 proteins covering multiple signaling pathways in human islets including integrin signaling and MAP kinase, NF-κβ, and JAK/STAT pathways (Metz et al. 2006). Combined genomic and proteomic techniques have been employed for profiling of glucagon-secreting α-cells (Maziarz et al. 2005). While a total of 5,945 gene products were detected in α-cells by the gene chips alone, only 1,651 proteins were identified with high confidence using shotgun proteomics and rigorous database searching. Seven hundred sixty-two cross-mapped gene product pairs (both the gene and corresponding protein) were jointly detected by both platforms. Conversely, 126 gene products were detected exclusively by proteomics, being somehow missed by the gene chip platform (Maziarz et al. 2005). In recent years, the growing number of islet proteomic data necessitates development of bioinformatics tools for easy data handling and data mining to assign subcellular location, functional properties, molecular networks, and known potential posttranslational modifications. It is becoming essential to create a common platform for islet proteomic users integrating molecular, cellular, phenotypic, and clinical information with experimental genetic and proteomic data.

An important feature of proteomics is that protein isoforms generated by posttranslational modifications can be separated by 2DGE. Among the hundreds of different types of protein modifications, reversible protein phosphorylation is a key regulatory mechanism of cellular signaling processes (Hunter and Karin 1992; Jones and Persaud 1998; Hunter 2007). To detect global phosphoproteome profiles of islets, the advantages of the fluorescent dye Pro-Q Diamond, which is suitable for the fluorescent detection of phosphoserine-, phosphothreonine-, and phosphotyrosine-containing proteins in 2D gels directly (Steinberg et al. 2003), have been exploited, and 90 different phosphorylated proteins were detected in the 2D map (unpublished data). However, vanishingly small amounts of phosphorylated proteins in cells and lack of robotic picker in our lab precluded spot cutting and identification of most of the spots. Only a few including ATP synthase α chain, elongation factor 1-α, actin, γ-aminobutyric acid receptor α-3 subunit, and α-2-HS-glycoprotein could be successfully identified. Further isolation and purification of phosphoproteins and increasing the loading amount by pooling islet samples will possibly increase the chances for better identification for comparative studies to elucidate how posttranslational modifications regulate insulin secretion. LC-MS/MS analysis for posttranslational modifications of mouse islet proteome identified relatively abundant secretion-regulatory proteins including chromogranin A and secretogranin-2 (Petyuk et al. 2008). Then again, it's just the very minute tip of the phosphoproteome iceberg.

In recent years, several proteomic strategies have been developed/optimized for the specific enrichment and fractionation of phosphoproteins and peptides (Fila and Honys 2012; Beltran and Cutillas 2012) such as immobilized metal affinity chromatography (IMAC), metal-oxide affinity chromatography (MOAC), hydrophilic interaction liquid chromatography (HILIC), polymer-based metal ion affinity capture (polyMAC), and strong anionic ion-exchange chromatography (SAX)/strong cationic ion-exchange chromatography (SCX). Using several strategies for sample preparation, phosphopeptide enrichment, and linear ion trap MS/MS, Han et al. (2012) have generated INS-1E phosphoproteome with 2,467 distinct phosphorylations on 1,419 phosphoproteins. They have also detected novel phosphorylation sites in several proteins including Rab proteins which may regulate insulin granule exocytosis as well as GLUT4 trafficking.

Comparative and Quantitative Islet Proteomics

The ability of the islet of Langerhans to respond with proper insulin release when the ambient glucose concentration is changed is of fundamental importance for glucose homeostasis (Rorsman 1997). In diabetes mellitus this ability is impaired with reduction in both first and second phase insulin secretion (van Haeften 2002; Scheen 2004) which leads to postprandial hyperglycemia. In the search for islet-derived factors responsible for the deranged insulin secretion, isolated islets have typically been cultured under different conditions, and it is well documented that elevated glucose concentrations (11 mM) during culture are essential for maintaining islet β-cell functions (Ling et al. 1994). Individual islets from the NOD and *ob/ob* mouse, which are animal models of type 1 and type 2 diabetes, respectively (Tochino 1987; Wolf 2001), have demonstrated improved glucose-stimulated insulin secretion (GSIS) after exposure to high glucose in culture medium (Bergsten and Hellman 1993; Lin et al. 1999). Such beneficial effects on GSIS have been correlated to changes in expression of individual proteins like glucokinase, glucose transporter 2, and uncoupling protein 2 (Heimberg et al. 1995; Liang et al. 1992; Chan et al. 2004). However, molecular details of the phenotypic shift in response to elevated glucose are to a large extent unknown. Since GSIS is a multifactorial event, approaches capable of determining multiple proteins simultaneously are essential for the elucidation of molecular mechanisms responsible for changes in GSIS. 2DGE and MS have been employed to characterize changes in global islet protein expressions related to exposing islets to high glucose (Ahmed and Bergsten 2005). In this proteomic study, the prohormone convertase 2 and cytokeratin 8 appeared as distinct spots on 2D gels of islets exposed to high glucose, but the proteins were barely visible on gels of freshly isolated islets (Ahmed and Bergsten 2005). The observed glucose-induced changes in global protein expression pattern suggested that enhanced insulin synthesis, restoration of insulin content and granule pools, and increased chaperone activity and antioxidants are important mechanisms underlying the augmented secretory effect of glucose in mouse islets. In comparison to other discrete hypothesis-driven studies, this report, for first time, showed orchestrated changes of multiple islet proteins that may contribute to the enhanced GSIS observed in these islets (Ahmed and Bergsten 2005). From this proteomic study, it is unclear how glucose-induced increase in cytokeratin interacts with kinesin-microtubule system and contributes, if any, in enhanced glucose responsiveness. However, it has been conjectured that kinesin-dependent interaction of cytokeratin with microtubules is mediated by the insulin granules where cytokeratins can interact with various lipids of the insulin granules, which are anchored to microtubules through kinesin interaction. In support of this view, oligonucleotide microarray studies showed an increase in cytokeratin 19 gene in pancreatic β-cells exposed to high (25 mM) glucose compared to low glucose (5.5 mM) for 24 h (Webb et al. 2000). An increase in cytokeratin level in different cultured cells has also been reported (Alge et al. 2003; Poland et al. 2002). This type II cytoskeletal 8 protein (KRT8) has been detected in 2D maps of glucose-responding mouse islets, INS-1E cells (Ahmed et al. 2005b), and human islets (Ahmed et al. 2005a). In the search for glucose-responsive proteins, a 65 kDa protein has been detected in 2D map of mouse islets (Collins et al. 1990), and glucose-induced synthesis of this protein was blocked by D-mannoheptulose, a specific blocker of glucose phosphorylation and metabolism. However, isolation and characterization of this protein has not been performed. Among the 2,000 different islet protein spots, 1.5 % was reported to be regulated by glucose in physiological concentration range (Collins et al. 1992). In another study, depolarization-induced Ca^{2+} influx and insulin release was found to be highly correlated with phosphorylation of a 60 kDa protein (Schubart 1982). Identification of this phosphoprotein revealed an intermediate filament protein of the keratin class in hamster insulinoma cells and in pancreatic islets (Schubart and Fields 1984). This cytokeratin protein exists in both a phosphorylated and unphosphorylated state and corresponds to the gel position of KRT8 detected by Ahmed and Bergsten (2005; Ahmed et al. 2005a, b). The gel position of the unidentified glucose-responsive 65 kDa protein also matches with the KRT8. In support of the suggestion that cytokeratin may be involved in the regulation of insulin release, cytokeratins 7, 8, 18, and 19 were localized to adult endocrine pancreas and insulinoma cells by immunohistochemistry and immunoblot analysis (Schubart and Fields 1984; Kasper et al. 1991; Farina and Zedda 1992; Francini et al. 2001), and it has been well documented that disturbances in cytoskeleton of the pancreatic β-cells drastically reduced their insulin secretory function and lifetime (Blessing et al. 1993).

Comparative proteomics of glucose-responsive and glucose-nonresponsive MIN6 cells using 2D differential in-gel electrophoresis (DIGE) (Dowling et al. 2006) also contributed to the understanding of the proteins involved in GSIS. Similar to the findings of Ahmed et al. (2005), they showed that glucose-nonresponsive cells have lower ER chaperone proteins (e.g., PDI, GRP78, endoplasmin, endoplasmic reticulum protein 29) and decreased antioxidative enzymes (e.g., carbonyl reductase 3, peroxiredoxin 4, and superoxide dismutase 1), suggesting proper protein folding and protection against oxidative stress are required for glucose-stimulated insulin release from pancreatic β-cells. To dissect the molecular events associated with β-cell dysfunction and development of diabetes, Lu et al. (2008) characterized global islet protein and gene expression changes in diabetic MKR mice and compared with nondiabetic control mice. Using iTRAQ, 159 proteins were found to be differentially expressed in MKR; marked upregulation of protein biosynthesis and endoplasmic reticulum stress pathways and parallel downregulation in insulin processing/secretion, energy utilization, and metabolism were observed. One hundred fifty-four of the differentially expressed proteins were able to be mapped to probe IDs on the microarray. In this study about 45.2 % of the differentiated proteins showed concordant changes (i.e., changes in the same direction) in mRNA, 0.6 % were discordant (i.e., having higher protein expression but lower mRNA expression), and notably 54.2 % showed changes in the proteome but not in the transcriptome. Similar approaches have been used for better understanding of the cellular and molecular functions of the signaling pathway of insulin synthesis and release in human β-cells (Jin et al. 2009). Of the 97 differentially expressed proteins involved in improved insulin release, the changes in protein and mRNA expression for 49 proteins (50.5 %) were in the same direction, while they moved oppositely for 14 proteins (14.4 %). Thirty-four of the 97 differentially expressed proteins were identified by protein expression but not by mRNA expression. The proteomic and genomic data indeed supplement each other and suggest a posttranscriptional and/or posttranslational regulation of a substantial number of differentially expressed proteins involved in islet function.

Imaging mass spectrometry (IMS) has been applied to identify differential expression of peptides in thin tissue section of pancreas of control and *ob/ob* mice (Minerva et al. 2008). Improvement and successful application of the IMS may lead to the discovery of new disorder-specific peptide biomarkers with potential applications in disease diagnosis. Protein expression profiling in fetal rat islets after protein restriction during gestation expanded our knowledge in the pathogenesis of type 1 and type 2 diabetes (Sparre et al. 2003).

Glucolipotoxicity and Islet Proteomics

Whereas glucose is the most important physiological stimulus for insulin secretion, chronic hyperglycemia causes desensitization and impairment of insulin release in response to glucose (Unger and Grundy 1985; Leahy et al. 1986; Eizirik et al. 1992; Purrello et al. 1996). Similarly, a high-fat intake, particularly if rich in saturated fatty acids, is associated with impaired insulin sensitivity and secretion and development of type 2 diabetes (Manco et al. 2004). It is commonly accepted that acute exposure (1–3 h) of pancreatic islets to free fatty acid leads to stimulation of GSIS both in vitro (Malaisse and Malaisse-Lagae 1968; Goberna et al. 1974; Campillo et al. 1979; Gravena et al. 2002) and in vivo (Paolisso et al. 1995; Carpentier et al. 1999; Boden and Chen 1999). However, the impact of long-term (>6 h) FFA exposure remains controversial (Ong et al. 2002; Park et al. 2009; Petricoin and Liotta 2004). The discrepancies may depend on the circulating free fatty acid levels and also on the percentage of unsaturation of the fatty acids (Dobbins et al. 2002; El-Assaad et al. 2003). It has been proposed that an increased FFA concentration alone is insufficient to induce β-cell failure and that an elevation of FFAs combined with high glucose is required to result in β-cell malfunction (Briaud et al. 2001;

Poitout and Robertson 2008), possibly as a result of accumulation of harmful lipid metabolites, e.g., ceramide in the cytoplasm (Prentki et al. 2002; Poitout 2008). This in turn is believed to interfere with the ability of the β-cells to respond to glucose with enhanced insulin secretion. Although the concept of glucolipotoxicity has become very popular and often debated, the underlying causes as well as functional consequences remain poorly defined. The main dietary fatty acids palmitate and oleate modulate the immediate early response genes, *c-fos* and *nur-77*, and a number of late genes of fatty acid metabolism including acetyl CoA carboxylase and fatty acid synthase (Roche et al. 1999). By analyzing global gene expression profiles in chronic fatty acid-treated MIN6 cells, it was found that the major groups of genes regulated by fatty acids are metabolic enzymes, transcription factors, and genes controlling distal secretory processes (Busch et al. 2002). However, in another study long-term lipid infusion in normal rats showed little influence on broad spectrum of islet-associated genes (Steil et al. 2001). A series of selected "candidate genes" have also been studied recently (Olofsson et al. 2007). The insulin (Ins1) and Glut2 transcript levels were significantly downregulated in the presence of both palmitate and oleate. Transcription of the mitochondrial acyl-CoA transporter carnitine palmitoyltransferase I (CPT1) was upregulated almost fourfold. In contrast to previous findings (Assimacopoulos-Jeannet et al. 1997; Lameloise et al. 2001; Joseph et al. 2004), the uncoupling protein UCP-2 was upregulated twofold in the presence of high glucose, but no additional effect by FFAs was detected (Olofsson et al. 2007). Therefore, it has been suggested that the failure of glucose to stimulate insulin secretion from FFA-pretreated islets is conceivably not due to increased uncoupling and reduced ATP generation (Olofsson et al. 2007). However, conflicting opinion also exists since Western blot analysis indicates that high glucose and fatty acid synergistically impaired the production of ATP in β-cells through reduction of ATP synthase β-subunit protein expression (Kohnke et al. 2007). Interestingly, we have found that the expression of ATP synthase subunit α (1.21-fold) and ATP synthase subunit β (1.16-fold) was significantly increased ($p < 0.05$) in islets isolated from high-fat-fed mice (unpublished data). In this gel-free LC-MS/MS-based proteomic study, compared to control mice, islets from high-fat-fed mice showed differential expression of 1,008 proteins. In accordance with the previous findings of fatty acid-induced inhibition of insulin gene transcription (Olofsson et al. 2007), insulin-degrading enzymes (Bhathena et al. 1985) are highly overexpressed in islets isolated from high-fat-fed mice, whereas both insulin 1 precursor and glucagon precursor are downregulated. Top 10 downregulated proteins in high-fat-diet islets include ARF (ADP ribosylation factor) GTPase-activating protein GIT1, flavin adenine dinucleotide (FAD) synthetase, CPT1, laminin subunit β2 precursor, γ-aminobutyric acid receptor subunit α-3, vesicle transport protein SEC20, reticulon 1, early endosome antigen 1, β-1,4-mannosyl-glycoprotein 4-β-N-acetylglucosaminyltransferase, and tudor domain-containing protein 5. The largely downregulated proteins include kelch-like protein 8, leucine-rich repeat-containing protein 8D, transcription factor E3, ras-related protein Rab 11B, Na^+-K^+ ATPase subunit α2 precursor, putative ATP-dependent RNA helicase DHX33, SCHAD,

F-actin capping protein subunit β, arylacetamide deacetylase, and type I inositol-3,4-bisphosphate 4-phosphatase. The vast amount of lipotoxicity proteomic data contains many novel proteins and opens new avenues for islet researchers. SELDI-TOF analysis of INS-1E cells and pancreatic islets exposed to fatty acids identified calmodulin, CPT1, and peptidylprolyl isomerase B as fatty acid-regulated proteins (Sol et al. 2009; Ortsäter et al. 2007). Using 2D-DIGE and MALDI-TOF/TOF, Maris et al. (2013) have recently demonstrated orchestrated regulation of chaperones, insulin processing and ubiquitin-related proteasomal degradation, vesicular transport and budding, as well as generation of toxic metabolites during triglyceride synthesis contribute to the fatty acid-induced β-cell dysfunction. They have shown that in the presence of high glucose, oleate and palmitate induce shunting of excess glucose and increase mitochondrial reactive oxygen species production which promote β-cell death (Schrimpe-Rutledge et al. 2012; Schubart 1982). To unravel the mechanism linking obesity to the development of T2D, Han et al. (2011) have applied iTRAQ proteomic strategy, and the differentially expressed proteins in Zucker diabetic fatty rats compared to Zucker fatty rats suggest that decreased mitochondrial oxidation, unbalance of TG/FFA cycling, and microvascular endothelial dysfunction may link obesity to T2D.

In a pioneering glucotoxicity proteomic study, Collins et al. (1992) used 2DGE of ^{35}S-methionine-labeled islet proteins that were exposed in vivo or in vitro to either low or high glucose. Approximately 2,000 protein spots were detected in 2D gels and 1.5 % and 1.6 % detectable proteins showed differential expression in response to prolonged glucose load in vitro and in vivo model, respectively. Lack of mass spectrometry did not allow protein identification of those glucose-responsive proteins. Schuit et al. (2002) purified rat β-cells and performed 2DGE of ^{35}S-methionine-labeled proteins synthesized over 4 h at 10 mM glucose after 10 days culture in low (6 mM) or high (20 mM) glucose. They distinguished two patterns of β-cell proteome change between 6 and 20 mM glucose. In one pattern two spots corresponding to proinsulin were increased almost ninefold in the presence of high glucose. Similar to this finding, on the 2DG map of INS-1E cells, proinsulin appeared as two spots. However, while one spot showed almost twofold upregulation in the presence of high glucose (25 mM), the other spot was fivefold downregulated by high glucose compared to exposure to low glucose (5.5 mM, unpublished data). The other pattern described by Schuit et al. (2002) showed suppression of translation of multiple spots close to pH 7 on 2D gels when the β-cells were exposed to 20 mM glucose. However, the identities of these protein spots were not determined. SELDI-TOF analysis of the different mitochondrial samples from INS-1E cells incubated for 5 days at 5.5, 11, 20, and 27 mM glucose showed 34 differentially expressed peaks among the samples (Nyblom et al. 2006). Such changes in expression of proteins were correlated with impairment of GSIS. Nevertheless, no identification of the differentially expressed peptides has been carried out. Comparison of INS-1E mitochondrial 2DG proteome revealed 75 spots showing twofold or more significant changes ($p < 0.05$) in relative abundance in the presence of 20 mM glucose compared to the cells exposed to 5.5 mM glucose.

Thirty-three protein spots appear only on the mitochondrial map of the INS-1E cells exposed to 5.5 mM glucose. Mitochondrial protein spots downregulated in glucotoxic conditions include ATP synthase α chain and δ chain, stress-70 protein, mitochondrial (75 kDa glucose-regulated protein; GRP75; HSPA9), malate dehydrogenase, aconitase, SCHAD, trifunctional enzyme β-subunit and NADH-cytochrome b5 reductase, and voltage-dependent anion-selective channel protein (VDAC) 2. There was upregulation of protein spots corresponding to VDAC3, trifunctional enzyme α subunit, heat shock protein 60, mitochondrial (HSP60), and 10 kDa heat shock protein, mitochondrial (HSP10). Typical to 2D map single protein appeared in multiple spots and several proteins co-migrated. For example, on the mitochondrial 2D map, five different spots corresponding to VDAC1 appeared at same molecular weight but having different pI. Three spots showed overexpression in response to high glucose and two other spots were downregulated. Changes in expression of a single isoform (spots) of a protein in 2DG do not necessarily signify alteration in total protein amount. Therefore, caution should be undertaken before concluding expression level of a protein on 2DG without validating the data with Western blot or other methods. In addition to the mitochondrial proteins, other differentially expressed proteins in glucotoxic condition include proinsulin; calreticulin; protein disulfide-isomerase A6 (PDIA6); PKC substrate 60.1 kDa protein; hypoxia upregulated protein 1 (ORP150); endoplasmin; heat shock cognate 71 kDa protein (HSPA8); heterogeneous nuclear ribonucleoproteins D0 and A2/B1; lamin B1; histones H2B, H3.3, and H4; and elongation factor 1-α-1. With label-free LC-MS/MS approach, 353 proteins were found to be differentially expressed in INS-1E cells exposed to 25 mM glucose compared to the cells cultured in the presence of 5.5 mM glucose (unpublished data). Ingenuity pathways analysis (IPA) revealed strong association of differentially expressed proteins with energy production, lipid metabolism, protein synthesis, DNA replication, recombination and repair, cell signaling, and metabolic disease. Using IPA we mapped biological networks affected by the differentially expressed proteins between 5.5 and 25 mM glucose-exposed INS-1E cells. Figure 3 shows the network involved in endocrine system development and function, lipid metabolism, and small-molecule biochemistry. In INS-1E cells exposed to 25 mM glucose, N-methylpurine DNA glycosylase (MPG) showed significant upregulation, while carboxypeptidase E (CPE) was fourfold downregulated. Other substantially downregulated proteins in response to high glucose exposure included chromogranin A (CGA), membrane-associated guanylate kinase (MAGI1), ubiquitin protein ligase E3 component n-recognin 5 (UBR5), and mitofusin (MFN). Although fold change is a commonly used criterion in quantitative proteomics, it does not provide an estimation of false-positive and false-negative rates that are often likely in a large-scale quantitative proteomic analysis such as in label-free LC-MS/MS quantitation. It is therefore essential for the islet researchers to adopt effective significance analysis of proteomic data which is particularly useful in the estimation of false discovery rates (Roxas and Li 2008). The proteomic data from 2DGE and LC-MS/MS analysis of the glucotoxic studies provide a comprehensive overview of the orchestrated changes in expression of multiple proteins involved in nutrient metabolism, energy production, nucleic acid metabolism, cellular defense, glycoprotein folding, molecular transport, protein trafficking, RNA damage and repair, DNA replication, apoptosis signaling, and mDNA stability. Farnandez et al. (2008) have correlated proteomic data with metabolomic findings in glucotoxic conditions in INS-1 β-cell line. While 75 proteins showed differential expression in the presence of high glucose, only 5 of those proteins were found to be involved in the observed metabolomic alterations, suggesting allosteric regulation and/or posttranslational modifications are more important determinants of metabolite levels than enzyme expression at the protein level (Fernandez et al. 2008). Combined SELDI-TOF and 2DGE approach identified 11 different proteins coupled to altered insulin release in response to high glucose (20 mM) (Sol EM, personal communication). A recent study on human islets using 2D fractionation and LC-MS/MS (Schrimpe-Rutledge et al. 2012) has identified 256 differentially proteins in response to high glucose. This study has identified a number of glucose-regulated proteins whose functions are presently unknown in β-cells, including pleiotropic regulator 1, retinoblastoma-binding protein 6, nuclear RNA export factor 1, Bax inhibitor 1, and synaptotagmin-17.

Type 1 Diabetes and Islet Proteomics

Type 1 diabetes (T1D) is an autoimmune disorder characterized by selective destruction of insulin-producing β-cells in the pancreas resulting from the action of environmental factors on genetically predisposed individuals (Kim and Lee 2009). The prevailing view for the pathogenesis of type 1 diabetes is that an autoimmune reaction, where cytokines play an important role, causes destruction of the β-cells (Eizirik and Mandrup-Poulsen 2001). Numerous reports have demonstrated both in rodent and human islets that interleukin-1β (IL-1β) alone or in combination with interferon-γ (IFN-γ) and tumor necrosis factor α (TNF-α) affects the transcription and translation of genes, which have been implicated in β-cell destruction (Mandrup-Poulsen 2001). To search for novel proteins involved in cytokine-induced destruction of β-cells, 2DGE has been used (Andersen et al. 1995). This approach has detected upregulation of 29 proteins on 2DG image of rat islets exposed to IL-1β compared to control islets, and addition of nicotinamide reduced the upregulation of 16 IL-1β-induced proteins (Andersen et al. 1995). In a subsequent study (Andersen et al. 1997), on 2D gels of ^{35}S-methionine-labeled rat islets, 52 spots were upregulated, 47 downregulated, and 6 synthesized de novo by IL-1β. Among these 105 differentially expressed proteins, 23 protein spots were found to be significantly affected when nitric oxide (NO) production was prevented, suggesting a major role of NO-independent IL-1β-mediated regulation of gene expression (John et al. 2000). Mass spectrometric analysis allowed identification of 15 proteins, which were most profoundly altered by cytokine treatment (John et al. 2000). Also, on the transcription level similar approaches have been made to search for genes involved in the

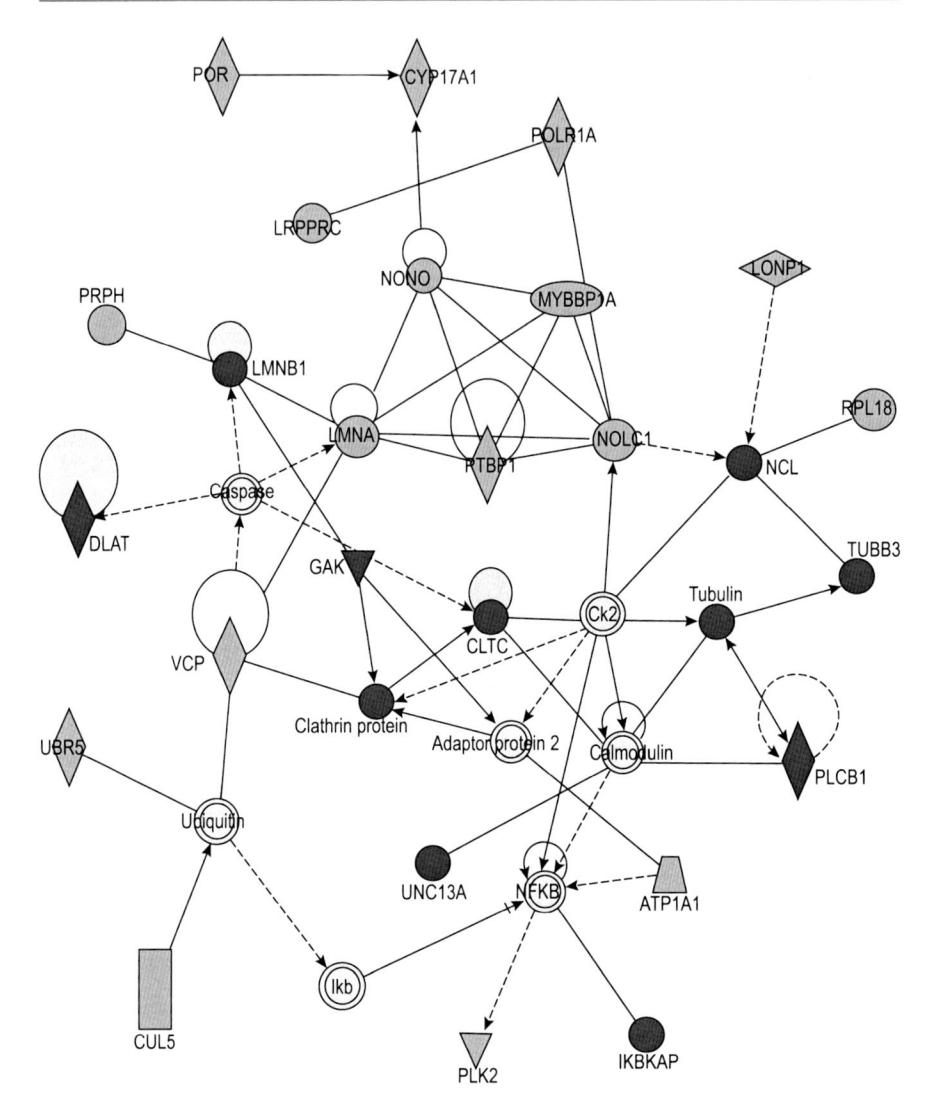

Fig. 3 Ingenuity pathway network obtained on a set of differentially regulated proteins detected in INS-1E cells exposed to 25 mM glucose compared to the cells cultured in the presence of 5.5 mM glucose. Proteins with a gray background were downregulated by high glucose, while other interacting proteins with a dark background were upregulated. *ATP1A1* ATPase, Na$^+$-K$^+$ transporting, α1 polypeptide, *CK2* casein kinase 2, *CLTC* clathrin, heavy chain, *CUL5* cullin 5, *CYP17A1* cytochrome P450, family 17, subfamily A, polypeptide 1, *DLAT* dihydrolipoamide S-acetyltransferase, *GAK* cyclin G-associated kinase, *Ikb* inhibitor of nuclear factor of κ light polypeptide gene enhancer in β-cells, β, *IKBKAP* inhibitor of kappa light polypeptide enhancer in β-cells, kinase complex-associated protein, *LMNA* lamin A/C, *LMNB1* lamin B1, *LONP1* lon peptidase 1, mitochondrial, *LRPPRC* leucine-rich PPR-motif containing, *MYBBP1A* MYB-binding protein (P160) 1a (p53-activated protein-2), *NCL* nucleolin, *NFKB* nuclear factor of κ light polypeptide gene enhancer in β-cells, *NOLC1* nucleolar and coiled-body phosphoprotein 1, *NONO* non-POU-domain-containing, octamer-binding protein, *PLCB1* phospholipase C, β

cytokine-induced alterations (Chen et al. 1999). Both these powerful approaches have yielded important information about putative genes/proteins involved in the development of the disease. Larsen et al. (2001) identified 57 different proteins from IL-1β-exposed rat islets and categorized them into several functional groups including energy transduction; glycolytic pathway; protein synthesis, chaperones, and protein folding; and signal transduction. Results of this differential expression analysis suggest that islet exposure to cytokines induces a complex pattern in β-cells comprising protective (e.g., upregulation of stress proteins) as well as deleterious (e.g., iNOS induction and NO production) events (Karlsen et al. 2001). The overall picture of the proteomic studies of type 1 diabetes is complex and does not allow us to predict which protein changes may be considered "primary" or "secondary" in importance, time, and sequence (Larsen et al. 2001). An integrative analysis method was developed combining genetic interactions using type 1 diabetes genome scan data and a high-confidence human protein interaction network (Bergholdt et al. 2007). Using this network analysis of the differentially expressed proteins in INS-1E cells exposed to cytokines, 42 of the differentially expressed proteins constituted a significant interaction network, suggesting extensive cross talk between the different proteins and the pathways in which they are involved with some proteins such as the chaperones GRP78, HSPA8, and GRP75 and the RNA synthesis/turnover proteins placed at the center of different networks. In fact all these islet proteomic studies strongly suggest a protective role of the chaperones in regulating β-cell dysfunction.

T1D is traditionally characterized by the presence of autoantibodies against β-cell proteins. However, up to 2–5 % of T1D patients exhibit no commonly detectable autoantibodies, and it has been speculated that the current T1D autoantigen panel is incomplete. To elucidate possible existence of other autoantigens in T1D, Massa et al. (2013) have applied the serological proteome analysis (SERPA) where they exploited the sera of T1D patients, who are negative for currently used T1D autoantibodies, for immunoblotting against 2DGE separated human islet proteins and detected 11 proteins by MS. Among the 11 proteins Rab GDP dissociation inhibitor β (GDIβ) is a possible novel candidate for T1D autoantigen. Proteomic approaches have also led to discover doublecortin as biomarker for β-cell injury in vitro (Jiang et al. 2013).

Fig. 3 (continued) 1, *PLK2* polo-like kinase 2, *POLR1A* polymerase (RNA) I polypeptide A, *POR* P450 (cytochrome) oxidoreductase, *PRPH* peripherin, *PTBP1* polypyrimidine tract-binding protein 1 (heterogeneous nuclear ribonucleoprotein I), *RPL18* ribosomal protein *L18*, *TUBB3* tubulin β-3, *UBR5* ubiquitin protein ligase E3 component n-recognin 5, *UNC13A* unc-13 homolog A, *VCP* valosin containing protein ◇ = enzyme; ◇ = peptidase; ⌂ = transporter; ☐ = ion channel, ⌾ = transcription regulator; ◉ = group or complex; ▽ = kinase; ○ = other. ——— = direct interaction; - - - - - = indirect interaction; ——— = binding only; ⟶ = acts on; ⟶⫯ = inhibits and acts on

Pharmacoproteomics and Pancreatic Islets

A potential application of proteomics in islet research is the detection of molecular alterations in diabetes and further characterization of existing or new drug (Chapal et al. 2004). One of the prime targets for the treatment of diabetes is to enhance the insulin sensitivity so that the tissues can precisely utilize glucose and keep its plasma level within physiological limit. Rosiglitazone, a member of the thiazolidinedione (TZD) class of antidiabetic agents, improves insulin sensitivity in both liver and peripheral tissues. TZDs bind to and activate the peroxisome proliferator-activated receptor (PPARγ) and regulate the coordinated expression of multiple genes that integrate the control of energy, glucose, and lipid homeostasis and therefore contribute to increased insulin sensitivity. Rosiglitazone has been shown to prevent islet cell hyperplasia and protect islets from toxic agents (Buckingham et al. 1998; Lin et al. 2005; Kim et al. 2007). In an elegant study using 2DGE, Sanchez et al. (2002) compared protein expression profiles of pancreatic islets from obese diabetic C57BL/6J *lep/lep* mice and their lean littermates treated with rosiglitazone. They identified nine differentially expressed proteins between lean and obese diabetic untreated mice. The expression levels of four of those nine proteins (tropomyosin 1, profilin, profilin fragment, and fatty acid-binding protein) were significantly modulated by rosiglitazone treatment of the obese mice. In a second set of experiments designed to identify proteins potentially associated with a low islet cell mass, they compared the islet protein expression between C57BL/6J and C57BL/Ks mice. The C57BL/Ks mice have a twofold less islet cell mass as compared with the C57BL/6J (Swenne and Andersson 1984) and, as a consequence, were more susceptible to diabetes (Kaku et al. 1989; Korsgren et al. 1990). Thirty-one proteins were found to be differentially expressed between the two mouse models and two of them, tropomyosin 1 and profilin, showed the same differential pattern between C57BL/Ks and obese diabetic C57BL/6J *lep/lep* mice. Taken together, these results suggest that actin-binding proteins could play an important role in defective islet function. We have a long way to go for the development of novel actin-modulating drugs for treatment of diabetes similar to microtubule-interacting or microtubule-stabilizing drugs developed for cancer treatment (Trivedi et al. 2008; Anchoori et al. 2008). In a recent study, the effects of imidazolines have been tested on rat islet proteome (Jagerbrink et al. 2007) with the optimism that if it were possible to develop one of the them into a drug, this compound may be effective without risk of hypoglycemic shock in subjects with low or normal blood glucose as imidazolines increase insulin release selectively at high glucose concentrations (Efanov et al. 2001). 2DG analysis revealed 53 differentially expressed proteins between imidazoline-treated and imidazoline-nontreated islets. Of special interest among the differentially expressed proteins are those involved in protecting cells from misfolded proteins (HSP60, PDI, and calreticulin), Ca^{2+} binding (calgizzarin, calcyclin, and annexin I), and metabolism or signaling (pyruvate kinase, α-enolase, and protein kinase C inhibitor 1). However, elucidation of exact mechanism of action of imidazolines and validation of targets require further studies.

Natural medicinal plant extracts and active components have antidiabetic activity (Jung et al. 2006), and the extracellular polysaccharides (EPS) obtained from mycelia culture of *Phellinus baumii* have strong hypoglycemic activity. Proteomic study provided insights into the mechanism of antidiabetic activity of the EPS in type 1 diabetes (Kim et al. 2008). 2DGE image analysis and mass spectrometry identified 10 downregulated and 16 upregulated proteins in streptozotocin-treated diabetic mice islets. The altered level of all these differentially expressed proteins was partially or fully restored to normal level by EPS treatment. The interesting downregulated proteins in diabetic model include cholesterol esterase, PDI, and islet regenerating protein, whereas the upregulated proteins are Cu-Zn superoxide dismutase, carbonyl reductase, GRP58, hydroxymethylglutaryl-CoA synthase, and putative human mitogen-activated protein kinase activator with WD repeat-binding protein. One advantage of this study is that the proteomic data was indeed supported by transcriptomics. It would be interesting to know how alteration of certain specific protein targets modulates the development and progress of type 1 diabetes. In a recent study, using proteomic approaches it has been demonstrated that RhoGDI-α/JNK pathway might be the focus of therapeutic target for the prevention of mycophenolic acid-induced islet apoptosis (Park et al. 2009).

Conclusion

During the last decade state-of-the-art proteomic technologies including the 2DGE and label-free LC-MS/MS quantitation have been applied to dissect the pathophysiology of islet function in an increasingly manner. A vast array of proteomic data has emerged from these studies providing molecular and comprehensive snapshot of complex disease process involving the pancreatic islet cells – but just like a trace of light through an age-old dark cave, coming from the ocean of bright light. Careful analysis and powerful bioinformatics tools are still required for functional summary of the datasets and generation of new hypothesis. These proteomic studies are indeed very early steps towards better understanding of the mechanism of pathophysiology of diabetes and providing new approaches for the prevention and treatment of the disease. Almost no functional proteomics has been performed in islet research. However, improvement and easy availability of high-throughput proteomic techniques will hopefully draw the attention of more islet biologist and generate more functional data. An important feature of diabetes is that it is a progressive condition. Pancreatic β-cell function, in particular, shows a progressive decline in the prediabetic phase and in established diabetes. To clearly define islet function, therefore, we need to measure it over a period of time amalgamating multiple platforms and involving cell biologists, physiologists, geneticists, and biochemists working together with proteomic specialists. A large-scale study will allow this, together with the detection of changes in islet protein patterns, and other metabolic traits will lead to a better understanding of how susceptible gene variants and their protein products predispose to diabetes. This will also help to explore novel biomarkers to predict future diabetes, for better understanding of the pathophysiology of diabetes, to reveal drug targets, as well as optimize the selection of molecules that interact with these targets.

Cross-References

▶ In vivo Biomarkers for Detection of β Cell Death

References

Abbott A (1999) A post-genomic challenge: learning to read patterns of protein synthesis. Nature 402:715–720
Adam BL, Vlahou A, Semmes OJ, Wright GL Jr (2001) Proteomic approaches to biomarker discovery in prostate and bladder cancers. Proteomics 1:1264–1270
Aebersold R, Goodlett DR (2001) Mass spectrometry in proteomics. Chem Rev 101:269–295
Aebersold R, Mann M (2003) Mass spectrometry-based proteomics. Nature 422:198–207
Aggarwal K, Choe LH, Lee KH (2006) Shotgun proteomics using the iTRAQ isobaric tags. Brief Funct Genomic Proteomic 5:112–120
Ahmed M, Bergsten P (2005) Glucose-induced changes of multiple mouse islet proteins analysed by two-dimensional gel electrophoresis and mass spectrometry. Diabetologia 48:477–485
Ahmed M, Forsberg J, Bergsten P (2005a) Protein profiling of human pancreatic islets by two-dimensional gel electrophoresis and mass spectrometry. J Proteome Res 4:931–940
Ahmed M, Bergsten P, McCarthy M, Rorsman P (2005b) Protein profiling of INS-1E cells by two-dimensional gel electrophoresis and mass spectrometry. Diabetologia 48(A):162
Alge CS, Suppmann S, Priglinger SG, Neubauer AS, May CA, Hauck S, Welge-Lussen U, Ueffing M, Kampik A (2003) Comparative proteome analysis of native differentiated and cultured dedifferentiated human RPE cells. Invest Ophthalmol Vis Sci 44:3629–3641
America AH, Cordewener JH (2008) Comparative LC-MS: a landscape of peaks and valleys. Proteomics 8:731–749
Anchoori RK, Kortenhorst MS, Hidalgo M, Sarkar T, Hallur G, Bai R, Diest PJ, Hamel E, Khan SR (2008) Novel microtubule-interacting phenoxy pyridine and phenyl sulfanyl pyridine analogues for cancer therapy. J Med Chem 51:5953–5957
Andersen HU, Larsen PM, Fey SJ, Karlsen AE, Mandrup-Poulsen T, Nerup J (1995) Two-dimensional gel electrophoresis of rat islet proteins. Interleukin 1 β-induced changes in protein expression are reduced by L-arginine depletion and nicotinamide. Diabetes 44:400–407
Andersen HU, Fey SJ, Larsen PM, Nawrocki A, Hejnaes KR, Mandrup-Poulsen T, Nerup J (1997) Interleukin-1β induced changes in the protein expression of rat islets: a computerized database. Electrophoresis 18:2091–2103
Anderson NL, Anderson NG (1998) Proteome and proteomics: new technologies, new concepts, and new words. Electrophoresis 19:1853–1861
Anderson NG, Matheson A, Anderson NL (2001) Back to the future: the human protein index (HPI) and the agenda for post-proteomic biology. Proteomics 1:3–12
Angel TE, Aryal UK, Hengel SM, Baker ES, Kelly RT, Robinson EW, Smith RD (2012) Mass spectrometry-based proteomics: existing capabilities and future directions. Chem Soc Rev 41:3912–3928
Assimacopoulos-Jeannet F, Thumelin S, Roche E, Esser V, McGarry JD, Prentki M (1997) Fatty acids rapidly induce the carnitine palmitoyltransferase I gene in the pancreatic β-cell line INS-1. J Biol Chem 272:1659–1664
Bartel DP (2009) MicroRNAs: target recognition and regulatory functions. Cell 136:215–233

Beltran L, Cutillas PR (2012) Advances in phosphopeptide enrichment techniques for phosphoproteomics. Amino Acids 43:1009–1024

Bergholdt R, Storling ZM, Lage K, Karlberg EO, Olason PI, Aalund M, Nerup J, Brunak S, Workman CT, Pociot F (2007) Integrative analysis for finding genes and networks involved in diabetes and other complex diseases. Genome Biol 8:R253

Bergsten P, Hellman B (1993) Glucose-induced cycles of insulin release can be resolved into distinct periods of secretory activity. Biochem Biophys Res Commun 192:1182–1188

Bertone P, Snyder M (2005) Advances in functional protein microarray technology. FEBS J 272:5400–5411

Bhathena SJ, Timmers KI, Oie HK, Voyles NR, Recant L (1985) Cytosolic insulin-degrading activity in islet-derived tumor cell lines and in normal rat islets. Diabetes 34:121–128

Binz PA, Muller M, Hoogland C, Zimmermann C, Pasquarello C, Corthals G, Sanchez JC, Hochstrasser DF, Appel RD (2004) The molecular scanner: concept and developments. Curr Opin Biotechnol 15:17–23

Blessing M, Ruther U, Franke WW (1993) Ectopic synthesis of epidermal cytokeratins in pancreatic islet cells of transgenic mice interferes with cytoskeletal order and insulin production. J Cell Biol 120:743–755

Boden G, Chen X (1999) Effects of fatty acids and ketone bodies on basal insulin secretion in type 2 diabetes. Diabetes 48:577–583

Boden G, Chen X, Rosner J, Barton M (1995) Effects of a 48-h fat infusion on insulin secretion and glucose utilization. Diabetes 44:1239–1242

Briaud I, Harmon JS, Kelpe CL, Segu VB, Poitout V (2001) Lipotoxicity of the pancreatic β-cell is associated with glucose-dependent esterification of fatty acids into neutral lipids. Diabetes 50:315–321

Brunner Y, Coute Y, Iezzi M, Foti M, Fukuda M, Hochstrasser DF, Wollheim CB, Sanchez JC (2007) Proteomics analysis of insulin secretory granules. Mol Cell Proteomics 6:1007–1017

Buckingham RE, Al-Barazanji KA, Toseland CD, Slaughter M, Connor SC, West A, Bond B, Turner NC, Clapham JC (1998) Peroxisome proliferator-activated receptor-γ agonist, rosiglitazone, protects against nephropathy and pancreatic islet abnormalities in Zucker fatty rats. Diabetes 47:1326–1334

Busch AK, Cordery D, Denyer GS, Biden TJ (2002) Expression profiling of palmitate- and oleate-regulated genes provides novel insights into the effects of chronic lipid exposure on pancreatic β-cell function. Diabetes 51:977–987

Cahill DJ, Nordhoff E, O'Brien J, Klose J, Eickhoff H, Lehrach H (2001) Bridging genomics and proteomics. In: Pennington SR, Dunn MJ (eds) Proteomics: from protein sequence to function. BIOS Scientific Publishers Ltd, Oxford, pp 1–22

Campillo JE, Luyckx AS, Torres MD, Lefebvre PJ (1979) Effect of oleic acid on insulin secretion by the isolated perfused rat pancreas. Diabetologia 16:267–273

Canas B, Lopez-Ferrer D, Ramos-Fernandez A, Camafeita E, Calvo E (2006) Mass spectrometry technologies for proteomics. Brief Funct Genomic Proteomic 4:295–320

Carpentier A, Mittelman SD, Lamarche B, Bergman RN, Giacca A, Lewis GF (1999) Acute enhancement of insulin secretion by FFA in humans is lost with prolonged FFA elevation. Am J Physiol 276:E1055–E1066

Celis JE, Ostergaard M, Jensen NA, Gromova I, Rasmussen HH, Gromov P (1998) Human and mouse proteomic databases: novel resources in the protein universe. FEBS Lett 430:64–72

Chan CB, Saleh MC, Koshkin V, Wheeler MB (2004) Uncoupling protein 2 and islet function. Diabetes 53(Suppl 1):S136–S142

Chao TC, Hansmeier N (2013) Microfluidic devices for high-throughput proteome analyses. Proteomics 13:467–479

Chapal N, Molina L, Molina F, Laplanche M, Pau B, Petit P (2004) Pharmacoproteomic approach to the study of drug mode of action, toxicity, and resistance: applications in diabetes and cancer. Fundam Clin Pharmacol 18:413–422

Chen MC, Schuit F, Eizirik DL (1999) Identification of IL-1β-induced messenger RNAs in rat pancreatic β cells by differential display of messenger RNA. Diabetologia 42:1199–1203

Chiu DT (2010) Interfacing droplet microfluidics with chemical separation for cellular analysis. Anal Bioanal Chem 397:3179–3183

Collins HW, Buettger C, Matschinsky F (1990) High-resolution two-dimensional polyacrylamide gel electrophoresis reveals a glucose-response protein of 65 kDa in pancreatic islet cells. Proc Natl Acad Sci U S A 87:5494–5498

Collins H, Najafi H, Buettger C, Rombeau J, Settle RG, Matschinsky FM (1992) Identification of glucose response proteins in two biological models of β-cell adaptation to chronic high glucose exposure. J Biol Chem 267:1357–1366

Cretich M, Damin F, Pirri G, Chiari M (2006) Protein and peptide arrays: recent trends and new directions. Biomol Eng 23:77–88

Dobbins RL, Szczepaniak LS, Myhill J, Tamura Y, Uchino H, Giacca A, McGarry JD (2002) The composition of dietary fat directly influences glucose-stimulated insulin secretion in rats. Diabetes 51:1825–1833

Domon B, Broder S (2004) Implications of new proteomics strategies for biology and medicine. J Proteome Res 3:253–260

Dowling P, O'Driscoll L, O'Sullivan F, Dowd A, Henry M, Jeppesen PB, Meleady P, Clynes M (2006) Proteomic screening of glucose-responsive and glucose non-responsive MIN-6 β cells reveals differential expression of proteins involved in protein folding, secretion and oxidative stress. Proteomics 6:6578–6587

Efanov AM, Zaitsev SV, Mest HJ, Raap A, Appelskog IB, Larsson O, Berggren PO, Efendic S (2001) The novel imidazoline compound BL11282 potentiates glucose-induced insulin secretion in pancreatic β-cells in the absence of modulation of K_{ATP} channel activity. Diabetes 50:797–802

Eizirik DL, Mandrup-Poulsen T (2001) A choice of death – the signal-transduction of immune-mediated β-cell apoptosis. Diabetologia 44:2115–2133

Eizirik DL, Korbutt GS, Hellerstrom C (1992) Prolonged exposure of human pancreatic islets to high glucose concentrations in vitro impairs the β-cell function. J Clin Invest 90:1263–1268

El-Assaad W, Buteau J, Peyot ML, Nolan C, Roduit R, Hardy S, Joly E, Dbaibo G, Rosenberg L, Prentki M (2003) Saturated fatty acids synergize with elevated glucose to cause pancreatic β-cell death. Endocrinology 144:4154–4163

Espina V, Mehta AI, Winters ME, Calvert V, Wulfkuhle J, Petricoin EF 3rd, Liotta LA (2003) Protein microarrays: molecular profiling technologies for clinical specimens. Proteomics 3:2091–2100

Ezzell C (2002) Proteins rule. Sci Am 286:40–47

Farina V, Zedda M (1992) On the expression of cytokeratins and their distribution in some rabbit gland tissues. Eur J Histochem 36:479–488

Fenselau C (2007) A review of quantitative methods for proteomic studies. J Chromatogr B Analyt Technol Biomed Life Sci 855:14–20

Fernandez C, Fransson U, Hallgard E, Spegel P, Holm C, Krogh M, Warell K, James P, Mulder H (2008) Metabolomic and proteomic analysis of a clonal insulin-producing β-cell line (INS-1 832/13). J Proteome Res 7:400–411

Fields S (2005) High-throughput two-hybrid analysis. The promise and the peril. FEBS J 272:5391–5399

Figeys D (2003) Proteomics in 2002: a year of technical development and wide-ranging applications. Anal Chem 75:2891–2905

Fila J, Honys D (2012) Enrichment techniques employed in phosphoproteomics. Amino Acids 43:1025–1047

Florens L, Washburn MP (2006) Proteomic analysis by multidimensional protein identification technology. Methods Mol Biol 328:159–175

Francini F, Del Zotto H, Gagliardino JJ (2001) Effect of an acute glucose overload on Islet cell morphology and secretory function in the toad. Gen Comp Endocrinol 122:130–138

Geho D, Lahar N, Gurnani P, Huebschman M, Herrmann P, Espina V, Shi A, Wulfkuhle J, Garner H, Petricoin E 3rd, Liotta LA, Rosenblatt KP (2005) Pegylated, steptavidin-conjugated quantum dots are effective detection elements for reverse-phase protein microarrays. Bioconjug Chem 16:559–566

Goberna R, Tamarit J Jr, Osorio J, Fussganger R, Tamarit J, Pfeiffer EF (1974) Action of B-hydroxy butyrate, acetoacetate and palmitate on the insulin release in the perfused isolated rat pancreas. Horm Metab Res 6:256–260

Gorg A, Obermaier C, Boguth G, Harder A, Scheibe B, Wildgruber R, Weiss W (2000) The current state of two-dimensional electrophoresis with immobilized pH gradients. Electrophoresis 21:1037–1053

Gorg A, Weiss W, Dunn MJ (2004) Current two-dimensional electrophoresis technology for proteomics. Proteomics 4:3665–3685

Gravena C, Mathias PC, Ashcroft SJ (2002) Acute effects of fatty acids on insulin secretion from rat and human islets of Langerhans. J Endocrinol 173:73–80

Graves PR, Haystead TA (2003) A functional proteomics approach to signal transduction. Recent Prog Horm Res 58:1–24

Guest PC, Bailyes EM, Rutherford NG, Hutton JC (1991) Insulin secretory granule biogenesis. Co-ordinate regulation of the biosynthesis of the majority of constituent proteins. Biochem J 274(Pt 1):73–78

Gygi SP, Rochon Y, Franza BR, Aebersold R (1999) Correlation between protein and mRNA abundance in yeast. Mol Cell Biol 19:1720–1730

Han D, Moon S, Kim H, Choi SE, Lee SJ, Park KS, Jun H, Kang Y, Kim Y (2011) Detection of differential proteomes associated with the development of type 2 diabetes in the Zucker rat model using the iTRAQ technique. J Proteome Res 10:564–577

Han D, Moon S, Kim Y, Ho WK, Kim K, Kang Y, Jun H, Kim Y (2012) Comprehensive phosphoproteome analysis of INS-1 pancreatic β-cells using various digestion strategies coupled with liquid chromatography-tandem mass spectrometry. J Proteome Res 11:2206–2223

Heimberg H, De Vos A, Pipeleers D, Thorens B, Schuit F (1995) Differences in glucose transporter gene expression between rat pancreatic α- and β-cells are correlated to differences in glucose transport but not in glucose utilization. J Biol Chem 270:8971–8975

Hoogland C, Sanchez JC, Walther D, Baujard V, Baujard O, Tonella L, Hochstrasser DF, Appel RD (1999) Two-dimensional electrophoresis resources available from ExPASy. Electrophoresis 20:3568–3571

Hunter T (2007) The age of crosstalk: phosphorylation, ubiquitination, and beyond. Mol Cell 28:730–738

Hunter T, Karin M (1992) The regulation of transcription by phosphorylation. Cell 70:375–387

Hutton JC, Penn EJ, Peshavaria M (1982) Isolation and characterisation of insulin secretory granules from a rat islet cell tumour. Diabetologia 23:365–373

Issaq HJ, Veenstra TD, Conrads TP, Felschow D (2002) The SELDI-TOF MS approach to proteomics: protein profiling and biomarker identification. Biochem Biophys Res Commun 292:587–592

Jagerbrink T, Lexander H, Palmberg C, Shafqat J, Sharoyko V, Berggren PO, Efendic S, Zaitsev S, Jornvall H (2007) Differential protein expression in pancreatic islets after treatment with an imidazoline compound. Cell Mol Life Sci 64:1310–1316

Janzi M, Odling J, Pan-Hammarstrom Q, Sundberg M, Lundeberg J, Uhlen M, Hammarstrom L, Nilsson P (2005) Serum microarrays for large scale screening of protein levels. Mol Cell Proteomics 4:1942–1947

Jestin JL (2008) Functional cloning by phage display. Biochimie 90:1273–1278

Jia L, Lu Y, Shao J, Liang XJ, Xu Y (2013) Nanoproteomics: a new sprout from emerging links between nanotechnology and proteomics. Trends Biotechnol 31:99–107

Jiang L, Brackeva B, Stange G, Verhaeghen K, Costa O, Couillard-Despres S, Rotheneichner P, Aigner L, Van Schravendijk C, Pipeleers D, Ling Z, Gorus F, Martens GA (2013) LC-MS/MS identification of doublecortin as abundant β cell-selective protein discharged by damaged β cells in vitro. J Proteomics 80:268–280

Jin J, Park J, Kim K, Kang Y, Park SG, Kim JH, Park KS, Jun H, Kim Y (2009) Detection of differential proteomes of human β-cells during islet-like differentiation using iTRAQ labeling. J Proteome Res 8:1393

John NE, Andersen HU, Fey SJ, Larsen PM, Roepstorff P, Larsen MR, Pociot F, Karlsen AE, Nerup J, Green IC, Mandrup-Poulsen T (2000) Cytokine- or chemically derived nitric oxide alters the expression of proteins detected by two-dimensional gel electrophoresis in neonatal rat islets of Langerhans. Diabetes 49:1819–1829

Jones PM, Persaud SJ (1998) Protein kinases, protein phosphorylation, and the regulation of insulin secretion from pancreatic β-cells. Endocrinol Rev 19:429–461

Joseph JW, Koshkin V, Saleh MC, Sivitz WI, Zhang CY, Lowell BB, Chan CB, Wheeler MB (2004) Free fatty acid-induced β-cell defects are dependent on uncoupling protein 2 expression. J Biol Chem 279:51049–51056

Jung M, Park M, Lee HC, Kang YH, Kang ES, Kim SK (2006) Antidiabetic agents from medicinal plants. Curr Med Chem 13:1203–1218

Kaku K, Province M, Permutt MA (1989) Genetic analysis of obesity-induced diabetes associated with a limited capacity to synthesize insulin in C57BL/KS mice: evidence for polygenic control. Diabetologia 32:636–643

Karlsen AE, Sparre T, Nielsen K, Nerup J, Pociot F (2001) Proteome analysis – a novel approach to understand the pathogenesis of type 1 diabetes mellitus. Dis Markers 17:205–216

Kasper M, von Dorsche H, Stosiek P (1991) Changes in the distribution of intermediate filament proteins and collagen IV in fetal and adult human pancreas. I. Localization of cytokeratin polypeptides. Histochemistry 96:271–277

Kelly RT, Tang K, Irimia D, Toner M, Smith RD (2008) Elastomeric microchip electrospray emitter for stable cone-jet mode operation in the nanoflow regime. Anal Chem 80:3824–3831

Kelly RT, Page JS, Marginean I, Tang K, Smith RD (2009) Dilution-free analysis from picoliter droplets by nano-electrospray ionization mass spectrometry. Angew Chem Int Ed Engl 48:6832–6835

Kiehntopf M, Siegmund R, Deufel T (2007) Use of SELDI-TOF mass spectrometry for identification of new biomarkers: potential and limitations. Clin Chem Lab Med 45:1435–1449

Kikuchi T, Carbone DP (2007) Proteomics analysis in lung cancer: challenges and opportunities. Respirology 12:22–28

Kim HS, Lee MS (2009) Role of innate immunity in triggering and tuning of autoimmune diabetes. Curr Mol Med 9:30–44

Kim EK, Kwon KB, Koo BS, Han MJ, Song MY, Song EK, Han MK, Park JW, Ryu DG, Park BH (2007) Activation of peroxisome proliferator-activated receptor-γ protects pancreatic β-cells from cytokine-induced cytotoxicity via NFκB pathway. Int J Biochem Cell Biol 39:1260–1275

Kim SW, Hwang HJ, Baek YM, Lee SH, Hwang HS, Yun JW (2008) Proteomic and transcriptomic analysis for streptozotocin-induced diabetic rat pancreas in response to fungal polysaccharide treatments. Proteomics 8:2344–2361

Klose J (1975) Protein mapping by combined isoelectric focusing and electrophoresis of mouse tissues. A novel approach to testing for induced point mutations in mammals. Humangenetik 26:231–243

Kohnke R, Mei J, Park M, York DA, Erlanson-Albertsson C (2007) Fatty acids and glucose in high concentration down-regulates ATP synthase β-subunit protein expression in INS-1 cells. Nutr Neurosci 10:273–278

Korsgren O, Jansson L, Sandler S, Andersson A (1990) Hyperglycemia-induced B cell toxicity. The fate of pancreatic islets transplanted into diabetic mice is dependent on their genetic background. J Clin Invest 86:2161–2168

Kusnezow W, Hoheisel JD (2002) Antibody microarrays: promises and problems. Biotechniques 33(Suppl):14–23

Lalonde S, Ehrhardt DW, Loque D, Chen J, Rhee SY, Frommer WB (2008) Molecular and cellular approaches for the detection of protein-protein interactions: latest techniques and current limitations. Plant J 53:610–635

Lameloise N, Muzzin P, Prentki M, Assimacopoulos-Jeannet F (2001) Uncoupling protein 2: a possible link between fatty acid excess and impaired glucose-induced insulin secretion. Diabetes 50:803–809

Larsen PM, Fey SJ, Larsen MR, Nawrocki A, Andersen HU, Kahler H, Heilmann C, Voss MC, Roepstorff P, Pociot F, Karlsen AE, Nerup J (2001) Proteome analysis of interleukin-1-β-induced changes in protein expression in rat islets of Langerhans. Diabetes 50:1056–1063

Leahy JL, Cooper HE, Deal DA, Weir GC (1986) Chronic hyperglycemia is associated with impaired glucose influence on insulin secretion. A study in normal rats using chronic in vivo glucose infusions. J Clin Invest 77:908–915

Liang Y, Najafi H, Smith RM, Zimmerman EC, Magnuson MA, Tal M, Matschinsky FM (1992) Concordant glucose induction of glucokinase, glucose usage, and glucose-stimulated insulin release in pancreatic islets maintained in organ culture. Diabetes 41:792–806

Lin JM, Sternesjo J, Sandler S, Bergsten P (1999) Preserved pulsatile insulin release from prediabetic mouse islets. Endocrinology 140:3999–4004

Lin CY, Gurlo T, Haataja L, Hsueh WA, Butler PC (2005) Activation of peroxisome proliferator-activated receptor-γ by rosiglitazone protects human islet cells against human islet amyloid polypeptide toxicity by a phosphatidylinositol 3'-kinase-dependent pathway. J Clin Endocrinol Metab 90:6678–6686

Ling Z, Hannaert JC, Pipeleers D (1994) Effect of nutrients, hormones and serum on survival of rat islet β cells in culture. Diabetologia 37:15–21

Lopez JL (2007) Two-dimensional electrophoresis in proteome expression analysis. J Chromatogr B Analyt Technol Biomed Life Sci 849:190–202

Lu H, Yang Y, Allister EM, Wijesekara N, Wheeler MB (2008) The identification of potential factors associated with the development of type 2 diabetes: a quantitative proteomics approach. Mol Cell Proteomics 7:1434–1451

Malaisse WJ, Malaisse-Lagae F (1968) Stimulation of insulin secretion by noncarbohydrate metabolites. J Lab Clin Med 72:438–448

Manco M, Calvani M, Mingrone G (2004) Effects of dietary fatty acids on insulin sensitivity and secretion. Diabetes Obes Metab 6:402–413

Mandrup-Poulsen T (2001) β-cell apoptosis: stimuli and signaling. Diabetes 50(Suppl 1):S58–S63

Mann M, Jensen ON (2003) Proteomic analysis of post-translational modifications. Nat Biotechnol 21:255–261

Maris M, Waelkens E, Cnop M, D'Hertog W, Cunha DA, Korf H, Koike T, Overbergh L, Mathieu C (2011) Oleate-induced β cell dysfunction and apoptosis: a proteomic approach to glucolipotoxicity by an unsaturated fatty acid. J Proteome Res 10:3372–3385

Maris M, Robert S, Waelkens E, Derua R, Hernangomez MH, D'Hertog W, Cnop M, Mathieu C, Overbergh L (2013) Role of the saturated nonesterified fatty acid palmitate in β cell dysfunction. J Proteome Res 12:347–362

Mason TM, Goh T, Tchipashvili V, Sandhu H, Gupta N, Lewis GF, Giacca A (1999) Prolonged elevation of plasma free fatty acids desensitizes the insulin secretory response to glucose in vivo in rats. Diabetes 48:524–530

Massa O, Alessio M, Russo L, Nardo G, Bonetto V, Bertuzzi F, Paladini A, Iafusco D, Patera P, Federici G, Not T, Tiberti C, Bonfanti R, Barbetti F (2013) Serological Proteome Analysis (SERPA) as a tool for the identification of new candidate autoantigens in type 1 diabetes. J Proteomics 82:263–273

Maurya P, Meleady P, Dowling P, Clynes M (2007) Proteomic approaches for serum biomarker discovery in cancer. Anticancer Res 27:1247–1255

Maziarz M, Chung C, Drucker DJ, Emili A (2005) Integrating global proteomic and genomic expression profiles generated from islet α cells: opportunities and challenges to deriving reliable biological inferences. Mol Cell Proteomics 4:458–474

McCafferty J, Griffiths AD, Winter G, Chiswell DJ (1990) Phage antibodies: filamentous phage displaying antibody variable domains. Nature 348:552–554

Merchant M, Weinberger SR (2000) Recent advancements in surface-enhanced laser desorption/ionization- time of flight-mass spectrometry. Electrophoresis 21:1164–1177

Messana I, Cabras T, Iavarone F, Vincenzoni F, Urbani A, Castagnola M (2013) Unraveling the different proteomic platforms. J Sep Sci 36:128–139

Metz TO, Jacobs JM, Gritsenko MA, Fontes G, Qian WJ, Camp DG 2nd, Poitout V, Smith RD (2006) Characterization of the human pancreatic islet proteome by two-dimensional LC/MS/MS. J Proteome Res 5:3345–3354

Minerva L, Clerens S, Baggerman G, Arckens L (2008) Direct profiling and identification of peptide expression differences in the pancreas of control and *ob/ob* mice by imaging mass spectrometry. Proteomics 8:3763–3774

Molloy MP, Brzezinski EE, Hang J, McDowell MT, VanBogelen RA (2003) Overcoming technical variation and biological variation in quantitative proteomics. Proteomics 3:1912–1919

Nicolls MR, D'Antonio JM, Hutton JC, Gill RG, Czwornog JL, Duncan MW (2003) Proteomics as a tool for discovery: proteins implicated in Alzheimer's disease are highly expressed in normal pancreatic islets. J Proteome Res 2:199–205

Nyblom HK, Thorn K, Ahmed M, Bergsten P (2006) Mitochondrial protein patterns correlating with impaired insulin secretion from INS-1E cells exposed to elevated glucose concentrations. Proteomics 6:5193–5198

O'Farrell PH (1975) High resolution two-dimensional electrophoresis of proteins. J Biol Chem 250:4007–4021

Old WM, Meyer-Arendt K, Aveline-Wolf L, Pierce KG, Mendoza A, Sevinsky JR, Resing KA, Ahn NG (2005) Comparison of label-free methods for quantifying human proteins by shotgun proteomics. Mol Cell Proteomics 4:1487–1502

Olofsson CS, Collins S, Bengtsson M, Eliasson L, Salehi A, Shimomura K, Tarasov A, Holm C, Ashcroft F, Rorsman P (2007) Long-term exposure to glucose and lipids inhibits glucose-induced insulin secretion downstream of granule fusion with plasma membrane. Diabetes 56:1888–1897

Olsen JV, Andersen JR, Nielsen PA, Nielsen ML, Figeys D, Mann M, Wisniewski JR (2004) HysTag – a novel proteomic quantification tool applied to differential display analysis of membrane proteins from distinct areas of mouse brain. Mol Cell Proteomics 3:82–92

Ong SE, Blagoev B, Kratchmarova I, Kristensen DB, Steen H, Pandey A, Mann M (2002) Stable isotope labeling by amino acids in cell culture, SILAC, as a simple and accurate approach to expression proteomics. Mol Cell Proteomics 1:376–386

Ortsäter H, Sundsten T, Lin JM, Bergsten P (2007) Evaluation of the SELDI-TOF MS technique for protein profiling of pancreatic islets exposed to glucose and oleate. Proteomics 7:3105–3115

Paolisso G, Gambardella A, Amato L, Tortoriello R, D'Amore A, Varricchio M, D'Onofrio F (1995) Opposite effects of short- and long-term fatty acid infusion on insulin secretion in healthy subjects. Diabetologia 38:1295–1299

Park YJ, Ahn HJ, Chang HK, Kim JY, Huh KH, Kim MS, Kim YS (2009) The RhoGDI-α/JNK signaling pathway plays a significant role in mycophenolic acid-induced apoptosis in an insulin-secreting cell line. Cell Signal 21:356–364

Petricoin EF, Liotta LA (2004) SELDI-TOF-based serum proteomic pattern diagnostics for early detection of cancer. Curr Opin Biotechnol 15:24–30

Petyuk VA, Qian WJ, Hinault C, Gritsenko MA, Singhal M, Monroe ME, Camp DG 2nd, Kulkarni RN, Smith RD (2008) Characterization of the mouse pancreatic islet proteome and comparative analysis with other mouse tissues. J Proteome Res 7:3114–3126

Poitout V (2008) Glucolipotoxicity of the pancreatic β-cell: myth or reality. Biochem Soc Trans 36:901–904

Poitout V, Robertson RP (2008) Glucolipotoxicity: fuel excess and β-cell dysfunction. Endocrinol Rev 29:351–366

Poland J, Sinha P, Siegert A, Schnolzer M, Korf U, Hauptmann S (2002) Comparison of protein expression profiles between monolayer and spheroid cell culture of HT-29 cells revealed fragmentation of CK18 in three-dimensional cell culture. Electrophoresis 23:1174–1184

Prentki M, Joly E, El-Assaad W, Roduit R (2002) Malonyl-CoA signaling, lipid partitioning, and glucolipotoxicity: role in β-cell adaptation and failure in the etiology of diabetes. Diabetes 51 (Suppl 3):S405–S413

Purrello F, Rabuazzo AM, Anello M, Patane G (1996) Effects of prolonged glucose stimulation on pancreatic β cells: from increased sensitivity to desensitization. Acta Diabetol 33:253–256

Rabilloud T (2012) The whereabouts of 2D gels in quantitative proteomics. Methods Mol Biol 893:25–35

Ray S, Chandra H, Srivastava S (2010) Nanotechniques in proteomics: current status, promises and challenges. Biosens Bioelectron 25:2389–2401

Resing KA, Ahn NG (2005) Proteomics strategies for protein identification. FEBS Lett 579:885–889

Roche E, Buteau J, Aniento I, Reig JA, Soria B, Prentki M (1999) Palmitate and oleate induce the immediate-early response genes *c-fos* and *nur-77* in the pancreatic β-cell line INS-1. Diabetes 48:2007–2014

Rogowska-Wrzesinska A, Le Bihan MC, Thaysen-Andersen M, Roepstorff P (2013) 2D gels still have a niche in proteomics. J Proteomics 88:4–13

Rorsman P (1997) The pancreatic β-cell as a fuel sensor: an electrophysiologist's viewpoint. Diabetologia 40:487–495

Roxas BA, Li Q (2008) Significance analysis of microarray for relative quantitation of LC/MS data in proteomics. BMC Bioinforma 9:187

Sali A, Glaeser R, Earnest T, Baumeister W (2003) From words to literature in structural proteomics. Nature 422:216–225

Sanchez JC, Chiappe D, Converset V, Hoogland C, Binz PA, Paesano S, Appel RD, Wang S, Sennitt M, Nolan A, Cawthorne MA, Hochstrasser DF (2001) The mouse SWISS-2D PAGE database: a tool for proteomics study of diabetes and obesity. Proteomics 1:136–163

Sanchez JC, Converset V, Nolan A, Schmid G, Wang S, Heller M, Sennitt MV, Hochstrasser DF, Cawthorne MA (2002) Effect of rosiglitazone on the differential expression of diabetes-associated proteins in pancreatic islets of C57Bl/6 *lep/lep* mice. Mol Cell Proteomics 1:509–516

Scheen AJ (2004) Pathophysiology of insulin secretion. Ann Endocrinol (Paris) 65:29–36

Schrimpe-Rutledge AC, Fontes G, Gritsenko MA, Norbeck AD, Anderson DJ, Waters KM, Adkins JN, Smith RD, Poitout V, Metz TO (2012) Discovery of novel glucose-regulated proteins in isolated human pancreatic islets using LC-MS/MS-based proteomics. J Proteome Res 11:3520–3532

Schubart UK (1982) Regulation of protein phosphorylation in hamster insulinoma cells. Identification of Ca^{2+}-regulated cytoskeletal and cAMP-regulated cytosolic phosphoproteins by two-dimensional electrophoresis. J Biol Chem 257:12231–12238

Schubart UK, Fields KL (1984) Identification of a calcium-regulated insulinoma cell phosphoprotein as an islet cell keratin. J Cell Biol 98:1001–1009

Schuit F, Flamez D, De Vos A, Pipeleers D (2002) Glucose-regulated gene expression maintaining the glucose-responsive state of β-cells. Diabetes 51(Suppl 3):S326–S332

Schvartz D, Brunner Y, Coute Y, Foti M, Wollheim CB, Sanchez JC (2012) Improved characterization of the insulin secretory granule proteomes. J Proteomics 75:4620–4631

Schweitzer B, Predki P, Snyder M (2003) Microarrays to characterize protein interactions on a whole-proteome scale. Proteomics 3:2190–2199

Sol ER, Hovsepyan M, Bergsten P (2009) Proteins altered by elevated levels of palmitate or glucose implicated in impaired glucose-stimulated insulin secretion. Proteome Sci 7:24

Sparre T, Reusens B, Cherif H, Larsen MR, Roepstorff P, Fey SJ, Mose Larsen P, Remacle C, Nerup J (2003) Intrauterine programming of fetal islet gene expression in rats – effects of maternal protein restriction during gestation revealed by proteome analysis. Diabetologia 46:1497–1511

Speer R, Wulfkuhle JD, Liotta LA, Petricoin EF 3rd (2005) Reverse-phase protein microarrays for tissue-based analysis. Curr Opin Mol Ther 7:240–245

Steil GM, Trivedi N, Jonas JC, Hasenkamp WM, Sharma A, Bonner-Weir S, Weir GC (2001) Adaptation of β-cell mass to substrate oversupply: enhanced function with normal gene expression. Am J Physiol Endocrinol Metab 280:E788–E796

Steinberg TH, Agnew BJ, Gee KR, Leung WY, Goodman T, Schulenberg B, Hendrickson J, Beechem JM, Haugland RP, Patton WF (2003) Global quantitative phosphoprotein analysis using multiplexed proteomics technology. Proteomics 3:1128–1144

Strohman R (1994) Epigenesis: the missing beat in biotechnology. Biotechnology (N Y) 12:156–164

Sun X, Kelly RT, Tang K, Smith RD (2010) Ultrasensitive nanoelectrospray ionization-mass spectrometry using poly (dimethylsiloxane) microchips with monolithically integrated emitters. Analyst (Cambridge, U K) 135:2296–2302

Sundsten T, Ortsater H (2008) Proteomics in diabetes research. Mol Cell Endocrinol 24:148

Swanson SK, Washburn MP (2005) The continuing evolution of shotgun proteomics. Drug Discov Today 10:719–725

Swenne I, Andersson A (1984) Effect of genetic background on the capacity for islet cell replication in mice. Diabetologia 27:464–467

Thiede B, Hohenwarter W, Krah A, Mattow J, Schmid M, Schmidt F, Jungblut PR (2005) Peptide mass fingerprinting. Methods 35:237–247

Tochino Y (1987) The NOD mouse as a model of type I diabetes. Crit Rev Immunol 8:49–81

Trivedi M, Budihardjo I, Loureiro K, Reid TR, Ma JD (2008) Epothilones: a novel class of microtubule-stabilizing drugs for the treatment of cancer. Futur Oncol 4:483–500

Unger RH, Grundy S (1985) Hyperglycaemia as an inducer as well as a consequence of impaired islet cell function and insulin resistance: implications for the management of diabetes. Diabetologia 28:119–121

Uttamchandani M, Yao SQ (2008) Peptide microarrays: next generation biochips for detection, diagnostics and high-throughput screening. Curr Pharm Des 14:2428–2438

van Haeften TW (2002) Early disturbances in insulin secretion in the development of type 2 diabetes mellitus. Mol Cell Endocrinol 197:197–204

Vercauteren FG, Arckens L, Quirion R (2006) Applications and current challenges of proteomic approaches, focusing on two-dimensional electrophoresis. Amino Acids 33:405

Wasinger VC, Cordwell SJ, Cerpa-Poljak A, Yan JX, Gooley AA, Wilkins MR, Duncan MW, Harris R, Williams KL, Humphery-Smith I (1995) Progress with gene-product mapping of the Mollicutes: *Mycoplasma genitalium*. Electrophoresis 16:1090–1094

Webb GC, Akbar MS, Zhao C, Steiner DF (2000) Expression profiling of pancreatic β cells: glucose regulation of secretory and metabolic pathway genes. Proc Natl Acad Sci USA 97:5773–5778

Wilkins MR, Pasquali C, Appel RD, Ou K, Golaz O, Sanchez JC, Yan JX, Gooley AA, Hughes G, Humphery-Smith I, Williams KL, Hochstrasser DF (1996) From proteins to proteomes: large scale protein identification by two-dimensional electrophoresis and amino acid analysis. Biotechnology (N Y) 14:61–65

Williams KL, Hochstrasser DF (1997) Introduction to the proteome. In: Wilkins MR, Williams KL, Appel RD, Hochstrasser DF (eds) Proteome research: new frontiers in functional genomics. Springer, New York, pp 1–12

Wingren C, Borrebaeck CA (2006) Antibody microarrays: current status and key technological advances. Omics 10:411–427

Wolf G (2001) Insulin resistance associated with leptin deficiency in mice: a possible model for noninsulin-dependent diabetes mellitus. Nutr Rev 59:177–179

Yates JR III (1998) Mass spectrometry and the age of the proteome. J Mass Spectrom 33:1–19

Yee A, Chang X, Pineda-Lucena A, Wu B, Semesi A, Le B, Ramelot T, Lee GM, Bhattacharyya S, Gutierrez P, Denisov A, Lee CH, Cort JR, Kozlov G, Liao J, Finak G, Chen L, Wishart D, Lee W, McIntosh LP, Gehring K, Kennedy MA, Edwards AM, Arrowsmith CH (2002) An NMR approach to structural proteomics. Proc Natl Acad Sci U S A 99:1825–1830

Zare RN, Kim S (2010) Microfluidic platforms for single-cell analysis. Annu Rev Biomed Eng 12:187–201

Zhou M, Veenstra T (2008) Mass spectrometry: m/z 1983–2008. Biotechniques 44:667–668, 670

Zhu H, Snyder M (2003) Protein chip technology. Curr Opin Chem Biol 7:55–63

Zhu H, Bilgin M, Bangham R, Hall D, Casamayor A, Bertone P, Lan N, Jansen R, Bidlingmaier S, Houfek T, Mitchell T, Miller P, Dean RA, Gerstein M, Snyder M (2001) Global analysis of protein activities using proteome chips. Science 293:2101–2105

Advances in Clinical Islet Isolation

Andrew R. Pepper, Boris Gala-Lopez, and Tatsuya Kin

Contents

Introduction	1166
Donor Selection	1168
Pancreas Preservation Prior to Islet Isolation	1170
Pancreas Dissociation and Enzyme	1172
Islet Purification	1175
Islet Culture	1177
Assessment of Islet Preparations	1181
Cytoprotective Strategies During Islet Isolation	1184
Conclusions	1185
Cross-References	1185
References	1186

Abstract

Currently, islet transplantation is continuing to emerge as a viable treatment strategy for selected patients with type 1 diabetes who suffer from severe hypoglycemia and glycemic instability. Subsequent to the initial report, in 2000 from Edmonton, of insulin independence in seven out of seven consecutive recipients, there has been an immense expansion in clinical islet transplantation in specialized centers worldwide. The challenge now is to avoid the observed islet graft attrition over time. Isolating high-quality human islets, which survive and function indefinitely, will undoubtedly contribute to the further improvements in long-term clinical outcome. This chapter outlines the criteria

---

A.R. Pepper • B. Gala-Lopez
Clinical Islet Transplant Program, University of Alberta, Edmonton, AB, Canada
e-mail: apepper@ualberta.ca; galalope@ualberta.ca

T. Kin (✉)
Clinical Islet Laboratory, University of Alberta, Edmonton, AB, Canada
e-mail: tkin@ualberta.ca

for selecting appropriate donors for islet isolation and transplantation, describes the processes involved in islet isolation, and discusses the scope for areas of further improvements.

Keywords

Culture • Islet purification • Organ preservation • Pancreas dissociation

Abbreviations

BMI	Body mass index
CA	Catalytic antioxidants
CBD	Collagen binding domains
CI	Class I collagenase
CII	Class II collagenase
DCD	Donation after cardiac death
FDA	Fluorescein diacetate
HbA_{1c}	Hemoglobin A_{1c}
HTK	Histidine-tryptophan-ketoglutarate
IEs	Islet equivalents
OCR	Oxygen consumption rate
PI	Propidium iodide
TLM	Two-layer method
UW	University of Wisconsin

Introduction

The transplantation of pancreatic tissue in either the form of whole organ or isolated islets of Langerhans has become an attractive alternative clinical option to daily insulin injection to achieve a more physiological means for precise restoration of glucose homeostasis in patients with type 1 diabetes. Successful β-cell replacement through transplantation can offer the recipient the advantage of achieving normal carbohydrate control, eliminate the need for exogenous insulin, restore normal physiological level of hemoglobin A_{1c} (HbA_{1c}), minimize micro- and macro-disease, reduce diabetes-induced organ dysfunction, and ameliorate the life-threatening risk of severe hypoglycemia often associated with exogenous insulin therapy. The abrogation of these primary and secondary complications is often accompanied by an overall improvement in the quality of life for the transplant recipient.

Pancreas transplantation is a highly successful and well-established treatment for selected cases of type 1 diabetes; however, it is associated with significant surgical morbidity. Conversely, islet transplantation offers advantages as stated above and has a low morbidity, but has historically been considered investigational and experimental because of limited success rates. Nevertheless, recent advances in islet isolation technology have opened the door for the reinstitution and development of new clinical islet transplantation programs around the globe, which have reported increasing success. In 2000, the Edmonton group attained insulin independence in all of seven patients by using freshly isolated islets from multiple donors while utilizing a steroid-free antirejection therapy, a procedure now known as the "Edmonton protocol" (Shapiro et al. 2000; Ryan et al. 2001). This protocol has set the standard worldwide for islet transplantation allowing other groups to achieve similar successes (Hering et al. 2004; Markmann et al. 2003; Warnock et al. 2005).

While it is clear that major advances have been made in achieving more consistent insulin independence following islet transplantation, it is also evident that the majority of islet recipients experience graft attrition over time, with an insulin independence rate of only 10 % at 5 years posttransplant (Ryan et al. 2005). The chronic decay in islet graft function is likely impacted by subclinical allograft rejection and recurrent autoimmunity. In light of the multiple pathways known to be involved in β-cell dysfunction as well as the alloresponse to foreign antigens, it is unlikely that a monotherapy will further optimize clinical islet transplantation and lead to single-donor recipients, indicating a need for refined immunosuppression protocols (Gala-Lopez et al. 2013). However, experimental studies in the absence of specific immunological destruction have indicated slowly progressive dysfunction of transplanted islets with time in nonhuman primates (Gray et al. 1986; Koulmanda et al. 2006), dogs (Alejandro et al. 1986; Kaufman et al. 1990; Levy et al. 2002; Warnock et al. 1988), and rodents (Hiller et al. 1991; Keymeulen et al. 1997; Orloff et al. 1988). In addition, clinical studies in autotransplantation show that patients experience a gradual decrease in islet graft function after a sustained period of graft function, despite the absence of graft-specific immunity (Oberholzer et al. 1999, 2000; Robertson et al. 2001). The autografts function well, with 85 % of patients retaining some function 2 years posttransplant. However, the insulin independence rate 1-year post-autotransplantation is approximately 32 %, indicating nonimmunological factors may hinder islet engraftment and function (Sutherland et al. 2008). Diminution of long-term autograft function may be attributed to a combination of low islet yield, suboptimal islet vascularization, engraftment site, and innate immune reactions such as the instant blood-mediated inflammatory reaction (Bennet et al. 2000). In contrast to allografts, autografts are routinely conducted in the absence of factors such as extended cold ischemia, alloimmunity, lack of diabetic immunity, and the toxicity associated with immunosuppressive drugs. Therefore, the gradual allograft attrition can be partially attributed to nonimmunological factors, emphasizing the need for further optimization of islet transplantation strategies to improve overall patient outcomes.

Islets are subjected to numerous stresses prior to transplantation. The process of islet attrition appears to begin at the time of donor brain death and continues during the islet isolation procedure. Despite many advances in technical aspects of human islet isolation, it still remains a technically demanding procedure, with several different factors influencing isolation outcome (Harlan et al. 2009). In addition, it is difficult to isolate a sufficient number of viable islets with any regularity. Providing high-quality human islets that survive and function for a longer period will no doubt contribute to further improvement of long-term clinical outcome.

The entire process of islet preparation comprises a number of steps. Among these steps are donor selection, pancreas preservation, enzymatic digestion of the pancreas, islet purification, islet culture, and islet assessment prior to transplantation. We outline each of these steps and provide the rationale for continued efforts in islet isolation. This chapter has been updated and partially adapted from the previously published version (Kin 2010).

Donor Selection

Identifying donor-based specific markers of islet isolation success may indeed provide a means of improving the success rates of the subsequent islet transplant. Previous single-center retrospective studies have identified several donor-related variables affecting islet isolation outcome, including but not limited to donor age, cause of death, body mass index (BMI), cold ischemia time, length of hospitalization, use of vasopressors, and blood glucose levels (Kin 2010; Goto et al. 2004; Ihm et al. 2006; Lakey et al. 1995, 1996; Matsumoto et al. 2004; Nano et al. 2005; Zeng et al. 1994). Pancreas weight has not been considered as a donor selection criterion because a value cannot be obtained prior to organ procurement. In most cases, a larger pancreas contains a larger islet mass (Nano et al. 2005; Kin et al. 2006a). Thus, pancreas size can serve as a surrogate parameter for donor islet mass on its own. One study developed a formula to predict pancreas weight, by analyzing data from 345 deceased donors (Kin et al. 2006a). Key observations made by this group are (1) males have a larger pancreas than females; (2) pancreas weight increases with age, reaching plateau in the fourth decade; and (3) BMI correlates with pancreas weight, but body surface area is a better predictor of pancreas weight than BMI. The finding that males have a larger pancreas size is consistent with the observations from other such studies (de la Grandmaison et al. 2001; Saisho et al. 2007). It has been reported that male donors provided a higher probability of yielding adequate islets (Hanley et al. 2008; Ponte et al. 2007). As for donor age, a positive correlation between age and islet yield is well documented (Lakey et al. 1996; Nano et al. 2005). A juvenile donor pancreas is often difficult to obtain an adequate number of islets (Balamurugan et al. 2006; Socci et al. 1993), which in part can be explained by its small size. Regarding BMI, several groups have indicated that BMI positively affects islet yield (Brandhorst et al. 1995a), leading many to consider BMI as an important donor factor influencing islet isolation outcome (Lakey et al. 1996; Matsumoto et al. 2004; Nano et al. 2005). However, this view has led to the misconception that an obese donor is a good candidate for successful islet isolation and transplantation. Supporting this notion is a recent report by Berney and Johnson who conclude that islet mass transplanted does not unequivocally correlate with islet graft function and therefore argue that donor selection criteria for islet transplantation, and hence allocation rules (pancreas for whole organ or islet transplant), may need to be redefined (Berney and Johnson 2010).

Donors with type 2 diabetes are considered unsuitable for islet isolation and transplantation because β-cell mass (Butler et al. 2003; Yoon et al. 2003) and function (Deng et al. 2004) may be decreased in type 2 diabetes. Type 2 diabetes is clinically insidious and can remain undiagnosed for many years. A negative medical history of diabetes obtained from the next of kin does not necessarily indicate the absence of glucose intolerance. Thus, it is not unexpected that a large proportion of organ donors for islets may have undiagnosed type 2 diabetes. In fact, a pancreas from an older donor with a higher BMI is not likely used for a whole organ transplant, but is preferred for islet isolation and transplantation (Ris et al. 2004; Stegall et al. 2007). Such a donor may be on the spectrum toward type 2 diabetes. Previous studies indicated that high glycemic values during donor management were detrimental to islet recovery after isolation (Lakey et al. 1996; Nano et al. 2005; Zeng et al. 1994). However, blood glucose levels are far too unreliable to use for the assessment of the donors' glucose metabolism in light of the pathophysiology of brain death and the pharmacology of drugs administered during the management of brain death. Although HbA_{1c} itself is not a diagnostic criterion for diabetes mellitus, its measurement in potential donors would provide useful information, since it has a high degree of specificity for detecting chronic hyperglycemia. Our islet isolation laboratory at the University of Alberta has implemented the routine measurement of donor HbA_{1c} levels prior to islet isolation. Our current practice is that donors with $HbA_{1c} > 7\,\%$ are excluded for clinical islet transplantation (Koh et al. 2008).

O'Gorman and colleagues developed a scoring system based on donor characteristics that can predict islet isolation outcomes (O'Gorman et al. 2005). This scoring system has proven to be effective in assessing whether a pancreas should be processed for islet isolation (Witkowski et al. 2006). It also allows for better management of the islet processing facility, as the cost of islet isolations is high. However, using a donor score of 79 as the most appropriate cutoff value, the sensitivity and specificity for predicting successful islet isolations were only 43 % and 82 %, respectively. Moreover, it is not clear about the actual impact of donor score on transplantation outcome, because the scoring system was developed solely based on islet isolation outcome. Similarly, other published studies dealing with donor factors do not take transplant outcome into consideration (Kin 2010; Goto et al. 2004; Lakey et al. 1995, 1996; Matsumoto et al. 2004; Nano et al. 2005; Zeng et al. 1994). An older donor with a higher BMI may be a better donor with respect to successful islet isolation, but probably is not ideal for islet transplantation when the biology of islets derived from such a donor is considered. An improved scoring system which takes both the islet isolation and transplantation outcomes into consideration would be more appropriate.

Donation after cardiac death (DCD) has also been used for islet transplantation. In experimental settings, islet yield and function derived from DCD pancreases seem to be comparable with those from their brain dead counterparts (Ris et al. 2004). However, in clinical settings, the results are not entirely promising. Japan has one of the most extensive experiences using DCD donors for organ transplant. They have been using optimized retrieval in these type of donors, as well as the Kyoto preservation solution and the two-layer method (TLM) (Nagata et al. 2006). Their most recent report for islet transplantation from this source shows that overall graft survival was 76.5 %, 47.1 %, and 33.6 % at 1, 2, and 3 years, respectively, whereas corresponding graft survival after multiple transplantations was 100 %, 80.0 %, and 57.1 %, respectively. All recipients remained free of severe hypoglycemia, while three achieved insulin independence for 14, 79, and 215 days (Saito et al. 2010). This is a clear indication of the potential benefits of DCD as an alternative source if used under strict releasing criteria, particularly in countries where heart-beating donors may not be readily available.

Pancreas Preservation Prior to Islet Isolation

According to a report from the National Islet Cell Resource Center Consortium in the USA, University of Wisconsin (UW) solution is currently the standard preservation solution prior to islet isolation. Recently, more pancreata are stored in histidine-tryptophan-ketoglutarate (HTK) solution, while TLM is decreasingly employed for pancreas preservation, at least in the USA.

HTK solution, originally developed for use as a cardioplegia solution, has long been used for abdominal organ preservation in Europe (de Boer et al. 1999). In 1995, Brandhorst and colleagues compared HTK and UW solutions in pancreas preservation for human islet isolation for the first time (Brandhorst et al. 1995b). They observed that both solutions were comparable in terms of islet yield, in vitro functional viability of islets, and in vivo islet function in a mouse transplant model. Similar findings were subsequently reported by Salehi and colleagues (Salehi et al. 2006). In an experimental model performed in pigs, Stadlbauer and colleagues did not find any differences in frequency of apoptotic islet cells between pancreata preserved in UW versus those in HTK (Stadlbauer et al. 2003). At the present time, there is no evidence that HTK solution is superior to UW regarding islet isolation outcome. However, cost advantages in utilization of HTK may see further increased use of this solution for organ preservation.

The reason for decreased utilization of TLM is not clear but might be explained by recent observations in 166 and 200 human pancreata indicating no beneficial effect of TLM (Caballero-Corbalan et al. 2007; Kin et al. 2006b). TLM was developed in the 1980s by Kuroda, who focused on organ protection from hypoxia by supplying oxygen via perfluorocarbon during cold preservation (Kuroda et al. 1988). Maintenance of adenosine triphosphate production in pancreata stored at the interface between perfluorocarbon and UW solution was observed as a result of oxygenation (Morita et al. 1993). Tanioka and colleagues applied for the first time TLM prior to islet isolation in a canine model (Tanioka et al. 1997). Subsequently many centers introduced TLM prior to clinical islet isolation and reported improvement in islet isolation outcomes for pancreata preserved in TLM, when compared with pancreata stored in UW alone (Hering et al. 2002; Ricordi et al. 2003; Tsujimura et al. 2002). Of note, most of the initial studies employed a short period of TLM with continuous oxygenation at the islet isolation facilities. In an attempt to enhance the beneficial effect of TLM, our center at the University of Alberta introduced TLM for an entire period of pancreas preservation using pre-oxygenated perfluorocarbon. However, in contrast to the expectation, no advantages of TLM over UW were observed in terms of pancreatic adenosine triphosphate level, islet yield, in vitro functional viability, and in vivo function after clinical transplantation (Kin et al. 2006b). These findings were subsequently confirmed by other groups (Ponte et al. 2007; Caballero-Corbalan et al. 2007). In a recent, meta-analysis study comparing pancreas preservation with the TLM versus UW for islet transplantation, the research concluded that the TLM was indeed beneficial for pancreata with prolonged cold ischemia prior to human islet isolation; however, benefits of the TLM for short-term preservation were inconclusive (Qin et al. 2011). Thus, there remains much work to be done to optimize pancreas preservation methods. One such strategy for optimization of pancreas preservation is the formulation of novel solution. The Matsumoto group has developed a Kyoto solution that contains trehalose and ulinastatin as distinct components (Noguchi et al. 2006). This solution and modification of it have demonstrated favorable results in human islet isolation outcomes, such that it has become the standard preservation solution for Baylor's clinical islet transplant program (Noguchi et al. 2010, 2012a).

Recently, hypothermic machine perfusion has been gaining increasing acceptance as a preservation method mainly for marginal donor kidneys (Sung et al. 2008). Hypothermic machine perfusion has several advantages over static cold storage. First, preservation solution can be continuously supplied directly to all cells. In addition, machine perfusion permits ex vivo pharmacologic manipulation of the graft. Moreover, real-time assessment of graft quality can be done by analysis of perfusate. Toledo-Pereyra and colleagues reported a canine islet autotransplantation study with 60 % and 40 % animal survival following hypothermic machine perfusion for 24 and 48 h, respectively (Toledo-Pereyra et al. 1980). In porcine islet isolation, Taylor and colleagues showed that machine perfusion improved islet isolation outcomes when compared with static UW preservation (Taylor et al. 2008). Our center at the University of Alberta performed machine perfusion in 12 human pancreata using a LifePort^{TM} Kidney Transporter (Organ Recovery Systems, Des Plaines, IL, USA) (Kin et al. 2006c). The first four pancreata were placed on the machine, after 10 h of static preservation in UW, for up to 24 h; metabolic and histologic changes of pancreata were assessed. It was found that tissue energy charge was maintained during the first 3 h in the machine perfusion and thereafter it gradually decreased. Histologic analysis revealed that tissue edema became evident at 24 h. The next eight pancreata were processed for islet isolation after 6 h of machine perfusion. Islet recovery and viability tended to be higher in pancreata preserved with the machine perfusion than in matched pancreata stored in static UW. These results are in accordance with the work of Leeser and colleagues who showed a feasibility of pump perfusion of human pancreata prior to islet isolation (Leeser et al. 2004).

Another recent advancement in organ preservation is the use of hypothermic persufflation, gaseous oxygen perfusion, as a surrogate to machine liquid perfusion (Suszynski et al. 2013, 2012; Nagai et al. 2013; Srinivasan et al. 2012; Tolba et al. 2006; Minor et al. 2000). In a preclinical study, persufflation of the pancreas prior to islet isolation demonstrated improved yields in viable islets and extended the duration of preservation compared to the TLM (Scott et al. 2010). Currently the utility and efficacy of persufflation for clinical islet transplantation is being actively investigated at the University of Alberta's Clinical Islet Transplant Program.

Pancreas Dissociation and Enzyme

A critical component to successful islet transplantation is the enzymatic dissociation of insulin-producing islets of Langerhans, from the surrounding pancreatic exocrine tissue. To facilitate the isolation of the islets, enzyme blends are delivered to the islet-exocrine interface, of which collagen is the major structural protein constituting this interface (Hughes et al. 2006; Van Deijnen et al. 1994). As a result of its dense structure and mechanical strength, collagen is not generally degraded by ordinary protease; however, it can be efficiently degraded with high specificity by collagenase (Watanabe 2004). As such, collagenase is a key component of an enzyme product for isolating islets; however, the use of collagenase alone results in an inadequate tissue digestion (Wolters et al. 1992, 1995). It is apparent that the presence of non-collagenase impurities is required for enhanced pancreas dissociation. Prior to the 1990s, crude collagenase, a fermentation product derived from *Clostridium histolyticum*, was exclusively used for isolating islets. This crude preparation contains two different classes of collagenase: class I collagenase (CI) and class II collagenase (CII) in addition to non-collagenolytic enzymes including but not limited to amylase, cellulase, pectinase, chitinase, sialidase, hyaluronidase, lipase, and phospholipase (Johnson et al. 1996; Matsushita and Okabe 2001; Soru and Zaharia 1972). Unfortunately, a major limitation to successful islet isolation is the exceedingly variable composition and activity of crude collagenase preparations that exists between lots of commercially available products (Johnson et al. 1996; Kin et al. 2007a). In the late 1990s a purified enzyme blend became available from Roche (Roche Applied Science, Indianapolis, IN). This enzyme blend, Liberase HI, is comprised of CI and CII, in addition to a non-collagenolytic enzyme, thermolysin, derived from *Bacillus thermoproteolyticus*. The introduction of Liberase HI had a profound impact on the field as it helped reduce a significant portion of the lot-to-lot enzyme variability. Subsequent to Liberase HI's inception into clinical and experimental islet isolations, enhanced islet yield and function in the human and animal models, compared to the historical use of crude collagenase, were immediately observed (Berney et al. 2001; Lakey et al. 1998; Linetsky et al. 1997; Olack et al. 1999). However several studies showed that Liberase is no more effective than crude collagenase in certain models, such as neonatal rat (Hyder 2005) and fetal pig (Georges et al. 2002) pancreata, and induces functional damage to rat (Vargas et al. 2001) and human (Balamurugan et al. 2005) islets. Moreover, this enzyme blend is not immune to a degree of lot-to-lot variations (Barnett et al. 2005; Kin et al. 2007b).

While the use of non-collagenolytic enzyme components has been shown to enhance pancreas dissociation, excessive exposure of these enzyme is found to decrease islet yields through islet fragmentation and disintegration (Wolters et al. 1992; Bucher et al. 2004) and to reduce islet viability (Brandhorst et al. 2005).

This is a major limitation of traditional products as fine tuning the narrow dose range is not possible by the user. A recently developed Collagenase NB1 (Serva Electrophoresis, Heidelberg, Germany) contains only CI and CII, which can be blended with separately packaged Neutral Protease NB (Serva Electrophoresis) as a non-collagenolytic component. This type of product has several potential advantages over traditional enzyme blends. First, the ratio between non-collagenolytic activity and collagenase activity can be adjusted accordingly to further optimize the isolation procedure. Moreover, separate storage of the individual enzyme components would improve the overall stability of each enzyme activity. Finally and importantly, the non-collagenolytic component can be sequentially administered after intraductal collagenase distention, in an attempt to avoid excessive exposure of islets to non-collagenolytic enzyme (Kin et al. 2009).

Clostridium histolyticum possesses two homologous but distinct genes, ColG and ColH. The former encodes CI and the later encodes CII (Matsushita et al. 1999; Yoshihara et al. 1994). It is important to know similarities and differences between the two enzymes. CI and CII are quite different in their primary and secondary structures, but the catalytic machinery of the two enzymes is essentially identical (Mookhtiar and Van Wart 1992). Both enzymes have a similar segmental structure consisting of three different segments: catalytic domain, spacing domain, and binding domain (Kin et al. 2007a; Matsushita et al. 2001). CI has tandem collagen binding domains (CBD) but CII possesses a single CBD (Matsushita et al. 2001). Tandem CBDs of CI may have advantages for binding to collagens in the pancreas; because tandem-repeated binding domains are generally considered useful for stabilization of bindings (Linder et al. 1996). Kinetic studies evaluating the hydrolysis of collagens by CI or CII indicate a higher catalytic efficiency of CI on collagen (Mallya et al. 1992). On the other hand, CII is characterized by the ability to attack synthetic peptide substrates at a much greater rate than CI (Bond and Van Wart 1984). Wolters and colleagues showed that rat pancreas digestion was more effective when both classes were used together instead of CI or CII alone (Wolters et al. 1995). van Wart and colleagues found a synergistic effect of the two enzymes on collagen degradation (Van Wart and Steinbrink 1985). Wolters and colleagues concluded that CII plays a predominant role in rat pancreas dissociation, whereas CI is minor in comparison (Wolters et al. 1995). Several investigators (Antonioli et al. 2007) and manufacturers have emphasized the view that CII is a key player in pancreas dissociation. Indeed, manufacturers have measured only CII activity in their product specification and CI activity has been ignored so far. However, the importance of CII has been challenged by a recent study demonstrating that neither CI nor CII alone is able to release islets from a rat pancreas (Brandhorst et al. 2008). Findings from human studies are in conflict with the classical view, too. Barnett and colleagues showed that the stability of intact CI is of great importance to the quality of the blend (Barnett et al. 2005). It is also demonstrated that a better enzyme performance is ascribed to a higher proportion of CI rather than a higher proportion of CII (Kin et al. 2007b). It is further shown that excessive CII is not effective to release islets from a human pancreas and rather a balanced CII/CI ratio is of paramount importance (Kin et al. 2008a).

Cross and colleagues performed extensive immunohistological studies on binding of collagenase to collagen (Cross et al. 2008), suggesting that collagenase perfused through the duct binds to collagen inside the pancreas. Their findings also suggest that collagenase can bind to collagen without help of non-collagenolytic enzyme activation and that low temperature does not inhibit binding of collagenase to collagen which is in line with a previous report (Matsushita et al. 1998). Another important finding from their studies is that collagenase binds to collagen located inside the islets as a result of intraductal perfusion with collagenase. This may result in islet fragmentation when the enzyme is activated.

Understanding of the structure of the islet-exocrine interface, and the nature of substrate at this interface, will be exploited to optimize pancreas dissociation. Previous studies have described the distribution of collagen types in the human pancreas. Type IV collagen is present in basement membranes associated with ducts and acini (Kadono et al. 2004). Collagen subtypes identified in the islet-exocrine interface are I, III, IV, V, and VI (Van Deijnen et al. 1994; Hughes et al. 2005). Recently, Hughes and colleagues found that type VI collagen is one of the major collagen subtypes at the islet-exocrine interface of the adult human pancreas (Hughes et al. 2006). Type VI collagen has a high disulfide content which serves to protect the molecule from bacterial collagenase digestion (Heller-Harrison and Carter 1984). It is also known that type VI collagen does not form banded collagen fibrils and is extensively glycosylated (Aumailley and Gayraud 1998). Regarding the amount of collagen, it is well known that the total collagen content increases with age in most tissues (Akamatsu et al. 1999; Clausen 1963; Gomes et al. 1997; Sobel and Marmorston 1956). In addition, pancreatic collagen is affected by the normal aging process. Bedossa and colleagues found significantly higher collagen content in pancreata from patients over the age of 50, as compared to younger subjects (Bedossa et al. 1989). The importance of pancreatic ultrastructure has been pointed out and discussed over the past two decades (van Deijnen et al. 1992); unfortunately, there has been little progress in this field. A better understanding of the differences in biomatrix among donor pancreata, for example, older versus younger donors, will help to improve pancreas dissociation.

In early 2007, the islet transplant community was notified of the potential risk of prion disease transmission, when using Liberase, due to the manufacturing of the product, which evolves a bovine neural component. As a result many islet isolation centers switched to Serva collagenase, as it was considered a safer option; however, this conversion had a considerable impact on the field in terms of islet yield and quality. In two recent retrospective studies, the researchers compared the human islet isolation outcomes between pancreata processed by Liberase HI and Collagenase NB1. Despite reporting no difference with respect to pre-purification, post-purification, or post-culture islet mass or percent recovery between the two groups, they concluded that islets prepared with Liberase HI were significantly more viable than those isolated with Collagenase NB1 (Iglesias et al. 2012; Misawa et al. 2012). In a biochemical study, the collagenases from Roche (Liberase MTF), Serva (Collagenase NB1), and VitaCyte (CIzyme Collagenase HA) (Indianapolis, IN) were analyzed by analytical high-performance liquid chromatography and collagen degradation activity, an assay that preferentially detects intact CI; the researchers demonstrate that the highest collagen degradation activity was found in the VitaCyte product followed by the Roche and Serva enzyme products. Furthermore, this group successfully used the VitaCyte product in 14 human islet isolations (Balamurugan et al. 2010). Of note, the biochemical analysis of the purified Collagenase NB1 consistently showed that these products contain primarily truncated CI, resulting in a molecular form that has lost one of the two CBDs (Balamurugan et al. 2010). This may in part explain the finding by O'Gorman and colleague, which demonstrated favorable clinical islet isolation outcomes with the new Liberase MTF, compared to Serva NB1 (O'Gorman et al. 2010).

In light of this significant setback, some centers successively adapted Serva enzyme blend with minimal modification of their preexisting islet isolation protocol (Szot et al. 2009; Sabek et al. 2008). The University of California San Francisco group achieved a high rate of islet isolation success using ~1,600 units of Serva collagenase and ~200 units of neutral protease for younger donor pancreata (Szot et al. 2009). A recent in-depth report by Balamurugan and colleagues from the University of Minnesota detailed the evaluation of eight different enzyme combinations in an attempt to improve islet yield for both autologous and allogeneic human islet transplantation. This comprehensive study included 249 clinical islet isolations consisting of enzyme combinations including purified, intact, or truncated CI and CII and thermolysin from *Bacillus thermoproteolyticus rokko* or neutral protease from *Clostridium histolyticum*. Based on the biochemical characteristics of enzymes and observations of enzyme behavior during isolations, they blended intact CI/CII and neutral protease from *Clostridium histolyticum* instead of thermolysin into a new enzyme mixture (Balamurugan et al. 2012). This new enzyme mixture consistently produced significantly higher islet yields from both pancreatitis and deceased donor pancreata compared to standard enzyme combinations, while retaining islet potency and function (Balamurugan et al. 2012).

Now considerable research efforts are being directed to experimentation with nonenzymatic pancreas dissociation methodologies. Some of these novel modalities include but are not limited to in situ cryopreservation of islets and selective destruction of acinar tissue (Taylor and Baicu 2011), as well as utilizing dielectrophoresis as a noncontact technique for isolating islets of Langerhans (Burgarella et al. 2013). Whether these and other enzymatic and nonenzymatic isolation strategies can be translated to clinical islet isolation and improve outcomes has yet to be elucidated.

Islet Purification

Following the enzymatic digestion of an average-size human pancreas (~90 g), the total packed volume of digested tissue is typically greater than 40 mL. While it is known that human liver has a capacity for adaptation and revascularization in the context of portal vein occlusion (Casey et al. 2002), the liver cannot accommodate 40 mL of tissue consisting of particles on the 100-μm scale. Furthermore, the substantial evidence of liver embolism, thrombosis, damage, and even death are documented in clinical settings immediately after intraportal infusion of a large amount of tissue (Mehigan et al. 1980; Mittal et al. 1981; Walsh et al. 1982; Toledo-Pereyra et al. 1984; Shapiro et al. 1995). Notably, there is the need to reduce tissue volume with minimum loss of islets for the safer intraportal infusion as well as to accommodate islet graft transplanted into extrahepatic sites. This can be achieved by a procedure called "islet purification."

Density-dependent separation of islets from exocrine tissue is the most simple and effective approach for islet purification. This methodology is based on the principle that, during centrifugation, tissue will migrate and settle to the density that is equal to its own density. Using this technique, separation can be achieved based on intrinsic differences in density between islet tissue (~1.059 g/mL) and exocrine tissue (1.059–1.074 g/mL) (London et al. 1992). Theoretically, a larger difference in cellular density between the two tissues could result more efficacious separation. The ideal separation would be expected when the islets are free from exocrine tissue and the density of exocrine tissue is well preserved. In contrast, the worst scenario would happen when the majority of islets are entrapped in the exocrine tissue (thereby a higher density of tissue) and the density of exocrine tissue is decreased due to exocrine enzyme discharge or tissue swelling. Thus, there is a trade-off between purity of islets and islet mass recovered. Obtaining an extremely high purity of islets requires sacrificing a less pure layer, which contains a considerable amount of islets. Nearly 100 % of islets can be recovered if a less pure layer with a large amount of exocrine tissue is included to the final preparation, but this turns in a lower purity and may be deleterious to the subsequent islet transplant.

Purification of large numbers of human islets has advanced rapidly with the introduction of the COBE 2991 (COBE Laboratories Inc., Lakewood, CO, USA) (Lake et al. 1989). The COBE 2991 cell processor, originally developed for producing blood cell concentrates, is an indispensable equipment in human islet processing facilities. It allows processing of a large volume of tissue in an enclosed sterile system. Moreover, it offers decreased operating time with an ease of generating continuous density gradients in conjunction with a two-chamber gradient maker.

Various gradient media have been developed and tested for islet purification. One of the most commonly used media is a synthetic polymer of sucrose (Ficoll; Amersham, Uppsala, Sweden)-based media. Scharp and colleagues reported that islet recovery was improved when Ficoll was dialyzed before centrifugation, to remove the low molecular weight osmotically active contaminants (Scharp et al. 1973). Olack and colleagues used Euro-Collins, an organ preservation solution, as the vehicle for dissolving the Ficoll powder (Olack et al. 1991). Hypertonic density solutions such as Euro-Collins/Ficoll prevent edema of the exocrine tissue at low temperature and result in improved separation of islets from the exocrine tissue, when compared with standard Ficoll solution (London et al. 1992). More recently, purification with iodixanol (OptiPrep) has been recently reported in islet transplant series with successful clinical outcomes in addition to significantly reducing the secretion of proinflammatory cytokine/chemokines from the islet, in comparison to Ficoll-based density gradients (Mita et al. 2010). Of note, in a comparative preclinical study of Ficoll-, histopaque-, dextran-, or iodixanol-based gradients, histopaque led to the isolation of viable and functional islets with a reduced cost as compared to a standard Ficoll gradient (McCall et al. 2011).

UW solution has been used for storing the pancreatic digest prior to density gradient centrifugation in an attempt to reduce acinar tissue swelling (Chadwick et al. 1994; Robertson et al. 1992; van der Burg et al. 1990). The beneficial effect of UW storage is ascribed to the presence of the osmotic effective substances lactobionate, raffinose, and hydroxyethyl starch. To extend this beneficial action, Huang and colleagues have introduced a mixture of UW and Ficoll-sodium-diatrizoate (Biocoll; Biochrom, Berlin, Germany) for density gradient separation (Huang et al. 2004). They showed that their new gradient medium improved post-purification islet yield when compared with the standard medium. The UW-Bicoll purification method has been further refined by Barbaro and colleagues, who recovered 85 % of islets after purification using a narrow range of density gradients (Barbaro et al. 2007).

Ichii and colleagues performed discontinuous density gradient purification to recover islets from the exocrine-enriched fraction obtained after the initial purification procedure (Ichii et al. 2005a). This supplemental purification, so-called rescue purification, contributed to the increase in the number of islet preparations suitable for transplantation. In a subsequent report, however, the same group showed no benefit of rescue purification on isolation success (Ponte et al. 2007).

Among islet processing laboratories, islet recovery rates vary from 50 % to 85 %, depending on purification methods and the quality of the pancreas, summarized in Table 1 (Ihm et al. 2006; Nagata et al. 2006; Kin et al. 2007b, 2008a; Mita et al. 2010; Barbaro et al. 2007; Brandhorst et al. 2003; Wang et al. 2007; Yamamoto et al. 2007; Noguchi et al. 2009), clearly indicating that more research is warranted. Some of the strategies include the avoidance of the high shear forces associated with COBE 2991 cell processor-based purification methods. Shimoda and colleagues reported such a method with their top loading large bottle purification technique in conjunction with low viscosity ET-Kyoto and iodixanol gradient solutions, resulting in improved islet yield and viability compared to traditional isolation techniques (Shimoda et al. 2012). Furthermore, others are investigating the utility of magnetic separation of either encapsulated (Mettler et al. 2013) or nonencapsulated islets, completely avoiding density gradient separation all together. Despite promising preclinical results using porcine pancreata, this technology translation to the clinical setting is yet to be demonstrated.

Islet Culture

Clinical islet transplantations rely on successful isolation of the human islets from donors followed by the in vitro culturing of the cells to maintain functionality until transplantation can be performed. Although there is debate as to whether freshly isolated islets are superior to cultured islets in experimental transplantation (Ihm et al. 2009; King et al. 2005; Olsson and Carlsson 2005), preservation of human

Table 1 Islet recovery rate after purification

Author [reference]	Density gradient	n	Pre-purification islet yield (IE)	Post-purification islet yield (IE)	Recovery rate (%)[a]
Brandhorst et al. (2003)	HBSS-Ficoll	76	463,872	245,889	53.0
Nagata et al. (2006)	Ficoll	8[b]	660,770	444,426	67.3
Ihm et al. (2006)	Iodixanol or Biocoll	110	356,745	244,034	68.4
Barbaro et al. (2007)	Biocoll	32	359,425	194,022	64.5[c]
	UW-Biocoll	132	370,682	310,607	84.9[c]
Yamamoto et al. (2007)	Ficoll or ficoll followed by rescue[d]	169	454,049	306,728	67.6
Kin et al. (2007b)	Ficoll	251	348,794	227,832	65.3
Wang et al. (2007)	Biocoll	23	373,350	184,284	49.4
Kin et al. (2008a)	UW-Biocoll	21	394,619	303,905	77.0
Mita et al. (2010)	Iodixanol	5[e]	264,533	120,268	52.9
	Ficoll	5[e]	270,361	132,560	49.4
Noguchi et al. (2009)	Iodixanol-Kyoto	11	699,780	594,136	84.9
	Ficoll	19	670,939	377,230	55.6

[a]Mean post-purification IE/mean pre-purification IE × 100
[b]Nonheart beating donors
[c]Mean of each isolation's recovery rate
[d]See Ichii et al. (2005a) regarding "rescue" purification
[e]Pancreatic digest equally divided into two group

islets for a certain period of time by means of culture provides many benefits to clinical islet transplantation. First, islet culture allows travel time for patients living away from transplant centers, as most of these procedures are conducted in specialized centers. Pre-transplant culture can moreover provide attainment of therapeutic levels of immunosuppression before islet infusion. During the culture period additional quality control testing can be undertaken, including microbiological and pyrogenic tests. In addition to these practical advantages, modification or treatment of islets through culture provides a strategic opportunity to promote islet survival after transplantation. Surface modification of islets by bioconjunction during culture is one of the examples (Cabric et al. 2007; Totani et al. 2008). Thus, a strong rationale exists for culturing islets prior to transplantation. A number of issues will be discussed below, including culture temperature, base media, and risk of islet loss during culture.

In 1977, Kedinger and colleagues transplanted cultured or fresh allo-islets into the liver of diabetic rats (Kedinger et al. 1977). Rats receiving fresh islets returned to diabetic state in 8 days after transplantation. In contrast, when islets were cultured for 3–4 days prior to transplantation, graft survival was prolonged to 90 days without immunosuppression. Culture temperature is not described in the report, but this was the first attempt to alter or reduce immunogenicity of islets by in vitro culture. A few years later, Lacy and colleagues adapted room temperature for culturing islets in an attempt to prolong allograft survival (Lacy et al. 1979a). They reported that culture of rat islets at 24 °C resulted in a prolonged islet allograft survival in immunosuppressed recipients. Their idea of culturing islets at 24 °C was based on a study performed by Opelz and Terasaki (Opelz and Terasaki 1974), who found that lymphocytes being placed in culture at room temperature lost their ability to stimulate allogeneic cells when tested in subsequent mixed lymphocyte culture. The results of Lacy's experiments support the theory that passenger leukocytes are involved in rejection of the allografts. However, the study did not demonstrate culture temperature at 24 °C per se contributed to altering immunogenicity of islets. To investigate the influence of temperature, they further compared allograft survival in immunosuppressed rats receiving islets cultured at 24 °C versus 37 °C (Lacy et al. 1979b). Culture of islets at 24 °C resulted in a longer allograft survival as compared to 37 °C culture. When recipients were not immunosuppressed, however, culture at 24 °C did not prolong graft survival as compared to fresh islets. Following the initial study performed by Lacy and colleagues (Lacy et al. 1979a), many investigators have set up a culture system at 22–24 °C prior to transplantation (Horcher et al. 1995; Kamei et al. 1989; Lacy and Finke 1991; Lacy et al. 1989; Mandel and Koulmanda 1985; Morsiani and Lacy 1990; Ryu and Yasunami 1991; Scharp et al. 1987; Yasunami et al. 1994). There are, however, little published studies specifically showing immunological superiority of 22–24 °C culture over 37 °C culture in an islet transplant model. A group from Germany reported a marked prolongation of rat islet allograft survival by culture at 22 °C comparing to 37 °C culture (Woehrle et al. 1989). Interestingly, the prolongation of graft survival was observed only when islets were transplanted under the renal subcapsular space; islets cultured at 22 °C were rapidly destroyed at the liver. They further confirmed a similar effect in a rat-to-mouse xenotransplantation model (Jaeger et al. 1994). It is uncertain if the strategy used in the animal models will be as satisfactory in the clinical situation.

Another possible benefit of low-temperature culture is that the structural and functional integrity of islets is well preserved, likely due to a lowered metabolic rate at temperatures below 37 °C. For instance, Ono and colleagues reported that rat islets cultured at 37 °C possessed a higher rate of central necrosis than islets cultured at 22 °C (Ono et al. 1979). Lakey and colleagues assessed human islet recovery after 24 h culture and they described a mean recovery rate of 73 % at 22 °C compared to 55 % at 37 °C (Lakey et al. 1994). Similarly, inadequate recovery rate at 37 °C was observed in pig islets after culture compared to 22 °C culture (Brandhorst et al. 1999). On the other hand, investigators have asserted that low-temperature culture results in impaired insulin production (Escolar et al. 1990) and more apoptotic cells in islets (Ilieva et al. 1999). In clinical settings, there is no consensus regarding culture temperature. Some centers have employed culture temperature at 37 °C followed by culture at 22 °C prior to clinical transplantation (Hering et al. 2004). Initially our center at the University of Alberta adapted this approach but consistently found significant loss of islets after culture (unpublished observation). Since 2003, our center has been using only 22 °C during the entire period prior to transplantation (Kin et al. 2008b).

Connaught Medical Research Laboratory 1066, originally designed for use with fibroblasts and kidney epithelial cells, appears to be the most widely used base medium for islet culture. Other base media used for clinical transplantation include Ham's F10 (Keymeulen et al. 2006) and M199 (Bertuzzi et al. 2002). Regardless of whichever base medium is employed, supplementation of medium seems to be a routine practice. Because serum contains many components that have a beneficial effect on cell survival, animal serum such as fetal calf serum is traditionally added to culture media in experimental settings. However, when islets are destined for clinical transplantation, use of animal sera has been considered unacceptable because of potential risk associated with viral or prion-related disease transmission (Will et al. 1996). Other potential problems of animal sera are evoking immune or inflammatory reactions in host against animal proteins (Johnson et al. 1991; Mackensen et al. 2000; Meyer et al. 1982), which cannot be diminished even by several washing steps (Spees et al. 2004). Therefore, adding human serum albumin as an alternative is the current standard in clinical islet culture.

One of the major concerns with culturing islets is the uncertainty of islet recovery rate after culture. There is ample evidence of a reduction in the islet mass during culture. Bottino and colleagues reported that there was at most 80 % reduction in DNA content in islet preparations following 24 h culture (Bottino et al. 2002). Another report showed only 18 % recovery rate in islet mass after 48 h culture (Zhang et al. 2004). A retrospective study on 104 islet preparations for clinical transplantation has identified several factors associated with risk of islet loss (Kin et al. 2008b). These include longer cold ischemia time prior to islet isolation, TLM for pancreas preservation, lower islet purity, and higher proportion of larger islets in the pre-cultured preparation. One may wonder if a longer culture period leads to a greater decrease in islet mass. The retrospective study does not support this caution probably because most of the islet preparations were cultured for a short period (20 h, median) with a narrow range (14 h, interquartile range). Most islet processing centers employ short period culture (up to 3 days) prior to transplantation (Hering et al. 2004; Warnock et al. 2005; Ryan et al. 2005; Kin et al. 2008b). A recent study from Baylor All Saints Medical Center concluded that fresh human islets (less than 6 h of culture) are more potent compared to cultured islets based on their in vitro and in vivo data. As such, this group's current clinical islet transplantation protocol implements fresh islet transplantation without culture (Noguchi et al. 2012b). Of note, the Brussels group cultured islets derived from several donors for up to 1 month until a critical islet mass is reached and then infused all islet preparations into a recipient as a single transplant procedure (Keymeulen et al. 1998). The impact of prolonged culture period on islet loss seems to be significant as islets were combined from as many as nine donors. Importantly, two of the seven recipients became insulin independent after transplantation of islets cultured for long periods.

Multiple experimentations have occurred during the design of an optimal culture media for islets. However, protocols for islet culture have not completely been standardized and practice may vary between centers. Optimal culture conditions for isolated human islet preparations should aim at providing sufficient oxygen and nutrients to allow for recovery from isolation-induced damage while maintaining tridimensional structure of the clusters and preventing islet mass loss (Ichii et al. 2007). The simplest and most investigated approach to islet preservation and culture in vitro is through alteration in culture conditions, including temperature and media composition (Daoud et al. 2010a). Several studies have also involved the administration of various growth factors and compounds in order to enhance the suspension culture of islets such as glutamine compounds, human albumin, insulin, and sericin (Daoud et al. 2010a). These investigations have been the focus of many laboratories for the purpose of short-term culture for transplantation as well as long-term culture in vitro testing. The use of slightly impure islet preparations and co-culture with extracellular matrix components such as collagen has shown to enhance the viability and function of isolated islets (Daoud et al. 2010b). In addition, supplementation of culture medium with small intestinal submucosa has also shown to improve islet functioning and viability (Lakey et al. 2001). Media composition, seeding density, and temperature play a significant role (Ichii et al. 2007).

A very important area of investigation is the use of compounds capable of improving the islet health during culture and enhancing their function after transplant. Very promising results have been obtained with antioxidants, which seem to successfully preserve islet integrity during culture. When using these agents the functional behavior and phenotypic cell characteristics of treated islets were preserved, as was the capacity to normalize diabetic mice, even when a marginal mass of islets was transplanted (Bottino et al. 2004). Similar results were obtained when using a catalytic antioxidant (CA) formulation with mice and human with clear benefits during transplant (Sklavos et al. 2010). Another area of investigation is methods for reducing anoxia during culture. Isolated islets are especially susceptible to damage from anoxia due to their large size relative to single cells, high oxygen consumption rate, and low levels of enzymes necessary for energy production under anaerobic conditions (Papas et al. 2005; Lau et al. 2009). Islets cultured at high surface densities in standard T-flasks also exhibit low viable tissue recovery, viability, and potency, due to anoxic conditions. These effects have been prevented by culturing islets in gas-permeable devices, which increase oxygen availability to islets with clear benefits for clinical islet culture and shipment (Papas et al. 2005). Finally, several new biomaterials have also been used to improve islet viability during culture. Some of them provide a surrogate extracellular matrix for isolated islets as a way to recreate the native islet microenvironment (Daoud et al. 2010b). Some of the most promising advances include the use of ECM-coated surfaces, islet encapsulation, scaffolding techniques, and bioreactors, among others.

Assessment of Islet Preparations

Unlike rodent pancreata, substantial heterogeneity in islet size exists within a human pancreas. In order to accurately and consistently quantify the islets within an isolation preparation, both the number of islets and size should be taken into consideration. Historically, it has been difficult to accurately establish the amount of total islet equivalents (IEs) (or total islet volume) in the human pancreas, namely, due to the fact that islets are scattered in a large exocrine gland of which they represent only small percent in volume. As such there is some controversy as to how many islets are present in the average human pancreas. Korsgern and colleagues (Korsgren et al. 2005) estimated the number of IEs in a normal pancreas is about 500,000 IEs based on seven autopsy cases reporting islet volume of 0.5–1.3 mL (Saito et al. 1978). Others estimated that the total islet volume is 2.4 mL in a normal pancreas, which is corresponding to 1,300,000 IEs (Colton et al. 2006). In Table 2, islet volume data from autopsy studies are listed (Maclean and Ogilvie 1955; Rahier et al. 1983; Sakuraba et al. 2002; Westermark and Wilander 1978). Calculated total IEs varies from 500,000 to 1,000,000 IEs, depending on the size of pancreas studied. One Japanese study (Sakuraba et al. 2002) reported a mean pancreas weight of 122 g, which is the highest among other studies. Consequently, islet mass reported in the study is remarkably large, resulting in a calculated IE of >1,000,000 IEs. In contrast, the pancreas weight in Maclean's study is only 50 % of that reported in the Japanese study (Maclean and Ogilvie 1955). Accordingly, the calculated IE from Maclean's data is almost half of the number from the Japanese study. Based on all data listed in the table, it would be reasonable to say that the average number of IEs in a normal 90-g human pancreas (Kin et al. 2006a) is about 800,000 IEs.

In addition to the quantity of islets, the functional viability of an islet preparation is critical in predicting the success of islet transplantation. To date, there lacks a consensus within the islet transplantation field as to which assays accurately assess islet potency prior to transplantation and predicts their subsequent function posttransplant. The viability of an islet preparation is currently assessed with the use of fluorescent stains based on dye exclusion polarity. For example, fluorescein diacetate (FDA) is a nonpolar dye and passes through the plasma membrane of living cells, whereas propidium iodide (PI) can only enter cells that have a compromised membrane. Using these two dyes together, the proportion of viable (green, FDA-positive) versus dead (red, PI-positive) cells can be assessed. FDA/PI is currently a widely used method for viability determination of the islet preparation prior to transplantation. These tests can be rapidly performed and are less labor intensive, making them attractive for use just prior to transplantation. However, there are several problems, making them of limited value. The main problem is that membrane integrity tests cannot distinguish between islets and non-islets. Another problem with the tests is the difficulty in assessing live/dead cells within a three-dimensional structure. In addition these tests fail to measure the metabolic capacity of the islet preparation. Nevertheless, it is important to acknowledge that viability estimated by membrane integrity tests is predictive of some outcome measurements in clinical transplantation, according to an annual report from Collaborative Islet Transplant Registry (2008).

A potentially more efficacious marker to determine the islet functional capacity is assaying for mitochondrial activity. Mitochondrial integrity is central to islet quality because mitochondria play a crucial role for glucose-stimulated insulin

Table 2 Estimation of total islet equivalents in a pancreas

Author [reference]	Mean age (range) year	n	Pancreas size (range) g or mL	Islet mass g	Islet volume mL	Calculated islet equivalents IE
Sakuraba et al. (2002)	51.7 (27–69)	15	122 g (75–170)	2.03	1.92[a]	1,085,295
Westermark and Wilander (1978)	74.9 (66–88)	15	76 mL	NR	1.60	905,874
Rahier et al. (1983)	54 (18–86)	8	83 g (67–110)	1.395	1.32[a]	745,806
Maclean and Ogilvie (1955)	56.1 (15–81)	30	61.7 g (38.9–99.2)	1.06	1.00[a]	566,706

Adapted from Kin (2010)
[a]Calculated assuming that islet density is 1.059 g/mL London et al. (1992)
NR not reported

secretion (Maechler 2002) and islet cell apoptosis (Aikin et al. 2004). Mitochondrial activity can be evaluated using a variety of methods. These include oxygen consumption rate (OCR), detection of mitochondrial membrane potential using dyes, release of cytochrome c, and measurement of redox state. Papas and colleagues assessed OCR of human islet preparations; they also measured DNA content of the preparations in order to normalize the OCR (Papas et al. 2007). They showed that OCR/DNA assay predicted efficacy of human islets grafted into mice. Furthermore, other groups have also demonstrated the utility of modified oxygen consumption rates as a predictor of islet in vivo function (Pepper et al. 2012). Despite beginning able to dynamically ascertain cellular potency, a drawback for this assay is that it cannot offer islet specificity, as all cells in the islet preparation will consume oxygen. Like membrane integrity test, the purity of the islet preparations significantly influences the assays precision. To circumvent this limitation, Sweet and colleagues developed a flow culture system (Sweet et al. 2002) that allows one to measure the OCR response, in human islets, against glucose stimulation. They demonstrated that glucose-stimulated changes in OCR were well correlated with in vivo function of human islet grafts (Sweet et al. 2005, 2008). They also showed glucose stimulation hardly increased OCR in non-islet tissue (Sweet et al. 2008). Given the fact that a clinical islet preparation contains a considerable amount of non-endocrine tissue, their approach would be logical and practical.

Since β cells within the islet are the only cells with the capacity to secrete insulin, it could be argued that the viability of β cells is probably most important to the outcome of transplantation. Ichii and colleagues reported a method for quantitating β-cell-specific viability (Ichii et al. 2005b). They dissociated islets into single cells and then stained the cells with a zinc specific dye, Newport Green (Molecular Probes, Eugene, OR, USA), and with a mitochondrial dye, tetramethylrhodamine ethyl ester. The double positive cells were quantified on a flow cytometer after dead cells were excluded using a DNA-binding dye. They showed that the β-cell-specific viability of human islet preparations was a useful marker of the outcome of a mouse transplant assay. The major limitation of this method is that the dispersed single cells are not likely representative of the original islets because a substantial fraction of cells is lost during dissociation. This dissociation of the islet also may contribute to β-cell death, resulting in a false-negative outcome. In addition, necrotic cells or late-stage apoptotic cells were not counted as nonviable cells, thereby leading to overestimation. Finally, several recent studies brought into question the use of Newport Green for detection of β cells because of its low quantum yield and poor selectivity to zinc (Gee et al. 2002).

Cytoprotective Strategies During Islet Isolation

During the isolation procedure the islets are exposed to numerous types of stress induced by nonphysiological stimuli. These include ischemic stress during organ procurement, preservation and islet isolation, mechanical and enzymatic stress during digestion, and osmotic stress during purification. A number of investigators have explored strategies to confer islet resistance to stress-induced damage. Most investigations have centered on treatment of the islets during culture. Some are focusing on modification in the isolation procedure to protect islets.

The use of antioxidants during islet isolation to protect islets from oxidative cell injury is a rational approach, because islet cells harbor poor endogenous antioxidant defense systems (Tiedge et al. 1997). Oxidative stress is a mechanism associated with disease states marked by inflammatory processes, including, but not limited to, autoimmune diseases such as type 1 diabetes and islet graft dysfunction (Tse et al. 2004; Toyokuni 1999; McCord 2000). Oxidative stress is initiated by the excessive production of reactive oxygen species, which are potent inducers of proinflammatory stress responses often marked by proinflammatory cytokines and chemokine synthesis (Tse et al. 2004; Adler et al. 1999).

Avila and colleagues delivered glutamine to the human pancreas via the duct prior to pancreas dissociation (Avila et al. 2005). They found that glutamine treatment reduced islet cell apoptosis and improved islet yield and function. Similarly, Bottino and colleagues perfused human pancreata with a mimetic superoxide dismutase, a novel class of chemical antioxidant compounds (Bottino et al. 2002). Islet yield immediately after isolation from a treated pancreas was similar to those from a non-treated pancreas. However, in vitro islet survival was significantly improved when islets were further treated with this compound during subsequent culture.

The ability to catalytically modulate oxidation-reduction reactions within a cell may control signaling cascades necessary for generating inflammation and provide therapeutic benefit targeted at downregulation of the immune response. The metalloporphyrin-based CAs have been demonstrated to scavenge a broad range of oxidants (Day et al. 1995, 1997, 1999; Ferrer-Sueta et al. 2003; Batinic-Haberle et al. 1998). The utility of these CAs to ameliorate other inflammatory-mediated disease processes has been demonstrated in a type 1 diabetes model of adoptive transfer, apoptosis, and blocking of hydrogen peroxide-induced mitochondrial DNA damage and partial rescue of a lethal phenotype in a manganese superoxide dismutase knockout mouse (Melov et al. 1998; Piganelli et al. 2002). Furthermore, Piganelli and colleagues have demonstrated that redox modulation protects islets from both the stresses involved in the isolation procedure and transplant-related injury (Bottino et al. 2004; Sklavos et al. 2010). It is conceivable that islet-sparing agents (i.e., CAs), which decrease the production of free radicals and, therefore, inflammatory cytokines may have a positive impact on islet function posttransplant and reduced the prevalence of primary nonfunction, increase the incidences of insulin independence from single islet infusions. In addition CA may allow for the utilization of islet isolated from expanded-criteria donors and DCD donors potentially increase the number of patients that can receive an islet transplant.

Conclusions

In recent years, the results of clinical islet transplantation have improved dramatically; such that the 5-year insulin independence rates now match the results of whole organ pancreas transplantation. This reality has become possible through the implementation of new more potent immunosuppressive agents while avoiding corticosteroids. Not to be forgotten are the substantial advancements in human islet isolation technology that have provided critical contributions to the steady evolution of this therapy. The goal now should be to sharply focus on routinely obtaining a large number of viable islets that provide full functional survival for the long term. Once met, this goal will undoubtedly enhance the long-term rates of insulin independence from single-donor recipients in the clinical setting. Indeed, much work remains to be done to achieve this goal, but it is clear that there is scope and tangible path for significant improvements that will permit islet transplantation to be a practical therapy for all patients with type 1 diabetes.

Acknowledgments To members of the Clinical Islet Laboratory at the University of Alberta for technical help in islet preparation, to the organ procurement organizations across Canada for identifying donors, and to our colleagues in the Human Organ Procurement and Exchange program in Edmonton for assistance in organ procurement. The Clinical Islet Transplant Program at the University of Alberta receives funding from the Juvenile Diabetes Research Foundation and the National Institute of Diabetes and Digestive and Kidney Diseases of the National Institutes of Health and charitable donations administered through the Diabetes Research Institute Foundation Canada.

Cross-References

- ▶ Human Islet Autotransplantation
- ▶ Islet Isolation from Pancreatitis Pancreas for Islet Autotransplantation
- ▶ Mouse Islet Isolation
- ▶ Successes and Disappointments with Clinical Islet Transplantation

References

Adler V, Yin Z, Tew KD, Ronai Z (1999) Role of redox potential and reactive oxygen species in stress signaling. Oncogene 18:6104–6111

Aikin R, Rosenberg L, Paraskevas S, Maysinger D (2004) Inhibition of caspase-mediated PARP-1 cleavage results in increased necrosis in isolated islets of Langerhans. J Mol Med (Berl) 82:389–397

Akamatsu FE, De-Souza RR, Liberti EA (1999) Fall in the number of intracardiac neurons in aging rats. Mech Ageing Dev 109:153–161

Alejandro R, Cutfield RG, Shienvold FL, Polonsky KS, Noel J, Olson L et al (1986) Natural history of intrahepatic canine islet cell autografts. J Clin Invest 78:1339–1348

Antonioli B, Fermo I, Cainarca S, Marzorati S, Nano R, Baldissera M et al (2007) Characterization of collagenase blend enzymes for human islet transplantation. Transplantation 84:1568–1575

Aumailley M, Gayraud B (1998) Structure and biological activity of the extracellular matrix. J Mol Med (Berl) 76:253–265

Avila J, Barbaro B, Gangemi A, Romagnoli T, Kuechle J, Hansen M et al (2005) Intra-ductal glutamine administration reduces oxidative injury during human pancreatic islet isolation. Am J Transplant 5:2830–2837

Balamurugan AN, He J, Guo F, Stolz DB, Bertera S, Geng X et al (2005) Harmful delayed effects of exogenous isolation enzymes on isolated human islets: relevance to clinical transplantation. Am J Transplant 5:2671–2681

Balamurugan AN, Chang Y, Bertera S, Sands A, Shankar V, Trucco M et al (2006) Suitability of human juvenile pancreatic islets for clinical use. Diabetologia 49:1845–1854

Balamurugan AN, Breite AG, Anazawa T, Loganathan G, Wilhelm JJ, Papas KK et al (2010) Successful human islet isolation and transplantation indicating the importance of class 1 collagenase and collagen degradation activity assay. Transplantation 89:954–961

Balamurugan AN, Loganathan G, Bellin MD, Wilhelm JJ, Harmon J, Anazawa T et al (2012) A new enzyme mixture to increase the yield and transplant rate of autologous and allogeneic human islet products. Transplantation 93:693–702

Barbaro B, Salehi P, Wang Y, Qi M, Gangemi A, Kuechle J et al (2007) Improved human pancreatic islet purification with the refined UIC-UB density gradient. Transplantation 84:1200–1203

Barnett MJ, Zhai X, LeGatt DF, Cheng SB, Shapiro AM, Lakey JR (2005) Quantitative assessment of collagenase blends for human islet isolation. Transplantation 80:723–728

Batinic-Haberle I, Benov L, Spasojevic I, Fridovich I (1998) The ortho effect makes manganese (III) meso-tetrakis(N-methylpyridinium-2-yl)porphyrin a powerful and potentially useful superoxide dismutase mimic. J Biol Chem 273:24521–24528

Bedossa P, Lemaigre G, Bacci J, Martin E (1989) Quantitative estimation of the collagen content in normal and pathologic pancreas tissue. Digestion 44:7–13

Bennet W, Groth CG, Larsson R, Nilsson B, Korsgren O (2000) Isolated human islets trigger an instant blood mediated inflammatory reaction: implications for intraportal islet transplantation as a treatment for patients with type 1 diabetes. Ups J Med Sci 105:125–133

Berney T, Johnson PR (2010) Donor pancreata: evolving approaches to organ allocation for whole pancreas versus islet transplantation. Transplantation 90:238–243

Berney T, Molano RD, Cattan P, Pileggi A, Vizzardelli C, Oliver R et al (2001) Endotoxin-mediated delayed islet graft function is associated with increased intra-islet cytokine production and islet cell apoptosis. Transplantation 71:125–132

Bertuzzi F, Grohovaz F, Maffi P, Caumo A, Aldrighetti L, Nano R et al (2002) Successful transplantation of human islets in recipients bearing a kidney graft. Diabetologia 45:77–84

Bond MD, Van Wart HE (1984) Characterization of the individual collagenases from Clostridium histolyticum. Biochemistry 23:3085–3091

Bottino R, Balamurugan AN, Bertera S, Pietropaolo M, Trucco M, Piganelli JD (2002) Preservation of human islet cell functional mass by anti-oxidative action of a novel SOD mimic compound. Diabetes 51:2561–2567

Bottino R, Balamurugan AN, Tse H, Thirunavukkarasu C, Ge X, Profozich J et al (2004) Response of human islets to isolation stress and the effect of antioxidant treatment. Diabetes 53:2559–2568

Brandhorst H, Brandhorst D, Hering BJ, Federlin K, Bretzel RG (1995a) Body mass index of pancreatic donors: a decisive factor for human islet isolation. Exp Clin Endocrinol Diabetes 103(Suppl 2):23–26

Brandhorst H, Hering BJ, Brandhorst D, Hiller WF, Gubernatis G, Federlin K et al (1995b) Comparison of Histidine-Tryptophane-Ketoglutarate (HTK) and University of Wisconsin (UW) solution for pancreas perfusion prior to islet isolation, culture and transplantation. Transplant Proc 27:3175–3176

Brandhorst D, Brandhorst H, Hering BJ, Bretzel RG (1999) Long-term survival, morphology and in vitro function of isolated pig islets under different culture conditions. Transplantation 67:1533–1541

Brandhorst H, Brandhorst D, Hesse F, Ambrosius D, Brendel M, Kawakami Y et al (2003) Successful human islet isolation utilizing recombinant collagenase. Diabetes 52:1143–1146

Brandhorst H, Brendel MD, Eckhard M, Bretzel RG, Brandhorst D (2005) Influence of neutral protease activity on human islet isolation outcome. Transplant Proc 37:241–242

Brandhorst H, Raemsch-Guenther N, Raemsch C, Friedrich O, Huettler S, Kurfuerst M, Korsgren O, Brandhorst D (2008) The ratio between collagenase class I and class II influences the efficient islet release from the rat pancreas. Transplantation 35:456–461

Bucher P, Bosco D, Mathe Z, Matthey-Doret D, Andres A, Kurfuerst M et al (2004) Optimization of neutral protease to collagenase activity ratio for islet of Langerhans isolation. Transplant Proc 36:1145–1146

Burgarella S, Merlo S, Figliuzzi M, Remuzzi A (2013) Isolation of Langerhans islets by dielectrophoresis. Electrophoresis 34:1068–1075

Butler AE, Janson J, Bonner-Weir S, Ritzel R, Rizza RA, Butler PC (2003) β-cell deficit and increased β-cell apoptosis in humans with type 2 diabetes. Diabetes 52:102–110

Caballero-Corbalan J, Eich T, Lundgren T, Foss A, Felldin M, Kallen R et al (2007) No beneficial effect of two-layer storage compared with UW-storage on human islet isolation and transplantation. Transplantation 84:864–869

Cabric S, Sanchez J, Lundgren T, Foss A, Felldin M, Kallen R et al (2007) Islet surface heparinization prevents the instant blood-mediated inflammatory reaction in islet transplantation. Diabetes 56:2008–2015

Casey JJ, Lakey JR, Ryan EA, Paty BW, Owen R, O'Kelly K et al (2002) Portal venous pressure changes after sequential clinical islet transplantation. Transplantation 74:913–915

Chadwick DR, Robertson GS, Contractor HH, Rose S, Johnson PR, James RF et al (1994) Storage of pancreatic digest before islet purification. The influence of colloids and the sodium to potassium ratio in University of Wisconsin-based preservation solutions. Transplantation 58:99–104

Clausen B (1963) Influence of age on chondroitin sulfates and collagen of human aorta, myocardium, and skin. Lab Invest 12:538–542

Collaborative Islet Transplant Registry: Annual Report (2008) Available from https://web.emmes.com/study/isl/reports/reports.htm. Accessed 20 Oct 2013

Colton CK, Papas KK, Pisania A, Rappel MJ, Powers DE, O'Neil JJ et al (2006) Characterization of islet preparations. In: Halberstadt CD, Emerich DF (eds) Cellular transplantation: from laboratory to clinic. Elsevier, New York, pp 85–133

Cross SE, Hughes SJ, Partridge CJ, Clark A, Gray DW, Johnson PR (2008) Collagenase penetrates human pancreatic islets following standard intraductal administration. Transplantation 86:907–911

Daoud J, Rosenberg L, Tabrizian M (2010a) Pancreatic islet culture and preservation strategies: advances, challenges, and future outlook. Cell Transplant 19:1523–1535

Daoud J, Petropavlovskaia M, Rosenberg L, Tabrizian M (2010b) The effect of extracellular matrix components on the preservation of human islet function in vitro. Biomaterials 31:1676–1682

Day BJ, Shawen S, Liochev SI, Crapo JD (1995) A metalloporphyrin superoxide dismutase mimetic protects against paraquat-induced endothelial cell injury, in vitro. J Pharmacol Exp Ther 275:1227–1232

Day BJ, Fridovich I, Crapo JD (1997) Manganic porphyrins possess catalase activity and protect endothelial cells against hydrogen peroxide-mediated injury. Arch Biochem Biophys 347:256–262

Day BJ, Batinic-Haberle I, Crapo JD (1999) Metalloporphyrins are potent inhibitors of lipid peroxidation. Free Radic Biol Med 26:730–736

de Boer J, De Meester J, Smits JM, Groenewoud AF, Bok A, van der Velde O et al (1999) Eurotransplant randomized multicenter kidney graft preservation study comparing HTK with UW and Euro-Collins. Transpl Int 12:447–453

de la Grandmaison GL, Clairand I, Durigon M (2001) Organ weight in 684 adult autopsies: new tables for a Caucasoid population. Forensic Sci Int 119:149–154

Deng S, Vatamaniuk M, Huang X, Doliba N, Lian MM, Frank A et al (2004) Structural and functional abnormalities in the islets isolated from type 2 diabetic subjects. Diabetes 53:624–632

Escolar JC, Hoo-Paris R, Castex C, Sutter BC (1990) Effect of low temperatures on glucose-induced insulin secretion and glucose metabolism in isolated pancreatic islets of the rat. J Endocrinol 125:45–51

Ferrer-Sueta G, Vitturi D, Batinic-Haberle I, Fridovich I, Goldstein S, Czapski G et al (2003) Reactions of manganese porphyrins with peroxynitrite and carbonate radical anion. J Biol Chem 278:27432–27438

Gala-Lopez B, Pepper AR, Shapiro AM (2013) Biologic agents in islet transplantation. Curr Diab Rep 13:713–722

Gee KR, Zhou ZL, Ton-That D, Sensi SL, Weiss JH (2002) Measuring zinc in living cells. A new generation of sensitive and selective fluorescent probes. Cell Calcium 31:245–251

Georges P, Muirhead RP, Williams L, Holman S, Tabiin MT, Dean SK et al (2002) Comparison of size, viability, and function of fetal pig islet-like cell clusters after digestion using collagenase or liberase. Cell Transplant 11:539–545

Gomes OA, de Souza RR, Liberti EA (1997) A preliminary investigation of the effects of aging on the nerve cell number in the myenteric ganglia of the human colon. Gerontology 43:210–217

Goto M, Eich TM, Felldin M, Foss A, Kallen R, Salmela K et al (2004) Refinement of the automated method for human islet isolation and presentation of a closed system for in vitro islet culture. Transplantation 78:1367–1375

Gray DW, Warnock GL, Sutton R, Peters M, McShane P, Morris PJ (1986) Successful autotransplantation of isolated islets of Langerhans in the cynomolgus monkey. Br J Surg 73:850–853

Hanley SC, Paraskevas S, Rosenberg L (2008) Donor and isolation variables predicting human islet isolation success. Transplantation 85:950–955

Harlan DM, Kenyon NS, Korsgren O, Roep BO (2009) Current advances and travails in islet transplantation. Diabetes 58:2175–2184

Heller-Harrison RA, Carter WG (1984) Pepsin-generated type VI collagen is a degradation product of GP140. J Biol Chem 259:6858–6864

Hering BJ, Matsumoto I, Sawada T, Nakano M, Sakai T, Kandaswamy R et al (2002) Impact of two-layer pancreas preservation on islet isolation and transplantation. Transplantation 74:1813–1816

Hering BJ, Kandaswamy R, Harmon JV, Ansite JD, Clemmings SM, Sakai T et al (2004) Transplantation of cultured islets from two-layer preserved pancreases in type 1 diabetes with anti-CD3 antibody. Am J Transplant 4:390–401

Hiller WF, Klempnauer J, Luck R, Steiniger B (1991) Progressive deterioration of endocrine function after intraportal but not kidney subcapsular rat islet transplantation. Diabetes 40:134–140

Horcher A, Siebers U, Bretzel RG, Federlin K, Zekorn T (1995) Transplantation of microencapsulated islets in rats: influence of low temperature culture before or after the encapsulation procedure on the graft function. Transplant Proc 27:3232–3233

Huang GC, Zhao M, Jones P, Persaud S, Ramracheya R, Lobner K et al (2004) The development of new density gradient media for purifying human islets and islet-quality assessments. Transplantation 77:143–145

Hughes SJ, McShane P, Contractor HH, Gray DW, Clark A, Johnson PR (2005) Comparison of the collagen VI content within the islet-exocrine interface of the head, body, and tail regions of the human pancreas. Transplant Proc 37:3444–3445

Hughes SJ, Clark A, McShane P, Contractor HH, Gray DW, Johnson PR (2006) Characterisation of collagen VI within the islet-exocrine interface of the human pancreas: implications for clinical islet isolation? Transplantation 81:423–426

Hyder A (2005) Effect of the pancreatic digestion with liberase versus collagenase on the yield, function and viability of neonatal rat pancreatic islets. Cell Biol Int 29:831–834

Ichii H, Pileggi A, Molano RD, Baidal DA, Khan A, Kuroda Y et al (2005a) Rescue purification maximizes the use of human islet preparations for transplantation. Am J Transplant 5:21–30

Ichii H, Inverardi L, Pileggi A, Molano RD, Cabrera O, Caicedo A et al (2005b) A novel method for the assessment of cellular composition and β-cell viability in human islet preparations. Am J Transplant 5:1635–1645

Ichii H, Pileggi A, Khan A, Fraker C, Ricordi C (2007) Culture and transportation of human islets between centers. In: Shapiro AMJ, Shaw J (eds) Islet transplantation and β cell replacement therapy. Informa Healthcare, New York, p 251

Iglesias I, Valiente L, Shiang KD, Ichii H, Kandeel F, Al-Abdullah IH (2012) The effects of digestion enzymes on islet viability and cellular composition. Cell Transplant 21:649–655

Ihm SH, Matsumoto I, Sawada T, Nakano M, Zhang HJ, Ansite JD et al (2006) Effect of donor age on function of isolated human islets. Diabetes 55:1361–1368

Ihm SH, Matsumoto I, Zhang HJ, Ansite JD, Hering BJ (2009) Effect of short-term culture on functional and stress-related parameters in isolated human islets. Transpl Int 22:207–216

Ilieva A, Yuan S, Wang R, Duguid WP, Rosenberg L (1999) The structural integrity of the islet in vitro: the effect of incubation temperature. Pancreas 19:297–303

Jaeger C, Wohrle M, Bretzel RG, Federlin K (1994) Effect of transplantation site and culture pretreatment on islet xenograft survival (rat to mouse) in experimental diabetes without immunosuppression of the host. Acta Diabetol 31:193–197

Johnson LF, deSerres S, Herzog SR, Peterson HD, Meyer AA (1991) Antigenic cross-reactivity between media supplements for cultured keratinocyte grafts. J Burn Care Rehabil 12:306–312

Johnson PR, White SA, London NJ (1996) Collagenase and human islet isolation. Cell Transplant 5:437–452

Kadono G, Ishihara T, Yamaguchi T, Kato K, Kondo F, Naito I et al (2004) Immunohistochemical localization of type IV collagen α chains in the basement membrane of the pancreatic duct in human normal pancreas and pancreatic diseases. Pancreas 29:61–66

Kamei T, Yasunami Y, Terasaka R, Konomi K (1989) Importance of transplant site for prolongation of islet xenograft survival (rat to mouse) after low-temperature culture and cyclosporin A. Transplant Proc 21:2693–2694

Kaufman DB, Morel P, Field MJ, Munn SR, Sutherland DE (1990) Purified canine islet autografts. Functional outcome as influenced by islet number and implantation site. Transplantation 50:385–391

Kedinger M, Haffen K, Grenier J, Eloy R (1977) In vitro culture reduces immunogenicity of pancreatic endocrine islets. Nature 270:736–738

Keymeulen B, Anselmo J, Pipeleers D (1997) Length of metabolic normalization after rat islet cell transplantation depends on endocrine cell composition of graft and on donor age. Diabetologia 40:1152–1158

Keymeulen B, Ling Z, Gorus FK, Delvaux G, Bouwens L, Grupping A et al (1998) Implantation of standardized β-cell grafts in a liver segment of IDDM patients: graft and recipients characteristics in two cases of insulin-independence under maintenance immunosuppression for prior kidney graft. Diabetologia 41:452–459

Keymeulen B, Gillard P, Mathieu C, Movahedi B, Maleux G, Delvaux G et al (2006) Correlation between β cell mass and glycemic control in type 1 diabetic recipients of islet cell graft. Proc Natl Acad Sci U S A 103:17444–17449

Kin T (2010) Islet isolation for clinical transplantation. Adv Exp Med Biol 654:683–710

Kin T, Murdoch TB, Shapiro AM, Lakey JR (2006a) Estimation of pancreas weight from donor variables. Cell Transplant 15:181–185

Kin T, Mirbolooki M, Salehi P, Tsukada M, O'Gorman D, Imes S et al (2006b) Islet isolation and transplantation outcomes of pancreas preserved with University of Wisconsin solution versus two-layer method using preoxygenated perfluorocarbon. Transplantation 82:1286–1290

Kin T, Mirbolooki M, Brassil J, Shapiro AM, Lakey J (2006) Machine perfusion for prolonged pancreas preservation prior to islet isolation (Abstract). In: Annual scientific meeting, Canadian Society of Transplantation

Kin T, Johnson PR, Shapiro AM, Lakey JR (2007a) Factors influencing the collagenase digestion phase of human islet isolation. Transplantation 83:7–12

Kin T, Zhai X, Murdoch TB, Salam A, Shapiro AM, Lakey JR (2007b) Enhancing the success of human islet isolation through optimization and characterization of pancreas dissociation enzyme. Am J Transplant 7:1233–1241

Kin T, Zhai X, O'Gorman D, Shapiro AM (2008a) Detrimental effect of excessive collagenase class II on human islet isolation outcome. Transpl Int 21:1059–1065

Kin T, Senior P, O'Gorman D, Richer B, Salam A, Shapiro AM (2008b) Risk factors for islet loss during culture prior to transplantation. Transpl Int 21:1029–1035

Kin T, O'Gorman D, Zhai X, Pawlick R, Imes S, Senior P, Shapiro AM (2009) Nonsimultaneous administration of pancreas dissociation enzymes during islet isolation. Transplantation 87:1700–1705

King A, Lock J, Xu G, Bonner-Weir S, Weir GC (2005) Islet transplantation outcomes in mice are better with fresh islets and exendin-4 treatment. Diabetologia 48:2074–2079

Koh A, Kin T, Imes S, Shapiro AM, Senior P (2008) Islets isolated from donors with elevated HbA1c can be successfully transplanted. Transplantation 86:1622–1624

Korsgren O, Nilsson B, Berne C, Felldin M, Foss A, Kallen R et al (2005) Current status of clinical islet transplantation. Transplantation 79:1289–1293

Koulmanda M, Smith RN, Qipo A, Weir G, Auchincloss H, Strom TB (2006) Prolonged survival of allogeneic islets in cynomolgus monkeys after short-term anti-CD154-based therapy: nonimmunologic graft failure? Am J Transplant 6:687–696

Kuroda Y, Kawamura T, Suzuki Y, Fujiwara H, Yamamoto K, Saitoh Y (1988) A new, simple method for cold storage of the pancreas using perfluorochemical. Transplantation 46:457–460

Lacy PE, Finke EH (1991) Activation of intraislet lymphoid cells causes destruction of islet cells. Am J Pathol 138:1183–1190

Lacy PE, Davie JM, Finke EH (1979a) Prolongation of islet allograft survival following in vitro culture (24 degrees C) and a single injection of ALS. Science 204:312–313

Lacy PE, Davie JM, Finke EH (1979b) Induction of rejection of successful allografts of rat islets by donor peritoneal exudate cells. Transplantation 28:415–420

Lacy PE, Ricordi C, Finke EH (1989) Effect of transplantation site and α L3T4 treatment on survival of rat, hamster, and rabbit islet xenografts in mice. Transplantation 47:761–766

Lake SP, Bassett PD, Larkins A, Revell J, Walczak K, Chamberlain J et al (1989) Large-scale purification of human islets utilizing discontinuous albumin gradient on IBM 2991 cell separator. Diabetes 38(Suppl 1):143–145

Lakey JR, Warnock GL, Kneteman NM, Ao Z, Rajotte RV (1994) Effects of pre-cryopreservation culture on human islet recovery and in vitro function. Transplant Proc 26:820

Lakey JR, Rajotte RV, Warnock GL, Kneteman NM (1995) Human pancreas preservation prior to islet isolation. Cold ischemic tolerance. Transplantation 59:689–694

Lakey JR, Warnock GL, Rajotte RV, Suarez-Alamazor ME, Ao Z, Shapiro AM et al (1996) Variables in organ donors that affect the recovery of human islets of Langerhans. Transplantation 61:1047–1053

Lakey JR, Cavanagh TJ, Zieger MA, Wright M (1998) Evaluation of a purified enzyme blend for the recovery and function of canine pancreatic islets. Cell Transplant 7:365–372

Lakey JR, Woods EJ, Zieger MA, Avila JG, Geary WA, Voytik-Harbin SL et al (2001) Improved islet survival and in vitro function using solubilized small intestinal submucosa. Cell Tissue Bank 2:217–224

Lau J, Henriksnas J, Scennson J, Carlsson PO (2009) Oxygenation of islets and its role in transplantation. Curr Opin Organ Transpl 14:688–693

Leeser DB, Bingaman AW, Poliakova L, Shi Q, Gage F, Bartlett ST et al (2004) Pulsatile pump perfusion of pancreata before human islet cell isolation. Transplant Proc 36:1050–1051

Levy MM, Ketchum RJ, Tomaszewski JE, Naji A, Barker CF, Brayman KL (2002) Intrathymic islet transplantation in the canine: I. Histological and functional evidence of autologous intrathymic islet engraftment and survival in pancreatectomized recipients. Transplantation 73:842–852

Linder M, Salovuori I, Ruohonen L, Teeri TT (1996) Characterization of a double cellulose-binding domain. Synergistic high affinity binding to crystalline cellulose. J Biol Chem 271:21268–21272

Linetsky E, Bottino R, Lehmann R, Alejandro R, Inverardi L, Ricordi C (1997) Improved human islet isolation using a new enzyme blend, liberase. Diabetes 46:1120–1123

London NJ, James RF, Bell PR (1992) Islet purification. In: Ricordi C (ed) Pancreatic islet transplantation. Landes, Austin, pp 113–123

Mackensen A, Drager R, Schlesier M, Mertelsmann R, Lindemann A (2000) Presence of IgE antibodies to bovine serum albumin in a patient developing anaphylaxis after vaccination with human peptide-pulsed dendritic cells. Cancer Immunol Immunother 49:152–156

Maclean N, Ogilvie RF (1955) Quantitative estimation of the pancreatic islet tissue in diabetic subjects. Diabetes 4:367–376

Maechler P (2002) Mitochondria as the conductor of metabolic signals for insulin exocytosis in pancreatic β-cells. Cell Mol Life Sci 59:1803–1818

Mallya SK, Mookhtiar KA, Van Wart HE (1992) Kinetics of hydrolysis of type I, II, and III collagens by the class I and II Clostridium histolyticum collagenases. J Protein Chem 11:99–107

Mandel TE, Koulmanda M (1985) Effect of culture conditions on fetal mouse pancreas in vitro and after transplantation in syngeneic and allogeneic recipients. Diabetes 34:1082–1087

Markmann JF, Deng S, Huang X, Desai NM, Velidedeoglu EH, Lui C et al (2003) Insulin independence following isolated islet transplantation and single islet infusions. Ann Surg 237:741–749

Matsumoto I, Sawada T, Nakano M, Sakai T, Liu B, Ansite JD et al (2004) Improvement in islet yield from obese donors for human islet transplants. Transplantation 78:880–885

Matsushita O, Okabe A (2001) Clostridial hydrolytic enzymes degrading extracellular components. Toxicon 39:1769–1780

Matsushita O, Jung CM, Minami J, Katayama S, Nishi N, Okabe A (1998) A study of the collagen-binding domain of a 116-kDa *Clostridium histolyticum* collagenase. J Biol Chem 273:3643–3648

Matsushita O, Jung CM, Katayama S, Minami J, Takahashi Y, Okabe A (1999) Gene duplication and multiplicity of collagenases in *Clostridium histolyticum*. J Bacteriol 181:923–933

Matsushita O, Koide T, Kobayashi R, Nagata K, Okabe A (2001) Substrate recognition by the collagen-binding domain of *Clostridium histolyticum* class I collagenase. J Biol Chem 276:8761–8770

McCall MD, Maciver AH, Pawlick R, Edgar R, Shapiro AM (2011) Histopaque provides optimal mouse islet purification kinetics: comparison study with Ficoll, iodixanol and dextran. Islets 3:144–149

McCord J (2000) The evolution of free radicals and oxidative stress. Am J Med 108:652–659

Mehigan DG, Bell WR, Zuidema GD, Eggleston JC, Cameron JL (1980) Disseminated intravascular coagulation and portal hypertension following pancreatic islet autotransplantation. Ann Surg 191:287–293

Melov S, Schneider JA, Day BJ, Hinerfeld D, Coskun P, Mirra SS et al (1998) A novel neurological phenotype in mice lacking mitochondrial manganese superoxide dismutase. Nat Genet 18:159–163

Mettler E, Trenkler A, Feilen PJ, Wiegand F, Fottner C, Ehrhart F et al (2013) Magnetic separation of encapsulated islet cells labeled with superparamagnetic iron oxide nano particles. Xenotransplantation 20:219–226

Meyer AA, Manktelow A, Johnson M, deSerres S, Herzog S, Peterson HD (1982) Antibody response to xenogeneic proteins in burned patients receiving cultured keratinocyte grafts. J Trauma 28:1054–1059

Minor T, Tolba R, Akbar S, Dombrowski T, Muller A (2000) The suboptimal donor: reduction of ischemic injury in fatty livers by gaseous oxygen persufflation. Transplant Proc 32:10

Misawa R, Ricordi C, Miki A, Barker S, Molano RD, Khan A et al (2012) Evaluation of viable β-cell mass is useful for selecting collagenase for human islet isolation: comparison of collagenase NB1 and liberase HI. Cell Transplant 21:39–47

Mita A, Ricordi C, Messinger S, Miki A, Misawa R, Barker S et al (2010) Antiproinflammatory effects of iodixanol (OptiPrep)-based density gradient purification on human islet preparations. Cell Transplant 19:1537–1546

Mittal VK, Toledo-Pereyra LH, Sharma M, Ramaswamy K, Puri VK, Cortez JA et al (1981) Acute portal hypertension and disseminated intravascular coagulation following pancreatic islet autotransplantation after subtotal pancreatectomy. Transplantation 31:302–304

Mookhtiar KA, Van Wart HE (1992) *Clostridium histolyticum* collagenases: a new look at some old enzymes. Matrix Suppl 1:116–126

Morita A, Kuroda Y, Fujino Y, Tanioka Y, Ku Y, Saitoh Y (1993) Assessment of pancreas graft viability preserved by a two-layer (University of Wisconsin solution/perfluorochemical) method after significant warm ischemia. Transplantation 55:667–669

Morsiani E, Lacy PE (1990) Effect of low-temperature culture and L3T4 antibody on the survival of pancreatic islets xenograft (fish to mouse). Eur Surg Res 22:78–85

Nagai K, Yagi S, Afify M, Bleilevens C, Uemoto S, Tolba RH (2013) Impact of venous-systemic oxygen persufflation with nitric oxide gas on steatotic grafts after partial orthotopic liver transplantation in rats. Transplantation 95:78–84

Nagata H, Matsumoto S, Okitsu T et al (2006) Procurement of the human pancreas for pancreatic islet transplantation from marginal cadaver donors. Transplantation 82:327–331

Nano R, Clissi B, Melzi R, Calori G, Maffi P, Antonioli B et al (2005) Islet isolation for allotransplantation: variables associated with successful islet yield and graft function. Diabetologia 48:906–912

Noguchi H, Ueda M, Nakai Y, Iwanaga Y, Okitsu T, Nagata H et al (2006) Modified two-layer preservation method (M-Kyoto/PFC) improves islet yields in islet isolation. Am J Transplant 6:496–504

Noguchi H, Ikemoto T, Naziruddin B, Jackson A, Shimoda M, Fujita Y et al (2009) Iodixanol-controlled density gradient during islet purification improves recovery rate in human islet isolation. Transplantation 87:1629–1635

Noguchi H, Naziruddin B, Onaca N, Jackson A, Shimoda M, Ikemoto T et al (2010) Comparison of modified Celsior solution and M-kyoto solution for pancreas preservation in human islet isolation. Cell Transplant 19:751–758

Noguchi H, Naziruddin B, Jackson A, Shimoda M, Fujita Y, Chujo D et al (2012a) Comparison of ulinastatin, gabexate mesilate, and nafamostat mesilate in preservation solution for islet isolation. Cell Transplant 21:509–516

Noguchi H, Naziruddin B, Jackson A, Shimoda M, Ikemoto T, Fujita Y et al (2012b) Fresh islets are more effective for islet transplantation than cultured islets. Cell Transplant 21:517–523

Oberholzer J, Triponez F, Lou J, Morel P (1999) Clinical islet transplantation: a review. Ann N Y Acad Sci 875:189–199

Oberholzer J, Triponez F, Mage R, Andereggen E, Buhler L, Cretin N et al (2000) Human islet transplantation: lessons from 13 autologous and 13 allogeneic transplantations. Transplantation 69:1115–1123

O'Gorman D, Kin T, Murdoch T, Richer B, McGhee-Wilson D, Ryan E et al (2005) The standardization of pancreatic donors for islet isolations. Transplantation 80:801–806

O'Gorman D, Kin T, Imes S, Pawlick R, Senior P, Shapiro AM (2010) Comparison of human islet isolation outcomes using a new mammalian tissue-free enzyme versus collagenase NB-1. Transplantation 90:255–259

Olack B, Swanson C, McLear M, Longwith J, Scharp D, Lacy PE (1991) Islet purification using Euro-Ficoll gradients. Transplant Proc 23:774–776

Olack BJ, Swanson CJ, Howard TK, Mohanakumar T (1999) Improved method for the isolation and purification of human islets of langerhans using liberase enzyme blend. Hum Immunol 60:1303–1309

Olsson R, Carlsson PO (2005) Better vascular engraftment and function in pancreatic islets transplanted without prior culture. Diabetologia 48:469–476

Ono J, Lacy PE, Michael HE, Greider MH (1979) Studies of the functional and morphologic status of islets maintained at 24 C for four weeks in vitro. Am J Pathol 97:489–503

Opelz G, Terasaki PI (1974) Lymphocyte antigenicity loss with retention of responsiveness. Science 184:464–466

Orloff MJ, Macedo A, Greenleaf GE, Girard B (1988) Comparison of the metabolic control of diabetes achieved by whole pancreas transplantation and pancreatic islet transplantation in rats. Transplantation 45:307–312

Papas KK, Avgoustiniatos ES, Tempelman LA, Weir GC, Colton CK, Pisania A et al (2005) High-density culture of human islets on top of silicone rubber membranes. Transplant Proc 37:3412–3414

Papas KK, Colton CK, Nelson RA, Rozak PR, Avgoustiniatos ES, Scott WE 3rd et al (2007) Human islet oxygen consumption rate and DNA measurements predict diabetes reversal in nude mice. Am J Transplant 7:707–713

Pepper AR, Hasilo CP, Melling CW, Mazzuca DM, Vilk G, Zou G et al (2012) The islet size to oxygen consumption ratio reliably predicts reversal of diabetes posttransplant. Cell Transplant 21:2797–2804

Piganelli JD, Flores SC, Cruz C, Koepp J, Batinic-Haberle I, Crapo J et al (2002) A metalloporphyrin-based superoxide dismutase mimic inhibits adoptive transfer of autoimmune diabetes by a diabetogenic T-cell clone. Diabetes 51:347–355

Ponte GM, Pileggi A, Messinger S, Alejandro A, Ichii H, Baidal DA et al (2007) Toward maximizing the success rates of human islet isolation: influence of donor and isolation factors. Cell Transplant 16:595–607

Qin H, Matsumoto S, Klintmalm GB, De vol EB (2011) A meta-analysis for comparison of the two-layer and university of Wisconsin pancreas preservation methods in islet transplantation. Cell Transplant 20:1127–1137

Rahier J, Goebbels RM, Henquin JC (1983) Cellular composition of the human diabetic pancreas. Diabetologia 24:366–371

Ricordi C, Fraker C, Szust J, Al-Abdullah I, Poggioli R, Kirlew T et al (2003) Improved human islet isolation outcome from marginal donors following addition of oxygenated perfluorocarbon to the cold-storage solution. Transplantation 75:1524–1527

Ris F, Toso C, Veith FU, Majno P, Morel P, Oberholzer J (2004) Are criteria for islet and pancreas donors sufficiently different to minimize competition? Am J Transplant 4:763–766

Robertson GS, Chadwick D, Contractor H, Rose S, Chamberlain R, Clayton H et al (1992) Storage of human pancreatic digest in University of Wisconsin solution significantly improves subsequent islet purification. Br J Surg 79:899–902

Robertson RP, Lanz KJ, Sutherland DE, Kendall DM (2001) Prevention of diabetes for up to 13 years by autoislet transplantation after pancreatectomy for chronic pancreatitis. Diabetes 50:47–50

Ryan EA, Lakey JR, Rajotte RV, Korbutt GS, Kin T, Imes S et al (2001) Clinical outcomes and insulin secretion after islet transplantation with the Edmonton protocol. Diabetes 50:710–719

Ryan EA, Paty BW, Senior PA, Bigam D, Alfadhli E, Kneteman NM et al (2005) Five-year follow-up after clinical islet transplantation. Diabetes 54:2060–2069

Ryu S, Yasunami Y (1991) The necessity of differential immunosuppression for prevention of immune rejection by FK506 in rat islet allografts transplanted into the liver or beneath the kidney capsule. Transplantation 52:599–605

Sabek OM, Cowan P, Fraga DW, Gaber AO (2008) The effect of isolation methods and the use of different enzymes on islet yield and in vivo function. Cell Transplant 17:785–792

Saisho Y, Butler AE, Meier JJ, Monchamp T, Allen-Auerbach M, Rizza RA et al (2007) Pancreas volumes in humans from birth to age one hundred taking into account sex, obesity, and presence of type-2 diabetes. Clin Anat 20:933–942

Saito K, Iwama N, Takahashi T (1978) Morphometrical analysis on topographical difference in size distribution, number and volume of islets in the human pancreas. Tohoku J Exp Med 124:177–186

Saito T, Gotoh M, Satomi S, Uemoto S, Kenmochi T, Itoh T et al (2010) Islet transplantation using donors after cardiac death: report of the Japan Islet Transplantation Registry. Transplantation 90:740–747

Sakuraba H, Mizukami H, Yagihashi N, Wada R, Hanyu C, Yagihashi S (2002) Reduced β-cell mass and expression of oxidative stress-related DNA damage in the islet of Japanese Type II diabetic patients. Diabetologia 45:85–96

Salehi P, Hansen MA, Avila JG, Barbaro B, Gangemi A, Romagnoli T et al (2006) Human islet isolation outcomes from pancreata preserved with Histidine-Tryptophan Ketoglutarate versus University of Wisconsin solution. Transplantation 82:983–985

Scharp DW, Kemp CB, Knight MJ, Ballinger WF, Lacy PE (1973) The use of ficoll in the preparation of viable islets of langerhans from the rat pancreas. Transplantation 16:686–689

Scharp DW, Lacy PE, Finke E, Olack B (1987) Low-temperature culture of human islets isolated by the distention method and purified with Ficoll or Percoll gradients. Surgery 102:869–879

Scott WE 3rd, O'Brien TD, Ferrer-Fabrega J, Avgoustiniatos ES, Weegman BP, Anazawa T et al (2010) Persufflation improves pancreas preservation when compared with the two-layer method. Transplant Proc 42:2016–2019

Shapiro AM, Lakey JR, Rajotte RV, Warnock GL, Friedlich MS, Jewell LD et al (1995) Portal vein thrombosis after transplantation of partially purified pancreatic islets in a combined human liver/islet allograft. Transplantation 59:1060–1063

Shapiro AM, Lakey JR, Ryan EA, Korbutt GS, Toth E, Warnock GL et al (2000) Islet transplantation in seven patients with type 1 diabetes mellitus using a glucocorticoid-free immunosuppressive regimen. N Engl J Med 343:230–238

Shimoda M, Itoh T, Iwahashi S, Takita M, Sugimoto K, Kanak MA et al (2012) An effective purification method using large bottles for human pancreatic islet isolation. Islets 4:398–404

Sklavos MM, Bertera S, Tse HM, Bottino R, He J, Beilke JN et al (2010) Redox modulation protects islets from transplant-related injury. Diabetes 59:1731–1738

Sobel H, Marmorston J (1956) The possible role of the gel-fiber ratio of connective tissue in the aging process. J Gerontol 11:2–7

Socci C, Davalli AM, Vignali A, Pontiroli AE, Maffi P, Magistretti P et al (1993) A significant increase of islet yield by early injection of collagenase into the pancreatic duct of young donors. Transplantation 55:661–663

Soru E, Zaharia O (1972) *Clostridium histolyticum* collagenase. II. Partial characterization. Enzymologia 43:45–55

Spees JL, Gregory CA, Singh H, Tucker HA, Peister A, Lynch PJ et al (2004) Internalized antigens must be removed to prepare hypoimmunogenic mesenchymal stem cells for cell and gene therapy. Mol Ther 9:747–756

Srinivasan PK, Yagi S, Doorschodt B, Nagai K, Afify M, Uemoto S et al (2012) Impact of venous systemic oxygen persufflation supplemented with nitric oxide gas on cold-stored, warm ischemia-damaged experimental liver grafts. Liver Transpl 18:219–225

Stadlbauer V, Schaffellner S, Iberer F, Lackner C, Liegl B, Zink B et al (2003) Occurrence of apoptosis during ischemia in porcine pancreas islet cells. Int J Artif Organs 26:205–210

Stegall MD, Dean PG, Sung R, Guidinger MK, McBride MA, Sommers C et al (2007) The rationale for the new deceased donor pancreas allocation schema. Transplantation 83:1156–1161

Sung RS, Christensen LL, Leichtman AB, Greenstein SM, Distant DA, Wynn JJ et al (2008) Determinants of discard of expanded criteria donor kidneys: impact of biopsy and machine perfusion. Am J Transplant 8:783–792

Suszynski TM, Rizzari MD, Scott WE 3rd, Tempelman LA, Taylor MJ, Papas KK (2012) Persufflation (or gaseous oxygen perfusion) as a method of organ preservation. Cryobiology 64:125–143

Suszynski TM, Rizzari MD, Scott WE, Eckman PM, Fonger JD, John R et al (2013) Persufflation (gaseous oxygen perfusion) as a method of heart preservation. J Cardiothorac Surg 8:105

Sutherland DE, Gruessner AC, Carlson AM, Blondet JJ, Balamurugan AN, Reigstad KF et al (2008) Islet autotransplant outcomes after total pancreatectomy: a contrast to islet allograft outcomes. Transplantation 86:1799–1802

Sweet IR, Khalil G, Wallen AR, Steedman M, Schenkman KA, Reems JA et al (2002) Continuous measurement of oxygen consumption by pancreatic islets. Diabetes Technol Ther 4:661–672

Sweet IR, Gilbert M, Jensen R, Sabek O, Fraga DW, Gaber AO et al (2005) Glucose stimulation of cytochrome C reduction and oxygen consumption as assessment of human islet quality. Transplantation 80:1003–1011

Sweet IR, Gilbert M, Scott S, Todorov I, Jensen R, Nair I et al (2008) Glucose-stimulated increment in oxygen consumption rate as a standardized test of human islet quality. Am J Transplant 8:183–192

Szot GL, Lee MR, Tavakol MM, Lang J, Dekovic F, Kerlan RK et al (2009) Successful clinical islet isolation using a GMP-manufactured collagenase and neutral protease. Transplantation 88:753–756

Tanioka Y, Sutherland DE, Kuroda Y, Gilmore TR, Asaheim TC, Kronson JW et al (1997) Excellence of the two-layer method (University of Wisconsin solution/perfluorochemical) in pancreas preservation before islet isolation. Surgery 122:435–441

Taylor MJ, Baicu S (2011) Cryo-isolation: a novel method for enzyme-free isolation of pancreatic islets involving in situ cryopreservation of islets and selective destruction of acinar tissue. Transplant Proc 43:3181–3183

Taylor MJ, Baicu S, Leman B, Greene E, Vazquez A, Brassil J (2008) Twenty-four hour hypothermic machine perfusion preservation of porcine pancreas facilitates processing for islet isolation. Transplant Proc 40:480–482

Tiedge M, Lortz S, Drinkgern J, Lenzen S (1997) Relation between antioxidant enzyme gene expression and antioxidative defense status of insulin-producing cells. Diabetes 46:1733–1742

Tolba RH, Schildberg FA, Schnurr C, Glatzel U, Decker D, Minor T (2006) Reduced liver apoptosis after venous systemic oxygen persufflation in non-heart-beating donors. J Invest Surg 19:219–227

Toledo-Pereyra LH, Valgee KD, Castellanos J, Chee M (1980) Hypothermic pulsatile perfusion: its use in the preservation of pancreases for 24 to 48 hours before islet cell transplantation. Arch Surg 115:95–98

Toledo-Pereyra LH, Rowlett AL, Cain W, Rosenberg JC, Gordon DA, MacKenzie GH (1984) Hepatic infarction following intraportal islet cell autotransplantation after near-total pancreatectomy. Transplantation 38:88–89

Totani T, Teramura Y, Iwata H (2008) Immobilization of urokinase on the islet surface by amphiphilic poly(vinyl alcohol) that carries alkyl side chains. Biomaterials 29:2878–2883

Toyokuni S (1999) Reactive oxygen species-induced molecular damage and its application in pathology. Pathol Int 49:91–102

Tse HM, Milton MJ, Piganelli JD (2004) Mechanistic analysis of the immunomodulatory effects of a catalytic antioxidant on antigen-presenting cells: implication for their use in targeting oxidation-reduction reactions in innate immunity. Free Radic Biol Med 36:233–247

Tsujimura T, Kuroda Y, Kin T, Avila JG, Rajotte RV, Korbutt GS et al (2002) Human islet transplantation from pancreases with prolonged cold ischemia using additional preservation by the two-layer (UW solution/perfluorochemical) cold-storage method. Transplantation 74:1687–1691

van Deijnen JH, Hulstaert CE, Wolters GH, van Schilfgaarde R (1992) Significance of the peri-insular extracellular matrix for islet isolation from the pancreas of rat, dog, pig, and man. Cell Tissue Res 267:139–146

Van Deijnen JH, Van Suylichem PT, Wolters GH, Van Schilfgaarde R (1994) Distribution of collagens type I, type III and type V in the pancreas of rat, dog, pig and man. Cell Tissue Res 277:115–121

van der Burg MP, Gooszen HG, Ploeg RJ, Guicherit OR, Scherft JP, Terpstra JL et al (1990) Pancreatic islet isolation with UW solution: a new concept. Transplant Proc 22:2050–2051

Van Wart HE, Steinbrink DR (1985) Complementary substrate specificities of class I and class II collagenases from *Clostridium histolyticum*. Biochemistry 24:6520–6526

Vargas F, Julian JF, Llamazares JF, Garcia-Cuyas F, Jimenez M, Pujol-Borrell R et al (2001) Engraftment of islets obtained by collagenase and liberase in diabetic rats: a comparative study. Pancreas 23:406–413

Walsh TJ, Eggleston JC, Cameron JL (1982) Portal hypertension, hepatic infarction, and liver failure complicating pancreatic islet autotransplantation. Surgery 91:485–487

Wang W, Upshaw L, Zhang G, Strong DM, Reems JA (2007) Adjustment of digestion enzyme composition improves islet isolation outcome from marginal grade human donor pancreata. Cell Tissue Bank 8:187–194

Warnock GL, Cattral MS, Rajotte RV (1988) Normoglycemia after implantation of purified islet cells in dogs. Can J Surg 31:421–426

Warnock GL, Meloche RM, Thompson D, Shapiro RJ, Fung M, Ao Z et al (2005) Improved human pancreatic islet isolation for a prospective cohort study of islet transplantation vs best medical therapy in type 1 diabetes mellitus. Arch Surg 140:735–744

Watanabe K (2004) Collagenolytic proteases from bacteria. Appl Microbiol Biotechnol 63:520–526

Westermark P, Wilander E (1978) The influence of amyloid deposits on the islet volume in maturity onset diabetes mellitus. Diabetologia 15:417–421

Will RG, Ironside JW, Zeidler M, Cousens SN, Estibeiro K, Alperovitch A et al (1996) A new variant of Creutzfeldt-Jakob disease in the UK. Lancet 347:921–925

Witkowski P, Liu Z, Cernea S, Guo Q, Poumian-Ruiz E, Herold K et al (2006) Validation of the scoring system for standardization of the pancreatic donor for islet isolation as used in a new islet isolation center. Transplant Proc 38:3039–3040

Woehrle M, Beyer K, Bretzel RG, Federlin K (1989) Prevention of islet allograft rejection in experimental diabetes of the rat without immunosuppression of the host. Transplant Proc 21:2705–2706

Wolters GH, Vos-Scheperkeuter GH, van Deijnen JH, van Schilfgaarde R (1992) An analysis of the role of collagenase and protease in the enzymatic dissociation of the rat pancreas for islet isolation. Diabetologia 35:735–742

Wolters GH, Vos-Scheperkeuter GH, Lin HC, van Schilfgaarde R (1995) Different roles of class I and class II *Clostridium histolyticum* collagenase in rat pancreatic islet isolation. Diabetes 44:227–233

Yamamoto T, Ricordi C, Messinger S, Sakuma Y, Miki A, Rodriguez R et al (2007) Deterioration and variability of highly purified collagenase blends used in clinical islet isolation. Transplantation 84:997–1002

Yasunami Y, Ryu S, Ueki M, Arima T, Kamei T, Tanaka M et al (1994) Donor-specific unresponsiveness induced by intraportal grafting and FK506 in rat islet allografts: importance of low temperature culture and transplant site on induction and maintenance. Cell Transplant 3:75–82

Yoon KH, Ko SH, Cho JH, Lee JM, Ahn YB, Song KH et al (2003) Selective β-cell loss and α-cell expansion in patients with type 2 diabetes mellitus in Korea. J Clin Endocrinol Metab 88:2300–2308

Yoshihara K, Matsushita O, Minami J, Okabe A (1994) Cloning and nucleotide sequence analysis of the colH gene from *Clostridium histolyticum* encoding a collagenase and a gelatinase. J Bacteriol 176:6489–6496

Zeng Y, Torre MA, Karrison T, Thistlethwaite JR (1994) The correlation between donor characteristics and the success of human islet isolation. Transplantation 57:954–958

Zhang G, Matsumoto S, Hyon SH, Qualley SA, Upshaw L, Strong DM et al (2004) Polyphenol, an extract of green tea, increases culture recovery rates of isolated islets from nonhuman primate pancreata and marginal grade human pancreata. Cell Transplant 13:145–152

Islet Isolation from Pancreatitis Pancreas for Islet Autotransplantation

42

A. N. Balamurugan, Gopalakrishnan Loganathan, Amber Lockridge, Sajjad M. Soltani, Joshua J. Wilhelm, Gregory J. Beilman, Bernhard J. Hering, and David E. R. Sutherland

Contents

Introduction	1201
Islet Cell Isolation	1204
Pancreatectomy and Pancreas Transport	1204
Preservation Solution	1205
Packaging	1205
Pancreas Trimming and Cannulation	1206
Trimming	1206
Cannulation	1207
Enzyme Selection	1208
Enzyme Combination	1208
Enzyme Dose	1208
Enzymatic Perfusion of Pancreas	1209
Perfusion Method	1209
Temperature, Pressure, and Flow Rate	1209
Interstitial Perfusion	1210
Post-distension Trimming	1210
Tissue Digestion	1211
Digest Sampling	1212
Digestion: Phase 1, Recirculation	1212
Digestion: Phase 2, Collection	1213
Digestion of the Severely Fibrotic Pancreas	1213
Other Common Complications	1214

A.N. Balamurugan (✉)
Islet Cell Laboratory, Cardiovascular Innovation Institute, Department of Surgery, University of Louisville, Louisville, KY, USA
e-mail: isletologist@hotmail.com; bala.appakalai@louisville.edu

G. Loganathan • A. Lockridge • S.M. Soltani • J.J. Wilhelm • G.J. Beilman • B.J. Hering • D.E.R. Sutherland
Department of Surgery, Schulze Diabetes Institute, University of Minnesota, Minneapolis, MN, USA
e-mail: gopal@umn.edu; lockr008@umn.edu; solt0040@umn.edu; wilhe010@umn.edu; beilm001@umn.edu; bhering@umn.edu; dsuther@umn.edu

M.S. Islam (ed.), *Islets of Langerhans*, DOI 10.1007/978-94-007-6686-0_48,
© Springer Science+Business Media Dordrecht 2015

Tissue Recombination 1215
    Fraction Collection 1215
    Recombination 1216
Purification Process 1217
    Standard Density Gradients 1218
    Analytical Test Gradient System 1219
    High-Density Gradients 1219
    COBE Purification Process 1220
Transplant Preparation 1221
Conclusion 1222
Cross-References 1223
References 1223

Abstract

For patients suffering from intractable chronic pancreatitis, surgical removal of the pancreas may be recommended. While pancreatectomy has the potential to alleviate suffering and prolong life, the induction of iatrogenic diabetes, through the loss of insulin-producing islet cells, becomes an immediate threat to the postoperative patient. Since the procedure was first performed in 1977 at the University of Minnesota, autologous islet transplantation has become the best treatment option to restore a patient's ability to endogenously regulate their blood sugar. Autologous islet isolation starts with a specific procurement and packaging of the pancreas, which is then transported to a specialized clean-room facility. The pancreas is distended with tissue dissociation enzymes that digest the extracellular matrix of the pancreas, freeing the embedded cells, which are combined into a single tube. If necessary, this tissue is purified by density gradient and the islets transferred to a transplant bag for intraportal infusion back into the patient. The most critical factor for a positive metabolic outcome from this procedure is the islet mass transplanted, making total isolation yield of particular concern. While the pancreas contains an abundance of islets, even the best isolation techniques capture only half of these. Furthermore, patient-donor characteristics and tissue conditions, like fibrosis and cell atrophy, can further diminish islet yields. Our goal at the University of Minnesota has been to research the underlying factors that influence islet yield and viability and to propose specific procedures designed to optimize isolation success regardless of tissue condition. The purpose of this review is to describe our basic protocol as well as to highlight our system of flexible techniques that can be adapted based on an ongoing evaluation of each individual pancreas and the procedural progress itself.

Keywords

Chronic pancreatitis • Pancreatectomy • Autograft • Allograft • Human islets • Autologous islet isolation • Transplantation • Insulin independence • Collagenase digestion

Abbreviations

ATGS	Analytical test gradient system
ATP	Adenosine triphosphate
cGMP	Current good manufacturing practices
CIT	Clinical islet transplantation
CP	Chronic pancreatitis
ECM	Extracellular matrix
FDA	Fluorescein diacetate
HBSS	Hanks balanced salt solution
HSA	Human serum albumin
HTK	Histidine-tryptophan-ketoglutarate
IAT	Islet autotransplantation
IEQ	Islet equivalents
NEM	New enzyme mixture
PI	Propidium iodide
TLM	Two-layer method
TP	Total pancreatectomy
TP/IAT	Total pancreatectomy and islet autotransplantation
UW	University of Wisconsin (solution)

Introduction

Chronic pancreatitis (CP) is a severe, debilitating, and progressive inflammatory disease that results in irreversible destruction of the pancreatic structure (DiMagno and DiMagno 2013; Braganza et al. 2011). CP is characterized by disabling abdominal pain and a gradual loss of both exocrine and endocrine function. Various therapies are available to manage pain or slow disease progression, but when a patient presents with refractory disease or signs of pancreatic cancer, surgery is often deemed necessary (D'Haese et al. 2013; Sah et al. 2013). Drainage/decompression and denervation operations are typically considered first while pancreatic resection, or pancreatectomy, is usually indicated after other procedures fail to alleviate pain (Gachago and Draganov 2008). Pancreas resection may be partial, involving the tail and body of the organ (distal pancreatectomy) or the head as well as the neighboring duodenum and gall bladder (pancreaticoduodenectomy, or Whipple's procedure). The most effective option for CP is a total or near-total pancreatectomy (TP) in which at least 95 % of the pancreas is removed (Blondet et al. 2007).

Total pancreatectomy necessarily results in surgical (iatrogenic) diabetes, due to a loss of insulin-producing pancreatic islet cells, but the risks of this condition can be effectively mitigated through simultaneous islet isolation and autograft transplantation via intraportal infusion (Fig. 1) (Sutherland et al. 2012). Unfortunately, while TP operations are relatively common, conjunctive islet autotransplantation (IAT) is not always available as the cost and difficulty of establishing an experienced and cGMPs (current good manufacturing practices) compliant islet isolation

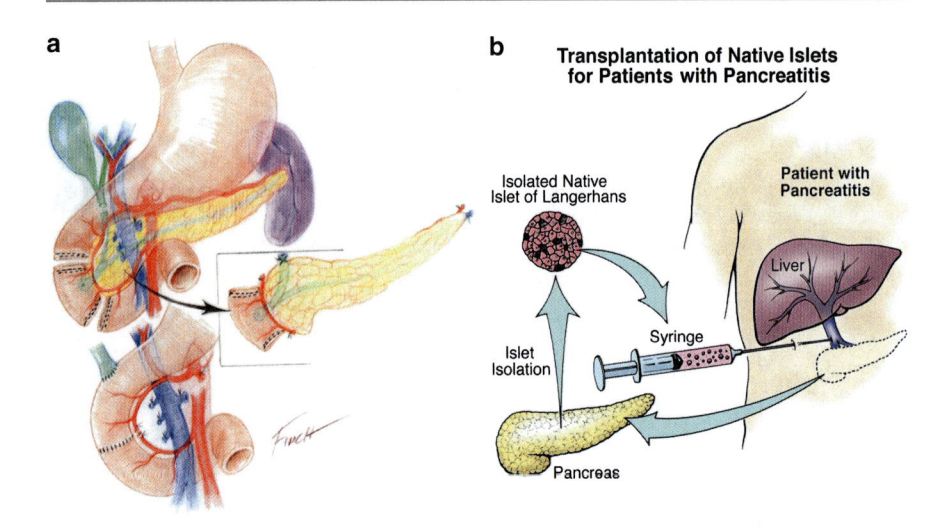

Fig. 1 (a) Surgical technique for total pancreatectomy. Total pancreatectomy and pylorus- and distal-sparing duodenectomy with orthotopic reconstruction by means of duodenostomy and choledochoduodenostomy (Adapted from Farney AC, Najarian JS, Nakhleh RE, et al. Autotransplantation of dispersed pancreatic islet tissue combined with total or near-total pancreatectomy for treatment of chronic pancreatitis. Surgery 1991; 110(2):427–37 [discussion: 437–9]; with permission. Copyright © 1991, Elsevier). (b) Sequence of events to preserve β cell mass in patients undergoing a total pancreatectomy for benign disease. The resected pancreas is dispersed by collagenase digestion followed by islet isolation. Autologous islets are then embolized to the patient's liver by means of the portal vein (From Blondet et al. 2007)

laboratory have limited the number of facilities capable of performing such a procedure (Rabkin et al. 1999a). For the best results, the resected pancreas should be processed immediately and may require 4–8 h for a trained team to isolate and package an islet preparation for transplant (Gruessner et al. 2004). Nevertheless, the importance of IAT for pancreatectomized patients is realized from the fact that iatrogenic diabetes is more brittle and difficult to manage than other forms of diabetes, with increased insulin sensitivity and more frequent episodes of severe hypoglycemia, which elevates the immediate danger to the postoperative patient (Sutherland et al. 2008; Morrison et al. 2002; Watkins et al. 2003).

The first IAT was performed in 1977 by Dr. Najarian and Dr. Sutherland at the University of Minnesota following a CP total pancreatectomy (Najarian et al. 1979). Apart from CP, however, IAT can be used in other scenarios (Berney et al. 2004; Forster et al. 2004; Alsaif et al. 2006; Garraway et al. 2009) where a pancreatic resection is performed, so long as the precipitating condition does not damage the islets themselves. Another indication for IAT is pancreatic trauma, following blunt or penetrating injury that requires a large (80 % or 90 %) pancreatic resection (Garraway et al. 2009). The basic principles of IAT can also be used to isolate islets from brain-dead or cadaver donors to transplant into patients with type I diabetes, whose islets have been lost. These allograft transplants come with their

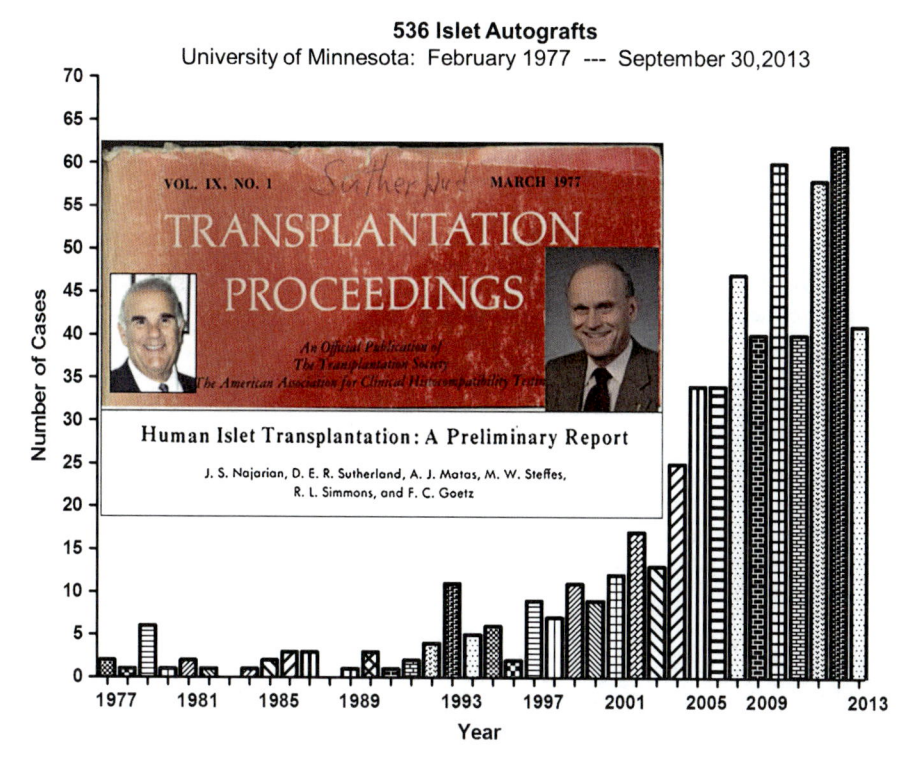

Fig. 2 Islet autograft experience at the University of Minnesota by year, in 536 patients (adults and children) from February 1977 to September 30, 2013. Dr. Najarian and Dr. Sutherland performed the world's first islet autotransplantation at the University of Minnesota, in 1977

own set of problems, however, namely, the need for prophylactic immunosuppression (Hering et al. 2005). In our institute, we have successfully completed more than 536 islet autotransplantations (Fig. 2).

The single most critical factor for optimal survival and function of transplanted islets is the size of the islet mass acquired from the isolation procedure. Sutherland et al. have reported up to 2 years of insulin independence in 71 % of patients who received an IAT of >300,000 islet equivalents (IEQ) (Sutherland et al. 2008, 2012; Bellin et al. 2008, 2011). A healthy human pancreas contains more than sufficient one to two million islets but a typical isolation recovers only 30–50 % of these (Lakey et al. 2003; Balamurugan et al. 2012). The efficacy of enzyme digestion is limited by the heterogeneous composition of the extracellular matrix (ECM), which is composed of a honeycomb of reticular fibers, adhesive proteins, large molecules, and multiple isoforms of collagen. A successful islet isolation protocol must break down the ECM to release the cells and separate endocrine from exocrine tissue without degrading the bonds that hold the individual islet cells together. Furthermore, disease pathology, acute trauma, previous surgical intervention, and each donor's unique biological characteristics result in variable tissue conditions that can dramatically impact islet yield and viability (Bellin et al. 2012a; Kobayashi et al. 2011).

To achieve consistent success in clinical outcome, an experienced isolation team must learn to recognize key variables, such as fibrosis and donor age, and to understand the principles of the isolation procedure sufficiently to adapt it for each individual organ. Thus, while a seemingly simple procedure, IAT requires highly specialized and flexible techniques in order to obtain an adequate number of healthy islets from a variety of donor types (Morrison et al. 2002). This report aims to describe the technical aspects of pancreas processing and islet isolation that are critical for achieving successful islet autotransplantations.

Islet Cell Isolation

Human islet isolation is a time-sensitive procedure that requires the trained attention and cooperation of, minimally, a three- to four-person team led by an experienced supervisor. The role of each individual must be carefully considered to avoid errors, identify and correct problems, and ensure the efficiency of the procedure to achieve the best possible yield of healthy islets for transplant (Lakey et al. 2003).

Adequate laboratory preparation prior to the isolation procedure is important so that the team is ready to begin as soon as the pancreas arrives to the clean-room facility. Using sterile technique, two people should set up the biological safety cabinets (or laminar flow hood) with materials needed for pancreas trimming, cannulation, distension, digestion, recombination, purification, and transplant bag preparation. Simultaneously, the remaining individuals prepare media and other in-use solutions that will be needed during isolation. Team members should also ensure that all necessary instrumentation, such as centrifuges and thermal probes, are turned on and functioning within normal parameters.

Pancreatectomy and Pancreas Transport

For autologous isolations, the pancreas is dissected and immersed immediately in a 1 L Nalgene container filled with cold preservation solution. Excess fat, connective tissue and duodenal tissue are removed before packing on ice, and pancreas is delivered to the islet isolation lab. During pancreatectomy, some transplant centers recommend flushing the ductal system with chilled preservation solution, containing a trypsin inhibitor, to minimize warm ischemia time and to maintain optimal ductal integrity for enzyme delivery and distension (Naziruddin et al. 2012; Shimoda et al. 2012a). A recent study, however, found that the more common vascular flush technique was just as effective as a ductal injection (Nakanishi et al. 2012). We have found these methods unnecessary for short transport of the pancreas, but for allograft isolations, where ischemia time is longer and donor conditions less ideal, we require en bloc pancreatic harvest with arterial flush (Kin 2010; Kin and Shapiro 2010).

Preservation Solution

Cold storage preservation relies on hypothermia and carefully tailored solutions to slow metabolism inhibit endogenous enzyme activity and support critical cellular processes despite the loss of an oxygenated blood supply. Organ packaging methods and solution ingredients have been designed to address several key problems associated with hypothermic ischemia followed by reperfusion including cellular swelling, ionic imbalance, acidosis, calcium accumulation, and the production of reactive oxygen species.

University of Wisconsin (UW) solution was developed in 1986, specifically for pancreas cold storage preservation (Wahlberg et al. 1986). UW contains phosphate as a pH buffer against anaerobic lactic acidosis, the large molecule saccharide raffinose and the anionic lactobionate as membrane-impermeant osmotic balancers, allopurinol and glutathione to scavenge free radicals, adenosine as an ATP substrate, and a high, intracellular-mimicking K^+/Na^+ ratio that was originally thought to retard cell-damaging Na^+ absorption and K^+ efflux ('tHart et al. 2002). Lactobionic acid is also a calcium chelator that prevents this cation from activating degenerative phopholipases during reperfusion ('tHart et al. 2002). While UW has become the standard organ transport solution, it is also costly with a short shelf life and many of the ingredients, designed to inhibit tissue degradation, interfere with the intentionally catabolic activity of collagenase and neutral protease (Contractor et al. 1995). Other cold storage solutions have been proposed including histidine-tryptophan-ketoglutarate (HTK), Celsior, and the Kyoto solutions, but UW remains the most common for pancreas hypothermic preservation. In trimming solution, a modified UW reverses the Na^+/K^+ ratio to mimic the natural extracellular environment and exchanges lactobionate for the less expensive but equally effective gluconate. Since we accept cadaveric donors from a variety of institutions, we recommend allograft pancreata be stored in either UW or HTK solution for transport to our lab (Barlow et al. 2013).

Packaging

For transporting a pancreas, we use simple cold storage in trimming solution – the organ placed in a 1 L Nalgene, filled to the top with solution, and packed on ice inside a cooler. For allograft isolations, we request organs to be packaged using the two-layer method (TLM) of oxygenated perfluorocarbon underneath a layer of either UW or HTK solution (Hering et al. 2002). Although the efficacy of TLM, hypothesized to promote aerobic respiration and ATP production, has been vigorously debated, a large-scale meta-study recently concluded that TLM did increase pancreatic islet yield and viability for long cold ischemia times, which are common for allogeneic isolations (Qin et al. 2011).

The Baylor group has published a series of studies demonstrating improvements in isolation yield using a combination of cold ductal flush during procurement and TLM packaging, even for short distances (Naziruddin et al. 2012). However, we prefer to minimize manipulation of the pancreas prior to the start of isolation. In our autotransplantation series, a small number of autologous isolation cases have also necessitated long transports due to the patient's inability to travel, in which the resected pancreas was flown by plane to our isolation center in Minnesota (Rabkin et al. 1997, 1999b) or in a similar case to another center (Jindal et al. 2010; Khan et al. 2012). The islet mass was then flown back to the patient and transplanted at the original site, in several instances resulting in a complete return to normoglycemia (Rabkin et al. 1999b; Jindal et al. 2010). Despite the long travel time, the pancreatic tissue did not appear to suffer any ill effects from the use of simple storage in cold UW (Rabkin et al. 1997; Khan et al. 2012), offering hope that this kind of procedure could make IAT available at hospitals and surgical sites where a full isolation facility is not feasible.

Pancreas Trimming and Cannulation

After receiving a pancreas at our isolation facility, the temperature of the transport preservation solution is recorded and a 3 mL sample volume taken to assay for microbial contamination (aerobic/fungal, anaerobic, and gram stain). It is critical that the pancreas be kept cold from the time of organ removal to the initiation of enzymatic digestion (Lakey et al. 2002). In conjunction with the preservation solution, a low-temperature environment slows cellular respiration and tissue degradation ('tHart et al. 2002). Initial procedures should be performed in a cold room with prechilled solutions, but if this is not available, an ice bath can be prepared for working with the pancreas. At least 1 L of sterile saline should be frozen solid ahead of time. During setup, wrap the saline bag in a towel or cover to protect the integrity of the outer plastic while breaking up the inner ice block with a mallet or hammer. The bag should be opened using sterile technique and the saline ice fragments poured into a large tray, which is then covered with a sterile drape. A smaller dissection tray is placed inside, resting on top of the drape and underlying slush.

Trimming

The first isolation step inside the biosafety hood is to immerse the pancreas in ~500 mL of trimming solution inside a cold trimming pan. After a visual inspection, the organ is photographed for formal documentation. The gross morphology of every pancreas is different and we assess each as mild, moderate, or severe to determine enzyme dose and digestion conditions (Fig. 3). A trained team member then performs an initial brief trim, removing excess fat and connective tissue from around the pancreas but keeping the pancreatic capsule intact to reduce the likelihood of enzyme leakage during ductal perfusion. After trimming, the pancreas is submerged for 5 min in Antibiotic solution (Fungizone + Gentamicin + Cefazolin) that has been checked against potential patient allergies before treating the pancreas. If the solution

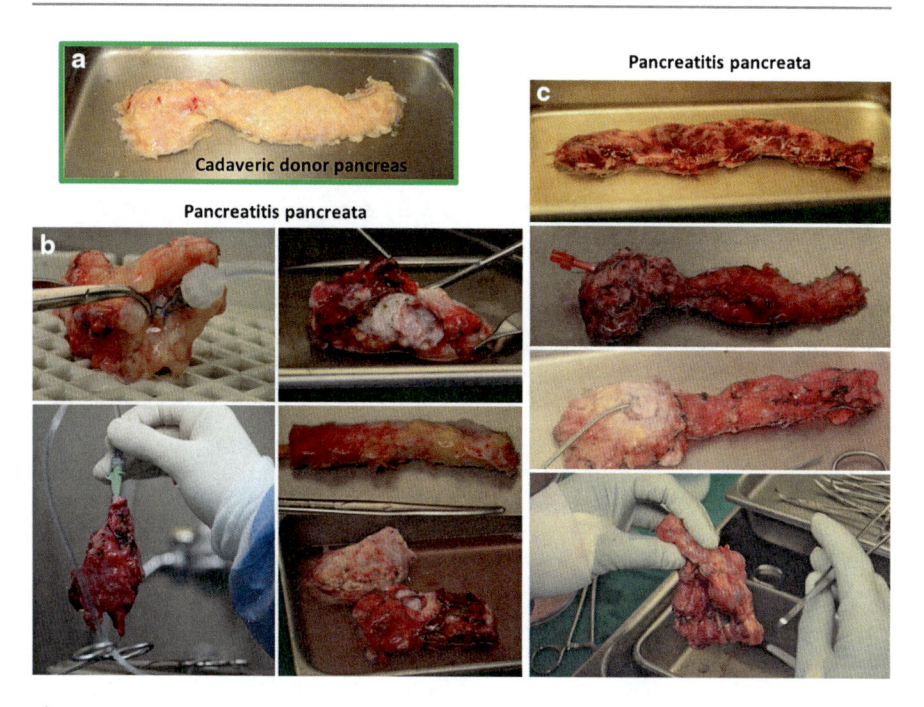

Fig. 3 Pancreas gross morphological differences between (**a**) normal cadaveric donor and (**b**), (**c**), chronic pancreatitis pancreata. For enzyme dosing, our lab categorizes the pancreas into mild, moderate, and severe categories

beaker is resting on a pre-tared scale, the organ can also be weighed during this time in order to calculate the per gram enzyme concentration that will be needed for perfusion/distension. The pancreas is then rinsed in two consecutive beakers, each containing 500 mL of sterile Hanks balanced salt solution (HBSS), and placed in a pan with trimming solution in preparation for cannulation.

Cannulation

Up to this point, we have referred to the pancreas as a single intact organ. However, depending on the severity of a patient's pancreatitis and the degree of near or total pancreatectomy, the organ may arrive whole, partial, or in multiple pieces. In the case of a whole pancreas, we have achieved better enzyme distention by dividing the pancreatic lobes and cannulating the head and the body-tail portions individually. The selection of a catheter should be based on the size of the pancreatic duct, which may vary between lobes. Typically, sizes range from 14 to 24 G but severe fibrosis or a dilated duct may require either a Christmas or metal catheter. The selected catheter should be sutured firmly in place to avoid enzyme backflow during perfusion. With the catheters secured, the pancreatic lobes are moved to the perfusion tray and the tray's basin filled with the prepared enzyme solution.

Enzyme Selection

Enzymatic tissue dissociation has been used to separate the exocrine and endocrine components of the pancreas since 1967, when crude collagenase enzyme, derived from the bacteria *Clostridium histolyticum*, was first used to isolate guinea pig and rat islets (Moskalewski 1965; Lacy and Kostianovsky 1967). Since then, researchers have observed the best tissue dissociation resulting after the ductal perfusion of a blend of collagenase and protease enzymes into the main pancreatic duct (Kin 2010; Caballero-Corbalan et al. 2009, 2010). We tailor our enzyme mixture according to pancreas weight, with the final enzyme solution prepared by diluting the reconstituted enzymes into approximately 350 mL HBSS + 10U/mL heparin solution (or up to 450 mL for larger pancreases) (Balamurugan et al. 2012).

Enzyme Combination

In the past, the universal adoption of Liberase HI for enzyme digestion provided a convenient formula in which one complete vial could be dissolved into the final desired solution volume (Hering et al. 2005). When Liberase was made clinically unavailable, we switched to SERVA collagenase and neutral protease (Anazawa et al. 2009). However, we have recently identified a new enzyme mixture (NEM) that has improved both the yield and viability of our islet products. We currently use VitaCyte collagenase HA, which contains a high proportion of intact C1 collagenase, to obtain greater islet numbers compared to SERVA's analogous product (Balamurugan et al. 2010). On the other hand, we observed SERVA neutral protease NB (GMP or premium grade) to produce better-quality islets with a more solid and intact structural morphology compared to VitaCyte's equivalent, thermolysin (Balamurugan et al. 2012). This novel combination (VitaCyte Collagenase + SERVA neutral protease) resulted in total islet yields of >200,000 IEQ in 90 % of our attempted autologous isolations and doubled the number of allograft isolations that reached transplantation threshold in our recently published study (Balamurugan et al. 2012).

Enzyme Dose

Initially, our laboratory used only a 1:1 ratio of whole enzyme vials, even with the NEM. We have since observed that customizing the enzyme dose based on pancreas weight and other donor/organ characteristics results in more consistent islet release. Pancreas weight is estimated based on the actual pre-distension weight minus the anticipated weight of non-pancreatic trimmed tissue, evaluated by an experienced team member based on visual and tactile observation. Following distension and final trimming, the actual pancreas weight is recorded and the real enzyme dose can be calculated. Improper enzyme concentration can then be compensated for by adjusting the temperature and timing of the digestion phase or by adding additional enzyme directly into the digest circuit. To produce our

NEM results, we used a dose formula of 22 W unit/g pancreas of collagenase and 1.5 DMC unit/g pancreas of neutral protease (Balamurugan et al. 2012). In practice, we vary the collagenase dose from 22 to 30 W unit/g and the neutral protease dose from 1.5 to 3.0 DMC unit/g pancreas depending on pancreatic characteristics.

Enzymatic Perfusion of Pancreas

During perfusion, enzyme solution is administered to the pancreatic tissue through a pressurized injection, manual or automated, into the cannulated pancreatic duct of either the whole or segmented pancreas (Lakey et al. 1999, 2003). Enzyme distension is the critical step in the islet isolation process (Fig. 4). Complete delivery of enzyme to the entire pancreatic parenchyma will reduce the remaining undigested tissue in the digestion chamber and maximize the final islet yield recovered and ultimately delivered to the patient.

Perfusion Method

Historically, enzyme solution was loaded directly into the ductal cannula with a handheld syringe, relying on retrograde perfusion to distend the pancreas. This approach diminishes the ability to address leaks as they occur and does not offer quantified values about pressure and flow rate. However, the manual method is still in common practice as either a primary or alternative perfusion technique at many clinical islet transplant (CIT) centers. The first use of a fully automated recirculating pump to perfuse a human pancreas through the pancreatic duct was published in 1999 and shown to improve islet yields over traditional syringe loading (Lakey et al. 1999). The superior yield produced by this method was confirmed using refined rather than crude collagenase (Linetsky et al. 1997a, b). In addition to improved distension and yield, automated pump perfusion provides precise control over injection pressure and enzyme solution temperature.

The modern automated perfusion system is equipped with peristaltic pumps, two pressure sensors, a heater, a touch screen, and data acquisition software (Bio-Rep) that combines the convenience of hands-free automation with the flexibility to make manual adjustments to a variety of programmable parameters. This is especially important to achieve even enzymatic distribution of cadaveric donor organs (Lakey et al. 1999) or to control pressure in a diseased pancreas with severe fibrosis or ductal alterations. This semiautomated system is used by several prominent CIT centers and is the primary perfusion method used in our laboratory.

Temperature, Pressure, and Flow Rate

Distension pressure, pump speed, flow rate, and temperature can all be monitored and controlled using a semiautomated perfusion system. Throughout the enzyme perfusion process, the temperature is kept between 6 °C and 16 °C while the desired injection pressure is maintained between 60 and 80 mmHg for the first 4 min and gradually increased to a perfusion pressure of 160–180 mmHg until completion (approximately 10–12 min total distention time). However, perfusion pressure can vary significantly depending on the condition of the organ. Distention pressures could be low for a severely damaged, leaking, pancreas or high for an organ with abnormal ductal anatomy (stricture or blockage) or severe fibrosis.

While it is important to maintain adequate perfusion pressure, vigilance is also required to maintain the pump speed and flow rate above 30 mL/min. A ductal occlusion or misplaced cannula can obstruct the flow of liquid, increasing the pressure and slowing the automated pump. However, enzyme injected at a low flow rate, even with a normal pressure reading, would result in an ineffective distention. In this case, the cannula may need to be repositioned or the team may decide that higher pressure or an interstitial perfusion method is required. Similarly an enzyme leak may cause the pump speed and flow rate to suddenly increase as the perfusion pressure drops and a stream of solution may be visible ejecting from the pancreas. Since leaks can significantly detriment the quality of the perfusion (Goto et al. 2004; Johnson 2010), they should be clamped or sutured immediately to ensure a complete and efficient distention.

Interstitial Perfusion

In some cases the extent of parenchymal fibrosis is so high that ductal enzyme perfusion is ineffective at delivering enzyme to the entire body of the pancreas (Fig. 4). In these cases, interstitial perfusion can be performed by repeated manual injections of cold enzyme solution into the tissue with a needle and syringe (Al-Abdullah et al. 1994) (Fig. 4d). On the other hand, the distal half of a pancreas, particularly if perfused intact, may fail to effectively distend with solution. This may be due to duct alterations caused by inflammation and fibrosis or intraductal calcification deposits, which have obstructed the flow of fluid through the duct. In these cases, it is possible to make a complete transverse cut before the distal section after the proximal end has finished distending and re-cannulate the distal end to attempt further distention in this area.

Post-distension Trimming

After adequate distention has been achieved, the pancreas is transferred to a trimming pan with fresh heparinized phase 1 solution. Final trimming of the pancreas should be performed as quickly as possible to reduce overall cold ischemia time. This includes removal of the pancreatic capsule, surface fat, vasculature tissue, and any sutures of staples used during the surgical procedure. Surface fat is a particular concern as it can clog the Ricordi chamber screen during digestion and obstruct the free flow of islets and solution. For normal pancreata, the trimmed

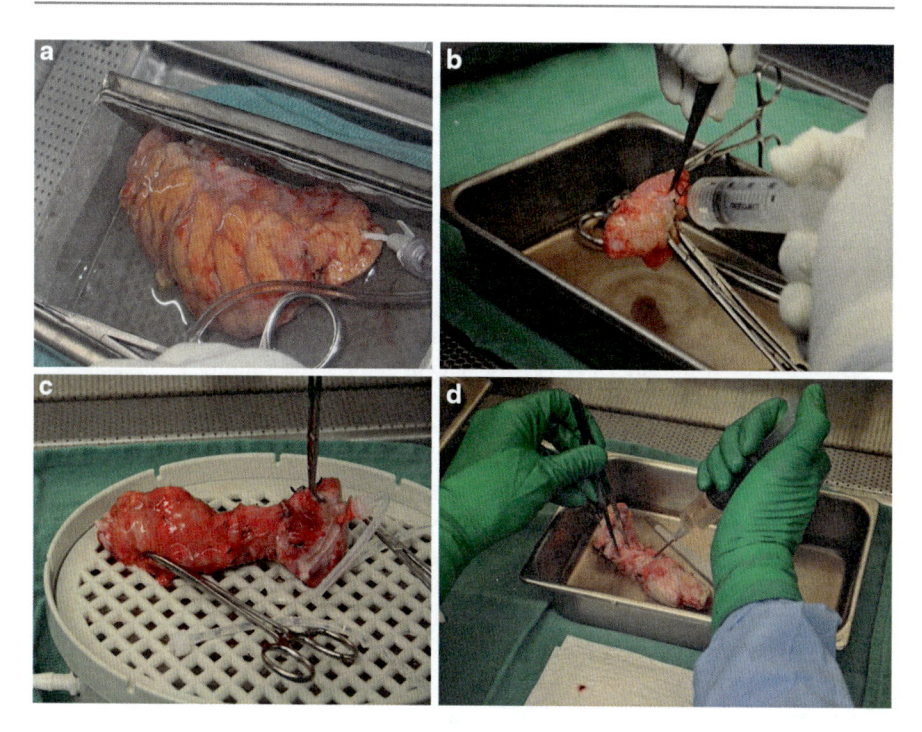

Fig. 4 Based on the nature of pancreas, the enzymatic distention will be performed using (**a**) a simple perfusion machine or (**b**) a hand syringe injection through the pancreatic duct or (**c**) an automated perfusion apparatus injection through the pancreatic duct or (**d**) in case of severely fibrosis or ductal collapsed pancreas, interstitial enzymatic distention (by using needle and syringe) should be performed

organ should be cut into large (3–5 cm diameter) pieces prior to digestion. For an inadequately distended or severely fibrotic pancreas, the tissue can be cut into many small pieces (up to 30 pieces, 2 cm in diameter) to enhance the exposed surface area available for mechanical and enzymatic digestion.

Tissue Digestion

The perfused enzyme begins to bind to the matrix tissue of the pancreas as soon as it is injected so we should not delay between the end of distension and the beginning of digestion. Currently, the semiautomated method for tissue digestion is employed by all centers isolating human islet preparations intended for clinical transplant programs (Ricordi 2003; Ricordi et al. 1988). Developed in 1988 by Dr. Camillo Ricordi, the digestion method utilizes a specialized "Ricordi" chamber to contain and collect the pancreatic digest as islets are released (Ricordi et al. 1988). Along with the pieces of pancreatic tissue, multiple stainless steel marbles are sealed inside the chamber to mechanically assist enzymatic dissociation as the chamber is gently shaken. This method has proven superior to manual methods in isolating high quantities of viable human islets for successful transplantation (Ricordi 2003; Ricordi et al. 1988; Paget et al. 2007).

Digest Sampling

To monitor the progress of tissue dissociation, it is critical to collect samples of the digest tissue and solution throughout the digestion process. We withdraw a 2 mL volume from a sterile syringe connected to an outlet port in the tubing circuit, which circulates solution through the Ricordi chamber lid. The sample is stained with dithizone inside a small petri dish and evaluated under a light microscope for quantity of acinar tissue, acinar diameter, number of islets, percent free islets, percent fragmented (over-digested) islets, and average islet score. These are recorded, along with the circuit temperature, at regular intervals. During phase I, sampling begins after 8 min of recirculation digestion and occurs every 90–120 s. After the switch to phase II collection, samples can be taken every 5–10 min.

Digestion: Phase 1, Recirculation

The distended and trimmed pancreas is aseptically transferred to a Ricordi chamber (usually 600 mL) connected to a digestion apparatus, along with all enzyme solution left in the perfusion tray or trimming pan. The digestion apparatus consists of a peristaltic pump that moves fluid through a closed circuit of tubing with a reservoir, a heat transfer coil, an inlet for diluent, and outlets for tissue collection or sampling. Once the tissue and bathing solution is inside the chamber, a 400–600 μm mesh screen (typically ~500 um) is placed across the opening before sealing the lid with a rubber gasket and screws or a large clamp. H phase 1 solution (+10 U/mL heparin) is added to fill the digestion circuit and purge air from the system. The start time of digestion is recorded when sufficient liquid has filled the reservoir to allow for sampling.

At the start of digestion, the pump speed is set to deliver a flow rate of 200 mL/min and the tubing clamps are adjusted to recirculate solution through the system. Over the first 5 min, the temperature is gradually increased to 34–37 °C while the chamber is gently rocked. When the target temperature has been reached (34–35 °C standard), the heating coil is removed from the heated water bath, the pump slowed to 100 mL/min, and more forceful shaking of the chamber begins. After 8 min (or when liberated tissue appears in the tubing), 2 mL samples of digest tissue are taken by syringe from the sampling port and viewed under a microscope after staining with dithizone. Successive samples with large quantities of free islets indicate the completion of phase 1 digestion and suggest a transition to phase 2 (switch time).

As mentioned above (see section "Enzyme Dose"), the actual enzyme dose, calculated by subtracting the real trimmed tissue weight from the original pancreas weight, can be a useful indicator to adjust the digestion settings. The target temperature set point can be raised or lowered to compensate for off-target enzyme concentrations, resulting from inaccurate estimation of the final trim or for a sluggish phase 1 progression. Additionally, in cases where phase 1 is proceeding exceedingly slow, extra collagenase or neutral protease can be added to the digest circuit directly to rapidly increase digestion rates. It is important to keep phase 1 as short as possible because prolonged exposure to active enzyme, and an increasingly basic chamber environment, can be harmful to islet yield and integrity (Gray and Morris 1987; Balamurugan et al. 2003; Tsukada et al. 2012). Temperature rate and set point, enzyme dose, and switch time should be determined by qualified and experienced personnel based on multiple factors such as the % dissociation of tissue in the Ricordi chamber and the progressive condition of the samples including amount of tissue, % of free islets, size/condition of acinar tissue, and morphology of the islets.

Digestion: Phase 2, Collection

Phase 2 of digestion starts by increasing the pump speed to a flow rate of 200 mL/min and switching the tubing clamps from recirculation of the digest material to collection. Cold phase 2 solution (RPMI 1640) is added to progressively weaken the enzyme concentration inside the digest circuit. The first 2 L of digest tissue is collected into four 1 L flasks in a series of increasing dilutions: first 25 % (250 mL of digest + 750 mL of cold RPMI 1640/2.5 % human serum albumin (HSA)) then 50 %, 50 %, and 75 % v/v. The prechilled collection media and the HSA both work to inhibit collagenase and neutral protease activity, which would otherwise continue to break down islet integrity during recombination. The decreasing dilution factors provide a greater inhibitory buffer during the initial collection fractions with the highest concentration of enzyme. Following the fourth flask, digest fractions are collected into 250 mL conical tubes, prefilled with 6.25 mL of chilled 25 % HSA. The stop point for phase II tissue collection is determined by the minimal presence or complete absence of islets and/or tissue in the dithizone-stained samples that have been under regular evaluation throughout digestion. As digestion comes to an end, the remaining RPMI is collected from the chamber and air introduced into the tubing to flush out the final solution. During this last wash, the chamber should be gently rotated to ensure that all fine tissue has been moved through the circuit. The weight of undigested tissue in the chamber can be used to ascertain the efficacy of the enzyme dose and the efficiency of the tissue dissociation.

Digestion of the Severely Fibrotic Pancreas

A common side effect of CP inflammation is the progressive development of fibrosis as a result of increased deposition and reduced degradation of ECM

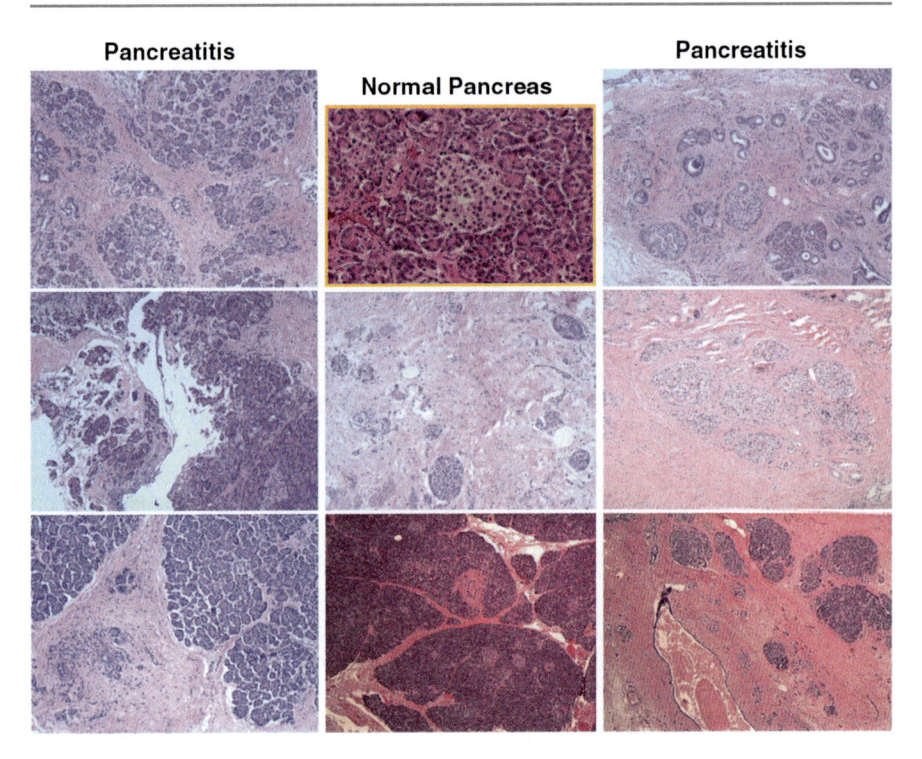

Fig. 5 Hematoxylin and eosin-stained histological images of normal (*middle*) and chronic pancreatitis pancreata. Cellular architecture differs in every case with the amount of fibrous tissue

materials (Fig. 5). This excess and hardened tissue is more resistant to enzyme digestion and can greatly lower islet yield if appropriate adjustments are not made to the procedure. Maintaining the solution temperature at the high end of the acceptable range (36.5–37) can intensify enzyme activity. If unusually little digestion is observed after 25 min, an enzyme recirculation protocol may be helpful (Balamurugan et al. 2003). This procedural variation involves the collection of free islets from the recirculating system early after release. The free islets are pelleted by a quick centrifugation step and transferred into fresh, cold media in the recombination container. The supernatant, containing active enzyme, is recycled back into the digestion system, which increases the effective enzyme dose for the undigested tissue remaining in the Ricordi chamber (Balamurugan et al. 2003). This recirculation can be repeated until nearly all tissue mass has dispersed or healthy islets are no longer apparent in the digestion samples.

Other Common Complications

In addition to the extent of fibrosis, other donor characteristics – especially age, pancreas weight, and fat infiltration – can lead to extreme discrepancies in islet release as the digestion phase progresses (Miki et al. 2009; Eckhard et al. 2004; Loganathan et al. 2013; Lake et al. 1989). A common observation in pancreas digestion from young donors is the release of mantle islets, which are embedded in a corona of acinar cells (Ricordi 2003; Balamurugan et al. 2006). These islets are difficult to recover after purification as the extra cells increase the islet density, settling them into a heavier and less differentiable density layer during COBE processing (Miki et al. 2009). Prolonged enzyme exposure or an extra chamber screen, to restrict the passage of larger particles, may be effective in reducing mantle islets collected during isolation of a young pancreas (Balamurugan et al. 2006). With very small pancreases, common with pediatric donors or near-total pancreatectomy, a smaller digest chamber (250 mL) is typically more effective by concentrating the enzyme activity and shortening the duration of the digestion phase. Excess liquid volume inside the chamber should be avoided since it can act as a cushion, making it difficult to mechanically agitate the tissue. As with the distention phase, experienced islet researchers and lab technologists can greatly enhance the success of the isolation by making slight modifications to the standard procedure.

Temperature, circulation speed, enzyme dose, apparatus setup, and/or the level of mechanical shaking can all be used to accommodate the inherent variation in pancreatic tissue condition caused by different disease pathologies. An understanding of how these digestion parameters affect the rate and quality of tissue dissociation is essential to minimize the amount of undigested tissue left in the Ricordi chamber and to maximize the number and quality of liberated islets obtained. This is particularly critical for autograft patients, who have invested their entire hope for insulin independence on the success of a single isolation, as opposed to allograft recipients, who may have the opportunity to receive islets from multiple donors and transplantations. Since the first autologous islet transplantation was performed at the University of Minnesota in 1977, international islet transplant registry data indicate that transplant recipients are achieving longer durations of insulin independence. As pancreas procurement, preservation, and islet isolation techniques have advanced, average islet yields and positive patient outcomes have increased, spurring support for the further expansion of islet autotransplantation.

Tissue Recombination

Fraction Collection

The recombination phase is closely associated with the digestion phase, beginning immediately after the digestion switch point. All digest solution collected in phase 2 contains a suspension of free islets that must be preserved from further enzyme degradation. The first four digest fractions, collected successively into four 1 L flasks of RPMI/HSA (see section "Digestion: Phase 2, Collection"), are each transferred to a separate biological safety cabinet hood and divided into 250 mL conicals then centrifuged at 140–$170 \times g$, 2–10 °C, for 3 min. The resulting cell pellets are separated by decanting, discarding the supernatant, and aspirating the remaining pellet into a 1 L flask prefilled with 750 mL of cold recombination solution (wash media). This cold storage/purification stock solution (Mediatech, Inc., Manassas, VA, USA) is supplemented with 10 U/mL of heparin and 2 % v/v penta starch. This latter ingredient prevents exocrine cells from swelling and becoming less dense, which would reduce the efficacy of the density gradient-dependent purification process later in the isolation (Eckhard et al. 2004). Subsequent digest fractions are drawn off directly into 250 mL conicals and can be centrifuged immediately after collection.

In some instances of CP, calcification deposits, up to 3 mm in diameter, may be observed in the collection conicals and should be removed before centrifugation. These deposits are much denser and larger than the observable digest tissue and can be mechanically separated by pipet aspiration. The calcifications should be gathered into a separate 250 mL conical where they can be rinsed with recombination solution to rescue any islets aspirated with the deposits.

Recombination

When phase 2 digestion has been completed and all tissue mass spun out and collected into the final recombination flask, 5 mL of tissue suspension is removed for density determination (see section "Analytical Test Gradient System"). The recombination flask is divided evenly into 4–5 conicals, rinsed thoroughly with wash media to scavenge all islets, and spun again. After decanting off the supernatant, these pellets are combined into a single conical and resuspended in fresh media up to 200 mL total volume. At this point, two samples of well-dispersed tissue suspension are taken to provide a "post-digest" islet count. This is done by gently rocking the capped conical, to evenly disrupt and distribute tissue aggregates, and then using a pipet to aspirate 100 μL of tissue suspension into 35 mm petri dishes containing 1 mL of wash media each. Packed pellet volume is estimated by bringing the recombination conical solution volume up to 250 mL before centrifuging at a higher speed, $220 \times g$, and reading the conical gradations. The count samples and total packed tissue volume are critical determinants for deciding whether a purification step is necessary. In all allograft cases and for autograft tissue volumes >20 mL, isopycnic gradient purification is advisable. Autograft pellet sizes smaller than 15 mL are usually not purified, to maximize the recovered islet mass, but tissue volume can be diminished by repeated washing: resuspending the pellet in fresh media, spinning at $140 \times g$, and decanting off the supernatant.

If islets are not purified, the final combined pellet is washed repeatedly with room-temperature transplant media (unsupplemented CMRL with 2.5 % HSA and 25 mM HEPES) until the supernatant is translucent and free of cell debris. The tissue pellet is then quantitatively transferred from the conical into a T-75 flask and brought up to a 100 mL volume with fresh transplant media. The islets are allowed to settle for 5 min so that samples of the supernatant may be aspirated for sterility testing and archival storage. The 100 μL sample counts from earlier are used to calculate the volume of islet-suspended solution that should be collected for islet viability/potency assays, minimally fluorescein diacetate (FDA)/propidium iodide (PI) (Loganathan et al. 2013), and others depending on total yield and patient consent. Density gradient purification necessitates additional recombination and sampling steps, which are described in the next section.

Purification Process

The goal of purification is to reduce tissue volume, particularly exocrine cell contamination, while minimizing islet loss. Several purification techniques have been reported but isopycnic density gradient centrifugation on the COBE 2991 cell processor is the only method that has been consistently successful and used clinically for large-scale human islet purification (Lake et al. 1989; Anazawa et al. 2011). This purification technique is accomplished by employing centrifugation through a density gradient to separate the less dense islets from the more dense exocrine tissue (Fig. 6).

The expediency of islet purification is debatable (Gores and Sutherland 1993) and should take into consideration the particular patient and potential for transplant complications. The decision to purify or not should be made by qualified and experienced personnel, often in consultation with the attending physician/surgeon. Avoiding density gradient purification is generally beneficial for islet viability as it avoids exposing the islets to harsh gradient solutions and additional mechanical stress (Mita et al. 2009). Some researchers also argue that exocrine-islet signaling plays an important role in islet survival and posttransplant function and that removing these communication pathways is detrimental to overall graft survival (Webb et al. 2012). On the other hand, significant evidence from transplant case reports describes the risks associated with transplanting large tissue volumes into the portal venous network, including portal vein thrombosis and portal hypertension (Bucher et al. 2004). Liver embolism, thrombosis, damage, and even death have been documented in clinical settings immediately after intraportal infusion of large amounts of tissue (Walsh et al. 1982; Kawahara et al. 2012). In addition, recent evidence from our laboratory indicates that activated proteolytic enzymes released by dying acinar cells and co-transplanted with islets may reduce in vivo islet function by degrading insulin (Loganathan et al. 2011).

Purification is also a risk when pellet size as small as 10–40 % of the total IEQ recovered from digest may be lost (Kin 2010) and it is well known that the total IEQ/kg of patient body weight is the current best predictor of a successful clinical outcome (Bellin et al. 2012b). As a general rule, we allow up to 0.25 cm^3 of tissue volume per kg of patient weight (in press, American Journal of Transplantation) for safe intraportal infusion, although we have safely infused a higher tissue content on several occasions when purification was anticipated to result in a particularly low yield (unpublished data).

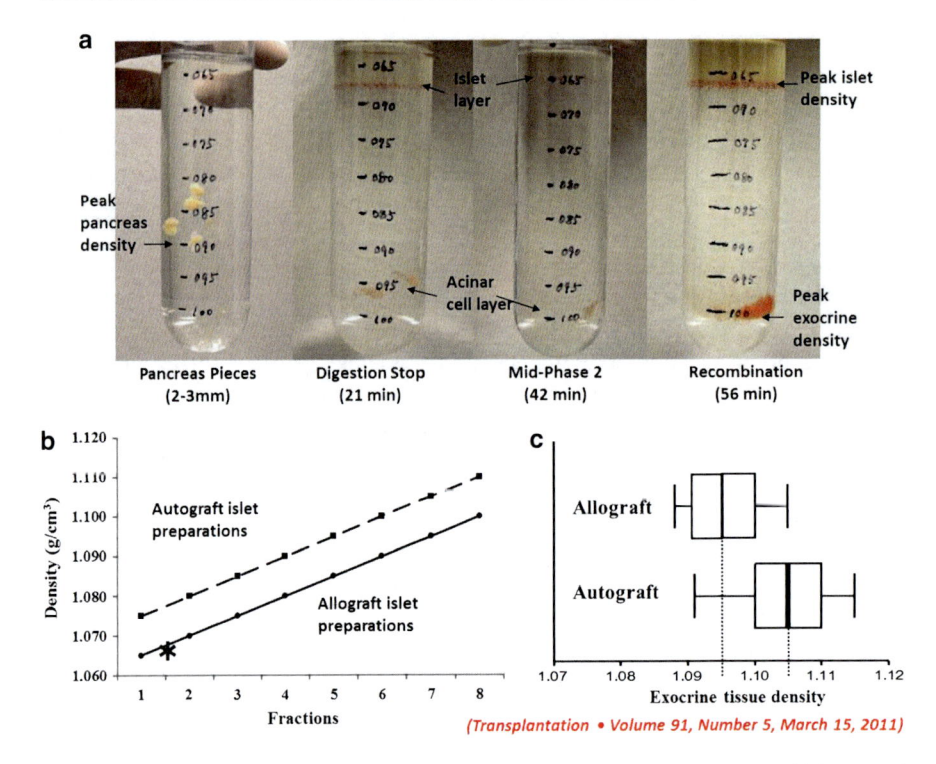

Fig. 6 (a) Examination of the density profile by prelabeled ATGS. The density at which most of the tissue floated is defined as the tissue density. Sequential photographs of digest tissue display a change in tissue density over time. Peak islet density is 1.068 g/cm^3, and exocrine tissue density is 1.101 g/cm^3. (b) The linearity of the analytical test gradient system (*ATGS*). For allograft purification (*solid line*), the ATGS is constructed of eight iodixanol-based density gradient levels, each increasing by 0.005 g/cm^3 at 5 mL intervals, running from heavy (1.105 g/cm^3) to light (1.065 g/cm^3) density. For autograft preparations (*dashed line*), the ATGS is constructed with a density gradient between 1.115 g/cm^3 (heavy) and 1.075 g/cm^3 (light). (c) Actual exocrine tissue density from all isolations indicated that autograft preparations were heavier than allograft preparations. The box-and-whiskers plot presents the median (*dark line*), the interquartile range (*box*), and the range between 10th percentile and 90th percentile (whiskers), *$P < 0.001$

Standard Density Gradients

When islets are purified with a COBE 2991 cell processor, the purification gradient used are often fixed between 1.060 and 1.100 g/cm^3, values currently utilized by the Clinical Islet Transplantation Consortium (Anazawa et al. 2011). A predefined, standard density gradient is less successful when exocrine density is unusually light (<1.100 g/cm^3), lowering the post-purification purity, or when islet density is unusually heavy (>1.100 g/cm^3, common for mantle islets), causing the islets to sediment into the COBE bag and reducing total recovery (Anazawa et al. 2011).

Traditionally, however, the exact density of pancreatic tissue components is unknown prior to purification, which enhances the difficulty of predicting or controlling the outcome of purification.

Analytical Test Gradient System

Predicting exocrine tissue and islet density is important for the selection of an optimal density gradient range for the COBE process, to maximize islet yield and purity (Fig. 6). Consequently, our center has developed an analytical test gradient system (ATGS) to determine the true density distribution of human pancreatic tissue components before purification (Anazawa et al. 2011). This method mimics the actual purification process but uses only a minute fraction of recombination tissue in a single tube. ATGS results can be quickly interpreted in order to customize the gradient for a full COBE purification.

The ATGS is prepared by using a small gradient maker (Hoefer SG30, GE Healthcare Bio-Sciences Corp, CA, USA) filled with heavy-density (1.115 g/cm^3) and light-density (1.065 g/cm^3) solutions in either of two chambers to create a continuous test gradient inside a 50 mL conical tube. Specific density solutions are made by changing the volumetric ratio of iodixanol (1.320 g/cm^3, OptiPrep, Axis-Shield, Oslo, Norway) and cold storage/purification stock solution (1.025 g/cm^3). The ATGS conical is first filled with 5 mL of pure heavy solution to create the bottom layer. A peristaltic pump (flow rate 2 mL/min) is then used to progressively mix the contents of the two machine chambers and force a gradually less dense solution into the conical. This procedure creates a 40 mL continuous density gradient from 1.115 to 1.065 g/cm^3 with 0.005 g/cm^3 increments every 5 mL (Fig. 6).

It is advisable to create the density test gradient during digestion so that it will be available as soon as the recombination product can be sampled. From the recombination flask, 5 mL of pancreatic digest (~0.1 mL of packed tissue) is carefully added to the top of the gradient. The ATGS conical is centrifuged at $400 \times g$ for 3 min at $4\ ^\circ C$ with no brake, similar to the actual COBE purification process. After centrifugation, the layers of acinar cells and islets are separated and settle at their respective tissue densities. Using the graduations on the side of the 50 mL conical, it is easy to determine the peak islet and acinar tissue density (Fig. 6a).

High-Density Gradients

Interpancreatic variations in exocrine and islet tissue density is influenced by the donor characteristics, the secretory status of exocrine cells, the pancreas procurement, the preservation protocols that affect cellular swelling and tissue edema, and the islet isolation procedure, which determines the extent of tissue dissociation and the size of aggregates formed (Anazawa et al. 2011; Chadwick et al. 1993;

London et al. 1998; Hering 2005). In particular, when isolating islets from CP pancreata for autograft transplant, many centers have observed frequent settling of islets in the COBE bag. Our data from use of the ATGS substantiates this observation, revealing a pattern of exocrine density discrepancy between living, chronic pancreatitis, donors (mean 1.105, range 1.085–1.115 g/cm^3), and brain-dead donors (mean 1.095, range 1.080–1.105 g/cm^3) (Anazawa et al. 2011) (Fig. 6b, c). To compensate for this effect, our center prefers to use a more dense gradient range for some autologous islet isolations, pushing the islets out of the COBE bag and into the collection fractions. As demonstrated, a more dense heavy solution effectively reduced tissue volume 63 % (from 30 ± 10.5 to 11 ± 9.2 ml) but also maintained a high post-purification islet recovery (84 ± 29.2 %). Following purification, autograft islets were found in all 12 collection fractions while allograft islets were primarily settled into the first 8, less dense layers.

COBE Purification Process

The COBE purification process begins after recombination has finished and the total tissue pellet has been consolidated into a single 250 mL conical. If the packed tissue volume is >25 mL, it may be split in half as a single COBE is likely to be less effective at segregating out exocrine tissue, which could result in significant islet sedimentation into the COBE bag. All purification steps should be done inside a cold room or custom-built refrigerator to ensure that the COBE apparatus and collection solutions are refrigerated throughout the process. The COBE 2991 cell processor is loaded with a larger gradient maker but in a similar manner as described earlier (see section "Analytical Test Gradient System"). We adjust our standard gradient range, 1.065–1.115 g/cm^3, according to the results of the analytical test gradient obtained during the recombination phase.

The centrifuged pellet is prepared for purification by decanting off the last of the wash media and resuspending the tissue in 20 mL of 25 % HSA. This suspension is transferred to a sterile 250 mL beaker; the conical is rinsed with cold storage solution and brought to a final volume of 120 mL by weight. The tissue is top loaded into the COBE bag by peristaltic pump (flow rate 20 mL/min), gently swirling the beaker for an even distribution of tissue, followed by an additional 30 mL of cold storage solution to rinse the beaker and tubing. After loading, the inlet tube is clamped and the COBE bag vented by carefully opening the outlet clamp and allowing the machinery to spin at 400 × g for 3 min. Subsequently, the first contents of the COBE bag are collected into an empty "waste" conical. When the tissue starts to appear in the tubing, the purification product is then distributed by 25 mL volumes into a consecutive series of 12 conicals, each prefilled with 225 mL of cold supplemented CMRL media.

The efficacy of the purification can be evaluated by taking a 1 mL sample from each well-mixed collection fraction, as well as a 200 µL sample of residual tissue remaining in the COBE bag. If a significant quantity of free islets is observed in the COBE bag, re-purification may be necessary. Assuming a successful purification, each collection conical is centrifuged at 220 × g, 2–10 °C, for 3 min. We use an estimation of each tube's pellet volume in conjunction with the sample counts to decide which fractions will be combined into a transplantable product. Ideally, the final pellet size will be <15 mL (<20 mL is acceptable) and contain the maximum number of islets with the minimum amount of contaminating tissue. After combining the selected fractions into a single 250 mL conical, the amassed pellet is washed in transplant media, under the same centrifugation parameters. The final pellet is transferred to a T-75 flask and resuspended in 100 mL of room-temperature transplant media, from which 100 μL samples are obtained for a "post-COBE" islet count and additional samples taken for potency and product sterility testing (see section "Recombination").

There have been some interesting innovations proposed in the last few years to improve and standardize this purification process. Friberg et al. have developed a computer-controlled closed system for loading the COBE 2991 that successfully reproduces a continuous density gradient but minimizes the potential for manual variation and contamination (Friberg et al. 2008). The Baylor research group found evidence that passage through the narrow neck of the COBE bag produces fluid shearing forces that encourage islet fragmentation (Shimoda et al. 2012b). They have shown some improvements in post-purification islet yield and size by using a wide plastic bottle with a lower-viscosity density gradient to reduce the necessary centrifugation speed (Shimoda et al. 2012c).

Transplant Preparation

Although quality control and sterility test results will not be available until after the product is released, the islet preparation should be immediately packaged for transplantation to avoid excess delay and operating time for the patient. Any subsequent failure to meet post-release criteria should be reported to the patient physician when assessed (especially sterility positives). Procedural deviations that have occurred during tissue processing should be reported to the physician before product release.

Once the tissue pellet, purified or unpurified, has been washed in transplant media and settled into 100 mL of fresh media for 5 min, the last sterility and retention samples can be taken from the supernatant to ensure the status of the final islet product immediately before packaging. From this point, extreme care should be taken to ensure the continued sterility of the product. All items in the biological safety cabinet where loading will take place should be sterile (if in contact with the product) or thoroughly disinfected and the operator should wear sterile gloves and sleeves during the loading process. If the packed tissue volume is greater than 10 mL, the pellet must be divided evenly and loaded into two 200 mL transplant bags, each labeled with the correct patient identification and FDA-required labeling. Each tissue load is suspended in 100 mL of transplant media and will require an additional 100 mL of media as a rinse solution. If the patient has no known allergy to ciprofloxacin, add 0.4 mL of Cipro® (1 % = 10,000 μg/mL) to each volume of rinse media.

After affixing a 60 mL syringe to the transplant bag, place the syringe upright in a clamp stand and transfer the 100 mL tissue suspension into the bag through the syringe. Rinse the tissue conical twice with 50 mL volumes of rinse solution to transfer any residual islets. Aseptically recap and clamp the bag's inlet tubing to ensure a thorough seal for transport. The sealed bag should be gently rocked to evenly suspend the islets. Repeat these steps for additional transplant bags if needed. Once the transplant physician at the operation room has been alerted, the islet preparation can be readied for transport in a room-temperature cooler equipped with temperature stabilizers (Kaddis et al. 2013).

Conclusion

Autologous islet isolation and transplantation has repeatedly demonstrated the ability to improve clinical outcomes by diminishing the impact of iatrogenic diabetes on patients undergoing pancreatectomy to alleviate CP or other disabling conditions. As practical experience has accumulated at an increasing number of qualified isolation centers, islet yield and viability, critical factors for achieving postoperative insulin independence, have progressively improved. Since performing the first IAT after CP pancreatectomy in 1977, our team at the University of Minnesota has been dedicated to understanding and improving the technical aspects of the isolation procedure that directly impact its immediate and posttransplant success. This includes the introduction of the simplified ATGS to improve purification yield (Anazawa et al. 2011) and the identification of post-isolation factors that detriment graft function (Loganathan et al. 2011). In the last few years, we have focused our research on the mechanics of enzyme digestion, proposing a new enzyme mixture and variable dose classes that have increased the flexibility of the procedure to respond to different donor and tissue characteristics (Balamurugan et al. 2010, 2012). Despite these and other advances, the best islet yields still average significantly less than the available store, indicating the need for more research.

There remains a clear need for better understanding enzyme mechanisms: the best role and timing for collagenase versus neutral protease, their functional ingredients, and how each interacts with different ECM components to define the length and efficiency of digestion. While good work has been done to identify factors that influence islet yield and viability, including donor and tissue characteristics, we need more specific techniques to be hypothesized and tested for overcoming these inevitable obstacles. Furthermore, there is currently a heavy cost burden to establish facilities and perform these procedures, which severely limits their availability, especially in developing countries. Ultimately, islet isolation should be considered a flexible and optimizable set of techniques, rather than a one-size-fits-all procedure. With well-trained teams and a deeper understanding of technical principles, it is our belief that IAT will continue to expand its reach all over the world, to prolong life and alleviate suffering among an increasingly diverse pool of potential recipients.

Acknowledgments The authors would like to thank Josh Wilhelm, Muhamad Abdulla, Mukesh Tiwari, Tom Gilmore, and Jeff Ansite.

Cross-References

▶ Advances in Clinical Islet Isolation
▶ Basement Membrane in Pancreatic Islet Function
▶ Human Islet Autotransplantation
▶ Successes and Disappointments with Clinical Islet Transplantation

References

Al-Abdullah IH, Kumar MS, Abouna GM (1994) Combined intraductal and interstitial distension of human and porcine pancreas with collagenase facilitates digestion of the pancreas and may improve islet yield. Transplant Proc 26:3384

Alsaif F, Molinari M, Al-Masloom A, Lakey JR, Kin T, Shapiro AM (2006) Pancreatic islet autotransplantation with completion pancreatectomy in the management of uncontrolled pancreatic fistula after whipple resection for ampullary adenocarcinoma. Pancreas 32:430–431

Anazawa T, Balamurugan AN, Bellin M, Zhang HJ, Matsumoto S, Yonekawa Y, Tanaka T, Loganathan G, Papas KK, Beilman GJ, Hering BJ, Sutherland DE (2009) Human islet isolation for autologous transplantation: comparison of yield and function using SERVA/Nordmark versus Roche enzymes. Am J Transplant 9:2383–2391

Anazawa T, Matsumoto S, Yonekawa Y, Loganathan G, Wilhelm JJ, Soltani SM, Papas KK, Sutherland DE, Hering BJ, Balamurugan AN (2011) Prediction of pancreatic tissue densities by an analytical test gradient system before purification maximizes human islet recovery for islet autotransplantation/allotransplantation. Transplantation 91:508–514

Balamurugan AN, Chang Y, Fung JJ, Trucco M, Bottino R (2003) Flexible management of enzymatic digestion improves human islet isolation outcome from sub-optimal donor pancreata. Am J Transplant 3:1135–1142

Balamurugan AN, Chang Y, Bertera S, Sands A, Shankar V, Trucco M, Bottino R (2006) Suitability of human juvenile pancreatic islets for clinical use. Diabetologia 49:1845–1854

Balamurugan AN, Breite AG, Anazawa T, Loganathan G, Wilhelm JJ, Papas KK, Dwulet FE, McCarthy RC, Hering BJ (2010) Successful human islet isolation and transplantation indicating the importance of class 1 collagenase and collagen degradation activity assay. Transplantation 89:954–961

Balamurugan AN, Loganathan G, Bellin MD, Wilhelm JJ, Harmon J, Anazawa T, Soltani SM, Radosevich DM, Yuasa T, Tiwari M, Papas KK, McCarthy R, Sutherland DE, Hering BJ (2012) A new enzyme mixture to increase the yield and transplant rate of autologous and allogeneic human islet products. Transplantation 93:693–702

Barlow AD, Hosgood SA, Nicholson ML (2013) Current state of pancreas preservation and implications for DCD pancreas transplantation. Transplantation 95:1419–1424

Bellin MD, Carlson AM, Kobayashi T, Gruessner AC, Hering BJ, Moran A, Sutherland DE (2008) Outcome after pancreatectomy and islet autotransplantation in a pediatric population. J Pediatr Gastroenterol Nutr 47:37–44

Bellin MD, Sutherland DE, Beilman GJ, Hong-McAtee I, Balamurugan AN, Hering BJ, Moran A (2011) Similar islet function in islet allotransplant and autotransplant recipients, despite lower islet mass in autotransplants. Transplantation 91:367–372

Bellin MD, Beilman GJ, Dunn TB, Pruett TL, Chinnakotla S, Wilhelm JJ, Ngo A, Radosevich DM, Freeman ML, Schwarzenberg SJ, Balamurugan AN, Hering BJ, Sutherland DE (2012a)

Islet autotransplantation to preserve β cell mass in selected patients with chronic pancreatitis and diabetes mellitus undergoing total pancreatectomy. Pancreas 42:317–321

Bellin MD, Balamurugan AN, Pruett TL, Sutherland DE (2012b) No islets left behind: islet autotransplantation for surgery-induced diabetes. Curr Diab Rep 12:580–586

Berney T, Mathe Z, Bucher P, Demuylder-Mischler S, Andres A, Bosco D, Oberholzer J, Majno P, Philippe J, Buhler L, Morel P (2004) Islet autotransplantation for the prevention of surgical diabetes after extended pancreatectomy for the resection of benign tumors of the pancreas. Transplant Proc 36:1123–1124

Blondet JJ, Carlson AM, Kobayashi T, Jie T, Bellin M, Hering BJ, Freeman ML, Beilman GJ, Sutherland DE (2007) The role of total pancreatectomy and islet autotransplantation for chronic pancreatitis. Surg Clin North Am 87:1477–1501, x

Braganza JM, Lee SH, McCloy RF, McMahon MJ (2011) Chronic pancreatitis. Lancet 377:1184–1197

Bucher P, Mathe Z, Bosco D, Becker C, Kessler L, Greget M, Benhamou PY, Andres A, Oberholzer J, Buhler L, Morel P, Berney T (2004) Morbidity associated with intraportal islet transplantation. Transplant Proc 36:1119–1120

Caballero-Corbalan J, Friberg AS, Brandhorst H, Nilsson B, Andersson HH, Felldin M, Foss A, Salmela K, Tibell A, Tufveson G, Korsgren O, Brandhorst D (2009) Vitacyte collagenase HA: a novel enzyme blend for efficient human islet isolation. Transplantation 88:1400–1402

Caballero-Corbalan J, Brandhorst H, Asif S, Korsgren O, Engelse M, de Koning E, Pattou F, Kerr-Conte J, Brandhorst D (2010) Mammalian tissue-free liberase: a new GMP-graded enzyme blend for human islet isolation. Transplantation 90:332–333

Chadwick DR, Robertson GS, Rose S, Contractor H, James RF, Bell PR, London NJ (1993) Storage of porcine pancreatic digest prior to islet purification. The benefits of UW solution and the roles of its individual components. Transplantation 56:288–293

Contractor HH, Johnson PR, Chadwick DR, Robertson GS, London NJ (1995) The effect of UW solution and its components on the collagenase digestion of human and porcine pancreas. Cell Transplant 4:615–619

D'Haese JG, Ceyhan GO, Demir IE, Tieftrunk E, Friess H (2013) Treatment options in painful chronic pancreatitis: a systematic review. HPB, Oxford

DiMagno MJ, DiMagno EP (2013) Chronic pancreatitis. Curr Opin Gastroenterol 29:531–536

Eckhard M, Brandhorst D, Brandhorst H, Brendel MD, Bretzel RG (2004) Optimization in osmolality and range of density of a continuous ficoll-sodium-diatrizoate gradient for isopycnic purification of isolated human islets. Transplant Proc 36:2849–2854

Forster S, Liu X, Adam U, Schareck WD, Hopt UT (2004) Islet autotransplantation combined with pancreatectomy for treatment of pancreatic adenocarcinoma: a case report. Transplant Proc 36:1125–1126

Friberg AS, Stahle M, Brandhorst H, Korsgren O, Brandhorst D (2008) Human islet separation utilizing a closed automated purification system. Cell Transplant 17:1305–1313

Gachago C, Draganov PV (2008) Pain management in chronic pancreatitis. World J Gastroenterol 14:3137–3148

Garraway NR, Dean S, Buczkowski A, Brown DR, Scudamore CH, Meloche M, Warnock G, Simons R (2009) Islet autotransplantation after distal pancreatectomy for pancreatic trauma. J Trauma 67:E187–E189

Gores PF, Sutherland DE (1993) Pancreatic islet transplantation: is purification necessary? Am J Surg 166:538–542

Goto M, Eich TM, Felldin M, Foss A, Kallen R, Salmela K, Tibell A, Tufveson G, Fujimori K, Engkvist M, Korsgren O (2004) Refinement of the automated method for human islet isolation and presentation of a closed system for in vitro islet culture. Transplantation 78:1367–1375

Gray DW, Morris PJ (1987) Developments in isolated pancreatic islet transplantation. Transplantation 43:321–331

Gruessner RW, Sutherland DE, Dunn DL, Najarian JS, Jie T, Hering BJ, Gruessner AC (2004) Transplant options for patients undergoing total pancreatectomy for chronic pancreatitis. J Am Coll Surg 198:559–567; discussion 568–559

Hering BJ (2005) Repurification: rescue rather than routine remedy. Am J Transplant 5:1–2

Hering BJ, Matsumoto I, Sawada T, Nakano M, Sakai T, Kandaswamy R, Sutherland DE (2002) Impact of two-layer pancreas preservation on islet isolation and transplantation. Transplantation 74:1813–1816

Hering BJ, Kandaswamy R, Ansite JD, Eckman PM, Nakano M, Sawada T, Matsumoto I, Ihm SH, Zhang HJ, Parkey J, Hunter DW, Sutherland DE (2005) Single-donor, marginal-dose islet transplantation in patients with type 1 diabetes. JAMA 293:830–835

Jindal RM, Ricordi C, Shriver CD (2010) Autologous pancreatic islet transplantation for severe trauma. New Engl J Med 362:1550

Johnson PR (2010) Suitable donor selection and techniques of islet isolation and purification for human islet-allograft transplantation. In: Hakim NS, Stratta RJ, Gray DW, Friend P, Colman A (eds) Pancreas, islet and stem cell transplantation for diabetes, 2nd edn. Oxford University Press, New York, pp 387–396

Kaddis JS, Hanson MS, Cravens J, Qian D, Olack B, Antler M, Papas KK, Iglesias I, Barbaro B, Fernandez L, Powers AC, Niland JC (2013) Standardized transportation of human islets: an islet cell resource center study of more than 2,000 shipments. Cell Transplant 22:1101–1111

Kawahara T, Kin T, Shapiro AM (2012) A comparison of islet autotransplantation with allotransplantation and factors elevating acute portal pressure in clinical islet transplantation. J Hepatobiliary Pancreat Sci 19:281–288

Khan A, Jindal RM, Shriver C, Guy SR, Vertrees AE, Wang X, Xu X, Szust J, Ricordi C (2012) Remote processing of pancreas can restore normal glucose homeostasis in autologous islet transplantation after traumatic whipple pancreatectomy: technical considerations. Cell Transplant 21:1261–1267

Kin T (2010) Islet isolation for clinical transplantation. Adv Exp Med Biol 654:683–710

Kin T, Shapiro AM (2010) Surgical aspects of human islet isolation. Islets 2:265–273

Kobayashi T, Manivel JC, Carlson AM, Bellin MD, Moran A, Freeman ML, Bielman GJ, Hering BJ, Dunn T, Sutherland DE (2011) Correlation of histopathology, islet yield, and islet graft function after islet autotransplantation in chronic pancreatitis. Pancreas 40:193–199

Lacy PE, Kostianovsky M (1967) Method for the isolation of intact islets of Langerhans from the rat pancreas. Diabetes 16:35–39

Lake SP, Bassett PD, Larkins A, Revell J, Walczak K, Chamberlain J, Rumford GM, London NJ, Veitch PS, Bell PR et al (1989) Large-scale purification of human islets utilizing discontinuous albumin gradient on IBM 2991 cell separator. Diabetes 38(Suppl 1):143–145

Lakey JR, Warnock GL, Shapiro AM, Korbutt GS, Ao Z, Kneteman NM, Rajotte RV (1999) Intraductal collagenase delivery into the human pancreas using syringe loading or controlled perfusion. Cell Transplant 8:285–292

Lakey JR, Kneteman NM, Rajotte RV, Wu DC, Bigam D, Shapiro AM (2002) Effect of core pancreas temperature during cadaveric procurement on human islet isolation and functional viability. Transplantation 73:1106–1110

Lakey JR, Burridge PW, Shapiro AM (2003) Technical aspects of islet preparation and transplantation. Transpl Int 16:613–632

Linetsky E, Bottino R, Lehmann R, Alejandro R, Inverardi L, Ricordi C (1997a) Improved human islet isolation using a new enzyme blend, liberase. Diabetes 46:1120–1123

Linetsky E, Lehmann R, Li H, Fernandez L, Bottino R, Selvaggi G, Alejandro R, Ricordi C (1997b) Human islet isolation using a new enzyme blend. Transplant Proc 29:1957–1958

Loganathan G, Dawra RK, Pugazhenthi S, Guo Z, Soltani SM, Wiseman A, Sanders MA, Papas KK, Velayutham K, Saluja AK, Sutherland DE, Hering BJ, Balamurugan AN (2011) Insulin degradation by acinar cell proteases creates a dysfunctional environment for human islets before/after transplantation: benefits of α-1 antitrypsin treatment. Transplantation 92:1222–1230

Loganathan G, Graham ML, Radosevich DM, Soltani SM, Tiwari M, Anazawa T, Papas KK, Sutherland DE, Hering BJ, Balamurugan AN (2013) Factors affecting transplant outcomes in diabetic nude mice receiving human, porcine, and nonhuman primate islets: analysis of 335 transplantations. Transplantation 95:1439–1447

London NJ, Swift SM, Clayton HA (1998) Isolation, culture and functional evaluation of islets of Langerhans. Diabetes Metab 24:200–207

Miki A, Ricordi C, Messinger S, Yamamoto T, Mita A, Barker S, Haetter R, Khan A, Alejandro R, Ichii H (2009) Toward improving human islet isolation from younger donors: rescue purification is efficient for trapped islets. Cell Transplant 18:13–22

Mita A, Ricordi C, Miki A, Barker S, Khan A, Alvarez A, Hashikura Y, Miyagawa S, Ichii H (2009) Purification method using iodixanol (OptiPrep)-based density gradient significantly reduces cytokine chemokine production from human islet preparations, leading to prolonged β-cell survival during pretransplantation culture. Transplant Proc 41:314–315

Morrison CP, Wemyss-Holden SA, Dennison AR, Maddern GJ (2002) Islet yield remains a problem in islet autotransplantation. Arch Surg 137:80–83

Moskalewski S (1965) Isolation and culture of the islets of Langerhans of the Guinea pig. Gen Comp Endocrinol 44:342–353

Najarian JS, Sutherland DE, Matas AJ, Goetz FC (1979) Human islet autotransplantation following pancreatectomy. Transplant Proc 11:336–340

Nakanishi W, Imura T, Inagaki A, Nakamura Y, Sekiguchi S, Fujimori K, Satomi S, Goto M (2012) Ductal injection does not increase the islet yield or function after cold storage in a vascular perfusion model. PLoS One 7:e42319

Naziruddin B, Matsumoto S, Noguchi H, Takita M, Shimoda M, Fujita Y, Chujo D, Tate C, Onaca N, Lamont J, Kobayashi N, Levy MF (2012) Improved pancreatic islet isolation outcome in autologous transplantation for chronic pancreatitis. Cell Transplant 21:553–558

Paget M, Murray H, Bailey CJ, Downing R (2007) Human islet isolation: semi-automated and manual methods. Diab Vasc Dis Res 4:7–12

Qin H, Matsumoto S, Klintmalm GB, De vol EB (2011) A meta-analysis for comparison of the two-layer and university of Wisconsin pancreas preservation methods in islet transplantation. Cell Transplant 20:1127–1137

Rabkin JM, Leone JP, Sutherland DE, Ahman A, Reed M, Papalois BE, Wahoff DC (1997) Transcontinental shipping of pancreatic islets for autotransplantation after total pancreatectomy. Pancreas 15:416–419

Rabkin JM, Olyaei AJ, Orloff SL, Geisler SM, Wahoff DC, Hering BJ, Sutherland DE (1999a) Distant processing of pancreas islets for autotransplantation following total pancreatectomy. Am J Surg 177:423–427

Rabkin JM, Olyaei AJ, Orloff SL, Geisler SM, Wahoff DC, Hering BJ, Sutherland DE (1999b) Distant processing of pancreas islets for autotransplantation following total pancreatectomy. Am J Surg 177:423–427

Ricordi C (2003) Islet transplantation: a brave new world. Diabetes 52:1595–1603

Ricordi C, Lacy PE, Finke EH, Olack BJ, Scharp DW (1988) Automated method for isolation of human pancreatic islets. Diabetes 37:413–420

Sah RP, Dawra RK, Saluja AK (2013) New insights into the pathogenesis of pancreatitis. Curr Opin Gastroenterol 29:523–530

Shimoda M, Itoh T, Sugimoto K, Iwahashi S, Takita M, Chujo D, Sorelle JA, Naziruddin B, Levy MF, Grayburn PA, Matsumoto S (2012a) Improvement of collagenase distribution with the ductal preservation for human islet isolation. Islets 4:130–137

Shimoda M, Itoh T, Iwahashi S, Takita M, Sugimoto K, Kanak MA, Chujo D, Naziruddin B, Levy MF, Grayburn PA, Matsumoto S (2012b) An effective purification method using large bottles for human pancreatic islet isolation. Islets 4:398–404

Shimoda M, Itoh T, Iwahashi S, Takita M, Sugimoto K, Kanak MA, Chujo D, Naziruddin B, Levy MF, Grayburn PA, Matsumoto S (2012c) An effective purification method using large bottles for human pancreatic islet isolation. Islets 4:398–404

Sutherland DE, Gruessner AC, Carlson AM, Blondet JJ, Balamurugan AN, Reigstad KF, Beilman GJ, Bellin MD, Hering BJ (2008) Islet autotransplant outcomes after total pancreatectomy: a contrast to islet allograft outcomes. Transplantation 86:1799–1802

Sutherland DE, Radosevich DM, Bellin MD, Hering BJ, Beilman GJ, Dunn TB, Chinnakotla S, Vickers SM, Bland B, Balamurugan AN, Freeman ML, Pruett TL (2012) Total pancreatectomy and islet autotransplantation for chronic pancreatitis. J Am Coll Surg 214:409–424; discussion 424–406

'tHart NA, Leuvenink HG, Ploeg RJ (2002) New solutions in organ preservation. Transplant Rev 16:131–141

Tsukada M, Saito T, Ise K, Kenjo A, Kimura T, Satoh Y, Saito T, Anazawa T, Oshibe I, Suzuki S, Hashimoto Y, Gotoh M (2012) A model to evaluate toxic factors influencing islets during collagenase digestion: the role of serine protease inhibitor in the protection of islets. Cell Transplant 21:473–482

Wahlberg JA, Southard JH, Belzer FO (1986) Development of a cold storage solution for pancreas preservation. Cryobiology 23:477–482

Walsh TJ, Eggleston JC, Cameron JL (1982) Portal hypertension, hepatic infarction, and liver failure complicating pancreatic islet autotransplantation. Surgery 91:485–487

Watkins JG, Krebs A, Rossi RL (2003) Pancreatic autotransplantation in chronic pancreatitis. World J Surg 27:1235–1240

Webb MA, Dennison AR, James RF (2012) The potential benefit of non-purified islets preparations for islet transplantation. Biotechnol Genet Eng Rev 28:101–114

Human Islet Autotransplantation

43

Martin Hermann, Raimund Margreiter, and Paul Hengster

Contents

Introduction	1230
Lessons from Islet Autotransplantations	1232
Still Open Issues in Islet Autotransplantation	1235
Islet Mass	1235
Islet Shipment	1235
Cell Death	1237
Which Are the Best Islets? Does Size Matter?	1237
The Role of the Surrounding Tissue: Site Matters	1238
Conclusion	1239
Cross-References	1240
References	1240

Abstract

Total pancreatectomy is a last option for patients in whom all other efforts for managing intractable pain caused by chronic pancreatitis have failed. When performed with a simultaneous islet autotransplantation, endogenous insulin production can be preserved to some extent. Although it does not necessarily prevent any future need for exogenous insulin, the diabetic state of the patient is less severe compared to pancreatectomy alone.

In contrast to islet allotransplantation, the patients do not require immunosuppressive drugs and are not at risk for autoimmune destruction or alloimmune rejection. Consequently, insulin independence is more likely to be preserved in

M. Hermann (✉)
Department of Anaesthesiology and Critical Care Medicine, Medical University of Innsbruck, Innsbruck, Austria
e-mail: martin.hermann@uki.at

R. Margreiter • P. Hengster
Daniel Swarovski Laboratory, Department of Visceral-, Transplant- and Thoracic Surgery, Center of Operative Medicine, Innsbruck Medical University, Innsbruck, Austria
e-mail: raimund.margreiter@uki.at; paul.hengster@uki.at

autoislet-transplanted patients than in those patients who received alloislet transplants. Therefore, the setting of islet autotransplantation allows the study of the transplanted islets without interference from medication or immunological variables.

Keywords

Human islet autotransplantation • Islet shipment • Real-time live confocal microscopy

Introduction

Chronic pancreatitis (CP) is a progressive inflammatory disease causing irreversible structural damage to the pancreatic parenchyma. Besides affecting the pancreatic exocrine function, in severe cases, the endocrine function may also be impaired leading to the onset of diabetes mellitus (Steer et al. 1995). As in many patients CP is clinically silent, its prevalence can only be estimated and ranges from 0.4 % to 5 % before the onset of clinically apparent disease. Besides heavy consumption of alcohol (150–170 g/day), pancreatic obstructions such as posttraumatic ductal strictures, pseudocysts, mechanical or structural changes of the pancreatic-duct sphincter, and periampullary tumors may result in chronic pancreatitis. Of high importance is the recent recognition of a set of genetic mutations such as the loss-of-function mutations of pancreatic secretory trypsin inhibitor (SPINK1), which were shown to be present in CP cases that previously had been considered idiopathic (for review see Naruse et al. 2007). Also, sphincter of Oddi dysfunction (SOD) has increasingly been recognized as being present in CP (McLoughlin and Mitchell 2007).

Due to the progress in imaging techniques such as endoscopic retrograde cholangiopancreatography, magnetic resonance imaging, and cross-sectional imaging, we now have a better understanding of the pathophysiology and origin of inflammation and pain in CP. Nevertheless, chronic pancreatitis still remains an inscrutable process of uncertain pathogenesis, unpredictable clinical course, and difficult treatment (Steer et al. 1995; Riediger et al. 2007). Chronic pancreatitis is associated with a mortality rate that approaches 50 % within 20–25 years. Approximately 15–20 % of patients die of complications associated with acute attacks of pancreatitis (Steer et al. 1995).

Complications such as biliary or duodenal stenosis, as well as intractable pain, are the current indications for surgery in patients with CP. Surgical drainage of the duct in CP has largely been replaced by endoscopic duct drainage procedures of sphincterotomy and stent placement in the duct. Patients with CP whose pain persists after endoscopic pancreatic-duct drainage are candidates for total pancreatectomy and islet autotransplantation (IAT) (Evans et al. 2008). Islet autotransplantation was also shown to be a successful option for patients suffering severe trauma requiring a complete removal of the pancreas (Khan et al. 2012; Jindal et al. 2010).

In the Cincinnati series of total pancreatectomy in combination with simultaneous IAT, unremitting abdominal pain refractory to high-dose narcotics was the indication for surgery (Ahmad et al. 2005; Rodriguez Rilo et al. 2003). Narcotic independence due to pain relief after total pancreatectomy and islet autotransplantation was achieved in 58–81 % of the patients (Ahmad et al. 2005; Blondet et al. 2007). Notably, in a retrospective survey, more than 95 % of the patients stated they would recommend total pancreatectomy in combination with islet autotransplantation (Blondet et al. 2007).

Mortality as well as morbidity associated with pancreatic resections in patients suffering from chronic pancreatitis was shown to be very low and normally leads to adequate pain control in the majority of CP patients. One drawback of surgical resection is the development of exo- and endocrine insufficiencies. Therefore, surgical resection of the pancreas is considered as a final option in the treatment of CP. Nevertheless, the addition of an islet autotransplant offers the possibility of a postoperative glucose control and should therefore always be a considerable option.

Besides being applicable to prevent surgical diabetes after extensive pancreatic resection for chronic pancreatitis, islet autotransplantation is additionally pertinent in benign tumors located at the neck of the pancreas. Even without pancreatic inflammation, extensive pancreatic resection of more than 70 % of the pancreas may cause diabetes (Slezak and Andersen 2001).

Islet autotransplantation, after extended pancreatectomy performed for the resection of benign tumors of the mid-segment of the pancreas, was shown to be a feasible option with excellent metabolic results and low morbidity. Due to the non-inflammatory nature of the pancreata, higher islet yields and, consequently, higher transplanted islet masses were achieved compared to those from organs resected for chronic pancreatitis. At a median follow-up of 5 years (range, 1–8 years), all patients ($n = 7$) had β;-cell function as assessed by a positive C-peptide level. Six out of the seven patients were insulin independent (Berney et al. 2004). Pivotal for such an approach is the unequivocal diagnosis of the benign nature of the tumor, before making the decision to perform the isolation and transplantation procedure.

The first total pancreatectomy in combination with islet autotransplantation to treat chronic pancreatitis (CP) in humans was performed 30 years ago at the University of Minnesota (Sutherland et al. 1978). Besides aiming to relieve the pain of the CP patient in whom other measures had failed, the additional goal was to preserve β-cell mass and insulin secretion in order to avoid the otherwise inevitable surgical diabetes. Since then, more than 300 islet autotransplantations have been performed and reported worldwide, most of them at the University of Minnesota. With a few exceptions, the intraportal site has been predominantly applied as an implantation site for the transplanted islets (Blondet et al. 2007; White et al. 2000). Since 1990 the results of autologous islet transplantation have been reported to the International Islet Transplant Registry (ITR) in Giessen, Germany (Bretzel et al. 2007).

Combined pancreatectomy and islet autotransplantation can be performed in adults, as well as in pediatric patients. For both patient populations, the procedures are identical and described in detail elsewhere (Bellin et al. 2008; Wahoff et al. 1995; White et al. 2000). Performing islet autotransplantations provides the possibility to compare the metabolic outcomes between islet autografts and islet allografts, the latter still being subject to declining function with time (Shapiro et al. 2006). Besides, and prior to, the outstanding results from the Edmonton study fuelling the whole field of islet transplantation with new energy, the "Minnesota islet autotransplantation" provided the pivotal biological "proof of principle" for the feasibility of a long-lasting successful glucose control after islet transplantation.

Islet allotransplantation shows a 5-year post-islet transplantation graft survival of approximately 80 %, and insulin independence around 10 % at 5 years (Ryan et al. 2005). Differences in the success of allogenic islet transplantation among different centers illustrate the complexity of the procedure (Shapiro et al. 2006). Therefore, the ultimate goal, defined by insulin independence in the long term being achieved on a regular basis, has still not been achieved.

Notably, the results from islet autotransplantation obtained so far clearly show that long-term insulin independence after islet transplantation is a goal which can be realized, although also here not on a regular basis (Blondet et al. 2007; Robertson et al. 1999, 2001). In a recently published study, the outcomes of islet function over time were compared between intraportal islet autotransplant recipients at the University of Minnesota and diabetic islet allograft recipients as reported by the Collaborative Islet Transplant Registry (CITR). With regard to insulin independence, 74 % of islet autotransplant recipients retained insulin independence at 2-year posttransplant versus only 45 % of the CITR allograft recipients who initially became insulin independent. Notably, 46 % of the islet autotransplant patients were still insulin independent at 5 years and 28 % at 10 years posttransplant (Sutherland et al. 2008).

Lessons from Islet Autotransplantations

Three metabolic states were described in patients after islet autotransplantations: one third of the patients after islet autotransplantation in the University of Minnesota series were long-term insulin independent, another third of the recipients became fully diabetic, and the last third achieved near normoglycemia and are therefore partially insulin independent requiring only one daily injection of insulin (Fig. 1) (Sutherland et al. 2012; Blondet et al. 2007).

A remarkable result when comparing islet allo- with islet autotransplantation is the generally higher long-term success rate of the latter (Robertson et al. 2001; Ryan et al. 2005). There are at least three known causes (Fig. 2) for organ/cell stress which are present in islet allotransplantation but not in autotransplantation, thereby possibly explaining the better long-term success rates of the latter:
1. Brain death: In islet allotransplantation, the organ is obtained from brain dead patients. In animal models, brain death was shown to negatively affect islet yield as well as function due to the activation of pro-inflammatory cytokines (Contreras et al. 2003).

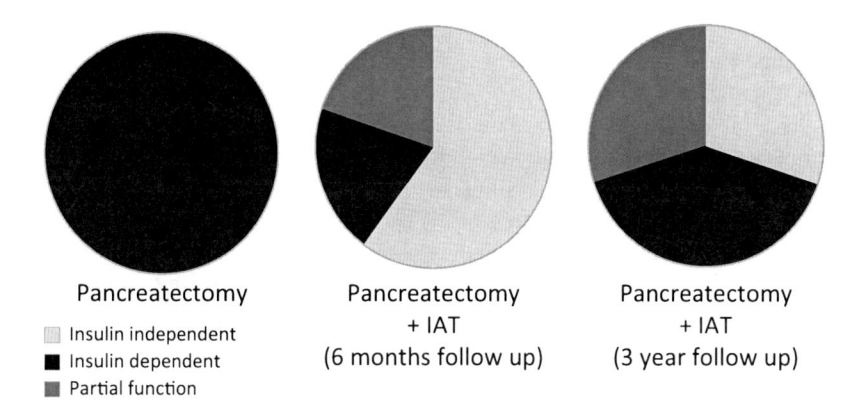

Fig. 1 A schematic representation of the three metabolic states described in patients after pancreatectomy with or without islet autotransplantation. Total pancreatectomy without a simultaneous islet autotransplantation unequivocally leads to insulin dependence of the patient. In contrast, approximately one third of autoislet transplant recipients become insulin independent and an additional one third require minimal insulin replacement (Bellin et al. 2011; Blondet et al. 2007)

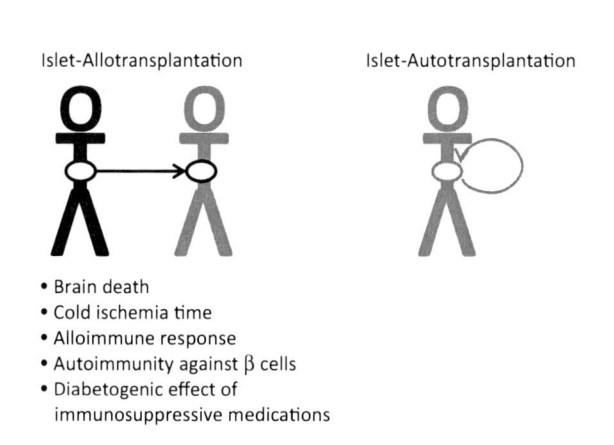

Fig. 2 In contrast to islet autografts, islet allografts are subject to several additional cell stress conditions. Brain death (Contreras et al. 2003), longer cold ischemia times before islet isolation from the donor pancreas (Hering et al. 2002), the patients' alloimmune response to the donor tissue, the autoimmunity against β cells (Huurman et al. 2008; Monti et al. 2008), and the diabetogenic effect of the immunosuppressive medications (Egidi et al. 2008) are the main reasons limiting long-term success of islet allotransplantation

2. Ischemia: In islet autotransplantation, the organ is not subjected to prolonged cold ischemia times which are normally present in islet allotransplantation due to the transport of the organ to the islet procurement center. Such cold ischemia times are known to damage the organ and impair cell viability, as well as function (Emamaullee and Shapiro 2007).

3. Immunosuppression: Besides ischemia-associated organ damage, the need for immunosuppression in islet allotransplantation is the third major limiting cause in the long-term success of islet allotransplantation (Emamaullee and Shapiro 2007). In human islet allotransplantation, immunosuppressive regimens are implemented in order to cope with both auto- and alloimmunity after transplantation. However, many of the immunosuppressive drugs are known to be directly β-cell toxic. Using a transgenic mouse model for conditional ablation of pancreatic β cells in vivo, Nir and coworkers elegantly demonstrated that β cells have a significant regenerative capacity which is prevented by the addition of the immunosuppressant drugs sirolimus and tracrolimus (Nir et al. 2007). As shown in humans, up to 15 % of nondiabetic patients who received solid organ transplantation were shown to develop posttransplant diabetes as a result of calcineurin inhibitor therapy (i.e., tacrolimus) (Emamaullee and Shapiro 2007). Therefore, the declining function of β cells after human allotransplantation may also be explained by the inhibition of β-cell turnover due to the administration of immunosuppressive drugs (Ruggenenti et al. 2008).

Allograft rejection and recurrent autoimmunity, both conditions not present in islet autotransplant recipients, may additionally contribute to the decreasing insulin independence over time observed in the allogeneic setting (Huurman et al. 2008; Monti et al. 2008). Recently it was shown that immunosuppression with FK506 and rapamycin after islet transplantation in patients with autoimmune diabetes induced homeostatic cytokines that expand autoreactive memory T cells. It was therefore proposed that such an increased production of cytokines might contribute to recurrent autoimmunity in transplanted patients with autoimmune disease and that a therapy that prevents the expansion of autoreactive T cells will improve the outcome of islet allotransplantation (Monti et al. 2008).

Another recently published study reports that cellular islet autoimmunity associates with the clinical outcome of islet allotransplantation. In this study, 21 type 1 diabetic patients received islet grafts prepared from multiple donors, while immunosuppression was maintained by means of antithymocyte globulin (ATG) induction and tacrolimus and mycophenolate treatment. Immunity against auto- and alloantigens was measured before and during 1 year after transplantation. Interestingly, cellular autoimmunity before and after transplantation was shown to be associated with delayed insulin independence and lower circulating C-peptide levels during the first year after islet allotransplantation. While seven out of eight patients without preexistent T-cell autoreactivity became insulin independent, none of the four patients reactive to both islet autoantigens GAD and IA-2 achieved insulin independence. Consequently, tailored immunotherapy regimens targeting cellular islet autoreactivity may be required (Huurman et al. 2008).

An additional explanation for the lack of long-term insulin independence after islet transplantation was suggested to be the detrimental effect of hyperglycemia on β-cell physiology. As shown in mice, increased apoptosis and reduced β-cell mass were found in islets exposed to chronic hyperglycemia (Biarnes et al. 2002). Consequently, both (auto- and allo-) human islet recipients usually receive insulin early on to maintain euglycemia as much as possible. However, no study in humans has been performed so far comparing islet engraftment with and without this measure.

Still Open Issues in Islet Autotransplantation

Islet Mass

The timing of the pancreatectomy and simultaneous islet allotransplantation has a direct impact on islet yield. The highest islet yields and insulin independence can be achieved when the islet autotransplantation is performed earlier in the disease course of CP (Ahmed et al. 2006; Rodriguez Rilo et al. 2003). The search for the optimal enzyme blend that maximizes human islet yield for transplantation is still ongoing (Balamurugan et al. 2012).

Although the islet yield is an important predictor of insulin independence (Gruessner et al. 2004), there are exceptions: one patient who received only 954 IEQ/kg remained insulin-free even 4 years after transplantation (Webb et al. 2006, 2008). Considering the scarcity of available organs, such results are a crucial proof of principle showing that even very low amounts of transplanted islets may be sufficient to provide long-term insulin independence. Interestingly islet autografts show durable function and, once established, are associated with a persisting high rate of insulin independence, although the β-cell mass transplanted is lesser than that used for islet allografts (Sutherland et al. 2008).

Evaluating and comparing the different outcomes after islet allo- versus autotransplantations may help clarify the extent to which different stress parameters account for islet damage resulting in limited success rates of islet allotransplantation. There are several causes for cellular stress in islet autotransplantation.

Islet Shipment

Exposure of islets to a series of damaging physicochemical stresses already during explantation of the pancreas may amplify the damage caused during cold storage as well as the following islet isolation procedure. There is consensus among the major islet transplantation centers that islet yields and quality can be improved with better pancreas procurement techniques such as in situ regional organ cooling which protects the pancreas from warm ischemic injury (for review see Iwanaga et al. 2008). In addition, the development of more sophisticated pancreas preservation protocols promises to translate into an improved islet yield as well as quality.

While pancreatectomy can be performed at most hospitals, only a few centers are able to perform islet isolations. Therefore, human islet autotransplantation is often limited due to the absence of an on-site islet-processing facility. The setup of an islet isolation facility, designed according to the rules of good manufacturing practice, is a technically challenging and cost- and time-intensive process (Hengster et al. 2005; Guignard et al. 2004). Consequently, several institutions have decided to perform transplantation of islets isolated at another center with already established expertise. Such an "outsourcing solution" was not only shown to be applicable in human islet allotransplantation (Guignard et al. 2004; Ichii et al. 2007; Yang et al. 2004) but also in human islet autotransplantation (Rabkin et al. 1997, 1999). In the latter, the resected pancreata were transferred to an islet-processing laboratory, which then sent back the freshly isolated islets that were transplanted into the same patient. All five patients experienced complete relief from pancreatic pain, and three of the five patients had minimal or no insulin requirement, thereby demonstrating the feasibility of islet shipment for autotransplantation (median follow-up of 23 months) (Rabkin et al. 1999). A recent case reports a successful emergency autologous islet transplantation after a traumatic Whipple operation with islets processed in a remote center. Although the number of islets was suboptimal, near-normal glucose tolerance was achieved, thereby showing that islets processed at a remote site are also suitable for transplantation (Khan et al. 2012; Jindal et al. 2010).

Although practicability as well as feasibility of islet transportation has already been proven, many questions such as the one addressing the optimal transport conditions for islets remain to be answered. While there is a worldwide consensus of how to isolate islets under GMP conditions, this is not the case for the transport of the freshly isolated islets. Many different media and transport devices have been used, ranging from 50 ml flasks, syringes, and gas permeable bags (Ichii et al. 2007). Other solutions such as rotary devices avoiding detrimental cell compaction (Merani et al. 2006) may be an alternative, especially when vitality parameters such as temperature, pH, or oxygen concentration are actively controlled (Wurm et al. 2007). Determining the optimal conditions for the transport of islets promises to yield better islet quality after the transport of islets and consequently an improved transplantation outcome. In addition, a gain of knowledge concerning the issues addressing the regeneration potential of freshly isolated islets may help not only to avoid unnecessary additional cellular stress but also counterbalance it in a preemptive way.

In this context, the topic of islet quality assessment has to be mentioned: similar to the transport conditions of human islets, this issue remains a matter of debate. Predicting the outcome of islet transplantation is still not possible due to the lack of reliable markers of islet potency, which might potentially be used to screen human islet preparations prior to transplantation. According to these pretransplant criteria, islet preparations that failed to reverse diabetes were indistinguishable from those that exhibited excellent function (Ichii et al. 2007).

Therefore, one of the primary challenges also in islet autotransplantation is to identify and understand the changes taking place in islets after the isolation, culture, and transport. Description of such changes in living islet cells offers insights not achievable by the use of fixed cell techniques. Combining real-time live confocal microscopy with three fluorescent dyes, dichlorodihydrofluorescein (DCF) diacetate, tetramethylrhodamine methyl ester (TMRM) perchlorate, and fluorescent wheat germ agglutinin (WGA), offers the possibility to assess overall oxidative stress, time-dependent mitochondrial membrane potentials, and cell morphology (Hermann et al. 2005, 2007). The advantage of such a method resides in the fast and accurate imaging at a cellular and even subcellular level. Taking into account the use of other fluorescent dyes which can be used to visualize additional cell viability parameters such as calcium concentrations (measured with rhod-2) or apoptosis (measured with annexin-V), such an approach promises to be of great value for a better future islet assessment, post-isolation, culture, and/or transport. For detecting islets at transplantation in a clinical setting, intraoperative ultrasound examination was shown to be useful (Sakata et al. 2012).

Cell Death

A significant proportion of the transplanted islet mass fails to engraft due to apoptotic cell death. Several strategies have been implemented to inhibit this process by blocking the extrinsic apoptosis-inducing signals (cFLIP or A20), although only with limited impact. More recently, investigations of downstream apoptosis inhibitors that block the final common pathway (i.e., X-linked inhibitor of apoptosis protein [XIAP]) have shown promising results, in human (Emamaullee et al. 2005a, b; Hui et al. 2005) as well as in rodent (Plesner et al. 2005) models of islet engraftment. XIAP-transduced human islets were significantly less apoptotic in an in vitro system that mimics hypoxia-induced injury. In addition, transplanting a series of marginal mass islet graft transplants in streptozotocin-induced diabetic NOD-RAG$^{-/-}$ mice resulted in 89 % of the animals becoming normoglycemic, with only 600 XIAP-transduced human islets (Emamaullee et al. 2005b). Moreover, XIAP overexpression has been shown to prevent the diabetogenicity of the immunosuppressive drugs tacrolimus and sirolimus in vitro (Hui et al. 2005).

Which Are the Best Islets? Does Size Matter?

In islet allo- as well as autotransplantation, it is still a matter of debate to define the features of an ideal islet able to ensure proper long-lasting glucose homeostasis after transplantation into the liver. The central question is whether bigger islets are better suited than smaller islets.

In the early phase after transplantation, the islets are supplied with oxygen and nutrients only by diffusion. In addition, data obtained from rat islet transplantations have shown that being in the portal vein, islets encounter a hypoxic state with an oxygen tension of 5 mmHg compared to 40 mmHg in the pancreas (Carlsson et al. 2001). In a study determining whether the size of the islets could influence the success rates of islet transplantations in rats, the small islets (<125 μm) were shown to be superior compared to their larger counterparts (>150 μm). The superiority of small islets was shown in vitro, via functional assays, as well as in vivo after transplanting them under the kidney capsule of diabetic rats. Using only marginal islet equivalencies for the renal subcapsular transplantation, large islets failed to produce euglycemia in any recipient rat, whereas small islets were successful in 80 % of the cases (MacGregor et al. 2006). A recent study analyzed the influence of islet size on insulin production in human islet transplantation. The results convincingly showed that small islets are superior to large islets with regard to in vitro insulin secretion and higher survival rates (Lehmann et al. 2007). Therefore, islet size seems to be of importance for the success of human islet transplantation, and at least regarding islets, it might be stated: "Small is beautiful!"

The question that remains to be answered is how to improve the transplantation outcome, when using large islets. Besides applying measures that promote islet engraftment, such as the addition of the iron chelator deferoxamine which increases vascular endothelial growth factor expression (Langlois et al. 2008), an alternative would be to customize large islets into small "pseudoislets" using the hanging drop technique (Cavallari et al. 2007).

The Role of the Surrounding Tissue: Site Matters

To what extent is the surrounding tissue necessary or beneficial for islet function?

Besides the long-lasting functionality of autologous transplanted islets, there are at least two additional findings in islet autotransplantation that merit attention: the relatively low amounts of islets needed to achieve normoglycemia and the impurity of transplanted islets.

In islet allotransplantation, about 850,000 islets, normally obtained from two to four pancreases, are needed to achieve insulin independence in a single type 1 diabetes patient. As a consequence, the available pool of pancreata for islet allotransplantation is limited and is therefore one of the foremost problems in islet transplantation. Interestingly islet autotransplantation has shown us that even low amounts of islets may result in long-lasting insulin independence (Robertson et al. 2001; Pyzdrowski et al. 1992).

Due to extensive fibrosis, which is often present in pancreata of pancreatitis patients, the digestion process is incomplete. Theoretically, such an incomplete digestion might result in lower success rates after islet transplantation. Surprisingly, in a recent study, 8 of 12 patients who showed insulin independence after islet autotransplantation had less than 40 % islet cleavage (Webb et al. 2008). Therefore, a protective role of the tissue surrounding the islets might be postulated. Besides postulating such a protective role of the surrounding tissue, one could speculate that the digestion process may also lead to the loss of the basement membrane surrounding the islets (Rosenberg et al. 1999) which might be detrimental as it is a well-recognized fact that the extracellular matrix provides the islets with biotrophic support (Ilieva et al. 1999; Pinkse et al. 2006; Rosenberg et al. 1999).

Besides the innate surrounding tissue of the islets, the ectopic site into which the islets are implanted also seems to exert an influence on their biology: while autoislet β-cell biology can be normal (as shown by fasting glucose and hemoglobin A1c levels and intravenous glucose disappearance rates) for up to 13 years

(Robertson et al. 2001), there seem to be abnormalities in α-cell responsiveness to insulin-induced hypoglycemia.

Although responses from intrahepatically autotransplanted islets to intravenous arginine were shown to be present, their responsiveness to insulin-induced hypoglycemia was absent (Kendall et al. 1997). Similar observations were also made in islet allotransplantation: two normoglycemic type 1 diabetic patients who had been successfully transplanted with alloislets into the liver also failed to secrete glucagon during hypoglycemia (Kendall et al. 1997). These findings led to a study comparing the α-cell function between autoislets transplanted either in the liver or in the peritoneal cavity of dogs. As expected from the situation in humans, the animals that received their islets transplanted into the liver did not have a glucagon response during hypoglycemic clamps. Interestingly, in the animals that received their autoislets transplanted into the peritoneal cavity, the glucagon response was present. Both groups showed similar responses to intravenous arginine (Gupta et al. 1997).

In a recent study, autologous pancreatic islets were successfully transplanted into a human bone marrow. In all four patients islet function was sustained up to the maximum follow-up of 944 days (Maffi et al. 2013).

Taking everything together it could be said, "site matters," and there is certainly not just one site which deserves a closer "sight-seeing."

Conclusion

The technical feasibility of islet autotransplantation has been demonstrated by several centers (Clayton et al. 2003; Gruessner et al. 2004; Rodriguez Rilo et al. 2003). In spite of the problems that autologous transplanted islets encounter in their new surrounding, pancreatic islet autotransplantation has prevented the onset of diabetes in pancreatectomized patients for more than two decades (Robertson 2004). Therefore, the biological proof of principle, for a long-lasting stable glucose control by islets transplanted into the liver, has already been established. This success is equally surprising as well as inspiring for the more difficult task of islet allotransplantation. Understanding how autotransplanted islets can sustain their homeostasis and function in the liver, even for decades, might help us to find answers for still open questions regarding the molecular and cellular basis necessary for a successful islet allotransplantation.

Islet autotransplantation can abrogate the onset of diabetes and may therefore be considered as a valuable addition to surgical resection of the pancreas. The results obtained after islet autotransplantation have definitively provided a significant proof of principle: islets are able to regulate glucose homeostasis over decades when transplanted into the liver.

In times like these, when the enthusiasm regarding clinical islet allotransplantation has been dampened by the inadequate long-term results, such a proof of principle is a vital beacon reminding us of the ultimate goal and prospects of islet transplantation.

Cross-References

▶ Advances in Clinical Islet Isolation
▶ Approaches for Imaging Islets: Recent Advances and Future Prospects
▶ Islet Isolation from Pancreatitis Pancreas for Islet Autotransplantation

References

Ahmad SA, Lowy AM, Wray CJ, D'Alessio D, Choe KA, James LE, Gelrud A, Matthews JB, Rilo HL (2005) Factors associated with insulin and narcotic independence after islet autotransplantation in patients with severe chronic pancreatitis. J Am Coll Surg 201:680–687

Ahmed SA, Wray C, Rilo HL, Choe KA, Gelrud A, Howington JA, Lowy AM, Matthews JB (2006) Chronic pancreatitis: recent advances and ongoing challenges. Curr Probl Surg 43:127–238

Balamurugan AN, Loganathan G, Bellin MD, Wilhelm JJ, Harmon J, Anazawa T, Soltani SM, Radosevich DM, Yuasa T, Tiwari M, Papas KK, McCarthy R, Sutherland DE, Hering BJ (2012) A new enzyme mixture to increase the yield and transplant rate of autologous and allogeneic human islet products 2. Transplantation 93:693–702

Bellin MD, Carlson AM, Kobayashi T, Gruessner AC, Hering BJ, Moran A, Sutherland DE (2008) Outcome after pancreatectomy and islet autotransplantation in a pediatric population. J Pediatr Gastroenterol Nutr 47:37–44

Bellin MD, Sutherland DE, Beilman GJ, Hong-McAtee I, Balamurugan AN, Hering BJ, Moran A (2011) Similar islet function in islet allotransplant and autotransplant recipients, despite lower islet mass in autotransplants 5. Transplantation 91:367–372

Berney T, Mathe Z, Bucher P, Demuylder-Mischler S, Andres A, Bosco D, Oberholzer J, Majno P, Philippe J, Buhler L, Morel P (2004) Islet autotransplantation for the prevention of surgical diabetes after extended pancreatectomy for the resection of benign tumors of the pancreas. Transplant Proc 36:1123–1124

Biarnes M, Montolio M, Nacher V, Raurell M, Soler J, Montanya E (2002) β-cell death and mass in syngeneically transplanted islets exposed to short- and long-term hyperglycemia. Diabetes 51:66–72

Blondet JJ, Carlson AM, Kobayashi T, Jie T, Bellin M, Hering BJ, Freeman ML, Beilman GJ, Sutherland DE (2007) The role of total pancreatectomy and islet autotransplantation for chronic pancreatitis. Surg Clin North Am 87:1477–1501, x

Bretzel RG, Jahr H, Eckhard M, Martin I, Winter D, Brendel MD (2007) Islet cell transplantation today. Langenbecks Arch Surg 392:239–253

Carlsson PO, Palm F, Andersson A, Liss P (2001) Markedly decreased oxygen tension in transplanted rat pancreatic islets irrespective of the implantation site. Diabetes 50:489–495

Cavallari G, Zuellig RA, Lehmann R, Weber M, Moritz W (2007) Rat pancreatic islet size standardization by the "hanging drop" technique. Transplant Proc 39:2018–2020

Clayton HA, Davies JE, Pollard CA, White SA, Musto PP, Dennison AR (2003) Pancreatectomy with islet autotransplantation for the treatment of severe chronic pancreatitis: the first 40 patients at the leicester general hospital. Transplantation 76:92–98

Contreras JL, Eckstein C, Smyth CA, Sellers MT, Vilatoba M, Bilbao G, Rahemtulla FG, Young CJ, Thompson JA, Chaudry IH, Eckhoff DE (2003) Brain death significantly reduces isolated pancreatic islet yields and functionality in vitro and in vivo after transplantation in rats. Diabetes 52:2935–2942

Egidi MF, Lin A, Bratton CF, Baliga PK (2008) Prevention and management of hyperglycemia after pancreas transplantation. Curr Opin Organ Transplant 13:72–78

Emamaullee JA, Shapiro AM (2007) Factors influencing the loss of β-cell mass in islet transplantation. Cell Transplant 16:1–8

Emamaullee J, Liston P, Korneluk RG, Shapiro AM, Elliott JF (2005a) XIAP overexpression in islet β-cells enhances engraftment and minimizes hypoxia-reperfusion injury. Am J Transplant 5:1297–1305

Emamaullee JA, Rajotte RV, Liston P, Korneluk RG, Lakey JR, Shapiro AM, Elliott JF (2005b) XIAP overexpression in human islets prevents early posttransplant apoptosis and reduces the islet mass needed to treat diabetes. Diabetes 54:2541–2548

Evans KA, Clark CW, Vogel SB, Behrns KE (2008) Surgical management of failed endoscopic treatment of pancreatic disease. J Gastrointest Surg 12:1924–1929

Gruessner RW, Sutherland DE, Dunn DL, Najarian JS, Jie T, Hering BJ, Gruessner AC (2004) Transplant options for patients undergoing total pancreatectomy for chronic pancreatitis. J Am Coll Surg 198:559–567

Guignard AP, Oberholzer J, Benhamou PY, Touzet S, Bucher P, Penfornis A, Bayle F, Kessler L, Thivolet C, Badet L, Morel P, Colin C (2004) Cost analysis of human islet transplantation for the treatment of type 1 diabetes in the Swiss-French Consortium GRAGIL. Diabetes Care 27:895–900

Gupta V, Wahoff DC, Rooney DP, Poitout V, Sutherland DE, Kendall DM, Robertson RP (1997) The defective glucagon response from transplanted intrahepatic pancreatic islets during hypoglycemia is transplantation site-determined. Diabetes 46:28–33

Hengster P, Hermann M, Pirkebner D, Draxl A, Margreiter R (2005) Islet isolation and GMP, ISO 9001:2000: what do we need – a 3-year experience. Transplant Proc 37:3407–3408

Hering BJ, Matsumoto I, Sawada T, Nakano M, Sakai T, Kandaswamy R, Sutherland DE (2002) Impact of two-layer pancreas preservation on islet isolation and transplantation. Transplantation 74:1813–1816

Hermann M, Pirkebner D, Draxl A, Margreiter R, Hengster P (2005) "Real-time" assessment of human islet preparations with confocal live cell imaging. Transplant Proc 37:3409–3411

Hermann M, Margreiter R, Hengster P (2007) Molecular and cellular key players in human islet transplantation. J Cell Mol Med 11:398–415

Hui H, Khoury N, Zhao X, Balkir L, D'Amico E, Bullotta A, Nguyen ED, Gambotto A, Perfetti R (2005) Adenovirus-mediated XIAP gene transfer reverses the negative effects of immunosuppressive drugs on insulin secretion and cell viability of isolated human islets. Diabetes 54:424–433

Huurman VA, Hilbrands R, Pinkse GG, Gillard P, Duinkerken G, van de Linde P, van der Meer-Prins PM, Versteeg-van der Voort Maarschalk MF, Verbeeck K, Alizadeh BZ, Mathieu C, Gorus FK, Roelen DL, Claas FH, Keymeulen B, Pipeleers DG, Roep BO (2008) Cellular islet autoimmunity associates with clinical outcome of islet cell transplantation. PLoS ONE 3:e2435

Ichii H, Sakuma Y, Pileggi A, Fraker C, Alvarez A, Montelongo J, Szust J, Khan A, Inverardi L, Naziruddin B, Levy MF, Klintmalm GB, Goss JA, Alejandro R, Ricordi C (2007) Shipment of human islets for transplantation. Am J Transplant 7:1010–1020

Ilieva A, Yuan S, Wang RN, Agapitos D, Hill DJ, Rosenberg L (1999) Pancreatic islet cell survival following islet isolation: the role of cellular interactions in the pancreas. J Endocrinol 161:357–364

Iwanaga Y, Sutherland DE, Harmon JV, Papas KK (2008) Pancreas preservation for pancreas and islet transplantation. Curr Opin Organ Transplant 13:445–451

Jindal RM, Ricordi C, Shriver CD (2010) Autologous pancreatic islet transplantation for severe trauma. N Engl J Med 362:1550

Kendall DM, Teuscher AU, Robertson RP (1997) Defective glucagon secretion during sustained hypoglycemia following successful islet allo- and autotransplantation in humans. Diabetes 46:23–27

Khan A, Jindal RM, Shriver C, Guy SR, Vertrees AE, Wang X, Xu X, Szust J, Ricordi C (2012) Remote processing of pancreas can restore normal glucose homeostasis in autologous islet transplantation after traumatic whipple pancreatectomy: technical considerations. Cell Transplant 21:1261–1267

Langlois A, Bietiger W, Mandes K, Maillard E, Belcourt A, Pinget M, Kessler L, Sigrist S (2008) Overexpression of vascular endothelial growth factor in vitro using deferoxamine: a new drug to increase islet vascularization during transplantation. Transplant Proc 40:473–476

Lehmann R, Zuellig RA, Kugelmeier P, Baenninger PB, Moritz W, Perren A, Clavien PA, Weber M, Spinas GA (2007) Superiority of small islets in human islet transplantation. Diabetes 56:594–603

MacGregor RR, Williams SJ, Tong PY, Kover K, Moore WV, Stehno-Bittel L (2006) Small rat islets are superior to large islets in in vitro function and in transplantation outcomes. Am J Physiol Endocrinol Metab 290:E771–E779

Maffi P, Balzano G, Ponzoni M, Nano R, Sordi V, Melzi R, Mercalli A, Scavini M, Esposito A, Peccatori J, Cantarelli E, Messina C, Bernardi M, Del Maschio A, Staudacher C, Doglioni C, Ciceri F, Secchi A, Piemonti L (2013) Autologous pancreatic islet transplantation in human bone marrow 1. Diabetes 62(10):3523–3531

McLoughlin MT, Mitchell RM (2007) Sphincter of Oddi dysfunction and pancreatitis. World J Gastroenterol 13:6333–6343

Merani S, Schur C, Truong W, Knutzen VK, Lakey JR, Anderson CC, Ricordi C, Shapiro AM (2006) Compaction of islets is detrimental to transplant outcome in mice. Transplantation 82:1472–1476

Monti P, Scirpoli M, Maffi P, Ghidoli N, De TF, Bertuzzi F, Piemonti L, Falcone M, Secchi A, Bonifacio E (2008) Islet transplantation in patients with autoimmune diabetes induces homeostatic cytokines that expand autoreactive memory T cells. J Clin Invest 118:1806–1814

Naruse S, Fujiki K, Ishiguro H (2007) Is genetic analysis helpful for diagnosing chronic pancreatitis in its early stage? J Gastroenterol 42(Suppl 17):60–65

Nir T, Melton DA, Dor Y (2007) Recovery from diabetes in mice by β cell regeneration. J Clin Invest 117:2553–2561

Pinkse GG, Bouwman WP, Jiawan-Lalai R, Terpstra OT, Bruijn JA, de Heer E (2006) Integrin signaling via RGD peptides and anti-β1 antibodies confers resistance to apoptosis in islets of Langerhans. Diabetes 55:312–317

Plesner A, Liston P, Tan R, Korneluk RG, Verchere CB (2005) The X-linked inhibitor of apoptosis protein enhances survival of murine islet allografts. Diabetes 54:2533–2540

Pyzdrowski KL, Kendall DM, Halter JB, Nakhleh RE, Sutherland DE, Robertson RP (1992) Preserved insulin secretion and insulin independence in recipients of islet autografts. N Engl J Med 327:220–226

Rabkin JM, Leone JP, Sutherland DE, Ahman A, Reed M, Papalois BE, Wahoff DC (1997) Transcontinental shipping of pancreatic islets for autotransplantation after total pancreatectomy. Pancreas 15:416–419

Rabkin JM, Olyaei AJ, Orloff SL, Geisler SM, Wahoff DC, Hering BJ, Sutherland DE (1999) Distant processing of pancreas islets for autotransplantation following total pancreatectomy. Am J Surg 177:423–427

Riediger H, Adam U, Fischer E, Keck T, Pfeffer F, Hopt UT, Makowiec F (2007) Long-term outcome after resection for chronic pancreatitis in 224 patients. J Gastrointest Surg 11:949–959

Robertson RP (2004) Islet transplantation as a treatment for diabetes – a work in progress. N Engl J Med 350:694–705

Robertson RP, Sutherland DE, Lanz KJ (1999) Normoglycemia and preserved insulin secretory reserve in diabetic patients 10–18 years after pancreas transplantation. Diabetes 48:1737–1740

Robertson RP, Lanz KJ, Sutherland DE, Kendall DM (2001) Prevention of diabetes for up to 13 years by autoislet transplantation after pancreatectomy for chronic pancreatitis. Diabetes 50:47–50

Rodriguez Rilo HL, Ahmad SA, D'Alessio D, Iwanaga Y, Kim J, Choe KA, Moulton JS, Martin J, Pennington LJ, Soldano DA, Biliter J, Martin SP, Ulrich CD, Somogyi L, Welge J, Matthews JB, Lowy AM (2003) Total pancreatectomy and autologous islet cell transplantation as a means to treat severe chronic pancreatitis. J Gastrointest Surg 7:978–989

Rosenberg L, Wang R, Paraskevas S, Maysinger D (1999) Structural and functional changes resulting from islet isolation lead to islet cell death. Surgery 126:393–398

Ruggenenti P, Remuzzi A, Remuzzi G (2008) Decision time for pancreatic islet-cell transplantation. Lancet 371:883–884

Ryan EA, Paty BW, Senior PA, Bigam D, Alfadhli E, Kneteman NM, Lakey JR, Shapiro AM (2005) Five-year follow-up after clinical islet transplantation. Diabetes 54:2060–2069

Sakata N, Goto M, Gumpei Y, Mizuma M, Motoi F, Satomi S, Unno M (2012) Intraoperative ultrasound examination is useful for monitoring transplanted islets: a case report 1. Islets 4:339–342

Shapiro AM, Ricordi C, Hering BJ, Auchincloss H, Lindblad R, Robertson RP, Secchi A, Brendel MD, Berney T, Brennan DC, Cagliero E, Alejandro R, Ryan EA, DiMercurio B, Morel P, Polonsky KS, Reems JA, Bretzel RG, Bertuzzi F, Froud T, Kandaswamy R, Sutherland DE, Eisenbarth G, Segal M, Preiksaitis J, Korbutt GS, Barton FB, Viviano L, Seyfert-Margolis V, Bluestone J, Lakey JR (2006) International trial of the Edmonton protocol for islet transplantation. N Engl J Med 355:1318–1330

Slezak LA, Andersen DK (2001) Pancreatic resection: effects on glucose metabolism. World J Surg 25:452–460

Steer ML, Waxman I, Freedman S (1995) Chronic pancreatitis. N Engl J Med 332:1482–1490

Sutherland DE, Matas AJ, Najarian JS (1978) Pancreatic islet cell transplantation. Surg Clin North Am 58:365–382

Sutherland DE, Gruessner AC, Carlson AM, Blondet JJ, Balamurugan AN, Reigstad KF, Beilman GJ, Bellin MD, Hering BJ (2008) Islet autotransplant outcomes after total pancreatectomy: a contrast to islet allograft outcomes. Transplantation 86:1799–1802

Sutherland DE, Radosevich DM, Bellin MD, Hering BJ, Beilman GJ, Dunn TB, Chinnakotla S, Vickers SM, Bland B, Balamurugan AN, Freeman ML, Pruett TL (2012) Total pancreatectomy and islet autotransplantation for chronic pancreatitis 1. J Am Coll Surg 214:409–424

Wahoff DC, Papalois BE, Najarian JS, Kendall DM, Farney AC, Leone JP, Jessurun J, Dunn DL, Robertson RP, Sutherland DE (1995) Autologous islet transplantation to prevent diabetes after pancreatic resection. Ann Surg 222:562–575

Webb MA, Illouz SC, Pollard CA, Musto PP, Berry D, Dennison AR (2006) Long-term maintenance of graft function after islet autotransplantation of less than 1000 IEQ/kg. Pancreas 33:433–434

Webb MA, Illouz SC, Pollard CA, Gregory R, Mayberry JF, Tordoff SG, Bone M, Cordle CJ, Berry DP, Nicholson ML, Musto PP, Dennison AR (2008) Islet auto transplantation following total pancreatectomy: a long-term assessment of graft function. Pancreas 37:282–287

White SA, Robertson GS, London NJ, Dennison AR (2000) Human islet autotransplantation to prevent diabetes after pancreas resection. Dig Surg 17:439–450

Wurm M, Lubei V, Caronna M, Hermann M, Margreiter R, Hengster P (2007) Development of a novel perfused rotary cell culture system. Tissue Eng 13:2761–2768

Yang Z, Chen M, Deng S, Ellett JD, Wu R, Langman L, Carter JD, Fialkow LB, Markmann J, Nadler JL, Brayman K (2004) Assessment of human pancreatic islets after long distance transportation. Transplant Proc 36:1532–1533

Successes and Disappointments with Clinical Islet Transplantation

44

Paolo Cravedi, Piero Ruggenenti, and Giuseppe Remuzzi

Contents

Introduction	1246
The Burden of Type 1 Diabetes Mellitus	1247
Pathophysiology of Type 1 Diabetes Mellitus	1247
Standard Management of Patients with Type 1 Diabetes	1249
Who May Benefit from Islet Transplantation?	1250
Islet Transplantation: A Historical Perspective	1250
Clinical Outcomes of Islet Transplantation	1251
Insulin Independence and Improved Glycemic Control	1251
Reduction of Hypoglycemic Episodes	1254
Long-Term Diabetic Complications	1254
Adverse Events in Islet Transplantation	1256
Immunosuppressive Regimens for Islet Transplantation	1257
Autologous Islet Transplantation	1259
Cost-Effectiveness of Islet Transplantation	1261
Future Developments	1262
New Sources of Islets	1262
Improving the Transplant Procedure	1263
Induction of Immune Tolerance	1265
Novel Therapeutic Perspectives for Type 1 Diabetes Mellitus	1265
Conclusion	1267
Cross-References	1267
References	1267

P. Cravedi (✉)
IRCCS – Istituto di Ricerche Farmacologiche Mario Negri, Bergamo, Italy
e-mail: paolo.cravedi@mssm.edu

P. Ruggenenti • G. Remuzzi
IRCCS – Istituto di Ricerche Farmacologiche Mario Negri, Bergamo, Italy

Unit of Nephrology, Azienda Ospedaliera Papa Giovanni XXIII, Bergamo, Italy
e-mail: pruggenenti@hpg23.it; gremuzzi@hpg23.it

M.S. Islam (ed.), *Islets of Langerhans*, DOI 10.1007/978-94-007-6686-0_23,
© Springer Science+Business Media Dordrecht 2015

Abstract

Islet transplantation is considered a therapeutic option for patients with type 1 diabetes who have life-threatening hypoglycemic episodes. After the procedure, the frequency and severity of hypoglycemic episodes generally decrease and the majority of patients have sustained graft function as indicated by detectable levels of C-peptide. However, true insulin independence is seldom achieved and generally not long-lasting. Apart from the low insulin-independence rates, reasons for concern regarding this procedure are the side effects of the immunosuppressive therapy, alloimmunization, and the high costs. Moreover, whether islet transplantation prevents the progression of diabetic micro- and macrovascular complications more effectively than standard insulin therapy is largely unknown. Areas of current research include the development of less toxic immunosuppressive regimens, the control of the inflammatory reaction immediately after transplantation, the identification of the optimal anatomical site for islet infusion, and the possibility to encapsulate transplanted islets to protect them from the alloimmune response. Nowadays, islet transplantation is still an experimental procedure, which is only indicated for a highly selected group of type 1 diabetic patients with life-threatening hypoglycemic episodes.

Keywords

Islet transplantation • Type 1 diabetes • Immunosuppression • Diabetic complications

Introduction

Type 1 diabetes is the clinical consequence of immune-mediated destruction of insulin-producing pancreatic β cells. Since from its first description, the main goal of treatment has been the identification of strategies to replace insulin deficiency by exogenous insulin and, later on, pancreas transplantation. More recently, islet transplantation has been proposed as an alternative, supposedly safer way to replace β-cell function compared to whole pancreas transplantation. However, despite early excitement and optimism, results have been unsatisfactory for many years.

In 2000, research in the field of pancreatic islet cell transplantation was boosted by a key paper reporting insulin independence in seven out of seven patients with type 1 diabetes mellitus over a median follow-up of 12 months (Shapiro et al. 2000). The two major novelties of this protocol were the administration of pancreatic islet doses higher than those previously used and a steroid-free immunosuppression. Until then, clinical outcomes had been disappointing. Of the 267 islet preparations transplanted since 1990, less than 10 % had resulted in insulin independence for more than 1 year (Brendel et al. 1999). With the new protocol, success rates have increased in parallel with significant improvements in the technical procedure and medical management of islet transplantation. However, true insulin-independence rates for a prolonged period of time are still very low, and patients are required to take immunosuppressive medication as long as there is evidence of residual graft function. Moreover, islet transplantation remains a highly complex procedure that typically requires the use of at least two donor pancreases and may compete with the number of organs available for whole organ transplantation. Thus, islet transplantation is still far from representing an effective and widely available cure for type 1 diabetes. This review describes the successes and disappointments of clinical islet transplantation programs.

The Burden of Type 1 Diabetes Mellitus

Type 1 diabetes is the most common metabolic disease in childhood with incidence rates ranging from 8 to >50 per 100,000 population per year in Western countries (Daneman 2009). For children aged 0–14 years, the prevalence of type 1 diabetes is estimated to be at least 1 million worldwide by the year 2025 (Green 2008). Onkamo et al. reported that the global incidence of type 1 diabetes is increasing by 3 % per year (Onkamo et al. 1999) and the Project Group estimated a global annual increase in incidence of 2.8 %, from 1990 to 1999 (DIAMOND Project Group 2006).

One of the most accredited theories to explain the increase in type 1 diabetes incidence is the hygiene hypothesis (Strachan 1989), suggesting that exposure to a variety of infectious agents during early childhood might protect against autoimmune diseases, including type 1 diabetes (Kolb and Elliott 1994). Consistently, the constant increase in the incidence of type 1 diabetes reported by a variety of epidemiological studies (Gale 2005; Soltesz et al. 2007) has been paralleled by a gradual decrease in infectious diseases, such as tuberculosis, mumps, measles, hepatitis A, and enteroviruses (Coppieters et al. 2012). Other factors, however, are probably involved in the increased type 1 diabetes incidence (reviewed in Egro (2013))

Children with type 1 diabetes usually present with a several day history of symptoms such as frequent urination, excessive thirst, and weight loss, which appear when about 80 % of the pancreatic β cells have already been lost. If those symptoms are misinterpreted, progressive insulin deficiency leads to a potentially life-threatening condition in the form of diabetic ketoacidosis. Patients with type 1 diabetes require daily subcutaneous injections of insulin in an effort to mimic the physiological release of insulin during meals and during fasting periods. Optimal glycemic control is crucial to reduce the incidence and slow the progression of microvascular and macrovascular complications (The Diabetes Control and Complications Trial Research Group 1993; Nathan et al. 2005).

Pathophysiology of Type 1 Diabetes Mellitus

Most people with type 1 diabetes do not have a family history of the disease; nonetheless, there is clearly a genetic predisposition for developing β-cell autoimmunity (Harjutsalo et al. 2006; Rich 1990; Tuomilehto et al. 1995) with first-degree relatives having a lifetime risk of developing type 1 diabetes of 6 % versus 0.4 % in the general population. In addition, there is a strong, approximately 50 %, concordance in identical twins for the development of type 1 diabetes (Kyvik et al. 1995; Redondo et al. 1999). Human leukocyte antigen (HLA)-related immunogenotype accounts for approximately 60 % of the genetic influence in type 1 diabetes. In individuals who are genetically at risk, an environmental trigger is thought to initiate an immune response targeting the insulin-secreting pancreatic islet β cells (Bluestone et al. 2010; Pugliese 2013). The initial immune response may initiate secondary and tertiary immune responses that contribute to further impair β-cell function and to their destruction (Peakman 2013).

The initial manifestation of β-cell injury is the appearance of diabetes-related autoantibodies. Whether they exert a direct pathogenic role or they are just markers of such injury is still debated. Regardless, with progressive impairment of β-cell function, metabolic abnormalities become measurable, initially either as loss of early insulin response to intravenous glucose (Chase et al. 2001) or as reduced β-cell sensitivity to glucose resulting in decreased insulin secretion (Ferrannini et al. 2010) and eventually as impaired glucose tolerance and hyperglycemia (Sosenko et al. 2009).

At the point of development of overt diabetes, there is still evidence of persistent β-cell function shown through measurement of C-peptide. After diagnosis of type 1 diabetes, however, there is a progressive decline in C-peptide as β-cell function decays (Greenbaum et al. 2012). Nonetheless, even many years after diagnosis, some patients with type 1 diabetes may have low detectable levels of C-peptide (Keenan et al. 2010).

In a prospective study, Ziegler et al. (2013) showed that among 585 children who developed two or more diabetes-related autoantibodies, nearly 70 % (280 of 401 available for follow-up) had developed type 1 diabetes within 10 years, and 84 % (299 of 355 available for follow-up) had developed type 1 diabetes within 15 years. Because the participants were recruited from both the general population and offspring of parents with type 1 diabetes, the similar findings take on added significance, suggesting that the same sequence of events occurs in individuals with so-called sporadic type 1 diabetes and in relatives of individuals with type 1 diabetes.

These data offer new opportunities for type 1 diabetes prevention in patients with autoantibodies. Importantly, evidence has been generated that β cells can regenerate, suggesting that even secondary prevention of type 1 diabetes could be possible. Importantly, evidence has been provided that mature β cells can proliferate not only in the presence of euglycemia (Dor et al. 2004) but also in the diabetic "milieu." Using genetically modified mice made diabetic by inducing 80 % β-cell ablation via endogenous diphtheria toxin production, Nir et al. (2007) showed that β-cell mass and glucose homeostasis can be fully restored in a few weeks after exhaustion of toxin production. Finding that this was accompanied by a dramatically increased proliferation rate even in face of severe hyperglycemia challenged the common belief that glucotoxicity is a major impediment to β-cell survival. Thus, provided the autoimmune process is inhibited, the possibility to induce surviving cells to proliferate and replenish the β-cell compartment might allow restoring normal glucose homeostasis and achieving freedom from exogenous insulin dependency for millions of type 1 diabetics worldwide.

Standard Management of Patients with Type 1 Diabetes

Glycemic control is the keystone of treatment for type 1 diabetes, which requires a meticulous balance of insulin replacement with diet and exercise. In 1993, the Diabetes Control and Complications Trial (DCCT) showed that a system of intensive diabetes management aimed at near-normal glycemic control dramatically reduces the risk of microvascular complications and favorably affects the risk of macrovascular complications compared to a less strict control approach (The Diabetes Control and Complications Trial Research Group 1993; Nathan et al. 2005; Lachin et al. 2008). Over a median follow-up period of 22 years, extension studies of the DCCT trial showed that intensive diabetes therapy significantly prevented reduction in glomerular filtration rate versus conventional diabetes therapy (de Boer et al. 2011). Unfortunately, the treatment regimens used by subjects randomized to the intensive treatment arm of the DCCT also significantly increased their risk of severe hypoglycemia and led to more weight gain (The Diabetes Control and Complications Trial Research Group 1993). Since the publication of the DCCT results, a variety of insulin analogues, better and more sophisticated insulin pumps, and faster and more accurate glucose meters have become widely used in the treatment of type 1 diabetes, making the prospects for patients with type 1 diabetes far better than they were in the past (Nordwall et al. 2004; Nathan et al. 2009).

In most type 1 diabetes patients, however, the goal of near normalization of glycated hemoglobin (HbA1c, a parameter of glucose control over the last 3 months) remains elusive. Several large, multicenter studies demonstrated a persistent gap between attained and target HbA1c levels. Successful implementation of intensive diabetes management in routine clinical practice continues to be a major challenge. The unremitting daily task of controlling blood glucose while avoiding hypoglycemia is arduous and often frustrating. A meta-analysis show that the use of insulin analogues and pump therapy, when compared with conventional insulins and injection-based regimens, respectively, has had only a modest impact on glycemic control and rates of adverse events (Yeh et al. 2012).

Glycemic control is particularly challenging in adolescent patients. In the DCCT the mean HbA1c for adolescents as compared to adults was 1–2 % higher in both the intensive and conventionally treated arms. Despite this, rates of hypoglycemia were higher in adolescents than in adults (Pescovitz et al. 2009). Studies published after DCCT have shown that mean levels of HbA1c have remained higher than current glycemic goals (Danne et al. 2001). Management of type 1 diabetes requires many life-long daily tasks that the child and/or family must perform to maintain a relatively healthy metabolism and glycemic control. Although in younger children these tasks are performed primarily by the care givers, in the teenage years the burden of diabetes management falls on the adolescents themselves. These patients more than others also require considerable psychosocial support, ongoing education, and guidance from cohesive diabetes team working with each patient to set and achieve individualized treatment goals.

Who May Benefit from Islet Transplantation?

Frequent and severe hypoglycemic events are most common indications for islet transplantation alone. Patients with so-called "brittle" diabetes may have an improvement in quality of life or may even be saved from fatal hypoglycemia when provided with functionally active β cells (Ryan et al. 2006). In addition, islet transplantation may be considered in patients with severe clinical and emotional problems with exogenous insulin therapy (Robertson et al. 2006).

The Edmonton group has proposed two scores to quantify the severity of labile diabetes. The HYPO score quantifies the extent of the problem of hypoglycemia by assigning scores to capillary glucose readings from a four-week observation period in combination with a score for self-reported hypoglycemic episodes in the previous year. The lability index (LI) quantifies the extent of glucose excursions over time and is calculated using the formula as described by this group (Ryan et al. 2004a).

The American Diabetes Association acknowledges the potential advantages of islet transplantation over whole pancreas transplantation in terms of morbidity and mortality associated with the operative procedure. However, they clearly state that islet transplantation is still an experimental procedure, only to be performed in the setting of controlled research studies. As for patients who will also be receiving a kidney transplantation, simultaneous pancreas transplantation is the treatment of choice, because it may improve kidney survival and will provide insulin independence in the majority of patients (Robertson et al. 2006). Islet after kidney transplantation is restricted to selected patients with end-stage renal disease affected by type 1 diabetes who underwent kidney transplantation alone or who rejected the pancreas after simultaneous pancreas-kidney transplantation.

Islet Transplantation: A Historical Perspective

The first evidence that islet transplantation might be considered a cure for type 1 diabetes emerged in 1972, when experiments in rodents showed that artificially induced diabetes mellitus could be reversed by transplanted pancreatic islets (Ballinger and Lacy 1972). In 1977, Paul Lacy discussed the feasibility of islet transplantation to treat type 1 diabetes in humans (Lacy 1978), which was subsequently attempted in patients. Success rates, however, were generally low, with less than 10 % of patients being insulin independent at 1 year after transplantation (Sutherland 1981). More encouraging results were obtained in patients who had already had a kidney transplant, with higher rates of insulin independency and graft function as defined by C-peptide secretion (Secchi et al. 1997; Benhamou et al. 2001). In 2000, a report was published describing seven type 1 diabetes patients with a history of severe hypoglycemia and poor metabolic control who underwent islet transplantation alone using a modified, steroid-free immunosuppressive protocol. In addition, each patient received at least two different islet transplantations; thus, the total transplanted islet mass per patient was remarkably higher than in previous series. Over a median follow-up of 11.9 months (range 4.4–14.9), all patients were insulin-independent (Shapiro et al. 2000). The so-called Edmonton protocol was subsequently adopted and modified by many centers. Results of a large multicenter trial using the Edmonton protocol were published in 2006 (Shapiro et al. 2006). Seventy percent of patients had an improved glycemic control after 2 years, but the insulin-independence rate was disappointingly low (14 %).

Clinical Outcomes of Islet Transplantation

Insulin Independence and Improved Glycemic Control

Many centers have published results obtained in their islet transplant programs (Hirshberg et al. 2003; Frank et al. 2004; Goss et al. 2004; Hering et al. 2004, 2005; Froud et al. 2005; Hafiz et al. 2005; Warnock et al. 2005; Keymeulen et al. 2006; O'Connell et al. 2006; Maffi et al. 2007; Gangemi et al. 2008; Gillard et al. 2008). Here, we present some of the largest reports from diverse geographic regions.

In 2005, single-center outcomes of 65 islet transplant recipients treated according to the Edmonton protocol were reported, showing that 44 (68 %) patients had become insulin independent, with a median duration of insulin independency of 15 months (IQR 6.2–25.5). Five of these subjects received only a single islet infusion, 33 received two infusions, and six received three infusions. Insulin independency after 5 years was 10 %. Nonetheless, after 5 years, some residual graft function could be demonstrated in about 80 % of patients on the basis of detectable serum C-peptide levels. Diabetic lability and the occurrence of severe hypoglycemia were effectively diminished (Ryan et al. 2005).

Following the initial Edmonton results in 2000, a large international trial in nine centers in the United States and Europe was initiated by the Immune Tolerance Network to examine the feasibility and reproducibility of islet transplantation using the Edmonton protocol. The primary end point, defined as insulin independency with adequate glycemic control 1 year after the final transplantation, was met by 16 out of 36 subjects (44 %). Only five of these patients were still insulin independent after 2 years (14 %). Of note, the considerable differences in results obtained by the various participating sites emphasize the need for concentration of this procedure in highly experienced centers. Again, graft function as defined by detectable C-peptide levels and associated improvements in diabetic control was preserved in a higher percentage of patients (70 % after 2 years) (Shapiro et al. 2006).

The Groupe de Recherche Rhin Rhone Alpes Geneve pour la transplantation d'Ilots de Langerhans (GRAGIL) reported results obtained in 10 patients who received one or two islet infusions. Only three out of ten patients had prolonged insulin independence after 1 year of follow-up. However, five more transplantations were considered successful, since after 1 year recipients fulfilled the predefined criteria of success consisting of a basal C-peptide ≥ 0.5 ng/ml, HbA1c ≤ 6.5 %, disappearance of hypoglycemic events, and ≥ 30 % reduction of insulin needs (Badet et al. 2007).

A recent report from the Japanese Trial of Islet Transplantation showed that only 3 out of 18 recipients of islet transplantation achieved insulin independency and only for a period of 2 weeks to 6 months. Graft function was preserved in 63 % after 2 years. As in the other reports, HbA1c levels decreased and blood glucose levels stabilized, with disappearance of hypoglycemia unawareness. In this report, no information was provided about the amount of islet equivalents (IEQ; number of islets in a preparation adjusted for size of the islet, one IEQ equals a single islet of 150 µm in diameter) per kg body weight infused. Of note, in Japan all pancreata are obtained from non-heart-beating donors, since pancreata from brain-dead donors are usually allocated to whole pancreas or pancreas-kidney transplantation. In addition, the presence of brain death is frequently not examined because of cultural reasons, and invasive procedures are usually not allowed even in brain-dead donors before cardiac arrest occurs. This may lead to decreased viability of pancreatic tissue when compared with pancreata from brain-dead donors (Kenmochi et al. 2009).

The largest registry of islet transplant data is the Collaborative Islet Transplant Registry (CITR), which retrieves its data mainly from US and Canadian medical institutions and two European centers. In their 2008 update considering 279 recipients of an islet transplantation reported between 1999 and 2007, the registry reported 24 % insulin independence after 3 years. Graft function as defined by detectable C-peptide levels after 3 years was 23–26 %. The prevalence of hypoglycemic events decreased dramatically, and mean HbA1c levels substantially improved. Predictors of better islet graft function were higher number, size, and viability of infused islets; older age and lower HbA1c levels in the recipient, whether the processing center was affiliated with the transplantation center; and the use of daclizumab, etanercept, or calcineurin inhibitors in the immunosuppressive regimens. In-hospital administration of steroids was associated with a negative outcome (Alejandro et al. 2008; Collaborative Islet Transplantation 2010).

Recently, the CITR reported the outcome data of 627 islet transplants performed between 1999 and 2010. Insulin independence at 3 years after transplant improved from 27 % in patients transplanted between 1999 and 2002 to 44 % in patients who received a transplant between 2007 and 2010 (Fig. 1). This success, however, was probably more the result of a careful selection of recipients (lower serum creatinine and donor-specific antibody titer) than of a real improvement in the transplant procedure. Consistently, the number of islet transplant performed each year did decline during the last era (Barton et al. 2012).

Fig. 1 Rate of insulin independence at different time points after islet infusion according to the transplant era (Data derived from Collaborative Islet Transplantation Registry (CITR) 2012, Barton et al. (2012))

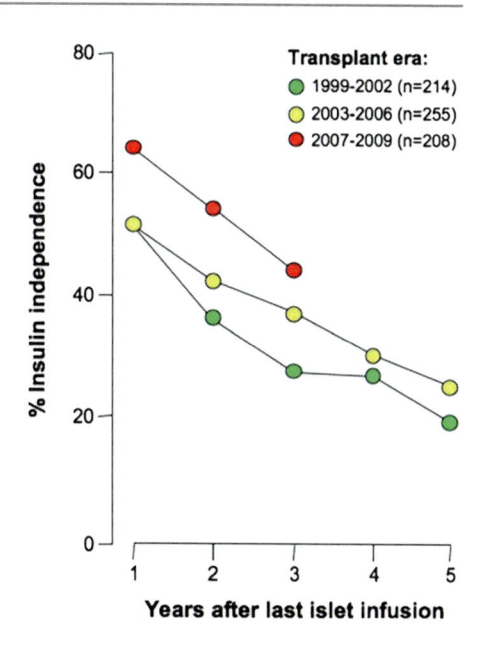

Table 1 Clinical outcomes of whole pancreas transplantation versus islet transplantation

Insulin independence	Whole pancreas	Islets
At 1 year	77 %	47 %
Long term	58 % (5 years)	24 % (3 years)

Data were derived from the International Pancreas Transplant Registry (until June 2004, $n = 1,008$ pancreas transplantation alone) and from the Collaborative Islet Transplantation Registry (until January 2008, $n = 279$ islet transplantation alone) (Alejandro et al. 2008; Gruessner and Sutherland 2005)

Table 1 shows success rates for pancreatic islet transplantation compared with whole pancreas transplantation alone as reported by the Collaborative Islet Transplantation Registry and the International Pancreas Transplant Registry, respectively (Collaborative Islet Transplantation 2010; Gruessner and Sutherland 2005). Indications for pancreas transplantation alone are similar to those for islet transplantation. However, whole pancreas transplantation is more frequently performed simultaneously with kidney transplantation or after kidney transplantation in type 1 diabetic patients with end-stage renal disease. For simultaneous whole pancreas-kidney transplantation, favorable effects on micro- and possibly macrovascular diabetic complications have consistently been described (White et al. 2009). For pancreas-after-kidney and for pancreas transplantation alone, data are less consistent, and mild or no benefits or even worsening of patient survival after these procedure have been reported compared to insulin-treated patients (Venstrom et al. 2003; Gruessner et al. 2005).

Reduction of Hypoglycemic Episodes

Intrahepatic islet transplantation restores physiologic islet cell hormonal responses to insulin-induced hypoglycemia in type 1 diabetes patients whereby endogenous insulin secretion is appropriately suppressed and glucagon secretion is partially restored (Rickels et al. 2005a).

Studies using continuous glucose monitoring systems in islet transplant recipients have shown significant decreases to abolition of time spent in the hypoglycemic range (<60 mg/dL) (Paty et al. 2006). Indeed, in addition to normalizing the glycemic threshold for counter-regulatory glucagon secretion, islet transplant recipients have normalization of the glycemic thresholds for counter-regulatory epinephrine, autonomic symptom, and growth hormone (GH) responses (Fig. 2). A study using paired hyperinsulinemic, hypoglycemic, and euglycemic clamps with stable isotope tracers in islet transplant recipients has preliminarily shown that the recovery of intact islet cell and sympathoadrenal responses is associated with a restored endogenous (primarily hepatic) glucose production response that is ultimately required to protect patients from the development of low blood glucose (Rickels et al. 2011).

Even patients with graft failure after islet transplantation showed significantly fewer hypoglycemic episodes compared with pretransplant (Leitao et al. 2008). Whether this result is due to the reduction in exogenous insulin requirement or to the restoration of glucose counter-regulation is a matter of active investigation (Rickels et al. 2005b).

Long-Term Diabetic Complications

Until now, it has not been sufficiently established whether pancreatic islet transplantation can prevent diabetic complications or halt their progression (Lee et al. 2006; Fiorina et al. 2008).

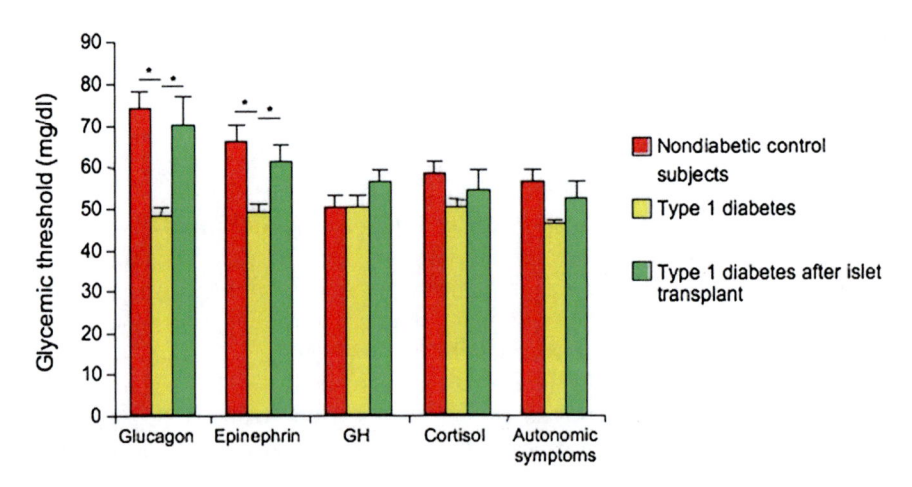

Fig. 2 Glycemic thresholds for counterregulatory responses in patients with type 1 diabetes, in islet transplant recipients, and in healthy controls. *$P < 0.05$ (Data from Rickels et al. 2007)

Cardiovascular Complications– In a retrospective study, cardiovascular function was compared between a group of 17 patients who received an islet-after-kidney transplantation and a group of 25 patients with previous kidney transplantation who were still on the waiting list for islet transplantation or who had experienced early islet graft failure. Baseline characteristics for both groups were similar. Islet transplantation was associated with an improvement in ejection fraction and left ventricular diastolic function compared to baseline. Moreover, arterial intima-media thickness was stable in the islet transplant group but worsened in the kidney-only group (Fiorina et al. 2005). Recently, data have been published showing that in a cohort of 15 consecutive islet transplant recipients who reached insulin independence carotid intima-media thickness did actually decrease after the procedure, suggesting that optimal glycemic control may lead to a regression of atherosclerotic lesions (Danielson et al. 2012).

Nephropathy– Increased kidney graft survival rates and stabilization of microalbuminuria have been reported in kidney transplant recipients with successful islet transplantation (fasting C-peptide levels of >0.5 ng/ml for >1 year) compared to kidney transplant patients with unsuccessful islet transplantation at 4 years after surgery (Fiorina et al. 2003). The effect of islet transplantation on kidney transplant survival compared with insulin therapy is still unknown.

An uncontrolled, observational study by the Edmonton group suggested an overall decline in estimated glomerular filtration rate during 4 years of follow-up after islet transplantation alone and an increase in albuminuria in a significant proportion of patients independently from insulin activity (Senior et al. 2007). Subsequently, Maffi et al. showed that even a mildly decreased renal function, before transplantations should be considered a contraindication for the currently used immunosuppressive regimen of sirolimus in combination with tacrolimus, since it was associated with progression to end-stage renal disease (Maffi et al. 2007). In patients with renal impairment, nephrotoxicity of immunosuppressive drugs like calcineurin and mTOR inhibitors might offset the benefits of improved metabolic control. Renal impairment progressively worsens even in those selected patients with type 1 diabetes who benefit of a 5-year normoglycemia period after a single pancreas transplantation, as a result of both immunosuppressive drug toxicity and, probably, the marginal effect of glycemia control on already damaged diabetic kidney (Fioretto et al. 1993, 1998). The same applies also to the rare subjects with prolonged normoglycemia after islet transplantation, which limits the indications for the procedure to those type 1 diabetes patients with normal renal function. In this cohort, aggressive treatment of risk factors for nephropathy, such as blood pressure and low-density lipoprotein cholesterol, together with careful tacrolimus level monitoring have been associated with preserved renal function after islet transplantation according to a retrospective series of 35 patients (Leitao et al. 2009).

Retinopathy – The Edmonton and the Miami series reported ocular problems posttransplantation in 8.5 % and 15 % of patients, respectively. Adverse events included retinal bleeds, tractional retinal detachment, and central retinal vein occlusion (Hafiz et al. 2005; Ryan et al. 2005). However, after 1–2 years, diabetic retinopathy seems to stabilize (Lee et al. 2005). Moreover, at 1 year after transplantation, arterial and venous retinal blood flow velocities are significantly increased, possibly indicating improved retinal microcirculation (Venturini et al. 2006). The acute adverse effects on retinopathy may be due to the sudden improvement in glycemic control after islet transplantation. The DCCT also reported initial deterioration of diabetic retinopathy in patients with preexisting disease who were treated in the intensive insulin treatment arm as compared to those in the conventional treatment arm; however, after 1 year differences between treatment arms disappeared and after 36 months of follow-up, intensive treatment was consistently associated with significantly less progression of diabetic retinopathy (The Diabetes Control and Complications Trial Research Group 1993). Consistently, a study on 44 patients showed that at 36 months after transplant progression of retinopathy was slower in subjects who received islet transplantation compared to those on intensive insulin therapy (Thompson et al. 2008). More studies are needed to assess whether the overall effect of islet transplantation on diabetic retinopathy is beneficial in the long term.

Neuropathy – Reports on the effect of islet transplantation on diabetic neuropathy suggest that the procedure has only marginal effect on this microvascular complication. Lee et al. performed nerve conduction studies in eight patients with at least 1 year of follow-up after transplantation. They concluded that peripheral neuropathy stabilized or maybe even improved, although no formal statistical analysis was provided and conclusions were based on clinical observations by a single neurologist (Lee et al. 2005). Del Carro et al. compared nerve conduction studies in patients who had received an islet-after-kidney transplantation to patients having received kidney transplantation only. In their interpretation of the results, they suggested that worsening of diabetic neuropathy seemed to be halted by islet transplantation, but no statistically significant differences between the two groups could be demonstrated (Del Carro et al. 2007). A prospective cohort study compared the progression of microvascular complications between 31 patients who received an islet transplant and 11 who remained on the waiting list. Despite an association with lower HbA1c levels and slower progression of retinopathy, islet transplantation did not lead to any benefit in neuropathy (Warnock et al. 2008).

By and large, the potential benefits of the islet transplant procedure have been identified only in the minority of patients who reach prolonged insulin independence. Moreover, lack of comparisons with control groups on insulin therapy prevents any consideration on real superiority of islet transplantation. Therefore, large, multicenter, randomized trials are needed to assess the role of islet transplantation in slowing the progression of diabetic complications over conventional supportive therapy.

Adverse Events in Islet Transplantation

Adverse events related to islet transplantation are principally related to the procedure itself and to the consequences of the immunosuppressive regimen. During the procedure, a large mass of β cells is percutaneously and transhepatically injected into the portal vein. This may lead to portal vein thrombosis or thrombosis of segmental branches. On the other hand, incidence rates of up to 14 % have been reported for intraperitoneal bleeding, which may require blood transfusion or even surgical intervention. This complication can be effectively prevented by sealing the catheter tract using thrombostatic coils and tissue fibrin glue (Villiger et al. 2005). Other relatively frequent procedure-related complications are abdominal pain from puncturation of the peritoneum or gallbladder and a transient rise of hepatic enzymes (Ryan et al. 2004b). Posttransplantation focal hepatic steatosis occurs in approximately 20 % of patients, possibly due to a local paracrine effect of insulin, but its significance with regard to graft function is not clear yet (Markmann et al. 2003; Bhargava et al. 2004).

Type 1 diabetes patients receiving pancreatic islet transplantation may need an additional kidney and/or whole pancreas transplantation later in life. Therefore, posttransplantation alloimmunization in roughly 10–30 % of patients using immunosuppression is a cause for concern (Campbell et al. 2007a; Cardani et al. 2007). Of note, up to 100 % of patients develop HLA alloreactivity, with 71 % having HLA panel-reactive antibodies (PRA) ≥50 %, after withdrawal of immunosuppression because of islet graft failure or side effects (Campbell et al. 2007a; Cardani et al. 2007). Pre- or posttransplantation alloreactivity against HLA class I and II may also be associated with reduced pancreatic islet graft survival itself (Lobo et al. 2005; Campbell et al. 2007b), although some authors suggested that increased PRA had no clinical significance under adequate immunosuppression (Cardani et al. 2007). As opposed to solid organ transplantation, pretransplantation testing of PRA is currently not performed in pancreatic islet transplantation. Thus, the impact of PRA positivity on clinical outcome after islet transplantation or on future whole organ transplantation has to be further investigated.

Immunosuppressive Regimens for Islet Transplantation

Following the publication by the Edmonton group in 2000 (Shapiro et al. 2000), the steroid-free immune-suppressive protocol this group used was adopted by many centers, although it was not the only change being introduced. Changes with regard to recipient and donor selection, the technical procedure, and the infusion of a large number of pancreatic islets from multiple donors will all have contributed to the favorable short-term results. The Edmonton immunosuppressive regimen consists of induction therapy with a monoclonal antibody against the interleukin-2 receptor (daclizumab) and maintenance therapy with a calcineurin inhibitor (tacrolimus) and a mammalian mTOR inhibitor (sirolimus). Sirolimus has been shown to display significant synergy with calcineurin inhibitors, control autoimmunity, induce apoptosis of T cells and other inflammatory cells, and induce generation of regulatory T cells. However, data have also emerged showing its potentially harmful effects on β-cell regeneration (Nir et al. 2007; Berney and Secchi 2009). The same applies for calcineurin inhibitors; although proven to be very effective in organ transplantation, they are toxic to β cells and cause insulin resistance and diabetes mellitus.

Moreover, sirolimus and tacrolimus exert direct nephrotoxic effects and they often induce the development of hyperlipidemia and hypertension, which may further increase the risk of micro- and macrovascular complications (Halloran 2004). Therefore, the combined use of sirolimus and tacrolimus to prevent acute rejection of transplanted pancreatic islets is certainly not ideal.

To increase islet transplantation success rates and diminish the often severe side effects associated with chronic use of immunosuppressive drugs (Hafiz et al. 2005), various centers are implementing new immunosuppressive regimens, both for the induction and for the maintenance phase (Hering et al. 2005; Gillard et al. 2008; Ghofaili et al. 2007; Froud et al. 2008a; Mineo et al. 2009). In an attempt to promote a pro-tolerogenic state, Froud et al. tested induction therapy with alemtuzumab in three islet transplant recipients (Froud et al. 2008a). Alemtuzumab is a humanized monoclonal antibody against CD-52, which is present on the surface of mature lymphocytes. Its administration leads to severe lymphocyte depletion and may favorably influence the regulatory T-cell versus effector T-cell ratio during T-cell repopulation (Weaver and Kirk 2007). Indeed, in these three patients, glucose metabolism seemed to be better than in historic controls, with no major infectious complications. However, other changes in the immunosuppressive regimen, such as the use of steroids on the day before islet infusion, the early switch from tacrolimus to mycophenolate mofetil (MMF) during the maintenance phase, and the use of etanercept (see below) may all have contributed to improved outcomes in this study.

Tumor necrosis factor (TNF)α is a regulator of the immune response, and its activity is inhibited by etanercept, a recombinant TNFα receptor protein. From the University of Minnesota came an interesting report of high success rates in eight patients using a protocol in which etanercept was administered as induction therapy, combined with prednisone, daclizumab, and rabbit antithymocyte globulin. Of the eight patients, five were still insulin independent after 1 year. Of note, patients received an islet graft from a single donor (Hering et al. 2005). More centers are now using etanercept as additional induction therapy, a strategy which is supported by the fact that the CITR found an association between etanercept use and graft survival (Gangemi et al. 2008; Alejandro et al. 2008; Faradji et al. 2008).

Some studies investigated the combination of etanercept induction with long-term use of subcutaneous exenatide, a glucagon-like peptide-1 (GLP-1) analogue. GLP-1 is a hormone derived from the gut, which stimulates insulin secretion, suppresses glucagon secretion, and inhibits gastric emptying (Gentilella et al. 2009). Combined treatment with etanercept and exenatide in addition to the Edmonton immunosuppressive protocol was shown to reduce the number of islets needed to achieve insulin independence (Gangemi et al. 2008). In addition, combined etanercept and exenatide use improved glucose control and graft survival in patients who needed a second transplantation because of progressive graft dysfunction (Faradji et al. 2008). In two studies with islet transplantation patients, exenatide reduced insulin requirements, although in one study they tended to rise again at the end of the 3-month study period, possibly due to exhaustion of β cells (Ghofaili et al. 2007; Froud et al. 2008a, b). However, these studies were very small and nonrandomized. Of note, exenatide use involves the administration of twice-daily subcutaneous injections, causes severe nausea, and may lead to hypoglycemia. Therefore, randomized controlled trials are needed to define whether its use confers additional benefit over immunosuppressive therapy alone in islet transplantation recipients (Rickels and Naji 2009).

An isolated case with more than 11 years of insulin independency after islet transplantation was described in 2009 (Berney et al. 2009). The intriguing question is which factors have contributed to the outcome in this particular patient. The patient had previously received a kidney transplant and was on an immunosuppressive regimen comprising antithymocyte globulin as induction therapy followed by prednisone (which was rapidly tapered), cyclosporine, and azathioprine, which was later switched to MMF. Interestingly, the authors investigated the cellular immune response and found that the patient was hyporesponsive toward donor antigens, possibly as a result of the expanded regulatory T-cell (Treg) pool. This may have contributed to the excellent long-term survival of the graft. Huurman et al. examined cytokine profiles and found that allograft-specific cytokine profiles were skewed toward a Treg phenotype in patients who achieved insulin independence and that expression of the Treg cytokine interleukin-10 was associated with low alloreactivity and superior islet function (Huurman et al. 2009). The role of Tregs in allograft tolerance has long been recognized in solid organ and bone marrow transplantation, and much research is devoted to translating this knowledge into therapeutic options, which may also benefit islet transplantation (Schiopu and Wood 2008).

Despite immunosuppressive therapy aimed at preventing rejection (i.e., alloimmunity), outcomes of islet transplantation may also be adversely influenced by autoimmune injury. A recent study showed delayed graft function in patients with pretransplant cellular autoreactivity to β-cell autoantigens; in 4 out of 10 patients with recurrence of autoreactivity posttransplantation, insulin independence was never achieved. Moreover, in 5 out of 8 patients in whom cellular autoreactivity occurred de novo after transplantation, time to insulin independence was prolonged (Huurman et al. 2008). In the international trial of the Edmonton protocol, patients with one or two autoantibodies in the serum before the final infusion had a significantly lower insulin-independence rate than those without autoantibodies (Shapiro et al. 2006).

Autologous Islet Transplantation

The concept of autologous islet transplantation after pancreatectomy arose at the University of Minnesota in the late 1970s (Sutherland et al. 1978). Pancreatectomy had already been performed as a treatment for patients with chronic, unrelentingly painful pancreatitis (Braasch et al. 1978). However, it was seen as an undesirable method, in part because removal of the gland inevitably causes insulin-dependent diabetes. This drawback led to the concept of not disposing of the resected pancreas but using it as a source of islet tissue that could be used for autologous transplant.

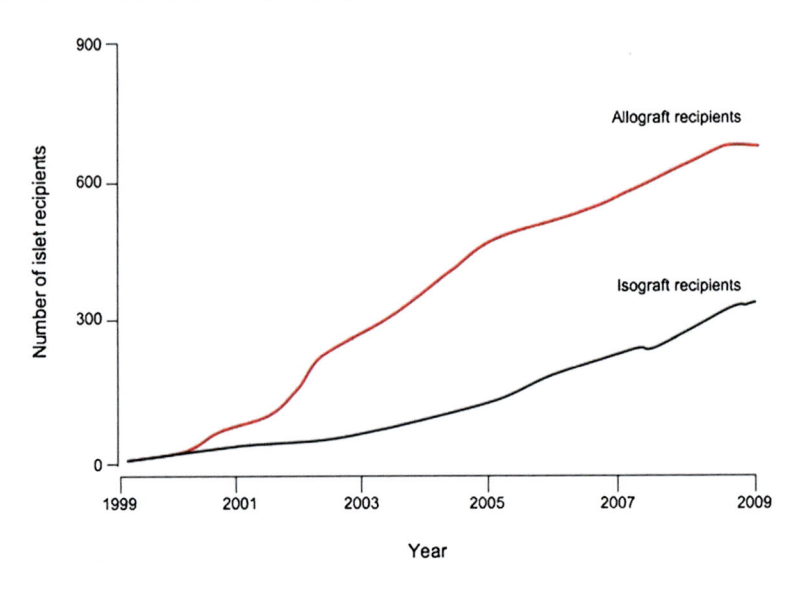

Fig. 3 Number of recipients of allograft and isograft islet transplantation, according to the Collaborative Islet Transplant Registry (CITR), 2011

In contrast to alloimmune islet transplantation in patients with type 1 diabetes mellitus, no immunosuppression is needed for patients receiving autologous islet infusion, due to the autologous source of the islets and lack of preexisting autoimmune reactivity. These considerations boosted the use of islet autotransplantation to prevent diabetes after pancreatectomy performed for chronic pancreatitis (Fig. 3). More recently, encouraging results have been published showing that islet autotransplantation can effectively prevent diabetes after 50–60 % distal partial pancreatectomy for benign pancreatic tumors (Jin et al. 2013).

A report of 85 total pancreatectomy patients from the United Kingdom showed that the group of 50 patients receiving concomitant autologous islet transplantation had a significantly lower median insulin requirement than those without concomitant transplantation, although only five patients remained insulin independent (Garcea et al. 2009). Of 173 recipients of autologous islet transplantations post-pancreatectomy at the University of Minnesota, 55 (32 %) were insulin independent, and 57 (33 %) had partial islet function as defined by the need of only once-daily long-acting insulin at some posttransplant point. Importantly, the rate of decline of insulin independence was remarkably limited, with 46 % insulin independence at 5 years follow-up and 28 % at 10 years (Sutherland et al. 2008). Despite the lower number of islets needed to provide insulin independence in autologous compared to allogeneic transplantation, islet cell mass is an important predictor of success in both transplant procedures (Shapiro et al. 2006; Wahoff et al. 1995). Therefore, improvements in islet yields from fibrotic and inflamed pancreata are expected to further improve outcomes of autologous islet transplantation (Naziruddin et al. 2012).

Cost-Effectiveness of Islet Transplantation

So far, no study has addressed the issue of the cost-effectiveness of islet transplantation in terms of the costs per quality-adjusted life year or per micro- or macrovascular diabetic complication prevented. The GRAGIL network has estimated the average cost of an islet transplantation in the year 2000 at €77,745.00. These costs even slightly exceed those for whole pancreas transplantation, mainly due to the high expenses of cell isolation (Guignard et al. 2004). A study by Frank et al. also found that pancreas processing-related costs led to higher total costs for isolated islet transplantation than for whole pancreas transplantation, even though the former was associated with less procedure-related morbidity and shorter hospital stays (Frank et al. 2004). These high costs may be justified in patients in whom islet transplantation is deemed to be lifesaving because of severe hypoglycemic episodes. However, in other settings, they will compare extremely unfavorably to the costs of current strategies to prevent diabetic complications, such as adequate glycemic control, blood pressure and lipid profile optimization, diet and weight loss, and angiotensin-converting enzyme inhibitor use.

According to the 2000 French National Cost Study (Guignard et al. 2004), the costs of hospitalization for pancreas transplantation (DRG 279) were €25,674. The processing of the 5.6 pancreata used for a single islet transplantation costs €23,755; with the hospitalization, the total costs are about €34,178. Since about 80 % of patients are still insulin independent at 3 years after a pancreas transplantation, compared to only 44 % of those who receive an islet transplantation (Barton et al. 2012; Boggi et al. 2012), it can be estimated that one patient achieving insulin independence at 3 year costs to health-care providers €29,175 or €77,677 if this target is achieved after pancreas or islet transplantation, respectively. Thus, achieving long-lasting insulin independency is almost threefold more expensive by islet than by pancreas transplantation. Based on these data, pancreas transplantation is likely to be remarkably more cost-effective compared to islet transplantation, even taking into account the higher risk of surgical complications related to whole organ grafting.

A recent cost-effectiveness exploratory analysis comparing islet transplantation with standard insulin therapy in type 1 diabetes adult patients affected by hypoglycemia unawareness showed that islet transplantation becomes cost saving at about 9–10 years after the procedure (Beckwith et al. 2012). However, since the vast majority of islet transplant recipients loose islet independence within 5 years after transplantation, this result in fact means that islet transplantation is never cost-effective compared to standard insulin therapy. Therefore, despite the absence of ad hoc studies formally addressing the issue of the cost-effectiveness of islet transplantation compared to other treatments for type 1 diabetes, available data suggest that such procedure is hardly superior to pancreas transplantation or insulin therapy.

A recent study showed that islet cell autotransplantation after total pancreatectomy for chronic pancreatitis resulted in improved survival over total pancreatectomy alone. Importantly, the cost of pancreatectomy plus islet cell autotransplantation with attendant admission and analgesia costs over the 16-year survival period was £110,445 compared with £101,608 estimated 16-year costs if no pancreatectomy was undertaken (Garcea et al. 2013). This suggests that islet autotransplantation after pancreatectomy might be a cost-effective strategy to treat chronic pancreatitis.

Future Developments

New Sources of Islets

Islet cell harvest from whole pancreas remains a limiting step, as the efficiency of the procedure and cell viability postharvest is relatively poor. There are several steps in the whole procedure of islet transplantation which may be targeted in order to improve islet recovery and posttransplantation protection. Pretransplantation procedures related to pancreas preservation, enzymatic digestion, purification, culture, and shipment may be further refined (Ichii and Ricordi 2009). While many laboratories are developing methods to improve these processes, given the severe shortage of donor pancreases, other investigators are exploring alternative sources of β cells.

Other sources of pancreatic β cells are mesenchymal stem cells (MSCs) that according to some studies might possibly display the capability to transdifferentiate into insulin-producing cells (Porat and Dor 2007; Claiborn and Stoffers 2008). Of even more importance are the immunomodulatory and anti-inflammatory properties of MSCs, which might control the autoimmune response preventing immune injury of newly proliferating cells (English 2012). However, before clinical use, substantial issues should be addressed regarding the safety, function, mode of isolation, and experimental handling of MSCs (Bianco et al. 2013).

Alternative sources for β-cell replacement include human embryonic stem cells (hESCs). Recently, it has been shown that the small molecule (-)-indolactam V induces differentiation of hESCs into pancreatic progenitor cells in vitro (Chen et al. 2009). The more plentiful pancreatic ductal cells isolated from human donor pancreases can be transdifferentiated into β cells (Bonner-Weir et al. 2000). Similarly, mouse experiments have shown that bile duct epithelial cells (Nagaya et al. 2009), acinar cells (Zhou et al. 2008), and hepatic cells (Aviv et al. 2009) can also be transdifferentiated into β cells. The differentiation of human fibroblast-derived induced pluripotent stem cells (iPSCs) into β cells provides another alternative that is particularly appealing due to its potential avoidance of allogeneic rejection (Tateishi et al. 2008).

Another possible source of pancreatic islets is xenotransplantation, with which some experience has been gained in humans. Six groups have independently reported that pig islets transplanted into nonhuman primates can maintain normoglycemia for periods in excess of 6 months, and more positive results are expected by genetic manipulation of transplanted islets (van der Windt et al. 2012).

In 1994, a Swedish group reported xenotransplantation with fetal porcine pancreatic islets in 10 diabetic patients. Although insulin requirements did not decrease, the procedure was well tolerated and there was no evidence of transmission of porcine endogenous retroviruses after 4–7 years of follow-up (Groth et al. 1994; Heneine et al. 1998). More recently, xenotransplantation has been performed in China, Russia, and Mexico (Valdes-Gonzalez et al. 2005; Wang 2007). In 2005, the group from Mexico reported a 4-year follow-up of 12 diabetic patients not taking immunosuppressive therapy who had received one to three subcutaneous implantations of a device containing porcine pancreatic islets and Sertoli cells. Sertoli cells, being immune-privileged, were added because they may confer immunoprotection to transplanted endocrine tissue. Follow-up showed a decreased insulin requirement in 50 % of patients, compared to pretransplant levels and non-transplanted controls. Porcine C-peptide was not detectable in the urine, and the significance of this study remains to be determined. Importantly, severe ethical issues have been raised with regard to xenotransplantation as it is currently being performed. The program in China was suspended, and the International Xenotransplantation Association has seriously objected to the Mexican and Russian studies, as they feel that the safety of the patient and of the general public (especially with regard to the spread of porcine endogenous retroviruses) is not sufficiently guaranteed (Sykes et al. 2006, 2007; Groth 2007). More experimental studies are needed before clinical trials in human can be initiated (Rajotte 2008).

In 2005, Matsumoto et al. performed the first islet transplantation from a living related donor in a patient who had brittle diabetes due to chronic pancreatitis. The procedure resulted in good glycemic control and no major complications in both the donor and the recipient (Matsumoto et al. 2005, 2006). However, results cannot be generalized to the type 1 diabetes population, as diabetic disease in the recipient did not result from an autoimmune process. Moreover, partial pancreatectomy in the donor implies major surgery with associated risks of morbidity and mortality. In the long term, donors may be at increased risk of developing diabetes mellitus themselves (Hirshberg 2006).

Improving the Transplant Procedure

Efficacy of islet transplant procedure is hampered by massive cell loss shortly after infusion because of an inflammatory reaction termed instant blood-mediated inflammatory reaction (IBMIR). This reaction involves activation of the complement and coagulation cascades, ultimately resulting in clot formation and infiltration of leukocytes into the islets, which leads to disruption of islet integrity and islet destruction (Fig. 4). Different strategies have been proposed to prevent this phenomenon. Since heparin can prevent clotting and decrease complement activation, peri-transplant heparinization of either the patient or, to prevent bleeding complications, the pancreatic islets themselves has been proposed as a strategy to improve outcomes (Johansson et al. 2005; Cabric et al. 2007; Koh et al. 2010). This may prevent the immediate and significant post-procedural islet loss. Moreover, it is now possible to visualize islets in the peri-transplantation phase using 18 - F-fluorodeoxyglucose positron-emission tomography combined with computed

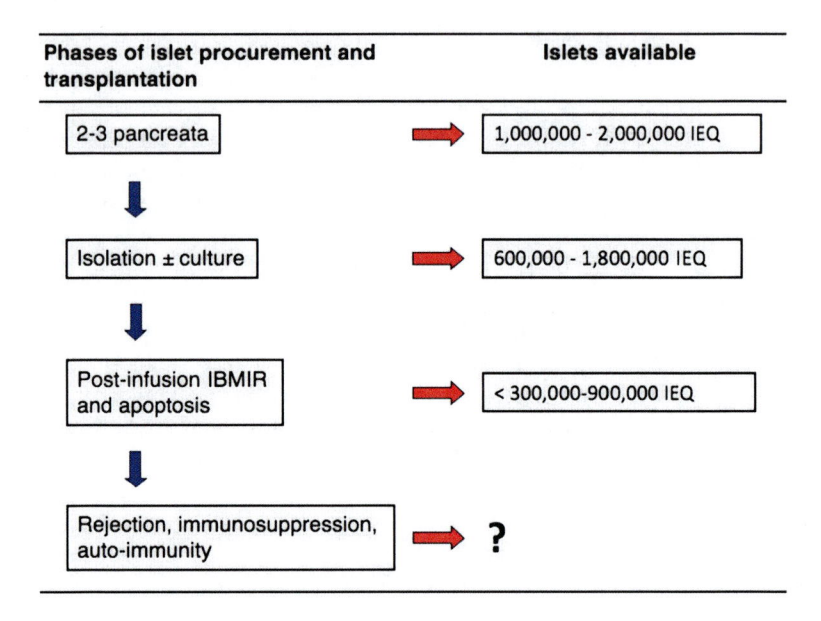

Phases of islet procurement and transplantation	Islets available
2-3 pancreata	1,000,000 - 2,000,000 IEQ
Isolation ± culture	600,000 - 1,800,000 IEQ
Post-infusion IBMIR and apoptosis	< 300,000-900,000 IEQ
Rejection, immunosuppression, auto-immunity	?

Fig. 4 Loss of pancreatic islet mass, from graft preparation to post-infusion degradation. *IEQ* islet equivalents, *IBMIR* instant blood-mediated inflammatory reaction. The IBMIR reduces islet mass by 50–70 % (Korsgren et al. 2005)

tomography in order to assess islet survival and distribution, which may also be used to evaluate alternative sites of implantation (Eich et al. 2007).

For the transplantation procedure itself, it has been recognized that the liver is not the ideal site for transplantation because of the procedure-related complications, the relatively low oxygen supply in the liver, the exposure to toxins absorbed from the gastrointestinal tract, and the IBMIR, which causes substantial islet loss shortly after infusion. Many alternative sites have been explored, including the omentum, pancreas, gastrointestinal tract, and muscular tissue. More recently, the bone marrow has been proposed as an alternative site of islet injection (Maffi et al. 2013). However, these alternative approaches have so far remained experimental, with none of them being convincingly superior to the currently used method (Merani et al. 2008; van der Windt et al. 2008).

Islet encapsulation as a strategy to improve graft survival is one of the main areas in experimental research. The use of semipermeable encapsulation material such as alginate gel or membrane devices should protect the islets against the alloimmune response while at the same time allowing them to sense glucose levels and secrete insulin (Figliuzzi et al. 2006; Beck et al. 2007). This technique must allow adequate diffusion of oxygen and nutrients to maintain islet viability and function but, at the same time, must be selective enough to prevent the permeation of host immune proteins. Islet encapsulation in alginate gel and in polysulfone hollow fibers allows adequate transport of nutrients to maintain islet function and viability, making the use of these immunoisolation strategies for transplantation a potentially important field of investigation to transplant also xenogeneic islets (Cornolti et al. 2009; O'Sullivan et al. 2011).

Recently, four patients received an intraperitoneal infusion of encapsulated allogeneic islets. Despite no immunosuppression, all patients turned positive for serum C-peptide response, both in basal and after stimulation, and anti-MHC class I–II and GAD65 antibodies all tested negative at 3 years after transplant. Daily mean blood glucose, as well as HbA1c levels, significantly improved after transplant, with daily exogenous insulin consumption declining in all cases, but with full insulin independence reached, just transiently, in one single patient (Basta et al. 2011).

Induction of Immune Tolerance

Over the past several decades, the generation of a large array of immunosuppressive agents has increased the number of therapeutic tools available to prevent acute rejection. However, as detailed above, toxicity of immunosuppressive drugs may offset their benefits; hence, the final goal of transplant medicine is to achieve T- and β-cell tolerance that is antigen specific without the need for long-term generalized immunosuppression. Many strategies have been proposed to promote transplant tolerance in rodent model of transplantation, including infusion of regulatory T cells, mesenchymal stem cells, or immature dendritic cells. Most strategies, however, failed when transferred to nonhuman primate models. Recently, a treatment that selectively depletes activated cytopathic donor reactive T cells while sparing resting and immunoregulatory T cells has been tested in a model of islet transplantation. Short-term sirolimus and IL-2/Ig plus mutant antagonist-type IL-15/Ig cytolytic fusion proteins posttransplantation resulted in prolonged, drug-free engraftment (Koulmanda et al. 2012).

Novel Therapeutic Perspectives for Type 1 Diabetes Mellitus

Other therapeutic approaches for patients with type 1 diabetes are also underway. Indeed, refinement of insulin pumps in combination with continuous glucose monitoring systems may lead to better glycemic control (The Juvenile Diabetes Research Foundation Continuous Glucose Monitoring Study Group 2008). In the future, patients will ideally be able to use a closed-loop system consisting of a glucose sensor and an insulin pump, as well as software to automatically translate measured glucose levels into appropriate insulin doses. A recent trial showed that over a 3-month period, the use of sensor-augmented insulin-pump therapy with the threshold-suspend feature reduced nocturnal hypoglycemia without increasing glycated hemoglobin values compared to insulin pumps without the threshold-suspend feature (Bergenstal et al. 2013). Promising results from two random-order crossover studies show that "bionic" pancreas improves glycemic control and reduces hypoglycemic episodes compared to the insulin pumps (Russel et al. 2014).

Intriguingly, attempts have also been made to protect pancreatic islets from autoimmunity, allowing regeneration of β cells in the early phases of type 1 diabetes. This appears to be a promising approach, since it is well documented that β cells can regenerate, as observed during pregnancy and in subjects with insulin resistance (Porat and Dor 2007; Claiborn and Stoffers 2008). Along this line, compelling evidence has accumulated suggesting that in addition to their immunosuppressive properties, CD3-specific antibodies can induce immune tolerance especially in the context of an ongoing immune response (Chatenoud and Bluestone 2007). Clinical studies have shown that this therapy may, at least partially, preserve β-cell mass in newly diagnosed type 1 diabetics (Chatenoud and Bluestone 2007; Keymeulen et al. 2005). A recent study showed that treatment with teplizumab, a nonactivating Fc-modified, anti-CD3 monoclonal antibody, significantly reduced C-peptide loss at 2 years after diagnosis, but its efficacy in delaying the need of insulin therapy was only partial (Hagopian et al. 2013). Modulation of T-cell costimulatory signals has been tested in a trial with 112 patients recently diagnosed with type 1 diabetes treated with CTLA4Ig fusion protein abatacept or vehicle control. Overall, abatacept slowed reduction in β-cell function (Orban et al. 2011). However, despite continued administration over 24 months, abatacept did not affect the decrease in β-cell function rate compared to placebo.

Starting from the rationale that interleukin-1 is involved in β-cell dysfunction and apoptosis, two recent randomized, controlled trials tested the effect of two interleukin-1 inhibitors, canakinumab and anakinra, on β-cell decline in recent-onset type 1 diabetes. Both drugs, however, failed to retard disease onset over placebo (Moran et al. 2013).

An alternative approach is targeting B lymphocytes, given the importance of the humoral response in the pathogenesis of type 1 diabetes and the fact that B lymphocytes also have a role as antigen-presenting cells. Promising results with a B lymphocyte depleting monoclonal antibody have been obtained in a mouse model of diabetes (Hu et al. 2007), and a clinical trial showed that a four-dose course of rituximab partially preserved β-cell function over a period of 12 months in patients with type 1 diabetes (Pescovitz et al. 2009).

A more drastic strategy to bypass autoimmunity is autologous non-myeloablative hematopoietic stem cell transplantation, which may reset autoreactive T cells and reverse the disease in new-onset type 1 diabetes (Couri et al. 2009). With this approach, persistent normoglycemia was achieved for a mean of 2.5 years in 60 % of patients. However, acute drug toxicity, risk of infections, and sterility may outweigh the benefits of this protocol.

Alternative approaches to the induction of tolerance include molecular biological strategies. In particular, evidence has been provided that immature dendritic cells (DCs) can promote tolerance. To this end, CD40, CD80, and CD86 cell surface molecules were specifically downregulated by ex vivo treating DCs from mice with a mixture of specific antisense oligonucleotides. This promoted the emergence of regulatory T cells that might possibly prevent the occurrence of diabetes (Machen et al. 2004). Intriguingly, to circumvent the technical issues of ex vivo DC manipulation, a recent study in mice showed that the same immature phenotype can be induced by using a microsphere-based vaccine injected subcutaneously (Phillips et al. 2008). This approach effectively prevented new-onset diabetes or even reversed it, providing the basis for testing this approach also in humans.

Conclusion

Islet transplantation is a dynamic field to which much time and resources are being devoted. If successful transplantation is defined as a transplant after which short-term quality of life and glycemic control improve, then success rates of this procedure are quite acceptable. However, if success is defined in terms of long-term insulin independence or prevention of diabetes-related complications, then outcomes are outright disappointing in the first, and largely unknown in the second.

At present, islet transplantation is far from representing a standard of care for the large majority of patients with type 1 diabetes (Cravedi et al. 2008; Ruggenenti et al. 2008). Only a highly selected group of patients with brittle diabetes may benefit from the procedure, which requires a high degree of expertise. In the absence of major advancements in islet transplantation, the 2006 recommendations by the American Diabetes Association (Robertson et al. 2006) that islet transplantation should be considered an experimental procedure appear still up to date.

Cross-References

▶ Advances in Clinical Islet Isolation
▶ Human Islet Autotransplantation
▶ Islet Isolation from Pancreatitis Pancreas for Islet Autotransplantation

References

Alejandro R, Barton FB, Hering BJ, Wease S (2008) 2008 update from the Collaborative Islet Transplant Registry. Transplantation 86:1783–1788

Aviv V, Meivar-Levy I, Rachmut IH, Rubinek T, Mor E, Ferber S (2009) Exendin-4 promotes liver cell proliferation and enhances the PDX-1-induced liver to pancreas transdifferentiation process. J Biol Chem 284:33509–33520

Badet L, Benhamou PY, Wojtusciszyn A et al (2007) Expectations and strategies regarding islet transplantation: metabolic data from the GRAGIL 2 trial. Transplantation 84:89–96

Ballinger WF, Lacy PE (1972) Transplantation of intact pancreatic islets in rats. Surgery 72:175–186

Barton FB, Rickels MR, Alejandro R et al (2012) Improvement in outcomes of clinical islet transplantation: 1999–2010. Diabetes Care 35:1436–1445

Basta G, Montanucci P, Luca G et al (2011) Long-term metabolic and immunological follow-up of nonimmunosuppressed patients with type 1 diabetes treated with microencapsulated islet allografts: four cases. Diabetes Care 34:2406–2409

Beck J, Angus R, Madsen B, Britt D, Vernon B, Nguyen KT (2007) Islet encapsulation: strategies to enhance islet cell functions. Tissue Eng 13:589–599

Beckwith J, Nyman JA, Flanagan B, Schrover R, Schuurman HJ (2012) A health economic analysis of clinical islet transplantation. Clin Transplant 26:23–33

Benhamou PY, Oberholzer J, Toso C et al (2001) Human islet transplantation network for the treatment of Type I diabetes: first data from the Swiss-French GRAGIL consortium (1999–2000). Groupe de Recherche Rhin Rhjne Alpes Geneve pour la transplantation d'Ilots de Langerhans. Diabetologia 44:859–864

Bergenstal RM, Klonoff DC, Garg SK et al (2013) Threshold-based insulin-pump interruption for reduction of hypoglycemia. N Engl J Med 369(3):224–232

Berney T, Secchi A (2009) Rapamycin in islet transplantation: friend or foe? Transpl Int 22:153–161

Berney T, Ferrari-Lacraz S, Buhler L et al (2009) Long-term insulin-independence after allogeneic islet transplantation for type 1 diabetes: over the 10-year mark. Am J Transplant 9:419–423

Bhargava R, Senior PA, Ackerman TE et al (2004) Prevalence of hepatic steatosis after islet transplantation and its relation to graft function. Diabetes 53:1311–1317

Bianco P, Cao X, Frenette PS et al (2013) The meaning, the sense and the significance: translating the science of mesenchymal stem cells into medicine. Nat Med 19:35–42

Bluestone JA, Herold K, Eisenbarth G (2010) Genetics, pathogenesis and clinical interventions in type 1 diabetes. Nature 464:1293–1300

Boggi U, Vistoli F, Egidi FM et al (2012) Transplantation of the pancreas. Curr Diab Rep 12:568–579

Bonner-Weir S, Taneja M, Weir GC et al (2000) In vitro cultivation of human islets from expanded ductal tissue. Proc Natl Acad Sci U S A 97:7999–8004

Braasch JW, Vito L, Nugent FW (1978) Total pancreatectomy of end-stage chronic pancreatitis. Ann Surg 188:317–322

Brendel M, Hering B, Schulz A, Bretzel R (1999) International islet transplant registry report. University of Giessen, Giessen, pp 1–20

Cabric S, Sanchez J, Lundgren T et al (2007) Islet surface heparinization prevents the instant blood-mediated inflammatory reaction in islet transplantation. Diabetes 56:2008–2015

Campbell PM, Senior PA, Salam A et al (2007a) High risk of sensitization after failed islet transplantation. Am J Transplant 7:2311–2317

Campbell PM, Salam A, Ryan EA et al (2007b) Pretransplant HLA antibodies are associated with reduced graft survival after clinical islet transplantation. Am J Transplant 7:1242–1248

Cardani R, Pileggi A, Ricordi C et al (2007) Allosensitization of islet allograft recipients. Transplantation 84:1413–1427

Chase HP, Cuthbertson DD, Dolan LM et al (2001) First-phase insulin release during the intravenous glucose tolerance test as a risk factor for type 1 diabetes. J Pediatr 138:244–249

Chatenoud L, Bluestone JA (2007) CD3-specific antibodies: a portal to the treatment of autoimmunity. Nat Rev Immunol 7:622–632

Chen S, Borowiak M, Fox JL et al (2009) A small molecule that directs differentiation of human ESCs into the pancreatic lineage. Nat Chem Biol 5:258–265

Claiborn KC, Stoffers DA (2008) Toward a cell-based cure for diabetes: advances in production and transplant of β cells. Mt Sinai J Med 75:362–371

Collaborative Islet Transplantation Registry (2010) Seventh annual report. http://citregistry.com Accessed on April 28, 2014

Coppieters KT, Wiberg A, Tracy SM, von Herrath MG (2012) Immunology in the clinic review series: focus on type 1 diabetes and viruses: the role of viruses in type 1 diabetes: a difficult dilemma. Clin Exp Immunol 168:5–11

Cornolti R, Figliuzzi M, Remuzzi A (2009) Effect of micro- and macroencapsulation on oxygen consumption by pancreatic islets. Cell Transplant 18:195–201

Couri CE, Oliveira MC, Stracieri AB et al (2009) C-peptide levels and insulin independence following autologous nonmyeloablative hematopoietic stem cell transplantation in newly diagnosed type 1 diabetes mellitus. JAMA 301:1573–1579

Cravedi P, Mannon RB, Ruggenenti P, Remuzzi A, Remuzzi G (2008) Islet transplantation: need for a time-out? Nat Clin Pract Nephrol 4:660–661

Daneman D (2009) State of the world's children with diabetes. Pediatr Diabetes 10:120–126

Danielson KK, Hatipoglu B, Kinzer K et al (2012) Reduction in carotid intima-media thickness after pancreatic islet transplantation in patients with type 1 diabetes. Diabetes Care 19:2012 (Published ahead of print November)

Danne T, Mortensen HB, Hougaard P et al (2001) Persistent differences among centers over 3 years in glycemic control and hypoglycemia in a study of 3,805 children and adolescents with type 1 diabetes from the Hvidore Study Group. Diabetes Care 24:1342–1347 de Boer IH, Sun W, Cleary PA et al (2011) Intensive diabetes therapy and glomerular filtration rate in type 1 diabetes. N Engl J Med 365:2366–2376

Del Carro U, Fiorina P, Amadio S et al (2007) Evaluation of polyneuropathy markers in type 1 diabetic kidney transplant patients and effects of islet transplantation: neurophysiological and skin biopsy longitudinal analysis. Diabetes Care 30:3063–3069

DIAMOND Project Group (2006) Incidence and trends of childhood type 1 diabetes worldwide 1990–1999. Diabet Med 23:857–866

Dor Y, Brown J, Martinez OI, Melton DA (2004) Adult pancreatic β-cells are formed by self-duplication rather than stem-cell differentiation. Nature 429:41–46

Egro FM (2013) Why is type 1 diabetes increasing? J Mol Endocrinol 51(1):R1–R13

Eich T, Eriksson O, Lundgren T (2007) Visualization of early engraftment in clinical islet transplantation by positron-emission tomography. N Engl J Med 356:2754–2755

English K (2012) Mechanisms of mesenchymal stromal cell immunomodulation. Immunol Cell Biol 91(1):19–26

Faradji RN, Tharavanij T, Messinger S et al (2008) Long-term insulin independence and improvement in insulin secretion after supplemental islet infusion under exenatide and etanercept. Transplantation 86:1658–1665

Ferrannini E, Mari A, Nofrate V, Sosenko JM, Skyler JS (2010) Progression to diabetes in relatives of type 1 diabetic patients: mechanisms and mode of onset. Diabetes 59:679–685

Figliuzzi M, Plati T, Cornolti R et al (2006) Biocompatibility and function of microencapsulated pancreatic islets. Acta Biomater 2:221–227

Fioretto P, Mauer SM, Bilous RW, Goetz FC, Sutherland DE, Steffes MW (1993) Effects of pancreas transplantation on glomerular structure in insulin-dependent diabetic patients with their own kidneys. Lancet 342:1193–1196

Fioretto P, Steffes MW, Sutherland DE, Goetz FC, Mauer M (1998) Reversal of lesions of diabetic nephropathy after pancreas transplantation. N Engl J Med 339:69–75

Fiorina P, Folli F, Zerbini G et al (2003) Islet transplantation is associated with improvement of renal function among uremic patients with type I diabetes mellitus and kidney transplants. J Am Soc Nephrol 14:2150–2158

Fiorina P, Gremizzi C, Maffi P et al (2005) Islet transplantation is associated with an improvement of cardiovascular function in type 1 diabetic kidney transplant patients. Diabetes Care 28:1358–1365

Fiorina P, Shapiro AM, Ricordi C, Secchi A (2008) The clinical impact of islet transplantation. Am J Transplant 8:1990–1997

Frank A, Deng S, Huang X et al (2004) Transplantation for type I diabetes: comparison of vascularized whole-organ pancreas with isolated pancreatic islets. Ann Surg 240:631–643

Froud T, Ricordi C, Baidal DA et al (2005) Islet transplantation in type 1 diabetes mellitus using cultured islets and steroid-free immunosuppression: Miami experience. Am J Transplant 5:2037–2046

Froud T, Baidal DA, Faradji R et al (2008a) Islet transplantation with alemtuzumab induction and calcineurin-free maintenance immunosuppression results in improved short- and long-term outcomes. Transplantation 86:1695–1701

Froud T, Faradji RN, Pileggi A et al (2008b) The use of exenatide in islet transplant recipients with chronic allograft dysfunction: safety, efficacy, and metabolic effects. Transplantation 86:36–45

Gale EA (2005) Spring harvest? Reflections on the rise of type 1 diabetes. Diabetologia 48:2445–2450

Gangemi A, Salehi P, Hatipoglu B et al (2008) Islet transplantation for brittle type 1 diabetes: the UIC protocol. Am J Transplant 8:1250–1261

Garcea G, Weaver J, Phillips J et al (2009) Total pancreatectomy with and without islet cell transplantation for chronic pancreatitis: a series of 85 consecutive patients. Pancreas 38:1–7

Garcea G, Pollard CA, Illouz S, Webb M, Metcalfe MS, Dennison AR (2013) Patient satisfaction and cost-effectiveness following total pancreatectomy with islet cell transplantation for chronic pancreatitis. Pancreas 42:322–328

Gentilella R, Bianchi C, Rossi A, Rotella CM (2009) Exenatide: a review from pharmacology to clinical practice. Diabetes Obes Metab 11:544–556

Ghofaili KA, Fung M, Ao Z et al (2007) Effect of exenatide on β cell function after islet transplantation in type 1 diabetes. Transplantation 83:24–28

Gillard P, Ling Z, Mathieu C et al (2008) Comparison of sirolimus alone with sirolimus plus tacrolimus in type 1 diabetic recipients of cultured islet cell grafts. Transplantation 85:256–263

Goss JA, Goodpastor SE, Brunicardi FC et al (2004) Development of a human pancreatic islet-transplant program through a collaborative relationship with a remote islet-isolation center. Transplantation 77:462–466

Green A (2008) Descriptive epidemiology of type 1 diabetes in youth: incidence, mortality, prevalence, and secular trends. Endocr Res 33:1–15

Greenbaum CJ, Beam CA, Boulware D et al (2012) Fall in C-peptide during first 2 years from diagnosis: evidence of at least two distinct phases from composite type 1 diabetes trialNet data. Diabetes 61:2066–2073

Groth C (2007) Towards developing guidelines on Xenotransplantation in China. Xenotransplantation 14:358–359

Groth CG, Korsgren O, Tibell A et al (1994) Transplantation of porcine fetal pancreas to diabetic patients. Lancet 344:1402–1404

Gruessner AC, Sutherland DE (2005) Pancreas transplant outcomes for United States (US) and non-US cases as reported to the United Network for Organ Sharing (UNOS) and the International Pancreas Transplant Registry (IPTR) as of June 2004. Clin Transplant 19:433–455

Gruessner RW, Sutherland DE, Gruessner AC (2005) Survival after pancreas transplantation. JAMA 293:675

Guignard AP, Oberholzer J, Benhamou PY et al (2004) Cost analysis of human islet transplantation for the treatment of type 1 diabetes in the Swiss-French Consortium GRAGIL. Diabetes Care 27:895–900

Hafiz MM, Faradji RN, Froud T et al (2005) Immunosuppression and procedure-related complications in 26 patients with type 1 diabetes mellitus receiving allogeneic islet cell transplantation. Transplantation 80:1718–1728

Hagopian W, Ferry RJ Jr, Sherry N et al (2013) Teplizumab preserves C-peptide in recent-onset type 1 diabetes: 2-year results from the randomized, placebo-controlled Protege trial. Diabetes 62(11):3901–3908

Halloran PF (2004) Immunosuppressive drugs for kidney transplantation. N Engl J Med 351:2715–2729

Harjutsalo V, Reunanen A, Tuomilehto J (2006) Differential transmission of type 1 diabetes from diabetic fathers and mothers to their offspring. Diabetes 55:1517–1524

Heneine W, Tibell A, Switzer WM et al (1998) No evidence of infection with porcine endogenous retrovirus in recipients of porcine islet-cell xenografts. Lancet 352:695–699

Hering BJ, Kandaswamy R, Harmon JV et al (2004) Transplantation of cultured islets from two-layer preserved pancreases in type 1 diabetes with anti-CD3 antibody. Am J Transplant 4:390–401

Hering BJ, Kandaswamy R, Ansite JD et al (2005) Single-donor, marginal-dose islet transplantation in patients with type 1 diabetes. JAMA 293:830–835

Hirshberg B (2006) Can we justify living donor islet transplantation? Curr Diab Rep 6:307–309

Hirshberg B, Rother KI, Digon BJ 3rd et al (2003) Benefits and risks of solitary islet transplantation for type 1 diabetes using steroid-sparing immunosuppression: the National Institutes of Health experience. Diabetes Care 26:3288–3295

Hu CY, Rodriguez-Pinto D, Du W et al (2007) Treatment with CD20-specific antibody prevents and reverses autoimmune diabetes in mice. J Clin Invest 117:3857–3867

Huurman VA, Hilbrands R, Pinkse GG et al (2008) Cellular islet autoimmunity associates with clinical outcome of islet cell transplantation. PLoS One 3:e2435

Huurman VA, Velthuis JH, Hilbrands R et al (2009) Allograft-specific cytokine profiles associate with clinical outcome after islet cell transplantation. Am J Transplant 9:382–388

Ichii H, Ricordi C (2009) Current status of islet cell transplantation. J Hepatobiliary Pancreat Surg 16:101–112

Jin SM, Oh SH, Kim SK et al (2013) Diabetes-free survival in patients who underwent islet autotransplantation after 50 % to 60 % distal partial pancreatectomy for benign pancreatic tumors. Transplantation 95:1396–1403

Johansson H, Lukinius A, Moberg L et al (2005) Tissue factor produced by the endocrine cells of the islets of Langerhans is associated with a negative outcome of clinical islet transplantation. Diabetes 54:1755–1762

Keenan HA, Sun JK, Levine J et al (2010) Residual insulin production and pancreatic β-cell turnover after 50 years of diabetes: Joslin Medalist Study. Diabetes 59:2846–2853

Kenmochi T, Asano T, Maruyama M et al (2009) Clinical islet transplantation in Japan. J Hepatobiliary Pancreat Surg 16:124–130

Keymeulen B, Vandemeulebroucke E, Ziegler AG et al (2005) Insulin needs after CD3-antibody therapy in new-onset type 1 diabetes. N Engl J Med 352:2598–2608

Keymeulen B, Gillard P, Mathieu C et al (2006) Correlation between β cell mass and glycemic control in type 1 diabetic recipients of islet cell graft. Proc Natl Acad Sci U S A 103:17444–17449

Koh A, Senior P, Salam A et al (2010) Insulin-heparin infusions peritransplant substantially improve single-donor clinical islet transplant success. Transplantation 89:465–471

Kolb H, Elliott RB (1994) Increasing incidence of IDDM a consequence of improved hygiene? Diabetologia 37:729

Korsgren O, Nilsson B, Berne C et al (2005) Current status of clinical islet transplantation. Transplantation 79:1289–1293

Koulmanda M, Qipo A, Fan Z et al (2012) Prolonged survival of allogeneic islets in cynomolgus monkeys after short-term triple therapy. Am J Transplant 12:1296–1302

Kyvik KO, Green A, Beck-Nielsen H (1995) Concordance rates of insulin dependent diabetes mellitus: a population based study of young Danish twins. BMJ 311:913–917

Lachin JM, Genuth S, Nathan DM, Zinman B, Rutledge BN (2008) Effect of glycemic exposure on the risk of microvascular complications in the diabetes control and complications trial – revisited. Diabetes 57:995–1001

Lacy PE (1978) Workshop on Pancreatic Islet Cell Transplantation in Diabetes sponsored by the National Institute of Arthritis, Metabolism, and Digestive Diseases and held at the National Institutes of Health in Bethesda, Maryland, on November 29 and 30, 1977. Diabetes 27:427–429

Lee TC, Barshes NR, O'Mahony CA et al (2005) The effect of pancreatic islet transplantation on progression of diabetic retinopathy and neuropathy. Transplant Proc 37:2263–2265

Lee TC, Barshes NR, Agee EE, O'Mahoney CA, Brunicardi FC, Goss JA (2006) The effect of whole organ pancreas transplantation and PIT on diabetic complications. Curr Diab Rep 6:323–327

Leitao CB, Tharavanij T, Cure P et al (2008) Restoration of hypoglycemia awareness after islet transplantation. Diabetes Care 31:2113–2115

Leitao CB, Cure P, Messinger S et al (2009) Stable renal function after islet transplantation: importance of patient selection and aggressive clinical management. Transplantation 87:681–688

Lobo PI, Spencer C, Simmons WD et al (2005) Development of anti-human leukocyte antigen class 1 antibodies following allogeneic islet cell transplantation. Transplant Proc 37:3438–3440

Machen J, Harnaha J, Lakomy R, Styche A, Trucco M, Giannoukakis N (2004) Antisense oligonucleotides down-regulating costimulation confer diabetes-preventive properties to nonobese diabetic mouse dendritic cells. J Immunol 173:4331–4341

Maffi P, Bertuzzi F, De Taddeo F et al (2007) Kidney function after islet transplant alone in type 1 diabetes: impact of immunosuppressive therapy on progression of diabetic nephropathy. Diabetes Care 30:1150–1155

Maffi P, Balzano G, Ponzoni M et al (2013) Autologous pancreatic islet transplantation in human bone marrow. Diabetes 62(10):3523–3531

Markmann JF, Rosen M, Siegelman ES et al (2003) Magnetic resonance-defined periportal steatosis following intraportal islet transplantation: a functional footprint of islet graft survival? Diabetes 52:1591–1594

Matsumoto S, Okitsu T, Iwanaga Y et al (2005) Insulin independence after living-donor distal pancreatectomy and islet allotransplantation. Lancet 365:1642–1644

Matsumoto S, Okitsu T, Iwanaga Y et al (2006) Follow-up study of the first successful living donor islet transplantation. Transplantation 82:1629–1633

Merani S, Toso C, Emamaullee J, Shapiro AM (2008) Optimal implantation site for pancreatic islet transplantation. Br J Surg 95:1449–1461

Mineo D, Sageshima J, Burke GW, Ricordi C (2009) Minimization and withdrawal of steroids in pancreas and islet transplantation. Transpl Int 22:20–37

Moran A, Bundy B, Becker DJ et al (2013) Interleukin-1 antagonism in type 1 diabetes of recent onset: two multicentre, randomised, double-blind, placebo-controlled trials. Lancet 381:1905–1915

Nagaya M, Katsuta H, Kaneto H, Bonner-Weir S, Weir GC (2009) Adult mouse intrahepatic biliary epithelial cells induced in vitro to become insulin-producing cells. J Endocrinol 201:37–47

Nathan DM, Cleary PA, Backlund JY et al (2005) Intensive diabetes treatment and cardiovascular disease in patients with type 1 diabetes. N Engl J Med 353:2643–2653

Nathan DM, Zinman B, Cleary PA et al (2009) Modern-day clinical course of type 1 diabetes mellitus after 30 years' duration: the diabetes control and complications trial/epidemiology of diabetes interventions and complications and Pittsburgh epidemiology of diabetes complications experience (1983–2005). Arch Intern Med 169:1307–1316

Naziruddin B, Matsumoto S, Noguchi H et al (2012) Improved pancreatic islet isolation outcome in autologous transplantation for chronic pancreatitis. Cell Transplant 21:553–558

Nir T, Melton DA, Dor Y (2007) Recovery from diabetes in mice by β cell regeneration. J Clin Invest 117:2553–2561

Nordwall M, Bojestig M, Arnqvist HJ, Ludvigsson J (2004) Declining incidence of severe retinopathy and persisting decrease of nephropathy in an unselected population of type 1 diabetes-the Linkoping Diabetes Complications Study. Diabetologia 47:1266–1272

O'Connell PJ, Hawthorne WJ, Holmes-Walker DJ et al (2006) Clinical islet transplantation in type 1 diabetes mellitus: results of Australia's first trial. Med J Aust 184:221–225

Onkamo P, Vaananen S, Karvonen M, Tuomilehto J (1999) Worldwide increase in incidence of Type I diabetes – the analysis of the data on published incidence trends. Diabetologia 42:1395–1403

Orban T, Bundy B, Becker DJ et al (2011) Co-stimulation modulation with abatacept in patients with recent-onset type 1 diabetes: a randomised, double-blind, placebo-controlled trial. Lancet 378:412–419

O'Sullivan ES, Vegas A, Anderson DG, Weir GC (2011) Islets transplanted in immunoisolation devices: a review of the progress and the challenges that remain. Endocr Rev 32:827–844

Paty BW, Senior PA, Lakey JR, Shapiro AM, Ryan EA (2006) Assessment of glycemic control after islet transplantation using the continuous glucose monitor in insulin-independent versus insulin-requiring type 1 diabetes subjects. Diabetes Technol Ther 8:165–173

Peakman M (2013) Immunological pathways to β-cell damage in type 1 diabetes. Diabet Med 30:147–154

Pescovitz MD, Greenbaum CJ, Krause-Steinrauf H et al (2009) Rituximab, B-lymphocyte depletion, and preservation of β-cell function. N Engl J Med 361:2143–2152

Phillips B, Nylander K, Harnaha J et al (2008) A microsphere-based vaccine prevents and reverses new-onset autoimmune diabetes. Diabetes 57:1544–1555

Porat S, Dor Y (2007) New sources of pancreatic β cells. Curr Diab Rep 7:304–308

Pugliese A (2013) The multiple origins of type 1 diabetes. Diabet Med 30:135–146

Rajotte RV (2008) Moving towards clinical application. Xenotransplantation 15:113–115

Redondo MJ, Rewers M, Yu L et al (1999) Genetic determination of islet cell autoimmunity in monozygotic twin, dizygotic twin, and non-twin siblings of patients with type 1 diabetes: prospective twin study. BMJ 318:698–702

Rich SS (1990) Mapping genes in diabetes. Genetic epidemiological perspective. Diabetes 39:1315–1319

Rickels MR, Naji A (2009) Exenatide use in islet transplantation: words of caution. Transplantation 87:153

Rickels MR, Schutta MH, Mueller R et al (2005a) Islet cell hormonal responses to hypoglycemia after human islet transplantation for type 1 diabetes. Diabetes 54:3205–3211

Rickels MR, Schutta MH, Markmann JF, Barker CF, Naji A, Teff KL (2005b) β-Cell function following human islet transplantation for type 1 diabetes. Diabetes 54:100–106

Rickels MR, Schutta MH, Mueller R et al (2007) Glycemic thresholds for activation of counterregulatory hormone and symptom responses in islet transplant recipients. J Clin Endocrinol Metab 92:873–879

Rickels M, Cullison K, Fuller C (2011) Improvement of glucose counter-regulation following human islet transplantation in long-standing type 1 diabetes: preliminary results [abstract]. Diabetes 60:293, OR

Robertson RP, Davis C, Larsen J, Stratta R, Sutherland DE (2006) Pancreas and islet transplantation in type 1 diabetes. Diabetes Care 29:935

Ruggenenti P, Remuzzi A, Remuzzi G (2008) Decision time for pancreatic islet-cell transplantation. Lancet 371:883–884

Russell SJ1, El-Khatib FH, Sinha M, Magyar KL, McKeon K, Goergen LG, Balliro C, Hillard MA, Nathan DM, Damiano ER (2014) Outpatient Glycemic Control with a Bionic Pancreas in Type 1 Diabetes. N Engl J Med. Jun 15. [Epub ahead of print]

Ryan EA, Shandro T, Green K et al (2004a) Assessment of the severity of hypoglycemia and glycemic lability in type 1 diabetic subjects undergoing islet transplantation. Diabetes 53:955–962

Ryan EA, Paty BW, Senior PA, Shapiro AM (2004b) Risks and side effects of islet transplantation. Curr Diab Rep 4:304–309

Ryan EA, Paty BW, Senior PA et al (2005) Five-year follow-up after clinical islet transplantation. Diabetes 54:2060–2069

Ryan EA, Bigam D, Shapiro AM (2006) Current indications for pancreas or islet transplant. Diabetes Obes Metab 8:1–7

Schiopu A, Wood KJ (2008) Regulatory T cells: hypes and limitations. Curr Opin Organ Transplant 13:333–338

Secchi A, Socci C, Maffi P et al (1997) Islet transplantation in IDDM patients. Diabetologia 40:225–231

Senior PA, Zeman M, Paty BW, Ryan EA, Shapiro AM (2007) Changes in renal function after clinical islet transplantation: four-year observational study. Am J Transplant 7:91–98

Shapiro AM, Lakey JR, Ryan EA et al (2000) Islet transplantation in seven patients with type 1 diabetes mellitus using a glucocorticoid-free immunosuppressive regimen. N Engl J Med 343:230–238

Shapiro AM, Ricordi C, Hering BJ et al (2006) International trial of the Edmonton protocol for islet transplantation. N Engl J Med 355:1318–1330

Soltesz G, Patterson CC, Dahlquist G (2007) Worldwide childhood type 1 diabetes incidence – what can we learn from epidemiology? Pediatr Diabetes 8(Suppl 6):6–14

Sosenko JM, Palmer JP, Rafkin-Mervis L et al (2009) Incident dysglycemia and progression to type 1 diabetes among participants in the Diabetes Prevention Trial-Type 1. Diabetes Care 32:1603–1607

Strachan DP (1989) Hay fever, hygiene, and household size. BMJ 299:1259–1260

Sutherland DE (1981) Pancreas and islet transplantation. II. Clinical trials. Diabetologia 20:435–450

Sutherland DE, Matas AJ, Najarian JS (1978) Pancreatic islet cell transplantation. Surg Clin North Am 58:365–382

Sutherland DE, Gruessner AC, Carlson AM et al (2008) Islet autotransplant outcomes after total pancreatectomy: a contrast to islet allograft outcomes. Transplantation 86:1799–1802

Sykes M, Cozzi E, d'Apice A et al (2006) Clinical trial of islet xenotransplantation in Mexico. Xenotransplantation 13:371–372

Sykes M, Pierson RN 3rd, O'Connell P et al (2007) Reply to "critics slam Russian trial to test pig pancreas for diabetes". Nat Med 13:662–663

Tateishi K, He J, Taranova O, Liang G, D'Alessio AC, Zhang Y (2008) Generation of insulin-secreting islet-like clusters from human skin fibroblasts. J Biol Chem 283:31601–31607

The Diabetes Control and Complications Trial Research Group (1993) The effect of intensive treatment of diabetes on the development and progression of long-term complications in insulin-dependent diabetes mellitus. N Engl J Med 329:977–986

The Juvenile Diabetes Research Foundation Continuous Glucose Monitoring Study Group (2008) Continuous glucose monitoring and intensive treatment of type 1 diabetes. N Engl J Med 359:1464–1476

Thompson DM, Begg IS, Harris C et al (2008) Reduced progression of diabetic retinopathy after islet cell transplantation compared with intensive medical therapy. Transplantation 85:1400–1405

Tuomilehto J, Podar T, Tuomilehto-Wolf E, Virtala E (1995) Evidence for importance of gender and birth cohort for risk of IDDM in offspring of IDDM parents. Diabetologia 38:975–982

Valdes-Gonzalez RA, Dorantes LM, Garibay GN et al (2005) Xenotransplantation of porcine neonatal islets of Langerhans and Sertoli cells: a 4-year study. Eur J Endocrinol 153:419–427

van der Windt DJ, Echeverri GJ, Ijzermans JN, Coopers DK (2008) The choice of anatomical site for islet transplantation. Cell Transplant 17:1005–1014

van der Windt DJ, Bottino R, Kumar G et al (2012) Clinical islet xenotransplantation: how close are we? Diabetes 61:3046–3055

Venstrom JM, McBride MA, Rother KI, Hirshberg B, Orchard TJ, Harlan DM (2003) Survival after pancreas transplantation in patients with diabetes and preserved kidney function. JAMA 290:2817–2823

Venturini M, Fiorina P, Maffi P et al (2006) Early increase of retinal arterial and venous blood flow velocities at color Doppler imaging in brittle type 1 diabetes after islet transplant alone. Transplantation 81:1274–1277

Villiger P, Ryan EA, Owen R et al (2005) Prevention of bleeding after islet transplantation: lessons learned from a multivariate analysis of 132 cases at a single institution. Am J Transplant 5:2992–2998

Wahoff DC, Papalois BE, Najarian JS et al (1995) Autologous islet transplantation to prevent diabetes after pancreatic resection. Ann Surg 222:562–575 (discussion 575–569, 1995)

Wang W (2007) A pilot trial with pig-to-man islet transplantation at the 3rd Xiang-Ya Hospital of the Central South University in Changsha. Xenotransplantation 14:358

Warnock GL, Meloche RM, Thompson D et al (2005) Improved human pancreatic islet isolation for a prospective cohort study of islet transplantation vs best medical therapy in type 1 diabetes mellitus. Arch Surg 140:735–744

Warnock GL, Thompson DM, Meloche RM et al (2008) A multi-year analysis of islet transplantation compared with intensive medical therapy on progression of complications in type 1 diabetes. Transplantation 86:1762–1766

Weaver TA, Kirk AD (2007) Alemtuzumab. Transplantation 84:1545–1547

White SA, Shaw JA, Sutherland DE (2009) Pancreas transplantation. Lancet 373:1808–1817

Yeh HC1, Brown TT, Maruthur N, Ranasinghe P, Berger Z, Suh YD, Wilson LM, Haberl EB, Brick J, Bass EB, Golden SH (2012) Comparative effectiveness and safety of methods of insulin delivery and glucose monitoring for diabetes mellitus: a systematic review and meta-analysis. Ann Intern Med. Sep 4;157(5):336–47

Zhou Q, Brown J, Kanarek A, Rajagopal J, Melton DA (2008) In vivo reprogramming of adult pancreatic exocrine cells to β-cells. Nature 455:627–632

Ziegler AG, Rewers M, Simell O et al (2013) Seroconversion to multiple islet autoantibodies and risk of progression to diabetes in children. JAMA 309:2473–2479

Islet Xenotransplantation: An Update on Recent Advances and Future Prospects

45

Rahul Krishnan, Morgan Lamb, Michael Alexander, David Chapman, David Imagawa, and Jonathan R. T. Lakey

Contents

Introduction to Type 1 Diabetes and Treatments	1276
Treatment of Type 1 Diabetes	1276
Limitations to Allotransplantation	1277
Xenotransplantation as a Solution	1278
What Is Xenotransplantation?	1278
Potential of Xenografts in Organ Transplantation	1278
Potential Sources of Islet Donors for Clinical Xenotransplantation	1280
Limitations of Xenotransplantation	1281
Pig to Human Infections	1282
Facilities and Microbial Safety	1282
Islet Xenotransplantation Transplant Studies: Large Animals and Clinical Trials	1283
Canine Islet Xenotransplantation	1283
Nonhuman Primate Islet Xenotransplantation	1284
Clinical Islet Xenotransplanation	1285
Recent Developments and Future Applications	1287
Genetically Modified Porcine Islets	1287
Gal-Deficient Galactosyl Transferase Knockout and Pathogen-Free Hybrids	1287

R. Krishnan • M. Lamb • M. Alexander • D. Imagawa
Department of Surgery, University of California Irvine, Orange, CA, USA
e-mail: rkrishn1@uci.edu; lamb.morgan@gmail.com; mlamb@uci.edu; michaela@uci.edu; dkimagaw@uci.edu

D. Chapman
Department of Experimental Surgery/Oncology, University of Alberta, Edmonton, AB, Canada
e-mail: dave.w.chapman@gmail.com

J.R.T. Lakey (✉)
Biomedical Engineering, Department of Surgery, University of California Irvine, Orange, CA, USA
e-mail: jlakey@uci.edu

M.S. Islam (ed.), *Islets of Langerhans*, DOI 10.1007/978-94-007-6686-0_28,
© Springer Science+Business Media Dordrecht 2015

Immunoisolation for Islet Xenotransplantation 1288
Future Direction of Islet Xenotransplantation 1288
Cross-References 1289
References 1289

Abstract

Type 1 diabetes (T1D) is an autoimmune disorder characterized by pancreatic β cell destruction, leading to a gradual loss of endogenous insulin secretion and ensuing hyperglycemia. Current treatment methods for T1D include the administration of exogenous insulin or the replacement of β-cell mass through either pancreas or islet transplantation. Current β-cell replacement trials have reported a lower incidence of potentially fatal hypoglycemic episodes when compared to exogenous insulin therapy. Islet allotransplants have also shown success in establishing insulin independence during the first year post transplant; however, transplant recipients progressively become increasingly dependent on insulin, and this therapy has been limited to a selected pool of patients due to a shortage of islet donors. The outcomes of the procedure have also been inconsistent, with the most success coming from highly experienced transplant centers.

Due to these inherent limitations of islet allotransplantation, xenotransplantation has been investigated as a possible alternative, since there are several potential advantages. Xenotransplantation has the potential of offering a reliable and consistent supply of islets. Having a readily available source of islets provides a significant advantage of being able to schedule the transplant procedure in advance. This is of great importance since most immunosuppression induction strategies require pre-dosing patients before the procedure. In addition, the development of specific pathogen-free (SPF) and designated disease pathogen-free (DPF) facilities and the propagation of animals with limited or no transmissible zoonotic diseases make xenotransplantation the treatment of choice for T1D patients.

This chapter aims to provide a comprehensive review of islet xenotransplantation for the treatment of type 1 diabetes, including a historical review, problems faced by xenotransplantation, current research on xenotransplantation outcomes and clinical trials, and future developments.

Keywords

Xenotransplantation • Type 1 diabetes • Islet transplant • Porcine islet • Piglet islet

Introduction to Type 1 Diabetes and Treatments

Treatment of Type 1 Diabetes

Type 1 diabetes (T1D) patients lack adequate endogenous insulin secretion due to the immune-mediated destruction of pancreatic β cells (Cnop et al. 2005). Current treatment regimens for T1DM patients aim to maintain blood glucose levels within the physiological range and can take the form of either exogenous insulin injections several times a day or replacement of β-cells by pancreas or islet transplantation. Long-term studies have suggested that such efforts to maintain blood glucose control may delay the development and progression of chronic complications of T1D (National Kidney Disease Education Program 2013; Dvornik 1978).

In contrast to exogenous insulin administration, replacement of β-cells has shown more reliability in achieving good glycemic control and preventing hypoglycemic episodes (Heller 2008). In 2000, the Edmonton islet transplant group at the University of Alberta, Canada, achieved insulin independence in seven consecutive patients, using islets isolated from cadaver organ donors under a specific steroid-free immunosuppression protocol (Shapiro et al. 2000). Although still in its early stages, islet transplantation results in fewer complications while maintaining comparable patient outcomes and graft survival (Frank et al. 2004), making it a suitable alternative to whole pancreas transplantation (Sutherland et al. 2001). In 2005 a study out of the University of Minnesota reported that out of the eight islet transplant recipients, all became insulin independent, with five out of eight remaining independent after 1 year. All eight patients also had improved HA1c levels and demonstrated sustained C-peptide production 5 years post transplant (Hering et al. 2005).

Limitations to Allotransplantation

The scarcity of suitable donors represents the largest hurdle for islet transplantation. Donor criteria for islet transplantation are similar to criteria established for whole pancreas transplant but may be more lenient on donor parameters (Ris et al. 2004).

Pancreases are first offered for whole pancreas transplantation because of its long-standing record of long-term function and patient survival (Ris et al. 2004). Pancreases are then offered to islet isolation centers. Current donor criteria associated with optimal whole pancreas transplants are young donors (<50 years old) with low body mass indexes (BMI <30), while older donors (>50 years old) with high BMI (>30) are factors associated with improved islet yield, viability, and overall ease of the islet isolation process (Vrochides et al. 2009; Wang et al. 2013).

Given the limitations associated with current islet isolation and allotransplantation techniques, there are three major factors that affect islet transplant. These factors include the associated adverse effects of current immunosuppressive protocols, the rejection associated with allotransplantation, and the limited supply of donor tissue (Titus et al. 2000). To overcome these limitations, current studies have turned their focus to xenotransplantation, with the porcine model being the most widely studied due to the genetic similarities of porcine and human insulin (Han and Tuch 2001). Porcine insulin has also been administered to diabetic patients for decades with successful patient outcomes (Richter and Neises 2002; Greene et al. 1983), and in theory, porcine islets should be able to provide a sustained and dynamic release of insulin required to maintain euglycemia in T1D patients.

Xenotransplantation as a Solution

What Is Xenotransplantation?

Xeno or "foreign" transplantation can be defined as the implantation, transplantation, or infusion of living cells, tissues, or organs from one species to another (National Health and Medical Research Council 2013; World Health Organization 2005).

Xenotransplantation includes a variety of different procedures that can be classified into three categories; external animal therapies, solid organ transplant, and tissue and cellular therapies, and transplants (HumanXenoTransplant 2013).

External animal therapies are procedures that are performed outside the body of the human patient, involving the use of animal cells, tissue, or organ. Liver perfusion involving the introduction of pig cells for the purification of blood and the growing of human skin over a layer of animal skin in a laboratory for later skin transplants into clinical patients are some examples of clinically relevant external animal therapies.

Solid organ transplantation can be defined as the transplantation of solid viscera into a human recipient in order to replace a diseased or damaged organ. Examples include the following: liver, heart, kidneys, pancreas, and even skin.

Tissue and cellular therapies and transplants include the implantation, infusion, or transplantation of animal cells into a human recipient to replenish or compensate for the patient's dysfunctional or diseased cells or tissue. Examples of tissue and cell therapies include the use of pancreatic islets, neurons, bone marrow, and stem cells.

Although xenotransplantation has the potential to resolve the problem of human organ shortage, it also has the potential for transmitting infectious diseases, both known and undiscovered, from animals to humans (U.S. Food and Drug Administration 2013).

Potential of Xenografts in Organ Transplantation

Due to the limited availability of suitable human organs and cells for clinical transplantation, pigs have been investigated as a potential alternative source. Xenotransplantation using porcine tissue could offer a scalable source of organs and cells, which have been shown to be physiological similar to humans (Soto-Gutierrez et al. 2012). Over the years there has been an increase in the use of porcine organs and cells in transplantation research, largely due to the increasing availability of specialized pig models (Cooper and Ayares 2011). In nonhuman primates, porcine heterotopic heart grafts were reported to have a survival rate of up to 8 months; primates that received porcine kidney transplants showed function for 3 months (Cooper and Ayares 2011). Although there has been significant advancement in this field, several barriers still exist, including immunity and organ rejection that must be overcome before this can become a standard therapy in the clinic.

One such barrier involving host immunity is the acute organ rejection that occurs after transplantation from pigs to baboons. It is related to the coagulation dysfunction triggered by activation of the host's humoral immune system that results in thrombotic microangiopathy (Kuwaki et al. 2004, 2005; Tseng et al. 2005; Ierino et al. 1998) and consumptive coagulopathy secondary to antibody deposition in the donor organs (Kozlowski et al. 1999; Buhler et al. 2000; Lin et al. 2010). This coagulation dysfunction may be related to molecular incompatibilities in the coagulation systems between pigs and humans (Bulato et al. 2012). Current efforts are focusing on the genetic modification of the cell source or of the complete organ itself (Ekser and Cooper 2010). Studies out of the Massachusetts General Hospital and Harvard Medical School have shown that the use of genetically modified pigs as a donor source for transplant can show improvements in graft function and survival (Kuwaki et al. 2004). Cowan et al. also showed that the transplantation of transgenic porcine kidneys that express human complement factors CD55 and CD59, along with the intravenous treatment of antithrombin III, has a significantly higher graft survival rate when compared to controls receiving normal porcine kidneys (Cowan et al. 2002). The transplantation of transgenic porcine islets has also shown encouraging results when transplanted into nonhuman primate models. Current studies have shown that diabetic nonhuman primates transplanted with Gal-deficient islets expressing human CD46, human CD39, and/or human TFPI can maintain graft survival and function and establish normoglycemia for up to a year post transplant (Ayares et al. 2013).

Xenotransplantation has also explored the use of encapsulation to prevent rejection of the graft by the recipient's immune system and protect recipient tissue by creating a biocompatible barrier around the transplanted tissue. Current cases using encapsulated islets have demonstrated long-term graft survival, without the use of harmful immunosuppression regimes (Chhabra and Brayman 2011). The most promising encapsulation technology in use involves creating an immunoisolation barrier around transplanted cells using sodium alginate. Current alginate encapsulated islet experiments using porcine islets transplanted into diabetic nonhuman primates have shown a significant reduction in daily insulin requirements when compared to un-transplanted controls (Elliott et al. 2005). A study by Dufrane et al. showed survival and function of encapsulated porcine islet grafts up to 6 months post transplant into Cynomolgus maccacus, with detectable C-peptide levels and anti-pig IgM/IgG 1 month after transplantation (Dufrane et al. 2006a). Unfortunately, even though the results were promising, the number of surviving functional islets was insufficient to make a significant impact on the required daily insulin dose after the first year. A major obstacle reported with transplant encapsulation of islets is opacified or cloudy capsules. Reports suggest that capsule opacification may affect the functional longevity of the islets, thus negatively impacting islet longevity (Elliott et al. 2007; Clayton et al. 1993).

Potential Sources of Islet Donors for Clinical Xenotransplantation

Piscine Islets

Although fish may not be the first thought when thinking of islet transplantation, but studies show that tilapia may be a perfect factory fish because it breeds easily, and studies have shown that tilapia islets, known as Brockmann bodies (BBs), can be easily and inexpensively harvested (Yang et al. 1997; Wright et al. 2004). They are currently being explored as an alternative islet source for xenotransplantation (Wright et al. 2004). BBs are separate organs located near the liver within a triangular region of adipose tissue (Wright and Yang 1997; Yang and Wright 1995; Dickson et al. 2003; Yang et al. 1997). According to reports by Wright et al., when BBs are transplanted into streptozotocin-induced diabetic mice under the kidney capsule, function is restored immediately and the animals are able to maintain long-term normoglycemia (>50 days) (Wright et al. 1992, 2004). However, BBs do have a few drawbacks. Tilapia insulin differs from human insulin by 17 amino acids; thus, it is required to be "humanized" prior to transplant in order to become biologically active in the human recipient (Alexander et al. 2006; Wright et al. 1992, 2004; Wright and Yang 1997). Furthermore, the xenotransplantation of fish BBs into humans would require the need for encapsulation to prevent recipient's immunological rejection of the foreign tissue.

Bovine Islets

Bovine islets can be isolated by a procedure similar to human islet isolation, using collagenase digestion and density gradient purification, and reports have shown graft function when they have been transplanted into diabetic nude mice (Lanza et al. 1995a). In addition, bovine islets can be cultured up to 4 weeks while maintaining glucose responsiveness (Marchetti et al. 1995). Encapsulated bovine islets have also been shown to restore euglycemia in non-immunosuppressed immunocompetent diabetic mice (Lanza et al. 1995b) and rats (Lanza et al. 1991). These results demonstrate the potential use of bovine islets as islet donors, because they are readily available and in high volume in the USA. However promising bovine islets are, they also possess a few drawbacks. Bovine islets are difficult and expensive to isolate and require an experienced team to be isolated (Figliuzzi et al. 2000; Friberg 2011). Studies have shown that bovine insulin differs from human insulin by three amino acids, which causes a different response to glucose levels compared to human insulin when used in exogenous therapies, and reports have shown that it can be more immunogenic (Derewenda and Dodson 1993).

Porcine Islets

Porcine islets have received the most attention as a possible source for islet transplantation. Porcine insulin differs from human insulin by only one amino acid (Storms et al. 1987), and current studies support the use of these islets in the clinical setting. In the early 1990s, studies by Lanza et al. showed that encapsulated islets that were isolated from adult pigs and then transplanted into the peritoneal cavity of diabetic Lewis rats could establish euglycemia within the first month and maintain it for a prolonged period of time (Lanza et al. 1991).

Another study by Sun et al. reported that euglycemia was maintained for up to 10 months post transplant in diabetic Balb/c mice that received a transplant of encapsulated adult porcine islets (Sun et al. 1992). Another study by Foster et al. used fetal pig islet-like clusters and reported 20 weeks of normoglycemia following transplantation into diabetic mice (Foster et al. 2007), while another study found that encapsulated neonatal porcine islets maintained euglycemia for more than 174 days (Omer et al. 2003). The study of porcine islets as a plausible islet source continues, and a variety of groups, using adult, fetal, and neonatal pig islets, also support the findings of Lanza, Sun, and Foster that porcine islets can maintain normal glycemia for extended periods of time (Calafiore et al. 2004; Cruise et al. 1999; Dufrane et al. 2006b; Duvivier-Kali et al. 2004; Foster et al. 2007; Lanza et al. 1993, 1999; Omer et al. 2003; Safley et al. 2008).

Although all of these show promising results and have stimulated interest in using porcine islets as a clinical source for transplant, several obstacles still remain. Adult porcine islets are shown to be difficult and expensive to isolate, and the islets tend to fragment during the isolation and purification processes (Falorni et al. 1996; Kuo et al. 1993). Conversely, neonatal and fetal islet isolation are much less difficult; however, studies have shown these islets require time to function in vivo (Mancuso et al. 2010). Even though these drawbacks can be addressed through changes in processes, all transplanted porcine islets need to be either encapsulated, within an immunoprotective layer to circumvent the species barrier, thus protect the islets from graft rejection, or the recipients receiving them must be given adequate immunosuppressive therapy, with numerous adverse effects (Hering et al. 2006; Buhler et al. 2002).

Limitations of Xenotransplantation

Xenotransplantation has a variety of limitations, most notably the immunologic and physiologic incompatibilities between animals and humans (Ibrahim et al. 2006). Another cause for concern is the potential of transplanted animal tissue to transmit endogenous retroviruses and other infectious agents across the species barrier, posing significant risk to the patient's health (Boneva et al. 2001; Fano et al. 1998).

According to the United States Federal Drug Administration's guidelines for xenotransplantation process, the potential health risk to the patient and community may include the possible transmission of infectious agents, which may not be detectable or infectious to the animal host. With the possibility of transmission, the FDA's report also outlines that there is an elevated risk of infection when transplanted into immunocompromised recipients, and there is a potential for recombination of animal and human genetic material, within the recipient, thus creating new pathogenic forms of microorganisms ((CBER) CfBEaR 2003).

Pig to Human Infections

Due to the similarities in overall physiology, there is a potential for pathogens to be transmitted from pigs to humans (Fano et al. 1998). There has been a special interest in a group of viruses known as porcine endogenous retroviruses, or PERVs. This interest is, namely, because of their potential to infect humans post transplant, shown by their ability to infect human cells in vitro and their continued presence in the donor animal's genome (Clark 1999; Denner 1998; Vanderpool 1999; Wilson 2008). Even though the potential of transmission exists, extensive investigations have uncovered no evidence of a human infection with PERV (Cozzi et al. 2005; Denner and Tonjes 2012).

Even though there has yet to be a recorded human infection with PERV or other porcine-transmitted microorganisms, there is still the possibility to infect and harm the transplant patient, which could extend additional risk to the transplant hospital personnel, family members, and to the wider community (Fishman 1997; Fishman and Patience 2004; Bach et al. 1998).

Due to the risk factors associated with the transmission of infectious agents, there must be well-outlined and regulated standards implemented long before any clinical trial can commence. In 2004 the International Xenotransplant Association (IXA) Ethics Committee released a position paper stating that there must be strict regulations in order to minimize the risks of contracting animal pathogens during and after transplantation, including the need to ensure appropriate animal husbandry, which entails the use of barrier-contained breeding facilities and stringent controls for surgical procedures and screening of cells or organs for transplant. All of these regulations must also be overseen by federal or national agencies to be effective in minimizing health risks (Sykes et al. 2003).

These standards will hopefully help minimize the use of xenotransplantation without clear governmental regulation and increase the understanding for national guidelines to minimize any potential health risk to the recipient and their community.

Facilities and Microbial Safety

Microbial safety is an important focus when it comes to xenotransplantation, and many facilities and transplant centers are planning to create standards for animal housing, surgery, islet isolations, and clinical transplant, to minimize the transmission of pathogens. Many centers have implemented current good manufacturing practices (cGMP) for islet processing facilities to provide a clean and safe environment for clinical transplants (Larijani et al. 2012; Korbutt 2009). Many studies have focused on improving these facilities and determining the microbial output they may have. One study performed reported that there was a 31 % microbial contamination in the transplant solution of human pancreases, but through the process of surface decontamination and islet processing, there was a 92 % clearance in microbial contamination (Kin et al. 2007). Even though the original solution was positive for contamination, the risk of a positive culture after processing is low but still possible within a cGMP facility. From these results in cadaveric human islet transplantation, many centers are trying to develop sterile animal housing facilities for breeding and pancreas procurement, along with established clean rooms used for islet processing and final product release (Schuurman 2009).

One of the first pathogen-free pig farms was implemented in New Zealand and led by Living Cell Technologies (LCT) (2013). The LCT farm holds a herd of rare pathogen-free Auckland Island pigs. The current facility, under the regulation of the New Zealand government, houses over 1,000 virus- and disease-free pigs for breeding and islet isolation (Living Cell Technologies 2013).

The Spring Point Project (SPP), in Minneapolis Minnesota, currently has a pathogen-free, medical grade porcine facility capable of breeding, housing, and pancreas procurement and has received the Class V medical license from the FDA in Aug 2012 (Bosserman 2010).

Facilities like LCT and SPP require extensive resources in order to develop. Stable farm sentinel pigs must be tested monthly for a variety of pathogens, including *Mycoplasma hyopneumoniae* and *Actinobacillus pleuropneumoniae*, and all products or animals leaving the facilities must be tested to confirm a non-disease state (Muirhead and Alexander 1997).

The Public Health Service (PHS) Guidelines state that extensive preclinical data must be obtained in order to progress to clinical trials and screening must be performed on all source herds and the final product for transplantation, along with stringent testing of clinical patients (Schuurman 2008).

Despite stringent testing being implemented and improved facilities being designed, the risk of infection still exists. To remove this risk, these pathogens must be removed from the source animal's genome. PHS states that a virus-free herd and final product are needed to ensure no pathogens are transferred from the source to the patient (Allan 1995), but this process will take time and is still in the early stages.

Islet Xenotransplantation Transplant Studies: Large Animals and Clinical Trials

Canine Islet Xenotransplantation

Large animal xenotransplantation trials help collect the requisite preclinical data necessary to justify proceeding into clinical trials because of the metabolic and immunological similarities between larger animals and humans. Canines have been used as large animal models because of the ease of training and also because they are a well-established diabetes model, either by chemical induction (streptozotocin or alloxan) or by surgical pancreatectomy (Anderson et al. 1993; Guptill et al. 2003). Many groups have shown positive results using encapsulated porcine islets transplanted into diabetic canines. In 2009 a group from Argentina microencapsulated adult pig islets and reported decreased insulin requirements and glycosylated hemoglobin (HbA1c) for up to 6–12 months post transplantation (Abalovich et al. 2009). Previous articles have reported that encapsulated porcine islets transplanted into diabetic canines can maintain euglycemia for >100 days with detectable porcine C-peptide levels in the serum (Kin et al. 2002; Monaco et al. 1991; Petruzzo et al. 1991).

With these reports and continued interest among researchers, canines offer a valid model for preclinical xeno-islet transplantation studies.

Nonhuman Primate Islet Xenotransplantation

Nonhuman primate (NHP) transplant models offer a stepping stone to the clinic by offering the closest model to a human for transplant, due to the similarities they share with humans, genetically, immunologically, and physically (Dufrane and Gianello 2012). Studies suggest that there is the potential for long-term survival and function of xenotransplanted, in contrast to studies performed with primate solid organ xenotransplantation (Cozzi 2009).

However, a study out of the University of Minnesota reported that when adult porcine islets were transplanted into NHP, the C-peptide levels were lower than NHPs transplanted with cadaveric human islets, thus suggesting lower glucose-stimulated insulin release by the transplanted porcine islets (Graham et al. 2011). Although encouraging, these results must be further addressed if porcine islets are to be used as a viable source for transplant and lead to the development of an improved, functional porcine islet type. A study in 2010 by Dufrane et al. reported that, without encapsulation, adult porcine islets are rejected after 7 days when transplanted under the kidney capsule of NHPs without immunosuppression, while the same islets can maintain viability and function in vivo for up to 6 months, if encapsulated first (Dufrane et al. 2010).

In 2012, a comprehensive review in Diabetes reported that several research centers have had successes with pig to NHP islet transplants, with normoglycemia maintained for up to 6 months post transplant. Out of the six reported studies, five reported the combination of transplants and immunosuppression with porcine donor tissue of varying ages ranging from adult to fetal islet tissue. Furthermore, one study even transplanted GTKO neonatal porcine islets with two out of the six groups within this study used encapsulation techniques (van der Windt et al. 2012).

It is still unknown which porcine islet model and age is "best" for clinical transplant and whether or not there is a need for encapsulation or immunosuppression regimes at this stage.

Clinical Islet Xenotransplanation

With all the current support for xenotransplantation and the NHP data to support it, this procedure has been implemented in many countries. However, efforts within the United States have yet to result in meaningful change. As of today, several countries all over the world are performing clinical trials using porcine islets (Table 1).

In 1994, Groth et al. found that isolated fetal porcine islets transplanted into diabetic kidney transplant patients, either intraportally or under the kidney capsule, were able to produce porcine C-peptide for up to 400 days post transplantation (Groth et al. 1994). This study, along with current findings, offers encouragement that the use of optimized islet preparations and encapsulation techniques will lead to improved clinical outcomes and result in long-term survival of transplanted islets (Elliott et al. 2007).

One of the most notable and well-documented clinical trials are the trials being performed by Living Cell Technologies (LCT). LCT is a company based in Australia that focuses on encapsulated cellular therapies for human therapeutics and currently has clinical trials in three countries for treating and curing T1D (Living Cell Technologies 2013). LCT's clinical trials use pathogen-free, medical grade fetal pigs that have been encapsulated with their "Diabecell" device. The first clinical trial lead by LCT was conducted in Australia. This trial was to demonstrate the safety and function of the "Diabecell" device (Living Cell Technologies 2012a). This trial reported that patients showed improvement in HbA1c levels and decreases in hypoglycemia unawareness (Living Cell Technologies 2012b). The results of the first trial from Australia led to a phase I/IIa clinical trial in New Zealand, where a total of 14 patients were transplanted. They were classified into groups based on the dose of islets they received: 5,000, 10,000, 15,000, and 20,000 IEQ/kg doses. A reduction in daily insulin requirements in all groups, with a statistically significant improvement in hypoglycemic unawareness events in the groups transplanted with 5,000 and 10,000IEQ was reported. In addition, there was no evidence of any transmissible zoonotic infection in the recipients (Living Cell Technologies 2012a).

Then in 2011 LCT commenced a third phase I/IIa clinical trial in Argentina. Patients enrolled in this trial received a dose of either 10,000 IEQ/kg (n = 4) or two implants (12 weeks apart) of 5,000 IEQ/kg each (n = 4). Reductions in hypoglycemic events, HbA1c levels, and daily insulin requirements were noted in both groups after 24 weeks post transplant (Living Cell Technologies 2012c). The results from all of these studies are still underway, but the results from all of them have been overwhelmingly positive. Patients demonstrated lower HbA1c levels and lower daily blood glucose levels.

With an increasing number of clinical trials being performed all over the world in the field of xenotransplantation, there has been increased interest in further developments and future applications. Even though the majority of patients have seen improvements in their diabetes management, without infection of evidence of PERV injection noted to date, long-term studies are still required to confirm improvements in recipient quality of life (Denner and Tonjes 2012; Argaw et al. 2004; Delwart and Li 2012; Garkavenko et al. 2004; Boneva et al. 2001).

Table 1 Xenotransplant clinical trials: pig to human transplant. Summary of clinical islet xenotransplantation for type 1 diabetes treatment (National Health and Medical Research Council 2013; Sgroi et al. 2010)

Tissue type	Country	Trial period	# Patients (age)	Encapsulation type	Company/author	Results
Fetal pig islets	New Zealand	2010–present	20+ (adult)	Alginate capsules	LCT (R.B Elliot) (Living Cell Technologies 2012b)	Improved HbA1c and reduction in hyperglycemia. Two patients were able to stop insulin after 4 months
Fetal pig islets	Russia	2007–present	8 (21–68)	Alginate capsules	LCT (R.B Elliot) (Living Cell Technologies 2010)	Six out of the eight patients had improved blood glucose control, with reduction in both insulin requirements and HbA1c. Two patients discontinued insulin for 8 months
Fetal pig islets	Argentina	2011–present	8 (adult), 20 recruited	Alginate capsules	LCT (R.B Elliot) (Living Cell Technologies 2012c)	No reports yet
Porcine islets	Mexico	2002 and 2005	12 (14.7 mean years)	Collagen devise w/2 steel mesh tubes	Valdes Gonzales (2002)	All patients infection free and two became insulin independent for several months; one continues to be insulin free after 2 years
Pig islets	Ukraine	1995, in progress	Unknown	None	I.S. Turchin (1995)	Unknown
Fetal pig islet like clusters	Sweden	1979–1993	10	None	C Gustav Groth (1994)	No C-peptide detected in plasma, but detected in urine after 460 days. Kidney biopsies showed viable cells
Pig islets	China	2005	22	None; with immunosupression	W. Wang (2011)	Reduction of insulin requirements and lowering of HbA1c. Detection of porcine C-peptide, immunosuppressants stopped after 1 year

Recent Developments and Future Applications

Genetically Modified Porcine Islets

Genetically modified porcine islets could offer advantages in graft function and limit the host's immune response to the transplant. Current studies focus on gene transfer techniques in order to decrease the host response to the graft. The genetic transduction of genes like antiapoptotic and antinecrotic Bcl-2 into porcine islets has shown to decrease the release of natural xenoreactive antibodies and cell lysis by the complement cascade (Contreras et al. 2001). Other molecules, like MSPEG (PEG-mono-succimidyl-succinate) and DSPEG (PEG-di-succimidyl-succinate), in combination with Bcl-2 transduction have shown no negative effect on islet viability, function, or morphology (Hui et al. 2005). Results from studies using these methods have also shown that these combinations have positive effects, with a reduction in lactate dehydrogenase release, a marker for cellular necrosis, when transplanted into animal recipients (Contreras et al. 2004). The use of T-cell inhibitors to limit the graft rejection by the adaptive immune system has also been studied. In a study by Klymiuk et al., neonatal porcine islets were isolated from transgenetic pigs that expressed LEA29Y, which inhibits T-cell costimulation. Isolated islets were then transplanted into diabetic immunocompetent mice and resulted in complete protection from being rejected by the host (Klymiuk et al. 2012).

These approaches and the development of genetically modified porcine islets may allow for transplantation without the need for toxic immunosuppressive regimens allowing for better survival of the islet graft and better overall health of the transplant recipient.

Gal-Deficient Galactosyl Transferase Knockout and Pathogen-Free Hybrids

The first target in antibody-mediated rejection in the pig to human transplant paradigm is the carbohydrate galactose-α1,3-galactose (α-gal). In porcine islets, expression levels varied depending on the size of the islet and not the age of the pig donor (Dufrane et al. 2005). Since nearly all porcine islets express α-gal, there have been recent developments targeted at the creation of a Gal knockout pig (GTKO) strain. GTKO pigs are deficient in α 1,3 galactosyl transferase, a strategy that addresses antibody-mediated α-Gal-based rejection. Studies characterizing GTKO pig islets have demonstrated insulin independence and establishment of normoglycemia (Thompson et al. 2011). A study in 2011 compared wild-type neonatal porcine islets (non-GTKO islets) with neonatal-GTKO islets and found an 80 % insulin independence with the GTKO islets compared to 20 % insulin independence with the wild type (Thompson et al. 2011). This experiment and the development of GTKO pigs allow for improved consistency and overall transplantation outcomes for the treatment of T1D.

Along with GTKO pigs, researchers have also been focusing on limiting or eradicating the occurrence of porcine-specific pathogens, thus creating pathogen-free pigs (Denner and Tonjes 2012; Vanrompay et al. 2005). Controlled environment breeding and housing facilities of porcine donors along with sterile rooms for pancreas procurement allow for a decreased risk of microbial contamination from outside sources and enhanced overall transplantation success (Gouin et al. 1997). Specific pathogen-free (SPF) pig farms have also created a suitable and reliable source for islet transplant that limit the risk of porcine microbe transmission but are limited because they do not limit PERVs because they are so well integrated into the pigs genome (Denner and Tonjes 2012). Without the development of a cross-species model, the study and validation of PERV transmission rates is nearly impossible (Denner et al. 2009). Thus, the creation of a PERV-free pig strain remains crucial to further research in this field.

Immunoisolation for Islet Xenotransplantation

The implementation of xenogeneically sourced islets for transplant therapy of T1D requires a solution to tackle the problem of acute rejection (Lanza et al. 1997; Platt 2000). While immunosuppression has previously been the standard therapy to prevent rejection in human islet allotransplantation, immunoisolation has been studied as an alternative for xenograft islets due to the presence of cell surface antigens such as α-gal (La Flamme et al. 2007; Wang et al. 1997).

A successful immunoisolation technique can allow islets to be transplanted without any immunosuppression. One such in vivo study utilizing porcine islets that have been immunoisolated using encapsulation in alginate showed remarkable longevity in the reversal of diabetes, showing blood glucose management up to 550 days when transplanted into diabetic rats (Meyer et al. 2008). There are several factors that influence this potential for success. One major factor involved in transplant success is the quality and purity of the islets that have been isolated because a purer preparation of islets reduces the chance of extraneous exocrine tissue to protrude out of the encapsulation layer (Narang and Mahato 2006; Omer et al. 2005). Another factor involved is the chemical composition, and purity of the alginate of greater purity has shown to improve the biocompatibility of the capsules, such that the immunoisolated islets are not recognized as foreign bodies (King et al. 2003; Hernandez et al. 2010).

Future Direction of Islet Xenotransplantation

The future direction of xenotransplantation may be headed in the development of a genetically engineered donor animal in order to reduce immunogenicity post transplant or through the development of immunoisolating materials such as alginate to allow the transplanted xenograft to remain undetected by the host immune system. Improvements in these technologies could make xenotransplantation a viable treatment option to circumvent the current donor tissue shortage that truly limits our ability to treat the diabetic population.

Cross-References

▶ Advances in Clinical Islet Isolation
▶ Islet Encapsulation
▶ Successes and Disappointments with Clinical Islet Transplantation
▶ The Comparative Anatomy of Islets

References

(CBER) CfBEaR (2003) Guidance for industry: source animal, product, preclinical, and clinical issues concerning the use of xenotransplantation products in humans. Food and Drug Administration. http://www.fda.gov/biologicsbloodvaccines/guidancecomplianceregulatoryinformation/guidances/xenotransplantation/ucm074354.htm#CLINICALISSUESINXENOTRANSPLANTATION. Accessed 27 Aug 2013

Abalovich AG, Bacque MC, Grana D, Milei J (2009) Pig pancreatic islet transplantation into spontaneously diabetic dogs. Transplant Proc 41(1):328–330

Alexander EL, Dooley KC, Pohajdak B, Xu BY, Wright JR Jr (2006) Things we have learned from tilapia islet xenotransplantation. Gen Comp Endocrinol 148(2):125–131

Allan JS (1995) Xenograft transplantation of the infectious disease conundrum. ILAR J/Natl Res Counc Inst Lab Anim Resour 37(1):37–48

Anderson HR, Stitt AW, Gardiner TA, Lloyd SJ, Archer DB (1993) Induction of alloxan/streptozotocin diabetes in dogs: a revised experimental technique. Lab Anim 27(3):281–285

Argaw T, Colon-Moran W, Wilson CA (2004) Limited infection without evidence of replication by porcine endogenous retrovirus in guinea pigs. J Gen Virol 85(Pt 1):15–19

Ayares D, Phelps C, Vaught T, Ball S, Monahan J, Walters A, Giraldo A, Bertera S, van der Windt D, Wijkstrom M, Cooper D, Bottino R, Trucco M (2013) Multi-transgenic pigs for xenoislet transplantation. Xenotransplantation 20(1):46

Bach FH, Fishman JA, Daniels N, Proimos J, Anderson B, Carpenter CB, Forrow L, Robson SC, Fineberg HV (1998) Uncertainty in xenotransplantation: individual benefit versus collective risk. Nat Med 4(2):141–144

Boneva RS, Folks TM, Chapman LE (2001) Infectious disease issues in xenotransplantation. Clin Microbiol Rev 14(1):1–14

Bosserman J (2010) Animal care facility receives highest accreditation. Spring Point Project. http://www.springpointproject.org/wp-content/uploads/2012/07/July-2010-Animal-care-facility-receives-highest-accreditation.pdf. Accessed 27 Aug 2013

Buhler L, Basker M, Alwayn IP, Goepfert C, Kitamura H, Kawai T, Gojo S, Kozlowski T, Ierino FL, Awwad M, Sachs DH, Sackstein R, Robson SC, Cooper DK (2000) Coagulation and thrombotic disorders associated with pig organ and hematopoietic cell transplantation in nonhuman primates. Transplantation 70(9):1323–1331

Buhler L, Deng S, O'Neil J, Kitamura H, Koulmanda M, Baldi A, Rahier J, Alwayn IP, Appel JZ, Awwad M, Sachs DH, Weir G, Squifflet JP, Cooper DK, Morel P (2002) Adult porcine islet transplantation in baboons treated with conventional immunosuppression or a non-myeloablative regimen and CD154 blockade. Xenotransplantation 9(1):3–13

Bulato C, Radu C, Simioni P (2012) Studies on coagulation incompatibilities for xenotransplantation. Methods Mol Biol 885:71–89

Calafiore R, Basta G, Luca G, Calvitti M, Calabrese G, Racanicchi L, Macchiarulo G, Mancuso F, Guido L, Brunetti P (2004) Grafts of microencapsulated pancreatic islet cells for the therapy of diabetes mellitus in non-immunosuppressed animals. Biotechnol Appl Biochem 39(Pt 2):159–164

Chhabra P, Brayman KL (2011) Current status of immunomodulatory and cellular therapies in preclinical and clinical islet transplantation. J Transplant 2011:637692

Clark MA (1999) This little piggy went to market: the xenotransplantation and xenozoonose debate. J Law Med Ethics 27(2):137–152

Clayton HA, James RF, London NJ (1993) Islet microencapsulation: a review. Acta Diabetol 30(4):181–189

Cnop M, Welsh N, Jonas JC, Jorns A, Lenzen S, Eizirik DL (2005) Mechanisms of pancreatic β-cell death in type 1 and type 2 diabetes: many differences, few similarities. Diabetes 54(Suppl 2):S97–S107

Contreras JL, Bilbao G, Smyth C, Eckhoff DE, Xiang XL, Jenkins S, Cartner S, Curiel DT, Thomas FT, Thomas JM (2001) Gene transfer of the Bcl-2 gene confers cytoprotection to isolated adult porcine pancreatic islets exposed to xenoreactive antibodies and complement. Surgery 130(2):166–174

Contreras JL, Xie D, Mays J, Smyth CA, Eckstein C, Rahemtulla FG, Young CJ, Anthony Thompson J, Bilbao G, Curiel DT, Eckhoff DE (2004) A novel approach to xenotransplantation combining surface engineering and genetic modification of isolated adult porcine islets. Surgery 136(3):537–547

Cooper DK, Ayares D (2011) The immense potential of xenotransplantation in surgery. Int J Surg 9(2):122–129

Cowan PJ, Aminian A, Barlow H, Brown AA, Dwyer K, Filshie RJ, Fisicaro N, Francis DM, Gock H, Goodman DJ, Katsoulis J, Robson SC, Salvaris E, Shinkel TA, Stewart AB, d'Apice AJ (2002) Protective effects of recombinant human antithrombin III in pig-to-primate renal xenotransplantation. Am J Transplant 2(6):520–525

Cozzi E (2009) On the road to clinical xenotransplantation. Transpl Immunol 21(2):57–59

Cozzi E, Bosio E, Seveso M, Vadori M, Ancona E (2005) Xenotransplantation-current status and future perspectives. Br Med Bull 75–76:99–114

Cruise GM, Hegre OD, Lamberti FV, Hager SR, Hill R, Scharp DS, Hubbell JA (1999) In vitro and in vivo performance of porcine islets encapsulated in interfacially photopolymerized poly(ethylene glycol) diacrylate membranes. Cell Transplant 8(3):293–306

Delwart E, Li L (2012) Rapidly expanding genetic diversity and host range of the Circoviridae viral family and other Rep encoding small circular ssDNA genomes. Virus Res 164(1–2):114–121

Denner J (1998) Immunosuppression by retroviruses: implications for xenotransplantation. Ann N Y Acad Sci 862:75–86

Denner J, Tonjes RR (2012) Infection barriers to successful xenotransplantation focusing on porcine endogenous retroviruses. Clin Microbiol Rev 25(2):318–343

Denner J, Schuurman HJ, Patience C (2009) The International Xenotransplantation Association consensus statement on conditions for undertaking clinical trials of porcine islet products in type 1 diabetes – chapter 5: strategies to prevent transmission of porcine endogenous retroviruses. Xenotransplantation 16(4):239–248

Derewenda U, Dodson GG (1993) Molecular structures in biology. Oxford University Press, Oxford, pp 260–277

Dickson BC, Yang H, Savelkoul HF, Rowden G, van Rooijen N, Wright JR Jr (2003) Islet transplantation in the discordant tilapia-to-mouse model: a novel application of alginate microencapsulation in the study of xenograft rejection. Transplantation 75(5):599–606

Dufrane D, Gianello P (2012) Pig islet for xenotransplantation in human: structural and physiological compatibility for human clinical application. Transplant Rev 26(3):183–188

Dufrane D, Goebbels RM, Guiot Y, Gianello P (2005) Is the expression of Gal-α1,3Gal on porcine pancreatic islets modified by isolation procedure? Transplant Proc 37(1):455–457

Dufrane D, Goebbels RM, Saliez A, Guiot Y, Gianello P (2006a) Six-month survival of microencapsulated pig islets and alginate biocompatibility in primates: proof of concept. Transplantation 81(9):1345–1353

Dufrane D, Steenberghe M, Goebbels RM, Saliez A, Guiot Y, Gianello P (2006b) The influence of implantation site on the biocompatibility and survival of alginate encapsulated pig islets in rats. Biomaterials 27(17):3201–3208

Dufrane D, Goebbels RM, Gianello P (2010) Alginate macroencapsulation of pig islets allows correction of streptozotocin-induced diabetes in primates up to 6 months without immunosuppression. Transplantation 90(10):1054–1062

Duvivier-Kali VF, Omer A, Lopez-Avalos MD, O'Neil JJ, Weir GC (2004) Survival of microencapsulated adult pig islets in mice in spite of an antibody response. Amer J Transplant 4(12):1991–2000

Dvornik D (1978) Chapter 17. Chronic complications of diabetes. In: Annual reports in medicinal chemistry, pp 159–166

Ekser B, Cooper DK (2010) Overcoming the barriers to xenotransplantation: prospects for the future. Expert Rev Clin Immunol 6(2):219–230

Elliott RB, Escobar L, Tan PL, Garkavenko O, Calafiore R, Basta P, Vasconcellos AV, Emerich DF, Thanos C, Bambra C (2005) Intraperitoneal alginate-encapsulated neonatal porcine islets in a placebo-controlled study with 16 diabetic cynomolgus primates. Transplant Proc 37(8):3505–3508

Elliott RB, Escobar L, Tan PL, Muzina M, Zwain S, Buchanan C (2007) Live encapsulated porcine islets from a type 1 diabetic patient 9.5 yr after xenotransplantation. Xenotransplantation 14(2):157–161

Falorni A, Basta G, Santeusanio F, Brunetti P, Calafiore R (1996) Culture maintenance of isolated adult porcine pancreatic islets in three-dimensional gel matrices: morphologic and functional results. Pancreas 12(3):221–229

Fano A, Cohen MJ, Cramer M, Greek R, Kaufman SR (1998) Of pigs, primates, and plagues: a layperson's guide to the problems with animal-to-human organ transplants. Med Res Modernization Comm, http://www.mrmcmed.org/pigs.html. Accessed August 27, 2013 2013

Figliuzzi M, Zappella S, Morigi M, Rossi P, Marchetti P, Remuzzi A (2000) Influence of donor age on bovine pancreatic islet isolation. Transplantation 70(7):1032–1037

Fishman JA (1997) Xenosis and xenotransplantation: addressing the infectious risks posed by an emerging technology. Kidney Int Suppl 58:S41–S45

Fishman JA, Patience C (2004) Xenotransplantation: infectious risk revisited. Ame J Transplant 4(9):1383–1390

Foster JL, Williams G, Williams LJ, Tuch BE (2007) Differentiation of transplanted microencapsulated fetal pancreatic cells. Transplantation 83(11):1440–1448

Frank A, Deng S, Huang X, Velidedeoglu E, Bae YS, Liu C, Abt P, Stephenson R, Mohiuddin M, Thambipillai T, Markmann E, Palanjian M, Sellers M, Naji A, Barker CF, Markmann JF (2004) Transplantation for type I diabetes: comparison of vascularized whole-organ pancreas with isolated pancreatic islets. Ann Surg 240(4):631–640; discussion 640–633

Friberg A (2011) Standardization of islet isolation and transplantation variables. Digit Compr Summ Upps Diss Fac Med 669:76

Garkavenko O, Croxson MC, Irgang M, Karlas A, Denner J, Elliott RB (2004) Monitoring for presence of potentially xenotic viruses in recipients of pig islet xenotransplantation. J Clin Microbiol 42(11):5353–5356

Gouin E, Rivereau AS, Darquy S, Cariolet R, Jestin A, Reach G, Sai P (1997) Minimisation of microbial contamination for potential islet xenografts using specific pathogen-free pigs and a protected environment during tissue preparation. Diabetes Metab 23(6):537–540

Graham ML, Bellin MD, Papas KK, Hering BJ, Schuurman HJ (2011) Species incompatibilities in the pig-to-macaque islet xenotransplant model affect transplant outcome: a comparison with allotransplantation. Xenotransplantation 18(6):328–342

Greene SA, Smith MA, Cartwright B, Baum JD (1983) Comparison of human versus porcine insulin in treatment of diabetes in children. British Med J 287(6405):1578–1579

Groth CG, Korsgren O, Tibell A, Tollemar J, Moller E, Bolinder J, Ostman J, Reinholt FP, Hellerstrom C, Andersson A (1994) Transplantation of porcine fetal pancreas to diabetic patients. Lancet 344(8934):1402–1404

Guptill L, Glickman L, Glickman N (2003) Time trends and risk factors for diabetes mellitus in dogs: analysis of veterinary medical data base records (1970–1999). Vet J 165(3):240–247

Han X, Tuch BE (2001) Cloning and characterization of porcine insulin gene. Comp Biochem Physiol B Biochem Mol Biol 129(1):87–95

Heller SR (2008) Minimizing hypoglycemia while maintaining glycemic control in diabetes. Diabetes 57(12):3177–3183

Hering BJ, Kandaswamy R, Ansite JD, Eckman PM, Nakano M, Sawada T, Matsumoto I, Ihm SH, Zhang HJ, Parkey J, Hunter DW, Sutherland DE (2005) Single-donor, marginal-dose islet transplantation in patients with type 1 diabetes. JAMA 293(7):830–835

Hering BJ, Wijkstrom M, Graham ML, Hardstedt M, Aasheim TC, Jie T, Ansite JD, Nakano M, Cheng J, Li W, Moran K, Christians U, Finnegan C, Mills CD, Sutherland DE, Bansal-Pakala P, Murtaugh MP, Kirchhof N, Schuurman HJ (2006) Prolonged diabetes reversal after intraportal xenotransplantation of wild-type porcine islets in immunosuppressed nonhuman primates. Nat Med 12(3):301–303

Hernandez RM, Orive G, Murua A, Pedraz JL (2010) Microcapsules and microcarriers for in situ cell delivery. Adv Drug Deliv Rev 62(7–8):711–730

Hui H, Khoury N, Zhao X, Balkir L, D'Amico E, Bullotta A, Nguyen ED, Gambotto A, Perfetti R (2005) Adenovirus-mediated XIAP gene transfer reverses the negative effects of immunosuppressive drugs on insulin secretion and cell viability of isolated human islets. Diabetes 54(2):424–433

HumanXenoTransplant (2013) Inventory of human xenotransplantation practices. http://www.humanxenotransplant.org/home/index.php. Accessed 26 Aug 2013

Ibrahim Z, Busch J, Awwad M, Wagner R, Wells K, Cooper DK (2006) Selected physiologic compatibilities and incompatibilities between human and porcine organ systems. Xenotransplantation 13(6):488–499

Ierino FL, Kozlowski T, Siegel JB, Shimizu A, Colvin RB, Banerjee PT, Cooper DK, Cosimi AB, Bach FH, Sachs DH, Robson SC (1998) Disseminated intravascular coagulation in association with the delayed rejection of pig-to-baboon renal xenografts. Transplantation 66(11):1439–1450

Kin T, Iwata H, Aomatsu Y, Ohyama T, Kanehiro H, Hisanaga M, Nakajima Y (2002) Xenotransplantation of pig islets in diabetic dogs with use of a microcapsule composed of agarose and polystyrene sulfonic acid mixed gel. Pancreas 25(1):94–100

Kin T, Rosichuk S, Shapiro AM, Lakey JR (2007) Detection of microbial contamination during human islet isolation. Cell Transplant 16(1):9–13

King A, Strand B, Rokstad AM, Kulseng B, Andersson A, Skjak-Braek G, Sandler S (2003) Improvement of the biocompatibility of alginate/poly-L-lysine/alginate microcapsules by the use of epimerized alginate as a coating. J Biomed Mater Res A 64(3):533–539

Klymiuk N, van Buerck L, Bahr A, Offers M, Kessler B, Wuensch A, Kurome M, Thormann M, Lochner K, Nagashima H, Herbach N, Wanke R, Seissler J, Wolf E (2012) Xenografted islet cell clusters from INSLEA29Y transgenic pigs rescue diabetes and prevent immune rejection in humanized mice. Diabetes 61(6):1527–1532

Korbutt GS (2009) The International Xenotransplantation Association consensus statement on conditions for undertaking clinical trials of porcine islet products in type 1 diabetes–chapter 3: pig islet product manufacturing and release testing. Xenotransplantation 16(4):223–228

Kozlowski T, Shimizu A, Lambrigts D, Yamada K, Fuchimoto Y, Glaser R, Monroy R, Xu Y, Awwad M, Colvin RB, Cosimi AB, Robson SC, Fishman J, Spitzer TR, Cooper DK, Sachs DH (1999) Porcine kidney and heart transplantation in baboons undergoing a tolerance induction regimen and antibody adsorption. Transplantation 67(1):18–30

Kuo CY, Myracle A, Burghen GA, Herrod HG (1993) Neonatal pig pancreatic islets for transplantation. In Vitro Cell Dev Biol Anim 29A(9):677–678

Kuwaki K, Knosalla C, Dor FJ, Gollackner B, Tseng YL, Houser S, Mueller N, Prabharasuth D, Alt A, Moran K, Cheng J, Behdad A, Sachs DH, Fishman JA, Schuurman HJ, Awwad M, Cooper DK (2004) Suppression of natural and elicited antibodies in pig-to-baboon heart transplantation using a human anti-human CD154 mAb-based regimen. Am J Transplant 4(3):363–372

Kuwaki K, Tseng YL, Dor FJ, Shimizu A, Houser SL, Sanderson TM, Lancos CJ, Prabharasuth DD, Cheng J, Moran K, Hisashi Y, Mueller N, Yamada K, Greenstein JL, Hawley RJ, Patience C, Awwad M, Fishman JA, Robson SC, Schuurman HJ, Sachs DH, Cooper DK (2005) Heart transplantation in baboons using α1,3-galactosyltransferase gene-knockout pigs as donors: initial experience. Nat Med 11(1):29–31

La Flamme KE, Popat KC, Leoni L, Markiewicz E, La Tempa TJ, Roman BB, Grimes CA, Desai TA (2007) Biocompatibility of nanoporous alumina membranes for immunoisolation. Biomaterials 28(16):2638–2645

Lanza RP, Butler DH, Borland KM, Staruk JE, Faustman DL, Solomon BA, Muller TE, Rupp RG, Maki T, Monaco AP et al (1991) Xenotransplantation of canine, bovine, and porcine islets in diabetic rats without immunosuppression. Proc Natl Acad Sci U S A 88(24):11100–11104

Lanza RP, Beyer AM, Staruk JE, Chick WL (1993) Biohybrid artificial pancreas. Long-term function of discordant islet xenografts in streptozotocin diabetic rats. Transplantation 56(5):1067–1072

Lanza RP, Kuhtreiber WM, Ecker DM, Marsh JP, Chick WL (1995a) Transplantation of porcine and bovine islets into mice without immunosuppression using uncoated alginate microspheres. Transplant Proc 27(6):3321

Lanza RP, Kuhtreiber WM, Ecker D, Staruk JE, Chick WL (1995b) Xenotransplantation of porcine and bovine islets without immunosuppression using uncoated alginate microspheres. Transplantation 59(10):1377–1384

Lanza RP, Cooper DK, Chick WL (1997) Xenotransplantation. Sci Am 277(1):54–59

Lanza RP, Jackson R, Sullivan A, Ringeling J, McGrath C, Kuhtreiber W, Chick WL (1999) Xenotransplantation of cells using biodegradable microcapsules. Transplantation 67(8):1105–1111

Larijani B, Arjmand B, Amoli MM, Ao Z, Jafarian A, Mahdavi-Mazdah M, Ghanaati H, Baradar-Jalili R, Sharghi S, Norouzi-Javidan A, Aghayan HR (2012) Establishing a cGMP pancreatic islet processing facility: the first experience in Iran. Cell Tissue Bank 13(4):569–575

Lin CC, Ezzelarab M, Shapiro R, Ekser B, Long C, Hara H, Echeverri G, Torres C, Watanabe H, Ayares D, Dorling A, Cooper DK (2010) Recipient tissue factor expression is associated with consumptive coagulopathy in pig-to-primate kidney xenotransplantation. Am J Transplant 10(7):1556–1568

Living Cell Technologies (2010) Report from LCT's Russian phase I/IIa diabetes trial. Living Cell Technologies, New Zealand

Living Cell Technologies (2012a) Diabecell clinical trials update. Living Cell Technologies, New Zealand

Living Cell Technologies (2012b) LCT announces positive DIABECELL Phase I/IIa trial. Living Cell Technologies, New Zealand

Living Cell Technologies (2012c) Strong interim results in Argentinian DIABECELL® trial. Living Cell Technologies, New Zealand

Living Cell Technologies (2013) LCT: the cell implant company. http://www.lctglobal.com. Accessed 27 Aug 2013

Mancuso F, Calvitti M, Luca G, Nastruzzi C, Baroni T, Mazzitelli S, Becchetti E, Arato I, Boselli C, Ngo Nselel MD, Calafiore R (2010) Acceleration of functional maturation and differentiation of neonatal porcine islet cell monolayers shortly in vitro cocultured with microencapsulated sertoli cells. Stem Cells Int 2010:587213

Marchetti P, Giannarelli R, Cosimi S, Masiello P, Coppelli A, Viacava P, Navalesi R (1995) Massive isolation, morphological and functional characterization, and xenotransplantation of bovine pancreatic islets. Diabetes 44(4):375–381

Meyer T, Hocht B, Ulrichs K (2008) Xenogeneic islet transplantation of microencapsulated porcine islets for therapy of type I diabetes: long-term normoglycemia in STZ-diabetic rats without immunosuppression. Pediatr Surg Int 24(12):1375–1378

Monaco AP, Maki T, Ozato H, Carretta M, Sullivan SJ, Borland KM, Mahoney MD, Chick WL, Muller TE, Wolfrum J et al (1991) Transplantation of islet allografts and xenografts in totally pancreatectomized diabetic dogs using the hybrid artificial pancreas. Ann Surg 214(3):339–360, discussion 361–332

Muirhead M, Alexander T (1997) The health status of the herd managing pig health and the treatment of disease

Narang AS, Mahato RI (2006) Biological and biomaterial approaches for improved islet transplantation. Pharmacol Rev 58(2):194–243

National Health and Medical Research Council (2013) Animal to human transplantation research (Xenotransplantation). http://www.nhmrc.gov.au/health-ethics/ethical-issues/animal-human-transplantation-research-xenotransplantation. Accessed 12 Apr 2013

National Kidney Disease Education Program (2013) Slow progression and reduce complications. http://nkdep.nih.gov/identify-manage/manage-patients/slow-progression.shtml. Accessed 26 Aug 2013

Omer A, Duvivier-Kali VF, Trivedi N, Wilmot K, Bonner-Weir S, Weir GC (2003) Survival and maturation of microencapsulated porcine neonatal pancreatic cell clusters transplanted into immunocompetent diabetic mice. Diabetes 52(1):69–75

Omer A, Duvivier-Kali V, Fernandes J, Tchipashvili V, Colton CK, Weir GC (2005) Long-term normoglycemia in rats receiving transplants with encapsulated islets. Transplantation 79(1):52–58

Petruzzo P, Pibiri L, De Giudici MA, Basta G, Calafiore R, Falorni A, Brunetti P, Brotzu G (1991) Xenotransplantation of microencapsulated pancreatic islets contained in a vascular prosthesis: preliminary results. Transpl Int 4(4):200–204

Platt JL (2000) Immunobiology of xenotransplantation. Transpl Int 13(Suppl 1):S7–S10

Richter B, Neises G (2002) 'Human' insulin versus animal insulin in people with diabetes mellitus. Cochrane Database Syst Rev (3):CD003816

Ris F, Toso C, Veith FU, Majno P, Morel P, Oberholzer J (2004) Are criteria for islet and pancreas donors sufficiently different to minimize competition? Amer J Transplant 4(5):763–766

Safley SA, Cui H, Cauffiel S, Tucker-Burden C, Weber CJ (2008) Biocompatibility and immune acceptance of adult porcine islets transplanted intraperitoneally in diabetic NOD mice in calcium alginate poly-L-lysine microcapsules versus barium alginate microcapsules without poly-L-lysine. J Diabetes Sci Technol 2(5):760–767

Schuurman HJ (2008) Regulatory aspects of pig-to-human islet transplantation. Xenotransplantation 15(2):116–120

Schuurman HJ (2009) The International Xenotransplantation Association consensus statement on conditions for undertaking clinical trials of porcine islet products in type 1 diabetes – chapter 2: source pigs. Xenotransplantation 16(4):215–222

Sgroi A, Buhler LH, Morel P, Sykes M, Noel L (2010) International human xenotransplantation inventory. Transplantation 90(6):597–603

Shapiro AM, Lakey JR, Ryan EA, Korbutt GS, Toth E, Warnock GL, Kneteman NM, Rajotte RV (2000) Islet transplantation in seven patients with type 1 diabetes mellitus using a glucocorticoid-free immunosuppressive regimen. N Engl J Med 343(4):230–238

Soto-Gutierrez A, Wertheim JA, Ott HC, Gilbert TW (2012) Perspectives on whole-organ assembly: moving toward transplantation on demand. J Clin Invest 122(11):3817–3823

Storms FE, Lutterman JA, van't Laar A (1987) Comparison of efficacy of human and porcine insulin in treatment of diabetic ketoacidosis. Diabetes Care 10(1):49–55

Sun AM, Vacek I, Sun YL, Ma X, Zhou D (1992) In vitro and in vivo evaluation of microencapsulated porcine islets. ASAIO J 38(2):125–127

Sutherland DE, Gruessner RW, Dunn DL, Matas AJ, Humar A, Kandaswamy R, Mauer SM, Kennedy WR, Goetz FC, Robertson RP, Gruessner AC, Najarian JS (2001) Lessons learned from more than 1,000 pancreas transplants at a single institution. Ann Sur 233(4):463–501

Sykes M, Dapice A, Sandrin M, Committee IXAE (2003) Position paper of the Ethics Committee of the International Xenotransplantation Association. Xenotransplantation 10(3):194–203

Thompson P, Badell IR, Lowe M, Cano J, Song M, Leopardi F, Avila J, Ruhil R, Strobert E, Korbutt G, Rayat G, Rajotte R, Iwakoshi N, Larsen CP, Kirk AD (2011) Islet xenotransplantation using gal-deficient neonatal donors improves engraftment and function. Am J Transplant 11(12):2593–2602

Titus T, Badet L, Gray DW (2000) Islet cell transplantation for insulin-dependent diabetes mellitus: perspectives from the present and prospects for the future. Expert Rev Mol Med 2(6):1–28

Tseng YL, Kuwaki K, Dor FJ, Shimizu A, Houser S, Hisashi Y, Yamada K, Robson SC, Awwad M, Schuurman HJ, Sachs DH, Cooper DK (2005) α1,3-Galactosyltransferase gene-knockout pig heart transplantation in baboons with survival approaching 6 months. Transplantation 80(10):1493–1500

Turchin IS, Tronko MD, Komissarenko VP (1995) Experience of 1.5 thousand transplantation of β-cells cultures to patients with diabetes mellitus. In: 5th IPITA Congress, Miami, 1995

U.S. Food and Drug Administration (2013) Guidance, compliance & regulatory information (Biologics). http://www.fda.gov/BiologicsBloodVaccines/GuidanceComplianceRegulatoryInformation/default.htm. Accessed 26 Aug 2013

Valdes R (2002) Xenotransplantation trials. Lancet 359(9325):2281

van der Windt DJ, Bottino R, Kumar G, Wijkstrom M, Hara H, Ezzelarab M, Ekser B, Phelps C, Murase N, Casu A, Ayares D, Lakkis FG, Trucco M, Cooper DK (2012) Clinical islet xenotransplantation: how close are we? Diabetes 61(12):3046–3055

Vanderpool HY (1999) Commentary: a critique of Clark's frightening xenotransplantation scenario. J Law Med Ethics 27(2):153–157

Vanrompay D, Hoang TQ, De Vos L, Verminnen K, Harkinezhad T, Chiers K, Morre SA, Cox E (2005) Specific-pathogen-free pigs as an animal model for studying Chlamydia trachomatis genital infection. Infect Immun 73(12):8317–8321

Vrochides D, Paraskevas S, Papanikolaou V (2009) Transplantation for type 1 diabetes mellitus. Whole organ or islets? Hippokratia 13(1):6–8

Wang T, Lacik I, Brissova M, Anilkumar AV, Prokop A, Hunkeler D, Green R, Shahrokhi K, Powers AC (1997) An encapsulation system for the immunoisolation of pancreatic islets. Nat Biotechnol 15(4):358–362

Wang W, Mo Z, Ye B, Hu P, Liu S, Yi S (2011) A clinical trial of xenotransplantation of neonatal pig islets for diabetic patients. Zhong Nan Da Xue Xue Bao Yi Xue Ban 36(12):1134–1140

Wang Y, Danelison KK, Ropski A, Harvat T, Barbar B, Paushter D, Qi M, Oberholzer J (2013) Systematic analysis of donor and isolation factor's impact on human islet yield and size distribution. Cell Transplant. doi:10.3727/096368912X662417

Wilson CA (2008) Porcine endogenous retroviruses and xenotransplantation. Cell Mol Life Sci 65(21):3399–3412

World Health Organization (2005) Xenotransplantation. http://www.who.int/transplantation/xeno/en/. Accessed 12 Apr 2013

Wright JR Jr, Yang H (1997) Tilapia Brockmann bodies: an inexpensive, simple model for discordant islet xenotransplantation. Ann Transplant 2(3):72–75

Wright JR Jr, Polvi S, MacLean H (1992) Experimental transplantation with principal islets of teleost fish (Brockmann bodies). Long-term function of tilapia islet tissue in diabetic nude mice. Diabetes 41(12):1528–1532

Wright JR Jr, Pohajdak B, Xu BY, Leventhal JR (2004) Piscine islet xenotransplantation. ILAR J/Natl Res Counc Inst Lab Anim Resour 45(3):314–323

Yang H, Wright JR Jr (1995) A method for mass harvesting islets (Brockmann bodies) from teleost fish. Cell Transplant 4(6):621–628

Yang H, O'Hali W, Kearns H, Wright JR Jr (1997) Long-term function of fish islet xenografts in mice by alginate encapsulation. Transplantation 64(1):28–32

Islet Encapsulation

46

Jonathan R. T. Lakey, Lourdes Robles, Morgan Lamb, Rahul Krishnan, Michael Alexander, Elliot Botvinick, and Clarence E. Foster

Contents

Introduction	1298
History	1298
Current Research	1299
Animal and Human Trials	1299
Biomaterials in Transplantation	1301
Future Applications	1304
Improved Capsule Engineering	1304
Stem Cells	1305
Conclusion	1306
References	1306

Abstract

Type 1 diabetes is an autoimmune disorder that destroys the insulin-producing cells of the pancreas. The mainstay of treatment is replacement of insulin through injectable exogenous insulin. Improvements in islet isolation techniques and immunosuppression regimens have made islet transplants a treatment option

---

J.R.T. Lakey (✉) • E. Botvinick
Department of Surgery, Biomedical Engineering, University of California Irvine, Orange, CA, USA
e-mail: jlakey@uci.edu; elliot.botvinick@gmail.com

L. Robles • M. Lamb • R. Krishnan • M. Alexander
Department of Surgery, University of California Irvine, Orange, CA, USA
e-mail: lyrobles@uci.edu; lamb.morgan@gmail.com; mlamb@uci.edu; michaela@uci.edu

C.E. Foster
Department of Surgery, Biomedical Engineering, University of California Irvine, Orange, CA, USA

Department of Transplantation, University of California Irvine, Orange, CA, USA
e-mail: fosterc@uci.edu

M.S. Islam (ed.), *Islets of Langerhans*, DOI 10.1007/978-94-007-6686-0_29,
© Springer Science+Business Media Dordrecht 2015

for select patients. Islet transplants have improved graft function over the years; however, graft function beyond year two is rare and notably these patients require immunosuppression to prevent rejection. Cell encapsulation has been proposed for numerous cell types, but it has found increasing enthusiasm for islets. Since islet transplants have experienced a myriad of success, the next step is to improve graft function and avoid systemically toxic immunosuppressive regimens. Cell encapsulation hopes to accomplish this goal. Encapsulation involves placing cells in a semipermeable biocompatible hydrogel that allows the passage of nutrients and oxygen and however blocks immune regulators from destroying the cell, thus avoiding systemic drugs. Several advances in encapsulation engineering and cell viability promise to make this a revolutionary discovery. This paper provides a comprehensive review of cell encapsulation of islets for the treatment of type 1 diabetes including a historical outlook, current research, and future studies.

Keywords
Cell encapsulation • Cell engineering • Islet transplants • Microcapsules • Type 1 diabetes

Introduction

Islet transplantation to treat type 1 diabetes has achieved vast improvements over the years, with more recipients able to achieve insulin independence for a longer period of time (Barton et al. 2012). Unfortunately, the lack of available donor organs and the use of antirejection medications continue to impede further progress. Encapsulation of islets for transplantation into diabetic recipients aims to provide a solution to these problems. Cell encapsulation is a revolutionary method of enveloping cells in a biocompatible matrix that provides a gradient for the diffusion of oxygen and nutrients but prevents large immune molecules from reaching the cell, thus avoiding rejection. This idea has been described since the 1930s but notable achievements have occurred over the last decade. This chapter aims to provide a review including a historical perspective, current research, and future applications of cell encapsulation of islets for the treatment of type 1 diabetes.

History

Over 25 million people in the United States (US) suffer from diabetes with approximately 5 % characterized as type 1. Diabetes is ranked as the seventh leading cause of death in the United States (CDC 2012). Type 1 diabetes (T1DM) is characterized as an autoimmune destruction of the β-cells of the pancreas with resultant insulin deficiency (ADA 2004). Currently, the mainstay of treatment for T1DM is glycemic control through injectable insulin. However, improvements in islet transplantation continue to occur, hopefully changing the treatment paradigm into cell replacement rather than supportive care. According to the Collaborative Islet Transplant Registry (CITR), there have been a total of 677 islet transplant recipients from 1999 to 2010. The percentage of recipients that achieve insulin independence for 3 years is 44 % between 2007 and 2010 compared to 27 % from 1999 to 2002 (Barton et al. 2012). Various immunosuppressive regimens have been implemented in islet transplant centers to maintain graft function. However, like other solid organ transplants, immune rejection medications are implicated in numerous adverse effects to the patient, as well as toxicity to the graft (Hafiz et al. 2005; Niclauss et al. 2011). To circumvent these issues, cell encapsulation has been proposed as the next treatment option for islet transplants with the goal of elimination of immunosuppression. Although cell encapsulation has been tested in other disease processes, such as neurodegenerative diseases, pain, and epilepsy to name a few, by far the greatest achievement using this technology has been in the encapsulation of islets for the treatment of T1DM (Bachoud-Lévi et al. 2000; Jeon 2011; Eriksdotter-Jönhagen et al. 2012; Fernández et al. 2012; Huber et al. 2012). It is clear that insulin independence can be achieved through the infusion of isolate islets, but improvements in graft viability and avoidance of systemically toxic drugs can be accomplished through encapsulation (Hearing and Ricordi 1999; Shapiro et al. 2003). The following paragraph will discuss advances in encapsulated islet technology.

Current Research

Animal and Human Trials

The first researcher to be accredited with transplantation of encapsulated tissue was Biscegeli in 1933. He placed mouse tumor cells in a polymer matrix into the abdomen of a guinea pig and was able to achieve survival without rejection (Bisceglie 1933). This idea was not replicated until approximately 50 years later when Lim and Sun were the first to use encapsulation for islet transplants in diabetic animals. They placed 2,000–3,000 islet equivalent (IEQ) in an alginate hydrogel for transplantation into the peritoneum of diabetic rats to achieve normoglycemia for up to 3 weeks compared to only 8 days in unencapsulated islets (Lim and Sun 1980). Currently, there are a number of achievements in encapsulating islets seen in small and large animal studies, as well as in early-phase clinical trials. In studies performed by Kobayashi et al. in 2003, NOD mice were used as both donor and recipient. The authors used a 5 % agarose microcapsule encasing 1,500–2,000 islet equivalents (IEQ) per mouse for intraperitoneal implantation as well as omental pouch transplants and witnessed prolonged euglycemia for a period of 100 days compared to 8 days for unencapsulated islet transplants (Kobayashi et al. 2003). The same authors repeated the study in 2006 and observed the same period of normoglycemia in the recipients; however, they also retrieved the devices after a period of 400 days and noted viable islets were recovered with a small percentage of necrotic cells (Kobayashi et al. 2006). In a more recent study, Dong et al. performed syngeneic transplants into streptozotocin (STZ)-induced BALB/c mice using polyethylene glycol-polylactic-co-glycolic acid (PEG-PLGA) nanoparticles with only 500–600 IEQ and revealed that over half of the mice achieved normal glucose levels for up to 100 days (Dong et al. 2012).

Less consistent but still notable results were accomplished with large animal models. Shoon-Shiong performed several encapsulated islet transplants into diabetic canines. In one publication in 1993, islets at a dose of 15,000–20,000 islets/kg were encapsulated in a microcapsule made of alginate into the intraperitoneal cavity and witnessed independence from exogenous insulin for a period of 110 days, as well as the presence of C-peptide in the blood for an average of 483 days (Soon-Shiong et al. 1993). In 2010, Dufrane used nonhuman primates as donor and recipients for subcutaneous and kidney capsule transplants of alginate micro- and macroencapsulated islets using 30,000 IEQ/kg. The authors noted correction of diabetes for a period of up to 28 weeks (Dufrane et al. 2010). In another study using cynomolgus monkeys as recipients by Elliott et al., neonatal pig islets were isolated (10,000 IEQ/Kg) and encapsulated in alginate microcapsules resulting in a more than 40 % reduction in injectable insulin dose compared to preimplantation (Elliott et al. 2005). Based on some noteworthy achievements in large animal studies, several researchers have been granted approval for stage one and two human clinical trials. Due to the previous success by Shoon-Shiong using a canine model, the authors were authorized for the first human clinical trial in 1994. A 38-year-old male, with type 1 diabetes and end-stage renal disease status postoperative kidney transplant, maintained on low-dose immunosuppression, became the first recipient of an encapsulated islet transplant. The patient initially received 10,000 IEQ/kg of cadaver islets encapsulated in an alginate microcapsule followed by a repeat infusion of 5,000 IEQ/kg 6 months later. The patients' insulin requirements were reduced to 1–2 insulin units per day, and eventually he was able to discontinue all exogenous insulin after 9 months (Soon-Shiong et al. 1994). In 2006 Calafiore et al. reported on their encapsulated islet transplants. Human cadavers (400,000–600,000 IEQ) were isolated for encapsulation into sodium alginate beads for intraperitoneal injection. The patients experienced improved daily glucose levels and a decline in daily exogenous insulin intake; however, neither patient became insulin independent (Calafiore et al. 2006).

Living Cell Technologies, a company out of Australia, has achieved the best outcomes for encapsulated islet transplants. In one arm of their studies, xenotransplants from fetal pigs were isolated from a pathogen-free farm in New Zealand. The islets were encapsulated in alginate microcapsules for intraperitoneal injections into human recipients. Several early-phase clinical trials have been performed from this group with promising results. The most significant achievement has been in the reduction of hypoglycemic episodes to around 40 %. Several patients achieved improvements in daily glucose levels and a reduction in exogenous insulin dosing; two patients became insulin independent after 4 months (Elliott et al. 2007, 2010, Elliott and Living cell technologies Ltd-LCT 2011). These are promising results and achievements; however, not all researchers have been able to achieve such results, and the lack of reproducibility is threatening to dampen enthusiasm. For example, a human clinical trial by Tuch et al. used alginate microcapsules for human cadaveric islet transplants and noted the presence of plasma C-peptide levels for up to 2.5 years; however, there was no change in insulin requirements for the recipients (Tuch et al. 2009). Likewise, in a follow-up publication by Elliot et al., from Living Cell Technologies, one recipient experienced early success with a 30 % reduction in insulin dosing; however, at follow-up at 49 weeks, the patient was back on his original insulin dosing regimen (Elliott et al. 2007). Although the purpose of these early-phase clinical trials is to assure safety and determine dosing, it is notable that most encapsulated islet recipients do not achieve sustainable insulin independence. Likewise, there is yet to be a standardized protocol for the type of biomaterial used and the dose of islets to be infused. It is clear, however, based on novel in vivo studies that the type of biomaterial does impact graft survival. King et al. tested several encapsulation methods using alginate with poly-L-lysine (PLL) or without as well as with high glucuronic acid (G) or high mannuronic acid (M) in mouse recipients and revealed that significant results were achieved with PLL-free high M microcapsules, with sustained normoglycemia for 8 weeks (King et al. 2003). Likewise, Lanza et al. revealed that improved capsule integrity and graft function could be achieved by altering the concentration of alginate in their xenotransplants into diabetic Lewis rats (Lanza et al. 1999). Although no consensus is achieved regarding the best encapsulation vehicles for islets transplants, no review would be complete without a discussion regarding biomaterials.

Biomaterials in Transplantation

Chang et al. were one of the first to describe the use of a semipermeable membrane for encapsulation. They postulated that liver enzymes or cells can be delivered in a polymer membrane for treatment of a disorder (Chang 1964). Now 50 years later, several types of encapsulation methods have been developed over the years however; currently the most common employed method are alginate microcapsules (King et al. 2003; Zimmermann et al. 2007; Krishnamurthy and Gimi 2011; Khanna et al. 2012). Capsule vehicles have taken the form of vascular shunts and macro-, micro-, and nanoscale devices. The original vascular device was developed over three decades ago as capillary fibers in culture-coated medium (Chick et al. 1975). Maki et al. performed several studies with vascular devices implanted as arteriovenous shunts into diabetic canines. These devices showed promising results with several canines achieving reduced insulin requirements (Maki et al. 1991, 1996). Ultimately, the difficulty with these devices was the ability to provide enough islets to coat the fibers. When attempts were made to elongate the fibers to include more islets, the devices resulted in clotting and fibrosis. This eventually led to its disuse as the higher dose of islets needed to achieve insulin independence in humans would result in the requirement for multiple devices to be employed Lanza et al. (1992). Macroscale devices have been used by a relative few, due to their

Fig. 1 Macroencapsulated islets. Islets were isolated from young piglets (average 20 days), matured for 8 days, and injected into a prefabricated islet sheet made of ultrapure high M alginate. Islets were placed in the center of the sheet with a scrim added for greater stability. Encapsulation design by Islet Science, LLC (Picture provided by Dr. Jonathan RT Lakey, UCI)

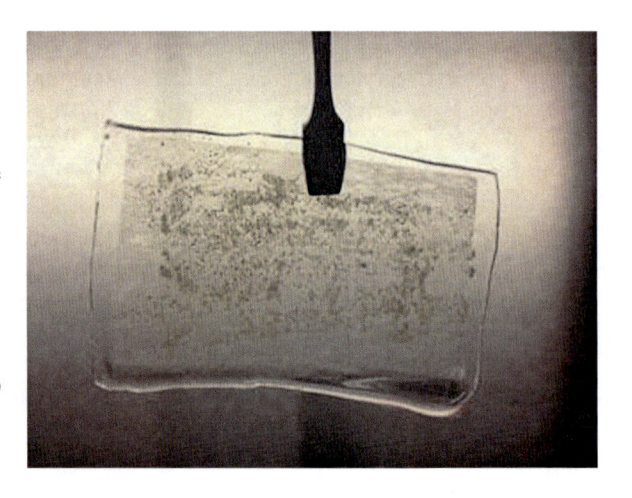

increased immunogenicity as well as the larger diffusion parameters required for oxygen and nutrients to reach the cell; however, they offer several advantages including implantation ease and retrievability. In a study using an alginate macroscaled sheet, diabetic canines achieved improved glucose levels for over 80 days (Storrs et al. 2001) (Fig. 1). Nanoencapsulation offers the advantage of improved diffusion parameters, improving the response of insulin to rising blood glucose levels. PEG has been used for nanoencapsulation devices, and when exposed to UV or visible light, it can be cross-linked with minimal damage to the inner cell. However, PEG is not as biocompatible as other hydrogels and does not provide complete protection from cytokines (Jang et al. 2004). Despite these concerns, some success has been attained with these gels (Dong et al. 2012). By far the most common encapsulation device is a microscale vehicle. These beads have mechanical stability, have an improved surface area to volume ratio, and have enhanced immunologic profiles (Krishnamurthy and Gimi 2011; Borg and Bonifacio 2011), but more importantly several companies make so-called encapsulators that can produce uniform capsules using air jet-driven droplet technology (Fiszman et al. 2002; Sun 1988) (Fig. 2). This is highly important when discussing using these capsules for clinical use as standardization, safety, and cost-effectiveness are going to be important aspects for its widespread clinical use. All of the clinical trials discussed in the previous paragraph used microcapsules for encapsulating islets. Along with the different capsule size, various materials have also been tested. Extracellular matrixes have included both synthetic and biologic materials. The most common synthetic agents employed in encapsulation engineering include polyethylene oxide, polyacrylic acid, polyvinyl alcohol, polyphosphazene, polypeptides, and derivatives. Naturally occurring hydrogels include gelatin, fibrin, agarose, hyaluronate, chitosan, and alginate (Nicodemus and Bryant 2008; Lee and Mooney 2000). Polyglycolic and lactic acid polymers continue to be the most commonly used synthetic material used in medical devices. However, several concerns are raised when using synthetic materials as a scaffold,

Fig. 2 Microencapsulated islets. Islets were isolated from young pigs (range 18–26 days), matured for 8 days, and then encapsulated in ultrapure high M (UPLVM) alginate microcapsule, using an electrostatic encapsulator. Encapsulated islets were then stained with dithizone. Sample image is an isolated islet, approximately 150 μm in diameter, encapsulated in a 500 μm microcapsule. Scale 100 μm (Picture provided by Dr. Jonathan RT Lakey, UCI)

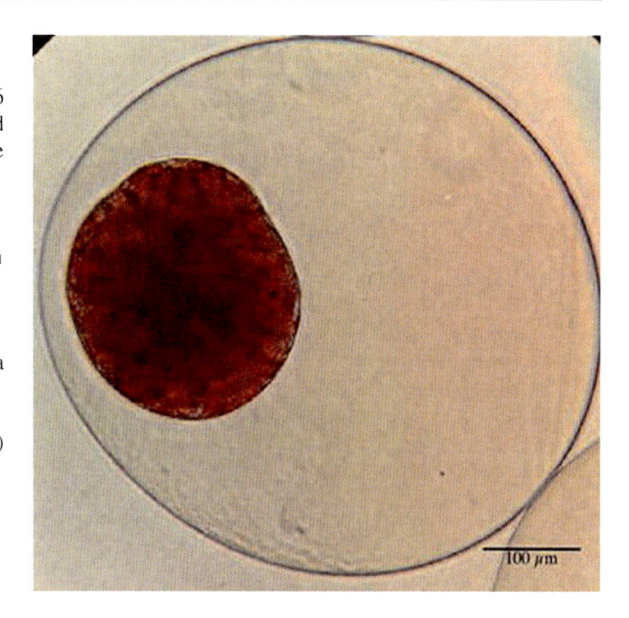

due to the possible potential that the foreign material will elicit a greater response by the body leading to fibrosis and loss of the encased cells. The production of these synthetic constituents would need to occur with nontoxic materials, and these materials would need to be purified in a method that is gentle enough on the cells and the transplant site. These capsules typically are heavily modified so that they can interact with the environment and degrade under physiologic conditions. Consequently, biologic materials are being used with an increasing incidence, with collagen being the most widely used naturally derived polymer in medical devices today. However, these gels exhibit poor strength, are expensive, and display significant variations between collagen batches making standardization of the process a problem (Lee and Mooney 2000). Therefore, alginate has become increasingly utilized for encapsulation due to its excellent biocompatibility, hydrophilic properties, easy gelation process, stable architecture, and relatively low cost. Alginate is a polysaccharide derived from seaweed which has to undergo extensive purification to improve its immunologic profile (O'Sullivan et al. 2010). Impure alginate has been implicated in islet cell necrosis and recruitment of inflammatory mediators (De Vos et al. 1997). Alginate is made up of chains of mannuronic (M) and guluronic (G) acid; the sequence of these units determines many of the encapsulation properties such as mechanical strength, durability, and permeability. For example, high G alginates form gels which are smaller and stronger than high M (O'Sullivan et al. 2010). Wang et al. tested over 200 capsule designs for their islet transplants finally deciding on a polymethyl coguanidine-cellulose sulfate/poly-L-lysine-sodium alginate (PMCG)-CS/PLL (Wang et al. 1997) for their canine syngeneic transplants resulting in normoglycemia achieved for approximately 160 days, with one canine achieving euglycemia for 214 days (Wang et al. 2008). The use of polycations and anions has been shown to provide improved permeability and mechanical strength; however, they also tend to result in an increase biologic response. Some counteract this response by adding another thin layer of alginate to provide a barrier (O'Sullivan et al. 2010). In broad terms, the process of gelation involves cross-linking by covalent, ionic, or physical bonds. For example, alginate converts into a gel form by ionic cross-linking with bivalent cations such as calcium, magnesium, and more commonly barium (King et al. 2001). The diffusion gradient that provides the bidirectional flow of materials is established by the degree of cross-linking. Mesh or pore size is typically much smaller than the encased cells however; hydrogels do not result in uniform pore size, and permselectability has never been clearly defined (O'Sullivan et al. 2010). What is apparent is that an increase in the degree of cross-linking results in gels that have superior mechanical strength but inversely reduces the size of the pores available for diffusion.

Several components need to be considered in engineering scaffolds that are safe for the encapsulated cells and the surrounding environment. The process of gelation needs to be mild; the capsule structure and chemistry should be nontoxic and reproducible. The degradation process must follow physiologic tissue growth, and its products must not adversely affect the coated cells or the body. Above all, the process of hydrogel engineering needs to be easily scaled up for industry, acceptable to surgeons, patients, and regulatory committees.

Clearly, capsule engineering is of vital importance for promoting graft survival and function, and many steps need to be considered in capsule construction to promote islet transplantations. Many advances have been achieved, but some studies have illustrated the difficulty with this technology. In animal studies performed by Suzuki et al., Tze, and Duvivier-Kali as well as the human clinical trial by Tuch, capsule fibrosis was a significant problem encountered (Suzuki et al. 1998; Tze et al. 1998; Duvivier-Kali et al. 2004; Tuch et al. 2009). Theoretically immune isolation is achieved by encapsulating cells, however, in these studies; clearly, some component of rejection was experienced. Likewise, oxygen and nutrients are able to freely diffuse across a matrix; however, studies illustrated by De Vos and Xin noted the absence of fibrosis, however, retrieved capsules containing necrotic islets, indicating the lack of appropriate oxygen to promote graft survival (De Vos et al. 1999; Xin et al. 2005). Due to these issues, some researchers have gone on to improve capsule engineering by means of co-encapsulation. These advances will be discussed in future applications as most of these undertakings are currently being performed in vitro.

Future Applications

Improved Capsule Engineering

Co-encapsulation is the process of adding additional molecules to the capsule to enhance the performance of the enveloped cell. In a novel study performed by Bunge, islets encapsulated with dexamethasone witnessed improved islet survival in mice recipients compared to those without the steroid (Bunger et al. 2005). In another study, Baruch encapsulated mouse monocyte macrophage cells and hamster kidney cells with ibuprofen and witnessed improved cell survival in vitro and in vivo (Baruch et al. 2009). Although cells are protected from large molecules such as antibodies, proinflammatory cytokines have a smaller molecular weight and can diffuse across most hydrogel gradients; thus, protection from these cytokines may promote capsule survival. In an in vitro study performed by Leung, capsules with anti-TNF α were able to remove active TNF α from a culture media (Leung et al. 2008). In order to improve oxygen supply to the cell, the access to a rich vascular bed is essential. As such, Khanna showed that the angiogenic factor, fibroblast growth factor 1, was able to be encapsulated and revealed continuous release for a 1 month period in vitro (Khanna et al. 2010). In another study, Pedraza et al. was able to encapsulate solid peroxide within polydimethylsiloxane resulting in sustained oxygen generation for approximately 6 weeks (Pedraza et al. 2012). Clearly, co-encapsulation is possible; however, most of these studies are still in the early phases of investigation, so we are yet to see whether this will improve graft function in vivo. The latest forefront for encapsulation cell technology for the treatment of diabetes involves the use of stem cells as a source of islets. Although some researchers have begun doing in vivo studies, the results have been varied.

Stem Cells

Stem cells are an attractive addition to cell encapsulation as the availability of human cadaveric donors for islet transplants continues to be a major problem. A company out of San Diego, Viacyte LLC, has performed the bulk of encapsulated stem cell-related transplants. Human embryonic stem cells directed down a pancreatic lineage were encapsulated in a device, called TheraCyte, which is made of a double membrane of polytetrafluoroethylene. In a study by Kroon et al., diabetic mice achieved normalization of glucose levels after 2 months (Kroon et al. 2008). Likewise, Lee et al. noted initially low glucose responses and plasma C-peptide levels after 12 weeks; however, after 5 months, improvements in glucose and C-peptide levels were evident, indicating differentiation continued to occur while encapsulated (Lee et al. 2009). In contrast, Matveyenko et al. did not achieve such outcomes, and in fact, their devices were noted to be encased in fibrotic tissue and upon retrieval did not stain positive for endocrine cells (Matveyenko et al. 2010). In a study using mesenchymal stem cells from human amniotic membranes, Kadam et al. were able to produce functional islet-like clusters which were encapsulated in polyurethane-polyvinylpyrrolidone microcapsules for transplantation into diabetic mice which resulted in euglycemia after day 15 until approximately day 30 (Kadam et al. 2010). Although, no study has been able to accomplish sustainable insulin independence using stem cells, improvements in stem cell differentiation are being accomplished and will hopefully improve upon this method (Blyszczuk et al. 2003).

Conclusion

Cell encapsulation of islets for the treatment of T1DM is a promising field that aims to revolutionize the treatment paradigm for diabetics. Although significant achievements have occurred, there are several obstacles that need to be addressed before achieving widespread use of this technology. Improvements in graft viability, biomaterial engineering, and islet isolation techniques will continue to promote success in this field.

References

American Diabetes Association (2004) Diagnosis and classification of diabetes mellitus. Diabetes Care 1:s5–s10

Bachoud-Lévi A, Déglon N, Nguyen JP, Bloch J, Bourdet C, Winkel L, Rémy P, Goddard M, Lefaucheur J, Brugières P, Baudic S, Cesaro P, Peschanski M, Aebischer P (2000) Neuroprotective gene therapy for Huntington's disease using a polymer encapsulated BHK cell line engineered to secrete human CNTF. Hum Gene Ther 11:1723–1729

Barton FB, Rickels MR, Alejandro R, Hering B, Wease S, Naziruddin B et al (2012) Improvement in outcomes of clinical islet transplantation. Diabetes Care 35:1436–1445

Baruch L, Benny O, Gilbert A, Ukobnik O (2009) Alginate-PLL cell encapsulation system co-entrapping PLGA-microspheres for the continuous release of anti-inflammatory drugs. Biomed Microdevices 11:1103–1113

Bisceglie V (1933) Uber die antineoplastische Immunität; heterologe Einpflanzung von Tumoren in Hühner-embryonen. Z Krebsforsch 40:122–140

Blyszczuk P, Czyz J, Kania G, Wagner M, Roll U, St-Onge L, Wobus AM (2003) Expression of Pax4 in embryonic stem cells promotes differentiation of nestin-positive progenitor and insulin-producing cells. Proc Natl Acad Sci U S A 100:998–1003

Borg DJ, Bonifacio E (2011) The use of biomaterials in islet transplantation. Curr Diab Rep 11:434–444

Bunger CM, Tiefenbach B, Jahnkea A, Gerlacha C, Freier T, Schmitz KP, Hopt UT, Schareck W, Klar E, De Vos P (2005) Deletion of the tissue response against alginate-pll capsules by temporary release of co-encapsulated steroids. Biomaterials 26:2353–2360

Calafiore R, Basta G, Luca G, Lemmi A, Montanucci P, Calabrese G, Racanicchi L, Mancuso F, Brunetti P (2006) Microencapsulated pancreatic islet allografts into non immunosuppressed patients with type diabetes: first two cases. Diabetes Care 29:1137–1139

Centers for Disease Control and Prevention (2012) Centers for Disease Control and Prevention: Diabetes Public Health Resources. http://www.cdc.gov/diabetes. Accessed 1 Aug 2012

Chang T (1964) Semipermeable microcapsules. Science 156:524–525

Chick WL, Like AA, Lauris V (1975)␤-cell culture on synthetic capillaries: an artificial endocrine pancreas. Science 187:847–849

De Vos P, De Haan BJ, Wolters GH, Strubbe JH, Van Schilfgaarde R (1997) Improved biocompatibility but limited graft survival after purification of alginate for microencapsulation of pancreatic islets. Diabetologia 40:262–270

De Vos P, Van Straaten JF, Nieuwenhuizen AG, De Groot M, Ploeg RJ, De Haan BJ, Van Schilfgaarde R (1999) Why do microencapsulated islet grafts fail in the absence of fibrotic overgrowth? Diabetes 48:1381–1388

Dong H, Fahmy TM, Metcalfe SM, Morton SL, Dong X, Inverardi L, Adams DB, Gao W, Wang H (2012) Immuno-isolation of pancreatic islet allografts using pegylated nanotherapy leads to long-term normoglycemia in full MHC mismatch recipient mice. PLoS One 7:e50265

Dufrane D, Goebbels RM, Gianello P (2010) Alginate macroencapsulation of pig islets allows correction of streptozotocin-induced diabetes in primates up to 6 months without immunosuppression. Transplantation 90:1054–1062

Duvivier-Kali VF, Omer A, Lopez-Avalos MD, O'Neil JJ, Weir GC (2004) Survival of microencapsulated adult pig islets in mice in spite of an antibody response. Am J Transplant 12:1991–2000

Elliott RB, Living cell technologies Ltd-LCT (2011) In: International pancreas and islet transplant association meeting, Prague

Elliott RB, Escobar L, Tan PL, Garkavenko O, Calafiore R, Basta P, Vasconcellos AV, Emerich DF, Thanos C, Bambra C (2005) Intraperitoneal alginate-encapsulated neonatal porcine islets in a placebo-controlled study with 16 diabetic cynomolgus primates. Transplant Proc 37:3505–3508

Elliott RB, Escobar L, Tan PL et al (2007) Live encapsulated porcine islets from a type 1 diabetic patient 9.5 yr after xenotransplantation. Xenotransplantation 14:157–161

Elliott RB, Garkavenko O, Tan P, Skaletsky NN, Guliev A, Draznin B (2010) Transplantation of microencapsulated neonatal porcine islets in patients with type 1 diabetes: safety and efficacy.70th scientific sessions. American Diabetes Association, Orlando

Eriksdotter-Jönhagen M, Linderoth B, Lind G, Aladellie L, Almkvist O, Andreasen N, Blennow K, Bogdanovic N, Jelic V, Kadir A, Nordberg A, Sundström E, Wahlund LO, Wall A, Wiberg M, Winblad B, Seiger A, Almqvist P, Wahlberg L (2012) Encapsulated cell biodelivery of nerve growth factor to the basal forebrain in patients with Alzheimer's disease. Dement Geriatr Cogn Disord 33:18–28

Fernández M, Barcia E, Fernández-Carballido A, Garcia L, Slowing K, Negro S (2012) Controlled release of rasagiline mesylate promotes neuroprotection in a rotenone-induced advanced model of Parkinson's disease. Int J Pharm 438:266–278

Fiszman GL, Karara AL, Finocchiaro LM, Glikin GC (2002) A laboratory scale device for microencapsulation of genetically engineered cells into alginate beads. Electron J Biotechnol 5:23–24

Hafiz MM, Faradji RN, Froud T, Pileggi A, Baidal DA, Cure P, Ponte G, Poggioli R, Cornejo A, Messinger S, Ricordi C, Alejandro R (2005) Immunosuppression and procedure-related complications in 26 patients with type 1 diabetes mellitus receiving allogeneic islet cell transplantation. Transplantation 80:1718–1728

Hearing B, Ricordi C (1999) Islet transplantation for patients for patients with type 1 diabetes; results, research priorities, and reasons for optimism. Graft 2:12–27

Huber A, Padrun V, Deglon N, Aebischer P, Hanns M, Boison D (2012) Grafts of adenosine-releasing cells suppress seizures in kindling epilepsy. Proc Natl Acad Sci U S A 98:7611–7616

Jang JY, Lee DY, Park SJ, Byun Y (2004) Immune reactions of lymphocytes and macrophages against PEG-grafted pancreatic islets. Biomaterials 25:3663–3669

Jeon Y (2011) Cell based therapy for the management of chronic pain. Korean J Anesthesiol 60:3–7

Kadam S, Muthyala S, Nair P, Bhonde R (2010) Human placenta-derived mesenchymal stem cells and islet-like cell clusters generated from these cells as a novel source for stem cell therapy in diabetes. Rev Diabet Stud 7:168–182

Khanna O, Moya EC, Opara EC, Brey EM (2010) Synthesis of multilayered alginate microcapsules for the sustained release of fibroblast growth factor-1. J Biomed Mater Res A 95:632–640

Khanna O, Larson JC, Moya ML, Opara EC, Brey EM (2012) Generation of alginate microspheres for biomedical applications. J Vis Exp 66:e3388

King S, Dorian R, Storrs R (2001) Requirements for encapsulation technology and the challenges for transplantation of islets of Langerhans. Graft 4:491–499

King A, Lau J, Nordin A, Sandler S, Andersson A (2003) The effect of capsule composition in the reversal of hyperglycemia in diabetic mice transplanted with microencapsulated allogeneic islets. Diabetes Technol Ther 5:653–663

Kobayashi T, Aomatsu Y, Iwata H, Kin T, Kanehiro H, Hisanaga M, Ko S, Nagao M, Nakajima Y (2003) Indefinite islet protection from autoimmune destruction in nonobese diabetic mice by agarose microencapsulation without immunosuppression. Transplantation 75:619–625

Kobayashi T, Aomatsu Y, Iwata H, Kin T, Kanehiro H, Hisanga M, Ko S, Nagao M, Harb G, Nakajima Y (2006) Survival of microencapsulated islets at 400 days posttransplantation in the omental pouch of NOD mice. Cell Transplant 15:359–365

Krishnamurthy NV, Gimi B (2011) Encapsulated cell grafts to treat cellular deficiencies and dysfunction. Crit Rev Biomed Eng 39:473–491

Kroon E, Martinson LA, Kadoya K, Bang AG, Kelly OG, Eliazer S, Young H, Richardson M, Smart NG, Cunningham J, Agulnick AD, D'Amour KA, Carpenter MK, Baetge EE (2008) Pancreatic endoderm derived from human embryonic stem cells generates glucose-responsive insulin-secreting cells in vivo. Nat Biotechnol 26:443–452

Lanza RP, Borland KM, Lodge P, Carretta M, Sullivan SJ, Muller TE, Solomon BA, Maki T, Monaco AP, Chick WL (1992) Treatment of severely diabetic, pancreatectomized dogs using a diffusion-based hybrid pancreas. Diabetes 41:886–889

Lanza RP, Jackson R, Sullivan A, Ringeling J, McGrath C, Kühtreiber W, Chick WL (1999) Xenotransplantation of cells using biodegradable microcapsules. Transplantation 67:1105–1111

Lee K, Mooney D (2000) Hydrogels for tissue engineering. Chem Rev 101:1869–1879

Lee SH, Hao E, Savinov AY, Geron I, Strongin AY, Itkin-Ansari P (2009) Human β-cell precursors mature into functional insulin-producing cells in an immunoisolation device: implications for diabetes cell therapies. Transplantation 87:983–991

Leung A, Lawrie G, Nielson LK, Trau M (2008) Synthesis and characterization of alginate/poly-L-ornithine/alginate microcapsules for local immunosuppression. J Microencapsul 25:387–398

Lim F, Sun AM (1980) Microencapsulated islets as bioartificial pancreas. Science 210:908

Maki T, Ubhi CS, Sanchez-Farpon H, Sullivan SJ, Borland K, Muller TE, Solomon BA, Chick WL, Monaco AP (1991) Successful treatment of diabetes with the biohybrid artificial pancreas in dogs. Transplantation 51:43–51

Maki T, Otsu I, O'Neil JJ, Dunleavy K, Mullon CJ, Solomon BA, Monaco AP (1996) Treatment of diabetes by xenogeneic islets without immunosuppression. Use of a vascularized bioartificial pancreas. Diabetes 45:342–347

Matveyenko AV, Georgia S, Bhushan A, Butler PC (2010) Inconsistent formation and nonfunction of insulin-positive cells from pancreatic endoderm derived from human embryonic stem cells in athymic nude rats. Am J Physiol Endocrinol Metab 299:13–20

Niclauss N, Bosco D, Morel P, Giovannoni L, Berney T, Parnaud G (2011) Rapamycin impairs proliferation of transplants islet β cells. Transplantation 91:714–722

Nicodemus G, Bryant S (2008) Cell encapsulation in biodegradable hydrogels for tissue engineering applications. Tissue Eng Part B Rev 14:149–165

O'Sullivan ES, Johnson AS, Omer A, Hollister-Lock J, Bonner-Weir S, Colton CK, Weir GC (2010) Rat islet cell aggregates are superior to islets for transplantation in microcapsules. Diabetologia 53:937–945

Pedraza E, Coronel MM, Fraker CA, Ricordi C, Stabler CL (2012) Preventing hypoxia-induced cell death in β cells and islets via hydrolytically activated, oxygen-generating biomaterials. Proc Natl Acad Sci U S A 109:4245–4250

Shapiro AM, Nanji SA, Lakey JR (2003) Clinical islet transplant: current and future directions towards tolerance. Immunol Rev 196:219–236

Soon-Shiong P, Feldman E, Nelson R, Heintz R, Yao Q, Yao Z, Zheng T, Merideth N, Skjak-Braek G, Espevik T et al (1993) Long-term reversal of diabetes by the injection of immunoprotected islets. Proc Natl Acad Sci U S A 90:5843–5847

Soon-Shiong P, Heintz RE, Merideth N, Yao QX, Yao Z, Zheng T, Murphy M, Moloney MK, Schmehl M, Harris M et al (1994) Insulin independence in a type 1 diabetic patient after encapsulated islet transplantation. Lancet 343:950–951

Storrs R, Dorian R, King SR, Lakey J, Rilo H (2001) Preclinical development of the islet sheet. Ann N Y Acad Sci 944:252–266

Sun AM (1988) Microencapsulation of pancreatic islet cells: a bioartificial endocrine pancreas. Methods Enzymol 137:575–580

Suzuki K, Bonner-Weir S, Trivedi N, Yoon KH, Hollister-Lock J, Colton CK, Weir GC (1998) Function and survival of macroencapsulated syngeneic islets transplanted into streptozocin-diabetic mice. Transplantation 66:21–28

Tuch BE, Keogh GW, Williams LJ et al (2009) Safety and viability of microencapsulated human islets transplanted into diabetic humans. Diabetes Care 32:1887–1889

Tze WJ, Cheung SC, Tai J, Ye H (1998) Assessment of the in vivo function of pig islets encapsulated in uncoated alginate microspheres. Transplant Proc 30:477–478

Wang T, Lacík I, Brissová M, Anilkumar AV, Prokop A, Hunkeler D, Green R, Shahrokhi K, Powers AC (1997) An encapsulation system for the immunoisolation of pancreatic islets. Nat Biotechnol 15:358–362

Wang T, Adcock J, Kühtreiber W, Qiang D, Salleng KJ, Trenary I, Williams P (2008) Successful allotransplantation of encapsulated islets in pancreatectomized canines for diabetic management without the use of immunosuppression. Transplantation 85:331–337

Xin ZL, Ge SL, Wu XK, Jia YJ, Hu HT (2005) Intracerebral xenotransplantation of semipermeable membrane-encapsuled pancreatic islets. World J Gastroenterol 11:5714–5717

Zimmermann H, Shirley SG, Zimmermann U (2007) Review Alginate-based encapsulation of cells: past, present, and future. Curr Diab Rep 7:314–320

Stem Cells in Pancreatic Islets

47

Erdal Karaöz and Gokhan Duruksu

Contents

Introduction	1312
The Type of Stem Cells in the Pancreatic Tissue	1313
Pancreatic Duct-Derived Stem/Progenitor Cells	1314
Acinar Cells	1316
Pancreatic Stellate Cells	1317
Nestin-Positive Islet-Derived Progenitor Cells	1318
β Cell Replication	1326
Concluding Remarks	1327
Cross-References	1328
References	1328

Abstract

Adult stem cells are unspecialized cells with the capacity to differentiate into many different cell types in the body. These cells together with the progenitor cells involve in the tissue repair processes and maintain the functionality of the tissue in which they are found. During the onset and progression of the diabetes mellitus type 1, the β cells in Langerhans islets undergo progressive and selective destruction. To restore the normal insulin levels to regulate glucose homeostasis, the dropping number of β cells is replaced by progenitor/stem cells in the environment of pancreatic islets. Due to the complexity of the organ organization, these cells could originate from different sources. In this chapter, the types of the stem cells derived from pancreatic environment with the differentiation capacity into pancreatic islets and their cells were mainly focused. Some of these cells do not only involve in the regeneration of insulin-producing cells, but they

---

E. Karaöz (✉) • G. Duruksu
Kocaeli University, Center for Stem Cell and Gene Therapies Research and Practice, Institute of Health Sciences, Stem Cell Department, Kocaeli, Turkey
e-mail: ekaraoz@hotmail.com

also function in the preservation of β cell viability. Besides providing an alternative renewable source for β cell replacement, the interaction of pancreatic stem cells with immune system in diabetes mellitus type 1 is also discussed.

Keywords
β cells • Differentiation • Immune regulatory • Regeneration • Stem cells

Abbreviations

ACTA2	Actin α 2 (smooth muscle actin)
CHIBs	Cultured human islet buds
ECFP	Cyan fluorescent protein
FACS	Fluorescence-activated cell sorting
FGF	Fibroblast growth factors
GLP-1	Glucagon-like peptide-1
HGF	Hepatocyte growth factor
ICAM-1	Intercellular adhesion molecule 1 (CD54)
IFN-γ	Interferon γ
IL-	Interleukin
MSCs	Mesenchymal stem cells
NG2	Neuron-glial antigen 2
NOD	Nonobese diabetic
PDGFRβ	Platelet-derived growth factor receptor, β polypeptide
PDX-1	Pancreatic duodenal homeobox-containing factor-1
PI-SCs	Pancreatic islet-derived stem cells
TGF-α	Transforming growth factor α
TGF-β	Transforming growth factor β
TNF-α	Tumor necrosis factor α
VCAM-1	Vascular cell adhesion molecule 1 (CD106)

Introduction

The imbalance between the generation and destruction of β cells has been seen as one of the probable causes of diabetes. According to this view, improving the regenerative potential of β cells might be an alternative approach for the treatment of type 1 diabetes by increasing the number of β cells in islets to compensate for the damage of autoimmunity (Banerjee and Bhonde 2003; Guz et al. 2001; Lechner et al. 2002). For this purpose, many experimental studies were carried out in diabetic animal models, and all the studies shared three common targets regardless of which experimental strategies they preferred:
- To demonstrate the continuation of existence of stem/progenitor cells in pancreas or islets of Langerhans in diabetes animal models
- To clarify the role of this stem/progenitor cells in β cell regeneration in pancreatic islets in response to changing conditions

- To determine the possible effect of β cell regeneration (or neogenesis) on manifestations of diabetes mellitus

Adult pancreatic β cells were previously thought to be the terminal cells (cells in last stage of differentiation) with limited regeneration/self-renewal capacity. However, the replication of adult islet cells was shown to induce in response to stimulators, like glucose, certain hormones, and growth factors, especially HGF which initiates the maturation of neonatal islet cells (Hayek et al. 1995; Peck and Ramiya 2004; Swenne 1992). In addition, the reversible dynamic alterations in the mass of pancreatic endocrine cells were commonly observed to meet the requirements during pregnancy and obesity throughout the life. Even during the growth and development of body, the increase in cell mass was reported (Bonner-Weir et al. 1989; Brelje et al. 1993; Marynissen et al. 1983; Peck and Ramiya 2004; Street et al. 2004).

Based on these data, Finegood et al. developed a mathematical model to evaluate the dynamics of pancreatic β cell transformation (Finegood et al. 1995). They suggested the existence of a balance between the division, development, and loss of β cells in healthy individuals. Despite the number of β cells closely linked with their replication rate, their average life expectancy was reported to be varying from 1 to 3 months. In another words, 1–4 % of β cells in adult pancreas were regenerated daily. These data demonstrate that the β cells significantly regenerated during the whole lifetime.

The Type of Stem Cells in the Pancreatic Tissue

In the experimental animal models of pancreatic damage, many evidences were collected for regeneration of mature islet cells by differentiation of pancreas stem or progenitor cells. The development of new islets from pancreatic duct epithelium or neighboring cells after partial pancreatectomy in rodents is a well-documented example (Bonner-Weir et al. 1993). In another damage model, alloxan and streptozotocin, which are known for their potential toxic effect on β cells, were observed to induce the regeneration of endocrine cells from intra-islet precursor cells (Korcakova 1971; Cantenys et al. 1981). In other studies, the role of pancreatic islet progenitor/stem cells in regeneration or neogenesis of islets was reported in the cases of partial duct obstruction (Rosenberg and Vinik 1992); copper deficiency (Rao et al. 1989); soybean trypsin inhibitor therapy (Weaver et al. 1985); caerulein-induced pancreatitis (Elsässer et al. 1986); transient hyperglycemia (Bonner-Weir et al. 1989); overexpression of IFN-γ (Gu and Sarvetnick 1993), Reg1 (Yamaoka et al. 2000), and TGF-α (Sandgren et al. 1990); and after injection of steroids (Kem and Logothetopoulos 1970), insulin antibodies (Logothetopoulos and Bell 1966), and growth factors (Otonkoski et al. 1994).

After the demonstration of the progenitor/stem cells' role in completion of insufficient pancreatic islet functions after damage to meet the requirements,

studies focused on the characterization of these cells with pancreatic islets' or islet cells' producing capacity. In many in vitro and in vivo studies, these cells could be classified with respect to their location, morphological, immunophenotypic, and gene expression profile characteristics as the following:
- Pancreatic duct stem cells
- Acinar cells
- Pancreatic stellate cells (fibroblast-like cells)
- Nestin-positive islet-derived progenitor cells

Pancreatic Duct-Derived Stem/Progenitor Cells

The duct epithelium-derived regeneration of islets under defined conditions was demonstrated for the first time in 1993 (Bonner-Weir et al. 1993). Two years later, Peck et al. published that the isolated islet-producing stem cells from pancreatic channel developed functional islets in long-term culture (Cornelius et al. 1997; Peck and Cornelius 1995; Peck and Ramiya 2004). These cells were obtained from duct epithelial cell cultures prepared by partial digestion of fresh pancreas tissue of prediabetic NOD (nonobese diabetic) mouse or human donors. With the recent improvements, this protocol consists of four main steps: first, duct epithelial cells and islets are isolated from digested pancreas, and they are cultured in the media with the supporting capacity for monolayer-forming epithelial cells with neuroendocrine cell-like phenotype. Epithelial-like islet precursor cells were observed to colonize in single layers, and as they continue to proliferate, these units turned into either sphenoid structures or single-cell suspensions depending on the type of culture and culture conditions. In the third step, islet-derived precursor cells were stimulated by high glucose concentrations and growth factors. By this procedure, it was aimed to support the cell maturation and to induce the generation of multicellular islet-like structures containing endocrine cells in various maturation stages. At the end, the formed islet-like structures were cultured in vivo (e.g., under the kidney capsule) to mature.

Many reports are available on the differentiation of duct cells into islet-like cells or islets by using different protocols (Wang et al. 1995; Fernandes et al. 1997; Rosenberg 1998; Finegood et al. 1999; Sharma et al. 1999; Xu et al. 1999; Yamamoto et al. 2000; Guz et al. 2001; Bonner-Weir et al. 2004; Ogata et al. 2004; Rooman and Bouwens 2004; Bouwens and Rooman 2005; Holland et al. 2005; Oyama et al. 2006; Wang et al. 2008). In the study by Ramiya et al. (2000), duct cells were derived from pancreas of prediabetic NOD mice by enzymatic digestion with collagenase, and they were cultured for 3 years (there are no information available for the culture period of duct cells from human sources). The islet number obtained from five mice pancreatic tissues in 3 years in secondary cultures was equivalent to the number of 10,000 pancreatic islets.

In the other study by Bonner-Weir et al., a monolayer of epithelial cells formed three-dimensional channel cystic structures in a diameter of 50–150 μm after the culturing of duct cells obtained by a similar protocol from human tissues

(Bonner-Weir et al. 2000), and from these structures, islet-like cell clusters comprising pancreas endocrine cells were budded. After 3–4 weeks of culture, the formation of islet-like structures (annotated as *cultured human islet buds* or *CHIBs*) was observed, and both duct and endocrine cells were characterized after detailed analyses. Moreover, uncharacterized cells with ungranulated cytoplasm were detected in culture. Significant increase in insulin secretion was measured, such that 10–15-fold increases in the amount of insulin were measured in each culture. Bonner-Weir later suggested that these duct cells could function as an important precursor cell source for both islet and acinar tissues and duct epithelia could be considered as *facultative stem cells*, consequently (Bonner-Weir et al. 2004).

Important issue in this approach is that every produced islet should be matured and checked for their functionality. For this purpose, the islet-like structures were generally transplanted under the kidney capsule (subcapsular) or subcutaneous region of diabetic or nondiabetic animals. To test the produced islets after the last in vivo maturation phases, Ramiya et al. (2000) transferred mouse islet-like clusters into the kidney subcapsular region of syngeneic female diabetic NOD mice (300 islets per subject). Within 1 week, the blood sugar levels of transplanted mice were decreased to 180–220 mg/dl, and they become insulin independent.

Another interesting result of syngeneic mouse islet implantation study was the lack of the reactive autoimmune response generation against in vitro-produced islets. During the histological examination of implant location, the absence of immune cells in this region was accepted as the evidence for this unresponsiveness (Ramiya et al. 2000; Peck and Ramiya 2004). Researchers usually connected these observations with:

- Loss of β cell autoantigens' expression in in vitro culture
- Development of peripheral tolerance by newly formed transplanted islets after re-induction of autoimmune response
- Lack of time required for the resumption of an autoimmune response
- Characteristics of the implant location

The origin of islets, which were derived from duct epithelial stem cells in this study, could be associated with the unresponsiveness of autoimmune system in transplanted diabetic mouse. Whatever the reason, the important conclusion of this study was that the duct stem/progenitor cells might be accepted as a potential islet/β cell source in cell replacement therapy of diabetes type 1 due to the lack of reactivation of autoimmune response in this system.

All of these studies indicate the existence of progenitor/stem cells with the differentiation capacity into both exocrine and endocrine cells within the population of pancreatic duct cells. Since the first hypothesis of islet regeneration from duct epithelium under certain conditions by Bonner-Weir et al. (1993), there have been many attempts to obtain functional islets and/or β cells from human or rodent duct cells by different techniques and approaches including gene transfer (Leng and Lu 2005; Kim et al. 2006). Considering the molecular and structural development stages of pancreas, all endocrine, acinar, and duct cells were developed from the same endodermal epithelial tissue derived from embryonic foregut (duodenal) region, and adult duct cells share some common features with embryonic primitive channels. As a result, pancreatic duct cells may contribute to the production of endocrine cells in adulthood.

Multipotent precursor/stem cells were isolated from islet and duct cell culture expressing markers of both neuronal and pancreatic precursor cells. These cells have the ability to differentiate into neuronal and glial cells as well as into pancreatic lineage cells (Seaberg et al. 2004). Because of their expression of distinctive markers excluding the mesoderm or neural crest characters, the existence of an intrinsic pancreatic precursor cell population was suggested. The argument about the origin of these cells being other tissue was eliminated by labeling the non-endocrine pancreas epithelial cells and eliminating contaminating mesenchymal cells with selection for drug resistance (Hao et al. 2006).

These progenitor/stem cells were isolated by FACS (fluorescence-activated cell sorting) according to the negative selection for CD45, Ter119, c-Kit (CD117), and Flk-1 (hematopoietic markers) and the positive selection for c-Met (hepatocyte growth factor receptor). Their differentiation into exocrine and endocrine pancreatic cell lineages was observed both in vitro and in vivo (Suzuki et al. 2004). In most of the studies on the recovery of pancreatic islets after the experimental damage, the number of β cells was increased. However, the origin of the β cells was not clearly defined, whether they differentiated from progenitor/stem cells into β cells or generated from β cells. To solve this problem, Ngn3-positive progenitor/stem cells in duct epithelium were studied. In the specific culture conditions, these cells were shown to differentiate into islet-/β cell-like cells. After pancreatic duct ligation, the number of Ngn3-positive cells was increased in mice (Xu et al. 2008). The silencing of Ngn3 in these cells resulted in the impairment of β/islet regeneration after duct ligation. On the other hand, the transplantation of these cells into Ngn3-knockout mice induced the generation of islet-like cells in embryonic pancreas. These evidences indicate the important role of Ngn3-positive multipotent cells in islet/β regeneration after duct ligation, but their origin is still unknown, whether they are duct cells or derived from a subpopulation of the duct epithelial cells.

Acinar Cells

There are many manuscripts that reported the differentiation of pancreatic exocrine/acinar cells into pancreatic duct and endocrine cells (transdifferentiation), the molecular marker expression of three important pancreas tissue components (exocrine, pancreas endocrine, and duct cells) of these stem/progenitor cells, and their supporting roles in pancreas regeneration and islet neogenesis.

Following the experimental pancreas damaged by various factors (duct ligation, pancreatectomy, and chemical agents), pancreatic acinar cells were induced to differentiate primarily into duct cell-like cells (*acino-ductal metaplasia*) and involved in the neogenesis of islets in some of these studies (Gu et al. 1994, 1997; Gmyr et al. 2000; De Haro-Hernandez et al. 2004; Holland et al. 2004;

Sphyris et al. 2005). These observations were supported by identification of *transition cells* expressing specific markers of endocrine/exocrine cells and duct cells.

Lipsett and Finegood (2002) reported that the acinar cells were directly responsible for increased β cell neogenesis in long-term experimental hyperglycemia model. Interestingly in the immunohistochemical examination of pancreatic tissues from dexamethasone-treated animals with pancreatic duct ligation, Lardon et al. detected the acino-insular structures (transition from acinar into islet) consisted of amylase and insulin-expressing transitional cells (Lardon et al. 2004). Some evidences were provided for the differentiation of adult mouse pancreatic cells into insulin-producing cells by cell line monitoring with Cre/Lox-based techniques (Minami et al. 2005). The newly formed cells possessed all the essential requirements for glucose-stimulated insulin secretion, which is the fundamental mechanism of insulin secretion. In this lineage monitoring by Cre-loxP recombination techniques, enhanced cyan fluorescent protein (ECFP) was expressed under the control of mouse amylase-2 and rat elastase-1 promoters. Pancreatic acinar cells, obtained from mice with R26R-ECFP reporter, were transdifferentiated into insulin-producing cells.

Similar observations were reported for the transdifferentiation of acinar cells into endocrine pancreas lineage cells (Baeyens et al. 2005; Minami et al. 2005; Okuno et al. 2007). Under defined conditions, acinar to duct cell transdifferentiation was detected (Means et al. 2005), whereas the acinar to β cell transdifferentiation was not (Desai et al. 2007). With the reprogramming of mouse acinar cells by inducing the expression of Ngn3, Pdx-1, and MafA, β cell transdifferentiation was induced in these cells (Zhou et al. 2008). The detailed analyses on transcription factors for β cell transdifferentiation showed that Pdx-1, Ngn3, and MafA expressions were critical for both differentiation and function. Indirect evidences were collected for acinar to β cells transdifferentiation in other studies (Lipsett and Finegood 2002; Rooman and Bouwens 2004). The capacity of pancreatic oval cells to differentiate into pancreatic lineages has not been proven yet.

Pancreatic Stellate Cells

The presence of stem cell-like cells in the pancreas, which was morphologically and functionally similar to Ito cells (known as fat-storing cells, vitamin A-storing fibroblast-like cells, or as myofibroblast-like cells) or stellate cells in the liver, was recently reported.

These cells, also called as pancreatic stellate cells, were first discovered in 1982 in mice (Watari et al. 1982) and 8 years later in human tissue (Ikejiri 1990). In the following studies, these cells were shown to be responsible for pancreatic fibrogenesis (Saotome et al. 1997). In the detailed examination, it was determined that they share many common features (like morphological and immunohistochemical properties, and synthesis capacity of extracellular matrix) with the hepatic stellate cells in the space of Disse. Hepatic stellate cells are localized in the interlobular septa and interacinar (periacinar) areas and have a pathogenic role in fibrosis (Haber et al. 1999). Pancreatic stellate cells were observed to support the proliferation of acinar cells and tubular structures in trypsin-induced necrohemorrhagic pancreatitis (Lechene et al. 1991). Recently, they were reported to involve in the regeneration of early stages of acute pancreatitis in humans (Zimmermann et al. 2002). All these findings point out that those pancreatic stellate cells not only are involved in fibrogenesis but also contribute in the tissue remodeling at the same time.

After reporting the role of pancreatic stellate cells in regeneration process to various pathogenic conditions, Kruse et al. (2004) isolated pancreas stellate cell-like cells from rodent exocrine pancreas, and these cells were cultured for a long time (at least 20 passages) without differentiation. These cells formed three-dimensional cell clusters, called *organoid bodies* (similar to the embryoid bodies), following culturing in droplets for 2 days in vitro. After incubation of these organoid bodies in tissue culture dishes for 7 weeks, they developed cell colonies, which formed new organoid bodies spontaneously. Without any induction, these bodies were observed to differentiate into various cell types of all three germ layers (smooth muscle cells, neurons, glial cells, epithelial cells, chondrocytes, and pancreatic exocrine and endocrine cells). The formation of embryonic cell line from organoid bodies of undifferentiated adult stem cells and their spontaneous differentiation into cell lineages of endoderm, mesoderm, and ectoderm lead some researchers to consider these cells as the new class of undifferentiated pluripotent adult stem cells.

Nestin-Positive Islet-Derived Progenitor Cells

Before the study by Zulewski et al., pancreatic islet precursor cells were believed to be found only in pancreatic ducts, and the related studies were focused consequently on this region (Zulewski et al. 2001). Due to the common features of β cells and neural cells in the early stages of mammalian embryonic development, and their similar characteristics, the expression of neural stem cell-specific marker, nestin, in rat and human pancreatic islet cells was explored. After all, these intermediate filament protein nestin-expressing cells were isolated from human and rat islets and propagated in culture conditions. These cells were named as nestin-positive islet-derived progenitor cells, and after induction for differentiation, they were transformed into islet cells with insulin, glucagon, and Pdx-1 gene expressions. Besides the expression of other neuroendocrine cells, hepatic and pancreatic exocrine genes were shown by RT-PCR.

Another study of this research group revealed the increase of the differentiation ratio into islet-like insulin-producing cell clusters in response to GLP-1 in vitro (Abraham et al. 2002). With this recent study, the researchers showed the presence of GLP-1 receptor in nestin-positive islet precursor cells. Therefore, their differentiation into insulin-producing cells was induced by binding of GLP-1 to appropriate receptors. The effect of GLP-1 in vivo is quite intriguing, if it significantly involved in β cell differentiation in vitro. Assuming the induction of β cell regeneration by local production of GLP-1 in pancreas, patients with type 2 diabetes were treated by GLP-1 injection. The upregulation in insulin secretion was observed afterwards, and the insulin-mediated sugar utilization was increased (Meneilly et al. 2001, 2003a). Similar study was also performed in patients with type 1 diabetes, but the upregulatory effect of GLP-1 on insulin-mediated glucose utilization was not observed in this case (Meneilly et al. 2003b). Despite the ineffectiveness in type 1diabetes, the positive results of GLP-1, with which clinical trials are still ongoing, are very important in type 2 diabetes. In addition, GLP-1 agonists have been shown to induce β cell development and differentiation besides their cell protective role and antiapoptotic effects on β cells. In accordance with these findings, pancreatic stem/progenitor cells might be used to produce β cells in vitro by stimulatory effect of GLP-1 and might be used in replacement therapy (List and Habener 2004).

One to three percent of these islet-derived nestin-positive precursor cells were observed to have similar immunophenotypic features of undifferentiated bone marrow stem cells and were defined as side population. In parallel with these findings, nestin-positive precursor cells were suggested to contain a subpopulation of immature stem cells with differentiation potential (Lechner et al. 2002). The detection of cells with immunophenotypic markers specific for bone marrow-derived cells in pancreatic islets was another important point in these studies. These cells, also possess specific immunophenotypic markers of pancreatic islet stem cells, were considered to migrate from bone marrow (Lechner and Habener 2003). On the other hand, some evidences were reported that these cells could not originate from hematopoietic cell lineages. For example, Poliakova et al. defined similar cells in the pancreases of human and monkey, but the vast majority of these cells (95 %) were negative for hematopoietic stem cell-specific markers, like CD34 and CD45 (Poliakova et al. 2004). In addition, a study was reported that no evidence was found on the development of endocrine cells from nestin-positive stem cells (Gao et al. 2003).

Other studies on the islet progenitor or stem cells mostly focused on demonstrating the presence of these adult pancreas-derived insulin-producing cells with the production capacity of βcell-like cells by using different culture conditions. These cells are called by different names, like intra-islet precursor cells (Guz et al. 2001; Banerjee and Bhonde 2003), pancreatic stem cells (Schmied et al. 2001; Suzuki et al. 2004), small cells (Petropavlovskaia and Rosenberg 2002), islet-derived progenitor cells (Wang et al. 2004; Linning et al. 2004; von Mach et al. 2004), multipotent stem cell (Choi et al. 2004), or β stem cells (Duvillie et al. 2003). Although these cells commonly differentiated into β-like insulin-producing cells in vitro and in vivo, they were also observed to form other types of cells, like central nervous system and multiple cell types with phenotypic characteristics of neural crista (Choi et al. 2004), exocrine cells (Schmied et al. 2001), and liver, stomach, and intestinal cells (Suzuki et al. 2004), under defined media in vitro. Their differentiation capacity into acinar (Suzuki et al. 2004) and liver cells (von Mach et al. 2004) was demonstrated in vivo.

In the culture of the postmortem adult human islets, mesenchymal-type cells exhibiting fibroblast morphology were characterized. Although these cells lack the ability of insulin secretion in response to glucose stimuli, they can be induced to differentiate into hormone-expressing islet-like cells (Gershengorn et al. 2004, 2005). Initially, they were thought to be the precursor cells of insulin-expressing cells undergone to mesenchymal-to-epithelial transition, but the human islet-derived precursor cells were later demonstrated to have the mesenchymal stem cell characteristics (Gershengorn et al. 2005; Davani et al. 2007). On the other hand, the inability of the epithelial-to-mesenchymal-to-epithelial transition by mouse β cells was revealed (Atouf et al. 2007; Morton et al. 2007; Meier et al. 2006). Some studies suggested the dedifferentiation of pancreatic islet cells into precursor cells, and the redifferentiation into the pancreatic lineage cells might be the source of the regenerated β cells (Lechner et al. 2005; Ouziel-Yahalom et al. 2006; Banerjee and Bhonde 2003).

Another clonogenic cell population, nestin-positive progenitor/stem cells, was isolated from islets showing bone marrow MSC phenotypic markers (Table 1) and differentiating into insulin-producing cells in vitro (Lechner et al. 2002; Gallo et al. 2007; Gershengorn et al. 2005; Zhang et al. 2005; Atouf et al. 2007; Davani et al. 2007; Karaoz et al. 2010a; Carlotti et al. 2010; Montanucci et al. 2011). These cells were mainly negative for c-peptide and Pdx-1 in the culture, but under serum-free conditions, these cells lost their MSC phenotype, formed islet-like clusters, and were able to secrete insulin. Although the endocrine hormones were not expressed in the normal culture conditions, these cells have the tendency to form clusters, which induce the expression of insulin, glucagon, and somatostatin genes in serum-free medium (Carlotti et al. 2010). Like mesenchymal stem cells (MSCs), they express lineage-specific markers, i.e., CD13, CD29, CD44, CD54, CD73, CD90, CD105, nestin, and vimentin, while they are negative for hematopoietic, endothelial, and ductal cell markers (Fig. 1) and differentiate into osteogenic, adipogenic, and chondrogenic cell lineages after induction (Fig. 2). The expression of ACTA2, CD146, NG2, and PDGFRβ by these cells indicates their pericyte characteristics. Besides, the expression of pluripotency markers by rat pancreatic islet-derived progenitor/stem cells, i.e., Oct-4, Rex-1, and Sox-2, was shown (Karaoz et al. 2010a). For that reason, these nestin-positive cells were called as pancreatic islet-derived stem cells (PI-SCs) (Karaoz et al. 2010a). Montanucci et al. also reported the existence of insulin-producing progenitor cells with the expression of Oct-4, Sox-2, Nanog, ABCG2, Klf-4, and CD117 in human islets (Montanucci et al. 2011). Similarly, they showed differentiation capacity like MSCs, including the differentiation ability into insulin-secreting endocrine cells.

The differentiation of nestin-positive progenitor/stem cells into insulin-secreting, β-like cells might be a promising approach to compensate for β cell loss in diabetes. Recently, the studies on MSCs were focused on their immunosuppressive activity as well. MSCs derived from bone marrow were demonstrated to have regulatory function on the components of immune system (Guo et al. 2006; Spaggiari et al. 2008; Comoli et al. 2008). These cells control the immune response by modifying their cytokine secretion profiles to show anti-inflammatory effect or tolerant phenotype (Chang et al. 2006). IFN-γ secreted by T and NK cells and IL-1 group cytokines by mononuclear cells were found to activate the suppressive

Table 1 Immunocytochemical properties of undifferentiated rPI-SCs in passage 3

Marker molecule	rPI-SCs[a]	Marker molecule	rPI-SCs[a]
β3Tubulin	+	Glut2	∅
Actin beta (ACTβ)	+	HNK-1ST	∅
ACTA2	+	IL-10	∅
BDNF	+	IL-1β	−/+
BMP-2	+	IL-1ra	+
BMP-4	+	IL-4	∅
BrdU	+	Insulin	∅
CD 31 (PECAM-1)	∅	Ki67	+
CD 34 (HPCA1)	∅	MAP 2a,b	+
CD 45 (PTP)	∅	MPO	+
CD 71 (TfR1)	∅	MyoD	∅
CD105 (Endoglin)	+	Myogenin	+
CD146 (MCAM)	+	Myosin IIa	+
c-Fos	+	Nestin	+
CNTF	∅	NSE (Eno2)	+
Collagen Ia1	−/+	Osteocalcin	+
Collagen II	+	Osteonectin	+
Connexin43	+	Osteopontin	+[b]
C-Peptide	∅	PCNA	+
Cytokeratin 18	∅	PDX-1	∅
Cytokeratin 19	∅	S100	+
Desmin	+	Somatostatin	∅
EP3	+	STAT3	+
Fibronectin	+	Vimentin	+
GFAP	+	vWF	∅
Glucagon	∅	β-tubulin	+

[a]Expression: +: positive, ∅: lack, −/+: weak
[b]Immunoreactivity was positive in 10–20 % of the cells

mechanisms in MSCs. Besides the soluble factors, like IL-10 (Krampera et al. 2003), IL-6 (Noel et al. 2007), TGF-β (Di Nicola et al. 2002; Aggarwal and Pittenger 2005), and HGF (Rasmusson 2006), cell-to-cell contact was shown to have role in regulation (Krampera et al. 2003; Di Nicola et al. 2002).

Similar to bone marrow-derived MSCs, pancreatic islet MSCs could also suppress proliferation of stimulated T cells. In the study by Kim et al. (2012), pancreatic islet MSCs showed higher IL-1β, IL-6, STAT3, and FGF9 expressions compared to bone marrow-derived MSC and less IL-10. As bone marrow-derived MSCs and pancreatic islet-derived stem cells share many common features, both types of cells were expected to have immune regulatory effect. For this aim, islet stem cells were cocultured with chemically stimulated T cells, and cellular responses were analyzed (unpublished data). After coculture, the proliferation of stimulated T cells was decreased, and the number of apoptotic cells was increased. The expressions of pro-inflammatory cytokines (IL-1a, IL-2, IL-12b, IFN-γ, and

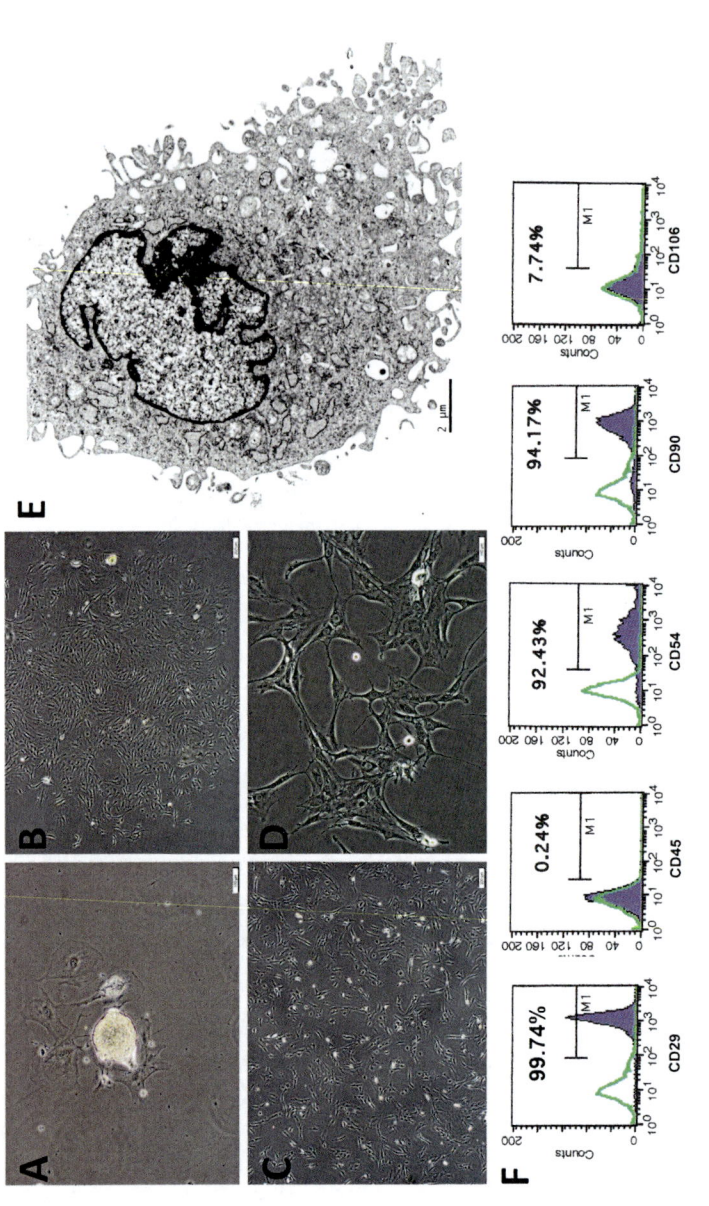

Fig. 1 Colony formation and characterization of rat pancreatic islet-derived stem cells. During the culture, cells with mesenchymal morphology expanded out from the pancreatic islet (**a**) and form colonies (**b**). Later, the cells exhibited large, flattened, or fibroblast-like morphology (**c**, **d**). The ultrastructural analysis of the cells revealed their eccentric and irregularly shaped large nuclei with multiple nucleoli (**e**). Chromatin-formed thin and dense layer inside the perinuclear cisternae. The cytoplasms had many rough endoplasmic reticulum cisternae. The plasma membrane of pancreatic islet stem cells has several thick pseudopodia-like structures (Scale Bars: 2 μm). The cell maker analysis showed typical mesenchymal marker expression for these cells, i.e., positive for CD29, CD54, and CD90 and negative for hematopoietic marker, CD45. CD106 (Vcam1) expression was partially positive

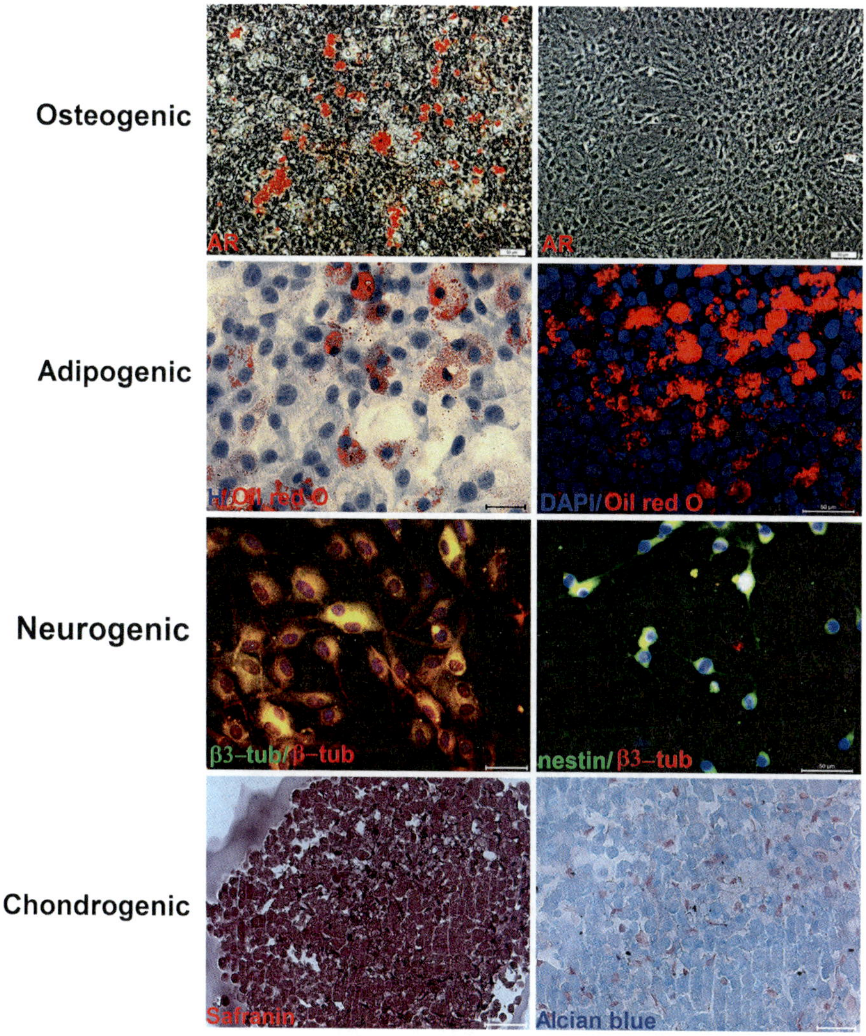

Fig. 2 Differentiation of pancreatic islet-derived stem cells into different cell lineages. After the culture in specific differentiation media, stem cells were characterized with respect to histochemical and immunofluorescence staining. Osteogenic differentiation was shown by staining of mineral deposits with *Alizarin red S*. Staining was not observed in the undifferentiated cells. Adipogenic differentiation was proven by staining of cytoplasmic oil deposits with *Oil Red O*. Neurogenic differentiation was determined both by alteration in cell morphology and staining for neurogenic markers (β-tubulin, β 3-tubulin, and nestin). Chondrogenic differentiation was shown by safranin and alcian blue staining

TNF-a) decreased while the level of anti-inflammatory cytokines (IL-4, IL-6, and IL-10) increased during the cocultures. Moreover, the number of T_{reg} significantly increased during the the presence of rPI-SCs, which is known to suppress the T-cell activity. The expression of IL-1ra in pancreatic islet stem cells blocks the inflammatory responses by inhibiting the activity of IL-1a and IL-1β (Jiang et al. 2005; Ye et al. 2008). The expression levels of ICAM-1, VCAM-1, IL-4, IL-6, IL-10, IL-13, TGF-β, and HGF increased in both direct and indirect cocultures. Their increased expression in the cocultures of rPI-SCs with T cells may be the sign for the immunosuppressive mechanism (Glennie et al. 2005; English et al. 2009; Zheng et al. 2004). Because of the immune regulatory effect, these cells were suggested to be used in organ transplants to decrease the rejection or in β cell protection from the immune attacks during the onset of diabetes. A very important question arouse consequently. If those stem cells in pancreatic islets were effectively modulating the immune response, then why can they not function effectively in the diabetic persons? The reasonable explanation for this might be the limited number of these cells for the regulation effects.

After stimulation with IFN-γ, mesenchymal stem cells were shown to act as antigen-presenting cells (Stagg et al. 2006; Chan et al. 2006; Morandi et al. 2008). This phenomenon is quite similar to the antigen-presenting cells in pancreatic islets that are considered to be the initiator of type 1 diabetes by presenting βcell-derived peptides to immune cells (Lacy et al. 1979; Nauta et al. 2006). In these studies, the antigen presentation was observed in the bone marrow-derived MSCs, but it is still unclear whether pancreatic islet-derived stem cells exhibit the similar activity. Indirect evidences were collected in the analyses of these cells. Pancreatic islet-derived stem cells showed CD40 and CD80 expressions (Klein et al. 2005) but not CD86. Similar case was also reported by Lei et al. (2005), in which murine keratinocyte stem cells expressed CD80 but not CD86, indicating that these cells could act as antigen-presenting cells (Lei et al. 2005). Furthermore, MHC II was also not expressed by pancreatic islet-derived stem cells, which makes them nonprofessional antigen-presenting cells. Remarkably, the expression of CX3CR1, which is typically expressed by monocytes, dendritic cells, natural killer cells, and helper T cells (type 1), was observed in pancreatic islet stem cells (Fraticelli et al. 2001; Nishimura et al. 2002; Dichmann et al. 2001). The similar characteristics of pancreatic islet-derived stem cells with antigen-presenting cells make them an important target in the early onset of the diabetes type 1.

In this context, the interaction of endocrine cells, mainly β cells, with islet-derived stem cells was analyzed at the microscopic level following the preparation of single-cell suspension of rat pancreatic islets (Fig. 3). Endocrine cells could be cultured for more than 30 days in appropriate culture conditions in the presence of islet-derived stem cells. However, the removal of these stem cells from the culture caused the decrease in the viability that made the long-term culture of endocrine cells impossible under defined condition (unpublished data). According to this observation, islet-derived stem cells might have similar protective effect like none marrow-derived stem cells (Karaoz et al. 2010b). Considering there could be similar interactions in vivo, the role of islet-derived stem cells was studied in the apoptosis of β cells. After coculture of damaged pancreatic islets with pancreatic islet-derived stem cells, the expression of regulatory proteins in apoptosis, like Bcl3, TNIP1 (TNFAIP3 interacting protein 1), and MAPKAPK2, was increased under stress in pancreatic islets (unpublished data). The number of viable cells and

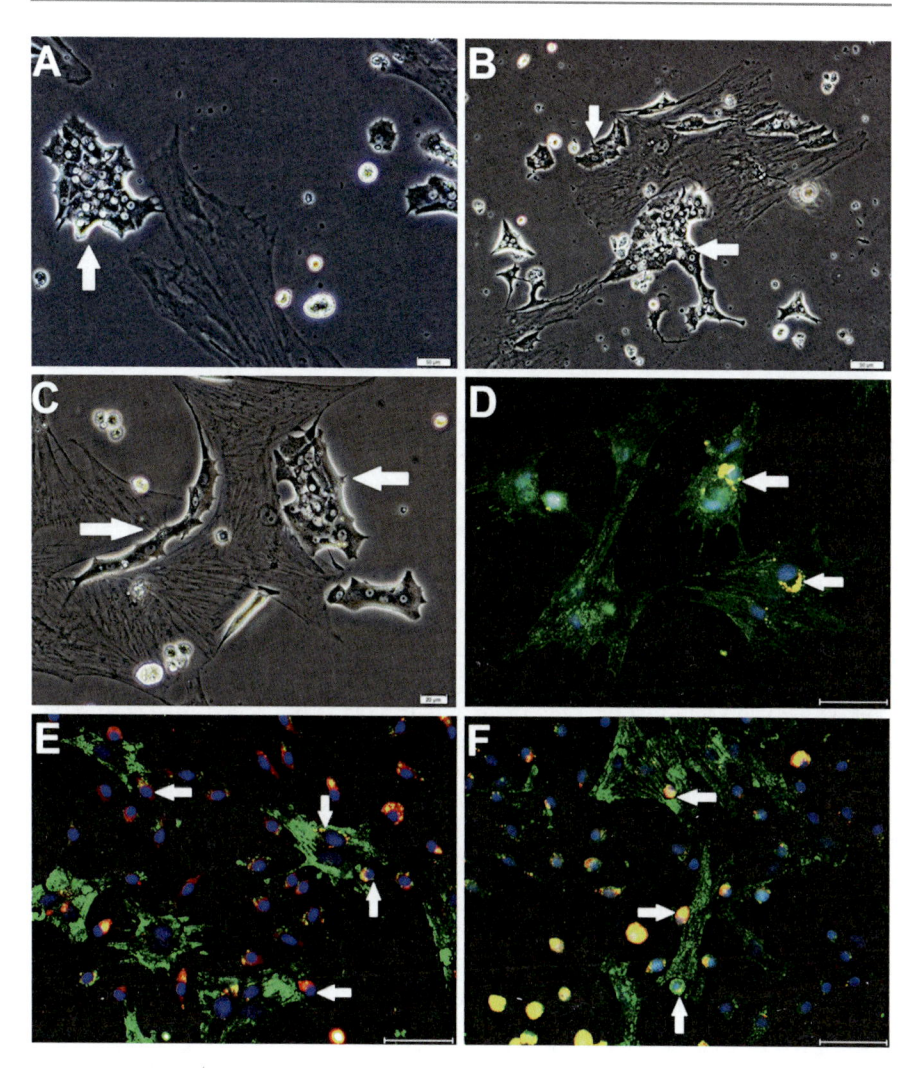

Fig. 3 Interaction of pancreatic islet endocrine cells with stem cells derived from pancreatic islets. Single-cell suspension of endocrine cells, prepared by enzymatic digestion of rat pancreatic islets, was cultured for 30 days without any significant loss of their viability together with stem cells derived from pancreatic islets. The endocrine cells, mostly β cells, interact closely with the stem cells in the culture (*white arrow*, **a–c**). Pancreatic islet-derived stem cells were identified by staining with antibodies against α-Actin-2 (**d**; *green*) and β-actin (**e, f**; *green*). The single-cell suspension of endocrine cells from islets with high number of insulin-positive cells (*red*) signifies that the suspension mostly composed of β cells. Stem cells in the environment of the islets support the β cell viability in vitro, and similar effect is also expected to exist

insulin-secretion capacity was preserved in the coculture with stem cells, whereas necrotic bodies were formed in the absence of the stem cells.

By direct and indirect mechanisms, islet stem cells have a protective role on pancreatic islet cells. Although a number of molecules involve in the antiapoptotic effect, the expression of IL-6 and TGF-β was significantly increased during the coculture (unpublished data). By the reverse experimental approach, in which the effect of IL-6 was neutralized by antibody, the pivotal role of IL-6 in the preserving the β-cell viability was shown. On the other hand, significant effect of IL-6 supplementation to the culture of damaged islets was not observed. The same experiment with TGF-β supplementation increased the cell viability, but the effect remained limited. These findings indicate that there are many key molecules in the regulation of apoptosis with complex interactions, although their expression levels are not as high as IL-6 or TGF-β. In the gene expression analysis of the cocultures after IL-6 neutralization, the expression levels of IL-2 and IL-9 were decreased considerably. The IL-2- and IL-9-mediated regulation of apoptosis in pancreatic islets was mainly mediated by JAK/STAT signaling, which controls numerous important biologic responses like immune function and cellular growth (Demoulin et al. 1996; Rabinovitch et al. 2002; Chentoufi et al. 2011; Tang et al. 2008).

The origin of progenitor/stem cells in pancreatic islets is still a controversial issue. Gong et al. (2012) reported that the nestin-positive cells involve in repair of damage in acinar tissues and islets. They take part only in the supporting of damaged islets, and functional differentiation of these cells was not observed. The expression of c-Kit (CD117) by these cells, and rarely by normal pancreatic tissues, becomes the main point of their suggestion that the origin of pancreatic islet stem cells might be the bone marrow stem cells. Those cells were reported to locate only in the islets and involve only in the repair process of islets damaged. For that reason, they were defined as progenitor cells rather stem cells (Gong et al. 2012). In another study, pancreatic islet-derived stem cells were pointed out to involve in endocrine pancreatic regeneration with a lower frequency (Joglekar et al. 2007).

β Cell Replication

Until recently, the β-cells regeneration was thought to be very limited. However, the report by a group of researchers from Harvard University published in 2004 changed this notion. The research group lead by Douglas Melton observed the self-renewal and regeneration of β cells in mice after partial pancreatectomy by the method of genetic lineage tracing (Dor et al. 2004; Dor and Melton 2004; Nir and Dor 2005). In the following studies, they demonstrated the replication of β-cells both in vivo and in vitro (Georgia and Bhushan 2004; Meier et al. 2006). Another evidence for β-cell regeneration by replication of mature β-cells was provided by Gershengorn et al. (2004, 2005). The newly formed cells were reported to be originated from nestin-positive mesenchymal-like progenitor cells that were generated by dedifferentiation of previously existing β cells, also called epithelial-mesenchymal transition process. Interestingly, the supporting studies came from two independent research groups later (Lechner et al. 2005; Ouziel-Yahalom et al. 2006). Both groups reported that the β-cell regeneration in human pancreatic islets occurs by de- and redifferentiation of these cells.

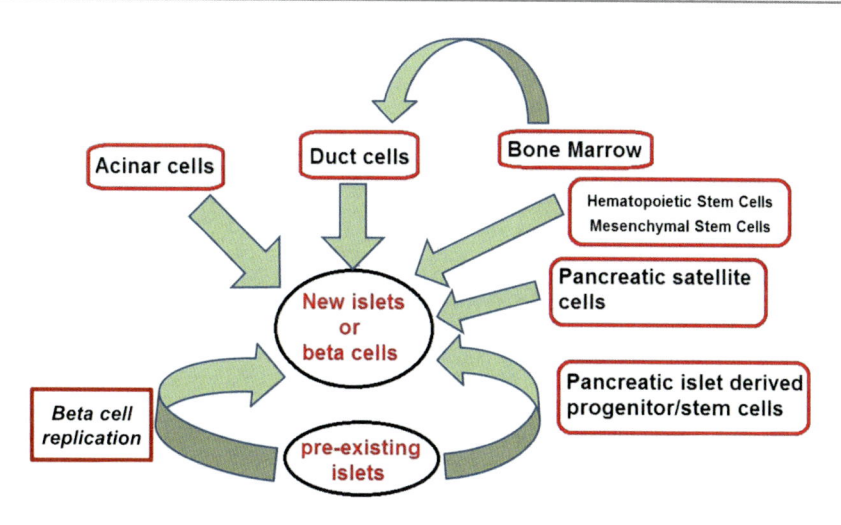

Fig. 4 Schematic drawing of the possible sources of regenerated pancreatic islets and β-cells

All these results indicated the generation of new islet or β cells occurred by the normal and pathological processes in adult pancreatic islets (Fig. 4). These cells might be the stem cells mobilized from bone marrow, cells originated from duct cells that homed in islets, undifferentiated stem cell cells from the early stage of development, or cells de- and redifferentiated in response to various physiological signals. Regardless of their source, the pancreatic islet stem cells promise by the demonstration of β-cell regeneration in adulthood the development of radical treatment procedures for type 1 diabetes in the future.

Concluding Remarks

As pancreatic islet-derived stem cells carry both immune suppressive and antiapoptotic features, they might have important roles in pathogenesis in diabetes type 1. Under normal circumstances, these cells are expected to function as protector of β-cells and other endocrine cells in islets maintaining both their viability and functionality. However, the weakness in their function during the onset of the diabetes is crucial in this case. Therefore, the interaction between stem cells and endocrine cells in pancreatic islets should be focused in the future more in detail for better understanding of the disease. In vitro and in vivo studies are required to explain why pancreatic islet-derived stem cells may not suppress attacks of autoreactive immune cells towards pancreatic islets in people with diabetes type 1. The comparison of stem cells from pancreatic islets of diseased people with healthy individuals at both genomic and proteomic levels would benefit to find alternative treatment approaches.

Cross-References

▶ Generating Pancreatic Endocrine Cells from Pluripotent Stem Cells
▶ Regulation of Pancreatic Islet Formation

References

Abraham EJ, Leech CA, Lin JC, Zulewski H, Habener JF (2002) Insulinotropic hormone glucagon-like peptide-1 differentiation of human pancreatic islet-derived progenitor cells into insulin-producing cells. Endocrinology 143:3152–3161

Aggarwal S, Pittenger MF (2005) Human mesenchymal stem cells modulate allogeneic immune cell responses. Blood 105:1815–1822

Atouf F, Park CH, Pechhold K, Ta M, Choi Y, Lumelsky NL (2007) No evidence for mouse pancreatic β-cell epithelial-mesenchymal transition in vitro. Diabetes 56:699–702

Baeyens L, De Breuck S, Lardon J, Mfopou JK, Rooman I, Bouwens L (2005) İn vitro generation of insulin-producing β cells from adult exocrine pancreatic cells. Diabetologia 48:49–57

Banerjee M, Bhonde RR (2003) Islet generation from intra islet precursor cells of diabetic pancreas: in vitro studies depicting in vivo differentiation. JOP 4:137–145

Bonner-Weir S, Deery D, Leahy JL, Weir GC (1989) Compensatory growth of pancreatic β-cells in adults after short-term glucose infusion. Diabetes 38:49–53

Bonner-Weir S, Baxter LA, Schuppin GT, Smith FE (1993) A second pathway for regeneration of adult exocrine and endocrine pancreas. A possible recapitulation of embryonic development. Diabetes 42:1715–1720

Bonner-Weir S, Taneja M, Weir GC, Tatarkiewicz K, Song KH, Sharma A, O'Neil JJ (2000) In vitro cultivation of human islets from expanded ductal tissue. Proc Natl Acad Sci USA 97:7999–8004

Bonner-Weir S, Toschi E, Inada A, Reitz P, Fonseca SY, Aye T, Sharma A (2004) The pancreatic ductal epithelium serves as a potential pool of progenitor cells. Pediatr Diabetes 5:16–22

Bouwens L, Rooman I (2005) Regulation of pancreatic β-cell mass. Physiol Rev 85:1255–1270

Brelje TC, Scharp DW, Lacy PE, Ogren L, Talamantes F, Robertson M, Friesen HG, Sorenson RL (1993) Effect of homologous placenta lactogens, prolactins, and growth hormones on islet β-cell division and insulin secretion in rat, mouse, and human islets: implication for placental lactogen regulation of islet function during pregnancy. Endocrinology 132:879–887

Cantenys D, Portha B, Dutrillaux MC, Hollande E, Roze C, Picon L (1981) Histogenesis of the endocrine pancreas in newborn rats after destruction by streptozotosin. An immunocytochemical study. Virchows Arch B Cell Pathol Incl Mol Pathol 35:109–122

Carlotti F, Zaldumbide A, Loomans CJ, van Rossenberg E, Engelse M, de Koning EJ, Hoeben RC (2010) Isolated human islets contain a distinct population of mesenchymal stem cells. Islets 2:164–173

Chan JL, Tang KC, Patel AP, Bonilla LM, Pierobon N, Ponzio NM, Rameshwar P (2006) Antigen-presenting property of mesenchymal stem cells occurs during a narrow window at low levels of interferon-γ. Blood 107:4817–4824

Chang CJ, Yen ML, Chen YC, Chien CC, Huang HI, Bai CH, Yen BL (2006) Placenta-derived multipotent cells exhibit immunosuppressive properties that are enhanced in the presence of interferon-γ. Stem Cells 24:2466–2477

Chentoufi AA, Gaudreau S, Nguyen A, Sabha M, Amrani A, Elghazali G (2011) Type I diabetes-associated tolerogenic properties of interleukin-2. Clin Dev Immunol 2011:289343

Choi Y, Ta M, Atouf F, Lumelsky N (2004) Adult pancreas generates multipotent stem cells and pancreatic and nonpancreatic progeny. Stem Cells 22:1070–1084

Comoli P, Ginevri F, Maccario R, Avanzini MA, Marconi M, Groff A, Cometa A, Cioni M, Porretti L, Barberi W, Frassoni F, Locatelli F (2008) Human mesenchymal stem cells inhibit antibody production induced in vitro by allostimulation. Nephrol Dial Transplant 23:1196–1202

Cornelius JG, Tchemev V, Kao KJ, Peck AB (1997) İn vitro generation of islets in long-term cultures of pluripotent stem cells from adult mouse pancreas. Horm Metab Res 29:271–277

Davani B, Ikonomou L, Raaka BM, Geras-Raaka E, Morton RA, Marcus-Samuels B, Gershengorn MC (2007) Human islet-derived precursor cells are mesenchymal stromal cells that differentiate and mature to hormone-expressing cells in vivo. Stem Cells 25:3215–3222

De Haro-Hernandez R, Cabrera-Munoz L, Mendez JD (2004) Regeneration of β-cells and neogenesis from small ducts or acinar cells promote recovery of endocrine pancreatic function in alloxan-treated rats. Arch Med Res 35:114–120

Demoulin JB, Uyttenhove C, Van Roost E, DeLestré B, Donckers D, Van Snick J, Renauld JC (1996) A single tyrosine of the interleukin-9 (IL-9) receptor is required for STAT activation, antiapoptotic activity, and growth regulation by IL-9. Mol Cell Biol 16:4710–4716

Desai BM, Oliver-Krasinski J, De Leon DD, Farzad C, Hong N, Leach SD, Stoffers DA (2007) Preexisting pancreatic acinar cells contribute to acinar cell, but not islet β cell, regeneration. J Clin Invest 117:971–977

Di Nicola M, Carlo-Stella C, Magni M, Milanesi M, Longoni PD, Matteucci P, Grisanti S, Gianni AM (2002) Human bone marrow stromal cells suppress T-lymphocyte proliferation induced by cellular or nonspecific mitogenic stimuli. Blood 99:3838–3843

Dichmann S, Herouy Y, Purlis D, Rheinen H, Gebicke-Harter P, Norgauer J (2001) Fractalkine induces chemotaxis and actin polymerization in human dendritic cells. Inflamm Res 50:529–533

Dor Y, Melton DA (2004) How important are adult stem cells for tissue maintenance? Cell Cycle 3:1104–1106

Dor Y, Brown J, Martinez OI, Melton DA (2004) Adult pancreatic β-cells are formed by self-duplication rather than stem-cell differentiation. Nature 429:41–46

Duvillie B, Attali M, Aiello V, Quemeneur E, Scharfmann R (2003) Label-retaining cells in the rat pancreas: location and differentiation potential in vitro. Diabetes 52:2035–2042

Elsässer HP, Adler G, Kern HF (1986) Time course and cellular source of pancreatic regeneration following acute pancreatitis in the rat. Pancreas 1:421–429

English K, Ryan JM, Tobin L, Murphy MJ, Barry FP, Mahon BP (2009) Cell contact, prostaglandin E 2 and transforming growth factor β 1 play non-redundant roles in human mesenchymal stem cell induction of CD4 CD25 high forkhead box P3 regulatory T cells. Clin Exp Immunol 156:149–160

Fernandes A, King LC, Guz Y, Stein R, Wright CV, Teitelman G (1997) Differentiation of new insulin-producing cells is induced by injury in adult pancreatic islets. Endocrinology 138:1750–1762

Finegood DT, Scaglia L, Bonner-Weir S (1995) Dynamics of β-cell mass in the growing rat pancreas. Estimation with a simple mathematical model. Diabetes 44:249–256

Finegood DT, Weir GC, Bonner-Weir S (1999) Prior streptozotocin treatment does not inhibit pancreas regeneration after 90% pancreatectomy in rats. Am J Physiol 276:E822–E827

Fraticelli P, Sironi M, Bianchi G, D'Ambrosio D, Albanesi C, Stoppacciaro A, Chieppa M, Allavena P, Ruco L, Girolomoni G, Sinigaglia F, Vecchi A, Mantovani A (2001) Fractalkine (CX3CL1) as an amplification circuit of polarized Th1 responses. J Clin Invest 107:1173–1181

Gallo R, Gambelli F, Gava B, Sasdelli F, Tellone V, Masini M, Marchetti P, Dotta F, Sorrentino V (2007) Generation and expansion of multipotent mesenchymal progenitor cells from cultured human pancreatic islets. Cell Death Differ 14:1860–1871

Gao R, Ustinov J, Pulkkinen MA, Lundin K, Korsgren O, Otonkoski T (2003) Characterization of endocrine progenitor cells and critical factors for their differentiation in human adult pancreatic cell culture. Diabetes 52:2007–2015

Georgia S, Bhushan A (2004) β cell replication is the primary mechanism for maintaining postnatal β cell mass. J Clin Invest 114:963–968

Gershengorn MC, Hardikar AA, Wei C, Geras-Raaka E, Marcus-Samuels B, Raaka BM (2004) Epithelial-to-mesenchymal transition generates proliferative human islet precursor cells. Science 306:2261–2264

Gershengorn MC, Geras-Raaka E, Hardikar AA, Raaka BM (2005) Are better islet cell precursors generated by epithelial-to-mesenchymal transition? Cell Cycle 4(3):380–382

Glennie S, Soeiro I, Dyson PJ, Lam EW, Dazzi F (2005) Bone marrow mesenchymal stem cells induce division arrest anergy of activated T cells. Blood 105:2821–2827

Gmyr V, Kerr-Conte J, Belaich S, Vandewalle B, Leteurtre E, Vantyghem MC, Lecomte-Houcke-M, Proye C, Lefebvre J, Pattou F (2000) Adult human cytokeratin 19-positive cells reexpress insulin promoter factor 1 in vitro: further evidence for pluripotent pancreatic stem cells in humans. Diabetes 49:1671–1680

Gong J, Zhang G, Tian F, Wang Y (2012) Islet-derived stem cells from adult rats participate in the repair of islet damage. J Mol Histol 43:745–750

Gu D, Sarvetnick N (1993) Epithelial cell proliferation and islet neogenesis in IFN-γ transgenic mice. Development 118:33–46

Gu D, Lee MS, Krahl T, Sarvetnick N (1994) Transitional cells in the regenerating pancreas. Development 120:1873–1881

Gu D, Arnush M, Sarvetnick N (1997) Endocrine/exocrine intermediate cells in streptozotocin-treated Ins-IFN-γ transgenic mice. Pancreas 15:246–250

Guo Z, Li H, Li X, Yu X, Wang H, Tang P, Mao N (2006) In vitro characteristics and in vivo immunosuppressive activity of compact bone-derived murine mesenchymal progenitor cells. Stem Cells 24:992–1000

Guz Y, Nasir I, Teitelman G (2001) Regeneration of pancreatic β cells from intra-islet precursor cells in an experimental model of diabetes. Endocrinology 142:4956–4968

Haber PS, Keogh GW, Apte MV, Moran CS, Stewart NL, Crawford DH, Pirola RC, McCaughan GW, Ramm GA, Wilson JS (1999) Activation of pancreatic stellate cells in human and experimental pancreatic fibrosis. Am J Pathol 155:1087–1095

Hao E, Tyrberg B, Itkin-Ansari P, Lakey JR, Geron I, Monosov EZ, Barcova M, Mercola M, Levine F (2006) β-cell differentiation from nonendocrine epithelial cells of the adult human pancreas. Nat Med 12:310–316

Hayek A, Beattie GM, Cirulli V, Lopez AD, Ricordi C, Rubin JS (1995) Growth factor/matrix-induced proliferation of human adult β cells. Diabetes 44:1458–1460

Holland AM, Gonez LJ, Harrison LC (2004) Progenitor cells in the adult pancreas. Diabetes Metab Res Rev 20:13–27

Holland AM, Góñez LJ, Naselli G, Macdonald RJ, Harrison LC (2005) Conditional expression demonstrates the role of the homeodomain transcription factor Pdx1 in maintenance and regeneration of β-cells in the adult pancreas. Diabetes 54:2586–2595

Ikejiri N (1990) The vitamin A-storing cells in the human and rat pancreas. Kurume Med J 37:67–81

Jiang XX, Zhang Y, Liu B, Zhang SX, Wu Y, Yu XD, Mao N (2005) Human mesenchymal stem cells inhibit differentiation and function of monocyte-derived dendritic cells. Blood 105:4120–4126

Joglekar MV, Parekh VS, Hardikar AA (2007) New pancreas from old: microregulators of pancreas regeneration. Trends Endocrinol Metab 18:393–400

Karaoz E, Ayhan S, Gacar G, Aksoy A, Duruksu G, Okçu A, Demircan PC, Sariboyaci AE, Kaymaz F, Kasap M (2010a) Isolation and characterization of stem cells from pancreatic islet: pluripotency, differentiation potential and ultrastructural characteristics. Cytotherapy 12:288–302

Karaoz E, Genç ZS, Demircan PÇ, Aksoy A, Duruksu G (2010b) Protection of rat pancreatic islet function and viability by coculture with rat bone marrow-derived mesenchymal stem cells. Cell Death Dis 1:e36

Kem H, Logothetopoulos J (1970) Steroid diabetes in the guinea pig. Studies on islet-cell ultrastructure and regeneration. Diabetes 19:145–154

Kim BM, Kim SY, Lee S, Shin YJ, Min BH, Bendayan M, Park IS (2006) Clusterin induces differentiation of pancreatic duct cells into insulin-secreting cells. Diabetologia 49:311–320

Kim J, Breunig MJ, Escalante LE, Bhatia N, Denu RA, Dollar BA, Stein AP, Hanson SE, Naderi N, Radek J, Haughy D, Bloom DD, Assadi-Porter FM, Hematti P (2012) Biologic and immunomodulatory properties of mesenchymal stromal cells derived from human pancreatic islets. Cytotherapy 14:925–935

Klein D, Barbé-Tuana F, Pugliese A, Ichii H, Garza D, Gonzalez M, Molano RD, Ricordi C, Pastori RL (2005) A functional CD40 receptor is expressed in pancreatic β cells. Diabetologia 48:268–276

Korcakova L (1971) Mitotic division and its significance for regeneration of granulated β-cells in the islets of Langerhans in allozan-diabetic rats. Folia Morphol (Praha) 19:24–30

Krampera M, Glennie S, Dyson J, Scott D, Laylor R, Simpson E, Dazzi F (2003) Bone marrow mesenchymal stem cells inhibit the response of naive and memory antigen-specific T cells to their cognate peptide. Blood 101:3722–3729

Kruse C, Birth M, Rohwedel J, Assmuth K, Goepel A, Wedel T (2004) Pluripotency of adult stem cells derived from human and rat pancreas. Appl Phys A 79:1617–1624

Lacy PE, Davie JM, Finke EH (1979) Prolongation of islet allograft survival following in vitro culture (24 degrees C) and a single injection of ALS. Science 204:312–313

Lardon J, Huyens N, Rooman I, Bouwens L (2004) Exocrine cell transdifferentiation in dexamethasone-treated rat pancreas. Virchows Arch 444:61–65

Lechene de la Porte P, Iovanna J, Odaira C, Choux R, Sarles H, Berger Z (1991) Involvement of tubular complexes in pancreatic regeneration after acute necrohemorrhagic pancreatitis. Pancreas 6:298–306

Lechner A, Habener JF (2003) Stem/progenitor cells derived from adult tissues: potential for the treatment of diabetes mellitus. Am J Physiol Endocrinol Metab 284:E259–E266

Lechner A, Leech CA, Abraham EJ, Nolan AL, Habener JF (2002) Nestin-positive progenitor cells derived from adult human pancreatic islets of Langerhans contain side population (SP) cells defined by expression of the ABCG2 (BCRP1) ATP-binding cassette transporter. Biochem Biophys Res Commun 293:670–674

Lechner A, Nolan AL, Blacken RA, Habener JF (2005) Redifferentiation of insulin-secreting cells after in vitro expansion of adult human pancreatic islet tissue. Biochem Biophys Res Commun 327:581–588

Lei J, Cheng J, Li Y, Li S, Zhang L (2005) CD80, but not CD86, express on cultured murine keratinocyte stem cells. Transplant Proc 37:289–291

Leng SH, Lu FE (2005) Induction of pancreatic duct cells of neonatal rats into insulin-producing cells with fetal bovine serum: a natural protocol and its use for patch clamp experiments. World J Gastroenterol 11:6968–6974

Linning KD, Tai MH, Madhukar BV, Chang CC, Reed DN Jr, Ferber S, Trosko JE, Olson LK (2004) Redox-mediated enrichment of self-renewing adult human pancreatic cells that possess endocrine differentiation potential. Pancreas 29:e64–e76

Lipsett M, Finegood DT (2002) β-cell neogenesis during prolonged hyperglycemia in rats. Diabetes 51:1834–1841

List JF, Habener JF (2004) Glucagon-like peptide 1 agonists and the development and growth of pancreatic β-cells. Am J Physiol Endocrinol Metab 286:E875–E881

Logothetopoulos J, Bell EG (1966) Histological and autoradiographic studies of the islets of mice injected with insulin antibody. Diabetes 15:205–211

Marynissen G, Aerts L, Van Assche FA (1983) The endocrine pancreas during pregnancy and lactation in the rat. J Dev Physiol 5:373–381

Means AL, Meszoely IM, Suzuki K, Miyamoto Y, Rustgi AK, Coffey RJ Jr, Wright CV, Stoffers DA, Leach SD (2005) Pancreatic epithelial plasticity mediated by acinar cell transdifferentiation and generation of nestin-positive intermediates. Development 132:3767–3776

Meier JJ, Ritzel RA, Maedler K, Gurlo T, Butler PC (2006) Increased vulnerability of newly forming β cells to cytokine-induced cell death. Diabetologia 49:83–89

Meneilly GS, McIntosh CH, Pederson RA, Habener JF, Gingerich R, Egan JM, Finegood DT, Elahi D (2001) Effect of glucagon-like peptide 1 on non-insulin-mediated glucose uptake in the elderly patient with diabetes. Diabetes Care 24:1951–1956

Meneilly GS, Greig N, Tildesley H, Habener JF, Egan JM, Elahi D (2003a) Effects of 3 months of continuous subcutaneous administration of glucagon-like peptide 1 in elderly patients with type 2 diabetes. Diabetes Care 26:2835–2841

Meneilly GS, McIntosh CH, Pederson RA, Habener JF, Ehlers MR, Egan JM, Elahi D (2003b) Effect of glucagon-like peptide 1 (7–36 amide) on insulin-mediated glucose uptake in patients with type 1 diabetes. Diabetes Care 26:837–842

Minami K, Okuno M, Miyawaki K, Okumachi A, Ishizaki K, Oyama K, Kawaguchi M, Ishizuka N, Iwanaga T, Seino S (2005) Lineage tracing and characterization of insulin-secreting cells generated from adult pancreatic acinar cells. Proc Natl Acad Sci USA 102:15116–15121

Montanucci P, Pennoni I, Pescara T, Blasi P, Bistoni G, Basta G, Calafiore R (2011) The functional performance of microencapsulated human pancreatic islet-derived precursor cells. Biomaterials 32:9254–9262

Morandi F, Raffaghello L, Bianchi G, Meloni F, Salis A, Millo E, Ferrone S, Barnaba V, Pistoia V (2008) Immunogenicity of human mesenchymal stem cells in HLA-class I-restricted T-cell responses against viral or tumor-associated antigens. Stem Cells 26:1275–1287

Morton RA, Geras-Raaka E, Wilson LM, Raaka BM, Gershengorn MC (2007) Endocrine precursor cells from mouse islets are not generated by epithelial-to-mesenchymal transition of mature β cells. Mol Cell Endocrinol 270:87–93

Nauta AJ, Kruisselbrink AB, Lurvink E, Willemze R, Fibbe WE (2006) Mesenchymal stem cells inhibit generation and function of both $CD34^+$-derived and monocyte-derived dendritic cells. J Immunol 177:2080–2087

Nir T, Dor Y (2005) How to make pancreatic β cells – prospects for cell therapy in diabetes. Curr Opin Biotechnol 16:524–529

Nishimura M, Umehara H, Nakayama T, Yoneda O, Hieshima K, Kakizaki M, Dohmae N, Yoshie O, Imai T (2002) Dual functions of fractalkine/CX3C ligand 1 in trafficking of perforin+/granzyme B+ cytotoxic effector lymphocytes that are defined by CX3CR1 expression. J Immunol 168:6173–6180

Noel D, Djouad F, Bouffi C, Mrugala D, Jorgensen C (2007) Multipotent mesenchymal stromal cells and immune tolerance. Leuk Lymphoma 48:1283–1289

Ogata T, Li L, Yamada S, Yamamoto Y, Tanaka Y, Takei I, Umezawa K, Kojima I (2004) Promotion of β-cell differentiation by conophylline in fetal and neonatal rat pancreas. Diabetes 53:2596–2602

Okuno M, Minami K, Okumachi A, Miyawaki K, Yokoi N, Toyokuni S, Seino S (2007) Generation of insulin-secreting cells from pancreatic acinar cells of animal models of type 1 diabetes. Am J Physiol Endocrinol Metab 292:E158–E165

Otonkoski T, Beattie GM, Rubin JS, Lopez AD, Baird A, Hayek A (1994) Hepatocyte growth factor/scatter factor has insulinotropic activity in human fetal pancreatic cells. Diabetes 43:947–953

Ouziel-Yahalom L, Zalzman M, Anker-Kitai L, Knoller S, Bar Y, Glandt M, Herold K, Efrat S (2006) Expansion and redifferentiation of adult human pancreatic islet cells. Biochem Biophys Res Commun 341:291–298

Oyama K, Minami K, Ishizaki K, Fuse M, Miki T, Seino S (2006) Spontaneous recovery from hyperglycemia by regeneration of pancreatic β-cells in Kir6.2G132S transgenic mice. Diabetes 55:1930–1938

Peck AB, Cornelius JG (1995) İn vitro growth of mature pancreatic islets of Langerhans from single, pluripotent stem cells isolated from pre-diabetic adult pancreas. Diabetes 44:10A

Peck AB, Ramiya V (2004) İn vitro-generation of surrogate islets from adult stem cells. Transpl Immunol 12:259–272

Petropavlovskaia M, Rosenberg L (2002) Identification and characterization of small cells in the adult pancreas: potential progenitor cells? Cell Tissue Res 310:51–58

Poliakova L, Pirone A, Farese A, MacVittie T, Farney A (2004) Presence of nonhematopoietic side population cells in the adult human and nonhuman primate pancreas. Transplant Proc 36:1166–1168

Rabinovitch A, Suarez-Pinzon WL, Shapiro AM, Rajotte RV, Power R (2002) Combination therapy with sirolimus and interleukin-2 prevents spontaneous and recurrent autoimmune diabetes in NOD mice. Diabetes 51:638–645

Ramiya VK, Maraist M, Arfors KE, Schatz DA, Peck AB, Cornelius JG (2000) Reversal of insulin dependent diabetes using islets generated in vitro from pancreatic stem cells. Nat Med 6:278–282

Rao MS, Dwivedi RS, Yeldandi AV, Subbarao V, Tan XD, Usman MI, Thangada S, Nemali MR, Kumar S, Scarpelli DG, Reddy JK (1989) Role of periductal and ductular epithelial cells of the adult rat pancreas in pancreatic hepatocyte lineage. A change in the differentiation commitment. Am J Pathol 134:1069–1086

Rasmusson I (2006) Immun modulation by mesenchymal stem cells. Exp Cell Res 312:2169–2179

Rooman I, Bouwens L (2004) Combined gastrin and epidermal growth factor treatment induces islet regeneration and restores normoglycaemia in C57Bl6/J mice treated with alloxan. Diabetologia 47:259–265

Rosenberg L (1998) Induction of islet cell neogenesis in the adult pancreas: the partial duct obstruction model. Microsc Res Tech 43:337–346

Rosenberg L, Vinik AL (1992) Trophic stimulation of the ductular-islet cell axis: a new approach to the treatment of diabetes. Adv Exp Med Biol 321:95–104

Sandgren EP, Luetteke NC, Palmiter RD, Brinster RL, Lee DC (1990) Overexpression of TGF α in transgenic mice: induction of epithelial hyperplasia, pancreatic metaplasia, and carcinoma of the breast. Cell 61:1121–1135

Saotome T, Inoue H, Fujimiya M, Fujiyama Y, Bamba T (1997) Morphological and immunocytochemical identification of periacinar fibroblast-like cells derived from human pancreatic acini. Pancreas 14:373–382

Schmied BM, Ulrich A, Matsuzaki H, Ding X, Ricordi C, Weide L, Moyer MP, Batra SK, Adrian TE, Pour PM (2001) Transdifferentiation of human islet cells in a long-term culture. Pancreas 23:157–171

Seaberg RM, Smukler SR, Kieffer TJ, Enikolopov G, Asghar Z, Wheeler MB, Korbutt G, van der Kooy D (2004) Clonal identification of multipotent precursors from adult mouse pancreas that generate neural and pancreatic lineages. Nat Biotechnol 22:1115–1124

Sharma A, Zangen DH, Reitz P, Taneja M, Lissauer ME, Miller CP, Weir GC, Habener JF, Bonner-Weir S (1999) The homeodomain protein IDX-1 increases after an early burst of proliferation during pancreatic regeneration. Diabetes 48:507–513

Spaggiari GM, Capobianco A, Abdelrazik H, Becchetti F, Mingari MC, Moretta L (2008) Mesenchymal stem cells inhibit natural killer-cell proliferation, cytotoxicity, and cytokine production: role of indoleamine 2,3-dioxygenase and prostaglandin E2. Blood 111:1327–1333

Sphyris N, Logsdon CD, Harrison DJ (2005) Improved retention of zymogen granules in cultured murine pancreatic acinar cells and induction of acinar-ductal transdifferentiation in vitro. Pancreas 30:148–157

Stagg J, Pommey S, Eliopoulos N, Galipeau J (2006) Interferon-γ-stimulated marrow stromal cells: a new type of nonhematopoietic antigen-presenting cell. Blood 107:2570–2577

Street CN, Sipione S, Helms L, Binette T, Rajotte RV, Bleackley RC, Korbutt GS (2004) Stem cell-based approaches to solving the problem of tissue supply for islet transplantation in type 1 diabetes. Int J Biochem Cell Biol 36:667–683

Suzuki A, Nakauchi H, Taniguchi H (2004) Prospective isolation of multipotent pancreatic progenitors using flow-cytometric cell sorting. Diabetes 53:2143–2152

Swenne I (1992) Pancreatic β-cell growth and diabetes mellitus. Diabetologia 35:193–201

Tang Q, Adams JY, Penaranda C, Melli K, Piaggio E, Sgouroudis E, Piccirillo CA, Salomon BL, Bluestone JA (2008) Central role of defective interleukin-2 production in the triggering of islet autoimmune destruction. Immunity 28:687–697

von Mach MA, Hengstler JG, Brulport M, Eberhardt M, Schormann W, Hermes M, Prawitt D, Zabel B, Grosche J, Reichenbach A, Muller B, Weilemann LS, Zulewski H (2004) İn vitro cultured islet-derived progenitor cells of human origin express human albumin in severe combined immunodeficiency mouse liver in vivo. Stem Cells 22:1134–1141

Wang RN, Kloppel G, Bouwens L (1995) Duct- to islet-cell differentiation and islet growth in the pancreas of duct-ligated adult rats. Diabetologia 38:1405–1411

Wang J, Song LJ, Gerber DA, Fair JH, Rice L, LaPaglia M, Andreoni KA (2004) A model utilizing adult murine stem cells for creation of personalized islets for transplantation. Transplant Proc 36:1188–1190

Wang ZV, Mu J, Schraw TD, Gautron L, Elmquist JK, Zhang BB, Brownlee M, Scherer PE (2008) PANIC-ATTAC: a mouse model for inducible and reversible β-cell ablation. Diabetes 57:2137–2148

Watari N, Hotta Y, Mabuchi Y (1982) Morphological studies on a vitamin A-storing cell and its complex with macrophage observed in mouse pancreatic tissues following excess vitamin A administration. Okajimas Folia Anat Jpn 58:837–858

Weaver CV, Sorenson RL, Kaung HC (1985) Immunocytochemical localization of insulin-immunoreactive cells in the pancreatic ducts of rats treated with trypsin inhibitor. Diabetologia 28:781–785

Xu G, Stoffers DA, Habener JF, Bonner-Weir S (1999) Exendin-4 stimulates both β-cell replication and neogenesis, resulting in increased β-cell mass and improved glucose tolerance in diabetic rats. Diabetes 48:2270–2276

Xu X, D'Hoker J, Stangé G, Bonné S, De Leu N, Xiao X, Van de Casteele M, Mellitzer G, Ling Z, Pipeleers D, Bouwens L, Scharfmann R, Gradwohl G, Heimberg H (2008) β cells can be generated from endogenous progenitors in injured adult mouse pancreas. Cell 132:197–207

Yamamoto K, Miyagawa J, Waguri M, Sasada R, Igarashi K, Li M, Nammo T, Moriwaki M, Imagawa A, Yamagata K, Nakajima H, Namba M, Tochino Y, Hanafusa T, Matsuzawa Y (2000) Recombinant human betacellulin promotes the neogenesis of β-cells and ameliorates glucose intolerance in mice with diabetes induced by selective alloxan perfusion. Diabetes 49:2021–2027

Yamaoka T, Yoshino K, Yamada T, Idehara C, Hoque MO, Moritani M, Yoshimoto K, Hata J, Itakura M (2000) Diabetes and tumor formation in transgenic mice expressing Reg I. Biochem Biophys Res Commun 278:368–376

Ye Z, Wang Y, Xie HY, Zheng SS (2008) Immunosuppressive effects of rat mesenchymal stem cells: involvement of $CD4^+$ $CD25^+$ regulatory T cells. Hepatobiliary Pancreat Dis Int 7:608–614

Zhang L, Hong TP, Hu J, Liu YN, Wu YH, Li LS (2005) Nestin-positive progenitor cells isolated from human fetal pancreas have phenotypic markers identical to mesenchymal stem cells. World J Gastroenterol 11:2906–2911

Zheng SG, Wang JH, Koss MN, Quismorio F Jr, Gray JD, Horwitz DA (2004) $CD4^+$ and $CD8^+$ regulatory T cells generated ex-vivo with IL-2 and TGF-β suppress a stimulatory graft-versus host disease with lupus-like syndrome. J Immunol 172:1531–1539

Zhou Q, Brown J, Kanarek A, Rajagopal J, Melton DA (2008) In vivo reprogramming of adult pancreatic exocrine cells to β-cells. Nature 455:627–632

Zimmermann A, Gloor B, Kappeler A, Uhl W, Friess H, Buchler MW (2002) Pancreatic stellate cells contribute to regeneration early after acute necrotising pancreatitis in humans. Gut 51:574–578

Zulewski H, Abraham EJ, Gerlach MJ, Daniel PB, Moritz W, Muller B, Vallejo M, Thomas MK, Habener JF (2001) Multipotential nestin-positive stem cells isolated from adult pancreatic islets differentiate ex vivo into pancreatic endocrine, exocrine, and hepatic phenotypes. Diabetes 50:521–533

Generating Pancreatic Endocrine Cells from Pluripotent Stem Cells

48

Blair K. Gage, Rhonda D. Wideman, and Timothy J. Kieffer

Contents

Introduction	1337
Origins and Utility of Pluripotent Stem Cells	1338
Pancreatic Differentiation of PSCs	1341
Pancreatic Development	1341
Definitive Endoderm, Foregut, and Pancreatic Endodermal Progenitors from PSCs	1344
Pancreatic Endocrine Development In Vitro	1346
Pancreatic Endocrine Development In Vivo	1348
Future Challenges	1352
Reproducibility	1352
Scaling Up	1354
Immunological Control and Encapsulation	1356
In Vitro Maturation and Models of Development	1359
Concluding Remarks	1362
Cross-References	1363
References	1363

B.K. Gage • R.D. Wideman
Department of Cellular and Physiological Sciences, Laboratory of Molecular and Cellular Medicine, University of British Columbia Vancouver, Vancouver, BC, Canada
e-mail: blairkgage@gmail.com; rdwideman@gmail.com

T.J. Kieffer (✉)
Department of Cellular and Physiological Sciences, Laboratory of Molecular and Cellular Medicine, University of British Columbia Vancouver, Vancouver, BC, Canada

Department of Surgery, Life Sciences Institute, University of British Columbia Vancouver, Vancouver, BC, Canada
e-mail: tim.kieffer@ubc.ca

M.S. Islam (ed.), *Islets of Langerhans*, DOI 10.1007/978-94-007-6686-0_49,
© Springer Science+Business Media Dordrecht 2015

Abstract

Human islet transplantation shows considerable promise as a physiological means of insulin replacement for type 1 diabetes, although donor availability and progressive loss of graft function continues to hamper more widespread implementation. Pluripotent stem cells (PSCs) by definition have the ability to form all tissues of the body including insulin-secreting pancreatic β-cells. This potential has led many academic and industry groups to examine methods for efficient production of functional insulin-producing cells from PSCs. Engineered in vitro differentiation protocols generally mimic known pancreatic developmental cascades, which convert undifferentiated cells, through germ-layer specification, to restricted pancreatic endodermal progenitors. The continued development of these progenitors in vivo results in the formation of functional pancreatic endocrine cells capable of preventing and/or reversing diabetic hyperglycemia in rodent models. While the insulin-producing, antidiabetic capacity of differentiated PSCs is very promising, key questions remain about optimizing differentiation processes for functional in vitro maturation, as well as whether production of pancreatic endocrine tissue is reproducible on a clinical scale and sufficiently safe. With additional research and development in these areas, the induction of differentiation processes to yield pancreatic endocrine-like cells could yield a potentially limitless supply of functional β-cells capable of replacing current human islet transplantation therapies for diabetes.

Keywords

Pluripotency • Embryonic stem cells • Pancreatic development • Pancreatic progenitor • Cell differentiation • Cellular therapy • Regenerative medicine • Diabetes

Abbreviations

BMP	Bone morphogenic protein
DMSO	Dimethyl sulfoxide
EC	Embryonal carcinoma
FGF	Fibroblast growth factor
GLP1	Glucagon-like peptide 1
HEPES	4-(2-hydroxyethyl)-1-piperazineethanesulfonic acid
hESC(s)	Human embryonic stem cell(s)
HGF	Hepatocyte growth factor
IGF	Insulin-like growth factor
iPSC(s)	Induced pluripotent stem cell(s)
LADA	Latent autoimmune diabetes of adults
mESC(s)	Mouse embryonic stem cell(s)
MODY	Maturity-onset diabetes of the young
PSC(s)	Pluripotent stem cell(s)
SCNT	Somatic cell nuclear transfer

Shh   Sonic hedgehog
TGF-β   Transforming growth factor β
VEGF   Vascular endothelial growth factor

Introduction

One of the many potential clinical applications of differentiated derivatives of pluripotent cells is a treatment for diabetes mellitus, a disease characterized by lack or insufficient activity of the glucose-lowering hormone insulin. Type 1 diabetes is an autoimmune disease in which insulin-producing pancreatic β-cells are destroyed by autoreactive $CD8^+$ T-cells (van Belle et al. 2011), while in type 2 diabetes, insulin's activity is insufficient due to progressive insulin resistance and eventual loss of pancreatic β-cells (Prentki and Nolan 2006). As β-cells are gradually lost, blood glucose levels rise due to insufficient levels of circulating insulin. While careful administration of insulin via injections can restore relatively normal blood glucose control in many patients temporarily, exogenous insulin administration does not fully match the sensitive and dynamic release kinetics achieved by endogenous β-cells, leading to numerous long-term complications including retinopathy, nephropathy, neuropathy, and cardiovascular disease (Melendez-Ramirez et al. 2010). Moreover, even when glucose levels are tightly controlled with insulin injections, patients are at an increased risk of potentially fatal hypoglycemic episodes due to an inability to mimic the normal decrease in insulin release in response to low blood glucose levels (The Diabetes Control and Complications Trial Research Group 1993; Fowler 2011).

Cadaveric islet transplantation, in conjunction with specific antirejection therapies, was developed as an alternative means of restoring physiological blood glucose control systems to patients with type 1 diabetes (Shapiro et al. 2000). The success of this trial where seven patients regained insulin independence and normal blood glucose control for up to 1 year following a relatively simple infusion of human islet cells into the hepatic circulation led to a great interest in islet transplantation as a curative approach for type 1 diabetes. Subsequent follow-up on this group of patients and others revealed that insulin independence upon islet transplantation was not sustainable for the majority of patients over a 5-year time period (Ryan et al. 2005). While insulin independence was maintained in only 10 % of these patients, 80 % had detectable C-peptide indicating that islet grafts were partially functional which may explain why these patients experienced fewer hypoglycemic events and had improved overall blood glucose control compared to patients treated by insulin injection alone (Ryan et al. 2005; Merani and Shapiro 2006; Shapiro 2011). Despite a poor rate of sustained insulin independence, the clinical benefits of islet transplant as well as the continuing goal of improving the durability of insulin independence has led to a number of islet transplantation programs being set up around the world (Shapiro et al. 2006). However, these programs continue to be constrained by a significant lack of cadaveric islet tissue available for transplant, particularly considering most patients require multiple independent islet preparations to achieve initial insulin independence. (For more information on this topic, see the chapter entitled "▶ Successes and Disappointments with Clinical Islet Transplantation.")

In an effort to produce a defined and vast supply of cells for transplant into patients with diabetes, many groups have explored the production of functional pancreatic endocrine cells from human embryonic stem cells (hESCs) both in vitro (Gage et al. 2013; D'Amour et al. 2006; Jiang et al. 2007a, b; Nostro et al. 2011; Mfopou et al. 2010; Rezania et al. 2011; Shim et al. 2007; Xu et al. 2011; Zhang et al. 2009; Gutteridge et al. 2013; Bose et al. 2012; Micallef et al. 2012; Kunisada et al. 2012; Cheng et al. 2012; Seguin et al. 2008; Leon-Quinto et al. 2004) and in vivo (Kelly et al. 2011; Kroon et al. 2008; Rezania et al. 2011, 2012, 2013; Basford et al. 2012; Bruin et al. 2013; Shim et al. 2007; Gutteridge et al. 2013; Bose et al. 2012; Soria et al. 2000). As reviewed from a number of perspectives (Fryer et al. 2013; Bruin and Kieffer 2012; Guo and Hebrok 2009; Van Hoof et al. 2009; Pagliuca and Melton 2013; Yang and Wright 2009; Nostro and Keller 2012; Champeris Tsaniras and Jones 2010; Stanley and Elefanty 2008; Seymour and Sander 2011; Hebrok 2012; Hosoya 2012; Roche et al. 2006), the generation of functional pancreatic endocrine cells either in vitro or in vivo from hESCs could result in an unlimited supply of cells for insulin replacement therapy while theoretically enabling physiological blood glucose control without the concomitant risk of hypoglycemia associated with exogenous insulin.

However, while in vivo maturation of in vitro-derived pancreatic progenitors can produce pancreatic endocrine cells capable of controlling blood glucose, strictly in vitro protocols have been far less successful at producing functional endocrine cells. In this chapter, we will briefly review the early study of pluripotent cells and more recent research on their utility. We then focus on the generation of pancreatic cells through normal pancreas development and in vitro differentiation strategies. Finally, we address future challenges associated with the generation of pancreatic endocrine cells from progenitors, including reproducibility, scaling up, and protection against immune attack for clinical applications, and generation of new models for further research.

Origins and Utility of Pluripotent Stem Cells

The study of pluripotency began in the early 1950s and 1960s when the first mouse teratocarcinoma cells were observed, isolated, and found to be capable of forming derivatives of all three embryonic germ layers from a single originating embryonal carcinoma (EC) cell (Stevens and Little 1954; Kleinsmith and Pierce 1964). These early EC cell studies revealed that a single cell could form any cell type in the body, but unfortunately their murine origin, considerable genetic rearrangements, and a host of other concerns associated with the tumorigenic EC cells prevented their use in regenerative medicine applications. The first karyotypically normal mouse embryonic stem cells (mESCs) were isolated from the inner cell mass of cultured blastocysts in 1981 (Evans and Kaufman 1981; Martin 1981). The development of these techniques allowed researchers to examine the factors which maintained the pluripotent state as well as signals which stimulate directed differentiation from the isolated cells.

Additionally, the isolation of mESCs capable of efficient chimera generation allowed for genetic targeting and the creation of modified mouse strains (Bradley et al. 1984). Seventeen years after the isolation of mESCs, the derivation of hESCs was first reported (Thomson et al. 1998), after considerable advances in non-human primate ESC derivation in the mid-1990s (Thomson et al. 1995, 1996). The lengthy delay in hESC isolation was presumably due to key differences between mESCs and hESCs. For example, mESCs are dependent on leukemia inhibitor factor for pluripotent growth, while hESCs rely on basic fibroblast growth factor to maintain a phenotype during extended in vitro expansion (Yu and Thomson 2008).

The generation of mESCs and hESCs has provided a benchmark definition of the properties that are required to define a cell as an embryonic stem cell and more importantly what experimental conditions must be met to define a cell as pluripotent. Associated with the pluripotent phenotype is the expression of key transcription factors such as OCT4, NANOG, and SOX2 as well as surface markers such as TRA-1-60, TRA-1-81, and SSEA3/4 in humans (SSEA1 in mice) (Adewumi et al. 2007; Smith et al. 2009). When combined with classical cell morphology of a high nuclear-to-cytoplasmic ratio and alkaline phosphatase activity, ESCs can be distinguished from other cell types in culture (O'Connor et al. 2008). Unfortunately, these assays are only correlative, and given the definition of a PSC is to be pluripotent, the standard assays of this trait are functional in nature, aiming to test the abilities of PSCs to develop into tissues of all three germ layers. This is most often tested through uncontrolled differentiation of PSCs in vitro in embryoid body differentiation assays and in vivo in teratoma formation assays. While the in vitro embryoid body assay provides some data regarding potency, the most widely accepted test for PSCs of all species remains the teratoma assay where undifferentiated PSCs are injected into immunocompromised mice and allowed to develop (Ungrin et al. 2007). A few months after transplant, the outgrowth is excised and examined for histological generation of differentiated derivatives of all three embryonic germ layers to demonstrate pluripotency. In the specific case of non-human PSCs, there is also the option of the generation of chimeras where the PSCs (most often of mouse origin) are injected into blastocysts to mix with the developing inner cell mass cells during subsequent in utero maturation. The resulting embryos are eventually examined to determine which tissues the PSCs were able to generate during in vivo development, and in the case of many genetically modified mouse strains, this process is continued to test for germline transmission to generate the entire embryo from a single PSC chimera-derived gamete (O'Connor et al. 2011; Ungrin et al. 2007).

The development of pluripotent stem cells (PSCs) took a dramatic step forward in 2006 with the report of induced pluripotent stem cells (iPSCs). In this work, combinatorial screening revealed that retrovirus-mediated expression of Oct4, Sox2, c-Myc, and Klf4 was sufficient to reprogram mouse fibroblasts into cells resembling mESCs (Takahashi and Yamanaka 2006). This approach was quickly extended to the reprogramming of human fibroblasts (Takahashi et al. 2007) and independently validated in another screen identifying OCT4, SOX2, NANOG, and LIN28 as being capable of reprogramming human somatic cells to iPSCs (Yu et al. 2007). Further studies have recapitulated this reprogramming process with even fewer factors delivered using a variety of technologies including adenoviruses, plasmids, transposons, mRNAs, proteins, and even solely with small molecules (Gonzalez et al. 2011; Hou et al. 2013). Assuming that efficient differentiation protocols exist, the eventual iPSCs derived from these methods make it theoretically possible to generate a cellular therapy for a specific disease state from PSC-derived cells that are immunologically specific to an individual patient (Maehr 2011). Whether such an immunological patient match provides any advantages for autoimmune diseases, such as type 1 diabetes, remains to be determined. Nevertheless, their utility for studying disease mechanisms and treatments is clearly evident.

Interestingly, the type of cell used to generate the iPSC line seems to have an effect on the resulting final product, most notably when mature specialized cell types are used as a starting material for iPSC generation. Reprogramming mature human cell types has been observed in a number of systems including both mouse and human adult β-cells (Bar-Nur et al. 2011; Stadtfeld et al. 2008). While the reprogramming process was able to induce pluripotency, the conversion was incomplete in some aspects, and the resulting iPSCs seemed to retain some legacy of their origin. Upon differentiation back to β-cells, iPSCs derived from insulin-positive cells expressed higher amounts of insulin than did iPSCs derived from non-insulin-positive pancreatic cells. This predisposition toward the somatic cell type of origin was attributed to the similar genomic DNA methylation patterns observed in human β-cells and β-cell-derived iPSCs, which were not fully converted to undifferentiated PSC patterns during the reprogramming process (Bar-Nur et al. 2011).

One of the most recent advances in human PSC generation is the application of somatic cell nuclear transfer (SCNT). In this process, an unfertilized oocyte has its nucleus removed and replaced with the nucleus of a differentiated diploid somatic cell. Upon microinjection into the oocyte and parthenogenetic activation, the maternal contents of the cell elicit epigenetic changes in the somatic nucleus resulting in a diploid zygote free from the fertilization process. The SCNT method has allowed the cloning of domestic livestock such as the famed sheep "Dolly" (Wilmut et al. 1997). After 10 years of additional research, the first non-human primate-derived ESCs were produced, albeit at very low efficiencies (Byrne et al. 2007). This process was finally extended to human cells by Tachibana et al. who were able to achieve SCNT with human oocytes and subsequently generate hESC lines from the developing blastocysts (Tachibana et al. 2013). This study represents a key milestone in efforts to generate specific cell types from progenitors, as SCNT may offer more complete reprogramming of somatic cells without significant epigenetic legacy marks compared to iPSCs. If so, SCNT could allow efficient, patient-specific hESC generation with the reproducible differentiation capacity required for regenerative medicine applications.

Pancreatic Differentiation of PSCs

Inducing differentiation of human PSCs into pancreatic cells has been pursued to achieve two primary goals: (1) to generate functional insulin-secreting cells capable of restoring euglycemia from a diabetic state and (2) to provide a model system for exploring the processes underlying the development of glucagon, somatostatin, pancreatic polypeptide, ghrelin, and insulin-positive cell types in healthy and diseased humans. While both goals seek to understand and exploit natural human development processes, the limited availability of human fetal tissue and the inability to apply advanced genetic tools to such tissue continue to present hurdles. Consequently, the majority of our understanding of pancreatic development is based on data collected from model systems such as frogs, mice, and zebrafish. Zebrafish are appropriate for rapid combinatorial genetic studies and offer the advantage of a rapid life cycle in an animal that contains a minimal pancreatic endocrine islet structure (Tiso et al. 2009; Kinkel and Prince 2009). While comparatively slower to reproduce and mature, mouse models are also appropriate for a considerable number of genetic tools, while also enabling whole-body physiology analysis through which the consequences of altered pancreatic development can be evaluated. Together, these models have provided a basic framework of pancreatic development (see Fig. 1) which continues to be modified as new genes involved in pancreas development and maintenance are identified from animal models and confirmed in a wide variety of human-based approaches including tissue taken from human fetal samples and from patients with monogenic forms of diabetes (e.g., maturity-onset diabetes of the young (MODY), latent autoimmune diabetes of adults (LADA), and neonatal diabetes). Once identified in human systems, these genes often require previously described model organisms to better understand the complex roles the genes play in mammalian pancreatic development. Ultimately, this iterative process refines our pancreatic developmental model, which serves as a road map for many fields of study.

Pancreatic Development

Efforts to induce hESC differentiation into pancreatic endocrine cells typically attempt to recapitulate the current understanding of the normal pancreatic developmental cascade (see Fig. 1). Induction protocols are generally derived from empirical testing of various signalling molecules and culture conditions identified in developmental model systems such as frogs, fish, and mice to optimize the stage-specific differentiation conditions required to convert undifferentiated PSCs to either progenitors or hormone-positive cells. Pancreatic development begins with the specification of embryonic germ layers via gastrulation in a process requiring signalling from TGF-β family members such as Nodal (Brennan et al. 2001). TGF-β signalling induces formation of SOX17- and FOXA2-copositive definitive endoderm cells, which are capable of further developing into all endoderm-derived tissues including the pharynx, thyroid, lung, stomach, liver, pancreas, and intestine

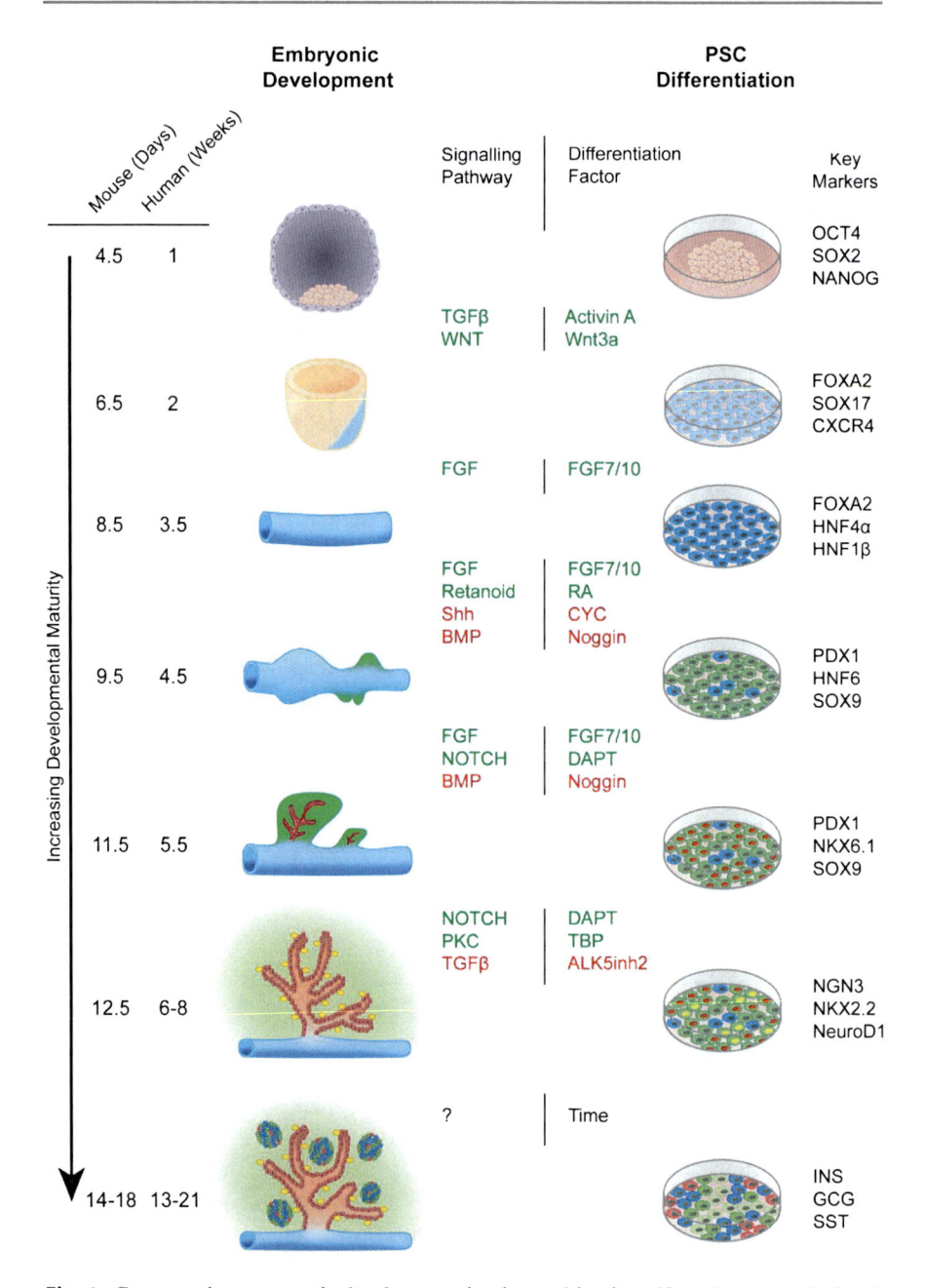

Fig. 1 Comparative pancreatic development in vivo and in vitro. Normal pancreatic development occurs through a complex series of morphogenic events that convert pluripotent cells into all potential cell types of the body. An approximate time line for mouse (days) and human (weeks) development is provided. Initially, pluripotent inner cell mass cells and their equivalent OCT4-positive hESCs (shown in *pink*) transition through a primitive streak intermediary stage to form

(Kanai-Azuma et al. 2002; Ang and Rossant 1994; Weinstein et al. 1994). The resulting sheet of endodermal cells invaginates into a tube formation in response to soluble factors including FGF4, which is released from the neighboring mesectodermal tissue (Wells and Melton 2000). This endoderm tube is next patterned along its length, making different portions permissive to the development of organ buds (Wells and Melton 1999). In mice, the earliest Pdx1-positive pancreas competent cells are derived from the transition zone between anterior Sox2-positive stomach progenitors and posterior Cdx2-positive intestinal progenitors (Jorgensen et al. 2007). The induction of Pdx1 positivity occurs in response to retinoic acid, BMP, and Shh signalling cascades (Stafford and Prince 2002; Tiso et al. 2002; Hebrok et al. 1998; Chen et al. 2004) and produces a cell population capable of generating the entire pancreatic organ in mice and humans (Jonsson et al. 1994; Stoffers et al. 1997). Similarly, available human developmental data suggest that these PDX1-positive cells subsequently gain expression of NKX6.1, identifying them as restricted pancreatic endodermal progenitors capable of differentiation to endocrine, exocrine, and ductal lineages (Jennings et al. 2013). In human development, it is this population which seems to form a complex tubular epithelial system which is regionalized based on location along the lengthening ductal tube into GATA4-positive "tip" progenitors which form PTF1a-positive exocrine cells and GATA4-negative "trunk" progenitors which can form both ductal and endocrine cells (Jennings et al. 2013). Commitment of PDX1-/NKX6.1-copositive, GATA4-negative "trunk" progenitors to the endocrine lineage is associated with transient expression of NGN3 to initiate endocrine fate specification (Jennings et al. 2013; Polak et al. 2000; Jeon et al. 2009; Slack 1995; Riedel et al. 2012; Piper et al. 2004). During this highly transcription factor-dependent cascade, within the pancreatic bud, some cells from the branched epithelial tree form early endocrine progenitor cells.

◂───

Fig. 1 (continued) cells committed to the endoderm lineage through TGF-β and WNT signalling pathways. Definitive endodermal cells (*light blue*) develop into endoderm-derived foregut cells (*dark blue*), which retain the ability to form any endoderm organ, in an FGF signalling-dependent process. Specification of the PDX1-positive dorsal and ventral pancreatic buds (shown in *green*) from the foregut tube occurs posterior to the developing stomach via high levels of FGF and retinoid signalling and inhibition of sonic hedgehog (*Shh*) and bone morphogenic protein (*BMP*) signals. Continued FGF and retinoid signalling specifies the NKX6.1-/PDX1-copositive pancreatic epithelial tree (*red* tree/nuclei within *green* bud/cells) within the expanding pancreatic buds, while the ventral bud rotates to fuse with the dorsal bud. NGN3-positive pancreatic endocrine precursor cells (shown as *yellow* buds/nuclei) form from pancreatic epithelial tree cells in a NOTCH signalling-dependent process enhanced by PKC activation and TGF-β inhibition. Over a considerable time frame and through processes that are incompletely understood, these endocrine precursor cells further mature through a number of fate specification stages (not described) into hormone-positive cells which coalesce into endocrine clusters (shown as clusters of *red*, *green*, and *blue* cells) within the surrounding pancreatic mesenchyme (*light green*). Signalling pathways are either activated (*green* text) or inhibited (*red* text) to drive development progression to the next stage. Differentiation factors (activators shown in *green* and inhibitors shown in *red*) utilized to modulate these signalling pathways are provided as examples and are not exhaustive. Key markers expressed at different developmental checkpoints are shown to the *right* of the figure

These cells continue to develop and eventually end up embedded in the surrounding pancreatic mesenchyme through incompletely understood mechanisms where they mature into functional endocrine cells (Pan and Wright 2011; Oliver-Krasinski and Stoffers 2008) (see Fig. 1).

Definitive Endoderm, Foregut, and Pancreatic Endodermal Progenitors from PSCs

The induced differentiation of PSCs into pancreatic cells generally follows known developmental stages as described above, but on an accelerated time line (see Fig. 1). While the formation of endocrine cells in humans takes approximately 7 weeks of development (Piper et al. 2004), early endocrine cells are first formed in differentiating hESCs by 2 weeks of culture (D'Amour et al. 2006). This considerably shortened culture time line of PSCs has been made possible by staging differentiating cells for efficient homogenous in vitro development without the need to enable morphogenesis of ectodermal and mesodermal lineages. To begin differentiation, pluripotent PSCs are seeded in cell culture plates. Once the cells have grown to a predetermined optimal density, which primes them to exit the replicative cell cycle (Chetty et al. 2013; Gage et al. 2013), the first inductive signals are provided via daily media changes which continue throughout the culture time line. Key marker genes and proteins are routinely monitored to ensure stage-specific differentiation and to identify cell homogeneity throughout the process. Primarily in response to activin A (a TGF-β family member) in low or no serum, the pluripotency program is repressed, and the formation of cells of the endoderm germ layer is stimulated (D'Amour et al. 2005; Kubo et al. 2004; Seguin et al. 2008; Brown et al. 2011). These signals induce the PSCs to move through an intermediate mesoderm/endoderm step that is developmentally similar to the primitive streak. Cells in this transient state can be identified via expression of the transcription factor Brachyury by 12–24 h after induction (D'Amour et al. 2005). Expression of FOXA2 and SOX17 follows approximately 48 h later and, together with the absence of extraembryonic primitive endoderm markers such as SOX7, demarks the formation of true definitive endoderm cells (D'Amour et al. 2005; Yasunaga et al. 2005; Seguin et al. 2008). The formation of this cell population was a key milestone in the pursuit of developing β-cells and other endoderm-derived tissues from PSCs. Remarkably, endodermal progenitors can also be isolated and expanded in culture producing a purified population free from pluripotent cells that maintains its ability to differentiate in an endoderm lineage-restricted manner (Cheng et al. 2012). Whether stimulated in a transient manner or from the endodermal progenitors which result from the in vitro expansion process, highly pure definitive endoderm cell populations are key to the success of later differentiation steps. Given the broad developmental potential of definitive endoderm progenitors, these cell populations are now being used to study the generation and continued differentiation of many tissue types including the lung (Mou et al. 2012), liver (Basma et al. 2009; Touboul et al. 2010;

Gouon-Evans et al. 2006; Hannan et al. 2013), intestine (Spence et al. 2011), as well as organs of the anterior foregut (Green et al. 2011; Kearns et al. 2013).

Following the generation of relatively pure definitive endoderm cells, the next challenge is to further pattern the sheet of cells to mimic the foregut stage of development. This is achieved with the addition of FGF signalling agonists, namely, FGF7 or FGF10, concomitant with prompt removal of the growth factors used to trigger previous differentiation stages. Expression of HNF4a and HNF1b marks the transition into foregut cells, which occurs over the next 72 h in culture (D'Amour et al. 2006; Kroon et al. 2008; Nostro et al. 2011; Schulz et al. 2012). This population of cells can form the gall bladder, hepatic, intestinal, and pancreatic cells, but requires further specification and repression of unwanted developmental programs. Retinoic acid plays a central role in the induction of pancreas formation from the foregut and has been found to be key in stimulating PDX1 expression in differentiating hESCs (D'Amour et al. 2006; Cai et al. 2010; Mfopou et al. 2010). At the same time, repression of hepatic and intestinal cell fates by the inhibition of BMP and Shh signalling is critical for proper specific pancreatic induction from hESCs (Spence et al. 2011; Green et al. 2011; Mfopou et al. 2007, 2010; Nostro et al. 2011; Sui et al. 2013; Gouon-Evans et al. 2006; Cho et al. 2012). Together, this mix of signalling cascades stimulates the formation of relatively homogenous PDX1-positive cell populations, which approach 95 % in purity in some reports, over the course of 3–5 days (Cai et al. 2010; Rezania et al. 2012). These PDX1-positive cells represent a key developmental step where the PSCs are partially restricted in cell fate but still retain the ability to form off-target tissues including extrahepatic biliary duct in humans depending on PDX1 expression levels (Jennings et al. 2013). Final maturation to PDX1-/NKX6.1-copositive pancreatic endodermal progenitors similar to those that predominate the pancreatic epithelium at 8–9 weeks of human fetal development (Riedel et al. 2012) and before the spatial regionalization associated with GATA4 expression to distinguish "tip" and "trunk" progenitors occurs over the next 72 h. The maturation of PDX1-positive cells to PDX1-/NKX6.1-copositive cells has been shown to occur in the absence of exogenous stimuli (Kroon et al. 2008) but can be enhanced by a mixture of BMP and ALK5 inhibition and PKC activation (Rezania et al. 2012, 2013). The state of the art in efficiency and homogeneity of PDX1-/NKX6.1-copositive progenitors is reported to be up to 86 % PDX1/NKX6.1 copositivity in 70 % of differentiation runs (Rezania et al. 2012). When transplanted into immunocompromised mouse models, these cells give rise to ductal cells and endocrine cells including functional insulin-positive cells, while PDX1-/NKX6.1-negative hormone-expressing cells appear to give rise predominantly to glucagon-positive α-cells (Rezania et al. 2011, 2012, 2013; Kelly et al. 2011; Kroon et al. 2008). The in vitro generation of a pancreatic progenitor pool from hESCs is an important checkpoint that has been achieved by many research groups and represents the second key milestone toward producing functional β-cells of sufficient quality and in quantities appropriate for future transplantation studies. However, these pancreatic endodermal progenitors have not yet specified which pancreatic cell type they will become.

Pancreatic Endocrine Development In Vitro

While the production of pancreatic endodermal progenitors has been relatively successful, the continued development of these cells in culture into fully functional endocrine cells remains poorly understood. To this end, two developmental routes are being explored. Both begin with the in vitro differentiation of PDX1-/NKX6.1-copositive progenitors over a 14–17-day culture period. Subsequently, these progenitors are either transplanted into immunocompromised mice to undergo relatively uncontrolled but highly successful development in vivo, toward functional endocrine cells, or alternatively the progenitors are cultured in vitro under more regulated conditions, again in an effort to elicit functional maturation of the cells.

The cumulative developmental literature on pancreatic endocrine induction, fate specification, and functional maturation suggests that temporally and spatially specific transcription factor expression is likely critical to efficient stimulation of β-cell formation. In particular, sequential expression of the endocrine restriction marker NGN3 followed by a number of fate-specifying factors (NKX2.2, PAX4, ARX, PAX6, and others) is thought to be critical for specification of pancreatic endodermal progenitors (Oliver-Krasinski and Stoffers 2008). After fate specification, endocrine cells begin to express maturation factors and eventually hormones, with MafA-driven insulin production in the β-cell being perhaps the most well-studied example (Oliver-Krasinski and Stoffers 2008; Zhang et al. 2005). During hESC differentiation, the induction of the endocrine cascade remains largely stochastic for many of the early differentiation protocols (D'Amour et al. 2006). This suggests that the process could be cell autonomous or more likely that the cultures themselves produce the signalling molecules required to activate endocrine development within the culture system.

In an effort to accelerate this endocrine induction process and improve its efficiency, a wide range of signalling molecules have been used, including but not limited to nicotinamide, exendin-4, IGF-1, HGF, noggin, bFGF, BMP4, VEGF, WNT, and various inhibitors of BMP, Shh, TGF-β, and Notch signalling pathways (Nostro et al. 2011; D'Amour et al. 2006; Rezania et al. 2011, 2012, 2013; Kelly et al. 2011; Kroon et al. 2008; Schulz et al. 2012; Jiang et al. 2007a, b; Xu et al. 2011; Sui et al. 2013). Some of these factors have a rational basis for testing as agents driving endocrine maturation. As one example, exendin-4 is a mimetic of the natural gut-derived hormone glucagon-like peptide 1 (GLP-1), which stimulates β-cell proliferation, decreases β-cell apoptosis, and renders β-cells glucose competent (Wideman and Kieffer 2009). In addition to rational factors, recently even seemingly innocuous factors have been found to have dramatic effects on differentiating hESCs. The organic solvent dimethyl sulfoxide (DMSO) decreased cell proliferation to a similar extent as high-density cell culture conditions, dramatically enhancing differentiation to definitive endoderm, PDX1-positive cells, and C-peptide-positive cells in more than 25 hESC and iPSC lines (Chetty et al. 2013). Even the buffering component HEPES was found to have significant inhibitory effects on endocrine maturation from pancreatic endodermal progenitors, with HEPES stimulating intestinal commitment (elevated CDX2 expression) at the expense of the pancreatic endocrine lineage (decreased NKX6.1, NGN3, NEUROD1, PDX1, and PTF1a expression) (Rezania et al. 2012).

Beyond these unexpected results, the modulation of TGF-β, WNT, BMP, and protein kinase C signalling has also resulted in considerable improvements in the efficiency of conversion to an endocrine fate. The inhibition of endogenous WNT signalling from foregut-stage hESC cultures impaired the eventual expression of insulin, and when optimally agonized by addition of WNT3a, a 15-fold increase in insulin expression was observed (Nostro et al. 2011). During the generation of pancreatic progenitors, TGF-β agonists were found to have a positive effect on hESC differentiation by increasing the number of PDX1-positive cells (Guo et al. 2013). Remarkably tight temporal regulation of this signalling pathway is required for further maturation toward endocrine cells as continued administration of TGF-β agonists repressed insulin expression (Guo et al. 2013). This idea was solidified as TGF-β inhibition with ALK5 inhibitor II caused a dose-dependent increase in NGN3-positive cells from progenitor cultures (Rezania et al. 2011). This effect continued down the cascade, promoting increased expression of NKX2.2, NEUROD1, and eventually insulin and glucagon without appreciably decreasing PDX1 expression (Rezania et al. 2011). Similarly, protein kinase C (PKC) activation has been identified as a potentially key pathway required to maintain PDX1 expression in pancreatic progenitors (Chen et al. 2009a). Indeed, when the PKC activator TPB was added to ALK5 inhibition and continued BMP inhibition by noggin, this three-factor mixture in the absence of HEPES buffering stimulated increased expression of NGN3, NEUROD1, and NKX6.1 without loss of PDX1 expression in pancreatic progenitors or induction of off-target differentiation which would be characterized by expression of albumin (liver) or CDX2 (intestine) (Rezania et al. 2012). Recently, one report suggested that BMP signalling is key to maintaining a proliferative PDX1-positive progenitor pool and that BMP antagonism is subsequently required to induce further pancreatic endocrine maturation (Sui et al. 2013).

The formation of pancreatic progenitors from hESCs that can generate functional endocrine cells including β-cells in vivo has been convincingly demonstrated (Rezania et al. 2012; Kroon et al. 2008). While this suggests that in vitro-derived pancreatic progenitors should have the capacity to produce functional endocrine cells, presently the majority of the pancreatic endocrine cells produced in vitro by various groups are still immature in function and typically express multiple hormones including insulin, glucagon, and somatostatin, with a bias toward glucagon-positive lineages (Rezania et al. 2011; Nostro et al. 2011; Jiang et al. 2007a; D'Amour et al. 2006; Segev et al. 2004). Of particular note, one of the highest in vitro endocrine differentiation efficiencies reported to date yielded up to 75 % endocrine cells (synaptophysin positive) (Rezania et al. 2011). While these cells were initially polyhormonal (insulin and glucagon positive), during extended culture or transplantation, they developed into functional endocrine cells expressing only glucagon and not other endocrine hormones. High expression of the key α-cell transcription factor ARX, along with low expression of PAX4, PDX1, and NKX6.1, may have caused this biased maturation to α-cells (Rezania et al. 2011).

One of the key objectives of in vitro differentiation of hESCs is the development of functional insulin-secreting cells. With this goal in mind, D'amour et al. (2006) examined the capacity of their differentiated endocrine cultures to responsively release insulin/C-peptide into the culture media. In this study, differentiated hESCs clusters contained approximately one third the amount of C-peptide per μg DNA found in human islets, with a high proportion of proinsulin remaining unprocessed. hESC clusters released C-peptide (two to sevenfold over basal) in response to depolarizing stimuli such as KCl, K_{ATP} channel blockade by tolbutamide, increased cAMP levels by IBMX addition, and nutrient supplementation by methyl-pyruvate α-ketoisocaproic acid, L-leucine, and L-glutamine (D'Amour et al. 2006). Importantly, these clusters were unable to reproducibly release C-peptide in response to glucose, with many experiments recording stable or even decreased C-peptide release in response to increased extracellular glucose concentrations. While some groups have shown modest insulin secretion (~2-fold) from differentiated mESCs in response to elevated glucose levels (Jiang et al. 2008), the majority of reports suggest that this key attribute is lacking under current in vitro culture systems which employ human cells. These immature hESC-derived endocrine cells share some characteristics with neonatal β-cells which have significantly elevated insulin release in low glucose conditions and blunted release in high glucose conditions (Martens et al. 2013). This poor glucose responsiveness in neonatal β-cells has been attributed to a deficit in mitochondrial energy shuttling associated with poor glucose-stimulated NAD(P)H generation (Martens et al. 2013; Jermendy et al. 2011). One notable exception to the unresponsive nature of in vitro PSC-derived insulin-positive cells is the differentiated progeny of in vitro purified endodermal progenitors (Cheng et al. 2012). These cells, despite being differentiated with protocols that typically generate polyhormonal cells lacking robust glucose responsiveness with other ESC lines, were found to express C-peptide without glucagon or somatostatin and release C-peptide in response to elevated glucose levels similarly to adult human islets (Cheng et al. 2012). Given that the endodermal progenitors used are lineage restricted, can rapidly self-renew yet are nontumorigenic, and could be effectively differentiated into glucose-responsive insulin-secreting cells, the authors revealed an alternative differentiation method which pauses at the definitive endoderm stage to improve directed differentiation purity and potential safety of the final product. Nevertheless, the generally limited responsiveness of most hESCs differentiated exclusively in vitro has led many groups to examine the development of pancreatic endodermal progenitors in vivo as an alternate strategy to yield functional endocrine cells with more reasonable efficiency.

Pancreatic Endocrine Development In Vivo

Given that current protocols for in vitro differentiation of pancreatic progenitors into endocrine cells tends to produce immature polyhormonal cells with poor glucose responsiveness, research efforts have turned to in vivo maturation strategies to elicit functional maturation of progenitor cells. This strategy is based on the success of functional maturation of human fetal pancreatic tissue upon transplantation in mice (Hayek and Beattie 1997; Castaing et al. 2001) and on the notion that in vivo maturation might enable exploitation of the full complexity of cellular interactions that drive normal pancreas development. Since our knowledge of pancreatic endocrine cell development, particularly the signals governing late endocrine maturation processes, remains incomplete, rational, literature-driven in vitro maturation is likely to remain challenging in the short term. However, if in vivo cell maturation is possible, it may provide key insights into the required signals that govern this process which are presumably deficient in the current in vitro culture systems. Moreover, in considering an eventual cell therapy product, the shorter time line associated with differentiation just to pancreatic progenitors is attractive, assuming adequate performance and safety following completion of maturation in vivo.

As previously reviewed, in vivo maturation protocols tend to begin from PDX1-/NKX6.1-copositive pancreatic endodermal progenitors generated through in vitro processes (see Figs. 1 and 2). These progenitor cells if differentiated as an adherent monolayer culture must be detached and prepared for transplantation in the form of a suspension of cell clusters (Kroon et al. 2008; Kelly et al. 2011; Rezania et al. 2012) or alternatively differentiated entirely in suspension prior to transplantation (Schulz et al. 2012). The composition and purity reported for these clusters vary among different groups, but they tend to be comprised predominantly of PDX1-/NKX6.1-copositive progenitors with lower numbers of pre-committed pancreatic endocrine cells (Rezania et al. 2012). Following harvest, progenitor cells are typically transplanted under the kidney capsule (Rezania et al. 2011, 2012; Kroon et al. 2008) or as part of a gel-foam disk transplanted into the epididymal fat pad (Kelly et al. 2011; Kroon et al. 2008; Schulz et al. 2012) of immunocompromised rodents. Initial engraftment of the progenitor mixture occurs over the next few weeks as blood vessels from the transplant recipient grow toward the transplanted tissue, likely in a VEGF-A-dependent process similar to islet engraftment and vascularization in mice (Vasir et al. 2001; Jansson and Carlsson 2002; Zhang et al. 2004; Brissova et al. 2006). The subsequent development of hormone-positive cells within the engrafted hESC origin tissue is initially rapid and results in the production of polyhormonal cells around 1 month posttransplantation (Rezania et al. 2012) (see Fig. 2). Over the next 2–3 months, the immature polyhormonal cell population decreases in number, and more mature cell types expressing a single major islet hormone predominate in the transplant tissue (see Fig. 2). This transition is also marked by the reorganization of endocrine cells within the grafts into endocrine clusters resembling islets and a significant increase in secretion of C-peptide from the graft (Rezania et al. 2012; Kroon et al. 2008). With extended in vivo maturation, glucose and/or meal-responsive C-peptide release continues to increase in grafts, along with nuclear NKX6.1 and MafA expression in insulin-positive cells (Kroon et al. 2008; Rezania et al. 2012). NKX6.1 has been shown to be both necessary and sufficient to maintain and specify the β-cell phenotype primarily due to repression of α-cell biasing factors such as ARX and PAX6 (Schaffer et al. 2013; Gauthier et al. 2007). MafA expression, which is known to mark maturation of insulin-producing cells into a glucose-responsive state

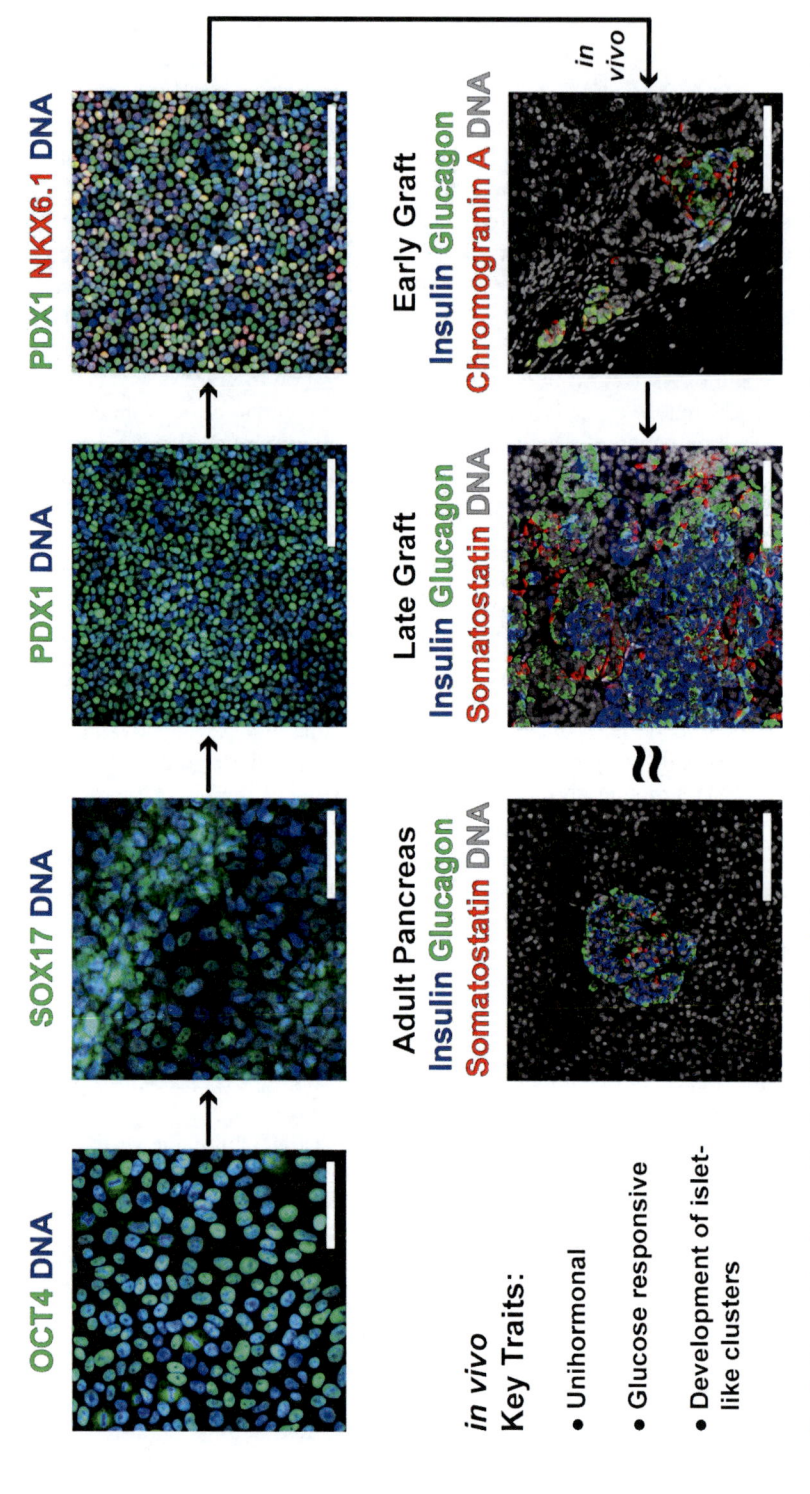

Fig. 2 Current in vitro and in vivo hESC differentiation mimicking pancreatic development. Representative immunohistochemical images are shown for various stages of in vitro generation of pancreatic progenitor cells and subsequent in vivo maturation to functional endocrine grafts. Pancreatic differentiation of hESCs converts OCT4-positive pluripotent cells into SOX17-positive definitive endoderm cells. These cells further develop into PDX1-positive foregut

(Zhang et al. 2005), is also associated with the point at which the hESC-derived grafts were able to restore normoglycemia in diabetic transplant recipients suggesting that a key functional transition had occurred within the graft (Rezania et al. 2012). Marking a key milestone in the field, this in vivo maturation process to yield glucose-responsive, insulin-producing cells has been independently shown to occur in normoglycemic (Kroon et al. 2008) and diabetic (Rezania et al. 2012) environments. However, the possibility that in vivo cell maturation in the host environment may be variable due to differing exposure to a variety of factors such as hormones and drugs remains a potential limitation of this approach.

The functional success of in vivo maturation strategies has led to intriguing questions about how to mimic the functional maturation of insulin-positive cells in culture and what cell population forms the final insulin-positive cell compartment in glucose-responsive grafts. Kelly et al. elegantly examined this question using a cell separation and transplantation strategy (Kelly et al. 2011). The authors followed an established in vitro differentiation protocol to produce a heterogeneous pancreatic cell population that was the basis for a flow-cytometry-based assay of 217 commercially available antibodies aimed at distinguishing endocrine cells from progenitors. Ultimately, CD142 was found to label a population of predominantly hormone-negative, NKX6.1-positive endodermal progenitors, while CD200 and CD318 preferentially labeled hormone-positive cells. The authors separated CD142-positive endodermal progenitor fractions (82 % progenitors) and CD318-positive hormone-positive fractions (84 % endocrine) by immunomagnetic cell separation methods and transplanted the cells into immunocompromised mice. Nine weeks after transplantation, CD318-enriched endocrine cells had developed mostly into glucagon-positive cells, while 13-week-old transplants of CD142-enriched pancreatic endodermal progenitors contained large numbers of cells expressing insulin, glucagon, or somatostatin, arranged in islet-like structures and surrounded by cells expressing markers of exocrine and ductal pancreatic cells. Taken together, this work suggests that the CD142-positive, NKX6.1-positive, hormone-negative population is the common progenitor for ductal, exocrine, and endocrine pancreatic cells including insulin, somatostatin, and glucagon lineages. In contrast, the in vitro-generated hormone-positive cells expressing CD318 and CD200 seemed to be predestined to form glucagon-positive cells (Kelly et al. 2011). Interestingly, the most functional grafts were generated from mixed cell populations which contained both the hormone-negative and hormone-positive populations (Kelly et al. 2011), the reasons for which are unclear. We compared maturation of hESC-derived pancreatic progenitors that contained high (~80 %) or

Fig. 2 (continued) endoderm cells that mature into pancreatic endodermal progenitors coexpressing NKX6.1 and PDX1. This population of cells is the basis for in vivo maturation where maturation of progenitors is achieved by transplantation into immunocompromised recipients. Transplanted hESC-derived cells mature from immature polyhormonal endocrine populations expressing insulin, glucagon, and the pan-endocrine marker chromogranin A into functional islet-like clusters resembling adult pancreatic islets and comprised of unihormonal cells. The colours of each marker are as indicated above each image. All scale bars are 100 μm

low (~25 %) fractions of NKX6.1-positive cell populations (Rezania et al. 2013). Upon transplantation and in vivo maturation of these cells, the NKX6.1-high grafts were found to have robust C-peptide release in response to physiological stimuli including meals, arginine, and glucose, which was not observed from the NKX6.1-low grafts. After 5 months of development, both NKX6.1-high and NKX6.1-low grafts generated pancreatic endocrine cells at high efficiencies, but the NKX6.1-high grafts contained increased numbers of insulin- and somatostatin-positive cells, while the NKX6.1-low grafts contained predominantly glucagon-positive cells (Rezania et al. 2013). Both of these studies support previous in vitro extended culture and transplantation studies in which glucagon-positive cells at the end of in vitro and in vivo differentiation protocols were found to arise from the glucagon-/insulin-copositive cells seen in the earlier in vitro differentiation stages (Rezania et al. 2011; Basford et al. 2012). Analysis of human fetal pancreas samples also supports the notion that polyhormonal endocrine cells are present during development and may give rise to mature, single hormone-producing cells (Riedel et al. 2012; Polak et al. 2000). Thus, the in vivo maturation of hESC-derived precursor cells presents a useful model for exploring the developmental capacity of cells initially produced in vitro. Moreover, in vivo maturation studies may help to further define the optimal cell population to produce functional hESC-derived pancreatic endocrine cells.

Future Challenges

Despite the success of in vitro differentiation of pancreatic endodermal progenitors following a pancreatic development model and the in vivo maturation of these progenitors to functional endocrine islet-like structures capable of reversing diabetic hyperglycemia, the cues that govern developmental conversion of these progenitors to functional endocrine cells are still incompletely understood. Moreover, while studies to date do solidify the notion that hESC-derived cells are able to control blood glucose in a diabetic recipient, the field must resolve a significant number of challenges ranging from how to reproducibly generate large amounts of functional, stable cells to how to prevent the recipient's immune system from destroying these cells once transplanted.

Reproducibility

If hESCs, or more broadly PSCs, are going to become a realistic source material for generating cellular therapies for diabetes, the reproducibility of differentiation and development must be addressed. In the undifferentiated state, hESC lines are generally considered to be quite uniform in terms of expression of pluripotency markers (Adewumi et al. 2007). Unfortunately, this uniformity does not seem to extend to the differentiation potential of hESCs. One report suggested that even among similarly derived hESC lines, more than 100-fold differences in lineage specification efficiency exist (Osafune et al. 2008), perhaps because of inherent DNA methylation patterns which predispose different hESC and iPSC lines toward certain lineages (Bock et al. 2011). This inherent variability between lines has contributed to the inability of investigators to repeat published protocols with different cells and obtain the same results. For example, some hESC lines are better able to generate pancreatic endocrine cells with the protocol developed by D'Amour et al. than others (D'Amour et al. 2006; Mfopou et al. 2010; Osafune et al. 2008). As a direct test of the pancreatic differentiation propensity of different hESC lines, Mfopou et al. applied the differentiation conditions optimized by D'Amour et al. for the CyT203 hESC line (D'Amour et al. 2006) to five in-house-generated hESC lines (VUB01, VUB02, VUB07, VUB14, and VUB17). In the VUB lines, the D'Amour protocol effectively generated definitive endoderm, gut tube, and foregut cells but eventually produced hepatocytes instead of pancreatic endocrine cells (Mfopou et al. 2010). Ultimately, cell line-specific alterations in the differentiation protocol, namely, adjustment of the timing and dosage of BMP and FGF signalling modulators, were required to restore pancreatic endocrine differentiation capacity to the VUB lines. These modifications were similar to those applied by other groups (Nostro et al. 2011). Line-to-line variation was also described by the ViaCyte, Inc. group (formerly Novocell, Inc.), which reported varying progenitor differentiation efficiencies among the CyT49, CyT203, and MEL1 hESC lines (Kelly et al. 2011). Administration of DMSO to hESC lines resistant to pancreatic lineages has recently been reported to significantly improve differentiation efficiency in more than 25 hESC and iPSC lines (Chetty et al. 2013). This effect of DMSO is likely mediated by cell cycle arrest in view of other studies demonstrating a requirement for transition from the G2/M cell cycle phases to G1/Go in order for hESCs to be capable of targeted differentiation (Calder et al. 2013; Sela et al. 2012; Pauklin and Vallier 2013; Gage et al. 2013) and can be mimicked by high cell seeding density (Gage et al. 2013). Taken together, these data suggest that while a single protocol is unlikely to be effective at inducing differentiation of multiple hESC lines, most lines are likely to be capable of efficient pancreatic development given the appropriate signals. However, it may be more effective to identify hESC (or PSC) lines which have inherently reproducible differentiation for widespread use than to develop multiple individually optimized protocols for each PSC line.

In addition to resolving the preferred PSC starting material(s) and the reproducibility of differentiation protocols, the methods and tools used to characterize and quantify nutrient-responsive hormone release from resulting cells should be harmonized to enable direct comparison among research groups. Previous efforts to repeat observations have revealed such things as the confounding effects of insulin uptake in differentiated mESCs (Lumelsky et al. 2001; Hansson et al. 2004; Rajagopal et al. 2003) and variation in hESC pancreatic differentiation propensity (Mfopou et al. 2010; Osafune et al. 2008). Rigorous cell characterization is also important since the standards for surrogate β-cells are very high, including the paramount importance of glucose-regulated insulin secretion (Halban et al. 2001). Within the field of pancreatic islet research, methods for assessing insulin release are well established, including normalizing secreted insulin amounts to total DNA or insulin content of the cell sample. These practices have not been uniformly applied to the task of testing PSC-derived pancreatic endocrine populations. Many in vitro studies simply report relative fold secretion of insulin (often as C-peptide) under static high glucose conditions versus low glucose conditions with different cell samples. This makes it hard to compare results between studies, and unfortunately, this approach fails to account for the number of endocrine cells in each sample, the pre-culture conditions (often higher glucose differentiation media), and the kinetics of secretion. Absolute hormone secretion in response to standardized glucose concentrations ideally in a kinetic secretion system with comparison to isolated islets would be preferable. By similarly assessing the function of differentiated PSCs, labs would be able to better compare methods and resultant cell populations, which will ultimately facilitate further improvements of PSC differentiation protocols toward developing robustly functional β-cells in vitro.

Scaling Up

Assuming that reproducible hESC differentiation can be achieved, the next major hurdle will be the production of clinically relevant quantities of pancreatic progenitors or functional endocrine cells in an economically feasible manner (Wong et al. 2012). One key aspect of this scale-up is determining what population of pancreatic endocrine cells will provide the most effective and safe treatment of diabetes. The functional capacity of β-cells seems to be significantly improved when islet cells, including α-cells, are clustered together with β-cells (Maes and Pipeleers 1984; Pipeleers et al. 1982). This may reflect the highly conserved natural arrangement of endocrine cells within islets and established paracrine signals between these cells (Bosco et al. 2010; Steiner et al. 2010; Hauge-Evans et al. 2009; Brunicardi et al. 2001). (For more information on this topic, please see the chapter entitled "▶ The Comparative Anatomy of Islets."). However, King et al. compared the ability of enriched β-cells and reaggregated islet cells to recover glycemic control in diabetic mice (King et al. 2007). The authors concluded that non-β-cell endocrine cells are not essential for transplantation success suggesting that a pure a β-cell product may be as effective as mixed islet cells, should protocols be successfully developed to produce pure a β-cells. Given the success in making relatively pure α-cells (Rezania et al. 2011), it should be possible to make highly enriched functional β-cells in the near future. While estimates vary, it is possible that one billion hESC-derived β-cells could be required to treat a single patient with diabetes (Docherty et al. 2007). In order to achieve this scale of production, considerable expansion of hESCs will be required. This is likely a reasonable goal given that hESCs are highly proliferative, doubling every 20 h (Chen et al. 2010) to allow an up to sixfold expansion in just 4–7 days in the undifferentiated state in stirred bioreactor systems (Zweigerdt et al. 2011). Once expanded, cultures can be differentiated toward pancreatic progenitor and or endocrine cells following the loosely established conversion ratio of 1:1 (undifferentiated hESC/differentiated progeny) (D'Amour et al. 2006).

Recently, the ViaCyte group reported a scalable production strategy for pancreatic progenitors (Schulz et al. 2012) which was subsequently reproduced and enhanced by researchers at Pfizer (Gutteridge et al. 2013). A large bank of frozen vials of undifferentiated hESCs is maintained, whereby a sample can be thawed, expanded over 2 weeks of adherent culture, and formed into suspension cell aggregates by dynamic rotation of dissociated undifferentiated cells. In the ViaCyte report, these aggregates were then differentiated into pancreatic progenitors in a rotating suspension format and transplanted in the epididymal fat pad of immunocompromised mice for final maturation. Graft maturation occurred over the next few months and ultimately resulted in glucose-stimulated C-peptide release at approximately 11–15 weeks posttransplant. Maturation continued until 4–5 months posttransplant, when graft tissue could maintain normal glycemic control in animals in which mouse pancreatic β-cells had been destroyed by injection of streptozotocin after graft maturation. In addition to the scalability of the rotational culture adaptation, the in vitro temporal expression patterns and the in vivo matured endocrine compartment were remarkably similar to previous reports, which were based on a minimally scalable two-dimensional adherent system (Kroon et al. 2008). While this bodes well for increased production of pancreatic progenitors using this potentially scalable differentiation method, some challenges remain. While the suspension differentiation methods significantly reduced the formation of non-endodermal origin tissues, approximately half of the grafts were considered cystic and thus were incompletely pancreatic endocrine cells (Schulz et al. 2012). The cultures did not uniformly express key pancreatic transcription factors, and ~2 % of the cells were unidentified, thereby carrying the risk of unknown developmental potential (Schulz et al. 2012). Presuming that all uniformity and safety concerns are addressed, the generation of these therapeutic cells is going to be costly. Indeed, relative to cadaveric and xenogeneic islet sources, there has been some debate as to whether a therapeutic product of this nature is ultimately economically viable (Wong et al. 2012; Wong and Nierras 2010).

One of the key aspects of clinical scale production which will continue to require further research and is likely to have a significant impact on the cost of the final product is the conversion of current differentiation protocols to ones relying on small molecules with fully defined composition rather than protein growth factors derived from animal products. As an example of such efforts, Borowiak et al. used high-throughput small-molecule screening to form ESC-derived cell types along the pancreatic developmental cascade. Using a fluorescent reporter mESC line in which expression of SOX17 was tracked by red fluorescent DsRED expression, ~4,000 compounds were screened to reveal that IDE1 and IDE2 significantly enhanced definitive endoderm induction from undifferentiated human and mouse ESCs (Borowiak et al. 2009). These two compounds could replace activin A, the recombinant protein widely used to activate TGF-β signalling via canonical phosphorylation of Smad2 (Borowiak et al. 2009). Further down the differentiation cascade, another screen was employed to identify small molecules capable of improving the induction of PDX1-positive pancreatic progenitor cells. Using an antibody-based high-content screen with dissociated and replated foregut progenitor cells treated with one of ~5,000 compounds or DMSO, (-)-indolactam V (ILV) was found to increase numbers of PDX1-positive cells (Chen et al. 2009b). When coadministered with FGF10, it also improved production of pancreatic progenitors capable of in vivo maturation to functional insulin-positive cells. Based on the inability of retinoic acid to synergize with ILV and the similar effects of protein kinase C activators with respect to PDX1-positive cell stimulation, ILV is believed to activate protein kinase C by direct binding, although this has yet to be proven explicitly (Chen et al. 2009b). Taken together, the high-efficiency formation of pancreatic progenitors under clinically amenable, defined, and scalable culture conditions seems feasible. While work is still required to merge these independent research efforts, the demand for transplantable tissue remains high, and progress in the field is expected to be swift.

Immunological Control and Encapsulation

With the efficient and scalable production of functional pancreatic progenitors getting closer to a reality, cellular therapy for diabetes nevertheless presents an immunological problem. Human PSC-derived grafts will face not only alloimmune attack but also the specific autoimmune-mediated attack of insulin-positive cells associated with the pathogenesis of type 1 diabetes. With just a few efficiently differentiating hESCs lines established, cells derived from an equally minimal number of human leukocyte antigen (HLA) types are expected to be available for immunological matching to patients. HLA matching is an important variable influencing the success of human islet transplantation (Mohanakumar et al. 2006), so the matching of hESC-derived pancreatic progenitors to recipient HLA types may dampen alloimmune graft rejection, as was observed in the early islet transplant experience (Scharp et al. 1990). With the adoption of glucocorticoid-free immunosuppressive regimens, survival and ongoing function of transplanted hESC-derived cells seems feasible, similar to the initial insulin independence observed in many islet recipients in the first year posttransplant (Shapiro et al. 2000). However, despite strong immunosuppressive regimes, graft function decreases over the next 4 years after transplant in islet recipients due to progressive immune rejection by allo and autoimmune mechanisms as well as apoptosis associated with high metabolic and insulin production demands on the engrafted islets (Shapiro 2011; Plesner and Verchere 2011). The immunosuppressive agents used to protect the grafts have also been found to be directly toxic to β-cells (Nir et al. 2007; Johnson et al. 2009). Over time, the loss of islet function requires that patients return to supplementation with exogenous insulin (Shapiro et al. 2006). While PSC-derived sources offer the potential to deliver larger amounts of pancreatic endocrine tissue, it remains to be seen whether such transplants will be able to sustain long-term insulin independence in recipients.

Upon transplantation, both PSC-derived grafts and cadaveric islet grafts will be faced with an activated recipient immune system. One way to prevent graft loss associated with host immune attack, without using immunosuppressant drugs, is by using a physical barrier to isolate the graft from the circulating immune system. This idea has taken form in a series of immunoisolation devices which range from thin cellular coatings to microencapsulation with thick cell cluster/islet coatings or macroencapsulation with engineered transplantable devices, as reviewed elsewhere (O'Sullivan et al. 2011; Weir 2013; Scharp and Marchetti 2013). These approaches are being actively developed and may be amenable to protecting PSC-based cell therapies.

Alginate microencapsulation forms a coating around islets or hESC-derived clusters which protects the cells from direct contact with host immune cells. This separation is presumed to be essential to preventing cytotoxic death of the transplanted cells, but due to the porous nature of the alginate gel, the graft can still secrete insulin in response to rising interstitial glucose concentrations (O'Sullivan et al. 2011). Depending on the chemical nature of the gel, this method may also protect the graft from antibody-mediated attack, but this defense typically comes at the cost of increased hypoxia-related necrosis (De Vos et al. 1999). Despite these considerable challenges, simple extrusion alginate encapsulation is remarkably effective in some mouse models, even with minimal surface modifications to restrict cytokine entry (Duvivier-Kali et al. 2001). In immunodeficient mice, alginate-encapsulated human islet cells delivered to the intraperitoneal cavity functioned better than free implants under the kidney capsule (Jacobs-Tulleneers-Thevissen et al. 2013). A pilot study was conducted in a patient with type 1 diabetes who received a peritoneal implant of the same encapsulated islet cells while on immunosuppression. While the transplant was without metabolic effect, likely due to the marginal transplant mass, functional cells within intact microcapsules were recovered 3 months posttransplant (Jacobs-Tulleneers-Thevissen et al. 2013). Clinical trials utilizing alginate-encapsulated porcine islets have begun with Living Cell Technologies reporting long-term graft survival, albeit in a single type 1 diabetic patient (Elliott et al. 2007). One key limitation of the alginate encapsulation system is that standard methods are minimally scalable due to processing capacities of current extrusion technologies (Hoesli et al. 2011). While scalable emulsification methods can effectively encapsulate β-cell lines at efficiencies adequate to reverse diabetes in mouse models (Hoesli et al. 2011, 2012), these methods have not yet been applied to larger animal models or at clinical scales. (For more information on this topic, please see the chapter entitled "▶ Islet Encapsulation.")

Macroencapsulation methods offer an alternate approach to immunoisolation of a transplanted graft with ease of retrieving representing a key advantage over microencapsulation methods. The TheraCyte™ device offers one example of this approach, whereby cells are loaded into the multilayered cell impermeable thin pouch via an access port. The loaded device may then be surgically implanted in a variety of places within the body, most simply subcutaneously. After its exterior surface becomes vascularized by the host, the graft gains function, ideally enabling effective blood glucose control without direct physical contact between the graft and host, thus providing an immunological barrier to protect the graft from the host immune system. One caveat of such an approach for diabetes therapy is that mature islets have a high demand for oxygen and thus are traditionally challenging to maintain in a macroencapsulation device in the absence of substantial vascularization or oxygen supplementation (Ludwig et al. 2012). Indeed, mature islets do not do particularly well within the TheraCyte™ unless the devices are preimplanted to allow some vascularization of the outer membranes before cells are loaded (Rafael et al. 2003). Notably, by directly oxygenating alginate-encapsulated human islet preparations within a multilayer transplantable device, functional glucose-responsive insulin secretion was maintained up to 10 months after implantation in one patient with long-standing type 1 diabetes (Ludwig et al. 2013). Remarkably, this functional islet mass was protected from attack by the recipient's immune system despite the absence of any immunosuppressive agents supporting the functionality of such a macroencapsulation device (Ludwig et al. 2013). More immature cells may have advantages surviving in a macroencapsulation device. For example, human fetal pancreas tissue appears better able to survive the transiently hypoxic transplantation environment (Rafael et al. 2003; Lee et al. 2009). Likewise, despite one report of inconsistent development of hESC-derived progenitor cells within this type of macroencapsulation device (Matveyenko et al. 2010), we recently showed that functional maturation of hESC-derived pancreatic progenitor populations is possible and efficient within the TheraCyte™ device including the ability to reverse diabetic hyperglycemia in mice (Bruin et al. 2013; Rezania et al. 2013). Thus, pancreatic precursor cells derived from hESCs may be more like fetal cells and more resilient to the hypoxic transplantation environment. If similar results are not obtained when more mature islet endocrine cells are developed from PSCs, progenitor cells may have a distinct advantage for macroencapsulation.

While a considerable amount of effort continues to be focused on the generation of a scalable transplantation product, some aspects of this strategy remain concerning. The continuing concern with transplanting hESC-derived cells is the notion that they might overgrow, enabling uncontrolled release of secreted products such as insulin, or that they might form teratomas. This risk is presumably greater with the transplant of progenitor cells than it would be with fully differentiated cells. Indeed, transplantation of progenitors has been associated with varying levels of teratoma or overgrowth formation, due to incomplete differentiation of cultures such that pluripotent or possibly multipotent cell types capable of ectodermal and mesodermal lineages remain at the time of transplantation (Kroon et al. 2008; Kelly et al. 2011; Rezania et al. 2012). The formation of teratomas and overgrowths seems considerably reduced following enrichment of the transplanted cell population with endodermal progenitors (Cheng et al. 2012) or CD142-positive cells (Kelly et al. 2011), high uniformity scaled-up progenitor differentiation (Schulz et al. 2012), and transplantation of an in vitro matured endocrine population (Rezania et al. 2011). These data suggest that further adjustments to the differentiation, selection, and/or transplantation protocols may reduce or eliminate the capacity for graft overgrowth and teratoma formation. Indeed, we have shown that simple modification of established differentiation protocols followed by macroencapsulation was able to nearly eliminate all off-target germ layer development to one device of 74 compared to 18 out of 40 devices with our previous protocol (Bruin et al. 2013). The physical constraints provided by transplantable macroencapsulation devices also serve an important risk reduction role in containing the graft. Thus, delivery of cells within a defined, growth limiting, and ultimately retrievable physical space provides considerable protection to the transplant recipient. If the device proves to be a completely effective immunoisolation system, the absence or at least minimization of immunosuppression requirements could provide an additional advantage as any escaping graft cells would presumably be quickly targeted by the host immune system given their foreign nature. Taken together, this suggests that a number of strategies may alleviate the concerns of tumor formation. When uniformly differentiated cells are combined with a sturdy graft encapsulation method allowing immunocompetent recipients, the safety profile of hESC-derived progenitor transplants may even rival current islet transplantation methods where immunosuppression poses significant risks.

In Vitro Maturation and Models of Development

As noted previously, in vitro hormone-positive cells generated by many current protocols seem to be developmentally biased to become glucagon-secreting cells, while in vivo maturation of the same progenitors can yield the full complement of mature endocrine cells albeit in a very uncontrolled and poorly defined manner. This suggests that our current understanding of in vitro pancreatic differentiation is deficient in critical stimuli which are required for the complete maturation observed in vivo. If we can correct this deficiency, we will be better able to produce a well-defined functional human pancreatic endocrine cell population that can be used as a platform for drug discovery and as a transplantation source that has reduced risk for formation of off-target cell types. Such a well-defined product free from contaminating non-endocrine cell types will have the advantage of functioning to control blood glucose levels in patients immediately following transplant and may have an improved safety profile over current progenitor populations that might respond unpredictably during their maturation in the uncontrolled transplantation environment of human patients. While previous studies have suggested that only β-cells are critical for successful reversal of diabetes and that non-β-cell islet endocrine cells are not required to ameliorate hyperglycemia during transplantation in mice (King et al. 2007), it remains to be seen whether this is true in hESC-derived endocrine cell transplants, as pure β-cell grafts have yet to be generated under any maturation or purification process reported to date. Similarly, whether the normal islet architecture seen in both endogenous human islets and in vivo matured hESC-derived grafts (Rezania et al. 2012; Kroon et al. 2008) is required for optimal graft function in terms of glycemic control is unknown. This issue may be particularly relevant to encapsulation technologies which may disrupt the normal islet architecture. To address these questions, the production of uniform functional endocrine cells from hESCs in vitro remains a key challenge in the field and at the same time positions hESC-derived pancreatic progenitor maturation as an interesting model with which to study human pancreas development.

Transcription factors play a key role in pancreatic development (Oliver-Krasinski and Stoffers 2008), and recently, researchers are turning to genetic modification of hESCs to allow targeted study of transcription factor-activated pathways and networks in an effort to understand and control pancreatic endocrine development from PSCs. Based on this concept, mouse and human ESC lines bearing forced overexpression of single or multiple transcription factors, including SOX17, FOXA2, NGN3, NKX2.2, NKX6.1, NEUROD1, PAX4, and PDX1 (Blyszczuk et al. 2003; Miyazaki et al. 2004; Lavon et al. 2006; Bernardo et al. 2009; Lin et al. 2007; Liew et al. 2008; Treff et al. 2006; Shiroi et al. 2005; Kubo et al. 2011; Raikwar and Zavazava 2012; Marchand et al. 2009; Raikwar and Zavazava 2011; Seguin et al. 2008), have been generated. The majority of these studies have expressed transgenes by random integration of plasmids or lentivirus. Such strategies suffer from transgene silencing and loss of expression, which does not seem to be the case when targeted homologous recombination approaches using safe harbor loci are employed (Hockemeyer et al. 2009; Smith et al. 2008; Liew et al. 2007). Despite these limitations and the difficulties involved with modifying ESCs, a considerable number of developmental insights have been gained by transcription factor overexpression studies. In almost all cases examined, the forced expression of (combinations of) these transcription factors stimulated transcription of endogenous genes, most notably insulin, glucagon, and somatostatin over the course of in vitro differentiation. However, studies with the transcription factor PDX1 highlight a key caveat of these types of studies. Constitutive overexpression of PDX1 in hESCs increased pancreatic endocrine and exocrine induction in an embryoid body model, although robust insulin expression was distinctly absent in vitro (Lavon et al. 2006). Given that Pdx1 expression is believed to be biphasic in nature over mouse embryonic development from E13.5 and out to adulthood (Jorgensen et al. 2007), further examinations of PDX1 expression in hESCs attempted to recreate this expression pattern. Using a tamoxifen-inducible PDX1-expressing hESC line, Bernardo et al. (2009) found that a specific expression pattern of one 5-day pulse after definitive endoderm followed by a 5-day delayed pulse most efficiently induced insulin expression while minimizing expression of exocrine (amylase) and the liver (AFP and albumin) lineages. This work suggests that increased understanding of pancreatic development and the dynamics of transcription factor expression may yet inform key improvements in hESC differentiation protocols.

Similar to the idea of using transcription factor expression to control hESC development, small 22-nucleotide microRNAs (miRNAs) may show similarly powerful effects in PSC differentiation given their dynamic expression patterns (Krichevsky et al. 2006; Ivey et al. 2008). The importance of miRNAs within pancreatic development was established with mice which are unable to generate miRNAs due to deletion of the Dicer gene within the Pdx1-positive-developing pancreas – these animals show dramatic defects in exocrine, endocrine, and ductal cell pancreatic development (Lynn et al. 2007). miRNAs have been found to be dynamically regulated during hESC differentiation in a similar manner to those seen in human fetal pancreatic development (Wei et al. 2013; Joglekar et al. 2009; Gutteridge et al. 2013; Chen et al. 2011; Liao et al. 2013). During pancreatic endocrine development, miRNA-7 has been implicated in controlling both β-cell proliferation as well as specification of hormone-positive cells (Wang et al. 2013;

Kredo-Russo et al. 2012). Following the murine work that revealed the key role of miRNA 375 to regulate α-cell and β-cell mass (Poy et al. 2009) and similar work in zebrafish (Kloosterman et al. 2007), overexpression of miRNAs in differentiating hESCs has been recently explored. Lentiviral gene delivery to hESCs which resulted in increased expression of miRNA 375 was associated with an increased commitment and maturation toward the pancreatic endocrine lineage including increased expression of insulin, PDX1, NKX6.1, and PAX6 (Lahmy et al. 2013). In a similar study, miRNA-375 overexpression resulted in reduced translation of the miRNA-375 targets SOX9 and HNF1β (Wei et al. 2013), further emphasizing the role of miRNAs to modify gene expression patterns and influence pancreatic endocrine development of hESCs. While the full impact of miRNAs on the many stages of hESC differentiation is not completely known, it seems clear that these small modifiers of gene expression are likely to have dramatic effects in the functional maturation of hESC and could serve as useful targets to control this process.

The idea of modeling human development in hESCs was recently applied to understanding the effects of mutations in the glucokinase gene (GCK), which are associated with MODY2. In an elegant study by Hua et al. (2013), skin biopsies were acquired from two patients with MODY2. Patient-specific iPSCs were generated from each biopsy and were found to be pluripotent while retaining the heterozygous deletion in GCK. Upon in vitro differentiation and in vivo maturation, GCK mutant grafts developed to contain insulin-producing cells which displayed an impaired functional response to elevated glucose levels similar to that commonly observed in individuals with MODY2. The authors then used homologous recombination to repair the genetic lesion in GCK in the undifferentiated iPSCs and found that this restored normal glucose responsiveness to insulin-producing cells upon in vitro and in vivo maturation (Hua et al. 2013). This work suggests that PSC differentiation can be used to recapitulate and understand the effects of human genetic phenotypes. Such approaches could ultimately allow for the generation of patient-specific cellular therapies to restore functionally normal cells in patients bearing genetic mutations in a particular cell type and for developing new drugs.

The strategy of retesting knowledge generated from mouse developmental and rare human models in hESCs has been relatively fruitful. However, discovery-based methods using hESC lines that allow live cell lineage tracing and prospective isolation are also enabling the identification of new factors that influence the development of human diabetes. One popular recent tool has been hESC cell lines which allow tracking of endogenous human insulin promoter activation in its native loci through expression of cytoplasmic eGFP (Nostro et al. 2011; Basford et al. 2012; Micallef et al. 2012). By using homologous recombination, this approach circumvents the problems associated with variable integration and expression of transgenes and the epigenetic silencing which has been observed with lentiviral and retroviral transgenesis (Pannell and Ellis 2001; Cockrell and Kafri 2007; Hockemeyer et al. 2009). The ability to illuminate and isolate the cell type of interest has allowed direct whole transcriptome analysis, as well as visualization of single-cell cytosolic calcium mobilization and single-whole-cell K_{ATP}, Ca_v, Na^+, and KV currents in hESC-derived insulin/eGFP-positive cells (Basford et al. 2012). As this kind of live cell labeling strategy expands to multiple colors and even to subcellularly localized reporters, the ability to specifically examine the characteristics of rare populations such as NKX6.1, PDX1, MafA, and insulin quadruple-positive, ARX, glucagon, and somatostatin triple-negative cells will become possible. If it is possible to understand how a single cell functionally develops, then we may be able to translate this knowledge to reproducibly guide the generation of glucose-responsive cells in vitro.

Concluding Remarks

With the debilitating complications of diabetes continuing to present a tremendous socioeconomic burden for patients, physicians, and health systems, the drive to develop novel curative approaches is high. While insulin injections continue to save the lives of patients with diabetes, cellular therapies with functional insulin-secreting cell populations should offer more dynamic and finely tuned glycemic control resulting in better quality of life. Human islet transplantation highlighted the potential of a cellular, transplantable source of β-cells for type 1 diabetes treatment but also the critical need for a readily available supply. To push toward this lofty goal, the stem cell field has approached and surpassed a number of key milestones. Research into the myriad of pluripotent cell types has shown that PSCs are broadly capable of producing the cells required for diabetes therapy, and the study of targeted differentiation strategies for PSCs has laid the groundwork for a PSC-derived β-cell source for transplantation. The formation of definitive endoderm and eventual pancreatic progenitors has become a key checkpoint in PSC development and maturation. With demonstration that in vivo maturation of these progenitors can ameliorate diabetic symptoms in multiple models, the capacity of PSC-derived cells to eventually release insulin is no longer a concern. Nevertheless, important questions remain about how this maturation occurs, and how to replicate it consistently and safely on a clinically appropriate scale, and ideally in vitro to generate cells that are fully differentiated and functional immediately upon transplantation. Given the fast pace of the field, progress in terms of fully in vitro hESC-derived β-cells is expected to be swift. While most research in this area has focused on exploiting the potential of hESCs for type 1 diabetes treatment, there remains a significant opportunity to apply these findings to the setting of insulin-dependent type 2 diabetes as well. While much is left to explore, optimize, and implement, the rewards of cell-based physiological glucose regulation will continue to drive innovation and ingenuity in this field of regenerative medicine.

Acknowledgments We thank Dr. J. Bruin and Dr. S. Erener for providing images used in this chapter. We are also indebted to Miss S. White for technical assistance in the generation of Fig. 1. BKG has received PhD funding from the Natural Sciences and Engineering Research Council, the Michael Smith Foundation for Health Research, and the University of British Columbia.

Cross-References

▶ Islet Encapsulation
▶ Regulation of Pancreatic Islet Formation
▶ Stem Cells in Pancreatic Islets
▶ Successes and Disappointments with Clinical Islet Transplantation
▶ The Comparative Anatomy of Islets

References

Adewumi O, Aflatoonian B, Ahrlund-Richter L, Amit M, Andrews PW, Beighton G, Bello PA, Benvenisty N (2007) Characterization of human embryonic stem cell lines by the International Stem Cell Initiative. Nat Biotechnol 25(7):803–816

Ang SL, Rossant J (1994) HNF-3 β is essential for node and notochord formation in mouse development. Cell 78(4):561–574

Bar-Nur O, Russ HA, Efrat S, Benvenisty N (2011) Epigenetic memory and preferential lineage-specific differentiation in induced pluripotent stem cells derived from human pancreatic islet β cells. Cell Stem Cell 9(1):17–23

Basford CL, Prentice KJ, Hardy AB, Sarangi F, Micallef SJ, Li X, Guo Q, Elefanty AG, Stanley EG, Keller G, Allister EM, Nostro MC, Wheeler MB (2012) The functional and molecular characterisation of human embryonic stem cell-derived insulin-positive cells compared with adult pancreatic β cells. Diabetologia 55(2):358–371

Basma H, Soto-Gutierrez A, Yannam GR, Liu L, Ito R, Yamamoto T, Ellis E, Carson SD, Sato S, Chen Y, Muirhead D, Navarro-Alvarez N, Wong RJ, Roy-Chowdhury J, Platt JL, Mercer DF, Miller JD, Strom SC, Kobayashi N, Fox IJ (2009) Differentiation and transplantation of human embryonic stem cell-derived hepatocytes. Gastroenterology 136(3):990–999

Bernardo AS, Cho CH, Mason S, Docherty HM, Pedersen RA, Vallier L, Docherty K (2009) Biphasic induction of Pdx1 in mouse and human embryonic stem cells can mimic development of pancreatic β-cells. Stem Cells 27(2):341–351

Blyszczuk P, Czyz J, Kania G, Wagner M, Roll U, St-Onge L, Wobus AM (2003) Expression of Pax4 in embryonic stem cells promotes differentiation of nestin-positive progenitor and insulin-producing cells. Proc Natl Acad Sci USA 100(3):998–1003

Bock C, Kiskinis E, Verstappen G, Gu H, Boulting G, Smith ZD, Ziller M, Croft GF, Amoroso MW, Oakley DH, Gnirke A, Eggan K, Meissner A (2011) Reference maps of human ES and iPS cell variation enable high-throughput characterization of pluripotent cell lines. Cell 144(3):439–452

Borowiak M, Maehr R, Chen S, Chen AE, Tang W, Fox JL, Schreiber SL, Melton DA (2009) Small molecules efficiently direct endodermal differentiation of mouse and human embryonic stem cells. Cell Stem Cell 4(4):348–358

Bosco D, Armanet M, Morel P, Niclauss N, Sgroi A, Muller YD, Giovannoni L, Parnaud G, Berney T (2010) Unique arrangement of α- and β-cells in human islets of Langerhans. Diabetes 59(5):1202–1210

Bose B, Shenoy SP, Konda S, Wangikar P (2012) Human embryonic stem cell differentiation into insulin secreting β-cells for diabetes. Cell Biol Int 36(11):1013–1020

Bradley A, Evans M, Kaufman MH, Robertson E (1984) Formation of germ-line chimaeras from embryo-derived teratocarcinoma cell lines. Nature 309(5965):255–256

Brennan J, Lu CC, Norris DP, Rodriguez TA, Beddington RS, Robertson EJ (2001) Nodal signalling in the epiblast patterns the early mouse embryo. Nature 411(6840):965–969

Brissova M, Shostak A, Shiota M, Wiebe PO, Poffenberger G, Kantz J, Chen Z, Carr C, Jerome WG, Chen J, Baldwin HS, Nicholson W, Bader DM, Jetton T, Gannon M, Powers AC (2006)

Pancreatic islet production of vascular endothelial growth factor–a is essential for islet vascularization, revascularization, and function. Diabetes 55(11):2974–2985

Brown S, Teo A, Pauklin S, Hannan N, Cho CH, Lim B, Vardy L, Dunn NR, Trotter M, Pedersen R, Vallier L (2011) Activin/Nodal signaling controls divergent transcriptional networks in human embryonic stem cells and in endoderm progenitors. Stem Cells 29(8):1176–1185

Bruin J, Kieffer T (2012) Differentiation of human embryonic stem cells into pancreatic endocrine cells. In: Hayat MA (ed) Stem cells and cancer stem cells, vol 8. Springer, Dordrecht, pp 191–206

Bruin JE, Rezania A, Xu J, Narayan K, Fox JK, O'Neil JJ, Kieffer TJ (2013) Maturation and function of human embryonic stem cell-derived pancreatic progenitors in macroencapsulation devices following transplant into mice. Diabetologia 56:1987–1998

Brunicardi FC, Kleinman R, Moldovan S, Nguyen TH, Watt PC, Walsh J, Gingerich R (2001) Immunoneutralization of somatostatin, insulin, and glucagon causes alterations in islet cell secretion in the isolated perfused human pancreas. Pancreas 23(3):302–308

Byrne JA, Pedersen DA, Clepper LL, Nelson M, Sanger WG, Gokhale S, Wolf DP, Mitalipov SM (2007) Producing primate embryonic stem cells by somatic cell nuclear transfer. Nature 450(7169):497–502

Cai J, Yu C, Liu Y, Chen S, Guo Y, Yong J, Lu W, Ding M, Deng H (2010) Generation of homogeneous PDX1$^+$ pancreatic progenitors from human ES cell-derived endoderm cells. J Mol Cell Biol 2(1):50–60

Calder A, Roth-Albin I, Bhatia S, Pilquil C, Lee JH, Bhatia M, Levadoux-Martin M, McNicol J, Russell J, Collins T, Draper JS (2013) Lengthened G1 phase indicates differentiation status in human embryonic stem cells. Stem Cells Dev 22(2):279–295

Castaing M, Peault B, Basmaciogullari A, Casal I, Czernichow P, Scharfmann R (2001) Blood glucose normalization upon transplantation of human embryonic pancreas into β-cell-deficient SCID mice. Diabetologia 44(11):2066–2076

Champeris Tsaniras S, Jones PM (2010) Generating pancreatic β-cells from embryonic stem cells by manipulating signaling pathways. J Endocrinol 206(1):13–26

Chen Y, Pan FC, Brandes N, Afelik S, Solter M, Pieler T (2004) Retinoic acid signaling is essential for pancreas development and promotes endocrine at the expense of exocrine cell differentiation in Xenopus. Dev Biol 271(1):144–160

Chen M, Song G, Wang C, Hu D, Ren J, Qu X (2009a) Small-molecule selectively recognizes human telomeric G-quadruplex DNA and regulates its conformational switch. Biophys J 97(7):2014–2023

Chen S, Borowiak M, Fox JL, Maehr R, Osafune K, Davidow L, Lam K, Peng LF, Schreiber SL, Rubin LL, Melton D (2009b) A small molecule that directs differentiation of human ESCs into the pancreatic lineage. Nat Chem Biol 5(4):258–265

Chen X, Chen A, Woo TL, Choo AB, Reuveny S, Oh SK (2010) Investigations into the metabolism of two-dimensional colony and suspended microcarrier cultures of human embryonic stem cells in serum-free media. Stem Cells Dev 19(11):1781–1792

Chen BZ, Yu SL, Singh S, Kao LP, Tsai ZY, Yang PC, Chen BH, Shoei-Lung Li S (2011) Identification of microRNAs expressed highly in pancreatic islet-like cell clusters differentiated from human embryonic stem cells. Cell Biol Int 35(1):29–37

Cheng X, Ying L, Lu L, Galvao AM, Mills JA, Lin HC, Kotton DN, Shen SS, Nostro MC, Choi JK, Weiss MJ, French DL, Gadue P (2012) Self-renewing endodermal progenitor lines generated from human pluripotent stem cells. Cell Stem Cell 10(4):371–384

Chetty S, Pagliuca FW, Honore C, Kweudjeu A, Rezania A, Melton DA (2013) A simple tool to improve pluripotent stem cell differentiation. Nat Methods 10(6):553–556

Cho CH, Hannan NR, Docherty FM, Docherty HM, Joao Lima M, Trotter MW, Docherty K, Vallier L (2012) Inhibition of activin/nodal signalling is necessary for pancreatic differentiation of human pluripotent stem cells. Diabetologia 55(12):3284–3295

Cockrell AS, Kafri T (2007) Gene delivery by lentivirus vectors. Mol Biotechnol 36(3):184–204, MB:36:3:184 [pii]

D'Amour KA, Agulnick AD, Eliazer S, Kelly OG, Kroon E, Baetge EE (2005) Efficient differentiation of human embryonic stem cells to definitive endoderm. Nat Biotechnol 23(12):1534–1541

D'Amour KA, Bang AG, Eliazer S, Kelly OG, Agulnick AD, Smart NG, Moorman MA, Kroon E, Carpenter MK, Baetge EE (2006) Production of pancreatic hormone-expressing endocrine cells from human embryonic stem cells. Nat Biotechnol 24(11):1392–1401

De Vos P, Van Straaten JF, Nieuwenhuizen AG, de Groot M, Ploeg RJ, De Haan BJ, Van Schilfgaarde R (1999) Why do microencapsulated islet grafts fail in the absence of fibrotic overgrowth? Diabetes 48(7):1381–1388

Docherty K, Bernardo AS, Vallier L (2007) Embryonic stem cell therapy for diabetes mellitus. Semin Cell Dev Biol 18(6):827–838

Duvivier-Kali VF, Omer A, Parent RJ, O'Neil JJ, Weir GC (2001) Complete protection of islets against allorejection and autoimmunity by a simple barium-alginate membrane. Diabetes 50(8):1698–1705

Elliott RB, Escobar L, Tan PL, Muzina M, Zwain S, Buchanan C (2007) Live encapsulated porcine islets from a type 1 diabetic patient 9.5 yr after xenotransplantation. Xenotransplantation 14(2):157–161

Evans MJ, Kaufman MH (1981) Establishment in culture of pluripotential cells from mouse embryos. Nature 292(5819):154–156

Fowler MJ (2011) The diabetes treatment trap: hypoglycemia. Clin Diabetes 29(1):36–39

Fryer BH, Rezania A, Zimmerman MC (2013) Generating β-cells in vitro: progress towards a Holy Grail. Curr Opin Endocrinol Diabetes Obes 20(2):112–117

Gage BK, Webber TD, Kieffer TJ (2013) Initial cell seeding density influences pancreatic endocrine development during in vitro differentiation of human embryonic stem cells. PLoS One 8:e82076

Gauthier BR, Gosmain Y, Mamin A, Philippe J (2007) The β-cell specific transcription factor Nkx6.1 inhibits glucagon gene transcription by interfering with Pax6. Biochem J 403(3):593–601

Gonzalez F, Boue S, Izpisua Belmonte JC (2011) Methods for making induced pluripotent stem cells: reprogramming a la carte. Nat Rev Genet 12(4):231–242

Gouon-Evans V, Boussemart L, Gadue P, Nierhoff D, Koehler CI, Kubo A, Shafritz DA, Keller G (2006) BMP-4 is required for hepatic specification of mouse embryonic stem cell-derived definitive endoderm. Nat Biotechnol 24(11):1402–1411

Green MD, Chen A, Nostro MC, d'Souza SL, Schaniel C, Lemischka IR, Gouon-Evans V, Keller G, Snoeck HW (2011) Generation of anterior foregut endoderm from human embryonic and induced pluripotent stem cells. Nat Biotechnol 29(3):267–272

Guo T, Hebrok M (2009) Stem cells to pancreatic β-cells: new sources for diabetes cell therapy. Endocr Rev 30(3):214–227

Guo T, Landsman L, Li N, Hebrok M (2013) Factors expressed by murine embryonic pancreatic mesenchyme enhance generation of insulin-producing cells from hESCs. Diabetes 62(5):1581–1592

Gutteridge A, Rukstalis JM, Ziemek D, Tie M, Ji L, Ramos-Zayas R, Nardone NA, Norquay LD, Brenner MB, Tang K, McNeish JD, Rowntree RK (2013) Novel pancreatic endocrine maturation pathways identified by genomic profiling and causal reasoning. PLoS One 8(2):e56024

Halban PA, Kahn SE, Lernmark A, Rhodes CJ (2001) Gene and cell-replacement therapy in the treatment of type 1 diabetes: how high must the standards be set? Diabetes 50(10):2181–2191

Hannan NR, Segeritz CP, Touboul T, Vallier L (2013) Production of hepatocyte-like cells from human pluripotent stem cells. Nat Protoc 8(2):430–437

Hansson M, Tonning A, Frandsen U, Petri A, Rajagopal J, Englund MC, Heller RS, Hakansson J, Fleckner J, Skold HN, Melton D, Semb H, Serup P (2004) Artifactual insulin release from differentiated embryonic stem cells. Diabetes 53(10):2603–2609, 53/10/2603 [pii]

Hauge-Evans AC, King AJ, Carmignac D, Richardson CC, Robinson IC, Low MJ, Christie MR, Persaud SJ, Jones PM (2009) Somatostatin secreted by islet δ-cells fulfills multiple roles as a paracrine regulator of islet function. Diabetes 58(2):403–411

Hayek A, Beattie GM (1997) Experimental transplantation of human fetal and adult pancreatic islets. J Clin Endocrinol Metab 82(8):2471–2475

Hebrok M (2012) Generating β cells from stem cells-the story so far. Cold Spring Harb Perspect Med 2(6):a007674

Hebrok M, Kim SK, Melton DA (1998) Notochord repression of endodermal Sonic hedgehog permits pancreas development. Genes Dev 12(11):1705–1713

Hockemeyer D, Soldner F, Beard C, Gao Q, Mitalipova M, DeKelver RC, Katibah GE, Amora R, Boydston EA, Zeitler B, Meng X, Miller JC, Zhang L, Rebar EJ, Gregory PD, Urnov FD, Jaenisch R (2009) Efficient targeting of expressed and silent genes in human ESCs and iPSCs using zinc-finger nucleases. Nat Biotechnol 27(9):851–857

Hoesli CA, Raghuram K, Kiang RL, Mocinecova D, Hu X, Johnson JD, Lacik I, Kieffer TJ, Piret JM (2011) Pancreatic cell immobilization in alginate beads produced by emulsion and internal gelation. Biotechnol Bioeng 108(2):424–434

Hoesli CA, Kiang RL, Mocinecova D, Speck M, Moskova DJ, Donald-Hague C, Lacik I, Kieffer TJ, Piret JM (2012) Reversal of diabetes by βTC3 cells encapsulated in alginate beads generated by emulsion and internal gelation. J Biomed Mater Res Part B, Appl Biomater 100(4):1017–1028

Hosoya M (2012) Preparation of pancreatic β-cells from human iPS cells with small molecules. Islets 4(3):249–252

Hou P, Li Y, Zhang X, Liu C, Guan J, Li H, Zhao T, Ye J, Yang W, Liu K, Ge J, Xu J, Zhang Q, Zhao Y, Deng H (2013) Pluripotent stem cells induced from mouse somatic cells by small-molecule compounds. Science 341:651–654

Hua H, Shang L, Martinez H, Freeby M, Gallagher MP, Ludwig T, Deng L, Greenberg E, Leduc C, Chung WK, Goland R, Leibel RL, Egli D (2013) iPSC-derived β cells model diabetes due to glucokinase deficiency. J Clin Invest 123:3146–3153

Ivey KN, Muth A, Arnold J, King FW, Yeh RF, Fish JE, Hsiao EC, Schwartz RJ, Conklin BR, Bernstein HS, Srivastava D (2008) MicroRNA regulation of cell lineages in mouse and human embryonic stem cells. Cell Stem Cell 2(3):219–229

Jacobs-Tulleneers-Thevissen D, Chintinne M, Ling Z, Gillard P, Schoonjans L, Delvaux G, Strand BL, Gorus F, Keymeulen B, Pipeleers D, β Cell Therapy Consortium E-F (2013) Sustained function of alginate-encapsulated human islet cell implants in the peritoneal cavity of mice leading to a pilot study in a type 1 diabetic patient. Diabetologia 56(7):1605–1614

Jansson L, Carlsson PO (2002) Graft vascular function after transplantation of pancreatic islets. Diabetologia 45(6):749–763

Jennings RE, Berry AA, Kirkwood-Wilson R, Roberts NA, Hearn T, Salisbury RJ, Blaylock J, Hanley KP, Hanley NA (2013) Development of the human pancreas from foregut to endocrine commitment. Diabetes 62:3514–3522

Jeon J, Correa-Medina M, Ricordi C, Edlund H, Diez JA (2009) Endocrine cell clustering during human pancreas development. J Histochem Cytochem 57(9):811–824

Jermendy A, Toschi E, Aye T, Koh A, Aguayo-Mazzucato C, Sharma A, Weir GC, Sgroi D, Bonner-Weir S (2011) Rat neonatal β cells lack the specialised metabolic phenotype of mature β cells. Diabetologia 54(3):594–604

Jiang J, Au M, Lu K, Eshpeter A, Korbutt G, Fisk G, Majumdar AS (2007a) Generation of insulin-producing islet-like clusters from human embryonic stem cells. Stem Cells 25(8):1940–1953

Jiang W, Shi Y, Zhao D, Chen S, Yong J, Zhang J, Qing T, Sun X, Zhang P, Ding M, Li D, Deng H (2007b) In vitro derivation of functional insulin-producing cells from human embryonic stem cells. Cell Res 17(4):333–344

Jiang W, Bai Z, Zhang D, Shi Y, Yong J, Chen S, Ding M, Deng H (2008) Differentiation of mouse nuclear transfer embryonic stem cells into functional pancreatic β cells. Diabetologia 51(9):1671–1679

Joglekar MV, Joglekar VM, Hardikar AA (2009) Expression of islet-specific microRNAs during human pancreatic development. Gene Expr Patterns 9(2):109–113

Johnson JD, Ao Z, Ao P, Li H, Dai LJ, He Z, Tee M, Potter KJ, Klimek AM, Meloche RM, Thompson DM, Verchere CB, Warnock GL (2009) Different effects of FK506, rapamycin, and mycophenolate mofetil on glucose-stimulated insulin release and apoptosis in human islets. Cell Transplant 18(8):833–845

Jonsson J, Carlsson L, Edlund T, Edlund H (1994) Insulin-promoter-factor 1 is required for pancreas development in mice. Nature 371(6498):606–609

Jorgensen MC, Ahnfelt-Ronne J, Hald J, Madsen OD, Serup P, Hecksher-Sorensen J (2007) An illustrated review of early pancreas development in the mouse. Endocr Rev 28(6):685–705

Kanai-Azuma M, Kanai Y, Gad JM, Tajima Y, Taya C, Kurohmaru M, Sanai Y, Yonekawa H, Yazaki K, Tam PP, Hayashi Y (2002) Depletion of definitive gut endoderm in Sox17-null mutant mice. Development 129(10):2367–2379

Kearns NA, Genga RM, Ziller M, Kapinas K, Peters H, Brehm MA, Meissner A, Maehr R (2013) Generation of organized anterior foregut epithelia from pluripotent stem cells using small molecules. Stem Cell Res 11(3):1003–1012

Kelly OG, Chan MY, Martinson LA, Kadoya K, Ostertag TM, Ross KG, Richardson M, Carpenter MK, D'Amour KA, Kroon E, Moorman M, Baetge EE, Bang AG (2011) Cell-surface markers for the isolation of pancreatic cell types derived from human embryonic stem cells. Nat Biotechnol 29(8):750–756

King AJ, Fernandes JR, Hollister-Lock J, Nienaber CE, Bonner-Weir S, Weir GC (2007) Normal relationship of β- and non-β-cells not needed for successful islet transplantation. Diabetes 56(9):2312–2318

Kinkel MD, Prince VE (2009) On the diabetic menu: zebrafish as a model for pancreas development and function. Bioessays 31(2):139–152

Kleinsmith LJ, Pierce GB Jr (1964) Multipotentiality of single embryonal carcinoma cells. Cancer Res 24:1544–1551

Kloosterman WP, Lagendijk AK, Ketting RF, Moulton JD, Plasterk RH (2007) Targeted inhibition of miRNA maturation with morpholinos reveals a role for miR-375 in pancreatic islet development. PLoS Biol 5(8):e203

Kredo-Russo S, Mandelbaum AD, Ness A, Alon I, Lennox KA, Behlke MA, Hornstein E (2012) Pancreas-enriched miRNA refines endocrine cell differentiation. Development 139(16):3021–3031

Krichevsky AM, Sonntag KC, Isacson O, Kosik KS (2006) Specific microRNAs modulate embryonic stem cell-derived neurogenesis. Stem Cells 24(4):857–864

Kroon E, Martinson LA, Kadoya K, Bang AG, Kelly OG, Eliazer S, Young H, Richardson M, Smart NG, Cunningham J, Agulnick AD, D'Amour KA, Carpenter MK, Baetge EE (2008) Pancreatic endoderm derived from human embryonic stem cells generates glucose-responsive insulin-secreting cells in vivo. Nat Biotechnol 26(4):443–452

Kubo A, Shinozaki K, Shannon JM, Kouskoff V, Kennedy M, Woo S, Fehling HJ, Keller G (2004) Development of definitive endoderm from embryonic stem cells in culture. Development 131(7):1651–1662

Kubo A, Stull R, Takeuchi M, Bonham K, Gouon-Evans V, Sho M, Iwano M, Saito Y, Keller G, Snodgrass R (2011) Pdx1 and Ngn3 overexpression enhances pancreatic differentiation of mouse ES cell-derived endoderm population. PLoS One 6(9):e24058

Kunisada Y, Tsubooka-Yamazoe N, Shoji M, Hosoya M (2012) Small molecules induce efficient differentiation into insulin-producing cells from human induced pluripotent stem cells. Stem Cell Res 8(2):274–284

Lahmy R, Soleimani M, Sanati MH, Behmanesh M, Kouhkan F, Mobarra N (2013) Pancreatic islet differentiation of human embryonic stem cells by microRNA overexpression. J Tissue Eng Regen Med. doi:10.1002/term.1787

Lavon N, Yanuka O, Benvenisty N (2006) The effect of overexpression of Pdx1 and Foxa2 on the differentiation of human embryonic stem cells into pancreatic cells. Stem Cells 24(8):1923–1930

Lee SH, Hao E, Savinov AY, Geron I, Strongin AY, Itkin-Ansari P (2009) Human β-cell precursors mature into functional insulin-producing cells in an immunoisolation device: implications for diabetes cell therapies. Transplantation 87(7):983–991

Leon-Quinto T, Jones J, Skoudy A, Burcin M, Soria B (2004) In vitro directed differentiation of mouse embryonic stem cells into insulin-producing cells. Diabetologia 47(8):1442–1451

Liao X, Xue H, Wang YC, Nazor KL, Guo S, Trivedi N, Peterson SE, Liu Y, Loring JF, Laurent LC (2013) Matched miRNA and mRNA signatures from a hESC-based in vitro model of pancreatic differentiation reveal novel regulatory interactions. J Cell Sci 126:3848–3861

Liew CG, Draper JS, Walsh J, Moore H, Andrews PW (2007) Transient and stable transgene expression in human embryonic stem cells. Stem Cells 25(6):1521–1528

Liew CG, Shah NN, Briston SJ, Shepherd RM, Khoo CP, Dunne MJ, Moore HD, Cosgrove KE, Andrews PW (2008) PAX4 enhances β-cell differentiation of human embryonic stem cells. PLoS One 3(3):e1783

Lin HT, Kao CL, Lee KH, Chang YL, Chiou SH, Tsai FT, Tsai TH, Sheu DC, Ho LL, Ku HH (2007) Enhancement of insulin-producing cell differentiation from embryonic stem cells using pax4-nucleofection method. World J Gastroenterol 13(11):1672–1679

Ludwig B, Rotem A, Schmid J, Weir GC, Colton CK, Brendel MD, Neufeld T, Block NL, Yavriyants K, Steffen A, Ludwig S, Chavakis T, Reichel A, Azarov D, Zimermann B, Maimon S, Balyura M, Rozenshtein T, Shabtay N, Vardi P, Bloch K, de Vos P, Schally AV, Bornstein SR, Barkai U (2012) Improvement of islet function in a bioartificial pancreas by enhanced oxygen supply and growth hormone releasing hormone agonist. Proc Natl Acad Sci USA 109(13):5022–5027

Ludwig B, Reichel A, Steffen A, Zimerman B, Schally AV, Block NL, Colton CK, Ludwig S, Kersting S, Bonifacio E, Solimena M, Gendler Z, Rotem A, Barkai U, Bornstein SR (2013) Transplantation of human islets without immunosuppression. Proc Natl Acad Sci USA 110(47):19054–19058

Lumelsky N, Blondel O, Laeng P, Velasco I, Ravin R, McKay R (2001) Differentiation of embryonic stem cells to insulin-secreting structures similar to pancreatic islets. Science 292(5520):1389–1394

Lynn FC, Skewes-Cox P, Kosaka Y, McManus MT, Harfe BD, German MS (2007) MicroRNA expression is required for pancreatic islet cell genesis in the mouse. Diabetes 56(12):2938–2945

Maehr R (2011) iPS cells in type 1 diabetes research and treatment. Clin Pharmacol Ther 89(5):750–753

Maes E, Pipeleers D (1984) Effects of glucose and 3',5'-cyclic adenosine monophosphate upon reaggregation of single pancreatic β-cells. Endocrinology 114(6):2205–2209

Marchand M, Schroeder IS, Markossian S, Skoudy A, Negre D, Cosset FL, Real P, Kaiser C, Wobus AM, Savatier P (2009) Mouse ES cells over-expressing the transcription factor NeuroD1 show increased differentiation towards endocrine lineages and insulin-expressing cells. Int J Dev Biol 53(4):569–578

Martens GA, Motte E, Kramer G, Gaarn LW, Stange G, Hellemans K, Aerts J, Nielsen JH, Ling Z, Pipeleers D (2013) Functional characteristics of neonatal rat β cells with distinct markers. J Mol Endocrinol 52:11–28

Martin GR (1981) Isolation of a pluripotent cell line from early mouse embryos cultured in medium conditioned by teratocarcinoma stem cells. Proc Natl Acad Sci USA 78(12):7634–7638

Matveyenko AV, Georgia S, Bhushan A, Butler PC (2010) Inconsistent formation and nonfunction of insulin-positive cells from pancreatic endoderm derived from human embryonic stem cells in athymic nude rats. Am J Physiol Endocrinol Metab 299(5):E713–E720

Melendez-Ramirez LY, Richards RJ, Cefalu WT (2010) Complications of type 1 diabetes. Endocrinol Metab Clin North Am 39(3):625–640

Merani S, Shapiro AM (2006) Current status of pancreatic islet transplantation. Clin Sci 110(6):611–625

Mfopou JK, De Groote V, Xu X, Heimberg H, Bouwens L (2007) Sonic hedgehog and other soluble factors from differentiating embryoid bodies inhibit pancreas development. Stem Cells 25(5):1156–1165

Mfopou JK, Chen B, Mateizel I, Sermon K, Bouwens L (2010) Noggin, retinoids, and fibroblast growth factor regulate hepatic or pancreatic fate of human embryonic stem cells. Gastroenterology 138(7): 2233–2245, 2245 e2231–e2214

Micallef SJ, Li X, Schiesser JV, Hirst CE, Yu QC, Lim SM, Nostro MC, Elliott DA, Sarangi F, Harrison LC, Keller G, Elefanty AG, Stanley EG (2012) INS(GFP/w) human embryonic stem cells facilitate isolation of in vitro derived insulin-producing cells. Diabetologia 55(3):694–706

Miyazaki S, Yamato E, Miyazaki J (2004) Regulated expression of pdx-1 promotes in vitro differentiation of insulin-producing cells from embryonic stem cells. Diabetes 53(4):1030–1037

Mohanakumar T, Narayanan K, Desai N, Ramachandran S, Shenoy S, Jendrisak M, Susskind BM, Olack B, Benshoff N, Phelan DL, Brennan DC, Fernandez LA, Odorico JS, Polonsky KS (2006) A significant role for histocompatibility in human islet transplantation. Transplantation 82(2):180–187

Mou H, Zhao R, Sherwood R, Ahfeldt T, Lapey A, Wain J, Sicilian L, Izvolsky K, Musunuru K, Cowan C, Rajagopal J (2012) Generation of multipotent lung and airway progenitors from mouse ESCs and patient-specific cystic fibrosis iPSCs. Cell Stem Cell 10(4):385–397

Nir T, Melton DA, Dor Y (2007) Recovery from diabetes in mice by β cell regeneration. J Clin Invest 117(9):2553–2561

Nostro MC, Keller G (2012) Generation of β cells from human pluripotent stem cells: potential for regenerative medicine. Semin Cell Dev Biol 23(6):701–710

Nostro MC, Sarangi F, Ogawa S, Holtzinger A, Corneo B, Li X, Micallef SJ, Park IH, Basford C, Wheeler MB, Daley GQ, Elefanty AG, Stanley EG, Keller G (2011) Stage-specific signaling through TGFβ family members and WNT regulates patterning and pancreatic specification of human pluripotent stem cells. Development 138(5):861–871

O'Connor MD, Kardel MD, Iosfina I, Youssef D, Lu M, Li MM, Vercauteren S, Nagy A, Eaves CJ (2008) Alkaline phosphatase-positive colony formation is a sensitive, specific, and quantitative indicator of undifferentiated human embryonic stem cells. Stem Cells 26(5):1109–1116

O'Connor MD, Kardel MD, Eaves CJ (2011) Functional assays for human embryonic stem cell pluripotency. Methods Mol Biol 690:67–80

O'Sullivan ES, Vegas A, Anderson DG, Weir GC (2011) Islets transplanted in immunoisolation devices: a review of the progress and the challenges that remain. Endocr Rev 32(6):827–844

Oliver-Krasinski JM, Stoffers DA (2008) On the origin of the β cell. Genes Dev 22(15):1998–2021

Osafune K, Caron L, Borowiak M, Martinez RJ, Fitz-Gerald CS, Sato Y, Cowan CA, Chien KR, Melton DA (2008) Marked differences in differentiation propensity among human embryonic stem cell lines. Nat Biotechnol 26(3):313–315

Pagliuca FW, Melton DA (2013) How to make a functional β-cell. Development 140(12):2472–2483

Pan FC, Wright C (2011) Pancreas organogenesis: from bud to plexus to gland. Dev Dyn 240(3):530–565

Pannell D, Ellis J (2001) Silencing of gene expression: implications for design of retrovirus vectors. Rev Med Virol 11(4):205–217

Pauklin S, Vallier L (2013) The cell-cycle state of stem cells determines cell fate propensity. Cell 155(1):135–147

Pipeleers D, in't Veld PI, Maes E, Van De Winkel M (1982) Glucose-induced insulin release depends on functional cooperation between islet cells. Proc Natl Acad Sci USA 79(23):7322–7325

Piper K, Brickwood S, Turnpenny LW, Cameron IT, Ball SG, Wilson DI, Hanley NA (2004) β cell differentiation during early human pancreas development. J Endocrinol 181(1):11–23

Plesner A, Verchere CB (2011) Advances and challenges in islet transplantation: islet procurement rates and lessons learned from suboptimal islet transplantation. J Transplant 2011:979527

Polak M, Bouchareb-Banaei L, Scharfmann R, Czernichow P (2000) Early pattern of differentiation in the human pancreas. Diabetes 49(2):225–232

Poy MN, Hausser J, Trajkovski M, Braun M, Collins S, Rorsman P, Zavolan M, Stoffel M (2009) miR-375 maintains normal pancreatic α- and β-cell mass. Proc Natl Acad Sci USA 106(14):5813–5818

Prentki M, Nolan CJ (2006) Islet β cell failure in type 2 diabetes. J Clin Invest 116(7):1802–1812

Rafael E, Wu GS, Hultenby K, Tibell A, Wernerson A (2003) Improved survival of macroencapsulated islets of Langerhans by preimplantation of the immunoisolating device: a morphometric study. Cell Transplant 12(4):407–412

Raikwar SP, Zavazava N (2011) Spontaneous in vivo differentiation of embryonic stem cell-derived pancreatic endoderm-like cells corrects hyperglycemia in diabetic mice. Transplantation 91(1):11–20

Raikwar SP, Zavazava N (2012) PDX1-engineered embryonic stem cell-derived insulin producing cells regulate hyperglycemia in diabetic mice. Transplant Res 1(1):19

Rajagopal J, Anderson WJ, Kume S, Martinez OI, Melton DA (2003) Insulin staining of ES cell progeny from insulin uptake. Science 299(5605):363

Rezania A, Riedel MJ, Wideman RD, Karanu F, Ao Z, Warnock GL, Kieffer TJ (2011) Production of functional glucagon-secreting α-cells from human embryonic stem cells. Diabetes 60(1):239–247

Rezania A, Bruin JE, Riedel MJ, Mojibian M, Asadi A, Xu J, Gauvin R, Narayan K, Karanu F, O'Neil JJ, Ao Z, Warnock GL, Kieffer TJ (2012) Maturation of human embryonic stem cell-derived pancreatic progenitors into functional islets capable of treating pre-existing diabetes in mice. Diabetes 61(8):2016–2029

Rezania A, Bruin JE, Xu J, Narayan K, Fox JK, O'Neil JJ, Kieffer TJ (2013) Enrichment of human embryonic stem cell-derived NKX6.1-expressing pancreatic progenitor cells accelerates the maturation of insulin-secreting cells in vivo. Stem Cells 31:2432–2442

Riedel MJ, Asadi A, Wang R, Ao Z, Warnock GL, Kieffer TJ (2012) Immunohistochemical characterisation of cells co-producing insulin and glucagon in the developing human pancreas. Diabetologia 55(2):372–381

Roche E, Ensenat-Waser R, Reig JA, Jones J, Leon-Quinto T, Soria B (2006) Therapeutic potential of stem cells in diabetes. Handb Exp Pharmacol 174:147–167

Ryan EA, Paty BW, Senior PA, Bigam D, Alfadhli E, Kneteman NM, Lakey JR, Shapiro AM (2005) Five-year follow-up after clinical islet transplantation. Diabetes 54(7):2060–2069

Schaffer AE, Taylor BL, Benthuysen JR, Liu J, Thorel F, Yuan W, Jiao Y, Kaestner KH, Herrera PL, Magnuson MA, May CL, Sander M (2013) Nkx6.1 controls a gene regulatory network required for establishing and maintaining pancreatic β cell identity. PLoS Genet 9(1):e1003274

Scharp DW, Marchetti P (2013) Encapsulated islets for diabetes therapy: history, current progress, and critical issues requiring solution. Adv Drug Deliv Rev. doi:10.1016/j.addr.2013.07.018

Scharp DW, Lacy PE, Santiago JV, McCullough CS, Weide LG, Falqui L, Marchetti P, Gingerich RL, Jaffe AS, Cryer PE et al (1990) Insulin independence after islet transplantation into type I diabetic patient. Diabetes 39(4):515–518

Schulz TC, Young HY, Agulnick AD, Babin MJ, Baetge EE, Bang AG, Bhoumik A, Cepa I, Cesario RM, Haakmeester C, Kadoya K, Kelly JR, Kerr J, Martinson LA, McLean AB, Moorman MA, Payne JK, Richardson M, Ross KG, Sherrer ES, Song X, Wilson AZ, Brandon EP, Green CE, Kroon EJ, Kelly OG, D'Amour KA, Robins AJ (2012) A scalable system for production of functional pancreatic progenitors from human embryonic stem cells. PLoS One 7(5):e37004

Segev H, Fishman B, Ziskind A, Shulman M, Itskovitz-Eldor J (2004) Differentiation of human embryonic stem cells into insulin-producing clusters. Stem Cells 22(3):265–274

Seguin CA, Draper JS, Nagy A, Rossant J (2008) Establishment of endoderm progenitors by SOX transcription factor expression in human embryonic stem cells. Cell Stem Cell 3(2):182–195

Sela Y, Molotski N, Golan S, Itskovitz-Eldor J, Soen Y (2012) Human embryonic stem cells exhibit increased propensity to differentiate during the G1 phase prior to phosphorylation of retinoblastoma protein. Stem Cells 30(6):1097–1108

Seymour PA, Sander M (2011) Historical perspective: beginnings of the β-cell: current perspectives in β-cell development. Diabetes 60(2):364–376

Shapiro AM (2011) Strategies toward single-donor islets of Langerhans transplantation. Curr Opin Organ Transplant 16(6):627–631

Shapiro AM, Lakey JR, Ryan EA, Korbutt GS, Toth E, Warnock GL, Kneteman NM, Rajotte RV (2000) Islet transplantation in seven patients with type 1 diabetes mellitus using a glucocorticoid-free immunosuppressive regimen. N Engl J Med 343(4):230–238

Shapiro AM, Ricordi C, Hering BJ, Auchincloss H, Lindblad R, Robertson RP, Secchi A, Brendel MD, Berney T, Brennan DC, Cagliero E, Alejandro R, Ryan EA, DiMercurio B, Morel P, Polonsky KS, Reems JA, Bretzel RG, Bertuzzi F, Froud T, Kandaswamy R, Sutherland DE, Eisenbarth G, Segal M, Preiksaitis J, Korbutt GS, Barton FB, Viviano L, Seyfert-Margolis V, Bluestone J, Lakey JR (2006) International trial of the Edmonton protocol for islet transplantation. N Engl J Med 355(13):1318–1330

Shim JH, Kim SE, Woo DH, Kim SK, Oh CH, McKay R, Kim JH (2007) Directed differentiation of human embryonic stem cells towards a pancreatic cell fate. Diabetologia 50(6):1228–1238

Shiroi A, Ueda S, Ouji Y, Saito K, Moriya K, Sugie Y, Fukui H, Ishizaka S, Yoshikawa M (2005) Differentiation of embryonic stem cells into insulin-producing cells promoted by Nkx2.2 gene transfer. World J Gastroenterol 11(27):4161–4166

Slack JM (1995) Developmental biology of the pancreas. Development 121(6):1569–1580

Smith JR, Maguire S, Davis LA, Alexander M, Yang F, Chandran S, ffrench-Constant C, Pedersen RA (2008) Robust, persistent transgene expression in human embryonic stem cells is achieved with AAVS1-targeted integration. Stem Cells 26(2):496–504

Smith KP, Luong MX, Stein GS (2009) Pluripotency: toward a gold standard for human ES and iPS cells. J Cell Physiol 220(1):21–29

Soria B, Roche E, Berna G, Leon-Quinto T, Reig JA, Martin F (2000) Insulin-secreting cells derived from embryonic stem cells normalize glycemia in streptozotocin-induced diabetic mice. Diabetes 49(2):157–162

Spence JR, Mayhew CN, Rankin SA, Kuhar MF, Vallance JE, Tolle K, Hoskins EE, Kalinichenko VV, Wells SI, Zorn AM, Shroyer NF, Wells JM (2011) Directed differentiation of human pluripotent stem cells into intestinal tissue in vitro. Nature 470(7332):105–109

Stadtfeld M, Brennand K, Hochedlinger K (2008) Reprogramming of pancreatic β cells into induced pluripotent stem cells. Curr Biol 18(12):890–894

Stafford D, Prince VE (2002) Retinoic acid signaling is required for a critical early step in zebrafish pancreatic development. Curr Biol 12(14):1215–1220

Stanley EG, Elefanty AG (2008) Building better β cells. Cell Stem Cell 2(4):300–301

Steiner DJ, Kim A, Miller K, Hara M (2010) Pancreatic islet plasticity: interspecies comparison of islet architecture and composition. Islets 2(3):135–145

Stevens LC, Little CC (1954) Spontaneous testicular teratomas in an inbred strain of mice. Proc Natl Acad Sci USA 40(11):1080–1087

Stoffers DA, Zinkin NT, Stanojevic V, Clarke WL, Habener JF (1997) Pancreatic agenesis attributable to a single nucleotide deletion in the human IPF1 gene coding sequence. Nat Genet 15(1):106–110

Sui L, Geens M, Sermon K, Bouwens L, Mfopou JK (2013) Role of BMP signaling in pancreatic progenitor differentiation from human embryonic stem cells. Stem Cell Rev 9:569–577

Tachibana M, Amato P, Sparman M, Gutierrez NM, Tippner-Hedges R, Ma H, Kang E, Fulati A, Lee HS, Sritanaudomchai H, Masterson K, Larson J, Eaton D, Sadler-Fredd K, Battaglia D, Lee D, Wu D, Jensen J, Patton P, Gokhale S, Stouffer RL, Wolf D, Mitalipov S (2013) Human embryonic stem cells derived by somatic cell nuclear transfer. Cell 153:1228–1238

Takahashi K, Yamanaka S (2006) Induction of pluripotent stem cells from mouse embryonic and adult fibroblast cultures by defined factors. Cell 126(4):663–676

Takahashi K, Tanabe K, Ohnuki M, Narita M, Ichisaka T, Tomoda K, Yamanaka S (2007) Induction of pluripotent stem cells from adult human fibroblasts by defined factors. Cell 131(5):861–872

The Diabetes Control and Complications Trial Research Group D (1993) The effect of intensive treatment of diabetes on the development and progression of long-term complications in insulin-dependent diabetes mellitus. The Diabetes Control and Complications Trial Research Group. N Engl J Med 329(14):977–986

Thomson JA, Kalishman J, Golos TG, Durning M, Harris CP, Becker RA, Hearn JP (1995) Isolation of a primate embryonic stem cell line. Proc Natl Acad Sci USA 92(17):7844–7848

Thomson JA, Kalishman J, Golos TG, Durning M, Harris CP, Hearn JP (1996) Pluripotent cell lines derived from common marmoset (*Callithrix jacchus*) blastocysts. Biol Reprod 55(2):254–259

Thomson JA, Itskovitz-Eldor J, Shapiro SS, Waknitz MA, Swiergiel JJ, Marshall VS, Jones JM (1998) Embryonic stem cell lines derived from human blastocysts. Science 282(5391):1145–1147

Tiso N, Filippi A, Pauls S, Bortolussi M, Argenton F (2002) BMP signalling regulates anteroposterior endoderm patterning in zebrafish. Mech Dev 118(1–2):29–37

Tiso N, Moro E, Argenton F (2009) Zebrafish pancreas development. Mol Cell Endocrinol 312(1–2):24–30

Touboul T, Hannan NR, Corbineau S, Martinez A, Martinet C, Branchereau S, Mainot S, Strick-Marchand H, Pedersen R, Di Santo J, Weber A, Vallier L (2010) Generation of functional hepatocytes from human embryonic stem cells under chemically defined conditions that recapitulate liver development. Hepatology 51(5):1754–1765

Treff NR, Vincent RK, Budde ML, Browning VL, Magliocca JF, Kapur V, Odorico JS (2006) Differentiation of embryonic stem cells conditionally expressing neurogenin 3. Stem Cells 24(11):2529–2537

Ungrin M, O'Connor M, Eaves C, Zandstra PW (2007) Phenotypic analysis of human embryonic stem cells. Curr Protoc Stem Cell Biol. Chapter 1:Unit 1B 3. doi:10.1002/9780470151808. sc01b03s2

van Belle TL, Coppieters KT, von Herrath MG (2011) Type 1 diabetes: etiology, immunology, and therapeutic strategies. Physiol Rev 91(1):79–118

Van Hoof D, D'Amour KA, German MS (2009) Derivation of insulin-producing cells from human embryonic stem cells. Stem Cell Res 3(2–3):73–87

Vasir B, Jonas JC, Steil GM, Hollister-Lock J, Hasenkamp W, Sharma A, Bonner-Weir S, Weir GC (2001) Gene expression of VEGF and its receptors Flk-1/KDR and Flt-1 in cultured and transplanted rat islets. Transplantation 71(7):924–935

Wang Y, Liu J, Liu C, Naji A, Stoffers DA (2013) MicroRNA-7 regulates the mTOR pathway and proliferation in adult pancreatic β-cells. Diabetes 62(3):887–895

Wei R, Yang J, Liu GQ, Gao MJ, Hou WF, Zhang L, Gao HW, Liu Y, Chen GA, Hong TP (2013) Dynamic expression of microRNAs during the differentiation of human embryonic stem cells into insulin-producing cells. Gene 518(2):246–255

Weinstein DC, Ruiz i Altaba A, Chen WS, Hoodless P, Prezioso VR, Jessell TM, Darnell JE Jr (1994) The winged-helix transcription factor HNF-3 β is required for notochord development in the mouse embryo. Cell 78(4):575–588

Weir GC (2013) Islet encapsulation: advances and obstacles. Diabetologia 56(7):1458–1461

Wells JM, Melton DA (1999) Vertebrate endoderm development. Annu Rev Cell Dev Biol 15:393–410

Wells JM, Melton DA (2000) Early mouse endoderm is patterned by soluble factors from adjacent germ layers. Development 127(8):1563–1572

Wideman RD, Kieffer TJ (2009) Mining incretin hormone pathways for novel therapies. Trends Endocrinol Metab 20(6):280–286

Wilmut I, Schnieke AE, McWhir J, Kind AJ, Campbell KH (1997) Viable offspring derived from fetal and adult mammalian cells. Nature 385(6619):810–813

Wong AL, Nierras CR (2010) Do stem cell-derived islets represent a commercially viable treatment for Type 1 and 2 diabetes? Regen Med 5(6):839–842

Wong AL, Hwa A, Hellman D, Greenstein JL (2012) Surrogate insulin-producing cells. F1000 Med Rep 4:15

Xu X, Browning VL, Odorico JS (2011) Activin, BMP and FGF pathways cooperate to promote endoderm and pancreatic lineage cell differentiation from human embryonic stem cells. Mech Dev 128(7–10):412–427

Yang YP, Wright C (2009) Chemicals turn human embryonic stem cells towards β cells. Nat Chem Biol 5(4):195–196

Yasunaga M, Tada S, Torikai-Nishikawa S, Nakano Y, Okada M, Jakt LM, Nishikawa S, Chiba T, Era T, Nishikawa S (2005) Induction and monitoring of definitive and visceral endoderm differentiation of mouse ES cells. Nat Biotechnol 23(12):1542–1550

Yu J, Thomson JA (2008) Pluripotent stem cell lines. Genes Dev 22(15):1987–1997

Yu J, Vodyanik MA, Smuga-Otto K, Antosiewicz-Bourget J, Frane JL, Tian S, Nie J, Jonsdottir GA, Ruotti V, Stewart R, Slukvin II, Thomson JA (2007) Induced pluripotent stem cell lines derived from human somatic cells. Science 318(5858):1917–1920

Zhang N, Richter A, Suriawinata J, Harbaran S, Altomonte J, Cong L, Zhang H, Song K, Meseck M, Bromberg J, Dong H (2004) Elevated vascular endothelial growth factor production in islets improves islet graft vascularization. Diabetes 53(4):963–970

Zhang C, Moriguchi T, Kajihara M, Esaki R, Harada A, Shimohata H, Oishi H, Hamada M, Morito N, Hasegawa K, Kudo T, Engel JD, Yamamoto M, Takahashi S (2005) MafA is a key regulator of glucose-stimulated insulin secretion. Mol Cell Biol 25(12):4969–4976

Zhang D, Jiang W, Liu M, Sui X, Yin X, Chen S, Shi Y, Deng H (2009) Highly efficient differentiation of human ES cells and iPS cells into mature pancreatic insulin-producing cells. Cell Res 19(4):429–438

Zweigerdt R, Olmer R, Singh H, Haverich A, Martin U (2011) Scalable expansion of human pluripotent stem cells in suspension culture. Nat Protoc 6(5):689–700

Pancreatic Neuroendocrine Tumors

49

Apostolos Tsolakis and George Kanakis

Contents

Introduction	1376
Epidemiology	1377
Classification and Staging	1377
Pathology	1379
Genetics	1381
Sporadic Pancreatic Neuroendocrine Tumors	1381
Hereditary Pancreatic Neuroendocrine Tumors	1381
Specific Tumor Types: Clinical Manifestations	1383
Well-Differentiated Pancreatic Neuroendocrine Tumors	1383
Neuroendocrine Carcinomas	1387
Biomarkers	1387
General Biomarkers	1387
Tumor-Specific Biomarkers	1388
Localization	1390
Ultrasonography	1390
Computerized Tomography and Magnetic Resonance Imaging	1391
Radionuclear Imaging	1391
Angiography	1392
Diagnostic Procedures	1393
EUS-Guided Fine Needle Aspiration and Core Needle Biopsy	1393

A. Tsolakis (✉)
Department of Medical Sciences, Section of Endocrine Oncology, Uppsala University, Uppsala, Sweden
e-mail: apostolos.tsolakis@medsci.uu.se

G. Kanakis
Department of Pathophysiology, University of Athens Medical School, Athens, Greece
e-mail: geokan@endo.gr

Treatment	1393
Surgery	1393
Cytoreductive Therapies of Liver Disease	1395
Medical Treatment	1396
Cross-References	1400
References	1400

Abstract

Pancreatic neuroendocrine tumors (pNETs) are rare neoplasms, with a prevalence of 1–2 per 100,000 people. Based on the presence or absence of a specific hormone-related clinical syndrome, they are divided into functioning and nonfunctioning. Among the former tumors, insulinomas are the most common. pNETs may be sporadic or associated with hereditary syndromes. The histopathology evaluation should include immunostaining with general (chromogranin A and synaptophysin) and specific neuroendocrine markers, as well as with the proliferation index Ki67. Chromogranin A is currently the most useful neuroendocrine biomarker for diagnosis and follow-up of pNETs. Other specific hormones released by the neoplastic cells can also be included in the biochemical evaluation. For tumor localization both noninvasive and invasive techniques may be used. Debulking procedures and medical therapy are the possible treatment options for pNETs, but surgery is the only modality that offers the possibility of cure.

Keywords

Pancreatic neuroendocrine tumors (pNETs) • Functioning/nonfunctioning pNETs • Well-differentiated pNETs • Neuroendocrine carcinomas • Sporadic/hereditary pNETs • Chromogranin A • Synaptophysin • Ki67 • Noninvasive/invasive tumor localization techniques • Debulking procedures • Medical treatment

Introduction

Pancreatic neuroendocrine tumors (pNETs) are rare neoplasms with a wide spectrum of clinical presentation. Their rarity in combination to the syndromic clinical scenario that accompanies a fraction of them has intensified fascination among the physicians that deal with these tumors. In contradiction to their infrequency, the relevant long survival of many patients with pNETs compared to that of their adenocarcinoma counterpart suggests that the point prevalence is not inconsiderable. However, pNETs can be difficult to be diagnosed, especially in their early stages. Nowadays, there have been a number of recent advances in various aspects of pNETs including diagnosis, molecular pathology, as well as in their treatment. The present book chapter reviews a number of these advances and the management of the pNETs.

Epidemiology

pNETs have a prevalence of 1–2 per 100,000 of the population and comprise 1–2 % of all pancreatic tumors (Kaltsas et al. 2004b; Halfdanarson et al. 2008). According to the Surveillance, Epidemiology, and End Results (SEER) Program (Modlin et al. 2008; Yao et al. 2008a), an increase in the frequency of pNETs has been observed, which might also be attributed to the employment of modern imaging modalities and the incidental detection of otherwise silent tumors (Vagefi et al. 2007). This notion is supported by data from autopsy series where the reported incidence of small (<1 cm) pNETs ranges from 0.8 % to 10 % (Kimura et al. 1991). No gender predisposition is observed, and these tumors may be diagnosed at all ages, with a peak incidence between 30 and 60 years (Halfdanarson et al. 2008).

Classification and Staging

Traditionally, pNETs have been classified in functioning and nonfunctioning (NF) according to their ability to secrete bioactive substances in sufficient amounts to evoke a relevant clinical syndrome (Klimstra et al. 2010). It should be noted that many pNETs may be accompanied by increased circulating hormone concentrations and/or show hormone immunoreactivity; however they do not give rise to a clinical syndrome, and thus they cannot be characterized as functioning (Tsolakis and Janson 2008). The majority of pNETs are NF (~60 %); among functioning tumors insulinomas and gastrinomas are the most frequently encountered, while the rest of functioning pNETs are quite uncommon (Falconi et al. 2012). The latter group is named rare functioning pNETs and includes VIPomas, glucagonomas, and somatostatinomas (SSomas), as well as tumors that ectopically secrete substances like adrenocorticotropic hormone (ACTH), corticotropin-releasing hormone (CRH), growth hormone (GH), growth hormone-releasing hormone (GHRH), parathyroid hormone-related protein (PTHrP), and calcitonin (Jensen et al. 2012).

According to the most recent classification system introduced by WHO, neuroendocrine neoplasms in general (pNETs included) are divided in three categories based on mitotic count or Ki67 proliferation index (Bosman et al. 2010):
- G1 with a mitotic count <2 per 10 high-power fields (HPF) and/or Ki67 index ≤2 %
- G2 with a mitotic count 2–20 per 10 HPF and/or Ki67 index 3–20 %
- G3 with a mitotic count <20 per 10 HPF and/or Ki67 index >20 %

G1 and G2 tumors comprise a subgroup of well-differentiated neuroendocrine tumors (NETs), whereas G3 tumors are characterized as poorly differentiated neuroendocrine carcinomas (NECs) of small or large cell type.

A summary of pNETs' characteristics is presented in Table 1.

Recently, the European Neuroendocrine Tumors Society (ENETS) introduced a staging system for NETs (Table 2) in line with the UICC (Union for International Cancer Control) TNM system (Rindi et al. 2006). According to a recent study, the

Table 1 Summary of pancreatic neuroendocrine tumors' characteristics

Tumor type (hormone)	Clinical presentation	Pancreatic localization	Incidence (cases/100,000/year)	Malignancy (%)	MEN-1 associated (%)
Insulinoma (insulin)	Whipple's triad[a]	Equal incidence in all parts	0.4	<10	4–6
Gastrinoma (gastrin)	Zollinger–Ellison syndrome[b]	Head	0.05–0.4	>60	20–30
Glucagonoma (glucagon)	Glucagonoma syndrome[c]	Tail	0.005	50–80	15
VIPoma (VIP)	Verner–Morrison syndrome[d]	Tail	0.01	50–60	<3
SSoma (somatostatin)	SSoma syndrome[e]	Head	<0.01	>70	<1
ACTHoma (ACTH)	Ectopic Cushing	No preferential	<0.01	95	Rare
CRHoma (CRH)	Ectopic Cushing	No preferential	Unknown (very rare)	>80	Rare
Calcitonin-secreting pNET	Diarrhea/hypocalcemic symptoms	No preferential	Unknown (very rare)	>80	10–20
GHRHoma	Acromegaly	Tail	Unknown (very rare)	>80	Unknown
PTHrPoma (PTHrP)	Hypercalcemic symptoms	No preferential	Unknown (very rare)	>80	Rare
Serotonin-secreting pNET	Typical carcinoid syndrome[f]	Head	Unknown (very rare)	>80	Unknown
NF-pNET (e.g., PP)	Mass effect	Head	1–3	60–80	10–20
NECs	Mass effect/syndromes (rare)	Head	<0.01	100	Unknown

Data obtained from Ehehalt et al. (2009) and Tsolakis and Janson (2008)

ACTH adrenocorticotropic hormone, *CRH* corticotropin-releasing hormone, *GHRH* growth hormone-releasing hormone, *NECs* neuroendocrine carcinomas, *NF* nonfunctioning, *pNET* pancreatic neuroendocrine tumor, *PP* pancreatic polypeptide, *PTHrP* parathyroid hormone-related peptide, *SS* somatostatin, *VIP* vasoactive intestinal polypeptide

[a]Hypoglycemic symptoms, low blood glucose levels, reversible upon glucose intake
[b]Diarrhea, hypergastrinemia, gastric acid hypersecretion, peptic ulcer diathesis
[c]Glucagonoma syndrome: necrolytic migrating erythema, diabetes mellitus, deep vein thrombosis, anemia, weight loss, neuropsychiatric symptoms
[d]WDHA syndrome: watery diarrhea, hypokalemia, achlorhydria
[e]SSoma syndrome: diabetes, hypochlorhydria, cholelithiasis, steatorrhea, anemia, weight loss
[f]Flushing, diarrhea, cardiac valvular diseases, bronchospasm

corresponding frequencies for each stage were stage I (5 %), stage II (15 %), stage III (22 %), and stage IV (55 %) (Pape et al. 2008).

Furthermore, various prognostic factors suggested to correlate to prognosis have been proposed for pNETs (Table 3) (Tsolakis and Janson 2008).

Table 2 Proposal for a TNM classification and disease staging for endocrine tumors of the pancreas according to Rindi et al. (2006)

TNM			
T – primary tumor			
TX	Primary tumor cannot be assessed		
T0	No evidence of primary tumor		
T1	Tumor limited to the pancreas and size <2 cm		
T2	Tumor limited to the pancreas and size 2–4 cm		
T3	Tumor limited to the pancreas and size >4 cm or invading duodenum or bile duct		
T4	Tumor invading adjacent organs (stomach, spleen, colon, adrenal gland) or the wall of large vessels (celiac axis or superior mesenteric artery)		
	For any T, add (m) for multiple tumors		
N – regional lymph nodes			
NX	Regional lymph node cannot be assessed		
N0	No regional lymph node metastasis		
N1	Regional lymph node metastasis		
M – distant metastases			
MX	Distant metastasis cannot be assessed		
M0	No distant metastases		
M1[a]	Distant metastasis		
Stage			
Disease stages			
Stage I	T1	N0	M0
Stage IIa	T2	N0	M0
IIb	T3	N0	M0
Stage IIIa	T4	N0	M0
IIIb	Any T	N1	M0
Stage IV	Any T	Any N	M1

[a]M1-specific sites defined according to Sobin and Wittekind (2002)

Pathology

pNETs are predominantly solid tumors; however they can rarely be cystic, being larger in size than solid ones. Cystic pNETs show a higher prevalence among Multipe Endocrine Neoplasia (MEN)-1 patients (Ligneau et al. 2001). The size of pNETs is quite variable, ranging from less than 1 cm to more than 15 cm. The majority of tumors are 1–5 cm, with NF tumors being larger at the time of diagnosis than their functioning counterparts. This is probably the result of delayed diagnosis due to the indolent course of NF tumors, rather than of more aggressive behavior (Falconi et al. 2012).

Table 3 Prognostic factors suggested to correlate to prognosis

Favorable prognosis	Unfavorable prognosis
1 Size <2 cm	1 Size >2 cm
2 Ki-67/MIB-1 <2 %	2 Ki-67/MIB-1 >2 %
3 Mitoses <2 per 10 HPF	3 Mitoses >2 per 10 HPF
4 No sign of angioinvasion	4 Sign of angioinvasion
5 No sign of perineural invasion	5 Sign of perineural invasion
6 No sign of metastases	6 Sign of metastases
7 No penetration through tumor capsule	7 Penetration through tumor capsule
8 Macroscopically radical surgery	8 No macroscopically radical surgery
9 No tumor necrosis	9 Tumor necrosis
10 Insulinoma	10 Functioning tumors (except insulinoma)

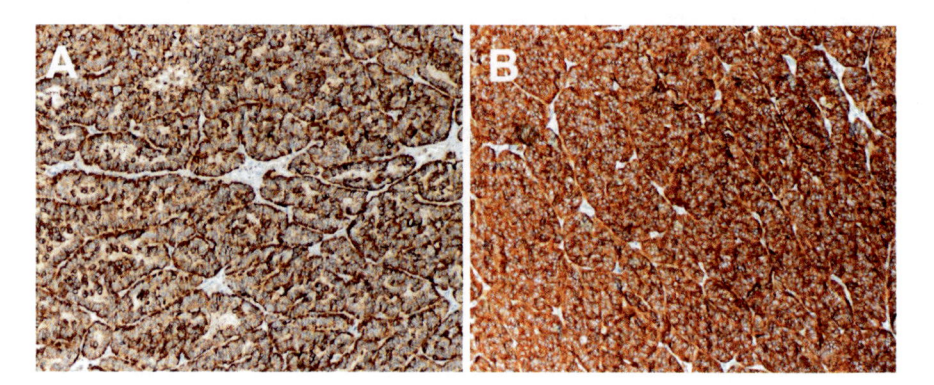

Fig. 1 A well-differentiated pancreatic neuroendocrine tumor immunostained with (**a**) chromogranin A and (**b**) synaptophysin; virtually all neoplastic cells are immunoreactive for both markers

The growth pattern of well-differentiated pNETs can vary (solid, trabecular, glandular, gyriform, tubuloacinar, pseudo rosette, and, rarely, rhabdoid), while different patterns can be observed in the same tumor (Perez-Montiel et al. 2003). Tumor cells are uniform with a centrally located nucleus, occasionally with oncocytic differentiation. In general, the histological growth pattern is not conclusive for the diagnosis, with the exception of amyloid deposits in insulinomas and psammoma bodies in SSomas (Kloppel et al. 2007b). NECs, on the other hand, display apparent atypia and high mitotic rate and/or Ki-67/MIB-1 proliferation index, as well as areas with necroses. In general it is difficult to recognize their neuroendocrine differentiation in the absence of immunohistochemical analysis (Rindi and Kloppel 2004).

The neuroendocrine nature of tumor cells is confirmed immunohistochemically with general neuroendocrine markers such as chromogranin A (CgA) (Fig. 1a) that is expressed in the large dense-core vesicles; synaptophysin (Fig. 1b) which is associated to small vesicles; cell membrane markers, such as the neural cell adhesion molecule (NCAM/CD56); and the cytosolic marker neuron-specific enolase (NSE) (Kloppel et al. 2007b). Further, in the case of functioning tumors, the relevant hormone might be sought by means of immunohistochemistry to confirm the diagnosis. However, this is not mandatory, as functioning pNETs may display weak (or even absent) immunoreactivity to the specific hormone because of its immediate release (Jensen et al. 2012).

According to the WHO guidelines, the diagnostic report should include (Bosman et al. 2010):
- The histological classification of the lesion (as NET or NEC, small or large cell type)
- The grade (G1, G2, or G3)
- The relevant TNM stage (according to ENETs and UICC 2009)
- Optionally, the expression of hormones, transcription factors, or somatostatin receptors

The observation in patients with MEN-1 syndrome that NETs less than 5 mm in diameter coexist with macro-tumors (>5 mm in diameter) has raised the hypothesis of precursor lesions and has been termed micro adenomatosis (Kloppel et al. 2007a). However, such precursor lesions have not been described as yet in sporadic pNETs.

Genetics

Sporadic Pancreatic Neuroendocrine Tumors

In the search of the genetic events in somatic cells that may lead to neoplasia, it is demonstrated that almost a third of pNET patients manifest allelic loss on chromosome 3p, a locus that harbors various tumor-suppressor genes (Chung et al. 1997). In addition, losses of chromosomes 1 and 11q, as well as gains of 9q, occur in tumors less than 2 cm in size, suggesting that they are early events in the development of pNETs (Zhao et al. 2001). Further, loss of heterozygosity (LOH) at chromosome 11q is encountered in functioning pNETs but not in NF ones. By contrast, LOH at chromosome 6q is associated with the development of NF tumors (Rigaud et al. 2001).

Hereditary Pancreatic Neuroendocrine Tumors

Germline mutations can give rise to genetic syndromes that manifest a predilection for NETs and pNETs in particular. Such syndromes are MEN-1, von Hippel–Lindau (vHL), neurofibromatosis (NF)-1, and tuberous sclerosis (TSCL). A summary of the hereditary pNETs is presented in Table 4.

MEN-1 is the result of a germline mutation in MENIN, a tumor-suppressor gene, located on chromosome 11q13. Some 20–60 % of MEN-1 patients have pancreatic lesions when assessed by clinical screening and the incidence increases with

Table 4 Inherited disorders associated with pancreatic neuroendocrine tumors

Syndrome	Associated clinical features	Chromosomal location	Pancreatic neuroendocrine tumor type
Multiple endocrine neoplasia-1	Primary hyperparathyroidism	11q13	Nonfunctioning
	Pituitary tumor		Gastrinoma
	Less commonly		Insulinoma
	Adrenocortical tumor		Various
	Non-medullary thyroid tumor		
Von Hippel–Lindau disease	CNS hemangioblastoma	3p25	Nonfunctioning
	Retinal angioma		Various, including cystic tumors
	Pheochromocytoma (often bilateral)		
	Renal cell carcinoma		
Neurofibromatosis-1	Neurofibromas	17q11.2	Various
	Café au lait spots		
	Axillary or inguinal freckling		
	Bone defects		
	Optic glioma		
	Iris Lisch nodule		
	Duodenal somatostatinoma		
	Gastrointestinal stromal tumor		
Tuberous sclerosis	CNS involvement	9q34 and 16p13	Nonfunctioning
	Epilepsy		
	Subependymal nodule		
	Subependymal giant cell astrocytoma		Insulinoma
	Mental retardation		
	Retinal hamartoma		
	Renal angiomyolipoma		
	Renal cysts		
	Cardiac rhabdomyoma		
	Angiofibroma		
	Lymphangiomyomatosis		

advancing age (3 %, 34 %, and 53 % of patients at age 20, 50, and 80 years, respectively). This figure, however, approaches 100 % in autopsy materials. The majority of tumors are NF, whereas gastrinomas are the most common functioning ones. pNETs usually appear in MEN-1 at an earlier age than the sporadic tumors do, and they represent the major mortality factor among other neoplasias of this syndrome (Tonelli et al. 2011).

The vHL syndrome is a rare autosomal-dominant disease associated with the mutation of a gene located on the short arm of chromosome 3p25.5. The frequency of pancreatic lesions in vHL patients ranges from 17 % to 56 %; however the majority represent benign cysts (Hough et al. 1994).

NF-1 is an inherited autosomal-dominant disorder evoked by mutations of the tumor-suppressor gene NF-1 on chromosome 17, and pNETs are a rare manifestation of this syndrome. TSCL is another autosomal-dominant genetic disorder which results from mutations in the TSCl or TSC2 genes. In rare cases, NF-pNETs or insulinomas can occur (Verhoef et al. 1999).

Specific Tumor Types: Clinical Manifestations

Well-Differentiated Pancreatic Neuroendocrine Tumors

Nonfunctioning Pancreatic Neuroendocrine Tumors

These tumors are the most common pNETs accounting for approximately 60 % in recent studies (Falconi et al. 2012). NF-pNETs can be truly nonfunctional without any capability of secretion (null secreting cells); however, they might also consist of pancreatic polypeptide- (PP-) secreting cells (PPomas), without being related to a distinct clinical syndrome. Due to their silent nature, they are often diagnosed at advanced stages, and 46–73 % of them present liver metastases at the time of diagnosis, while only 14 % have localized disease (Garcia-Carbonero et al. 2010). As a consequence, the age at presentation is advanced and peaks in the sixth and seventh decade of life. Nevertheless, up to 38 % of patients remain asymptomatic and are diagnosed incidentally during abdominal surgery or imaging for an unrelated disease (Zerbi et al. 2011).

Whenever symptoms are present, the most common are abdominal pain (35–78 %), weight loss (20–35 %), and anorexia and nausea (45 %). Less frequent are intra-abdominal hemorrhage (4–20 %), jaundice (17–50 %), or a palpable mass (7–40 %). In rare cases, the activation of an NF tumor to become functioning has been reported (Madura et al. 1997; Phan et al. 1998; Matthews et al. 2000).

Functioning Pancreatic Neuroendocrine Tumors
Insulinoma
Insulinomas are the most common type of functioning pNETs; they have an annual incidence of around four cases per one million people (Tsolakis and Janson 2008). The median age at presentation is 47 years (range 8–82 years), and women seem to be slightly more frequently affected than men (1.5:1) (Service et al. 1991). These patients can have symptoms for several years and present a long history of seeking medical attention before the right diagnosis is set (mean duration varying between 26 and 50 months).

Insulinomas are located almost exclusively in the pancreas and affect all of its parts with equal frequency (Vagefi et al. 2007). The vast majority of insulinomas (90–95 %) is benign at presentation and has an excellent prognosis, with an overall

5-year survival rate of 97 %. The rate of malignancy is less than 10 %; however in such case, especially in the presence of liver metastases, the median survival is less than 2 years according to older data (Soga et al. 1998). In contrast to the sporadic cases that are in the vast majority (91 %), solitary tumors in MEN-1 patients are often multiple (59 %), and they present earlier in life (15–45 years of age) (Norton et al. 2006).

The clinical hallmark of insulinoma are episodes of hypoglycemia characterized by adrenergic symptoms (tachycardia, tremor, sweating) and as glucose levels decrease neurological symptoms (visual disturbances, confusion, comma). These episodes often occur several hours after the last meal (fasting, typically during the night or before breakfast) and can be triggered or aggravated by physical exercise or the omission of meals. Eventually, patients often overeat to compensate and develop obesity (de Herder et al. 2006).

Gastrinoma (Zollinger–Ellison Syndrome)

Gastrinoma is the most common type of malignant functioning pNETs and the second most common functioning pNET in general after insulinoma, with an overall annual incidence between 0.5 and 4.0 per million people. This figure rises to approximately 0.1 % among patients with duodenal ulcer disease. The male to female ratio is 3:2, and the mean age at diagnosis is 48 years (range 8–94 years), with a delay from onset of symptoms of 5.5 years (Anlauf et al. 2006).

Almost 90 % of gastrinomas develop in the anatomical region underlying the right upper abdominal quadrant, which includes the head of the pancreas, the superior and descending portion of the duodenum, and the associated lymph nodes (gastrinoma triangle) (Stabile et al. 1984). Sporadic tumors were thought till recently more common to develop in the pancreas; however advances in endoscopy have revealed an increased incidence of duodenal gastrinomas that now account for 50–88 % in sporadic Zollinger–Ellison syndrome (ZES) patients. At the time of diagnosis, over 60 % of gastrinomas are metastatic, but in general, these are slowly progressive tumors with 10-year survival rate estimated at 90–100 % of patients with radical resection, 46 % with lymph node metastasis, and 40 % in patients with hepatic metastases. Pancreatic gastrinomas are larger in size and present liver metastases (22–35 %) more frequently than their duodenal counterparts (Jensen et al. 2006). Among ZES patients, 20–30 % of the cases present in the context of MEN-1 syndrome, and in such case they develop at an earlier age than sporadic (mean 32–35 years). Consequently, the diagnosis of a gastrinoma should be followed by screening for other components of MEN-1 syndrome (Anlauf et al. 2006).

The clinical symptoms result from gastric acid hypersecretion (ZES) and consist of abdominal pain from developing peptic ulcer disease (PUD) or gastroesophageal reflux disease, as well as diarrhea and malabsorption due to inactivation of pancreatic enzymes (Roy et al. 2000). Atypical or multiple ulcers strongly suggest the diagnosis even though the use of proton pump inhibitors (PPIs) has made this manifestation less frequent than in the past. Moreover, successful therapy with PPIs may mask the diagnosis, which is often delayed (Corleto et al. 2001; Banasch and Schmitz 2007). ZES should be suspected in cases of recurrent, severe, or familial PUD; *H. pylori*-negative PUD without other risk factors (gastrotoxic drugs); PUD and hypergastrinemia concomitantly with prominent gastric folds; or PUD with hypercalcemia. Patients with MEN-1 should be screened for the presence of ZES as gastrinomas are quite frequent in this group (20–30 %) (Gibril et al. 2004).

Rare Functioning Pancreatic Neuroendocrine Tumors
The majority of patients with rare functioning pNETs have advanced disease at the time of diagnosis, and their prognosis is determined mainly by the size of the primary tumor. Five-year survival for these patients is 29–45 % (O'Toole et al. 2006).

Glucagonoma
Glucagonomas represent approximately 5 % of all pNETs and 8–13 % of functioning tumors. A male to female ratio of 0.8:1 with a mean age of 52.5 years (range 11–88) at diagnosis has been reported, while 80 % are already metastatic. The 10-year survival in patients without metastases is 64 % and in those with metastatic disease does not exceed 51 % (Soga and Yakuwa 1998a). Approximately 15 % of patients with glucagonomas have MEN-1. The pancreatic tail is the site where these tumors are more frequently found (Levy-Bohbot et al. 2004).

Glucagon excess is associated with catabolic action, and patients with glucagonomas present a distinct clinical syndrome (glucagonoma syndrome) consisting of diabetes mellitus, normochromic and normocytic anemia, and a predilection to venous thrombosis. The most characteristic clinical finding is necrolytic migratory erythema (NME), described in 65–80 % of the patients (Fig. 2). NME is a pruritic skin rash that usually begins in the groin and perineum, subsequently spreading to the extremities and complicated with secondary infections of the skin by *Candida* and *Staphylococcus aureus*. The cause of NME is still unknown (Kindmark et al. 2007). Interestingly, clinical cases of hyperglucagonemia and pancreatic α-cell hyperplasia/islet cell tumors that lack the manifestations of the relevant clinical syndrome have been reported. This entity has been attributed to inactivating mutations of the human glucagon receptor (Zhou et al. 2009).

VIPoma
These rare pancreatic tumors secreting vasoactive intestinal polypeptide (VIP) represent 3–8 % of all pNETs (O'Toole et al. 2006). Men and women are equally affected and the overall median age at diagnosis is 42 years (range 1–82 years). VIPomas are located mainly in the tail, followed by the head and the body. Apart from pancreatic VIPomas (75–80 %), exopancreatic neurogenic tumors located in the retroperitoneum and in the mediastinum have been reported. Approximately half of all VIPomas are metastatic, the majority affecting the liver. The 5-year survival is estimated at 94.4 % for patients without metastatic disease and 59.6 % for patients with metastases (Soga and Yakuwa 1998b).

Fig. 2 Necrolytic migratory erythema in a patient with glucagonoma

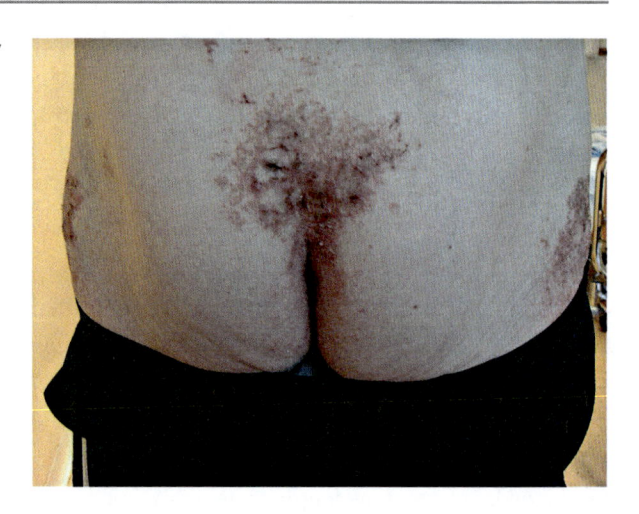

VIP stimulates the secretion and inhibits the absorption of sodium, chloride, potassium, and water from the mucosa of the small bowel. It also stimulates bowel motility. These effects lead to the syndrome of diarrhea/hypokalemia/achlorhydria (watery diarrhea hypokalemia achlorhydria – WDHA) or Verner-Morrison syndrome. Diarrheas are devastating and can range from 0.5 to 15.0 l/day, with consequent dehydration.

Somatostatinoma

Pancreatic SSomas are usually large, with an average diameter of 5 cm, and show a predilection for the head of the pancreas. Duodenal SSomas have also been reported that tend to be smaller (O'Brien et al. 1993). The estimated annual incidence is less than 1 in 10 million people. Men and women are equally affected (Soga and Yakuwa 1999).

The clinical features of the SSoma syndrome are quite insidious and result from the inhibitory properties of somatostatin: hyperglycemia caused by insulin release inhibition, cholelithiasis as a result of decreased cholecystokinin release, and consequently diminished gallbladder contractility. Steatorrhea is also observed due to inhibition of pancreatic enzyme/bicarbonate secretion and diminished intestinal absorption. Finally, hypochlorhydria/achlorhydria is often caused by the inhibitory effect of SS on gastrin release and acid secretion (O'Toole et al. 2006).

Rare Ectopic Hormone Syndromes

Due to the pluripotent secretory capacity of neuroendocrine cells, pNETs may secrete substances not related to the endocrine pancreas (ectopic secretion), leading to distinct clinical syndromes (Kaltsas et al. 2010). The clinical syndrome may herald the presence of the neoplasm and lead to early diagnosis; however, it may sometimes dominate the clinical picture and confuse the clinician. About 10–20 % of total cases presenting with Cushing's syndrome (CS) are due to ectopic secretion of ACTH (Newell-Price et al. 2006). Moreover, CS occurs in 5 % of cases with sporadic ZES (Maton et al. 1986). Similarly, acromegaly has been reported in patients with pNETs, being the second most frequent cause of paraneoplastic syndrome related to acromegaly after lung carcinoids (Gola et al. 2006). In this setting, the humoral syndrome is mainly related to GHRH hypersecretion rather than to GH itself (Biermasz et al. 2007). There are also case reports in the literature of humoral hypercalcemia of malignancy due to PTH-related peptide and syndrome of inappropriate antidiuretic hormone secretion (Kanakis et al. 2012).

Neuroendocrine Carcinomas

NECs are highly malignant epithelial neoplasms that constitute approximately 2–3 % of pNETs. By means of morphology, they can be divided in small- and large-cell variants based on the criteria used for the classification of the pulmonary counterparts (Travis et al. 2004). The tumors are already large at diagnosis, with an average diameter of 4 cm, and present frequently metastases involving the liver and regional lymph nodes, as well as distant organs. The prognosis is poor, with survival times ranging from 1 month to up to approximately 1 year (Nilsson et al. 2006).

Biomarkers

Several serum tumor markers are used for the diagnosis and management of pNETs, and they are divided in general (nonspecific that are found in all NETs) and tumor-specific markers (O'Toole et al. 2009; Oberg 2011).

General Biomarkers

The most widely used general markers of neuroendocrine differentiation are CgA, PP, and NSE. CgA is currently the most useful general biomarker available for the diagnosis of pNETs as it is co-secreted by the majority of neuroendocrine cells and persists after malignant transformation. The overall sensitivity of the assay ranges from 67 % to 93 % (Stridsberg et al. 2003). However, there are conditions in which CgA measurements could be misleading: the most common being those associated with achlorhydria. Chronic atrophic gastritis (type A), and treatment with antisecretory medications, especially PPIs cause falsely elevated CgA levels. Therefore, proper discontinuation of such agents should be applied prior to CgA assessment (Marotta et al. 2012). Apart from diagnosis, CgA serves as a marker for the evaluation of the response to therapy of patients with liver metastases as it correlates well with tumor burden (Massironi et al. 2010).

PP is a 36 amino acid linear oligopeptide mostly useful in the diagnosis of NF-pNETs. It is often elevated in metastatic disease and has an overall sensitivity of about 50–80 % (Lonovics et al. 1981). The diagnostic accuracy is further improved by combining serum CgA and PP levels offering an increase in sensitivity to 90 %

(Walter et al. 2012). In order to maximize the sensitivity of PP assay in the diagnosis of pNETs, a specific meal stimulatory test (mixed meal) has been developed; however its value has been challenged in recent guidelines (Langer et al. 2001; O'Toole et al. 2009). The assessment of PP may be particularly useful for early detection of pancreatic tumors in MEN-1 patients. NSE is elevated in 30–50 % of patients with pNETs, particularly those of poor differentiation. In such patients, NSE might be elevated despite normal CgA levels (Nobels et al. 1997).

Tumor-Specific Biomarkers

To establish the diagnosis of functioning pNETs, inappropriate elevation of the relevant, tumor-specific marker (i.e., gastrin in ZES) should be sought, as well as clinical/laboratory evidence of the relevant clinical syndrome (e.g., high gastric output).

Insulinoma
In order to diagnose insulinomas, the presence of endogenous hyperinsulinemia (EH) (e.g., inappropriately high levels of insulin) during hypoglycemia should be demonstrated (Jensen et al. 2012). This can be established using the following criteria:
- Blood glucose level ≤ 2.2 mmol/l (≤ 40 mg/dl)
- Concomitant serum insulin level ≥ 6 μU/l (≥ 36 pmol/l) or ≥ 3 U/l by ICMA
- Plasma/serum C-peptide level ≥ 200 pmol/l
- Serum proinsulin level of at least 5 pmol/l
- Plasma β-hydroxybutyrate levels ≤ 2.7 mmol/l
- Absence of sulfonylurea (metabolites) in patient's plasma and/or urine

These might be documented during a spontaneous episode, providing that the triad suggested by Whipple is fulfilled: (1) symptoms of hypoglycemia, (2) blood glucose ≤ 2.2 mmol/l (≤ 40 mg/dl), and (3) relief of symptoms with administration of glucose (Service 1995). The diagnosis is supported by the absence of plasma/urine ketones during fasting as a consequence of suppression of β-oxidation by endogenous hyperinsulinemia. Furthermore, exogenous insulin administration or sulfonylurea abuse should be excluded. The work-up is complete with the measurement of proinsulin that might reveal insulinomas that secrete immature forms of insulin due to defective hormone processing (Kao et al. 1994).

In case of difficulty to document such episodes, the patient should undergo fasting for 72 h. This test is considered the gold standard in the diagnosis of insulinomas with sensitivity reaching 100 %. The test should be ended when the patient develops symptoms and plasma glucose levels are ≤ 2.2 mmol/l (≤ 40 mg/dl) or when the 72 h are completed (Service and Natt 2000). Usually, insulinoma patients develop hypoglycemia early after fasting (one third of patients at 12 h and 80 % within 24 h). The assessment of the results relies on the criteria mentioned for spontaneous hypoglycemia. If the results are still equivocal, 1 mg of glucagon is administered immediately after the 72-h fasting. In healthy patients glucose levels after glucagon administration should not rise more than 25 mg/dl above basal due to depletion of hepatic glycogen stores (Hoff and Vassilopoulou-Sellin 1998). An alternative diagnostic test is the calcium infusion test, during which hypoglycemia is achieved 2 h after intravenous calcium infusion.

Gastrinoma

The diagnosis of gastrinoma is set by elevated fasting gastrin levels in the setting of high gastric acid output (Arnold 2007). The criteria used to establish the diagnosis are:
- Gastric juice pH <2.0
- Fasting serum gastrin (FSG) > tenfold the upper limit of normal (ULN)

In cases of marginal FSG values (two to tenfold ULN), performing a secretin test and assessment of basal acid output (BAO) is suggested. During the test, 2 U/kg secretin is given as an intravenous bolus after overnight fasting, and serum gastrin is measured before and 2, 5, 10, 15, 20, and 30 min after infusion. A delta gastrin >120 pg/ml any time during the test is indicative of the presence of a gastrinoma with a sensitivity of 94 %. If the secretin test is negative but the suspicion for the presence of ZES remains high, the calcium provocation test may be helpful as it may be positive in 5–10 % of such cases (Frucht et al. 1989). An alternative diagnostic test is the glucagon provocative test, during which glucagon (20 mg/kg) is given intravenously, followed by 20 mg/kg/h for the next 30 min. A gastrinoma is suggested if plasma gastrin levels peak within 10 min after glucagon administration, with an increase of greater than 200 pg/ml and greater than 35 % of the basal value (Shibata et al. 2012). Regarding BAO, a rate of more than 15 mmol/h is highly suggestive of ZES (O'Toole et al. 2009).

Elevated FSG is not specific for ZES, as it can be encountered in other conditions with hyperchlorhydria as in *H. pylori*-positive gastritis, gastric outlet obstruction, renal insufficiency, and antral G-cell syndromes. Moreover, elevated FSG can also be found in states of achlorhydria, such as atrophic gastritis, and chronic PPI use (Rehfeld et al. 2012). Consequently, PPIs should be interrupted at least 1 week prior to FSG testing and replaced for this period by histamine receptor 2 (H2) blockers, which in turn have to be discontinued 48 h before the test (Banasch and Schmitz 2007).

Rare Functioning Tumors

In patients with rare functioning tumors, the diagnosis can be established by demonstrating elevated levels of the relevant hormone combined with general markers of neuroendocrine differentiation (O'Toole et al. 2006). In glucagonomas, fasting glucagon levels are usually increased 10– 20-fold. In the rare case of equivocal values, tolbutamide or arginine stimulation test may be used to establish the diagnosis. In VIPomas, levels of VIP exceeding 60 pmol/L are diagnostic. Similarly, diagnosis of SSomas should be based on high SS levels; however, SS assays are cumbersome and not routinely available. The diagnosis of ectopic Cushing's syndrome comprises initially the documentation of hypercortisolemia and afterward the detection of its source. Once the diagnosis is established, ACTH or CRH may serve as follow-up markers (O'Toole et al. 2009).

Localization

The morphology and imaging properties of different pNETs vary according to the specific tumor type. Thus, there is no general consensus regarding the most appropriate imaging modality for the detection of pNETs. ENETS has recently proposed an algorithm for the localization and staging of NF-pNETs implicating abdominal ultrasound (US), abdominal computerized tomography (CT), magnetic resonance imaging (MRI), somatostatin receptor scintigraphy (SRS), positron emission tomography (PET), and endoscopic/intraoperative US (EUS/IOUS) (Sundin et al. 2009; Falconi et al. 2012). The technique which proves to depict better the individual tumor should be preferred for follow-up.

Ultrasonography

Abdominal US (B-mode) is a widely available, low-cost modality, which however has a low detection rate for pNETs that do not exceed 40 %. Besides, it is an operator-sensitive modality. Several techniques have been developed to improve the diagnostic accuracy of US (Sundin et al. 2009):

Contrast-enhanced ultrasonography (CEUS) allows the depiction of the micro-/macrovasculature and enhancement patterns of pNETs. A correlation between CEUS enhancement pattern and the Ki67 index has been reported (D'Onofrio et al. 2004; Malago et al. 2009).

EUS has enhanced sensitivity (overall >90 %), especially for small pancreatic lesions (even smaller than 0.5 cm), due to the proximity between the transducer and the pancreas (Fig. 3). pNETs are generally depicted as round lesions with well-defined margins that tend to be hypoechoic, but they may also be isoechoic or anechoic. The vascular nature of these tumors can be appreciated using the color Doppler. In addition, EUS can estimate the distance between the tumor and the pancreatic duct, which is essential when deciding if tumor enucleation is feasible.

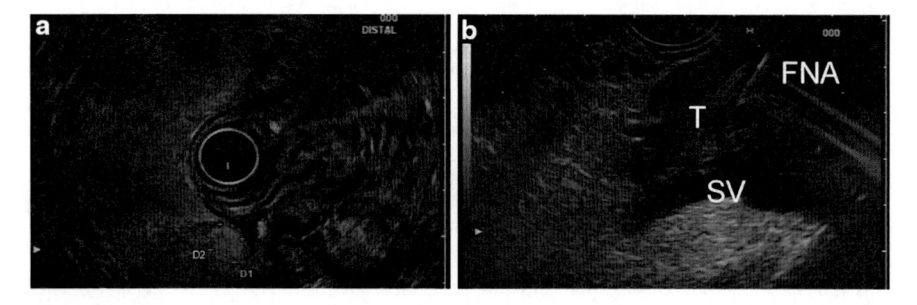

Fig. 3 (a) Endoscopic ultrasound of an insulinoma located at the tail of the pancreas. The tumor is isoechoic and measures D1:9 × D2:7 mm. (b) Fine needle aspiration (FNA) under endoscopic ultrasound guidance of a nonfunctioning pancreatic neuroendocrine tumor (T). The tumor is anechoic and in close proximity to the splenic vein (SV)

Thus, EUS is the modality of choice for small insulinomas that cannot be detected otherwise.

Furthermore, preoperative EUS-guided injection of India ink has been shown to aid in intraoperative localization of insulinomas. Nevertheless, EUS's accuracy is decreased for tumors located at the pancreatic tail (McLean and Fairclough 2005).

IOUS, when combined with palpation of the pancreas, has sensitivity rates that reach 98 %. This technique is essential in MEN-1 patients as it allows the detection of small multiple pNETs or duodenal tumors and depending on the findings may alter the surgical procedure (Norton et al. 1988; Jensen et al. 2008).

Computerized Tomography and Magnetic Resonance Imaging

Abdominal CT usually depicts pNETs as lesions isodense to the surrounding pancreatic parenchyma and can be distinguished by the strong enhancement they show after the infusion of contrast medium (Nino-Murcia et al. 2003). This is a result of the increased vascularity that characterizes such lesions and is better observed during the arterial phase images. The presence of calcification, invasion of adjacent organs, and central necrosis all suggest malignancy. In order to reveal possible liver metastases, the combination of arterial phase with portal venous phase is recommended (Ichikawa et al. 2000). The sensitivity and specificity of CT scan in the diagnosis of pNETs are 73 % and 96 %, respectively (Sundin et al. 2009).

MRI has similar sensitivity rates (85–94 %). pNETs show low signal intensity on T1-weighted images and high signal on T2-weighted images and also exhibit strong enhancement after infusion of paramagnetic contrasts. Fat-suppressed T1-weighted images are preferred for detecting pancreatic lesions, whereas T2-weighted images are useful in distinguishing pNETs from adenocarcinomas that have low signal (Owen et al. 2001).

Radionuclear Imaging

SRS is a functional imaging technique, using the ability of somatostatin analogs (SSAs) to bind to somatostatin receptor (SSTR) subtypes. Given the fact that 80–90 % of NETs express SSTRs (Zamora et al. 2010), the sensitivity and specificity rates of SRS for pNETs are similarly high (90 % and 80 %, respectively) (Lamberts et al. 1992). An exception is benign insulinomas that do not sufficiently express SSTR2, the main target of SRS, and the corresponding sensitivity rates fall to 50–60 % (Portela-Gomes et al. 2000). Interestingly, insulinomas have been shown to overexpress glucagon-like peptide (GLP) receptors; therefore, radiolabeled GLP-1 analogs could serve for depicting such tumors (Christ et al. 2009). Eventually, SRS is the modality of choice for staging of pNETs, especially for the detection of extrahepatic metastases, since it allows whole-body imaging. Routine

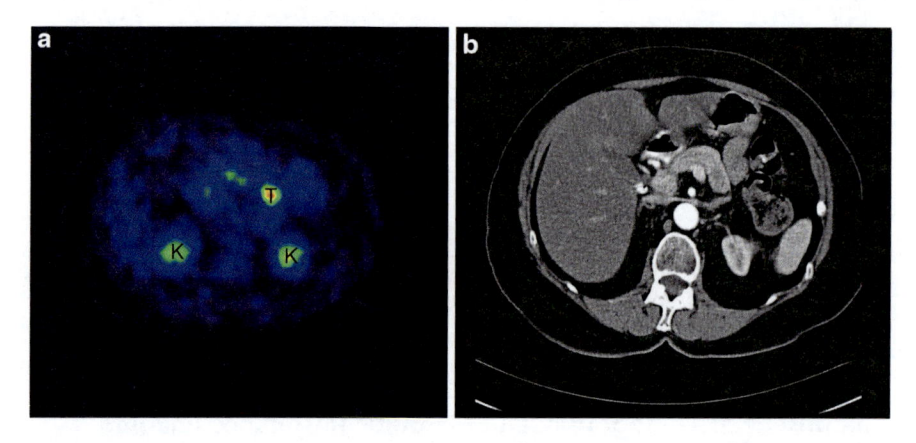

Fig. 4 (a) Positron emission tomography using ^{11}C-5-hydroxy-L-tryptophan as a tracer (transaxial view). Distinct uptake by the tumor (T) localized in cauda pancreatis and normal uptake by the kidneys (K) are evident. (b) The corresponding computerized tomography scan (arterial phase) did not depict the tumor

SRS should not be employed if other imaging studies for NETs are negative. The most commonly used tracer is 111In-DTPA-octreotide.

PET is a more recent technique with better spatial resolution than scintigraphy that allows the quantification of tracer uptake. Several specific tracers have been used for the detection of pNETs, among which the 68Ga-labeled SSA DOTA-D-Phe1-Tyr3-octreotide (DOTATOC) is the most widely employed (Buchmann et al. 2007). ^{11}C-5-hydroxy-L-tryptophan (5-HTP) (Fig. 4) seems to have the optimal properties for staging pNETs; however it is not widely available (Orlefors et al. 2005). Standard PET with ^{18}F-labeled deoxyglucose (FDG) depicts tumors with rapid proliferation; thus it may not be accurate in detecting well-differentiated pNETs. However, it is useful in the detection of aggressive poorly differentiated tumors with a proliferation index around or greater than 15 % (Pasquali et al. 1998).

Angiography

This is an invasive technique that allows visualization of pNETs due to their increased vascularity, especially during the early arterial phase (Jackson 2005). However, it has a detection rate of approximately 60 %, being superseded by noninvasive cross-sectional imaging techniques (MRI/CT). Thus, the main utility of this technique is to determine the localization of multiple gastrinomas/insulinomas in MEN-1 patients by combining angiography with hepatic venous sampling for the relevant hormone after intra-arterial secretin/calcium administration. In such case, the diagnostic accuracy reaches 88–100 % (Doppman et al. 1995).

Diagnostic Procedures

Tissue diagnosis is the cornerstone of designing the clinical management strategy and predicting the outcome of an underlying malignancy. Several procedures of tissue sampling have been developed and applied in pancreatic tumors. In general, if the pancreatic lesion is clearly resectable and the patient is fit for surgery, it is suggested not to perform an attempt to obtain neoplastic tissue preoperatively. In that way the possible risk of needle tract seeding is avoided. Imaging-guided fine needle aspiration and core needle biopsy CT and abdominal US may be used to guide percutaneous pancreatic fine needle aspiration (FNA). FNA or core needle biopsy (CNB) reaches a sensitivity and accuracy of 93.9 % and 94.4 %, respectively (Paulsen et al. 2006). CNB may have an advantage over FNA, in which false-positive results have been reported; however it has been associated with higher complication rates. The initial enthusiasm over the use of percutaneous FNA has been decreased as a result of needle tract seeding and the development of EUS FNA.

EUS-Guided Fine Needle Aspiration and Core Needle Biopsy

With the guidance of EUS, it is possible to obtain cytologic material from pancreatic or liver lesions via FNA (Fig. 3b), and the results have been shown to be superior to conventional computerized tomography-guided FNA (Horwhat et al. 2006; Chatzipantelis et al. 2008). EUS FNA is performed using a linear array echoendoscope, while the patient is under conscious sedation and cardiorespiratory monitoring. Apart from diagnosis, FNA can provide prognostic information by the estimation of the Ki-67 proliferation index; however an experienced cytopathologist may be needed and the results are still equivocal (Piani et al. 2008; Kaklamanos et al. 2011). The diagnostic yield of EUS may be increased by employing CNB instead of FNA; however the safety profile of CNB is yet to be established (Thomas et al. 2009).

Treatment

Surgery

Complete surgical resection represents the only potential curative therapy for pNETs, providing that the diagnosis has been set at early stages (Hill et al. 2009). The extent of the surgery depends on the biological behavior (grading), the metastatic extent (staging), the secretory status (functioning or not), and the possible presence of the tumor in the context of a hereditary syndrome.

The type of resection of the primary pNETs depends largely on the location of the tumor, the tumor size, as well as the histological type. For any tumor larger than

2 cm, a standard pancreatic resection should be preferred: pancreaticoduodenectomy for tumors located at the head and distal pancreatectomy with or without spleen preservation for lesions of the body and tail (Falconi et al. 2010). For smaller well-differentiated tumors with clear margins, parenchyma-sparing procedures may be an option such as middle pancreatectomy for tumors of the body. For small tumors located no closer than 2–3 mm from the pancreatic duct, especially insulinomas, enucleation might be considered.

Laparoscopic removal of pNETs, either by resection or enucleation, is a surgical option that is associated with less morbidity and mortality than standard resections. The overall surgical morbidity rate ranges from 31 % to 42 %, with pancreatic fistula being the most common complication (Fernández-Cruz et al. 2008). Although laparoscopic tumor enucleation is a parenchyma-sparing procedure, it is associated with a higher rate of development of pancreatic fistulas. It is thus recommended that laparoscopic tumor enucleation may be reserved for tumors of the pancreatic head that are considered noninvasive, whereas tumors of the pancreatic body and/or tail should be treated with laparoscopic resection (Haugvik et al. 2013). The results between G1 and G2 tumors are comparable; however it should be noted that in case of malignant tumors, laparoscopic techniques cannot ensure complete lymphadenectomy and negative surgical margins (Al-Taan et al. 2010).

Gastrinomas should undergo resection even in the case of metastatic tumors as debulking is helpful for the control of hormone secretion (Lorenz and Dralle 2007). The general properties mentioned above for pNET surgery should be followed. Additionally, duodenotomy with transillumination should be routinely performed to detect small duodenal tumors as liver exploration and regional lymphadenectomy to exclude abdominal metastases (Norton et al. 2004). Consequently, laparoscopic surgery is not indicated for this type of tumors.

Insulinomas, on the other hand, due to their excellent prognosis are treated laparoscopically with parenchyma-sparing techniques (Richards et al. 2011). No standard lymphadenectomy is recommended. In case the tumor cannot be localized preoperatively, a laparotomy is indicated as well as pancreas exploration by both palpation and IOUS (Espana-Gomez et al. 2009). If this is also negative, intraoperative insulin sampling may be helpful (Gimm et al. 2007).

MEN-1 patients represent a distinct subgroup of patients with pNETs whose pancreas may harbor multiple tumors; therefore, careful intraoperative exploration of the pancreas is mandatory (Anlauf et al. 2006). Regarding NF-pNETs, all tumors accompanied by metastases, larger than 2 cm, and presenting a yearly increased size >0.5 cm should be resected (Triponez et al. 2006). When multiple tumors are found throughout the pancreas, a reasonable approach is distal subtotal pancreatectomy and enucleation of the tumors located in the head. Tumors smaller than 2 cm show a malignant potential less than 6 %, and their excision is debated; an intensive follow-up is recommended instead (Kouvaraki et al. 2006). Regarding gastrinomas, pancreaticoduodenectomy is the treatment of choice; otherwise the risk of recurrence is high (Fendrich et al. 2007). If insulin hypersecretion is detected in association with multiple pancreatic tumors, it is mandatory to define which of the tumors are insulinomas (Norton et al. 2006). This can be obtained preoperatively with venous sampling following arterial calcium stimulation or with intraoperative insulin sampling. Tumor resection follows the gradient indicated by these procedures (Doppman et al. 1995; Gimm et al. 2007).

In the setting of locally advanced or metastatic disease, the minimal requirements of the surgery include the resection of the primary tumor, the regional lymph nodes, and/or regional and distant metastases, providing that at least 90 % of the disease is resectable (Joseph et al. 2011). This is particularly important in the case of functioning tumors since reducing the tumor burden ameliorates the hormonal syndrome (Jensen et al. 2012). On the contrary, debulking is debated in the case of NF tumors, unless surgery is indicated for compassionate reasons to treat acute bleeding, obstruction, or perforation of the adjacent organs. In all cases a standard pancreatectomy is performed. Conditions that may exclude a patient from surgery are the invasion of portal vein and the entrapment of superior mesenteric artery (Boudreaux 2011).

Particularly for metastatic liver disease, ENETS has proposed guidelines based on the magnitude of liver involvement (Steinmuller et al. 2008): for localized disease involving one or two contiguous segments, an anatomic resection is adequate, whereas for "complex" disease where the contralateral lobe is involved, a multistep approach is recommended. Initially, the primary tumor is resected along with part of liver metastases, and a second operation is performed after adequate restoration of the liver parenchyma has occurred (Kianmanesh et al. 2008). Patients with well-differentiated (G1 or G2) tumors when treated with hepatic resection exhibit 5-year survival rates of 47–76 % compared to 30–40 % of untreated patients, providing that no extra-abdominal metastases are present (Sarmiento et al. 2003). The most advanced stage of liver involvement is the "diffuse" pattern, where more than 75 % of the liver, including both lobes, is affected. In such case surgery is not recommended and other cytoreductive therapies should be sought (Steinmuller et al. 2008).

Cytoreductive Therapies of Liver Disease

Radiofrequency Ablation

In patients with unresectable liver disease, radiofrequency ablation (RFA) is an option, providing that the lesions do not exceed 5 cm in diameter (Gillams et al. 2005). The heat produced locally by high-frequency alternative current destroys malignant cells and the transducer can be inserted percutaneously or laparoscopically, depending on the lesions' location. Furthermore, RFA may serve as neoadjuvant therapy, in order to convert an unresectable disease into a resectable one (Ahlman et al. 2000).

Embolization, Chemoembolization, and Radioembolization

Selective embolization of branches of the hepatic arteries with particles (microspheres) is an alternative ablation modality for patients with unresectable liver disease, particularly if the lesions are not susceptible to RFA. This technique is based on the property that most of the tumor vasculature originates from the hepatic arteries, whereas the normal parenchyma is irrigated mainly from the portal vein system (Gupta et al. 2005). Nevertheless, liver function enzymes should be monitored closely after the embolization, as necrosis of great areas of the liver may occur.

Chemoembolization (transarterial chemoembolization/TACE) is a variation of this technique where chemotherapeutic drugs (5-FU, doxorubicin, and mitomycin C) are used instead of particles (Drougas et al. 1998). There are no data comparing neither the efficacy of transarterial embolization and TACE in pNETs nor guidelines regarding the regiment of drugs and dosages administered. The most recent achievement in this type of therapy is radioembolization with the use of radioactive yttrium-90 microspheres, a β-emitter with long-range tissue penetration of up to 11 mm which is as effective as and potentially less toxic than TACE (Kalinowski et al. 2009).

Liver Transplantation

An alternative option in patients with diffuse liver involvement may be orthotopic liver transplantation (OLT); however strict criteria should be set as high recurrence rates (up to 76 %) have been reported (Lehnert 1998). Moreover, the use of immunosuppressive drugs required after OLT may even accelerate the development of new metastases at the transplanted organ or elsewhere. Thus, patients younger than 50 year without extra-abdominal disease are generally preferred, while some centers may not accept patients with liver involvement exceeding 50 % (Olausson et al. 2002). Additionally, patients with poorly differentiated tumors are not good candidates due to their dismal prognosis. A proliferation index of Ki-67 less than 5 % is suggested (Rosenau et al. 2002).

Medical Treatment

Biotherapy

SSAs and interferon α (IFNα) have been both used successfully in the control of symptoms in functioning pNETs (Oberg et al. 2010).

Currently available SSAs (octreotide and lanreotide) target SSTRs 2, 3, and 5. Octreotide controls diarrhea in over 75 % of patients with VIPomas and reduces VIP concentrations in 80 % of the patients. Similarly it cures NME in 30 % of cases and reduces circulating glucagon in 80 % (Maton 1993). The corresponding rates for biochemical response in insulinomas are somewhat lower, since these tumors do not express SSTR2 avidly (Vezzosi et al. 2005). Interestingly, the symptomatic relief caused by SSAs is not always accompanied by a reduction in circulating hormone levels, implying a direct effect of these compounds on the peripheral target organ. However, in some cases, an escape from the initial response occurs that is attributed to tachyphylaxis and managed by increasing the therapeutic dose (Oberg et al. 2010).

Regarding metastatic NF-pNETs, a single center phase 4 study using octreotide acetate LAR as a first-line treatment resulted in stable disease in 38 % of the patients after a median follow-up of 49.5 months. Sixty-two percent of the patients developed disease progression after a median follow-up of 18 months (Butturini et al. 2006). The authors concluded that a Ki67 index >5 % may justify more aggressive therapy without waiting for radiological progression of the disease. Apart from their antisecretory action, SSAs were also considered to have an antiproliferative effect; however hitherto this has been demonstrated only for well-differentiated small intestine NETs (PROMID study) (Rinke et al. 2009). This benefit was mostly evident in patients who had less than 10 % tumor liver disease and in those in whom the primary tumor was removed. Moreover, a novel multivalent SSA has been developed, SOM 230 (pasireotide) that has high affinity for all SSTRs except SSTR4 and has been shown to control the symptoms of carcinoid syndrome in 25 % of patients resistant to octreotide (Schmid and Schoeffter 2004). A trial assessing the antiproliferative effects of SOM-230 in gastroenteropancreatic NETs is currently conducted. Furthermore, an ongoing phase 3 study of lanreotide (NCT00353496) in patients with NF-NETs is enrolling patients with pNETs and may provide more information about the antitumor efficacy of SSAs in pNETs.

IFNa is an alternative therapeutic choice that has been shown to elicit an objective biochemical response in 77 %, symptomatic response in 40–60 %, and a reduction of tumor size in 10–15 % of patients with advanced pNETs (Fazio et al. 2007). It can also be added to chemotherapy regiments; however, it seems to be less effective than SSAs and with more side effects.

Chemotherapy

Patients with inoperable metastatic disease and G2 tumors are candidates for systemic chemotherapy. Till recently, the only licensed chemotherapy for well-differentiated pNETs was streptozotocin (STZ) in combination with doxorubicin (DOX) and/or 5-fluorouracil (5-FU), based on the results reported by Moertel et al. (1980). In more recent studies, the combination of STZ and 5-FU proved to be more effective than STZ and DOX with less serious adverse effects (Moertel et al. 1992). The response rates of the above combinations range from 30 % to 40 %. Regarding G3 tumors, the therapy is similar to that used in small-cell lung cancer, based on the combination of etoposide (VpI6) with cisplatin (CDDP) (Moertel et al. 1991).

In addition to registered therapies, several other compounds have been used in the treatment of pNETs either as monotherapy or as combinations (Eriksson 2010). Temozolomide (TMZ), the active metabolite of dacarbazine taken orally, was reported to elicit a partial response of 8 % and stabilize the disease in 67 % of patients with pNETs resistant to other chemotherapeutic regimens (Ekeblad et al. 2007). In this study the expression of O-6-methylguanine DNA methyltransferase (MGMT) was a negative predictive factor for the response to therapy. In another study of TMZ in patients with progressive pNETs, responses up to 53 % were reported; however, response was not related to MGMT expression

(Kulke et al. 2006). Based on in vitro data, TMZ has been used in combinations with capecitabine and thalidomide, achieving impressive response rates as high as 45 % and 74 %, respectively (Strosberg et al. 2010).

Novel Molecular Therapies

Bevacizumab is a monoclonal antibody that targets vascular endothelial growth factor and is used in combination with other chemotherapeutic agents in the treatment of pNETs (Yao 2007). Combined with TMZ partial response up to 24 % has been reported, whereas a randomized phase III trial comparing octreotide plus IFNα or bevacizumab in advanced gastroenteropancreatic NETs is ongoing (Yao et al. 2008b).

Sunitinib is a tyrosine kinase inhibitor showing antiproliferative and antiangiogenic properties. It was recently licensed for the treatment of advanced pNETs based on the results of a phase III study (sunitinib 37.5 mg/d vs. placebo) that demonstrated a significantly higher progression-free survival for the patients treated with sunitinib (Raymond et al. 2011). Regarding the safety profile of this drug, grade 3 and 4 neutropenia, hypertension, hand-foot syndrome, abdominal pain, diarrhea, and fatigue have been reported. Nevertheless, as yet sunitinib should not be considered as a first-line therapy, but only in selected cases.

An alternative treatment option for pNETs progressing on conventional regimens is everolimus, an inhibitor of the mammalian target of rapamycin (mTOR) pathway. The RADIANT-3 study, a phase III randomized trial of octreotide plus everolimus 10 mg or placebo in patient with metastatic pNETs, showed promising results with increased progression-free survival and relatively minor side effects (Yao et al. 2011). These results were equally positive in treatment naïve, as well as in patients already treated with other agents. Thus everolimus could be used as first-line therapy in selected cases.

Radiolabeled Somatostatin Analogs and Peptide Receptor Radionuclide Therapy

Radiolabeled SSAs use high-energy β-emitters including indium 111, yttrium 90, and lutetium 177 (a combined β and γ emitter) in order to stop the growth of SSTR2 baring tumor cells (Kaltsas et al. 2004a; Kwekkeboom et al. 2005; Kalinowski et al. 2009). Thus, the expression of SSTR2 should be sought by means of SRS, before the use of PPRT. PRRT is applicable for both functioning and NF-pNETs, and the most promising results refer to 177Lu-DOTA0,Tyr3 for which partial remission rates up to 33 % have been reported. The most important side effects are related to hematological and renal toxicity (Kwekkeboom et al. 2008). Nevertheless, PRRT should not be used as first-line therapy, but after failure of medical therapy.

External-Beam Radiotherapy

External-beam radiotherapy treatment has been considered ineffective for pNETs, except for patients with bone and brain metastases (Alexandraki and Kaltsas 2012).

Supportive Therapy

Occasionally, patients with advanced disease, particularly with functioning tumors, become refractory to the effect of previously mentioned treatment alternatives. In such cases the administration of supportive therapy is mandatory.

In patients with insulinomas, frequent small meals can be used for small periods (e.g., preoperatively) to prevent prolonged fasting. If hypoglycemia symptoms are more frequent, diazoxide that inhibits the release of insulin may be helpful. In refractory cases glucocorticoids and intravenous infusion of glucose may support to ameliorate hypoglycemia (Service 1995). In case of malignant insulinomas, mTOR inhibitors (everolimus) may be an option as besides their antitumoral action they have been shown to impede insulin secretion and thus control hypoglycemia (Kulke et al. 2009).

Regarding gastrinomas, the employment of PPIs in the therapy of ZES has almost changed the natural history of the disease, as they are very effective in the treatment of PUD, as yet no resistance or tachyphylaxis has been reported and the safety profile is quite favorable (Banasch and Schmitz 2007). In MEN1/ZES patients, the correction of primary hyperparathyroidism can ameliorate gastrin secretion and facilitate the control of ZES.

In patients with VIPomas, fluid and electrolytes should be monitored and replaced appropriately due to gross intestinal losses. Medical agents reducing intestinal motility such as loperamide and in advanced stages morphine may also be helpful. Glucagonoma patients may suffer from hyperglycemia and antidiabetic drugs should be then employed (Jensen et al. 2012). Patients with CS should promptly receive adrenostatic (ketoconazole, metyrapone) or adrenolytic agents (mitotane). Newer agents, such as the novel SSA pasireotide and glucocorticoid receptor antagonists, like mifepristone, have shown promising results. In refractory metastatic disease, bilateral adrenalectomy should be considered (Feelders et al. 2010).

Treatment Algorithm

Since the primary treatment goal in pNETs is the eradication of the neoplastic cells, surgery is still the first-line therapy. In patients who present metastases and when curative surgery is not feasible, debulking surgery with additional measures is suggested, in order to achieve symptom relief and suppress further tumor progression. If the disease is confined locoregionally and involves a limited part of the liver, cytoreductive therapies like RFA and TACE are usually sufficient to reduce tumor burden. By contrast, in patients presenting disseminated disease or when patients are unfit for surgery, medical management with systematically administered agents is required. SSAs or chemotherapy represents the first-line medical treatment. In cases of metastatic well-differentiated pNETs (particularly those with a Ki67 proliferation index <5 %), the tumors may respond to biotherapy (usually disease stabilization) with SSAs or IFNα. Biotherapy can also improve the symptoms of secretory syndromes. Using the same criteria as for SSA treatment, systemic chemotherapy is used, which may be combined with biotherapy if needed (e.g., functioning tumors). If the above regimens are not effective, a treatment among everolimus, receptor tyrosine kinase inhibitors, other anti-angiogenic factors, or PRRT should be sought. In patients with poorly differentiated tumors, surgical intervention is usually obsolete and cytotoxic chemotherapy should be the initial choice. The above algorithm is usually adjusted depending on the expertise and possible available treatment in the respective NET center.

Cross-References

▸ Basement Membrane in Pancreatic Islet Function
▸ Microscopic Anatomy of the Human Islet of Langerhans
▸ Regulation of Pancreatic Islet Formation
▸ Stem Cells in Pancreatic Islets

References

Ahlman H, Wangberg B et al (2000) Interventional treatment of gastrointestinal neuroendocrine tumours. Digestion 62(Suppl 1):59–68

Alexandraki KI, Kaltsas G (2012) Gastroenteropancreatic neuroendocrine tumors: new insights in the diagnosis and therapy. Endocrine 41(1):40–52

Al-Taan OS, Stephenson JA et al (2010) Laparoscopic pancreatic surgery: a review of present results and future prospects. HPB (Oxford) 12(4):239–243

Anlauf M, Garbrecht N et al (2006) Sporadic versus hereditary gastrinomas of the duodenum and pancreas: distinct clinico-pathological and epidemiological features. World J Gastroenterol 12 (34):5440–5446

Arnold R (2007) Diagnosis and differential diagnosis of hypergastrinemia. Wien Klin Wochenschr 119(19–20):564–569

Banasch M, Schmitz F (2007) Diagnosis and treatment of gastrinoma in the era of proton pump inhibitors. Wien Klin Wochenschr 119(19–20):573–578

Biermasz NR, Smit JW et al (2007) Acromegaly caused by growth hormone-releasing hormone-producing tumors: long-term observational studies in three patients. Pituitary 10(3):237–249

Bosman F, Carneiro F et al (eds) (2010) WHO classification of tumours of the digestive system. IARC Press, Lyon

Boudreaux JP (2011) Surgery for gastroenteropancreatic neuroendocrine tumors (GEPNETS). Endocrinol Metab Clin North Am 40(1):163–171, ix

Buchmann I, Henze M et al (2007) Comparison of 68Ga-DOTATOC PET and 111In-DTPAOC (Octreoscan) SPECT in patients with neuroendocrine tumours. Eur J Nucl Med Mol Imaging 34(10):1617–1626

Butturini G, Bettini R et al (2006) Predictive factors of efficacy of the somatostatin analogue octreotide as first line therapy for advanced pancreatic endocrine carcinoma. Endocr Relat Cancer 13(4):1213–1221

Chatzipantelis P, Salla C et al (2008) Endoscopic ultrasound-guided fine-needle aspiration cytology of pancreatic neuroendocrine tumors: a study of 48 cases. Cancer 114(4):255–262

Christ E, Wild D et al (2009) Glucagon-like peptide-1 receptor imaging for localization of insulinomas. J Clin Endocrinol Metab 94(11):4398–4405

Chung DC, Smith AP et al (1997) A novel pancreatic endocrine tumor suppressor gene locus on chromosome 3p with clinical prognostic implications. J Clin Invest 100(2):404–410

Corleto VD, Annibale B et al (2001) Does the widespread use of proton pump inhibitors mask, complicate and/or delay the diagnosis of Zollinger-Ellison syndrome? Aliment Pharmacol Ther 15(10):1555–1561

D'Onofrio M, Mansueto G et al (2004) Neuroendocrine pancreatic tumor: value of contrast enhanced ultrasonography. Abdom Imaging 29(2):246–258

de Herder WW, Niederle B et al (2006) Well-differentiated pancreatic tumor/carcinoma: insulinoma. Neuroendocrinology 84(3):183–188

Doppman JL, Chang R et al (1995) Localization of insulinomas to regions of the pancreas by intra-arterial stimulation with calcium. Ann Intern Med 123(4):269–273

Drougas JG, Anthony LB et al (1998) Hepatic artery chemoembolization for management of patients with advanced metastatic carcinoid tumors. Am J Surg 175(5):408–412

Ehehalt F, Saeger HD et al (2009) Neuroendocrine tumors of the pancreas. Oncologist 14:456–467

Ekeblad S, Sundin A et al (2007) Temozolomide as monotherapy is effective in treatment of advanced malignant neuroendocrine tumors. Clin Cancer Res 13(10):2986–2991

Eriksson B (2010) New drugs in neuroendocrine tumors: rising of new therapeutic philosophies? Curr Opin Oncol 22(4):381–386

Espana-Gomez MN, Velazquez-Fernandez D et al (2009) Pancreatic insulinoma: a surgical experience. World J Surg 33:1966–1970

Falconi M, Zerbi A et al (2010) Parenchyma-preserving resections for small nonfunctioning pancreatic endocrine tumors. Ann Surg Oncol 17(6):1621–1627

Falconi M, Bartsch DK et al (2012) ENETS Consensus Guidelines for the management of patients with digestive neuroendocrine neoplasms of the digestive system: well-differentiated pancreatic non-functioning tumors. Neuroendocrinology 95(2):120–134

Fazio N, de Braud F et al (2007) Interferon-α and somatostatin analog in patients with gastroenteropancreatic neuroendocrine carcinoma: single agent or combination? Ann Oncol 18(1):13–19

Feelders RA, Hofland LJ et al (2010) Medical treatment of cushing's syndrome: adrenal-blocking drugs and ketoconazole. Neuroendocrinology 92(Suppl 1):111–115

Fendrich V, Langer P et al (2007) Management of sporadic and multiple endocrine neoplasia type 1 gastrinomas. Br J Surg 94(11):1331–1341

Fernández-Cruz L, Blanco L, Cosa R, Rendón H. (2008) Is laparoscopic resection adequate in patients with neuroendocrine pancreatic tumors? World J Surg 32(5):904–17.

Frucht H, Howard JM et al (1989) Secretin and calcium provocative tests in the Zollinger-Ellison syndrome. A prospective study. Ann Intern Med 111(9):713–722

Garcia-Carbonero R, Capdevila J et al (2010) Incidence, patterns of care and prognostic factors for outcome of gastroenteropancreatic neuroendocrine tumors (GEP-NETs): results from the National Cancer Registry of Spain (RGETNE). Ann Oncol 21(9):1794–1803

Gibril F, Schumann M et al (2004) Multiple endocrine neoplasia type 1 and Zollinger-Ellison syndrome: a prospective study of 107 cases and comparison with 1009 cases from the literature. Medicine (Baltimore) 83(1):43–83

Gillams A, Cassoni A et al (2005) Radiofrequency ablation of neuroendocrine liver metastases: the Middlesex experience. Abdom Imaging 30(4):435–441

Gimm O, Konig E et al (2007) Intra-operative quick insulin assay to confirm complete resection of insulinomas guided by selective arterial calcium injection (SACI). Langenbecks Arch Surg 392(6):679–684

Gola M, Doga M et al (2006) Neuroendocrine tumors secreting growth hormone-releasing hormone: pathophysiological and clinical aspects. Pituitary 9(3):221–229

Gupta S, Johnson MM et al (2005) Hepatic arterial embolization and chemoembolization for the treatment of patients with metastatic neuroendocrine tumors: variables affecting response rates and survival. Cancer 104(8):1590–1602

Halfdanarson TR, Rabe KG et al (2008) Pancreatic neuroendocrine tumors (PNETs): incidence, prognosis and recent trend toward improved survival. Ann Oncol 19(10):1727–1733

Haugvik SP, Marangos IP et al (2013) Long-term outcome of laparoscopic surgery for pancreatic neuroendocrine tumors. World J Surg 37(3):582–590

Hill JS, McPhee JT et al (2009) Pancreatic neuroendocrine tumors: the impact of surgical resection on survival. Cancer 115(4):741–751

Hoff AO, Vassilopoulou-Sellin R (1998) The role of glucagon administration in the diagnosis and treatment of patients with tumor hypoglycemia. Cancer 82(8):1585–1592

Horwhat JD, Paulson EK et al (2006) A randomized comparison of EUS-guided FNA versus CT or US-guided FNA for the evaluation of pancreatic mass lesions. Gastrointest Endosc 63(7):966–975

Hough DM, Stephens DH et al (1994) Pancreatic lesions in von Hippel-Lindau disease: prevalence, clinical significance, and CT findings. AJR Am J Roentgenol 162(5):1091–1094

Ichikawa T, Peterson MS et al (2000) Islet cell tumor of the pancreas: biphasic CT versus MR imaging in tumor detection. Radiology 216(1):163–171

Jackson JE (2005) Angiography and arterial stimulation venous sampling in the localization of pancreatic neuroendocrine tumours. Best Pract Res Clin Endocrinol Metab 19(2):229–239

Jensen RT, Niederle B et al (2006) Gastrinoma (duodenal and pancreatic). Neuroendocrinology 84(3):173–182

Jensen RT, Berna MJ et al (2008) Inherited pancreatic endocrine tumor syndromes: advances in molecular pathogenesis, diagnosis, management, and controversies. Cancer 113(7 Suppl):1807–1843

Jensen RT, Cadiot G et al (2012) ENETS Consensus Guidelines for the management of patients with digestive neuroendocrine neoplasms: functional pancreatic endocrine tumor syndromes. Neuroendocrinology 95(2):98–119

Joseph S, Wang YZ et al (2011) Neuroendocrine tumors: current recommendations for diagnosis and surgical management. Endocrinol Metab Clin North Am 40(1):205–231, x

Kaklamanos M, Karoumpalis I et al (2011) Diagnostic accuracy and clinical significance of the fine needle aspiration Ki-67 labelling index in pancreatic endocrine tumours. Endocr Relat Cancer 18(6):L1–L3

Kalinowski M, Dressler M et al (2009) Selective internal radiotherapy with Yttrium-90 microspheres for hepatic metastatic neuroendocrine tumors: a prospective single center study. Digestion 79(3):137–142

Kaltsas G, Rockall A et al (2004a) Recent advances in radiological and radionuclide imaging and therapy of neuroendocrine tumours. Eur J Endocrinol 151(1):15–27

Kaltsas GA, Besser GM et al (2004b) The diagnosis and medical management of advanced neuroendocrine tumors. Endocr Rev 25(3):458–511

Kaltsas G, Androulakis II et al (2010) Paraneoplastic syndromes secondary to neuroendocrine tumours. Endocr Relat Cancer 17(3):R173–R193

Kanakis G, Kaltsas G et al (2012) Unusual complication of a pancreatic neuroendocrine tumor presenting with malignant hypercalcemia. J Clin Endocrinol Metab 97(4):E627–E631

Kao PC, Taylor RL et al (1994) Proinsulin by immunochemiluminometric assay for the diagnosis of insulinoma. J Clin Endocrinol Metab 78(5):1048–1051

Kianmanesh R, Sauvanet A et al (2008) Two-step surgery for synchronous bilobar liver metastases from digestive endocrine tumors: a safe approach for radical resection. Ann Surg 247(4):659–665

Kimura W, Kuroda A et al (1991) Clinical pathology of endocrine tumors of the pancreas. Analysis of autopsy cases. Dig Dis Sci 36(7):933–942

Kindmark H, Sundin A et al (2007) Endocrine pancreatic tumors with glucagon hypersecretion: a retrospective study of 23 cases during 20 years. Med Oncol 24(3):330–337

Klimstra DS, Modlin IR et al (2010) The pathologic classification of neuroendocrine tumors: a review of nomenclature, grading, and staging systems. Pancreas 39(6):707–712

Kloppel G, Anlauf M et al (2007a) Endocrine precursor lesions of gastroenteropancreatic neuroendocrine tumors. Endocr Pathol 18(3):150–155

Kloppel G, Rindi G et al (2007b) Site-specific biology and pathology of gastroenteropancreatic neuroendocrine tumors. Virchows Arch 451(Suppl 1):S9–S27

Kouvaraki MA, Shapiro SE et al (2006) Management of pancreatic endocrine tumors in multiple endocrine neoplasia type 1. World J Surg 30(5):643–653

Kulke MH, Stuart K et al (2006) Phase II study of temozolomide and thalidomide in patients with metastatic neuroendocrine tumors. J Clin Oncol 24(3):401–406

Kulke MH, Bergsland EK et al (2009) Glycemic control in patients with insulinoma treated with everolimus. N Engl J Med 360(2):195–197

Kwekkeboom DJ, Mueller-Brand J et al (2005) Overview of results of peptide receptor radionuclide therapy with 3 radiolabeled somatostatin analogs. J Nucl Med 46(Suppl 1):62S–66S

Kwekkeboom DJ, de Herder WW et al (2008) Treatment with the radiolabeled somatostatin analog [177 Lu-DOTA 0, Tyr3]octreotate: toxicity, efficacy, and survival. J Clin Oncol 26(13):2124–2130

Lamberts SW, Reubi JC et al (1992) Somatostatin receptor imaging in the diagnosis and treatment of neuroendocrine tumors. J Steroid Biochem Mol Biol 43(1–3):185–188

Langer P, Wild A et al (2001) Prospective controlled trial of a standardized meal stimulation test in the detection of pancreaticoduodenal endocrine tumours in patients with multiple endocrine neoplasia type 1. Br J Surg 88(10):1403–1407

Lehnert T (1998) Liver transplantation for metastatic neuroendocrine carcinoma: an analysis of 103 patients. Transplantation 66(10):1307–1312

Levy-Bohbot N, Merle C et al (2004) Prevalence, characteristics and prognosis of MEN 1-associated glucagonomas, VIPomas, and somatostatinomas: study from the GTE (Groupe des Tumeurs Endocrines) registry. Gastroenterol Clin Biol 28(11):1075–1081

Ligneau B, Lombard-Bohas C et al (2001) Cystic endocrine tumors of the pancreas: clinical, radiologic, and histopathologic features in 13 cases. Am J Surg Pathol 25(6):752–760

Lonovics J, Devitt P et al (1981) Pancreatic polypeptide. A review. Arch Surg 116(10):1256–1264

Lorenz K, Dralle H (2007) Surgical treatment of sporadic gastrinoma. Wien Klin Wochenschr 119:597–601

Madura JA, Cummings OW et al (1997) Nonfunctioning islet cell tumors of the pancreas: a difficult diagnosis but one worth the effort. Am Surg 63(7):573–577, discussion 577–578

Malago R, D'Onofrio M et al (2009) Contrast-enhanced sonography of nonfunctioning pancreatic neuroendocrine tumors. AJR Am J Roentgenol 192(2):424–430

Marotta V, Nuzzo V et al (2012) Limitations of Chromogranin A in clinical practice. Biomarkers 17(2):186–191

Massironi S, Conte D et al (2010) Plasma chromogranin A response to octreotide test: prognostic value for clinical outcome in endocrine digestive tumors. Am J Gastroenterol 105(9):2072–2078

Maton PN (1993) Use of octreotide acetate for control of symptoms in patients with islet cell tumors. World J Surg 17(4):504–510

Maton PN, Gardner JD et al (1986) Cushing's syndrome in patients with the Zollinger-Ellison syndrome. N Engl J Med 315(1):1–5

Matthews BD, Heniford BT et al (2000) Surgical experience with nonfunctioning neuroendocrine tumors of the pancreas. Am Surg 66(12):1116–1122, discussion 1122–1113

McLean AM, Fairclough PD (2005) Endoscopic ultrasound in the localisation of pancreatic islet cell tumours. Best Pract Res Clin Endocrinol Metab 19(2):177–193

Modlin IM, Oberg K et al (2008) Gastroenteropancreatic neuroendocrine tumours. Lancet Oncol 9(1):61–72

Moertel CG, Hanley JA et al (1980) Streptozocin alone compared with streptozocin plus fluorouracil in the treatment of advanced islet-cell carcinoma. N Engl J Med 303 (21):1189–1194

Moertel CG, Kvols LK et al (1991) Treatment of neuroendocrine carcinomas with combined etoposide and cisplatin. Evidence of major therapeutic activity in the anaplastic variants of these neoplasms. Cancer 68(2):227–232

Moertel CG, Lefkopoulo M et al (1992) Streptozocin-doxorubicin, streptozocin-fluorouracil or chlorozotocin in the treatment of advanced islet-cell carcinoma. N Engl J Med 326(8):519–523

Newell-Price J, Bertagna X et al (2006) Cushing's syndrome. Lancet 367(9522):1605–1617

Nilsson O, Van Cutsem E et al (2006) Poorly differentiated carcinomas of the foregut (gastric, duodenal and pancreatic). Neuroendocrinology 84(3):212–215

Nino-Murcia M, Tamm EP et al (2003) Multidetector-row helical CT and advanced postprocessing techniques for the evaluation of pancreatic neoplasms. Abdom Imaging 28(3):366–377

Nobels FR, Kwekkeboom DJ et al (1997) Chromogranin A as serum marker for neuroendocrine neoplasia: comparison with neuron-specific enolase and the α-subunit of glycoprotein hormones. J Clin Endocrinol Metab 82(8):2622–2628

Norton JA, Cromack DT et al (1988) Intraoperative ultrasonographic localization of islet cell tumors. A prospective comparison to palpation. Ann Surg 207(2):160–168

Norton JA, Alexander HR et al (2004) Does the use of routine duodenotomy (DUODX) affect rate of cure, development of liver metastases or survival in patients with Zollinger-Ellison syndrome? Ann Surg 239:617–626

Norton JA, Fang TD et al (2006) Surgery for gastrinoma and insulinoma in multiple endocrine neoplasia type 1. J Natl Compr Canc Netw 4(2):148–153

O'Brien TD, Chejfec G et al (1993) Clinical features of duodenal somatostatinomas. Surgery 114(6):1144–1147

O'Toole D, Salazar R et al (2006) Rare functioning pancreatic endocrine tumors. Neuroendocrinology 84(3):189–195

O'Toole D, Grossman A et al (2009) ENETS Consensus Guidelines for the standards of care in neuroendocrine tumors: biochemical markers. Neuroendocrinology 90(2):194–202

Oberg K (2011) Circulating biomarkers in gastroenteropancreatic neuroendocrine tumours. Endocr Relat Cancer 18(Suppl 1):S17–S25

Oberg KE, Reubi JC et al (2010) Role of somatostatins in gastroenteropancreatic neuroendocrine tumor development and therapy. Gastroenterology 139(3):742–753, 753 e741

Olausson M, Friman S et al (2002) Indications and results of liver transplantation in patients with neuroendocrine tumors. World J Surg 26(8):998–1004

Orlefors H, Sundin A et al (2005) Whole-body (11)C-5-hydroxytryptophan positron emission tomography as a universal imaging technique for neuroendocrine tumors: comparison with somatostatin receptor scintigraphy and computed tomography. J Clin Endocrinol Metab 90(6):3392–3400

Owen NJ, Sohaib SA et al (2001) MRI of pancreatic neuroendocrine tumours. Br J Radiol 74(886):968–973

Pape UF, Jann H et al (2008) Prognostic relevance of a novel TNM classification system for upper gastroenteropancreatic neuroendocrine tumors. Cancer 113(2):256–265

Pasquali C, Rubello D et al (1998) Neuroendocrine tumor imaging: can ^{18}F-fluorodeoxyglucose positron emission tomography detect tumors with poor prognosis and aggressive behavior? World J Surg 22(6):588–592

Paulsen SD, Nghiem HV et al (2006) Evaluation of imaging-guided core biopsy of pancreatic masses. AJR Am J Roentgenol 187(3):769–772

Perez-Montiel MD, Frankel WL et al (2003) Neuroendocrine carcinomas of the pancreas with 'Rhabdoid' features. Am J Surg Pathol 27(5):642–649

Phan GQ, Yeo CJ et al (1998) Surgical experience with pancreatic and peripancreatic neuroendocrine tumors: review of 125 patients. J Gastrointest Surg 2(5):472–482

Piani C, Franchi GM et al (2008) Cytological Ki-67 in pancreatic endocrine tumours: an opportunity for pre-operative grading. Endocr Relat Cancer 15(1):175–181

Portela-Gomes GM, Stridsberg M et al (2000) Expression of the five different somatostatin receptor subtypes in endocrine cells of the pancreas. Appl Immunohistochem Mol Morphol 8(2):126–132

Raymond E, Dahan L et al (2011) Sunitinib malate for the treatment of pancreatic neuroendocrine tumors. N Engl J Med 364(6):501–513

Rehfeld JF, Bardram L et al (2012) Pitfalls in diagnostic gastrin measurements. Clin Chem 58(5):831–836

Richards ML, Thompson GB et al (2011) Setting the bar for laparoscopic resection of sporadic insulinoma. World J Surg 35:785–789

Rigaud G, Missiaglia E et al (2001) High resolution allelotype of nonfunctional pancreatic endocrine tumors: identification of two molecular subgroups with clinical implications. Cancer Res 61(1):285–292

Rindi G, Kloppel G (2004) Endocrine tumors of the gut and pancreas tumor biology and classification. Neuroendocrinology 80(Suppl 1):12–15

Rindi G, Kloppel G et al (2006) TNM staging of foregut (neuro)endocrine tumors: a consensus proposal including a grading system. Virchows Arch 449(4):395–401

Rinke A, Muller HH et al (2009) Placebo-controlled, double-blind, prospective, randomized study on the effect of octreotide LAR in the control of tumor growth in patients with metastatic neuroendocrine midgut tumors: a report from the PROMID Study Group. J Clin Oncol 27(28):4656–4663

Rosenau J, Bahr MJ et al (2002) Ki67, E-cadherin, and p53 as prognostic indicators of long-term outcome after liver transplantation for metastatic neuroendocrine tumors. Transplantation 73(3):386–394

Roy PK, Venzon DJ et al (2000) Zollinger-Ellison syndrome. Clinical presentation in 261 patients. Medicine (Baltimore) 79(6):379–411

Sarmiento JM, Heywood G et al (2003) Surgical treatment of neuroendocrine metastases to the liver: a plea for resection to increase survival. J Am Coll Surg 197(1):29–37

Schmid HA, Schoeffter P (2004) Functional activity of the multiligand analog SOM230 at human recombinant somatostatin receptor subtypes supports its usefulness in neuroendocrine tumors. Neuroendocrinology 80(Suppl 1):47–50

Service FJ (1995) Hypoglycemic disorders. N Engl J Med 332(17):1144–1152

Service FJ, Natt N (2000) The prolonged fast. J Clin Endocrinol Metab 85(11):3973–3974

Service FJ, McMahon MM et al (1991) Functioning insulinoma–incidence, recurrence, and long-term survival of patients: a 60-year study. Mayo Clin Proc 66(7):711–719

Shibata C, Kakyo M, et al (2012) Criteria for the glucagon provocative test in the diagnosis of gastrinoma. Surg Today 43(11):1281–5

Sobin LH, Wittekind C (2002) TNM classification of malignant tumours. Wiley-Liss, New York

Soga J, Yakuwa Y (1998a) Glucagonomas/diabetico-dermatogenic syndrome (DDS): a statistical evaluation of 407 reported cases. J Hepatobiliary Pancreat Surg 5(3):312–319

Soga J, Yakuwa Y (1998b) Vipoma/diarrheogenic syndrome: a statistical evaluation of 241 reported cases. J Exp Clin Cancer Res 17(4):389–400

Soga J, Yakuwa Y (1999) Somatostatinoma/inhibitory syndrome: a statistical evaluation of 173 reported cases as compared to other pancreatic endocrinomas. J Exp Clin Cancer Res 18(1):13–22

Soga J, Yakuwa Y et al (1998) Insulinoma/hypoglycemic syndrome: a statistical evaluation of 1085 reported cases of a Japanese series. J Exp Clin Cancer Res 17(4):379–388

Stabile BE, Morrow DJ et al (1984) The gastrinoma triangle: operative implications. Am J Surg 147(1):25–31

Steinmuller T, Kianmanesh R et al (2008) Consensus guidelines for the management of patients with liver metastases from digestive (neuro)endocrine tumors: foregut, midgut, hindgut, and unknown primary. Neuroendocrinology 87(1):47–62

Stridsberg M, Eriksson B et al (2003) A comparison between three commercial kits for chromogranin A measurements. J Endocrinol 177(2):337–341

Strosberg JR, Fine RL et al (2010) First-line chemotherapy with capecitabine and temozolomide in patients with metastatic pancreatic endocrine carcinomas. Cancer 117(2):268–275

Sundin A, Vullierme MP et al (2009) ENETS Consensus Guidelines for the standards of care in neuroendocrine tumors: radiological examinations. Neuroendocrinology 90(2):167–183

Thomas T, Kaye PV et al (2009) Efficacy, safety, and predictive factors for a positive yield of EUS-guided Trucut biopsy: a large tertiary referral center experience. Am J Gastroenterol 104(3):584–591

Tonelli F, Giudici F et al (2011) Pancreatic endocrine tumors in multiple endocrine neoplasia type 1 syndrome: review of literature. Endocr Pract 17(Suppl 3):33–40

Travis WD, World Health Organization et al (2004) Pathology and genetics of tumours of the lung, pleura, thymus and heart. Oxford University Press, Oxford (distributor)

Triponez F, Goudet P et al (2006) Is surgery beneficial for MEN1 patients with small ($<$ or $=2$ cm), nonfunctioning pancreaticoduodenal endocrine tumor? An analysis of 65 patients from the GTE. World J Surg 30(5):654–662, discussion 663–654

Tsolakis AV, Janson ET (2008) Endocrine pancreatic tumors: diagnosis and treatment. Expert Rev Endocrinol Metab 3(2):187–205

Vagefi PA, Razo O et al (2007) Evolving patterns in the detection and outcomes of pancreatic neuroendocrine neoplasms: the Massachusetts General Hospital experience from 1977 to 2005. Arch Surg 142(4):347–354

Verhoef S, van Diemen-Steenvoorde R et al (1999) Malignant pancreatic tumour within the spectrum of tuberous sclerosis complex in childhood. Eur J Pediatr 158(4):284–287

Vezzosi D, Bennet A et al (2005) Octreotide in insulinoma patients: efficacy on hypoglycemia, relationships with Octreoscan scintigraphy and immunostaining with anti-sst2A and anti-sst5 antibodies. Eur J Endocrinol 152(5):757–767

Walter T, Chardon L et al (2012) Is the combination of chromogranin A and pancreatic polypeptide serum determinations of interest in the diagnosis and follow-up of gastro-entero-pancreatic neuroendocrine tumours? Eur J Cancer 48(12):1766–1773

Yao JC (2007) Neuroendocrine tumors. Molecular targeted therapy for carcinoid and islet-cell carcinoma. Best Pract Res Clin Endocrinol Metab 21(1):163–172

Yao JC, Hassan M et al (2008a) One hundred years after "carcinoid": epidemiology of and prognostic factors for neuroendocrine tumors in 35,825 cases in the United States. J Clin Oncol 26(18):3063–3072

Yao JC, Phan A et al (2008b) Targeting vascular endothelial growth factor in advanced carcinoid tumor: a random assignment phase II study of depot octreotide with bevacizumab and pegylated interferon α-2b. J Clin Oncol 26(8):1316–1323

Yao JC, Shah MH et al (2011) Everolimus for advanced pancreatic neuroendocrine tumors. N Engl J Med 364(6):514–523

Zamora V, Cabanne A et al (2010) Immunohistochemical expression of somatostatin receptors in digestive endocrine tumours. Dig Liver Dis 42(3):220–225

Zerbi A, Capitanio V et al (2011) Surgical treatment of pancreatic endocrine tumours in Italy: results of a prospective multicentre study of 262 cases. Langenbecks Arch Surg 396(3):313–321

Zhao J, Moch H et al (2001) Genomic imbalances in the progression of endocrine pancreatic tumors. Genes Chromosomes Cancer 32(4):364–372

Zhou C, Dhall D et al (2009) Homozygous P86S mutation of the human glucagon receptor is associated with hyperglucagonemia, α cell hyperplasia, and islet cell tumor. Pancreas 38(8):941–946

Index

A
ABCC8, 7, 313–314
Abscisic acid, 618
Acetylcholine, 277
Acinar cells, 12, 29, 32, 113, 419, 420,
    612, 1175
Adaptive immune responses, 1056
Adenylyl cyclase, 572, 753
Adult stem cell, 1318
A-kinase anchoring proteins (AKAPs), 584
AKT, 668–669, 671–673
Allograft, 1202, 1205, 1216
Allotransplantation, 41,1232–1239
Alloxan, 208, 209, 613, 1061
α-cell/alpha cell, 22–23, 192–202, 278–282,
    514, 519, 520, 722, 768, 925
Amino acids, 202
Amplifying pathway, 641
Amylin/islet amyloid polypeptide (IAPP), 222
Amyloid, 856–857
Anatomy, of islets, 1–14, 27–28
Androgen, 790
Animal models, 84–85
Anoctamin 1, 383
Anti-apoptotic genes and proteins, 862, 866
Antigen-presenting cells, 1050, 1056–1058,
    1061, 1064
Anti-glucagon therapies in diabetes, 220
Apoptosis, 804, 809, 846–847, 849–852,
    862–863, 867, 1116
Aquaporins, 390
Arachidonic acid, 135–136, 613
ARNT/HIF1beta, 806
Aryl hydrocarbon receptor nuclear
    translocator like (Arntl). *See* Brain and
    muscle aryl hydrocarbon receptor
    nuclear translocator like protein-1
ATP/ADP ratio, 696
ATP-sensitive potassium (K_{ATP}) channel, 305

Autoantibodies, 31, 1066–1067, 1083–1085
Autoantigens, 1051–1052, 1056, 1058, 1067
Autograft, 1216, 1218
Autoimmune diabetes, 874
Autoimmunity, 25
Autologous islet isolation, 1220, 1222
Autonomic regulation of glucagon
    secretion, 204
Autotransplantation, 1199–1222, 1229–1239

B
Basal calcium, 607
Basement membrane, 40
Beta-catenin, 710–718
β-cell / beta cell, 23–25, 252, 514, 846,
    849–850, 852, 857, 859, 863, 865, 1313,
    1318, 1324, 1326
β-cell clock, 694
  development, 749
  differentiation, 668, 672, 1341, 1344, 1348,
    1353, 1358
  dysfunction, 837
  exhaustion, 834–835, 837
  function, 551
  imaging, 62, 65–66
  mass, 667, 672, 803–805, 897–898, 909,
    1116, 1125
  regeneration, 727, 733
  survival, 672, 787
  volume of, 802, 803
Beta cell proliferation, 541, 717–718, 809, 905,
    1360
β1-integrin, 44, 49
β-oxidation, 185
Bcl family, 849, 860, 863, 866–867
Big-K channels, 198
Bioluminescence imaging, 66–67
Biosynthesis, 177, 222

Biphasic insulin secretion, 135, 491–493, 608–609
BK channels (Big K), *aka* Maxi-K or slo1, *See* Big-K channels
Blood glucose signal transduction, 548
Blood vessels, 47
Body mass index (BMI), 803, 904, 1168
Brain and muscle aryl hydrocarbon receptor nuclear translocator like protein-1 (Bmal1, Mop3), 689
BRIN-BD11 cells, 375
Bursting, 454, 461

C
Ca^{2+}-activated Cl channels, 418
Ca^{2+} handling, 148
Ca^{2+} channel, 860, 867
cADPR, 619–619
Calcium dynamics, 456
Ca^{2+}-induced Ca^{2+} release (CICR), 619
Calcium oscillations, 617, 621
Calcium-release activated calcium channel, 353
Calcium-release activated non-selective cation channel, 340, 353
Calcium signaling, 338
Calpain, 880
CAMP response element binding protein (CREB), 582
Cancer, 717, 1133, 1397
Canonical Wnt signaling, 711–713
Caspase, 847, 849, 851, 862, 867
Ca_v Channels, 199
CCK, 11
$CD4^+$ T cells, 1052, 1058, 1061, 1063, 1065
$CD8^+$ T cells, 1052, 1062, 1065, 1067
CD20 antibody, 1090
CD3 antibody, 1088–1089
Cell/basement membrane interaction, 39–52
Cell engineering, 1305
Cell fate, 117, 1345
Cell proliferation, 710, 712, 715, 718, 729, 789
Cellular therapy, 1340, 1352, 1361–1362
Central clock, 691
CFTR. *See* Cystic fibrosis transmembrane conductance regulator
Chemoosmotic hypothesis, 371
Chemotherapy, 1397–1398
CHI. *See* Congenital hyperinsulinism
Chromogranin A, 1380
Chronic pancreatitis (CP), 1201, 1213–1214, 1216

CICR. *See* Ca^{2+}-induced Ca^{2+} release
Circadian clock, 688–689
Circadian disruption, 700
Circadian rhythm, 688
$[Cl^-]i$, 405, 431
Clinical islet transplantation, 888, 1177–1178, 1245–1267
Clinical outcome, 31, 1234, 1251–1257
Clostridium histolyticum, 1172
COBE, 1220–1221
Cold ischemia time, 1180
Collagen, 41, 49
Collagenase, 751, 1172–1174
Collagenase digestion, 1202, 1213
Combinatorial therapies for treatment of obesity and diabetes, 223
Comparative, 1–14, 279, 749, 906, 1143–1145, 1342
Comparative species, 3
Complexin, 478
Complication, 51, 316, 608, 647, 708, 877–878, 911, 1084, 1230, 1254–1256, 1337
Computerized tomography (CT), 70
Congenital hyperinsulinism (CHI), 260, 277, 312
Continuous density gradient, 1176
Control of glucagon secretion by glucose, 203, 211
Cost-efficacy, 1261
Counterregulation, 183–184, 187
Coupling factor, 139, 641–642
CPT-1, 140
CT. *See* Computerized tomography
Culture, 1177–1181
Cyclic AMP, 589
Cystic fibrosis transmembrane conductance regulator (CFTR), 422
Cytokines, 833, 835, 999–1001, 1010–1012
Cytoprotection, 720
Cytomegalovirus

D
Dark cycle, 695
δ-cell/delta cell, 25–26, 282
Debulking procedures, 1394–1395, 1399
definition, 1277
(de)differentiation, 22, 46, 534, 715, 757, 758, 806, 811, 1056, 1062, 1320, 1341
DEND, 261, 315–316

Index   1409

Dendritic cells, 1056, 1059, 1068
Depolarization, 427, 610
Desensitization, 149
Development, 112, 115–116, 120
2DGE. *See* Two-dimensional gel
    electrophoresis
D-glucose anomeric specificity, 158, 160–163,
    166–167, 169
Diabetes, 186–188, 573, 582, 646, 937, 1337,
    1341, 1356–1357
Diabetic complications, 1255, 1261
Differentiation, 111–112, 120, 1313, 1315,
    1317–1320
Direct effects of glucose on glucagon
    secretion, 216
Doublecortin, 1119
DPP-4 inhibitors, 910

E
Effects of glucagon, 223
Electrical activity, 194, 264–266, 269–271,
    307, 454, 456, 615, 621
Electrophysiology of α-cells, 194
Embryonic stem cells, 1338–1339
Endocrine
    cells, 4, 12, 13, 22–28, 41, 110, 113, 177,
    191, 213, 478, 486, 714, 745, 748, 1263
    tumours, 1375–1400
Endocrine progenitor cells, 117
Endoderm, 2, 112, 715, 1344–1352
Endoplasmic reticulum, 804, 807, 809
Endoplasmic reticulum stress, 879
Endothelial cells, 28, 47–48, 646, 713,
    747, 1057
Enzymatic pancreas dissociation,
    1172–1175
Epac2, 578, 587
ER stress, 539, 612,774, 774, 888, 898
Estrogen, 787, 789
Evanescent-field microscopy, 481
Evolution, 4–6, 180, 1185
Exenatide, 905
Exendin-4, 568, 586, 811, 905, 1089, 1346
Exercise, 16, 547
Exocytosis, 134–135, 201, 264–266, 307,
    475–503, 574–578, 615, 641–644, 723,
    758, 806, 1139
Expansion, 115, 147, 715, 749, 773–774, 878,
    905, 1063, 1102,1234, 1354
EXTL3/EXTR1, 963
Extracellular hypotonicity, 378
Extrapancreatic glucagon, 181, 191

F
Fatty acids, 185, 203
    synthesis, 185–186
Fibrosis, 20, 745–747, 1238
Foxa2, 115, 183, 496
Foxo-1, 806
FoxP3, 1057, 1094
Free fatty acids (FFA), 148, 536, 643, 775,
    1145–1147
Functioning tumours, 1389

G
GABA, 211
γ-Hydroxybutyrate, 211
GAD. *See* Glutamic acid decarboxylase
GAD vaccine, 1086–1087
Galanin, 276
Gastrin, 10, 182, 885, 1389, 1399
Gastrinoma, 1384–1385, 1389
Gastrointestinal peptide, 325
GEFs, 574
Gene-environment interactions, 874, 878
Gene expressed in hepatocellular carcinoma-
    intestine-pancreas, 959
Gene expression, 147, 806–808
Gene polymorphism, 808
Gene regulatory networks, 115
Genetic polymorphisms, 1055
Genetics, 32, 42, 150, 263, 268, 269, 315,
    327, 502, 539, 540, 639, 715, 729,
    745, 768, 808–810, 881–883, 898,
    1050, 1053–1055, 1100, 1132,
    1142, 1230, 1247–1248, 1262,
    1381–1383
Genome-wide association studies, 722–727,
    808, 876, 882
Genomics and proteomics, 1142
Gestation, 21, 113, 532, 749, 1100, 1145
GIP, 131–133, 182, 267, 325, 572, 775, 810,
    884, 903–904
GIRK. *See* G protein-gated inwardly rectifying
    K^+ channels
GK rat, 745, 748, 752, 757
Glicentin, 178
Glicentin-related pancreatic polypeptide
    (GRPP), 178
Glinides, 258
Glitazones, 902
GLP 1. *See* Glucagon-like peptide 1
GLP 2. *See* Glucagon-like peptide 2
Glucagon, 4, 175, 183
    and diabetes, 187

Glucagon-like peptide 1 (GLP-1), 178, 221, 572, 585, 710, 717–718, 720
Glucagon-like peptide 2 (GLP-2), 178, 180
Glucagonoma 186, 1385, 1386
Glucagon receptor, 183, 221, 223
Glucagon secretion, 158, 164, 167, 169, 393
Glucokinase, 859
Glucolipotoxicity, 535–538
Gluconeogenesis, 184, 186
Glucose-induced/stimulated insulin secretion, 696, 856, 859, 863, 867
Glucose toxicity/glucotoxicity, 809, 846, 859–860, 867, 877, 1147
Glucotoxicity and lipotoxicity, 148, 774–775
Glutamate, 638, 643
Glutamate dehydrogenase, 139, 313
Glutamic acid decarboxylase (GAD), 65, 1118
Glyburide, 903
Glycemic control, 771, 902, 1066
Glycogen synthase kinase-3, 711
Glycogenolysis, 184
Glycogen synthesis, 184
Glycolysis, 184
GPR40, 141, 643
GPR49, 730
GPR119, 569
G protein-gated inwardly rectifying K^+ channels (GIRK), 198
Growth factor, 8, 43–44, 142, 716–717, 884, 1055, 1089, 1238, 1313, 1339, 1355
GRPP. *See* Glicentin-related pancreatic polypeptide
Guanine nucleotide exchange factors, 500, 577, 585, 621

H
HCN Channels, 198
HIP, 959
Histocompatibility antigens (HLA) 1050, 1082, 1257, 1356
HNF1α, 495, 882
HNF1β, 120
HNF6, 115
H_2O_2, 136, 259, 260, 613, 645, 756
Homeostasis model assessment (HOMA), 902
Hormone, 2, 11
Human islet allotransplantation, 1234
Human islets, 1204, 1217
autotransplantation, 1093, 1230, 1232, 1235
2-Hydroxyesteriol, 389
Hyperglucagonemia, 187, 190, 203
Hyperglycaemic

Hyperglycemia, 110, 177, 220, 531, 538, 646, 708, 772, 775, 877–878, 897, 1088, 1169
Hyperinsulinemia, 828, 833, 835, 838
Hyperinsulinemic, 210, 313, 768
Hypertrophy, 535, 775
Hypoglycemia, 177, 187, 610, 883, 909, 1084

I
IA-2 antibody, 1067
IAPP. *See* Amylin/islet amyloid polypeptide
IGF-1, 8, 10, 659–675, 749, 885
IGF-1 receptor, 661–662, 664, 666, 668, 673
IGF2, 729, 749
IGF2BP2, 729
Immune cells, 993, 1009
Immune regulatory, 1321, 1324
Immunocytochemistry, 849, 851, 856
Immunoisolation, 1288
Immunology of β-cell destruction, 1084
Immunosuppression, 1257
Impaired glucagon response to hypoglycemia, 187
Incretin mimetics, 132, 903–906
Incretins, 132–133, 903
Inflammation, 809
Inflammatory pathways linked to β cell demise in diabetes, 1084
INGAP. *See* Islets neogenesis-associated protein
Ingenuity pathway analysis, 1148
Inhibitors, 942
Innervation, 29
Insulin, 4, 6, 7, 158, 160, 162–163, 165, 167–169, 207, 277, 846, 850, 856, 860, 863, 867
autoantibodies, 1051
biosynthesis, 130, 750–751, 770, 905
granules, 484–487, 723, 803, 1051, 1144
hypersecretion, 833
independence, 1203, 1215
receptor, 660, 662, 664, 666
release, 751, 753–754, 758
resistance, 829, 831, 834, 838
secreting cells, 660, 662–663, 669, 672
secretion, 137, 308, 324, 345–347, 353–354, 404, 458, 478, 491, 495, 548, 639, 641, 663–666, 670, 787, 792, 802, 805, 806, 899, 914
signaling, 149, 208, 272, 538, 770, 885
transcription, 49
Insulinoma, 23, 116, 131, 259, 612, 641, 1051, 1144, 1383–1384, 1388–1389

Insulitis, 31, 1049, 1057, 1060, 1062–1063
Integrin, 44, 49–50, 1062, 1142
Invasive tumor localization techniques, 1392
Invertebrates, 2–5, 42
In vivo mouse models, 1125
Ion channels, 826
IRS-1, 670–671, 675
IRS-2, 670–672
Islet, 1–14, 786, 934, 940, 1138
    amyloid polypeptide, 856–857
    architecture, 11, 323, 745–747, 878
    autoantibodies, 1050–1052, 1060, 1067
    autoimmunity, 1049–1050, 1065, 1067
    cell, 748
    donors, 1279
    encapsulation, 1093
    equivalent, 1182, 1252
    function, 39–52, 65, 460, 710, 743–759, 770, 1152, 1181, 1232, 1260
    isolation, 84–85, 101
    morphology, 20–21
    pancreatic, 454
    purification, 91, 93, 1175–1177
    shipment, 1235
    structure, 1–14
    transplantation, 888, 1122, 1246, 1251, 1267
    transplants, 1093, 1276–1277, 1282, 1298, 1303, 1305
    vasculature, 28–29
    xenotransplantation, 1093, 1277, 1279, 1281, 1283, 1288
Islets neogenesis-associated protein (INGAP, Reg3 δ), 971, 974–975

J
JC-1 assay, 696

K
K_{ATP} channel, 194, 204, 214, 216–217, 610
K_{ATP} Channel-Dependent Model, 216
K_{Ca} channel, 266–271
KCNJ11, 261, 314–316, 539, 723, 808, 881
Ketone bodies, 182, 186, 189
Ki67, 1377, 1390, 1397
Ki67 index, 1377, 1397
$K_{IR}6.1$, 252
$K_{IR}6.2$, 252, 261, 309–311, 316–321
K_{IR}, 252
K_v channels, 196
    lactogen, 792

L
Label-free proteomics, 1137
Lactation, 531
Lactogens, 791
Laminins, 42, 49
    large animals and clinical trials, 1283
Laser scanning microscopy (LSM)
Latent autoimmune diabetes in adults (LADA), 1082
LC-MS/MS, 1142
Leptin, 222, 769
Light cycle, 695
    limitations of, 1281
Lipofuscin, 25
Lipotoxicity, 809, 880, 1147
12-Lipoxygenase, 999–1001, 1024–1027
Liraglutide, 906
Liver, 7, 65, 149, 184, 454, 476, 540, 623, 642, 768, 883, 1152, 1317, 1344
Long chain acyl CoA (LC acyl CoA), 140–141, 255, 617, 643
Low-temperature culture, 1179
Lutheran, 47

M
Macrophages, 28, 747, 881, 1056, 1097
Magnetic resonance imaging (MRI), 72
Major proglucagon fragment (MPGF), 178
Making islets from human embryonic stem cells, 1094
Malate–aspartate shuttle, 139
Malic enzyme, 140, 641
MAPK. *See* Mitogen-activated protein kinase
Mass spectrometry, 1135, 1139, 1141, 1153
Maternal, 532, 647, 1050
Maturity onset diabetes of the young (MODY), 111, 732, 874
Maxi-K channels. *See* Big-K channels
Mechanisms of Action, 183
Medical treatment, 1396–1399
Membrane integrity test, 1182
Membrane potential, 608, 621
MEN1, 1399
Mesenchymal stem cell (MSC), 1262, 1320, 1324
Metabolic oscillations, 453–469, 772
Metabolic stress, 786
Metabolism in α-cells, 192
Metformin, 903, 906, 909
Methylated insulin gene, 1122
Microanatomy, of the islets of Langerhans, 191
Microcapsules, 1299–1300, 1303

microRNA, 495–496, 1133
Misfolded proteins, 879, 1059, 1152
Mitochondria, 634, 806
Mitochondrial death pathway, 877
Mitochondrial DNA (mtDNA), 647–648
Mitogen-activated protein kinase (MAPK), 668–669, 673–674
  inhibitors, 388
Monocytes, 1063, 1324
Mop3. *See* Brain and muscle aryl hydrocarbon receptor nuclear translocator like protein-1
Morphology, 12, 22, 23,69, 324, 532, 640, 773, 806, 1237, 1319, 1339
Mouse, 768
MSG-obesity model, 558
Multiple endocrine neoplasia, 1382
Munc13-1 478, 755
Munc18-1 478, 755

N
NAD(P)H oxidase-derived H2O2, 380
NADPH oxidase (NOX), 925, 927
Na$_v$ channels, 198
NBCe1 Na$^+$/HCO$_3^-$ cotransporters, 373
Necrosis, 1121
Neonatal diabetes, 314
Neonates, 22, 324, 538–539
Nephrin
Nervous systems, 180
NeuroD, 121, 1347
Neuroendocrine carcinomas, 1377, 1387
Neurogenin-3 (ngn3), 22,115–116, 731–732, 883
Neuropeptide, 29, 771, 1052
Neurotransmitters, 13, 134,182, 276–278, 480, 623, 634, 770
Neutral protease, 1175
Nidogen, 43
Nitric oxide, 143, 259, 613, 747, 1149
NKCC1, 407, 409, 433
NKCC2, 407, 409, 412
Nkx2.2, 116, 1347
Nkx6.1, 115, 116, 1343
Nkx6.2, 116
NOD mouse, 875, 1086, 1314
Non-endocrine islet cells, 28
Non-invasive tumor localization techniques, 1392, 1394
Noradrenaline, 276
NOX. *See* NADPH oxidase

NPPB, 388
Nrf2, 697
Nuclear imaging, 4
Nutrient metabolism, 136, 148, 257, 1149
Nutrition, 531, 539

O
ob/ob, 1143, 1145, 767–776
Obesity, 820, 831, 833
Offspring, 749
Optical coherence tomography (OCT), 65–67
Optical imaging, 62
Orai, 342–343, 346–347
Organ preservation, 1170, 1171
Orthology, 959
Oscillations, 453
Oxidative stress, 697
OXPHOS, 696

P
Pancreas, 2, 6, 8, 10, 12, 535
Pancreas
  development, 2, 21,65, 111–112, 116, 119, 713–716
  weight, 1168
Pancreas dissociation, 1172–1175
Pancreatectomized diabetic rats, 556
Pancreatectomy, 1201–1202, 1204
Pancreatic β-cell, 133, 307, 633
Pancreatic development, 1341–1342, 1346, 1349, 1352, 1359
Pancreatic islet, 16, 476, 547, 771
Pancreatic islets α-cells, 158, 161, 165
Pancreatic islets beta-cells, 307, 771–774
Pancreatic neuroendocrine tumors (pNETs), 1376–1379, 1381–1382, 1387, 1390–1392, 1394, 1398
  Functioning pNETs, 1377, 1381, 1383–1385, 1388, 1396, 1398
  Hereditary pNETs, 1381
  Non-functioning pNETs, 1383, 1387, 1390, 1397–1398
  Sporadic pNETs, 1381
  Well differentiated pNETs, 1380, 1383, 1392, 1397, 1399
Pancreatic polypeptide, 11, 13, 21, 178, 180, 1341, 1383
Pancreatic progenitor, 1347, 1354, 1356
Pancreatic stone protein (PSP), 959, 963
Pancreatic thread protein (PTP), 959
Pancreatitis, 906, 907, 1238, 1259

Pancreatitis-associated protein (PAP), 959, 962
PAP. See Pancreatitis-associated protein
Paracrine regulation, of glucagon secretion, 204
PC1/3, 180
PC2, 178, 180, 220
8-pCPT-2′-O-Me-cAMP 575, 582
Pdx-1, 21, 114–115, 147, 538–539, 571, 715–716, 882, 883, 808, 1317, 1345
Peripheral clocks, 691, 694
Perlecan, 43–44
PERV
PET-CT, 67, 69, 77
Pharmacoproteomics, 1152–1153
Phosphodiesterase, 567, 753
Phosphodiesterase inhibitors, 387
Phosphoproteome, 1142
Phosphotransfer, 255–257
PI3K, 663, 665, 669, 671
Piglet islet, 1280
PKB, 48
PLA2, 136, 267
Pluripotency, 1338–1340
Polymorphism, 512, 523
pNETs. See Pancreatic neuroendocrine tumors
Porcine islets, 1277, 1279–1280
Positron Emission Tomography (PET), 67–69, 1263, 1392
PPARγ, 881
PP cells, 26
Prediction, 1052, 1067
Prevention, 1064
Progenitor, 22, 114–115, 709, 774
Progenitor cell 709
Progesterone, 791
Proglucagon, 178, 180
Programmed cell death, 883, 1089
Programming, 120, 531–542, 749, 758
Prohormone convertases, 178
Pro-inflammatory cytokines, 936
"Proinsulin", 23–25, 645, 731, 751, 812, 882, 897, 1056, 1348, 1388
Proliferation, 50, 261, 534, 582–583, 642, 717–718, 874, 885, 1052
Protein expression, 806–808
Protein kinase A, 578, 585
Protein kinase C (PKC), 135, 577, 713, 754, 1148
Protein microarray, 1138
Protein phosphatase inhibitor, 1, 1121
Proteome, 1132
Proteomics, 18, 1131
PSP. See Pancreatic stone protein

PTP. See Pancreatic thread protein
Pulsatility, 454
Pulsatility of Insulin, Glucagon, and Somatostatin Secretion, 205

Q
Quality assessment, 97, 100

R
Rab3A 478
Radioactive tracer molecules, 67–68
Radiotracer, 67–68
Rapamycin, 883, 1234, 1398
Reactive oxygen species (ROS), 645, 697, 925, 934, 936, 939
Real time live confocal microscopy, 1236
Reg1, 964
Reg2, 966
Reg3, 967
Reg3 δ. See Islets neogenesis-associated protein
Reg4. See Regenerating protein-like protein
Regenerating protein-like protein (RELP, Reg4), 959, 972
Regeneration, 804, 957, 967, 1313, 1316, 1318, 1326
Regenerative medicine, 1338, 1340
Regulatory T cell, 1057–1058, 1093, 1259
RELP, 959
Reproductive hormones, 786
Retinopathy, 1255–1256
Retrovirus, 1263
Rev-erb, 690
Ryanodine receptor, 621
RyR1, 617
RyR2, 616
RyR3, 616

S
Saturated fat, 530
SELDI-TOF, 1135
Selective plane illumination microscopy, 48
Sensitization, 273, 575, 617
Sequential exocytosis, 483, 487, 489
SERCA, 456, 614, 753, 881
Short-chain L-3-hydroxyacyl-CoAdehydrogenase(SCHAD), 313, 314, 1141, 1148
Signaling pathways, 112

Signal transduction, 133, 148
Sik1, 700
Single Photon Emission Computed
    Tomography (SPECT), 67–69
SIRT4, 639
siRNA, 49, 137, 140, 584, 617, 716, 809
Small G Protein, 495, 712
SNAP-25, 264, 490–491, 755
SNAREs, 489–491
Soluble adenylyl cyclase, 386
Somatostatin, 213, 277
Somatostatin secretion, 161
Species, 2, 4, 5
Stem cells, 1313, 1315, 1317, 1319–1320, 1324, 1326
Stem cells in pancreatic islets, 1094
Stimulus-secretion coupling, 133, 351, 353
Store-operated calcium entry, 338, 341, 344–345, 355
Store-operated channels, 201
Store-operated ion channels, 16, 337
Streptozotocin (STZ), 788, 1118, 1121, 1125
Streptozotocin (STZ)-induced diabetic rats, 555
Stromal cell-derived factor-1, 710, 718
Stromal interaction molecule, 341, 349
Structure, 6–8, 27, 43, 72, 121, 133, 179, 252–255, 309–311, 458, 488, 622, 743–759, 1134, 1174, 1314–1315, 1341, 1351
Sulfonylureas, 258
Supra chiasmatic nucleus (SCN), 691
Surgery, 1230, 1231, 1263, 1393–1395
Survival, 48–49, 569, 716, 718–721, 1359
Survival and growth factors, 716, 720–721, 884–885
Synaptic-like microvesicles, 496–502, 1051
Synaptophysin, 1380
Synchronization, 15
Syntaxin-1, 478, 755

T

Tannic acid, 384
TCF7L2, 712, 720–721
Tenidap, 374, 388
TGF-beta, 1326, 1341
Thermodynamic equilibrium, 405
Thioredoxin, 136, 1141
Tolerance, 47, 4, 220, 266, 323, 531, 572, 583, 622, 648, 728, 750, 769, 802, 875, 887, 898, 914, 1055–1056, 1085
Toll like receptors, 24, 25
Total pancreatectomy, 314, 324, 21, 1231, 1260

Toxin, 200, 263, 618, 1050, 1061, 1085–1086, 1248
Transcription factors, 111, 114–115, 117, 120
Transdifferentiation, 974
Transgenic mouse, 33, 1234
Transient receptor potential (TRP) channels, 342, 349, 351, 611
TRPM2, 613
TRPM3, 613
TRPV1, 612
TRPV2, 612
TRPV4, 612
Transplantation, 1221
T- regulatory cells (Tregs), 1053, 1056–1059, 1061, 1066
TRP channels. See Transient receptor potential (TRP) channels
TRPM4, 272, 613–614
TRPM5, 272, 614
TRPV5, 612–613
TUNEL assay, 851–852, 859
Two-dimensional gel electrophoresis (2DGE), 1134
Two-photon microscopy, 481
Two-pore channel, 353
Type 1 diabetes (T1DM), 29–32, 846, 993–1001, 1015–1016, 1017–1022, 1247, 1261, 1265, 1276, 1288, 1298
Type 2 diabetes (T2DM), 32–33, 515, 539–542, 744, 747, 846, 897, 1001–1014, 1016–1019, 1023–1027

U

Ucp2. See Uncoupling Protein 2
Ultrasound (US), 69
Uncoupling Protein 2 (Ucp2), 697
Uptake of amino acids, 186

V

Vascular basement membrane, 47–48
Vascularization, 21, 747, 1349
Vasculature, 4, 28, 47
Vasopressin receptors, 700
Vertebrates, 2–5
Vesicle-associated membrane protein-2 (VAMP2), 478, 1052
Vitamin D, 1096–1097
Volume-regulated anion channel (VRAC), 377, 393, 416, 426, 435
Volume-sensitive anion channels (VSACs), 271

W
Weanlings, 532
Wnt signaling, 571, 707–733

X
Xenopus, 8, 317, 715
Xenotransplantation, 1262–1263

Z
ZDF rats. *See* Zucker diabetic fatty (ZDF) rats
Zinc, 515
Zinc-specific dye, 1183
Zinc transporters, 521
Zn^{2+}, 210
ZnT8, 25, 1067
Zucker diabetic fatty (ZDF) rats, 557